Handbook of Theoretical Computer Science

Volume A

ALGORITHMS AND COMPLEXITY

ELSEVIER
AMSTERDAM • NEW YORK • OXFORD • TOKYO

THE MIT PRESS
CAMBRIDGE, MASSACHUSETTS

Handbook of Theoretical Computer Science

Volume A

ALGORITHMS AND COMPLEXITY

edited by

JAN VAN LEEUWEN,
Utrecht University, The Netherlands

1990

ELSEVIER
AMSTERDAM • NEW YORK • OXFORD • TOKYO

1990

THE MIT PRESS
CAMBRIDGE, MASSACHUSETTS

ELSEVIER SCIENCE PUBLISHERS B.V.
P.O. Box 211,
1000 AE Amsterdam, The Netherlands

Co-publishers for the United States, Canada, and Japan:

The MIT Press
55 Hayward Street
Cambridge, MA 02142, U.S.A.

Library of Congress Cataloging-in-Publication Data

Handbook of theoretical computer science/editor, Jan van Leeuwen.
p. cm.
Includes bibliographical references and indexes.
Contents: v. A. Algorithms and complexity -- v. B. Formal models and semantics.
ISBN 0-444-88075-5 (U.S. : set). -- ISBN 0-444-88071-2 (U.S. : v. A). -- ISBN 0-444-88074-7 (U.S. : v. B). --ISBN 0-262-22040-7 (MIT Press : set). --ISBN 0-262-22038-5 (MIT Press : v. A). -- ISBN 0-262-22039-3 (MIT Press : v. B).
1. Computer science. I. Leeuwen, J. van (Jan)
QA76.H279 1990
004--dc20 90-3485
CIP

Elsevier Science Publishers B.V.
ISBN: 0 444 88071 2 (Volume A)
ISBN: 0 444 88075 5 (Set of Vols A and B)

The MIT Press
ISBN: 0 262 22038 5 (Volume A)
ISBN: 0 262 22040 7 (Set of Vols A and B)

Printed in The Netherlands

Printed on acid-free paper

Preface

Modern developments in computer and software systems have raised many challenging issues concerning the design and efficiency of complex programming applications. There is an increasing need for "advanced theory", to understand and exploit basic concepts and mechanisms in computing and information processing. The *Handbook of Theoretical Computer Science* is designed to provide a wide audience of professionals and students in Computer Science and related disciplines with an overview of the major results and developments in the theoretical exploration of these issues to date.

There are many different roles for "theory" in Computer Science. On the one hand it provides the necessary mathematical foundations for studying formal systems and algorithms that are needed. On the other hand, it provides concepts and languages to capture the essence, in algorithmic and descriptive terms, of any system from specification to efficient implementation. But the mathematical frameworks that have proved to be invaluable for Computer Science are used increasingly in many other disciplines as well. Wherever notions of information or information processing are identified, questions of representation and computation can be formalized in computer science terms. Theoretical Computer Science concerns itself with all formal models and methods and all techniques of description and analysis that are required in this domain.

As a consequence there are many facets to Theoretical Computer Science. As a discipline it employs advanced techniques from Mathematics and Logic, but at the same time it has established its own formal models and fundamental results in which the original motivations have remained visible. The *Handbook of Theoretical Computer Science* attempts to offer an in-depth view of the field of Theoretical Computer Science as a whole, by a comprehensive exposition of the scientific advances in this area.

In order to keep the Handbook within manageable limits, it was decided to restrict the material to the recognized core areas of Theoretical Computer Science. Even with this restriction imposed, further choices had to be made in the subjects covered and the extent of the material provided on each subject. The current version of the Handbook is presented in two volumes:

Vol. A: Algorithms and Complexity
Vol. B: Formal Models and Semantics

This more or less reflects the division between algorithm-oriented and description-oriented research that can be witnessed in Theoretical Computer Science, and it seemed natural to follow it for these books. Whereas the volumes can be used independently, there are many interesting connections that show that the two areas really are highly intertwined. Together, the volumes give a unique impression of research in Theoretical Computer Science as it is practised today. If not for reference purposes, we hope you will use these books to get a feeling for the theoretical work that is going on in the many fields of Computer Science or to simply satisfy your curiosity.

Each volume consists of close to twenty chapters, with each chapter devoted to a representative subject of current research. Volume A presents the basic material on models of computation, complexity theory, data structures and efficient computation in many recognized subdisciplines of Theoretical Computer Science. Volume B presents a choice of material on the theory of automata and rewriting systems, the foundations of modern programming languages, logics for program specification and verification, and a number of studies aimed at the theoretical modeling of advanced computing applications. Both volumes have been organized to reflect the development of Theoretical Computer Science from its classical roots to the modern complexity-theoretic and logic approaches to, for example, parallel and distributed computing. In most cases an extensive bibliography has been provided to assist in further study of the field and finding specific source documents for further reference. Specific attention has been given to a structured development of the material in the various chapters in order to make the Handbook largely self-contained. However, some familiarity with Computer Science at an undergraduate level is assumed.

The writing of this Handbook was started by an initiative of the publisher, who identified the need for a comprehensive and encyclopedic treatise on the fundamental areas of Theoretical Computer Science. The advisory board consisting of A.R. Meyer (Cambridge, MA), M. Nivat (Paris), M.S. Paterson (Coventry) and D. Perrin (Paris) was instrumental in defining the early version of the Handbook project. The complete project that has now resulted in the present volumes has grown far beyond the limits originally foreseen for the Handbook, and only because of the loyal support of all contributing colleagues to the project has it been possible to complete the gigantic task of presenting the current overview of the core areas of Theoretical Computer Science. I would like to thank all authors, the advisory board and the editorial staff at Elsevier Science Publishers for their invaluable support and contributions to the Handbook project.

J. van Leeuwen
Managing Editor
Utrecht, 1990

List of Contributors to Volume A

A.V. Aho, *AT&T Bell Laboratories, Murray Hill* (Ch. 5)
R.B. Boppana, *New York University* (Ch. 14)
Ph. Flajolet, *INRIA, Rocquencourt* (Ch. 9)
D.S. Johnson, *AT&T Bell Laboratories, Murray Hill* (Ch. 2)
R.M. Karp, *University of California, Berkeley* (Ch. 17)
Th. Lengauer, *Universität Paderborn* (Ch. 16)
A.K. Lenstra, *AT&T Bell Laboratories, Morristown* (Ch. 12)
H.W. Lenstra, Jr, *University of California, Berkeley* (Ch. 12)
M. Li, *University of Waterloo* (Ch. 4)
K. Mehlhorn, *Universität des Saarlandes, Saarbrücken* (Ch. 6)
N. Pippenger, *University of British Columbia, Vancouver* (Ch. 15)
V. Ramachandran, *University of Texas, Austin* (Ch. 17)
R. Rivest, *Massachusetts Institute of Technology, Cambridge, MA* (Ch. 13)
J. Seiferas, *University of Rochester* (Ch. 3)
J.T. Schwartz, *Courant Institute, New York* (Ch. 8)
M. Sharir, *Courant Institute, New York & Tel Aviv University* (Ch. 8)
M. Sipser, *Massachusetts Institute of Technology, Cambridge, MA* (Ch. 14)
V. Strassen, *Universität Konstanz* (Ch. 11)
A. Tsakalidis, *University of Patras* (Ch. 6)
L.G. Valiant, *Harvard University, Cambridge, MA* (Ch. 18)
P. van Emde Boas, *Universiteit van Amsterdam* (Ch. 1)
J. van Leeuwen, *Rijksuniversiteit Utrecht* (Ch. 10)
P.M.B. Vitányi, *CWI Amsterdam & Universiteit van Amsterdam* (Ch. 4)
J.S. Vitter, *Brown University, Providence, RI* (Ch. 9)
F.F. Yao, *Xerox PARC, Palo Alto, CA* (Ch. 7)

Contents

CHAPTER 1

Machine Models and Simulations

Peter van EMDE BOAS

Department of Mathematics and Computer Science, University of Amsterdam, Plantage Muidergracht 24, 1018 TV Amsterdam, Netherlands, and
Centre for Mathematics and Computer Science, Kruislaan 413, 1098 SJ Amsterdam, Netherlands

Contents

HANDBOOK OF THEORETICAL COMPUTER SCIENCE
Edited by J. van Leeuwen

1. Introduction

The study of computational complexity requires that one agrees on a model of computation, normally called a *machine model*, for effectuating algorithms. Unfortunately many different machine models have been proposed in the past, ranging from theoretical devices like the Turing machine to more or less realistic models of random access machines and parallel computers. Consequently there are many notions of computational complexity, depending on the machine model one likes to adopt, and there seem to be no uniform notions of *time* or *space* complexity. In this chapter we will explore the more fundamental aspects of complexity in machine models and address the question to what extent the machine-based complexity measures for the various models are related or different. We will identify the unifying principles that bind machine models together or set them apart, by criteria of efficiency for mutual simulations.

1.1. The invariance of computational complexity theory

If one wants to reason about complexity measures such as the time and space consumed by an algorithm, then one must specify precisely what notions of time and space are meant. The conventional notions of time and space complexity within theoretical computer science are based on the implementation of algorithms on abstract machines, called *machine models*. The study of complexity measures in real computers is not pursued in theory, because it would make the results dependent on existing technology and hardware peculiarities rather than on insight and mathematical experience. Rather, through suitable machine models one attempts to provide a reasonable approximation of what one might expect if a real computer were used for one's computations.

Still, even if we base complexity theory on abstract instead of concrete machines, the arbitrariness of the choice of a model remains. It is at this point that the notion of *simulation* enters. If we present mutual simulations between two models and give estimates for the time and space overheads incurred by performing these simulations, we show in fact that the time and space complexities in the two models do not differ by more than the given *overhead factors*. The size of the overhead factors tells us to what extent assertions about complexity are machine-based and to what extent they are machine-independent.

The existence of many different machine models does not necessarily mean that there are no uniform underlying notions of computational complexity. In the theory of computation a large variety of models and formal calculi has been proposed to capture the notion of effective computation. This has not lead to a proliferation of computation theories, due to the basic fact that the proposed formalisms have all been shown to be equivalent in the following sense: each computation in one formalism can be simulated by a computation in the other formalism. From the equivalence one concludes that if a problem is unsolvable in one particular model, then it is also unsolvable for all other formalized computing devices to which this particular model is related by mutual simulation. Consequently the notion of (un)solvability does not really depend on the model used. It has allowed us to return to an informal and intuitive style of working, relying on what has become known as the "inessential use" of CHURCH'S THESIS: *whatever is felt to be effectively computable can be brought within the scope of one of the formal models.* The basic models of computation themselves remain on the shelf, to be used if and when needed: we teach these models to our students and their discovery has become history [23, 59]. The validity of Church's Thesis is something which cannot be proven mathematically. At best the thesis could be invalidated by some convincing counterexample, but so far no counterexample has been discovered. All formalisms for computability have been found to satisfy the thesis. We will restrict therefore ourselves to models for which Church's Thesis holds.

When the theory of computational complexity was founded in the early nineteen-sixties [43] the study of machine models was revived. Models which had been shown to be computationally equivalent became different again, since they had different properties with respect to time and space complexity measures. Also a large collection of new models was introduced, based on the experience and developments in the world of practical computing. There no longer seemed to be a uniform relationship between the various models like there was at the level of general computability theory. Also, classical machine models like the unary Turing machine [22, 130] and Minsky's multicounter machine [81, 119] turned out to be too unwieldy for modeling realistic computations and were gradually discarded as viable machine models.

Depending on the objects one likes to manipulate in the computations (numbers like nonnegative integers, or alphanumeric strings), two models have obtained a dominant position in machine-based complexity theory:

- the *off-line multitape Turing machine* [1], which represents the standard model for string-oriented computation, and
- the *random access machine* (*RAM*) as introduced by Cook and Reckhow [21], which models the idealized Von Neumann style computer.

But other models were proposed as well and were accepted without any difficulty. At the same time, during the early nineteen-seventies, one became aware of the fact that some complexity notions seemed to have a *model-independent* meaning. The most important of these notions concerned the characterization of *computational feasibility* or *tractability*.

A problem is said to be *feasible* if it can be solved in polynomial time (as stated for the first time by Edmonds [26]). But this very notion seems to require that one identifies a machine model on which this time is measured. The crucial fact is now this: *if it can be shown that reasonable machines simulate each other within polynomial-time bounded overhead, it follows that the particular choice of a model in the definition of feasibility is irrelevant, as long as one remains within the realm of reasonable machine models.*

A fundamental problem, discovered in the early nineteen-seventies as well, is the $P \stackrel{?}{=} NP$ problem. Here P denotes the class of problems solvable by deterministic Turing machines in polynomial time, and NP denotes the class of problems solvable by *non*deterministic Turing machines in polynomial time. The $P \stackrel{?}{=} NP$ question asks for the power of nondeterminism: are there problems which a nondeterministic machine can solve in polynomial time, which cannot be solved in polynomial time by a reasonable deterministic machine? Stated differently: are nondeterministic machines feasible for practical computing?

The fundamental complexity classes P and NP became part of a *fundamental hierarchy:* LOGSPACE, NLOGSPACE, P, NP, PSPACE, EXPTIME, ... And again theory faced the problem that each of these classes has a *machine-dependent* definition, and that efficient simulations are needed before one can claim that these classes are in fact *machine-independent* and represent fundamental concepts of computational complexity. It seems therefore that complexity theory, as we know it today, is based on the following assumption.

INVARIANCE THESIS: *"Reasonable" machines can simulate each other within a polynomially bounded overhead in time and a constant-factor overhead in space.*

A further development arose from the study of parallelism. It turned out that there was a deep connection between "time" in parallel machines and "space" in sequential models. This relation is known as the Parallel Computation Thesis.

PARALLEL COMPUTATION THESIS: *Whatever can be solved in polynomially bounded space on a reasonable sequential machine model can be solved in polynomially bounded time on a reasonable parallel machine, and vice versa.*

This chapter deals with the evidence that exists in support of the validity of the above two theses. The study of simulations with polynomial-time bounded overhead is a traditional subject in complexity theory (see [1, 75, 138]). The additional requirement of constant-factor overhead in space complexity seems to have been largely ignored. However, as we will see in Section 2, this problem is not as easy as it may seem. The Parallel Computation Thesis has received quite some attention, after it was first formulated by Goldschlager [39].

The Invariance Thesis can be interpreted in two ways:

- the *orthodox interpretation* requires that a single simulation achieves both bounds on the overheads involved simultaneously;
- the *liberal interpretation* allows for a time-efficient simulation and an entirely different space-efficient simulation (which may turn out to require a much larger, e.g. exponential, overhead in time).

We believe that the orthodox interpretation is the correct one to adhere to for computational complexity theory. Clearly, if one investigates space or time bounded complexity classes independently, the liberal interpretation suffices in order to make the classes in the fundamental hierarchy machine-independent. Almost all simulations one finds in the literature achieve both overheads at the same time. Also it is useful to study classes defined by simultaneous time and space bounds, and the orthodox interpretation is needed in order to make these multiple-resource bounded classes machine-invariant. Stockmeyer [126] has argued that one reason for using the Turing machine model in the foundation of complexity theory is precisely this invariance that it provides.

The escape in defending the Invariance Thesis and the Parallel Computation Thesis is clearly present in the word *reasonable*. This is easily inferred from the standard practice of the theoretical computer science community when it is faced with evidence against the validity of the Invariance Thesis. For example, when in 1974 it was found that a RAM model with unit-time multiplication and addition instructions (together with bitwise Boolean operations) is as powerful as a parallel machine, this model (the MBRAM) was thrown out of the realm of reasonable (sequential) machines and was considered to be a "parallel machine" instead. The standard strategy seems to be to adjust the definition of "reasonable" when needed. The theses become a guiding rule for specifying the right classes of models rather than absolute truths and, once accepted, the theses will never be invalidated. This strategy is made explicit if we replace the word *reasonable* by some more neutral phrase. We have proposed for this purpose the use of the notions of *a machine class* [131, 132]:

- the *first machine class* consists of those sequential models which satisfy the Invariance Thesis with respect to the traditional Turing machine model;
- the *second machine class* consists of those (parallel or sequential) devices which satisfy the Parallel Computation Thesis with respect to the traditional, sequential Turing machine model.

Not all machine models encountered in the literature belong to one of these two machine classes. The classical mathematical machine models like the unary Turing machine and the multicounter machine are too weak to keep pace with the members of the first machine class. There are intermediate models which could be located in between the two classes, and there are parallel models which are even more powerful than the members of the second machine class, up to the models which can compute everything in constant time. But so far we have not found reasons for introducing a *third* machine class at this level.

In the course of this chapter various machine models will be introduced and their computational power will be explained and clarified. We will do so in an informal manner. The primary focus will be on understanding the fundamental relationships between the complexities in the various machine models. For an impression of the complete field of machine models and related subjects, the reader is referred to the book of Wagner and Wechsung [138].

1.2. Formalization of machine models

In this section we summarize the common ingredients found in many machine models. Any machine model can be obtained by choosing the appropriate interpretation of the basic components. We also introduce the basic concepts of simulation and complexity.

1.2.1. Machines and computations

A *machine model M* is a class of similarly structured devices M_i $(i \geqslant 0)$ called *machines*, which can be described as mathematical objects in the language of set theory. It is common to define these objects as tuples of finite sets. For example, the standard single-tape Turing machine can be defined as a seven-tuple consisting of three finite sets, one finite relation and three elements in these sets (cf. Section 2.1).

The set-theoretical object provides only partial information on how the machine will behave and what its computations will look like. Common in these definitions is the presence of a finite object, called *program* or *finite control*, which is to operate on a structure called a *memory*, which can be unbounded. Sometimes "finite control" refers to some notion of a *processor* for the execution of the program. At any moment in time, finite control is in a particular *state*, and the number of different states that are possible is finite. Memory could just be a repository for finite chunks of information composed of symbols from one or more finite sets called *alphabets*; the alphabets are specified in the tuple which defines the machine. In another possible structure, memory is filled with finite sets of numbers, usually taken from the set of nonnegative integers ω.

Memory is always modeled as a regular structure of *cells*, where the information is stored. The memory structures are usually linear and discrete, but other memory structures like higher-dimensional grids and trees also occur. Although the memory of a machine model is infinite in principle, only a finite part of the memory is *used* in any realistic computation. Formally this is achieved by assuming that a special value (a *blank symbol* or the number 0) is stored in those locations where nothing has happened so far.

Programs are written in some type of "programming language". Typically there will be instructions for fetching information from or storing information in memory, for modifying the information in reachable cells in memory, for testing of conditions and performing conditional or unconditional jumps in the program, and for performing input and output. In the Turing machine model all possible types of instructions have been fused into a single read-test-write-move-goto instruction; for models like the RAM the instruction set resembles a primitive assembly code.

For the purpose of input and output, two special sections of memory are distinguished: the *input section* from which information is read one symbol at a time, and the *output section* through which information is communicated to the outside world, again one symbol at a time. The program communicates with the input and output sections through explicit *channels* that can scan the sections in some programmed fashion. For the input section and its use two interpretations offer themselves:

- *On-line computation*: the input characters are regarded as entities which offer themselves for processing only once, like signals received from an external source. If one wants to read an input symbol for a second time, one must store it in some (other) part of the memory.
- *Off-line computation*: the input symbols are permanently available for inspection and can be read as often as one likes (possibly in another order than the sequential order of the input).

Some models, like the single-tape Turing machine, have no separate input or output sections at all; in this case the input is stored in memory at the start of the computation and the result must be retrieved from memory when the computation has ended.

In order to formalize the notion of computation we need the concept of a *configuration* of the machine. A configuration is a full description of the state of a machine, including e.g. the complete state of its memory. Typically a configuration consists of a pointer to the "current instruction" in the program, the structure and contents of the finite part of the memory that has participated in the computation so far, and the states of the input and output sections of memory. Moreover, for all components of memory the configuration must indicate on which cells the interaction between finite control and memory is currently taking place (the state of the channels).

Next one must define the *transition relation* between two configurations of the machine. Configuration C_2 is obtained by a transition from configuration C_1, notation: $C_1 \vdash C_2$, if the instruction of the program that is imminent in C_1 takes the machine from C_1 to C_2. Computations can now be described by taking the reflexive, transitive closure $\vdash^*$ of this relation $\vdash$. The word *computation* refers primarily to sequences of configurations, connected by $\vdash$, which establish the presence of a pair consisting of two configurations in $\vdash^*$. It can also refer to the pairs in $\vdash^*$ themselves.

At this point the distinction between *deterministic* and *nondeterministic* machines must be made. For a deterministic machine there exists for every configuration C_1 at most one configuration C_2 such that $C_1 \vdash C_2$, whereas in a nondeterministic machine there may exist several such configurations C_2. In virtually all machine models that exist, the program itself is the source of the nondeterminism. The program typically indicates that more than one instruction can be performed in the given configuration, which leads to *bounded nondeterminism*: the number of possible configurations C_2 that follow from C_1 is finite and bounded by a number dependent on the machine's program only. *Unbounded nondeterminism* is obtained if also the interaction with the memory can be nondeterministic. For example, in a RAM one could consider an instruction which loads a random integer in a memory cell. In this chapter we will restrict ourselves to bounded nondeterminism.

Given the transition relation, one must define what is meant by *initial, final, accepting* and *rejecting* configurations next. *Initial* configurations consist of an initial state in the finite control (indicated in the tuple describing the machine), a memory which is blank except possibly for the input section, an empty output section, and all channels connecting program and memory are initialized to some standard position. An input x thus completely determines the initial configuration for machine M_i, denoted by $C_i(x)$. Every configuration that can be obtained by some computation from an initial configuration is called *reachable*. A configuration C is called *final* if there exists no configuration C' such that $C \vdash C'$. If the state in the finite control in a final configuration equals a specific accepting state—designated as such in the tuple describing the machine—then the final configuration is called *accepting*, otherwise it is called *rejecting*. There exist other ways of defining configurations to be accepting or rejecting: for example, sometimes it is required that in an accepting configuration, memory is reset to some "clean" situation.

A *full computation* is a computation which starts in an initial configuration and which does not end in a nonfinal one. Thus, a full computation is either infinite, in which case it is called a *divergent computation*, or it *terminates* in a final configuration. A terminating computation is called *accepting* or *rejecting* depending on whether its final configuration is accepting or rejecting, respectively.

Machine computations can be used for several purposes:

- Machine M_i *accepts the language* $D(M_i)$ consisting of those inputs x for which a terminating computation starting with $C_i(x)$ exists.
- Machine M_i *recognizes* the language $L(M_i)$ consisting of those inputs x for which an accepting computation starting with $C_i(x)$ can be constructed. Moreover all computations of the machine M_i must be terminating.

- Machine M_i *computes* the relation $R(M_i)$ consisting of those pairs $\langle x, y\rangle$ of inputs x and outputs y such that there exists an accepting computation which starts in $C_i(x)$ and which terminates in a configuration where y denotes the contents of the output memory.

A nondeterministic machine model can also be used as a *probabilistic machine*. In this case inputs are considered as "recognized" or "accepted" only if more than 50% of all possible computations (rather than just "at least one") on the input are terminating or accepting, respectively.

In this formalization there exists only one finite control—or program—which is connected to the memory by possibly more than one channel. As a consequence the machine will execute only one instruction at a time (even in the case of a nondeterministic machine). We emphasize this feature by calling the given type of machine model a *sequential machine model*. A *parallel machine model* is obtained if we drop the condition that there exists just one finite control. In a parallel machine finite control is replaced by a set of programs or *processors* which may be infinite in the same way as memory can be infinite: the number of processors has no fixed bound, but in every reachable configuration only a finite number of them will have been activated in the computation. In some models large amounts of memory are accessible to all processors (*shared memory*), in addition to the own *local memory* that the processors have. Each processor has its private channels for interacting with memory. Processors can communicate either directly (e.g. through some *interconnection network*) or through shared memory (or both).

The condition that only one symbol at a time is communicated between the in- or output section of memory and the finite control is also relaxed in parallel machine models. In parallel models the processors can exchange information directly by communication channels without the use of intermediate passive memory. In an extreme case like the cellular automata model [128] there are only processors, and their finite controls have taken over the role of the memory entirely.

Transitions in a parallel machine are the combined result of transitions within each of the active processors. In one formalization a global transition is obtained as the cumulative effect of all local transitions performed by the active processors which operate at the same time (*synchronous computation*). An alternative formalization is obtained if the processors proceed at their own speed in some unrelated way (*asynchronous computation*). In both cases the processors can interfere with each other while acting on the shared memory; this holds in particular when one processor wants to write at a location where another processor wants to read or write at the same time. It depends on the precise definition of a parallel model to stipulate what must happen when read/write conflicts arise, if these conflicts may occur at all during a legal computation.

1.2.2. Simulations

Intuitively, a simulation of M by M' is some construction which shows that everything a machine $M_i \in M$ can do on inputs x can be performed by some machine $M'_i \in M'$ on the same inputs as well. The evidence that the behavior on inputs x is preserved should be retrievable from the computations themselves rather than from the input/output relations established by the devices M_i and M'_i.

It is difficult to provide a more specific formal definition of the notion of simulation which is sufficiently general at the same time; for every definition one is likely to find examples of "simulations" which are not covered by it. A first problem is that M' may not be able to process the input x at all, e.g. because the structure of the objects which can be stored in the input section of the memories of the two machine models may be quite different. If M_i is a Turing machine which operates on strings and M'_i is a RAM operating on numbers, then the sets of possible inputs for the two machines are disjoint. In order to overcome this problem, one allows that M' operates on a suitable "standard" encoding of the input x rather than on x itself. The encoding has to be of a simple and effective nature, in order to prevent spurious interpretations which would allow for the recognition of nonrecursive sets using simulations.

One could introduce a relation between configurations C_1 of M_i and C_2 of M'_i to express that C_2 "represents" or simulates the configuration C_1. Denote such a relation by $C_1 \approx C_2$. For a simulation one could now require a condition like

$$\forall C_1, C_2, C_3((C_1 \vdash C_3 \wedge C_1 \approx C_2) \Rightarrow \exists C_4[(C_2 \vdash C_4 \wedge C_3 \approx C_4)])$$

or the weaker condition

$$\forall C_1, C_2, C_3((C_1 \vdash C_3 \wedge C_1 \approx C_2) \Rightarrow \exists C_4[(C_2 \vdash^* C_4 \wedge C_3 \approx C_4)]).$$

These conditions are far too restrictive. For example, the first condition implies that the running time of a computation is preserved by a simulation. The second condition requires that each configuration of M_i has its analogue in the computation of M'_i and that the general order of the computation by M_i is preserved by the simulation.

A next attempt is to consider an entire computation of M_i and require that for each configuration C_1 occurring in its computation there exists a "corresponding" configuration C_2 in the simulating computation of M'_i. This extension allows e.g. recursive simulations like [104] or simulations where the computation runs in reverse order [120], but still would exclude simulations in which the entire transition relation of M_i becomes an object on which the computation of M'_i operates, and the effect of M_i *on some input* x is simulated by letting M'_i evaluate the reflexive transitive closure of this transition relation. As we will see in Sections 3 and 4, this type of simulation is quite common in the theory of parallel machine models.

The above attempts show how hard it is to define simulation as a mathematical object and still remain sufficiently general. The initial definition seems to be the best one can provide, and we leave it to the reader to develop a feeling for what simulations can do by looking at the examples later in this chapter and elsewhere in the literature.

1.2.3. Complexity measures and complexity classes

Given the intuitive description of a computation, it is easy to define the time measure for a terminating computation: simply take the number of configurations in the sequence which describes the computation. What we do here implicitly is to assign one unit of time to every transition, the *uniform time* or *cost measure*. This measure applies both to sequential and parallel models.

In models where a single transition may involve the manipulation of objects from an infinite domain (like in the case of a RAM) this measure may be unrealistic, and one often uses a refined measure, in which each transition is weighted according to the amount of information manipulated in the transition. For example, in the *logarithmic time measure* for the RAM an instruction is charged an amount of time proportional to the number of bits involved in the execution of the instruction.

A crude measure for the *space* consumed by a computation is the number of cells in memory which have been affected by the computation. A more refined measure also takes into account how much information is stored in every cell. Moreover, it is tradition not to charge for the memory used by the input and the output, in case the corresponding sections of memory are clearly separated from the rest of the memory. This convention is required in order to distinguish e.g. functions which can be computed in sublinear space. However, in structural complexity theory [3] the alternative, in which output is charged, is used as well; it leads to a theory which differs from the one obtained in the usual interpretation.

Having defined the notions of time and space for each individual computation, we can define these measures for the machine as a whole. The way these measures are defined depends on what the machine is intended to do with the input: recognize, accept or compute. Moreover, one is usually interested not so much in the time and space requirement for specific inputs x, but in the global relation between the time and space measure of a computation and the length of an input x, denoted by $|x|$. The length $|x|$ of the input depends on the encoding of x for the particular machine at hand. Let $f(n)$ be a function from the set of nonnegative integers ω to itself.

1.1. DEFINITION. Machine M_i is said to terminate in time $f(n)$ (space $f(n)$) in case every computation on an input x of length n consumes time (space) $\leqslant f(n)$. A deterministic machine M_i accepts in time $f(n)$ (space $f(n)$) if every terminating computation on an input of length n consumes time (space) $\leqslant f(n)$. A nondeterministic machine M_i accepts in time $f(n)$ (space $f(n)$) if for every accepted input of length n there exists a terminating computation which consumes time (space) $\leqslant f(n)$.

1.2. DEFINITION. A language L is recognized in time $f(n)$ (space $f(n)$) by machine M_i if $L = L(M_i)$ and M_i terminates in time $f(n)$ (space $f(n)$). A language L is accepted in time $f(n)$ (space $f(n)$) by machine M_i if $L = D(M_i)$ and M_i accepts in time $f(n)$ (space $f(n)$).

One could also define that L is recognized by M_i in time $f(n)$ (space $f(n)$) if $L = L(M_i)$ and M_i *accepts* in time $f(n)$ (space $f(n)$). This definition turns out to be equivalent to the one given, for nice models and nice functions $f(n)$ for which it is possible to shut off computations which exceed the amount of resources allowed. For such resource bounds in fact the notions of accepting and recognizing become equivalent. In this chapter we will restrict ourselves to these nice resource bounds. In the general case, however, the three definitions are different.

We can now define the important machine-dependent *complexity classes*.

1.3. Definition. The set of languages recognized by a machine model M in time $f(n)$ (space $f(n)$) consists of all languages L which are recognized in time $f(n)$ (space $f(n)$) by some member M_i of M. We denote this class by M-TIME($f(n)$) (M-SPACE($f(n)$)). The class of languages recognized simultaneously in time $f(n)$ and space $g(n)$ will be denoted M-TIME&SPACE($f(n), g(n)$). The intersection of the classes M-TIME($f(n)$) and M-SPACE($g(n)$) consisting of the languages recognized both in time $f(n)$ and space $g(n)$ will be denoted M-TIME,SPACE($f(n), g(n)$).

We will especially look at the following classes $\mathscr{F}$ of resource-bounding functions:

$$Log = \{k \cdot \log n \mid k \in \omega\}, \qquad Lin = \{k \cdot n \mid k \in \omega\}, \qquad Poly = \{k \cdot n^k + k \mid k \in \omega\},$$
$$Expl = \{k \cdot 2^{k \cdot n} \mid k \in \omega\}, \qquad Exp = \{k \cdot 2^{n^k} \mid k \in \omega\}.$$

A set L is said to be recognized (accepted) in time (space) $\mathscr{F}$ if it is recognized (accepted) in time (space) f for some $f \in \mathscr{F}$. For any specific machine model M this leads to the following basic complexity classes.

1.4. Definition

M-LOGSPACE $= \{L \mid L$ is recognized by $M_i \in M$ in space $Log\}$,
M-PTIME $= \{L \mid L$ is recognized by $M_i \in M$ in time $Poly\}$,
M-PSPACE $= \{L \mid L$ is recognized by $M_i \in M$ in space $Poly\}$,
M-EXPLTIME $= \{L \mid L$ is recognized by $M_i \in M$ in time $Expl\}$,
M-EXPTIME $= \{L \mid L$ is recognized by $M_i \in M$ in time $Exp\}$,
M-EXPLSPACE $= \{L \mid L$ is recognized by $M_i \in M$ in space $Expl\}$,
M-EXPSPACE $= \{L \mid L$ is recognized by $M_i \in M$ in space $Exp\}$.

The above classes clearly are machine-dependent. It is possible, however, to obtain the hierarchy of fundamental complexity classes as indicated in Section 1.1 by selecting for M the standard model of off-line multitape Turing machines (cf. [1]). Denoting deterministic Turing machines by T and nondeterministic Turing machines by NT we obtain the following definition.

1.5. Definition

LOGSPACE = T-LOGSPACE,	NLOGSPACE = T-NLOGSPACE,
P = T-PTIME,	NP = T-NPTIME,
PSPACE = T-PSPACE,	NPSPACE = T-NPSPACE,
EXPTIME = T-EXPTIME,	NEXPTIME = T-NEXPTIME,
EXPSPACE = T-EXPSPACE,	NEXPSPACE = T-NEXPSPACE.

Here we use a notation where the resource is prefixed by N rather than the machine model itself. So for example T-NPTIME and NT-PTIME denote the same class.

From the elementary properties of the Turing machine model we obtain the hierarchy

$$\begin{aligned}\text{LOGSPACE} \subseteq \text{NLOGSPACE} \subseteq \text{P} \subseteq \text{NP} \subseteq \text{PSPACE} = \text{NPSPACE}\\ \subseteq \text{EXPTIME} \subseteq \text{NEXPTIME} \subseteq \text{EXPSPACE}\\ = \text{NEXPSPACE}\ldots\end{aligned}$$

Moreover, all fundamental complexity classes mentioned above have obtained their standard meaning. Note the notational distinction between the classes described by exponential bounds with linear functions and polynomials in the exponent, and compare the definitions used in [56]. The strictness question for every inclusion shown represents a major open problem in complexity theory, except for the equalities PSPACE = NPSPACE and EXPSPACE = NEXPSPACE which follow from Savitch's Theorem [104]. The $\text{P} \stackrel{?}{=} \text{NP}$ problem is perhaps the most notorious of these open problems.

1.2.4. Simulation overheads and the machine classes

We now return to the quantitative aspects of simulations. The unanswered question from Section 1.2.2 of what is formally meant by a simulation remains open, but we can now at least quantify the overhead costs of simulations as follows.

1.6. DEFINITION. For machine models M and M' we say that M' simulates M with time (space) overhead $f(n)$ if there exists a function s with the following property: for every machine M_i in M the machine $M'_{s(i)}$ in M' simulates M_i, and for every input x of M if $c(x)$ is the encoded input of $M'_{s(i)}$ which represents x and if $t(x)$ is the time (space) bound needed by M_i for processing x, then the time (space) required by $M'_{s(i)}$ for processing $c(x)$ is bounded by $f(t(x))$.

The fact that a model M' simulates a model M with time or space overhead $f(n)$ will be denoted by $M \leqslant M'$ (*time* $f(n)$) or $M \leqslant M'$ (*space* $f(n)$) respectively.

In this definition, processing stands for either recognizing, accepting, rejecting the input, or evaluating some (possibly partial and/or multivalued) function of the input. The encoding $c(x)$ must be recursive and of a reasonably low complexity. If $s(i)$ is recursive as well, the simulation is said to be *effective*. In abstract complexity theory [10], it is shown that simulations can be made effective and that recursive overheads always exist.

For classes of functions $\mathcal{F}$, we say that M' simulates M with time (space) overhead $\mathcal{F}$ in case the simulation holds for some overhead $f \in \mathcal{F}$. This will be denoted by $M \leqslant M'$ (*time* $\mathcal{F}$)) or $M \leqslant M'$ (*space* $\mathcal{F}$).

This enables us to identify some important classes of simulations:

(1) $M \leqslant M'$ (*time Poly*): *polynomial-time simulation*,

(2) $M \leqslant M'$ (*time Lin*): *linear-time simulation*,

(3) $M \leqslant M'$ (*space Lin*): *constant-factor space overhead simulation.*

Special cases of linear-time simulation are the so-called *real-time* and *constant-delay* simulation. These simulations behave according to one of the attempted definitions discussed in Section 1.2.2, in which the configurations of M are simulated by corresponding configurations of M' in an order-preserving way; moreover, the number of configurations of M' in the simulation of two successive configurations of M is bounded by a constant in the case of a constant-delay simulation; for the real-time case this constant equals 1. For machine models like the Turing machine, which have a constant-factor speed-up property, it is possible to transform any constant-delay simulation into a real-time simulation. We denote the existence of a real-time simulation by $M \leqslant M'$ (*real-time*).

If a single simulation achieves both a time overhead $f(n)$ and a space overhead $g(n)$ we denote this by $M \leqslant M'$ (*time* $f(n)$ & *space* $g(n)$).

If both overheads can be achieved but not necessarily in the same simulation, we express this by the notation $M \leqslant M'$ (*time* $f(n)$, *space* $g(n)$).

Again these notations are extended to classes of functions where needed. If simulations exist in both directions, achieving similar overheads, then we replace the symbol $\leqslant$ by the equivalence symbol $\approx$. For example, $M \approx M'$ (*time Poly*) expresses the fact that the machine models M and M' simulate each other within polynomial-time bounded overhead.

The given notions enable us to define the *first machine class* and the *second machine class*.

1.7. Definition. Let T be the Turing machine model and let M be an arbitrary machine model. Then M is a *first class machine model* if $\mathrm{T} \approx M$ (*time Poly* & *space Lin*). We say that M is a *second class machine model* if M-PTIME $= M$-NPTIME $=$ PSPACE.

Hence, first class machine models are exactly those models which satisfy the Invariance Thesis under the orthodox interpretation, whereas second class machines are those machines which satisfy the Parallel Computation Thesis both for their deterministic and nondeterministic version. In case a machine model does not have a nondeterministic version the latter condition is omitted.

The first machine class represents the class of reasonable sequential models. However, for the machine models in the second machine class it is not clear at all that they can be considered to be reasonable. It seems that the marvelous speed-ups obtained by the parallel models of the second machine class require severe violations of basic laws of nature [136, 137]. Stated differently: if physical constraints are taken into account, all gains of parallelism seem to be lost [17, 117]. Hence, being a "natural" model or not can never be a criterion for selecting a particular class of parallel machines. The models are of interest rather because of the surprising implications that certain computational assumptions of parallelism or some extensions of the arithmetic power in the sequential case have in the standard models.

1.2.5. Reductions and completeness

We said that one machine model M' simulates another model M as soon as M'-machines can do everything that M-machines can do. We stipulated that this property should be inferred from the behavior of the machines and not just from the input-output relationship. If this condition is omitted, we obtain a notion which is crucial in complexity theory as well: the notion of a *reduction*. Reductions are a valuable tool for establishing the existence of simulations. In this section we will look at this technique in detail.

1.8. DEFINITION. Let A and B be subsets of Σ^* and T^* respectively, and let f be some function from Σ^* to T^*. We say that *f reduces A to B*, notation $A \leqslant_m B$ *by* f, in case $f^{-1}(B) = A$.

The above notion of a reduction is the well-known *many-one reduction* from recursive function theory. There exist many types of reductions (see e.g. [138]) but for our purposes this notion suffices. In complexity theory it is usually required that f can be evaluated in polynomial time (with respect to the length of the input). This is expressed by the notation $A \leqslant_m^P B$ and we speak of a *polynomial-time reduction* or *p-reduction*. If f satisfies the even stronger condition that it can be evaluated in logarithmic space, then we speak of a *logspace reduction*.

In recursion theory the reduction $A \leqslant_m B$ establishes that A is recursive provided B is and, consequently, if A is undecidable then so is B. In complexity theory a reduction $A \leqslant_m^P B$ establishes the fact that A is easy in case B is, and consequently that B is hard provided A is hard.

One can prove that almost all fundamental complexity classes in the hierarchy

$$\begin{aligned} \text{LOGSPACE} &\subseteq \text{NLOGSPACE} \subseteq \text{P} \subseteq \text{NP} \subseteq \text{PSPACE} = \text{NPSPACE} \\ &\subseteq \text{EXPTIME} \subseteq \text{NEXPTIME} \subseteq \text{EXPSPACE} = \\ &= \text{NEXPSPACE} \ldots \end{aligned}$$

are closed under polynomial-time reductions. Thus we have, for example, that for all languages A and B: $(B \in \text{P} \wedge A \leqslant_m^P B) \Rightarrow A \in \text{P}$. Exceptions are the classes EXPLTIME and NEXPLTIME which are not closed under polynomial-time reductions; this is due to the fact that the composition of 2^n with a polynomial $p(n)$ yields a function $2^{p(n)}$ which is a member of *Exp* rather than *Expl*. At the level of LOGSPACE and NLOGSPACE one needs the stricter logspace reductions.

1.9. DEFINITION. Let B be some set in Σ^* and $\mathscr{X}$ a family of sets. B is called $\leqslant_m^P$*-hard for* $\mathscr{X}$ (or simply *$\mathscr{X}$-hard*) if $A \leqslant_m^P B$ for every $A \in \mathscr{X}$. B is called $\leqslant_m^P$*-complete* for $\mathscr{X}$ (or *$\mathscr{X}$-complete*) if $B \in \mathscr{X}$ and B is $\mathscr{X}$-hard.

The best-known instances of these notions are the concepts of NP-completeness and PSPACE-completeness. Cook [18] proved in 1971 that NP-complete problems exist, by establishing the NP-completeness of the set SATISFIABILITY of all Boolean formulae for which a satisfying truth-value assignment exists. (The similar set in which the Boolean formulae are required to be in conjunctive normal form is NP-complete as well.) The set MAJORITY, consisting of all Boolean formulae for which more than half of all possible truth-value assignments are satisfying, is complete for the class of languages accepted by probabilistic polynomial-time bounded machines. The existence of PSPACE-complete problems follows from [124]. Nowadays many NP-complete and PSPACE-complete problems are known [56].

Problems that are hard or complete for a machine-based family of sets $\mathscr{X}$ somehow encode the computational behavior of all machines for $\mathscr{X}$. This implies that some machine model M' that can solve instances of a complete problem in polynomial time can simulate all machines in the corresponding family of machines. In order to understand this we observe that the existence of complete problems for the fundamental classes in the fundamental hierarchy is easy to obtain. For example, the *universal* language $L_{\mathrm{NP}}=\{\langle M_i, x, w\rangle \mid$ nondeterministic Turing machine M_i accepts x in time $\leqslant |w|\}$ is NP-complete. If we replace time by space and consider deterministic in stead of nondeterministic machines we similarly obtain a universal language L_{PSPACE} that is PSPACE-complete.

The universal languages can often be encoded by so-called *master reductions* into combinatorial problems. For example, in the proof of Cook's Theorem [18] one constructs for every triple $\langle M_i, x, w\rangle$ a propositional formula $\Phi(M_i, x, w)$ such that satisfying truth-value assignments for its propositional variables correspond to accepting computations of $M_i(x)$ of length $\leqslant |w|$. Other master reductions are based on e.g. the use of tiling problems [65, 103] (a construction that works for both NP and PSPACE). See also Reif's ingenious construction for PSPACE-bounded computations [96].

Virtually all reductions which are used in proofs establishing the hardness or completeness of specific problems have the property that they are *parsimonious*: they transform not only solvable instances into solvable instances but also solutions into solutions. The property of master reductions that solutions encode accepting computations is therefore preserved under parsimonious reductions. We note that many complete problems have the property that one can efficiently compute a solution as soon as one can efficiently establish the existence of solutions.

All these specific properties of the traditional complete problems together yield the observation that *solving a complete problem for a machine model M by means of a machine model M′ establishes the existence of simulations*. From deciding the solvability of some combinatorial encoding of a machine computation, one first computes the solution to the combinatorial problem and subsequently obtains the original computation by decoding this solution. Thus we obtain for example that if M solves a PSPACE-complete problem in polynomial time, then all polynomial-space bounded machines can be simulated by M with polynomial-time overhead. This observation will be used frequently in Sections 3 and 4. The resulting proofs are also much shorter than the proofs by direct simulation of the machine models involved.

2. Sequential machine models

2.1. *Turing machines*

Turing machines have established themselves as the standard machine model in contemporary complexity theory. Only in the related area of analysis of algorithms this role is taken over by the RAM model. In this section we describe the basic version of the Turing machine model. Additional features could be added that deal e.g. with the structure, dimension and number of tapes, the numbers of heads on a tape, and restrictions on the use of tapes. Every selection of some particular set of features leads to another version of the Turing machine model.

This leads to the following important question that we address in this section: *Is the Invariance Thesis true for all Turing machine models?*

2.1.1. Description of the Turing machine models

In the Turing machine model, the memory consists of a finite collection of tapes divided into tape cells. Each cell is capable of storing a single symbol from the worktape alphabet. For each tape there exist one or more channels (*read-write heads*) connecting the finite control to the tape. Each head is positioned at some tape cell. Several heads may be located at the same cell.

In the program of the Turing machine, various types of instruction have been integrated into a single type: the machine reads the ordered set of symbols from the worktapes which are currently scanned by the heads. Depending on these symbols and the internal state of the finite control, the appropriate instruction will cause the machine to overwrite some or all of the scanned symbols by new ones, move some or all of the heads to an adjacent cell on the corresponding tape and enter a new state in the finite control.

The simplest version is the single-tape Turing machine which can be described by a tuple $M = \langle K, \Sigma, P, q_0, q_f, b, \Delta \rangle$, in which

- K is a finite set of internal states;
- Σ is the tape alphabet;
- P is the program stored in the finite control, consisting of a finite set of quintuples $\langle q, s, q', s', m \rangle \in K \times \Sigma \times K \times \Sigma \times \Delta$; here $\Delta = \{L, 0, R\}$ is the set of possible head moves: Left, No Move or Right; the meaning of a quintuple $\langle q, s, q', s', m \rangle$ is: if the finite control is in state q and the head is scanning symbol s, then write symbol s', perform move m and let finite control enter state q';
- q_0 and q_f are two special elements in K denoting the initial and the final state, respectively;
- b is a special tape symbol called *blank* which represents the contents of a tape cell which has never been scanned by the head.

In the single-tape model there is no special input or output tape. In the initial configuration the input is written on the one available tape and the unique head is positioned over the leftmost input symbol. If one wants the model to produce output, one normally adopts an ad hoc convention for extracting the output from the final configuration. (For example, the output may consist of all nonblank tape symbols written to the left of the head in the final configuration).

If we denote configurations of M in the format $\$\Sigma^*K\Sigma^*\$$, with the state symbol written in front of the currently scanned tape symbol, the transitions between two successive configurations can be described by a simple *context-sensitive grammar*. For example, for every instruction $\langle q, s, q', s', R \rangle$ one includes the production rules $(qsa, s'q'a)$ for every $a \in \Sigma$, together with the rule $(qs\$, s'q'b\$)$ for the blank symbol b. Similar rules encode the behavior of left-moving instructions or instructions where the head does not move. This transformation of the program into a grammar establishes a one-one correspondence between computations of the machine and derivations in the grammar. This syntactic representation of computations opens the way to encode Turing machine computations in a large variety of combinatorial structures, like tilings

[42, 65, 103] and regular expressions [124], leading to a large collection of master reductions in computation theory. The single-tape model is very popular for its use in reductions and other simulations.

In the more complicated Turing machine models, Σ is replaced by a power Σ^k where k denotes the number of tapes. The set of moves becomes also a k-fold Cartesian product where, moreover, each set of moves is adjusted to the dimension of the tape on which the corresponding head moves. In case several heads are moving on a single tape, a new coordinate must be added in order to let the heads "feel" that they are positioned at the same square.

The components of an instruction can always be split in an observation part and an action part: in the above case the first state and symbol represent the observation and the rest represents the action. If a single observation can lead to at most one action, then the machine is said to be *deterministic*, otherwise it is *nondeterministic*. Nondeterminism of Turing machines is bounded by definition since the set of possible actions is bounded by the size of the program, which is finite.

Since it serves no purpose to read left of the first input symbol on a read-only input tape and since the output tape (if present) is one-way, it can be presumed that in- and output tapes are semi-infinite; their cells are indexed by the nonnegative integers. The worktapes can be taken to be semi-infinite as well. There exist several restricted types of worktapes which have obtained special names:

- A *stack* is a semi-infinite worktape with the special property that, whenever the head makes a left move, the previous contents of the cell are erased (*popping the stack*). After a write instruction the head can move right (*pushing a symbol on the stack*). The head cannot move left of the origin but it can feel the beginning (the *bottom*) of the stack and in this way test whether the stack is empty.
- A *queue* is a semi-infinite tape with two right-only heads. The first head is a write-only head, whereas the second one is a read-only head. The second head therefore can read (only once) everything the first head has written before. Also a queue can be tested for emptyness.
- A *counter* is a stack with a single-letter alphabet. Its purpose is to count a number n by storing a string of n copies of the one tape symbol; the counter can test whether the number equals zero by checking whether the stack is empty. The push and the pop move of a counter correspond to incrementing and decrementing the integer represented in the counter.

For higher-dimensional tapes there are several possible choices for the set of possible head moves. It makes a difference whether the heads can move only along the edges in the grid or whether they can proceed in diagonal steps as well. Other moves which have been considered are *fast rewinds* (a head proceeds in one step to its original position) and *head-to-head jumps* (one head moving in one step to the position of another).

2.1.2. Simulation overheads

In this section we discuss the major results on simulations of enhanced Turing machines on simpler ones. We use the notation $\leqslant$ introduced in Definition 1.8 for relating machine models. We omit in this list the trivial real-time constant-factor space simulations which exist between a Turing machine model and a more enhanced version: adding new features will never increase time or space consumption as long as the additional features are not used. In quoting the results we will refer to a Turing machine model by just specifying its relevant features.

2.1. THEOREM. *The following extensions of the Turing machine model can be simulated by more elementary features with the following overheads:*

(1) *1-tape* $\leqslant$ *2-stacks* (*real-time & space Lin*),
(2) *1-stack* $\leqslant$ *2-counters* (*time Expl & space Expl*),
(3) *1-tape* $\leqslant$ *2-counters* (*time* $2^{2^{k \cdot n}}$ *& space* $2^{2^{k \cdot n}}$),
(4) *m-counters* $\leqslant$ *1-tape* (*real-time & space Log*),
(5) *m-tapes* $\leqslant$ *1-tape* (*time* $k \cdot n^2$ *& space Lin*),
(6) *m-tapes* $\leqslant$ *2-tapes* (*time* $k \cdot n \cdot \log n$ *& space Lin*),
(7) *multihead tapes* $\leqslant$ *single-head tapes* (*real-time & space Lin*),
(8) *multihead tapes + jumps* $\leqslant$ *single-head tapes* (*real-time & space Lin*),
(9) *2-dimtapes* $\leqslant$ *1-tape* (*time Poly & space* $k \cdot n^2$),
(10) *2-dimtapes* $\leqslant$ *1-tape* (*time Poly & space* $k \cdot n \cdot \log n$),
(11) *2-dimtapes* $\leqslant$ *2-tapes* (*time* $k \cdot n^{3/2}$ *& space* $k \cdot \log n$),
(12) *2-dimtapes* $\leqslant$ *1-tape* (*time Poly & space Lin*).

PROOF (*sketch*). Results (1) and (2) together imply (3) with four counters and a single exponential increase in time and space. But in fact two counters suffice (at the price of one more exponential increase). These results are due to Minsky and imply that two-counter machines are universal [81].

Result (4) is less well-known and due to Vitányi [135]; see also [118]. It is based on an oblivious simulation of a single counter on a single tape, where oblivious means that the position of the heads of the simulator depends on the number of steps in the computation only and not on the contents of the tape. The simulation uses a redundant number representation and an ingenious method of simulating a recursive procedure without a stack. It is easy to see that—given an oblivious simulation of one counter—one can simulate k counters as well on the same tape, without loss of space or time.

Result (5) can be found in most textbooks [51] whereas (6), which is due to Hennie and Stearns [48], again is based on an oblivious simulation of a single tape on two tapes. This simulation uses a technique of tape segmentation, where a tape is decomposed into regions of exponentially increasing sizes, which are located farther and farther away from the current head position and which are serviced with an exponentially decreasing frequency.

Result (7) was originally given by Fischer, Meyer and Rosenberg [30]. Later improvements required fewer tapes for the simulation [64]. Extension (8) with jumps was obtained by Kosaraju [61].

As far as the Invariance Thesis is concerned, the simulations of multidimensional tapes on single-dimensional ones seem to have received too little attention. The standard simulations of two-dimensional tapes on single-dimensional tapes ((9) and (10)) require more than constant-factor space overheads, in combination with time overheads $k \cdot n^3$ and $k \cdot n^2 \cdot \log n$. The same is true for the time-efficient simulation (11) by Stoss [127] (also presented in [13]) which again is based on tape segmentation. The only reference known to the author in which simulation (12) with a constant-factor space overhead is given is Hemmerling's report [47]; see also [138].

Simulation (12) is based on the so-called *history method.* The idea is to store address-contents records using relative rather than absolute addresses related to the adjacent records on the simulated tape. This requires that the entire visited region of the higher-dimensional tape is traced by a path which visits every cell, proceeding from every cell to one of its neighbors on the grid, and which finally is not longer than a constant multiple of the number of cells visited. It should therefore not visit a cell more than a constant number of times. It turns out that a full traversal of a depth-first search tree in the subgraph of the grid which represents the visited area of the tape has these properties.

Simulations (9)–(12) all extend to the case of higher-dimensional tapes. The time overheads in (9), (10) and (11) become $k \cdot n^{d+1}$, $k \cdot n^2 \cdot \log n$ and $k \cdot n^{2-1/d} \cdot \log n$ respectively, whereas the space overhead in (9) becomes n^d. The space bounds in (10) and (11) and both bounds in Hemmerling's simulation are independent of the dimension. □

2.1.3. Other Turing machine simulations

An important property of the Turing machine model is the possibility of constant-factor speed-up both in space and time. By changing the tape alphabet from Σ to Σ^k one can compress k symbols into a single square. This makes it possible to design a machine which works k times as fast and uses k times fewer space. The anomaly of the constant-factor speed-up clearly disappears if one assigns a weight to every step or tape square of a Turing machine proportional to the amount of information processed. After compressing k symbols into one, the manipulation of the symbol would become k times as expensive in a weighted time measure. Also the amount of information stored in a single symbol would be multiplied by the same factor k. A space measure called *capacity* has been proposed which accounts for this anomaly [15]. On the other hand, the constant-factor speed-up is very convenient: one can ignore all constant factors implied by $O(f)$ terms in complexity estimates.

Most proofs of the constant-factor speed-up result start with a preprocessing stage in which the input is condensed (k symbols into one); this preprocessing requires linear time provided the machine has two tapes; otherwise preprocessing requires quadratic time, which would lead to a constant-factor speed-up in time only for time bounds larger than n^2. Next the original computation is simulated, but in order to be certain that one can simulate k steps of the original computation in a single move, one must scan the neighboring cells first and subsequently perform some updates to this neighborhood as well. This leads to a speed-up in which k moves are simulated by six moves. Hence, in order to achieve a speed-up by a factor k one must condense the tape by a factor $6 \cdot k$ rather than k. In the original proof of Hartmanis and Stearns [46] one block of symbols of the tapes is kept in the finite control, thus allowing k moves for one move in the simulation.

An important question is whether the overheads as indicated in the above theorem are optimal or not. Folklore has it that the simulation of two tapes on one requires quadratic time overhead, since the set of palindromes can be recognized in linear time on two tapes but requires time $\Omega(n^2)$ on a single-tape machine [51]. But note that this result does not imply anything for the optimality of the square-time overhead for simulating two worktapes by one worktape. Such problems have been solved only recently using the technique of Kolmogorov complexity; see for example [66, 67, 73, 74, 87].

The following result expresses some well-known relations.

2.2. THEOREM. *The complexity classes of the Turing machine model* T *satisfy the following inclusions:*

(1) $\text{T-TIME}(f(n)) \subseteq \text{NT-TIME}(f(n)) \subseteq \text{T-TIME}(2^{k \cdot f(n)})$,
(2) $\text{T-SPACE}(f(n)) \subseteq \text{NT-SPACE}(f(n)) \subseteq \text{T-SPACE}(f(n)^2)$,
(3) $\text{T-TIME}(f(n)) \subseteq \text{T-SPACE}(f(n)/\log f(n))$.

PROOF (*sketch*). Relations (1) and (2) provide the known bounds on the relation between determinism and nondeterminism. For the time bounded classes the bounds are essentially trivial. The second inclusion in (2), which requires $f(n) \geqslant \log n$, represents Savitch's Theorem [104]. For the case of linear time it is known that the first inclusion in (1) is a proper one [88] if beside a time bound $f(n)$ also a space bound $s(n)$ is known. Wiedermann [141] has recently improved the upper bound in (1) to

$$\text{NT-TIME \& SPACE}(f(n), s(n)) \subseteq \text{T-TIME}(f(n) \cdot s(n)^2 \cdot 2^{k \cdot s(n)}).$$

This also improves the time bound which results from the proof of Savitch's Theorem:

$$\text{NT-TIME \& SPACE}(f(n), s(n)) \subseteq \text{T-TIME}(2^{k \cdot s(n) \cdot \log f(n)}).$$

Result (3) represents the important fact that space can be a more powerful resource than time, but beware: the proof given by Hopcroft, Paul and Valiant [50] is valid for one-dimensional tapes only. The result extends to higher-dimensional Turing machine s but then it requires a new proof [91]. There exist also versions for the RAM model [89] and for the Storage Modification Machine [41].

The proof of (3) involves several techniques: first the Turing machine computation is made block-respecting with a block size $k(n) \approx f(n)^{2/3}$; this means that both time and tapes are divided into segments of size $k(n)$ such that segment boundaries are only crossed at the end of a time slot. The overhead for this part of the simulation is linear in time and space. Computing onwards for one block of time now only requires knowledge of the tape segments scanned at the beginning of a block period. The initial configurations of these blocks are collected into a computation graph which turns out to be a directed graph with in-degree $m+1$, where m equals the number of tapes. A correspondence is made between solutions of a pebble game on this directed graph and space efficient simulations of the original computation. The fact that any graph of $m := f(n)^{1/3}$ nodes can be pebbled with $m/\log m$ pebbles now yields the logarithmic gain in space efficiency in (3). □

For the single-tape Turing machine model, denoted by T1, a better bound is known:

$$\text{T1-TIME}(f(n)) \subseteq \text{T1-SPACE}(\sqrt{f(n)})$$

provided $f(n) \geqslant n^2$; see [86]. The result requires that the time bound $f(n)$ is constructible. Contrary to the theorem of Hopcroft, Paul and Valiant the space-efficient simulation for single-tape Turing machines can be made polynomial time as well. Ibarra and Moran [52] have shown that

$$\text{T1-TIME}(f(n)) \subseteq \text{T1-TIME \& SPACE}(f(n)^2, \sqrt{f(n)}).$$

The above result extends also to Turing machines with a two-way read-only input tape and one one-dimensional worktape. The simulation overheads become slightly different, due to the fact that the position of the input head must be stored. The space overhead becomes $\sqrt{f(n)} \cdot \log n$, but the time overhead remains at $f(n)^2$, where the constructibility condition is still required. The simulation extends to nondeterministic models as well. Ibarra and Moran [52] have shown that

$$\text{T1-NTIME}(f(n)) \subseteq \text{T1-NTIME \& SPACE}(f(n)^{3/2}, \sqrt{f(n)}).$$

This result was further improved to [70, 129]

$$\text{T1-NTIME}(f(n)) \subseteq \text{T1-NTIME \& SPACE}(f(n), \sqrt{f(n)}).$$

The modified proof uses nondeterminism for a guess-and-certify strategy comparable to the strategies used in several efficient simulations of parallel models. It is interesting to combine the above result with the breadth-first simulation proposed by Wiedermann. One obtains a result with a sublinear function in the exponent [141]

$$\text{T1-NTIME}(f(n)) \subseteq \text{T1-TIME}(f(n)^2 \cdot 2^{k \cdot \sqrt{f(n)}}).$$

2.2. *Register machines*

2.2.1. *The variety of models of register machines*

Register-based machines [21] have become the standard machine model for the analysis of concrete algorithms. On the one hand, register machines can be recognized as the model of a real computer, reduced to its minimal essential instruction repertoire. On the other hand, register machines do not have some essential characteristics of all real computers: they neither have a finite word length nor a finite address space. All models are equipped with a "compiler" for some ALGOL-like programming language, and one is quite free to use whatever high-level language features one likes in the description of RAM algorithms. It is clear that, in order to obtain realistic estimates on the complexity of the algorithms dealt with, some knowledge of the efficiency of the compiled code is presupposed.

The basic RAM model consists of a finite control where a program is stored, one or more accumulator registers *Acc*, an instruction counter, and an infinite collection of memory registers $R[0], R[1], \ldots$ The accumulator and the memory registers have an unbounded word length, but in any reachable configuration only a finite value will be stored in these registers.

The instruction repertoire of the RAM (and that of other machines we will meet in the sequel) can be divided into four categories:

(1) instructions which influence the flow of control: this includes the unconditional instructions **goto**, *Accept*, *Reject*, *Halt* and conditional jumps like **if** condition **then goto** and **if** condition **then skip**; the "conditions" are simple tests like $Acc = 0?$, $Acc \geqslant 0?$, and $Acc = R[i]?$;

(2) instructions for input and output: *Read* and *Print*;

(3) instructions for transport of data between accumulator and memory;

(4) instructions for performing arithmetic;

RAM programs are labeled sequences of instructions taken from the instruction repertoire.

The meaning of the instructions is self-explanatory and indicates in each case how the instruction pointer is adjusted in the program. In all instructions that are not jumps the instruction pointer is simply set to the instruction that literally follows the one that is completed in the program. Nondeterminism in RAM programs is obtained if each "line" of the program actually lists two instructions, from which the machine must choose one upon every execution of this "line". More common is the multiple use of labels: if the machine executes a jump to a label that occurs twice in the program, then one of the occurrences of the label is chosen as the place where the execution of the program proceeds.

Depending on the precise model chosen, the transput instructions either transfer entire integers between arbitrary registers, or require the accumulator as an intermediate storage location for these transfers (*Read Acc*, *Print Acc*). An even more restricted form of transput occurs in the model in which a read instruction inputs a single bit from the input channel and performs a conditional jump to a subsequent instruction label, depending on whether the bit equals 0 or 1.

The name RAM (*random access machine*) indicates the main feature in the instructions of the third type: the use of *indirect addressing*. There exist three types of load instructions and two types of store instructions:

$LOADD\ i$	$(Acc := i)$,		
$LOAD\ i$	$(Acc := R[i])$,	$STORE\ i$	$(R[i] := Acc)$,
$LOADI\ i$	$(Acc := R[R[i]])$	$STOREI\ i$	$(R[R[i]] := Acc)$.

In the weakest model the only arithmetical instruction is the increment-by-one instruction: $Acc := Acc + 1$. The resulting model is called the "Successor RAM". The standard model has both addition and subtraction, in which the second argument is to be fetched from memory: $ADD\ j\ (Acc := Acc + R[j])$. More powerful models extend the set of arithmetic instructions with multiplication and division. For these powerful models it also becomes possible to treat the register contents as bit strings rather than as numbers; in this interpretation we have instructions for concatenation of bit strings and bitwise Boolean operations.

In the sequel we will use the following notations:

- SRAM denotes the Successor RAM, the model which only has the successor instruction;
- RAM denotes the standard model with addition and subtraction;
- MRAM denotes the model that has addition and subtraction, and multiplication and division;
- if the bitwise Boolean instructions are available, we add a B in the name of a model, (for example MBRAM); and
- if a model is nondeterministic, an N will be added in front of its "name".

For every one of the different models there are at least two different methods of measuring the time and space consumption of a RAM computation. In the *uniform measure* every executed instruction is counted as one step, regardless of the size of the values operated on. In the *logarithmic measure* every executed instruction is given a cost equal to the sum of the logarithms of all quantities involved in the instruction, even including the addresses involved in direct or indirect addressing.

Von Neumann's idea to store a program in the memory rather than in the finite control of a computer brings us to the RASP model [21] (*Random Access Stored Program machine*). Here the memory has been divided into registers with even and with odd addresses. Two adjacent registers together store an instruction, with the operation code in the even addressed register and the operand address in its odd neighbor. By writing in an odd location the machine can obtain the effect of indirect addressing without having it in its instruction repertoire. It has been shown that for the purposes of complexity theory the RAM and the RASP are fully equivalent: they simulate each other in real-time with constant-factor space overhead.

Even the weakest model (the SRAM) is universal as a computing device and it remains so even when the instructions for indirect addressing are removed. In this case all addresses accessed during a computation are represented explicitly in the program. This model is equal to the *register machine*, used as a model for basic recursion theory by Shepherdson and Sturgis [119]. The machine uses only a fixed finite set of its registers during the computation, and can be described as a finite control equipped with a fixed number of counters. Note that Minsky has shown that two registers suffice to obtain universal computing power [81].

The use of indirect addressing leads to two, less desirable effects. It becomes possible to use registers in a very sparse way, leaving big gaps in the memory where nothing has happened. It also becomes less reasonable to presume that at the start of the computation all registers are properly initialized at zero; what if a preceding program has left memory full of scattered values? If one has to initialize memory, who is going to pay for this? Both these problems have been considered and solved [1] but, as we will see below, it is not clear that the solutions that show how to get around these problems with constant-factor overheads in time and space are measure-independent.

2.2.2. Time complexity measures for RAMs

Let us consider the two time measures for the RAM model in more detail. In the uniform measure every instruction takes one unit of time. In the logarithmic measure every instruction is charged for the sum of the lengths of all data manipulated implicitly or explicitly by the instruction. This length is based on a size function on numbers based on the binary logarithm:

$$size(n) = \textbf{if } n \leqslant 1 \textbf{ then } 1 \textbf{ else } \lceil \log n \rceil + 1 \textbf{ fi}.$$

For example, the cost of an $ADDI\,j$ instruction, with the effect $Acc := Acc + R[R[j]]$ will be something like $size(x) + size(j) + size(R[j]) + size(R[R[j]])$, where x denotes the current value of the accumulator Acc. In the sequel the use of the uniform or logarithmic measure will be indicated by prefixing the word TIME or SPACE in the denotation of complexity classes with U and L, respectively.

It is clear that a computation costs "less" in the uniform time measure than in the logarithmic measure. The gap between the two measures may become as large as a multiplicative factor equal to the size of the largest value involved in the computation. How large this value can grow depends on the available arithmetic instructions.

2.3. THEOREM. *The uniform and the logarithmic time measure are related as follows:*

(1) SRAM-UTIME $\leqslant$ SRAM-LTIME (*time* $n \cdot \log n$),
(2) RAM-UTIME $\leqslant$ RAM-LTIME (*time* n^2),
(3) MRAM-UTIME $\leqslant$ MRAM-LTIME (*time Exp*).

The same relations hold for the corresponding nondeterministic classes. As can be seen, the gap between uniform and logarithmic measure for the MRAM is so large that no simulation with polynomial-time bounded overhead can be guaranteed. Indeed, the MRAM with the uniform time measure has to be discarded from the realm of reasonable models: it is a member of the second machine class [9].

It is easy to see that the bitwise Boolean instructions can be simulated by models that do not have them in time polynomial in the length of the operands. This leads to the following result.

2.4. THEOREM. *The simulation overheads related to the bitwise Boolean instructions are given by*

(1) SBRAM-UTIME $\leqslant$ SRAM-UTIME (*time* $n \cdot \log n$),
(2) SBRAM-LTIME $\leqslant$ SRAM-LTIME (*time* $n \cdot \log n$),
(3) BRAM-UTIME $\leqslant$ RAM-UTIME (*time Poly*),
(4) BRAM-LTIME $\leqslant$ RAM-LTIME (*time Poly*),
(5) MBRAM-UTIME $\leqslant$ MRAM-UTIME (*time Poly*),
(6) MBRAM-LTIME $\leqslant$ MRAM-LTIME (*time Poly*).

PROOF (*sketch*). For (1) and (2) the standard technique of translating between numbers and their bit patterns does not work; one must explicitly store and manipulate the binary strings in arrays. Simulation (5) is a rather deep result about the second machine class [9]: bitwise Boolean instructions are not needed for the power of parallelism in the presence of both addition and multiplication. The remaining simulations are straightforward and are left to the reader. Again the results hold also for nondeterministic models. □

The absence or presence of multiplicative and parallel bit manipulation operations is of relevance for the correct understanding of some results in the analysis of algorithms. It is frequently assumed that the machine used can perform innocent looking multiplicative instructions on "small" values, since this is possible in real-world computers. It is also common to use the uniform time measure. At the same time one is intersted in obtaining complexity bounds which are precise up to logarithmic factors. It is therefore relevant to know to what extent the complexity bounds depend on the precise instruction repertoire that is used.

Assume by way of example that the instructions for performing left and right shifts of bit patterns (multiplications and divisions by powers of (2) are used. Assume moreover that we are allowed to perform these operations on arguments $\leqslant n^k$, where n denotes the largest value in the input and k is a fixed constant. It turns out that one can obtain a large collection of other multiplicative or bit-manipulation instructions with a time overhead O(1) in the uniform time measure by the basic technique of table look-up. The results of any such operation can be stored in a table of size n^{2k}; the restricted multiplication allows us to simulate indexing in an n^k by n^k array, and therefore the operator can be performed in time O(1), once the table has been computed. But for many specific operations of a sufficiently "local" nature the same result can be obtained using a table of size $O(n)$ (or even size $O(\sqrt{n})$) by splitting the operands in pieces of length $\frac{1}{2}\log n$ or even shorter and reconstructing in constant time the entire result from the piecewise results, by some suitable combining function which depends on the instruction. Since the table size becomes negligible compared to n, so does the time required for precomputing its values. The parallel bitwise logical operators and ordinary multiplication have this locality property, and so does the instruction which counts the number of 1-bits in a binary number.

This observation is a generalization of a strategy used by Schmidt and Siegel in [110]. It shows that there hardly exists such a thing as an "innocent" extension of the standard RAM model in the uniform time measure; either one only has additive arithmetic, or one might as well include all reasonable multiplicative and/or bitwise Boolean instructions on small operands.

The relation between the various RAM models and Turing machines is expressed in the following results.

2.5. THEOREM. *The various RAM models can simulate the Turing machine model* T *with polynomial time overhead as follows:*

(1) T $\leqslant$ SRAM-UTIME (*time real-time*),
(2) T $\leqslant$ SRAM-LTIME (*time* $k\cdot n\cdot\log n$),
(3) T $\leqslant$ RAM-UTIME (*time* $k\cdot n/\log n$),
(4) T $\leqslant$ RAM-LTIME (*time* $n\cdot\log\log n$).

PROOF (*sketch*). The first of these simulations is due to Schönhage and follows as a corollary of his results on Storage Modification Machines [114]. The second simulation can be found in e.g. [1]. Result (3) is an observation by Hopcroft, Paul and Valiant [49]. The fourth simulation was given by Katajainen, Penttonen and Van Leeuwen [58]. □

2.6. THEOREM. *The Turing machine model* T *can simulate the various RAM models with the following overheads*:

(1) SRAM-UTIME $\leqslant$ T (*time* $n^2 \cdot \log n$),
(2) RAM-UTIME $\leqslant$ T (*time* n^3),
(3) MRAM-UTIME $\leqslant$ T (*time Exp*),
(4) SRAM-LTIME $\leqslant$ T (*time* n^2),
(5) RAM-LTIME $\leqslant$ T (*time* n^2),
(6) RAM-LTIME $\leqslant$ T (*time* $n^2/\log US$),
(7) MRAM-LTIME $\leqslant$ T (*time Poly*).

PROOF (*sketch*). The above results are obtained by allocating address-value pairs of the RAM on Turing machine worktapes and estimating the time needed for processing these structures. Clearly, most time in the simulation is spent on searching for a record on the linear tape. The first two simulations are due to Cook and Reckhow [21]. For simulation (3) there seems to be no alternative for an exponential time overhead simulation due to the possible growth of the length of the values involved. Results (4) and (5) again are due to Cook and Reckhow [21]. Simulation (6), where *US* denotes uniform space, is obtained by slightly rearranging the information contained in the RAM registers on the tape such that the shorter data are closer to the left end of the tape and are therefore easier to access. This result is due to Wiedermann [139]. Result (7) is easy. □

If higher-dimensional tapes are used, the overheads are reduced to the effect that the factor n^2 in the simulation overhead for a linear tape can be replaced by $n^{1+1/d}$ where d denotes the dimension of the worktapes used [138].

2.2.3. Space complexity measures for the RAM

It is well-known that space on a RAM should not be measured by in terms of the number of registers used, as two register already provide universal computing power [81]. Yet this uniform space measure is commonly used in the analysis of algorithms. This can be justified on the basis that in concrete algorithms the intermediate results stored in registers are always bounded in terms of the input values: either by a polynomial, or by a simple exponential function. Estimating the sizes of intermediate results therefore represents an essential part of the analysis of algorithms in algebraic and number-theoretical computations, since the above assumptions need justification.

In the theory of machine models every RAM register is charged a cost corresponding to the size of its contents. This regime is used in the definition of the logarithmic time measure, so why not use it for the space measure as well. But even now there are several ways to proceed. A rather crude way is to charge every register used for the size of the largest value produced during the entire computation. A more refined method is to charge every register for the largest value stored there during the entire computation. The latter heuristics leads to an expression

$$space = \sum_{j=0}^{maxaddr} size_2(j, \max(j)) \tag{2.1}$$

where *maxaddr* is the index of the highest address accessed during the computation, and max(j) is the largest integer ever stored in $R[j]$ during the computation. The parameter j is added in the size function $size_2$ in order to make it possible to consider size functions which depend on the address of the register.

An even more refined method would be to look at individual RAM configurations. Charge every register used in some configuration for the size of the value currently stored there and compute the sum of these register costs for the registers which are currently used. The maximum of this sum over all reachable configurations would become the space measure. In this measure a copy of a large value which is repeatedly transported from one register to another one, after which the first register gets cleared is counted only twice, whereas in the measure expressed in (2.1) it is charged for every register it visits. This difference, however, turns out to be nonproblematic, and therefore we will restrict ourselves to space measures described by the above expression (2.1). See [138] for several other ways of defining space measures for the RAM.

The traditional RAM model supports the use of initialized storage: registers which have never been accessed before store the value *0*. It is quite possible that there are registers with an index in the range *0*, ..., *maxaddr* which are never accessed during the computation, and therefore in the above formula something must be said about the space consumed by those unused registers. Savitch [106] considers the following size function:

$$size_w(i,x) = \textbf{if } x \leq 1 \textbf{ then } 1 \textbf{ else } \lceil \log x + 1 \rceil \textbf{ fi} \tag{2.2}$$

(This is the same as *size*(x).) The resulting measure charges unused registers for an amount of one bit for the value 0 stored there. As a consequence it becomes possible to consume exponential space during polynomial time by performing a program like

```
addr := 1;
for i from 1 to n do
   addr := addr + addr;
   R[addr] := 1
od;
```

A more generally accepted measure (see for example [1]) uses the same size function for used registers but gives the unused registers for free:

$$size_s(i,x) = \textbf{if } R[i] \text{ is unused } \textbf{then } 0 \textbf{ else } size(x) \textbf{ fi}. \tag{2.3}$$

This measure solves the anomaly stated above but introduces a new problem which seems to have been overlooked in the literature: how to simulate a RAM on a Turing machine with constant-factor overhead in space?

The standard trick of storing address-value records on a worktape requires additional space for the addresses, whereas a sequential allocation of all registers in the range 0, ..., *maxaddr* requires space proportional to Savitch's measure (2.2). So none of the two proposed measures leads to an invariance of space in an evident way. Hence, taking the standard simulation of a RAM on a Turing machine as a point of departure, we arrive at the following size measure:

$$size_b(i, x) = \textbf{if } R[i] \text{ is unused } \textbf{then } 0 \textbf{ else } size(i) + size(x) \textbf{ fi}. \tag{2.4}$$

We claim that $size_b$ represents the intuitively correct way of measuring space on a RAM. If the purpose of the space measure were just to define the logarithmic time measure by the relation *time for some instruction = total space affected by performing this instruction* then it would make no difference whether $size_b$ or $size_s$ is used. The address space charged by $size_b$ to a register accessed by an indirect addressing is charged to the address anyhow, since this address itself is affected by the instruction as well. The difference between the two measures only becomes visible when the space consumption for a sequence of instructions has to be measured.

Basing the RAM space measure on $size_b$ (2.4) has several advantages. The constant-factor space overhead for simulation of a Turing machine on a RAM remains intact, although the standard simulation which stores tape cells in consecutive registers has to be abandoned. This simulation would introduce an $\Omega(S \cdot \log S)$ space overhead due to the lengths of the addresses of S registers. But by using the standard trick of "one tape = two stacks", and by storing a single stack in a single register, a simulation with constant-factor space overhead is obtained (at the price of increasing the time overhead by a factor S or S^2 depending on the time measure used). Hence the proposed size measure validates the Invariance Thesis.

Another advantage of the use of $size_b$ is connected with the simulation of uninitialized storage as suggested in [1, Exercise 2.12]. When using $size_s$ the space overhead becomes $\Omega(S \cdot \log S)$, whereas it is a constant-factor space overhead when using $size_b$. A similar observation can be made about the standard method of compacting sparsely used registers into a dense set by the creation of address-value pairs on the RAM itself.

Having chosen the right size measure, the space complexity of the RAM becomes fully equivalent to that of the Turing machine.

2.2.4. *The problematic simulation of a RAM on a Turing machine*

The reader might ask at this point whether the difference between $size_s$ and $size_b$ is really important. Surprisingly it is. As shown in [121, 122], the Invariance Thesis, which is evidently valid if we use the measure based on $size_b$, becomes problematic if the definition based on $size_s$ (2.3) is used. Using the extremely complicated simulation sketched below, it can be shown that for deterministic off-line computations a Turing machine can simulate a RAM with the space measure $size_s$ with constant-factor overhead in space. However, since this simulation requires exponential overhead in time, it is not sufficient for showing that the RAM model (again with the standard $size_s$ space measure) is a first class machine model! It can be shown that for on-line computations a constant-factor overhead simulation does not exist, and in the case of nondeterministic computations the problem is wide open.

We now sketch the simulation. The straightforward simulation does not work, since no space is available for the address parts of the address-value records which must be stored on the worktape of the Turing machine. On the other hand, the problem evaporates as soon as the size of the data stored in RAM memory exceeds or is just proportional to the space needed by these addresses. Therefore the problem will only arise in the unlikely situation that small chunks of data (say characters) are stored in registers with large addresses, encoding in this manner information about the address rather than about the data itself. This situation is illustrated by the following recognition problem:

$$L_0 = \{w_1 \# w_2 \# \ldots \# w_k \# w_0 \mid w_i \in \{0,1\}^* \wedge w_0 \in \{w_i \mid 1 \leqslant i \leqslant k\}\}. \tag{2.5}$$

It is not hard to see that on a RAM the above language can be recognized on-line by reading the words w_i as addresses, and storing a 1 in the corresponding registers. If we consider the typical case where the k words w_i in the input have length $|w_i| = m$, the above algorithm will consume space $O(m+k)$ on a RAM with space measure based on $size_s$, whereas it requires space $O(m \cdot k)$ if the measure based on $size_b$ is used. The latter amount of space is a provable lower bound for Turing machine on-line recognition, since the Turing machine on-line acceptor must write a full description of its input on its worktapes before it can process the last word w_0. Clearly this lower bound argument collapses if off-line processing is allowed, and it is not difficult to construct an off-line Turing machine acceptor which runs in space $O(m + \log k)$.

The space measure based on $size_s$ seems to underestimate the true space consumption when one uses a sparse table. If such a situation arises in actual computing, the problem may be solved by hashing techniques. Hashing maps a sparse set of logical addresses on a much denser set of physical addresses. Hence, rather than storing the address-value pairs directly, we represent the addresses by their hash codes for a suitable hash function, and reconsider the resulting space requirements. Two observations can now be made:

- In order for the computations not to be disrupted by confusion of registers, the hash function used should be perfect with respect to the addresses used during the computation; it should be a 1-1 function if restricted to those arguments on which it will be evaluated.
- If the actual number of addresses used during the computation equals k, and if the hash function has a range $c \cdot k$ for some constant c, the physical address-value records can be allocated sequentially on a worktape, so the hash codes no longer need to be stored. Using a variable-size format for the remaining value records, it can be achieved that inside this hash table an amount of space is consumed proportional to the space as measured by the $size_s$ space measure.

This idea leads to the following problem on perfect hashing which must be solved: *given a set A of k elements of a universe $U = \{0, \ldots, u-1\}$ of size u, find a perfect hash function f which scatters this set A completely into a hash table of size $c \cdot k$ for some fixed constant c.*

Functions satisfying the above requirements evidently exist. But for the problem at hand, these functions themselves should satisfy a space requirement as well. In the above situation, the RAM to be simulated may quite well consume no more than

$O(k+\log u)$ space, and consequently, the manipulation of the hash function should not require more than $O(k+\log u)$ space as well. So the final constraint on f reads:

- the hash function f should be of program size $O(k+\log u)$ and require space $O(k+\log u)$ for its evaluation.

Hash functions satisfying the first requirement were constructed explicitly by Fredman, Komlós and Szemerédi [34], but their construction did not satisfy the last requirement. Mehlhorn [78] showed that for the program size the upper bound could be met and improved to a tight $\Theta(k+\log\log u)$ bound, but his function required far more space for evaluation. In [121, 122] Slot and the author succeeded in matching the Mehlhorn bound in combination with an $O(k+\log u)$ evaluation space. The required space-efficient logical-to-physical address translation therefore exists. See also [35, 53, 79] and the paper by Schmidt and Siegel [110], where perfect hash functions are constructed which achieve the Mehlhorn bound and can be evaluated in unit time on an MRAM.

The next question is how the simulator can ever hope to find a suitable perfect hash function. Since we are looking for a deterministic simulation, the method of guessing a hash function and using it is not allowed. One has to try out a large collection of hash functions and select a good one, i.e. a perfect hash function with respect to the set of addresses in the simulated computation. Whether a hash function is good must be certified. In the process of certification it is not possible to make a list of all addresses encountered and keeping track of possible collisions under the hash function, since this will reintroduce the space consumption we are trying to eliminate. However, it is possible within the available space bounds to check for every slot in the range of the hash function that there exists at most one address which is hashed onto this slot. Since the computation is deterministic, this process can be repeated for every slot individually. This explains why the simulation breaks down for both probabilistic and nondeterministic modes of computation. Under these modes of computation one never knows whether a second run of the computation will access the same registers as the first run.

As long as the hash function has not obtained its certificate of being perfect, the simulation has to cope with the possibility that the hash function is in fact not perfect and that the simulator therefore may confuse registers. It may provide wrong answers, it may consume far more space than it should, and it may even diverge on bounded memory. The first problem is not a serious problem since the answer of the simulation will not be trusted before the hash function has been certified, and the second problem can be solved using Savitch's trick of incremental space [104]. It is the third problem which causes the greatest trouble. Since the space consumption $O(k+\log u)$ may turn out to be $o(\log n)$ where n denotes the input length, detection of loops on bounded storage by counting is not allowed. Luckily this problem has been solved already by Sipser [120], who proposed a simulation by backward search from the accepting configuration. By this simulation loops are prevented and no extra space is consumed.

Details of the above simulation and the theory on perfect hashing on which it is based can be found in [122]. The simulation requires a lot of recomputations, causes an exponential time overhead, and fails to solve the problem for other modes of computation. As such it provides no evidence that the Invariance Thesis holds if the

RAM space measure is based on $size_s$ (2.3). This measure should therefore be replaced in the literature by the measure based on $size_b$ (2.4).

2.3. *Storage Modification Machines*

The Storage Modification Machine (SMM) was introduced by Schönhage in 1970 (see [114]). The model is similar to the model proposed by the Kolmogorov and Uspenskii (*KUM*) in 1958 [60]. For reasons that will become apparent, the model is also called a *pointer machine.*

The SMM model resembles the RAM in having a stored program and a similar flow of control. However, instead of operating on registers in memory it operates on a single storage structure, called a *Δ-structure*, where Δ is a finite alphabet of at least two symbols. A Δ-structure S is a finite directed graph in which each node has $k=\#\Delta$ outgoing edges, which are labeled by the k elements of Δ. A KUM is like an SMM but it operates on undirected instead of directed graphs.

There exists a designated node a in S, called the *center* of S. There exists a map p^* from Δ^* to S defined as follows: for the empty string ε: $p^*(\varepsilon)=a$, and for all strings $w\in\Delta^*$ and $a\in\Delta$: $p^*(wa)$="the endpoint of the edge labeled a starting in $p^*(w)$". The map p^* does not have to be surjective; however, nodes which can not be reached by tracing a word w in Δ^* starting from the center a will play no subsequent role during the computations of the SMM, and therefore nodes may be assumed to have disappeared when they become unreachable.

Similar to RAMs, the program of the Storage Modification Machine consists of flow of control instructions (**goto**, *Accept*, *Halt*, ...), transput instructions (*Read* and *Print*, where a *Read* will input a single bit and act like a conditional jump depending on the value of the bit read), and instructions which operate on memory, in this case a Δ-structure S. There exist three types of instructions of the latter type:

(1) **new** w: creates a new node which will be located at the end of the path traced by w; if $w=\varepsilon$, the new node will become the center; otherwise the last edge on the path labeled w will be directed towards the new node. All outgoing edges of the new node will be directed to the former node $p^*(w)$

(2) **set** w **to** v: redirects the last pointer on the path labeled by w to the former node $p^*(v)$; if $w=\varepsilon$, this simply means that $p^*(v)$ becomes the new center; otherwise the structure of the graph is modified.

(3) **if** $v=w$ (**if** $v\neq w$) **then** ...: the conditional instruction (conditional jump suffices); here it is tested whether the nodes $p^*(v)$ and $p^*(w)$ coincide or not.

Simple as it may seem, a computation by a Storage Modification Machine is extremely hard to trace. It is as complicated as any pointer-based algorithm. Note that e.g. after "**new** w" it is not necessarily true that $p^*(w)$ denotes the new node. Also, the "Hoare formula" $p^*(v)=x$ {**set** w **to** v} $p^*(w)=x$ is invalid. These two anomalies represent the so-called *paradoxes of assignment*.

2.3.1. *Complexity measures for the SMM*

What should the time and space complexity measures for an *SMM* be. For the time measure there exists only one candidate: the uniform time measure. One might consider charging an instruction according to the length of the paths traced during the instruction but, since the arguments of the instructions are denoted explicitly in the program, this length is a constant independent of the Δ-structure currently in memory. Therefore such a weighted measure will differ from the uniform measure by no more than a constant factor which is fully determined by the program.

The reasonable space measure for the SMM seems to be the number of nodes in the current Δ-structure which corresponds, in fact, to the definition given in [41]. This definition has, however, some problematic aspects. There simply exist too many Δ-structures of n nodes. For the number $\mathscr{X}(k,n)$ of Δ-structures of n nodes over an alphabet Δ of size k, Schönhage presents the following estimates:

$$n^{n\cdot k-n+1} \leqslant \mathscr{X}(k,n) \leqslant \binom{(k\cdot n)}{n} \cdot \frac{n^{n\cdot k-n+1}}{n\cdot k+1}.$$

From the above estimate it follows that in space n one can encode $O(n\cdot k\cdot \log n)$ bits of information, rather than $O(n\cdot \log k)$ bits as on n squares of a Turing machine tape with alphabet Δ. This leads to a similar situation as we have seen in Section 2.2.4: consider the following on-line recognition problem:

$L_1 = \{I_1 \# \ldots \# I_k \ \&\ v = w \mid I_1 \# \ldots \# I_k$ encodes a straight-line SMM program such that after performing the program $p^*(v) = p^*(w)\}$.

It is evident that an SMM can recognize this language in $O(n)$ space, where n denotes the number of **new** instructions in the straight-line program $I_1 \# \ldots \# I_k$. On the other hand, a Turing machine on-line recognizer must have written a full description of the Δ-structure generated by program $I_1 \# \ldots \# I_k$ when it reads the &-symbol and therefore, by an information-theoretic argument, it must have consumed space $\Omega(n\cdot \log n)$ at this moment.

This example leaves the situation open for the case of off-line computations, but it is not difficult to see that also in this case $O(n)$ nodes on an SMM are capable of storing as much information as $O(n\cdot \log n)$ tape squares on a Turing machine. The basic technique consists of a Δ-structure which represents a cycle of $O(n)$ nodes with in each node a pointer to some other node in the cycle. The latter pointer stores in this way $O(\log n)$ bits of information, and it is a matter of simple programming to show that this representation can be used and updated, using no more than $O(\log n)$ additional nodes. A similar technique works also for the KUM. For more details see [133].

More recently it has been shown by Luginbuhl and Loui [71] that the above gain in space can actually be obtained with a constant-factor overhead in time by using a complete binary tree in stead of a cycle. The paths toward the leaves encode the $O(\log n)$ bits stored in this node and it is easy to see that they can be read out with constant-factor time overhead, provided that after crossing the border between two tape blocks of size $O(\log n)$ the simulated Turing machine can compute onwards for at least $O(\log n)$ steps. This additional property can be obtained with constant-factor overhead in time as well; compare with the idea of making a Turing machine computation block-respecting in the proof of the theorem of Hopcroft, Paul and Valiant [50].

Based on the above observation one obtains the space measure $n \cdot \log n$ for a Δ-structure of n nodes. Under this measure the SMM can be simulated with constant-factor space overhead on a Turing machine and vice versa. An alternative is to make the dependence on k explicit by assigning to an n-node Δ-structure over an alphabet Δ of size k the measure $n \cdot k \cdot \log n$. This measure has been proposed by Borodin et al. [15], where it is called the *capacity*. Introducing the factor k in this space measure is similar to charging each tape cell on a Turing machine a space of $\log k$, where k is the size of the tape alphabet. Note, however, that this choice for a space measure would destroy the constant-factor speed-up property for Turing machine space.

2.3.2. Simulations for the SMM model

Schönhage [114] has compared the SMM model both with (multidimensional) Turing machines and with some RAM models in the uniform time measure.

2.7. THEOREM. *The SMM is related to other sequential machine models as follows:*

(1) SMM $\approx$ SRAM-UTIME (*real-time*),

(2) T $\leqslant$ SMM (*real-time*).

As a consequence, the SMM can be simulated on a standard RAM in the logarithmic time measure with $n \cdot \log n$ time overhead. Schönhage [116] also proved a corresponding lower bound which shows that there exists a gap between the SMM model and the RAM in the logarithmic time measure. He proves that storing a bitstring in a RAM requires nonlinear time, even if reading is for free and if arbitrary arithmetic instructions are available. An advantage claimed for the SMM model is that it supports an implementation of n-bit integer multiplication which runs in linear time. For a discussion of the connection between theoretical fast multiplication algorithms and practical machines, see Schönhage [115].

There are a few theoretical results known about the SMM model. In [41] Halpern et al. prove a version of the result of Hopcroft, Paul, and Valiant that time $T(n)$ can be simulated in space $O(T(n)/\log T(n))$. For this purpose they use the space measure determined by the number of nodes which, as we indicated above, differs from Turing machine space by a logarithmic factor. Since the time measure seems to differ from Turing machine time by the same logarithm, one can still maintain that it is a similar result. More recent is a claim by Schnitger [111] which suggests that there exist a nonlinear gap in time between the SMM and the KUM. Since the main distinction between the two models is that Schönhage uses a directed graph as a storage structure, where Kolmogorov and Uspenskii use an undirected graph, the implicit bound of $\#\Delta$ on the out-degree for the SMM becomes a bound on the in-degree for the KUM as well. Therefore there cannot exist too many short paths leading to some particular node. Schnitger introduces a well-designed on-line real-time recognition problem which exploits this feature, and establishes a nonlinear lower bound for this problem on the KUM, depending however on some unproven conjecture about communication complexity.

The SMM represents an interesting theoretical model but, in the context of the above observations, its attractiveness as a fundamental model for complexity theory is questionable. Its time measure is based on uniform time in a context where this measure is known to underestimate the true time complexity. The same observation holds for the space measure for the machine; under the plausible definition of the space measure the SMM does not belong to the first machine class as defined in Section 1.2.4.

2.4. *Circuits and nonuniform models*

All machine models discussed so far have the property that a single machine or a single program operates on inputs of arbitrary length. Occasionally the machine, in performing some computation, will first determine the length of the input and then, after having set up the proper parameters, the "real" computation will start.

Circuits represent an alternative approach to the complexity of decision problems and function evaluation. Given some problem, say deciding membership for the language L, one first reduces this problem to a family of finite problems $L_n = L \cap \Sigma^n$. These finite problems are easy with respect to machine computations since their solution can be programmed using table look-up. Instead one investigates the size s_n of the particular devices which solve these finite problems. The quantity s_n as a function of the input size n now becomes a measure of the complexity of the problem L.

For inputs x of arbitrary length the process of computation thus decomposes in two steps: from the input length $n = |x|$ one determines the nth device M_n; next x is processed by M_n in order to produce the required answer. Intuitively the time complexity of L is bounded by the complexity of the two stages together. But note that neither of these two stages has a time complexity which is measured by s_n as a function of n. For the first stage one obtains at best a lower bound: retrieving an object of size s_n requires time $\Omega(s_n)$. For the second stage s_n seems not to be related at all to the time required to operate M_n on x.

At this point some restrictions are enforced in order to make the theory meaningful. The first restriction is that the mapping $n \Rightarrow M_n$ is *uniform* (it should be computable in polynomial time and/or logspace). The second restriction is that the devices M_n are extremely easy to evaluate, as is the case e.g. with formulas and circuits.

Without the first condition it is possible to have undecidable problems with trivial circuit complexity: take any undecidable problem $U \subseteq \omega$, and consider the language L_U defined by $x \in L_U \Leftrightarrow |x| \in U$. For every length n there exist very simple devices which accept or reject all inputs of length n, but there is no effective way to select the right device for each n. There exists a meaningful alternative for making the mapping $n \Rightarrow M_n$ uniform: consider Turing machine computability relative to some suitable oracle (see [112, 57] for more details).

Without the (second) restriction that the devices are extremely simple to evaluate, the theory cannot provide any complexity bounds above $\log n$: take a universal Turing machine with a program for L; this is an object of constant size. If one adds a description of the length n in $\log n$ bits, one can obtain a finite device prepared for dealing with inputs of length n. Asymptotically, the $\log n$ bits of n dominate in the description of the size of this object.

We will discuss a few results which connect this theory to machine models (see also

[102].) Information on the role of nonuniform complexity in structural complexity theory can be found in [3].

2.4.1. The circuit model

A *logical network* or *Boolean circuit* is a finite, labeled, directed acyclic graph. Input nodes (output nodes) are nodes without ancestors (successors) in the graph. Input nodes are labeled with the names of input variables $x_1, \ldots, x_n$. The internal nodes in the graph are labeled with functions from a finite collection, called the basis of the circuit. Here the condition is enforced that the number of ancestors of an internal node is equal to the number of arguments of its label; moreover a suitable edge labeling establishes a 1-1 correspondence between the ancestors and the arguments of the label.

By induction one defines for every node in the circuit a function represented by this node: for an input variable labeled by x_j this is the projection function $(x_1, \ldots, x_n) \to x_j$; an internal node with label $g(y_1, \ldots, y_n)$ for which the k ancestors represent the functions $h_1(x_1, \ldots, x_n), \ldots, h_k(x_1, \ldots, x_n)$ represents the function $g(h_1(x_1, \ldots, x_n), \ldots, h_k(x_1, \ldots, x_n))$. A function $f(x_1, \ldots, x_n)$ is computed by the circuit iff it is represented by an output node.

The most important size functions for circuits are

- *depth*: the length of a longest path in the graph from an input node to an output node;
- *size*: the total number of nodes in the graph.

It is common to restrict oneself to circuits over a basis of logical functions consisting of monadic and binary functions only. There exist four monadic functions (of which only the *not* is important), and there are sixteen binary logical functions like *and*, *or*, and *xor*. It is well known that there exist single binary functions like the *nand* function, which form a complete basis by themselves. Still the most common circuit basis consists of three functions: *and*, *or*, and *not*. If the *not* is omitted, the circuits only represent *monotone functions* and will be called *monotone circuits*. Networks with unbounded fan-in *and* and *or* gates are also considered. If the number of successors of every node is restricted to 1 the circuit becomes a tree, and one speaks of a formula rather than a circuit.

The complexity measures introduced below depend on the logical basis, and on whether one deals with circuits or formulas. In the latter case one speaks about formula complexity rather than circuit complexity. It is not hard to see that for circuit complexity the effect of the choice of a basis is at most a constant factor in depth and size; for formulas the story is different, see [63, 93]. If unbounded fan-in is disallowed in the basis, the constant factor in size remains, but a logarithmic penalty factor in depth may occur.

2.8. DEFINITION. For a given finite collection of logical functions $f_1(x_1, \ldots, x_n), \ldots, f_k(x_1, \ldots, x_n)$ the circuit complexities $size(f_1, \ldots, f_k)$ and $depth(f_1, \ldots, f_k)$ are defined to be the minimal size and depth of a circuit such that the entire set is computed by this circuit. If $\mathscr{F}_n$ for $n \in \omega$ is a family of n-argument functions, one likewise defines $size(\mathscr{F}_n)$ and $depth(\mathscr{F}_n)$ as a function of n. If L is a language over $\{0,1\}^*$, then the circuit complexities of L are defined as the circuit complexities of the family $\mathscr{F}_n$ where $\mathscr{F}_n$ for every n consists of the characteristic function L_n of the set $L \cap \{0,1\}^n$.

2.9. THEOREM. *For arbitrary languages L the following estimates hold:*

(1) $size(\mathcal{F}_n) \leqslant \#\mathcal{F}_n \cdot 2^n$,
(2) $size(\mathcal{F}_n) \leqslant 2^n/(n \cdot (1+\varepsilon))$ *for* $\#\mathcal{F}_n = 1$, $n \geqslant N(\varepsilon)$,
(3) $depth(\mathcal{F}_n) \leqslant n+1$,
(4) $depth(\mathcal{F}_n) \leqslant size(\mathcal{F}_n)$,
(5) $size(\mathcal{F}_n) \leqslant \#\mathcal{F}_n \cdot 2^{depth(\mathcal{F}_n)}$.

PROOF (*sketch*). Result (1) is easy and results (4) and (5) are trivial. Result (2) is the upper bound established by Lupanov [72], and result (3) has been obtained by McColl and Paterson [76]. It is known that the bound in (2) is asymptotically tight; by enumeration of all possible circuits one shows that for every $\delta > 0$ only a fraction of all functions can be computed by a circuit of size $\leqslant 2^n(1-\delta)/n$ since there are not enough of those small circuits. It is a much harder problem to prove lower bounds in circuit complexity for explicit functions. There the largest lower bound in the literature for circuit complexity shown for an individual explicit function equals $3 \cdot n + o(n)$ [12]. For formula complexity one has an $\Omega(n^2/\log n)$ lower bound [82]. For monotone circuits exponential lower bounds for specific functions have been established in 1985 [2, 95]. □

2.4.2. Relation with Turing machine complexity

The finite control of a Turing machine can be considered as a finite automaton which can be implemented using a Boolean circuit. The size of this circuit tells something of the complexity of the program of the machine, whereas the depth indicates how fast the machine can perform a single step of its computation, assuming that the machine has to be implemented using logical gates.

Consider the time-space diagram which encodes the complete computation of some Turing machine M_i on some input of length n. Without loss of generality we can assume that the M_i consumes both maximal time and space on every input of length n, so the size of this diagram does not depend on the precise input. The structure of the diagram is such that it is filled with characters and that each character is determined completely by the characters located at a small local set of neighbors. These dependencies are directed forward in time. It is not hard to design a Boolean circuit of size proportional to the size of the time-space diagram which evaluates (an encoding of) the character at position (s, t) in the space-time diagram by an amount of circuitry for this location which depends on the program only. In this way one obtains the following result from [102].

2.10. THEOREM. *$size(L_n) \leqslant P \cdot T(n) \cdot S(n)$, where $T(n)$ and $S(n)$ denote the time and space complexity of the Turing machine and P denotes its size.*

In the above simulation a large amount of circuitry is wasted for calculations needed to copy the contents of an unvisited tape cell from one configuration to the next. These local circuits could be eliminated, provided one would know in advance where the heads reside on the tapes. This problem was solved by the Hennie–Stearns simulation in which a k-tape Turing machine was obliviously simulated on a two-tape machine.

Exploiting this idea one obtains the following improvement due to Pippenger and Fischer [92], see also [112].

2.11. THEOREM. *size*$(L_n) \leqslant P \cdot T(n) \cdot \log T(n)$.

A tighter bound was given in [90]. There exists also a tight relation between the depth measure and the space complexity of Turing machines. For space bounds $s(n) \geqslant \log n$ which are constructible in space $O(\max(\log n, \log s(n)))$ Borodin [14] has shown the following theorem.

2.12. THEOREM. *depth*$(L_n) \leqslant O(s(n)^2)$, *when* $L \in \mathrm{NSPACE}(s(n))$.

Moreover, for these functions $s(n)$, circuits of depth $s(n)$ can be evaluated in space $O(s(n))$.

More recently circuits are used as a tool for characterizing the complexity of parallel computation as well. This connection has lead to the introduction of three hierarchies of complexity classes:

- NC_k = the class of languages L recognized by logspace uniform families of circuits of polynomial size and depth $O(\log(n)^k)$,
- AC_k = the class of languages L recognized by logspace uniform families of unbounded fan-in circuits of polynomial size and depth $O(\log(n)^k)$,
- SC_k = T-TIME & SPACE(*Poly*, $O(\log(n)^k)$).

The families NC_k, AC_k and SC_k form a hierarchy within P. Also, the first two hierarchies can be combined to a single hierarchy:

$$NC_0 \subseteq AC_0 \subseteq \cdots \subseteq NC_k \subseteq AC_k \subseteq NC_{k+1} \subseteq \cdots$$

(see e.g. [19, 20] or later chapters in this Handbook for further details).

3. The second machine class

In this section we explore the second machine class, which consists of the machine models which satisfy the Parallel Computation Thesis. Machine models do not have to use parallelism in order to become a member of the second machine class, as we will see. The models of "real" parallel devices first appear in Section 3.4.

There is a great variety of parallel machine models and related issues that have proved useful for some reason or another: models consisting of RAMs or finite-state devices in an arbitrary interconnection pattern [39], models for systolic computation [40, 77], cellular automata [128], the subject of relative quality of one interconnection pattern with respect to another [36, 62, 80], the relation between parallel models and network resource bounds [19, 20], the write-conflict resolution methods [25, 28, 29, 68] and the study of additional resource bounds on parallel devices, like polynomial bounds on the number of processors used or other hardware bounds [24]. The reader interested in such subjects is referred to the indicated literature and later chapters in this Handbook.

For the class of models discussed here we will establish a tight connection between simulations of sequential devices and transitive closure algorithms.

3.1. PSPACE *and transitive closure*

There exists a generic translation between space bounded computations and paths in a suitably constructed configuration graph. For simplicity we will assume from now on that all space bounds are $\Omega(\log n)$. For a given input x, a given Turing machine M and a given space bound S we can form the graph $G(x, M, S)$ of all configurations C of machine M which use space $\leqslant S$; an edge connects two configurations C_1 and C_2 if there is a one-step transition from C_1 to C_2. This graph has the following properties:

(1) for every input x and space bound S there exists a unique node corresponding to the initial configuration on input x;

(2) assuming that a suitable notion of acceptance has been chosen, the accepting configuration is unique;

(3) the number of nodes in the graph $G(x, M, S)$ is bounded by some exponential function $2^{c \cdot S}$, where the constant c depends on M but not on x;

(4) the graph $G(x, M, S)$ can be encoded in such a way that an S-space bounded Turing machine on input x and a description of M can write the encoding of $G(x, M, S)$ on some write-only output tape;

(5) if M is deterministic, then every node in $G(x, M, S)$ has out-degree $\leqslant 1$; if M is nondeterministic then the out-degree of some nodes can be $\geqslant 2$, but for a suitable restriction of the Turing machine model the out-degree can be assumed to be $\leqslant 2$ as well;

(6) the input x is accepted in space S by M if there exists a path from the unique initial configuration on input x in $G(x, M, S)$ to an (or, with a suitable restriction on the model, the unique) accepting configuration. This path can be assumed to be loop-free, and therefore its length can be assumed to be $2^{c \cdot S}$.

These properties suggest the following universal algorithm for testing membership for languages in PSPACE: assume that L is recognized by M in space n^k; then, in order to test whether $x \in L$, we first construct $G(x, M, |x|^k)$, compute the reflexive transitive closure of this graph, and investigate whether in this transitive closure there exists an edge between the unique initial configuration on x and the unique accepting configuration. Although it seems that this is a rather expensive method for simulating a single computation, it is exactly this transitive closure algorithm which is hidden in almost all proofs that some particular parallel machine obeys the Parallel Computation Thesis. This transitive closure algorithm provides the unified perspective on the various parallel machine models we will discuss in this section. The fact that it represents a brute-force try-out-everything simulation gives an indication for the massive parallelism required in order to have parallel models belonging to the second machine class.

3.1.1. *Transitive closure algorithms and* PSPACE-*complete problems*

Various designs of a transitive closure algorithm lead to various insights. In the first place we can represent the graph $G(x, M, S)$ by a Boolean matrix $A(x, M, S)$ where a 1 denotes the presence of an edge and a 0 denotes the absence of an edge. The row and column indices denote configurations. Clearly, these configurations can be written down in space S; the total number (and therefore also the size N of the matrix $A(x, M, S)$) is exponential in S: $N = 2^{c \cdot S}$. If we let $A(x, M, S)[i, i] = 1$ for all $i \leqslant N$, then the transitive closure of M can be computed by $c \cdot S$ squarings of $A(x, M, S)$ using Boolean matrix multiplication. If N^3 processors are available, each squaring can be performed in time $O(S)$ (needed for adding N Boolean values), so the entire transitive closure algorithm takes time $O(S^2)$. If a time bound T on the computation is given, this time is reduced to $O(S \cdot \log T)$, due to the fact that no paths longer than T edges have to be investigated. Finally, if the parallel model has a concurrent-write feature, the time needed for adding the N Boolean values can be reduced to $O(1)$, and in this case the time for the transitive closure algorithm becomes $O(\log T)$. Note however that, in order to perform this algorithm, the matrix $A(x, M, S)$ must be constructed first.

Next we investigate the following recursive function *path*(w, i, j) which evaluates to **true** in case there exists a path from node i to node j of length 2^w:

```
proc path=(int w, node i,j) bool:
  if w=0 then i=j or edge(i,j)
  else
    bool found:= false;
    forall node n while not found
      found:=found or (path(order-1, i, n) and path(order-1, n, j))
    od;
  found
  fi;
```

Existence of an accepting computation can be evaluated by the call *path*$(c \cdot S$, *init*, *final*), where *init* and *final* are the unique initial and final configurations in $G(x, M, S)$; given a time bound T the initial parameter w can also be chosen to be $\log T$. With recursion depth $c \cdot S$ (respectively $\log T$) and parameter size $O(S)$, this recursive procedure can be evaluated in space $O(S^2)$ (respectively $O(S \cdot \log T)$). This is the main ingredient of the proof of Savitch's Theorem [104] which implies that PSPACE = NPSPACE.

3.1. THEOREM. *If a language L is recognized by some nondeterministic Turing machine in space* $S(n) \geqslant \log n$, *then it can be recognized by some deterministic Turing machine in space* $S(n)^2$.

A third formulation of the transitive closure algorithm yields the PSPACE-completeness of the problem QUANTIFIED BOOLEAN FORMULAS (QBF) [125].

QUANTIFIED BOOLEAN FORMULAS

Instance: A formula of the form $Q_1 x_1 \ldots Q_n x_n [P(x_1, \ldots, x_n)]$, where each Q_i equals $\forall$ or $\exists$, and $P(x_1, \ldots, x_n)$ is a propositional formula in the Boolean variables $x_1, \ldots, x_n$.

Question: Does this formula evaluate to **true**?

3.2. THEOREM. *The problem* QUANTIFIED BOOLEAN FORMULAS *is* PSPACE-*complete.*

PROOF (*sketch*). The idea is to encode nodes in $G(x, M, S)$ by a Boolean valuation to a sequence of $k := c' \cdot S$ Boolean variables. From the proof of Cook's Theorem which establishes the NP-completeness of SATISFIABILITY (see [18, 37]), one obtains the existence of a propositional formula P_0 in $2 \cdot k$ variables $P_0(x_1, \ldots, x_k, y_1, \ldots, y_k)$, where the variables $x_1, \ldots, x_k$ encode node i, the variables $y_1, \ldots, y_k$ encode node j and P_0 expresses that $i = j$ or there exists an edge from i to j.

One can now define by induction a sequence of quantified propositional formulas P_d such that $P_d(x_1, \ldots, x_k, y_1, \ldots, y_k)$ expresses the presence of a path of length $\leqslant 2^d$ between node i and node j. In a naive approach the formula P_d would include only existential quantifiers, and P_d would include two copies of P_{d-1}; by a standard trick from complexity theory we reduce the number of occurrences of P_{d-1} in P_d to one; this trick, however, introduces universal quantifiers:

$$\begin{aligned} P_d(x_1, \ldots, x_k, y_1, \ldots, y_k) = \exists z_1, \ldots, z_k [\forall u_1, \ldots, u_k [\forall v_1, \ldots, v_k \\ [((u_1, \ldots, u_k = x_1, \ldots, x_k \\ \wedge v_1, \ldots, v_k = z_1, \ldots, z_k) \\ \vee (u_1, \ldots, u_k = z_1, \ldots, z_k \\ \wedge v_1, \ldots, v_k = y_1, \ldots, y_k)) \\ \Rightarrow P_{d-1}(u_1, \ldots, u_k, v_1, \ldots, v_k)]]]. \end{aligned}$$

Substituting for $x_1, \ldots, x_k$ and $y_1, \ldots, y_k$ the codes of the initial and final node in $G(x, M, S)$ in $P_K(x_1, \ldots, x_k, y_1, \ldots, y_k)$, where $K = c \cdot S$ denotes the logarithm of the size of $G(x, M, S)$, we obtain a closed quantified Boolean formula, the truth of which expresses the existence of an accepting computation. It is not difficult to see that P_K is a formula of length $O(k^2)$ in $O(k^2)$ variables, since $K = \Theta(k)$. Here we have counted each variable as a single symbol; clearly a representation in a finite alphabet will introduce another factor $\log k$ in the length of the formula. From this one concludes that QBF is PSPACE-complete. □

The "alternation of quantifiers" in QBF leads to an interesting game-theoretic interpretation. SATISFIABILITY is the prototype of a solitaire game, where the player has to look for some configuration with a particular property or for a sequence of simple moves leading to some particular goal state.

The alternating quantifiers turn QBF into a two-person game. Two players Alice and Emily in turn choose the truth values to be assigned to existentially or universally quantified variables in the order of their nesting inside the formula. Emily tries to establish the truth of the formula whereas Alice tries to show that the given formula is false. The truth of the entire formula is equivalent with the existence of a winning strategy for Emily in this game.

Starting with this game-theoretic interpretation of QBF, several authors have investigated the endgame analysis of real games. A useful intermediate game is GENERALIZED GEOGRAPHY [109]; from there one can reach HEX on arbitrary graphs and even HEX on the traditional hexagonal board; see [27, 97]. CHECKERS, CHESS and GO are not on the list because, after earlier PSPACE-hardness results [32, 69], they turn out to be even more difficult than PSPACE; see [33, 98, 99]. There also are examples of group-theoretic problems which have been shown to be PSPACE-hard [54, 55]. Together with problems which encode PSPACE-bounded computations in a more direct way (like Reif's GENERALIZED MOVER'S PROBLEM [96]) this has led to an interesting range of PSPACE-complete problems (cf. [38]). It should, however, be no longer a surprise that alternation forms a fundamental concept in one of the machine models in the second machine class.

3.1.2. Establishing membership in the second machine class

Suppose that we have some parallel or otherwise powerful machine model $\mathscr{X}$ and want to prove that it is a member of the second machine class: $\mathscr{X}$-PTIME = $\mathscr{X}$-NPTIME = PSPACE. The above observations provide us with some tools for proving this result.

The inclusion PSPACE $\subseteq \mathscr{X}$-PTIME can be shown by inventing a polynomial-time algorithm on $\mathscr{X}$ which solves a PSPACE-complete problem like QBF. An alternative is to show that on $\mathscr{X}$ one can implement one of the transitive closure algorithms from the previous section in polynomial time. For the inclusion $\mathscr{X}$-NPTIME $\subseteq$ PSPACE one uses in most cases a guess-and-verify method. A nondeterministic machine is used to guess a trace of an accepting $\mathscr{X}$-computation on the given input. Such a trace consists of instructions executed and memory values stored during this computation but in general not all information stored in memory during the $\mathscr{X}$-computation can be written down in polynomial space. Therefore one writes down enough information from which the remaining memory contents can be reconstructed. The reconstruction is done by a recursive procedure which reflects the machine architecture of model $\mathscr{X}$; the recursive procedure tells how the memory contents at time t depend on those at time $t-1$. Care has to be taken that all arguments for this procedure and also its values can be written down in polynomial space. Using the trace and the recursive procedure it becomes possible to certify the trace as being consistent: for every conditional instruction the correct branch has been selected.

If one has established for two models $\mathscr{X}$ and $\mathscr{X}'$ that they are both members of the second machine class, then one has indirectly shown that $\mathscr{X}$ and $\mathscr{X}'$ simulate each other with polynomially bounded overhead in time. In the literature only a few instances of direct simulations establishing such polynomial time overheads are given. One example is the refined analysis of the power of various models of vector machines by Ruzzo [100], where explicit overheads for the simulation of alternating Turing machines and vector machines are given. Another example can be found in the paper by van Leeuwen and Wiedermann on array processing machines [134], where time overheads for simulation on the SIMDAG and for a reverse simulation are determined.

3.2. The alternation model

3.2.1. The concept of alternation

The concept of alternation [16] leads to machine models which obey the Parallel Computation Thesis without providing any intrinsic parallelism at all. As a computational device an *alternating Turing machine* is very similar to a standard nondeterministic sequential Turing machine; only the definition of acceptance for inputs is different.

Since the machine is nondeterministic, the computation can be represented as a computation tree in which the branches represent possible transitions. The leaves—the terminal configurations where the machine halts—are either accepting or rejecting as usual. For a standard nondeterministic machine a computation tree is considered to represent an accepting computation as soon as a single accepting leaf can be found. But for the alternating machine the notion of acceptance is slightly more complicated.

The main idea is to equip states in the Turing machine program with labels *existential* and *universal*. Configurations inherit the label of the state included in this configuration. Next one assigns a quality *Accept*, *Reject* or *Undef* to every node in the computation tree according to the following rules:

(1) the quality of an accepting (rejecting) leaf equals *Accept* (*Reject*);

(2) the quality of an internal node representing an *existential* configuration is *Accept* if one of its successor configurations has quality *Accept*;

(3) the quality of an internal node representing an *existential* configuration is *Reject* if all of its successor configurations have quality *Reject*;

(4) the quality of an internal node representing a *universal* configuration is *Reject* if one of its successor configurations has quality *Reject*;

(5) the quality of an internal node representing a *universal* configuration is *Accept* if all of its successor configurations have quality *Accept*;

(6) the quality of a node with one successor equals the quality of its successor;

(7) the quality of any node the quality of which is not determined by application of the above rules is *Undef*,

Clearly the quality *Undef* arises only if the computation tree contains infinite branches, but even nodes which have infinite offspring can obtain a definite quality since, for example, an accepting son of an *existential* node overrides the *Undef* label of another son. By definition, an alternating device accepts its input in case the root node of the computation tree, representing the initial configuration on that input, obtains the quality *Accept*. The above treatment is a minor simplification of the presentation in [16] in so far as the feature of negating states is not included.

It should be clear that the notion of an alternating mode of computation makes sense for virtually every machine model and is not restricted to Turing machines. For example, it makes sense to consider alternating finite automata, alternating PDAs etc. For every device some power may be gained by proceeding to the level of alternation but whether the gain is substantial depends on the specific model considered. For example, in the case of finite automata, the languages recognized by alternating finite automata are still regular languages; the gain is a potentially doubly exponential reduction of the size of the automation.

The time (space) consumed by an alternating computation is measured as the maximal time (space) consumption along any branch in the computation tree. Another complexity measure considered by Ruzzo [101] for the alternating Turing machine is the minimal size of an accepting computation tree. An accepting tree is a subtree of the full computation tree which yields a certificate that the input is accepted; it includes all successors of a *universal* node, but for an *existential* node only one successor with quality *Accept* must be included in the accepting computation tree. This measure is related both to nondeterministic time in the sequential model and parallel time, and leads to various interesting classes with simultaneous resource bounds.

3.2.2. Relation between alternating Turing machines and sequential models

The alternating machine, which is just a standard machine in disguise, clearly inherits the simulation results for the first machine class models. As a consequence there exists a machine-independent hierarchy for alternating classes:

$$\text{ALOGSPACE} \subseteq \text{APTIME} \subseteq \text{APSPACE} \subseteq \text{AEXPTIME}.$$

Its behavior like a parallel machine model is now expressed by the following result.

3.3. THEOREM. *The alternating Turing machine model is related to the sequential complexity hierarchy by the equality* APTIME = PSPACE. *The other complexity classes are shifted versions in the sequential hierarchy as well:*

$$\text{ALOGSPACE} = \text{P}, \qquad \text{APSPACE} = \text{EXPTIME},$$
$$\text{AEXPTIME} = \text{EXPSPACE}.$$

PROOF (*sketch*). Note that the alternating devices have no nondeterministic mode of computation. We give a short indication why the above equalities are true. First consider the inclusion APTIME ⊆ PSPACE. It suffices to show that the quality of the initial configuration in a polynomial-time bounded computation tree can be evaluated in polynomial space. Clearly, this quality can be evaluated by a recursive procedure which traverses the nodes of the computation tree. This procedure has a recursion depth proportional to the running time of the alternating device, whereas each recursive call requires an amount of space proportional to the space consumed by the alternating device. Hence the space needed by the deterministic simulator is proportional to the space-time product of the alternating device, which in turn is bounded by the square of the running time.

The reverse inclusion PSPACE ⊆ APTIME follows as soon as we show how to evaluate QBE in polynomial time on an alternating machine. This is almost trivial: let a machine guess the valuation for the quantified variables where the universally (existentially) quantified variables are guessed in a universal (existential) state; these values are guessed in the order of the nesting in the formula. Next the formula is evaluated in a deterministic mode and the machine accepts (rejects) if the result becomes true (false).

The equality AEXPTIME = EXPSPACE is obtained by a standard padding argument from the equality just obtained.

The inclusion ALOGSPACE ⊆ P is shown as follows: note that a logspace bounded alternating machine for a given input has only a polynomial number of configurations. These configurations can be written on a worktape. The terminal configurations obtain a quality based on the included state. Next, by repeatedly scanning the list of configurations, the quality of intermediate configurations can be determined by application of the rules (2)–(6). This scanning process terminates if during a scan no new quality can be determined. Since during each sweep either at least one quality is determined or the process terminates, and since the time needed for a single sweep is bounded by the square of the size of the list of configurations, the running time for this procedure is polynomial.

The reverse inclusion P$\subseteq$ALOGSPACE is shown as follows. Assume that the language L is recognized in time $T(n)$ by a standard single-tape Turing machine M. It suffices to show how an alternating device can recognize L in space $\log T(n)$. Consider therefore the standard computation diagram of the computation of M on input x. This diagram can be represented in the form of a $K \times K$ table of symbols, where $K = T(|x|)$. The top row of this table describes the initial configuration on input x, and the bottom row describes the final configuration which should be an accepting one. Each intermediate symbol is completely determined by the three symbols in the row directly above it, since the machine M is deterministic.

The alternating device now guesses the position in the bottom row of the occurrence of an accepting state, and certifies this symbol by generating in a universal state three offspring machines which guess in an existential state the symbols in the three squares above it. These guesses are certified in the same way, all the way up to the top row, where guesses are certified by comparison with the input x. The space required by this procedure is proportional to the space required for writing down the position of the square considered in the diagram, which is $O(\log T(|x|))$. Since the machine M is deterministic, only those guesses which are correct can be certified (shown by induction on the row number) and therefore the guesses are globally consistent. This last observation breaks down for nondeterministic devices M and therefore we cannot obtain the inclusion NP$\subseteq$ALOGSPACE in this way.

Again by a simple padding argument one obtains the equality APSPACE = EXPTIME, as an easy consequence of the equality ALOGSPACE = P. □

3.3. *Sequential machines operating on huge objects in unit time*

The first machine model for which the validity of the Parallel Computation Thesis was established, was the *vector machine model* of Pratt and Stockmeyer [94], shortly later succeeded by the MRAM of Hartmanis and Simon [44, 45]. These models have in common that their power originates from the possibility to operate on objects of exponential size in unit time.

All these models are derived from the RAM model with the uniform time measure, by extending it with new powerful arithmetic instructions. In the vector machine this extension consists of the introduction of a new type of registers, called vectors, which can be shifted by amounts stored in the arithmetical registers of the RAM. The contents of the vector registers can also be subjected to parallel bitwise Boolean operations like *and*, *or*, or *xor*. This makes it possible to program the concatenation of the contents of two vectors and to perform various masking operations. In the MRAM model it was realized that shifting a vector amounts to multiplication or division by a suitable power of 2. Hence the separation between vectors and arithmetic registers is an inessential feature in the model; the same power can be achieved by introducing multiplication and division in unit time, preserving the bitwise Boolean operations.

Restrictions of the model have been investigated. For example one can forsake one of the two shift directions (right shift of one register is simulated by left shifting all the others); as a consequence one can drop the division instruction, which yields the MRAM model as proposed in [44]. More recently it has been established that the combination of multiplication and division, in the absence of the bitwise Boolean instructions, suffices as well [9, 113]. This result shows the power of a purely arithmetical model.

Below we will illustrate the power of these models by a model which has been obtained by moving into the other direction: stressing the pure symbol manipulation instructions and dropping the powerful arithmetic. This is the EDITRAM model, proposed in [123] as a model of a text editor. A proof will be outlined for the validity of the Parallel Computation Thesis for the EDITRAM and we will indicate the connection with the proofs for other models.

3.3.1. The EDITRAM model

In the EDITRAM we extend the standard RAM with a fixed finite set of text files. Standard arithmetic registers can be used as cursors in a text file. Beside the standard instructions on the arithmetic registers the EDITRAM has instructions for

(1) reading a symbol from a file via a cursor,
(2) writing a symbol into a file via a cursor,
(3) positioning a cursor at the end of a file (thus computing its length),
(4) positioning a cursor into a file by loading an arithmetic value into the cursor,
(5) systematic replacement of *string*1 by *string*2 in a text file,
(6) concatenation of text files,
(7) copying of segments of text files as indicated by cursor positions,
(8) deletion of segments of text files as indicated by cursor positions.

In the systematic string replacement instruction (5), arguments *string*1 and *string*2 are to be presented by literals in the program; substitution of the contents of an entire text file for a single character would allow a doubly exponential growth of the size of text files, which is more than we are aiming for.

Time complexity in the model is defined by using the uniform time measure for the arithmetic registers. Thus an edit instruction is charged one unit of time. A reasonable alternative would be to use the logarithmic time measure with respect to the arithmetic registers. Then an edit instruction would be charged according to the logarithm of the values of the involved cursors, which would be proportional to the logarithm of the length of the (affected portion of) the text file. Since in our model the growth of a text file is at most exponential, the two measures will be polynomially related. This observation no longer holds if we allow that entire text files are substituted for strings by the replace operation; the doubly exponential growth of the text files will require exponential time in the logarithmic measure and hence, the uniform and the logarithmic measure are no longer polynomially related.

3.3.2. *The EDITRAM is a second machine class device*

3.4. THEOREM. *The EDITRAM belongs to the second machine class.*

PROOF (*sketch*). In order to verify that the EDITRAM obeys the Parallel Computation Thesis we must prove the two inclusions EDITRAM-NPTIME ⊆ PSPACE and PSPACE ⊆ EDITRAM-PTIME.

The proof of the first inclusion is characteristic for this type of models. Given an input we must test in polynomial space whether the given EDITRAM machine accepts this input or not. But by Theorem 3.1 our simulation may be nondeterministic. Therefore we first guess the trace of some accepting computation and write it down on a worktape. The accepting computation being polynomial-time bounded, we can write down the sequence of instructions in the program of the EDITRAM which are executed. Moreover, since the length of the values of the arithmetic registers is linearly bounded by the time, we can also maintain a log on the register values in polynomial space.

We cannot maintain a log on the contents of the text files, since their length may grow exponentially. Instead we introduce a recursive procedure *char*(*time*, *position*, *textfile*) which evaluates to the character located at the given position in the given text file after performing the instruction at the given time. The arguments of this procedure can be written down in polynomial space, due to the fact that the growth of the length of a text file is bounded by a simple exponential function in the time (both systematic string replacement and concatenation will at most multiply the length of a text file by a constant). Given this procedure it is possible to certify that the trace written on the worktape indeed represents an accepting computation.

From the meaning of the individual instructions one can obtain a recursive procedure which expresses the value of *char*(*time*, *position*, *textfile*) in terms of similar values after the previous instruction at time *time* − 1. In the case where the present instruction is a systematic string replacement, we face the problem to figure out where the character at the given position was located before the replacement. Since this requires information on the number of occurrences of the replaced pattern preceding this position in the given text file, the entire text file, up to the given position, must be recomputed by recursive calls. This is the most complicated case in the description of this recursive procedure. For details see [123].

The total space required by the evaluation of this procedure is bounded by the product of the size of an individual call (which we indicated to be polynomial) and the recursion depth (which is bounded by the running time of the EDITRAM computation being simulated, which was also assumed to be polynomial). This completes the proof of the first inclusion.

Next we consider the inclusion PSPACE ⊆ EDITRAM-PTIME. It suffices to show how to solve the PSPACE-complete problem QBF in polynomial time on a deterministic EDITRAM. Consider a given instance $Q_1x_1 \ldots Q_nx_n[P(x_1, \ldots, x_n)]$ of QBF. Our algorithm is performed in three stages:

Step 1: Remove the quantifiers in the order of their nesting from inside to outside by programming the transformations

$$\forall x_i[P(\ldots,x_i,\ldots)] \quad\Rightarrow\quad (P(\ldots,0,\ldots)\wedge P(\ldots,1,\ldots)),$$
$$\exists x_i[P(\ldots,x_i,\ldots)] \quad\Rightarrow\quad (P(\ldots,0,\ldots)\vee P(\ldots,1,\ldots)).$$

Clearly, each transformation preserves the truth of the involved formula; the involved formula P is a quantifier-free formula, due to the order of the quantifier eliminations. After elimination of all quantifiers a formula of exponential size is obtained which still is equivalent to the given instance of QBF.

Step 2: Evaluate the resulting formula by systematic string replacements of the type

$$\begin{array}{llll} (0\vee 0)\Rightarrow 0, & (0\vee 1)\Rightarrow 1, & (1\vee 0)\Rightarrow 1, & (1\vee 1)\Rightarrow 1,\\ (0\wedge 0)\Rightarrow 0, & (0\wedge 1)\Rightarrow 0, & (1\wedge 0)\Rightarrow 0, & (1\wedge 1)\Rightarrow 1,\\ (\neg 0)\Rightarrow 1, & (\neg 1)\Rightarrow 0, & (0)\Rightarrow 0, & (1)\Rightarrow 1. \end{array}$$

These transformations can be produced by local systematic string replacements.

Step 3: Check whether the resulting literal equals 0 or 1.

Note that after a single cycle through the replacements in Step 2, the depth of the involved propositional expression has been decreased by at least 1. If the depth of the propositional kernel of the given instance was k, then after the transformations of Step 1, the depth of the intermediate formula is $k+n$ (n being the number of quantifiers eliminated). Therefore the number of iterations in Step 2 is polynomial.

It remains to show how to perform the transformations in Step 1. Clearly it suffices to locate and read the innermost quantifier and to form the conjunction or disjunction of two copies of the propositional kernel, provided the quantified variable has been replaced by 0 and 1 respectively in these copies. But since our EDITRAM program allows only literal strings as arguments in systematic replacement instructions, we must program the later substitutions. We design therefore a subroutine which copies the string of characters representing variable x_i into a special-purpose text file (this encoding will include some binary representation of its index i), and which next subjects all occurrences of variables in the propositional kernel P to a treatment of systematic replacements which will turn all occurrences of x_i into a special pattern, and which will leave all other variables undisturbed. Then by substituting 0 or 1 for the special pattern, the required substitutions are obtained. For details of this subroutine, see [123]. This completes the proof of the second inclusion. □

In order to prove the first inclusion for the cases of the vector machine and the MRAM, a similar recursive procedure can be defined which evaluates the contents of a specified bit of a vector or a specified bit of an arithmetic register at some given time. In the vector machine the size of a vector grows exponentially but not worse, whereas for the MRAM all arithmetic registers may grow exponentially in length. Due to the presence of carries, the simulation of a multiplication becomes as complicated as the case of a string replacement in the EDITRAM. Divisions have been reduced to multiplications inside the MRAM model itself at an earlier stage of the proof. Also the length of register addresses remains bounded by the standard trick enabling the machine to use consecutive registers in its memory. In general these simulations achieve the required space bound at the price of a huge consumption of time; the same values are computed over and over again by the recursion.

The proofs of the corresponding second inclusion for the cases of vector machines and MRAMs invoke a direct simulation of the transitive closure algorithm, by subroutines which build the matrix $A(x, M, S)$ into a register and which compute its transitive closure by iterated squarings. A main ingredient is the programming of a routine which builds a bit string consisting of the 2^K bit strings representing the first 2^K integers, separated by markers, and of bit strings to be used as masks for extracting a given bit position from these integers in parallel in a single instruction. The idea of simplyfying these proofs by invoking QBF as a PSPACE-complete problem was also used in [9].

3.4. Machines with true parallelism

In this section we consider models which provide observable parallelism by having multiple processors that operate on shared data and/or shared channels. All will become clear from the next section, for these models it really matters how the model is defined in details. Some models which belong to the second machine class turn out to become more powerful when a minor feature of their definition is changed.

In these models a computation can proceed either synchronously (all processors perform a step of the computation at the same time, driven by a local clock) or asynchronously (each processor computes at its own speed).

There exist a number of possible strategies for resolving the write conflicts which arise when several processors attempt to write in the same location of shared memory. In the *priority-write* strategy the processor with the lowest index will succeed in writing in a shared location and the other values will be lost. Other strategies which have been investigated are *exclusive-write* (no two processors can write in the same global register at all), *common-write* (if two processors try to write different values at the same time in the same register then the computation jams but writing the same value is permitted), and *arbitrary-write* (one of the writers wins but it is nondeterministically determined which one). For yet another approach to multiple writes, see [25]. The computational power of the parallel RAM models based on these resolution strategies has been compared in [28, 29, 68] under the (unrealistic) assumption that the individual processors have arbitrary computing power. In these investigations the priority-write model has established itself as the most powerful one.

There are several methods of controlling the creation of parallel processors. Some models have a large or even infinite collection of identical processors which operate in parallel. In other models processors can create, by their own action, a finite number of new processors running in parallel. In this way an arbitrarily large tree of active processors can be activated as time proceeds.

Curiously enough there exists no proposal for a parallel version of the SMM. It seems that this possibility has never been investigated from a complexity point of view. A parallel version of the KUM has been introduced in the Soviet literature already twenty-five years ago; this *Kolmogorov–Barzdin' machine* has been described by Barzdin' [4, 5, 6] but mainly for its computational properties like the existence of a universal machine and the problem of giving a precise formulation of the involved parallel transformation of a graph (predating the literature on graph grammars). In two papers with Kalnin'sh [7, 8] which appeared ten years later, these computational investigations were continued. There is one paper which deals with complexity issues: in [6] it is established that a higher-dimensional grid cannot be simulated on a lower-dimensional one with constant-factor overhead in space (see also [24]). As such the construction of a parallel version of the SMM which behaves like a second machine class member remains an interesting exercise.

3.4.1. The SIMDAG model

In the SIMDAG model [39] (*Single Instruction, Multiple Data AGregate*), there exists a single global processor which can broadcast instructions to a potentially infinite sequence of local processors, in such a way that only a finite number of them are activated. The mechanism to keep finite the number of processors activated in a single step uses the signature of the local processors. Each local processor has a read-only register, called the signature, containing a number which uniquely identifies this local processor. In broadcasting an instruction, the global processor includes a threshold value, and any local processor with a signature less than the threshold value transmitted performs the instruction while the others remain inactive.

Since the global processor can at most double the value of its threshold during a single step, it follows that the number of subprocessors activated is bounded by an exponent in the running time of the SIMDAG computation. The local processors operate both on local memory and on the shared memory of the global processor, where write conflicts are resolved by priority; the local processor with the lower index becomes the winner in case of a write conflict.

For a device like the SIMDAG it is necessary to restrict the power of the arithmetic instructions involved. Otherwise—as we will see in the sequel—the machine may become too powerful. In the SIMDAG model the instruction repertoire for local and global processors involves additive arithmetic and parallel Boolean operations, combined with restricted shifting (division by 2). The writing of data in global storage by local processors is made conditional by stipulating that writing the value 0 is suppressed. In this way the *or* of a list of Boolean values computed by the local processors can be computed in a single write instruction by letting each local processor write a 1 in a fixed register in global memory if its bit equals 1, whereas the value 0 is not transmitted to the global memory.

3.5. THEOREM. *The SIMDAG model is a member of the second machine class.*

PROOF (*sketch*). The inclusion SIMDAG-NPTIME$\subseteq$PSPACE is shown by an argument similar to that used in the case of the EDITRAM.

The trace of a nondeterministic SIMDAG computation can be guessed and written down on a worktape in polynomial space. Next one defines a pair of recursive procedures *global*(*time*, *register*) and *local*(*time*, *register*, *signature*) which evaluate to the value stored at the given time in the given register of the global and given local processor respectively. The arguments of these recursive procedures can be written down in polynomial space (due to the restrictions on the arithmetics of the SIMDAG) and the recursion depth is bounded by the running time of the SIMDAG computation. Using these recursive procedures the guessed trace of the SIMDAG computation can be certified to be a correct accepting computation. As before, the time needed for the simulation is very large due to the recomputation of intermediate results.

The converse inclusion PSPACE $\subseteq$ SIMDAG-PTIME is shown by presenting an implementation of the transitive closure algorithm which runs in polynomial time. This algorithm first loads the $K \times K$ matrix $A(x, M, S)$ which was introduced in Section 3.1 in global memory. For convenience assume that K is a power of 2. The matrix is constructed by letting processor $i+K \cdot j$ evaluate the entry $A(x, M, S)[i, j]$. By inspecting its signature, the processor (using the Boolean operations and the division by 2 as a shift operator) can determine the values of i and j first and unravel these values as bit patterns next in order to see whether the two encoded Turing machine configurations are equal or are connected by a single step. By a global write the matrix is loaded into global memory.

After formation of the matrix the transitive closure is computed by iterated squaring. Each squaring is computed by letting local processor $i+K \cdot j+K \cdot K \cdot k$ read the values of $A(x, M, S)[i, k]$ and $A(x, M, S)[k, j]$, form the *and* of these two values and write the result (conditionally) in $A(x, M, S)[i, j]$. This requires a constant number of steps. After these squarings the existence of an accepting computation is determined by the global processor by inspecting the proper matrix entry of $A(x, M, S)$. □

3.4.2. *The array processing machine*

Our next model, the *array processing machine* (APM), was proposed by van Leeuwen and Wiedermann [134]. It was inspired by the contemporary vectorized supercomputers. This machine has the storage structure of an ordinary RAM but, besides the traditional accumulator, it also has a *vector accumulator* which consists of a potentially unbounded linear array of standard accumulators.

The array processing machine combines the instruction set of a standard RAM with a new repertoire of vector instructions which operate on the vector accumulator. These instructions allow for reading, writing, transfer of data and arithmetic on vectors of matching size which consist of consecutive locations in storage and/or an initial segment of the vector accumulator.

Each operation on the vector accumulator destroys its previous content. Conditional control on a vector operation is possible by the use of a mask which consists of an array of Boolean values (0 or 1) of the same size as the vector operands; the vector instruction now is performed only at those locations corresponding to occurrences of 1 in the mask. A complete address for a vector operation therefore may consist of four integers: lower and upper bounds of the vector argument and the mask respectively.

The power of parallelism is provided to the model by the time measure used: uniform time or logarithmic time, where every vector instruction is charged according to its most expensive scalar component. So in a vector-*LOAD* the logarithmic time complexity is proportional to the logarithm of the upper bound of the operand and/or mask plus the logarithm of the largest value loaded into the vector accumulator.

In [134] it is proved that the array processing machine is a member of the second machine class, by providing mutual simulations with respect to the SIMDAG. The simulations require polynomial n^4 overhead. In both directions the simulations require nontrivial programming techniques, and a parallel $O(\log^2 n)$ implementation of Batcher's sort is an essential element of the simulation.

Inspection of the model shows that a proof that the APM obeys the Parallel Computation Thesis can be easily obtained by the techniques used for other devices in this chapter.

3.6. THEOREM. *The APM model is a member of the second machine class.*

PROOF (*sketch*). To prove PSPACE$\subseteq$APM-PTIME one can show that QBF can be solved in polynomial time on an APM; here the main ingredient is the construction of n vectors in storage of length 2^n where vector j contains the value of bit j in the binary representation of the numbers $0, \ldots, 2^n-1$. Using those vectors one can evaluate a given propositional kernel in linear time using the vector instructions, and the resulting vector can be folded together according to the quantifiers in order to provide the final result.

For the converse, APM-NPTIME$\subseteq$PSPACE, the usual technique of writing down a computation trace and certifying it by means of a recursive procedure will work. The detour via PSPACE implies the existence of mutual polynomial time overhead simulations of SIMDAG and APM but it does not provide explicit simulation overheads as indicated above. □

3.4.3. Models with recursive parallelism

As an example of a parallel machine of a different character we mention the recursive Turing machine introduced by Savitch [105]. In this model every copy of the device can spawn new copies which start computing in their own environment of worktapes, and which communicate with their originator by means of channels shared by two copies of the machine.

A similar model, based on the RAM, is the k-PRAM described by Savitch and Stimson [108]. In this model a RAM-like device can create up to k copies of itself, which start computing in their local environment, while their creator is continues computing. These copies themselves can create new offspring as well. Data are transmitted from parent to child at generation time by loading parameters in the registers of the offspring. Upon termination a child can return a result to its parent by writing a value in a special register of the parent which is read-only for the parent. The parent can inspect the register to see whether the child has already written into it; if he reads the register before the child has assigned a value to it, the parent's computation is suspended. Using these features a parent can activate a number of children and consume the first value which is returned, aborting the computations of the remaining children which have not yet terminated.

Clearly, models based on such local communication channels are much slower in broadcasting information to a collection of subprocessors and in gathering the answers. However, the validity of the Parallel Computation Thesis is not disturbed by this delay, due to the fact that the slowdown is at worst polynomial: in order to activate an exponential number of processors by spawning subprocessors, polynomial time suffices in case a complete binary tree is formed; the time consumed for activating this tree is usually polynomial in terms of the time consumed by writing down the data to be processed by the subprocessors.

It is not difficult to see that the above models support an implementation of the recursive version of the transitive closure algorithm from Section 3.1, which runs in polynomial time. This explains why PSPACE is contained in the parallel-PTIME class for these models. The time overhead, which used to be n^2 before, now becomes n^3 due to the fact that the recursive algorithm contains also a loop over all intermediate nodes. This loop would require exponential time if performed sequentially, and therefore it is replaced by a recursive expansion where a call at order w produces an exponential number of calls at order $w-1$. The recursion depth of the procedure is $\log^2 N$ instead of $\log N$ where N is the size of the transition matrix.

The same recursive expansion trick is used in order to show that for these models parallel-NPTIME $\subseteq$ parallel-PTIME. A single nondeterministic computation of length T is replaced by 2^T deterministic computations relative to a choice string which is different for every copy. Transforming this idea into a complete proof exposes a number of complications. The final proof for this inclusion yields an n^6 time overhead for elimination of nondeterminism, see [105, 108].

A crucial difference between the SIMDAG model and the recursive models is that in the SIMDAG model the local processors, if they are active at all at some time, all execute the same instruction on data which may be different. As a consequence it is possible to write down the trace of executed instructions of a SIMDAG computation in polynomial space. In the recursive models each processor, once being activated, performs its own program except for the impact of communication with its parent or its offspring. It becomes therefore impossible to write down the complete computation trace in polynomial space if an exponential number of processors is activated.

This has consequences for the proof of the inclusion parallel-PTIME $\subseteq$ PSPACE for these models. Rather than writing down the entire trace of the computation and certifying it by a recursive procedure, the entire genealogical tree of machine incarnations is searched by a recursive procedure. Recursion depth is bounded by the polynomial bound on the running time of the parallel machine. A polynomial bound on the size of the recursive stack frames is obtained from the fact that the machine has no powerful arithmetic. It is crucial at this place that the communication between parent and offspring is entirely local; if offspring can write in global memory, the proof breaks down. A further complication results from the fact that a parent can abort some of its children before they have terminated; evaluating the result produced by such a child may lead to a divergent computation. Therefore all devices are extended with a counter, to keep track of global synchronous time. Such counters are needed anyhow in order to let a parent know which one of its two children was the first to terminate. For more details, see [105, 108].

4. Parallel machine models outside the second machine class

In this section we consider several models which can not be classified as members of either the first or the second machine class. One model is much weaker than the models in the previous section, other models are more powerful than the members of the second machine class, including a model claimed to perform all computations in constant time.

4.1. A weak parallel machine

The *parallel Turing machine* (PTM), introduced by Wiedermann [140], should not be confused with the models introduced by Savitch under the name recursive Turing machines [105]. In both cases one considers a Turing machine with a nondeterministic program, where a choice of possible successor states leads to the creation of several devices, each continuing in one of the possible configurations. Where in Savitch's model the entire configuration is multiplied, in Wiedermann's model only the finite control and the heads are multiplied, thus leading to a proliferation of Turing automata all operating on the same collection of tapes.

The PTM consists of a finite control with k d-dimensional work-tapes, the first of which contains the input at the start of the computation. Each control has one head on every tape. The program of the device is a standard nondeterministic Turing program for a machine with k d-dimensional single-head tapes; however, instead of choosing a next state when facing a nondeterministic move, the machine creates new copies of its control and heads, which go on computing on the same tapes. There are no read conflicts; write conflicts are resolved by the *common-write strategy*: if two heads try to write different symbols at the same square, the computation aborts and rejects; if two heads try to write the same symbol, this symbol is written and the heads move on. An accepting computation is a computation in which every finite control that is created during the computation halts in an accepting state.

The crucial observation which makes this model weaker than the members of the second machine class we have met, is related to the achievable degree of parallelism. Although the machine can activate an exponential number of copies of itself in polynomial time, these copies operate on the same tapes and therefore only a polynomial number of essentially different copies will be active at any moment in time. If two finite controls are in the same internal state and have their heads positioned on the same tape squares, their behavior will be equal from that time onwards; hence the two controls actually merge into a single unit. This leads to an upperbound of $q \cdot S^k$ on the number of different controls being active at the same time, where q denotes the number of states in the program and S denotes the space used by the device. Since space is bounded by time this leads to a polynomial bound on the number of different copies.

Based on this observation it becomes possible to simulate the PTM by a standard deterministic Turing machine with polynomial time overhead: the simulator maintains on some additional worktape a list of all active finite controls with their head positions, and by maintaining a pair of old and new worktapes the machine can process the updates of each control in sequence, taking care of the needed multiplication of controls and checking for write conflicts.

It cannot be inferred from this simulation that the PTM is a member of the first machine class, since it is not clear that the above simulation can be modified in such a way that the space overhead becomes a constant factor. For the special case of a single-tape PTM a constant-factor space overhead simulation is achieved, by storing with each tape cell the set of states achieved by heads scanning this cell; this set (a subset of the fixed set of states of the machine) can be written down in an amount of space which is independent of the length of the input. But for the case of more tapes or more heads on a single tape, it is not sufficient to mark the tape cells by the states in which they are scanned, since one must also know which heads belong together to a single finite control. This requires the encoding of head positions for each device. Therefore the naive simulation requires space $O(S^k \cdot \log S)$.

Wiedermann observes that one can recognize the language of palindromes with a PTM with two one-dimensional tapes, where the first tape is a read-only input tape, in space O(1) if the space of the input tape is not counted. But the heads on the read-only input tape may multiply, thus storing information on this tape by their positions. It is therefore unlikely that a constant-factor space overhead simulation is possible, since palindromes cannot be recognized in constant space by a standard machine. This particular example, however, seems to depend on the particular interpretation that the space on the input tape is not counted, and an example where the input head is common for all copies of the finite control seems to be required for a more convincing separation result.

The simulations show that P = PTM-PTIME and that PSPACE = PTM-PSPACE; the PTM model therefore does not obey the Parallel Computation Thesis, unless P = PSPACE. The PTM model has the interesting property that for several practical problems in P an impressive speed-up by pipelining can be achieved, even though the device is not a member of the second machine class. For details, see [140].

4.2. *Beyond the second machine class*

In this section we present several models which seem to be more powerful than the members of the second machine class. We also discuss a critique of the Parallel Computation Thesis.

4.2.1. *The* MIMD-RAM

The MIMD-RAM which was introduced by Fortune and Wyllie [31] as a hybrid of the SIMDAG and the k-PRAM described in Section 3.4. In the MIMD-RAM a machine can create offspring by forking, where the offspring will perform the instructions of its own program. Upon creation the offspring machine will start executing at the first instruction in its program. The subsequent course of the computation is influenced by the initial value of the accumulator, which is set by the creator. The machine has standard additive RAM arithmetic.

Each offspring processor has its own local memory. The machines communicate through global memory. Global and local memories are standard RAM memories. Each processor can read and write in its local memory using the standard RAM instructions. For passing information upwards, processors write in global memory. Simultaneous reads from global memory are allowed, but simultaneous writes are prohibited. The machine accepts if the oldest copy of a processor accepts by halting with a 1 in its accumulator. In the model a special input convention is used which supports the reading of an input of length n in time $O(\log n)$, so sublinear running times become meaningful.

The MIMD-RAM has the property that its deterministic version is a member of the second machine class, whereas its nondeterministic version provides us with a full exponential speed-up in time. This is expressed by the following result.

4.1. THEOREM. *The MIMD-RAM model satisfies the following equations:*

(1) MIMD-RAM-PTIME = PSPACE, *and*

(2) MIMD-RAM-NPTIME = NEXPTIME.

Before we outline the result, we need an NP-complete problem that will be used in the proof: BOUNDED TILING [103], called SQUARE TILING (GP13) by Garey and Johnson [37].

BOUNDED TILING

Instance: A finite set W of tiles (squares with colors given on their edges), and an $N \times N$ square V with a given coloring on the $4 \cdot N$ unit-edge segments on the border of V.

Question: Is it possible to tile the square V with copies of the tiles in W (without rotations or reflections) such that each pair of adjacent tiles have matching colors on their common edge, and such that the tiles adjacent to the border of the square V have colors matching the given coloring of the border on their exterior edges?

PROOF (*sketch*). PSPACE $\subseteq$ MIMD-RAM-PTIME: One proof consists of the simulation of a 2-PRAM where the channels have been replaced by global registers. Another proof consists of a version of the transitive closure algorithm for the computation graph. Simultaneous writes are prevented by designing a fan-in algorithm for evaluation of the *and* or *or* of $O(S)$ bits. In both cases the resulting program is a deterministic MIMD-RAM program.

MIMD-RAM-PTIME $\subseteq$ PSPACE: Due to the restrictions on the arithmetic, in time T at most 2^T machines are created, each operating on at most 2^T registers and operating on values bounded by 2^T. This implies that all arguments for a recursive procedure $val(i, j, t)$, which yields the value of register j in device i at time t, can be written down in polynomial space. A similar procedure can be obtained for the contents of the global memory. However, it is no longer possible to write down in polynomial space the entire trace of the computation, since each device performs its own instructions. Instead the trace of the computation is built together with the tree of recursive calls; a certain value is present in some register as a result of the execution of previous instructions which have been executed, because of the presence of certain values in other registers at some earlier time etc. It is not hard to see that an alternating RAM can evaluate this recursion, guessing and certifying the trace at the same time. From the fact that the MIMD-RAM involved is deterministic, it follows that only the true values and instructions performed can be certified. It also follows that this alternating RAM consumes time polynomially bounded by T. Hence the above inclusion follows.

MIMD-RAM-NPTIME $\subseteq$ NEXPTIME and MIMD-RAM-NLOGTIME $\subseteq$ NP: These inclusions can be shown by brute-force simulations: the overhead in simulating a nondeterministic MIMD-RAM computation by writing down an entire record of its computation is at most exponential in time.

NP $\subseteq$ MIMD-RAM-NLOGTIME and NEXPTIME $\subseteq$ MIMD-RAM-NPTIME: These inclusions are shown by designing a nondeterministic MIMD-RAM algorithm which recognizes an NP-complete problem in logarithmic time. This yields the first inclusion, and the second one follows by a similar argument or by a padding argument. For the NP-complete problem we take BOUNDED TILING. An instance of BOUNDED TILING can be solved by a nondeterministic MIMD-RAM which operates as follows: first it creates N^2 copies, each covering a single unit square inside V. This requires time $O(\log N)$. Next, in a nondeterministic move, each processor guesses the tile to be placed on its square. This is the unique nondeterministic move in the entire program. Subsequently, using the global memory, each processor exchanges its tile with the processors representing its neighbors, and certifies whether the choices match. The result is obtained by a standard fan-in communication of the bits computed in this way. Evaluation of the match requires constant time; the fan-in procedure requires $O(\log N)$ steps. □

4.2.2. The LPRAM

The LPRAM is a model introduced by Savitch [107]. It is a hybrid of the k-PRAM and the MRAM, since it combines the recursive parallelism of the k-PRAM with the powerful arithmetic of the MRAM. Communication is by channels only, as in the k-PRAM.

4.2. THEOREM. *The LPRAM satisfies the following equations:*

(1) LPRAM-PTIME = PSPACE, *and*

(2) LPRAM-NPTIME = NEXPTIME.

PROOF (*sketch*). PSPACE $\subseteq$ LPRAM-PTIME: trivial; do not use the vector instructions.

LPRAM-PTIME $\subseteq$ PSPACE: by a standard guess-and-verify method. The same alternating verifier can be used as in Section 4.2.1 but, since the vector instructions allow for exponential growth of values, the certification of values should proceed at the bit level rather than at the register level. This is the same trick as is used for the inclusion MRAM-NPTIME $\subseteq$ PSPACE by Pratt and Stockmeyer [94] and Hartmanis and Simon [44, 45].

LPRAM-NPTIME $\subseteq$ NEXPTIME and LPRAM-NLOGTIME $\subseteq$ NP: These inclusions are again obtained by a brute-force simulation.

NP $\subseteq$ LPRAM-NLOGTIME and NEXPTIME $\subseteq$ LPRAM-NPTIME: These inclusions are again proved by implementing a nondeterministic LPRAM-algorithm for BOUNDED TILING which runs in logarithmic time: the machine first creates a tree of N^2 offspring processors which guess a tile for some unit square inside V. These guesses are communicated to the oldest processor by a fan-in procedure, where along the way the exponential growing information is stored using the vector instructions. Next the same information is distributed once more over N^2 processors, in such a way that each processor obtains a cell with its four neighbors. These N^2 processors then check whether the tiles match, and finally the results are communicated to the oldest processor by a bounded fan-in procedure. □

4.2.3. Extending the SIMDAG with powerful arithmetic

We now consider a hybrid of the SIMDAG and the MRAM. The starting point is a more refined analysis of the running time of the transitive closure algorithm in Section 3.3.1. Such an analysis shows that the total running time consists of three parts:

(1) $O(\log K)=O(S)$ for the evaluation of the matrix size K,

(2) $O(S)$ for unraveling the configurations and computing $A(x, M, S)$, and

(3) $O(\log T)=O(S)$ for computing the transitive closure of $A(x, M, S)$.

This more refined analysis explains why the power of the arithmetic in the model is a crucial factor in establishing whether the model belongs to the second machine class or not. Assuming that we can use multiplication in unit time, the first contribution is reduced to $O(\log\log K)=O(\log S)$; given more powerful parallel Boolean instructions, the unraveling can be distributed over $O(S)$ processors, and the contribution of step (2) becomes $O(\log S)$ as well. As a consequence, for the resulting PSIMDAG model (for "powerful SIMDAG") one obtains NEXPTIME$\subseteq$PSIMDAG-PTIME. Hence it is unlikely that such a model obeys the Parallel Computation Thesis, unless PSPACE$=$ NEXPTIME.

These ideas are the basis for the objections against the Parallel Computation Thesis as put forward by N. Blum [11]. However, in his paper he only considers the third phase of the above algorithm: the computation of the transitive closure itself in $\log T$ steps. It should be clear that in a model in which addition is the only arithmetic instruction available, steps (1) and (2) will still require time $O(S)$ But, given a suitable set of powerful arithmetical instructions, the required speed-ups of steps (1) and (2) from $O(S)$ to $O(\log S)$ become possible. It therefore seems that Blum's model requires such strong arithmetical instructions, but this makes his claims much weaker, given the fact that models showing this kind of behavior have been proposed before by Fortune and Wyllie and by Savitch.

Blum's machine models resemble the SIMDAG but, instead of the priority-strategy for write conflicts, the exclusive-write strategy is used in the PRAM and the common-write strategy in the WRAM. For these models Blum claims the following results.

4.3. THEOREM. *When equipped with powerful arithmetic the models satisfy*

(1) NEXPTIME$\subseteq$WRAM-PTIME,

(2) EXPTIME$\subseteq$PRAM-PTIME.

PROOF (*sketch*). Taking the computation of the transition matrix for granted, given the powerful arithmetic, the first inclusion, NEXPTIME$\subseteq$WRAM-PTIME, follows from the fact that in the transitive closure algorithm only 0s and 1s need to be written, and the writing of a 0 can be suppressed. Hence, in the case of a multiple-write, the writes will be consistent. Since a bound T on the running time of the simulated machine is given, we know that instead of the transitive closure the Tth power of the transition matrix can be computed.

The second inclusion, EXPTIME $\subseteq$ PRAM-PTIME, follows by realizing that, if the machine is deterministic, all powers of the transition matrix have the property that every row contains at most a single non-zero entry. From this it follows that during a squaring every matrix element $A(x, M, S)[i, j]$ will be written by at most one processor. Note that also step (2) of the algorithm must be modified in order to have the required fan-in of $O(S)$ bits of local information into a single matrix element: this can be obtained at the price of an additional time of $O(\log S)$. So Blum's construction is valid on the PRAM for deterministic computations. □

4.2.4. Arbitrary computations in constant time

Intuition in computation theory states that, whatever machine model one selects, it should always remain true that computation time for the model behaves like a complexity measure [10]. In particular, it should never be possible to perform arbitrary complex computations in constant time. A model which violates this very mild restriction has been described in the literature. We find the model in the context of an analysis of the correct interpretation of the Parallel Computation Thesis by Parberry [83, 84]. For similar results with slightly improved resource bounds, see [85].

Parberry considers a parallel RAM model which runs fully synchronously. The number of active processors is determined at the start of the computation. Beside the standard additive arithmetic, each processor has some shifts or multiplication-like instructions which enable processors to extract bits from bit strings at arbitrary locations. The processors communicate via a global memory. Each processor has a read-only register initialized at its index. Resources considered are time T, storage S (counted in RAM words), word size W (both in local and global storage), and the number of processors P as functions of the input size n: $T(n)$, $S(n)$, $W(n)$ and $P(n)$.

The model resembles the SIMDAG, with two major differences. Rather than having all processors activated by a single central processor, all processors with index $\leqslant P(n)$ are activated at time $t=0$ and remain active until all have halted. The result of the computation is found by inspecting register 0 in global memory. The second difference is that the model is nonuniform: some quantities like $P(n)$ and derived quantities are given to all processors in advance.

In this way the machine has sufficient power to perform the transitive closure algorithm from Section 3.3.1 in constant time. Remember the three parts of the running time analysis introduced in Section 4.2.3; if we let $T'(n)$ and $S'(n)$ denote the time and space bounds of the Turing machine computations on input x with $|x| = n$, and if $K(n)$, the size of the transition matrix $A(x, M, S'(n))$, equals $2^{c' \cdot S'(n)}$ for a suitable constant c', then we obtain

(1) $O(\log K(n)) = O(S'(n))$ for the evaluation of the matrix size $K(n)$,

(2) $O(S'(n))$ for unraveling the configurations and computing $A(x, M, S'(n))$, and

(3) $O(\log T'(n)) = O(S'(n))$ for computing the transitive closure of $A(x, M, S'(n))$.

Step (1) becomes $O(1)$ due to the nonuniformity involved. Quantities like space $S'(n)$

and time $T'(n)$ for the Turing machine to be simulated are given in advance, as are the number of processors $P(n)$ which will be invoked.

Steps (2) and (3) are combined. In the simulation by Blum a team of $O(S'(n))$ processors is called upon in order to decompose the index of processor $i \cdot K(n)+j$ into bits and pieces to determine whether $A(x, M, S'(n))[i,j]$ equals 0 or 1. So the index of a processor is analyzed whether it represents a legal transition between two successive transitions of the machine which is simulated. In Parberry's simulation this idea is carried to the extreme: the index of a processor is analyzed to see whether it represents the coding of an entire accepting computation. The length of a string encoding of such a computation becomes $O(S'(n) \cdot T'(n))$, hence the total number of indices analyzed in this way becomes $2^{c \cdot T'(n) \cdot S'(n)}$ for some constant c. The analysis of one such index can be performed by breaking it into bits and pieces: given an encoding like the one used in time-space diagrams of Turing machines, a team of $T'(n) \cdot S'(n)$ processors can perform the comparisons of the coded configurations at time t and time $t+1$ in time $O(1)$; as Parberry indicates, T processors in a single team already suffice. One must check furthermore that the initial configuration at time $t=0$ represents a proper configuration on the real input, but this can be done in a similar way.

The result obtained in this way is that the simulation time becomes $O(1)$, at the price of using $P(n)=2^{c \cdot T'(n) \cdot S'(n)} \cdot T'(n)$ processors and word size $W(n)=O(T'(n) \cdot S'(n))$.

By a similar proof Parberry shows that if one can simulate with word size $W(n)$ a $B(n)$ time bounded deterministic Turing machine in time $B'(n)$ on his model, then one can use this simulation as a tool for speeding up a $T'(n)$ time bounded deterministic Turing machine computation with word size $O(W(n)+B(n)+\log T(n))$ to time $O(T(n)/B(n)+B'(n))$; the second term in the time analysis of the simulation is the time required to set up a transition matrix structure for computing $B(n)$ steps in time $O(1)$ by table look-up, whereas the first term measures the time needed for simulation of $T(n)$ steps in blocks of $B(n)$ steps each.

This brings Parberry to his analysis of how the above simulations support the Parallel Computation Thesis: a "reasonable" parallel device should satisfy the constraint that $W(n)$ is polynomially bounded by $T(n)$ and that (as a consequence) $P(n)$ is bounded by some expression $2^{c \cdot T(n)^k}$ for some constant k. The result above shows that, within these bounds, arbitrarily large polynomial speed-ups can be obtained with "reasonable" devices, but the speed-up to constant time is possible only at the price of violating this "reasonability" criterion. Parberry's result shows once more the hideous power of shifts and multiplicative instructions. Also the nonuniformity of the model makes it suspect. If quantities like $T'(n)$, $S'(n)$ or $P(n)$ must be computed from the input, the $O(1)$ time bound is invalidated.

Acknowledgment

The idea of the first machine class originated in a discussion in 1982 with L. Torenvliet, when the author tried to convince him of the self-evidence of what would later become the Invariance Thesis. The results for the traditional space measure for RAMs were discovered in close cooperation with C. Slot. Discussions with J.P.

Schmidt in the Summer of 1988 were very valuable for understanding the true power of the restricted unit-time multiplication instruction in RAMs. I am grateful to A. Schönhage, M.C. Loui and D.R. Luginbuhl for valuable discussions about the SMM model. The difference between the orthodox and the liberal interpretation of the Invariance Thesis was pointed out to me by J. Wiedermann. I am grateful to R. Freivals for providing the references to the work on the Kolmogorov–Barzdin' machine model.

The interest in parallel machines which led to the introduction of the second machine class (which pre-dates the first machine class by about one year) was provoked by the invitation by J. van Leeuwen and P.M.B. Vitányi to present an overview on models of parallelism in a seminar at the University of Utrecht [131]. This work underwent several revisions during presentations at the Banach Seminar on Foundations of Computation Theory (Warsaw) in 1985.

Finally, I thank L. Stockmeyer, L. Torenvliet, J. van Leeuwen, P.M.B. Vitányi, J. Wiedermann, and the students of my classes on machine models for valuable comments on this chapter.

References

[1] AHO, A.V., J.E. HOPCROFT and J.D. ULLMAN, *The Design and Analysis of Computer Algorithms* (Addison-Wesley, Reading, MA, 1974).

[2] ANDREEV, A.E., On a method for obtaining lower bounds for the complexity of individual monotone functions, *Soviet Math. Dokl.* **31** (1985) 530–534.

[3] BALCÁZAR, J.L., J. DÍAZ and J. GABARRÓ, *Structural Complexity I*, EATCS Monographs on Theoretical Computer Science, Vol. 11 (Springer, Berlin, 1988).

[4] BARZDIN', Ya.M., Universal pulsing elements, *Soviet Phys. Dokl.* **9** (1965) 523–525.

[5] BARZDIN', Ya.M., Universality problems in the theory of growing automata, *Soviet Phys. Dokl.* **9** (1965) 535–537.

[6] BARZDIN', Ya.M., Capacity of the medium and behavior of automata, *Soviet Phys. Dokl.* **10** (1966) 8–11.

[7] BARZDIN', Ya.M. and Ya.Ya. KALNIN'SH, On a language for the transformation of graphs intended for the specification of automata, *Automat. i Vychisl. Tekhn.* **7**(5) (1973) 22–28.

[8] BARZDIN', Ya.M and Ya.Ya. KALNIN'SH, A universal automaton with variable structure, *Autom. i Vychisl. Tekhn.* **8**(2) (1974) 9–17.

[9] BERTONI, A., G. MAURI and N. SABADINI, Simulations among classes of random access machines and equivalence among numbers succinctly represented, *Ann. Discrete Math.* **25** (1985) 65–90.

[10] BLUM, M., A machine-independent theory of the complexity of recursive functions, *J. Assoc. Comput. Mach.* **14** (1967) 322–336.

[11] BLUM, N., A note on the "Parallel Computation Thesis", *Inform. Process. Lett.* **17** (1983) 203–205.

[12] BLUM, N., A Boolean function requiring 3*n* network size, *Theoret. Comput. Sci.* **28** (1984) 337–345.

[13] BÖHLING, K.H. and B. VON BRAUNMÜHL, *Komplexität bei Turingmaschinen*, Reihe Informatik, Vol. 14 (B.I. Wissenschaftsverlag, Zürich, 1974).

[14] BORODIN, A., On relating time and space to size and depth, *SIAM J. Comput.* **6** (1977) 733–744.

[15] BORODIN, A., M. FISCHER, D.G. KIRKPATRICK, N.A. LYNCH and M. TOMPA, A time-space tradeoff for sorting on non-oblivious machines, *J. Comput. System Sci.* **22** (1981) 351–364.

[16] CHANDRA, A.K., D.C. KOZEN and L.J. STOCKMEYER, Alternation, *J. Assoc. Comput. Mach.* **28** (1981) 114–133.

[17] CHAZELLE, B. and L. MONIER, Unbounded hardware is equivalent to deterministic Turing machines, *Theoret. Comput. Sci.* **24** (1983) 120–123.

[18] COOK, S.A., The complexity of theorem proving procedures, in: *Proc. 3rd Ann. ACM Symp. on Theory of Computing* (1971) 151–158.

[19] COOK, S.A., Towards a complexity theory of synchronous parallel computation, *Enseign. Math.* **27** (1980) 99–124.

[20] COOK, S.A., The classification of problems which have fast parallel algorithms, in: M. Karpinski, ed., *Proc. Internat. Conf. on Fundamentals of Computation Theory*, Lecture Notes in Computer Science, Vol. 158 (Springer, Berlin, 1983) 78–93.

[21] COOK, S.A. and R.A. RECKHOW, Time bounded random access machines, *J. Comput. System Sci.* **7** (1973) 354–375.

[22] DAVIS, M., *Computability and Unsolvability* (McGraw-Hill, New York, 1958).

[23] DAVIS, M., Why Gödel didn't have Church's thesis, *Inform. and Control* **54** (1982) 3–24.

[24] DYMOND, P.W. and S.A. COOK, Hardware complexity and parallel computation, in: *Proc. 21st Ann. IEEE Symp. on Foundations of Computer Science* (1980) 360–372.

[25] DYMOND, P.W. and W.L. RUZZO, Parallel RAMs with owned global memory and deterministic context-free language recognition, in: L. Kott, ed., *Proc. 13th Internat. Coll. on Automata, Languages and Programming*, Lecture Notes in Computer Science, Vol. 226 (Springer, Berlin, 1986) 95–104.

[26] EDMONDS, J., Paths, trees, and flowers, *Canad. J. Math.* **17** (1965) 449–467.

[27] EVEN, S. and R.E. TARJAN, A combinatorial problem which is complete in polynomial space, *J. Assoc. Comput. Mach.* **23** (1976) 710–719.

[28] FICH, F.E., P.L. RAGDE and A. WIGDERSON, Relations between concurrent-write models of parallel computation, *SIAM J. Comput.* **17** (1988) 606–627.

[29] FICH, F.E., F. MEYER AUF DER HEIDE, P.L. RAGDE and A. WIGDERSON, One, two, three ... infinity: lower bounds for parallel computation, in: *Proc. 17th Ann. ACM Symp. on Theory of Computing* (1985) 48–58.

[30] FISCHER, P.C., A.R. MEYER and A.L. ROSENBERG, Real-time simulation of multihead tape units, *J. Assoc. Comput. Mach.* **19** (1972) 590–607.

[31] FORTUNE, S. and J. WYLLIE, Parallelism in random access machines, in: *Proc. 10th Ann. ACM Symp. on Theory of Computing* (1978) 114–118.

[32] FRAENKEL, A.S., M.R. GAREY, D.S. JOHNSON, T. SCHAEFER and Y. YESHA, The complexity of checkers on an $N \times N$ board (preliminary report), in: *Proc. 19th Ann. IEEE Symp. on Foundations of Computer Science* (1978) 55–64.

[33] FRAENKEL, A.S. and D. LICHTENSTEIN, Computing a perfect strategy for $n \times n$ chess requires time exponential in n, *J. Combin. Theory* **31** (1981) 199–214.

[34] FREDMAN, F.L., J. KOMLÓS and E. SZEMERÉDI, Storing a sparse table with O(1) worst case access time, in: *Proc. 23rd Ann. IEEE Symp. on Foundations of Computer Science* (1982) 165–169.

[35] FREDMAN, F.L., J. KOMLÓS and E. SZEMERÉDI, Storing a sparse table with O(1) worst case access time, *J. Assoc. Comput. Mach.* **31** (1984) 538–544.

[36] GALIL, Z. and W. PAUL, An efficient general-purpose parallel computer, *J. Assoc. Comput. Mach.* **30** (1983) 360–387.

[37] GAREY, M.S. and D.S. JOHNSON, *Computers and Intractability: a Guide to the Theory of NP-completeness* (Freeman, San Francisco, CA, 1979).

[38] GILBERT, J.R., T. LENGAUER and R.E. TARJAN, The pebbling problem is complete in polynomial space, *SIAM J. Comput.* **9** (1980) 513–524.

[39] GOLDSCHLAGER, L.M., A universal interconnection pattern for parallel computers, *J. Assoc. Comput. Mach.* **30** (1982) 1073–1086.

[40] GRUSKA, J., Systolic automata—power, characterizations, nonhomogeneity, in: M.P. Chytil and V. Koubek, eds., *Proc. Mathematical Foundations of Computer Science '84*, Lecture Notes in Computer Science, Vol. 176 (Springer, Berlin, 1984) 32–49.

[41] HALPERN, J.Y., M.C. LOUI, A.R. MEYER and D. WEISE, On time versus space III, *Math. Systems Theory* **19** (1986) 13–28.

[42] HAREL, D., Recurring dominos: making the highly undecidable highly understandable, *Ann. Discrete Math.* **24** (1985) 51–72.

[43] HARTMANIS, J., Observations about the developmentof theoretical computer science, *Ann. Hist. Comput.* **3** (1981) 42–51.

[44] HARTMANIS, J. and J. SIMON, On the power of multiplication in random access machines, in: *Proc. 15th Ann. IEEE Conf. on Switching and Automata Theory* (1974) 13–23.

[45] HARTMANIS, J. and J. SIMON, On the structure of feasible computations, in: M. Rubinoff and M.C. Yovits, eds., *Advances in Computers, Vol. 14* (Academic Press, New York, 1976) 1–43.

[46] HARTMANIS, J. and R.E. STEARNS, On the computational complexity of algorithms, *Trans. Amer. Math. Soc.* **117** (1966) 285–306.

[47] HEMMERLING, A., On the space complexity of multidimensional Turing automata, Preprint 2, E.-M. Arndt Univ., Greifswald, 1979.

[48] HENNIE, F.C. and R.E. STEARNS, Two-way simulation of multi-tape Turing machines, *J. Assoc. Comput. Mach.* **13** (1966) 533–546.

[49] HOPCROFT, J., W. PAUL and L. VALIANT, On time versus space and related problems, in: *Proc. 16th Ann. IEEE Symp. on Foundations of Computer Science* (1975) 57–64.

[50] HOPCROFT, J., W. PAUL and L. VALIANT, On time versus space, *J. Assoc. Comput. Mach.* **24** (1977) 332–337.

[51] HOPCROFT, J.E. and J.D. ULLMAN, *Introduction to Automata Theory, Languages, and Computation* (Addison-Wesley, Reading, MA, 1979).

[52] IBARRA, O.H. and S. MORAN, Some time-space trade-off results concerning single-tape and offline TM's, *SIAM J. Comput.* **12** (1983) 388–394.

[53] JACOBS, C.T.M. and P. VAN EMDE BOAS, Two results on tables, *Inform. Process. Lett.* **22** (1986) 43–48.

[54] JERRUM, M., The complexity of finding minimum-length generator sequences (extended abstract), in: J. Paredaens, ed., *Proc. 11th Internat. Coll. on Automata, Languages and Programming*, Lecture Notes in Computer Science, Vol. 172 (Springer, Berlin, 1984) 270–280.

[55] JERRUM, M., The complexity of finding minimum-length generator sequences, *Theoret. Comput. Sci.* **36** (1985) 265–290.

[56] JOHNSON, D.S., A catalog of complexity classes, in: J. van Leeuwen, ed., *Handbook of Theoretical Computer Science, Vol. A* (North-Holland, Amsterdam, 1990) 67–161 (this volume).

[57] KARP, R.M. and R.J. LIPTON, Some connections between nonuniform and uniform complexity classes, in: *Proc. 12th Internat. ACM Symp. on Theory of Computing* (1980) 302–309.

[58] KATAJAINEN, J., M. PENTTONEN and J. VAN LEEUWEN, Fast simulation of Turing machines by random access machines, *SIAM J. Comput.* **17** (1988) 77–88.

[59] KLEENE, S.C., Origins of recursive function theory, *Ann. Hist. Comput.* **3** (1981) 52–67.

[60] KOLMOGOROV, A.N. and V.A. USPENSKII, On the definition of an algorithm, *Uspekhi Mat. Nauk* **13** (1958) 3–28; English translation in: *Russian Math. Surveys* **30** (1963) 217–245.

[61] KOSARAJU, S.R., Real-time simulation of concatenable double-ended queues, in: *Proc. 11th Ann. ACM Symp. on Theory of Computing* (1979) 346–351.

[62] KOSARAJU, S.R. and M.J. ATALLAH, Optimal simulations between mesh-connected arrays of processors, *J. Assoc. Comput. Mach.* **35** (1988) 635–650.

[63] KRAPCHENKO, V.M., Complexity of the realization of a linear function in the class of P-circuits, *Mat. Zametki* **9**(1) 35–40; English translation in: *Math. Notes Acad. Sci. USSR* (1971) 21–23.

[64] LEONG, B.L., and J.I. SEIFERAS, New real-time simulations of multihead tape units, *J. Assoc. Comput. Mach.* **28** (1981) 166–180.

[65] LEWIS, H.R. and C.H. PAPADIMITRIOU, *Elements of the Theory of Computation* (Prentice Hall, Englewood Cliffs, NJ, 1981).

[66] LI, M., L. LONGPRÉ and P.M.B. VITÁNYI, The power of the queue, in: A.L. Selman, ed., *Structure in Complexity Theory*, Lecture Notes in Computer Science, Vol. 223 (Springer, Berlin, 1986) 219–233.

[67] LI, M. and P.M.B. VITÁNYI, Tape versus queue and stacks: the lower bounds, *Inform. and Comput.* **78** (1988) 56–85.

[68] LI, M. and Y. YESHA, Separation and lower bounds for ROM and nondeterministic models of computation, *Inform. and Comput.* **73** (1987) 102–128.

[69] LICHTENSTEIN, D. and M. SIPSER, Go is polynomial-space hard, *J. Assoc. Comput. Mach.* **27** (1980) 393–401.

[70] LORYS, K. and M. LISKIEWICZ, Two applications of Fürer's counter to one-tape non-deterministic TMs, in: M.P. Chytil, L. Janiga and V. Koubek, eds., *Proc. Mathematical Foundations on Computer Science '88*, Lecture Notes in computer Science, Vol. 324 (Springer, Berlin, 1988) 445–453.

[71] LUGINBUHL, D.R. and M.C. LOUI, Hierarchies and space measures for pointer machines, Report UILU-ENG-88-2245, Dept. of Electrical Engineering, Univ. of Illinois at Urbana-Champaign, 1988.

[72] LUPANOV, O.B., On the asymptotic bounds of complexities of formulas which realize logic algebra functions, *Dokl. Akad. Nauk SSSR* **128** (1959) 464–467; English translation in: *Autom. Expr.* **2** (1960) 12–14.

[73] MAASS, W., Combinatorial lower bound arguments for deterministic and nondeterministic Turing machines, *Trans. Amer. Math. Soc.* **302** (1985) 675–693.

[74] MAASS, W. and G. SCHNITGER, An optimal lower bound for Turing machines with one work tape and a two-way input tape, in: A.L. Selman, ed., *Structure in Complexity Theory*, Lecture Notes in Computer Science, Vol. 223 (Springer, Berlin, 1986) 249–264.

[75] MACHTEY, M. and P. YOUNG, *An Introduction to the General Theory of Algorithms*, Theory of Computation Series, Vol. 2 (North-Holland, New York, 1978).

[76] MCCOLL, W.F. and M.S. PATERSON, The depth of all Boolean functions, *SIAM J. Comput.* **6** (1977) 373–380.

[77] MEAD, C. and L. CONWAY, *Introduction to VLSI Systems* (Addison-Wesley, Reading MA, 1980).

[78] MEHLHORN, K., On the program size of perfect universal hash functions, in: *Proc. 23rd Ann. IEEE Symp. on Foundations of Computer Science* (1982) 170–175.

[79] MEHLHORN, K., *Data Structures and Algorithms 1: Sorting and Searching*, EATCS Monographs on Theoretical Computer Science, Vol. 1 (Springer, Berlin, 1984).

[80] MEYER AUF DER HEIDE, F., Efficiency of universal parallel computers, *Acta Inform.* **19** (1983) 269–296.

[81] MINSKY, M., *Computation, Finite and Infinite Machines* (Prentice Hall, Englewood Cliffs, NJ, 1967).

[82] NEČIPORUK, È.I., A Boolean function, *Soviet Math. Dokl.* **7** (1966) 999–1000.

[83] PARBERRY, I., A complexity theory for parallel computation, Ph.D. Thesis, Dept. of Computer Science, Univ. of Warwick, Coventry, 1984.

[84] PARBERRY, I., Parallel speedup of sequential machines: a defense of the parallel computation thesis, *SIGACT News* **18**(1) (1986) 54–67.

[85] PARBERRY, I. and G. SCHNITGER, Parallel computation with threshold functions (preliminary version), in: A.L. Selman, ed., *Structure in Complexity Theory*, Lecture Notes in Computer Science, Vol. 223 (Springer, Berlin, 1986) 272–290.

[86] PATERSON, M.S., Tape bounds for time bounded Turing machines, *J. Comput. System Sci.* **6** (1972) 116–124.

[87] PAUL, W.J., On-line simulation of $k+1$ tapes by k tapes requires nonlinear time, *Inform. and Control* **53** (1982) 1–8.

[88] PAUL, W.J., N. PIPPENGER, E. SZEMERÉDI and W.T. TROTTER, On determinism versus non-determinism and related problems, in: *Proc. 24th Ann. IEEE Symp. on Foundations of Computer Science* (1983) 429–438.

[89] PAUL, W.J. and R. REISCHUK, On time versus space II, *J. Comput. System. Sci.* **23** (1981) 108–126.

[90] PIPPENGER, N., On simultaneous resource bounds, in: *Proc. 20th Ann. IEEE Symp. on Foundations of Computer Science* (1979) 307–311.

[91] PIPPENGER, N., Probabilistic simulations, in: *Proc. 14th Ann. ACM Symp. on Theory of Computing* (1982) 17–26.

[92] PIPPENGER, N. and M.J. FISCHER, Relations among complexity measures, *J. Assoc. Comput. Mach.* **26** (1979) 361–381.

[93] PRATT, V.R., The effect of basis on size of Boolean expressions, in: *Proc. 16th Ann. IEEE Symp. on Foundations of Computer Science* (1975) 119–121; also: *Kibernet. Sb. Nov. Ser.* **17** (1980) 114–123 (in Russian).

[94] PRATT, V.R. and L.J. STOCKMEYER, A characterization of the power of vector machines, *J. Comput. System Sci.* **12** (1976) 198–221.

[95] RAZBOROV, A.A., Lower bounds for the monotone complexity of some Boolean functions, *Soviet Math. Dokl.* **31** (1985) 354–357.

[96] REIF, J.H., Complexity of the mover's problem and generalizations (extended abstract), in: *Proc. 20th Ann. IEEE Symp. on Foundations of Computer Science* (1979) 421–427.

[97] REISCH, S., HEX ist PSPACE-volständig, *Acta Inform.* **15** (1981) 167–191.

[98] ROBSON, J.M., The complexity of go, in: R.E.A. Mason, ed., *Information Processing '83, Proc. 9th IFIP World Computer Congress* (North-Holland, Amsterdam, 1983) 413–418.

[99] ROBSON, J.M., *N* by *N* checkers is exptime complete, *SIAM J. Comput.* **13** (1984) 252–267.
[100] RUZZO, W.L., An improved characterisation of the power of vector machines, Report DCS-TR-78-10-01, Dept. of Computer Science, Univ. of Washington, Seattle, 1978.
[101] RUZZO, W.L., Tree-size bounded alternation, *J. Comput. System Sci.* **21** (1980) 218–235.
[102] SAVAGE, J.E., *The Complexity of Computing* (Wiley, New York, 1976).
[103] SAVELSBERG, M. and P. VAN EMDE BOAS, BOUNDED TILING: an alternative to SATISFIABILITY?, in: G. Wechsung, ed., *Proc. 2nd Frege Conf.* (Akademie Verlag, Berlin (GDR), 1984) 354–365.
[104] SAVITCH, W.J., Relations between deterministic and nondeterministic tape complexities, *J. Comput. System Sci.* **4** (1970) 177–192.
[105] SAVITCH, W.J., Recursive Turing machines, *Internat. J. Comput. Math.* **6** (1977) 3–31.
[106] SAVITCH, W.J., The influence of the machine model on computational complexity, in: J.K. Lenstra, A.H.G. Rinnooy Kan and P. van Emde Boas, eds., *Interfaces between Computer Science and Operations Research*, Mathematical Centre Tracts, Vol. 99 (Centre for Mathematics and Computer Science, Amsterdam, 1978) 1–32.
[107] SAVITCH, W.J., Parallel random access machines with powerful instruction sets, *Math. Systems Theory* **15** (1982) 191–210.
[108] SAVITCH, W.J. and M.J. STIMSON, Time-bounded random access machines with parallel processing, *J. Assoc. Comput. Mach.* **26** (1979) 103–118.
[109] SCHAEFER, T.J., Complexity of some two-person perfect-information games, *J. Comput. System Sci.* **16** (1978) 185–225.
[110] SCHMIDT, J.P. and A. SIEGEL, The spatial complexity of oblivious *k*-probe hash functions, Manuscript, Courant Institute, New York, 1988.
[111] SCHNITGER, G., Storage Modification Machines versus Kolmogorov–Uspenskii Machines (an information flow analysis), Manuscript, Dept. of Computer Science, Pennsylvania State Univ., College Park, PA, 1987.
[112] SCHNORR, C.P., The network complexity and the Turing machine complexity of finite functions, *Acta Inform.* **7** (1976) 95–107.
[113] SCHÖNHAGE, A., On the power of random access machines, in: H.A. Maurer, ed., *Proc. 6th Internat. Coll. on Automata, Languages and Programming*, Lecture Notes in Computer Science, Vol. 71 (Springer, Berlin, 1979) 520–529.
[114] SCHÖNHAGE, A., Storage modification machines, *SIAM J. Comput.* **9** (1980) 490–508.
[115] SCHÖNHAGE, A., Tapes versus pointers, a study in implementing fast algorithms, *EATCS Bull.* **30** (1986) 23–32.
[116] SCHÖNHAGE, A., A nonlinear lower bound for Random-Access Machines under logarithmic cost, *J. Assoc. Comput. Mach.* **35** (1988) 748–754.
[117] SCHORR, A., Physical parallel devices are not much faster than sequential ones, *Inform. Process. Lett.* **17** (1983) 103–106.
[118] SEIFERAS, J. and P.M.B. VITÁNYI, Counting is easy, *J. Assoc. Comput. Mach.* **35** (1988) 985–1000.
[119] SHEPHERDSON, J.C. and H.E. STURGIS, Computability of recursive functions, *J. Assoc. Comput. Mach.* **10** (1963) 217–255.
[120] SIPSER, N., Halting space-bounded computations, *Theoret. Comput. Sci.* **10** (1980) 335–337.
[121] SLOT, C. and P. VAN EMDE BOAS, On tape versus core; an application of space efficient hash functions to the invariance of space, in: *Proc. 16th Ann. ACM Symp. on Theory of Computing* (1984) 391–400.
[122] SLOT, C. and P. VAN EMDE BOAS, The problem of space invariance for sequential machines, *Inform. and Comput.* **77** (1988) 93–122.
[123] STEGWEE, R.A., L. TORENVLIET and P. VAN EMDE BOAS, The power of your editor, Report RJ 4711 (50179), IBM Research Lab., San Jose, CA, 1985.
[124] STOCKMEYER, L., The complexity of decision problems in automata theory and logic, Report MAC-TR-133, Lab. for Computer Science, Massachusetts Institute of Technology, Cambridge, MA, 1974.
[125] STOCKMEYER, L., The polynomial time hierarchy, *Theoret. Comput. Sci.* **3** (1977) 1–22.
[126] STOCKMEYER, L., Classifying the computational complexity of problems, *J. Symbolic Logic* **52** (1987) 1–43.
[127] STOSS, H.J., Zwei-Band-Simulation von Turingmaschinen, *Comput.* **7** (1971) 222–235.

[128] TOFFOLI, T. and N. MARGOLUS, *Cellular Automata Machines, a New Environment for Modeling* (MIT Press, Cambridge, MA, 1987).

[129] TORENVLIET, L. and P. VAN EMDE BOAS, A note on time and space, in: *Computing Science in the Netherlands, Conf. Papers CSN 87* (Centre for Mathematics and Computer Science, Amsterdam, 1987) 225–234.

[130] TURING, A.M., On computable numbers, with an application to the Entscheidungsproblem, *Proc. London Math. Soc.* (2) **42** (1936) 230–265; correction, *ibidem* **43** (1937) 544–546.

[131] VAN EMDE BOAS, P., The second machine class: models of parallelism, in: J.K. Lenstra and J. van Leeuwen, eds., *Parallel Computers and Computations*, CWI Syllabus, Vol. 9 (Centre for Mathematics and Computer Science, Amsterdam, 1985) 133–161.

[132] VAN EMDE BOAS, P., The second machine class 2: an encyclopaedic view on the Parallel Computation Thesis, in: H. Rasiowa, ed., *Mathematical Problems in Computation Theory*, Banach Center Publications, Vol. 21 (PWN, Warsaw, 1988) 235–256.

[133] VAN EMDE BOAS, P., Space measures for storage modification machines, *Inform. Process Lett.* **30** (1989) 103–110.

[134] VAN LEEUWEN, J. and J. WIEDERMANN, Array processing machines, *BIT* **27** (1987) 25–43.

[135] VITÁNYI, P.M.B., An optimal simulation of counter machines, *SIAM J. Comput.* **14** (1985) 1–33.

[136] VITÁNYI, P.M.B., Non-sequential computation and laws of nature, in: F. Makedon et al., eds., *VLSI Algorithms and Architectures, Aegean Workshop AWOC 86*, Lecture Notes in Computer Science, Vol. 227 (Springer, Berlin, 1986) 108–120.

[137] VITÁNYI, P.M.B., Locality, communication, and interconnect length in multicomputers, *SIAM J. Comput.* **17** (1988) 659–672.

[138] WAGNER, K. and G. WECHSUNG, *Computational Complexity*, Mathematische Monographien, Vol. 19 (VEB Deutscher Verlag der Wissenschaften, Berlin (GDR), 1986; also: Reidel, Dordrecht, 1986).

[139] WIEDERMANN, J., Deterministic and nondeterministic simulation of the RAM by the Turing machine, in: R.E.A. Mason, ed., *Information Processing '83, Proc. 9th IFIP World Computer Congress* (North-Holland, Amsterdam, 1983) 163–168.

[140] WIEDERMANN, J., Parallel Turing machines, Tech. Report RUU-CS-84-11, Dept. of Computer Science, Univ. of Utrecht, 1984.

[141] WIEDERMANN, J., Fast simulation of nondeterministic Turing machines with application to the knapsack problem, Report VUSEI-AR4/1986, VUSEI-AR, Bratislava, 1986.

CHAPTER 2

A Catalog of Complexity Classes

David S. JOHNSON

AT&T Bell Laboratories, Murray Hill, NJ 07974, USA

Contents

HANDBOOK OF THEORETICAL COMPUTER SCIENCE
Edited by J. van Leeuwen

1. Preliminaries

One of the goals of complexity theory is to classify problems as to their intrinsic computational difficulty. Given a problem, how much computing power and/or resources do we need in order to solve it? To date, we have not made much progress toward finding precise answers to such questions. We have, however, made a great deal of progress in classifying problems into general "complexity classes", which characterize, at least in a rough way, something of their inherent difficulties. This chapter will survey the most popular such classes, and the types of problems they contain.

This first section contains the preliminary definitions and concepts needed for defining the complexity classes. Sections 2 through 5 can be viewed as a catalog of complexity classes, with Section 6 devoted to last-minute developments, and tables and figures for use as ready-reference material. In addition to providing definitions of the classes, we will discuss what is known about the relationships between them. We shall also, where possible, provide examples of typical problems in each class, and describe something of what is known of the "structure" of the class.

1.1. Problems and instances

We begin by defining our terms. For our purposes, a "problem" is a total relation on strings, say over the alphabet $\{0,1\}$. More precisely, we have the following definition (where "$\{0,1\}^*$" represents the set of all finite strings made up of 0s and 1s).

DEFINITION. A *problem* is a set X of ordered pairs (I, A) of strings in $\{0,1\}^*$, where I is called the *instance*. A is called an *answer* for that instance, and every string in $\{0,1\}^*$ occurs as the first component of at least one pair.

This abstract and technical definition is required because the complexity classes we shall be discussing are all defined in terms of machine models like those covered in [153, 246]. Such machines are only equipped to handle inputs and outputs that are strings, rather than the more interesting combinatorial objects that we normally think of as the subjects of "problems", such as graphs, equations, or logical expressions.

Note, however, that such combinatorial objects are not all that different from strings, at least in the ways we use them. As a cogent example, consider the following problem.

GRAPH ISOMORPHISM

Instance: Two undirected graphs $G=(V,E)$ and $G'=(V',E')$, where V and V' are finite sets of vertices, and E and E' are finite sets of edges (unordered pairs of vertices from V and V' respectively).

Answer: "Yes" if there is a one-one onto function $f: V \rightarrow V'$ such that for all pairs $\{u,v\} \subseteq V$, $\{u,v\} \in E$ if and only if $\{f(u), f(v)\} \in E'$. Otherwise, "no".

As defined, this problem makes sense only as a question about *descriptions* of graphs, asking in effect whether two descriptions describe the same graph. In other words, it is a question about symbolic representations of the combinatorial objects called graphs, rather than the graphs themselves. Typically, we will present such symbolic representations in a linear fashion, i.e., as strings. It is thus not too much of a jump to think of the GRAPH ISOMORPHISM problem as a string relation. Let a_Y and a_N be strings chosen to represent "yes" and "no" respectively. The string relation would then be

$$\{(x, a_Y)\text{: string } x \text{ consists of two representations of the same graph } G\}$$
$$\cup\{(x, a_N)\text{: string } x \text{ does not consist of two representations of the same graph } G\}.$$

Similar translations will work for any other problem we might consider solving by digital computer, since we must have some way to represent instances and answers as strings if we hope to represent them in computer memory, which itself can be viewed as a string.

The above definition of the string relation corresponding to the GRAPH ISOMORPHISM problem is of course incomplete, for we have failed to specify precisely how strings are to represent graphs. As pointed out in [247], various schemes are available, from adjacency lists to incidence matrices. And even if we choose one basic scheme, say adjacency lists, there are still many fine details of punctuation and syntax to be filled in. Each choice would give a different string relation.

Fortunately, all the "reasonable" representation schemes are relatively interchangeable. Given one form of representation, one can quickly translate to any other. As we shall explain more fully in Section 1.5, it is normally safe to describe a problem in abstract terms, leaving the representational details to be filled in in the standard ways. The results we quote will hold no matter which string relation would result, assuming the details are "reasonable". When representational issues do make a difference, we shall say so.

We conclude this section with several technical points. First, recall that we require that a string relation be total if it is to be considered a "problem". Thus in our formulation of GRAPH ISOMORPHISM as a string relation, the answer is "no" both when x consists of representations of different graphs and when x does not represent graphs at all. Alternatively, one might wish to have a third "answer string" a_M ("M" for "meaningless input"). In either case, our definition requires that the problem be completely specified, even for meaningless input.

Second, note that our definition of problem does not require that the relation be a function, i.e., there can be more than one pair in X that have the same first component. For instance, in the GRAPH ISOMORPHISM problem, each pair (x, a_Y) could be replaced by the set of pairs of the form (x, w), where w is the representation of a function f that provides an isomorphism between the two graph representations included in x (there can be more than one such f). In this new problem, *any* w such that $(x, w) \in X$ would be a satisfactory answer for instance x. If one desired *all* such w, one could define a different problem X' where $(x, w) \in X'$ if x consists of two graph representations and w is the (possible empty) list contaning all the possible isomorphisms between the two graphs represented by x.

As the above example illustrates, our definition of problem is sufficiently general to meet most needs. To emphasize this generality, string relations are sometimes referred to as generalized "search problems" [87]. This is in opposition to such important special cases as *functions*, *decision problems*, and *counting problems*, defined below.

DEFINITION. A *function* is a string relation in which each string $x \in \{0, 1\}^*$ is the first component of precisely one pair.

DEFINITION. A *decision probem* is a function in which the only possible answers are "yes" and "no".

DEFINITION. A *counting problem* is a function in which all answers are nonnegative integers.

Decision problems are a particularly important special case. Most of the complexity classes we shall be considering are restricted to such problems, or, more precisely, to the *languages* derived from them.

DEFINITION. A *language* is any subset of $\{0, 1\}^*$.

There is a natural correspondence between languages and decision problems:

DEFINITION. If L is a language, then the decision problem R_L corresponding to L is $\{(x, \text{yes}): x \in L\} \cup \{(x, \text{no}): x \notin L\}$.

DEFINITION. Given a decision problem R, the language $L(R)$ corresponding to it is simply $L(R) = \{x \in \{0, 1\}^*: (x, \text{yes}) \in R\}$.

Note that there is an asymmetry between "yes" and "no" in this latter definition (i.e., "no" does not appear although "yes" does). Interchanging "yes" and "no" would yield a different and "complementary" language, defined as follows.

DEFINITION. If L is a language, then its *complementary language* is co-$L = \{0, 1\}^* - L$.

This asymmetry in the definition of $L(R)$ is of substantial theoretical importance. As we shall see in the next section, many of our models of computation, in particular the nondeterministic ones, are similarly asymmetrical, and hence $L(R)$ and co-$L(R)$ need not always belong to the same complexity class. Indeed, a common question we shall be asking is whether a given complexity class C is "closed under complement", i.e., whether $L \in C$ implies co-$L \in C$ for all languages L.

1.2. *How to solve a problem*

In everyday speech we often talk about "solving" particular instances of problems, as in solving a crossword puzzle. Complexity theorists, however, do not consider a problem X solved unless they have a general method that will work for *any* instance (assuming enough time, memory, and other resources are provided). In practice, such methods may be usable only for a small finite set of instances, given physical bounds on the resources available. Because the methods are general, however, they will automatically let us solve larger instances should the amount of available resources ever increase. The key question about such methods is thus how their resource requirements increase with instance size. As we shall see, there are many different notions of what a "method" can be, some rather abstract, but the following definitions will be relevant to all.

DEFINITION. The *size* of an instance I, written $|I|$, is its length, i.e., the number of symbols it contains. (Recall that formally our problems are string relations and our instances are strings.)

DEFINITION. If M is a method for solving problem X and R is a resource used by that method, then $R_M: Z^+ \rightarrow Z^+$ is the function defined by letting $R_M(n)$ be the maximum, over all strings x of length n, of the amount of resource R used when M is applied to input x.

Note that $R_M(n)$ is a worst-case measure. For a particular instance x with $|x| = n$, and even for most such instances, M may use far less than $R_M(n)$ of resource R. However, $R_M(n)$ provides the best possible overall guarantee, and thus provides a certainty that an average-case measure cannot.

For the purpose of defining complexity classes, it is useful to be able to talk about the resource requirements of problems themselves, rather than just of the particular methods used to solve them. It is difficult to define this notion of "requirement" directly, however, due to the fact that there may be no "best" method for a given problem X. For example, it could be the case that there are methods $M[\alpha]$ with $R_{M[\alpha]}(n) = n^{1+\alpha}$ for all $\alpha > 0$, but no method M with $R_M(n) = O(n)$. We shall thus settle for an indirect definition in terms of upper and lower bounds. Suppose X is a problem, R is a resource, and C is a class of methods.

DEFINITION. The requirements of problem X for resource R under methods from C is *upper bounded by* $T(n)$ if and only if there is a method $M \in C$ for solving X that has $R_M(n) = O(T(n))$.

DEFINITION. The requirement of problem X for resource R under methods from C is *lower bounded by* $T(n)$ if and only if all methods $M \in C$ for solving X have $R_M(n) = \Omega(T(n))$.

The above definitions are in terms of "classes" of methods. So far the term "method" has been used to represent the thing that does the "solving", but has otherwise been left unspecified. There are many notions of "method" available, each with its own uses and applicability. The goal of any method that solves problem X is, given an instance x, to produce an a such that $(x, a) \in X$. How that a is produced, however, can vary substantially. Depending on the class of methods in use and the type of problem, the answer a might be represented by anything from the contents of an output tape to the result of applying some logical or arithmetic function to the structure of an abstract computation tree.

Our choice of "method class" can also yield very different notions of the resource requirements for a problem. This choice, and the bounds we place on the relevant resource usage functions, are what defines a complexity class. In the next section, we survey the main classes of methods that we shall consider.

1.3. Machine models

Each class of methods that we discuss is most easily described in terms of a model of computation. A standard variant on the deterministic Turing machine can serve as our first model and a natural point of reference.

Deterministic Turning machines (DTMs)

For a general introduction to Turing machines, see [246]. In our variant, the machine has three tapes, one semi-infinite read-only tape for input, one semi-infinite write-only tape for output, and a read-write worktape. Such a machine "solves" a problem, if, whenever it is started with a string written in the leftmost cells of its input tape (all other cells blank), it eventually halts with an acceptable answer written in the leftmost cells of the output tape (all other cells blank). We will be principally interested in two resources for the basic machine, "time" and "space". The *time* for a computation is simply the number of steps made before the machine halts. The *space* is the number of cells of the worktape that ever were visited by the worktape head during the computation. Note that, for this machine, the space used can be much smaller than the input size (and we shall see complexity classes defined by placing such a "small" bound on the space usage allowed). The running time, on the other hand, must be at least as large as the input size unless answers depend only on some initial substring of the input.

Nondeterministic Turing machines (NDTMs)

This is a variant in which at each step the Turing machine has several choices as to its next move. The set of all possible computations can thus be viewed as a tree, with each reachable configuration of the Turing machine and its tapes having as its children those configurations that can be reached from it in one legal move. This machine can only yield the answers "yes" and "no", and hence is applicable only to decision problems. It answers "yes" if the tree of reachable configurations contains *any* configuration in which the machine is halted and the output tape contains the string representing yes. The answer is "no" if no such configuration is reachable. For the kinds of problems and algorithms in which we are interested, we may assume that the machine is designed in such a way that the tree is finite and all halting configurations (leaves) are at the same depth in the tree. In this case, the time used is the depth of the tree, and the space is the maximum, over all configurations in the tree, of the number of worktape cells in use.

Alternating Turing machines (ATMs)

This variant, proposed in [52], is again restricted to decision problems. In it, we again have a finite tree of computations with all halting configurations at the same level. Now, however, each configuration in the tree is labelled as either *universal* (∀) or as *existential* (∃). We inductively determine if the answer is "yes" as follows: A halting configuration is a yes-configuration if the output tape contains the string representing yes. A nonhalting ∃-configuration is a yes-configuration if at least one of its children is a yes-configuration. A nonhalting ∀-configuration is a yes-configuration if *all* its children are yes-configurations. The answer is "yes" if and only if the initial configuration is a yes-configuration.

The resources of time and space are defind as for NDTMs, but now we have an additional resource, which we call "alternations". This is the maximum, over all paths from an initial configuration to a final one, of the number of times a configuration has a different label from its parent (the root has no parent, but is assumed to contribute the first alternation). Thus, for instance, NDTMs can be viewed as ATMs with one alternation (and all configurations labelled by ∃).

Turing machines with other acceptance criteria

A variety of other output conventions besides those given above are possible for Turing machines that produce computation trees. A *probabilistic Turing machine* yields answer "yes" if and only if more than half the halting configurations are accepting configurations. A *random Turing machine* says "yes" if at least half the halting configurations are yes-configurations, says "no" if none of the halting configurations are yes-configurations, and otherwise says nothing. (It solves a decision problem only if it always says something, and that "something" is the correct answer.) A *counting Turing machine* yields as answer the *number* of halting configurations that are yes-configurations (and hence is applicable only to counting problems). Still other options are possible, but these will be discussed in the context of the complexity classes they are used to define (see Section 4).

Oracle Turing machines (OTMs)

These are Turing machines (DTMs, NDTMs, ATMs) with an additional semi-infinite "oracle tape", which alternates between write-only and read-only modes. Associated with any oracle machine is a particular problem (string relation) Y which the oracle can solve for free. While the oracle tape is in write-only mode, the DTM can at any time enter a query state, in which case at the next step the contents y of the oracle tape will be automatically replaced by a string b such that $(y, b) \in Y$, and the tape will become read-only. (Once the result has been read, the machine can in one step erase the oracle tape and return it to write-only mode.) Querying the oracle is thus like invoking a subroutine for solving Y, except that we do not count the time required by the subroutine.

Time and space requirements are the same as before. There are technical questions having to do with whether the number of cells used on the oracle tape should be counted (as with input and output tapes, this number can greatly exceed the number needed for the worktape). Unless otherwise stated, we shall assume that oracle tape space *is* counted in determining total space usage. One additional resource we may wish to count is oracle queries, the maximum, over all computation paths, of the number of times the machine enters its query state.

Parallel random access machines (PRAMs)

For a general introduction to random access machines and parallel random access machines, see [153, 246]. We need only consider the parallel version here, first introduced in [82]. We shall assume that the machine, in addition to having an arbitrary number of processors, is equipped with three sets of memory registers, one of read-only cells for input, one of write-only cells for output, and one of read-write memory cells for work. The input-output requirements are analogous to those for DTMs. For the sake of specificity, we assume that PRAM programs satisfy the concurrent-read, exclusive-write (CREW) restriction. (Two processors can read the contents of the same cell at the same time, but they cannot both write something in the cell at the same time.)

The space resource is measured in the standard way: the memory cells are indexed in sequence $c_1, c_2, \ldots$, and the space used equals the maximum k such that cell c_k is accessed by some processor during the computation. Time is a bit more complicated. We assume that all the processors work synchronously, one step at a time. The time for a given step is the maximum, over all processors, of the "semilogarithmic" cost measure for the operation performed by that processor. (The semilogarithmic cost for an operation that places a number N in register i when the input is of size n is $\lceil 1+(\log N+\log i)/\log n \rceil$). The third resource is the number of processors used. (The familiar "random access machine" or RAM is simply a PRAM wth only one processor.)

Families of Boolean circuits

Again, for a general introduction, see [153]. A family F of Boolean circuits is an infinite collection of acyclic Boolean circuits $\{B_1, B_2, \ldots\}$, made up of AND, OR, and NOT gates. Such families are applicable only to problems X satisfying the property that for any two pairs $(x, a), (y, b)$ in X, $|x|=|y|$ implies $|a|=|b|$. (Decision problems obey this requirement, and most other search problems can be reformulated as string relations satisfying that property.) In a family that solves a problem X, circuit B_n has an ordered sequence of n inputs and an ordered sequence of outputs whose number equals the length of the answers for instances of size n. Given the bits of a string x of length n as input, it produces the bits of an answer for x at its outputs.

Families of circuits constitute what is called a "nonuniform" method for solving a problem. There need be no commonality between the circuits, no way to build circuit B_n given n. A *uniform* circuit family is one in which B_n can be generated automatically, given n (say by an appropriately resource-bounded DTM).

As with PRAMs, circuit families can be viewed as a model for parallel computation. For circuit families (uniform or nonuniform), the resource analogous to (parallel) time is *depth*, the number of gates in the longest path from an input to an output in a circuit. The resource analogous to sequential time is *size*, the total number of gates in the circuit. The resource most closely analogous to space is *width*. To define this, we assign levels to all the gates (with input gates having level 0 and each other gate having a level one greater than the maximum level among those gates that provide its input). We call a gate "live at level i" if the level of the gate is i or less, and an output wire from the gate is an input wire for some gate at level greater than i. The width of a circuit is the maximum, over all i, of the number of gates (other than input gates) that are live at level i. Two final resources, these having no obvious correlate among the Turing machine resources, are *fan-in*, the maximum number of inputs an AND or OR gate can have, and *fan-out*, the maximum number of outputs a gate can have.

1.4. Resource bounds

In the previous section, we saw a wide variety of resources. In restricting these resources, we shall normally choose from a short list of types of bounds:

- *Constant*: there exists a constant k such that $R_M(n) \leqslant k$ for all $n \geqslant 0$.
- *Logarithmic*: $R_M(n) = O(\log n)$.
- *Polylogarithmic*: there exists a constant k such that $R_M(n) = O(\log^k n)$.
- *Linear*: $R_M(n) = O(n)$.
- *Polynomial*: there exists a constant k such that $R_M(n) = O(n^k)$.
- *Exponential*: there exists a constant k such that $R_M(n) = O(2^{n^k})$.
- *Unbounded*: no constant at all is imposed on $R_M(n)$, beyond the fact that only a finite amount of resource R is used for any particular instance.

Note that these alternatives do not allow us to make as precise distinctions as are theoretically possible. In particular, the following "hierarchy" theorems show that very fine distinctions can be made. We state them here since, as a corollary, they also imply that each of the above restrictions is stronger than its successors, at least as far as DTM and NDTM space and time are concerned.

H1. THEOREM (Hartmanis and Stearns [113]). *If $F_1(n)$ and $F_2(n)$ are "time constructible functions" and if*

$$\liminf_{n\to\infty} \frac{F_1(n)\log_2(F_2(n))}{F_2(n)} = 0,$$

then there exists a language L that can be recognized by a DTM in time bounded by $F_2(n)$, but not by any DTM with time bounded by $F_1(n)$.

H2. THEOREM (Hartmanis, Lewis and Stearns [112]). *If $F_1(n)$ and $F_2(n)$ are "space constructible functions" with $F_2(n) \geqslant \log_2 n$ for all $n \geqslant 1$, and if*

$$\liminf_{n\to\infty} \frac{F_1(n)}{F_2(n)} = 0,$$

then there exists a language L that can be recognized by a DTM in space bounded by $F_2(n)$, but not by any DTM with space bounded by $F_1(n)$.

H3. THEOREM (Seiferas, Fischer and Meyer [223]). *If $F_1(n)$ and $F_2(n)$ are "time constructible functions" and if*

$$\liminf_{n\to\infty} \frac{F_1(n)}{F_2(n)} = 0,$$

then there exists a language L that can be recognized by an NDTM with time complexity bounded by $F_2(n)$, but not by any NDTM with time complexity bounded by $F_1(n)$.

(Separation for NDTM space classes, at least the coarse separation we require, follows from Theorem H2 and a result of [216] that we shall discuss in more detail in Section 2.6. More refined separations of NDTM space classes can be found, for example in [222].)

The reason we settle for coarse distinctions, rather than the fine ones offered by Theorems H1 through H3, is that, currently, coarse distinctions are the only ones we know how to make. (And even when we make them, their validity may depend on unproven conjectures.) Theorems H1 through H3 are proved using diagonalization arguments that construct problems with the desired properties, but give no insight into how a natural problem might be shown to have them. Thus such important algorithmic questions as whether, for example, the bipartite matching problem can be solved in time $O(n^2)$ as well as $O(n^{2.5})$ remain out of reach.

Fortunately, there is a theoretical advantage to the consideration of such coarse bounds: They often free us from concern about the precise details of our models of computation. Similarly, they allow us much latitude in the way we represent problems as string relations, as we shall see in the next section.

1.5. *A first example: the class P (and the class FP)*

As an illustrative example, let us consider the perhaps most famous of all complexity classes, the class "P" of decision problems solvable (or languages recognizable) by DTMs obeying a polynomial bound on running time. Informally, one often sees "P" used to refer to the class of *all* search problems solvable in polynomial time, but we shall use the notation "FP" to denote this more general class. This is not a standard usage (there is none), but will help clarify the issues in what follows. By the simulation results reported in [246], we know that the class P (and the class FP) will not be altered if we add extra worktapes to our machine model, or even if we replace the DTM by a RAM. Furthermore, on the assumption that any "reasonable" representation of a problem as a string relation can be translated into any other in polynomial time, P (and FP) can be viewed as representation-independent. (For example, note that the adjacency list representation for a graph can be translated into an adjacency matrix representation in $O(n^2)$ time if n is the length of the former representation.)

The significance of P (and FP) was first pointed out in [56, 72]. In particular, Edmonds was the first to propose polynomial-time solvability as a theoretical equivalent to the informal notion of "efficiently solvable", an identification that has held up over the years, despite its obvious exceptions. (For more on the motivation of this identification and its drawbacks, see for example [87, 135].)

A wide range of problems are known to be in P (and FP), and researchers are continually attempting to identify more members. Perhaps the most significant addition to the membership list in the 1980s is LINEAR PROGRAMMING [149, 157]. This is formulated as a decision problem as follows.

LINEAR PROGRAMMING

Instance: Integer valued vectors $V_i = (v_i[1], \ldots, v_i[n])$, $1 \leqslant i \leqslant m$, and $D = (d[1], \ldots, d[m])$, and $C = (c[1], \ldots, c[n])$, and an integer B.

Answer: "Yes" if and only if there is a vector $X = (x[1], \ldots, x[n])$ of rational numbers such that $C \cdot X \geqslant B$ and $V_i \cdot X \leqslant d[i]$ for all i, $1 \leqslant i \leqslant m$.

LINEAR PROGRAMMING is of particular significance for P, for, in a sense to be explained in Section 1.7, it is an example of the *hardest* kind of problem in the class.

1.6. *Reductions*

In the main body of this chapter, we shall be defining complexity classes and illustrating them by giving examples of problems they do and do not contain. To show that a particular problem is in a given class, we need only exhibit a "method" for solving the problem that meets the requirements of the class's definition (as to machine model and resource bounds). Proofs of non-membership must be more indirect. One key technique, however, is applicable to both sorts of proof: the "reduction".

One common programming trick that can be incorporated into most of our models of computation is the "subroutine". For instance, an algorithm for solving the network flow problem (see [247]) might proceed by solving a sequence of shortest path problems, i.e., by using a *subroutine* for the shortest path problem. If we restrict ourselves to Turing machine models, the concept of subroutine can be formalized in terms of oracles. Suppose for example that we have a polynomial-time deterministic oracle Turing machine that solves problem X under the assumption that the oracle solves problem Y. In this case, the oracle acts just like a subroutine for solving Y. If it should turn out that problem Y is in P, one could replace the oracle by an appropriate version of a DTM program for solving Y (the subroutine), thus obtaining a polynomial-time oracle-free DTM that solves X. (We use here the fact that if p and q are polynomials, then so is the composition $p \circ q$ of p and q.)

DEFINITION. If X and Y are problems (string relations), a (*Turing*) *reduction* from X to Y is any OTM that solves X given an oracle for Y.

When restricting attention to decision problems, we often do not need to use the full power of this definiton, with its ability to ask an unbounded number of queries. Indeed, the following much restricted version often suffices.

DEFINITION. If X and Y are decision problems, a *transformation* from X to Y is a (DTM-computable) function $f: \{0,1\}^* \rightarrow \{0,1\}^*$ such that x has answer "yes" under X if and only if $f(x)$ has answer "yes" under Y.

Note that this is the same thing as a Turing machine reduction in which the OTM can ask but one query of its oracle, and must give as is own answer the answer it receives from the oracle. This type of reduction is also called a "many-one" reduction in the literature, given that it can conceivably map many x to the same string y.

As indicated above, reductions become useful tools in dealing with complexity classes when they themselves obey resource bounds. If R is a restricted class of reduction, let us write "$X \leqslant_R Y$" to signify that there is a reduction in R from X to Y.

DEFINITION. If C is a complexity class and R is a class of resource bounded reductions, we say that R is *compatible* with C if for any problems X and Y, $X \leqslant_R Y$ and $Y \in C$ imply that $X \in C$.

For the class P, the most general compatible class of reductions that is commonly considered consists of Turing reductions that obey polynomial time bounds (*polynomial-time Turing reductions*, sometimes abbreviated as "$\leqslant_T$" reductions). Also compatible, of course, is the more restricted class of *polynomial transformations* (transformations f with $f \in$ FP, abbreviated as "$\leqslant_p$" reductions). Another popular class of reductions is obtained by a further restriction, this time to *log-space transformations* (or "$\leqslant_{\text{log-space}}$" reductions), i.e., those transformations that are computable in logarithmic space. (Note that such transformations are also polynomial transformations, since the logarithmic bound on workspace means that there are only a polynomial number of distinct states for the worktape). In addition to being compatible with P, these three classes of reductions all have one more important property: they are "transitive".

DEFINITION. A class R of reductions is *transitive* if for all problems A, B, and C, $A \leqslant_R B$ and $B \leqslant_R C$ together imply $A \leqslant_R C$.

(Proving that log-space transformations are transitive is an interesting exercise, as we do not necessarily have enough space on our logarithmic worktape to write down the entire output of the first transformation, and so must generate it bit by bit, as needed.)

In Section 4.4, we shall introduce some additional forms of reduction, more complex than those given here, such as "γ-reductions" and "random" reductions of various types. In the meantime, however, the ones given here will suffice for our discussions.

1.7. *Completeness*

Reductions can clearly simplify the job of proving membership in a class. More importantly, however, they can also be used to prove non-membership. Suppose we know that problem X is not in class C. Then if we can exhibit a reduction from X to Y that is compatible with C, we will know that Y cannot be in C either, by the definition of "compatible". The difficult task, of course, is finding that first problem X that is not in C. Reduction can help here too.

We know by Theorems H1 and H3 of Section 1.4 that certain class containments $C' \subseteq C$ are proper, i.e., although every X in C' is in C, there is an X in C that is not in C'. As already pointed out, the diagonalization arguments used to prove H1 through H3 unfortunately do not make examples of problems in $C - C'$ readily available to us. Suppose, however, that we could show, for some class R of reductions compatible with

C', that *all* problems in C were reducible to a given problem X. Then if X were in C', we would have $C \subseteq C'$, a contradiction. This motivates the following definition.

DEFINITION. Suppose X is a problem, C is a class of problems, and R is a class of reductions. If $Y \leqslant_R X$ for all $Y \in C$, then we say that X is *hard* for C (*under R-reductions*), or simply *R-hard for C*. If also $X \in C$, then we say that X is *complete for C* (*under R-reductions*), or *R-complete for C*.

Note that if X is R-complete for a class C, it can be viewed as typical of the "hardest" problems in C (at least from the viewpoint of any class $C' \subseteq C$ with which R is compatible). As an example, consider the LINEAR PROGRAMMING problem introduced in Section 1.6. When we said there that this was in a sense the "hardest problem" in P, we were referring to the fact that LINEAR PROGRAMMING is complete for P under log-space reductions ("log-space complete for P") [68]. Log-space reductions are compatible not only with P, but also with the class $L \subseteq P$ of decision problems solvable by DTMs obeying a logarithmic space bound. Since LINEAR PROGRAMMING is complete for P under a reduction that is compatible with L, we conclude that it is in L if and only if $P = L$.

The situation here is somewhat different from that envisioned above, however. Although we strongly suspect that L is properly contained in P, we do not yet know for sure that this is the case. Thus we cannot at this point prove that a problem is in $P - L$ simply by showing that it is log-space complete for P; all that we can currently conclude from such a result is that the problem is in $P - L$ unless a widely believed conjecture is false. Such conditional complexity results abound in the current theory of complexity classes, as so many of the relations between classes are still unresolved.

There remains the question of how we show a problem X to be hard for a class under a given type of reduction. Given that the classes of interest are all of infinite size, we clearly cannot exhibit distinct reductions to X from each member. For many of our complexity classes, however, there will be one "generic" problem for which all these reductions follow almost by definition. In the case of the class P and log-space transformations, that problem is as follows.

DTM ACCEPTANCE

Instance: Description of a DTM M, string x, and an integer n written in unary.

Answer: "Yes" if and only if M, when started with input x, halts with answer "yes" in n or fewer steps.

Note that if a problem Y is in P, there must be a DTM M that accepts the set of strings representing its yes-instances, and integers c and k such that M never takes more than $c|y|^k$ steps on any input y. If one has such a c, k, and M in mind, one can, given any instance y of problem Y, construct an instance of DTM ACCEPTANCE that has the same answer using only logarithmic space. Simply write down the description of M (constant effort, independent of y), copy y (requires constant workspace), and compute $n = c|y|^k$ (logarithmic space). Thus DTM ACCEPTANCE is trivially log-space hard for P.

In fact, DTM ACCEPTANCE is log-space *complete* for P, as it is easy to see that it is itself in P: we can simply simulate the running of the Turing machine M on the input x, and the number of steps n that we have to simulate is polynomially bounded in the size of the input. (Note that here our requirement that n be written in unary notation is crucial; if n were written in the more standard binary notation, the simulation might take exponential time in the "size" of the input. Representing numbers in unary fashion is not "reasonable" in any ordinary sense, but it has its technical uses, and we shall see more of them in Section 2.1.)

Once we have our first R-complete problem Y for a class C, obtaining others is conceptually much more straightforward, assuming our class of reductions is transitive. All we need do to show that problem X is complete is prove $Y \leqslant_R X$; the transitivity of R does the rest. In the descriptions of complexity classes that follow, we shall when possible give examples of complete problems for each that are a bit more interesting than the generic one. We shall not, however, give the reductions used to prove completeness. Readers unfamiliar with the techniques involved in constructing such reductions are referred to [87] or any of the standard textbooks on algorithms and complexity, such as [5, 192].

1.8. Relativized worlds

We will be discussing many open problems concerning the relationship between classes in what follows. One common measure of the potential obstacles to resolving such problems is derived by considering how they "relativize". If X is a class defined in a particular way, and A is any subset of $\{0, 1\}^*$, then X^A is the analogous class defined using the same resource bounds but augmenting the machine model with a (perhaps additional) oracle tape for asking questions about membership in A. For instance, P^A would be the set of all languages that can be recognized in polynomial time by an oracle Turing machine with oracle A.

The question of whether class X equals class Y can usually be judged "difficult" if it relativizes both ways, i.e., if there exists an oracle A such that $X^A = Y^A$ and an oracle B such that $X^B \neq Y^B$. This is because the standard proof techniques for proving classes equal or unequal, such as simulation and diagonalization, continue to work even if relativized. If X is shown to equal Y by one of these techniques, then X^A will equal Y^A for all A, and similarly if we can use them to show $X \neq Y$, then we will have $X^A \neq Y^A$ for all A. Hence if a question relativizes both ways, the standard techniques cannot be applied to it. Unfortunately, most of the interesting open questions have this property.

A few closed problems also are known to have the property, but these tend to be of three well understood basic types, and do not appear relevant to the types of open questions we shall be discussing:

(1) Questions defined in terms of specially constructed oracles, so that their "relativized" versions deal with *doubly* relativized classes (i.e., classes defined in terms of machines with two separate oracle tapes and two separate oracles). For example, one can construct languages A, B, and C such that $\mathrm{P}^A = \mathrm{NP}^A$, and yet both $(\mathrm{P}^A)^B = (\mathrm{NP}^A)^B$, and $(\mathrm{P}^A)^C \neq (\mathrm{NP}^A)^C$ [104, 126].

(2) Problems discussing running time distinctions that are so fine as to be machine-dependent, such as the result of [197] concerning deterministic and non-deterministic linear time on a particular Turing machine model.

(3) Problems concerning sublinear space bounds, when the oracle tape is not required to obey the space bound. These last will be discussed further in Section 5.1, but see also [107].

Sometimes, even though a particular open question relativizes both ways, there is an additional property that at least hints at what the correct answer might be. Each oracle set A can be viewed as establishing an "alternate universe", with its own properties and relation between classes. For some questions, even though there are alternate universes where each possible outcome occurs, one of those outcomes is vastly preferred. More precisely, if one takes the natural measure over the space of all subsets of $\{0, 1\}^*$, the set of all oracles A for which a given outcome occurs has measure 1, i.e., the outcome occurs for almost all A. We then say that the given outcome holds "for a random oracle". There is unfortunately no a priori reason why this should imply that the outcome occurs for the particular oracle $A=\emptyset$, i.e., the one which yields the base case and hence the world in which we are normally most interested. (A "random oracle hypothesis", asserting that if a property holds for a random oracle then it holds for the empty oracle, was wishfully proposed in [29], but counterexamples, albeit technical ones, were quickly found [165].) Nevertheless, such results at least add an extra aura of believability to conjectures to which we already subscribe for other reasons, and we shall mention them where they are known. (For more on relativization, see Section 6.1.)

1.9. The organization of the catalog

The next four sections constitute a "catalog" of complexity classes. We have attempted to be moderately complete, covering all the major and many of the minor classes. Inevitably, however, some of the more obscure classes had to be omitted for space reasons. Normally, however, papers that discuss such classes will begin by relating them to more famous classes, so even in such cases the current survey may help in putting things in perspective. In discussing each class, we try to summarize some of the most important and interesting results about its structure and its relation to other classes. Here our choice of what to present is of necessity somewhat more selective, but we have attempted to present those results that the non-specialist might find the most intriguing.

Note that while the presentation must be sequential, the relationships between classes are not. Thus the presentation will unavoidably contain many pointers back and forth. If a property of class X does not make sense until we have discussed class Y, it cannot be fully explained until that later class has been introduced, and hence can only be alluded to by a forward reference when X is discussed. Our intention is that the non-expert should be able to read the presentation straight through as a tutorial, and subsequently be able to use it as a reference, with the aid of the tables and figures in Section 6 and the pointers in the text.

One additional note about the presentation. It is not our purpose here to present an historical picture of the development of the field, but rather to survey the current "state of the art" for complexity classes. We shall, of course, attempt to cite all the fundamental papers in the field. There will be cases, however, where a paper that was instrumental in pointing the way to a particular result will be ignored in favor of the paper that actually first presented the result itself. For readers interested in the development of ideas, the papers we do cite should provide an adequate point of departure for the needed literature search. Other general surveys and tutorials related to the subject of complexity classes can be found in [20, 87, 224, 236].

2. Presumably intractable problems

2.1. *The class NP, NP-complete problems, and structural issues*

The class "NP" is defined to be the set of all decision problems solvable (languages recognizable) by NDTMs in polynomial-bounded time. It trivially contains P, since DTMs are special cases of NDTMs. It would appear to contain much more, however, as nondeterminism seems to add significant power to time-bounded computations.

To see this, it is perhaps easier to consider a more restricted, though no less powerful version of the polynomial-time NDTM: the polynomial time "guess-and-check" procedure for verifying yes-answers (language membership). On input x, such a procedure first (nondeterministically) guesses a string y such that $|y| \leqslant p(|x|)$ for some fixed polynomial p, and then runs a polynomial-time algorithm on input (x, y). The answer is "yes" (x is in L) if and only if there is a guessable y such that the algorithm will answer "yes". It is easy to see that a decision problem (language) is in NP if and only if such a scheme exists.

Moreover, it is easy to see that a wide variety of seemingly intractable problems are susceptible to polynomial-time guess-and-check procedures, with the "guess" offering a shortcut to what would otherwise seem to be an unavoidably exponential-time exhaustive search. For instance, consider the following famous problem.

SATISFIABILITY

Instance: List of *literals* $U=(u_1, \bar{u}_1, u_2, \bar{u}_2, \ldots, u_n, \bar{u}_n)$, sequence of *clauses* $C=(c_1, c_2, \ldots, c_m)$, where each clause c_i is a subset of U.

Answer: "Yes" if there is a *truth assignment* for the variables $u_1, \ldots, u_n$ that satisfies all the clauses in C, i.e., a subset $U' \subseteq U$ such that $|U' \cap \{u_i, \bar{u}_i\}| = 1$, $1 \leqslant i \leqslant n$, and such that $|U' \cap c_i| \geqslant 1$, $1 \leqslant i \leqslant m$.

Note that there are 2^n possible truth assignments, and at present we know of no foolproof way to determine in subexponential time whether there exists one that satisfies all the clauses. A guess-and-check algorithm, however, need only guess a satisfying truth assignment (if one exists); verifying that all the clauses are satisfied is then trivial.

A second example is the following problem.

CLIQUE

Instance: Graph $G=(V, E)$, positive integer K.

Answer: "Yes" if G contains a complete subgraph of size K, i.e., if there is a subset $V' \subseteq V$ with $|V| = K$ such that for all $u, v \in V'$, $u \neq v$, the pair $\{u, v\}$ is an edge in E.

Note that there are roughly $|V|^K$ possible subsets, and at present we know of no way to determine if one with the desired property exists in subexponential time. A guess-and-check algorithm, however, need only guess V' and verify that all the required edges are present in E.

A third example is the following.

TRAVELLING SALESMAN PROBLEM (TSP)

Instance: List $c_1, c_2, \ldots, c_n$ of *cities*, a positive integer *distance* $d(c_i, c_j) = d(c_j, c_i)$ for each pair $\{c_i, c_j\}$ of distinct cities, and an integer bound B.

Answer: "Yes" if there is a *tour* through all the cities of total length B or less, i.e., a permutation π of indices $1, \ldots, n$ such that

$$\sum_{i=1}^{n-1} d(c_{\pi(i)}, c_{\pi(i+1)}) + d(c_{\pi(n)}, c_{\pi(1)}) \leqslant B.$$

Note that there are $n! = 2^{\Theta(n \log n)})$ possible tours, and again there is no known way of finding the answer deterministically in subexponential time. A guess-and-check algorithm, however, need only guess the desired tour and verify its length.

These three examples, and many more like them, have led most theoretical computer scientists to the belief that P is a strict subclass of NP, i.e., that P$\neq$NP. Moreover, although the question of P versus NP relativizes both ways (there exist oracles A and B such that $\mathrm{P}^A = \mathrm{NP}^A$ and $\mathrm{P}^B \neq \mathrm{NP}^B$ [17]), the two classes are unequal for a "random" oracle [29].

Given the likelihood that P$\neq$NP, there can be considerable practical significance in identifying a problem as "complete" for NP under an appropriate notion of reducibility. If the reducibility is compatible with P in the sense described in Section 1.6, such problems will be polynomial-time solvable if and only if P$=$NP. Such a result thus provides strong theoretical support for the belief that a problem is intractable, and can direct us to more productive approaches to the problem, such as settling for near-optimal rather than optimal solutions in the case of optimization problems.

Section 1.6 listed three types of reduction that were compatible with P. In order of apparently decreasing power, they were polynomial-time Turing reductions, polynomial transformations, and log-space transformations. By tradition, we reserve the unadorned adjective "NP-complete" for just one of these:

DEFINITION. A problem (language) is NP-*complete* if it is complete for NP under polynomial transformations.

If completeness under a different type of reduction is intended, most authors will normally point this out explicitly, as in "complete for NP under polynomial-time Turing reductions". It is not yet clear whether such distinctions are meaningful, however. Although polynomial-time Turing reductions are known to be more powerful than polynomial transformations for classes larger than NP [170], this distinction is not known to hold within NP. Indeed, it can only hold there if P$\neq$NP, which remains to be proved. Similary, polynomial-time transformations cannot be more powerful than log-space transformations unless $P \neq L$, another unproved conjecture. All three problems introduced above as members of NP are complete for NP under log-space reductions, the weakest of the three, and there are few, if any, examples of problems that are complete for NP but are not known to be complete under such reductions.

The first NP-complete problem to be identified (other than the generic problem of "NDTM ACCEPTANCE", defined analogously to the DTM ACCEPTANCE problem of Section 1.8) was SATISFIABILITY. The result that identified it is now called "Cook's Theorem" in honor of its author, Steven Cook.

1. THEOREM (Cook [58]). SATISFIABILITY *is* NP-*complete.*

In the same paper, Cook also proved that CLIQUE was NP-complete. (At roughly the same time, Leonid Levin in the Soviet Union independently came up with an idea equivalent to NP-completeness and proved a version of Theorem 1 in which SATISFIABILITY was replaced by a variant on the problem of "TILING" that we will cover in Section 3.2 [174].) Shortly thereafter, the class of known NP-complete problems was expanded to include the TSP and a wide variety of other problems by Richard Karp in the landmark paper [150] (see also [151]). By 1979, over 300 problems could be listed [87]. NP-complete problems are now known to permeate all areas of computer science, operations research, and mathematics, and are not unknown in such disparate areas as biology, physics, and political science. Readers are referred to the above book to obtain a better feel for the variety of results that have been obtained, as well as a much more thorough treatment of the subject and its history. More recent surveys can be found in the current author's "*NP-completeness Column*", starting with the first edition in [129].

An interesting observation is that it often takes very little to change a problem from polynomial-time solvable to NP-complete. Table 1 lists several pairs of such related problems, one in P and one NP-complete. (All the NP-completeness results in Table 1 are from [150], except the result for QUADRATIC DIOPHANTINE EQUATIONS, which is from [182].

The distinction illustrated in the last row of Table 1 is worthy of special note. Many a false proof that $P = NP$ has been based on the mistaken impression that UNARY PARTITION is NP-complete, whereas in fact it can be solved in polynomial time by a straightforward dynamic programming algorithm. Such an algorithm requires time exponential in the input size for BINARY PARTITION, even though it is polynomial in the perhaps exponentially larger) size for the corresponding UNARY PARTITION. This distinction is worth formalizing.

DEFINITION. A *pseudopolynomial-time* algorithm is one whose running time would be polynomial if all input numbers were expressed in unary notation.

The existence of a pseudopolynomial-time algorithm for a given NP-complete problem may mean that the problem is not so "intractable" after all. (Pseudopolynomial time becomes polynomial time if one restricts attention to instances in which the maximum number obeys a constant or polynomial bound in terms of the input size, and such restrictions may well hold in practice.) Thus it may be important to determine

Table 1
Problems on the frontier

POLYNOMIAL TIME	NP-COMPLETE
EDGE COVER *Instance*: Graph $G=(V, E)$, integer k. *Answer*: Yes if there is a subset $E' \subseteq E$ with $\|E'\| \leqslant k$ such that every vertex is the endpoint of an edge in E'.	VERTEX COVER *Instance*: Graph $G=(V, E)$, integer k. *Answer*: Yes if there is a subset $V' \subseteq V$ with $\|V'\| \leqslant k$ such that every edge has an endpoint in V'.
FEEDBACK EDGE SET *Instance*: Graph $G=(V, E)$ integer k. *Answer*: Yes if there is a subset $E' \subseteq E$ with $\|E'\| \leqslant k$ such that every cycle in G contains an edge in E'.	FEEDBACK ARC SET *Instance*: Directed graph $G=(V, A)$ integer k. *Answer*: Yes if there is a subset $A' \subseteq A$ with $\|A'\| \leqslant k$ such that every (directed) cycle in G contains an arc in E'.
EULER CYCLE *Instance*: Graph $G=(V, E)$ with $m=\|E\|$. *Answer*: Yes if there is an ordering $e_1, e_2, \ldots, e_m$ such that e_1 and e_m share an endpoint, as do all pairs $\{e_i, e_{i+1}\}$, $1 \leqslant i < m$.	HAMILTONIAN CYCLE *Instance*: Graph $G=(V, E)$ with $n=\|V\|$. *Answer*: Yes if there is an ordering $v_1, v_2, \ldots, v_n$ such that $\{v_1, v_n\}$ forms an edge, as do all pairs $\{v_i, v_{i+1}\}$, $1 \leqslant i < n$.
2-SATISFIABILITY (2-SAT) *Instance*: Instance of SATISFIABILITY in which no clause contains more than 2 literals. *Answer*: Yes if the clauses are satisfiable.	3-SATISFIABILITY (3-SAT) *Instance*: Instance of SATISFIABILITY in which no clause contains more than 3 literals. *Answer*: Yes if the clauses are satisfiable.
LINEAR DIOPHANTINE EQUATIONS *Instance*: Positive integers a, b, and c *Answer*: Yes if there are positive integers x and y such that $ax+by=c$.	QUADRATIC DIOPHANTINE EQUATIONS *Instance*: Positive integers a, b, and c. *Answer*: Yes if there are positive integers x and y such that $ax^2+by=c$.
UNARY PARTITION *Instance*: Set $A=\{a_1, \ldots, a_n\}$ of integers written in unary notation. *Answer*: Yes if there is a subset $A' \subset A$ such that $\Sigma_{a \in A'} a = \Sigma_{a \in A-A'} a$.	BINARY PARTITION *Instance*: Set $A=\{a_1, \ldots, a_n\}$ of integers written in binary notation. *Answer*: Yes if there is a subset $A' \subset A$ such that $\Sigma_{a \in A'} a = \Sigma_{a \in A-A'} a$.

whether a problem is merely NP-complete in the ordinary sense, or whether it has the following stronger property.

DEFINITION. A problem is said to be NP-*complete in the strong sense* if the variant of it in which all input numbers are written in unary notation is NP-complete.

Note that problems like CLIQUE are trivially NP-complete in the strong sense, since the only number in the input is by definition bounded by the input size (in this case, the input bound K is bounded by the number of vertices $|V|$). A more meaningful example of a "strongly NP-complete" problem is the following variant of the "PARTITION" problems of Table 1.

3-PARTITION

Instance: Sequence a_i, $1 \leqslant i \leqslant 3n$, of positive integers (in binary notation).

Answer: "Yes" if there is a partition of these integers into disjoint 3-element sets A_j, $1 \leqslant j \leqslant n$, such that all n sets A_i have exactly the same sum.

This problem is proved NP-complete in the strong sense by exhibiting a polynomial p and showing that the problem remains NP-complete (in the ordinary sense) even if we require $a_i \leqslant p(n)$ for $1 \leqslant i \leqslant 3n$. (See [86] for the proof and a more rigorous development of these concepts.)

Distinctions like that provided by "strong NP-completeness" are only possible if one imposes some form of semantics on the language in NP, i.e., identifies certain parts of the input string as "numbers", "graphs", etc. There is much one can say, however, without at all considering what the languages in NP "mean", but by simply examining their structural, set-theoretic properties. A first question along these lines concerns what are called "sparse" languages, defined as follows.

DEFINITION. A language L is *sparse* if there is a polynomial p such that L contains no more than $p(n)$ strings of length n for all $n > 1$.

As simple examples of sparse languages, consider languages over a one-letter alphabet, i.e., subsets of $\{1\}^*$. These are sometimes called "tally languages". For such languages, there can be at most one string of length n for any n.

A natural "structural" question to ask is whether there can be a sparse NP-complete language. Unfortunately, this possibility seems ruled out. In [179] it was proved that a sparse language can be NP-complete under polynomial transformations only if P = NP. Weaker, but not more believable conclusions follow if a sparse language is complete for NP under polynomial-time Turing reductions [146, 179] (see Section 2.4). Indeed, there are interesting consequences if there is a sparse language anywhere in NP − P [111] (see Section 3.2). (For more detailed tutorials on results concerning sparsity, see [105, 180].

Thus, assuming P ≠ NP, all NP-complete sets have a certain gross structural similarity. Could the similarity actually be much closer? Could they, for instance, all be isomorphic?

DEFINITION. A polynomial transformation is a *polynomial-time isomorphism* if it is a one-one onto function and its inverse is also polynomial-time computable.

In [31] Berman and Hartmanis was conjectured that if a problem is NP-complete under polynomial-time transformations then it is polynomial-time isomorphic to SATISFIABILITY (and hence all NP-complete problems are isomorphic in this sense). Berman and Hartmanis were able to show that this was true for all the NP-complete problems known to them at the time, although of course this did not constitute a proof of the conjecture. (Indeed a complete proof would have implied that P ≠ NP: If P = NP, then all languages in NP would be NP-complete, including finite and infinite ones, and no finite language can be isomorphic to an infinite one.)

A further obstacle to proving this "isomorphism conjecture" is the fact that it may well be false even if $P \neq NP$. In particular, a class of specially constructed "k-creative" languages, introduced in [145, 258], seems to be rich source of candidates for NP-complete languages *not* isomorphic to the more normal NP-complete languages or to each other. In [166] it is shown that the desired nonisomorphic NP-complete k-creative languages exist so long as a particular type of polynomial-time "scrambling" function exists, and that, although no such function has yet been found, they do exist with respect to a random oracle. Thus the isomorphism conjecture fails with respect to a random oracle (and no oracle is yet known for which the conjecture succeeds; the best we have is an oracle for which the analogous conjecture succeeds for the larger class Σ_2^P, to be defined in Section 2.5 [120]).

In this light, it is interesting to note that if there exist nonisomorphic NP-complete languages, then there exist infinitely many distinct isomorphism classes, and these have a rich structure, as shown in [181]. A similar statement can be made about the structure of $NP - P$ under polynomial transformations, assuming $P \neq NP$. In [167], it is shown that under this assumption and these reductions, $NP - P$ must consist of an infinite collection of equivalence classes, of which the NP-complete problems are only one, albeit the hardest. Two famous problems have long been considered candidates for membership in such intermediate equivalence classes: GRAPH ISOMORPHISM (as described in Section 1.1) and COMPOSITE NUMBER (Given an integer n written in binary notation, are there positive integers p, q, $1 < p, q < n$, such that $n = pq$?). We shall have more to say about these two in later sections (COMPOSITE NUMBER in Sections 2.2 and 4.3, GRAPH ISOMORPHISM in Sections 2.5 ad 4.1). As we shall see, there are strong arguments why neither is likely to be NP-complete.

The equivalence class containing GRAPH ISOMORPHISM, in particular, has been so well-studied that a name for it (or rather its generalization to arbitrary search problems) has been introduced:

DEFINITION. The class of *GRAPH ISOMORPHISM-complete* problems consists of all those search problems X such that X and GRAPH ISOMORPHISM are polynomial-time Turing reducible to each other.

Included in this class are various special cases of graph isomorphism (e.g., GRAPH ISOMORPHISM restricted to regular graphs [38]), as well as related problems (e.g. "Given G, what is the order of its automorphism group?" [183]) and at first glance unrelated ones (e.g., a special case of CLIQUE [162]). These are to be contrasted with the many special cases of GRAPH ISOMORPHISM that are now known to be in P, for surveys of which, see for instance [87, 129].

We conclude this section with a more fundamental issue. So far, in considering the question of P versus NP, we have assumed that there were only two possible answers: $P = NP$ and $P \neq NP$. There is the possibility, however, that the question is *independent* of the standard proof systems on which we rely. This would mean that there exist models for both possible outcomes, with neither model offering an affront to our fundamental assumptions. This possibility was first raised explicitly in [109], which showed that for any formal system F there exists an oracle A such that $P^A = NP^A$ is independent of F.

The result as proved in [109] was limited, since it held only for a particular description A. That is, independence was proved only for the statement "$P^{L(M)} = NP^{L(M)}$", where M was a particular DTM and "$L(M)$" denotes the language accepted by M. (Moreover, in this case, $L(M)$ was in fact the empty set, and it was only the opaque choice for its representation that yielded the independence result.) Subsequently, however, the restriction to a particular representation has been removed. In particular, given F, there exists a recursive set A such that for *any* provably total DTM M that recognizes A, $P^{L(M)} = NP^{L(M)}$ is independent of F [103, 204]. (A DTM is "total" if it halts on all inputs.)

The question thus arises, could this more powerful conclusion hold for the empty oracle and for a proof system we commonly use, such as Peano Arithmetic? Certain computer science questions about the termination of programs in particular typed languages *have* been proved to be independent of Peano Arithmetic (even of second-order Peano Arithmetic) in [81]. It does not appear, however, that the techniques used could be applied to the P versus NP question. So far, the only results for P = NP have been consistency results for substantially weaker systems, as in [67]. Unfortunately, these systems are so weak that many standard, known results cannot be proved in them either [141, 142, 143, 173], and so independence of such systems tells us little. Moreover, it can be argued that even Peano Arithmetic is too weak a theory for independence of it to be significant, and that at the very least one would want the question to be independent of full Zermelo–Fraenkel set theory before one would be willing to give up the quest for an answer. Although such independence is believed unlikely, it is a possibility that cannot be totally ignored. We will nonetheless ignore it for the remainder of this chapter. Readers wishing more background on the issue are referred to [144]. For a detailed discussion of the logical theories mentioned above, see for instance [25].

2.2. Co-NP, NP∩co-NP, and nondeterministic reducibilities

One aspect of NP and the NP-complete problems that we have failed so far to emphasize is their one-sidedness. Consider the complementary problem to SATISFIABILITY, in which we are asked if it is the case that every truth assignment fails to satisfy at least one clause. This is simply SATISFIABILITY with the answers "yes" and "no" reversed, and hence in practical terms is computationally equivalent to SATISFIABILITY. However, it is not at all clear that this "CO-SATISFIABILITY" problem is NP-complete or even in NP. (What would be a "short proof" that all truth assignments have the desired property?)

In general, if we let "co-NP" denote the set of all languages $\{0, 1\}^* - L$, where $L \in$ NP, we are left with the open question of whether NP = co-NP, and the suspicion that it is not. This suspicion is supported by the fact that the two classes are distinct for a random oracle [29], although there do exist oracles A such that $NP^A = \text{co-}NP^A$ even though $P^A \neq NP^A$ [17], and so the question seems unlikely to be resolved soon. Note that the two classes *must* be identical if P = NP, and indeed that P = NP if and only if P = co-NP. Note also that NP must equal co-NP if any NP-complete problem, such as SATISFIABILITY, should prove to be in co-NP, and that if NP ≠ co-NP, then neither set can contain the other.

Assuming that NP and co-NP are distinct, one wonders if there is any structural, rather than computational, way in which they differ. If one takes "structural property" in a suitably broad sense, there is at least one such difference, albeit a technical one. In [111] it is shown that there is an oracle for which co-NP $-$ P contains a sparse language but no tally (i.e., one-symbol) languages, in contrast to the fact that for all oracles, NP $-$ P contains a sparse language if and only if it contains a tally language.

A more important consequence of the presumed inequality of NP and co-NP is that it would provide us with yet another potentially interesting class, NP$\cap$co-NP. This class would be a proper subclass of both and would contain P, although not necessarily properly. The question then becomes, does P $=$ NP$\cap$co-NP? This question relativizes both ways for oracles that yield P $\neq$ NP [17], although it is not currently known whether either result is preferred by a random oracle (inequality holds with probability 1 for a "random permutation") [29]. The one current candidate of importance for NP$\cap$co-NP $-$ P is the above-mentioned COMPOSITE NUMBER problem. As remarked in the previous secton, this is in NP. By a result of [199] that there exist short proofs of primality, it is also in co-NP. Unfortunately, the status of COMPOSITE NUMBER as a candidate is somewhat weakened by the fact that many researchers think it is in P, and that it is only a matter of time before the required polynomial-time algorithm is discovered. For instance, it is shown in [188] that COMPOSITE NUMBER is in P if the Extended Riemann Hypothesis is true. A better candidate would be a problem that is complete for NP$\cap$co-NP, but unfortunately, no such problem is known. Indeed, it is strongly expected that none exists, since completeness results proved by known methods relativize and there exist oracles A such that $\mathrm{NP}^A \cap \text{co-NP}^A$ has no complete sets under polynomial transformations [228]. Moreover, it will be no easier finding complete sets under the more general notion of polynomial-time Turing reducibility: it is shown in [110] that either both types of complete sets exist, or neither.

The class NP$\cap$co-NP may still have its uses, however. Consider the concept of "γ-reducibility," introduced in [3]. This reducibility, while not necessarily compatible with P, is compatible with NP$\cap$co-NP. Thus, although a problem that is complete for NP under γ-reductions could conceivably be in P even if P $\neq$ NP, it could not be in P (or even in co-NP) if NP $\neq$ co-NP. A γ-reduction is an example of a "nondeterministic" reduction, in that it is defined in terms of an NDTM rather than a DTM:

DEFINITION. A language X is *γ-reducible* to a problem Y if there is a polynomial-time NDTM M such that, on any input string x, M has at least one accepting computation, and such that, if y is the contents of the tape at the end of any such accepting computation, then $x \in X$ if and only if $y \in Y$.

Several problems are proved to be complete for NP under γ-reductions in [3], none of which has yet been proved NP-complete under the more restrictive reductions of the previous section. All were of a number-theoretic nature, with a typical one being LINEAR DIVISIBILITY (Given positive integers a and c in binary representation, is there a positive integer x such that $ax+1$ divides c?).

This notion of nondeterministic reduction was modified and generalized in [177] to what is called "strong nondeterministic polynomial-time Turing reducibility".

DEFINITION. A language X is *strongly nondeterministically polynomial-time Turing reducible* to a problem Y if there is a nondeterministic polynomial-time OTM M with Y as oracle such that

(i) all computations halt with one of the three outcomes {yes, no, don't know},

(ii) for any string x, M has at least one computation that yields the correct answer, and

(iii) all computations that do not result in "don't know" yield the correct answer.

If a problem Y is complete for NP under this more general notion of reducibility, it can be shown that $Y \in$ co-NP if and only if NP = co-NP, the same conclusion that holds for γ-reducibility. In [55] strong nondeterministic Turing reducibility is used to prove the (presumed) intractability of a broad class of network testing problems, only a few of which are currently known to be NP-complete.

If one drops from the above definition the requirement that (ii) hold for nonmembers of X, one obtains (ordinary) polynomial-time nondeterministic Turing reductions, as introduced in [187]. Although these do not have the nice sorts of consequences discussed here, they have their uses, e.g., see [17, 170, 187]. We shall be introducing still further types of reducibilities and of "completeness" for NP once we introduce the concept of "randomized" computation in Section 4.

2.3. *NP-hard, NP-easy, and NP-equivalent problems (the class $F\Delta_2^P$)*

In the previous section we mentioned that NP was limited as a class because of its one-sided nature. A second apparent limitation is due to the fact that NP is restricted to decision problems, whereas in practice the problems that interest us are often function computations or more general search problems. For instance, the TRAVELLING SALESMAN (TSP) decision problem defined above is not really the problem one wants to solve in practice. In practice, one wants to find a tour (permutation) that has the shortest possible length, not simply to tell whether a tour exists that obeys a given length bound. The latter problem was simply chosen as a decision problem in NP that could serve as a "stand-in" for its optimization counterpart, based on the observation that the optimization problem could be no easier than its stand-in. (Finding an optimal tour would trivially enable us to tell whether any tour exists whose length obeys the given bound.) The term "NP-hard" has been introduced for use in this context.

DEFINITION. A search problem X is NP-*hard* if for some NP-complete problem Y there is a polynomial-time Turing reduction from Y to X.

Note the obvious consequence that an NP-hard problem cannot be solvable in polynomial time unless P = NP.

The optimization version of the TSP is clearly NP-hard, as are all other search problems that contain NP-complete problems as special cases. Moreover, the term "NP-hard" serves a valuable function even in the realm of decision problems, where it provides a simple way of announcing the presumed intractability of problems that may not be in NP, but are still no easier than the NP-complete problems. For instance, the complement co-X of an NP-complete problem X, although just as intractable as X, will not be NP-complete unless NP = co-NP. It can be called NP-hard, however, if all we want to do is emphasize its presumed intractability. (When more precise language is required, the term "co-NP-complete" is available for this case.)

Unlike the NP-complete problems, the NP-hard problems do not form an equivalence class. Although no NP-hard problem can be in P unless P = NP, the converse need not be true. (Even undecidable problems can be NP-hard.) To get an equivalence class, we need to be able to impose an upper bound on the complexity of the problems. For this purpose, we introduce the following definition.

DEFINITION. A search problem X is NP-*easy* if for some problem Y in NP there is a polynomial-time Turing reduction from X to Y.

Note that all NP-easy problems will be solvable in polynomial time if P = NP.

Perhaps surprisingly, the optimization problem version of the TSP is NP-easy as well as NP-hard. It can be solved by binary search using an oracle for the following problem in NP: Given a sequence $c_1, c_2, \ldots, c_n$ of cities with specified intercity distances, a bound B, and an integer k, is there a tour that has length B or less whose first k cities are c_1 through c_k, in order? Similar arguments apply to many other search problems.

The set of search problems that are both NP-hard and NP-easy, an equivalence class under polynomial-time Turing reductions, thus constitutes a natural extension of the class of NP-complete problems to search problems in general, an extension that is captured in the following definition (from [87]).

DEFINITION. A search problem is NP-*equivalent* if it is both NP-hard and NP-easy.

Some additional notation will prove useful in what follows.

DEFINITION. The class $\mathrm{F}\Delta_2^{\mathrm{P}}$ (Δ_2^{P}) consists of the set of all NP-easy search problems (decision problems).

Commonly used pseudonyms for $\mathrm{F}\Delta_2^{\mathrm{P}}$ and Δ_2^{P} that are perhaps more informative are "$\mathrm{FP}^{\mathrm{NP}}$" and "$\mathrm{P}^{\mathrm{NP}}$". The symbolism of the "Δ" notation will become clearer when we discuss the polynomial hierarchy in Section 2.5. Note that the NP-equivalent problems are simply those that are complete under polynomial-time Turing reductions for $\mathrm{F}\Delta_2^{\mathrm{P}}$.

2.4. *Between NP and Δ_2^{P}: the class D^{P} and the Boolean hierarchy*

For practical purposes there appears to be little point in distinguishing amongst the NP-equivalent problems. As with NP-complete problems, each can be solved in single exponential time ($\mathrm{O}(2^{p(n)})$ for some polynomial p), but none can be solved in polynomial time unless P = NP (in which case they all can). Furthermore, if any problem in NP can be solved in time $\mathrm{O}(n^{c \log n})$ for some c, so can all NP-equivalent problems, and similar statements hold for time $2^{c \log^k n}$ and any fixed k. Nevertheless, there are significant theoretical distinctions to be made, and interesting classes of decision problems that lie midway between NP and Δ_2^P, classes that are seemingly distinct from each other with respect to polynomial transformations (although not with respect to polynomial-time Turing reductions).

Of particular interest is the following class, introduced in [194].

DEFINITION. The class D^P consists of all those languages that can be expressed in the form $X \cap Y$, where $X \in$ NP and $Y \in$ co-NP.

Note that this is far different from the class NP∩co-NP. It in fact contains all of NP∪co-NP, and does not equal that union unless NP = co-NP. Indeed, the three statements "D^P = NP∪co-NP", "Δ_2^P = NP∪co-NP", and "NP = co-NP" are all equivalent [194]. As we have already seen, oracles exist for which both outcomes occur. A variety of interesting types of problems appear to be in D^P − (NP∪co-NP).

A first example is the *exact answer* problem. For instance, in the exact answer version of the TSP, we are given a list of cities together with their intercity distances and a bound B, and are asked whether the optimal tour length is *precisely* B. This is the intersection of the TRAVELLING SALESMAN decision problem given in Section 2.1 (in NP) and the (co-NP) question that asks whether all tours are of length B or more. This problem is in fact complete for D^P under polynomial transformations (or "D^P-complete") [194] and hence cannot be in either NP or co-NP unless the two classes are equal. Another example of an exact answer problem that is complete for D^P is EXACT CLIQUE: Given a graph G and an integer K, is the maximum clique size precisely K? [194].

A second type of problem in D^P is the *criticality* problem. For instance, consider CRITICAL SATISFIABILITY: Given an instance of SATISFIABILITY, is it the case that the set of clauses is unsatisfiable, but deleting any single clause is enough to yield a subset that is satisfiable? Here the "unsatisfiability" restriction determines a co-NP language, and the m satisfiability restrictions, where m is the number of clauses, can be combined into a single instance of SATISFIABILITY, which is in NP. This problem is also D^P-complete [193].

A third type of problem in D^P is the *uniqueness* problem, as in UNIQUE SATISFIABILITY: Given an instance of SATISFIABILITY, is it the case that there is one, and only one, satisfying truth assignment? This is once again clearly in D^P and it is not difficult to show that it is NP-hard (e.g., see [33]). In this case, however, we do not know whether it is D^P-complete. There are oracles for which it is and for which it is not [33]. (D^P properly contains NP∪co-NP for both oracles.) What we do know is that UNIQUE SATISFIABILITY cannot be in NP unless NP = co-NP [33]. The possibility that UNIQUE SATISFIABILITY is in co-NP has yet to be similarly limited. (For a slightly more thorough treatment of D^P, see [133].)

Related to the class D^P are two intermingled hierarchies of classes within Δ_2^P: the *Boolean hierarchy* and the *query hierarchy* (e.g., see [48, 49, 106, 250, 252]).

DEFINITION. The *Boolean hierarchy* consists of the classes BH_k, $k \geqslant 0$, as follows:

(1) $BH_0 = P$.

(2) If $k>0$, BH_k is the set of all languages expressible as $X-Y$, where $X \in \mathrm{NP}$ and $Y \in \mathrm{BH}_{k-1}$.

At present there is no consistently adopted notation for these classes; "BH(k)" is used for BH_k in [53, 147] and "NP(k)" is preferred in [48]. The notation used here is chosen for its mnemonic value, especially in relation to the following notation, which *is* consistently used.

DEFINITION. The class BH is equal to the union $\bigcup_{k=0}^{\infty} \mathrm{BH}_k$.

See Fig. 1 (Section 6) for a schematic view of this hierarchy and how it relates to the other classes we have seen so far. Note that $\mathrm{BH}_1 = \mathrm{NP}$ and $\mathrm{BH}_2 = \mathrm{D}^P$. There exist oracles for which this hierarchy is infinite, i.e., for which $\mathrm{BH}_k \neq \mathrm{BH}_{k+1}$ for all $k \geqslant 0$, as well as oracles for which it collapses, i.e., for which $\mathrm{BH} = \mathrm{BH}_k$ for some $k \geqslant 0$. Indeed, for any $k \geqslant 0$ there is an oracle such that all classes BH_i, $0 \leqslant i < k$, are distinct, but for which $\mathrm{BH}_k = \mathrm{BH} = \Delta_2^{\mathrm{P}}$ (in fact, it will equal PSPACE, but we have not defined PSPACE yet; see Section 2.6) [48]. Note that it need not be the case that $\mathrm{BH} = \Delta_2^{\mathrm{P}}$. Indeed, there are oracles that separate the two classes. In particular, there are oracles for which BH does not have any complete problems under polynomial transformations, even though Δ_2^{P} always has such complete problems [48]. The individual classes BH_k also have complete problems, although for classes above $\mathrm{BH}_2 = \mathrm{D}^{\mathrm{P}}$ the currently known complete problems are somewhat contrived. For instance, for each even $k>0$, the following problem is complete for BH_k: "Given a graph G, is $\chi(G)$ an odd number lying in the interval $[3k, 4k]$?" [48].

Let us now turn to the related *query hierarchy*. In defining it, we shall make use of the following formalism:

DEFINITION. For each function $f: Z^+ \to Z^+$, the class $\mathrm{P}^{\mathrm{NP}[f(n)]}$ is the set of all languages that can be recognized by an oracle Turing machine with an oracle for SATISFIABILITY that makes no more than $f(n)$ queries of the oracle, where n is the input size.

(Note that $\mathrm{P}^{\mathrm{NP}} = \Delta_2^{\mathrm{P}}$ is the same as $\mathrm{P}^{\mathrm{NP}[n^{O(1)}]} = \bigcup_{k=1}^{\infty} \mathrm{P}^{\mathrm{NP}[n^k]}$.)

DEFINITION. The *query hierarchy* consists of the classes QH_k, $k \geqslant 0$, where for each such k, $\mathrm{QH}_k = \mathrm{P}^{\mathrm{NP}[k]}$. (Here k represents the constant function $f(n)=k$.)

DEFINITION. The class QH is equal to the union $\bigcup_{k=0}^{\infty} \mathrm{QH}_k$.

It is easy to see that $\mathrm{BH}_k \subseteq \mathrm{QH}_k$; QH_k can in turn be shown to reside in a class of the Boolean hierarchy, albeit one that is somewhat higher up [50] (for a proof, see [250]). Thus $\mathrm{QH} = \mathrm{BH}$, and we can conclude that either both hierarchies are infinite, or both collapse to some finite level [106, 250]. If they do collapse, there are some interesting consequences for a more famous hierarchy, and we shall discuss these in the next section.

The above results do not carry over to the analogs of these hierarchies for arbitrary search problems. Let $\mathrm{FP}^{\mathrm{NP}[f(n)]}$ be the search problem analog of the class $\mathrm{P}^{\mathrm{NP}[f(n)]}$. The hierarchy of classes $\mathrm{FP}^{\mathrm{NP}[k]}$, $k \geqslant 0$, cannot collapse unless $\mathrm{P}=\mathrm{NP}$, a much stronger result than is known for the query hierarchy. Indeed, the precise number of queries asked makes a difference (assuming $\mathrm{P} \neq \mathrm{NP}$) up to at least $\frac{1}{2} \log n$ queries [163].

This raises the question of classes defined using a nonconstant bound on the number of queries, in particular $\mathrm{P}^{\mathrm{NP}[\mathrm{O}(\log n)]}$ and $\mathrm{FP}^{\mathrm{NP}[\mathrm{O}(\log n)]}$. Note that the former contains (and may properly contain) all of $\mathrm{QH}=\mathrm{BH}$. With these more general classes, we once again can obtain interesting complete problems. The derivation is somewhat more straightforward in the case of the search problem classes. For instance, with an appropriate definition of reduction (the "metric" reduction of [163]), the problem of determining (rather than merely verifying) the size of the largest clique in a graph G is complete for the class $\mathrm{FP}^{\mathrm{NP}[\mathrm{O}(\log n)]}$. The $\mathrm{O}(\log n)$ here comes from the number of questions that need to be asked when performing a binary search to find the answer, given that the maximum clique size is no more than the number of vertices in G. If the optimal value one is asking to determine can be significantly larger than the input size, one can get much higher complexities. In particular, determining the length of an optimal travelling salesman tour is complete for all of $\mathrm{F}\Delta_2^{\mathrm{P}}$ [163].

Obtaining problems that are complete for the decision problem classes $\mathrm{P}^{\mathrm{PN}[\mathrm{O}(\log n)]}$ and Δ_2^{P} under polynomial transformations is a bit more of a challenge. The natural plan of attack would be to find appropriate decision versions of the maximum clique search problem and the TSP, but note that our previous attempts to convert such problems to decision problems forced the problems to lose complexity. "Is there a tour of length B or less?" dropped the TSP from Δ_2^{P} to NP, and "Is the optimal tour length exactly B?" dropped it to D^{P}, almost as far. To obtain completeness for $\mathrm{P}^{\mathrm{NP}[\mathrm{O}(\log n)]}$ and Δ_2^{P}, it turns out that one must find ways to provide instances with less in the way of hints as to the optimal clique size or tour length. Two questions that do the trick for the TSP and Δ_2^{P} are as follows: "Is there exactly one optimal tour?" [191] and "Is the optimal tour length divisible by k?" (given k as part of the input) [163]. Similarly, the problem of determining whether the maximum clique size for graph G is divisible by k (given G and k) is complete for $\mathrm{P}^{\mathrm{NP}[\mathrm{O}(\log n)]}$ [163].

A final question to consider is whether $\mathrm{P}^{\mathrm{NP}[\mathrm{O}(\log n)]}=\Delta_2^{\mathrm{P}}$ or $\mathrm{FP}^{\mathrm{NP}[\mathrm{O}(\log n)]}=\mathrm{F}\Delta_2^{\mathrm{P}}$. Here it is once again easier to prove separation results for the function classes: It is shown in [163] that the $\mathrm{F}\Delta_2^{\mathrm{P}}$ equality can hold only if $\mathrm{P}=\mathrm{NP}$, whereas it is not known whether the consequences of the Δ_2^{P} equality would be so severe. Indeed, oracles exist for which the Δ_2^{P} equality holds but not the one for $\mathrm{F}\Delta_2^{\mathrm{P}}$ [163].

2.5. *The polynomial hierarchy*

The class Δ_2^{P} gets its name from its membership in a hierarchy that is far more famous than those of the previous section. The *polynomial hierarchy*, introduced in [187] as a computational analog to the Kleene arithmetic hierarchy of recursion theory, consists of classes Δ_k^{P}, Σ_k^{P}, and Π_k^{P}, $k \geqslant 0$, defined as follows:

$$\Delta_0^{\mathrm{P}}=\Sigma_0^{\mathrm{P}}=\Pi_0^{\mathrm{P}}=\mathrm{P},$$

and, for all $i \geqslant 0$,

$$\Delta_{k+1}^{\mathrm{P}} = \mathrm{P}^{\Sigma_k^{\mathrm{P}}}, \qquad \Sigma_{k+1}^{\mathrm{P}} = \mathrm{NP}^{\Sigma_k^{\mathrm{P}}}, \qquad \Pi_{k+1}^{\mathrm{P}} = \text{co-}\Sigma_{k+1}^{\mathrm{P}}.$$

In other words, $\Delta_{k+1}^{\mathrm{P}}$ ($\Sigma_{k+1}^{\mathrm{P}}$) is the set of all languages recognizable in polynomial time (nondeterministic polynomial time) with an oracle to a problem in Σ_k^{P}, and Π_{k+1}^{P} consists of the complements of all languages in $\Sigma_{k+1}^{\mathrm{P}}$. In particular, $\Delta_1^{\mathrm{P}} = \mathrm{P}$, $\Sigma_1^{\mathrm{P}} = \mathrm{NP}$, and $\Pi_1^{\mathrm{P}} = \text{co-NP}$, while, as stated earlier, $\Delta_2^{\mathrm{P}} = \mathrm{P}^{\mathrm{NP}}$. Note that $\Delta_k^{\mathrm{P}} \subseteq \Sigma_k^{\mathrm{P}} \cap \Pi_k^{\mathrm{P}}$. It is an open problem whether the containment is proper, just as it was open for the case of $k=1$ ($\mathrm{P} \subseteq \mathrm{NP} \cap \text{co-NP}$). Similarly, $\Delta_{k+1}^{\mathrm{P}}$ contains $\Sigma_k^{\mathrm{P}} \cup \Pi_k^{\mathrm{P}}$ and we do not know whether the containment is proper for any $k \geqslant 1$. Given these containment relationships, we can capture the set of all languages in the polynomial hierarchy with the following definition.

Definition. The class PH is equal to the union $\bigcup_{k=0}^{\infty} \Sigma_k^{\mathrm{P}}$.

For a schematic illustration of PH and the classes of the polynomial hierarchy, see Fig. 2. There is an interesting alternative method for defining the classes Σ_k^{P} and Π_k^{P}, based on alternations of quantifiers. Note that we can extend the notion of "string relation" from the binary relations we discussed in Section 1.1 to k-ary relations for arbitrary $k \geqslant 2$. Σ_k^{P} is then the set of languages that can be expressed in the following format:

$$\begin{aligned}\{x\colon &(\exists y_1 \text{ with } |y_1| \leqslant p(|x|))\\ &\quad(\forall y_2 \text{ with } |y_2| \leqslant p(|x|))\\ &\quad\quad\ldots\\ &\quad\quad\quad(Q_k y_k \text{ with } |y_k| \leqslant p(|x|))\\ &\quad\quad\quad\quad[\langle y_1, y_2, \ldots, y_k, x\rangle \in R]\}\end{aligned}$$

where R is a polynomial-time recognizable k-ary relation, p is a polynomial, and Q_k is $\exists$ if k is odd, $\forall$ otherwise, and in general the quantifiers alternate [234]. (Note that, in the terminology of Section 1.3, this is equivalent to saying that Σ_k^{P} is the set of languages recognized by polynomial-time alternating Turing machines where the root of the computation tree is always labelled by "$\exists$" and the number of alternations is bounded by k.) The definition of Π_k^{P} is obtained by replacing all $\exists$s by $\forall$s, and vice versa.

Using this formalism, it is easier to identify problems as belonging in particular classes of the hierarchy. For instance, consider sentences of the form

$$(Q_1 x_1)(Q_2 x_2)\ldots(Q_j x_j)[F(x_1, x_2, \ldots, x_j)]$$

where each Q_i is a quantifier ($\exists$ or $\forall$) and F is a Boolean expression in the given variables. Let us say that such a sentence has a "quantifier alternation" for each $i > 1$ such that $Q_i \neq Q_{i-1}$ and for Q_1 itself. Then the set $\mathrm{QBF}_{k,\exists}$ of true sentences of this form with k quantifier alternations and with $Q_1 = \exists$ is a language in Σ_k^{P}. In fact, not surprisingly, it is complete for that class [255] under polynomial transformations. Similarly, the set $\mathrm{QBF}_{k,\forall}$ of true sentences with k quantifier alternations and with $Q_1 = \forall$ is complete for Π_k^{P}.

A more interesting example that is complete for a level of the polynomial hierarchy above Δ_2^P is the following problem, which is complete for Σ_2^P [237].

INTEGER EXPRESSION INEQUIVALENCE

Instance: Two *integer expressions* g and h, where the syntax and semantics of integer expressions are defined inductively as follows: the binary representation of a positive integer n is an integer expression representing the singleton set $\{n\}$; if e and f are integer expressions representing sets E and F, then $(e \cup f)$ is an expression representing the set $E \cup F$ and $(e+f)$ is an expression representing the set $\{m+n: m \in E \text{ and } n \in F\}$.

Answer: "Yes" if g and h represent different sets.

One level up, we have the following problem, complete for Δ_3^P [164].

DOUBLE KNAPSACK

Instance: Positive integers $x_1, \ldots, x_m, y_1, \ldots, y_n, N, k$.

Answer: "Yes" if the kth bit of the number M (to be defined below) is 1. To define M, let Z be the set of integers I such that (a) there exists an $S \subseteq \{1, 2, \ldots, m\}$ with $\sum_{i \in S} x_i = I$, and (b) there is no $T \subseteq \{1, 2, \ldots, n\}$ such that $\sum_{i \in T} y_i = I$. M is the largest element of $Z \cap \{1, 2, \ldots, N\}$ (or 0, if the set is empty).

For other interesting natural problems that appear to be located above Δ_2^P in the hierarchy, see [172]. Note that a problem that is complete for Σ_k^P cannot be in a class that is lower in the hierarchy unless the two classes are equal. This raises the question of whether the polynomial hierarchy might collapse to some fixed level.

As a first and most important observation along this line, note that if P = NP, then $\Sigma_{k+1}^P = \Sigma_k^P = \text{P}$ for all $k \geq 0$, and so the hierarchy collapses to P. In fact P = NP if and only if P = PH. Observe that this in a sense makes the fine structure of the polynomial hierarchy about as "academic" as that of $\Delta_2^P - \text{NP}$. If P ≠ NP, the hardest problems in PH will be intractable, but no more so than NP-complete problems seem to be (all problems in PH can be solved straightforwardly in single exponential time). If P = NP, all problems in PH will be in P. There might be a practical distinction, however, if NP ended up in some class intermediate between P and single exponential time. For instance, if the NP-complete problems could be solved in time $2^{c \log^2 n}$, then although this would extend to Δ_2^P, it would not seem to extend to all of PH. The best we can say a priori is that each problem in PH would be solvable in time $2^{c \log^k n}$ for some k, i.e., in time exponential in "polylog n". (The proof is by straightforward padding arguments.)

The fact that PH collapses to P if P = NP, i.e., if $\Sigma_0^P = \Sigma_1^P$, generalizes. It is not difficult to see that for any $k > 0$, $\Sigma_k^P = \Sigma_{k+1}^P$ implies $\text{PH} = \Sigma_k^P$. Similar statements hold for Δ_k^P and Π_k^P, as well as for $\Sigma_k^P = \Pi_k^P$ [234]. Another interesting and somewhat surprising result, alluded to in the previous section, links the collapse of the polynomial hierarchy to that of the Boolean and query hierarchies within Δ_2^P. In [53, 147] it is shown that if the Boolean hierarchy (or equivalently the query hierarchy) is finite, i.e. if it collapses to some fixed level, then $\text{PH} \subseteq \Delta_3^P$. A result to which we alluded in Section 2.1 is also relevant: If there is a sparse language that is complete for NP under Turing reductions, then PH collapses into Δ_2^P [179] (in fact to $\text{P}^{\text{NP}[O(\log n)]}$, as defined in the previous section

[146]). Finally, it has recently been shown that the polynomial hierarchy will collapse to Π_2^P if GRAPH ISOMORPHISM turns out to be NP-complete [39, 220].

Despite all these intriguing results, the consensus of the experts currently is that the hierarchy does not collapse. If so, the result is likely to be hard to prove, as each of the important alternatives for PH is now known to hold in at least one relativized world. Oracles for which PH is an infinite hierarchy and does not collapse at all are presented in [257] and Ko has recently shown that for each $k \geqslant 0$, there is an oracle for which $\Sigma_k^P \neq \Sigma_{k+1}^P = \text{PH}$ [159]. (With respect to a random oracle, all we currently know is that PH does not collapse below NP$\cup$co-NP [29].) The story continues in the next section.

2.6. *PSPACE and its subclasses*

The polynomial hierarchy consists of all those languages recognizable by polynomial-time bounded alternating Turing machines obeying fixed bounds on the number of alternations. One might reasonably ask what happens when the number of alternations is not so bounded. Let "APTIME" denote the class of languages accepted by polynomial-time alternating Turing machines (with no constraints on the number of alternations). We are thus asking whether PH = APTIME.

This is an even more interesting question than it might at first seem, for APTIME is a more interesting class than it might first seem. In fact it equals "PSPACE", the set of all languages recognizable by polynomial-*space* bounded (deterministic) Turing machines. This is a consequence of a more general result proved by Chandra, Kozen and Stockmeyer. Before stating that theorem, we introduce some notation that will be useful both here and later:

DEFINITION. If $T(n)$ is a function from the positive integers to themselves, then DTIME[$T(n)$] (NTIME[$T(n)$], ATIME[$T(n)$]) is the set of all languages recognized by DTMs (NDTMs, ATMs) in time bounded by $T(n)$, where n is the size of the input.

DEFINITION. If $S(n)$ is a function from the positive integers to themselves, then DSPACE[$S(n)$] (NSPACE[$S(n)$], ASPACE[$S(n)$]) is the set of all languages recognized by DTMs (NDTMs, ATMs) in space bounded by $S(n)$, where n is the size of the input.

2. THEOREM (Chandra, Kozen and Stockmeyer [52]). *For any function* $T(n) \geqslant n$, $\text{ATIME}[T(n)] \subseteq \text{DSPACE}[T(n)] \subseteq \bigcup_{c>0} \text{ATIME}[c \cdot T(n)^2]$.

Taking a union with $T(n)$ ranging over all polynomials yields the claimed result that $\text{APTIME} = \bigcup_{k>0} \text{ATIME}[n^k] = \bigcup_{k>0} \text{DSPACE}[n^k] = \text{PSPACE}$.

With PSPACE we reach our first complexity class containing P that is not known to collapse to P if P = NP. There are oracles such that P = NP and yet P $\neq$ PSPACE [159] (as well, of course, as oracles for which P = NP = PSPACE and such that P $\neq$ NP = PSPACE) [17]. Indeed, the range of oracular possibilities with respect to the entire polynomial hierarchy is now quite comprehensive. Among oracles for which PH does not collapse, there are ones both for PH = PSPACE and for which PH $\neq$ PSPACE [114, 115, 257]. Similarly, for any $k \geqslant 0$, among those oracles for which PH collapses to

Σ_k^P, there are also ones both for PH = PSPACE and for which PH ≠ PSPACE [159]. (With respect to a random oracle, however, PH ≠ PSPACE [47, 13].)

Thus it would seem that APTIME = PSPACE is a new and distinct class, and that a problem complete for it under polynomial transformations ("PSPACE-complete") is not likely to be in PH (much less P). There are interesting PSPACE-complete languages. An easy first example is the following generalization of the languages $\mathrm{QBF}_{k,\exists}$ and $\mathrm{QBF}_{k,\forall}$ as defined in the previous section.

QUANTIFIED BOOLEAN FORMULAS (QBF)

Instance: A sentence of the form $S=(Q_1x_1)(Q_2x_2)\ldots(Q_jx_j)[F(x_1,x_2,\ldots,x_j)]$ where each Q_i is a quantifier (∃ or ∀) and F is a Boolean expression in the given variables. (Note that there is no restriction on the number of alternations.)

Answer: "Yes" if S is a true sentence.

Starting with QBF, researchers have identified a wide variety of other PSPACE-complete problems. First among them are problems about games. Note that the alternation inherent in the APTIME definition of PSPACE is analogous to the thought processes one must go through in determining whether one has a forced win in a two-person game. ("Do I have a move such that for all possible next moves of my opponent there is a second move for me such that for all second moves of my opponent there is a third move for me such that. . . ") Thus it is perhaps not surprising that many problems about games are PSPACE-complete. Here is a simple one from [219].

GENERALIZED GEOGRAPHY

Instance: Directed graph $G=(V, A)$, specified vertex v_0.

Answer: "Yes" if the Player 1 has a forced win in the following game. Players alternate choosing new arcs from the set A. Player 1 starts by choosing an arc whose tail is v_0, and thereafter each player must choose an arc whose tail equals the head of the previously chosen arc. The first player unable to choose a new arc loses.

Note that this generalizes the "geography" game in which players alternate choosing the names of countries, with the first letter of each name having to agree with the last letter of its predecessor. Other PSPACE-complete games include a generalization of the game Hex from the Hex board to an arbitrary graph [74] and the endgame problem for a generalization of the game GO to arbitrarily large grids [176]. (For surveys of further recreational examples, see [87, 131].)

There are also many PSPACE-complete problems in which the alternation is more circumspect, such as FINITE-STATE AUTOMATA EQUIVALENCE (given the state diagrams of two nondeterministic finite automata, do they recognize the same language? [237]), REGISTER MINIMIZATION (given a straight-line program P and a bound B, can the output of P be computed using just B registers if recomputation is allowed? [89]), and the GENERALIZED MOVER'S PROBLEM (given a collection of three-dimensional "jointed robots", represented by polyhedra joined at their vertices, an assignment of locations to these robots in a 3-dimensional space with fixed polyhedral obstacles, and a set of desired final locations for the robots, can they be moved to their desired locations without ever causing an overlap between separate robots or a robot and an obstacle? [206]).

With PSPACE, we have reached the top of the tower of classes for which the question of containment in P remains open. The obvious candidate for a next step, nondeterministic polynomial space or "NPSPACE", has a rather serious drawback: like APTIME, it equals PSPACE itself. This follows from "Savitch's Theorem", proved in 1970.

3. THEOREM (Savitch [216]). *For any "space constructible" function* $T(n) \geqslant \log_2 n$, $\text{NSPACE}[T(n)] \subseteq \text{PSPACE}[T(n)^2]$.

Thus any dream of an alternating hierarchy above PSPACE in analogy to the polynomial hierarchy above P collapses back into PSPACE itself.

Savitch's result does not, however, dispose of the question of nondeterminism for the interesting subclass of those problems solvable in *linear* space.

DEFINITION. The class LIN-SPACE (NLIN-SPACE) equals $\bigcup_{c>0} \text{DSPACE}[cn]$ ($\bigcup_{c>0} \text{NSPACE}[cn]$).

These classes are especially interesting since NLIN-SPACE consists precisely of those languages that can be generated by "context-sensitive grammars" or, equivalently, recognized by "(nondeterministic) linear bounded automata" or "LBAs" (see [122, 214]). The question of whether LIN-SPACE = NLIN-SPACE is thus equivalent to the long open problem of whether deterministic LBAs are as powerful as general (nondeterministic) ones. Savitch's Theorem does not resolve the question, since it only shows that $\text{NLIN-SPACE} \subseteq \bigcup_{c>0} \text{DSPACE}[cn^2]$.

It should be noted that, as complexity classes, LIN-SPACE and NLIN-SPACE differ somewhat from the ones that we have seen so far. They are machine-independent in much the same sense that P and PSPACE are, but there is one valuable property of P and PSPACE that they do not share: they are *not* closed under polynomial transformations (as can be shown by padding arguments using Theorem H2 of Section 1.4 [35]). One consequence of this is that membership in LIN-SPACE (NLIN-SPACE) can be strongly dependent on the way in which problem instances are represented. A graph problem that is in LIN-SPACE if instances are given by adjacency matrices, might conceivably fail to be in it if instances are given by adjacency lists, even though both representations are "reasonable" in the sense of Section 1.1. A second consequence concerns the relationship of the two classes to P and NP.

By Theorem H2 of Section 1.4, LIN-SPACE and NLIN-SPACE are proper subclasses of PSPACE. Unlike PSPACE, however, they are not known to contain NP, or even P. Indeed, all we know for certain about their relationships to the latter two classes is that LIN-SPACE (NLIN-SPACE) does not equal either P or NP. (LIN-SPACE and NLIN-SPACE are not closed under polynomial transformations whereas P and NP both are.) The most natural assumption is that LIN-SPACE and NLIN-SPACE are incomparable to P and NP, although strict containment one way or the other is not ruled out. For instance, NLIN-SPACE $\subset$ P if P = PSPACE. This is actually an "if and only" statement, since there are PSPACE-complete problems in NLIN-SPACE, for instance, CONTEXT-SENSITIVE LANGUAGE MEMBERSHIP: given a context-sensitive grammar G and a string s, is s in the language generated by G? Indeed, there is a *fixed* context-sensitive language whose membership problem is PSPACE-complete [37].

Although the question of LIN-SPACE versus NLIN-SPACE remains open, a related problem of almost equal antiquity has recently been resolved. This one involves co-NLIN-SPACE, the set of all languages whose complements are in NLIN-SPACE, and its resolution puts an end to any hopes of building an alternation hierarchy above LIN-SPACE. The key result here is the following result.

4. THEOREM (Immerman [125] and Szelepcsényi [239]). *For any function* $S(n) \geqslant \log n$, NSPACE$[S(n)]$ = co-NSPACE$[S(n)]$.

As a consequence, NLIN-SPACE = co-NLIN-SPACE, and the alternation hierarchy above LIN-SPACE collapses to NLIN-SPACE, even if we assume LIN-SPACE $\neq$ NLIN-SPACE.

We shall see further applications of Theorems 3 and 4 later in this catalog.

3. Provably intractable problems

None of the complexity classes considered in the previous section contains a problem that has been *proved* to be intractable, no matter how hard some of those problems may have looked. If P = PSPACE, a possibility that has not yet been ruled out, then all problems in the classes discussed would be solvable in polynomial time. In this section we consider complexity classes that *are* known to contain intractable problems. A schema of the classes to be covered is given in Fig. 3.

First, it should be noted that for any well-behaved functional bound f that grows faster than any polynomial (for instance, $f(n) = n^{\log\log\log\log n}$), the class DTIME$(f(n))$ of decision problems solvable by $f(n)$ time-bounded DTMs must, by Theorem H1 of Section 1.4, contain problems that are not in P and hence are intractable by our definition. We shall not be interested in such "subexponential" classes however, as all the interesting theoretical work has concerned classes with exponential (or worse) resource bounds, and these are the ones we shall discuss.

3.1. *The class EXPTIME and its variants*

There are currently two distinct notions of "exponential time" in use. In what follows, we shall introduce a terminological distinction between the two notions, but the reader should be aware that the terminology we use is not completely standard. In perusing the literature, one should thus take care to ascertain the definition being used. The first and more natural notion of exponential time gives rise to the following class.

DEFINITION. The class EXPTIME equals $\bigcup_{k>0}$ DTIME[2^{n^k}], i.e., is the set of all decision problems solvable in time bounded by $2^{p(n)}$, where p is a polynomial.

As defined, EXPTIME equals the class APSPACE of all languages recognizable by alternating Turing machines (ATMs) using polynomial-bounded space. This equivalence follows from a theorem of Chandra, Kozen and Stockmeyer that complements Theorem 2 above.

5. THEOREM (Chandra, Kozen and Stockmeyer [52]). *For any function $S(n) \geqslant \log n$,* ASPACE[$S(n)$] $= \bigcup_{c>0}$ DTIME[$2^{c \cdot S(n)}$].

EXPTIME thus contains both PSPACE and NP. (There are oracles that separate EXPTIME from PSPACE, as well as oracles for which NP = PSPACE = EXPTIME [66].)

The second notion of "exponential time" restricts attention to linear exponents and gives rise to the following class:

DEFINITION. The class ETIME equals $\bigcup_{c>0}$ DTIME[2^{cn}].

Note that, by Theorem H1, ETIME is properly contained in EXPTIME. Like EXPTIME, it is independent of machine model (since a function $p(2^{cn})$, p a polynomial, is simply $O(2^{c'n})$ for a possibly larger constant c'). ETIME has several theoretical drawbacks, however. Like LIN-SPACE, it is not closed under polynomial transformations, and so a problem that is in ETIME under one choice of "reasonable" input representation could conceivably *not* be in ETIME if another "reasonable" choice was made. Furthermore, the computational dominance of ETIME over P is perhaps not quite as strong as one would like. In particular, ETIME $\neq$ $\mathrm{P}^{\mathrm{ETIME}}$, i.e., one can do more in polynomial time with an oracle for a problem in ETIME than one can do in ETIME alone. This is bcause $\mathrm{P}^{\mathrm{ETIME}}$ = EXPTIME, as can be shown by simple padding arguments (e.g., see [246]). In contrast, note that EXPTIME = $\mathrm{P}^{\mathrm{EXPTIME}}$. Finally, unlike EXPTIME, ETIME is not known to contain NP, although it is known to be unequal to NP (because it is not closed under polynomial transformations [34, 35]). There exist oracles yielding each of the three possibilities (ETIME incomparable to NP, ETIME $\subset$ NP, and NP $\subset$ ETIME) [66].

Because EXPTIME contains problems that require exponential time, any problem that can be shown to be complete for EXPTIME under polynomial transformations, i.e., "EXPTIME-complete", is provably intractable. (By padding arguments we can show that any problem that is ETIME-complete is EXPTIME-complete, so we restrict ourselves to the latter definition.) Some examples make explicit use of the aternation implicit in the definition of EXPTIME as APSPACE, such as GENERALIZED CHECKERS and GENERALIZED CHESS (given an endgame position in one of these games, as generalized from the standard 8×8 board to an $N \times N$ one, does white have a forced win?) [83, 210]. Others are less obvious, such as determining whether a given attribute grammar has the "circularity" property [128].

3.2. *The class NEXPTIME*

The nondeterministic analogs of EXPTIME and ETIME are as follows.

DEFINITION. The class NEXPTIME equals $\bigcup_{k>0}$NTIME[2^{n^k}].

DEFINITION. The class NETIME equals $\bigcup_{c>0}$NTIME[2^{cn}].

It is not difficult to see that P=NP implies both that EXPTIME=NEXPTIME and that ETIME=NETIME, and that ETIME=NETIME implies EXPTIME=NEXPTIME. None of the three converses is known to hold, however, and there exist oracles for which each is violated [66]. (There also exist oracles such that EXPTIME ≠ NEXPTIME, such that PSPACE ≠ EXPTIME, and such that PSPACE = EXPTIME = NEXPTIME [66].) Although P ≠ NP does not by itself seem to imply ETIME ≠ NETIME, we can get that conclusion if we also assume something about the structure of NP − P. In particular, ETIME ≠ NETIME if and only if NP − P contains a sparse language [111]. Whether this argues that NP − P contains sparse languages or that ETIME = NETIME is left to the biases of the reader. A final connection between NETIME and the P-versus-NP question (also of unclear significance) is the fact that $\mathrm{P}^{\mathrm{NETIME}} = \mathrm{NP}^{\mathrm{NETIME}}$ [116].

Since Theorem H3 of Section 1.4 implies that NEXPTIME must properly contain NP, any problem that is NEXPTIME-complete (under polynomial transformations) is not only not in P, it is not even in NP. Thus such results not only prove intractability, they prove the nonexistence of "short proofs" for yes-answers. (Once again, NETIME-completeness implies NEXPTIME-completeness and so can be ignored.) Examples of NEXPTIME-complete problems come from a variety of fields. A first example is worth highlighting, as we shall see variants on it in later sections. It is due to Stockmeyer and Meyer [237] and is reminiscent of the INTEGER EXPRESSION INEQUIVALENCE problem of Section 2.5, although this time we are talking about *regular* expressions and different operators.

INEQUIVALENCE OF REGULAR EXPRESSIONS OVER $(\cup, \cdot, {}^2)$

Instance: Two regular expressions e_1 and e_2 involving only the operations of union, composition, and squaring, where such an expression has the following syntax and semantics: A single symbol (0 or 1) is an expression representing the set consisting of the 1-character string (0 or 1). If e and f are expressions representing sets E and F, then "$e \cup f$" is an expression representing $E \cup F$, "$e \cdot f$" is an expression representing $\{xy: x \in E \text{ and } y \in F\}$, and e^2 is an expression that serves as a shorthand representation for the expression $e \cdot e$.

Answer: "Yes" if e_1 and e_2 represent different sets.

As a second example of a NEXPTIME-complete problem, consider the following TILING problem. A *tile* is an ordered 4-tuple $T = (N_T, S_T, E_T, W_T)$ of integers. Suppose we are given a finite collection C of tiles, a specific sequence $T_1, \ldots, T_m$ of tiles, satisfying $E_{Ti} = W_{T_{i+1}}, 1 \leqslant i < m$, and an integer n (written in binary notation). The question we ask is whether there is a tiling of the $n \times n$ square, i.e., a mapping $f:\{1, 2, \ldots, n\} \times \{1, 2, \ldots, n\} \to C$, such that $f(1, i) = T_i$, $1 \leqslant i \leqslant m$, and such that, for all i, j, $1 \leqslant i, j \leqslant n$,

(a) $j < n$ implies that $E_{f(i,j)} = W_{f(i,j+1)}$, and

(b) $i < n$ implies that $N_{f(i,j)} = S_{f(i+1,j)}$.

(Note the key fact that n can be exponentially larger than m and $|C|$. If we restrict ourselves to instances with $m = n$, the problem is only NP-complete [174].)

The idea of converting NP-complete problems into NEXPTIME-complete problems by using succinct representations of instances is also the key to our final example. Let us say that a Boolean circuit C *represents* a graph $G = (V, E)$ if, when given as input two integers i, j, $1 \leqslant i < j \leqslant |V|$, C outputs 1 if the graph contains an edge between vertex i and vertex j, and outputs 0 otherwise. Not all graphs have exponentially succinct Boolean circuit representations, but enough of them do that the following problem is NEXPTIME-complete [195]: given the Boolean circuit representation of a graph $G = (V, E)$, does G contain a clique of size $|V|/2$? (Note that for ordinary graph representations, this problem is "only" NP-complete.) Similar tricks can be pulled with other graph problems that are NP-complete under the standard representation [195]. Moreover, even if one starts with a problem that is trivial under the standard representation, conversion to the Boolean circuit representation is often enough at least to yield NP-completeness [85].

3.3. *On beyond NEXPTIME*

In this section we complete our climb up the ladder of intractability, with brief stops along the way to look at a few more interesting classes. A likely choice for the first stop would be the following.

DEFINITION. The class EXPSPACE equals $\bigcup_{k>0} \mathrm{DSPACE}[2^{n^k}]$.

There is, however, an intermediate class between NEXPTIME and EXPSPACE that captures the precise complexity of an interesting problem. This class is defined by simultaneously bounding both time and alternations, rather than just a single one as in the definitions of NEXPTIME and EXPSPACE. (Note that by Theorem 2, EXPSPACE $= \bigcup_{k>0} \mathrm{ATIME}[2^{n^k}]$.) To facilitate definitions based on simultaneous bounds, we introduce some additional notation.

DEFINITION. The class TA$[t(n), a(n)]$ is the set of all problems solvable by ATMs using at most $t(n)$ time and $a(n)$ alternations on inputs of length n.

Note that we have already seen classes for which this type of notation would have been appropriate. For example, $\Sigma_2^{\mathrm{P}} \cup \Pi_2^{\mathrm{P}} = \bigcup_{k>0} \mathrm{TA}[n^k, 2]$, and the polynomial hierarchy PH $= \bigcup_{k>0, j>0} \mathrm{TA}[n^k, j]$. An "exponential hierarchy" (EH), including EXPTIME and NEXPTIME as its first two classes, can be obtained by analogy, simply by replacing "n^k" by "2^{n^k}" in the above formulation for PH. The class of interest here, however, is not EH but a class that appears to lie somewhere between EH and EXPSPACE. Specifically, it is $\bigcup_{k>0}\text{TA}[2^{n^k}, n]$. (No less formal name has yet been proposed.) Note that this class is presumably larger than EH, but (also presumably) smaller than EXPSPACE (for which $a(n)$ should be 2^{n^k} rather than simply n).

The problem whose complexity is captured by $\bigcup_{k>0}\text{TA}[2^{n^k}, n]$ is the THEORY OF REAL ADDITION: given a first-order sentence S involving the logical connectives, the "$<$" and "$=$" relational symbols, binary constants for all the integers, and variables (with binary indices) representing real numbers, is S true? The completeness of this problem for $\bigcup_{k>0}\text{TA}[2^{n^k}, n]$ follows from results in [30, 76]. (Note that, by being complete for this class, the problem is at least as hard as hard as any problem in NEXPTIME. Thus, for any axiomatization of the theory of real addition, there must be true theorems whose proofs are exponentially long.)

Stepping up the rest of the way to EXPSPACE, we can get even stronger intractability results, as problems complete for this class provably require exponential space as well as time. To obtain an example of an EXPSPACE-complete problem, all we need do is generalize the problem highlighted in the previous section, obtaining INEQUIVALENCE OF REGULAR EXPRESSIONS OVER $(\cup, \cdot, {}^2, *)$ [237]. This is the version of the problem in which the Kleene "*" operator is added, i.e., if e is an expression representing the set E, we also allow the expression "$e*$", which represents the set of all strings of the form $x_1 x_2 \ldots x_n$, where $n \geqslant 0$ ($n=0$ means we have the empty string) and $x_i \in E$, $1 \leqslant i \leqslant n$.

As might be expected from the alternative definition of EXPSPACE in terms of alternating exponential time, there are also games that are complete for EXPSPACE. These are quite complicated however, involving teams of players, incomplete information, and rules sufficiently convoluted that it is perhaps best simply to refer the interested reader to the original source [205] and to [131].

Next up the ladder we come to what we shall denote as "2-EXPTIME" and "2-NEXPTIME", the classes of decision problems solvable by DTMs (NDTMs) operating in time bounded by $2^{2^{p(n)}}$ for some polynomial p. However, we must go a bit further (halfway to 2-EXPSPACE) to obtain our next interesting completeness result. The following problem is complete for the class $\bigcup_{k>0}\text{TA}[2^{2^{n^k}}, n]$: PRESBURGER ARITHMETIC (the analog of the theory of real addition in which the variables are to be interpreted as natural numbers rather than arbitrary reals) [76, 30]. Thus, contrary to what we learned in school, integers are more difficult than real numbers (for which the theory of addition was exponentially easier, as mentioned above).

Continuing up the ladder we now have 2-EXPSPACE, and then k-EXPTIME, k-NEXPTIME, and k-NEXPSPACE, $k \geqslant 3$, where k refers to the number of levels of exponentiation in the appropriate resource bound. The next class of real interest, however, is the union of this hierarchy: the class of "elementary" decision problems.

DEFINITION. The class ELEMENTARY $= \bigcup_{k>1} k$-EXPTIME.

Even those unimpressed with the difficulty of problems in 2- or 3-EXPTIME will have to admit that if a problem is decidable but not in ELEMENTARY, it might as well not be decidable at all. Examples of such nonelementary problems are the "weak monadic second-order theory of successor" [186] and yet another generalization of the regular expression inequivalence problem, this time to INEQUIVALENCE OF REGULAR EXPRESSIONS OVER ($\cup, \cdot, \neg$). Here we replace "squaring" and the Kleene "*" operation by negation: if e is an expression representing the set E, then "$\neg e$" is an expression representing $\{0,1\}^*-E$. Although determining whether two such expressions are equivalent is decidable, it requires running time that grows with a tower of 2s that is at least $\log n$ levels tall [237]. (The currently best upper bound involves a tower that is n levels tall [237].)

One could of course define complexity classes that are even larger than the ones so far discussed, but at this point it seems best to skip ahead to the class representing the supreme form of intractability: the undecidable (or nonrecursive) problems. These are traditionally the domain of recursion theorists, and can themselves be grouped into more and more complex classes, but we shall leave a detailed discussion of these to the Handbook on Logic [65] and other surveys, such as those contained in [102, 122]. Suffice it to point out that it is almost as easy to change a problem from decidable to undecidable as it is to change one from polynomial-time solvable to NP-complete. For example, if in the TILING problem mentioned above one asks that the tiling extend to all of $Z \times Z$ instead of a simple $n \times n$ subsquare, the problem becomes undecidable [251], as does PRESBURGER ARITHMETIC if one augments it to include multiplication [184].

4. Classes that count

In this section we return to the world of the only *presumably* intractable, to examine complexity classes that do not fit in the standard format of the previous sections. The classes we shall study are defined by variants of the NDTM in which the answer depends, not just on the existence or absence of computations ending in accept states, but on the *number* of such computations. We begin with classes in which the *precise* number is significant, and conclude with a whole panoply of "probabilistic" classes where the important criterion is that the number exceed a given threshold.

4.1. *Counting Turing machines and the class #P*

The first class we consider is restricted to functions $f:\{0,1\}^* \to Z_0^+$, i.e., from strings to the nonnegative integers. It was introduced by Valiant in [243], and is defined in terms of what Valiant calls *counting Turing machines*:

DEFINITION. A *counting Turing machine* (*CTM*) is an NDTM whose "output" for a given input string x is the number of accepting computations for that input.

DEFINITION. The class #P is the set of all functions that are computable by polynomial-time CTMs.

There is some debate about how "#P" should be pronounced, as the "#"-sign variously abbreviates "sharp", "pound", and "number", depending on context. The context here would seem to favor the last of the three alternatives, and that was advocated by [87] although many authors prefer the first.

A typical member of #P is the following HAMILTONIAN CIRCUIT ENUMERATION problem: given a graph G, how many Hamiltonian circuits does it contain? Note that this problem is NP-hard in the sense of Section 1.3, since we can use an oracle for it to solve the NP-complete HAMILTONIAN CIRCUIT problem. (Trivially, G has such a circuit if and only if the number of such circuits that it has is greater than 0.) Note, however, that counting the number of Hamiltonian circuits might well be harder than telling if one exists. To begin with, the decision problem that simply asks, given G and k, whether G has k or more Hamiltonian circuits, is not known to be in NP. The number of Hamiltonian circuits that a graph with n vertices can have need not be polynomially bounded in n, and so the approach of simply guessing the requisite number of circuits and checking their Hamiltonicity may well take more than polynomial time. No clever tricks for avoiding this difficulty are known. Indeed, this decision problem is not known to lie anywhere within the polynomial hierarchy, although it is clearly in PSPACE.

Based on the presumed unlikelihood that #P⊆FP, Valiant [243] introduced a new class of (presumably) intractable problems: those that are "#P-complete".

DEFINITION. A problem X is #P-*hard* if there are polynomial-time Turing reductions to it from all problems in #P. If in addition $X \in$ #P, we say that X is #P-*complete*.

Note the corollary that a #P-hard problem can be in FP only if #P⊆FP. In practice, it turns out that a more restricted form of reduction, first introduced in [225] often suffices to prove #P-hardness, especially for enumeration problems whose underlying existence question is in NP.

DEFINITION. A *parsimonious transformation* is a polynomial transformation f from problem X to problem Y such that, if $\#(X,x)$ is defined to be the number of solutions that instance x has in problem X, then $\#(X,x) = \#(Y,f(x))$.

By appropriately modifying the standard transformations for showing NP-completeness, one can often obtain transformations that are parsimonious, or at least "weakly parsimonious" ($\#(X,x)$ can be computed in polynomial time from x and $\#(Y,f(x))$). In particular, Cook's Theorem can be modified to show that "#SAT", the problem of counting the number of satisfying truth assignments for an instance of SATISFIABILITY, is #P-complete. From this it follows that the enumeration versions of most NP-complete problems, including HAMILTONIAN CIRCUIT ENUMERATION, are #P-complete.

What is perhaps surprising is that enumeration versions of problems in P can also be #P-complete. The following is the most famous such #P-complete problem, first identified in [243].

PERMANENT COMPUTATION

Instance: An $n \times n$ 0-1 matrix A.

Answer: The value $perm(A)=\sum_\sigma\prod_{i=1}^n A_{i,\sigma(i)}$ of the *permanent* of A, where the summation is over all $n!$ permutations σ of $\{1,2,\ldots,n\}$.

Note that the permanent of a matrix is simply the variant on the determinant in which all summands are given positive signs. The analogous DETERMINANT COMPUTATION problem appears to be much easier however, since the determinant of a matrix A is computable in polynomial time, even when the entries of A are allowed to be arbitrary rationals (e.g., see [5]).

To see that PERMANENT COMPUTATION can be viewed as an enumeration problem, simply observe that $\prod_{i=1}^n A_{i,\sigma(i)}$ must be either 0 or 1. Hence $perm(A)$ is simply the number of permutations σ for which this product equals 1. Moreover, note that the problem of whether there *exists* a σ for which the product equals 1 is in P. Indeed, it is equivalent to the well-known polynomial-time solvable problem of telling whether a bipartite graph contains a perfect matching. Recall that for a given graph G, a *perfect matching* is a set M of edges such that every vertex of G is an endpoint of precisely one edge in M. Suppose we consider A to be the adjacency matrix for bipartite graph G_A on $2n$ vertices $v_1,\ldots,v_n$ and $u_1,\ldots,u_n$, with (u_i,v_j) an edge of G_A if and only if $A_{ij}=1$. Then it is easy to verify that $\prod_{i=1}^n A_{i,\sigma(i)}=1$ if and only if $M=\{(u_i,v_{\sigma(i)})\colon 1\leqslant i\leqslant n\}$ is a perfect matching for G_A. Thus $perm(A)$ is simply the number of distinct perfect matchings in G_A. Consequently, counting the number of perfect matchings in a bipartite graph is #P-complete, in direct contrast to such positive results as the fact that, given a graph G, one can in polynomial time determine both the number of distinct spanning trees it contains and the number of Euler circuits it contains [101].

For more examples of #P-complete problems, see [70, 200, 244]. One particularly interesting type of problem covered by these references is the *reliability* problem. As an example, consider the following problem.

GRAPH RELIABILITY

Instance: Directed graph $G(V,A)$ with specified source and sink vertices s and t. For each arc a, a rational number $p(a)\in[0,1]$.

Answer: The probability that there exists a path in G from s to t consisting entirely of "nonfailed" arcs, given that an arc a fails with probability $p(a)$, independently for all a.

Although not formally in #P, this problem is essentially the same as the problem of counting the number of distinct subgraphs of a certain type, and the latter problem *can* be shown to be #P-complete. (Technically, we should say that the reliability problems are #P-*equivalent*, in analogy with the NP-equivalent problems, as defined in Section 2.3. That is, they are #P-hard, and yet solvable in polynomial time with an oracle to a problem in #P, i.e., they are in the class $FP^{\#P}$. Another interesting #P-equivalent problem is that of computing the volume of the convex hull of a set of rational coordinate points in Euclidean n-space [70].)

One important enumeration problem that appears *not* to be #P-complete is the problem, given two graphs G and H, of counting the number of distinct isomorphisms between G and H. For this problem, the enumeration problem is polynomial-equivalent to the decision problem, and indeed, verifying the number of isomorphisms is known to be in NP, even though there may be an exponential number of them [183].

Another interesting set of enumeration problems that are not known to be #P-hard are such problems as, given an integer n (written in unary so that "polynomial time" means "polynomial in n"), how many distinct labelled graphs with n vertices contain Hamiltonian circuits? This is the form for many classical enumeration problems in combinatorial mathematics. Some such problems are known to be in P; for instance the much simpler problem, given n, of determining how many distinct labelled graphs with n vertices exist (there are precisely $2^{n(n-1)/2}$). Others, like the former, remain open. For problems like these, a new class becomes relevant.

DEFINITION. The class $\#P_1$ consists of all those problems in #P whose instances are restricted to strings over a single-letter alphabet.

The question of whether $\#P_1 \subseteq FP$ is addressed in [244]. It is perhaps easier than #P versus FP, but no less open. Furthermore, there are graph problems, albeit convoluted ones, that are $\#P_1$-complete under parsimonious transformations and hence in FP if and only if $\#P_1 \subseteq FP$. For details, see [244].

Returning to the class #P, let us briefly consider its relationship to some of the other classes we have been considering, in particular the polynomial hierarchy. In discussing this relationship, we must first deal with the fact that #P is a class of functions, while PH and its subclasses are classes of languages. One option here is to consider a functional analog of PH. Suppose we replace $\Delta_1^P = P$ by FP and, for $k > 1$, replace $\Delta_k^P = P^{\Sigma_{k-1}^P}$ by $F\Delta_k^P = FP^{\Sigma_{k-1}^P}$, the set of all functions computable in polynomial time with an oracle for a set in Σ_{k-1}^P. We could then ask whether there exists a k such that $\#P \subseteq F\Delta_k^P$, or more generally, whether $\#P \subseteq FPH = \bigcup_{k=1}^{\infty} F\Delta_k^P$?

This problem remains open, although it is known that there exist oracles for which #P is not contained in $F\Delta_2^P$[235]. Indeed, something much stronger can be said: there are relativized worlds in which there exist problems in #P that cannot even be *well-approximated* within $F\Delta_2^P$, in the sense of being computed to within a constant factor r for some r. This result does not extend any farther up in the hierarchy however. For any fixed ε and d, and any function f in #P, there is a function in $F\Delta_3^P$ that approximates f to within a factor of $1+\varepsilon|x|^{-d}$, where $|x|$ is the input length [235]. Note, however, that this is a far cry from saying that #P is actually contained in $F\Delta_3^P$, and current betting is that in fact #P is not even contained in FPH.

The other possibility, that #P in a sense dominates the polynomial hierarchy, is more appealing, and a recent result indicates a precise sense in which it is true. This result, due to Toda [240] concerns languages rather than functions, and so requires an appropriate stand -in for #P rather than one for PH. The stand-in chosen is the natural one of the "#P-easy" languages, i.e., the class $P^{\#P}$ of languages recognizable in polynomial time with an oracle to a problem in #P.

6. THEOREM (Toda [240]). $PH \subseteq P^{\#P}$.

Equality between PH and $P^{\#P}$ seems unlikely. The two classes are unequal in almost all relativized worlds, as a technical consequence of the proofs mentioned in Section 2.6

that PH $\neq$ PSPACE relative to a random oracle [13, 47]. Furthermore, an immediate consequence of Theorem 6 is that $P^{\#P}$ cannot be contained in PH unless the polynomial hierarchy collapses. The same statement holds for an apparently much simpler class than #P, independently introduced by [93] and [196] and defined as follows.

DEFINITION. A *parity Turing machine* ($\oplus$TM) is an NDTM that accepts a given input string x if and only if the number of accepting computations for that input is odd.

DEFINITION. The class $\oplus$P ("*parity*-P") is the set of all languages accepted by polynomial-time $\oplus$TMs.

As one might expect, a typical example of a $\oplus$P-complete problem is $\oplus$SAT (given an instance I of satisfiability, is the number of satisfying truth assignments for I even?). Superficially, this would seem possibly to be a simpler problem than actually counting the number of satisfying assignments (#SAT), and so we might assume that $\oplus$P is a less powerful class than #P. However, just as determining the parity of the optimal travelling salesman tour's length proved to be just as hard as computing that length (both problems are complete for $F\Delta_2^P$, as seen in Section 2.4), here too one does not lose much by settling for the parity of x rather than x itself. In [240], it is shown that there is a randomized sense in which PH is $\oplus$P-easy, and that consequently $\oplus$P cannot be in PH unless PH collapses. We shall cover the concept of randomized reduction in Section 4.4, and will have more to say about $\oplus$P both there and in Section 4.2.

We conclude this section with a brief discussion of two more classes whose definitions can be related to that of #P. Suppose our NDTMs are augmented with output tapes, the contents of which are interpreted as (binary) integers. Such machines can be interpreted as computing functions in a variety of ways. It is not difficult to see that #P is simply the set of functions computable by taking the sum of the output values over all accepting computations of a polynomial-time NDTM. Two other natural options are to take the maximum value or to take the number of distinct values. These give rise to the following two classes, defined respectively in [163] and [161].

DEFINITION. The class OptP is the set of all functions computable by taking the maximum of the output values over all accepting computations of a polynomial-time NDTM.

DEFINITION. The class span-P is the set of all functions computable as $|S|$, where S is the set of output values generated by the accepting computations of a polynomial-time NDTM.

It is not difficult to see that $\mathrm{OptP} \subseteq \mathrm{FP}^{\#P}$ (and indeed that $\mathrm{OptP} \subseteq \mathrm{FP}^{NP} = F\Delta_2^P$), given that the maximum can be determined by binary search. Thus it is likely that OptP is strictly weaker than #P. It does, however, capture essentially the full power of $F\Delta_2^P$. Although it is a proper subclass, given that it can only contain functions rather than arbitrary search problems, any problem complete for OptP (under the "metric reductions" to which we alluded in Section 2.4) is complete for all of $F\Delta_2^P$ [163]. An example (also mentioned in Section 2.4) is the problem of computing the length of an optimal travelling salesman tour.

Whereas OptP seems to be less powerful than #P, this is not the case for span-P. It is not difficult to show that both #P and OptP are contained in span-P[161]. Moreover, there is evidence supporting the contention that span-P *properly* contains #P, but for that we will need the definitions of the next section.

4.2. *Unambiguous Turing machines and the classes FUP and UP*

In this section we consider yet another way in which NDTMs with output tapes can be considered to compute functions. Rather than worry about how to combine the results of multiple accepting computations, let us require that there be only one! The following definition is from [242].

DEFINITION. An *unambiguous Turing machine* (UTM) is an NDTM that, for each possible input string, has at most one accepting computation.

Such a machine, if allowed an output tape, provides a unique output for every string it accepts, and hence can be viewed as computing a function over the domain of the language that it recognizes.

DEFINITION. FUP (UP) is the class of all partial functions computable (all languages recognizable) by polynomial-time UTMs.

(This class is called "UPSV" in [99].) There are a variety of functions that, although not at present known to be in FP, are known to be in FUP. For a primary example, consider the *discrete logarithm*. In its simplest formulation, this problem can be given as follows.

DISCRETE LOG

Instance: Given a prime p, a primitive root a modulo p, and an integer b, $0<b<p$.

Answer: The discrete logarithm of b with respect to p and a, i.e., the (unique) integer c, $0\leqslant c<p$, such that $a^c=b \bmod p$.

This problem, as stated, is almost, but not quite, in FUP. The answer is unique, but unfortunately the domain is not known to be in UP (which is required by definition). Although one can guess short proofs that substantiate primality and primitivity [199], no way of guessing *unique* proofs is known. Fortunately, the following augmented version of DISCRETE LOG, also not known to be in FP, *can* be computed by a polynomial-time UTM: given p, a, and b as before, together with short proofs that p is prime and a is a primitive root, compute the discrete log of b with respect to p and a.

Functions like (augmented) DISCRETE LOG that potentially lie in FUP − FP have cryptographic significance. The existence of such functions can be shown to be equivalent to that holy grail of modern cryptography theory, the "one-way function".

There are a variety of definitions of this concept, some stronger than others. The following captures about the minimum one would wish to have.

DEFINITION. A (partial) function f is (*weakly*) *one-way* if it satisfies the following three properties:

(1) f is "honest" (i.e., for all x in the domain of f, the string $f(x)$ can be no more than polynomially smaller than the string x),
(2) $f \in \mathrm{FP}$, and
(3) f^{-1} is not in FP.

Note that this definition of "one-way" requires only that the inverse be difficult in the worst-case. We have introduced the qualifier "weakly" so as to contrast this notion with the stronger one of [256] that we shall discuss briefly at the end of Section 4.3.

The discrete logarithm (or more precisely, its inverse) is currently one of the more popular candidates for the role of one-way function (according to both weak and strong definitions). Many current cryptographic schemes are based on presumed difficulty of the problem of computing c given p, a, and b (see [209]). At present, however, the only evidence we have of this difficulty is the fact that no one has as yet found an efficient algorithm for the problem.

A proof that a one-way function exists would be a major event, of course, as it would imply $\mathrm{P} \neq \mathrm{NP}$. The question on one-way function existence, however, can be even more tightly tied to a simple question about UTMs. Note that by definition $\mathrm{P} \subseteq \mathrm{UP} \subseteq \mathrm{NP}$, and it is not difficult to see that $\mathrm{FUP} = \mathrm{FP}$ if and only if $\mathrm{UP} = \mathrm{P}$. It has been shown in [99] that one-way functions exist if and only if $\mathrm{P} \neq \mathrm{UP}$. (One-way functions whose range is in P exist if and only if $\mathrm{P} \neq \mathrm{UP} \cap \text{co-UP}$.) Thus we might still hope to get strong candidates for one-way functions if we could identify complete problems for UP. Unfortunately, this seems unlikely: there are oracles for which UP has no such complete problems, either under polynomial transformations [108] or the more general polynomial-time Turing reductions [117]. The question of whether $\mathrm{UP} = \mathrm{NP}$ also has interesting connections to other areas, in this case to the classes of the previous section. In [161] it is shown that span-$\mathrm{P} = \#\mathrm{P}$ if and only if $\mathrm{UP} = \mathrm{NP}$.

As might be expected, there exist oracles for all four possible relations between P, UP and NP: $\mathrm{P} = \mathrm{UP} = \mathrm{NP}$, $\mathrm{P} \neq \mathrm{UP} \neq \mathrm{NP}$, $\mathrm{P} = \mathrm{UP} \neq \mathrm{NP}$, and $\mathrm{P} \neq \mathrm{UP} = \mathrm{NP}$ [202, 88]. There is also an oracle such that $\mathrm{P} \neq \mathrm{UP} \cap \text{co-UP}$ [108]. Relative to a random oracle the containments $\mathrm{P} \subseteq \mathrm{UP} \subseteq \mathrm{NP}$ are all strict [28, 211].

Reference [28] actually is addressed toward more detailed considerations. It addresses the interesting question of whether one gets a bigger class of languages every time one increases the allowable number of accepting computations.

DEFINITION. For each $k > 0$, the class UP_k is the set of languages recognizable by NDTMs that always have k or fewer accepting computations.

Note that $\mathrm{UP} = \mathrm{UP}_1$. It is not known whether there is any k such that UP_k is strictly contained in UP_{k+1}. (If there were, then we would have $\mathrm{P} \neq \mathrm{NP}$.) It is shown in [28] however, that with respect to a random oracle, strict containment holds for all $k \geqslant 1$.

Moreover, for random oracles we also have that $\bigcup_{k\geqslant 1} \mathrm{UP}_k$ is strictly contained in the following natural class, first introduced, under a slightly different name, in [9].

DEFINITION. The class FewP is the set of all languages recognizable by polynomial-time NDTMs for which the number of accepting computations is bounded by a fixed polynomial in the size of the input.

Note that FewP is contained in NP by definition. A less obvious (and more interesting) containment is the following. Recall the class ⊕P introduced in the previous section, defined in terms of NDTMs whose answers are determined by the parity of the number of accepting computations. Although there are indications that this is a very powerful class, at present we do not even know if ⊕P contains NP. (Oracles exist for which the two classes are incomparable [241].) In [51] it is proved that ⊕P does contain FewP.

See Fig. 4 for a schema of the relations between the classes of decision problems introduced in this and the previous section. For more on the classes UP and FewP, and the relation of the former to cryptography, see the above references, and [133, 160].

4.3. Random Turing machines and the classes R, co-R, and ZPP

In the previous section, we considered NDTMs that had either no accepting computations or exactly one. The restriction that yields the classes highlighted in this section is somewhat different, but has a similar flavor. In the definitions that follow, we continue to assume, as in Section 1.3, that our NDTMs are normalized so that all computations are finite and have the same length. Now, however, we also assume (without loss of generality) that every internal node in the computation tree has either one or two successors, and that every computation path has the same number of "branch points", i.e., nodes with two successors.

DEFINITION. A *random Turing machine* (RTM) is an NDTM such that, for each possible input string, either there are no accepting computations or else at least half of all computations are accepting.

DEFINITION. The class R consists of all decision problems solved by polynomial-time RTMs.

Here "R" stands for "random polynomial time", a terminology introduced in [3]. (Some writers use "RP" to denote the same class.) Note that $\mathrm{P}\subseteq\mathrm{R}\subseteq\mathrm{NP}$. The term "random" comes from the following observation. Suppose we attempted to simulate the operation of a given NDTM on a given input as follows: Starting at the initial configuration we proceed deterministically until we reach a branch point. At each branch point, we randomly pick one of the two alternatives for the next move, and then proceed. Given our assumption that all computations of the NDTM contain the same number of branch points, the probability we will end up in an accept state is simply the ratio of the number of accepting computations to the total number of computations.

Thus for a problem in R, the probability that such a simulation will mistakenly answer "no" (i.e., fail to accept) when the answer is "yes" is 0.5 or less. Note that, as with the definition of NP, this definition is one-sided; the probability that we will answer "yes" when the answer is "no" is 0.

A fortunate side-effect of this one-sidedness is that we can arbitrarily increase the probability of correctness by repeating the experiment. If we perform k independent random simulations and ever find an accepting computation, we know the answer is "yes", otherwise the probability that the answer is "yes" is at most $1/2^k$. Even for moderate values of k, say $k=50$, this can be less than the probability of a machine error in our computer, and so we will be fairly safe in asserting that the answer is "no".

Thus, computationally speaking, it is almost as good for a problem to be in R as for it to be in P. Moreover, the possibility that R = P has not been ruled out. There are oracles for which the two classes are distinct and either R ≠ NP or R = UP = NP [202]. There are also oracles for which all of P, UP, R, and NP are all distinct [88]. Relative to a random oracle however, P = R ≠ NP [29].

Several important problems, not known to be in P, are known either to be in R or co-R (the complements of decision problems in R, where it is the "yes" answers that may be false, although only with probability 0.5 or less). For a first example, consider the following.

PRODUCT POLYNOMIAL INEQUIVALENCE

Instance: Two collections $\boldsymbol{P}=\{P_1,\ldots,P_n\}$ and $\boldsymbol{Q}=\{Q_1,\ldots,Q_m\}$ of multivariate polynomials over the rationals, each polynomial represented by listing its terms with non-zero coefficients.

Answer: "Yes" if $\prod_{i=1}^{n} P_i$ and $\prod_{j=1}^{m} Q_j$ are different polynomials.

This problem is not known to be in P. We cannot simply multiply together each collection and compare the results, as such multiplications may result in an exponential blow-up in the number of terms. Nor can we factor the individual polynomials and compare the two lists of irreducible factors, since the irreducible factors of a polynomial can also have exponentially more terms than the original. There is, however, a simple randomized test for inequivalence of the two products, suggested by Schwartz in [221], that runs in polynomial time. One merely chooses a set of values for the arguments in an appropriate random fashion, *evaluates* each of the P_is and Q_is with these values substituted in, and then multiplies the two resulting collections of rational numbers together, thus obtaining the values of the two product polynomials at the given arguments. Schwartz shows that the evaluated products must differ at least half the time if the product polynomials are inequivalent. (They of course cannot differ if the product polynomials are equivalent.) The above "randomized algorithm" thus can serve as a basis for a polynomial-time RTM that solves PRODUCT POLYNOMIAL INEQUIVALENCE, which consequently is in R.

A second, although as we shall see possible weaker, candidate for membership in R − P is COMPOSITE NUMBER. As has been observed in [201, 233], a positive integer n is composite if and only if more than half of the integers b, $1 \leqslant b < n$, are "witnesses" to this in a technical sense that can be verified in polynomial time given just n and b. The following "randomized algorithm" can thus serve as the basis for a polynomial-time RTM that solves COMPOSITE NUMBER: Pick a random integer b between 1 and n and say yes if and only if b is a witness to the compositeness of n. The total time is $O(\log^3 n)$, i.e., polynomial in the size of the input number n. Thus COMPOSITE NUMBER is in R (and its complementary problem, PRIME NUMBER, is in co-R). COMPOSITE NUMBER is a weaker candidate for membership in R − P than PRODUCT POLYNOMIAL INEQUIVALENCE for two reasons. The first is the already cited result of [188] that, if the Extended Riemann Hypothesis holds, then COMPOSITE NUMBER $\in$ P. The second is the recent discovery claimed in [2] (and building strongly on results in [95]) that PRIME NUMBER is also in R, thus placing both the PRIME and COMPOSITE NUMBER problems in the elite class "ZPP".

DEFINITION. ZPP = R$\cap$co-R.

We refer to ZPP as an "elite class" because it also equals the set of those decision problems that can be solved by randomized algorithms that *always* give the correct answer and run in expected polynomial time. Note that every problem in ZPP is solvable by such an algorithm: One simply runs both the R and co-R algorithms for the problem repeatedly until one discovers a witness to the answer. The expected number of iterations is so small that the expected running time is proportional to that of the slower of the R and co-R algorithms. (In the case of COMPOSITE NUMBER, this unfortunately may be as bad as $\Omega(n^{100})$; the Adleman and Huang result [2] is of mostly theoretical interest.) Conversely, if a decision problem can be solved by a randomized algorithm that runs in expected polynomial time and always gives the correct answer, that problem must be in R$\cap$co-R: Associated with any such expected polynomial-time algorithm is a polynomial p such that, for any input x, the probability that the algorithm will halt before $p(|x|)$ steps have been taken exceeds 0.5. The required RTMs can be constructed by simulating the algorithm for $p(|x|)$ steps.

Another characterization of ZPP is that it consists of all those problems solvable by polynomial-time "Las Vegas" algorithms. This term, introduced in [262], refers to randomized algorithms which either give the correct answer or no answer at all (and for which the latter option occurs less than half the time). This is as opposed to "Monte Carlo" algorithms, where "Monte Carlo" is typically used as a generic term for "randomized". (See [132] for a more extended discussion of these terms, as well as more on probabilistic classes in general.)

See Fig. 5 for a schema relating R, co-R, and ZPP to the major nonrandomized classes presented earlier (as well as to some additional randomized classes that will be introduced in Section 4.5). There are many who believe that membership in ZPP is a strong hint of membership in P. Indeed, relative to a random oracle, P = ZPP = R, with all properly contained in NP [29]. Oracles do exist, however, for which P $\neq$ ZPP [123] and ZPP $\neq$ R [21].

Returning to the question of R versus P, observe that although we have presented potential examples of problems in R − P, we have provided no evidence that these are the most likely examples. That is, we have not identified any "R-complete problems". The fact is, we do not know of any, and it is unlikely that we shall find any soon. As is the case for NP$\cap$co-NP, there are oracles such that R contains no problems that are complete for it under polynomial transformations [228] or under polynomial-time Turing reductions [119].

We conclude this section by discussing ways in which R appears to differ from NP. As we have seen, although there are oracles such that R = NP, the two classes are distinct for a random oracle. Further evidence in favor of the proper containment of R in NP comes from the fact, pointed out in [1], that R has "small circuits". More precisely, R is contained in the following class, defined in terms of the Boolean circuit families of Section 1.3.

DEFINITION. The class P/poly consists of all those languages recognizable by (not necessarily uniform) families of polynomial-size circuits.

In other words, for any problem X in P/poly, there is a polynomial p_X with the following property: for any instance size n, there is a Boolean circuit B_n with n inputs and $O(p_X(n))$ gates that correctly outputs X's answer for all instances of size n. According to the definition, there need be no uniform way of generating the circuits B_n in time polynomial in n, and the proof in [1] that $R \subseteq P/poly$ takes advantage of this. That is, it proves that the circuits exist, but does not show how to construct them. Thus it does not also imply that R = P. It does, however, provide strong evidence that $R \neq NP$. Although nontrivial oracles exist for which $NP \subseteq P/poly$ (i.e., ones for which we also have $P \neq NP$) [126], such a containment would imply that the polynomial hierarchy collapses into Σ_2^P [152].

As a digression, we should point out that P/poly has an alternative definition, this one in terms of Turing machines with "advice".

DEFINITION. An *advice-taking Turing machine* is a Turing machine that has associated with it a special "advice oracle", one that is a (not necessarily recursive) function $A: Z^+ \to \{0,1\}^*$. On input x, a special "advice tape" is automatically loaded with $A(|x|)$ and from then on the computation proceeds as normal, based on the two inputs, x and $A(|x|)$.

DEFINITION. An advice-taking TM uses *polynomial advice* if its advice oracle A satisfies $|A(n)| \leqslant p(n)$ for some fixed polynomial p and all nonnegative integers n.

DEFINITION. If X is a class of languages defined in terms of resource-bounded TMs, then X/poly is the class of languages defined by TMs with the same resource bounds but augmented by polynomial advice.

Since the "advice" for an instance of size n can be the description of a polynomial size, n-input Boolean circuit, it is clear that a language in P/poly according to the circuit family definition is also in it according to the advice definition; the converse is almost as immediate.

Other results about polynomial advice from [152] include

(a) $PSPACE \subseteq P/poly$ implies $PSPACE \subseteq \Sigma_2^P \cap \Pi_2^P$ (and hence the polynomial hierarchy collapses and equals PSPACE);

(b) EXPTIME $\subseteq$ P/poly implies EXPTIME $= \Sigma_2^{\mathrm{P}}$ (and hence P $\neq$ NP); and
(c) EXPTIME $\subseteq$ PSPACE/poly implies EXPTIME = PSPACE.
The proof techniques used in these results are elaborated upon in [18, 19].

We conclude this section by alluding to an additional result that can be viewed as supporting the conjecture that R is strictly contained in NP. This is a result of A.C. Yao in [256] linking the complexity of R to the existence of "strong" one-way functions. We omit the technical details of the latter definition here, but suffice it to say that, whereas the "one-way functions" defined in the previous section were only difficult to invert in a worst-case sense, strong one-way functions are also difficult to invert "on average" (and even for nonuniform circuits). This is obviously a more useful cryptographic property, and many theoretical encryption schemes are based on the assumption that such functions exist (and indeed, that the DISCRETE LOG problem of the previous section provides one). However, if we assume that such functions exist, and also that NP-complete problems take exponential time (a slightly stronger assumption than P $\neq$ NP, but a common one), then Yao's result implies that R is strictly contained in NP. He shows that, assuming that strong one-way functions exist, we must have $\mathrm{R} \subseteq \bigcap_{\varepsilon>0} \mathrm{DTIME}(2^{n^{\varepsilon}})$.

4.4. *Randomized reductions and NP*

When we discussed the class NP previously in Section 2.1, we mentioned several different classes of transformations with respect to which problems could be proved "complete" for NP, and indicated that there would be more in a later section, once the appropriate groundwork had been laid. The introduction of random Turing machines in the previous section has laid that groundwork, and in this section we discuss three new forms of reduction, based on variants of random Turing machines, and the kinds of "completeness for NP" that they yield. Although other variants are possible, these three have to date been the most successful.

Each can be viewed as a variant to the γ-reduction described in Section 2.2, and all include polynomial transformations as a special case. With each, we reduce a decision problem X to decision problem Y by means of a polynomial-time NDTM that yields an output for each accepting computation and satisfies certain properties. The property for γ-reductions was that there be at least one accepting computation for each string x, and for each output y, y had the same answer in Y as x had in X. Here are properties that our three randomized reductions must satisfy for any given input string x:

(1) *Random reduction* (R-reduction) [3]: (a) At least half the computations are accepting, and (b) for all outputs y, y has the same answer in Y as x has in X. (Note that every R-reduction is thus also a γ-reduction, something that cannot be said for the following two types of reduction.)

(2) *Unfaithful random reduction* (UR-reduction) [4]: (a) All computations are accepting, (b) the outputs must be "faithful" for yes-instances, i.e., if the answer for x in X is "yes", then all outputs y have answer "yes" in Y, and (c) correct outputs must be "abundant" for no-instances, i.e., if the answer for x in X is "no", then at least $1/p(|x|)$ of the outputs y have answer "no" in Y, where p is a fixed polynomial.

(This definition is generalized somewhat from that given in [4], for which "abundant" meant "at least half". The generalization allows us to include that UR-reductions, like the other two types, are transitive.)

(3) *Reverse unfaithful random reduction* (RUR-reduction) [245]: Same as for UR-reductions, except now the outputs must be faithful for no-instances and correct outputs must be abundant for yes-instances. (This reduction goes unnamed in [245]; the above name was the best we could come up with on short notice.)

The properties that make these reductions useful are as follows:

(1) If X is hard for NP under R-reductions, then $X \in \mathrm{ZPP}$ implies $\mathrm{NP} = \mathrm{ZPP}$ and $X \in \mathrm{R}$ implies $\mathrm{NP} = \mathrm{R}$.

(2) If X is hard for NP under UR-reductions, then $X \in \text{co-R}$ implies $\mathrm{NP} = \mathrm{ZPP}$.

(3) If X is hard for NP under RUR-reductions, then $X \in \mathrm{R}$ implies $\mathrm{NP} = \mathrm{R}$.

Note that in all three cases we have at the very least that if X is in ZPP then $\mathrm{NP} = \mathrm{R}$. Thus, although proving that a problem is complete (or simply hard) for NP under any of the above three types of reduction is not as strong an argument for intractability as proving it NP-complete, such a proof still provides believable evidence that the problem cannot be solved in either polynomial time or polynomial expected time. We conclude this section with examples of complete problems of each type.

For a problem that is complete for NP under R-reductions, consider the following: given positive integers a, b, and c, where a is a power of 2, are there positive integers x and y such that $axy + by = c$? In [3] this problem is shown to be complete for NP under R-reductions, assuming the Extended Riemann Hypothesis. The latter assumption was used only for certifying primality however, so the new result that PRIME NUMBER is in R allows us to dispense with that hypothesis. This problem also turns out to be complete for NP under UR-reductions [4], and so it can be in neither R nor co-R without dire consequences.

A second, less number-theoretic example of a problem complete for NP under UR-reductions is the following problem, from [248]: ENCODING BY DTM: Given two strings x, y over $\{0, 1\}$ and an integer K, is there a DTM with K or fewer states that, when started with x on its worktape and its read-write head located at the leftmost symbol of x, writes y on its output tape in just $|y|$ steps?

RUR-reductions were introduced in [245] for the purpose of showing that the following variant on SATISFIABILITY, called "UPSAT" in [133] is hard: If there are no satisfying truth assignments, the answer is "no". If there is exactly one, the answer is "yes". If there are more than one satisfying truth assignments, then both "yes" and "no" are valid answers. (Note that this is not the same as the UNIQUE SATISFIABILITY problem described in Section 2.4.) Although UPSAT is not a decision problem as defined, it can be turned into one by specifying a particular answer, "yes" or "no", for each instance with more than one satisfying truth assignment. In [245] it was shown that all such restrictions are RUR-hard for NP. As a consequence (assuming $\mathrm{R} \neq \mathrm{NP}$), SATISFIABILITY would remain hard even if one could somehow be "promised" that no instances with more than one satisfying truth assignment would ever arise. (For a fuller discussion of such "promise" problems and how to reason about them, see [73, 133, 245].)

An important second example of the use of RUR-reductions concerns the $\oplus$SAT problem of Section 4.1, which is also shown to be RUR-hard for NP in [245]. It was this result that was generalized in [240] to prove that PH is ⊕P-easy in a randomized sense, or equivalently, that ⊕SAT is hard for PH under randomized reductions. (The actual reductions used in [240] are of yet a new type, and might be called "BPP-reductions". We will not define them here, but the interested reader should be able to deduce the definition after reading the next section.)

4.5. Probabilistic Turing machines and the classes PP and BPP

In the previous two sections we were for the most part concerned with "randomized" computations that used different criteria for the answers "yes" and "no". In this section we consider the situation when the two answers are treated symmetrically. For the definitions we are about to provide, we continue to assume as before that our NDTMs are normalized so that all branch points have outdegree 2, and all computations terminate in the same number of steps and contain the same number of branch points. We also assume that all computations terminate in either a "yes" or a "no" state.

DEFINITION. A *probabilistic Turing machine* (a PTM) is an NDTM whose output for a given input string x is "yes" if more than half of the computations terminate in "yes" states, and is "no" if more than half of the computations terminate in "no" states. (If the number of yes-computations equals the number of no-computations, the output is "don't know".)

DEFINITION. A PTM *solves* a problem X if and only if the PTM outputs the correct answer for each instance of the problem (and never claims ignorance).

Note that if, as before, we imagine ourselves as simulating such a PTM by making random choices at each branch point, the probability that we obtain the correct answer will by definition always exceed 0.5.

The first complexity class to be derived using these definitions is due to Gill [90].

DEFINITION. The class PP is the set of all decision problems that can be solved by polynomial-time PTMs.

Note that this class does not have the positive practical advantages of R, since a polynomial number of iterations of a PP algorithm may not be able to increase our confidence in the answer to worthwhile levels. For instance, it may be that the answer is correct only with probability $\frac{1}{2}+(\frac{1}{2})^n$, in which case we cannot reduce the error probability below $\frac{1}{4}$ without an exponential number of iterations. Indeed, it is not difficult to show that PP contains $\mathrm{P}^{\mathrm{NP}[O(\log n)]}$, and hence NP, co-NP, and many presumably intractable problems [28]. PP is, however, contained in PSPACE. (Relative to a random oracle, both containments PP$\subseteq$PSPACE and NP$\cup$co-NP$\subseteq$PP are proper [29].)

Unlike R, the class PP is known to have complete problems under polynomial transformations; in the canonical one we are given an instance of SATISFIABILITY and asked if more than half of the possible truth assignments satisfy all the clauses [90, 225]. Note the close relationship between this problem and that of actually counting the number of satisfying truth assignments, the problem that was observed to be complete for #P in Section 4.1. Indeed, it is easy to show that $P^{PP}=P^{\#P}$, i.e. an oracle for a problem in PP is just as good as an oracle for a problem in #P [10]. In light of Toda's result, mentioned in Section 4.1, that $PH \subseteq P^{\#P}$ [240], this implies that PH is also contained in P^{PP}, a corollary of which is that PP cannot lie in the polynomial hierarchy unless the hierarchy collapses.

Although PP does not of itself seem to guarantee useful randomized algorithms, there is an important subclass that maintains its symmetric nature and is capable of providing such algorithms. This class contains R∪co-R and can be viewed as the most general class of "efficiently solvable" problems.

DEFINITION. The class **BPP** is the set of all decision problems solvable by polynomial-time PTMs in which the answer always has probability at least $\frac{1}{2}+\delta$ of being correct, for some fixed $\delta>0$.

The "B" in BPP stands for "bounded away from $\frac{1}{2}$". Note that for randomized algorithms based on such PTMs, the probability of correctness can be rapidly increased by iteration; ineed we can simplify the definition without loss of generality by replacing "$\frac{1}{2}+\delta$" by $\frac{2}{3}$.

BPP is an interesting class. By definition we have $R \cup \text{co-}R \subseteq BPP \subseteq PP$, but currently we know nothing definite about the inclusion relations, if any, between NP and BPP. What we do know is the following: If $NP \subseteq BPP$, then $R=NP$ and, in addition, the polynomial hierarchy must collapse to BPP [158, 259]. One might thus conjecture that $BPP \subseteq NP$, especially since for a random oracle we have $P=ZPP=R=BPP \subseteq NP$, with the latter containment proper [29]. Moreover, like R (and for much the same reason, e.g., see [20]), BPP has "small circuits", i.e., $BPP \subseteq P/\text{poly}$ as defined in Section 4.3. There are oracles, however, for which BPP is not contained in NP, indeed, is not even contained in Δ_2^P [235]. BPP can, however, be shown to lie within the polynomial hierarchy, in fact in $\Sigma_2^P \cap \Pi_2^P$ [171, 230] (see Fig. 5).

The current consensus seems to be that the two sets NP and BPP are incomparable (as well as the two sets NP∪co-NP and BPP). Current candidates for membership in NP − BPP include the NP-complete problems. There are also copious candidates for BPP − NP, since BPP, being symmetric, contains both R and co-R. (The class co-R does not appear to be contained in NP, although it is contained in co-NP.) Thus we can get a candidate for BPP − NP simply by choosing a problem in co-R that is not known to be in NP, such as PRODUCT POLYNOMIAL EQUIVALENCE (the complement of the inequivalence problem introduced in the previous section). Analogous candidates for BPP − (NP∪co-NP) are harder to come by. Indeed all problems that have to date been identified as members of BPP are actually members of NP∪co-NP, although not necessarily in R∪co-R. (See [16] for examples of number-theoretic problems in NP that are candidates for BPP − (R∪co-R).) The hope that complete problems for BPP might offer candidates for BPP − (NP∪co-NP) is somewhat dim, given that there are oracles for which BPP has no complete problems (under polynomial transformations [108] or polynomial-time Turing reductions [117]). For more on BPP, see aso [260, 261], the former reference providing further insights into the classes of Sections 4.3 and 4.6 as well.

As a final class definable by symmetric radomized computations, let us briefly consider the class PPSPACE of decision problems solvable by polynomial *space* bounded PTMs. This clearly contains PSPACE. Surprisingly, as proved in [226], it is also contained in PSPACE, and hence is identical to it!

4.6. Stochastic Turing machines, interactive proofs, and the classes they define

So far in Section 4 we have seen counting Turing machines (CTMs), unambiguous Turing machines (UTMs), random Turing machines (RTMs) and probabilistic Turing machines (PTMs). In this section we introduce one final variant, the *stochastic* Turing machine of [190]. This will be a hybrid between a probabilistic Turing machine and an alternating Turing machine. As usual, we shall define it in terms of an NDTM computation tree. For simplicity in our definition, we assume that all configurations (except the leaves) have outdegree 2 in the computation tree, that all computation paths are of the same length, and that they all contain the same number of non-leaf nodes.

DEFINITION. A *stochastic Turing machine* (STM) is an NDTM whose output for a given input string x is specified as follows: As in an ATM, each configuration in the computation tree is identified as one of two types; this time, however, the types are "existential" and "random", rather than "existential" and "universal". We assume the types alternate from level to level. An *admissible computation* for such a tree is a subtree obtained by deleting one of the two subtrees hanging from each existential node. The output for input string x is "yes" if and only if the resulting computation tree contains an admissible subtree in which more than half the leaves are accepting.

As with ATMs we can view the computation of an STM as corresponding to a game, but here the game will be between a normal player and an "indifferent opponent", one who simply makes random moves, choosing with equal probability between the potential immediate successors. The answer for input x is "yes" if and only if there is a strategy under which the existential player has a probability greater than 0.5 of winning against his random opponent. We say that an STM solves a given decision problem X if the existential player has such a strategy for every yes-instance of X, but no such strategy for any no-instance of X.

The name "PPSPACE" was recycled in [190] to denote the class of decision problems solvable by polynomial-time STMs (the "time" in the resource restriction became "SPACE" in the class name because of the alternation involved in the machine model). As with the PPSPACE we saw in the previous section however, the name is not important. This PPSPACE turns out to be identical to ordinary PSPACE, as did its predecessor. Its main advantage is hence as a means for showing interesting new types of problems to be PSPACE-complete, such as certain scheduling problems where the task lengths vary according to a Poisson process, as well as problems from control theory [190]. One simple-to-understand example is the following variant on the #P-equivalent GRAPH RELIABILITY problem of Section 4.1.

DYNAMIC GRAPH RELIABILITY

Instance: Directed graph $G(V, A)$ with specified source and sink vertices s and t. For each pair v, a, where v is a vertex and a an arc, a rational number $p(v, a) \in [0, 1]$.

Answer: "Yes" if there is a strategy by which you can, starting at s, reach the destination t with probability exceeding $\frac{1}{2}$, assuming your travels are subject to the following rules: Your traversal proceeds in steps, where in each step you leave your current vertex and travel along a (still-existent) outgoing arc to an adjacent vertex. Initially all arcs in A exist and are traversable, but the arcs fail (disappear permanently) according to a random process that occurs while your traversal proceeds, with the probability that a given arc a disappears during a step that you started at vertex v being $p(v, a)$.

Here the "random opponent" is the process generating the arc failures. Note how much more game-theoretic this problem is than the original GRAPH RELIABILITY problem, in which one could view the random arc failures as all happening at the beginning of the process, after which one could count on the remaining arcs to persist, and thus could easily walk from s to t, assuming a path still existed.

Although the above first attempt at using polynomial-time STMs to define a new complexity class failed, subsequent attempts have been substantially more productive. These have generated new classes by imposing further restrictions on the STM computations, just as we restricted PTMs in order to define BPP. A first such class is the following (its name will be explained below).

DEFINITION. The class AM[poly] consists of all decision problems solvable by polynomial-time STMs satisfying the restriction that, for any input x, the existential player's best strategy yields a winning probability that either exceeds $\frac{2}{3}$ or is less than $\frac{1}{3}$.

This class (and its name) was introduced in [12], which provided an alternative definition in terms of conversations between an all-knowing wizard "Merlin" and a skeptical listener "Arthur", who has polynomial-time bounded computational resources and the ability to flip unbiased coins. Merlin's goal is to convince Arthur "beyond a reasonable doubt" that a given string x is a yes-instance of decision problem X. When Arthur speaks, he is limited to simply telling Merlin the outcome of some number of coin flips (polynomial in $|x|$). When Merlin speaks, his message (also polynomial in $|x|$) can depend on x and all the previous messages. The conversation is allowed to run for a polynomial number of interchanges, after which Arthur inputs x and a transcript of the conversation to a deterministic polynomial-time algorithm, which tells him whether or not to believe that x is a yes-instance. Protocols for conversations of this sort correspond to STMs, and X is in AM[poly] if there is a protocol such that, for any yes-instance x, Merlin can with probability greater than $\frac{2}{3}$ convince Arthur of this fact, and for each no-instance the probability is less than $\frac{1}{3}$ that

Merlin can fool Arthur, no matter what Merlin does. (In what follows, we will say such a protocol "solves" X.) Note that, given this interpretation of AM[poly] in terms of STMs, we have AM[poly] $\subseteq$ PPSPACE = PSPACE, and Merlin need not in fact have arbitrary computing power, but can settle for polynomial space.

Of particular interest are the subclasses of AM[poly] in which only a bounded number of messages are sent.

DEFINITION. For each $k > 0$, the class MA[k] consists of all those decision problems solvable by Arthur–Merlin protocols in which Merlin goes first and there are exactly k messages sent.

DEFINITION. For each $k > 0$, the class AM[k] consists of all those decision problems solvable by Arthur–Merlin protocols in which Arthur goes first and there are exactly k messages sent.

Note that MA[1] is simply the set of decision problems solvable by protocols in which the conversation begins and ends with Merlin's first transmission. This is the same as STMs all of whose configurations are existential, and clearly equals NP. Analogously, the class AM[1] consists of those decision problems solvable by protocols in which only Arthur speaks, and hence equals BPP. Things become a bit more interesting once we begin to allow some true interaction between Arthur and Merlin. As shorthands, we shall use the following notation, introduced in [12], for what turn out to be the two most important classes:

DEFINITION. MA = MA[2]; AM = AM[2].

The class MA, in which both parties speak once with Merlin first, can be viewed as a randomized version of NP, consisting of those problems for which all answers have short *probabilistic* proofs (i.e., proofs for which the validity problem is in BPP). The class AM, where Arthur speaks and Merlin responds, is also a generalization of NP, in that it contains precisely those languages that are in NP^B for almost all oracles B [189]. (The analogous result, that BPP consists of precisely those languages that are in P^B for almost all oracles B, was proved in [29].)

In [12], it is shown that NP $\subseteq$ MA $\subseteq$ AM. An oracle exists for which AM properly includes MA (and hence NP) [215]. A potential member of AM − NP is given in [12]. This is the problem MATRIX GROUP EXACT ORDER: given a prime power $q = p^n$, integers k and m, and a collection C of $k \times k$ matrices over GF(q), does the matrix group generated by C have order equal to m, i.e., are there precisely m distinct matrices that can be obtained by multiplying together sequences of members of C (where repetitions are allowed and the product of the empty sequence is taken to be the identity matrix)? Although telling whether the order is divisible by a given integer is in NP [15], the question for the EXACT ORDER problem remains open, even though it is shown in [12] to be in AM (indeed, in AM $\cap$ co-AM).

Turning to the classes MA[k] and AM[k], $k > 2$, we have yet another potential hierarchy. In this case, however, something unexpected happens: the hierarchy collapses! For all $k>2$, $MA[k]=AM[k]=AM$ ($=AM[2]$) [12]. The question of whether all of AM[poly] collapses to AM remains open, however, and their exist oracles for which it does not, indeed for which AM[poly] − PH is nonempty [6]. As to AM and MA, it can be shown that $AM \subseteq \Pi_2^p$ and that $MA \subseteq \Pi_2^p \cap \Sigma_2^p$ [12]. A final question is whether AM, which contains NP, also contains co-NP. This now seems unlikely, as it is shown in [39] that this would imply that the polynomial hierarchy collapses into AM. The class co-NP may not be in AM[poly] either: although no one has yet shown that PH would collapse if this occurred, there does exist an oracle under which containment fails to hold [80]. AM (and hence AM[poly]) does, however, contain one interesting problem in co-NP that is not known to be in NP: GRAPH NONISOMORPHISM [91]. (This is the complementary problem to the GRAPH ISOMORPHISM problem introduced in Section 1.1, with the answer being "yes" only if the two graph representation are *not* isomorphic.) (For new developments on AM[poly], see Section 6.1.)

Another surprising result compares the power of Arthur–Merlin protocols to the apparently much more general notion of an "interactive proof system" introduced in [96]. Such a system is also based on two players, one all powerful, one with a source of random bits and polynomial-time computational power. The distinction is that, while Merlin (now called "the prover") continues to be operating under the same constraints, Arthur (now called "the verifyer" and *always* sending the first message), can be cleverer. He no longer need simply send the random bits he generated. He may send the result of an arbitrary polynomial-time computation based on the input x, some new random bits, the transcript of the conversation so far, and the list of all random bits he previously generated.

The definition of "solving" a decision problem is the same for interactive proof systems as it was for Arthur–Merlin protocols. Note, however, that the fact that the verifyer can in effect keep "secrets" from the prover means that interactive proof systems cannot be modelled directly by STMs, as could Arthur–Merlin protocols. Let us thus define for interactive proof systems the analogs of the classes we had for Arthur–Merlin games.

DEFINITION. The class IP consists of all those decision problems solvable by interactive proof systems in which the total computation time of the verifyer is polynomially bounded, but there is otherwise no limit on the amount of interaction.

DEFINITION. For each $k>0$, the class $IP[k]$ is the set of all decision problems solvable by interactive proof systems in which the total computation time of the verifyer is polynomially bounded and neither the prover nor the verifyer sends more than k messages.

Note that, by definition, $AM[poly] \subseteq IP$ and, for all $k \geq 1$, $AM[2k] \cup MA[2k] \subseteq IP[k]$ (and so $AM \subseteq IP[1]$). Surprisingly, all the extra power that the verifyer has in an interactive proof system (and all the work we just went to making the above definitions) is for naught. As shown in [97], $IP=AM[poly]$ and $IP[k] \subseteq AM$, for all $k \geq 1$ (so yet another potential hierarchy collapses to AM).

Also contained in AM (indeed, in AM∩co-AM) is the set of all decision problems solvable by "perfect zero-knowledge" polynomial-time interactive proof systems, even when the number of rounds is unbounded [7, 79]. (The concept of a "zero-knowledge" proof system, which can convince a verifyer of a statement without revealing anything beyond the fact that the statement is true, was introduced in [96]. It is too involved to describe here, and so interested readers are directed to [137] and the references therein.)

For a schema of the classes introduced in this section, See Fig. 6.

5. Inside P

In Sections 2 through 4, we have concentrated on complexity classes that *contain* P, viewing equality with P as the most desirable possible outcome. It is not the case, however, that membership in P by itself ensures tractability in any practical sense. Thus researchers have of late devoted equal, if not more time to investigating classes that are in various senses "easier" than P (or at least incomparable). These classes are the subject of this final section. In contrast to the previous sections, our general order of traversal will be downwards rather than upwards, from larger classes to smaller ones.

5.1. *Classes defined by sublinear space bounds: POLYLOG-SPACE, L, NL, and SC*

With much of today's computing being done on personal computers that have limited main memory (and do not have sophisticated paging systems), the amount of memory an algorithm requires can often be more crucial than its running time. Consequently, there has been much interest in algorithms that use significantly less workspace than the size of their input (which can be read as needed off the floppy disk or input tape on which it is delivered). This ideal is captured by complexity theorists in the concept of the "log-space" and, less restrictively, the "polylog-space" DTM.

DEFINITION. The class L consists of all decision problems solvable by DTMs with workspace bounded by $O(\log|x|)$ on input x.

DEFINITION. The class POLYLOG-SPACE consists of all decision problems solvable with workspace bounded by $O(\log^k|x|)$ for some fixed k.

In the terminology of Section 2.6, L = DSPACE[$O(\log n)$] and POLYLOG-SPACE = $\bigcup_{k>1}$DSPACE[$\log^k n$] (or DSPACE[$\log^{O(1)} n$] for short). It is immediate that L ⊆ POLYLOG-SPACE, with the containment proper because of Theorem H2 of Section 1.4. What is more interesting are the relations between these two classes and P. Clearly L ⊆ P, since no log-space DTM can have more than a polynomial number of distinct memory states. (Indeed, for any log-space bounded DTM, there will be a polynomial p such that for any input x, if the DTM runs longer than time $p(|x|)$ on input x, it will never halt.) For POLYLOG-SPACE and P, however, the situation is more complicated. In a result analogous to the one for NP and ETIME in Section 3.1, it can be shown [36] that POLYLOG-SPACE ≠ P, although we do not know whether the two classes are incomparable or one properly contains the other. This inequality follows from the answer to a question we have previously asked about other classes: can POLYLOG-SPACE have a complete problem (in this case under log-space transformations)? Surprisingly, the answer is "no", and not just in special relativized worlds, but for (unrelativized) POLYLOG-SPACE itself. Essentially, it can be shown that should such a complete problem exist, we would have POLYLOG-SPACE $\subseteq$ LOG^k-SPACE (the set of decision problems solvable by $O(\log^k |x|)$-space DTMs) for some fixed k, an impossibility given Theorem H2. But this means that POLYLOG-SPACE cannot equal P, since there are problems that *are* log-space complete for P, as we saw in Section 1.5.

Of the three possibilities for the relationship between POLYLOG-SPACE and P, most researchers would probably favor incomparability. There exist likely candidates for P − POLYLOG-SPACE, for instance the problems that are log-space complete for P, since if any such problem is in POLYLOG-SPACE, then P $\subseteq$ POLYLOG-SPACE and in fact is contained in LOG^k-SPACE for some fixed k, an unlikely prospect. We have already seen two examples of such "P-complete" problems in Section 1.7: DTM ACCEPTANCE and LINEAR PROGRAMMING. Here are two more well-known examples. The first, proved P-complete in [168], can be viewed simply as a restatement of DTM ACCEPTANCE in terms of the Boolean circuit model of computation. The second, proved P-complete in [57], has an interesting graph-theoretic flavor and has been the setting for some intriguing lower bound arguments about algorithmic complexity.

CIRCUIT VALUE

Instance: A description of an n-input, 1-output Boolean circuit, together with an input value (0 or 1) for each of the input gates.

Answer: "Yes", if the output of the circuit on the given input is 1.

PATH SYSTEM ACCESSIBILITY

Instance: A finite set X of *nodes*, a relation $R \subseteq X \times X \times X$, and two sets $S, T \subseteq X$ of *source* and *terminal* nodes.

Answer: "Yes", if there is an "accessible" terminal node, where all sources are accessible, and if $(x, u, v) \in R$ and u and v are accessible, then x is accessible. (Note that this can be viewed as the question of whether a given straight-line program computes its claimed outputs, and is trivially in P.)

Although no one has yet been able to prove that P-complete problems require more than polylogarithmic space, there are some partial results in that direction for the latter problem. In particular, it has been shown to require $\Omega(|x|^{1/4})$ space under a fairly wide class of methods [60, 64].

Although POLYLOG-SPACE has no complete problems, one can come up with plausible candidates for POLYLOG-SPACE − P by considering problems that are complete for subclasses of POLYLOG-SPACE, such as DSPACE[$O(\log^k n)$] for a fixed $k > 1$, under log-space transformations. Note that the simple proof that L $\subseteq$ P does not carry over to DSPACE[$O(\log^k n)$] for any $k > 1$ since for each such k a $\log^k |x|$-space DTM has a potentially superpolynomial number of memory states.

Complete problems for DSPACE[O($\log^k n$)], $k>1$, do exist, but unfortunately, at present we do not have any "natural" candidates. Instead, we must settle for the relevant analog of the DTM ACCEPTANCE problem of Section 1.7. In this case, the question is simply: Given a DTM M, an integer c and a string x, does M, if given input x, halt in an accept state after a computation that never uses more than $c \log^k n$ space?

Assuming that both polynomial time and polylogarithmic space are important properties for an algorithm to have, but that neither appears to guarantee the other, a natural class to consider is the class "SC" consisting of all decision problems solvable by algorithms that *simultaneously* obey both types of bounds. To define this class formally, let us first introduce a notation analogous to the "TA[$t(n)$, $a(n)$]" used in Section 3.3 to capture the notion of simultaneous time and alternation bounds:

DEFINITION. The class TS[$t(n)$, $s(n)$] is the set of all problems solvable by DTMs that use at most $t(n)$ time and $s(n)$ space on inputs of size n.

DEFINITION. The class SC = TS[$n^{O(1)}$, $log^{O(1)} n$], i.e., it is the class of all decision problems solvable by DTMs that simultaneously obey polynomial time bounds and polylogarithmic space bounds.

The name "SC" stands for "Steve's Class", in honor of Steven Cook, who proved the first deep result about the class [61]. To motivate Cook's result, let us first observe that although SC $\subseteq$ P $\cap$ POLYLOG-SPACE, It may not *equal* that intersection. It might well be the case that a problem X can be solved by a polynomial-time algorithm and by a second algorithm that uses only polylog space, but by no algorithm that *simultaneously obeys* both resource bounds. Cook's result, mentioned above, addressed a class of apparent candidates for P $\cap$ POLYLOG-SPACE $-$ SC, the deterministic context-free languages ("DCFLs", see [32, 122]). These were known to be recognizable by linear-time algorithms and by $\log^2 n$ space algorithms, but did not in general seem to be recognizable by "SC-type" algorithms. Cook effectively dried up this potential source of candidates, by showing that all DCFLs can be recognized in simultaneous $\log^2 n$ space and polynomial (albeit not necessarily linear) time. Thus they are all in SC. To be more precise, they are in the subclass SC^2 of SC, where SC^k is the class of decision problems solvable in simultaneous $\log^k n$ space and polynomial time. (Note that $SC^1 = L$.) There are other sources of candidates for P $\cap$ POLYLOG-SPACE $-$ SC however, as we shall see below.

To conclude this section, let us look briefly at the important nondeterministic analogs of the classes we have seen in this section. The major candidate here is the following one.

DEFINITION. The class NL consists of all those decision problems solvable by NDTMs that use space bounded by O($\log|x|$), given input x.

Other natural possibilities for nondeterministic classes have significantly less interest: There seems little need for a nondeterministic version of SC; the class NPOLYLOG-SPACE, defined analogously to NL, simply equals POLYLOG-SPACE, as a consequence of Theorem 3 of Section 2.6; and the class co-NL simply equals NL, as a consequence of Theorem 4 of Section 2.6. (That theorem also implies, incidentally, that there is no point in defining analogs for log-space of the probabilistic classes R and ZPP: these turn out also to equal NL [43]. The analogs of BPP and PP may well be distinct, however. We shall have more to say about them in Section 5.3.)

By definition and [216], we have L$\subseteq$NL$\subseteq$LOG^2-SPACE$\subseteq$POLYLOG-SPACE, with the last containment proper by Theorem H2. Somewhat surprisingly, we also have NL$\subseteq$P [59]. Making this more surprising is the fact that there is an oracle A such that NL^A is not contained in P^A [169]. (NL$\subseteq$LOG^2-SPACE also fails to relativize.) Ladner and Lynch [169] argue that these failures are due to the non-step-by-step nature of the simulations that prove the two nonrelativizing results. Their oracle Turing machine model, however, does not include the cells of the oracle tape among the space to be bounded, and this omission can also be viewed as the source of the difficulty. Unfortunately, it is not clear that there is *any* reasonable way to construct relativizations of classes with sublinear space bounds, and so we will ignore such results for the remainder of this survey. Readers who are nonetheless interested in such results are referred to the above cited paper, and to [44, 45, 178, 218, 253, 254] for more of them.

Given that NL$\subseteq$P$\cap$POLYLOG-SPACE, the obvious question to ask is whether NL$\subseteq$SC. We suspect the answer is no. Containment does not follow from the simulation that shows NL$\subseteq$P, as that simulation requires polynomial space. Moreover, there are serious candidates for NL$-$SC (which are hence also the candidates for P$\cap$POLYLOG-SPACE$-$SC promised above). These are the problems that are log-space complete for NL. They are also candidates for NL$-$L, another presumably nonempty set. They can be in L (SC) if and only if NL$=$L (NL$\subseteq$SC). For a variety of examples, see [140]. We shall give just one, from [138, 217] that can be viewed as a simpler version of the PATH SYSTEM ACCESSIBILITY problem above.

GRAPH ACCESSIBILITY

Instance: A directed graph G with specified vertices s and t.
Answer: "Yes", if there is a directed path in G from s to t.

There are no similarly natural candidates for SC$-$NL, but once again an appropriate analog of DTM ACCEPTANCE will suffice. This is one that is log-space complete for TS[O(n^2), O($\log^2 n$)]: Given a DTM M, constants c and d, and a string x, does M, when given x as input, halt after cn^2 or fewer steps, having used space bounded by $d\log^2 n$?

Before leaving this section, there is one last nondeterministic class worth mentioning, this one inspired by the undirected version of the above GRAPH ACCESSIBILITY problem, also studied (from a randomized point of view) in [8]. Clearly this is a special case of (directed) GRAPH ACCESSIBILITY. Could it be easier? This may well be the case, as it can be shown to be log-space complete for a class that appears to be intermediate between L and NL. To define this class, we need yet one more variant on the NDTM, introduced in [175]. This is the "symmetric Turing machine", an NDTM in which the move relation is symmetric. (If it is possible to move from configuration c to configuration c' in one step, then it is also possible to more from c' to c in one step, for any c and c'.)

DEFINITION. The class SL consists of all those decision problems solvable by log-space bounded symmetric Turing machines.

Another problem complete for SL is the problem, given a graph G, of determining whether G contains an odd cycle. For more on SL and symmetric space-bounded computation, see [175]. It should be pointed out that, although the class is based on one sort of symmetry, it is not known to possess a more standard sort. That is, even though we now know that NL is closed under complement, no one has yet been able to prove that SL and co-SL are identical [43].

5.2. *Parallel computing and the classes NC and RNC*

In the previous section we used the personal computer, a relatively recent computing phenomenon, as a motivation. In this section we turn to another type of computer that is just now coming into its own: the massively parallel computer. Despite our arguments in the previous section as to why space could be more important than time for practical computing, there are situations where polynomial time, even linear time, may be too much, and practitioners would be more than willing to apply more processors to a task if doing so would substantially reduce the running time.

In addressing this issue from their traditional asymptotic point of view, theoretical computer scientists have focused on the PRAM model of computation from Section 1.3, and a class defined in terms of simultaneous time and processor bounds:

DEFINITION. The class TP[$t(n)$, $p(n)$] is the set of all problems solvable by PRAMs that use at most $t(n)$ time and $p(n)$ processors on inputs of size n.

DEFINITION. The class NC = TP[$\log^{O(1)} n$, $n^{O(1)}$], i.e., it consists of all those decision problems that are solvable on a PRAM that simultaneously obeys a polylogarithmic bound on the running time and a polynomial bound on the number of processors used.

More informally, we might say that NC consists of those problems solvable with a polynomial-bounded amount of hardware in polylog time. As with the sequential complexity classes of previous sections, this class is substantially model-independent. There are equivalent definitions in terms of uniform Boolean circuits of polynomial size and polylog depth, in terms of ATMs obeying simultaneous polylog time and log-space bounds, and in terms of a variety of other models and variants [62, 198, 213]. The name "NC" stands for "Nicks's Class", in honor of Nicholas Pippenger, the first researcher to study the class seriously.

A first observation about NC is that it lies in P, since a single processor can simulate a $\log^j n$ time PRAM computation that uses n^k processors in time $O(n^k \log^j n)$, a polynomial in n. NC also lies inside POLYLOG-SPACE, since by a result of [41] POLYLOG-SPACE is contained in the set of decision problems solvable in polylog time on PRAMs with an *unbounded* number of processors. Only slightly more difficult is the result that NL $\subseteq$ NC. (There is a straightforward NC algorithm for GRAPH ACCESSIBILITY which is log-space complete for NL.)

It is interesting to compare these results with those for the analogously named class SC. There are superficial reasons why one might thing that NC and SC would be identical. Both are contained in P∩POLYLOG-SPACE. Moreover, each class is defined in terms of two simultaneous resource constraints, and there is a one-to-one correspondence between these constraints if they are considered individually. To see this, let us take the definition of NC in terms of polynomial-size, polylog-depth Boolean circuits that are "log-space uniform", where this uniformity condition is defined as follows.

DEFINITION. A family $\{B_n: n \geqslant 1\}$ of Boolean circuits is *log-space uniform* if there is a DTM that, given n, constructs B_n using space $O(\log n)$.

Note that under this definition, the construction time must be polynomially bounded and hence so must be the size of the circuits. In what follows, we follow [61] in using "uniform" to mean log-space uniform unless we specifically say otherwise.

For the claimed individual correspondences, we use two results from [41]. First, polylog-space DTMs have precisely the same power as polylog-depth uniform Boolean circuits (if one ignores running time for the DTM and circuit size for the Boolean circuits). Second, polynomial-time DTMs have the same power as polynomial-size uniform Boolean circuits (if one ignores the space used by the DTM and the depth of the circuits).

Unfortunately, as we saw when we compared SC and P∩POLYLOG-SPACE, the fact that resource constraints must be obeyed simultaneously can substantially change their effects, and it appears that in fact SC and NC are incomparable. We have already noted one difference: NL is contained in NC but is not known be contained in SC. Thus log-space complete problems for NL, such as graph accessibility, are prime candidates for NC − SC. In [61] it was proposed that deterministic context-free languages were prime candidates for membership in SC − NC, but in [213] it was subsequently shown that all context-free languages are in NC, even nondeterministic ones. In their place, an alternative candidate for SC − NC was proposed. This candidate is a restricted version of the P-complete CIRCUIT VALUE problem mentioned in the previous section. The version of this problem restricted to circuits of polylog "width", as defined in Section 1.3, is a currently viable candidate for SC − NC (the version restricted to circuits of polylog depth is likewise a candidate for NC–SC) [213]. (For a alternative discussion of NC and SC, see [130].)

In considering the above examples, we must not lose sight of the fact that a far more important class comparison problem has yet to be resolved. This is the question of whether NC = P, a question widely viewed as the natural analog (for parallel computing) of the question of P versus NP. As with P versus NP, there has been little progress toward the resolution of NC versus P, but a flowering of research devoted to classifying the complexities of problems on the assumption that the two classes differ. Here the key lower bound technique is, of course, the completeness result, in this case log-space completeness for P. Since L ⊆ NC, a problem that is log-space complete for P cannot be in NC unless NC = P. We have already seen this to be the case with "NC" replaced by "POLYLOG-SPACE", "SC", "NL", or "L", but it was only when the connection to parallel computing was noticed and such completeness results could be interpreted as implying the "inherently sequential nature" of a problem, that the question of log-space completeness for P began to attract widespread attention. In this survey we shall make do with the four examples of problems log-space complete for P already exhibited in Sections 1.7 and 5.1, but industrious readers can find many more, for example in [69, 94, 139, 153], and the extensive, although as yet unpublished survey [121].

Balancing the above negative results, we also have had a substantial outpouring of results showing that important problems *are* in NC, and a rapidly developing body of expertise in parallel algorithm design, a survey of which appears elsewhere in this Handbook [153]. Much of this work has been devoted to functions and more general search problems. According to the above definition of NC as a class of decision problems, which corresponds to that given in the original papers that described the class such as [61], such problems are technically ineligible for membership in NC. This has not, however, prevented researchers from ascribing membership to them, either by abusing notation, or simply by redefining NC to be a class of search problems. There has been little consistency in this. For instance, in a sequence of papers authored by Cook [61, 63, 43], NC has switched from a class of decision problems to a class of search problems and back again. Indeed, even this Handbook is not consistent; the choice made here disagrees with the one made in at least one other chapter [153]. For consistency with other class definitions, however, it seems more appropriate to leave NC as a class of decision problems and introduce a new name for the corresponding search problem class. The new class name is chosen by analogy with the distinction we have already made between P and FP:

DEFINITION. The class FNC consists of all those search problems solvable in polylog time by PRAMs with a polynomially bounded number of processors.

Note that NC$\subseteq$FNC by definition. Among the more general search problems in FNC are matrix multiplication (by straightforward techniques), finding minimum spanning trees and Euler tours [11, 54], and a wide variety of algebraic problems, e.g., see [14, 207]. We shall see further examples in the next section.

As was the case for sequential computation, we can often finesse the difference between NC and FNC by showing that there is a decision problem X that has essentially the same complexity as the search problem Y that we are really interested in, i.e., X is in NC if and only if Y is in FNC. This is the case, for instance, when Y is a function whose output size is polynomially bounded. In this case X can simply be: "Given an instance I of Y and an integer k, is the kth bit of Y's answer for I equal to 1?" If X is in NC, one can simply combine a polynomial number of copies of an NC circuit for X (one for each output bit of Y) to obtain an FNC circuit for Y.

When Y is a more general search problem, however, with the possibility of a variety of answers, the correspondence between search and decision problems becomes much less clear than it was in the sequential case. As in the sequential case, the answer to the search problem may be easy to determine given a polynomial number of calls to a subroutine for the decision problem, but if those calls cannot be made in parallel (as they could when Y was a function), we may be unable to satisfy an overall polylog time bound even if the subroutine itself does. For more on this issue, see [155]. In light of the issue, it is often crucial that we perform our complexity analyses on search problems directly, rather than simply on decision problems that would traditionally have been their stand-ins.

Let us conclude this section by mentioning two interesting equivalence classes of problems that so far do not appear to be either in FNC or to be log-space complete for FP. These classes can be viewed as analogs of the class of "GRAPH-ISOMORPHISM equivalent" problems mentioned in Section 2.1, and will contain both decision problems and more general search problems.

For our notion of "equivalence" here, we shall introduce yet another type of reduction, one presumably more powerful than the log-space reduction, but one that is still compatible with NC and FNC (in the sense of Section 1.6). In the definition of this reduction, we assume that the PRAM model is augmented by a special shared memory for the (parallel) construction of oracle queries and for receiving the oracle's answers (which are presumed to arrive one step after the query construction is signalled to be complete). The oracle may process multiple queries in parallel.

DEFINITION. An NC (*Turing*) *reduction* from a search problem X to a search problem Y is an oracle PRAM program that, given an oracle for Y, solves X in polylog time using at most a polynomial number of processors.

These reductions are presumed to be more powerful than log-space reductions both because they allow multiple calls to the oracle, and because NC is presumably more powerful than L. We shall say two problems X and Y are "NC-equivalent" if there are NC reductions from X to Y and from Y to X.

The first class we shall describe is a generalization to search problems of the class of "CC-complete" problems introduced in [185]. Its first member is the "COMPARATOR CIRCUIT VALUE" problem (CCV), that variant of the CIRCUIT VALUE problem in which all circuit elements are "comparators", i.e., two-input gates with two outputs, one yielding $x \vee y$ and one yielding $x \wedge y$, where x and y are the (binary) values of the two inputs. (This problem is "CC-complete" by definition: CC, as defined in [185], is the class of decision problems that are log-space reducible to CCV.) Although no NC-style algorithms are known for CCV, it seems unlikely to be P-complete, given the lack of "fan-out" in the circuit elements. (In P-completeness reductions, as in NP-completeness reductions, one seems always to need a method for transmitting information from one location in the construction to many other locations, and fan-out, in one guise or other, seems to be what is necessary to do the trick.)

The class of CCV-equivalent search problems has surprising variety. A second member is LEXICOGRAPHIC MAXIMAL MATCHING, the problem of finding the lexicographically first maximal matching in a graph G, under some given naming of the graph's edges, where a matching M is *maximal* if all edges in G are either in M or share an endpoint with some edge in M. Note that without the lexicographic restriction, this problem is in FNC, as follows from the result of [156] that finding a maximal independent set is in FNC. (Interestingly enough, the lexicographic version of the latter problem appears to be harder than LEXICOGRAPHIC MAXIMAL MATCHING: finding the lexicographically first maximal independent set *is* known to be log-space complete for FP [63].)

A final example of a CCV-equivalent problem is the STABLE ROOMMATES problem. In this problem, we are given a set S of $2n$ people who are to be assigned to n 2-person rooms, together with a preference list $l(p)$ for each person p (p's rank-ordering of all the other people as potential roommates). We ask whether there is a partition of S into n roommate pairs such that no two non-roommates prefer each other to their current roommates. (That this problem was even in P was only discovered in 1985 [127].)

By altering the STABLE ROOMMATES problem slightly, one obtains the problem that forms the basis for our second (and presumably incomparable) class. Suppose that, instead of preference lists, we specify for each person p a set $a(v)$ of "acceptable" roommates (with p being acceptable to q only if q is acceptable to p). The question of whether there is a partition of S into n pairs of mutually acceptable roommates is the simply the question of whether an undirected graph contains a perfect matching (a problem whose restriction to bipartite graphs was discussed in Section 4.1). Let us call this the PERFECT MATCHING problem.

Among the problems NC-equivalent to PERFECT MATCHING are such problems as computing a maximum *weight* perfect matching (assuming edge weights are written in unary), constructing a maximum cardinality matching, and finding the maximum source-to-sink flow in a directed graph (with unary edge capacities) [154]. (The last of these problems is log-space complete for P if the edge capacities are written in binary [94]. It is not yet known whether maximum weight perfect matching with binary weights shares this property.)

All of these problems are in P or FP (e.g., see [247]), but none are known to be in NC (FNC) or to be log-space complete for P (FP). What is known, however, makes it unlikely that the latter is the case. It has been shown in [154], that PERFECT MATCHING is in the class "RNC", which is the randomized counterpart to NC in the same sense that R is the randomized counterpart to P. RNC could be defined in terms of coin-flipping PRAMs with certain probabilities of obtaining the correct answer, but it will be quicker to define it as follows.

DEFINITION. A decision problem X is in RNC if and only if there is a polynomial p and a problem Y in NC such that for all input strings x the following two properties hold:

(A) If x is a yes-instance of X, then for over half of the possible strings y with $|y| = p(|x|)$, the string xy is a yes-instance of Y.

(B) If x is a no-instance of X, then for all of the possible strings y with $|y| = p(|x|)$, the string xy is a no-instance of Y.

Note that if one replaces NC by P in the above definition, one obtains a definition of the class R, as claimed. As with NC, there is confusion in the literature over whether RNC is allowed to contain arbitrary search problems or not. Once again, we choose to resolve this confusion by giving the more general class a different name.

DEFINITION. A decision problem X is in FRNC if and only if there is a polynomial p and a search problem Y in FNC such that for all input strings x the following two properties hold:

(A) If y is such that $|y| = p(|x|)$, then any answer for xy in problem Y is an answer for x in problem X.

(B) If x has an answer in X, then for over half the possible strings y with $|y| = p(|x|)$, the string xy has an answer in Y.

Under this definition, all the search problems mentioned above as reducible to PERFECT MATCHING are in FRNC [154].

As a final comment on these problems, we note that, due to a result of [148], PERFECT MATCHING is in fact in RNC$\cap$co-RNC, and there is a randomized algorithm for it that uses a polynomially bounded number of processors, that always gives the right answer, and that runs in expected polylog time for every instance. (Just as R$\cap$co-R$=$ZPP, RNC$\cap$co-RNC equals the analogously defined class ZPNC.) We still do not know whether PERFECT MATCHING is in NC, however. The more general question of whether NC$=$RNC is also open. For more on NC and RNC, see [153] elsewhere in this Handbook.

5.3. *Inside NC*

In this and the next two sections, we shall briefly consider the structure of NC and survey the prominent complexity classes inside NC. In doing so, we must make a fundamental shift. Once inside NC, it is no longer possible to talk about complexity classes as "model-independent" in the sense we have used before. Precise details of the model in question often need to be specified if the class is to be well-defined.

For instance, the first classes we shall consider are the classes NC^k, $k \geqslant 1$, of decision problems solvable with polynomial hardware in time $O(\log^k|x|)$. (These classes are analogous to the subclasses SC^k of SC mentioned in the previous section.) For the PRAM definition of NC, the precise nature of the classes NC^k can depend heavily on the assumptions one makes about how algorithms behave when two processors want to access the same memory location at the same time. For instance, one might allow concurrent "reads" but disallow concurrent "writes", as in the "CREW" PRAM (concurrent-read, exclusive-write), allow both operations to take place concurrently, as in the "CRCW" PRAM, or allow neither, as in the "EREW" PRAM. In the case of concurrent writes, there is the additional choice as to which processor actually succeeds in writing to a cell when many attempt to do so concurrently, with the options being "random" (one writer will succeed, but you cannot predict which), "priority" (the processor with lowest index succeeds), or "common" (algorithms are constrained so that no two processors ever attempt to write different things to the same memory cell at the same time). For discussions of the relative power provided by the different choices, see for instance [77, 78, 153].

To avoid these issues, we shall follow most authors and use a definition in terms of uniform Boolean circuits, although even here there are choices to be made. In particular, the class NC^1 can depend significantly on the precise "uniformity" condition imposed on the circuits [213]. Here we make the most common choice and once again use log-space uniformity. We shall also continue our distinction between classes of decision problems and classes of search problems.

DEFINITION. For each $k \geqslant 1$, the class NC^k (FNC^k) consists of all languages recognizable (search problems solvable) by log-space uniform classes of Boolean circuits having polynomial size and depth $O(\log^k n)$.

A first observation about the classes NC^k is that $NC^k \subseteq LOG^k$-SPACE for all $k \geqslant 1$ [41]. A second observation concerns the relation of the NC^k to another sequence of classes that stratifies NC. This second sequence is defined in terms of "unbounded fan-in circuits", a variant on our standard Boolean circuit model in which AND-gates and OR-gates are allowed to have arbitrarily large fan-in. Note that something very much like unbounded fan-in occurs in programmable logic arrays (PLAs) and indeed any time one uses a bus to distribute data. Thus this concept is not simply a theoretical construct. Nevertheless, it does offer surprising powers. For instance the decision problem: "Is the input string made up of all 0s?" can be solved by a depth-1 unbounded fan-in circuit, whereas in an ordinary circuit this problem would require depth at least $\Omega(\log n)$, just so all the input bits could be *communicated* to the output gate. The classes based on this model are defined analogously to the classes NC^k.

DEFINITION. For each $k \geqslant 1$, the class AC^k consists of all languages recognizable by log-space uniform classes of unbounded fan-in circuits having polynomial size and depth $O(\log^k n)$.

It is not difficult to show that for all $k \geqslant 0$, $AC^k \subseteq NC^{k+1} \subseteq AC^{k+1}$, and hence the union of all the classes AC^k is simply NC. We shall not have anything more to say about the AC^k for now, but will have a lot to say about the especially interesting class AC^0 of bounded-depth circuits in Section 5.4.

So far, not many of the classes NC^k have proved individually interesting. Thus there has, however, been considerable research effort (and complexity class generation) inside the lowest classes of this hierarchy (NC^1 and NC^2). We shall cover the contents of NC^1 in the next section and will conclude this section by examining classes of problems that appear to lie *between* NC^1 and NC^2.

Before embarking on this discussion, however, we should admit that there is some controversy among theoreticians about the significance of the NC^k hierarchy. Many argue that the divison of NC into subclasses based on running times obscures the real bottleneck for parallel computing, which is the number of processors required. An algorithm that requires $|x|$ processors and $\log^2|x|$ running time is likely to be far more useful than one that requires $|x|^2$ processors and takes $\log|x|$ time, but the latter is in NC^1, the more restrictive (and hence presumably better) class. In the real world, where processors are limited, it is the *time* $\times$ *processor* product that may be the relevant complexity measure for parallel computation, as it provides the basis for a time–processor trade-off as more processors become available.

For this trade-off to be beneficial with relatively small numbers of processors, we would ideally like the product to equal the best running time known for a sequential algorithm that solves the problem, and indeed, parallel algorithm designers strive to meet this goal, at least to within a low-order polylog factor. Note, however, that useful near-optimal time–processor trade-offs can hold even for problems not in NC. A P-complete problem that can be solved by a parallel algorithm that runs in time $O(n)$ using n processors may well be much easier in practice than a problem in NC that can be solved in time $O(\log n)$ using n^3 processors, given that in the real world even obtaining n processors may be impossible for realistic values of n. Thus, one should be careful about taking at face value all claims (such as the one made in the previous section) that the P-complete problems are "inherently sequential".

Despite the above limitations, there is still much theoretical interest in determining where certain important problems lie in the NC^k hierarchy. Before we go on to consider this issue in detail, let us address two final general questions. First, how far up does the hierarchy extend? Is it infinite, or does it collapse to some level? (As with the polynomial hierarchy, if for some k, $NC^k = NC^{k+1}$, than $NC = NC^k$.) A collapse seems unlikely, but at present we know of no diagonalization argument that precludes it (as Theorem H2 precluded the collapse of the LOG^k-SPACE hierarchy). Indeed, at present we cannot prove even that $NC^1 \neq NP$!

A related question is the following: Can NC have complete problems? We can ask this question both for log-space transformations (which are in FNC^2 for the right choice of machine model) or for NC^1 reductions. The latter are defined in terms of the following machine model.

Definition. An *oracle-augmented Boolean circuit* is a Boolean circuit with an additional class of "oracle" gates allowed, where the latter can have any number of inputs and outputs. The input string for such a gate is the sequence of the values on its input gates; the output string is the sequence of values on its output gates. Given a search problem X as oracle, the output string of an oracle gate with input string x is any y that is an answer for x in X. (If no answer exists, the circuit containing the gate fails.)

Definition. An NC^1 reduction from problem X to problem Y is a log-space uniform family of oracle-augmented Boolean circuits that

(1) solves X given Y as oracle,

(2) contains at most a polynomial number of gates, and

(3) has $O(\log n)$ depth where, for each oracle gate g, the contribution of that gate to the length of the circuit paths containing it is counted as $\log(g_{in} + g_{out})$, where g_{in} and g_{out} are the number of input and output gates of g respectively (e.g., see [63]).

It is not difficult to see that NC^1 reductions are compatible with all the classes NC^k, $k \geqslant 1$, and that log-space reductions are compatible with all NC^k, $k \geqslant 2$. Consequently, if NC has complete problems under either log-space or NC^1 reductions, the NC^k hierarchy will collapse. It would thus be a major result if any such problems were to be identified, but they are at present not strictly ruled out.

Let us now turn to the promised discussion of classes of problems that are contained in NC^2 and contain NC^1. We have already seen three such classes; it can be shown that

$NC^1 \subseteq L \subseteq SL \subseteq NL \subseteq NC^2$. Recall that the GRAPH ACCESSIBILITY problems for directed and undirected graphs were complete for NL and SL respectively under log-space transformations. To obtain a complete problem for L (under NC^1 reductions), we need only restrict GRAPH ACCESSIBILITY to directed (or undirected) forests [266]. (For other examples, see [62, 264, 266].) Beyond L, SL, and NL, perhaps the most famous class in the range from NC^1 to NC^2 is the following.

DEFINITION. The class LOGCFL consists of all those decision problems that are log-space reducible to a context-free language.

This class has several alternative characterization. We will mention three, but see also [249]. First, LOGCFL is the set of decision problems solvable by nondeterministic auxiliary pushdown automata in log space and polynomial time (see [238] for definitions and proof). This can be shown to imply that $NL \subseteq LOGCFL$. Second, LOGCFL consists of those decision problems solvable by alternating Turing machines obeying an $O(\log n)$ space bound and a polynomial bound on the total size of the computation tree [212]. Using this characterization, one can conclude that LOGCFL $\subseteq AC^1$ and hence is contained in NC^2 [212]. Third, LOGCFL consists of precisely those decision problems solvable by AC^1 circuits in which no AND-gate has fan-in exceeding 2 (i.e., the fan-in is "semiunbounded" [43]). A final structural result about LOGCFL, proved using techniques similar to those used for proving Theorem 4 (Section 2.6), is that LOGCFL is closed under complement [43]. (NC^k and AC^k, being deterministic classes, are automatically closed under complement for all k.)

Examples of problems that are complete for LOGCFL include the "hardest context-free language" of [98] and the CIRCUIT VALUE problem restricted to monotone circuits having "degree" at most n [63, 265]. (The notion of "degree" used here is defined inductively: the degree of a constant and of an input variable or its negation is 1, the degree of an OR-gate is the maximum of the degrees of its inputs, and the degree of an AND-gate is the sum of the degrees of its inputs [231].) An example of a problem that is in LOGCFL but may not be complete for it is the CIRCUIT VALUE problem for monotone planar circuits [71], where a circuit is monotone if it contains no NOT-gates. (If either one of the two restrictions "planar" or "monotone" is applied by itself, the problem becomes log-space complete for P [92].) Also in LOGCFL are all decision problems log-space reducible to *deterministic* context-free languages, a class we might call "LOGDCFL", and one that is in $NC \cap SC$, by the result of [61] mentioned in the previous section. For more on LOGCFL and the problems in it, see [43, 63, 212, 213, 238].

A second class that has attracted attention, and appears to be incomparable to LOGCFL (as well as AC^1), although it contains L, SL, and NL [43] is the class "DET", first introduced in [63]. This is another class that has been defined both as a class of search problems [63] and as one of decision problems [43]. Since both versions have their usefulness, we shall once again introduce more precise terminology below.

DEFINITION. The class DET (FDET) consists of all those decision problems (search problems) that are log-space reducible to INTEGER DETERMINANT (the problem of computing the determinant of an n by n matrix of n-bit integers).

INTEGER DETERMINANT can be shown to be in FNC^2, and consequently $FDET \subseteq FNC^2$ (and $DET \subseteq NC^2$) [42]. The interesting complete problems here are all search problems, and hence technically complete only for FDET. The precise nature of reductions involved in the completeness results, and indeed in the definitions of DET and FDET, are not spelled out in [63], but presumably something like log-space Turing reductions with at most a constant number of calls to the oracle will do. Examples of complete problems from [63] are, in addition to computing the INTEGER DETERMINANT, the problems of computing the inverse of an n by n integer matrix as above (DETERMINANT INVERSE), and of computing the product of n such matrices (ITERATED MATRIX PRODUCT).

The class DET is also of note because it contains two probabilistic complexity classes [42, 43]. The most inclusive of these is the "unbounded two-sided error" class PL, whose relation to L is the same as was that of PP to P in Section 4.5. Both PL and its subclass BPL (the bounded two-sided error analog of BPP) may well be strictly larger than NL (unlike the analogs of R and ZPP, which can be shown to equal NL [43].) As far as we now know, PL and BPL are incomparable to LOGCFL. Further probabilistic classes, based on requiring simultaneous log-space and expected polynomial time can also be defined and placed inside DET (see [43]). For a schema of the classes discussed in this and the previous two sections, see Fig. 7.

5.4. *Inside* NC^1

We conclude Section 5 with a brief look inside NC^1. Before we can begin, however, we have to deal with the fact that there are many NC^1s. As mentioned in Section 5.3, even when restricting oneself to the uniform Boolean circuit definition of this class, the precise class defined can depend on the uniformity condition imposed. So far we have restricted attention to the standard log-space uniformity, but other possibilities exist and may be more appropriate.

For instance, one might ask why in practice one should require that the circuits be computable in log-space, rather than allowing full use of the power of polynomial time (in which case we call the circuits "P-uniform"). If in fact one were going to manufacture many copies of each circuit, one might well be able to amortize the polynomial design cost. We did not raise this issue earlier, as we know of no examples higher up in the NC^k hierarchy that suggest that the two types of uniformity differ. There *are* examples, however, when one compares (log-space uniform) NC^1 to P-uniform NC^1, or more precisely, when one compares the corresponding classes of search problem.

For example, consider the ITERATED PRODUCT problem, in which one is given n n-bit integers and asked simply to compute their product. This problem is in P-uniform FNC^1 [26], but the best log-space uniform circuit family known for it has depth $O(\log n \log\log n)$ [207], and hence is just slightly too deep to qualify. For further examples, see [26].

Alternatively, instead of asking for laxer definitions of uniformity, one might argue the conditions should be more stringent than log-space uniformity. As we saw in the previous section, L may properly contain NC^1. Thus in assuming log-space uniformity in the definition of NC^1, we are allowing the machine that constructs the circuits to have more power than the circuits themselves, a bothersome property when one is trying to make fine distinctions about the computational power of such circuits. For this and for other more technical reasons, some researchers feel that the uniformity condition used in defining NC^1 should be no stronger than NC^1-computability itself. Advocates of this position, including the authors of [43, 63], now usually suggest that we use the notion of "U_{E^*}-uniformity" originally proposed in [213].

This notion of uniformity is as technical as its name would suggest, and we shall not describe it in detail here. An essential point, however, is that the machine involved in the definition does not have to construct the circuits; it merely has to recognize a language describing the interconnections of their gates. This makes the machine's task easier, and thus makes it possible for us to get by with reduced computing power. The "reduced power" machine chosen for the definition of U_{E^*}-uniformity is a specially modified alternating Turing machine whose running time is $O(\log n)$, where n is the length of the input. The modification replaces the standard input tape with a random access mechanism: The Turing machine writes the address of an input bit it wishes to obtain on a length $O(\log n)$ indexing tape, and then receives that bit in a special read-only register on the next step. In this way, although no single $O(\log n)$ computation path can look at all the bits of an n-bit input, the entire computation tree *can* take all the input bits into account. (Had we used the standard linear input tape, the computation tree would have been restricted to the first $O(\log n)$ bits of the input.)

It should be pointed out that, in the definition of U_{E^*}-uniformity, these $O(\log n)$ time "random access" ATMs are given a significant boost. The strings in the "extended connection language" E^* that the ATMs must recognize, although of length n where n is the number of inputs to the circuit, contain only $O(\log n)$ relevant bits. (These are padded out to the required length with additional, meaningless bits.) In comparing the power of these ATMs to that of NC^1 machines, however, the relevant question is how well the ATMs can do without such help. More precisely, if the ATMs are to be no more powerful, the following class must be contained in U_{E^*}-uniform NC^1.

DEFINITION. The class ALOGTIME consists of all those decision problems solvable by $O(\log n)$ time bounded ATMs, where n is the length of the input.

Surprisingly, not only is ALOGTIME contained in U_{E^*}-uniform NC^1, the two classes are equal [213]! There is currently, however, no proof of equality between U_{E^*}-uniform NC^1 and log-space uniform NC^1. Thus all we know at present is that

$$\text{ALOGTIME} = U_{E^*}\text{-uniform } NC^1 \subseteq \text{log-space uniform } NC^1.$$

Note that, as suggested above, the distinctions between these uniformity conditions disappears for NC^k, $k>1$. For such k, log-space uniform NC^k equals U_{E^*}-uniform NC^k, and both equal the appropriate generalization of ALOGTIME, i.e., the class of languages recognized by ATMs with time and space bounded by $O(\log^k n)$ and $O(\log n)$ respectively [213].

Even if ALOGTIME does not equal (log-space uniform) NC^1, we can consider it to be the largest interesting class contained therein. The *smallest* nontrivial class that we shall consider is the analog of ALOGTIME for deterministic Turing machines.

DEFINITION. The class DLOGTIME consists of all those decision problems solvable by a "random access" DTM in O(log n) time, where n is the length of the input.

Needless to say, the O(log n) time DTM is a *very* weak model of computation, as its answers must ignore all but log n bits of the input (although *which* O(log n) bits it is may depend on the values of the bits seen). DLOGTIME will be contained in every significant class we examine from now on. It contains little in the way of interesting problems itself, unless one considers such tasks as using binary search to verify the length of your input as interesting. There are important uses for O(log n) time DTMs, however.

For one thing, such machines can be used to define a uniformity condition that is in no danger of providing hidden power to the "uniform" classes of machines it defines (as we worried that log-space uniformity might do to NC^1). As with U_{E^*} uniformity, this new "DLOGTIME-uniformity" condition is defined in terms of recognizing a language that describes the interconnection patterns of the gates in the "uniform" circuits. Again the strings in the language contain only O(log n) significant bits, but are padded out to length n. (The language itself is slightly different: a "direct" rather than "extended" connection language.) Adding to the theoretical appeal of DLOGTIME-uniformity is the fact that it is equivalent to a seemingly very different notion of uniformity proposed in [124] and based on definability by first-order formulae of mathematical logic [23]. Also appealing is the fact that DLOGTIME-uniform NC^1 remains equal to U_{E^*}-uniform NC^1.

A second use for O(log n) time DTMs is in reductions. Again the very weakness of DLOGTIME will ensure that the resulting class of transformations is compatible with all the classes in which we will be interested.

DEFINITION. A polynomial transformation f from a problem X to a problem y is a DLOGTIME *transformation* if the language $\{(x, i, c)$: the ith bit of $f(x)$ is $c\}$ is in DLOGTIME.

We shall return to these reductions at the end of this section, where we shall give some examples of problems complete for ALOGTIME under them. First, however, let us examine more of the subclasses of ALOGTIME that are presumably proper subclasses, and hence incapable of containing such ALOGTIME-complete problems.

Perhaps the most studied of these subclasses is AC^0, the class of all decision problems solvable by constant-depth, polynomial-size, unbounded fan-in circuits (as previously defined in Section 5.3), a class that is easily seen to be a strict superset of DLOGTIME. What makes constant-depth, unbounded fan-in circuits so interesting is the lower bound results we can prove for them. First, one can prove that AC^0 contains a noncollapsing hierarchy, based on alternation depth (which in this case is simply depth, since unbounded fan-in means that there is no need for two OR-gates or two

AND-gates in a row). Let the *depth* of an unbounded fan-in circuit be the maximum number of AND- and OR-gates in any path from an input to the output. (Depth 0 would be a circuit that simply hooked some single input gate directly to the output.)

DEFINITION. For all $k \geqslant 0$, the class (uniform) AC_k^0 consists of all problems solvable by DLOGTIME-uniform, depth-k, polynomial-size, unbounded fan-in circuits.

Note that $AC_0^0 \subset$ DLOGTIME $\subset AC_2^0$, while DLOGTIME and AC_1^0 are incomparable. With classes *this* simple, it is easy to get separations. Higher up in the AC_k^0 hierarchy, things become more interesting (and nontrivial). As shown in [229] however, there are problems in $AC_k^0 - AC_{k-1}^0$ for each $k > 0$. These problems are artificial ones designed to make maximal use of the alternation inherent in AC_k^0. Finding interesting "natural" problems inside AC^0 is more of a challenge. Indeed the class is more interesting for the problems it has been shown *not* to contain, such as PARITY (is the total number of 1s in the input odd?) and MAJORITY (are more than half the input bits 1s?), and a variety of others [40, 75, 84, 232].

Note that in order to prove non-membership in AC^0, one must in fact prove superpolynomial lower bounds on circuit size. Constant-depth, unbounded fan-in circuits are at present the most powerful model in which we have been able to prove such results for problems as "easy" as the ones in NP. The model is so weak, however, that one can prove these bounds for problems that are in NC^1, as are all the above examples. This is not to say that the results do not have any practical implications; they confirm the popular wisdom among VLSI designers that functions like parity and majority cannot be computed by reasonably sized fixed-depth PLAs[229]. Moreover, the results are in a sense "stronger" than necessary for this, as the lower bounds also hold for *nonuniform* circuit families. (Indeed, when the class AC^0 is mentioned in the literature, it is typically viewed in nonuniform terms, and compared to nonuniform versions of the NC^k classes. One also sees mention of the nonuniform classes AC^k, $k > 0$, which are the unbounded fan-in analogs of the NC^k. For our purposes here, we shall mean the DLOGTIME-uniform version of a class unless we state otherwise.)

As an aside, we note that the above lower bounds have recently been strengthened to be truly exponential (rather than merely superpolynomial), and this has important theoretical corollaries, given the formal analogies between AC^0 and the polynomial hierarchy of Section 2.4, elaborated in [40, 84, 229]. (Note that the levels of each can be viewed as offering the opportunity for bounded alternation.) Nonuniform exponential lower bounds on size in the cases of PARITY, proved in [257] and tightened in [114, 115], yield the oracle set for which PSPACE $\neq$ PH. Similarly, exponential lower bounds on size for depth-$(k-1)$ nonuniform circuit families solving problems in AC_k^0, announced in [257] and spelled out in [114, 115], yield an oracle for which PH does not collapse.

Given that bounded-depth, unbounded fan-in circuits made up of AND-, OR- and NOT-gates cannot solve everything in NC^1, a natural question to ask is whether the addition of more powerful gates might help. For example, if you add a "MOD(2)-gate", i.e., one that outputs a 1 if and only if an even number of its input gates are non-zero, then bounded-depth, unbounded fan-in circuits *can* solve the PARITY problem. They cannot, however, solve all problems in NC^1 [203]. More generally, let us consider the following classes of extensions to AC^0, all easily seen to be in NC^1.

DEFINITION. For any positive integer $k>1$, let $AC^0(k)$ consist of those languages recognized by polynomial-size, bounded-depth, unbounded fan-in Boolean circuits, augmented by "MOD(k)-gates", i.e., unbounded fan-in gates that output "1" if and only if the number of their non-zero inputs is congruent to 0 MOD(k).

For all primes p, $AC^0(p)$ is *strictly* contained in NC^1. In particular, for any primes $p \neq q$, the problem of determining congruence to 0 MOD(q) is not in $AC^0(p)$ [232]. This is as far as this line of research has gone, however: the classes $AC^0(p)$, p prime, are the largest classes known to be properly contained in NC^1. So far, no one has been able to prove even that $AC^0(6)$ does not equal NP! [24].

Should the $AC^0(6)$ challenge fall, there are two more classes of augmented AC^0 circuits waiting in the wings to be the next candidates for proper inclusion in NC^1. They are the following.

DEFINITION. The class ACC consists of all those languages recognized by polynomial-size, bounded-depth, unbounded fan-in Boolean circuits in which any MOD(k)-gate, $k>1$, may be used.

DEFINITION. The class TC^0 consists of all those languages recognized by polynomial-size, bounded-depth, unbounded fan-in Boolean circuits augmented by "threshold" gates, i.e., unbounded fan-in gates that output "1" if and only if more than half their inputs are non-zero.

For more on these classes, see [22, 24, 100, 208]. It is conjectured that the latter class contains a hierarchy of bounded-depth classes TC_k^0 analogous to the AC_k^0, although so far the highest level separation proved is between classes TC_2^0 and TC_3^0 [100]. The relation between these and our previous classes is summarized as

$$AC^0 \subset ACC \subseteq TC^0 \subseteq NC^1.$$

We conclude our coverage of the classes inside NC^1 with a brief discussion of hopes for an alternative hierarchy, one that would in a sense be perpendicular to the one provided by AC^0. This hierarchy is defined by placing constant bounds on the *width* of circuits (as defined in Section 1.3), rather than on their depth. We also restrict ourselves once more to circuits with bounded fan-in, say, fan-in 2. Define BW_k^0 to be the set of all problems solvable by polynomial-size, bounded fan-in circuits of width k or less, and $BW^0 = \bigcup_{k>1} BW_k^0$. It is not difficult to show that for all $k \geqslant 0$, $AC_k^0 \subseteq BW_k^0$. Moreover, AC^0 is properly contained in BW^0, since PARITY is in BW^0. Are there any candidates for $NC^1 - BW^0$? Here the answer is, surprisingly, no. Unlike the case with AC^0, we have $NC^1 = BW^0$. Moreover, the BW_k^0 hierarchy collapses to its fourth level and we in fact have $NC^1 = BW_4^0$ [22]. As with the above results for AC^0 and its variants, these last results hold in both the uniform and nonuniform case (DLOGTIME-uniformity will suffice [23].) For a somewhat expanded coverage of AC^0 and BW^0, see [134, 136]. For a summary of the inclusion relations between the classes inside NC^1 that we have presented, see Fig. 8.

This last result brings us to the final topic of this section: complete problems for NC^1 (or more precisely, ALOGTIME = U_{E^*}-uniform NC^1). Although the NC^k hierarchy would collapse if NC had complete problems, there is no such technical difficulty for NC^1, and indeed problems complete for it under DLOGTIME reductions have already been identified. (One can also get completeness results using the more powerful "AC^0-reduction", based on bounded-depth unbounded fan-in circuits with oracle gates, but so far the extra power seems unnecessary.)

The proof that BW_4^0 equals NC^1 follows directly from the observation that the former contains a problem that is complete for the latter under DLOGTIME-reductions [23, 22]. The problem in question is yet another "product" problem, following in line with ITERATED MATRIX PRODUCT of Section 5.4 and ITERATED PRODUCT of this section. This is PERMUTATION GROUP PRODUCT: "Given a sequence of elements from the permutation group S_5, does their product equal the identity?" A corollary of this completeness result that may be of interest to formal language theorists is that, whereas the hardest context-sensitive languages are complete for PSPACE (see Section 2.6), and the hardest context-free languages are complete for LOGCFL (see the previous section), the hardest regular languages are only complete for ALOGTIME [22]. (It is easy to see that all regular languages are contained in ALOGTIME.)

For our final example of an ALOGTIME-complete problem, we have the BOOLEAN FORMULA VALUE problem, where formulas are strings with syntax and semantics defined inductively as follows: "1" is a formula with value 1, "0" is a formula of value 0, and if f and g are formulas with values $v(f)$ and $v(g)$ respectively, then "$(f \wedge g)$" is a formula whose value is the logical AND of $v(f)$ and $v(g)$ and "$(f \vee g)$" is a formula whose value is the logical OR of $v(f)$ and $v(g)$. The hard part here is showing that the BOOLEAN FORMULA VALUE problem is even *in* ALOGTIME; see [46]. Its presence in the class indicates that NC^1 circuits, unlike the classes of circuits determining DLOGTIME, AC^0, ACC, and TC^0, can perform tasks that are far from rudimentary. Thus, although proving that $NC^1 \neq NP$ would only be a small first step along the way to a proof that $P \neq NP$, it could be the first important one.

6. New developments, tables and figures

We begin in this section with a brief look at an important new result that was obtained since the body of the chapter was written and that has cast some doubts on the comments made in Section 1.8 about the significance of relativized results. We then conclude with a collection of tables and figures that summarize the material in our "Catalog of Complexity Classes" and thus can be used for "ready reference".

6.1. New developments

In Section 4.6 we introduced the concept of "interactive proof system" and the class IP (also known as AM[poly]) of decision problems solvable by interactive proof systems with a polynomial number of rounds. We also observed that $IP \subseteq PSPACE$.

Furthermore, we noted that there is an oracle for which IP does not contain co-NP [80], thus seeming to imply, by the remarks of Section 1.8 on relativization, that co-NP$\subseteq$IP (and hence IP = PSPACE) would be hard to prove.

Surprisingly, both results have now been proved in rapid succession. First came the result that co-NP$\subseteq$IP and in fact PH$\subseteq$IP [267]. Adi Shamir then strengthened this result to show that all of PSPACE was in IP and hence IP = PSPACE [268]. Not only does this give us a precise determination of the power of interactive proofs, it also raises questions about the proper interpretation of relativization results.

In Section 1.8 we indicated that the standard proof techniques for answering questions about complexity classes were "all" known to relativize. This turns out to have been incorrect. It remains true that *most* standard techniques, such as simulation and diagonalization, do relativize, and so the existence of a relativized world in which a conjecture does not hold rules out the use of such techniques in proving it. Shamir's proof, however, is relatively simple and can itself be viewed as an instance of a "standard technique", albeit one that has been remarkable until now only for its failures. This is the approach commonly taken in the many false "proofs" that P = NP: show that a complete (or "hard") problem for the supposedly larger class (for example HAMILTONIAN CIRCUIT in the case of NP) is a member of the supposedly smaller class (i.e., in the case of P, can be solved in polynomial time). Such a proof would not relativize, since individual problems do not relativize. (Oracles can be attached to machines, but there is no natural concept of "relativized HAMILTONIAN CIRCUIT".) What Shamir has done is show that the PSPACE-complete QUANTIFIED BOOLEAN FORMULA problem of Section 2.6 can be solved by a polynomial round interactive proof system. (The problem used to show PH$\subseteq$IP in [267] was the PERMANENT COMPUTATION problem of Section 4.1, which is "hard" for PH based on the results of [240, 243], as discussed in that section.)

It is still not clear why this "specific hard problem" technique should have proved successful with IP while not elsewhere, but one should note that the interactive nature of the computation and the fact that information is hidden from one of the parties is crucial to the details of the proof. Thus it is not clear that this first major success of the specific hard problem approach will signal further successes in disparate domains. (In a similar domain, however, another success of the approach has recently been announced [263]. This time the result is that the class MIP of decision problems solvable by polynomially bounded interactive proof systems in which there are *two* provers is exactly equal to NEXPTIME. See [263] for full definitions and details.)

6.2. *Tables and figures*

This section is devoted to tables and figures that help summarize the material in this "Catalog of Complexity Classes". Tables 2(a) and 2(b) provide an index to the complexity classes we have defined, from AC_0 to ZPP. For each class, we indicate the section in which it is defined, any additional sections in which it is mentioned, and any figures in which it is represented. The list is intended to provide pointers, at least indirectly, for all the classes mentioned in the text. (We omit a few relatives of the included classes, but each of these can be located by following the pointers for the corresponding included class.) Table 3 provides similar indexing information for the

Table 2(a)
Index to the classes

CLASS NAME	DEFINED	MENTIONED/ILLUSTRATED
AC^0	5.3	5.4, Fig. 8
$AC^0(k)$	5.4	
AC_k^0	5.4	Fig. 7
AC^k	5.3	Fig. 7
ACC	5.4	Fig. 8
ALOGTIME	5.4	Fig. 8
AM	4.6	6.1, Fig. 6
AM[poly]	4.6	Fig. 6
BH	2.4	Fig. 1
BH_k	2.4	Fig. 1
BPL	5.3	Fig. 7
BPP	4.5	4.6, 5.1, 5.2, Figs. 5, 6
BW^0	5.4	Fig. 8
BW_k^0	5.4	Fig. 8
CC	5.2	
co-NP	2.2	2.3–2.5, 4.3, 4.5, Figs. 1, 2, 5, 6
co-R	4.3	4.4, 4.5, Fig. 5
D^P	2.4	Fig. 1
DET	5.3	Fig. 7
Δ_2^P	2.3	2.4, 2.5, 4.5, Figs. 1, 2
Δ_k^P	2.5	Fig. 2
DLOGTIME	5.4	Fig. 8
EH	3.3	Fig. 3
ELEMENTARY	3.3	Fig. 3
ETIME	3.1	5.1, Fig. 3
EXPSPACE	3.3	Fig. 3
EXPTIME	3.1	4.3, Fig. 3
$F\Delta_2^P$	2.3	2.4, 5.1
$F\Delta_k^P$	4.1	
FewP	4.2	Fig. 4
FDET	5.3	
FNC	5.2	
FNC^k	5.3	
FP	1.5	1.6, 1.7, 2.4, 4.2, 5.2
FP^{NP}	2.3	2.4
$FP^{NP[k]}$	2.4	
$FP^{NP[O(\log n)]}$	2.4	
FPH	4.1	
FRNC	5.2	
FUP	4.2	
IP	4.6	6.1, Fig. 6
IP[k]	4.6	
L	5.1	5.2, 5.3, Fig. 7
LIN-SPACE	2.6	
LOG^k-SPACE	5.1	5.3, Fig. 7
LOGCFL	5.3	Fig. 8
LOGDCFL	5.3	Fig. 8

Table 2(b)
Index to the classes (*continued*)

Table 3
Index to reducibilities and models of computation

REDUCIBILITY	DEFINED	MENTIONED
AC^0-reduction	5.4	
BPP-reduction	4.4	
DLOGTIME transformation	5.4	
γ-reduction	2.2	4.4
log-space transformation ($\leqslant_{\text{log-space}}$)	1.6	1.7, 5.1–5.3
metric reduction	2.4	4.1
NC-reduction	5.2	
NC^1-reduction	5.3	
nondeterministic polynomial-time Turing reduction	2.2	
parsimonious transformation	4.1	
polynomial transformation ($\leqslant_p$)	1.6	2.1, 2.2, 4.2, 4.3
polynomial-time isomorphism	2.1	
polynomial-time Turing reduction ($\leqslant_T$)	1.6	2.2, 2.3, 4.2, 4.3
R-reduction	4.4	
RUR-reduction	4.4	
strong nondeterministic polynomial-time Turing reduction	2.2	
UR-reduction	4.4	

MODEL OF COMPUTATION	DEFINED	MENTIONED
"advice-taking" Turing machine	4.3	4.5
Arthur–Merlin game	4.6	
alternating Turing machine (ATM)	1.3	2.6, 3.3, 4.6, 5.4
Boolean circuit family	1.3	
nonuniform	1.3	4.3, 5.4
log-space uniform	5.2	5.3, 5.4
P-uniform	5.4	
U_{E^*}-uniform	5.4	
DLOGTIME-uniform	5.4	
counting Turing machine (CTM)	4.1	
deterministic Turing machine (DTM)	1.3	1.5, 1.6, 2.6, 3.1, 3.3, 5.1, 5.2, 5.4
interactive proof	4.6	
nondeterministic Turing machine (NDTM)	1.3	2.1, 3.2, 3.3
oracle-augmented Boolean circuit family	5.3	1.8, 5.4
oracle Turing machine(OTM)	1.3	1.8, 2.3, 2.4, 2.5, 3.2, 4.1, 4.5
parallel random access machine (PRAM)	1.3	5.2, 5.3
parity Turing machine ($\oplus$TM)	4.1	
probabilistic Turing machine (PTM)	4.5	
"random access" Turing machine	5.4	
random Turing machine (RTM)	4.3	
stochastic Turing machine (STM)	4.6	
symmetric Turing machine	5.1	
unambiguous Turing machine (UTM)	4.2	

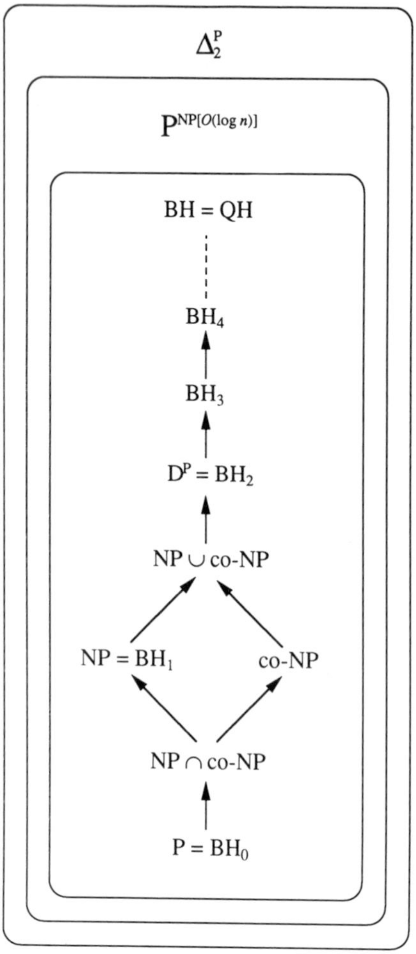

Fig. 1. The Boolean hierarchy.

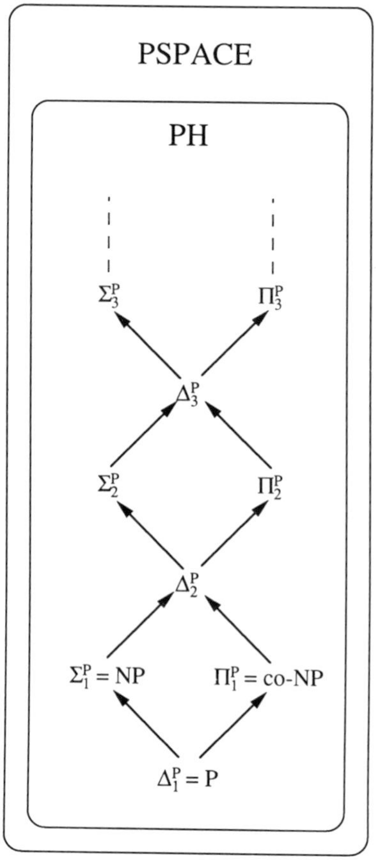

Fig. 2. The polynomial hierarchy.

various notions of "reducibility" and for the various models of computation that we have discussed. For less specialized directories, see the outline at the beginning of the chapter and the overall index to this Handbook.

We also include in this section the eight figures that have already been mentioned in the text. In these figures, an arrow from class A to class B indicates that $A \subseteq B$, and all arrows are drawn upwards. All currently known containment relations are indicated, although some are only present implicitly. (If there is an arrow from A to B and an arrow from B to C, there will be no arrow from A to C, even though the containment relation $A \subseteq C$ is implied.) The figures do not indicate which of the containments are proper. For what is known on this account, consult the appropriate sections of the text. (Because of space constraints, some of the arrows are omitted in Fig. 3. In this case, class A contains class B if the name of class A is written immediatey above that of class B.)

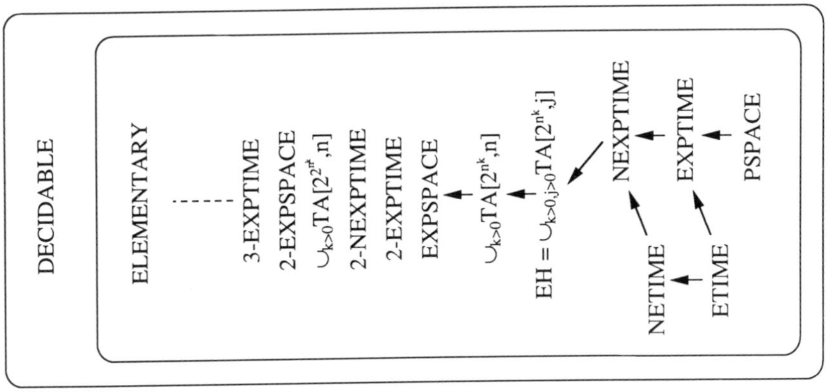

Fig. 3. Provably intractable problems.

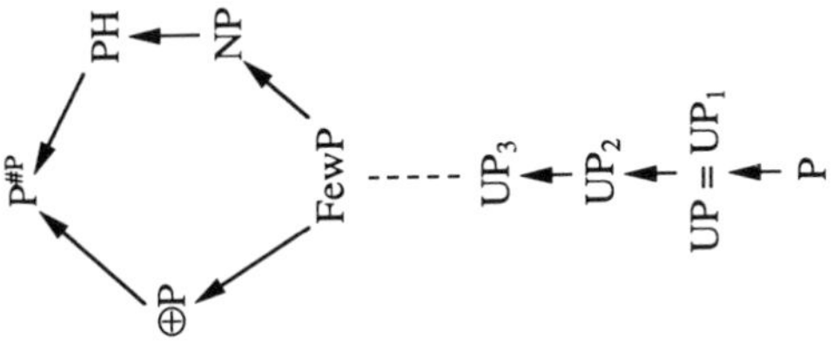

Fig. 4. Counting classes.

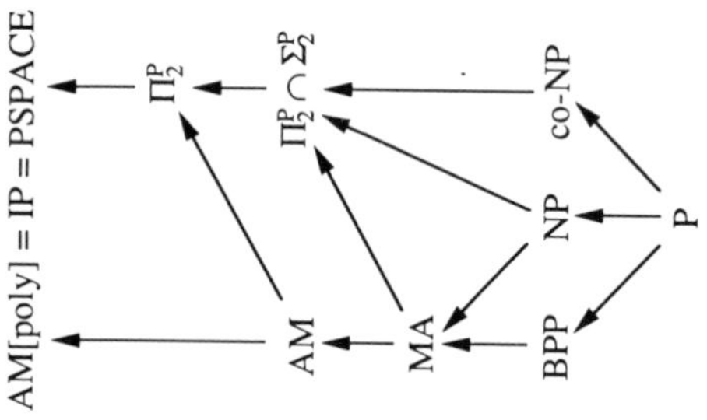

Fig. 6. Interactive complexity classes.

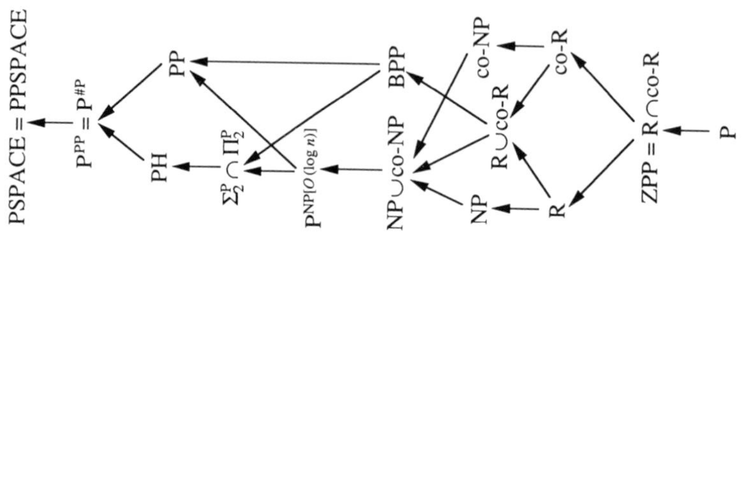

Fig. 5. Randomized complexity classes.

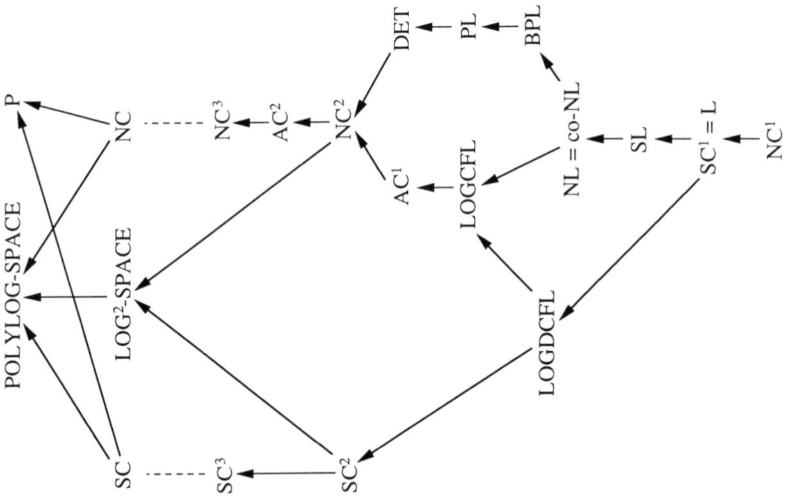

Fig. 7. Classes inside P.

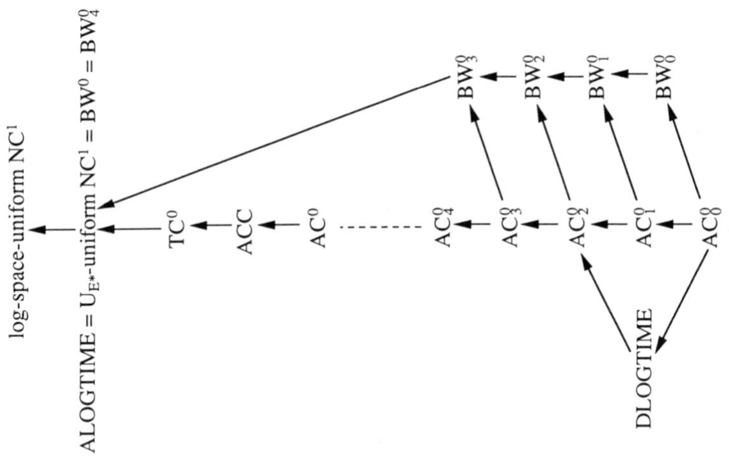

Fig. 8. Classes inside NC^1.

References

[1] ADLEMAN, L., Two theorems on random polynomial time, in: *Proc. 19th Ann. IEEE Symp on Foundations of Computer Science* (1978) 75–83.

[2] ADLEMAN, L. and M. HUANG, Recognizing primes in random polynomial time, in: *Proc. 19th Ann. ACM Symp. on Theory of Computing* (1987) 462–470.

[3] ADLEMAN, L. and K. MANDERS, Reducibility, randomness, and intractability, in: *Proc. 9th Ann. ACM Symp. on Theory of Computing* (1977) 151–163.

[4] ADLEMAN, L.M. and K. MANDERS, Reductions that lie, in: *Proc. 20th Ann. IEEE Symp. on Foundations of Computer Science* (1979) 397–410.

[5] AHO, A.V., J.E. HOPCROFT and J.D. ULLMAN, *The Design and Analysis of Computer Algorithms* (Addison-Wesley, Reading, MA, 1974).

[6] AIELLO, W., S. GOLDWASSER and J. HASTAD, On the power of interaction, in: *Proc. 27th Ann. IEEE Symp. on Foundations of Computer Science* (1986) 368–379.

[7] AIELLO, W. and J. HASTAD, Perfect zero-knowledge languages can be recognized in two rounds, in: *Proc. 28th Ann. IEEE Symp. on Foundations of Computer Science* (1987) 439–448.

[8] ALELIUNAS, R., R.M. KARP, R.J. LIPTON, L. LOVÁSZ and C. RACKOFF, Random walks, traversal sequences and the complexity of maze problems, in: *Proc. 20th. Ann. IEEE Symp. on Foundations of Computer Science* (1979) 218–223.

[9] ALLENDER, E., The complexity of sparse sets in P, in: A.L. Selman, ed., *Structure in Complexity Theory*, Lecture Notes in Computer Science, Vol. 223 (Springer, Berlin, 1986) 1–11.

[10] ANGLUIN, D., On counting problems and the polynomial-time hierarchy, *Theoret. Comput. Sci.* **12** (1980) 161–173.

[11] AWERBUCH, B., A. ISRAELI and Y. SHILOACH, Finding Euler circuits in logarithmic parallel time, in: *Proc. 16th Ann. ACM Symp. on Theory of Computing* (1984) 249–257.

[12] BABAI, L., Trading group theory for randomness, in: *Proc. 17th Ann. ACM Symp. on Theory of Computing* (1985) 421–429; a subsequent journal paper covering some of this material is: L. Babai and S. Moran, Arthur–Merlin games: A randomized proof system, and a hierarchy of complexity classes, *J. Comput. System Sci.* **36** (1988) 254–276.

[13] BABAI, L., Random oracles separate PSPACE from the polynomial-time hierarchy, *Inform. Process. Lett.* **26** (1987) 51–53.

[14] BABAI, L., E. LUKS and A. SERESS, Permutation groups in NC, in: *Proc. 19th Ann. ACM Symp. on Theory of Computing* (1987) 409–420.

[15] BABAI, L. and E. SZEMERÉDI, On the complexity of matrix group problems, in: *Proc. 25th Ann. IEEE Symp. on Foundations of Computer Science* (1984) 229–240.

[16] BACH, E., G. MILLER and J. SHALLIT, Sums of divisors, perfect numbers and factoring, *SIAM J. Comput.* **15** (1986) 1143–1154.

[17] BAKER, T., J. GILL and R. SOLOVAY, Relativizations of the $P \stackrel{?}{=} NP$ question, *SIAM J. Comput.* **4** (1975) 431–442.

[18] BALCÁZAR, J. L., Logspace self-reducibility, in: *Proc. Structure in Complexity Theory* (*3rd Ann. IEEE Conf.*) (1988) 40–46.

[19] BALCÁZAR, J.L., Nondeterministic witnesses and nonuniform advice, in: *Proc. Structure in Complexity Theory* (*4th Ann. IEEE Conf.*) (1989) 259–269.

[20] BALCÁZAR, J.L., J. DIAZ and J. GABARRÓ, *Structural Complexity I* (Springer, Berlin, 1988).

[21] BALCÁZAR, J.L. and D.A. RUSSO, Immunity and simplicity in relativizations of probabilistic complexity classes, *Theoret. Inform. Appl.* **22** (1988) 227–244.

[22] BARRINGTON, D.A., Bounded-width polynomial-size branching programs can recognize exactly those languages in NC^1, *J. Comput. System Sci.* **38** (1989) 150–164.

[23] BARRINGTON, D.A.M., N. IMMERMAN and H. STRAUBING, On uniformity conditions within NC^1, in: *Proc. Structure in Complexity Theory* (*3rd Ann. IEEE Conf.* (1988) 47–59.

[24] BARRINGTON, D.A.M. and D. THÉRIEN, Finite monoids and the fine structure of NC^1, *J. Assoc. Comput. Mach.* **35** (1988) 941–952.

[25] BARWISE, J., *Handbook of Mathematical Logic* (North-Holland, Amsterdam, 1977).

[26] BEAME, P.W., S.A. COOK and H.J. HOOVER, Log depth circuits for division and related problems, *SIAM J. Comput.* **15** (1986) 994–1003.

[27] BEIGEL, R., On the relativized power of additional accepting paths, in: *Proc. Structure in Complexity Theory (4th Ann. IEEE Conf.)* (1989) 216–224.

[28] BEIGEL, R., L.A. HEMACHANDRA and G. WECHSUNG, On the power of probabilistic polynomial time: $P^{NP[\log]} \subseteq PP$, in: *Proc. Structure in Complexity Theory (4th Ann. IEEE Conf.)* (1989) 225–227.

[29] BENNETT, C.H. and J. GILL, Relative to a random oracle A, $P^A \neq NP^A \neq co\text{-}NP^A$ with probability 1, *SIAM J. Comput.* **10** (1981) 96–113.

[30] BERMAN, L., The complexity of logical theories, *Theoret. Comput. Sci.* **11** (1980) 71–78.

[31] BERMAN, L. and J. HARTMANIS, On isomorphisms and density of NP and other complete sets, *SIAM J. Comput.* **6** (1977) 305–322.

[32] BERSTEL, J. and L. BOASSON, Context-free languages, in: J. van Leeuwen, ed., *Handbook of Theoretical Computer Science, Vol. B* (North-Holland, Amsterdam, 1990) Chapter 2.

[33] BLASS, A. and Y. GUREVICH, On the unique satisfiability problem, *Inform. and Control* **55** (1982) 80–88.

[34] BOOK, R.V., On languages accepted in polynomial time, *SIAM J. Comput.* **1** (1972) 281–287.

[35] BOOK, R.V., Comparing complexity classes, *J. Comput. System Sci.* **9** (1974) 213–229.

[36] BOOK, R.V., Translational lemmas, polynomial time, and $(\log n)^j$-space, *Theoret. Comput. Sci.* **1** (1976) 215–226.

[37] BOOK, R.V. On the complexity of formal grammars, *Acta Inform.* **9** (1978) 171–182.

[38] BOOTH, K.S., Isomorphism testing for graphs, semigroups, and finite automata are polynomially equivalent problems, *SIAM J. Comput.* **7** (1978) 273–279.

[39] BOPPANA, R.B., J. HASTAD and S. ZACHOS, Does co-NP have short interactive proofs?, *Inform. Process. Lett.* **25** (1987) 127–133.

[40] BOPPANA, R. and M. SIPSER, The complexity of finite functions, in: J. van Leeuwen, ed., *Handbook of Theoretical Computer Science, Vol. A* (North-Holland, Amsterdam, 1990) 757–804.

[41] BORODIN, A., On relating time and space to size and depth, *SIAM J. Comput* **6** (1977) 733–744.

[42] BORODIN, A., S.A. COOK and N. PIPPENGER, Parallel computations for well-endowed rings and space-bounded probablistic machines, *Inform. and Control.* **58** (1983) 113–136.

[43] BORODIN, A., S.A. COOK, P.W. DYMOND, W.L. RUZZO and M.L. TOMPA, Two applications of inductive counting for complementation problems, *SIAM J. Comput.* **18** (1989) 559–578.

[44] BUSS, J.F., A theory of oracle machines, in: *Proc. Structure in Complexity Theory (2nd Ann. IEEE Conf.)* (1987) 175–181.

[45] BUSS, J.F., Relativized alternation and space bounded computation, *J. Comput. System Sci.* **36** (1988) 351–378.

[46] BUSS, S.R., The Boolean formula value problem is in ALOGTIME, in: *Proc. 19th Ann. ACM Symp. on Theory of Computing* (1987) 123–131.

[47] CAI, J.-Y., With probability one, a random oracle separates PSPACE from the polynomial-time hierarchy, in: *Proc. 18th Ann. ACM Symp. on Theory of Computing* (1986) 21–29; journal version in: *J. Comput. System Sci.* **38** (1988) 68–85.

[48] CAI, J.-Y., T. GUNDERMANN, J. HARTMANIS, L.A. HEMACHANDRA, V. SEWELSON, K. WAGNER and G. WECHSUNG, The Boolean hierarchy I: Structural properties, *SIAM J. Comput.* **17** (1988) 1232–1252.

[49] CAI, J.-Y., T. GUNDERMANN, J. HARTMANIS, L.A. HEMACHANDRA, V. SEWELSON, K. WAGNER and G. WECHSUNG, The Boolean hierarchy II: Applications, *SIAM J. Comput.* **18** (1989) 95–111.

[50] CAI, J.-Y. and L.A. HEMACHANDRA, The Boolean hierarchy: Hardware over NP, Report No. TR 85-724, Dept. Computer Science, Cornell Univ., Ithaca, NY, 1985.

[51] CAI, J.-Y. and L.A. HEMACHANDRA, On the power of parity polynomial time, in: *Proc. 6th Ann. Symp. on Theoretical Aspects of Computing*, Lecture Notes in Computer Science, Vol. 349 (Springer, Berlin, 1989) 229–239.

[52] CHANDRA, A.K., D.C. KOZEN and L.J. STOCKMEYER, Alternation, *J. Assoc. Comput. Mach.* **28** (1981) 114–133.

[53] CHANG, R. and J. KADIN, The Boolean hierarchy and the polynomial hierarchy: a closer connection, Report No. TR 89-1008, Dept. Computer Science, Cornell Univ., Ithaca, NY, 1989.

[54] CHIN, F.-Y., J. LAM and I.-N. CHEN, Efficient parallel algorithms for some graph problems, *Comm. ACM* **25** (1982) 659–665.

[55] CHUNG, M.J. and B. RAVIKUMAR, Strong nondeterministic Turing reduction — a technique for proving intractability, in: *Proc. Structure in Complexity Theory* (*2nd Ann. IEEE Conf.*) (1987) 132–137.

[56] COBHAM, A., The intrinsic computational difficulty of functions, in: Y. Bar-Hillel, ed., *Proc. 1964 Internat. Congress for Logic Methodology and Philosophy of Science* (North-Holland, Amsterdam, 1964) 24–30.

[57] COOK, S.A., Path systems and language recognition, in: *Proc. 2nd Ann. ACM Symp. on Theory of Computing* (1970) 70–72.

[58] COOK, S.A., The complexity of theorem-proving procedures, in: *Proc. 3rd Ann. ACM Symp. on Theory of Computing* (1971) 151–158.

[59] COOK, S.A., Characterizations of pushdown machines in terms of time-bounded computers, *J. Assoc. Comput. Mach.* **18** (1971) 4–18.

[60] COOK, S.A., An observation on time-storage trade-off, *J. Comput. System Sci.* **9** (1974) 308–316.

[61] COOK, S.A., Deterministic CFL's are accepted simultaneously in polynomial time and log squared space, in: *Proc. 11th Ann. ACM Symp. on Theory of Computing* (1979) 338–345.

[62] COOK, S.A., Towards a complexity theory of synchronous parallel computation, *Enseign. Math.* **27** (1981) 99–124.

[63] COOK, S.A., A taxonomy of problems with fast parallel algorithms, *Inform. and Control* **64** (1985) 2–22.

[64] COOK, S. and R. SETHI, Storage requirements for deterministic polynomial time recognizable languages, *J. Comput. System Sci.* **13** (1976) 25–37.

[65] DAVIS, M., Unsolvable problems, in: J. Barwise, ed., *Handbook of Mathematical Logic* (North-Holland, Amsterdam, 1977) 567–594.

[66] DEKHTYAR, M.I., On the relativation of deterministic and nondeterministic complexity classes, in: *Mathematical Foundations of Computer Science*, Lecture Notes in Computer Science, Vol. 45 (Springer, Berlin, 1976) 255–259.

[67] DEMILLO, R.A. and R.J. LIPTON, The consistency of "P = NP" and related problems with fragments of number, theory in: *Proc. 12th Ann. ACM Symp. on Theory of Computing* (1980) 45–57.

[68] DOBKIN, D., R.J. LIPTON and S. REISS, Linear programming is log-space hard for P, *Inform. Process. Lett.* **8** (1979) 96–97.

[69] DWORK, C., P.C. KANELLAKIS and J.C. MITCHELL, On the sequential nature of unification, *J. Logic Programming* **1** (1984) 35–50.

[70] DYER, M.E. and A.M. FRIEZE, On the complexity of computing the volume of a polyhedron, *SIAM J. Comput.* **80** (1989) 205–226.

[71] DYMOND, P.W. and S.A. COOK, Complexity theory of parallel time and hardward, *Inform. and Comput.* **80** (1989) 205–226.

[72] EDMONDS, J., Paths, trees, and flowers, *Canad. J. Math.* **17** (1965) 449–467.

[73] EVEN, S., A.L. SELMAN and Y. YACOBI, The complexity of promise problems with applications to cryptography, *Inform. and Control* **61** (1984) 159–173.

[74] EVEN, S. and R.E. TARJAN, A combinatorial game which is complete in polynomial space, *J. Assoc. Comput. Mach.* **23** (1976) 710–719.

[75] FAGIN, R., M.M. KLAWE, N.J. PIPPENGER and L. STOCKMEYER, Bounded-depth, polynomial size circuits for symmetric functions, *Theoret. Comput. Sci.* **36** (1985) 239–250.

[76] FERRANTE, J. and C. RACKOFF, A decision procedure for the first order theory of real addition with order, *SIAM J. Comput.* **4** (1975) 69–76.

[77] FICH, F.E., P. RAGDE and A. WIGDERSON, Relations between concurrent-write models of parallel computation, *SIAM J. Comput.* **17** (1988) 606–627.

[78] FICH, F.E., P. RAGDE and A. WIGDERSON, Simulations among concurrent-write PRAMs, *Algorithmica* **3** (1988) 43–52.

[79] FORTNOW, L., The complexity of perfect zero-knowledge, in: *Proc. 19th Ann. ACM Symp. on Theory of Computing* (1987) 204–209.

[80] FORTNOW, L. and M. SIPSER, Are there interactive protocols for co-NP languages?, *Inform. Process. Lett.* **28** (1988) 249–251.

[81] FORTUNE, S., D. LEIVANT and M. O'DONNELL, The expressiveness of simple and second-order type structures, *J. Assoc. Comput. Mach.* **30** (1983) 151–185.

[82] FORTUNE, S. and J. WYLLIE, Parallelism in random access machines, in: *Proc. 10th Ann. ACM Symp. on Theory of Computing* (1978) 114–118.

[83] FRAENKEL, A.S. and D. LICHTENSTEIN, Computing a perfect strategy for $n \times n$ chess requires time exponential in n, *J. Combin. Theory Ser. A.* **31** (1981) 199–213.

[84] FURST, M., J. SAXE and M. SIPSER, Parity, circuits, and the polynomial time hierarchy, *Math. Systems Theory.* **17** (1984) 13–27.

[85] GALPERIN, H. and A. WIGDERSON, Succinct representation of graphs, *Inform. and Control* **56** (1983) 183–198.

[86] GAREY, M.R. and D.S. JOHNSON, Strong NP-completeness results: motivation, examples, and implications, *J. Assoc. Comput. Mach.* **25** (1978) 499–508.

[87] GAREY, M.R. and D.S. JOHNSON, *Computers and Intractability: A Guide to the Theory of NP-Completeness* (Freeman, San Francisco, 1979).

[88] GESKE, J. and J. GROLLMAN, Relativizations of unambiguous and random polynomial time classes, *SIAM J. Comput.* **15** (1986) 511–519.

[89] GILBERT, J.R., T. LENGAUER and R.E. TARJAN, The pebbling problem is complete for polynomial space, *SIAM J. Comput.* **9** (1980) 513–524.

[90] GILL, J., Computational complexity of probabilistic Turing machines, *SIAM J. Comput.* **6** (1977) 675–695.

[91] GOLDREICH, O., S. MICALI and A. WIGDERSON, Proofs that yield nothing but their validity and a methodology of cryptographic protocol design, in: *Proc. 27th Ann. IEEE Symp. on Foundations of Computer Science* (1986) 174–187.

[92] GOLDSCHLAGER, L.M., The monotone and planar circuit value problems are log space complete for P, *SIGACT News* **9**(2) (1977) 25–29.

[93] GOLDSCHLAGER, L. and I. PARBERRY, On the construction of parallel computers from various bases of Boolean functions, *Theoret. Comput. Sci.* **43** (1986) 43–58.

[94] GOLDSCHLAGER, L., R. SHAW and J. STAPLES, The maximum flow problem is log space complete for P, *Theoret. Comput. Sci.* **21** (1982) 105–111.

[95] GOLDWASSER, S. and J. KILIAN, Almost all primes can be quickly certified, in: *Proc. 18th Ann. ACM Symp. on Theory of Computing* (1986) 316–330.

[96] GOLDWASSER, S., S. MICALI and C. RACKOFF, The knowledge complexity of interactive proof-systems, in: *Proc. 17th Ann. ACM Symp. on Theory of Computing* (1985) 291–304; a journal version under the same title appears in: *SIAM J. Comput.* **18** (1989) 186–208.

[97] GOLDWASSER, S. and M. SIPSER, Private coins versus public coins in interactive proof systems, in: *Proc. 18th Ann. ACM Symp. on Theory of Computing* (1986) 59–68.

[98] GREIBACH, S.A., The hardest context-free language, *SIAM J. Comput.* **2** (1973) 304–310.

[99] GROLLMAN, J. and A.L. SELMAN, Complexity measures for public-key cryptosystems, *SIAM J. Comput.* **17** (1988) 309–335.

[100] HAJNAL, A., W. MAASS, P. PUDLÁK, M. SZEGEDY and G. TURÁN, Threshold circuits of bounded depth, in: *Proc. 28th Ann. IEEE Symp. on Foundations of Computer Science* (1987) 99–110.

[101] HARARY, F. and E.M. PALMER, *Graphical Enumeration* (Academic Press, New York, 1973).

[102] HAREL, D., *Algorithmics: The Spirit of Computing* (Addison-Wesley, Reading, MA, 1987).

[103] HARTMANIS, J., Independence results about context-free languages and lower bounds, *Inform. Process. Lett.* **20** (1985) 241–248.

[104] HARTMANIS, J., Solving problems with conflicting relativizations, *Bull. EATCS* **27** (1985) 40–49.

[105] HARTMANIS, J., Structural complexity column: Sparse complete sets for NP and the optimal collapse of the polynomial hierarchy, *Bull. EATCS* **32** (1987) 73–81.

[106] HARTMANIS, J., Structural complexity column: The collapsing hierarchies, *Bull. EATCS* **33** (1987) 26–39.

[107] HARTMANIS, J., Structural complexity column: Some observations about relativization of space bounded computations, *Bull. EATCS* **35** (1988) 82–92.

[108] HARTMANIS, J. and L. HEMACHANDRA, Complexity classes without machines: On complete languages for UP, *Theoret. Comput. Sci.* **58** (1988) 129–142.

[109] HARTMANIS, J. and J.E. HOPCROFT, Independence results in computer science, *SIGACT News* **8**(4) (1976) 13–24.

[110] HARTMANIS, J. and N. IMMERMAN, On complete problems for NP∩co-NP, in: *Proc. Internat. Coll. on Automata, Languages, and Programming*, Lecture Notes in Computer Science, Vol. 194 (Springer, Berlin, 1985) 250–259.

[111] HARTMANIS, J., N. IMMERMAN and V. SEWELSON, Sparse sets in NP − P: EXPTIME versus NEXPTIME, *Inform. and Control* **65** (1985) 159–181.

[112] HARTMANIS, J., P.M. LEWIS and R.E. STEARNS, Classification of computations by time and memory requirements, in: *Proc. IFIP Congress 1965* (Spartan, New York, 1965) 31–35.

[113] HARTMANIS, J. and R.E. STEARNS, On the computational complexity of algorithms, *Trans. Amer. Math. Soc.* **117** (1965) 285–306.

[114] HASTAD, J., Improved lower bounds for small depth circuits, in: *Proc. 18th Ann. ACM Symp. on Theory of Computing* (1986) 6–20.

[115] HASTAD, J., *Computational Limitations for Small-Depth Circuits* (MIT Press, Cambridge, MA, 1987).

[116] HEMACHANDRA, L.A., The strong exponential hierarchy collapses, in: *Proc. 19th Ann. ACM Symp. on Theory of Computing* (1987) 110–122; also: *J. Comput. System Sci.* **39** (1989) 299–322.

[117] HEMACHANDRA, L.A., Structure of complexity classes: separations, collapses, and completeness, in: *Mathematical Foundations of Computer Science*, Lecture Notes in Computer Science, Vol. 324 (Springer, Berlin, 1988) 59–73.

[118] HEMACHANDRA, L.A., Private communication, 1989.

[119] HEMACHANDRA, L.A. and S. JAIN, On relativization and the existence of Turing complete sets, Report No. TR-297, Computer Science Dept., Univ. of Rochester, Rochester, NY, 1989.

[120] HOMER, S. and A.L. SELMAN, Oracles for structural properties: the isomorphism problem and public-key cryptography, in: *Proc. Structure in Complexity Theory* (*4th Ann. IEEE Conf.*) (1989) 3–14.

[121] HOOVER, H.J. and W.L. RUZZO, A compendium of problems complete for P, Manuscript, 1985.

[122] HOPCROFT, J.E. and J.D. ULLMAN, *Introduction to Automata Theory, Languages, and Computation* (Addison-Wesley, Reading, MA, 1979).

[123] HUNT, J.W., Topics in probabilistic complexity, Ph.D. Dissertation, Dept. of Electrical Engineering. Stanford Univ., Stanford, CA, 1978.

[124] IMMERMAN, N., Expressibility as a complexity measure: results and directions, in: *Proc. Structure in Complexity Theory* (*2nd Ann. IEEE Conf.*) (1987) 194–202.

[125] IMMERMAN, N., Nondeterministic space is closed under complementation, *SIAM J. Comput.* **17** (1988) 935–938.

[126] IMMERMAN, N. and S.R. MAHANEY, Oracles for which NP has polynomial size circuits, in: *Proc. Conf. on Complexity Theory* (1983) 89–93.

[127] IRVING, R.W., An efficient algorithm for the stable room-mates problem, *J. Algorithms* **6** (1985) 577–595.

[128] JAZAYERI, M., W.F. OGDEN and W.C. ROUNDS, The intrinsically exponential complexity of the circularity problem for attribute grammars, *Comm. ACM* **18** (1975) 697–706.

[129] JOHNSON, D.S., The NP-completeness column: an ongoing guide (1st edition), *J. Algorithms* **2** (1981) 393–405.

[130] JOHNSON, D.S., The NP-completeness column: an ongoing guide (7th edition), *J. Algorithms* **4** (1983) 189–203.

[131] JOHNSON, D.S., The NP-completeness column: an ongoing guide (9th edition), *J. Algorithms* **4** (1983) 397–411.

[132] JOHNSON, D.S., The NP-completeness column: an ongoing guide (12th edition), *J. Algorithms* **5** (1984) 433–447.

[133] JOHNSON, D.S., The NP-completeness column: an ongoing guide (15th edition), *J. Algorithms* **6** (1985) 291–305.

[134] JOHNSON, D.S., The NP-completeness column: an ongoing guide (17th edition): Computing with one hand tied behind your back, *J. Algorithms* **7** (1986) 289–305.

[135] JOHNSON, D.S., The NP-completeness column: an ongoing guide (19th edition): The many faces of polynomial time, *J. Algorithms* **8** (1987) 285–303.

[136] JOHNSON, D.S., The NP-completeness column: an ongoing guide (20th edition): Announcements, updates, and greatest hits, *J. Algorithms* **8** (1987) 438–448.

[137] JOHNSON, D.S., The NP-completeness column: an ongoing guide (21st edition): Interactive proof systems for fun and profit, *J. Algorithms* **9** (1988) 426–444.

[138] JONES, N.D., Space-bouned reducibility among combinatorial problems, *J. Comput. System Sci.* **11** (1975) 68–85.

[139] JONES, N.D. and W.T. LAASER, Complete problems for deterministic polynomial time, *Theoret. Comput. Sci.* **3** (1976) 105–117.

[140] JONES, N.D., Y.E. LIEN and W.T. LAASER, New problems complete for nondeterministic log space, *Math. Systems Theory* **10** (1976) 1–17.

[141] JOSEPH, D., Polynomial time computations in models of ET, *J. Comput. System Sci.* **26** (1983) 311–338.

[142] JOSEPH, D. and P. YOUNG, Independence results in computer science?, *J. Comput. System Sci.* **23** (1981) 205–222.

[143] JOSEPH, D. and P. YOUNG, Corrigendum, *J. Comput. System Sci.* **24** (1982) 378.

[144] JOSEPH, D. and P. YOUNG, A survey of some recent results on computational complexity in weak theories of arithmetic, Report No. 83-10-01, Computer Science Dept., Univ. of Wisconsin, Madison, WI, 1983.

[145] JOSEPH, D. and P. YOUNG, Some remarks on witness functions for nonpolynomial and noncomplete sets in NP, *Theoret. Comput. Sci.* **39** (1985) 225–237.

[146] KADIN, J., $P^{NP[\log n]}$ and sparse Turing-complete sets for NP, in: *Proc. Structure in Complexity Theory (2nd Ann. IEEE Conf.)* (1987) 33–40.

[147] KADIN, J., The polynomial hierarchy collapses if the Boolean hierarchy collapses, *SIAM J. Comput.* **17** (1988) 1263–1282 (errors in the proof in this paper are corrected in [53]).

[148] KARLOFF, H., A Las Vegas RNC algorithm for maximum matching, *Combinatorica* **6** (1986) 387–392.

[149] KARMARKAR, N., A new polynomial-time algorithm for linear programming, *Combinatorica* **4** (1984) 373–395.

[150] KARP, R.M., Reducibility among combinatorial problems, in: R.E. Miller and J.W. Thatcher, eds., *Complexity of Computer Computations* (Plenum Press, New York, 1972) 85–103.

[151] KARP, R.M., On the complexity of combinatorial problems, *Networks* **5** (1975) 45–68.

[152] KARP, R.M. and R.J. LIPTON, Some connections between nonuniform and uniform complexity classes, in: *Proc. 12th Ann. ACM Symp. on Theory of Computing* (1980) 302–309; appeared in journal form as: R.M. KARP and R.J. LIPTON, Turing machines that take advice, *Enseign. Math.* **28** (1982) 191–209.

[153] KARP, R.M. and V. RAMACHANDRAN, Parallel algorithms for shared-memory machines, in: J. van Leeuwen, ed., *Handbook of Theoretical Computer Science, Vol. A* (North-Holland, Amsterdam, 1990) 869–941.

[154] KARP, R.M., E. UPFAL and A. WIGDERSON, Constructing a maximum matching is in random NC, *Combinatorica* **6** (1986) 35–48.

[155] KARP, R.M., E. UPFAL and A. WIGDERSON, The complexity of parallel search, *J. Comput. System Sci.* **36** (1988) 225–253.

[156] KARP, R.M. and A. WIGDERSON, A fast parallel algorithm for the maximal independent set problem, *J. Assoc. Comput. Mach.* **32** (1986) 762–773.

[157] KHACHIYAN, L.G., A polynomial algorithm in linear programming, *Dokl. Akad. Nauk. SSSR* **244** (1979) 1093–1096 (in Russian); English translation in: *Soviet Math. Dokl.* **20** (1979) 191–194.

[158] KO, K.-I., Some observations on the probabilistic algorithms and NP-hard problems, *Inform. Process. Lett.* **14** (1982) 39–43.

[159] KO, K.-I., Relativized polynomial time hierarchies having exactly *K* levels, *SIAM J. Comput.* **18** (1989) 392–408.

[160] KÖBLER, J., U. SCHÖNING, S. TODA and J. TORÁN, Turing machines with few accepting computations and low sets for PP, in: *Proc. Structure in Complexity Theory (4th Ann. IEEE Conf.)* (1989) 208–215.

[161] KÖBLER, J., U. SCHÖNING and J. TORÁN, On counting and approximation, *Acta Inform.* **26** (1989) 363–379.

[162] KOZEN, D., A clique problem equivalent to graph isomorphism, Manuscript, 1978.

[163] KRENTEL, M., The complexity of optimization problems, *J. Comput. System Sci.* **36** (1988) 490–509.

[164] KRENTEL, M., Generalizations of OptP to the polynomial hierarchy, Report No. TR88-79, Dept. of Computer Science, Rice Univ., Houston, TX, 1988.

[165] KURTZ, S.A., On the random oracle hypothesis, *Inform. and Control* **57** (1983) 40–47.

[166] KURTZ, S.A., S.R. MAHANEY and J.S. ROYER, The isomorphism conjecture fails relative to a random oracle, in: *Proc. 21st Ann. ACM Symp. on Theory of Computing* (1989) 157–166.

[167] LADNER, R.E., On the structure of polynomial time reducibility, *J. Assoc. Comput. Mach.* **22** (1975) 155–171.

[168] LADNER, R.E., The circuit value problem is log space complete for P, *SIGACT News* **7**(1) (1975) 18–20.

[169] LADNER, R.E. and N.A. LYNCH, Relativizations about questions of log space computability, *J. Comput. System Sci.* **10** (1976) 19–32.

[170] LADNER, R.E., N.A. LYNCH and A.L. SELMAN, A comparison of polynomial time reducibilities, *Theoret. Comput. Sci.* **1** (1975) 103–124.

[171] LAUTEMANN, C., BPP and the polynomial hierarchy, *Inform. Process. Lett.* **17** (1983) 215–218.

[172] LEGGETT JR, E.W. and D.J. MOORE, Optimization problems and the polynomial hierarchy, *Theoret. Comput. Sci.* 15 (1981) 279–289.

[173] LEIVANT, D., Unprovability of theorems of complexity theory in weak number theories, *Theoret. Comput. Sci.* **18** (1982) 259–268.

[174] LEVIN, L.A., Universal sorting problems, *Problemy Peredaci Informacii* **9** (1973) 115–116 (in Russian); English translation in: *Problems of Information Transmission* **9** (1973) 265–266.

[175] LEWIS, H.R. and C.H. PAPADIMITRIOU, Symmetric space-bounded computation, *Theoret. Comput. Sci.* **19** (1982) 161–187.

[176] LICHTENSTEIN, D. and M. SIPSER, GO is polynomial-space hard, *J. Assoc. Comput. Mach.* **27** (1980) 393–401.

[177] LONG, T.J., Strong nondeterministic polynomial-time reducibilities, *Theoret. Comput. Sci.* **21** (1982) 1–25.

[178] LYNCH, N., Log space machines with multiple oracle tapes, *Theoret. Comput. Sci.* **6** (1978) 25–39.

[179] MAHANEY, S.R., Sparse complete sets for NP: solution of a conjecture of Berman and Hartmanis, *J. Comput. System Sci.* **25** (1982) 130–143.

[180] MAHANEY, S.R., Sparse sets and reducibilities, in: R.V. Book, ed., *Studies in Complexity Theory* (Wiley, New York, 1986) 63–118.

[181] MAHANEY, S.R. and P. YOUNG, Reductions among polynomial isomorphism types, *Theoret. Comput. Sci.* **39** (1985) 207–224.

[182] MANDERS, K. and L. ADLEMAN, NP-complete decision problems for binary quadratics, *J. Comput. System Sci.* **16** (1978) 168–184.

[183] MATHON, R., A note on the graph isomorphism counting problem, *Inform. Process. Lett.* **8** (1979) 131–132.

[184] MATIJACEVIC, Y., Enumerable sets are Diophantine, *Dokl. Akad. Nauk. SSSR* **191** (1970) 279–282 (in Russian); English translation in: *Soviet Math. Doklady* **11** (1970) 354–357.

[185] MAYR, E.W. and A.S. SUBRAMANIAN, The complexity of circuit value and network stability, in: *Proc. Structure in Complexity Theory* (*4th Ann. IEEE Conf.*) (1989) 114–123.

[186] MEYER, A.R., Weak monadic second order theory of successor is not elementary recursive, Manuscript, 1973.

[187] MEYER, A.R. and L.J. STOCKMEYER, The equivalence problem for regular expressions with squaring requires exponential time, in: *Proc. 13th Ann. IEEE Symp. on Switching and Automata Theory* (1972) 125–129.

[188] MILLER, G.L., Riemann's hypothesis and tests for primality, *J. Comput. System Sci.* **13** (1976) 300–317.

[189] NISAN, N. and A. WIGDERSON, Hardness vs. randomness, in: *Proc. 29th Ann. IEEE Symp. on Foundations of Computer Science* (1988) 2–11.

[190] PAPADIMITRIOU, C.H., Games against nature, in: *Proc. 24th Ann. IEEE Symp. on Foundations of Computer Science* (1983) 446–450; revised version appeared as: C.H. PAPADIMITRIOU, Games against nature, *J. Comput. System Sci.* **31** (1985) 288–301.

[191] PAPADIMITRIOU, C.H., On the complexity of unique solutions, *J. Assoc. Comput. Mach.* **31** (1984) 392–400.

[192] PAPADIMITRIOU, C.H. and K. STEIGLITZ, *Combinatorial Optimization: Algorithms and Complexity* (Prentice-Hall, Englewood Cliffs, NJ, 1982).

[193] PAPADIMITRIOU, C.H. and D. WOLFE, The complexity of facets resolved, *J. Comput. System Sci.* **37** (1988) 2–13.

[194] PAPADIMITRIOU, C.H. and M. YANNAKAKIS, The complexity of facets (and some facets of complexity), *J. Comput. System Sci.* **28** (1984) 244–259.

[195] PAPADIMITRIOU, C.H. and M. YANNAKAKIS, A note on succinct representations of graphs, *Inform. and Control* **71** (1986) 181–185.

[196] PAPADIMITRIOU, C.H. and S. ZACHOS, Two remarks on the power of counting, in: *Proc. 6th GI Conf. on Theoretical Computer Science*, Lecture Notes in Computer Science, Vol. 145 (Springer, Berlin, 1983) 269–276.

[197] PAUL, W.J., N. PIPPENGER, E. SZEMERÉDI and W.T. TROTTER, On determinism versus non-determinism and related problems, in: *Proc. 24th Ann. IEEE Symp. on Foundations of Computer Science* (1983) 429–438.

[198] PIPPENGER, N., On simultaneous resource bounds, in: *Proc. 20th Ann. IEEE Symp. on Foundations of Computer Science* (1979) 307–311.

[199] PRATT, V., Every prime has a succinct certificate, *SIAM J. Comput.* **4** (1975) 214–220.

[200] PROVAN, J.S., The complexity of reliability computations in planar and acyclic graphs, *SIAM J. Comput.* **15** (1986) 694–702.

[201] RABIN, M.O., Probabilistic algorithm for testing primality, *J. Number Theory* **12** (1980) 128–138.

[202] RACKOFF, C., Relativized questions involving probabilistic algorithms, *J. Assoc. Comput. Mach.* **29** (1982) 261–268.

[203] RAZBOROV, A.A., Lower bounds for the size of bounded-depth networks over a complete basis with logical addition, *Mat. Zametki* **41** (1987) 598–607 (in Russian); English translation in: *Math. Notes Acad. Sci. USSR* **41** (1987) 333–338.

[204] REGAN, K., The topology of provability in complexity theory, *J. Comput. System Sci.* **36** (1988) 384–432.

[205] REIF, J.H., Universal games of incomplete information, in: *Proc. 11th Ann. ACM Symp. on Theory of Computing* (1979) 288–308.

[206] REIF, J.H., Complexity of the mover's problem and generalizations, in: *Proc. 20th Ann. IEEE Symp. on Foundations of Computer Science* (1979) 421–427; a formal version of this paper appeared as: Complexity of the generalized movers problem, in: J. Schwartz, ed., *Planning Geometry and Complexity of Robot Motion* (Ablex, Norwood, NJ, 1985) 421–453.

[207] REIF, J.H., Logarithmic depth circuits for algebraic functions, *SIAM J. Comput.* **15** (1986) 231–282.

[208] REIF, J.H., On threshold circuits and polynomial computation, in: *Proc. Structure in Complexity Theory* (*2nd Ann. IEEE Conf.*) (1987) 118–123.

[209] RIVEST, R., Cryptography, in: J. van Leeuwen, ed., *Handbook of Theoretical Computer Science, Vol. A* (North-Holland, Amsterdam, 1990) 717–755.

[210] ROBSON, J.M., *N* by *N* checkers is Exptime complete, *SIAM J. Comput.* **13** (1984) 252–267.

[211] RUDICH, S., Limits on provable consequences of one-way functions, Doctoral Dissertation, Univ. of California, Berkeley, CA, 1988.

[212] RUZZO, W.L., Tree-size bounded alternation, *J. Comput. System Sci.* **21** (1980) 218–235.

[213] RUZZO, W.L., On uniform circuit complexity, *J. Comput. System Sci.* **22** (1981) 365–383.

[214] SALOMAA, A., Formal languages and power series, in: J. van Leeuwen, ed., *Handbook of Theoretical Computer Science, Vol. B* (North-Holland, Amsterdam, 1990).

[215] SANTHA, M., Relativized Arthur–Merlin versus Merlin–Arthur games, *Inform. and Comput.* **80** (1989) 44–49.

[216] SAVITCH, W.J., Relationship between nondeterministic and deterministic tape classes, *J. Comput. System Sci.* **4** (1970) 177–192.

[217] SAVITCH, W.J., Nondeterministic $\log n$ space, in: *Proc. 8th Ann. Princeton Conf. on Information Sciences and Systems*, Dept. of Electrical Engineering, Princeton Univ., Princeton, NJ (1974) 21–23.

[218] SAVITCH, W.J., A note on relativized log space, *Math. Systems Theory* **16** (1983) 229–235.

[219] SCHAEFER, T.J., Complexity of some two-person perfect-information games, *J. Comput. System Sci.* **16** (1978) 185–225.

[220] SCHÖNING, U., Graph isomorphism is in the low hierarchy, in: *Proc. 4th Symp. on Theoretical Aspects of Computing*, Lecture Notes in Computer Science, Vol. 247 (Springer, Berlin, 1986) 114–124.

[221] SCHWARTZ, J.T., Fast probabilistic algorithms for verification of polynomial identities, *J. Assoc. Comput. Mach.* **27** (1980) 710–717.

[222] SEIFERAS, J.I., Relating refined space complexity classes, *J. Comput. System Sci.* **14** (1977) 100–129.

[223] SEIFERAS, J.I., M.J. FISCHER and A.R. MEYER, Separating nondeterministic time complexity classes, *J. Assoc. Comput. Mach.* **25** (1978) 146–167.

[224] SHMOYS, D.B. and E. TARDOS, Computational complexity, in: *Handbook of Combinatorics* (North-Holland, Amsterdam, to appear).

[225] SIMON, J., On some central problems in computational complexity, Doctoral Thesis, Dept. of Computer Science, Cornell Univ., Ithaca, NY, 1975.

[226] SIMON, J., Space-bounded probabilistic Turing machine complexity classes are closed under complement, in: *Proc. 13th Ann. ACM Symp. on Theory of Computing* (1981) 158–167.

[227] SIMON, J., On tape-bounded probabilistic Turing machine acceptors, *Theoret. Comput. Sci.* **16** (1981) 75–91.

[228] SIPSER, M., On relativization and the existence of complete sets, in: *Proc. Internat. Coll. on Automata, Languages, and Programming*, Lecture Notes in Computer Science, Vol. 140 (Springer, Berlin, 1982) 523–531.

[229] SIPSER, M., Borel sets and circuit complexity, in: *Proc. 15th Ann. ACM Symp. on Theory of Computing* (1983) 61–69.

[230] SIPSER, M., A complexity theoretic approach to randomness, in: *Proc. 15th Ann. ACM Symp. on Theory of Computing* (1983) 330–335.

[231] SKYUM, S., and L.G. VALIANT, A complexity theory based on Boolean algebra, *J. Assoc. Comput. Mach.* **32** (1985) 484–502.

[232] SMOLENSKI, R., Algebraic methods in the theory of lower bounds for Boolean circuit complexity, in: *Proc. 19th Ann. ACM Symp. on Theory of Computing* (1987) 77–82.

[233] SOLOVAY, R. and V. STRASSEN, A fast Monte-Carlo test for primality, *SIAM J. Comput.* **6** (1977) 84–85.

[234] STOCKMEYER, L., The polynomial time hierarchy, *Theoret. Comput. Sci.* **3** (1976) 1–22.

[235] STOCKMEYER, L., On approximation algorithms for #P, *SIAM J. Comput.* **14** (1985) 849–861.

[236] STOCKMEYER, L., Classifying the computational complexity of problems, *J. Symbolic Logic* **52** (1987) 1–43.

[237] STOCKMEYER, L.J. and A.R. MEYER, Word problems requiring exponential time, in: *Proc. 5th Ann. ACM Symp. on Theory of Computing* (1973) 1–9.

[238] SUDBOROUGH, I.H., On the tape complexity of deterministic context-free languages, *J. Assoc. Comput. Mach.* **25** (1978) 405–414.

[239] SZELEPCSÉNYI, R., The method of forcing for nondeterministic automata, *Bull. EATCS* **33** (1987) 96–100.

[240] TODA, S., On the computational power of PP and $\oplus$P, in: *Proc. 30th Ann. IEEE Symp. on Foundations of Computer Science* (1989) 514–519.

[241] TORÁN, J., Structural properties of the counting hierarchies, Doctoral Dissertation, Facultat d'Informatica, UPC Barcelona, 1988.

[242] VALIANT, L.G., Relative complexity of checking and evaluating, *Inform. Process. Lett.* **5** (1976) 20–23.

[243] VALIANT, L.G., The complexity of computing the permanent, *Theoret. Comput. Sci.* **8** (1979) 189–201.

[244] VALIANT, L.G., The complexity of enumeration and reliability problems, *SIAM J. Comput.* **8** (1979) 410–421.

[245] VALIANT, L.G. and V.V. VAZIRANI, NP is as easy as detecting unique solutions, *Theoret. Comput. Sci.* **47** (1986) 85–93.

[246] VAN EMDE BOAS, P., Machine models and simulations, in: J. van Leeuwen, ed., *Handbook of Theoretical Computer Science, Vol. A* (North-Holland, Amsterdam, 1990) 1–66.

[247] VAN LEEUWEN, J., Graph algorithms, in: J. van Leeuwen, ed., *Handbook of Theoretical Computer Science, Vol. A* (North-Holland, Amsterdam, 1990) 525–631.

[248] VAZIRANI, U.V. and V.V. VAZIRANI, A natural encoding scheme proved probabilistic polynomial complete, *Theoret. Comput. Sci.* **24** (1983) 291–300.

[249] VENKATESWARAN, H. and M. TOMPA, A new pebble game that characterizes parallel complexity classes, *SIAM J. Comput.* **18** (1989) 533–549.

[250] WAGNER, K.W., Bounded query computation, in: *Proc. Structure in Complexity Theory (3rd Ann. IEEE Conf.)* (1988) 260–277.

[251] WANG, H., Proving theorems by pattern recognition, *Bell Systems Tech. J.* **40** (1961) 1–42.

[252] WECHSUNG, G., On the Boolean closure of NP, in: *Proc. Internat. Conf. on Fundamentals of Computation Theory*, Lecture Notes in Computer Science, Vol. 199 (Springer, Berlin, 1985) 485–493.

[253] WILSON, C.B., Relativized circuit complexity, in: *Proc. 24th Ann. IEEE Symp. on Foundations of Computer Science* (1983) 329–334.

[254] WILSON, C.B., A measure of relativized space which is faithful with respect to depth, *J. Comput. System Sci.* **36** (1988) 303–312.

[255] WRATHALL, C., Complete sets for the polynomial time hierarchy, *Theoret. Comput. Sci.* **3** (1976) 23–34.

[256] YAO, A.C., Theory and applications of trapdoor functions, in: *Proc. 23rd Ann. IEEE Symp. on Foundations of Computer Science* (1982) 80–91.

[257] YAO, A.C.-C., Separating the polynomial-time hierarchy by oracles, in: *Proc. 26th Ann. IEEE Symp. on Foundations of Computer Science* (1985) 1–10.

[258] YOUNG, P., Some structural properties of polynomial reducibilities and sets in NP, in: *Proc. 15th Ann. ACM Symp. on Theory of Computing* (1983) 392–401.

[259] ZACHOS, S., Collapsing probabilistic polynomial hierarchies, in: *Proc. Conf. on Complexity Theory* (1983) 75–81.

[260] ZACHOS, S., Probabilistic quantifiers and games, *J. Comput. System Sci.* **36** (1988) 433–451.

[261] ZACHOS, S. and H. HELLER, A decisive characterization of BPP, *Inform. and Control* **69** (1986) 125–135.

[262] BABAI, L., Talk presented at the 21st Annual Symposium on Foundation of Computer Science, San Juan, Puerto Rico, October 29, 1979 (not in the Proceedings).

[263] BABAI, L., L. FORTNOW and C. LUND, Non-deterministic exponential time has two-prover interactive protocols, Manuscript, 1990.

[264] CHEN, J., A new complete problem for DSPACE(log n), *Discrete Applied Math.* **25** (1989) 19–26.

[265] COOK, S.A., Personal communication correcting typographical error in [63].

[266] COOK, S.A. and P. MCKENZIE, Problems complete for deterministic logarithmic space, *J. Algorithms* **8** (1987) 385–394.

[267] LUND, C., L. FORTNOW, H. KARLOFF and N. NISAN, The polynomial time hierarchy has interactive proofs, Announcement by electronic mail, December 13, 1989.

[268] SHAMIR, A., IP = PSPACE, Manuscript, 1989.

CHAPTER 3

Machine-Independent Complexity Theory

Joel I. SEIFERAS

Computer Science Department, University of Rochester, Rochester, NY 14627, USA

Contents

HANDBOOK OF THEORETICAL COMPUTER SCIENCE
Edited by J. van Leeuwen

1. Introduction

The familiar measures of computational complexity are *time* and *space*. Generally speaking, time is the number of discrete steps in a computation, and space is the number of distinct storage locations accessed by the instructions of the computation.

What we would like to classify is the computational complexity of an entire problem; but a problem has a countably infinite number of "instances", and may be solved in many different ways, on many different computer architectures. Since an overall solution has a time complexity and a space complexity for each instance, we can specify its complexity as a *function* on problem instances, or, for more gross purposes, on some measure of the "size" of problem instances. To specify the computational time or space used by a *partial* solution, we can use a *partial* function.

Each separate solution to a problem will have its own computational complexity. What should be *the* time or space complexity of the problem? The natural desire is to choose the *smallest* such complexity as the inherent one. A set of complexities, however, since they are functions and not mere members of the set N of natural numbers, need not necessarily have a smallest member; there might be "trade-offs", or even infinite descending chains. If no solution for a problem has the least complexity, we say that the problem "has speed-up", basing our terminology on the special case of *time* complexity. We show in this chapter that there really are computational problems with arbitrarily dramatic speed-up; and we show, on the other hand, that there are problems without it (i.e., that do have essentially optimum solutions). Moreover, we show that the existence of these phenomena is "machine-independent", depending neither on the choice of complexity measure nor on the choice of computer architecture, within reason.

Another natural, but very different, notion of complexity is *program size* [4]. Program size *is* just a single natural number, so a problem does have a well-defined best complexity of this static variety, at least for a particular, fixed computer architecture. It turns out that the essential features of this notion of complexity are again "machine-independent" [4, 13]. What is more remarkable, however, is the drastic inherent *trade-offs* that can exist between program size and dynamic computational complexity [24].

Although we are interested in machine-independent phenomena, we will start in a very specific, concrete setting. Our results in this setting can be viewed initially as mere illuminating examples. In the end, however, we will derive most of our machine-independent results as corollaries of these machine-dependent ones, a method first used by Hartmanis and Hopcroft [10]. There is an alternative approach, based on M. Blum's two simple axioms [3] (see below), that can be viewed as machine-independent from the outset. For more complete coverage of both approaches, and for many additional research directions, the reader is encouraged to consult additional surveys, overviews, and research literature. The contemporary presentations in [7, 29] are additional good sources of pointers into the literature to date.

2. Simple Turing machines and space complexity

The machines we consider each consist of a *finite-state program* with access to an *input tape* and a single *storage tape*. Recorded on the input tape is an *input word*, a nonnull, finite string of characters from some finite *input alphabet*. (If Σ is the alphabet, then Σ^+ denotes the set of all such strings.) Recorded on the storage tape is a string of characters from the fixed, binary alphabet $\{0, 1\}$. The initial content of the storage tape is the trivial word "0" of length 1. (In general, we denote the length of a word x by $|x|$.) A separate *tape head* is maintained in some position, initially the leftmost one, on each of the two tapes. The finite-state program consists of a finite set of *(control) states*, one of which is designated as the initial one, and a finite function indicating how each next computational "action" depends on the visible "display" of the current total state of the entire machine. (In this chapter, we consider only *deterministic* automata; so we assume each next action is fully determined by the current display.) The *display* consists of the current control state of the program, the symbol being "scanned" by each tape head, and indication whether each tape head is scanning its tape's leftmost or rightmost symbol. Each *action* can change the control state and also do any of the following: write 0 or 1 over the scanned symbol on the storage tape; shift a tape head left or right respectively, if it is not already at the left or right end of its tape; and append an additional 0 onto the right end of the storage tape, if the head is currently at that end. (Because of this special role for 0, we sometimes refer to it as the "blank" storage tape symbol.) Note that the machine cannot modify or extend its input tape, but that it can modify its storage tape and extend it arbitrarily far to the right. Let us call such a machine an *STM* (for "simple Turing machine").

From the informal description above, it should be clear how an STM starts and runs on an arbitrary word over its input alphabet. To view it as computing some function or partial function of its input, we must adopt some output convention. For a finite-valued function, we could specify a distinct "final" control state for each possible value, and say that the computed value is the one (if any) whose corresponding final state is first entered. For a more general function, the convention could be to leave the output value encoded somehow at the left end of the storage tape if and when a designated final state is first entered.

Under the output conventions suggested above, we can view computations as *ending* when they first enter final states, since subsequent steps do not affect their output. This finally leads to a definition of space complexity: *the space* $\mathrm{Space}_M(x)$ *used by STM M on input x* is the length of M's storage tape if and when its computation on that input ends. $\mathrm{Space}_M(x)$ is undefined if the computation by M on input x does not end. We denote this lack of definition by $\mathrm{Space}_M(x)=\infty$, even though this makes it look like $\mathrm{Space}_M(x)$ is defined, and even though a nonterminating computation might not ever use much space (if it is in a "loop"). (Again the terminology follows *time* complexity more closely: A nonterminating computation does consume infinite time.) A benefit is that our implicit conventions regarding arithmetic and comparisons when some of the operands might be undefined are consistent with viewing ∞ as "the value infinity".

Actually, it is safe to ignore looping (nonterminating computation on bounded space). From each STM that might loop, we can construct an *equivalent* one that does not loop (equivalent in the sense that it computes the same partial function, in exactly the same space). Moreover, the nonlooping STMs obtained in this way are easily recognizable, syntactically. For a description of this construction, see [26].

Our first example of the speed-up phenomenon is a simple but universal one:

the space complexity of any solution to *any* problem *can be reduced by any constant* c; i.e., whatever an STM can do in space $S(x)$, another STM can do in space $S'(x) = \max(1, S(x) - c)$. By induction, it suffices to prove this for $c = 1$.

PROPOSITION (additive-constant speed-up). *Whatever an STM M can do in space $S(x)$, another STM M' can do in space* $S'(x) = \max(1, S(x) - 1)$.

PROOF. The idea of the simulation is for M' to maintain the one leftmost bit of M's storage tape in its finite control, and to maintain only the rest of M's storage tape, including head position, on its own storage tape. Of course, exceptions must be made until the length of M's storage tape reaches 2, and whenever that tape's head scans its leftmost symbol. At each successive instant, the control state of M' will consist of the current control state of M, the leftmost bit of M's storage tape, indication whether M's storage tape has yet reached length 2, and indication whether M's storage tape head currently scans the leftmost symbol on its tape. It is inductively straightforward for M' to maintain all this and simultaneously to determine and simulate the successive transitions of M. □

3. Recursion, padding and compression

The second phenomenon is sometimes referred to as "compression". Since the additive-constant speed-up phenomenon above is universal, we cannot expect to find a problem whose optimum complexity is completely compressed down to a single, precise space bound. It turns out, however, that additive-constant speed-up is the only such concession we have to make.

In order to conveniently ignore additive-constant speed-up, we define a slightly "blurred" partial order $\preccurlyeq$ on functions and partial functions, so that $f(x) \preccurlyeq g(x)$ holds if and only if there is some additive constant c such that, for every x, $f(x) \leqslant c + g(x)$. In these terms, we have seen that we cannot hope to avoid speed-up from $S(x)$ to $S'(x)$ if $S'(x) \succcurlyeq S(x)$.

COMPRESSION THEOREM FOR STM SPACE. *Above any computable function, there is a computable space bound $S: \Sigma^+ \to N$ for which there is a predicate $P: \Sigma^+ \to \{0, 1\}$, computable within space $S(x)$, but not computable within space $S'(x)$* unless $S'(x) \succcurlyeq S(x)$.

Before we turn to the proof itself, let us introduce some additional technical machinery, notation, and terminology. The proof will involve diagonalization by an STM over other STMs, so we will need a way for an STM to represent other STMs on its tapes. Straightforward formalization and binary encoding of our informal definition of STMs computing partial functions from Σ^+ to $\{0, 1\}$ yields an easily decodable binary string, or "program", describing each such machine. For each binary string e, let M_e be the machine described by e. (Not every binary string will describe an (explicitly nonlooping) STM via our formalization; but the ones that are not "well-formed" in this sense will be easily recognizable, and we adopt the convention that each of these does implicitly "describe" some one fixed machine of our choice.) Let $\varphi_e(x)$ denote the partial function computed by M_e, and let $S_e(x) = \mathrm{Space}_{M_e}(x)$. (Recall that the latter is defined) on precisely the same domain as the former.) Finally, say that a total space bound $S: \Sigma^+ \to N$ is *constructible* if there is an STM that, on each input string x, extends its storage tape to length exactly $S(x)$, zeros it out, and halts, having "constructed" or "laid out" space $S(x)$. For example, n^3, 2^n, $n!$, $n^{2/3}$, $\log_2 n \log_2 \log_2 n$, and all other common superlogarithmic functions of $n = |x|$ turn out to be constructible (within an additive constant), including all the ones in Grzegorczyk's class $\mathcal{E}^2$ [21, Section 4].

Several of our proofs will involve procedures that "invoke themselves recursively". Implementation of such procedures is easiest and most transparent on "stored program" computers. Now that we do see how to develop the notion of a machine's "program", let us go back and slightly revise our STM model to include a "stored program" feature. First, we syntactically add one new "action" to the original definition. Our *intention* is for the new action to involve *writing the next bit of the machine's own program code*, if there is a next one, over the scanned symbol on the storage tape. With just the syntactic definition, we can already proceed to assign program codes as outlined above; and, having done so, we are finally able to complete the semantic part of the definition in the intended way. Noting that we can easily transform an STM of either variety into an equivalent one of the other variety, we may as well adopt the enhanced version as our standard.

To facilitate the implementation of even *mutual* recursion, we place one more, explicit but harmless, constraint on the assignment of program codes: at the end of every well-formed code, there should be a clearly delimited block of "unreachable code", more appropriately viewed as a "data segment". To design a pair of mutually recursive STMs then, we can simply write programs that differ only in the first bit of this data. Each of the two machines can do its own thing, according to the value of this bit; and each can also easily convert its own, directly available, program code to the other's, for the sake of mutual recursion.

To illustrate the utility of our newly enhanced STM model in the simplest possible setting, we give a transparent alternative to the usual obscure proof of the "fixed-point" version of Kleene's recursion theorem [23]. The result itself will turn out to be machine-independent.

RECURSION THEOREM. *For each total, computable program transformation* $\tau: \{0,1\}^+ \to \{0,1\}^+$, *there is a fixed point e, itself computable from a program code for* τ, *such that* $\varphi_e = \varphi_{\tau(e)}$.

PROOF. For the desired fixed point, just take a code e_0 for a program that retrieves its own code (e_0), transforms it (to $\tau(e_0)$), and then simulates the resulting machine ($M_{\tau(e_0)}$) on the given input. $\square$

One last useful tool lets us transform programs into "equivalent", "padded" ones that are longer. Codes e_1 and e_2 are *equivalent* if their STMs are equivalent (i.e., if $\varphi_{e_1} = \varphi_{e_2}$ and

$S_{e_1}=S_{e_2}$). To facilitate padding that is this innocuous even in the presence of recursion, we make one last simple change to our scheme for assignment of program codes and the semantics of the program-retrieval instruction: We add a *second* clearly delimited block of "unreachable code", this one *not even visible to the program-retrieval instruction.*

PADDING LEMMA. *There is a computable, one-to-one "padding" transformation* $p: \{0,1\}^+ \to \{0,1\}^+$ *such that, for each STM code e, p(e) is a longer code equivalent to e. Moreover, the corresponding* unpadding *transformation, which transforms e to the shortest code u(e) from which e can be obtained by repeated application of p, is so simple that an STM with access to e can simulate access to u(e) at no additional cost in space.*

PROOF. If e is a well-formed program code with empty padding block, then denote by $e(x)$ the equivalent program code with padding block x. The idea is to pad by adding one more 0 to the padding block ($p(e(x))=e(x0)$), and to unpad by removing all trailing 0s from the padding block. The only problem is that this ignores all the ill-formed program codes.

We pad ill-formed program codes to well-formed ones that are not otherwise further unpaddable, as follows: Fix some explicit, well-formed code e_0, with empty padding block, for the default STM that each ill-formed code implicitly describes. Then let $p(e)=e_0(e1)$ when e is ill-formed. The unpadding rule is now modified so that $u(e_0(e10^i))=e$ (rather than the originally proposed $e_0(e1)$) in the case that e is ill-formed. □

PROOF OF THE COMPRESSION THEOREM. Any constructible S will do. There is a constructible space bound above any computable function, since an STM can compute the function and then go on to extend its storage tape according to the computed value.

We define $P(x)$ for one x at a time, considering input strings x in their natural order $<$. (The natural order is by length, with strings of the same length ordered lexicographically.) To keep P hard enough, we aim to build in an exception to $P=\varphi_e$ whenever $S_e \not\succeq S$, by "canceling" e on some input x. To cancel e on input x, we aim simply to choose $P(x)$ to be the least member of $\{0,1\}$ not equal to $\varphi_e(x)$. On input x, we aim to cancel the least e (in that same natural order of binary strings) that should be, but has not yet been, canceled.

Naively, we might try, on input x, to cancel $c(x)=\min\{e \mid S_e \not\succeq S \text{ and } e \text{ is not } c(x') \text{ for any earlier } x'\}$. Since P has to be computable and not *too* complex, however, this does not work. To salvage the plan, we must exploit its slack:

(1) There is no harm in canceling a few *extra* e's, perhaps based on mere "suspicion" that $S_e \not\succeq S$.

(2) There is no harm in canceling some e's *more than once*, perhaps saving much of the space necessary for a full check into past accomplishments.

(3) There is no harm in canceling some e's *earlier* or *later* than we had planned, perhaps saving much of the space necessary to determine exactly the originally intended ordering.

(4) There is no harm in *never* explicitly canceling some of those e's for which φ_e is not total, since P will differ from these merely by being total.

Taking advantage of all these observations, we decide, on input x, to cancel

$$c(x) = \min\{e \mid S(x) \text{ is enough space to simulate } M_e \text{ on input } x, \\ \text{but } \textit{not} \text{ enough to discover that } e = c(x') \text{ for some } x' < x\}.$$

(When this set is empty, arbitrarily cancel some fixed e with S_e identically equal to 1). Note that it is the first condition that raises suspicion that maybe $S_e \not\succ S$. The second condition might be satisfied either because space $S(x)$ is inadequate for the review, or simply because e has not yet been canceled; our choice of $c(x)$ is based on the latter hope.

The key to the tightness of our compression theorem is that a single "universal" STM, given an arbitrary STM program code e on its storage tape, can simulate M_e in space equal to just $S_e(x)$ plus some constant c_e that depends only on e. Assuming an appropriate (but straightforward) formalization and encoding, in fact, $c_e = |e|$ will suffice; the idea is for the universal STM to maintain on its storage tape an up-to-date replica of M_e's storage tape, but with a copy of a slightly modified version of the self-delimiting program code e inserted at the current head position. The main reason for modification of the code is to indicate which is the current control state of M_e; redundant, and hence temporarily changeable, copies of each bit facilitate such marking. Such marking also makes it possible to compare state labels, in order to find each *next* control state of M_e.

In terms of the simulation we have outlined, the condition that "$S(x)$ is enough space to simulate M_e on input x" becomes simply "$S(x) \geqslant S_e(x) + |e|$". Once $e = c(x)$ is known, it will therefore be straightforward by simulation within space $S(x)$ to compute $\varphi_e(x)$ for the sake of diagonalization.

To compute $c(x)$ within space $S(x)$, we use a recursive procedure. The first step is to lay out space exactly $S(x)$. Within that space, never again extending the storage tape, we write down each successive binary string e until we find one that qualifies to be $c(x)$, or until the next one would not fit within the allocated space. For each such e, we check whether universal simulation of M_e on input x overflows the allocated space. (Except when the simulation fills the entire tape, this requires recognition of a simulated right end that is not the actual right end of the storage tape. A suitable convention under such circumstances is to maintain a 1 in the first unused position, and 0s in all the rest.)

If the universal simulation does overflow, then e does not satisfy the condition $S(x) \geqslant S_e(x) + |e|$, and we go on to the next value of e. If it does not overflow, then we go on to check the second condition, leaving the self-delimited string e written at the beginning of the storage tape. Within the remaining space $S(x) - |e|$, we write down a self-delimited version of each successive binary string x' that precedes x, until we find one with $c(x') = e$, or until the next x' would not fit within the allocated space. To check each x', we try recursively to calculate $c(x')$ within the remaining available space. This involves retrieval of the code d for the STM we are designing and universal simulation of M_d on x'; it is easy for the simulation to recognize e and the representation of x' as a bound, because they are both self-delimiting. If we run out of space before finding $c(x')$, or if $c(x') \neq e$, then we go on to the next value of x'; but, if we discover that $c(x') = e$, then e does not satisfy the second condition, and we go on to *its* next value.

Having thus shown that an STM can compute P within space S, we turn to the converse, that M_e *cannot* compute P if $S_e \not\succ S$. We have already noted that M_e cannot compute P unless φ_e is total; and, by explicit design, M_e does not compute P if e gets canceled. Therefore, it suffices to show that e gets canceled if φ_e is total and $S_e \not\succcurlyeq S$. Noting that the latter condition implies that both $S(x)-S_e(x)$ and $S(x)$ itself are simultaneously unbounded, just consider the first x so large that every $e'<e$ that ever gets canceled is $c(x')$ for some $x'<x$, and such that $S(x)$ is enough space both to simulate M_e on input x ($S(x)-S_e(x) \geqslant |e|$) and to discover an earlier cancellation of each such e'. For that x, $e=c(x)$. □

Before leaving the compression phenomenon, we should note that restricting space can have a drastic effect even on predicates that do remain computable. The effect is in terms of *minimum program size.* If we define

$$K(P,S)=\min\{|e| \mid M_e \text{ computes predicate } P \text{ within space } S\},$$

for example, then we can prove the following abrupt trade-off between program size and space complexity [17].

THEOREM. *Above any computable function, and for any "limit-computable" function* $h: N \to N$ *("h" for "huge"), there is a computable space bound* $S: \Sigma^+ \to N$ *for which there is a predicate* $P: \Sigma^+ \to \{0,1\}$, *computable within every space bound (even a constant one), but such that* $K(P,S') > h(K(P,S))$ *unless* $S'(x) \succcurlyeq S(x)$.

(h is defined to be *limit-computable* if $h(n)=\lim_{m\to\infty} g(m,n)$ for each n, for some computable function g.)

PROOF (*idea*). Adapt the proof of the Compression Theorem by adding the following additional requirement for membership of e in the set of candidates for $c(x)$:

it "looks like" $|e|$ does *not* exceed $h(|e_0|)$,
where e_0 is the program code for the STM being designed.

To implement the "looks like" aspect, make use of the approximation $g(m,|e_0|)$, for m as large as possible within the available space. If $S(x)$ tends to infinity, then the set will be empty when x gets sufficiently large, so that we are free to make $P(x)=0$ almost everywhere (i.e., with only finitely many exceptions). □

Through mutual recursion, one can work out *multi*step versions of the above trade-off. See [24] for the details, and for many more results on trade-offs between program size and computational complexity, in some very general settings.

4. Gaps and arbitrary speed-up

Before we turn explicitly to the speed-up phenomenon, let us re-examine the specification of the predicate P in the proof of the Compression Theorem. The mere *definition* of the cancellation function, and hence of P itself, does not depend on the constructibility of S. If S is merely *computable*, we still get a computable predicate P that is not computable within space S_e unless $S_e \succcurlyeq S$. On the other hand, for P to *be* computable within space $S_e \geqslant S$, it suffices to be able to delimit space $S(x)$ within already constructed, or "preallocated", space $S_e(x)$. If computable S is delimitable in this sense within *every* $S_e \geqslant S$, then let us call S *subconstructible*. From each subconstructible S, then, our specification gives us a predicate P that is computable within space S_e if and only if $S_e \succcurlyeq S$.

4.1. LEMMA. *If $S: \Sigma^+ \to N$ is subconstructible, then there is a predicate $P: \Sigma^+ \to \{0, 1\}$, computable within space S_e if and only if $S_e \succcurlyeq S$.*

We will see below that this slight generalization of the *Compression* Theorem is a major step in the proof of the *Speed-up* Theorem.

REMARK. It follows from the Fundamental Theorem [11, 12, 19] (discussed below) that the subconstructibility hypothesis in Lemma 4.1 can be weakened all the way to mere computability. For the *Compression Theorem*, on the other hand, it is *not* enough for S to be computable. The disparity is explained by the "gap" phenomenon.

The weak version of the gap phenomenon is that there are computable space bounds $S: \Sigma^+ \to N$ such that whatever an STM can do within space S, it can actually do in much less space, so that there is a "complexity gap" below S, in which no function's inherent space complexity lies. In fact it is easy to design S with the even stronger property that not even one particular STM can *use* space that lies in the gap infinitely often [28, 6]; thus there are computable space bounds with no constructible space bounds anywhere nearby, explaining the disparity described above. For a mundane "doubly exponential" gap, for example, one can design S as follows: for each x, choose $S(x)$ so that the interval from $S(x)$ down to $\log\log S(x)$ *misses* $S_e(x)$ for every $e \leqslant x$. It is straightforward to check whether a prospective choice is satisfactory; and there are only a finite number of spoiling values $S_e(x)$ to "dodge". With more effort, one can obtain a slightly (but necessarily) weaker gap as big as any "computable operator that preserves totalness" [8, 30]. We state this result below, without proof.

A *computable operator* is a mapping from computable partial functions to computable partial functions that can be specified by a computable program transformation. More formally, using STM codes as our programs, F is a computable operator if and only if there is a computable transformation $\tau: \{0, 1\}^+ \to \{0, 1\}^+$ for which $F(\varphi_e) = \varphi_{\tau(e)}$. (Note that not every computable transformation τ specifies an operator: we must have $\varphi_{\tau(e)} = \varphi_{\tau(e')}$ whenever $\varphi_e = \varphi_{e'}$.) If the image of each total function is total, then we say that *F preserves totalness*.

GAP THEOREM FOR STM SPACE. *For each computable operator F that preserves totalness, there are computable space bounds $S: \Sigma^+ \to N$ and $S' \leqslant S$ so much smaller that $F(S') \leqslant S$, but such that, for each e, if $S_e \leqslant S$ holds almost everywhere, then so does $S_e \leqslant S'$.*

Finally, we state our Speed-up Theorem, also in terms of computable operators [18].

"SPEED-UP" THEOREM FOR STM SPACE. *For each computable operator F that preserves totalness, there is a computable predicate $P:\Sigma^+\to\{0,1\}$ with "F-speed-up": Whenever P is computable within a space bound S, it is also computable within a space bound $S'\leqslant S$ so much smaller that $F(S')\leqslant S$ holds almost everywhere.*

By induction, of course, the Speed-up Theorem will yield an entire infinite, rapidly descending sequence $S, S', S'', \ldots$ Note that the finitely many exceptions to each "descent" $F(S^{(i+1)})\leqslant S^{(i)}$ cannot be avoided: even if $F(f)$ were just $f+1$, for each particular argument x we would reach a contradiction to *perfect* speed-up after just $S(x)$ successive applications. On the other hand, one major step in the proof is showing that, with only finitely many exceptions at each step, such rapid descent is possible.

4.2. LEMMA. *For each computable operator F that preserves totalness, there is a sequence of functions $g_1\geqslant g_2\geqslant g_3\geqslant\cdots$ $(g_i:\Sigma^+\to N)$ such that, for each i, $F(g_{i+1})\leqslant g_i$ holds almost everywhere. Moreover, the sequence $\{g_i\}$ is uniformly computable, and even "uniformly subconstructible", as defined below.*

A sequence $\{g_i\}$ *is uniformly subconstructible* if some single STM, given x on its input tape and i on its otherwise blank preallocated storage tape of length $m\geqslant|i|$, can halt with the storage tape head at position $\min\{m, g_i(x)\}$ and with no net change to the *content* of the storage tape. The need for some such condition is dictated by Lemma 4.3, the one missing link to complete a proof of the Speed-up Theorem based on Lemmas 4.1 and 4.2.

4.3. LEMMA. *For each uniformly subconstructible decreasing sequence $g_1\geqslant g_2\geqslant g_3\geqslant\cdots$ $(g_i:\Sigma^+\to N)$, there is a single subconstructible space bound $S:\Sigma^+\to N$ such that each e satisfies $S_e\succcurlyeq S$ if and only if it satisfies $S_e\succcurlyeq g_i$ for some i.*

It remains only to prove Lemmas 4.2 and 4.3. Since the demands of the latter dictate the appropriate notion of uniform subconstructibility, we give its proof first.

PROOF OF LEMMA 4.3. The argument is reminiscent of the proof of the Compression Theorem. We define $S(x)$ for one x at a time, again considering input strings x in their natural order. Again the specification involves diagonalization, but the task seems more difficult this time: if there is no i for which $S_e\succcurlyeq g_i$, then we must design S so that $S_e\not\succcurlyeq S$; but setting no single value of S can guarantee this, thereby decisively "canceling" e. All we can do on a particular input x is to "attack" some $e=a(x)$. The problem turns out not to be a real one, however, since, by the Padding Lemma, there are so many other STM codes e' with $S_{e'}$ identical to S_e.

To attack e on input x, we simply define $S(x)$ to be $g_e(x)$ (identifying e with its position in the standard ordering of binary strings). On the assumption that there is no i for which $S_e\succcurlyeq g_i$, we have in particular that $S_e\not\succcurlyeq g_e$, so that the choice of $S(x)$ *is* a modest step toward arranging $S_e\not\succcurlyeq S$. In terms of the "attack function" a, our definition is simply

$$S(x)=g_{a(x)}(x);$$

to complete the definition of S, it remains only to define a.

On input x, we attack STM code

$$a(x)=\min\{e \mid g_e(x) \text{ is enough space to simulate } M_e \text{ on input } x, \text{ but } \textit{not} \text{ enough to discover that } e=a(x') \text{ for some } x'<x\}.$$

(When this set is empty, we conservatively take $a(x)=1$ and attack the very first STM code, thus defining $S(x)$ to be $g_1(x)$.) Note that the first condition raises suspicion that $S_e \not\succcurlyeq g_e$, and hence at least a faint suspicion that there is no i for which $S_e \succcurlyeq g_i$.

Based only on some fixed algorithm for computing $g_i(x)$ from i and x, we can implement the recursive review as in the proof of the Compression Theorem, to get a total attack function $a(x)$ and a resulting computable space bound $S(x)$. It remains only to show, for each e, that S satisfies $S \preccurlyeq S_e$ if and only if some g_i satisfies $g_i \preccurlyeq S_e$, and to show that S is subconstructible.

First, note that e does get attacked (i.e., $e=a(x)$ for some x) if it satisfies $S_e \not\succcurlyeq g_e$. The argument is familiar from the proof of the Compression Theorem: Just consider the first x so large that every $e'<e$ that ever gets attacked is $a(x')$ for some $x'<x$, and such that $g_e(x)$ is enough space both to simulate M_e on input x ($g_e(x)-S_e(x) \geqslant |e|$) and to discover an earlier attack on each such e'. For that x, $e=a(x)$.

By the Padding Lemma, there is an infinite sequence of STM codes $e_1<e_2<e_3<\cdots$ such that $S_e=S_{e_1}=S_{e_2}=S_{e_3}=\cdots$. If there is no i for which $g_{e_i} \preccurlyeq S_e$, then it follows that $S_{e_i} \not\succcurlyeq g_{e_i}$ holds for every i, so that each e_i gets attacked. If $e_i=a(x)$, then $S(x)=g_{e_i}(x) \geqslant S_e(x)+|e_i|$. Therefore, $S(x)-S_e(x)$ gets arbitrarily large, so that $S \not\preccurlyeq S_e$.

Conversely, suppose $g_i \preccurlyeq S_e$. To show that $S \preccurlyeq S_e$, it suffices to show that $S(x)$ is bounded by $g_i(x)$ whenever both x and $S(x)$ are sufficiently large. We do this by showing that $a(x)>i$ holds under the latter circumstances, so that $S(x)=g_{a(x)}(x) \leqslant g_i(x)$.

The argument that $a(x)$ exceeds i whenever x and $S(x)$ are sufficiently large is again familiar. Just consider x so large that every $e' \leqslant i$ that ever gets attacked is $a(x')$ for some $x'<x$, and $S(x)$ so large that it is enough space to discover the earliest attack on each such e'. Then, since $g_{a(x)}(x)=S(x)$, $a(x) \leqslant i$ would imply choice of $a(x)$ violating the second requirement.

Finally, we must show how to "subconstruct" S. The idea is to try, within the preallocated space, for each successive STM code e, to lay out space $g_e(x)$ and to check within that space whether e satisfies the two qualification conditions for $a(x)$. (It is the successive subconstructions that dictate the need for uniform subconstructibility. Note the importance of not losing track of e in the subconstruction of $g_e(x)$, since the process must be able to continue later with the correct *next* value.) It might be necessary to skip the first few values of e, since, for those values, $g_e(x)$ might exceed the preallocated space; but, if one of these skipped values is $a(x)$, then $S(x)=g_{a(x)}(x)$ itself exceeds the preallocated space, so that subconstructibility places no constraint at all on the outcome. So when the first e qualifies, we can assume that $a(x)$ is that e. Similarly, we can assume $a(x)=1$ if we discover that no e will qualify. It is straightforward in either case to subconstruct space $g_{a(x)}(x)=S(x)$, preallocation permitting. □

PROOF OF LEMMA 4.2. The obvious idea is to design a recursive STM to uniformly subconstruct an appropriate sequence $g_1 \geqslant g_2 \geqslant g_3 \geqslant \cdots$. Given inputs i and x, and a space preallocation, the machine should try to calculate $F(g_{i+1})(x)$ in unary within the preallocated space, saving i and subconstructing the "high-water mark", the rightmost position needed for the calculation. This will guarantee that $g_i \geqslant F(g_{i+1})$ (everywhere!), but it will not guarantee that the g_i's are total. (We have already observed that they cannot all be total if there are no exceptions to the inequalities $g_i \geqslant F(g_{i+1})$.)

Because F preserves totalness, we do know that each g_i will be total *if* g_{i+1} *is*. We modify our design to include one auxiliary partial function g_0 such that every other g_i is total if g_0 is *not* total, and also such that g_0 is *total* if and only if every other g_i is total, all while introducing only finitely many exceptions to each inequality $g_i \geqslant F(g_{i+1})$ $(i \geqslant 1)$: As before, given inputs i and x, and a space preallocation, the machine works within the preallocated space, saving i and finally subconstructing the high-water mark. If $i=0$, it tries to calculate $g_j(y)$ in unary for all j, $y<x$ $(j \neq 0)$, finally concluding its subconstruction task by restoring the contents of the storage tape and positioning the storage head at the high-water mark. (This guarantees that g_0 will be total if and only if g_j is total for every $j>0$.) If $i>0$, it first tries for $|x|$ steps to calculate $g_0(z)$ for all $z<i$. If it fails to complete the calculation of any of these values (for lack of either space or time), then it concludes its subconstruction at this point. (If g_0 is not total, it will fail whenever i is sufficiently large, thus guaranteeing that g_i is total for each such i. Since we will continue to have g_i $(i \geqslant 1)$ total whenever g_{i+1} is, this will guarantee, in turn, that g_i is total for every $i \geqslant 1$.) On the other hand, if it successfully calculates $g_0(z)$ for every $z<i$, as it is in fact bound to do for every sufficiently large preallocation and input string x, then it goes on with the original plan, trying to calculate $F(g_{i+1})(x)$ in unary before concluding the subconstruction of the high-water mark. □

5. Effective speed-up

We say that f has *effective* F-speed-up if there is a computable (possibly partial) program transformation σ such that, if M_e computes f, then ($\sigma(e)$ is defined and) so does $M_{\sigma(e)}$, with $F(S_{\sigma(e)}) \leqslant S_e$ almost everywhere. The additive-constant speed-up of our first proposition, for example, is effective for every f. (The implicit operator there maps S to S', where $S'(x)=1$ if $S(x)=1$, and $S'(x)=S(x)+1$ otherwise.) The arbitrarily dramatic speed-up of our Speed-up Theorem, on the other hand, cannot be effective.

THEOREM (noneffectiveness of speed-up). *For no sufficiently large computable operator F that preserves totalness is there a computable function with* effective *F-speed-up.*

PROOF (*cf.* [5]). Suppose F is large (enough so that the difference $F(S)-S$ is superlinear say) and f has effective F-speed-up via σ. Fix some STM M that computes f, say within space S (necessarily large, of course). Without loss of generality, we can assume that σ is total, and that $S_{\sigma(e)} \leqslant S$ for every e. (If not, then transform instead to $\sigma'(e)$ such that, on input x, $M_{\sigma'(e)}$ first tries, within relatively insignificant space $|x|$, say, to calculate $\sigma(e)$. If $M_{\sigma'(e)}$ does not find $\sigma(e)$, then it goes on to behave just like M. If it does find $\sigma(e)$, then it behaves, by alternating simulation, like whichever of M and $M_{\sigma(e)}$ requires less space for simulation on input x. (The respective requirements are at most $S(x)$ and $S_{\sigma(e)}(x)+|\sigma(e)|$.) The result is that $S_{\sigma'(e)} \leqslant S$ for every e, and that f has F'-speed-up via σ', for F' only slightly smaller than F.)

The strategy is somehow to design an STM M_e that computes f by computing and simulating its own speed-up $M_{\sigma(e)}$. Such an M_e would compute f nearly as efficiently as $M_{\sigma(e)}$, a contradiction.

It is easy to design M_e to compute f, and it is easy to design M_e to simulate $M_{\sigma(e)}$ but it seems hard to design M_e to do *both*. The trick is to design *two* STMs, M_{e_0} and M_{e_1}, interdependent through mutual recursion. On a particular input, one of the two could make sure to compute f, and the other could behave like $M_{\sigma(e_0)}$ or $M_{\sigma(e_1)}$, whichever requires less space. If we assign the two roles appropriately, then the machine not explicitly computing f will be behaving like the (would-be) speed-up of the other, and hence hopefully computing f for that reason. Because our tentative argument is circular, however, with $\varphi_{e_0}=\varphi_{e_1}=f$ and $\varphi_{e_i}=\varphi_{\sigma(e_i)}$ depending on each other, we need an additional delicate "recursive review" patch.

On input x, within (relatively insignificant, but monotonically growing and unbounded) space $|x|$, for as long an initial segment of inputs $x'<x$ as possible, M_{e_i} checks the three-way equality $\varphi_{\sigma(e_0)}(x')=\varphi_{\sigma(e_1)}(x')=f(x')$ and whether or not x' is "special". It declares x itself to be special if it is the very first input or if it is the first input on which the allocated space is sufficient to reach the last successfully checked special x'. (By design, then, x will be the *very next* special input after x' in the latter case, and the sequence of special inputs will be infinite.) If x is not special, or if the three-way equalities did not all hold, then M_{e_i} just goes on to behave like M on x, explicitly computing $f(x)$. Otherwise (i.e., if x is special and all the successfully checked three-way equalities did hold), M_{e_i} determines by alternating simulation which of $M_{\sigma(e_0)}$ and $M_{\sigma(e_1)}$ requires less space on the current input x. (Arbitrarily break ties in favor of $M_{\sigma(e_0)}$, say.) If $M_{\sigma(e_i)}$ requires less space, then M_{e_i} again goes on to behave like M on x. Otherwise, M_{e_i} goes on to behave like $M_{\sigma(e_{1-i})}$ on x.

CLAIM. *Both M_{e_0} and M_{e_1} do compute f.*

PROOF. If not, then consider the least x for which $\varphi_{e_0}(x)$ or $\varphi_{e_1}(x)$ differs from $f(x)$. The two cases are symmetric, so assume it is $\varphi_{e_0}(x)$ that differs from $f(x)$. By design, x must be special, and we must have $\varphi_{e_0}(x)=\varphi_{\sigma(e_1)}(x)$ and $\varphi_{e_1}(x)=f(x)$, with $\varphi_{\sigma(e_1)}(x)\neq f(x)$. For each $x'<x$, $\varphi_{e_1}(x')=f(x')$ by assumption; and, for each $x'>x$, we will have $\varphi_{e_1}(x')=f(x')$ by design. (On each special $x'>x$, M_{e_1} will discover that $\varphi_{\sigma(e_1)}(x)\neq f(x)$.) Therefore, φ_{e_1} is f but $\varphi_{\sigma(e_1)}$ is not, a contradiction. $\square$

It follows that all the three-way equalities will always check out. Either for $i=0$ or for $i=1$, therefore, there will be an infinite sequence of special inputs x on which M_{e_i} ends up behaving like $M_{\sigma(e_{1-i})}$ on x and using space little more than $S_{\sigma(e_{1-i})}(x) \leqslant S_{\sigma(e_i)}(x)$.

Since F is large, this contradicts the original supposition that $F(S_{\sigma(e_i)}) \leqslant S_{e_i}$ holds almost everywhere. □

Effective speed-up of a weaker sort *is* possible. It is possible, for example, if we weaken our requirement from "$F(S') \leqslant S$ almost everywhere" to just "$F(S') \leqslant S$ *infinitely often* (*i.o.*)" (i.e., $F(S')(x) \leqslant S(x)$ for infinitely many values of x). In fact, it is easy to arrange such "i.o.-speed-up" all the way down to a trivial complexity level:

"LEVELING" THEOREM FOR STM SPACE. *For each computable space bound* $S: \Sigma^+ \to N$, *there is a computable predicate* $P: \Sigma^+ \to \{0, 1\}$, *not computable within space* S, *but with "i.o.-speed-up triviality", in the following sense: whenever* P *is computable within a space bound* S', *it is also computable within a space bound* $S'' \leqslant S'$ *so often so much smaller that the following set is infinite*:

$$\{x \mid S'(x) \geqslant S(x) \text{ and } S''(x) = 1\}.$$

Moreover, the speed-up can be effective.

PROOF (*cf.* [5]). We make P different from every φ_e with $S_e \leqslant S$ by straightforward diagonalization: For each e, if $S_e(e) \leqslant S(e)$, then we choose $P(e)$ to be the least member of $\{0, 1\}$ not equal to $\varphi_e(e)$. (So $S_e \leqslant S \Rightarrow P(e) \neq \varphi_e(e) \Rightarrow P \neq \varphi_e$.) If $S_e(e) > S(e)$, then we choose $P(e)$ to be 0, say. (Note, incidentally, that $P(x)$ *is* straightforwardly computable in just $|x|$ more space than $S(x)$ is.)

Now suppose M_{e_1} computes P. Letting p and u be the padding and unpadding transformations discussed earlier, consider the input strings $u(e_1)$, $p(u(e_1))$, $p^2(u(e_1))$, ... For each such string $e = p^i(u(e_1))$, $P(e)$ does equal $\varphi_{e_1}(e)$, so that we must have $S_e(e) > S(e)$ and $P(e) = 0$. Because e and e_1 are equivalent, we also have $S_{e_1}(e) > S(e)$. So, for the desired i.o.-speed-up to triviality, design M_{e_2} to output 0 without extending its storage tape if its input is a padded version of $u(e_1)$, and to behave exactly like M_{e_1} otherwise. □

6. Fundamental Theorem for STM space

Before we go on to develop the notion of machine independence, and to derive machine-independent results from the machine-dependent ones we have proven so far, we should mention that our compression and speed-up results for STM space complexity can be combined and strengthened to yield an *exact characterization* of those "prospective STM space complexities" that are "realizable".

A *prospective* (*STM space*) *complexity* is just a set $\mathscr{C}$ of STM codes. We write $\mathscr{C} \preccurlyeq S$ (or $S \succcurlyeq \mathscr{C}$) to indicate that $S_e \preccurlyeq S$ holds for some e in $\mathscr{C}$. A computable function f *realizes* prospective complexity $\mathscr{C}$ if, for each STM code e, f is computable by an STM within space S_e if and only if $S_e \succcurlyeq \mathscr{C}$.

We say that $\mathscr{C}$ and $\mathscr{C}'$ are *equivalent* if, for each STM code e, $\mathscr{C} \preccurlyeq S_e$ holds if and only if $\mathscr{C}' \preccurlyeq S_e$ does. Note that, if f realizes $\mathscr{C}$, and if $\mathscr{C}'$ is equivalent to C, then f realizes $\mathscr{C}'$, too.

It is easy to see that each realizable prospective complexity, or some equivalent one, must necessarily satisfy the following three conditions:

(1) For some e in $\mathscr{C}$, S_e is total.

(2) For each e_1 and e_2 in $\mathscr{C}$, there is some e_3 in $\mathscr{C}$ for which $S_{e_3} \preccurlyeq \min(S_{e_1}, S_{e_2})$.

(3) For some computable predicate $P(x, y, e)$, $\mathscr{C} = \{e \mid \exists x \forall y\, P(x, y, e)\}$.

The characterization theorem [11, 12, 19], which Meyer and Winklmann call "the Fundamental Theorem of complexity theory", adds that these three conditions also *suffice* for realizability, even by a function that assumes only two values, so that they *exactly characterize* the realizable prospective STM space complexities.

FUNDAMENTAL THEOREM FOR STM SPACE. *The following are equivalent for a prospective complexity* $\mathscr{C}$:

(i) $\mathscr{C}$ *is realizable.*

(ii) $\mathscr{C}$ *is realizable by a predicate.*

(iii) $\mathscr{C}$ *is equivalent to a prospective complexity that satisfies conditions* 1–3 *above.*

To complete a proof of the Fundamental Theorem, it suffices to prove an alternative to our Lemma 4.2, saying that, for every prospective complexity $\mathscr{C}$ that satisfies the three conditions, there is a uniformly subconstructible decreasing sequence $\{g_i\}$, essentially as above, such that, for each STM code e, $S_e \succcurlyeq \mathscr{C}$ if and only if $S_e \succcurlyeq g_i$ for some i.

From the Fundamental Theorem, one can prove our versions of the Compression and Speed-up Theorems as easy corollaries. One can also conclude that, for every computable function, there is a *predicate* with exactly the same STM space complexity, strongly justifying the emphasis in complexity theory on computing predicates and recognizing languages.

7. Machine independence

Let us turn now to machine independence. A theorem on "machine-independent computational complexity" will have to be true for every "machine model" and "measure of computational complexity". To make this precise, we need rigorous but broad notions of what can be a machine model and what can be a complexity measure.

We consider the issue of machine model first. Recall that, in the case of the STM, we formalized and encoded machine descriptions to pin down the details of the model. The result was what we now call a *programming system*, in which each binary "program" code e specifies a partial function ϕ_e. Rather than generalize the informal notion of "machine model", we focus on the resulting programming systems, viewing each code e as a "program" for, or particular instance of, some implicit "machine" that computes the partial function ϕ_e.

Not every programming system is reasonable. One expectation by modern programmers and those familiar with the Church–Turing thesis [23] is that programs written in any feasible programming language can be translated, or "compiled", into any other serious programming language. Although there might be other reasonable expectations, this one weak requirement turns out to be enough for a useful definition [22]: A programming system $\{\phi_e\}$ is *reasonable* (or perhaps we should say "as reasonable as the STM system") if there exist computable translations (or "cross compilers") $f, g: \{0,1\}^+ \to \{0,1\}^+$ such that, for each program code e, $\phi_e = \varphi_{f(e)}$ and $\varphi_e = \phi_{g(e)}$. For any *truly* reasonable programming system, of course, the translations f and g would be *efficiently* computable; but, by adopting the more liberal definition, we will be able to see that the results we have been using and proving for STMs are machine-independent in the most robust sense.

For example, it already follows from our liberal definition that every reasonable programming system $\{\phi_e\}$ has all of the following properties [22, 15]:

(1) (*Computability*) $\{\phi_e\}$ is the set of all computable partial functions.

(2) (*Uniform computability*) $\phi_e(x)$ is a computable partial function of e and x.

(3) (*Computable composition*) There is a computable function c such that, for each e_1 and e_2, $\phi_{c(e_1,e_2)}$ is the composition of ϕ_{e_1} and ϕ_{e_2} (i.e., $\phi_{c(e_1,e_2)}(x) = \phi_{e_1}(\phi_{e_2}(x))$ for every x).

(4) (*s-1-1 Theorem*) If we alternatively view each ϕ_e as a computable partial function of *two* variables, by some computable pairing convention, then there is a computable function s such that, for each e, x, and y, $\phi_{s(e,x)}(y) = \phi_e(x, y)$.

(5) (*Recursion Theorem*) For each total, computable program transformation $\tau: \{0,1\}^+ \to \{0,1\}^+$, there is a fixed point e, itself computable from a program code for τ, such that $\phi_e = \phi_{\tau(e)}$.

(6) (*Padding Lemma*) There is a computable, one-to-one "padding" function p such that, for each program code e, $p(e)$ is a longer code with $\phi_{p(e)} = \phi_e$.

(7) (*Isomorphism Theorem*) There is a *bijective* (i.e., one-to-one and onto) computable translation f such that $\phi_e = \varphi_{f(e)}$ for every program code e.

To see that every reasonable programming system $\{\phi_e\}$ has these seven properties, first note that the STM system does, and then try to translate the properties to $\{\phi_e\}$. For properties (1)–(4), this is completely straightforward. To find the fixed point for property (5), first find a fixed point e' for the composite transformation $f \circ \tau \circ g$ in the STM system, and then use its translation $g(e')$. This does yield

$$\phi_{g(e')} = \varphi_{e'} = \varphi_{f(\tau(g(e')))} = \phi_{\tau(g(e'))}.$$

Assuming property (6), it is not hard to obtain the bijection for property (7): Starting with an empty partial function f, extend both the domain and the range of f to include each successive e. To assign $f(e)$ or $f^{-1}(e)$, translate e and pad enough times to obtain an equivalent program that has not yet been assigned.

To prove property (6), we would like to translate to the STM programming system, pad there, and translate back. There is a problem, however, because the translations might not be one-to-one. Property (7) would solve our problem, but we have already used property (6) in its proof. The solution is to settle first for a *weakened* version of property (7), the proof of which is the only tricky step in the whole sequence.

INJECTION LEMMA. *Between any pair of reasonable programming systems, there is a* one-to-one *computable translation.*

PROOF. Start with any computable translation h from $\{\phi'_e\}$ to $\{\phi_e\}$, and aim to replace the images $h(e)$ in the unprimed system by a sequence of respectively equivalent but *distinct* programs there.

In the unprimed programming system, let s be the computable s-1-1 function of property (4). We will choose d_0 so that, for each e, $s(d_0, e)$ will be usable as the desired replacement for $h(e)$.

We obtain d_0 as the fixed point of a computable program transformation τ that transforms program d in such a way that

$$\phi_{\tau(d)}(e,x)=\begin{cases}0 & \text{if } s(d,e)\in\{s(d,e')\mid e'<e\};\\ 1 & \text{if not, but if } s(d,e)\in\{s(d,e')\mid e<e'\leqslant x\};\\ \phi_{h(e)}(x) & \text{otherwise.}\end{cases}$$

(To get such a transformation, straightforwardly design an STM to produce the desired output, and then translate the result into the unprimed abstract programming system.) Since $\phi_{s(d_0,e)}=\phi_{d_0}(e,x)=\phi_{\tau(d_0)}(e,x)$, this yields

$$\phi_{s(d_0,e)}(x)=\begin{cases}0 & \text{if } s(d_0,e)\in\{s(d_0,e')\mid e'<e\};\\ 1 & \text{if not, but if } s(d_0,e)\in\{s(d_0,e')\mid e<e'\leqslant x\};\\ \phi_{h(e)}(x) & \text{otherwise.}\end{cases}$$

To complete our proof, we must show that $s(d_0,e)$ is a one-to-one function of e, and that $\phi_{s(d_0,e)}=\phi_{h(e)}$ for every e. If the former is true, then certainly the latter will be true, since $s(d_0,e)$ will never belong to $\{s(d_0,e')\mid e'<e\}$ or $\{s(d_0,e')\mid e<e'\leqslant x\}$. Therefore, it remains only to show that $s\{d_0,e)$ is one-to-one.

For the sake of argument, suppose that $s(d_0,e)$ is not a one-to-one function of e. Let $s(d_0,e)=s(d_0,e')$ be a counterexample with $e<e'$, and with e as small as possible. Examining the cases, we see that $\phi_{s(d_0,e')}(x)=0$ for every x, and that $\phi_{s(d_0,e)}(x)=1$ for every $x\geqslant e'$, contradicting $s(d_0,e)=s(d_0,e')$. □

Our notion of a *complexity measure* on an arbitrary programming system $\{\phi_e\}$ should generalize our space measure $\{S_e\}$ on the STM programming system $\{\varphi_e\}$. To each partial function ϕ_e, it should assign an appropriate "resource bound" Φ_e. Since neither the "machine model" nor its "resource" is explicit, the constraints will have to be abstract axioms. In our liberal definition, there will be only two such axioms [3]:

(A1) For each e, Φ_e is defined on precisely the same domain as ϕ_e.

(A2) It is decidable from e, x, and b whether $\Phi_e(x)=b$.

Note that the prototype measure $\{S_e\}$ on $\{\varphi_e\}$ does satisfy these axioms. The natural notions of time also satisfy them.

Our notion of complexity measure is so general that it actually obviates the need to look at more than one reasonable programming system $\{\phi_e\}$. If $\{\Phi'_e\}$ is any complexity measure on any other reasonable programming system $\{\phi'_e\}$, then there is a complexity measure $\{\Phi_e\}$ on $\{\phi_e\}$ that is equivalent in the following strong sense: for each e, there is an e' such that $\phi'_{e'}=\phi_e$ and $\Phi'_{e'}=\Phi_e$; and, for each e', there is an e such that $\phi_e=\phi'_{e'}$ and $\Phi_e=\Phi'_{e'}$. In other words, each partial function is computable in exactly the same resource bounds in the two systems. Design of Φ_e is easy: just take it to be $\Phi'_{f(e)}$, where f is a bijective computable translation from the unprimed programming system to the primed one. When it is convenient, therefore, we can restrict attention to any one particular programming system—to our formalization of STMs, for example. Of course, complexity measures that seem natural in other programming systems might appear quite obscure when translated to STMs.

Through "cross compilation", it turns out that many complexity results carry over from one reasonable programming system and complexity measure to all others. This includes even the Fundamental Theorem, although the published versions [14, 25] are proved directly from the axioms. (In either case, the results are somewhat "blurred" and relatively awkward to state.) The key tool is the following consequence of our definitions:

RECURSIVE-RELATEDNESS THEOREM (cf. [3]). *If $\{\Phi_e\}$ and $\{\Phi'_e\}$ are complexity measures on the respective reasonable programming systems $\{\phi_e\}$ and $\{\phi'_e\}$, then there is a total, computable function r such that, for each e, there is an e′ such that $\phi'_{e'}=\phi_e$ and $\Phi'_{e'}=_r\Phi_e$, where the binary relation $=_r$ is the "r-blurred" version of equality that we define below.*

For $r:\Sigma^+\times N\to N$, the binary relation $=_r$ is defined in terms of the binary relation $\leqslant_r$. If ψ_1 and ψ_2 are partial functions from Σ^+ to N, then we say that $\psi_1\leqslant_r\psi_2$ holds if $\psi_1(x)\leqslant r(x,\psi_2(x))$ holds for almost every x, and we say that $\psi_1=_r\psi_2$ holds if both $\psi_1\leqslant_r\psi_2$ and $\psi_2\leqslant_r\psi_1$ hold. Note that, notation notwithstanding, neither $\leqslant_r$ nor $=_r$ is transitive, in general.

REMARKS. (i) *Machine-based* complexity theory dwells on questions such as how small the function r can be for specific programming systems and complexity measures of interest. (In general, of course, there might not be one best r; the recursive relation itself might happen to have speed-up.) The issues include the relative efficiency of various machine models and the relationship between the time and space measures on any particular model.

(ii) Our development provides no good reason, except for technical need, to tolerate finitely many exceptions to the inequality defining $\leqslant_r$. In the concrete case of STMs, we can modify the behavior of any machine in an arbitrary way on any particular finite set of input strings, without affecting its behavior elsewhere; but this need not be the case in an abstract setting. Some have proposed the imposition of an additional "finite-patching" axiom, to better capture the notion of "naturalness" for a complexity measure [27].

PROOF OF THE RECURSIVE-RELATEDNESS THEOREM. We can take e' to be $f(e)$ for any fixed computable translation f from the unprimed programming system to the primed one. Certainly, then, $\phi'_{e'}=\phi_e$. If we take

$$r_1(x,b)=\max\{\Phi'_{f(e)}(x)\mid e\leqslant x \text{ and } \Phi_e(x)=b\},$$

then we get $\Phi'_{f(e)}(x)\leqslant r_1(x,\Phi_e(x))$ for every $x\geqslant e$, so that $\Phi'_{f(e)}\leqslant_{r_1}\Phi_e$. (Note that all four of Φ_e, ϕ_e, $\phi'_{f(e)}$, and $\Phi'_{f(e)}$ share the same domain of definition. Therefore, the condition $\Phi_e(x)=b$ guarantees that $\Phi'_{f(e)}(x)$ is defined and can be found by exhaustive search.) Similarly, if we take

$$r_2(x,b)=\max\{\Phi_e(x)\mid e\leqslant x \text{ and } \Phi'_{f(e)}(x)=b\},$$

then we get $\Phi_e \leqslant_{r_2} \Phi'_{f(e)}$. If we define r to be the pointwise maximum of r_1 and r_2, therefore, then we do get $\Phi'_{f(e)} =_r \Phi_e$. □

For Corollaries 7.1–7.4 below, let $\{\phi_e\}$ be any reasonable programming system, and let $\{\Phi_e\}$ be any complexity measure on $\{\phi_e\}$. For each resource bound $B: \Sigma^+ \to N$, the *complexity class* defined or "named" by B, denoted here by

$$\mathrm{CCLASS}_{\{\phi_e\},\{\Phi_e\}}(B),$$

is the class of functions computable in a resource bound bounded by B almost everywhere. We will omit the subscripts from the notation when they are clear from context (but not in Corollary 7.5, where they are not).

7.1. COROLLARY ("blurred" compression). *There is a computable "blurring" function r such that, for each total resource bound Φ_e, there is a predicate P computable within some resource bound $\Phi_{e'} =_r \Phi_e$, but not computable within a resource bound $\Phi_{e''}$* unless *$\Phi_{e''} \geqslant_r \Phi_e$.*

REMARKS. (i) Another issue in machine-based complexity is how tight the compression can be for a particular complexity measure on a particular programming system of interest. (Here, again, however, there might not be one best tightness r.)

(ii) There are similarly blurred (less abrupt) versions of the trade-offs between program size and computational complexity.

PROOF OF COROLLARY 7.1. Let r_0 recursively relate $\{\Phi_e\}$ on $\{\phi_e\}$ to $\{S_e\}$ on $\{\varphi_e\}$, where we already have a compression result, and take r to be somewhat larger. (If we take r_0 to be monotonic in and at least equal to each argument, then $r(x,b)=r_0(x, r_0(x,b)+|x|)$ will do.) If Φ_e is total, then recursive relatedness gives us a d for which $S_d =_{r_0} \Phi_e$. Compression for STM space gives us a predicate P computable within space $S_{d'}$ if and only if $S_{d'} \succcurlyeq S_d$. We show that P satisfies the conclusion of the asserted corollary.

Since P is computable within space S_d, recursive relatedness guarantees that it is computable in the abstract system within some resource bound $\Phi_{e'} =_{r_0} S_d =_{r_0} \Phi_e$, which implies by composition that $\Phi_{e'} =_r \Phi_e$. On the other hand, if P is computable in the abstract system within resource bound $\Phi_{e''}$, then it is computable on an STM within a space bound $S_{d''}$ that must satisfy both $S_{d''} =_{r_0} \Phi_{e''}$ (by recursive relatedness) and $S_{d''} \succcurlyeq S_d$ (by the inherent complexity of P). Therefore, $\Phi_{e''} =_{r_0} S_{d''} \succcurlyeq S_d =_{r_0} \Phi_e$, so that $\Phi_{e''} \geqslant_r \Phi_e$. □

7.2. COROLLARY. *There is a total resource bound Φ_e almost everywhere above each computable function.*

7.3. COROLLARY (speed-up). *For each computable operator F that preserves totalness, there is a computable predicate P with "F-speed-up": Whenever P is computable within a resource bound* Φ, *it is also computable within a resource bound* Φ' *so small that* $F(\Phi') \leqslant \Phi$ *holds almost everywhere.*

7.4. COROLLARY (gap). *For each computable operator F that preserves totalness, there are computable resource bounds* Φ *and* $\Phi' \leqslant \Phi$ *so much smaller that* $F(\Phi') \leqslant \Phi$, *but such that, for each e, if* $\Phi_e \leqslant \Phi$ *holds almost everywhere, then so does* $\Phi_e \leqslant \Phi'$.

PROOF IDEA FOR COROLLARIES 7.2–4. As in the proof of Corollary 7.1, exploit the corresponding fact for STM space complexity. The "blurring" does not show up in any of the final results, because we can simply cite the STM result for a larger threshold or operator than is our real goal. □

7.5. COROLLARY. *For any pair of complexity measures, on any pair of reasonable programming systems, there is a computable resource bound* $B: \Sigma^+ \to N$ *that defines the same complexity class in both systems. (If the respective systems and complexity measures are given unprimed and primed versions of the usual names, then the conclusion is that there is always a bound B for which* $\mathrm{CCLASS}_{\{\phi_e\},\{\Phi_e\}}(B) = \mathrm{CCLASS}_{\{\phi'_e\},\{\Phi'_e\}}(B)$.)

PROOF (*idea*). Take B in the middle of a gap that is sufficiently bigger than the recursive relationship between the two complexity measures. □

REMARK. Yet another pursuit of machine-based complexity is to show that particular complexity measures of interest can *differ*.

As an example of an interesting machine-independent theorem that apparently does not generalize by recursive relatedness, we give the "Union Theorem" [16]. The theorem states that the union of any uniform hierarchy of complexity classes is itself a complexity class. (Note that Lemma 4.3 above can be viewed as a sort of *intersection* theorem. It follows from the Speed-up Theorem, however, that the exact analogue cannot hold [1].)

UNION THEOREM. *Consider any complexity measure on any reasonable programming system. For each uniformly computable nondecreasing sequence of resource bounds* $B_1 \leqslant B_2 \leqslant B_3 \leqslant \cdots$ $(B_i: \Sigma^+ \to N)$, *there is a single computable resource bound* $B_\infty: \Sigma^+ \to N$ *for which*

$$\mathrm{CCLASS}(B_\infty) = \bigcup_{i=1}^{\infty} \mathrm{CCLASS}(B_i).$$

PROOF. What we really prove is even stronger, and has very little to do with complexity classes: Given any nondecreasing sequence of functions $\{B_i\}$ and any sequence of functions $\{\Phi_e\}$ to "dodge", there is a function B_∞ that bounds every B_i almost everywhere, and such that every Φ_e that is bounded by B_∞ almost everywhere is already bounded by some B_i almost everywhere. Moreover, B_∞ is computable if $\{B_i\}$ and $\{\Phi_e\}$ are uniformly computable and $\{\Phi_e\}$ satisfies axiom (A2) for complexity measures.

We define $B_\infty(x)$ for one x at a time, considering arguments x in their usual natural order. To ensure that B_∞ bounds every B_i almost everywhere, we must take care that it should eventually stop dipping below each B_i; we will check this explicitly at the end of the proof. To dodge the functions we should dodge, we aim to "attack" each *pair* (e, i) for which B_i is not bounded almost everywhere by Φ_e, and for which $i > e$ holds. An attack will consist of a distinct dip by B_∞ below Φ_e, so that, if (e, i) gets attacked for all $i > e$, B_∞ will dip below Φ_e infinitely often, as required.

Unsurprisingly, our strategy on argument x is to attack the smallest pair (e, i), if there is one, that satisfies $i < x$, that was not attacked on any earlier argument $x' < x$, and that satisfies $\Phi_e(x) > B_i(x)$ (i.e., $\Phi_e(x) = b$ for no $b < B_i(x)$, which is decidable if axiom (A2) holds). Since we attack no pair more than once, we do eventually attack every pair (e, i) for which $\Phi_e > B_i$ holds infinitely often. The method of attack is simply to set $B_\infty(x)$ to $B_i(x) < \Phi_e(x)$. In the exceptional case that no pair is eligible for attack, set $B_\infty(x)$ to $B_x(x)$.

Finally, to see that our B_∞ does eventually stop dipping below B_i, note that $B_\infty(x) < B_i(x)$ holds only if some pair (e, i') with $e < i' < i$ gets attacked on argument x, or if $x < i$. Since no pair is attacked twice, the number of such arguments x is some finite number that depends only on i. □

As a final example, we describe a version of the "honesty" or "naming" theorem of McCreight and Meyer [16]. Informally, the theorem states that every complexity class with a computable name also has an "honest" name, by which we mean a naming resource bound whose arguments and values are an "honest indication" of its complexity. More formally, call a function $B: \Sigma^+ \to N$ *h-honest* if there is some program code e for which ϕ_e is B, and for which $\Phi_e(x) \leqslant h(x, B(x))$ holds for almost every x. For example, in the context of STM space complexity, "constructibility" can be viewed as h-honesty for $h(x, B(x)) = B(x)$; and, in the more general context, each "total resource bound" Φ_e can be viewed as h-honest for this same h.

Honesty Theorem. *For each computable resource bound $B: \Sigma^+ \to N$, there is an h-honest resource bound $B': \Sigma^+ \to N$ such that* CCLASS(B') = CCLASS(B), *where $h: \Sigma^+ \times N \to N$ is a computable function that depends only on the reasonable programming system and the complexity measure (and not on B).*

Proof (*idea*) (*cf.* [20]). The construction, like the one for the union theorem, actually has very little to do with complexity classes. (In fact both results can be viewed as consequences of a single closure property for classes of *program codes* [9].) From $B = \phi_{e_0}$, we construct a possibly partial honest bound B' in such a way that each Φ_e lies below B' almost everywhere if and only if it lies below B almost everywhere. To get a total bound with these properties, it suffices to replace B' with the pointwise minimum of B' and $\phi_{e_0} + \Phi_{e_0}$.

For this construction, we do *not* define $B'(x)$ strictly according to the natural order of its arguments x. Instead, we "dovetail" through an infinite number of "opportunities" for each of the infinitely many arguments x, until we discover the first "appropriate" opportunity to define each $B'(x)$. If stage n of the dovetailing is the ith opportunity to define B' at argument x, then we consider defining $B'(x)$ to be i.

The decision whether to define $B'(x)=i$ at this stage n is based on a growing, ordered agenda that we maintain. At stage n there is one agenda item for each program code $e<n$. The agenda item for e indicates whether or not we are looking currently for an(other) opportunity for B' to dip below Φ_e. The decision is made as follows:

(1) Do not redefine $B'(x)$ if it has been defined already at some stage $n'<n$.

(2) Otherwise, do define $B'(x)$ if the definition would fulfill a positive agenda item for some e ($i<\Phi_e(x)$ for that e) without violating any negative item for an e' higher on the agenda than e ($i \geqslant \Phi_{e'}(x)$ for each such e').

The agenda is updated as follows:

(3) Add a *negative* item to the bottom of the agenda for program code n.

(4) If $\Phi_{e_0}(x)=i$, then convert to *positive* the agenda item for each e such that $\Phi_e(x)>\phi_{e_0}(x)$. (The latter indicates a dip of $B=\phi_{e_0}$ below Φ_e. The prerequisite $\Phi_{e_0}(x)=i$ ensures that this dip is new.)

(5) If in (2) the decision was made to define $B'(x)=i$, then convert to *negative* the highest positive responsible agenda item, and move that item to the bottom of the agenda.

We leave it as an exercise to show that B' meets our specifications and is h-honest for some computable h that depends only on $\{\phi_e\}$ and $\{\Phi_e\}$. The latter is easiest seen as a consequence of the simple observation that it is decidable from e_0, x, and i whether $B'(x)=i$. For a careful proof, see [20]. □

Finally, we should note that machine independence is known not to be generally decidable. In particular, it is not decidable from a property of computable partial functions whether the property is measure-independent [2].

Acknowledgment

I thank George Hauser for his reactions to the earliest, least readable versions of this chapter, Jan van Leeuwen for his help and encouragement to finally finish, and Jun Tarui for constructive criticism of what I thought was the very last version.

References

[1] BASS, L.J., A note on the intersection of complexity classes of functions, *SIAM J. Comput.* **1**(4) (1972) 288–289.

[2] BENNISON, V.L., Recursively enumerable complexity sequences and measure independence, *J. Symbolic Logic* **45**(3) (1980) 417–438.

[3] BLUM, M., A machine-independent theory of the complexity of recursive functions, *J. Assoc. Comput. Mach.* **14**(2) (1967) 322–336.

[4] BLUM, M., On the size of machines, *Inform. and Control* **11**(3) (1967) 257–265.

[5] BLUM, M., On effective procedures for speeding up algorithms, *J. Assoc. Comput. Mach.* **18**(2) (1971) 290–305.

[6] BORODIN, A., Computational complexity and the existence of complexity gaps, *J. Assoc. Comput. Mach.* **19**(1) (1972) 158–174.

[7] CALUDE, C., *Theories of Computational Complexity* (Elsevier North-Holland, New York, 1988).

[8] CONSTABLE, R.L., The operator gap, *J. Assoc. Comput. Mach.* **19**(1) (1972) 175–183.

[9] EMDE BOAS, P. VAN, Some applications of the McCreight–Meyer algorithm in abstract complexity theory, *Theoret. Comput. Sci.* **7**(1) (1978) 79–98.

[10] HARTMANIS, J. and J.E. HOPCROFT, An overview of the theory of computational complexity, *J. Assoc. Comput. Mach.* **18**(3) (1971) 444–475.

[11] LEVIN, L.A., On storage capacity for algorithms, *Soviet Math. Dokl.* **14**(5) (1973) 1464–1466.

[12] LEVIN, L.A., Complexity of computation of computable functions (in Russian), in: V.A. Kosmidiadi, N.A. Maslov and N.V. Petri, eds., *Complexity of Computations and Algorithms* (Mir, Moscow, 1974) 174–185.

[13] LI, M. and P.M.B. VITÁNYI, Kolmogorov complexity and its applications, in: J. van Leeuwen, ed., *Handbook of Theoretical Computer Science, Vol. A* (North-Holland, Amsterdam, 1990) 187–254.

[14] LYNCH, N., "Helping": several formalizations, *J. Symbolic Logic* **40**(4) (1975) 555–566.

[15] MACHTEY, M. and P. YOUNG, *An Introduction to the General Theory of Algorithms* (Elsevier North-Holland, New York, 1978).

[16] MCCREIGHT, E.M. and A.R. MEYER, Classes of computable functions defined by bounds on computation (preliminary report), in: *Conf. Record ACM Symp. on Theory of Computing*, Marina del Rey, CA (1969) 79–88.

[17] MEYER, A.R., Program size in restricted programming languages, *Inform. and Control* **21**(4) (1972) 382–394.

[18] MEYER, A.R. and P.C. FISCHER, Computational speed-up by effective operators, *J. Symbolic Logic* **37**(1) (1972) 55–68.

[19] MEYER, A.R. and K. WINKLMANN, The fundamental theorem of complexity theory (preliminary version), in: J.W. de Bakker and J. van Leeuwen, eds., *Foundations of Computer Science III, Part 1: Automata, Data Structures, Complexity* Mathematical Centre Tracts, Vol. 108 (Centre for Mathematics and Computer Science, Amsterdam, 1979) 97–112.

[20] MOLL, R. and A.R. MEYER, Honest bounds for complexity classes of recursive functions, *J. Symbolic Logic* **39**(1) (1974) 127–138.

[21] RITCHIE, R.W., Classes of predictably computable functions, *Trans. Amer. Math. Soc.* **106** (1963) 139–173.

[22] ROGERS JR., H., Gödel numberings of partial recursive functions, *J. Symbolic Logic* **23**(3) (1958) 331–341.

[23] ROGERS JR., H., *Theory of Recursive Functions and Effective Computability* (McGraw-Hill, New York, 1967).

[24] ROYER, J.S. and J. CASE, Intensional subrecursion and complexity theory, Manuscript, 1988.

[25] SCHNORR, C.P. and G. STUMPF, A characterization of complexity sequences, *Z. Math. Logik Grundlag. Math.* **21**(1) (1975) 47–56.

[26] SIPSER, M., Halting space-bounded computations, *Theoret. Comput. Sci.* **10**(3) (1980) 335–338.

[27] SMITH, C.H., A note on arbitrarily complex recursive functions, *Notre Dame J. Formal Logic* **29**(2) (1988) 198–207.

[28] TRAKHTENBROT, B.A., Complexity of algorithms and computations (in Russian), Course Notes (Novosibirsk University, Novosibirsk, 1967).

[29] WAGNER, K. and G. WECHSUNG, *Computational Complexity* (VEB Deutscher Verlag der Wissenschaften, Berlin, GDR, 1986).

[30] YOUNG, P., Easy constructions in complexity theory: gap and speed-up theorems, in: *Proc. Amer. Math. Soc.* **37**(2) (1973) 555–563.

CHAPTER 4

Kolmogorov Complexity and its Applications

Ming LI

Aiken Computation Laboratory, Harvard University, Cambridge, MA 02138, USA

Paul M.B. VITÁNYI

Centrum voor Wiskunde en Informatica, Kruislaan 413, 1098 SJ Amsterdam, Netherlands, and Faculteit Wiskunde en Informatica, Universiteit van Amsterdam, Amsterdam, Netherlands

Contents

HANDBOOK OF THEORETICAL COMPUTER SCIENCE
Edited by J. van Leeuwen

1. Introduction

In everyday language we identify the information in an individual object with the essentials of a description for it. We can formalize this by defining the amount of information in a finite object (like a string) as the size (i.e., number of bits) of the smallest program that, starting with a blank memory, outputs the string and then terminates. A similar definition can be given for infinite strings, but in this case the program produces element after element forever. Thus, 1^n (a string of n ones) contains little information because a program of size about $\log n$ outputs it (like "print n 1s"). Likewise, the transcendental number $\pi = 3.1415\ldots$, an infinite sequence of seemingly "random" decimal digits, contains O(1) information. (There is a short program that produces the consecutive digits of π forever.) Such a definition would appear to make the amount of information in a string depend on the particular programming language used. Fortunately, it can be shown that all choices of programming languages (that make sense) lead to quantification of the amount of information that is invariant up to an additive constant.

The theory dealing with the quantity of information in individual objects goes by names such as "algorithmic information theory", "Kolmogorov complexity", "K-complexity", "Kolmogorov–Chaitin randomness", "algorithmic complexity", "descriptive complexity", "program-size complexity", and others. Although there is a case to be made for "Solomonoff–Kolmogorov–Chaitin complexity" as the most appropriate name, we regard "Kolmogorov complexity" as well entrenched and commonly understood, and use it hereafter.

At the outset we wanted to survey the applications of this theory to the theory of computation, primarily in connection with the analysis and synthesis of algorithms in relation to the resources in time and space such algorithms require. But we were dealing with deep notions and general principles arising from, and having an impact to, many more disciplines. Gradually, the subject acquired a sophisticated mathematical theory and applications in an increasingly large number of astoundingly different areas. This chapter attempts to grasp the mass of fragmented knowledge of this fascinating theory.

The mathematical theory of Kolmogorov complexity contains deep and sophisticated mathematics. Yet the amount of this mathematics one needs to know to apply the notions fruitfully in widely divergent areas, from recursive function theory to chip technology, is very little. However, formal knowledge does not necessarily imply the wherewithal to apply it, perhaps especially so in the case of Kolmogorov complexity. It is the purpose of this chapter to develop the minimum amount of theory needed, and briefly outline a scala of illustrative applications. In fact, while the pure theory of the subject will have its appeal to the select few, the surprisingly large field of its applications will, we hope, delight the multitude.

One can distinguish three application areas, according to the way Kolmogorov complexity is used. That is, we can use the fact that some strings are extremely compressible; that many strings are not compressible at all; and that some strings may be compressed but that it takes a lot of effort to do so.

Kolmogorov complexity has its roots in probability theory, combinatorics, and philosophical notions of randomness, and came to fruition using the recent development of the theory of algorithms. Consider Shannon's classical information theory [144] that assigns a quantity of information to an ensemble of possible messages. All messages in the ensemble being equally probable, this quantity is the number of bits needed to count all possibilities. This expresses the fact that each message in the ensemble can be communicated using this number of bits. However, it does not say anything about the number of bits needed to convey any individual message in the ensemble. To illustrate this, consider the ensemble consisting of all binary strings of length 9999999999999999. By Shannon's measure, we require 9999999999999999 bits on the average to encode such a string. However, the string consisting of 9999999999999999 ones can be encoded in about 55 bits by expressing 9999999999999999 in binary and adding the repeated pattern "1". A requirement for this to work is that we have agreed on an algorithm that decodes the encoded string. We can compress the string still further when we note that 9999999999999999 equals $3^2 \times 1111111111111111$, and that 1111111111111111 consists of 2^4 ones.

Thus, we have discovered an interesting phenomenon: the description of some strings can be compressed considerably. In fact, there is no limit to the amount to which strings can be compressed, provided they exhibit enough regularity. This observation, of course, is the basis of all systems to express very large numbers, and was exploited early on by Archimedes in "The Sand Reckoner". However, if regularity is lacking, it becomes more cumbersome to express large numbers. For instance, it seems easier to compress the number "one billion," than the number "one billion seven-hundred thirty-five million two-hundred sixty-eight thousand and three-hundred ninety-four," even though they are of the same order of magnitude.

This brings us to a related root of Kolmogorov complexity, the notion of randomness. In the context of the above discussion, random strings are strings that cannot be compressed. Now let us compare this with the common notions of mathematical randomness. To measure randomness, criteria have been developed which certify this quality. Yet, in recognition that they do not measure "true" randomness, we call these criteria "pseudo" random tests [71]. For instance, statistical survey of initial sequences of decimal digits of π have failed to disclose any significant deviations of randomness [71, 113, 147]. But clearly, this sequence is so regular that it can be described by a simple program to compute it, and this program can be expressed in a few bits. Von Neumann [120]:

> "Any one who considers arithmetical methods of producing random digits is, of course, in a state of sin. For, as has been pointed out several times, there is no such thing as a random number—there are only methods to produce random numbers, and a strict arithmetical procedure is of course not such a method. (It is true that a problem we suspect of being solvable by random methods may be solvable by some rigorously defined sequence, but this is a deeper mathematical question than we can go into now.)"

This fact prompts more sophisticated definitions of randomness. Notably R. Von Mises [115] proposed notions that approach the very essence of true randomness. In the early nineteenhundreds, Von Mises aimed at an axiomatic foundation of a calculus of probabilities. With the axioms validated by empirical evidence, in the manner of thermodynamics or mechanics, this would form a basis for real applications. However, while the ultimate justification of proposals for a proper theory of probabilities must lie in its applicability to real phenomena, this aspect was largely ignored in favor of the mathematical elegance of Kolmogorov's classic treatment of the set-theoretic axioms of his calculus of probability in 1933 [74].

> "This theory was so successful, that the problem of finding the basis of real applications of the results of the mathematical theory of probability became rather secondary to many investigators. ... [however] the basis for the applicability of the results of the mathematical theory of probability to real 'random phenomena' must depend in some form on the *frequency concept of probability*, the unavoidable nature of which has been established by Von Mises in a spirited manner." [75].

Let us go into some more detail. The usual treatment of probability theory is designed so that abstract probabilities can be computed, but nothing is said about what probability really means, or how the concept can be applied meaningfully to the actual world. In [115] Von Mises analyzes the situation in detail, and suggests that a proper definition of probability depends on obtaining a proper definition of a random sequence.

The *frequency theory* to interpret probability says roughly that if we perform an experiment many times, then the ratio of favorable outcomes to the total number n of experiments will, *with certainty*, tend to a limit, p say, as $n \to \infty$. This tells us something about the *meaning* of probability, namely the measure of the positive outcomes is p. But suppose we throw a coin 1000 times and wish to know what to expect. Is 1000 enough for convergence to happen? The statement above does not say. So we have to add something about the rate of convergence. But we cannot assert a *certainty* about a particular number of n throws, such as "the proportion of heads will be $p \pm \varepsilon$ for large enough n (with ε depending on n)". We can at best say "the proportion will lie between $p \pm \varepsilon$ with at least such and such probability (depending on ε and n_0) whenever $n > n_0$." But now we defined probability in an obviously circular fashion.

In 1919 Von Mises proposed to eliminate the problem by postulating that a sequence of outcomes of independent repetitions of random events in nature, like a sequence of tosses with a coin, satisfies certain properties. The properties selected were claimed to be validated by abundant empirical evidence. (We discuss the actual properties he suggests below.) The analogy Von Mises uses is with a physical science as thermodynamics, where, apart from the assumption of the basic laws like the law of conservation of energy, or the law of increasing entropy, the remainder is derived in a purely mathematical way. These laws have no other justification than a long history of failures of inventors of perpetuum mobiles (in contrast to the idea that perpetuum mobiles are impossible because of the laws of thermodynamics). Coming back to probability theory, granted the Von Mises axioms, the remainder of the calculus is developed in a purely mathematical way, and the mathematical laws of probability result. This approach actually satisfied Von Mises, and solves the problem noticed above because one property of a random sequence will be that the relative frequency limit exists. Other philosophers of science insist that additionally the random sequences defined form a set of full measure, and without exception do satisfy all laws of probability, because then it seems physically justifiable to assume that as a result of an (infinite) experiment only (or rather with probability one) random sequences appear (see [71, 100, 115, 173]).

Von Mises' particular interpretation of a notion of infinite random sequence of zeros and ones designated by the special name of *collective* (*Kollektiv* in German) is as follows. An infinite sequence $a_1 a_2 \ldots$ of zeros and ones is a random sequence in the special meaning of *collective* if the following two conditions are satisfied:

(1) Firstly, if f_n is the number of ones among the first n terms of the sequence, then

$$\lim_{n \to \infty} \frac{f_n}{n} = p,$$

for some p, $0 < p < 1$.

(2) Secondly, (1) is not only required for the original sequence, but (with the same limit p) also for every infinite subsequence $a_{n_1} a_{n_2} \ldots$ obtained by some *admissible* partial function ϕ, which is defined for all finite binary sequences and takes the values 0 and 1, and selecting one after the other those indices n for which $\phi(a_1 a_2 \ldots a_{n-1}) = 1$.

The existence of a relative frequency limit, condition (1), is a strong assumption. Empirical evidence from long runs of dice throws, in gambling houses, or with death statistics in insurance mathematics, suggests that the relative frequencies are *apparently convergent*. But clearly, no empirical evidence can be given for the existence of a definite limit for the relative frequency. However long the test run, in practice it will always be finite, and whatever the apparent behavior in the observed initial segment of the run, it is always possible that the relative frequencies keep oscillating forever if we continue.

Condition (2) says that, for any "admissible" strategy of successively selecting infinitely many elements from the sequence, the frequency of ones in the selection goes to the same limit as in condition (1). Put in other words, considering the sequence as fair coin tosses, condition (2) says there is no *strategy* ϕ (*principle of excluded gambling system*) that assures a player, betting at fixed odds and in fixed amounts on the tosses of the coin, to make infinite gain. That is, no advantage is gained in the long run by following some system, such as betting "head" after each run of seven consecutive tails, or (more plausibly) by placing the nth bet "head" after the appearance of $n+7$ tails in succession. According to Von Mises, the above conditions are sufficiently familiar and form an uncontroverted empirical generalization to serve as a basis of an applicable calculus of probabilities. The problem with this definition is that Von Mises was unable to give a rigorous definition of what is the admissibility criterion. He essentially appeals to the familiar notion that no gambler, making a fixed number of wagers of "heads", at fixed odds and in fixed amounts, on the flips of a coin, has profit in the long run from betting according to a system instead of betting at random. Says Church: "this definition... while clear as to general intent, is too inexact in form to serve satisfactorily as the basis of a mathematical theory."

It turns out that the naive mathematical approach to a concrete formulation comes to grief as follows. We completely ignore the clear intention of Von Mises concerning a nontrivial restriction implied by the phrase "admissible place selection functions" by admitting simply *all* partial functions. Since arbitrary functions are allowed as a strategy, this definition is too restrictive, and no sequence exists that satisfies it with probability p other than 0.

EXAMPLE. Let $a=a_1a_2\dots$ be any infinite string satisfying (1). Define ϕ_1 as $\phi_1(a_1\dots a_{i-1})=1$ if $a_i=1$, and undefined otherwise. But then $p=1$. However, this is not all of the story. Defining ϕ_0 by $\phi_0(a_1\dots a_{i-1})=b_i$, b_i the complement of a_i for all i, we obtain by (2) that $p=0$. Consequently, if we allow functions like ϕ_1 and ϕ_0 as strategies, then Von Mises' definition cannot be satisfied at all.

This counterexample was not recognized as such by Von Mises, because it apparently violates the admissibility condition that a_i is not used in the definition of $\phi(a_1\dots a_{i-1})$. Here Von Mises' position is succinctly expressed by "first the collective, then the probability." Each collective, a physical object, determines what the admissible place selection functions for it are. While we can generalize from experience that all lawlike place selection functions are admissible, a place selection function pulled out of the blue with reference to a particular collective is inadmissible for it (but may be admissible for other collectives). In the example it just happens that *after* a criterion for admissible ϕ has been fixed too widely, it turns out that for any sequence there is an admissible ϕ that coincides with a ψ that is defined in a clearly inadmissible fashion. Here we cannot go into the various arguments put forward by the contestants in the ensuing discussion, but note that several attempts to resolve this problem turned out to be unsatisfactory one way or the other.

A. Wald [167] showed that the restriction of the set of admissible ϕ to a countable set eliminates the contradiction above. For any countable set of admissible selection functions, almost all sequences are random. Can we meaningfully fix some countable set of functions? A. Church proposed to fix it [37] to the formal notion of *effectively computable functions*, or *recursive functions*, as developed by A.M. Turing and himself (gamblers use computable strategies). He points out that, with a total recursive ϕ, not only is the definition completely rigorous and do corresponding random sequences exist, but moreover they are abundant since the infinite random sequences with $p=\frac{1}{2}$ form a set of measure 1; and from the existence of random sequences with probability $\frac{1}{2}$, the existence of random sequences associated with other probabilities is readily derived. Let us call sequences satisfying (1) and (2) with computable ϕ *Mises–Wald–Church* random. Appealing to a theorem by Wald yields as a corollary that the set of Mises–Wald–Church random sequences associated with any fixed probability has the cardinality of the continuum. Moreover, each Mises–Wald–Church random sequence qualifies as a normal number. (A number is *normal* if each digit of the base, and each block of digits of any length, occurs with equal asymptotic frequency.) Note however, that not every normal number is Mises–Wald–Church random. This follows, for instance, from Champernowne's number

$$0.1234567891011121314151617181920\dots$$

that is normal and where the ith digit is easily calculated from i. The definition of a Mises–Wald–Church random sequence implies that its consecutive digits cannot be effectively computed. (Namely, existence of an effective ϕ_1 as above contradicts $0<p<1$ in (2).) Thus, an existence proof for Mises–Wald–Church random sequences is necessarily nonconstructive.

Unfortunately, the Von Mises–Wald–Church definition is not yet good enough, since it was discovered by Ville [160] that even standard properties such as the Law of the Iterated Logarithm do not follow from it. In 1965, P. Martin-Löf, visiting Kolmogorov, succeeded in defining random sequences in a manner that is free of such difficulties [109]. His notion of infinite random sequences is related to infinite sequences of which all finite initial segments have high Kolmogorov complexity (cf. Section 2.4; for a survey of the work on infinite random sequences, see [81, 82]).

Until now the discussion has centered on infinite random sequences where the randomness is defined in terms of limits of relative frequencies. However,

> "The frequency concept, based on the notion of *limiting frequency* as the number of trials increases to infinity, does not contribute anything to substantiate the application of the results of probability theory to real practical problems where we always have to deal with a finite number of trials," [75].

It seems more appealing to try to define randomness for finite strings first and only then define random infinite strings in terms of randomness of initial segments. The aim is to obtain a theory in which the existence of frequency limits follows from the randomness of the sequence, rather than the other way around [129]. However, properly defining random *finite* strings appeared to be an even more hopeless affair than such a definition for infinite strings. But the essence of the solution had already been discovered before. For instance, P.S. Laplace [83] and also Kolmogorov [75] observed that "randomness" consists in lack of "regularity", and that, if some regularity can be caused by a simple law, then the chance that it is caused by this law is far greater than that it arose spontaneously. Moreover, it can be noted that there cannot be a very large number of simple laws. Identifying "laws" with "algorithms" brings us to our topic proper.

1.1. The inventors

We feel it is important to give a careful treatment of the genesis of the ideas in this area. Kolmogorov complexity originated with the discovery of universal descriptions, and a recursively invariant approach to the concepts of complexity of description, randomness, and a priori probability. Historically, it is firmly rooted in R. Von Mises' notion of random infinite sequences [115] as discussed above.

With the advent of electronic computers in the 1950s, a new emphasis on computer algorithms and a maturing general recursive function theory, ideas tantamount to Kolmogorov complexity, came to many people's minds, because "when the time is ripe for certain things, these things appear in different places in the manner of violets coming to light in early spring" (Wolfgang Bolyai to his son Johann in urging him to claim the invention of non-Euclidean geometry without delay). Thus, with Kolmogorov complexity one can associate three inventors: R.J. Solomonoff in Cambridge, MA

[149], close in time but far away in geography followed by A.N. Kolmogorov in Moscow [75, 76], and then G.J. Chaitin in New York [25].

R.J. Solomonoff had been a student of R. Carnap at the University of Chicago in the fifties. His objective was to formulate a completely general theory of inductive inference that would overcome shortcomings of previous methods like [22]. Already in November 1960 Solomonoff had published a Zator Company technical report on the subject of "Kolmogorov" complexity [148]. In March 1964 he published a long paper [149] introducing a version of universal a priori probability, namely the forerunner of the Solomonoff–Levin distribution, through the intermediate definition of what we have termed "Kolmogorov complexity", and proved the Invariance Theorem. This paper received little attention until Kolmogorov started to refer to it from 1968 onward. It is interesting to note that Solomonoff also discusses informally the ideas of randomness of finite strings, noncomputability of Kolmogorov complexity, computability of approximations to Kolmogorov complexity, and resource-bounded Kolmogorov complexity. A paragraph referring to Solomonoff's work occurs in [114]. To our knowledge, these are evidently the earliest documents outlining an algorithmic theory of descriptions.

In 1933 the great Soviet mathematician A.N. Kolmogorov[1] supplied probability theory with a powerful mathematical foundation [74]. Following a four-decades long controversy on Von Mises' concept of randomness, Kolmogorov finally introduced complexity of description of finite individual objects, as a measure of individual information content and randomness, and proved the Invariance Theorem in his paper of spring 1965 [76]. Kolmogorov's invention of the complexity of description was in no way a haphazard occurrence, but on the contrary the inevitable confluence of several of his major research threads: the foundations of probability and random sequences, information theory, and the theory of algorithms. *Uspekhi Mat. Nauk* announced Kolmogorov's lectures on related subjects in 1961 and following years, and, says Kolmogorov: "I came to similar conclusions [as Solomonoff], before becoming aware of Solomonoff's work, in 1963–1964" [77].

G.J. Chaitin had finished the Bronx High School of Science, and was an 18-years old undergraduate student at the City College of the City University of New York when he submitted the original versions of [24, 25] for publication, in October and November 1965 respectively. Published in 1966, [24] investigated "state/symbol" complexities relative to arbitrary algorithms. In this work Chaitin extended C.E. Shannon's earlier work on coding concepts [145], and did not introduce any invariant notion of complexity. However, at the end of his 1969 publication [25], Chaitin apparently independently puts forward the proper notion of Kolmogorov complexity, proves the Invariance Theorem, and studies infinite random binary sequences (in the sense of having maximally random finite initial segments) and their complexity oscillations. According to Chaitin: "this definition [of Kolmogorov complexity] was independently proposed about 1965 by A.N. Kolmogorov and me ... Both Kolmogorov and I were then unaware of related proposals made in 1960 by Ray Solomonoff" [27].

[1]Andrei N. Kolmogorov, born 25 April 1903 in Tambov, USSR, died 20 October 1987 in Moscow. For biographical details see [3, 17, 52], or [166], and the obituary in the *Times* [121].

The Swedish mathematician P. Martin-Löf, visiting Kolmogorov in Moscow during 1964–1965, investigated complexity oscillations of infinite sequences and proposed a new definition of infinite random sequences which is based on constructive measure theory [109, 111]. L.A. Levin, then a student of Kolmogorov, found a definition of a priori probability (the Solomonoff–Levin distribution) as a maximal semicomputable measure [173], and introduced the quantity corresponding to the self-delimiting variant of Kolmogorov complexity as the negative logarithm of a priori probability. In 1974 he more explicitly introduced Kolmogorov complexity based on self-delimiting programs [86]. This work relates to P. Gács' results concerning the differences between symmetric and asymmetric expressions for information [45]. In 1975 Chaitin also discovered and investigated this Kolmogorov complexity based on self-delimiting programs [28]. Another variant of Kolmogorov complexity, viz., the length of the shortest program p that computes a function f such that $f(i)$ is the ith bit of the target string, was found by D.W. Loveland [103, 104] and used extensively in [173]. Other variants and results were given by Willis [169], Levin [85], Schnorr [140], and Cover [38]. Apart from Martin-Löf's work, we mention that of C.P. Schnorr [138, 139] on the relation between Kolmogorov complexity and randomness of infinite sequences. This chapter of the Handbook is a precursor of our forthcoming textbook [97].

2. Mathematical theory

Kolmogorov gives a cursory but fundamental and elegant exposition of the basic ideas in [78]. Currently, the most complete treatment of the fundamental notions and results in Kolmogorov complexity is Levin and Zvonkin's 1970 survey [173]. Since this survey is not up to date, it should be complemented by Schnorr's [139, 140] and Chaitin's more recent monograph [30]. Kolmogorov and Uspenskii present a survey covering research in the Soviet Union [80], and Kolmogorov's selected works in the area are contained in [79] (see also [174]). For the advanced reader we mention Levin's important work [88]. An introductory but complete treatment of Kolmogorov complexity and its applications will be given in [97].

There are several variants of Kolmogorov complexity; here we focus on the original version in [24, 76, 78, 173]. (Later, we also define the more refined "self-delimiting" version.)

NOTATION. It is useful to fix some notation first. All through this paper it is convenient to identify the positive integers with the finite binary strings as follows:

$$(0, \varepsilon), (1, 0), (2, 1), (3, 00), (4, 01), (5, 10), (6, 11), (7, 000), \ldots$$

That is, the natural number n corresponds with the nth binary string in lexicographic length-increasing order. We call such an n a *finite object*, and whether we view it as a natural number or a binary string will be apparent from the context. If x is a binary string (natural number) then $|x|$ denotes the *length* or number of zeros and ones in x. The *number of elements* in a finite set A is denoted by $d(A)$. Hence, with $A=\{1,2,\ldots,n\}$ we have $|d(A)|$ is about $\log n$. A few times we need to denote the *absolute value* of a number as in $|a-b|$, the absolute value of the difference of a and b. We feel that in each such case the context clearly indicates that we mean an absolute value and not a length, and refrain from introducing special notation.

First take a general viewpoint, as in [78], in which one assumes some domain D of objects with some standard enumeration of objects x by numbers $n(x)$. We are interested in the fact that $n(x)$ may not be the most economical way to specify x. To compare methods of specification, we agree to view such a method as a function S from natural numbers p written in binary notation to natural numbers n, $n = S(p)$. We do not yet assume that S is computable, but maintain full generality to show to what extent such a theory can also be developed with noneffective notions, and at which point effectiveness is required. For each object x in D we call the length $|p|$ of the smallest p that gives rise to it the *complexity of object x with respect to the specifying method S*:

$$K_S(x) = \min\{|p| : S(p) = n(x)\},$$

and $K_S(x) = \infty$ if there are no such p. In computer science terminology we can call p a program and S a programming method (or language). Then one can say that $K_S(x)$ is the minimal length of a program to obtain x under programming method S. Considering distinct methods $S_1, S_2, \ldots, S_r$ of specifying the objects of D, it is easy to construct a new method S that gives for each object x in D a complexity $K_S(x)$ that exceeds only by c, c less than about $\log r$, the original minimum of the complexities $K_{S_1}(x), K_{S_2}(x), \ldots, K_{S_r}(x)$. The only thing we have to do is to reserve the first $\log r$ bits of p to identify the method S_i that should be followed, using as a program the remaining bits of p. We say that a method S "absorbs a method S' with precision up to c" if for all x

$$K_S(x) \leqslant K_{S'}(x) + c.$$

Above we have shown how to construct a method S that absorbs any of the methods $S_1, \ldots, S_r$ with precision up to c, where $c \sim \log r$. Two methods S_1 and S_2 are called "c-equivalent" if each of them c-absorbs the other. As Kolmogorov remarks, this construction would be fruitless if the hierarchy of methods with respect to absorption were odd, for instance, if there is no bottom element. However, under relatively natural restriction on S this is not so. Namely, among the partial recursive functions (in the sense of Turing [155]), there exist *optimal* ones, say S, such that for any other computable function S'

$$K_S(x) \leqslant K_{S'}(x) + c_{S,S'}.$$

Clearly, all optimal methods S, S' of specifying objects in D are equivalent in the following way: the absolute value of the difference satisfies

$$|K_S(x) - K_{S'}(x)| \leqslant c_{S,S'}.$$

Thus, from an asymptotic point of view, the complexity $K(x)$ of an object x, when we restrict ourselves to optimal methods of specification, does not depend on accidental peculiarities of the chosen optimal method.

To fix thoughts, w.l.o.g., consider the problem of describing a finite object x. It is useful to develop the idea that the complexity of specifying an object can be facilitated when another object is already specified. Thus, we define the complexity of an object x, given an object y. Let $p \in \{0,1\}^*$, and we call p a *program*. Any computable function f together with strings p and y such that $f(p,y)=x$ is a description of x. We call f the *interpreter* or *decoding* function. The (descriptional) complexity K_f of x, with respect to f, conditional to y, is defined by

$$K_f(x|y)=\min\{|p|: p\in\{0,1\}^* \ \& \ f(p,y)=x\},$$

and $K_f(x|y)=\infty$ if there are no such p. The following theorem asserts that each finite object has an intrinsic complexity which is independent from the means of description. Namely, there exist asymptotically optimal functions such that the description length with respect to them minorizes the description length with respect to any other function, apart from an additive constant, for all finite objects. This important fact is what makes the theory work.

INVARIANCE THEOREM (Solomonoff [149], Kolmogorov [76], Chaitin [25]). *There exists a partial recursive function f_0, such that, for any other partial recursive function f, there is a constant c_f such that for all strings x, y, $K_{f_0}(x|y) \leqslant K_f(x|y)+c_f$.*

PROOF. Fix some standard enumeration of Turing machines, with an ordinary input tape, an extra input tape to contain the conditional information, a worktape, and an output tape. Let $n(T)$ be the number associated with Turing machine T. Assume that the conditional input y is contained on the extra input tape. Let f_0 be the *universal* partial recursive function computed by a universal Turing machine U. That is, U starting with input 0^n1p, $p\in\{0,1\}^*$, on the ordinary input tape and y on the extra input tape halts with output x on the output tape iff T starting with input p on the ordinary input tape and y on the extra input tape halts with x on its output tape, for $n(T)=n$. Choosing $c_f=n+1$ finishes the proof. □

Clearly, any function f_0 that satisfies the Invariance Theorem is *optimal* in the sense discussed above. Therefore, we are justified to fix a particular *reference machine U* as in the proof of the theorem and its associated partial recursive function f_0, and drop the subscripts on K. We define the *conditional Kolmogorov* complexity $K(x|y)$ of x under condition of y to be equal to $K_{f_0}(x|y)$ for this fixed optimal f_0. Define the *unconditional Kolmogorov* complexity of x as $K(x)=K(x|\varepsilon)$, where ε denotes the empty string ($|\varepsilon|=0$).

In his talks Kolmogorov used to credit A.M. Turing [155] for the universal Turing machine, which is the substance of the Invariance Theorem. Before we continue, we recall the definitions of the big-O notation.

NOTATION (*order of magnitude*). We use the *order of magnitude* symbols O, o, Ω and Θ. If f and g are functions on the real numbers, then

(i) $f(x)=O(g(x))$ if there are positive constants c, x_0, such that $|f(x)|\leqslant c\,|g(x)|$ for all $x\geqslant x_0$;

(ii) $f(x)=o(g(x))$ if $\lim_{x\to\infty} f(x)/g(x)=0$;

(iii) $f(x)=\Omega(g(x))$ if $f(x)\neq o(g(x))$; and

(iv) $f(x)=\Theta(g(x))$ if both $f(x)=O(g(x))$ and $f(x)=\Omega(g(x))$.

The relevant properties are extensively discussed in [72, 161]. This use of Ω was introduced first by Hardy and Littlewood in 1914, and must not be confused by Chaitin's real number Ω we meet in a later section. (Some computer scientists use the order of magnitude symbol Ω such that $f(x)=\Omega(g(x))$ iff there is a positive constant c such that $f(x)\geqslant cg(x)$ from some x onwards. This is different from our use of the traditional definition of Ω as the complement of o, as in (ii), which says: $f(x)=\Omega(g(x))$ iff there is a positive constant c such that $f(x)\geqslant cg(x)$ for infinitely many x.)

EXAMPLE. For each finite binary string x we have $K(xx)\leqslant K(x)+O(1)$. (The constant implied by the big-O notation is fixed by the choice of reference machine U.) Namely, let T compute x from program p. Now fix a universal machine V which, on input $0^{n(T)}1p$, simulates T just like the reference machine U in the proof of the Invariance Theorem, but additionally V doubles T's output before halting. Now V starting on $0^{n(T)}1p$ computes xx, and therefore U starting on $0^{n(V)}10^{n(T)}1p$ computes xx. Hence, for all x, $K(xx)\leqslant K(x)+n(V)+1$.

EXAMPLE. Let us define $K(x, y)=K(\langle x, y\rangle)$ with $\langle\cdot,\cdot\rangle$ a standard one-one mapping (pairing function) of pairs of natural numbers to natural numbers. That is, $K(x, y)$ is the length of a shortest program that outputs x and y and a way to tell them apart. It is seductive to conjecture $K(x, y)\leqslant K(x)+K(y)+O(1)$, the obvious (but false) argument running as follows. Suppose we have a shortest program p to produce x, and a shortest program q to produce y. Then with O(1) extra bits to account for some Turing machine T that schedules the two programs, we have a program to produce x followed by y. However, any such T will have to know where to divide its input to identify p and q. We can separate p and q by prefixing pq by a clearly distinguishable encoding r of the length $|p|$ in $O(\log|p|)$ bits (see Section 2.2 on self-delimiting strings). Consequently, we have at best established

$$K(x, y)\leqslant K(x)+K(y)+O(\log(\min(K(x), K(y)))).$$

In general this cannot be improved.

2.1. *Incompressibility*

Apart from showing that complexity is an attribute of the finite object alone, the Invariance Theorem has also another most important consequence: it gives an upper bound on the complexity. Namely, there is a fixed constant c such that for all x of length n we have

$$K(x)\leqslant n+c.$$

This is easy to see. If T is a machine that just copies its input to its output, then $p=0^{n(T)}1x$ is a program for the reference machine U to output x.

This says that $K(x)$ is bounded above by the length of x modulo an additive constant. The obvious question to ask further is: "how many x can be compressed how far?". Since there are 2^n binary strings of length n, but only 2^n-1 possible shorter descriptions, it follows that, for all n, there is a binary string x of length n such that $K(x)\geqslant n$. We call such strings *incompressible*. It also follows that, for any length n and any binary string y, there is a binary string x of length n such that $K(x|y) \geqslant n$.

EXAMPLE. Is a substring of an incompressible string also incompressible? A string $x = uvw$ can be specified by a short description for v of length $K(v)$, a description of $|u|$, and the literal description of uw. Moreover, we need information to tell these three items apart. Such information can be provided by prefixing each item with a self-delimiting description of its length, as explained in the section on self-delimitation. Together this takes $K(v) + |uw| + O(\log|x|)$ bits. Hence,

$$K(x) \leqslant K(v) + O(\log|x|) + |uw|.$$

Thus, if we choose x incompressible, $K(x) \geqslant |x|$, then we obtain

$$K(v) \geqslant |v| - O(\log|x|).$$

It can be shown that this is optimal—a substring of an incompressible string can be compressible. This conforms to a fact we know from probability theory: every sufficiently long random string must contain long runs of zeros.

EXAMPLE. Define $p(x)$ as a shortest program for x. We show that $p(x)$ is incompressible, in the sense that there is a constant $c > 0$ such that for all strings x, we have $K(p(x)) \geqslant |p(x)| - c$. Suppose the contrary. Define a universal machine V that works just like the reference machine U, except that V first simulates U on its input to obtain an output, and then uses this output as input on which to simulate U once more. But then, U with input $0^{n(V)}1p(p(x))$ computes x, and therefore $K(x) \leqslant |p(x)| - c + n(V) + 1$, for all $c > 0$, some x, which is impossible.

EXAMPLE. It is easy to see that $K(x|x) \leqslant n(T) + 1$, where T is a machine that just copies the input to the output. However, it is more interesting that, for *some* shortest program $p(x)$ of (x), $K(p(x)|x) \leqslant \log K(x) + O(1)$, which cannot be improved in general. Hint: later we show that K is a noncomputable function. This rules out that we can compute any shortest program $p(x)$ from x. However, we can dovetail the computation of all programs shorter than $|x| + 1$: run the first program one step, run the first program one step and the second program one step, and so on. This way we will eventually enumerate all programs that output x. However, since some computations may not halt, and the halting problem is undecidable, we need to know the length of a shortest program $p(x)$ to recognize any such program when it is found.

A natural question to ask is: how many strings are incompressible? It turns out that virtually all strings of given length n are incompressible. Namely, there is at least one x of length n that cannot be compressed to length $< n$ since there are 2^n strings of length n and but $2^n - 1$ programs of length less than n; at least $\frac{1}{2}$ of all strings of length n cannot be compressed to length $< n - 1$ since there are but $2^{n-1} - 1$ programs of length less than $n - 1$; at least $\frac{3}{4}$th of all strings of length n cannot be compressed to length $< n - 2$, and so on.

Generally, let $g(n)$ be an integer function. Call a string x of length n *g-incompressible* if

$K(x) \geqslant n-g(n)$. There are 2^n binary strings of length n, and only $2^{n-g(n)}-1$ possible descriptions shorter than $n-g(n)$. Thus, the ratio between the number of strings x of length n with $K(x)<n-g(n)$ and the total number of strings of length n is at most $2^{-g(n)}$, a *vanishing fraction* when $g(n)$ increases unboundedly with n. In general we loosely call a finite string x of length n *random* if $K(x) \geqslant n-O(\log n)$.

Intuitively, incompressibility implies the absence of regularities, since regularities can be used to compress descriptions. Accordingly, we like to identify incompressibility with absence of regularities or *randomness*. In the context of finite strings randomness like incompressibility is a matter of degree: it is obviously absurd to call a given string random and call nonrandom the string resulting from changing a bit in the string to its opposite value. Thus, we identify c-incompressible strings with c-*random* strings.

However, with infinite strings we may a priori hope to be able to use Kolmogorov complexity to sharply distinguish the random strings from the nonrandom ones, to finish the task set by Von Mises (see the Introduction). Let us call an infinite string x g-*incompressible* if each initial string $x_{1:n}$ of length n has $K(x_{1:n}) \geqslant n-g(n)$, from some n onward. In [109], Martin-Löf has defined a satisfactory notion for randomness of infinite strings. It turns out that Martin-Löf random strings are $(2\log n)$-incompressible, but not $(\log n)$-incompressible (cf. Section 2.4). We call finite or infinite $O(\log n)$-incompressible strings loosely "random" or "Kolmogorov random", but want to stress here that randomness for infinite strings according to Martin-Löf has a stricter and more profound definition. We return in somewhat more detail to this matter below.

Curiously, though most strings are random, it is impossible to effectively prove them random. The fact that almost all finite strings are random but cannot be proved to be random amounts to an information-theoretic version of Gödel's Theorem below. Strings that are not incompressible are *compressible* or *nonrandom*. The nonrandom infinite binary strings are very scarce: they have measure zero in the set of all infinite binary strings.

2.2. *Self-delimiting descriptions*

In previous sections we formalized the concept of a greatest lower bound on the length of a description. Now we look at feasibility. Let the variables x, y, x_i, $y_i, \ldots$ denote strings in $\{0,1\}^*$. A *description* of x, $|x|=n$, can be given as follows:

(1) A piece of text containing several formal parameters $p_1, \ldots, p_m$. Think of this piece of text as a formal parametrized procedure in an algorithmic language like Pascal.

It is followed by

(2) an ordered list of the actual values of the parameters.

The piece of text of (1) can be thought of as being encoded over a given finite alphabet, each symbol of which is coded in bits. Therefore, the encoding of (1) as prefix of the binary description of x requires $O(1)$ bits. This prefix is followed by the ordered list (2) of the actual values of $p_1, \ldots, p_m$ in binary. To distinguish one from the other, we encode (1) and the different items in (2) as self-delimiting strings, an idea used already by C.E. Shannon.

For each string $x \in \{0,1\}^*$, the string $\bar{x}$ is obtained by inserting a "0" in between each pair of adjacent letters in x, and adding a "1" at the end. That is,

$$\overline{01011} = 0010001011.$$

Let $x' = \overline{|x|}\, x$ (an encoding of the length of x in binary followed by x in binary). The string x' is called the *self-delimiting* version of x. So "100101011" is the self-delimiting version of "01011". (According to our convention "10" is the fifth binary string.) The self-delimiting binary version of a positive integer n requires $\log n + 2 \log \log n$ bits, and the self-delimiting version of a binary string w requires $|w| + 2 \log |w|$ bits. For convenience, we denote the length $|n|$ of a natural number n by "$\log n$".

EXAMPLE (*generalization*). More generally, for $x \in \{0,1\}^* - \{\varepsilon\}$, $d_0(x) = \bar{x}$ is the self-delimiting version of order 0 of x using $2|x|$ bits. Above we defined the "standard" self-delimiting version $d_1(x) = x'$ of order 1. In general, for $i \geqslant 1$, $d_i(x) = \bar{x}_i x_{i-1} \ldots x_1 x$, with $x_1 = |x|$ and $x_j = |x_{j-1}|$ $(1 < j \leqslant i)$, is the self-delimiting version of order i of x. Define $\log^{(1)} = \log$, and $\log^{(j+1)} = \log \log^{(j)}$ for $j \geqslant 1$. Then,

$$|d_i(x)| = |x| + \log^{(1)}|x| + \cdots + \log^{(i-1)}|x| + 2 \log^{(i)}|x|.$$

Obviously, further improvements are possible.

EXAMPLE. Self-delimiting descriptions were used in the proof of the Invariance Theorem (namely, in the encoding $0^{n(T)}1$). Using it explicitly, we can define Kolmogorov complexity as follows. Fix an effective coding c of all Turing machines as binary strings such that no code is a prefix of any other code. Denote the code of Turing machine M by $c(M)$. Then the Kolmogorov complexity of $x \in \{0,1\}^*$, with respect to c, is defined by $K_c(x) = \min\{|c(M)y| \colon M \text{ on input } y \text{ halts with output } x\}$.

EXAMPLE (*self-delimiting Kolmogorov complexity*). A code c such that $c(x)$ is not a prefix of $c(y)$ if $x \neq y$ is called a *prefix code*. We can define a variant of Kolmogorov complexity by requiring at the outset that we only consider Turing machines for which the set of programs is a prefix code. The resulting variant, called *self-delimiting* Kolmogorov complexity, has nicer mathematical properties than the original one, and has therefore become something of a standard in the field. This complexity is variously denoted in the literature by KP, I, H, or simply by K which results in confusion with the original notion. We treat it in Section 2.7 and denote it (only) there by K'. For most applications it does not matter whether we use $K'(x)$ or $K(x)$ since they coincide to within an additive term of $O(\log |x|)$. In the case of inductive inference, however, we need to use the self-delimiting version of complexity. We denote both versions indiscriminately by $K(x)$, and point out which version we mean if it matters.

EXAMPLE (*Li, Maass, and Vitányi*). In proving lower bounds in the theory of computation it is sometimes useful to give an efficient description of an incompressible string with "holes" in it. The reconstruction of the complete string is then achieved using an additional description. In such an application we aim for a contradiction where these two descriptions together have significantly smaller length than the incompressible string they describe. Formally, let $x_1 \ldots x_k$ be a binary string of length n with the x_i's

$(1 \leqslant i \leqslant k)$ blocks of equal length c. Suppose that d of these blocks are deleted and the relative distances in between deleted blocks are known. We can describe this information by

(1) a formalization of this discussion in O(1) bits, and

(2) the actual values of $c, m, p_1, d_1, p_2, d_2, \ldots, p_m, d_m$, where m $(m \leqslant d)$ is the number of "holes" in the string, and the literal representation of $\hat{x} = \hat{x}_1 \hat{x}_2 \ldots \hat{x}_k$.

Here $\hat{x}_i$ is x_i if it is not deleted, and is the empty string otherwise; p_j, d_j indicates that the next p_j consecutive x_i's (of length c each) are one contiguous group followed by a gap of $d_j c$ bits long. Therefore, $k-d$ is the number of (nonempty) $\hat{x}_i$'s, with

$$k = \sum_{i=1}^{m} (p_i + d_i) \quad \text{and} \quad d = \sum_{i=1}^{m} d_i.$$

The actual values of the parameters and $\hat{x}$ are coded in a self-delimiting manner. Then, by the convexity of the logarithm function, the total number of bits needed to describe the above information is no more than

$$(k-d)c + 3d \log(k/d) + \mathrm{O}(\log n).$$

We then proceed by showing that we can describe x by this description plus some description of the deleted x_i's, so that the total requires considerably less than n bits. Choosing x such that $K(x) \geqslant n$ then gives a contradiction (see [94]).

This finishes the Application Toolkit. We have now formalized the essence of what we need for most applications in the sequel. Having made our notions precise, many applications can be described informally yet rigorously. The remainder of the theory of Kolmogorov complexity we treat below is not always required for the later applications. But, for instance, for the proper definition of the Solomonoff–Levin distribution, as needed in the application to inductive inference, it is required to use the self-delimiting version of complexity we briefly discussed in an example above (see also Section 2.7).

2.3. *Quantitative estimate of K*

We want to get some insight in the quantitative behavior of K. We follow [173]. We start this section with a useful property. Consider the conditional complexity of a string x, with x an element of a given finite set M, given some string y. Let $d(M)$ denote the number of elements in M. Then the fraction of $x \in M$ for which $K(x|y) < |d(M)| - m$, does not exceed 2^{-m}, by a counting argument similar to that in Section 2.1. Hence we have shown that the conditional complexity of the majority of elements in a finite set cannot be significantly less than the complexity of the size of that set. The following lemma says that it cannot be significantly more either.

LEMMA (Kolmogorov). *Let A be an r.e. set of pairs (x, y), and let $M_y = \{x: (x, y) \in A\}$. Then, up to a constant depending only on A, $K(x|y) \leqslant |d(M_y)|$.*

PROOF. Let A be enumerated by a Turing machine T. Using y, modify T to T_y such that T_y enumerates all pairs (x, y) in A, without repetition. In order of enumeration we select the pth pair (x, y), and output the first element, i.e. x. Then we find $p<d(M_y)$, such that $T_y(p)=x$. Therefore, we have by the Invariance Theorem $K(x|y)\leqslant K_{T_y}(x)\leqslant|d(M_y)|$, as required. □

EXAMPLE. Let A be a subset of $\{0,1\}^*$. Let $A^{\leqslant n}$ equal $\{x\in A: |x|\leqslant n\}$. If the limit of $d(A^{\leqslant n})/2^n$ goes to zero for n going to infinity, then we call A *sparse*. For example, the set of all finite strings that have twice as many zeros as ones is sparse. This has as a consequence that all but finitely many of these strings have short programs.

CLAIM. (a) (Sipser) *If A is recursive and sparse, then for all constant c there are only finitely many x in A with $K(x)\geqslant|x|-c$. Using Kolmogorov's Lemma we can extend this result as follows.*

(b) *If A is r.e. and $d(A^{\leqslant n})/n^{-(1+\varepsilon)}2^n$, $\varepsilon>0$, goes to zero for n going to infinity, then, for all constant c, there are only finitely many x in A with $K(x)\geqslant|x|-c$.*

(c) *If A is r.e. and $d(A^{\leqslant n})\leqslant p(n)$ with p a polynomial, then, for all constant $c>0$, there are only finitely many x in A with $K(x)\geqslant|x|/c$.*

PROOF. (a) Consider the lexicographic enumeration of all elements of A. There is a constant d, such that the ith element x of A has $K(x)\leqslant K(i)+d$. If x has length n, then the sparseness of A implies that $K(i)\leqslant n-g(n)$, with $g(n)$ unbounded. Therefore, for each constant c and all n, if x in A is of length n, then $K(x)<n-c$ from some n onward.

(b) Fix c. Consider an enumeration of n-length elements of A. For all such x, the lemma above in combination with the sparseness of A implies that

$$K(x|n)\leqslant n-(1+\varepsilon)\log n+O(1).$$

Therefore, $K(x)\leqslant n-\varepsilon\log n+O(1)$, for some other fixed positive ε, and the right-hand side of the inequality is less than $n-c$ from some n onward.

(c) Similarly as above. □

We now look at unconditional complexity. We identify the binary string x with the natural number x, as in the correspondence mentioned at the outset of Section 2. This way, $K(x)$ can be considered as an integer function.

LEMMA (Kolmogorov). *For any binary string x, the following hold:*

(a) *$K(x)\leqslant|x|$, up to some constant not depending on x.*

(b) *The fraction of x for which $K(x)<l-m$ and $|x|=l$ does not exceed 2^{-m}, so that equality holds in* (a) *for the majority of words.*

(c) *$\lim_{x\to\infty}K(x)=\infty$, and*

(d) *for $m(x)$ being the largest monotonic increasing integer function bounding $K(x)$ from below, $m(x)=\min_{y\geqslant x}K(y)$, we have $\lim_{x\to\infty}m(x)=\infty$.*

(e) *For any partial recursive function $\phi(x)$ tending monotonically to ∞ from some x_0 onwards, we have $m(x)<\phi(x)$, for all large enough x.*

(f) *The absolute difference $|K(x+h)-K(x)|\leqslant2|h|$, up to some constant independent of x,h. That is, although $K(x)$ varies all the time between $|x|$ and $m(x)$, it does so fairly smoothly.*

PROOF. (a)–(d) have been argued above or are easy. For (e) see [173]. We prove (f). Let p be a minimal-length description of x so that $K(x)=|p|$. Then we can describe $x+h$ by $\bar{h}p$, $\bar{h}$ the (order 0) self-delimiting description of h, and a description of this discussion in a constant number of bits. Since $|\bar{h}|\leqslant 2|h|$ (see previous section), this proves (f). □

EXAMPLE. One effect of the information quantity associated with the length of strings is that $K(x)$ is *nonmonotonic on prefixes*. This can be due to the information contained in the length of x. That is, clearly for $m<n$ we can still have $K(m)>K(n)$. But then $K(x)<K(y)$ for $x=0^n$ and $y=0^m$, notwithstanding that y is a proper prefix of x. For example, if $n=2^k$ then $K(0^n)\leqslant \log\log n+O(1)$, while we have shown above that there are $m<n$ for which $K(m)\geqslant \log n-O(1)$. Therefore, the complexity of a part can turn out to be greater than the complexity of the whole. In an initial attempt to solve this problem we may try to eliminate the effect of the length of the string on the complexity measure by treating the length as given. However, this does not help as the next example shows.

EXAMPLE. For any binary string x, $|x|=n$, we have $K(x|n)\leqslant K(x)$, but usually the length of x does not give too much information about x. But sometimes it does. For instance for a string $x=0^n$. Then, $K(x)=K(n)+O(1)$, but $K(x|n)=O(1)$. Moreover, it is easy to find m such that $m<n$ and $K(0^m|n)=\Omega(\log n)$. But of course $K(0^m|m)=O(1)$ again. Thus, our next try is to look at the complexity $K(x|n)$ where $n=|x|$. But now we get nonmonotonicity in another way. Consider $x=n00\ldots0$ with $|x|=n$, that is, the nth binary string padded with zeros up to length n. These strings are called *n-strings* by Loveland [104]. Now always $K(x|n)=O(1)$, but by choosing the prefix n of x random we have $K(n|m)=\Omega(\log n)$ with $m=|n|$.

2.4. *Infinite random strings*

If $x=x_1x_2\ldots$ is a finite or infinite string of binary digits $x_i, |x|\geqslant n$, then $x_{m:n}$ denotes the substring $x_m x_{m+1}\ldots x_n$.

In connection with the task set by Von Mises (cf. Introduction) we would like to express the notion of randomness of infinite binary strings in terms of Kolmogorov complexity. The obvious way is to use the definition of finite random strings. That is, call an infinite binary string x random if there is a constant c such that, for all n, $K(x_{1:n})\geqslant n-c$. But in 1965 Martin-Löf found that such strings do not exist [109–111]:

THEOREM (Martin-Löf). *If $f(n)$ is a recursive function such that $\Sigma 2^{-f(n)}=\infty$, then for any infinite binary sequence x there are infinitely many n for which $K(x_{1:n})<n-f(n)$.*

EXAMPLE. $f(n)=\log n$ satisfies the condition of the theorem. Let $x=x_1x_2\ldots$ be any infinite binary string, and $x_{1:m}$ any m-length prefix of x. If $n-m$ is the natural number corresponding to $x_{1:m}$, so m is about $\log n$, then $K(x_{1:n})=K(x_{m+1}\ldots x_n)+O(1)$. This is easy to see, since we can uniquely reconstruct $x_{1:m}$ from the length $n-m$ of $x_{m+1}\ldots x_n$ with $O(1)$ additional bits of information.

However, it was observed by Martin-Löf that if $f(n)$ is such that the series

$$\sum 2^{-f(n)} \tag{2.1}$$

converges recursively (there is a recursive set of integers n_r such that $\sum_{n \geq n_r} 2^{-f(n)} \leq 2^{-r}$, for example $f(n) = \log n + 2 \log \log n$), then almost all strings x (in the sense of binary measure) have the property

$$K(x_{1:n}) \geq n - f(n), \tag{2.2}$$

from some n onwards. In a less precise form these phenomena were also presented by Chaitin [25]. Due to these complexity oscillations the idea of identifying infinite random sequences with those such that $K(x_{1:n}) \geq n - c$ does not work. These problems caused Martin-Löf to try another track and proceed directly to the heart of the matter. Namely, to justify any proposed definition of randomness, one will have to show that the sequences that are random in the stated sense satisfy the several properties of stochasticity we know from the theory of probability. So why not, instead of proving that each such property separately is satisfied by a proposed definition, formalize the property that the random sequences introduced possess, in an appropriate sense, all possible properties of stochasticity.

It turns out that the notion of infinite binary strings satisfying all properties of randomness, in the sense of all properties that hold with probability 1, is contradictory. However, if we restrict ourselves to only those properties that are effectively verifiable, and statistical tests for randomness invariably are effective, then the resulting notion of random infinite string is noncontradictory. Pursuing this approach through constructive measure theory, Martin-Löf [109] develops the notion of random binary sequences as having all "effectively verifiable" properties that from the point of view of the usual probability theory are satisfied with "probability 1". That is to say, they pass all effective statistical tests for randomness in the form of a "universal" test, where the bits represent the outcome of independent experiments with outcomes 0 or 1 with probability $\frac{1}{2}$. Not only do such random strings exist, indeed, it turns out that these random strings have measure 1 in the set of all strings. Using this definition of randomness he shows the following theorems [111].

Theorem (Martin-Löf). *Random binary sequences satisfy* (2.2) *from some n onwards, provided* (2.1) *converges recursively.*

Theorem (Martin-Löf). *If $K(x_{1:n}|n) \geq n - c$ for some constant c and infinitely many n, then x is a random binary sequence.*

For related work see also [103, 138, 139, 141, 173]. We mention that for the self-delimiting version (Section 2.7) of complexity K' it holds that x is random iff there is a constant $c > 0$ such that $K'(x_{1:n}|n) \geq n - c$, for all n.

2.5. *Algorithmic properties of K*

We select some results from Zvonkin and Levin's survey [173]. Again, we consider $K(x)$ as a function that maps a positive integer x to a positive integer $K(x)$.

THEOREM (Kolmogorov). (a) *The function $K(x)$ is not partial recursive. Moreover, no partial recursive function $\varphi(x)$, defined on an infinite set of points, can coincide with $K(x)$ over the whole of its domain of definition.*

(b) *There is a (total) recursive function $H(t, x)$, monotonically decreasing in t, such that $\lim_{t\to\infty} H(t, x) = K(x)$. That is, we can obtain arbitrary good estimates for $K(x)$ (but not uniformly).*

PROOF. (a) Every infinite r.e. set contains an infinite recursive subset [135, Theorem 5-IV]. Select an infinite recursive set A in the domain of definition of $\varphi(x)$. The function $f(m) = \min\{x: K(x) \geqslant m, x \in A\}$ is (total) recursive (since $K(x) = \varphi(x)$ on A), and takes arbitrarily large values. Also, by construction, $K(f(m)) \geqslant m$. On the other hand, $K(f(m)) \leqslant K_f(f(m)) + c_f$ by definition of K, and obviously $K_f(f(m)) \leqslant |m|$. Hence, $m \leqslant \log m$ up to a constant independent of m, which is false.

(b) Let c be a constant such that $K(x) \leqslant |x| + c$ for all x. Define $H(t, x)$ as the length of the smallest program p, with $|p| \leqslant |x| + c$, such that the reference machine U with input p halts with output x within t steps. □

EXAMPLE (Barzdin'). It is not too difficult to show by similar reasoning that if $f(x) < |x|$ is a total recursive function with $\lim_{x\to\infty} f(x) = \infty$, then the set $B = \{x: K(x) \leqslant f(x)\}$ is *simple* in the recursive-theoretic sense of Post. That is, B is recursively enumerable and the complement of B is infinite but does not contain an infinite recursively enumerable subset. It is then straightforward that for every axiomatized theory F (that is consistent and sound) there are only finitely many n for which the statement "$n \notin B$" is both true and provable in F. However, from the definition of B it follows that all x with $K(x) \geqslant |x|$ do not belong to B, and there are infinitely many of those. Hence, if F is strong enough to express statements of the form "$n \notin B$", for instance F contains arithmetic, then infinitely many true statements can be expressed in F but are not provable. This is a version of Gödel's famous incompleteness result. It is different from Gödel's original proof in the fact that our undecidable statements are not constructive. This result is attributed to Barzdin' in Levin and Zvonkin's 1970 survey [173].

It turns out that with Kolmogorov complexity one can quantify the distinction between r.e. sets and recursive sets. Let $x = x_1 x_2 \ldots$ be an infinite binary sequence such that the set of numbers n with $x_n = 1$ is r.e. That is, x is the *characteristic sequence* of the set $M = \{n: x_n = 1\}$. If the complementary set with the $x_n = 0$ were also r.e., then $f(n) = x_n$ would be computable, and the relative complexity $K(x_{1:n}|n)$ bounded. But in the general case, when the set of ones is r.e., $K(x_{1:n})$ can grow unboundedly.

THEOREM (Barzdin', Loveland). *For any binary sequence x with the set $M = \{n: x_n = 1\}$ being r.e., it holds that $K(x_{1:n}|n) \leqslant \log n + c_M$, where c_M is a constant dependent on M (but not dependent on n). Moreover, there are sequences such that for any n it holds that $K(x_{1:n}) \geqslant \log n$.*

PROOF. Let the number of ones in $x_{1:n}$ be $m \leqslant n$. Since M is r.e. we can recursively enumerate all of its elements without repetition. Given m, we know that after having enumerated m elements in M that are less or equal to n, we have found them all. Since

$K(m) \leqslant \log n + c$ for some fixed constant c, this proves the upper bound. The lower bound holds for universal sets like $K_0 = \{\langle x, y\rangle: T_x \text{ halts on input } y\}$ (see [9, 103, 173]). □

In [78] Kolmogorov gives the following interesting interpretation with respect to investigations in the foundations of mathematics: Label all Diophantine equations by natural numbers. Y.V. Mateyasevich has proved that there is no general algorithm to answer the question whether the equation D_n is soluble in integers (the answer to Hilbert's Tenth Problem is negative). Suppose we weaken the problem by asking for the existence of an algorithm that enables us to answer the question of the existence or nonexistence of solutions for the first n Diophantine equations with the help of some supplementary information of size related to n. The theorem above shows that this size can be as small as $\log n + O(1)$. Such information is in fact contained in the $\sim \log n$ length prefix of the mythical number Ω that encodes the solution to the halting problem for the first n Turing machines (cf. Section 3.6).

In the same 1968 paper [9] Barzdin' derives one of the first results in "time-limited" Kolmogorov complexity. It shows that by imposing recursive time limits on the decoding procedure, the length of the shortest description of a string can sharply increase. Let t be an integer function and T be a Turing machine. Define $K^t_T(x_{1:n}|n)$ as the minimum length of a program p such that T starting with conditional n on its extra input tape computes the n-length prefix of x within $t(n)$ steps, and then halts.

THEOREM (Barzdin'). *Let T be any Turing machine. For any binary sequence x with an r.e. set $M = \{n: x_n = 1\}$ and any constant $c > 0$, there exists a (total) recursive function t such that for infinitely many n, $K^t_T(x_{1:n}|n) \leqslant cn$ holds. Moreover, there are such sequences x such that for any (total) recursive t and any n, $K^t_T(x_{1:n}) \geqslant c_t n$ holds, with c_t a constant independent of n (but dependent on t.)*

2.6. *Information*

If the conditional complexity $K(x|y)$ is much less than the unconditional complexity $K(x)$, then we may interpret this as an indication that y contains much information about x. Consequently, up to an additive constant, we can regard the difference

$$I(x:y) = K(y) - K(y|x)$$

as a quantitative measure of the information about y contained in x. If we choose f_0, in the Invariance Theorem, such that $f_0(\varepsilon, x) = x$, then

$$K(x|x) = 0, \qquad I(x:x) = K(x).$$

In this way we can view the complexity $K(x)$ as the information contained in an object about itself. For applications, this definition of the quantity of information has the advantage that it refers to individual objects, and not to objects treated as elements of a set of objects with a probability distribution given on it, as in [144]. Does the new definition have the desirable properties that hold for the analogous quantities in classic information theory? We know that equality and inequality can hold only up to additive constants, according to the indeterminacy in the Invariance Theorem. For example, the equality $I(x:y)=I(y:x)$ cannot be expected to hold exactly, but a priori it can be expected to hold up to a constant related to the choice of reference function f_0. However, with the current definitions, information turns out to be symmetric only up to a logarithmic factor. Define $K(x, y)$ as the complexity of x and y together (see the examples *before* Section 2.1). That is, the length of the least program of U that prints out x and y and a way to tell them apart. The following lemma is due to Kolmogorov and Levin.

LEMMA (symmetry). *To within an additive term of* $O(\log K(x, y))$,

$$K(x, y)=K(x)+K(y|x).$$

In the general case it has been proved that equality up to a logarithmic error term is the best possible. From the lemma it follows immediately that, to within an additive term of $O(\log K(x, y))$,

$$K(x)-K(x|y)=K(y)-K(y|x),$$

and therefore the absolute value of the difference of the information quantities satisfies

$$|I(x:y)-I(y:x)|=O(\log K(x, y)).$$

It has been established that the difference can be of this order (see [173]).

2.7. *Self-delimiting Kolmogorov complexity*

This more refined version of complexity is, in a sense, implicit in Solomonoff's original a priori probability [148, 149]. A definition (the universal semicomputable semimeasure $\boldsymbol{m}(x)$, corresponding to the Solomonoff–Levin distribution we study later) was supplied in the 1970 survey of Levin and Zvonkin [173]. The quantity corresponding to the self-delimiting Kolmogorov complexity $K'(x)$ occurs already in the form of the negative logarithm of the a priori probability $\boldsymbol{m}(x)$. It is explicitly defined as below and studied by Levin and Gács in 1974 [45, 86]. It was also discovered in 1975 by Chaitin [28]. For the development of the theory $K'(x)$ is often a more useful complexity measure than the $K(x)$, but for many applications one can use both equally well because they coincide to within a logarithmic factor.

NOTE. With some abuse of notation, after this section we simply drop the prime of $K'(x)$ and denote all types of Kolmogorov complexity simply by $K(x)$. If it is important whether we intend the self-delimiting version or the non-self-delimiting one, then we will explicitly state which version we mean.

There are two ways to define $K'(x)$. First, consider a class of Turing machines $T_1, T_2, \ldots$ with a one-way input tape, a one-way output tape, and a two-way worktape. Let the infinite input tape contain only zeros or ones (no blanks). These are the self-delimiting Turing machines, because the set of inputs (programs) for which each machine halts is prefix-free. We call a binary string p a *program* for T if T starts scanning the leftmost bit of p and halts scanning the rightmost bit of p. Just as before, we can prove an Invariance Theorem concerning the complexities associated with the different machines, where the optimal complexity is provided by the universal self-delimiting machine. We fix one such machine U and call it the reference self-delimiting machine. The *self-delimiting complexity* $K'(x)$ is the length of the shortest program p of U that outputs x.

FACT. We have defined programs so that no program is the prefix of another one. Each program is *self-delimiting* with respect to T. This allows an alternative approach to define $K'(x)$.

Second approach: For a self-delimiting machine T and each binary string x, define $P(x)$ as the *probability* that T eventually halts with x written on the output tape. (Solomonoff has called P the a priori probability, cf. Section 2.9.) The *entropy* is defined as $H(x) = -\log P(x)$.

Let the symbols on the input tape be provided by independent tosses of an unbiased coin. This enables us to give a natural probability distribution over programs: the probability of program p is simply $2^{-|p|}$. We can now easily compose programs from self-delimiting subprograms by prefixing a sequence of n self-delimiting programs with a self-delimiting description of n. Choosing the method in the previous section, we can encode a binary string x by a program of length $|x| + 2\log|x|$. Namely, define Turing machine T such that it outputs a binary string x iff it first reads the self-delimiting binary encoding of the length of x, and then the usual binary representation of x. Thus, with respect to T,

$$P(x) \geqslant 2^{-|x|-2\log|x|}, \qquad H(x) \leqslant |x| + 2\log|x|, \qquad K'(x) \leqslant |x| + 2\log|x|.$$

To make the definitions meaningful, we normalize these measures with respect to an optimal *universal* machine. This choice maximizes P and minimizes H and K'. It can be shown that the optimal self-delimiting machine selected this way can be set equal to the reference machine U of the earlier approach! Hence the two approaches define the same K' (as usual, up to a fixed additive constant). Let U be such a machine. Then, the a priori probability of x is

$$P(x) = \sum_{U(p)=x} 2^{-|p|}.$$

(This is the *Solomonoff–Levin* distribution which we denote below by the special notation $\boldsymbol{m}(x)$. Thus the entropy $H(x) = -\log \boldsymbol{m}(x)$.) It follows immediately from the definitions that the relation between the different notions is $K(x) \leqslant K'(x) \leqslant H(x)$. Levin has shown the significant result that $K'(x) = -\log \boldsymbol{m}(x) + O(1)$. That is, if x has many *long* programs, it must also have a *short* program.

EXAMPLE. To within an additive constant, for all finite binary strings x, y we have $K'(x, y) \leqslant K'(x) + K'(y)$. Namely, let p and q be self-delimiting programs for x and y respectively. Let V be a universal machine just like the reference machine U, except that it simulates U first on p to produce x, then on q to produce y, and subsequently outputs x, y. Presented with input pq, V can tell p apart from q because p is self-delimiting. Hence, U with program $0^{n(V)}1pq$, computes xy. Therefore, for all finite binary strings x, y, $K'(x, y) \leqslant K'(x) + K'(y) + n(V) + 1$.

The self-delimiting complexity K' satisfies many laws without a logarithmic fudge term. Infinite random sequences can be more naturally defined using K' complexity. In fact, it turns out that an infinite binary string x is random in the sense of Martin-Löf (see before and [109]) iff there exists a constant c such that $K'(x_{1:n}) \geqslant n - c$ for all n. This interesting characterization was proposed by Chaitin [28], and proved by Schnorr. The following lemma, due to Levin and Gács [45], and also Chaitin [28], shows that $K'(x)$ is a symmetric measure of the information in x.

LEMMA (strong symmetry). *To within an additive constant,*

$$K'(x, K'(x)) = K'(x), \qquad K'(x, y) = K'(x) + K'(y \mid (x, K'(x))).$$

Therefore, the exact symmetry of information holds up to an additive constant in the sense that

$$K'(y) - K'(y \mid (x, K'(x))) = K'(x) - K'(x \mid (y, K'(y))).$$

REMARK. It is a fundamental result due to Gács [45] that if we delete $K'(x)$ and $K'(y)$ from the condition, the equalities in the lemma in general can only hold to within a term logarithmic in $K'(x, y)$. In Chaitin's formulation [28] of conditional complexity, say denoted by $Kc(x|y)$, he means the complexity of x given the lexicographically least program p for y. It is straightforward to verify that $Kc(x|y) = K'(x|p) = K'(x|(y, K'(y)))$. Furthermore, if simply $Kc(x) = K'(x)$ and $Kc(x, y) = K'(x, y)$, then we have $Kc(x, y) = Kc(x) + Kc(y|x)$ up to an additive constant.

2.8. *Probability theory*

Laplace [83] has pointed out the following conflict between our intuition and the classical theory of probability:

> "In the game of heads and tails, if head comes up a hundred times in a row then this appears to us extraordinary, because the nearly infinite number of combinations that can arise in a hundred throws are divided in regular sequences, or those in which we observe a rule that is easy to grasp, and in irregular sequences, that are incomparably more numerous."

Yet, one-hundred heads are just as probable as any other equal-length sequence of heads and tails, even though we feel that it is less "random" than some others. We can formalize this. (We follow Gács's insightful treatment [46] in this section). Let us call a *pay-off* function (or *martingale*) with respect to distribution P any nonnegative function $t(x)$ with $\sum_x P(x)t(x) \leqslant 1$. Suppose our favorite nonprofit casino asks 1 dollar for a game that consists of a sequence of flips of a fair coin, and claims that each outcome x has probability $P(x) = 2^{-|x|}$. To back up this claim, it ought to agree to pay

$t(x)$ dollars on outcome x. Accordingly, we propose a pay-off t_0 with respect to P_n (P restricted to sequences of n coin flips): put $t_0 = 2^{n/2}$ for all x whose even digits are 0 (head) and 0 otherwise. This bet will cost the casino $2^{50} - 1$ dollars for the outcome ($n = 100$) above. Since we must propose the pay-off function beforehand, it is unlikely that we define precisely the one that detects this particular fraud. However, fraud implies regularity, and, as Laplace suggests, the number of regular bets is so small that we can afford to make all of them in advance.

> "If we seek a cause wherever we perceive symmetry, it is not that we regard the symmetrical event as less possible than the others, but, since this event ought to be the effect of a regular cause or that of chance, the first of these suppositions is more probable than the second."

Let us make this formal, using some original ideas of Solomonoff as developed by Levin. We need the *Kraft inequality*: for each set S of finite binary strings such that no string in S is a proper prefix of another string in S, we have

$$\sum_{x \in S} 2^{-|x|} \leqslant 1.$$

We call S a *prefix-code*. With notation $K(x|y)$ we mean the self-delimiting complexity of the previous section. Then for each fixed y, the set of $K(x|y)$'s is the length set of a prefix-code, so Kraft's inequality applies. For most binary strings of length n, no significantly shorter description exists, since the number of short descriptions is small. We can sharpen this observation by, instead of counting the number of simple sequences, measuring their probability. By Kraft's inequality, for each fixed y,

$$\sum_{x} 2^{-K(x|y)} \leqslant 1, \tag{2.3}$$

so that only a few objects can have small complexity. Conversely, let μ be a *computable* probability distribution, i.e., such that there is an effective procedure that, given x, computes $\mu(x)$ to any degree of accuracy. Let $K(\mu)$ be the length of the smallest such program. Then

$$K(x) \leqslant -\log \mu(x) + K(\mu) + c, \tag{2.4}$$

with c a universal constant. Put $d(x|\mu) = -\log \mu(x) - K(x)$. By (2.3), $t(x|\mu) = 2^{d(x|\mu)}$ is a pay-off function. We can now beat any fraudulent casino. We propose the pay-off function $t(x) = 2^{-\log P_n(x) - K(x|n)}$. (We use conditional complexity $K(x|n)$ because the uniform distribution P_n depends on n.) If every other coin flip comes up heads, then $K(x|n) \leqslant (\frac{1}{2}n) + c_0$, and hence we win $2^{t(x)} \geqslant c_1 2^{n/2}$ from the casino ($c_0, c_1 > 0$), even though the bet does not refer to "heads".

The fact that $t(x|\mu)$ is a pay-off function, implies by Chebychev's First Inequality that for any $k > 0$,

$$\mu\{x : K(x) < -\log \mu(x) - k\} < 2^{-k}. \tag{2.5}$$

Together, (2.4) and (2.5) say that with large probability the complexity $K(x)$ of a random outcome x is close to its upper bound $-\log \mu(x) + K(\mu)$. If an outcome x violates any "laws of probability", then the complexity $K(x)$ falls far below the upper bound. Indeed, a proof of some law of probability (like the law of large numbers, the law of iterated logarithm, etc.) always gives rise to some simple computable pay-off function $t(x)$ taking large values on the outcomes violating the law. In general, the pay-off function $t(x|\mu)$ is *maximal* (up to a multiplicative constant) among all pay-off functions that are semicomputable (from below). Hence the quantity $d(x|\mu)$ constitutes a universal test of randomness—it measures the *deficiency* of randomness in the outcome x with respect to distribution μ, or the extend of justified suspicion against hypothesis μ given the outcome x.

2.9. *A priori probability: the Solomonoff–Levin distribution*

Let $K(x)$ be the self-delimiting variant of complexity. The incomputable *Solomonoff–Levin distribution* $\boldsymbol{m}(x)$ can be defined as

$$\boldsymbol{m}(x) = \sum_{U(p)=x} 2^{-|p|},$$

with U the reference universal self-delimiting Turing machine. Now $\boldsymbol{m}(x)$ can be interpreted as the probability that U halts with output x if we generate the input p by an indefinitely long sequence of random coin flips (U will use only the self-delimiting prefix p of this sequence as its program) (see also the section on self-delimiting complexity). This distribution was conceived by Solomonoff in [148, 149], but in different form. A mathematically natural form, in terms of a *universal semicomputable semimeasure* that dominates all semicomputable semimeasures, which can be shown to coincide with the above definition of $\boldsymbol{m}(x)$ up to a multiplicative constant, was given by Levin in [173]. A discrete semimeasure μ is a real-valued function satisfying $\sum_x \mu(x) \leqslant 1$. A function μ is semicomputable (from below) if the set $\{\langle p, q, x\rangle : p/q \leqslant \mu(x)\}$, p, q and x natural numbers, is recursively enumerable.

It can be shown that $\boldsymbol{m}$ is a universal semicomputable semimeasure in the sense that, for each semicomputable semimeasure μ, there exists a constant $c > 0$ such that, for all x, $\boldsymbol{m}(x) > c\mu(x)$. That is, $\boldsymbol{m}$ dominates μ multiplicatively.

It can be shown that $\boldsymbol{m}(x) = 2^{-K(x) \pm O(1)}$. It turns out that $\boldsymbol{m}(x)$ has the remarkable property that the test $d(x|\boldsymbol{m})$ shows all outcomes x random with respect to it. We can interpret (2.4) and (2.5) as saying that if the real distribution is μ, then $\mu(x)$ and $\boldsymbol{m}(x)$ are close to each other with large probability. Therefore, if x comes from some unknown computable distribution μ, then we can use $\boldsymbol{m}(x)$ as an estimate for $\mu(x)$. Accordingly, Solomonoff has called $\boldsymbol{m}$ "a priori probability". The randomness test $d(x|\mu)$ can be interpreted in the framework of hypothesis testing as the likelihood ratio between hypothesis μ and the fixed alternative hypothesis $\boldsymbol{m}$. In ordinary statistical hypothesis testing, some properties of an unknown distribution μ are taken for granted, and the role of the universal test can probably be reduced to some tests that are used in statistical practice. However, such conditions do not hold in general as is witnessed by prediction of time series in economics, pattern recognition or inductive inference (see Section 3.2).

Since the a priori probability $\boldsymbol{m}$ is a good estimate for the actual probability, we can use the conditional a priori probability for prediction—without reference to the unknown distribution μ. For this purpose, we first define a priori probability M for the set of infinite binary sequences as in [173]. For any finite sequence x, $M(x)$ is the a priori probability that the outcome is some extension of x. Let x, y be finite sequences. Then

$$\frac{M(xy)}{M(x)} \tag{2.6}$$

is an estimate of the conditional probability that the next terms of the outcome will be given by y provided that the first terms are given by x. It converges to the actual conditional probability $\mu(xy)/\mu(x)$ with μ-probability 1 for any computable distribution μ [150]. Inductive inference formula (2.6) can be viewed as a mathematical formulation of *Occam's razor*: predict by the simplest rule fitting the data. The a priori distribution M is incomputable, and the main problem of inductive inference can perhaps be stated as "finding efficiently computable optimal approximations to M" [46].

3. Applications of compressibility

It is not surprising that some strings can be compressed arbitrary far. Easy examples are the decimal expansions for some transcendental numbers like $\pi = 3.1415\ldots$ and $e = 2.7182\ldots$. These strings can be described in $O(1)$ bits, and have therefore constant Kolmogorov complexity. A moment's reflection suggests that the set of computable numbers, i.e., the real numbers computable by Turing machines which start with a blank tape, coincides *precisely* with the set of real numbers of Kolmogorov complexity $O(1)$. A nice application of what we may call extreme compressibility of some strings is a new version of Gödel's celebrated incompleteness theorem.

3.1. *A version of Gödels Theorem*

Recall Gödel's famous incompleteness result that each formal mathematical system which contains arithmetic is either inconsistent or contains theorems which cannot be proved in the system. Barzdin' has first formulated a new form of Gödel's Incompleteness Theorem in terms of simple sets (cf. Section 2.5). This was also treated by Chaitin in a sequence of papers [26, 27, 32], as follows. Let us view a theorem—a true statement—together with the description of the formal system, as a description of a proof of that theorem. Just as certain numbers can be really far compressed, like π or 10^{100}, in their descriptions, in a formal mathematical system the ratio between the length of the theorems and the length of their shortest proofs can be enormous. In a sense, the argument below shows that the worst-case such ratio expressed as a function of the length of the theorem increases faster than any computable function.

In Bennett's [11] phrase: *although most numbers are random, only finitely many of them can be proved random within a given consistent axiomatic system.* If T is an axiomatized system (which is sound, i.e. all theorems provable in T are true), whose axioms and rules of inference require about k bits to describe, then T cannot be used to prove the randomness of any number much longer than k bits. If the system could prove randomness for a number much longer than k bits, then the *first* such proof (first in an unending enumeration of all proofs obtainable by repeated application of axioms and rules of inference) could be used to derive a contradiction: an approximately k-bit program to find and print out the specific random number mentioned in this proof, a number whose smallest program is by assumption considerably larger than k bits. Therefore, even though most strings are random, we will never be able to explicitly exhibit a string of reasonable size which demonstrably possesses this property. (This formulation is due to Bennett.)

EXAMPLE (*Chaitin*). We use an approach that is slightly different from Chaitin's approach.

(a) Let axiomatized theory T be describable in k bits: $K(T) \leqslant k$. Assume that T is contradiction-free.

(b) Assume that all true formulas in T can be proved in T.

(c) Let $S_c(x)$ be a formula in T with the meaning: "x is the lexicographically least binary string of length c with $K(x) \geqslant c$." Here x is a formal parameter and c an explicit constant, so $K(S_c) \leqslant \log c$ up to a fixed constant independent of T and c.

For each c, there exists an x such that $S_c(x) = \mathbf{true}$ is a true statement by a simple counting argument. Moreover, S_c expresses that this x is unique. It is easy to see that combining the descriptions of T, S_c, we obtain a description of this x. Namely, by (b), for each candidate string y of length c, we can decide $S_c(y) = \mathbf{true}$ (holds for $y = x$) or $\mathrm{not}(S_c(y)) = \mathbf{true}$ (holds for $y \neq x$), by simple enumeration of all proofs in T. We need to distinguish the descriptions of T and S_c, so we code T's description self-delimiting in not more than $2k$ bits. Hence, for some fixed constant c' independent of x, T and c, we find $K(x) \leqslant 2k + \log c + c'$, which contradicts $K(x) > c$ for all $c > c_T$, where $c_T = 3k + c'$ for another constant c'.

As Chaitin expresses it: "... if one has ten pounds of axioms and a twenty-pound theorem, then that theorem cannot be derived from those axioms."

EXAMPLE (*Levin*). Levin has tried to gauge the implications of algorithmic complexity arguments to probe depths beyond the scope of Gödel's arguments. He derives what he calls "Information Conservation Inequalities". Without going into the subtle details, Levin's argument in [86] takes the form that the information $I(\alpha : \beta)$ in a string α about a string β cannot be significantly increased by either algorithmic or probabilistic means. In other terms, any sequence that may arise in nature contains only a finite amount of information about any sequence defined mathematically. Levin says: "Our thesis contradicts the assertion of some mathematicians that the truth of any valid proposition can be verified in the course of scientific progress by means of nonformal methods (to do so by formal methods is impossible by Gödel's Theorem)." (For a continuation of this research, see [88].)

3.2. *Inductive inference in theory formation*

This application stood at the cradle of Kolmogorov complexity proper. It led Solomonoff to formulate the important notion of a universal a priori probability, as described previously. Solomonoff's proposal [149], i.e., the Solomonoff–Levin distribution, is a synthesis of Occam's principle and Turing's theory of effective computability applied to inductive inference. His idea is to view a theory as a compact description of past observations together with predictions of future ones. The problem of theory formation in science is formulated as follows. The investigator observes increasingly larger and larger initial segments of an infinite binary sequence. We can consider the infinite binary sequence as the outcome of an infinite sequence of experiments on some aspect X of nature. To describe the underlying regularity of this sequence, the investigator tries to formulate a theory that governs X, on the basis of the outcome of past experiments. Candidate theories are identified with computer programs that compute infinite binary sequences starting with the observed initial segment.

To make the discussion precise, we give a simplified Solomonoff inductive inference theory. Given a previously observed data string S over $\{0, 1\}^*$, the inference problem is to predict the next symbol in the output sequence, i.e., extrapolating the sequence S. By Bayes' rule, for $a=0$ or 1,

$$P(Sa|S)=\frac{P(S|Sa)P(Sa)}{P(S)}, \tag{3.1}$$

where $P(S)=P(S|S0)P(S0)+P(S|S1)P(S1)$. Since $P(S|Sa)=1$ for any a, we have,

$$P(Sa|S)=\frac{P(Sa)}{P(S0)+P(S1)}. \tag{3.2}$$

In terms of inductive inference or machine learning, the final probability $P(Sa|S)$ is the probability of the next symbol being a given the initial sequence S. The goal of inductive inference in general is to be able to infer the underlying machinery that generated S, and hence be able to predict (extrapolate) the next symbol. Obviously we now only need the prior probability $P(Sa)$ to evaluate $P(Sa|S)$. Let K denote the self-delimiting Kolmogorov complexity. Using the Solomonoff–Levin distribution, assign $2^{-K(Sa)}$ as the prior probability to $P(Sa)$. This corresponds to Occam's razor principle of choosing the simplest theory and has two very nice properties:

(1) $\sum_x 2^{-K(x)} \leqslant 1$ by Kraft's inequality. Hence this is a proper probability assignment.

(2) For any computable probability P, there is a constant c such that $2^{-K(x)} \geqslant c\,P(x)$, for all x. In fact, $\boldsymbol{m}(x)=2^{-K(x)\pm O(1)}$ [47, 173].

This approach is guaranteed to converge in the limit to the true solution and in fact it converges *faster than* any other method up to a constant multiplicative factor. It turns out that Gold's idea of identification by enumeration [53] can also be derived from this approach [96].

The notion of complexity of equivalent theories as the length of a shortest program that computes a given string emerges forthwith, and also the invariance of this measure under changes of computers that execute them. The metaphor of natural law being a compressed description of observations is singularly appealing. Among others, it gives substance to the view that a natural law is better if it "explains" more, that is, describes more observations. On the other hand, if the sequence of observations is sufficiently random, then it is subject to no law but its own description. This metaphorical observation was also made by Chaitin [24].

3.3. *Rissanen's Minimum Description Length Principle*

Scientists formulate their theories in two steps: first a scientist must, based on scientific observations or given data, formulate alternative hypotheses, and second he selects one definite hypothesis. This was done by many ad hoc principles, among the most dominant, Occam's razor principle, the maximum likelihood principle, various ways of using Bayesian formula with different prior distributions. However, no single principle is satisfactory in all situations. But in an ideal sense, the Solomonoff approach we have discussed presents a *perfect* way of solving all induction problems using Bayes' rule with the universal prior distribution. However, due to the noncomputability of the universal prior function, such a theory cannot be directly used in practice. Inspired by the Kolmogorov complexity research, in 1978 Rissanen proposed the *Minimum Description Length Principle* (*MDLP*) [132]. See also [177] for a related approach pioneered by C.S. Wallace and D.M. Boulton in 1968. Quinlan and Rivest used this principle to construct an algorithm for constructing decision trees and the result was quite satisfactory compared to existing algorithms [130]. Using MDLP, Gao and Li [51] have implemented a system (on IBM PC) which on-line recognizes/learns hand-written English and Chinese characters. The *Minimum Description Length Principle* can be intuitively stated as follows.

MINIMUM DESCRIPTION LENGTH PRINCIPLE. *The best theory to explain a set of data is the one which minimizes the sum of*

(1) *the length (encoded in binary bits) of the theory;*
(2) *the length (in binary bits) of data when encoded with the help of the theory.*

EXAMPLE (*Kemeny*). The importance of "simplicity" for inductive inference was already exploited in an elegant paper by Kemeny [70]. This paper clearly anticipates the ideas developed in this section, and we cite one example. We are given n points in the plane. Which polynomial fits these points best? One extreme is to put a straight line through the cluster such that the χ^2 measure is minimized. The other extreme is to simply fit an $n-1$ degree polynomial through the n points. Neither choice seems very satisfactory, and it is customary to think that the problem is not formulated precisely enough. But MDLP says that we look for the mth degree polynomial, $m \leqslant n-1$, such that the description of the m-vector of coefficients of the polynomial, together with the description of the points relative to the polynomial, is minimized.

It is remarkable that, using Kolmogorov complexity, we can formally derive a version of the MDL principle, and explain how and why it works [96]. From Bayes' rule,

$$P(H|D) = \frac{P(D|H)P(H)}{P(D)},$$

we need to choose the hypothesis H such that $P(H|D)$ is maximized, where D denotes the data. Now taking the negative logarithm on both sides of Bayes' formula, we get

$$-\log P(H|D) = -\log P(D|H) - \log P(H) + \log P(D).$$

Since D is fixed, maximizing the term $P(H|D)$ is equivalent to minimizing

$$-\log P(D|H) - \log P(H)$$

Now to get the minimum description length principle, we only need to explain the above two terms properly. The term $-\log P(H)$ is straightforward. Assuming the Solomonoff–Levin distribution, $P(H) = 2^{-K(H)}$ where $K(H)$ is the self-delimiting Kolmogorov complexity of H. Then the term $-\log P(H)$ is precisely the *length* of a minimum-length *prefix-free encoding*, or shortest program, for the hypothesis H. Similarly for the term $-\log P(D|H)$.

Now just assume that P is computable. It can be shown that the universal probability distribution $\boldsymbol{m}(x)$ can approximate $P(x)$ in the sense that

(1) there is a constant c, such that $\boldsymbol{m}(x) \geqslant c\,P(x)$, and

(2) the probability that $\boldsymbol{m}(x) \leqslant k\,P(x)$ is at least $1 - 1/k$.

Since $\boldsymbol{m}(H) = 2^{-K(H) \pm O(1)}$, the quantity $2^{-K(H)}$ is a reasonable estimate for $P(H)$. Similarly, $2^{-K(D|H)}$ is an estimate for $P(D|H)$. Hence we must minimize $K(D|H) + K(H)$, i.e., find an H such that the sum of the description lengths is minimized.

In the original Solomonoff approach, H in general is a Turing machine. In practice we must avoid such a too general approach in order to keep things computable. In different applications, the hypothesis H can mean many different things. For example, H may refer to decision trees, finite automata, Boolean formulae, or polynomials of certain degree. Thus, Rissanen suggested the following approach. First convert (or encode) H to an integer. Then we try to assign a prior probability to each integer. Assigning $2^{-K(n)}$ to integer n would be perfect but it is not computable. Jeffreys [67] suggested to assign probability $1/n$ to integer n. But this results in an improper distribution since $\sum_{n=1}^{\infty} 1/n$ diverges. Rissanen defined the following length function: let

$$l^*(n) = \log n + \log\log n + \log\log\log n + \cdots$$

all positive terms, and let $L(n) = l^*(n) + \log c$ where $c = 2.865064\ldots$. It has been shown that $\sum_n 2^{-L(n)} = 1$, see [133].

3.4. *Learnability in the Valiant learning model*

This section is rather closely related to the previous two sections. There we did not consider the number of steps involved in making an inference, or the number of examples needed to learn a concept. Solomonoff's principle shows that we learn something perfectly in the limit, but how fast this converges is not prescribed at the outset. For instance, the well-known principle of Gold [53] of inference by enumeration can be viewed as a particular case of Solomonoff's principle (cf. [96]) and we note here without further explanation, and as immediately clear to those familiar with it, that the Gold paradigm of inductive inference is aimed at *precise* inference. But what if we want to learn a concept using a number of examples that is bounded a priori? Obviously if we are to *precisely* infer a law of nature, an infinite (or exponential) behavior is inherent. However, for the purpose of machine learning, it is sufficient to just *learn* such a law *approximately*: if a human child (or a computer) would recognize, with 0.99 *probability*, the next apple after seeing three apples, we consider that the concept of apple is learned. In 1983, Valiant [157] introduced such a learning model. Valiant emphasizes polynomial-time learnability. For simplicity and convenience, we consider the problem of learning Boolean formulae of n variables.

According to Valiant, a concept F is, say, a Boolean formula. Those vectors v such that $F(v)=1$ are called positive examples, the rest are negative examples of F. For any F, there are many possible Boolean formulae f such that f is consistent with the concept F. Let $|f|$ denote the least number of symbols needed to write such a representation f. The learning algorithm has available two buttons labeled POS and NEG. If POS (NEG) is pushed, a positive (negative) example is generated according to some fixed but unknown probability distribution D^+ (D^-). We assume nothing about the distributions D^+ and D^- except that $\sum_{f(v)=1} D^+(v)=1$ and $\sum_{f(v)=0} D^-(v)=1$. Let A be a class of concepts. Then A is learnable from examples iff there exists a polynomial p and a (possibly randomized) learning algorithm L such that, for f in A and $\varepsilon>0$, algorithm L halts in $p(n, |f|, 1/\varepsilon)$ time and examples, and outputs a formula $g\in A$ that, with probability at least $1-\varepsilon$, has the following properties: $\sum_{g(v)=0} D^+(v)<\varepsilon$ and $\sum_{g(v)=1} D^-(v)<\varepsilon$.

Various classes of concepts are shown to be learnable in Valiant's sense [16, 58, 68, 99, 134, 157, 158]. Many Valiant learnable classes are NP-complete to learn in the Gold sense (see [69] for a survey). In [16], again by Occam's principle, it was shown that given a set of positive and negative data, any consistent concept of size "reasonably" less than the size of data is an "approximately" correct concept. That is, if one can find a shorter representation of data, then one learns. The shorter the conjecture is, the more and better it explains with higher probability (see also [96]). Many concepts turn out to be too hard (like NP-complete) to learn in Valiant's sense. A further refinement to learning "simple" concepts under all "simple" distributions, using the universal distribution $\boldsymbol{m}$, is developed in [176].

3.5. Computable numbers are not random

One can make precise the off-hand claim made in the Introduction that the set of computable numbers coincides with the set of reals which have Kolmogorov complexity O(1). (We view a real number as an infinite sequence of digits.) Let $N=\{0,1,\ldots\}$ be the set of natural numbers, let $S=\{\varepsilon,0,1,01,10,11,\ldots\}$ be the set of finite binary strings, and let X be the set of infinite binary strings. We denote by $|s|$ the length of a string s, and by $s_{1:n}$ the prefix of length n of a string s. (If $x\in X$ then $|x|=\infty$.) An infinite string x is *recursive* iff there is a recursive function $f: N\to S$ such that $x_{1:n}=f(n)$ for all n. Let K denote the plain Kolmogorov complexity as in Section 2. It can be shown [29] that x is recursive iff there exists a constant $c>0$ such that for all $n\in N$ we have $K(x_{1:n})\leqslant K(n)+c$.

3.6. The number of wisdom Ω

This Cabalistic exercise follows Chaitin [26, 27] and Bennett [11]. A real is *normal* if each digit from 0 to 9, and each block of digits of equal length, occurs with equal asymptotic frequency. No rational number is normal to any base, and almost all irrational numbers are normal to every base. But for particular ones, like π and e, it is not known whether they are normal, although statistical evidence suggests they are. In contrast to the randomly appearing sequence of the decimal representation of π, the digit sequence of Champernowne's number 012345678910111213... is very nonrandom yet provably normal [34]. Once we know the law that governs π's sequence, we can make a fortune betting at fair odds on the continuation of a given initial segment, and most gamblers would eventually win against Champernowne's number because they will discover its law.

Almost all real numbers are Kolmogorov random, which implies that no possible betting strategy, betting against fair odds on the consecutive bits, can win infinite gain. Can we exhibit a specific such number? One can define an uncomputable number $k = 0.k_1 k_2 \ldots$ such that $k_i = 1$ if the ith program in a fixed enumeration of programs for some fixed universal machine halts, else $k_i = 0$. By the unsolvability of the halting problem, k is noncomputable. However, by Barzdin's Theorem (Section 2.5) k is *not* incompressible: each n-length prefix $k_{1:n}$ of k can be compressed to a string of length not more than $2 \log n$ (since $K(k_{1:n}|n) \leqslant \log n$), from some n onwards. It is also easy to see that a gambler can still make infinite profit, by betting only on solvable cases of the halting problem, of which there are infinitely many. Chaitin [28] has found a number that is random in the strong sense needed.

Ω equals the probability that the reference self-delimiting universal Turing machine halts when its program is generated by fair coin tosses. That is, $\Omega = \sum_x \boldsymbol{m}(x)$, with $\boldsymbol{m}(x)$ the Solomonoff–Levin distribution. Then, Ω is a number between 0 and 1. It is greater than 0 since some programs do halt, and it is less than one since some programs do not halt (use the Kraft inequality). It is Kolmogorov random, it is noncomputable, and no gambling scheme can make an infinite profit against it. It has the curious property that it encodes the halting problem very compactly. Namely, suppose we want to determine whether a program p halts or not. Let program p have length n. Its probability in terms of coin tosses is 2^{-n}. If we know the first n bits $\Omega_{1:n}$ of Ω, then $\Omega_{1:n} < \Omega < \Omega_{1:n} + 2^{-n}$. However, dovetailing (execute phases $1, 2, \ldots$. where phase i consists of executing one step of each of the first i programs) the running of all programs sufficiently long must yield eventually an approximation Ω' of Ω with $\Omega' > \Omega_{1:n}$. If p is not among the halted programs which contributed to Ω', then p will never halt, since otherwise its contribution would yield $\Omega \geqslant \Omega' + 2^{-n}$, which is a contradiction. (That is, $\Omega_{1:n}$ is a short program to obtain $k_{1:m}$ with $m \approx 2^n$.)

The argument suffices, via Barzdin's Theorem in Section 2.5, to establish that Ω is a random infinite sequence in the sense of Martin-Löf (Section 2.4).

Knowing the first 10,000 bits of Ω enables us to solve the halting of all programs of less than 10,000 bits. This includes programs looking for counterexamples to Fermat's Last Theorem, Riemann's Hypothesis and most other conjectures in mathematics that can be refuted by single finite counterexamples. Moreover, for all axiomatic mathematical theories which can be expressed compactly enough to be conceivably interesting to human beings, say in less than 10,000 bits, $\Omega_{1:10,000}$ can be used to decide for every statement in the theory whether it is true, false or independent. Finally, knowledge of $\Omega_{1:n}$ suffices to determine whether $K(x) \leqslant n$ for each finite binary string x. Thus, Ω is truly the number of wisdom, and "can be known of, but not known, through human reason" [11].

EXAMPLE (*Chaitin*). Recall that the Barzdin'–Loveland Lemma states that for all r.e. sets each n-length initial segment of their characteristic sequence has Kolmogorov complexity $O(\log n)$. Kolmogorov has remarked that this implies that the solubility of the first n Diophantine equations in an effective enumeration can be decided using at most $O(\log n)$ bits extra information. Namely, given the number $m \leqslant n$ of soluble equations in the first n equations, we can find them all effectively in the obvious way. Chaitin observed that this is not the case if we replace the question of mere solubility by the question of whether there are finitely many or infinitely many nontrivially different solutions. Namely, no matter how many solutions we find for a given equation, by itself this can give no information on the question to be decided. It turns out that the set of indices of the Diophantine equations with infinitely many different solutions is not r.e. In particular, in the characteristic sequence each initial segment of length n has Kolmogorov complexity of about n. Chaitin says that this shows that randomness is inherent not only in natural phenomena (e.g., related to quantum mechanics), but also occurs in mathematics [32, 33]. More precisely, we have the following claim.

CLAIM. *There is an (exponential) Diophantine equation* $A(n, x_1, x_2, \ldots, x_m) = 0$ *which has infinitely many solutions* $x_1, x_2, \ldots, x_m$ *iff the n-th bit of* Ω *is* 1.

PROOF. By dovetailing the running of all programs of the reference self-delimiting machine U in the obvious way we find a computable sequence of rational numbers $r_1 \leqslant r_2 \leqslant \cdots$ such that $\Omega = \lim_{n\to\infty} r_n$. The set $R = \{(n, k)$: the nth bit of r_k is a $1\}$ is a recursively enumerable (even recursive) set. The main step is to use a theorem due to J.P. Jones and Y.V. Mateyasevich (*J. Symbol. Logic* **49** (1984), pp. 818–829) to the effect that "every recursively enumerable set R has a singlefold exponential Diophantine representation $A(p, y)$". That is, $A(p, y) = 0$ is an exponential Diophantine equation, and the singlefoldedness consists in the property that $p \in R$ iff there is a y such that $A(p, y) = 0$ is satisfied and, moreover, there is only a single such y. (Here both p and y can be multituples of integers; in our case p represents $\langle n, x_1 \rangle$, and y represents $\langle x_2, \ldots, x_m \rangle$. For technical reasons we consider as proper solutions only solutions x involving no negative integers.) Representing R this way, there is a Diophantine equation $A(n, k, x_2, \ldots, x_m) = 0$ which has exactly one solution $x_2, \ldots, x_m$ if the nth bit of the binary expansion of r_k is a one, and it has no solution $x_2, \ldots, x_m$ otherwise. Consequently, the number of different m-tuples $x_1, x_2, \ldots, x_m$ which are solutions to $A(n, x_1, x_2, \ldots, x_m) = 0$ is infinite if the nth bit of the binary expansion of Ω is a 1, and this number is finite otherwise. □

4. Example of an application in mathematics: weak prime number theorems

Using Kolmogorov complexity, it is easy to derive a weak version of the prime number theorem. An adaptation of the proof that Chaitin gives of Euclid's theorem that the number of primes is infinite [31] yields a very simple proof of a weak prime number theorem. Let $\pi(n)$ denote the number of prime numbers less than n. (Recall that $\pi(n)$ is asymptotically $n/\log n$). We prove that $\pi(n)$ is $\Omega(\log n(\log\log n)^{-1})$. Let n be a random number with $K(n) \geqslant \log n - O(1)$. Consider a prime factorization

$$n = p_1^{e_1} \cdot p_2^{e_2} \cdot \ldots \cdot p_m^{e_m},$$

with $p_1, p_2, \ldots$ the sequence of primes in increasing order. With $m = \pi(n)$, we can describe n by the $\pi(n)$-length vector of exponents $(e_1, \ldots, e_{\pi(n)})$. Since $p_i \geqslant p_1 = 2$, it holds that $e_i \leqslant \log n$ and, bounding $K(e_i)$ by the length of self-delimiting descriptions of e_i,

$$K(e_i) \leqslant \log \log n + 2 \log \log \log n,$$

for all $i \leqslant m$. Therefore,

$$K(n) \leqslant \pi(n)(\log \log n + 2 \log \log \log n).$$

Substituting the lower bound on $K(n)$, we obtain the claimed lower bound on $\pi(n)$ for the special sequence of random n.

Recently, P. Berman [Personal communication] obtained the stronger result that the number of primes below n is $\Omega(n/\log^2 n)$, by an elementary Kolmogorov complexity argument. It is interesting because it shows a relation between primality and prefix codes. Recall that we identify the positive integer n with the nth binary string. Assume that we have a function $c: N \rightarrow N$ with the following property: for every two integers m, n, $c(m)$ is not a prefix of $c(n)$. Then c is called a *prefix code*. Consider only prefix codes c such that $c(n) = o(n^2)$. (For instance, choose $c(n) = \overline{|n|}n$, the self-delimiting description of n with binary length $|c(n)| = \log n + 2 \log \log n$ bits.)

LEMMA (Berman). *For an infinite subsequence of positive integers n, $p_n = O(c(n))$, where p_n is the n-th prime.*

Choosing $c(n)$ as above we have $c(n) \leqslant n \log^2 n$. Therefore, by the lemma, p_n is $O(n \log^2 n)$. Straightforward manipulation of the order-of-magnitude symbols then shows that $\pi(n)$ is $\Omega(n/\log^2 n)$. This can be strengthened by choosing more efficient codes, but not all the way to obtain $\pi(n) = \Omega(n/\log n)$.

5. Applications of incompressibility: proving lower bounds

It was observed in [126] that the static, descriptional (program size) complexity of a *single* random string can be used to obtain lower bounds on dynamic, computational (running time) complexity. The power of the static, descriptional Kolmogorov complexity in the dynamic, computational lower bound proofs rests on one single idea: there are incompressible (or Kolmogorov random) strings. A traditional lower bound proof by counting usually involves *all* inputs (or all strings of certain length) and one shows that the lower bound has to hold for *some* of these ("typical") inputs. Since a particular "typical" input is *hard* to construct, the proof has to involve all the inputs. Now we understand that a "typical input" can be constructed via a Kolmogorov random string. However, as we have shown in relation with Gödel's Theorem, we will never be able to put our hands on one of those strings or inputs and claim that it is random or "typical". No wonder the old counting arguments had to involve all inputs:

it was because a particular typical input cannot be *proved* to be "typical" or random. In a Kolmogorov complexity proof, we choose a random string that *exists*. That it cannot be exhibited is no problem, since we only need existence. As a routine, the way one proves a lower bound by Kolmogorov complexity is as follows: Fix a Kolmogorov random string which we know exists. Prove the lower bound with respect to this particular *fixed* string: show that if the lower bound does not hold, then this string can be compressed. Because we are dealing with only one fixed string, the lower bound proof usually becomes quite easy and natural.

In the next subsection, we give three examples to illustrate the basic methodology. In the following subsections, we survey the lower bound results obtained using Kolmogorov complexity of the past ten years (1979–1988). Many of these results resolve old or new, some of them well-known, open questions; some of these results greatly simplify and improve the existing proofs. The questions addressed in the next few subsections often deal with simulating one machine model by another, e.g., as treated in P. van Emde Boas' Chapter 1 "Machine Models and Simulations" in this Handbook.

5.1. *Three examples of proving lower bounds*

In this section, we illustrate how Kolmogorov complexity is used to prove lower bounds by three concrete examples.

5.1.1. *Example 1: one-tape Turing machines*

Consider a most basic Turing machine model with only one tape, with a two-way read/write head, which serves as both input and worktape. The input is initially put onto the first n cells of the only tape. We refer a reader not familiar with Turing machines to [62] for a detailed definition. The following theorem was first proved by Hennie and a proof by counting, for comparison, can be found in [62, page 318]. In [124] the following elegant proof is presented. Historically, this was the first lower bound obtained by Kolmogorov complexity.

THEOREM. *It requires $\Omega(n^2)$ steps for the above single-tape TM to recognize $L=\{ww^{\mathrm{R}}: w\in\{0,1\}^*\}$ (the palindromes). Similarly for $L'=\{w2w^{\mathrm{R}}: w\in\{0,1\}^*\}$.*

PROOF (*cf.* [124]). Assume on the contrary that M accepts L in $o(n^2)$ time. Let $|M|$ denote the length of the description of M. Fix a Kolmogorov random string w of length n for a large enough n. Consider the computation of M on ww^{R}. A *crossing sequence* associated with a tape square consists of the sequence of states the finite control is in when the tape head crosses the intersquare boundary between this square and its left neighbor. If cs is a crossing sequence, then $|cs|$ denotes the length of its description. Consider an input of $w0^nw^{\mathrm{R}}$ of length $3n$. Divide the tape segment containing the input into three equal-length segments of size n. If each crossing sequence associated with a square in the middle segment is longer than $n/(10|M|)$ then M spent $\Omega(n^2)$ time on this input. Otherwise there is a crossing sequence of length less than $n/(10|M|)$. Assume that this occurs at c_0. Now this crossing sequence requires at most $n/10$ bits to encode. Using this crossing sequence, we reconstruct w as follows. For every string $x0^nx^R$ of length $3n$, put it on the input tape and start to simulate M. Each time when the head reaches c_0 from the left, we take the next element in the crossing sequence to skip the computation of M when the head is on the right of c_0 and resume the simulation starting from the time when the head moves back to the left of (or on) c_0 again. If the simulation ends consistently, i.e. every time the head moves to c_0, the current status of M is consistent with that specified in the crossing sequence, then $w=x$. Otherwise, if $w \neq x$ and the crossing sequences are consistent in both computations, then M accepts a wrong input $x0^nw^R$. However, this implies

$$K(w) < |cs| + O(\log n) < n,$$

contradicting $K(w) \geqslant n$. □

5.1.2. Example 2: parallel addition

Consider the following widely used and most general parallel computation model, the priority PRAM. A priority PRAM consists of processors $P(i)$, $i = 1, 2, \ldots, n^{O(1)}$, and an infinite number of shared memory cells $C(i)$, $i = 1, 2, \ldots$ Each step of the computation consists of three parallel phases as follows. Each processor (1) reads from a shared memory cell, (2) performs a computation, and (3) may attempt writing into some shared memory cell. At each step each processor is in some *state*. The actions and the next state of each processor at each step depend on the current state and the value read. In case of *write conflicts*, the processor with the minimum index succeeds in writing.

Theorem. *Adding n integers, each of polynomial number of bits, requires $\Omega(\log n)$ parallel steps on a priority PRAM.*

Remark. A weaker version than the one above was first proved in [59] using a Ramsey theorem, and in [66, 122]. In these references one needs to assume that the integers have arbitrarily many bits, or exponentially many bits. A more precise version of the above theorem was proved in [93]. Beame [10] obtained a different proof, independently.

Proof (*cf.* [93]). Suppose that a priority PRAM M with $n^{O(1)}$ processors adds n integers in $o(\log n)$ parallel steps for infinitely many n's. The programs (maybe infinite) of M can be encoded into an oracle A. The oracle, when queried about (i, l), returns the initial section of length l of the program for $P(i)$. Fix a string $X \in \{0,1\}^{n^3}$ such that $K^A(X) \geqslant |X|$. Divide X equally into n parts $x_1, x_2, \ldots, x_n$. Then consider the (*fixed*) computation of M on input $(x_1, \ldots, x_n)$. We inductively define (with respect to X) a processor to be *alive* at step t in this computation if

(1) it writes the output; or

(2) it succeeds in writing something at some step $t' \geqslant t$ which is read at some step $t'' \geqslant t'$ by a processor who is alive at step t''.

An input is *useful* if it is read at some step t by a processor alive at step t. By simple induction on the step number we have that for a T-step computation, the number of useful inputs and the number of processors ever alive are both $O(2^T)$.

It is not difficult to see that, given all the useful inputs and the set $ALIVE=\{(P(i),t_i)$: $P(i)$ was alive until step $t_i>0\}$, we can simulate M to uniquely reconstruct the output $\sum_{i=1}^n x_i$. Since $T=o(\log n)$, we know $2^T=o(n)$. Hence there is an input x_{i_0} which is not useful. We need $O(2^T \log n^{O(1)})=o(n\log n)$ bits to represent $ALIVE$. To represent $\{x_i: i\neq i_0\}$ we need $n^3-n^2+\log n$ bits, where $\log n$ bits are needed to indicate the index i_0 of the missing input. The total number of bits needed in the simulation is less than

$$J=n^3-n^2+O(n\log n)+O(\log n)<n^3.$$

But from these J bits we can find $\sum_{i=1}^n x_i$ by simulating M using the oracle A, and then reconstruct x_{i_0} from $\sum_{i=1}^n x_i$ and $\{x_i: i\neq i_0\}$. But then $K^A(X)\leqslant J<n^3$. This contradicts the randomness of X. □

5.1.3. Example 3: Boolean matrix rank (Seiferas–Yesha)

For all n, there is an n by n matrix over GF(2) (a matrix with zero-one entries with the usual Boolean multiplication and addition) such that every submatrix of s rows and $n-r$ columns $(r,s\leqslant\frac{1}{4}n)$ has at least $\frac{1}{2}s$ linear independent rows.

REMARK. Combined with the results in [20, 172] this example implies that $TS=\Omega(n^3)$ is an optimal lower bound for Boolean matrix multiplication on any general random-access machines, where T stands for time and S stands for space.

PROOF. Fix a random sequence x of elements in GF(2) (zeros and ones) of length n^2, so $K(x)\geqslant|x|$. Arrange the bits of x into a matrix M, one bit per entry in, say, the rowmajor order. We claim that this matrix M satisfies the requirement. To prove this, suppose this is not true. Then consider a submatrix of M of s rows and $n-r$ columns, $r,s\leqslant\frac{1}{4}n$. Suppose that there are at most $\frac{1}{2}s-1$ linearly independent rows. Then $1+\frac{1}{2}s$ rows can be expressed by the linear combination of the other $\frac{1}{2}s-1$ rows. Thus we can describe this submatrix using

- The $\frac{1}{2}s-1$ linear independent rows, in $(\frac{1}{2}s-1)(n-r)$ bits;
- for each of the other $\frac{1}{2}s+1$ rows, use $(\frac{1}{2}s-1)$ bits.

Then to specify x, we only need to specify, in addition to the above, (i) M without the bits of the submatrix, and (ii) the indices of the columns and rows of this submatrix. When we list the indices of the rows of this submatrix, we list the $\frac{1}{2}s-1$ linearly independent rows first. Hence we only use

$$n^2-(n-r)s+(n-r)\log n+s\log n+(\tfrac{1}{2}s-1)(n-r)+(\tfrac{1}{2}s-1)(\tfrac{1}{2}s+1)<n^2$$

bits, for large n's. This contradicts the fact $K(x)\geqslant|x|$. □

REMARK. A lower bound obtained by Kolmogorov complexity usually implies that the lower bound holds for "almost all strings". This is the case for all three examples. In this sense the lower bounds obtained by Kolmogorov complexity are usually stronger than those obtained by its counting counterpart, since it usually also implies directly the lower bounds for nondeterministic or probabilistic versions of the considered machine. We will discuss this in Section 5.7.

5.2. *Lower bounds: more tapes versus fewer tapes*

Although Barzdin [9] and Paul [124] are the pioneers of using Kolmogorov complexity to prove lower bounds, the most influential paper is probably the one by Paul, Seiferas and Simon [126], which was presented at the 1980 STOC. This was partly because [124] was not widely circulated and, apparently, the paper by Barzdin [9] did not even reach this community. The major goal of [126] was "to promote the approach" of applying Kolmogorov complexity to obtain lower bounds. In [126], apart from other results, the authors with the aid of Kolmogorov complexity, remarkably simplified the proof of a well-known theorem of Aanderaa [1]: *real-time* simulation of k tapes by $k-1$ tapes is impossible for deterministic Turing machines.

In this model the Turing machine has k (work)tapes, apart from a separate input tape and (possibly) a separate output tape. This makes the machine for each $k \geqslant 1$ far more powerful than the model of Example 1, where the single tape is both input tape and worktape. For instance, a one-(work)tape Turing machine can recognize the marked palindromes of Example 1 in real time $T(n)=n$ in contrast with $T(n)=\Omega(n^2)$ required in Example 1.

In 1982 Paul [127], using Kolmogorov complexity, extended the results in [126] to: on-line simulation of real-time $(k+1)$-tape Turing machines by k-tape Turing machines requires $\Omega(n(\log n)^{1/(k+1)})$ time. Duriš, Galil, Paul and Reischuk [42] then improved the lower bound for the one- versus two-tape case to $\Omega(n \log n)$.

To simulate k tapes with 1 tape, the known (and trivial) upper bound on the simulation time was $O(n^2)$. The above lower bound decreased the gap with this upper bound only slightly. But in later developments w.r.t. this problem, Kolmogorov complexity has been very successful. The second author, not using Kolmogorov complexity, reported in [163] an $\Omega(n^{1.5})$ lower bound on the time to simulate a single pushdown store on-line by one *oblivious* tape unit. However, using Kolmogorov complexity the technique worked also without the oblivious restriction, and yielded in quick succession papers [164, 165] and the optimal results cited hereafter. Around 1983/1984, independently and in chronological order[2], Wolfgang Maass at UC Berkeley, the first author of the present chapter at Cornell and the second author at CWI Amsterdam, obtained a square lower bound on the time to simulate two tapes by one tape (deterministically), and thereby closed the gap between one tape versus

[2] *Historical note.* A claim for an $\Omega(n^{2-\varepsilon})$ lower bound for simulation of two tapes by both one deterministic tape and one nondeterministic tape was first circulated by W. Maass in August 1983, but did not reach Li and Vitányi. Maass submitted his extended abstract containing this result to STOC by November 1983, and this did not reach the others either. The final STOC paper of May 1984 (submitted February 1984) contained the optimal $\Omega(n^2)$ lower bound for the deterministic simulation of two tapes by one tape. In: M. Li, "On 1 tape versus 2 stacks", Tech. Rept. TR-84-591, Dept. Computer Science, Cornell University, January 1984, the $\Omega(n^2)$ lower bound was obtained for the simulation of two pushdown stores by one deterministic tape. In: P.M.B. Vitányi, "One queue or two pushdown stores take square time on a one-head tape unit", Tech. Rept. CS-R8406, Centre for Mathematics and Computer Science, Amsterdam, March 1984, the $\Omega(n^2)$ lower bound was obtained for the simulation of two pushdown stores (or the simulation of *one* queue) by one deterministic tape. Maass' and Li's result were for off-line computation with one-way input, while Vitányi's result was for on-line computation. Li and Vitányi combined these and other results in [94], while Maass published in [105].

k (w.l.o.g. two) tapes. These lower bounds, and the following ones, were proven with as simulator an *off-line* machine with *one-way* input. All three relied on Kolmogorov complexity, and actually proved more in various ways[2]. Thus, Maass also obtained a nearly optimal result for nondeterministic simulation: [105] exhibits a language that can be accepted by two deterministic one-head tape units in real time but for which a one-head tape unit requires $\Omega(n^2)$ time in the deterministic case, and

$$\Omega(n^2/(\log n)^2 \log\log n)$$

time in the nondeterministic case. This lower bound was later improved by [94] to

$$\Omega(n^2/\log n \log\log n)$$

time using Maass' language, and by Galil, Kannan and Szemeredi [49] to $\Omega(n^2/\log^{(k)} n)$ (for any k, with $\log^{(k)}$ the k-fold iterated logarithm) by an ingenious construction of a language whose computation graph does not have small separators. This almost closed the gap in the nondeterministic case. In their final combined paper, Li and Vitányi [94] presented the following lower bounds, all by Kolmogorov complexity. To simulate two pushdown stores, or only one queue, by one deterministic tape requires $\Omega(n^2)$ time. Both bounds are tight. (Note that the two-pushdown-stores result implies the two-tape result. However, the one-queue result is incomparable with either of them.) Further, one-tape nondeterministic simulation of two pushdown stores requires $\Omega(n^{1.5}/\sqrt{\log n})$ time. This is almost tight because of [89]. Finally, one-tape nondeterministic simulation of one queue requires $\Omega(n^{4/3}/\log^{2/3} n)$ time. The corresponding upper bound of the last two simulations is $O(n^{1.5}\sqrt{\log n})$ in [89]. In a successor paper, together with Longpré, we have extended the above work with a comprehensive study stressing queues in comparison to stacks and tapes [91]. There it was shown that a queue and a tape are not comparable, i.e. neither can simulate the other in linear time. Namely, simulating one pushdown store (and hence one tape) by one queue requires $\Omega(n^{4/3}/\log n)$, in both the deterministic and nondeterministic cases. Simulation of one queue by one tape was resolved above, and simulation of one queue by one pushdown store is trivially impossible. Nondeterministic simulation of two queues (or two tapes) by one queue requires $\Omega(n^2/(\log^2 n \log\log n))$ time, and deterministic simulation of two queues (or two tapes) by one queue requires quadratic time. All these results would be formidable without Kolmogorov complexity.

A next step is to attack the similar problem with a *two-way input* tape. Maass and Schnitger [106] proved that when the input tape is two-way, two worktapes are better than one for *computing a function* (in contrast to recognizing a language). The model is a Turing machine with no output tape; the function value is written on the worktape(s) when the machine halts. It is interesting to note that they considered a matrix transposition problem, as considered in Paul's original paper. Apparently, in order to transpose a matrix, a lot of information needs to be shifted around which is hard for a single tape. In [106] it is shown that transposing a matrix (with element size $O(\log n)$) requires $\Omega(n^{3/2}(\log n)^{-1/2})$ time on a one-tape off-line Turing machine with an extra two-way read-only input tape. The first version of this paper (authored by Maass alone) does not actually depend on Kolmogorov complexity, but has a cumbersome proof. The final Kolmogorov complexity proof was much easier and clearer. (This lower bound is also optimal [106].) This gives the desired separation of two tapes versus one, because with two worktapes, one can sort in $O(n \log n)$ time and hence do matrix transposition in $O(n \log n)$ time. Recently, Maass, Schnitger and Szemeredi [107] in 1987 finally resolved the question of whether two tapes are better than one with two-way input tape, for *language recognition*, with an ingenious proof. The separation language they used is again related to matrix transposition except that the matrices are Boolean and sparse (only $\log^{-2} n$ portion of non-zeros): $\{A^{**}B: A=B^t$ and $a_{ij} \neq 0$ only when $i,j=0 \bmod \log m$ where m is the size of matrices$\}$. The proof techniques used combinatorial arguments rather than Kolmogorov complexity. There is still a wide open gap between the $\Omega(n \log n)$ lower bound of [107] and the $O(n^2)$ upper bound. In [106] it was observed that if the Turing machine has a one-way output tape on which the transposed matrix can be written, transposition of Boolean matrices takes only $O(n^{5/4})$. Namely, with only one worktape and no output tape, once some bits have been written they can later be moved only by time-wasting sweeps of the worktape head. In contrast, with an output tape, as long as the output data are computed in the correct order, they can be output and do not have to be moved again. Using Kolmogorov complexity, in [41] Dietzfelbinger shows that transposition of Boolean matrices by Turing machines with two-way input tape, one worktape, and a one-way output tape requires $\Omega(n^{5/4})$ time, thus matching the upper bound for matrix transposition.

It turns out that the similar lower bound results for higher-dimensional tapes are also tractable, and sometimes easier to obtain. The original paper [126] contains such lower bounds. M. Loui proved the following results by Kolmogorov complexity. A *tree worktape* is a complete infinite rooted binary tree as storage medium (instead of a two-way infinite linear tape). A worktape head starts at the origin (the root) and in each step can move to the direct ancestor of the currently scanned node (if it is not the root) or to either one of the direct descendants. A *multihead tree machine* is a Turing machine with a tree worktape with $k \geqslant 1$ tree worktape heads. We assume that the finite control knows whether two worktape heads are on the same node or not. A *d-dimensional worktape* consists of nodes corresponding to d-tuples of integers, and in each step a worktape head can move from its current node to a node with each coordinate ± 1 of the current coordinates. Each worktape head starts at the origin which is the d-tuple with all zeros. A *multihead d-dimensional machine* is a Turing machine with a d-dimensional worktape with $k \geqslant 1$ worktape heads. M. Loui [102] has shown that a multihead d-dimensional machine simulating a multihead tree machine on-line (both machines have a one-way input tape and one-way output tape) requires time $\Omega(n^{1+1/d}/\log n)$ in the worst case, and he proved the same lower bound for the case where a multihead d-dimensional machine is made more powerful by allowing the worktape heads also to move from their current node to the current node of any other worktape head in a single step. The lower bound is optimal.

5.3. *Lower bounds: more heads versus fewer heads*

Again applying Kolmogorov complexity, Paul [125] showed that two-dimensional two-tape (with one head on each tape) Turing machines cannot simulate on-line two-dimensional Turing machines with two heads on one tape in real time. He was not able to resolve this problem for one-dimensional tapes, and, despite quite some effort, the following problem is open and believed to be difficult: Are two heads on one (one-dimensional) tape better than two (one-dimensional) tapes, each with one head? The following result, proved using Kolmogorov complexity, is intended to be helpful in separating these classes. A Turing machine with two one-head storage tapes cannot simulate a queue in both real time and with at least one storage head always within $o(n)$ squares from the start square [162]. (Thus, most prefixes of the stored string need to be shifted all the time, while storing larger and larger strings in the simulator, because the simulator must always be ready to reproduce the stored string in real time. It would seem that this costs too much time, but this has not been proved yet.) To eventually exploit this observation to obtain the desired separation, Seiferas [142] proved the following "equal information distribution" property. For no c (no matter how large) is there a function $f(n)=o(n)$, such that every sufficiently long string x has a description y with the properties: $|y|=c|x|$ and if x' is a prefix of x and y' is any subword of y with $|y'|=c|x'|$ then $K(x'|y')<f(K(x))$.

Multihead finite automata and pushdown automata were studied in parallel with the field of computational complexity in the 1960s and 1970s. One of the major problems on the interface of the theory of automata and complexity is to determine whether additional computational resources (heads, stacks, tapes, etc.) increase the computational power of the investigated machine. In the case of multihead machines it is natural to ask whether $k+1$ heads are better than k. A k-head finite (pushdown) automaton is just like a finite (pushdown) automaton except having k *one-way* heads on the input tape. Two rather basic questions were left open from the automata and formal language theory of the 1960s:

(1) Rosenberg Conjecture (1965): $(k+1)$-head finite automata are better than k-head finite automata [136, 137].

(2) Harrison–Ibarra Conjecture (1968): $(k+1)$-head pushdown automata are better than k-head pushdown automata. Or, there are languages accepted by $(k+1)$-DPDA but not k-PDA [55].

In 1965, Rosenberg [137] claimed a solution to problem (1), but Floyd [44] pointed out that Rosenberg's informal proof was incomplete. In 1971 Sudborough [152, 153], and later Ibarra and Kim [65] obtained a partial solution to problem (1) for the case of two heads versus three heads, with difficult proofs. In 1976 Yao and Rivest [171] finally presented a full solution to problem (1). A different proof was also obtained by Nelson [119]. Recently it was noted by several people, including Seiferas and the present authors, that the Yao–Rivest proof can be done very naturally and easily by Kolmogorov complexity: Let

$$L_b=\{w_1\#\ldots\#w_b\$w_b\#\ldots\#w_1: w_i\in\{0,1\}^*\}.$$

as defined by Rosenberg and Yao–Rivest. Let $b=\binom{k}{2}+1$. So L_b can be accepted by a $(k+1)$-DFA. Assume that a k-FA M also accepts L_b. Let W be a long enough Kolmogorov random string and W be equally partitioned into $w_1w_2\ldots w_b$. We say that the two w_i's in L_b are matched if there is a time such that two heads of M are in the two w_i's concurrently. Hence there is an i such that w_i is not matched. Then apparently, this w_i can be generated from $W-w_i$ and from the positions of heads and states for M when a head comes in/out w_i; $K(w_i|W-w_i)=O(k\log n)<\frac{1}{2}|w_i|$, a contradiction.

The Harrison–Ibarra Conjecture, however, was open until the time of applied Kolmogorov complexity. Several authors tried to generalize the Yao–Rivest method [116, 117] or the Ibarra–Kim method [35] to the k-PDA case, but only partial results were obtained. For the complete odyssey of these efforts see the survey in [36]. With the help of Kolmogorov complexity, [36] presented a complete solution to the Harrison–Ibarra Conjecture for the general case. The proof was constructive, and quite simple compared to the partial solutions. The basic idea, ignoring the technical details, was generalized from the above proof we gave for the Rosenberg Conjecture.

A related problem of whether a k-DFA can do string matching was raised by Galil and Seiferas [48]. They proved that a six-head *two-way* DFA can do string (pattern) matching, i.e., accept $L=\{x\#y: x \text{ is a substring of } y\}$. In 1982, when the first author and Yaacov Yesha, then at Cornell, tried to solve the problem, we achieved a difficult and tediously long proof (many pages), by counting, that 2-DFA cannot do string matching. Later Seiferas suggested the use of Kolmogorov complexity, which shortened the proof to less than a page [92]! By similar methods a proof that 3-DFA cannot do string matching was also obtained [90].

5.4. *Lower bounds: parallel computation and branching programs*

In Example 2 we have seen that the remarkable concept of Kolmogorov complexity does not only apply to lower bounds in restricted Turing machines, it also applies to lower bounds in other general models, like parallel computing models.

Fast addition or multiplication of n numbers in parallel is obviously important. In 1985 Meyer auf der Heide and Wigderson [59] proved, using Ramsey theorems, that on a priority PRAM, the most powerful parallel computing model, addition (and multiplication) requires $\Omega(\log n)$ parallel steps. Independently, a similar lower bound on addition was obtained by Israeli and Moran [66] and Parberry [122]. All these lower bounds depend on inputs from infinite (or exponentially large) domains. However, in practice, we are often interested in small inputs. For example, addition of n numbers of $n^{1/\log\log n}$ bits each can be done in $O(\log n/\log\log n)$ time with $n^{O(1)}$ processors which is *less* than the $\Omega(\log n)$ lower bound of [59]. In [93] Kolmogorov complexity is applied to obtain parallel lower bounds (and trade-offs) for a large class of functions with arguments in small domains (including addition, multiplication...) on priority PRAM. As a corollary, for example, we show that for numbers of polynomial size, it takes $\Omega(\log n)$ parallel steps for addition. This improved the results of [59, 66, 122]. Furthermore the proof is really natural and intuitive, rather than the complicated counting as before. Independently, Paul Beame at the same meeting also obtained similar results, but using a different partition method. A proof of the above result was given in Example 2.

As another example, we prove a depth-2 unbounded fan-in circuit requires $\Omega(2^n)$ gates from $\{\text{AND}, \text{OR}, \text{NOT}\}$ to compute the parity function. Assume the contrary. Let C be a binary encoding of integer n and such a circuit with $o(2^n)$ gates. Without loss of generality, let the first level of C be AND gates and the second level be an OR gate.

Consider an $x = x_1 \ldots x_n$ such that $K(x|C) \geqslant |x| = n$ and $PARITY(x) = 1$. Now, any AND gate of fan-in at least n must be 0 since otherwise we can specify x by the index of that gate which is $\log_2(o(2^n))$. Therefore, since $PARITY(x) = 1$ some AND gate G, of fan-in less than n, must be 1. Then G includes neither x_i nor $\overline{x_i}$ for some i. Hence changing only the value of x_i in x does not change the output (value 1) of G and C, a contradiction. (Note that more careful calculation on the constants can result in a more precise bound.)

Sorting is one of the most studied problems in computer science, due to its great practical importance. (As we have seen, it was also studied by Paul in [124].) In 1979 Borodin, Fischer, Kirkpatrick, Lynch and Tompa proved a time–space trade-off for comparison-based sorting algorithms [19]. This was improved and generalized to a very wide class of sequential sorting algorithms by Borodin and Cook [20] defined as "branching programs". The proof involved difficult counting. In [131] Reisch and Schnitger used Kolmogorov complexity, in one of their three applications, to simplify the well-known $\Omega(n^2/\log n)$ bound of Borodin and Cook [20] for the time–space trade-off in sorting with branching programs. They also improved the lower bound in [20] to $\Omega(n^2 \log\log n/\log n)$.

5.5. *Lower bounds: time–program size trade-off for searching a table*

"Is x in the table?" Let the table contain n keys. You can sort the table and do binary search on the table; then your program can be as short as $\log n$ bits, but you use about $\log n$ time (probes). Or you can do hashing; you can use a perfect hashing function $h(x) = [A/(Bx + C)]$ [23, 112]; then your program can be as long as $\Omega(n)$ bits since A, B, C need to have $\Omega(n)$ bits to make $h(x)$ perfect, but the search time is $O(1)$ probes. What is the size of the program? It is nothing but the Kolmogorov complexity.

Searching a table is one of the most fundamental issues in computer science. In a beautiful paper [108] Mairson literally studied *the program complexity* of table-searching procedures in terms of the number of bits that is required to write down such programs. In particular, he proved that a perfect hashing function of n keys needs $\Theta(n)$ bits to implement. He also provided the trade-offs between the time needed to search a table and the size of the searching program.

5.6. *Lower bounds: very large scale integration*

It should not be surprising that Kolmogorov complexity can be applied to VLSI lower bounds. Many VLSI lower bounds were based on the crossing sequence type of arguments similar to that of Turing machines [98]. This sort of arguments can be readily converted to much more natural and easier Kolmogorov complexity arguments like the one used in Example 1.

We use the model of Lipton and Sedgewick [98], which is a generalization of Thompson's model [154]. All lower bounds proved here also apply to the Thompson model. Roughly speaking, there are three main components in the model:

(a) the (n-input, 1-output) Boolean function $f(x_1, x_2, \ldots, x_n)$ which is to be computed;

(b) a synchronous circuit C, computing f, which contains AND, OR, NOT gates of arbitrary fan-in and fan-out and with n fixed input gates (i.e., what is called where-oblivious) that are not necessarily on the boundary of the layout of C (the time an input arrives may depend on the data value); and

(c) a VLSI (for convenience, rectangle) layout V that realizes C, where wires are of unit width and processors occupy unit squares.

A central problem facing the VLSI designers is to find C that computes a given f in time T, and a VLSI layout of C with area A, minimizing say AT^2 as introduced by Thompson [154] and later generalized in [98].

We prove $AT^2 = \Omega(n^2)$ lower bounds for many problems roughly as follows. Draw a line to divide the layout into two parts, with about half the inputs on each part. Suppose the line cuts through ω wires, then $A = \Omega(\omega^2)$. Further, since for each time unit only one bit of information can flow through a wire, $T > I/\omega$ where I is the *amount* of information that has to be passed between the two parts. Then for each specific problem one only needs to show that $I = \Omega(n)$ for any division. Lipton and Sedgewick defined a *crossing sequence* to be, roughly, the sequence of T tuples $(v_1, \ldots, v_\omega)$ where the ith tuple contains the values appearing at the cut of width ω at step i.

Now it is trivial to apply our Kolmogorov complexity to simplify the proofs of *all* VLSI lower bounds obtained this way. Instead of complicated and nonintuitive counting arguments which involves *all* inputs, we now demonstrate how easy one can use one single Kolmogorov random string instead. The lower bounds before the work of [98] were for n-input and n-output functions; the Kolmogorov complexity can be even more trivially applied there. We only look at the harder n-input 1-output problems stated in [98]. A sample question was formulated in [98].

Example (*pattern matching*). Given a binary text string of $(1-\alpha)n$ bits and a pattern of αn bits, with $\alpha < 1$, determine if the pattern occurs in the text.

Proof (*sketch*). Let C implement pattern matching with layout V. Consider any cut of V of width ω which divides inputs into two halves. Now it is trivial that $I = \Omega(n)$ since for a properly arranged Kolmogorov random text and pattern this much information must pass the cut. This finishes the proof of $AT^2 = \Omega(n^2)$. □

All other problems, selection/equality testing, DCFL, factor verification, listed in [98] can be done similarly, even under the nondeterministic, or randomized, circuits as defined in [98].

Some general considerations on VLSI lower bounds using Kolmogorov complexity were given by R. Cuykendall [39]. L.A. Levin and G. Itkis have investigated the VLSI computation model under different information transmission assumptions [87, 175]. In their model, if the speed of information transmission is superlinear, namely $\max(K(d) - \log f(d)) < \infty$ with $f(d)$ the time for a signal to traverse a wire of length d, then a chip can be simulated by a chip in which all long wires have been deleted (which results in considerable savings in required area). Note that $f(d) = \Omega(d \log^2 d)$ satisfies the requirements for f, but not $f(d) = O(d)$.

5.7. *Lower bounds: randomized algorithms*

We have seen that Kolmogorov complexity can be naturally applied to nondeterministic Turing machines. Hence, it is likely that it is useful for analyzing randomized algorithms. Indeed this is the case. In their paper about three applications of Kolmogorov complexity [131] Reisch and Schnitger analyzed, using Kolmogorov complexity, the probabilistic routing algorithm in the n-dimensional binary cubes of Valiant and Brebner [156].

In 1983 Paturi and Simon generalized the deterministic lower bounds previously proved in [1, 125–127, etc.] to probabilistic machines. This is based on the following elegant idea (based on a note of, and discussions with, R. Paturi). As we mentioned before, all Kolmogorov complexity proofs depend on only a fixed Kolmogorov random string α. If the lower bound fails, then this incompressible string can be compressed, hence a contradiction. A version of the Symmetry of Information Lemma stated in Sections 2.6 and 2.7 is proved in [123]. They show that for a sequence of random coin tossing, the probability that this sequence β of random coin tossing bits contains much information about α is vanishingly small. Observe that if α is Kolmogorov random relative to the coin tossing sequence β, then the old deterministic argument would just fall through with β as an extra useless input (or oracle as in Example 2). Note that many such α's exists. Hence, (ignoring technical details) using this idea and careful construction of the input for the probabilistic simulator, it was shown that, on the average, the probabilistic simulator would not give any advantage in reducing the computation time.

REMARK. Similar ideas were expressed earlier in 1974 by Levin who called the general principle involved "Law of Information Conservation" [86]; see for later developments also [88].

5.8. *Lower bounds: formal language theory*

The classic introduction to formal language theory is [62]. An important part of formal language theory is deriving a hierarchy of language families. The main division is the Chomsky hierarchy with regular languages, context-free languages, context-sensitive languages and recursively enumerable languages. The common way to prove that certain languages are not regular (not context-free) is by using "pumping" lemmas, i.e., the uvw-lemma ($uvwxy$-lemma respectively). However, these lemmas are complicated to state and cumbersome to prove or use. In contrast, below we show how to replace such arguments by simple, intuitive and yet rigorous Kolmogorov complexity arguments. We present some material from our paper [95]. Without loss of generality, languages are infinite sets of strings over a finite alphabet.

Regular languages coincide with the languages accepted by finite automata (FA). Another way of stating this is by the Myhill–Nerode Theorem: each regular language over an alphabet V consists of the union of some equivalence classes of a right-invariant equivalence relation on V^* $(=\bigcup_{i\geq 0} V^i)$ of finite index. Let us give an example of how to use Kolmogorov complexity to prove nonregularity. We prove that $\{0^k1^k: k\geq 1\}$

is not regular. To derive a contradiction, suppose it is regular. Fix k with $K(k) \geqslant \log k$, with k large enough to derive the contradiction below. The state q of the accepting FA after processing 0^k is, up to a constant, a description of k. Namely, by running the FA, starting from state q, on a string consisting of ones, it reaches its first accepting state precisely after k ones. Hence, there is a constant c, depending only on FA, such that $\log k < c$, which is a contradiction. We generalize this observation, actually a Kolmogorov-complexity interpretation of the Myhill–Nerode Theorem, as follows. (In lexicographic order, short strings precede long strings.)

Lemma (KC-regularity). *Let L be regular. Then for some constant c depending only on L and for each string x, if y is the n-th string in the lexicographical order in $L_x = \{y : xy \in L\}$ (or in the complement of L_x) then $K(y) \leqslant K(n) + c$.*

Proof. Let L be a regular language. A string y such that $xy \in L$, for some x and n as in the lemma, can be described by

(a) this discussion, and a description of the FA that accepts L,

(b) the state of the FA after processing x, and the number n. □

The KC-regularity lemma can be applied whenever the pumping lemma can be applied. It turns out that the converse of our lemma also holds and gives a Kolmogorov complexity characterization of regular languages [95]. Therefore, the above lemma also applies to situations when the normal pumping lemma(s) do(es) not apply. Further it is easier and more intuitive than pumping lemmas, as shown in the following example.

Example. We prove that $\{1^p : p \text{ is prime}\}$ is not regular. Consider the string xy consisting of p ones, with p the $(k+1)$st prime. In the lemma set x equal to $1^{p'}$ with p' the kth prime, so $y = 1^{p-p'}$ and $n = 1$. It follows that $K(p-p') = O(1)$. Since the differences between the consecutive primes rise unboundedly, this implies that there is an unbounded number of integers of Kolmogorov complexity $O(1)$. Since there are only $O(1)$ descriptions of length $O(1)$, we have a contradiction. (A simple way to argue that $p-p'$ rises unboundedly is as follows. Let P be the product of the first j primes. Clearly, no $P+i$, $1 < i \leqslant j$, is prime.)

Example. (*cf.* [62, *Exercise* 3.1(h*)]). Prove that $L = \{xx^{R}w : x, w \in \{0,1\}^* - \{\varepsilon\}\}$ is not regular. Set $x = (01)^n$, where $K(n) \geqslant \log n$. Then the lexicographically first word in L_x is $y = (10)^n 0$. Hence, $K(y) = \Omega(\log n)$, contradicting the KC-regularity Lemma.

Example (*cf.* [62, *Exercise* 3.6*]). Prove that $L = \{0^i 1^j : GCD(i,j) = 1\}$ is not regular. For each prime p, the string $0^p 1^p$ is the second word with prefix 0^p. Hence by the KC-regularity Lemma there is a constant c such that for all p we have $K(p) < c$, which is a contradiction.

Similar general lemmas can also be proved to separate DCFLs from CFLs. Previous proofs that a CFL is not a DCFL often use ad hoc methods. We refer the interested readers to [95].

Concerning formal language theory we must mention a beautiful proof due to Seiferas. It is known that linear context-free languages can be recognized on-line by a one-worktape Turing machine in $O(n^2)$ time. This result is due to Kasami. Gallaire, using a very complicated counting argument and *de Bruijn sequences* [50], proved that a multitape Turing machine requires $\Omega(n^2/\log n)$ time to recognize on-line linear context-free languages. In [143] Seiferas presented an elegant and short proof of the same bound using Kolomogorov complexity, significantly simplifying Gallaire's proof.

5.9. *Lower bounds: which method to use?*

Instead of attempting to answer this difficult question, we present a problem with three proofs: one by counting, one by probabilistic argument and one by Kolmogorov complexity. The question and first two proofs are taken from a beautiful book by Erdös and Spencer [43].

A *tournament* T is a complete directed graph, i.e., for each pair of vertices u and v in T, exactly one of the edges (u, v), (v, u) is in the graph. Given a tournament T of n nodes $\{1, \ldots, n\}$, fix any standard effective coding, denoted by $c(T)$, using $\frac{1}{2}n(n-1)$ binary bits, one bit for each edge. The bit of edge (u, v) is set to 1 iff $u<v$. The next theorem and the first two proofs are from the first example in [43].

THEOREM. *If $v(n)$ is the largest integer such that every tournament on $\{1, \ldots, n\}$ contains a transitive subtournament on $v(n)$ players, then $v(n) \leqslant 1+[2\log_2 n]$.*

REMARK. This theorem was proved first by Erdös and Moser in 1964. Stearns showed by induction that $v(n) \geqslant 1+[\log_2 n]$.

PROOF (*by counting*). Let $v=2+[2\log_2 n]$. Let $\Gamma=\Gamma_n$ be the class of all tournaments on $\{1, \ldots, n\}$, and Γ' the class of tournaments on $\{1, \ldots, n\}$ that do contain a transitive subtournament on v players. Then

$$\Gamma' = \bigcup_A \bigcup_\sigma \Gamma_{A,\sigma} \tag{5.1}$$

where $A \subseteq \{1, \ldots, n\}$, $|A|=v$, σ is a permutation on A, and $\Gamma_{A,\sigma}$ is the set of T such that $T|A$ is generated by σ. If $T \in \Gamma_{A,\sigma}$, the $\binom{v}{2}$ games of $T|A$ are determined. Thus

$$|\Gamma_{A,\sigma}| = 2^{\binom{n}{2}-\binom{v}{2}} \tag{5.2}$$

and by elementary estimates

$$|\Gamma'| \leqslant \sum_{A,\sigma} 2^{\binom{n}{2}-\binom{v}{2}} = \binom{n}{v} v! 2^{\binom{n}{2}-\binom{v}{2}} < 2^{\binom{n}{2}} = |\Gamma|. \tag{5.3}$$

Thus $\Gamma - \Gamma' \neq \emptyset$. That is, there exists $T \in \Gamma - \Gamma'$ not containing a transitive subtournament on v players. □

PROOF (*by probabilistic argument*). Let $v=2+[2\log_2 n]$. Let $\Gamma=\Gamma_n$ be the class of all tournaments on $\{1, \ldots, n\}$. Let also $A \subseteq \{1, \ldots, n\}$, $|A|=v$, and let σ be a permutation on A. Let $\boldsymbol{T}=T_n$ be a random variable. Its values are the members of Γ where, for each $T \in \Gamma$, $\Pr(\boldsymbol{T}=T)=2^{-\binom{n}{2}}$. That is, all members of Γ are equally probable values of $\boldsymbol{T}$.

Then

$$\begin{aligned}&\Pr(\boldsymbol{T} \text{ contains a transitive subtournament on } v \text{ players})\\ &\quad \leqslant \sum_A \sum_\sigma \Pr(\boldsymbol{T} \mid A \text{ generated by } \sigma)\\ &\quad = \binom{n}{v} v! 2^{-\binom{v}{2}} < 1.\end{aligned}$$

Thus some value T of $\boldsymbol{T}$ does not contain a transitive subtournament on v players. □

PROOF (*by Kolmogorov complexity*). Fix $T \in \Gamma_n$ such that

$$K(c(T) \mid n, v(n)) \geqslant |c(T)| = \tfrac{1}{2}n(n-1).$$

Suppose $v(n) = 2 + [2\log_2 n]$ and let S be the transitive tournament of $v(n)$ nodes. We effectively recode $c(T)$ as follows in less than $|c(T)|$ bits, and hence we obtain a contradiction, by

- listing in order of dominance the index of each node in S in front of $c(T)$, using $2(\lceil \log_2 n \rceil)^2 + 2\lceil \log_2 n \rceil + |c(T)|$ bits;
- deleting all bits from $c(T)$ for edges in between nodes in S to save $2(\lceil \log_2 n \rceil)^2 + 3\lceil \log_2 n \rceil + 1$ bits. □

5.10. *Lower bounds: open questions*

This section summarizes the open questions that we consider to be interesting and that may be solvable by Kolmogorov complexity.

(1) Can k-DFA do string matching [48]?

(2) Are two heads on one (one-dimensional) tape better than two (one-dimensional) tapes each with one head?

(3) Prove a tight, or $\Omega(n^{1+\varepsilon})$, lower bound for simulating two tapes by one for off-line Turing machines with an extra two-way input tape.

6. Resource-bounded Kolmogorov complexity and its applications

Here we treat several notions of resource-bounded Kolmogorov complexity, with applications ranging from the P = NP question to factoring integers and cryptography. Several authors suggested early on the possibility of restricting the power of devices used to compress strings. Says Kolmogorov [76] in 1965:

> "The concept discussed ... does not allow for the 'difficulty' of preparing a program p for passing from an object x to an object y.... [some] object permitting a very simple program, i.e. with very small complexity $K(x)$, can be restored by short programs only as the result of computations of a thoroughly unreal nature.... [this concerns] the relationship between the necessary complexity of a program and its permissible difficulty t. The complexity $K(x)$ that was obtained [before] is, in this case, the minimum of $K^t(x)$ on the removal of the constraints on t."

The earliest use of resource-bounded Kolmogorov complexity we know of is Barzdin's 1968 result [9] cited earlier. Time-limited Kolmogorov complexity was applied by Levin [84] in relation with his independent work on NP-completeness, and further studied in [88]. Adleman investigated such notions [2], in relation to factoring large numbers. Resource-bounded Kolmogorov complexity was extensively investigated by Daley [40], Hartmanis [56], and Ko [73]. Sipser [146] used time-limited Kolmogorov complexity to show that the class BPP (problems which can be solved in polynomial time with high probability) is contained in the polynomial time hierarchy: $\mathrm{BPP} \leqslant \Sigma_4 \cap \Pi_4$. (Gács improved this to $\mathrm{BPP} \leqslant \Sigma_2 \cap \Pi_2$.) We treat the more influential approaches of Adleman, Bennett, Hartmanis and Sipser in more detail below. Let us note here that there is some relation between the approaches to resource-bounded Kolmogorov complexity by Adleman [2], Levin [88] and Bennett [12].

6.1. *Potential*

In an elegant paper [2], Adleman formulates the notion of *potential* as the amount of time that needs to be pumped into a number by the computation that finds it. That is, while constructing a large composite number from two primes we spend only a small amount of time. However, to find the primes back may be difficult and take a lot of time. Is there a notion of storing potential in numbers with the result that high-potent primes have relatively low-potent products? Such products would be hard to factor, because all methods must take the time to pump the potential back. Defining the appropriate notion, Adleman shows that if factoring is not in P (the class of problems that can be solved by deterministic algorithms in time polynomial in the input length) then this is the reason why. Formally, we use the following definition.

DEFINITION. For all integers $k \geqslant 0$, for all $x \in \{0,1\}^*$ (for all $y \in \{0,1\}^*$), x *is k-potent* (*with respect to y*) iff there is a program p of size $\leqslant k \log|x|$ which with blanks (y) as input halts with output x in less than or equal to $|x|^k$ steps. (Recall that $|x|$ is the length of x and x can mean the positive integer x or the xth binary string.)

EXAMPLE. For almost all $n \in N$, 1^n is 2-potent. Namely, $|1^n| = n$ and $|n| \sim \log n$. Then it is not difficult to see that, for each large enough n, there is a program p, $|p| < 2 \log n$, that computes 1^n in less than n^2 steps.

EXAMPLE. For all k, for almost all incompressible x, x is not k-potent. This follows straightaway from the definitions.

EXAMPLE. Let u be incompressible. If $v = u + 1^{666}$, where "+" denotes "exclusive or", then v is incompressible, but also v is 1-potent with respect to u.

LEMMA (Adleman). *For all k, the function*

$$f_k(x, 1^n) = \{\, y \colon |y| \leqslant n \text{ and } y \text{ is } k\text{-potent w.r.t. } x \}$$

is computable in polynomial time.

Proof. There are at most $2 \cdot 2^{k|n|} \sim 2n^k$ programs of length $\leqslant k|n|$. By simulating all such programs (one after the other) on input x for at most n^k steps, the result is obtained. □

We informally state two results proved by Adleman.

Theorem (Adleman). *Factoring is difficult iff multiplication infinitely often takes highly potent numbers and produces relatively low-potent products.*

Theorem (Adleman). *With respect to the* $\mathrm{P}=\mathrm{NP}$ *question:* $SAT \in \mathrm{NP}-\mathrm{P}$ *iff for all* k *there exist infinitely many* $\phi \in SAT$ *such that, for all* T*, if truth assignment* T *satisfies* ϕ*, then* T *is not* k*-potent w.r.t.* ϕ*.*

6.2. *Logical depth*

Bennett has formulated an intriguing notion of logical depth [12–14]. Kolmogorov complexity helps to define individual information and individual randomness. It can also help to define a notion of "individual computational complexity" of a finite object. Some objects are the result of long development (=computation) and are extremely unlikely to arise by any probabilistic algorithm in a small number of steps. Logical depth is the necessary number of steps in the deductive or causal path connecting an object with its plausible origin. Concretely, the time required by a universal computer to compute an object from its maximally compressed description. Formally (in Gács' reformulation, using the Solomonoff–Levin approach to a priori probability $\boldsymbol{m}$, cf. Section 2.9),

$$depth_{\varepsilon}(x) = \min\{t : \boldsymbol{m}_t(x)/\boldsymbol{m}(x) \geqslant \varepsilon\}.$$

Here, $\boldsymbol{m}_t$ is the t-bounded analogue of $\boldsymbol{m}$. Thus, the depth of a string x is at least t with confidence $1-\varepsilon$ if the conditional probability that x arises in t steps *provided it arises at all* is less than ε. (One can also formulate logical depth in terms of shortest programs and running times, see [12] or the example below.) According to Bennett, quoted in [30]: "A structure is deep, if it is superficially random but subtly redundant, in other words, if almost all its algorithmic probability is contributed by slow-running programs. . . . A priori the most probable explanation of 'organized information' such as the sequence of bases in a naturally occurring DNA molecule is that it is the product of an extremely long biological process."

Example (*Bennett*). Bennett's original definition: fix, as usual, an optimal universal machine U; a string $x \in \{0, 1\}^*$ is logical (d, b)-deep, or "d-deep at confidence level 2^{-b}", if every program to compute x in time $\leqslant d$ is compressible by at least b bits.

The notion is intended to formalize the idea of a string for which the null hypothesis that it originated by an effective process of fewer than d steps is as implausible as tossing b consecutive heads. Depth should be stable, i.e., no trivial computation should be able to transform a shallow object into a deep one.

THEOREM (Bennett). *Deep strings cannot be quickly computed from shallow ones. More precisely, there is a polynomial $p(t)$ and a constant c, both depending on U, such that if x is a program to compute y in time t and if x is less than (d, b)-deep, then y is less than $(d+p(t), b+c)$-deep.*

EXAMPLE (*Bennett*). Similarly, depth is reasonably machine-independent. If U, U' are two optimal universal machines, then there exists a polynomial $p(t)$ and a constant c, both depending on U, U', such that $(p(d), b+c)$-depth on either machine is a sufficient condition for (d, b)-depth on the other.

EXAMPLE (*Bennett*). This example concerns the distinction between depth and information: consider the numbers k and Ω (see Section 3.6). k and Ω encode the same information, viz. solution to the halting problem. But k is deep and Ω shallow. Because Ω encodes the halting problem with maximal density (the first 2^n bits of k can be computed from the first $n+O(\log n)$ bits of Ω), it is recursively indistinguishable from random noise and practically useless: the time required to compute an initial segment of k from an initial segment of Ω increases faster than any computable function. That is, Barzdin' [9] showed that the initial segments of k are compressible to the logarithm of their length if unlimited time is allowed for decoding, but they can only be compressed by a constant factor if any recursive bound is imposed on the decoding time. The precise statement of this is given at the end of Section 2.5.

6.3. *Generalized Kolmogorov complexity*

Below we partly follow [56]. Assume that we have fixed a universal Turing machine U with an input tape, worktapes and an output tape. "A string x is computed from a string z" (z is a *description* of x) means that U starting with z on its input tape halts with x on its output tape.

REMARK. In order to be accurate in the reformulations of notions in the examples below, we shall assume w.l.o.g. that the set of programs for which U halts is an effective prefix code: no such program is the prefix of any other such program, i.e., we use self-delimiting Kolmogorov complexity as described in Section 27.

In the following we distinguish the main parameters we have been able to think of: compression factor, time, space, and whether the computation is inflating or deflating. A string x has *resource-bounded* Kolmogorov complexity ${}^{\mathrm{UP}}_{U}(K, T, S)$ if x can be computed from a string z, $|z| \leqslant K \leqslant |x|$, in $\leqslant T$ steps by U using $\leqslant S$ space on its worktape. A string x of length n is in complexity class $K_U^{\mathrm{UP}}[k(n), t(n), s(n)]$ if $K \leqslant k(n)$, $T \leqslant t(n)$ and $S \leqslant s(n)$. Thus, we consider a computation that *inflates* z to x. A string x has *resource-bounded* Kolmogorov complexity ${}^{\mathrm{DOWN}}_{U}(K, T, S)$ if some description z of x can be computed from x, $|z| \leqslant K \leqslant |x|$, in $\leqslant T$ steps by U using $\leqslant S$ space on its worktape. Here we consider a computation that *deflates* x to z. A string x of length n is in complexity class $K_U^{\mathrm{DOWN}}[k(n), t(n), s(n)]$ if $K \leqslant k(n)$, $T \leqslant t(n)$ and $S \leqslant s(n)$. Clearly,

$$K_U^{\mathrm{DOWN}}[k(n), \infty, \infty] = K_U^{\mathrm{UP}}[k(n), \infty, \infty] = K[k(n)]$$

with $k(n)$ fixed up to a constant and $K[k(n)]$ (with some abuse of notation) the class of binary strings x such that $K(x) \leqslant k(n)$. (Here we denote by K the self-delimiting Kolmogorov complexity).

It follows immediately by the Hennie–Stearns simulation of many worktapes by two worktapes (cf. [62]) that there is a U with two worktapes such that, for any multitape universal Turing machine V, there is a constant c such that

$$K_V^{\text{UP}}[k(n), t(n), s(n)] \subseteq K_U^{\text{UP}}[k(n)+c, c \cdot t(n) \log t(n)+c,\ c\, s(n)+c].$$

Thus, henceforth we drop the subscripts because the results we derive are invariant up to such small perturbations. It is not difficult to prove however that larger perturbations of the parameters separate classes. For instance,

$$K^{\text{UP}}[\log n, \infty, n^2] \subset K^{\text{UP}}[\log n, \infty, n^2 \log n],$$

$$K^{\text{UP}}[\log n, \infty, n^2] \subset K^{\text{UP}}[2 \log n, \infty, n^2].$$

The obvious relation between inflation and deflation is

$$K^{\text{UP}}[k(n), t(n), \infty] \subseteq K^{\text{DOWN}}[k(n), t(n)2^{k(n)}, \infty],$$

$$K^{\text{DOWN}}[k(n), t(n), \infty] \subseteq K^{\text{UP}}[k(n), t(n)2^n, \infty]$$

(there are at most $2^{k(n)}$ (resp. 2^n) possibilities to try). Our $K^{\text{UP}}[f(n), g(n), h(n)]$ will also be written as $K[f(n), g(n), h(n)]$ to be consistent with literature as in [56].

In his Ph.D. Thesis [101], Longpré analyzed the structure of the different generalized Kolmogorov complexity sets, with different time and space bounds (the UP version). Longpré builds the resource hierarchies for Kolmogorov complexity in the spirit of classical time and space complexity hierarchies. He related further structural properties to classical complexity. He also extended Martin-Löf's results to generalized Kolmogorov complexity: the space-bounded Kolmogorov complexity random strings pass all statistical tests which use less space than the space bound. Finally, he shows how to use Kolmogorov randomness to build a pseudorandom number generator that passes Yao's test [170].

Example (*Potency*). Adleman's potency [2] can now be reformulated as follows: $x \in \{0,1\}^*$, $|x| = n$, is k-potent if $x \in K[k \log n, n^k, \infty]$.

Example (*Computation time*). Related to the notions potential and logical depth is Levin's concept of *time of computation complexity Kt* [88]. In this framework we formulate it as follows: $x \in \{0,1\}^*$ has Kt-complexity $Kt(x) = m$ if $x \in K[m - \log t, t, \infty]$, m minimal.

Example (*Hartmanis*). The sparse set

$$\text{SAT} \cap K[\log n, n^2, \infty]$$

is a Cook-complete set for all other sparse sets in NP.

In [56] these and similar results are derived for PSPACE and sets of other densities.

It is also used to give new interpretations to oracle constructions, and to simplify previous oracle constructions. This leads to conditions in terms of Kolmogorov complexity under which there exist NP complete sets that are not polynomial-time isomorphic, as formulated in [15]. In [57] a characterization of the P = NP question is given in terms of time-bounded Kolmogorov complexity and relativization. Earlier, Adleman in [2] established a connection, namely NP ≠ P exactly when NP machines can "manufacture" randomness. Following this approach, Hemachandra [60] obtains unrelativized connections in the spirit of [57].

EXAMPLE (*Hartmanis*). Hartmanis noticed the following interesting fact: a polynomial machine cannot from simple input compute complicated strings and hence cannot ask complicated questions to an oracle A. Using this idea, he constructed several very elegant oracles. As an example, we construct the Baker–Gill–Solovay oracle A such that $P^A \neq NP^A$: By diagonalization, choose $C \subseteq \{1^{2^n}: n \geqslant 1\}$ and $C \in \mathrm{DTIME}(n^{\log n}) - \mathrm{P}$. For every n such that $1^{2^n} \in C$, put the first string of length 2^n from

$$K[\log n, n^{\log n}, \infty] - K[\log n, n^{\log\log n}, \infty]$$

in A. Clearly, $C \in NP^A$. But C cannot be in P^A since in polynomial time, a P^A-machine cannot ask any question about any string in A. Hartmanis also constructed two others including a random sparse oracle A such that $NP^A \neq P^A$ with probability 1.

EXAMPLE (*Longpré, Natarajan* [101, 118]). It was noticed that Kolmogorov complexity can be used to obtain space complexity hierarchies in Turing machines. Also it can be used to prove certain immunity properties. For example, one can prove that if $\lim_{n\to\infty} S(n)/S'(n) = 0$, then, for any universal machine U, if $S'(n) \geqslant n$ is a nondecreasing function and if $f(n)$ is a function not bounded by any constant and computable in space $S(n)$ by U, then we have that the complement of $K_U[f(n), \infty, S'(n)]$ is DSPACE($S(n)$)-immune for large n.

6.4. Generalized Kolmogorov complexity applied to structural proofs

Generalized Kolmogorov complexity turns out to be an elegant tool for studying the structure in complexity classes. The first such applications are probably due to Hartmanis, as we discussed in a previous section. Other work in this area includes [5, 8, 128]. In this section we try to present some highlights of the continuing research in this direction. We will present several excellent constructions, and describe some constructions in detail.

EXAMPLE (*an exponentially low set not in* P). A set A is exponentially low if $E^A = E$, where $E = \mathrm{DTIME}(2^{cn})$. Book, Orponen, Russo, and Watanabe [18] constructed an exponentially low set A which is not in P. We give this elegant construction in detail. Let $K = K[\frac{1}{2}n, 2^{3^n}, \infty]$ and $\bar{K}$ its complement. Let $A = \{x: x$ is the lexicographically least element of $\bar{K}$ of length $2^{2^{\cdots^2}}$ (stack of m 2s), for some $m > 0\}$. Obviously, $A \in E$. Further A is not in P since otherwise we let $A = L(M)$ and for $|x| \gg |M|$ and $x \in A$ we would have that $x \in K$, a contradiction. We also need to show that $E^A = E$. To simulate a computation of E^A by an E machine we do the following: If a query to A is of correct length (stack of 2s) and shorter than cn for a small constant c, then just do exhaustive search to decide. Otherwise, the answer is "no" since

(1) a string of wrong length (no stack of 2s) is not in A and

(2) a string of length greater than cn is not even in $\bar{K}$.

(2) is true since the query string can be calculated from the input of length n and the exponentially shorter previous queries, which can be encoded in, say, $\frac{1}{4}cn$ bits assuming c is chosen properly; therefore the query string is in K.

In [168], Watanabe used time/space-bounded Kolmogorov complexity to construct a more sophisticated set D which is polynomial Turing complete for E but not complete for E under polynomial truth-table reduction. Allender and Watanabe [6] used Kolmogorov complexity to characterize the class of sets which are polynomial many-one equivalent to tally sets, in order to study the question of whether $E^{\mathrm{P}}_{\mathrm{m}}(Tally) = E^{\mathrm{P}}_{\mathrm{btt}}(Tally)$ is true, where $E^{\mathrm{P}}_{-}(Tally) = \{L\text{: for some tally set } T,\ L =^{\mathrm{P}}_{-} T\}$. In [63, 64], Huynh started a series of studies on the concept of resource-bounded Kolmogorov complexity of languages. He defined the (time/space-bounded) Kolmogorov complexity of a language to be the (time/space-bounded) Kolmogorov complexity of

$$Seq(L^{<n}) = C_L(w_1)C_L(w_2)\ldots C_L(w_{2^n-1}),$$

where w_i is lexicographically the ith word and $C_L(w_i)=1$ iff $w_i \in L$. In particular, he shows that there is a language $L \in \mathrm{DTIME}(2^{2^{O(n)}})$ (any hard set for this class) such that the 2^{Poly}-time-bounded Kolmogorov complexity of L is exponential almost everywhere. That is, the sequence $Seq(L^{<n})$ cannot be compressed to a subexponentially short string within 2^{Poly} time for all but finitely many n's. Similar results were also obtained for space-bounded classes. He used these results to classify exponential-size circuits. Compare this with Barzdin's result, cited at the end of Section 2.5.

Allender and Rubinstein studied the relation between small resource-bounded Kolmogorov complexity and P-printability. Sets like $K[k \log n, n^k, \infty]$ for some constant k are said to have small time-bounded Kolmogorov complexity. A set S is said to be polynomial-time printable (P-printable) if there is a k such that all the elements of S up to size n can be printed by a deterministic machine in time $n^k + k$. Clearly, every P-printable set is a sparse set in P. Define the *ranking function* for a language L, $r_L : \Sigma^* \to N$, given by $r_L(x) = d(\{w \in L : w < x\})$ [54]. Allender and Rubinstein [7] proved that the following are equivalent:

(1) S is P-printable.

(2) S is sparse and has a ranking function computable in polynomial time.

(3) S is P-isomorphic to some tally set in P.

(4) $S \subseteq K[k \log n, n^k, \infty]$ for some constant k and $S \in \mathrm{P}$.

Note: The equivalence of (1) and (4) is due to Balcazar, and Book [8] and Hartmanis and Hemachandra [57].

6.5. *Time-bounded Kolmogorov complexity and language compression*

If A is a recursive set and x is lexicographically the ith element in A, then we know

$K(x) \leqslant \log i + c_A$ for some constant c_A not depending on x. Here we use plain Kolmogorov complexity as in the introduction of Section 2.

NOTATION. In this section let us write $K^t(x|y)$ to denote the conditional t-time-bounded Kolmogorov complexity of x, given y. Define the unconditional complexity of x as $K^t(x) = K^t(x|\varepsilon)$.

Further let $A \in \mathrm{P}$, where P is the class of problems decidable by deterministic Turing machines in polynomial time. It is seductive to think the following:

CONJECTURE. $\exists c \forall s \in A^n [K^p(s) \leqslant \log d(A^n) + c_A]$, *where* A^n *is the set of elements in* A *of length* n, *and* p *is a polynomial.*

However in polynomial time a Turing machine cannot search through 2^n strings as is assumed with A a recursive set as above. Whether or not the above conjecture is true is still an important open problem in time-bounded Kolmogorov complexity which we deal with exclusively in this section. It also has important consequences in language compression.

DEFINITION (*Goldberg and Sipser* [51]). (1) A function $f: \Sigma^* \to \Sigma^*$ is a *compression* of language L if f is one-to-one on L and, for all except finitely many $x \in L$, $|f(x)| < |x|$.

(2) A language L is *compressible in time* T if there is a compression function f for L which can be computed in time T, and also the inverse f^{-1} of f with domain $f(L)$, such that for any $x \in L$, $f^{-1}(f(x)) = x$, can be computed in time T.

(3) Compression function f *optimally compresses* a language L if, for any $x \in L$ of length n,

$$|f(x)| \leqslant \left\lceil \log\left(\sum_{i=0}^{n} d(L^i) \right) \right\rceil .$$

(4) One natural and optimal compression is *ranking*. The ranking function $r_L : L \to N$ maps $x \in L$ to its index in a lexicographical ordering of L.

Obviously, language compression is closely related to the Kolmogorov complexity of the elements in the language. *Efficient* language compression is closely related to the *time-bounded* Kolmogorov complexity of the elements of the language. By using a ranking function, we can obtain the optimal Kolmogorov complexity of any element in a recursive set, and hence, optimally compress the total recursive set. That was trivial. Our purpose is studying the polynomial-time setting of the problem. This is far from trivial.

6.5.1. *Language compression with the help of an oracle*

Let $\psi_1, \psi_2, \ldots$ be an effective enumeration of partial recursive predicates. Let T_ψ be a multitape Turing machine which computes ψ. $T_\psi(x)$ outputs 0 or 1. If T_ψ accepts x in t steps (time), then we also write $\psi^t(x) = 1$.

DEFINITION (*Sipser*). Let x, y, p be strings in $\{0,1\}^*$. Fixing ψ, we can define KD^t_ψ of x, *conditional to* ψ and y, by

$$KD^t_\psi(x|y) = \min\{|p|: \forall v, \psi^t(v, p, y) = 1 \text{ iff } v = x\},$$

and $KD^t_\psi(x|y) = \infty$ if there are no such p.

REMARK. One can prove an invariance theorem similar to that of the K^t version; we can hence drop the index ψ in KD^t_ψ.

The intuition of the above definition is that while $K^t(x)$ is the length of the shortest program *generating* x in $t(|x|)$ time, $KD^t(x)$ is the length of the shortest program *accepting* only x in $t(|x|)$ time. In pure Kolmogorov complexity, these two measures differ by only an additive constant. In the resource-bounded Kolmogorov complexity, they appear quite different. The KD version appears to be somewhat simpler, and the following were proved by Sipser [146]. Let p, q be polynomials, c be a constant, and NP be an NP-complete oracle.

(1) $\forall p \exists q [KD^q(s) \leqslant K^p(s) + O(1)]$.

(2) $\forall p \exists q [K^q(s \mid NP) \leqslant KD^p(s) + O(1)]$.

(3) $\forall c \exists d$, if $A \subseteq \Sigma^n$ and A is accepted by a circuit of size n^c, then for each $s \in A$,

$$KD^d(s \mid A, i_A) \leqslant \log d(A) + \log\log d(A) + O(1),$$

where i_A depends on A and has length about $n \log d(A)$.

(4) $\forall c \exists d$, if $A \subseteq \Sigma^n$ is accepted by a circuit of size n^c and there is a string i_A such that for each $s \in A$,

$$K^d(s \mid A, i_A, NP) \leqslant \log d(A) + \log\log d(A) + O(1),$$

then

$$K^d(s \mid A, \Sigma_2) \leqslant \log d(A) + \log\log d(A) + O(1).$$

In order to prove the above results, Sipser needed an important coding lemma which will be proved again below using Kolmogorov complexity. Let $A \subseteq \Sigma^n$, $k = d(A)$ and $m = 1 + \lceil \log k \rceil$. Let $h: \Sigma^n \to \Sigma^m$ be a linear transformation given by a randomly chosen $m \times n$ binary matrix $R = \{r_{ij}\}$, i.e. for $x \in \Sigma^n$, Rx is a string $y \in \Sigma^m$ where $y_i = (\Sigma_j r_{ij} \times x_j) \bmod 2$. Let H be a collection of such functions. Let $A, B \subseteq \Sigma^n$ and $x \in \Sigma^n$. *h separates x within A* if for every $y \in A$, different from x, $h(y) \neq h(x)$. *h separates B within A* if it separates each $x \in B$ within A. *H separates B within A* if for each $x \in B$ some $h \in H$ separates x within A. In order to give each element in A a (logarithmic) short code, we randomly hash elements of A into short codes. If collision can be avoided, then elements of A can be described by short programs.

CODING LEMMA (Sipser). *Let $A \subseteq \Sigma^n$, where $d(A) = k$. Let $m = 1 + \lceil \log k \rceil$. There is a collection H of m linear transformations $\Sigma^n \to \Sigma^m$ such that H separates A within A.*

PROOF. We give the main idea ignoring relevant issues as self-delimiting descriptions, etc. Fix a random string s of length nm^2 such that $K(s|A) \geqslant |s|$. Cut x into m equal pieces.

Use the nm bits from each piece to form an $n \times m$ binary matrix in the obvious way. Thus we have constructed a set H of m random matrices. We claim that H separates A within A.

Assume this is not true. That is, for some $x \in A$, no $h \in H$ separates x within A. Hence there exist $y_1, \ldots, y_m \in A$ such that $h_i(x) = h_i(y_i)$. Hence $h_i(x - y_i) = 0$. Since $x - y_i \neq 0$, the first column of h_i corresponding to a 1 in $x - y_i$ can be expressed by the rest of the columns using $x - y_i$. Now we can describe s using the following:

- index of x in A, using $\lceil \log k \rceil$ bits,
- indices of $y_1, \ldots, y_m$, in at most $m \lceil \log k \rceil$ bits,
- matrices $h_1, \ldots, h_m$ each minus the redundant column, in $m^2 n - m^2$ bits.

From the above information, given A, a short program will reconstruct h_i by the rest columns of h_i and x, y_i. The total length is only

$$m^2 n - m(\log k + 1) + \log k + m(\log k) \leqslant nm^2 - 1.$$

Hence, $K(s|A) < |s|$, a contradiction. □

From this lemma, Sipser also proved $\mathrm{BPP} \subseteq \Sigma_4 \cap \Pi_4$. Gács improved this to $\mathrm{BPP} \subseteq \Sigma_2 \cap \Pi_2$. We provide Gács' proof: Let $B \in \mathrm{BPP}$ be accepted by a probabilistic algorithm with error probability at most 2^{-n} on inputs of length n, which uses $m = n^k$ random bits. Let $E_x \subset \Sigma^m$ be the collection of random inputs on which M rejects x. For $x \in B$, $|E_x| \leqslant 2^{m-n}$. Letting $l = 1 + m - n$, the Coding Lemma states that there is a collection H of l linear transformations from Σ^m to Σ^l separating E_x within E_x. If x is not in B, $|E_x| > 2^{m-1}$ and by the pidgeon hole principle, no such collection exists. Hence $x \in B$ iff such an H exists. The latter can be expressed as

$$\exists H \, \forall e \in E_x \, \exists h \in H \, \forall e' \in E_x \, [e \neq e' \text{ implies } h(e) \neq h(e')].$$

The second existential quantifier has polynomial range, hence can be eliminated. Hence $\mathrm{BPP} \subseteq \Sigma_2$. Since BPP is closed under complement, $\mathrm{BPP} \subseteq \Pi_2$. Hence $\mathrm{BPP} \subseteq \Sigma_2 \cap \Pi_2$.

6.5.2. Language compression without oracle

Without the help of oracles, Sipser and Goldberg [51] obtained much weaker and more difficult results. For a given language L, define the density of L to be $\mu_L = \max\{\mu_L(n)\}$, where $\mu_L(n) = d(L^n)/2^n$. Goldberg and Sipser proved that if $L \in \mathrm{P}$, $k > 3$, and $\mu_L \leqslant n^{-k}$, then L can be compressed in probabilistic polynomial time; the compression function f maps strings of length n to strings of length $n - (k-3)\log n + c$ with probability approaching 1.

The above result is weak in two senses:

(1) If a language L is very sparse, say $\mu_L \leqslant 2^{-n/2}$, then one expects to compress $\frac{1}{2}n$ bits instead of only $O(\log n)$ bits given by the theorem; can this be improved?

(2) The current compression algorithm is probabilistic; can this be made deterministic?

In computational complexity, oracles sometimes help us to understand the possibility of proving a new theorem. Goldberg and Sipser show that when S, the language to be compressed, does not need to be in P and the membership query of S is given by an oracle, then the above result is optimal. Specifically, we have the following:

(1) There is a sparse language S which cannot be compressed by more than $O(\log n)$ bits by a probabilistic polynomial-time machine with an oracle for S.

(2) There is a language S, of density $\mu_S < 2^{-n/2}$, which cannot be compressed by any deterministic polynomial-time machine that uses the oracle for S.

See [151] for practical data compression techniques.

6.5.3. *Ranking: optimally compressible languages*

Ranking is a special and optimal case of compression. The ranking function r_L maps the strings in L to their indices in the lexicographical ordering of L. If $r_L: L \to N$ is polynomial-time computable, then so is $r_L^{-1}: N \to L$ in terms of the length of its output. We are only interested in polynomial-time computable ranking functions. In fact, there are natural language classes that are easy to compress. Goldberg and Sipser [54], and Allender [4] show that if a language L is accepted by a one-way logspace Turing machine, then r_L can be computed in polynomial time. Goldberg and Sipser also prove by diagonalization that

(a) there is an exponential-time language that cannot be compressed in deterministic polynomial time; and

(b) there is a double exponential-time language that cannot be compressed in probabilistic polynomial time.

Call $\boldsymbol{C}$ P-rankable if for all $L \in \boldsymbol{C}$, r_L is polynomial-time computable. Hemachandra in [61] proved that P is P-rankable iff NP is P-rankable, and P is P-rankable iff $P = P^{\#P}$, and PSPACE is P-rankable iff P = PSPACE. Say a set A is *k-enumeratively-rankable* if there is a polynomial-time computable function f so that for every x, $f(x)$ prints a set of k numbers, one of which is the rank of x with respect to A. Cai and Hemachandra [21] proved $P = P^{\#P}$ iff each set $A \in P$ for some k_A has a k_A-enumerative-ranker.

6.6. *A Kolmogorov random reduction*

The original ideas of this section belong to U. Vazirani and V. Vazirani [159]. We reformulate their results in terms of Kolmogorov complexity.

In 1979 Adleman and Manders defined a probabilistic reduction, called UR-reduction, and showed several number-theoretic problems to be hard for NP under UR-reductions but not known to be NP-hard under Turing reductions. In [159] the notion is refined as follows.

Definition. A is *PR-reducible* to B, denoted by $A \leqslant_{PR} B$, iff there is probabilistic polynomial-time TM T and $\delta > 0$ such that

(1) $x \in A$ implies $T(x) \in B$, and

(2) x not in A implies $\Pr(T(x)$ not in $B) \geqslant \delta$.

A problem is *PR-complete* if every NP problem can be PR-reduced to it.

Vazirani and Vazirani obtained a nonnumber-theoretic PR-complete problem, which is still not known to be NP-complete up to today.

ENCODING BY TM

Instance: Two strings $x, y \in \{0, 1, 2, \alpha, \beta\}^*$, integer k.

Question: Is there a TM M with k or fewer states that on input x generates y in $|y|$ steps. (M has one read-write tape initially containing x and a write-only tape to write y. M must write one symbol of y at each step, i.e. real time.)

PR-COMPLETENESS PROOF. We reduce ENCODING BY FST to our problem, where the former is NP-complete and is defined as follows.

ENCODING BY FST

Instance: Two strings $x, y \in \{0, 1, 2\}^*$, $|x| = |y|$, and integer k.

Question: Is there a finite-state transducer M with k or less states that outputs y on input x. (Each step, M must read a symbol and output a symbol.)

The reduction is now as follows: any instance (x, y, k) of ENCODING BY FST is transformed to (xr, yr, k) for ENCODING BY TM, where $K(r|x, y) \geqslant |r| - c$ and $r \in \{\alpha, \beta\}^*$. For $\delta = 2^{-c}$, Pr(generate such an r) $\geqslant 1 - \delta$. Clearly, if there is an FST F of at most k states that outputs y on input x, then we can construct a TM which outputs yr on input xr by simply adding two new transitions from each state back to itself on α, β and outputing what it reads. If there is no such FST, then the k-state TM must reverse its read head on prefix x, or halt, when producing y. Hence it produces r without seeing r (notice the real-time requirement). Hence $K(r|x, y) = O(1)$, a contradiction. □

7. Conclusion

The opinion has sometimes been voiced that Kolmogorov complexity has only very abstract use. We are convinced that Kolmogorov complexity is immensely useful in a plethora of applications ranging from very theoretic to quite practical. We believe that we have given conclusive evidence for that conviction by this collection of applications.

In our view the covered material represents only the onset of a potentially enormous number of applications of Kolmogorov complexity in mathematics and the sciences. By the examples we have discussed, readers may get the feel how to use this general-purpose tool in their own applications, thus starting the golden spring of Kolmogorov complexity.

Acknowledgment

We are grateful to Greg Chaitin, Peter Gács, Leonid Levin and Ray Solomonoff for taking lots of time to tell us about the early history of our subject, and introducing us to many exciting applications. Additional comments were provided by Donald Loveland and Albert Meyer. Juris Hartmanis and Joel Seiferas introduced us in various ways to Kolmogorov complexity. P. Berman, R. Paturi, and J. Seiferas and Y. Yesha kindly supplied to us (and gave us permission to use) their unpublished material about the prime number theorem, lower bounds for probabilistic machines and Boolean matrix rank respectively. O. Watanabe and L. Longpré supplied to us interesting material in connection with the structure of complexity classes. A.Kh. Shen' and A. Verashagin translated an initial draft version of this article into Russian, and pointed out several errors. Comments of Charles Bennett, Peter van Emde Boas, Jan Heering, Evangelos Kranakis, Ker-I Ko, Danny Krizanc, Michiel van Lambalgen, Lambert Meertens, John Tromp, Umesh Vazirani, and Mati Wax are gratefully acknowledged. Th. Tsantilas and J. Rothstein supplied many useful references. We dedicate this work to A.N. Kolmogorov, who unexpectedly died while this paper was being written.

The work of the first author was supported in part by National Science Foundation Grant DCR-8606366, Office of Naval Research Grant N00014-85-k-0445, Army Research Office Grant DAAL03-86-K-0171, and NSERC Operating Grant OGP0036747. His current affiliation is: Department of Computer Science, University of Waterloo, Ontario, Canada N2L 3G1.

Preliminary versions of parts of this article appeared as: "Two decades of applied Kolmogorov complexity; In memoriam A.N. Kolmogorov 1903–1987", in: *Proc. 3rd IEEE Structure in Complexity Theory Conference* (1988) 80–102; and: "Kolmogorovskaya swozhnost': dvadsat' let spustia", *Uspekhi Mat. Nauk* **43**(6) (1988) 128–166 (in Russian) (= *Russian Mathematical Surveys*).

References

[1] AANDERAA, S.O., On k-tape versus $(k-1)$-tape real-time computation, in: R.M. Karp, ed., *Complexity of Computation* (Amer. Mathematical Soc., Providence, RI, 1974) 75–96.

[2] ADLEMAN, L., Time, space, and randomness, Report MIT/LCS/79/TM-131, Laboratory for Computer Science, Massachusetts Institute of Technology, 1979.

[3] ALEKSANDROV, P.S., A few words on A.N. Kolmogorov, *Russian Math. Surveys* **38** (1983) 5–7.

[4] ALLENDER, E., Invertible functions, Ph.D. Thesis, Georgia Institute of Technology, Atlanta 1985.

[5] ALLENDER, E., Some consequences of the existence of pseudorandom generators, *J. Comput. System Sci.* **39** (1989) 101–124.

[6] ALLENDER, E. and O. WATANABE, Kolmogorov complexity and degrees of tally sets, in: *Proc. 3rd Ann. Conf. IEEE on Structure in Complexity Theory* (1988) 102–111.

[7] ALLENDER, E.A. and R.S. RUBINSTEIN, P-printable sets, *SIAM J. Comput.* **17** (1988) 1193–1202.

[8] BALCAZAR, J. and R. BOOK, On generalized Kolmogorov complexity, in: A.L. Selman, ed., *Structure in Complexity Theory*, Lecture Notes in Computer Science, Vol. 223 (Springer, Berlin, 1986) 334–340.

[9] BARZDIN', Y.M., Complexity of programs to determine whether natural numbers not greater than n belong to a recursively enumerable set, *Soviet Math. Dokl.* **9** (1968) 1251–1254.

[10] BEAME, P., Limits on the power of concurrent-write parallel machines, *Inform. and Comput.* **76** (1988) 13–28.

[11] BENNETT, C.H., On random and hard to describe numbers, Mathematics Tech. Report RC 7483 (#32272), IBM T.J. Watson Research Center, Yorktown Heights, 1979; also in: M. Gardner, Mathematical Games, *Scientific American* (November 1979) 20–34.

[12] BENNETT, C.H., *On the logical "depth" of sequences and their reducibilities to random sequences*, Unpublished manuscript, IBM T.J. Watson Research Center, Yorktown Heights (1981/2).

[13] BENNETT, C.H., Dissipation, information, computational complexity and the definition of organization, in: D. Pines, ed., *Emerging Syntheses in Science* (*Proc. Founding Workshops of the Santa Fe Institute, 1985*) (Addison-Wesley, Reading, MA, 1987) 297–313.

[14] BENNETT, C.H., Logical depth and physical complexity, in: R. Herken, ed., *The Universal Turing Machine; A Half-Century Survey* (Oxford University Press, Oxford and Kammerer & Unverzagt, Hamburg, 1988) 227–258.

[15] BERMAN, L. and J. HARTMANIS, On isomorphisms and density of NP and other complete sets, *SIAM J. Comput.* **6** (1977) 305–327.
[16] BLUMER, A., A. EHRENFEUCHT, D. HAUSSLER and M. WARMUTH, Classifying learnable geometric concepts with the Vapnik–Chervonenkis dimension, in: *Proc. 18th Ann. ACM Symp. on Theory of Computing* (1986) 273–282.
[17] BOGOLYUBOV, N.N., B.V. GNEDENKO and S.L. SOBOLEV, Andrei Nikolaevich Kolmogorov (on his eightieth birthday), *Russian Math. Surveys* **38** (1983) 9–27.
[18] BOOK, R., P. ORPONEN, D. RUSSO and O. WATANABE, Lowness properties of sets in the exponential-time hierarchy, *SIAM J. Comput.* **17** (1988) 504–516.
[19] BORODIN, A., M.J. FISCHER, D.G. KIRKPATRICK, N.A. LYNCH and M. TOMPA, A time-space tradeoff for sorting and related non-oblivious computations, in: *Proc. 20th Ann. IEEE Symp. on Foundations of Computer Science* (1979) 319–328.
[20] BORODIN, A. and S. COOK, A time-space tradeoff for sorting on a general sequential model of computation, in: *Proc. 12th Ann. ACM Symp. on Theory of Computing* (1980) 294–301.
[21] CAI, J. and L. HEMACHANDRA, Enumerative counting is hard, *Inform. and Comput.* **82** (1989) 34–44.
[22] CARNAP, R., *Logical Foundations of Probability* (Univ. of Chicago Press, Chicago, IL, 1950).
[23] CARTER, J. and M. WEGMAN, Universal classes of hashing functions, *J. Comput. System Sci.* **18** (1979) 143–154.
[24] CHAITIN, G.J., On the length of programs for computing finite binary sequences, *J. Assoc. Comput. Mach.* **13** (1966) 547–569.
[25] CHAITIN, G.J., On the length of programs for computing finite binary sequences: statistical considerations, *J. Assoc. Comput. Mach.* **16** (1969) 145–159.
[26] CHAITIN, G.J., Information-theoretic limitations of formal systems, *J. Assoc. Comput. Mach.* **21** (1974) 403–424.
[27] CHAITIN, G.J., Randomness and mathematical proof, *Scientific American* **232** (May 1975) 47–52.
[28] CHAITIN, G.J., A theory of program size formally identical to information theory, *J. Assoc. Comput. Mach.* **22** (1975) 329–340.
[29] CHAITIN, G.J., Information-theoretic characterizations of recursive infinite strings, *Theoret. Comput. Sci.* **2** (1976) 45–48.
[30] CHAITIN, G.J., Algorithmic information theory, *IBM J. Res. Develop.* **21** (1977) 350–359.
[31] CHAITIN, G.J., Toward a mathematical definition of "life", in: M. Tribus, ed., *The Maximal Entropy Formalism* (MIT Press, Cambridge, MA, 1979) 477–498.
[32] CHAITIN, G.J., Gödel's theorem and information, *Internat. J. Theoret. Phys.* **22** (1982) 941–954; reprinted in: T. Tymoczko, ed., *New Directions in the Philosophy of Mathematics* (Birkhäuser, Boston, 1986).
[33] CHAITIN, G.J., *Algorithmic Information Theory* (Cambridge Univ. Press, Cambridge, 1987).
[34] CHAMPERNOWNE, D.G., The construction of decimals normal in the scale of ten, *J. London Math. Soc.* (2) **8** (1933) 254–260.
[35] CHROBAK, M., Hierarches of one-way multihead automata languages, *Theoret. Comput. Sci.* **48** (1986) 153–181.
[36] CHROBAK, M. and M. LI, $k+1$ heads are better than k for PDAs, *J. Comput. System Sci.* **37** (1988) 144–155.
[37] CHURCH, A., On the concept of a random sequence, *Bull. Amer. Math. Soc.* **46** (1940) 130–135.
[38] COVER, T.M., Universal gambling schemes and the complexity measures of Kolmogorov and Chaitin, Tech. Report 12, Statistics Dept., Stanford Univ., Palo Alto, CA, 1974.
[39] CUYKENDALL, R.R., Kolmogorov information and VLSI lower bounds, Ph.D. Thesis, Univ. of California, Los Angeles, CA, 1984.
[40] DALEY, R.P., On the inference of optimal descriptions, *Theoret. Comput. Sci.* **4** (1977) 301–309.
[41] DIETZFELBINGER, M., Lower bounds on computation time for various models in computational complexity theory, Ph.D. Thesis, Dept. of Computer Science, Univ. of Illinois at Chicago, 1987.
[42] DURIŠ, P., Z. GALIL, W. PAUL and R. REISCHUK, Two nonlinear lower bounds for online computations, *Inform. and Control* **60** (1984) 1–11.
[43] ERDŐS, P. and J. SPENCER, *Probabilistic Methods in Combinatorics* (Academic Press, New York, 1974).
[44] FLOYD, R., Review 14, *Comput. Rev.* **9** (1968) 280.
[45] GÁCS, P., On the symmetry of algorithmic information, *Soviet Math. Dokl.* **15** (1974) 1477–1480; Correction, *Ibidem* **15** (1974) 1480.

[46] GÁCS, P., Randomness and probability—complexity of description, in: Kotz-Johnson, ed., *Encyclopedia of Statistical Sciences* (Wiley, New York, 1986) 551–555.

[47] GÁCS, P., Lecture notes on descriptional complexity and randomness, Unpublished manuscript, Boston University, Boston, MA, 1987.

[48] GALIL, Z. and J. SEIFERAS, Time-space optimal matching, in: *Proc. 13th Ann. ACM Symp. on Theory of Computing* (1981) 106–113.

[49] GALIL, Z., R. KANNAN, and E. SZEMEREDI, On nontrivial separators for k-page graphs and simulations by nondeterministic one-tape Turing machines, *J. Comput. System Sci.* **38** (1989) 134–149.

[50] GALLAIRE, H., Recognition time of context-free languages by on-line Turing machines, *Inform. and Control* **15** (1969) 288–295.

[51] GAO, Q. and M. LI, An application of minimum description length principle to online recognition of handprinted alphanumerals, in: *Proc. 11th Internat. Joint Conf. on Artificial Intelligence*, Detroit, MI (1989).

[52] GNEDENKO, B.V., Andrei Nikolaevich Kolmogorov (on the occasion of his seventieth birthday), *Russian Math. Surveys* **28** (1973) 5–16.

[53] GOLD, E.M., Language identification in the limit, *Inform. and Control* **10** (1967) 447–474.

[54] GOLDBERG, Y. and M. SIPSER, Compression and ranking, *SIAM J. Comput.* **19** (1990).

[55] HARRISON, M.A. and O.H. IBARRA, Multi-head and multi-tape pushdown automata, *Inform. and Control* **13** (1968) 433–470.

[56] HARTMANIS, J., Generalized Kolmogorov complexity and the structure of feasible computations, in: *Proc. 24th Ann. IEEE Symp. on Foundations of Computer Science* (1983) 439–445.

[57] HARTMANIS, J. and L. HEMACHANDRA, On sparse oracles separating feasible complexity classes, in: *Proc. 3rd Symp. on Theoretical Aspects of Computer Science (STACS '86)*, Lecture Notes in Computer Science, Vol. 210 (Springer, Berlin, 1986) 321–333.

[58] HAUSSLER, D., N. LITTLESTONE and M. WARMUTH, Expected mistake bounds for on-line learning algorithms, Manuscript, 1988.

[59] HEIDE, F. MEYER AUF DER and A. WIGDERSON, The complexity of parallel sorting, in: *Proc. 17th Ann. ACM Symp. on Theory of Computing* (1985) 532–540.

[60] HEMACHANDRA, L., Can P and NP manufacture randomness?, Tech. Report TR86-795, Dept. of Computer Science, Cornell Univ., Ithaca, NY, 1986.

[61] HEMACHANDRA, L., On ranking, in: *Proc. 2nd Ann. IEEE Conf. on Structure in Complexity Theory* (1987) 103–117.

[62] HOPCROFT, J.E. and J.D. ULLMAN, *Introduction to Automata Theory, Languages, and Computation* (Addison-Wesley, Reading, MA, 1979).

[63] HUYNH, D.T., Non-uniform complexity and the randomness of certain complete languages, Tech. Report TR 85-34, Iowa State Univ., 1985.

[64] HUYNH, D.T., Resource-bounded Kolmogorov complexity of hard languages, in: *Structure in Complexity Theory*, Lecture Notes in Computer Science, Vol. 223 (Springer, Berlin, 1986) 184–195.

[65] IBARRA, O.H. and C.E. KIM, On 3-head versus 2-head finite automata, *Acta Inform.* **4** (1975) 193–200.

[66] ISRAELI, A. and S. MORAN, Private communication.

[67] JEFFREYS, Z., *Theory of Probability* (Oxford at the Clarendon Press, Oxford, 3rd ed., 1961).

[68] KEARNS, M., M. LI, L. PITT and L. VALIANT, On the learnability of Boolean formulae, in: *Proc. 19th Ann. ACM Symp. on Theory of Computing* (1987) 285–295.

[69] KEARNS, M., M. LI, L. PITT and L. VALIANT, Recent results on Boolean concept learning, in: T. Mitchell, ed., *Proc. 4th Workshop on Machine Learning*, (Morgan Kaufmann, Los Altos, CA, 1987) 337–352.

[70] KEMENY, J.G., The use of simplicity in induction, *Philos. Rev.* **62** (1953) 391–408.

[71] KNUTH, D., *Semi-numerical Algorithms* (Addison-Wesley, Reading, MA, 2nd ed., 1981).

[72] KNUTH, D.E., Big Omicron and big Omega and big Theta, *SIGACT News* **8** (2) (1976) 18–24.

[73] KO, K.-I, Resource-bounded program-size complexity and pseudorandom sequences, Tech. Report, Dept. of Computer Science, Univ. of Houston, Houston, TX, 1983.

[74] KOLMOGOROV, A.N., *Grundbegriffe der Wahrscheinlichkeitsrechnung* (Springer, Berlin, 1933); *Osnovnye Poniatija Teorii Verojatnostej* (Nauka, Moscow, 2nd Russian ed., 1974).

[75] KOLMOGOROV, A.N., On tables of random numbers, *Sankhyā Ser. A* **25** (1963) 369–376.
[76] KOLMOGOROV, A.N., Three approaches to the quantitative definition of information, *Problems Inform. Transmission* **1** (1) (1965) 1–7.
[77] KOLMOGOROV, A.N., Logical basis for information theory and probability theory, *IEEE Trans. on Inform. Theory* **14** (5) (1968) 662–664.
[78] KOLMOGOROV, A.N., Combinatorial foundations of information theory and the calculus of probabilities, *Russian Math. Surveys* **38** (1983) 29–40.
[79] KOLMOGOROV, A.N., *Information Theory and Theory of Algorithms, Selected Works Vol. 3* (Nauka, Moscow, 1987) (in Russian).
[80] KOLMOGOROV, A.N. and V.A. USPENSKII, Algorithms and randomness, *SIAM J. Theory Probab. Appl.* **32** (1987) 389–412.
[81] LAMBALGEN, M. VAN, Von Mises' definition of random sequences reconsidered, *J. Symbolic Logic* **52** (1987) 725–755.
[82] LAMBALGEN, M. VAN, Random sequences, Ph.D. Thesis, Faculty of Mathematics and Computer Science, Univ. van Amsterdam, Amsterdam, 1987.
[83] LAPLACE, P.S., *A Philosophical Essay on Probabilities* (Dover, New York; original publication, 1819).
[84] LEVIN, L.A., Universal search problems, *Problems Inform. Transmission* **9** (1973) 265–266.
[85] LEVIN, L.A., On the notion of a random sequence, *Soviet Math. Dokl.* **14** (1973) 1413–1416.
[86] LEVIN, L.A., Laws of information conservation (non-growth) and aspects of the foundation of probability theory, *Problems Inform. Transmission* **10** (1974) 206–210.
[87] LEVIN, L.A., Do chips need wires?, Manuscript/NSF proposal MCS-8304498, Computer Science Dept., Boston Univ., 1983.
[88] LEVIN, L.A., Randomness conservation inequalities; information and independence in mathematical theories, *Inform. and Control* **61** (1984) 15–37.
[89] LI, M., Simulating two pushdowns by one tape in $O(n^{1.5}(\log n)^{0.5})$ time, *J. Comput. System Sci.* **37** (1988) 101–116.
[90] LI, M., Lower bounds in computational complexity, Ph.D. Thesis, Report TR-85-663, Computer Science Dept., Cornell Univ., Ithaca, NY, 1985.
[91] LI, M., L. LONGPRÉ and P.M.B. VITÁNYI, On the power of the queue, in: *Structure in Complexity Theory*, Lecture Notes in Computer Science, Vol. 223 (Springer, Berlin, 1986) 219–233.
[92] LI, M. and Y. YESHA, String-matching cannot be done by 2-head 1-way deterministic finite automata, *Inform. Process. Lett.* **22** (1986) 231–235.
[93] LI, M. and Y. YESHA, New lower bounds for parallel computation, *J. Assoc. Comput. Mach.* **36** (1989) 671–680.
[94] LI, M. and P.M.B. VITÁNYI, Tape versus queue and stacks: the lower bounds, *Inform. and Comput.* **78** (1988) 56–85.
[95] LI, M. and P.M.B. VITÁNYI, A new approach to formal language theory by Kolmogorov complexity, in: *Proc. 16th Internat. Coll. on Automata, Languages and Programming*, Lecture Notes in Computer Science, Vol. 372 (Springer, Berlin, 1989) 506–520.
[96] LI, M. and P.M.B. VITÁNYI, Inductive reasoning and Kolmogorov complexity, in: *Proc. 4th Ann. IEEE Structure in Complexity Theory Conf.* (1989) 165–185.
[97] LI, M. and P.M.B. VITÁNYI, *An Introduction to Kolmogorov Complexity and Its Applications* (Addison-Wesley, Reading, MA, to appear).
[98] LIPTON, R. and R. SEDGEWICK, Lower bounds for VLSI, in: *Proc. 13th Ann. ACM Symp. on Theory of Computing* (1981) 300–307.
[99] LITTLESTONE, N., Learning quickly when irrelevant attributes abound: a new linear threshold algorithm, in: *Proc. 28th Ann. IEEE Symp. on Foundations of Computer Science* (1987) 68–77.
[100] LITTLEWOOD, J.E., The dilemma of probability theory, in: B. Bollobás, ed., *Littlewood's Miscellany* (Cambridge Univ. Press, Cambridge, revised ed., 1986) 71–73.
[101] LONGPRÉ, L., Resource bounded Kolmogorov complexity, a link between computational complexity and information theory, Ph.D. Thesis, Tech. Report TR-86-776, Computer Science Dept., Cornell Univ., Ithaca, NY, 1986.
[102] LOUI, M., Optimal dynamic embedding of trees into arrays, *SIAM J. Comput.* **12** (1983) 463–472.

[103] LOVELAND, D.W., On minimal-program complexity measures, in: *Proc. ACM Symp. on Theory of Computing* (1969) 61–65.
[104] LOVELAND, D.W., A variant of the Kolmogorov concept of complexity, *Inform. and Control* **15** (1969) 510–526.
[105] MAASS, W., Combinatorial lower bound arguments for deterministic and nondeterministic Turing machines, *Trans. Amer. Math. Soc.* **292** (1985) 675–693.
[106] MAASS, W. and G. SCHNITGER, An optimal lower bound for Turing machines with one work tape and a two-way input tape, in: *Structure in Complexity Theory*, Lecture Notes in Computer Science, Vol. 223 (Springer, Berlin, 1986) 249–264.
[107] MAASS, W., G. SCHNITGER and E. SZEMEREDI, Two tapes are better than one for off-line Turing machines, in: *Proc. 19th Ann. ACM Symp. on Theory of Computing* (1987) 94–100.
[108] MAIRSON, H.G., The program complexity of searching a table, in: *Proc. 24th IEEE Symp. on Foundations of Computer Science* (1983) 40–47.
[109] MARTIN-LÖF, P., The definition of random sequences, *Inform. and Control* **9** (1966) 602–619.
[110] MARTIN-LÖF, P., Algorithmen und zufällige Folgen, Lecture notes, Univ. of Erlangen, 1966.
[111] MARTIN-LÖF, P., Complexity oscillations in infinite binary sequences, *Z. Wahrsch. Verw. Gebiete* **19** (1971) 225–230.
[112] MEHLHORN, K., On the program-size of perfect and universal hash functions, in: *Proc. 23rd Ann. IEEE Symp. on Foundations of Computer Science* (1982) 170–175.
[113] METROPOLIS, N.C., G. REITWEISER and J. VON NEUMANN, Statistical treatment of values of the first 2,000 decimal digits of e and π calculated on the ENIAC, in: A.H. Traub, ed., *John von Neumann, Collected Works, Vol. V* (MacMillan, New York, 1963).
[114] MINSKY, M.L., Steps towards Artificial Intelligence, *Proc. I.R.E.* (January 1961) 8–30.
[115] MISES, R. VON, *Probability, Statistics and Truth* (MacMillan, New York, 1939; reprint, Dover, New York, 1981).
[116] MIYANO, S., A hierarchy theorem for multihead stack-counter automata, *Acta Inform.* **17** (1982) 63–67.
[117] MIYANO, S., Remarks on multihead pushdown automata and multihead stack automata, *J. Comput. System Sci.* **27** (1983) 116–124.
[118] NATARAJAN, B.K., Personal communication, 1988.
[119] NELSON, C.G., One-way automata on bounded languages, Tech. Report TR14-76, Aiken Computer Lab., Harvard Univ., 1976.
[120] NEUMANN, J. VON, Various techniques used in connection with random digits, in: A.H. Traub, ed., *John von Neumann, Collected Works, Vol. V* (MacMillan, New York, 1963).
[121] Obituary, Mr. Andrei Kolmogorov—giant of mathematics, *Times* (October 26, 1987).
[122] PARBERRY, I., A complexity theory of parallel computation, Ph.D. Thesis, Dept. of Computer Science, Warwick Univ., Coventry, UK, 1984.
[123] PATURI, R. and J. SIMON, Lower bounds on the time of probabilistic on-line simulations, in: *Proc. 24th Ann. IEEE Symp. on Foundations of Computer Science* (1983) 343–350.
[124] PAUL, W., Kolmogorov's complexity and lower bounds, in: L. Budach, ed., *Proc. 2nd Internat. Conf. on Fundamentals of Computation Theory* (Akademie Verlag, Berlin, 1979) 325–334.
[125] PAUL, W., On heads versus tapes, *Theoret. Comput. Sci.* **28** (1984) 1–12.
[126] PAUL, W.J., J.I. SEIFERAS and J. SIMON, An information theoretic approach to time bounds for on-line computation, *J. Comput. System Sci.* **23** (1981) 108–126.
[127] PAUL, W.J., On-line simulation of $k+1$ tapes by k tapes requires nonlinear time, *Inform. and Control* (1982) 1–8.
[128] PETERSON, G., Succinct representations, random strings and complexity classes, in: *Proc. 21st Ann. IEEE Symp. on Foundations of Computer Science* (1980) 86–95.
[129] POPPER, K.R., *The Logic of Scientific Discovery* (Univ. of Toronto Press, Toronto, 1959).
[130] QUINLAN, J. and R. RIVEST, Inferring decision trees using the minimum description length principle, *Inform. and Comput.* **80** (1989) 227–248.
[131] REISCH, S. and G. SCHNITGER, Three applications of Kolmogorov-complexity, in: *Proc. 23rd Ann. IEEE Symp. on Foundations of Computer Science* (1982) 45–52.
[132] RISSANEN, J., Modeling by the shortest data description, *Automatica—J. IFAC* **14** (1978) 465–471.

[133] RISSANEN, J., A universal prior for integers and estimation by minimum description length, *Ann. Statist.* **11** (1982) 416–431.
[134] RIVEST, R., Learning decision-lists, Unpublished manuscript, Lab. for Computer Science, Massachusetts Institute of Technology, Cambridge, MA, 1986.
[135] ROGERS JR., H., *Theory of Recursive Functions and Effective Computability* (McGraw Hill, New York, 1967).
[136] ROSENBERG, A., Nonwriting extensions of finite automata, Ph.D. Thesis, Aiken Computer Lab., Harvard Univ., Cambridge, MA, 1965.
[137] ROSENBERG, A., On multihead finite automata, *IBM J. Res. Develop.* **10** (1966) 388–394.
[138] SCHNORR, C.P., Eine Bemerkung zum Begriff der zufälligen Folge, *Z. Wahrsch. Verw. Gebiete* **14** (1969/70) 27–35.
[139] SCHNORR, C.P., *Zufälligkeit und Wahrscheinlichkeit; Eine algorithmische Begründung der Wahrscheinlichkeitstheorie*, Lecture Notes in Mathematics, Vol. 218 (Springer, Berlin, 1971).
[140] SCHNORR, C.P., Process complexity and effective random tests, *J. Comput. System Sci.* **7** (1973) 376–388.
[141] SCHNORR, C.P., A survey of the theory of random sequences, in: R.E. Butts and J. Hintikka, eds., *Basic Problems in Methodology and Linguistics* (Reidel, Dordrecht, 1977) 193–210.
[142] SEIFERAS, J., The symmetry of information, and an application of the symmetry of information, Notes, Computer Science Dept, Univ. of Rochester, 1985.
[143] SEIFERAS, J., A simplified lower bound for context-free-language recognition, *Inform. and Control* **69** (1986) 255–260.
[144] SHANNON, C.E. and W. WEAVER, *The Mathematical Theory of Communication* (Univ. of Illinois Press, Urbana, IL, 1949).
[145] SHANNON, C.E., A universal Turing machine with two internal states, in: C.E. Shannon and J. McCarthy, eds., *Automata Studies* (Princeton Univ. Press, Princeton, NJ, 1956).
[146] SIPSER, M., A complexity theoretic approach to randomness, in: *Proc. 15th Ann. ACM Symp. on Theory of Computing* (1983) 330–335.
[147] SKIENA, S.S., Further evidence for randomness in π, *Complex Systems* **1** (1987) 361–366.
[148] SOLOMONOFF, R.J., A preliminary report on a general theory of inductive inference, Tech. Report ZTB-138, Zator Company, Cambridge, MA, 1960.
[149] SOLOMONOFF, R.J., A formal theory of inductive inference, Part 1 and Part 2, *Inform. and Control* **7** (1964) 1–22 and 224–254.
[150] SOLOMONOFF, R.J., Complexity-based induction systems: comparisons and convergence theorems, *IEEE Trans. Inform. Theory* **24** (1978) 422–432.
[151] STORER, J., *Data Compression: Method and Theory* (Computer Science Press, Rockville, MD, 1988).
[152] SUDBOROUGH, I.H., Computation by multi-head writing finite automata, Ph.D. Thesis, Pennsylvania State Univ., University Park, PA, 1974.
[153] SUDBOROUGH, I.H., One-way multihead writing finite automata, *Inform. and Control* **30** (1976) 1–20.
[154] THOMPSON, C.D., Area–time complexity for VLSI, in: *Proc. 11th Ann. ACM Symp. on Theory of Computing* (1979) 81–88.
[155] TURING, A.M., On computable numbers with an application to the Entscheidungsproblem, *Proc. London Math. Soc.* **42** (1936) 230–265; Correction, *Ibidem* **43** (1937) 544–546.
[156] VALIANT, L. and G. BREBNER, Universal schemes for parallel communication, in: *Proc. 13th Ann. ACM Symp. on Theory of Computing* (1981) 263–277.
[157] VALIANT, L.G., A theory of the learnable, *Comm. ACM* **27** (1984) 1134–1142.
[158] VALIANT, L.G., Deductive learning, *Philos. Trans. Royal Soc. Lond. Ser. A* **312** (1984) 441–446.
[159] VAZIRANI, U. and V. VAZIRANI, A natural encoding scheme proved probabilistic polynomial complete, *Theoret. Comput. Sci.* **24** (1983) 291–300.
[160] VILLE, J., *Etude Critique du Concept de Collectif* (Gauthier-Villars, Paris, 1939).
[161] VITÁNYI, P.M.B. and L. MEERTENS, Big Omega versus the wild functions, *SIGACT News* **16** (4) (1985) 56–59.
[162] VITÁNYI, P.M.B., On two-tape real-time computation and queues, *J. Comput. System Sci.* **29** (1984) 303–311.

[163] VITÁNYI, P.M.B., On the simulation of many storage heads by one, *Theoret. Comput. Sci.* **34** (1984) 157–168.

[164] VITÁNYI, P.M.B., Square time is optimal for the simulation of a pushdown store by an oblivious one-head tape unit, *Inform. Process. Lett.* **21** (1985) 87–91.

[165] VITÁNYI, P.M.B., An $N^{1.618}$ lower bound on the time to simulate one queue or two pushdown stores by one tape, *Inform. Process. Lett.* **21** (1985) 147–152.

[166] VITÁNYI, P.M.B., Andrei Nikolaevich Kolmogorov, *CWI Quarterly* **1** (2) (June 1988) 3–18.

[167] WALD, A., Sur la notion de collectif dans le calcul des probabilités, *C.R. Acad. Sci.* **202** (1936) 1080–1083.

[168] WATANABE, O., Comparison of polynomial time completeness notions, *Theoret. Comput. Sci.* **53** (1987) 249–265.

[169] WILLIS, D.G., Computational complexity and probability constructions, *J. Assoc. Comput. Mach.* **17** (1970) 241–259.

[170] YAO, A., Theory and application of trapdoor functions, in: *Proc. 23rd Ann. IEEE Symp. on Foundations of Computer Science* (1982) 80–91.

[171] YAO, A.C.-C. and R.L. RIVEST, $k+1$ heads are better than k, *J. Assoc. Comput. Mach.* **25** (1978) 337–340.

[172] YESHA, Y., Time-space tradeoffs for matrix multiplication and discrete Fourier transform on any general random access computer, *J. Comput. System Sci.* **29** (1984) 183–197.

[173] ZVONKIN, A.K. and L.A. LEVIN, The complexity of finite objects and the development of the concepts of information and randomness by means of the theory of algorithms, *Russ. Math. Surveys* **25** (1970) 83–124.

[174] COVER, T.M., P. GÁCS and R.M. GRAY, Kolmogorov's contributions to information theory and algorithmic complexity, *Ann. Probab.* **17** (1989) 840–865.

[175] ITKIS, G. and L.A. LEVIN, Power of fast VLSI models is insensitive to wires' thinness, in: *Proc. 30th Ann. IEEE Symp. on Foundations of Computer Science* (1989) 402–409.

[176] LI, M. and P.M.B. VITÁNYI, A theory of learning simple concepts under simple distributions and average case complexity for the universal distribution, in: *Proc. 30th Ann. IEEE Symp. on Foundations of Computer Science* (1989) 34–39.

[177] WALLACE, C.S. and P.R. FREEMAN, Estimation and inference by compact coding, *J. Royal Statist. Soc. Ser. B* **49** (1987) 240–251; Discussion, *Ibidem* **49** (1987) 252–265.

CHAPTER 5

Algorithms for Finding Patterns in Strings

Alfred V. AHO

AT & T Bell Laboratories, Murray Hill, NJ 07974, USA

Contents

HANDBOOK OF THEORETICAL COMPUTER SCIENCE
Edited by J. van Leeuwen

1. Introduction

String pattern matching is an important problem that occurs in many areas of science and information processing. In computing, it occurs naturally as part of data processing, text editing, term rewriting, lexical analysis, and information retrieval. Many text editors and programming languages have facilities for matching strings. In biology, string-matching problems arise in the analysis of nucleic acids and protein sequences, and in the investigation of molecular phylogeny. String matching is also one of the central and most widely studied problems in theoretical computer science.

The simplest form of the problem is to locate an occurrence of a keyword as a substring in a sequence of characters, which we will call the *input string*. For example, the input string *queueing* contains the "keyword" *ueuei* as a substring. Even for this problem, several innovative, theoretically interesting algorithms have been devised that run significantly faster than the obvious brute-force method. The problem becomes richer as we enlarge the class of patterns to include sets of keywords and regular expressions. This article examines the time–space trade-offs inherent in searching for occurrences of such patterns in text strings.

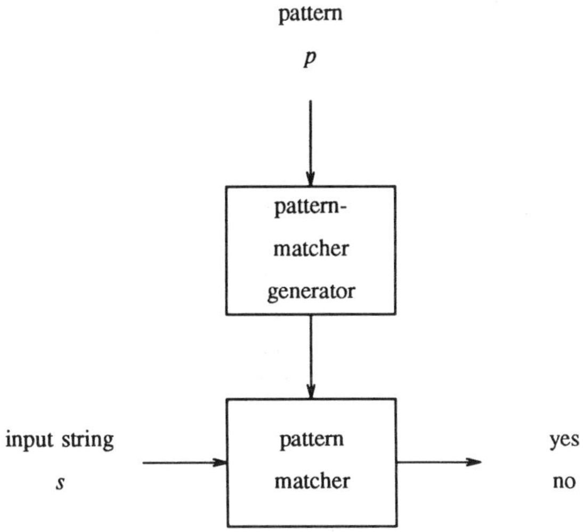

Fig. 1. Model for pattern-matching problems.

We will treat pattern-matching problems in the general setting shown in Fig. 1. The input consists of a pair (p, s) where p is the pattern and s is the input string. The pattern is transformed by the pattern-matcher generator into a pattern matcher, which is used to look for an occurrence of the pattern in the input string. The pattern matcher reports "yes" if s contains a substring matched by p, "no" otherwise.

The actual output of a pattern-matching algorithm depends on the application. In information retrieval, the input string is often a file consisting of lines of text (such as a dictionary, a program listing, or a manuscript), and we are interested in all lines matched by the pattern. For example, we might be interested in searching a dictionary for all words that contain the five vowels in order. In programming language compilation, the input string would be the sequence of characters making up the source program and we would be interested in partitioning the input into a sequence of lexical tokens such as comments, identifiers, operators, and so on, where the structure of the tokens is specified by the pattern. In text editing, we might want to identify the longest nonoverlapping substrings of the input string denoted by the pattern; for example, in a manuscript we might want to change all occurrences of *color* into *colour*. In this paper, we will simply consider the output of the pattern matcher to be "yes" if the input string contains a substring matched by the pattern, "no" otherwise.

We will measure the overall performance of a pattern-matching algorithm by the time and space taken to answer yes or no measured as a function of the lengths of p and s using the random-access machine (RAM) as the model of computation [6]. We will assume the pattern is given before the text string. In this way an algorithm can preprocess the pattern, constructing from it whatever kind of pattern-matching machine it needs before scanning any of the input string. In the setting of Fig. 1, we will analyze the time and space taken by the pattern-matcher generator as well as the pattern matcher itself.

In practice, a number of issues need to be resolved in the design of a patternmatching program. Decisions need to be made on what class of patterns to use, what notation to describe the patterns, and what distribution to expect on patterns and text strings. In some applications it may not be possible to store all of the input string in memory, so there may be a limit on how much backtracking over the input is possible. For applications such as text editing it is desirable to be able to construct the recognizer quickly since the input strings are likely to be short. For applications such as textual search we may be willing to spend more time constructing the recognizer if it can be made to run faster on long text files. In many applications the pattern-matching process is I/O bound, so it is important to be able to read the input string as quickly as possible. In the 1980s the performance of many pattern-matching programs improved by an order of magnitude or more due to algorithmic improvements of the kind discussed in this chapter [73].

2. Notations for patterns

In our pattern-matching model, the pattern matcher reports "yes" if the input string contains a substring matched by the pattern, "no" otherwise. In effect, the pattern is a notation for describing a set of substrings. The simplest patterns are single keywords that match themselves. For example, if we specified *dous* as a pattern, then we would report success on input strings such as *hazardous* and *horrendously*. A somewhat broader class of patterns would be sets of keywords. Since many text-processing systems use variants of regular expressions to describe patterns, we will use regular expressions as they are defined in language theory as our third notation for specifying patterns. We will also discuss the effect of some of the embellishments that have been added to regular expressions to make them more descriptive for practical use.

2.1. *Regular expressions*

Regular expressions, regular sets, and finite automata are central concepts in automata and formal language theory. As we shall see, these concepts are also central in a study of string pattern-matching algorithms.

DEFINITION. We define *regular expressions* and the strings they match recursively as follows:

(1) The following characters are metacharacters: | ( ) *

(2) A non-metacharacter a is a regular expression that matches the string a.

(3) If r_1 and r_2 are regular expressions, then $(r_1|r_2)$ is a regular expression that matches any string matched by either r_1 or r_2.

(4) If r_1 and r_2 are regular expressions, then $(r_1)(r_2)$ is a regular expression that matches any string of the form xy, where r_1 matches x and r_2 matches y.

(5) If r is a regular expression, then $(r)^*$ is a regular expression that matches any string of the form $x_1x_2 \ldots x_n$, $n \geqslant 0$, where r matches x_i for $1 \leqslant i \leqslant n$. In particular, $(r)^*$ matches the empty string, which we denote by ε.

(6) If r is a regular expression, then (r) is a regular expression that matches the same strings as r.

Many parentheses in regular expressions can be avoided by adopting the convention that the Kleene closure operator * has the highest precedence, then concatenation, then |. The two binary operators, concatenation and |, are left-associative. Under these conventions the regular expressions $(a|((b)^*)(c))$ and $a|b^*c$ are equivalent, in the sense that they match the same strings, namely, an a, or a sequence of zero or more b's followed by a c.

2.1. EXAMPLE. The regular expression

(hot|cold) (apple|blueberry|cherry) (pie|tart)

matches any of the twelve delicacies ranging from *hot apple pie* to *cold cherry tart*. The regular expression

*the (very,)*very hot cherry pie*

matches the strings *the very hot cherry pie*; *the very, very hot cherry pie*; *the very, very, very hot cherry pie*; and so on. The regular expression

$$(aa|bb)^* ((ab|ba)(aa|bb)^* (ab|ba)(aa|bb)^*)^*$$

matches all strings of a's and b's having both an even number of a's and an even number of b's.

A *regular set* is a set of strings matched by a regular expression. (Some authors use the terms *rational set* and *rational expression* for regular set and regular expression.) The origins of regular sets go back to the work of McCulloch and Pitts [94] who devised finite-state automata as a model for the behavior of neural nets. As a formalism for specifying strings, regular expressions and finite automata are equivalent in that they both describe the same sets of strings [110]. The notation of regular expressions arises naturally from the mathematical result of Kleene [79] that characterizes the regular sets as the smallest class of sets of strings which contains all finite sets of strings and which is closed under the operations of union, concatenation, and "Kleene closure".

2.2. Extensions to the regular expression notation

Many text-editing and searching programs add abbreviations and new operators to the basic regular expression notation above to make it easier to specify patterns. Here we mention some of the extensions used by the popular regular expression matching programs *awk* [7], *egrep* [96], and *lex* [84] on the UNIX® operating system. One useful extension is the quoting metacharacter, \, that permits metacharacters to be matched. In the definition above, the symbols |, (,), and * are metacharacters considered not to be part of the alphabet of input characters. With \, we can include these symbols, as well as \ itself, as targets for matches, by writing \| to match |, * to match *, and so on; \\ matches \.

We often want to match substrings at the beginning or end of the input string. To do this, *awk* adds the metacharacters ^ and $ to match the null string at the left and right ends of an input line. Thus *dous*$ matches the substring *dous* only when it is at the right end of the input line.

Another convenience is a succinct way to specify a match for any character in a set of characters. In *awk*, the symbol . is used to match any single character. We can think of . as a "wild card" or "don't care" symbol. The notation [*abc*] is a character class that matches an *a*, *b*, or *c*, and the notation [^*abc*] is a complemented character class that matches any single character that is not an *a*, *b*, or *c*. Thus the *awk* expression

^[^*aeiou*]**a* [^*aeiou*]**e* [^*aeiou*]**i* [^*aeiou*]**o* [^*aeiou*]**u* [^*aeiou*]*$

will match an input line in which the five vowels appear in lexicographic order, such as a line consisting of the word *abstemious* or *facetious*. A range of characters may be specified with an abbreviated character class; for example, [*a–z*] matches any lower-case letter, [*A–Za–z*] matches any upper- or lower-case letter, and [^0–9] matches any non-digit. None of these extensions, however, extends the descriptive power of regular expressions beyond regular sets.

Many pattern-matching tools allow Boolean combinations of patterns. Since the class of regular sets is closed under the operations of union, intersection, and complement, the Boolean operations *or*, *and*, and *not* do not add more descriptive power to regular expressions as a notation for describing string patterns, but in practice, they offer considerable convenience and, in both theory and practice, succinctness to the specification of patterns. For example, try writing a regular expression pattern for all words with no repeated letter (such as *ambidextrously* or *dermatoglyphics*). The general question concerning the relative succinctness of different notations for regular sets has been of considerable theoretical interest [65, 99].

2.3. *Regular expressions with back referencing*

In some applications it is desirable to be able to specify repeating structures in matches. For example, we might be interested in words of the form xx, that is, words consisting of a repeated substring such as *killeekillee* or *tangantangan*. Patterns such as these are sometimes called *squares* and they cannot be specified by regular expressions. An assignment operator called *back referencing* can be added to regular expressions to allow repeating patterns to be specified. Back referencing appeared in the first version of the SNOBOL programming language [48] and has been implemented in several commands on the UNIX system, notably the text editor *ed* and the pattern-matching program *grep* [96].

DEFINITION. The following rules define *regular expressions with back referencing* (rewbrs for short) and the strings they match. In these rules we assume the alphabet of characters is distinct from $\{v_1, v_2, \ldots\}$, the set of variable names. In a rewbr, a variable defines a string whose initial value is undefined.

(1) The following characters are metacharacters: | ( ) * %

(2) Each non-metacharacter a is a rewbr that matches the string a.

(3) Each variable v_i is a rewbr that matches the string defined by v_i.

(4) If r_1 and r_2 are rewbrs, then $(r_1|r_2)$ is a rewbr that matches any string matched by either r_1 or r_2.

(5) If r_1 and r_2 are rewbrs, then $(r_1)(r_2)$ is a rewbr that matches any string of the form xy, where r_1 matches x and r_2 matches y.

(6) If r is a rewbr, then $(r)^*$ is a rewbr that matches any string of the form $x_1x_2\ldots x_n$, $n \geqslant 0$, where r matches x_i for $1 \leqslant i \leqslant n$.

(7) If r is a rewbr that matches a string x, then $(r)\%v_i$ is a rewbr that matches x and v_i is assigned the value x.

(8) If r is a rewbr, then (r) is a rewbr that matches the same strings as r.

Rule (7) defines the back-referencing operator %. Its properties are similar to those of the conditional value assignment operator in SNOBOL4 [60]. As before, redundant parentheses can be avoided using the same precedences and associativities as in regular expressions. The back-referencing operator is left-associative and has the highest precedence.

2.2. EXAMPLE. A few examples should clarify this definition.

(1) The rewbr $(a|b|c)^*(a|b|c)\%v_1(a|b|c)^*v_1(a|b|c)^*$ matches any string of a's, b's or c's with at least one repeated character. To see this, note that $(a|b|c)^*$ matches any string of characters and that $(a|b|c)\%v_1$ will match any single character and assign to v_1 the value of that character. The second v_1 in the rewbr will match a second occurrence of that character in the input string.

(2) The rewbr $(a|b|c)^*\%v_1v_1$ matches any string of the form xx where x is any string of a's, b's or c's.

(3) The rewbr $(((a|b|c)^*\%v_1)v_1)^*$ matches any string of the form $x_1x_1x_2x_2\ldots x_nx_n$ where each x_i is a string of a's, b's or c's.

These examples illustrate some of the definitional power of rewbrs. Pattern (1) can be denoted by a somewhat longer ordinary regular expression. The complement of pattern (1) suggests a concise specification for the no-repeated-letter pattern mentioned above. Pattern (2), however, does not denote a regular set or even context-free language so it cannot be denoted by any regular expression or context-free grammar [70]. Pattern (3), likewise, cannot be expressed by a regular expression or context-free grammar. Patterns denoted by regular expressions with back referencing, only one variable name, and no alternation have been studied by Angluin [11].

3. Matching keywords

We are now ready to consider the first of our pattern-matching problems.

3.1. PROBLEM. Given a pattern p consisting of a single keyword and an input string s, answer "yes" if p occurs as a substring of s, that is, if $s=xpy$, for some x and y; "no" otherwise. For convenience, we will assume $p=p_1p_2\ldots p_m$ and $s=s_1s_2\ldots s_n$ where p_i represents the ith character of the pattern and s_j the jth character of the input string.

This section presents and compares four algorithms for this task. It is straightforward to generalize these algorithms to locate all occurrences of the pattern in the input string.

3.1. The brute-force algorithm

Our initial algorithm is the obvious one. First, the pattern-matcher generator reads and stores the pattern; this is its only function. Then, the pattern matcher reads the input string using a buffer, since it may have to backtrack if a partial match fails. It looks for a match by comparing the pattern $p_1p_2\ldots p_m$ with the m-character substring $s_ks_{k+1}\ldots s_{k+m-1}$ for each successive value of k from 1 to $n-m+1$. It compares the keyword with the substring of the input from left to right until it either matches all of the keyword or finds that $p_i\neq s_j$, for some $1\leqslant i\leqslant m$ and $k\leqslant j\leqslant k+m-1$:

$$\begin{array}{l} \qquad\qquad\qquad\quad i \\ \qquad\qquad\qquad\quad \downarrow \\ \qquad\quad p_1\ldots p_i\ldots p_m \\ s_1\ldots s_k\ldots s_j\ldots s_{k+m-1}\ldots s_n \\ \qquad\qquad\qquad\quad \uparrow \\ \qquad\qquad\qquad\quad j \end{array}$$

```
begin
  i:=1 /* pointer into the pattern keyword p=p1p2...pm */
  j:=1 /* pointer into the input string s=s1s2...sn */
  while
    i≤m and j≤n do
      if pi=sj then
        begin i:=i+1; j:=j+1 end
      else
        begin i:= 1; j:=j−i+2 end
  if i>m then return "yes" else return "no"
end
```

Fig. 2. The brute-force algorithm for matching a single keyword.

At this point it "slides" the pattern one character to the right and starts looking for a match by comparing p_1 with s_{k+1}. The pseudo-code in Fig. 2 details this behavior.

3.2. THEOREM. *The brute-force algorithm solves Problem* 3.1 *in* O(mn) *time and* O(m) *space.*

PROOF. The algorithm may take the maximum O(mn) time. For example, when $p=a^{m-1}b$ and $x=a^n$, the algorithm makes $mn-m^2+m-1$ character comparisons. The algorithm requires a buffer of size O(m) to hold the pattern and the m-character substring of the input. □

In practical situations the expected performance of the brute-force algorithm is usually O($m+n$), but a precise characterization depends on the statistical properties of the pattern and text string.

3.2. *The Karp–Rabin algorithm*

The brute-force algorithm looks for a match by comparing the pattern with the m-character substring $s_k s_{k+1} \ldots s_{k+m-1}$ for each successive value of k from 1 to $n-m+1$. Karp and Rabin [78] suggested that we use a judiciously designed hash function h, which they call a *fingerprint*, to lower the cost of comparing the pattern with each successive m-character substring.

Let h be a hash function that maps each m-character string to an integer. If $h(p_1 p_2 \ldots p_m) \neq h(s_k s_{k+1} \ldots s_{k+m-1})$, then we know $p_1 p_2 \ldots p_m$ cannot possibly match $s_k s_{k+1} \ldots s_{k+m-1}$. If, however, $h(p_1 p_2 \ldots p_m) = h(s_k s_{k+1} \ldots s_{k+m-1})$, then we must still compare $p_1 p_2 \ldots p_m$ with $s_k s_{k+1} \ldots s_{k+m-1}$ character by character to make sure we do not have a false match.

The hash function reduces m character comparisons into a single integer comparison. But we have not gained very much if it takes a long time to compute the hash value or if we get a lot of false matches. Karp and Rabin have suggested using the hash function $h(x) = x \bmod q$ where q is an appropriately large prime. We can transform the

m-character string $s_k s_{k+1} \ldots s_{k+m-1}$ into the integer

$$x_k = s_k b^{m-1} + s_{k+1} b^{m-2} + \cdots + s_{k+m-1}$$

by treating each character s_i as an integer and choosing b to be an appropriate radix. The value of x_{k+1} can be simply computed from the value of x_k:

$$x_{k+1} = (x_k - s_k b^{m-1})b + s_{k+m}.$$

Figure 3 contains the details of the Karp–Rabin algorithm as presented by Sedgewick [120].

```
begin
  d := b^{m-1} mod q
  h_p := (p_1*b^{m-1} + p_2*b^{m-2} + ··· + p_m) mod q
  h_s := (s_1*b^{m-1} + s_2*b^{m-2} + ··· + s_m) mod q
  for k := 1 to n-m+1 do
  begin
    if h_p = h_s and p_1 p_2 ... p_m = s_k s_{k+1} ... s_{k+m-1} then return "yes"
    h_s := (h_s + b*q - s_k*d) mod q /* the additive factor b*q is to keep the rhs positive */
    h_s := (h_s*b + s_{k+m}) mod q
    k := k+1
  end
  return "no"
end
```

Fig. 3. The Karp–Rabin algorithm for matching a single keyword.

Since the mod function is associative, we can apply it after each arithmetic operation to keep the numbers small and we will still end up with the same result as though we had performed all of the operations first and then applied the mod function.

In the worst case, at each iteration we might get a false match that results in m character comparisons to determine $p_1 p_2 \ldots p_k \neq s_k s_{k+1} \ldots s_{k+m-1}$. Most of the time, however, we would expect $h_p \neq h_s$ and thus each iteration would be done in constant time.

3.3. THEOREM. *The Karp–Rabin algorithm solves Problem* 3.1 *in* O(mn) *time in the worst case and in* O($m+n$) *time in the expected case. It requires* O(m) *space.*

One other advantage of the Karp–Rabin algorithm is that it can be used to do two-dimensional pattern matching [78].

3.3. The Knuth–Morris–Pratt algorithm

When a mismatch occurs at the jth input character in the brute-force algorithm, we do not need to reset the input pointer to position $j-i+2$ because the previous $i-1$

input characters are already known—they are the first $i-1$ characters of the pattern. Taking advantage of this observation, Knuth, Morris, and Pratt [80] published an elegant nonbacktracking algorithm that requires only $O(m+n)$ time in the worst case to determine whether p is a substring of s. The algorithm first reads the pattern and in $O(m)$ time constructs a table h, called the *next* function, that determines how many characters to slide the pattern to the right in case of a mismatch during the pattern-matching process. The pattern matcher then processes the input string. Since no backtracking is required, the input can be read one character at a time. The pattern matcher uses the table h in the way shown in Fig. 4 to locate a match.

```
begin
  i:=1 /* pointer into the pattern p=p1p2...pm */
  j:=1 /* pointer into the input string s=s1s2...sn */
  while i≤m and j≤n do
  begin
    while i>0 cand pi≠sj do i:=hi
    i:=i+1; j:=j+1
  end
  if i>m then return "yes" else return "no"
end
```

Fig. 4. The Knuth–Morris–Pratt algorithm for matching a single keyword.

The **cand** in the inner loop is a "conditional and" that compares p_i and s_j only if $i>0$. The key idea is that if we have successfully matched the prefix $p_1p_2\ldots p_{i-1}$ of the keyword with the substring $s_{j-i+1}s_{j-i+2}\ldots s_{j-1}$ of the input string and $p_i \neq s_j$, then we do not need to reprocess any of $s_{j-i+1}s_{j-i+2}\ldots s_{j-1}$ since we know this portion of the text string is the prefix of the keyword that we have just matched.

Instead, each iteration of the inner **while**-loop slides the pattern a certain number of characters to the right as determined by the next table h. In particular, the pattern is shifted $i-h_i$ positions to the right and i is set to h_i. The algorithm repeats this step until i becomes zero (in which case none of the pattern matches any of substring of the input string ending at character s_j) or until $p_i=s_j$ (in which case $p_1\ldots p_{i-1}p_i$ matches $s_{j-i+1}\ldots s_{j-1}s_j$ for the new value of i). The outer **while**-loop increments the pointers to the pattern and the input string.

The essence of the algorithm is the next function that is stored in the table h. To make the algorithm run correctly and in linear time it has the property that h_i is the largest k less than i such that $p_1p_2\ldots p_{k-1}$ is a suffix of $p_1p_2\ldots p_{i-1}$ (i.e., $p_1\ldots p_{k-1}=p_{i-k+1}\ldots p_{i-1}$) and $p_i \neq p_k$. If there is no such k, then $h_i=0$. One easy way to compute h is by the program in Fig. 5 that is virtually identical to the matching program itself.

The assignment statement $j:=h_j$ in the inner loop (the second **while**-statement in Fig. 5) is never executed more often than the statement $i:=i+1$ in the outer loop. Therefore, the computation of h is done in $O(m)$ time. Similarly, the assignment $i:=h_i$ in the inner

```
begin
  i:=1; j:=0; h_1:=0
  while i<m do
  begin
    while j>0 cand p_i≠p_j do j:=h_j
    i:=i+1; j:=j+1
    if p_i=p_j then h_i:=h_j else h_i:=j
  end
end
```

Fig. 5. Algorithm to compute the next function h for pattern $p=p_1p_2\ldots p_m$.

loop of Fig. 4 is never executed more often than the statement $j:=j+1$ in the outer loop. What this means is that the pattern is shifted to the right a total of at most n times by the inner loop. Therefore, the pattern matcher runs in $O(n)$ time. Consequently, we have the following theorem.

3.4. THEOREM. *The Knuth–Morris–Pratt algorithm takes* $O(m+n)$ *time and* $O(m)$ *space to solve Problem* 3.1.

Note that the running time of the Knuth–Morris–Pratt algorithm is independent of the size of alphabet. As another measure of performance, let us briefly examine how many times the inner loop of the Knuth–Morris–Pratt algorithm can be executed while scanning the same input character.

DEFINITION. The Fibonacci strings are defined as follows:

$$F_1=b,\qquad F_2=a,\quad \text{and}\quad F_n=F_{n-1}F_{n-2}\quad \text{for } n>2.$$

3.5. EXAMPLE. Consider the Fibonacci string $F_7=abaababaabaab$. The algorithm in Fig. 5 produces the following next function for F_7:

i	1	2	3	4	5	6	7	8	9	10	11	12	13
p_i	a	b	a	a	b	a	b	a	a	b	a	a	b
h_i	0	1	0	2	1	0	4	0	2	1	0	7	1

Let us apply the Knuth–Morris–Pratt algorithm to look for the pattern F_7 in the input string *abaababaabacabaababaabaab*. After matching the first 11 characters successfully, the algorithm finds a mismatch at the input character c. At this point $i=j=12$:

```
                        i
                        ↓
a b a a b a b a a b a a b
a b a a b a b a a b a c a b a a b a b a a b a a b
                        ↑
                        j
```

The first iteration of the inner loop of Fig. 4 sets $i=h_{12}=7$. This has the effect of shifting the pattern 5 characters to the right, so position 7 of the pattern is now aligned above position 12 of the input string:

$$\begin{array}{l} \qquad\qquad\qquad\qquad\qquad\quad i \\ \qquad\qquad\qquad\qquad\qquad\quad \downarrow \\ \qquad\qquad a\ b\ a\ a\ b\ a\ b\ a\ a\ b\ a\ a\ b \\ a\ b\ a\ a\ b\ a\ b\ a\ a\ b\ a\ c\ a\ b\ a\ a\ b\ a\ b\ a\ a\ b\ a\ a\ b \\ \qquad\qquad\qquad\qquad\qquad\quad \uparrow \\ \qquad\qquad\qquad\qquad\qquad\quad j \end{array}$$

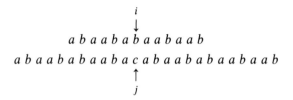

At this point, p_i still does not match s_j, so i is set to $h_7=4$. Mismatches continue for $i=4,2,1,0$. At this point the inner loop is exhausted without finding a match and the outer loop then sets i to 1 and j to 13:

$$\begin{array}{l} \qquad\qquad\qquad\qquad\qquad\qquad i \\ \qquad\qquad\qquad\qquad\qquad\qquad \downarrow \\ \qquad\qquad\qquad\qquad\qquad\qquad a\ b\ a\ a\ b\ a\ b\ a\ a\ b\ a\ a\ b \\ a\ b\ a\ a\ b\ a\ b\ a\ a\ b\ a\ c\ a\ b\ a\ a\ b\ a\ b\ a\ a\ b\ a\ a\ b \\ \qquad\qquad\qquad\qquad\qquad\qquad \uparrow \\ \qquad\qquad\qquad\qquad\qquad\qquad j \end{array}$$

From this point, the algorithm finds a successful match.

Matching a Fibonacci string of length m presents a worst case for the Knuth–Morris–Pratt algorithm. In one execution of the inner loop of Fig. 4 the pattern may be shifted $\log_\varphi m$ times, where $\varphi=\frac{1}{2}(1+\sqrt{5})$, the golden ratio. This is the maximum that is possible [80].

3.6. THEOREM. *The maximum number of times the inner loop of the Knuth–Morris–Pratt algorithm can shift the pattern right while scanning the same input character is at most* $1+\log_\varphi m$.

Note, however, that the total number of shifts of the pattern made by the algorithm in processing the complete input string is at most $n-1$.

3.4. The Boyer–Moore algorithm

The Boyer–Moore [32] algorithm for string matching is the fastest-known single-keyword pattern-matching algorithm in both theory and practice. It achieves its great speed by skipping over portions of the input string that cannot possibly contribute to a match. When the alphabet size is large, the algorithm determines whether a match occurs comparing only about n/m input string characters on the average.

The basic idea of the algorithm is to superpose the keyword on top of the input string and to look for a match by comparing characters in the keyword from right to left against the input string. Consequently, like the brute-force algorithm, it needs a buffer in which to store the current portion of the input. Initially, we compare p_m with s_m. If s_m occurs nowhere in the keyword, then there cannot be a match for the pattern beginning at any of the first m characters of the input. We can therefore safely slide the keyword m characters to the right and try matching p_m with s_{2m}, thus avoiding $m-1$ unnecessary character comparisons.

Consider the general case. We have just shifted the pattern to the right and are about to compare p_m with s_k:

$$\begin{array}{l} \qquad\qquad p_1 \qquad \cdots \overset{\downarrow}{p_m} \\ s_1 \cdots s_{k-m+1} \cdots \underset{\uparrow j}{s_k} \cdots s_n \end{array}$$

(1) We discover that p_m and s_k do not match. If the rightmost occurrence of s_k in the keyword is p_{m-g}, we can shift the pattern g positions to the right to align p_{m-g} and s_k, and then resume matching by comparing p_m with s_{k+g}:

$$\begin{array}{l} \qquad\qquad p_1 \qquad \cdots p_{m-g} \cdots \overset{\downarrow}{p_m} \\ s_1 \cdots s_{k-m+g+1} \cdots s_k \quad \cdots \underset{\uparrow j}{s_{k+g}} \cdots s_n \end{array}$$

As a special case, if s_k did not occur in the keyword, we would shift the pattern m positions to the right and resume matching by comparing p_m with s_{k+m}.

(2) Suppose the last $m-i$ characters of the keyword agree with the last $m-i$ characters of input string ending at position k; that is,

$$p_{i+1}p_{i+2}\cdots p_m = s_{k-m+i+1}s_{k-m+i+2}\cdots s_k.$$

If $i=0$, we have found a match. On the other hand, suppose $i>0$ and $p_i \neq s_{k-m+i}$. Two cases now arise.

(a) If the rightmost occurrence of the character s_{k-m+i} in the keyword is p_{i-g}, then as in Case 1 we can simply shift the keyword g positions to the right so that p_{i-g} and s_{k-m+i} are aligned and resume matching by comparing p_m with s_{k+g}:

$$\begin{array}{l} \qquad\qquad p_1 \qquad \cdots p_{i-g} \quad \cdots \overset{\downarrow}{p_m} \\ s_1 \cdots s_{k-m+g+1} \cdots s_{k-m+i} \cdots \underset{\uparrow j}{s_{k+g}} \cdots s_n \end{array}$$

If p_{i-g} is to the right of p_i, i.e., $g<0$, then we would instead shift the pattern one position to the right and resume matching by comparing p_m with s_{k+1}.

(b) A longer shift than that obtained in Case 2(a) may be possible. Suppose the suffix $p_{i+1}p_{i+2}\ldots p_m$ reoccurs as the substring $p_{i+1-g}p_{i+2-g}\ldots p_{m-g}$ in the keyword and $p_i \neq p_{i-g}$. (If there is more than one such reoccurrence, take the rightmost one.) Then we may get more of a shift than in Case 2(a) by aligning $p_{i+1-g}p_{i+2-g}\ldots p_{m-g}$ above $s_{k-m+i+1}s_{k-m+i+2}\ldots s_k$ and recommencing the search by comparing p_m with s_{k+g}:

$$\begin{array}{ccccccc} & & & & & & \downarrow \\ & p_1 & & \ldots p_{i+1-g} & & \ldots p_{m-g} \cdot \ldots & p_m \\ s_1 \ldots s_{k-m+g+1} & & \ldots s_{k-m+i+1} & & \ldots s_k & \ldots & s_{k+g} \ldots s_n \\ & & & & & & \uparrow \\ & & & & & & j \end{array}$$

The details of this process are given in Fig. 6.

```
begin
  j:=m
  while j≤n do
  begin
    i:=m
    while i>0 cand p_i=s_j do
      begin i:=i−1; j:=j−1 end
    if i=0 then return "yes"
    else j:=j+max(d_1[s_j], d_2[i])
  end
  return "no"
end
```

Fig. 6. The Boyer–Moore algorithm for matching a single keyword.

This algorithm uses two tables d_1 and d_2 to determine how far to slide the keyword to the right when $p_i \neq s_j$. The first table is indexed by characters. For every character c, $d_1[c]$ is the largest i such that $c=p_i$, or $c=m$ if the character c does not occur in the keyword. Table d_1 covers Cases 1 and 2(a) in the discussion above.

Case 2(b) is handled by the second table, which is indexed by positions in the keyword. For every $1 \leqslant i \leqslant m$, $d_2[i]$ gives the minimum shift g such that when we align p_m above s_{k+g}, the substring $p_{i+1-g}p_{i+2-g}\ldots p_{m-g}$ of the pattern agrees with the substring $s_{k-m+i+1}s_{k-m+i+2}\ldots s_k$ of input string, assuming p_i did not match s_{k-m+i}. Formally,

$$\begin{aligned} d_2[i]=\min\{g+m-i \mid g \geqslant 1 \text{ and } (g \geqslant i \text{ or } p_{i-g} \neq p_i) \\ \text{and } ((g \geqslant k \text{ or } p_{k-g}=p_k) \text{ for } i<k \leqslant m)\}. \end{aligned}$$

This table can be computed using the algorithm in Fig. 7.

```
begin
  for i:=1 to m do d2[i]:=2*m-i
  j:=m; k:=m+1
  while j>0 do
  begin
    f[j]:=k
    while k≤m cand p_j≠p_k do
    begin
      d2[k]:=min(d2[k],m-j)
      k:=f[k]
    end
    j:=j-1; k:=k-1
  end
  for i:=1 to k do d2[i]:=min(d2[i],m+k-i)
  j:=f[k]
  while k≤m do
  begin
    while k≤j do
    begin
      d2[k]:=min(d2[k], j-k+m)
      k:=k+1
    end
    j:=f[j]
  end
end
```

Fig. 7. Algorithm to compute shift table d_2.

Boyer and Moore originally had a different way of computing the second shift table d_2. The technique given above is due to Knuth [80] with a modification provided by Mehlhorn [124]. The intermediate function f computed by the algorithm in Fig. 7 has the property that $f[m]=m+1$ and for $1\leqslant j<m$, $f[j]=\min\{i \mid j<i\leqslant m$ and $p_{i+1}p_{i+2}\cdots p_m=p_{j+1}p_{j+2}\cdots p_{m+j-i}\}$.

3.7. Example. The algorithm in Fig. 7 produces the following values for d_2 and f for the pattern *abaababaabaab*:

i	1	2	3	4	5	6	7	8	9	10	11	12	13
p_i	*a*	*b*	*a*	*a*	*b*	*a*	*b*	*a*	*a*	*b*	*a*	*a*	*b*
$d_2[i]$	20	19	18	17	16	15	14	8	15	14	8	14	1
$f[i]$	9	10	11	12	8	9	10	11	12	13	13	13	14

The minimum number of character comparisons needed to determine all occurrences of a keyword of length m in an input string of length n is an interesting theoretical question that has been considered by Knuth [80], Guibas and Odlyzko [61], and Galil [53]. The most recent work by Apostolico and Giancarlo [18] has resulted in a variant of the Boyer–Moore algorithm in which the number of character comparisons is at most $2n$, regardless of the number of occurrences of the pattern in the input string. The following theorem summarizes the worst-case behavior of the Boyer–Moore algorithm.

3.8. THEOREM. *The Boyer–Moore algorithm solves Problem* 3.1 *in* $O(m+n)$ *time and* $O(m)$ *space.*

Rivest [114] has shown that any algorithm for finding a keyword in an input string must examine at least $n-m+1$ of the characters in the input string in the worst case. This result implies that there do not exist pattern-matching algorithms whose worst-case behavior is sublinear in n, providing a sharp contrast with the sublinear average behavior of the Boyer–Moore algorithm. Yao [139] has shown that the minimum average number of characters that need to be examined in looking for a pattern in a random text string is $\Omega(n\lceil\log_A m\rceil/m)$ for $n>2m$, where A is the alphabet size.

3.5. *Expected performance*

Problem 3.1 has been one of the most intensely studied string-matching questions in both theory and practice. The quintessential question is how well does an algorithm solve Problem 3.1 in practice. Several authors have conducted experiments evaluating the relative performance of the four algorithms presented in this section [44, 71, 124]. The primary conclusion is that the Boyer–Moore algorithm is noticeably faster for longer patterns especially in text-processing applications. Horspool observed that for pattern lengths of six or more the Boyer–Moore algorithm even outperformed a naive algorithm that was implemented using a special-purpose machine instruction to scan the input string for the character with the lowest frequency in the pattern [71]. He also noted that the performance of the simplified form of the algorithm in Fig. 8 with a single shift table indexed by characters is comparable to the original formulation. The comparison of $s_{j-m+1}s_{j-m+2}\ldots s_j$ with $p_1p_2\ldots p_m$ is done from right to left.

Several authors have used Markov-chain theory to derive analytical results on the

```
begin
  for each character c in the input alphabet do d[c]:=m
  for j:=1 to m−1 do d[p_j]:=m−j
  j:=m
  while j≤n do
  begin
    if s_j=p_m then
      if s_{j−m+1}s_{j−m+2}...s_j=p_1p_2...p_m then return "yes"
    j:=j+d[s_j]
  end
  return "no"
end
```

Fig. 8. Horspool's simplified Boyer–Moore algorithm.

expected number of character comparisons made by these algorithms on random strings [21, 23]. Baeza-Yates [21] and Schaback [119] have analyzed the expected performance of some variants of the Boyer–Moore algorithm.

3.6. *Theoretical considerations*

In 1971, Cook [38] published the following surprising result that has had a significant influence on the theory of string pattern matching.

3.9. THEOREM. *Every two-way deterministic pushdown automaton (2DPDA) language can be recognized in linear time on a random-access machine.*

This result states that there exists a linear-time pattern-matching algorithm for any set of strings that can be recognized by a 2DPDA, even though the 2DPDA may spend more than linear time recognizing the set of strings. For example, the existence of an $O(m+n)$ keyword-matching algorithm follows directly from Cook's result because the set of strings

$$\{p\#s \mid s = xpy \text{ for some } x \text{ and } y\}$$

can be recognized by a 2DPDA, and hence can be recognized in linear time on a random-access machine [6, 98]. The string-matching capabilities of other classes of automata, especially k-head finite automata, have also been of theoretical interest [16, 39, 58, 85].

Knuth, Morris, and Pratt give a fascinating account about the influence of Cook's theoretical result on their algorithm [80]. Knuth traced out the simulation used in Cook's constructive proof to derive a linear-time pattern-matching algorithm for the keyword-recognition problem. Pratt modified this algorithm to make its running time independent of the input alphabet size. The resulting algorithm was one which Morris had discovered and implemented independently, without the benefit of Cook's theorem.

The string-matching problem has added new vigor to the study of periods and overlaps in strings and to the study of the combinatorics of patterns in strings [24, 43, 46, 62, 63, 80]. We say that an integer k is a *period* of $s = s_1 s_2 \ldots s_n$ if $s_i = s_{i+k}$ for $1 \leqslant i \leqslant n-k$. The following two statements are equivalent:

(1) k is a period of s if and only if $s = (uv)^j u$ for some $j \geqslant 0$, where $|uv| = k$ and v is nonempty ($|x|$ denotes the length of a string x).

(2) k is a period of s if and only if $st = rs$ for some equal-length strings r and t.

When there is a mismatch in the Knuth–Morris–Pratt algorithm at symbol p_i in the pattern, the length of the ensuing shift is a period of $p_1 \ldots p_{i-1}$. In general, $h_i = i - k$, where k is the smallest period of $p_1 \ldots p_{i-1}$ that is not a period of $p_1 \ldots p_i$. Note that the Knuth–Morris–Pratt algorithm can be used to compute the period of a string in linear time.

An important mathematical property of periods is that if i and j are periods of s, and $i + j \leqslant |s| + gcd(i, j)$, then the $gcd(i, j)$ is also a period of s [88].

Galil and Seiferas [58] and Crochemore and Perrin [43] have developed keyword-matching algorithms that use only a constant amount of additional space and are intermediate in performance between the Knuth–Morris–Pratt and Boyer–Moore algorithms.

4. Matching sets of keywords

We now generalize the single-keyword pattern-matching problem to sets of keywords.

4.1. PROBLEM. Given a pattern p consisting of a set of keywords $\{w_1, w_2, \ldots, w_k\}$ and an input string $s = s_1 s_2 \ldots s_n$, answer "yes" if some keyword w_i occurs as a substring of s; "no" otherwise. We assume the sum of the lengths of the keywords is m.

In this section we describe two algorithms to solve this problem. The first constructs an automaton from the keywords to look for the matches in parallel rather than one at a time. The second adds Boyer–Moore-like techniques to the first algorithm. Both algorithms can also be used to locate all occurrences of the keywords in the input string.

4.1. The Aho–Corasick algorithm

The straightforward way to look for multiple keywords in an input string is to apply the fastest single-keyword pattern-matching algorithm once for each keyword. This would give a solution whose running time is $O(m + kn)$. In this section we describe an algorithm due to Aho and Corasick [3] that solves this problem in $O(m + n)$ time, which yields a significant improvement when the number of keywords is large.

As for a single keyword, the existence of an $O(m+n)$ time algorithm for the multiple-keyword pattern-matching problem is evident from Cook's Theorem. The language

$$L = \{w_1 \# w_2 \# \ldots \# w_k \# \# s \mid s = xw_iy \text{ for some } x \text{ and } y, \text{ and } 1 \leqslant i \leqslant k\}$$

can be recognized by a 2DPDA, and consequently Cook's result implies that there is an $O(m + n)$ algorithm to do the matching.

The Aho–Corasick algorithm generalizes the Knuth–Morris–Pratt algorithm to multiple-keyword patterns. It first constructs from the set of keywords a pattern-matching automaton in $O(m)$ time; the automaton then reads the input and looks for all the keywords simultaneously in $O(n)$ time. The pattern-matching automaton is like a deterministic finite automaton except that it has two transition functions, a forward transition function and a failure transition function. The failure transition function is used only when the forward transition fails.

DEFINITION. An Aho–Corasick pattern-matching automaton consists of the following components:

(1) Q, a finite set of states,
(2) Σ, a finite input alphabet,
(3) $g: Q \times \Sigma \rightarrow Q \cup \{fail\}$, a forward transition function,
(4) $h: Q \rightarrow Q$, a failure transition function,
(5) q_0, an initial state, and
(6) F, a set of accepting states.

Figure 9 shows how this automaton is used to determine whether an input string $s = s_1 s_2 \ldots s_n$ contains a keyword from the pattern set.

```
begin
  q := q0
  for j := 1 to n do
  begin
    while g[q, s_j] = fail do q := h[q]
    q := g[q, s_j]
    if q is in F then return "yes"
  end
  return "no"
end
```

Fig. 9. The Aho–Corasick algorithm for matching multiple keywords.

Initially, the pattern-matching automaton is in state q_0 scanning s_1, the first character of s. It then executes a sequence of moves. During a move on input symbol s_j the automaton makes zero or more failure transitions until it reaches a state q for which $g[q, s_j] \neq fail$. To complete the move, the automaton makes one forward transition to state $g[q, s_j]$. If this state is an accepting state, the automaton returns "yes" and halts; otherwise, the automaton makes a move on the next input character s_{j+1}.

The forward and failure transition functions of the pattern-matching automaton will have the following two properties:

(1) $g[q_0, a] \neq fail$ for all a in Σ.

(2) If $h[q] = q'$, then the depth of q' is less than the depth of q, where the *depth* of a state is the length of the shortest sequence of forward transitions from the initial state to that state.

The first property makes sure no failure transitions occur in the initial state. The second property ensures that the total number of failure transitions needed to process an input string will be less than the total number of forward transitions. Since exactly one forward transition is made on each input character, fewer than $2n$ transitions of both kinds will be made in processing an input string of length n. Thus, an Aho–Corasick pattern-matching automaton runs in $O(n)$ time.

We now show how to construct the automaton from the set of keywords $p = \{w_1, w_2, \ldots, w_k\}$.

(1) From the set of keywords, construct a trie in which each node represents a prefix of some keyword. The trie can be constructed in $O(m)$ time. The nodes of the trie are the

states of the automaton and the root is the initial state q_0, which represents the empty prefix. Each node corresponding to a complete keyword is an accepting state. The transition function g is defined so that $g[q, p_i] = q'$ when q corresponds to the prefix $p_1 \ldots p_{i-1}$ of some keyword and q' corresponds to $p_1 \ldots p_{i-1} p_i$.

(2) For state q_0, set $g[q_0, a] = q_0$ for each character a for which $g[q_0, a]$ was not defined in step (1).

(3) Set $g[q, a] = \mathit{fail}$ for all q and a for which $g[q, a]$ was not defined in steps (1) and (2).

The three steps above define the forward transition function. Note that state q_0 has the property that $g[q_0, a] \neq \mathit{fail}$ for any a in Σ.

DEFINITION. We define a *failure function* $f: Q \rightarrow Q$ on the nodes of the trie with the following property:

> if states q_u and q_v represent the prefixes u and v of some keywords in p, then $f[q_u] = q_v$ if and only if v is the longest proper suffix of u that is also the prefix of some keyword in p.

The failure function can be computed in O(m) time by making a traversal over the trie. We use a breadth-first traversal so that all states of depth d are visited before those of depth $d+1$. During the traversal, we compute the failure function at each state as follows:

(1) Set $f[q_0] = q_0$ and for all states q of depth 1 set $f[q] = q_0$.

(2) Assume that f has been defined for all states of depth less than $d \geqslant 2$ and suppose $g[q, a] = q'$ where q' is a state of depth d. Set $f[q'] = g[r, a]$, where r is the state determined by the following program:

> $r := f[q]$
> **while** $g[r, a] = \mathit{fail}$ **do** $r \leftarrow f[r]$

Note that since the depth of $f[r]$ is always less than that of r for all states r except the root, and $g[q_0, a] \neq \mathit{fail}$, the **while**-loop will always terminate.

4.2. EXAMPLE. Figure 10 gives the trie for the Fibonacci string *abaababa* and the values of the failure function at each state.

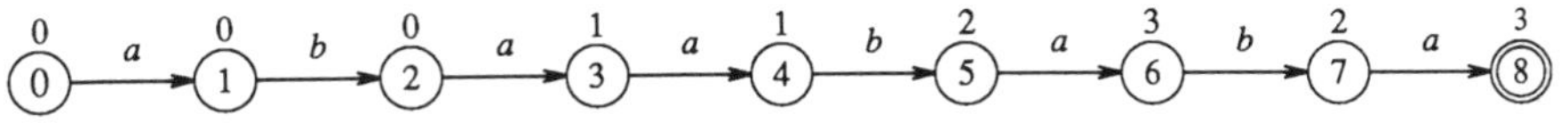

Fig. 10.

The failure function itself can be used as the failure transition function. However, the failure function may sometimes make more failure transitions than necessary. For example, if we try to match the input string *abaabb* with the failure function above, the pattern-matching automaton would make the following state transitions:

	a		b		a		a		b		b	
0		1		2		3		4		5		0
										2		
										0		

On each of the first five characters, the automaton makes a single forward transition. On the last character, the automaton makes a failure transition from state 5 to state 2, and then another failure transition to state 0, before making the final forward transition to state 0.

An "optimized" failure transition function h that avoids these unnecessary failure transitions can be constructed from f as follows:

(1) Let $h[q_0] = q_0$.

(2) Assume that h has been defined for all states of depth less than d and let q be a state of depth d. If the set of characters for which there is a forward non-*fail* transition in state $f[q]$ is a subset of the set of characters for which there is a forward non-*fail* transition in state q, then set $h[q] = h[f[q]]$; otherwise, set $h[q] = f[q]$.

The function h here is a generalization of the next function in the Knuth–Morris–Pratt algorithm. It can be computed in O(m) time by making another breadth-first traversal over the trie.

4.3. Example. Figure 11 shows the h function for the Fibonacci string *abaababa*.

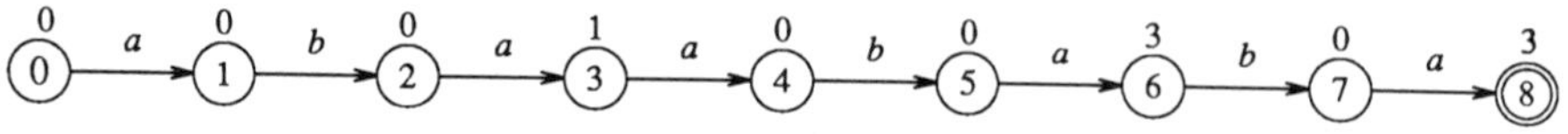

Fig. 11.

With h as the failure transition function, the automaton would make the following moves:

	a		b		a		a		b		b	
0		1		2		3		4		5		0
										0		

At the last character, this automaton makes one failure transition rather than two.

4.4. Example. For a more complete example, let us construct the Aho–Corasick pattern-matching automaton for the set of keywords {*cacbaa*, *acb*, *aba*, *acbab*, *ccbab*}. Figure 12 shows the trie. The forward transition function g, the failure function f, and the failure transition function h for this set of patterns are given in Table 1.

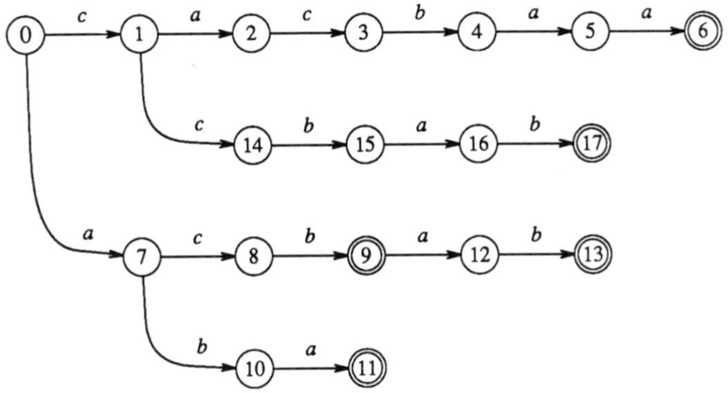

Fig. 12. Trie for keywords *cacbaa*, *acb*, *aba*, *acbab*, *ccbab*.

Table 1
Transition functions for the pattern-matching automaton of Example 4.5

	g			*f*	*h*
	a	*b*	*c*		
0	7	0	1	0	0
1	2	*fail*	14	0	0
2	*fail*	*fail*	3	7	7
3	*fail*	4	*fail*	8	1
4	5	*fail*	*fail*	9	0
5	6	*fail*	*fail*	12	12
6	*fail*	*fail*	*fail*	7	7
7	*fail*	10	8	0	0
8	*fail*	9	*fail*	1	1
9	12	*fail*	*fail*	0	0
10	11	*fail*	*fail*	0	0
11	*fail*	*fail*	*fail*	7	7
12	*fail*	13	*fail*	7	7
13	*fail*	*fail*	*fail*	10	10
14	*fail*	15	*fail*	1	1
15	16	*fail*	*fail*	0	0
16	*fail*	17	*fail*	7	7
17	*fail*	*fail*	*fail*	10	10

To summarize, we can construct the trie, and the forward and failure transition functions in $O(m)$ time and $O(m)$ space. The pattern-matching automaton processes the input string with no backtracking, making n forward transitions and at most $n-1$ failure transitions on input string of length n. Thus, the pattern matcher runs in $O(n)$ time.

4.5. THEOREM. *The Aho–Corasick algorithm solves Problem* 4.1 *in* $O(m+n)$ *time and* $O(m)$ *space.*

Let us examine the number of failure transitions that can be made while scanning one input character. If the pattern consists of only one keyword of length m, then the Aho–Corasick algorithm is essentially the Knuth–Morris–Pratt algorithm, and as we noted, $\log_\varphi m$ failure transitions are necessary and sufficient in any one move. If the pattern is a set of keywords, then $O(m)$ failure transitions may be necessary in a single move, where m is the sum of the lengths of the keywords. The total number of failure transitions in processing an input string of length n, however, is at most $n-1$.

In the early 1970s Aho and Corasick used this algorithm in a bibliographic search system in which a user could specify documents by prescribing Boolean combinations of keywords and phrases. When this algorithm was used in place of a straightforward multiple-keyword algorithm (the analog of the brute-force method for single keywords), the system typically ran 4 to 12 times faster [3]. Aho also implemented the original version of the UNIX system keyword-matching program *fgrep* using this algorithm.

The Aho–Corasick algorithm can also be used to match subtrees in trees by noting that if we label the branches of each node with the numbers 1, 2, 3, ..., from left to right, then a tree is uniquely characterized by the set of paths from the root to the leaves [4, 69]. Ben-Yehuda and Pinter [25] have extended the Aho–Corasick algorithm to do two-dimensional pattern matching.

4.2. *The Commentz-Walter algorithm*

Commentz-Walter has described an approach in which the ideas in the Boyer–Moore algorithm are combined with the Aho–Corasick algorithm to look for patterns consisting of sets of keywords [37]. The basic idea is to construct an Aho–Corasick style pattern-matching automaton, but for the keywords reversed. Let $k_{\min}$ be the length of a shortest keyword. Matching begins with the automaton in its initial state scanning s_i, the ith character of the input string, where $i = k_{\min}$. We can think of the automaton being superposed on top of the input string with the start state above the character s_i. The automaton then makes state transitions, reading the input string from right to left, until it either finds a match for a keyword or enters a state for which there is no transition on the current input symbol.

In the latter case, the automaton has read the input characters $s_{i-j+1}s_{i-j+2}\ldots s_i$ (from right to left), the automaton is in some state q, and there is no transition from state q on s_{i-j}. At this point, the automaton shifts its start state to a character to the right of s_i and recommences matching from the start state. The amount of the shift is determined as follows:

(1) Let $d_1[q]$ be a minimal shift so that $s_{i-j+1}s_{i-j+2}\ldots s_i$ will be aligned with a matching substring of some keyword.

(2) Let $d_2[q]$ be a minimal shift so that a suffix of $s_{i-j+1}s_{i-j+2}\ldots s_i$ will be aligned with a prefix of some keyword.

(3) Let $d_3[s_{i-j}, j]$ be a minimal shift so that s_{i-j} will be aligned with a matching character in some keyword.

The amount of shift at state q on input character s_{i-j} is then given by

$$shift[q, s_{i-j}, j] = \min\begin{cases}\max(d_1[q], d_3[s_{i-j}, j]),\\ d_2[q].\end{cases}$$

We can visualize each shift as moving the start state of the automaton to the right along the input string.

We will now describe how to construct the Commentz-Walter pattern-matching automaton from the set of keywords. We first construct a trie for the set of keywords reversed. As in the Aho–Corasick automaton, the states are the nodes of the trie, the root is the start state, and each node corresponding to a complete keyword in reverse is an accepting state. Let *path*(q) be the string spelled out by the characters on the path from the start state q_0 to the node q. The transition function g is defined so that $g[q, p_i] = q'$ where *path*(q) $= p_1 p_2 \ldots p_{i-1}$ for some reversed keyword $p_1 p_2 \ldots p_k, k \geqslant i$.

4.6. EXAMPLE. The Commentz-Walter trie for the set of keywords {*cacbaa*, *aba*, *acbab*, *ccbab*, *acb*} is shown in Fig. 13.

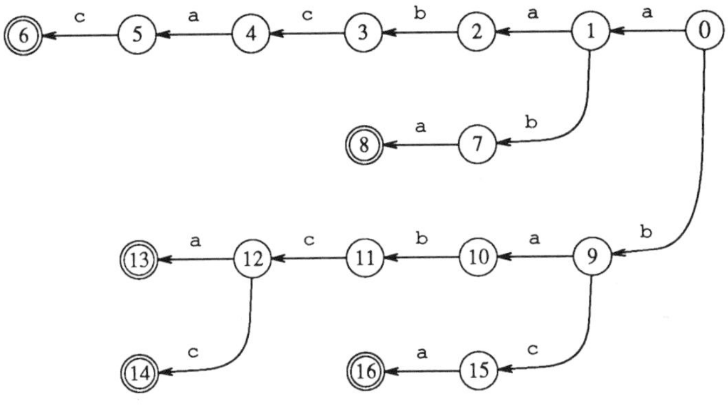

Fig. 13. Reversed trie for keywords *cacbaa, aba, acbab, ccbab, acb*.

We will now compute the two intermediate shift functions d_1 and d_2 for each state. Let *depth*(q) be the number of edges along the path from the root to node q. We compute two intermediate sets for each state other than the start state:

$$set_1(q) = \{r \mid path(q) \text{ is a proper suffix of } path(r)\},$$
$$set_2(q) = \{r \mid r \text{ is in } set_1(q) \text{ and } r \text{ is an accepting state}\}.$$

From these sets we compute the two shift functions d_1 and d_2. For the start state q_0, we define $d_1[q_0]=1$. For all other states q, we define

$$d_1[q]=\min\begin{cases} depth(r)-depth(q) & \text{where } r \text{ is in } set_1(q),\\ k_{\min}. \end{cases}$$

The function $d_1[q]$ is the smallest difference in depth between q and any state r for which $path(r)$ is the string $u\ path(q)$ for some nonnull u. ($d_1[q]$ is in fact the length of u.) In other words, if some reversed keyword contains another reoccurrence of substring $path(q)$ at the point at which a mismatch occurs, it is safe to move the start state to the right by $d_1[q]$ positions.

The second shift function is defined to align a suffix of $s_{i-j+1}s_{i-j+2}\ldots s_i$ (which is $path(q)$ reversed) with a prefix of some keyword:

$$d_2[q_0]=k_{\min},$$

$$d_2[q]=\min\begin{cases} depth(r)-depth(q) & \text{where } r \text{ is in } set_2(q),\\ d_2[parent(q)]. \end{cases}$$

The value of $d_2[q]$ is determined by those keywords w such that $path(q)$, or a prefix thereof, is a suffix of w^R, the reversal of w. Table 2 summarizes this information for the trie above.

Matching proceeds somewhat like in the Boyer–Moore algorithm with the trie playing the role of the single keyword. Define

$$char(c)=\min\begin{cases} depth(q) & \text{where the transition into } q \text{ is labeled } c,\\ k_{\min}+1. \end{cases}$$

Table 2
Shift functions for the trie in Fig. 13

node	d_1	d_2	set_1	set_2
0	1	3		
1	1	2	{2, 5, 8, 10, 13, 16}	{8, 13, 16}
2	3	2	∅	∅
3	3	2	∅	∅
4	3	2	∅	∅
5	3	2	∅	∅
6	3	2	∅	∅
7	1	2	{3, 11}	∅
8	3	2	∅	∅
9	1	3	{3, 7, 11}	∅
10	1	1	{8}	{8}
11	3	1	∅	∅
12	3	1	∅	∅
13	3	1	∅	∅
14	3	1	∅	∅
15	2	3	{4, 12}	∅
16	2	2	{5, 13}	{13}

Finally, let

$$shift[q, s_{i-j}, j] = \min\{\max(d_1[q], char(s_{i-j}) - j - 1), d_2[q]\}.$$

This function is used if there is a mismatch in state q at input character s_{i-j}. The $char(s_{i-j}) - j - 1$ term sometimes allows us to shift more than $d_1[q]$ positions by observing that d_1 is at most $k_{\min}$ and that the shortest distance from the root to a node whose incoming edge is labeled by s_{i-j} may be greater than this. In this case, it is safe to advance by $char(s_{i-j}) - j - 1$ positions (the term $-j-1$ adjusts for the relative position of the start state).

Note that we use the minimum of d_2 and the maximum of d_1 and *char* in computing *shift*. At node 14 in the trie of Fig. 13, for example, $d_1[14] = 3 = k_{\min}$ because $set_1(14)$ is empty. When we reach node 14, $s_{i-j+1} s_{i-j+2} \ldots s_i = ccbab$ with the start state over the rightmost *b*. It might appear that we could shift the start state 3 positions to the right once we reach node 14. But $d_2[14] = 1$ because of node 10 where we shift by 1 to check for a match of *aba*; thus, at any descendant of node 10 we cannot shift by more than 1 in case we miss this possible match.

In general, a shift to the right by $shift[q, s_{i-j}, j]$ positions is guaranteed not go past an occurrence of a keyword. The entire algorithm is summarized in Fig. 14.

```
begin
  q:=q0
  i:=k_min
  j:=0
  while i≤n do
  begin
    while j<i cand g[q, s_{i-j}]≠fail do
    begin
      q:=g[q, s_{i-j}]
      j:=j+1
      if q is accepting then return "yes"
    end
    i:=i+shift[q, s_{i-j}, j]
    q:=q0
    j:=0
  end
  return "no"
end
```

Fig. 14. The Commentz-Walter algorithm for matching multiple keywords.

The Commentz-Walter pattern-matching automaton can be constructed in $O(m)$ time using techniques similar to those in Section 4.1, but its worst-case running time on an input string of length n is $\Theta(mn)$. In practice, with small numbers of keywords, the Boyer–Moore aspect of the Commentz-Walter algorithm can make it faster than the Aho–Corasick algorithm, but with larger numbers of keywords the Aho–Corasick algorithm has a slight edge. It is possible at a greater than linear-time preprocessing cost to produce a version of the Commentz-Walter algorithm whose worst-case behavior is linear in n [37].

5. Matching regular expressions

We now generalize the pattern-matching problem to full regular expressions.

5.1. Problem. Given a pattern consisting of a regular expression r and an input string $s=s_1s_2\ldots s_n$ answer "yes" if r matches a substring of s; "no" otherwise. We assume the length of r is m.

This section describes two algorithms to solve this problem. In the first the pattern-matcher generator constructs a nondeterministic finite automaton, which is then used as the matcher to process the input string. In the second the generator constructs a deterministic finite automaton. Both approaches have been used in regular-expression pattern-matching programs.

5.1. Nondeterministic recognizers for regular expressions

This section outlines a regular-expression pattern-matching technique originally due to Thompson [127] that was used in the text editor *qed* and in a simplified form in the UNIX system command *grep*. We will prepend to the given regular expression the expression $(a_1|a_2|\ldots|a_f)^*$ where a_1 through a_f are the symbols of the input alphabet. This prefix allows matching to begin at any position in the input string. For the remainder of this section, we will assume r contains this prefix.

We begin by constructing a nondeterministic finite automaton (NDFA) from r. An NDFA is a directed graph in which the nodes are the states and each edge is labeled by a single character or the symbol ε, which stands for the empty string. One state is designated as an *initial* state, and some states as *accepting* states. An NDFA accepts (matches) a string if there is a path from the start state to an accepting state whose edge labels spell out the string.

Once we have constructed an NDFA for r, we run it on the input string s. If the NDFA enters an accepting state while processing s, we report that r matches s, otherwise, we report "no". The NDFA can take the form of an executable program or a state-transition table that is interpreted.

The recursive procedure below can be used to construct an NDFA for the regular expression. We first parse the regular expression into its constituent subexpressions according to the formation rules used to define a regular expression in Section 2.1. Using rule (1), we construct an NDFA for a non-metacharacter. Rules (2)–(5) show how to combine the NDFAs constructed from the constituent subexpressions.

(1) For a non-metacharacter c, construct the NDFA in Fig. 15(a) where i is a new initial state and a a new accepting state. This automaton clearly accepts exactly the string c.

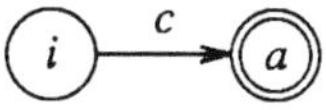

Fig. 15(a).

(2) Suppose N_{r_1} and N_{r_2} are NDFAs for r_1 and r_2. For the regular expression $r=r_1|r_2$, construct the NDFA N in Fig. 15(b) where i is a new initial state and a a new accepting state. There is an ε-transition from i to the start states of N_{r_1} and N_{r_2}. There is an ε-transition from the accepting states of N_{r_1} and N_{r_2} to the new accepting state a. (The initial and accepting states of N_{r_1} and N_{r_2} are not considered start or accepting states of N_r.) Note that any path from i to a must pass through either N_{r_1} or N_{r_2}. Thus, N accepts any string accepted by N_{r_1} or N_{r_2}.

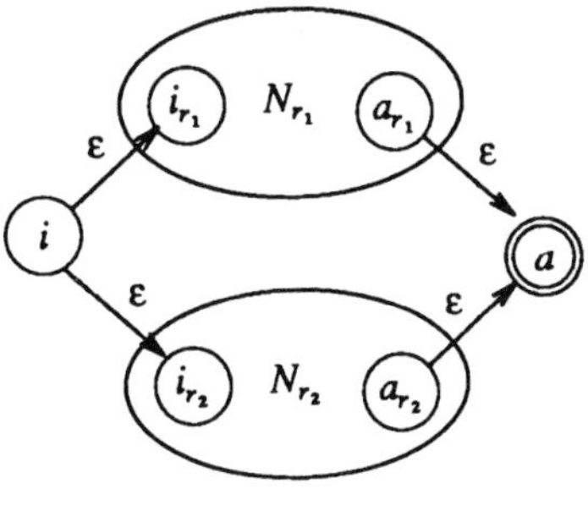

Fig. 15(b).

(3) Suppose N_{r_1} and N_{r_2} are NDFAs for r_1 and r_2. For the regular expression $r=r_1r_2$, construct the NDFA N in Fig. 15(c) where the start state of N_{r_1} becomes the start state of N and the accepting state of N_{r_2} becomes the accepting state of N. The accepting state of N_{r_1} is merged with the start state of N_{r_2}; that is, all transitions from the start state of N_{r_2} become transitions from the accepting state of N_{r_1}. The new merged state loses its status as a start or accepting state in N. A path from i_{r_1} to a_{r_2} must go first through N_{r_1} and then through N_{r_2}, so N accepts any string of the form xy where N_{r_1} accepts x and N_{r_2} accepts y.

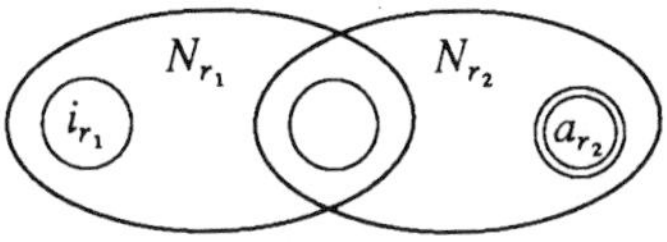

Fig. 15(c).

(4) Suppose N_{r_1} is an NDFA for r_1. For the regular expression $r=r_1^*$, construct the NDFA N in Fig. 15(d) where i is a new initial state and a a new accepting state. In N, we can go from i to a directly, along an edge labeled ε, representing the fact that s^* matches the empty string, or we can go from i to a passing through N_{r_1} one or more times. Thus, N accepts any string matched by r_1^*.

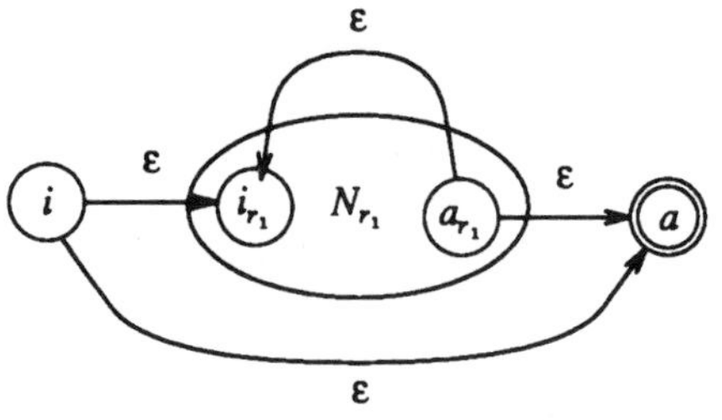

Fig. 15(d).

(5) For the regular expression (r) use the NDFA for r.

Figure 16 shows the NDFA that results from this construction for the regular expression $(a|b)^*aba$. This construction produces an NDFA N for r with the following properties:

(1) N has at most twice as many states as the length of r, since each step of the construction creates at most two new states.

(2) N has exactly one start state and one accepting state, and the accepting state has no outgoing transitions. This property holds for each of the constituent NDFAs as well.

(3) Each state has either one outgoing edge labeled by a character or at most two outgoing ε-edges.

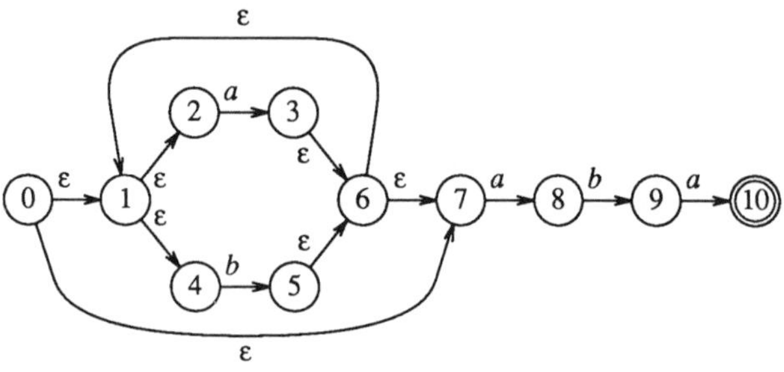

Fig. 16. NDFA for $(a|b)^*aba$.

Figure 17 contains an algorithm to determine whether an NDFA N with initial state i and accepting state a matches a substring of an input string $s = s_1 s_2 \ldots s_n$. The algorithm uses the function *epsilon*(Q) to compute all states that can be reached from a set of states Q by following only ε-edges. After reading each character, the algorithm determines Q, the current set of states for N. It computes the next set of states from Q in two stages. First, it determines *goto*(Q, c), all states that can be reached from a state in Q by a transition on c, the current input character. Then, it computes *epsilon*(*goto*(Q, c)), all states that can be reached from *goto*(Q, c) by following only ε-edges. The algorithm returns "yes" if the accepting state is in the set of current states. If no accepting state is ever encountered, "no" is returned.

```
begin
  Q := epsilon({i})
  if Q contains a then return "yes"
  for j := 1 to n do
  begin
    Q := epsilon(goto(Q, s_j))
    if Q contains a then return "yes"
  end
  return "no"
end
```

Fig. 17. Algorithm to simulate an NDFA on an input string $s_1 \ldots s_n$.

The NDFA can be simulated in time proportional to $|N| \times |s|$, where $|N|$ is the number of states in N and $|s|$ is the length of s, by taking advantage of the special properties of N. The set of states reachable after each input character can be efficiently computed using two stacks and a bit vector indexed by states. One stack is used to store Q, the current set of states, and the other stack to determine the next set of states. Since each state has at most two out-transitions, each state on the first stack can add at most two new states to the second stack. The bit vector is used to quickly determine whether a state is already on the second stack so that we do not add it twice. Once we have put the states in $goto(Q, c)$ onto the second stack, we can use a simple reachability algorithm to compute $epsilon(goto(Q, c))$. When we have computed all the reachable states on the second stack, we read the next input character and interchange the roles of the two stacks.

Since there can be at most $|N|$ states on a stack, the computation of the next set of states from the current set of states can be done in time proportional to $|N|$. Thus, the time needed to run N on input s is proportional to $|N| \times |s|$. Since the number of states in N is at most twice the length of r, the running time of this algorithm is $O(|r| \times |s|)$. Thus, we have the following theorem.

5.2. THEOREM. *Problem* 5.1 *can be solved with an NDFA in* $O(mn)$ *time and* $O(m)$ *space.*

Myers [105] has shown how the use of node listings and the "Four Russians" trick [6] can be applied to this algorithm to derive an $O(mn/\log n)$ time and space solution to Problem 5.1.

5.2. Deterministic recognizers for regular expressions

Although the algorithm in the previous section produces a compact NDFA, its running time is proportional to the product of the size of the regular expression and the length of the input. A deterministic finite automaton (DFA) is an NDFA in which there are no ε-transitions and in which every state has at most one transition on any input character. DFAs are well suited for regular-expression pattern matching because they are capable of recognizing all regular-expression patterns. Moreover, a DFA can be simulated in real time because an input character causes at most one state transition.

This section describes a DFA-based regular-expression pattern-matching algorithm. The algorithm first constructs a syntax tree for the regular expression, such as the one in Fig. 18 for the regular expression $(a|b)^*aba\#$. The symbol # is an endmarker appended to the expression; its function will be explained shortly. A dot is used to represent the concatenation operator.

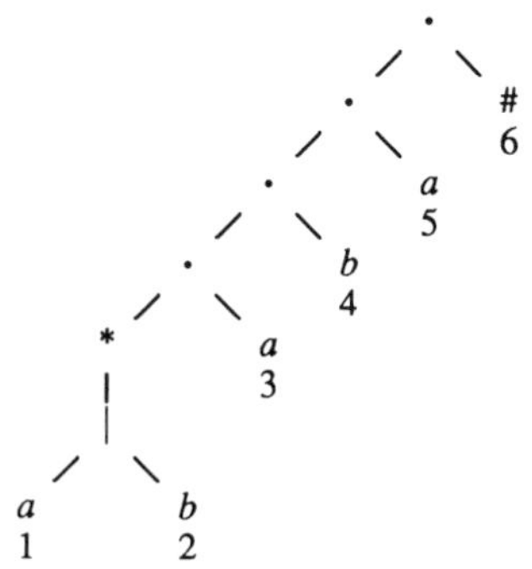

Fig. 18. Syntax tree for $(a|b)^*aba\#$.

The leaves of the syntax tree are labeled by the non-metacharacters in the regular expression. Using a technique suggested by McNaughton and Yamada [97] we associate with each leaf a unique integer called the *position* of the leaf. Positions are shown below the leaves in the syntax tree of Fig. 18.

The transitions of the DFA are constructed directly from the syntax tree. Each state of the DFA is the set of positions corresponding to the leaves that are active after having read some prefix of the input string. Initially, leaves 1, 2, and 3 are active so the initial state of the DFA is the set $\{1, 2, 3\}$.

The next state Q' representing the transition from state Q on an input character c is computed as follows. For each position i in Q whose leaf-symbol matches c, we add *follow*(i) to T, where *follow*(i) is the set of positions that can "follow" the leaf labeled by i in the syntax tree. These two rules define *follow*(i):

(1) If *left* and *right* are the two children of a node labeled by a concatenation operator in the syntax tree and i is a position that can last be active in the subtree rooted at *left*, then all positions initially active in the subtree rooted at *right* are in *follow*(i).

(2) If i is a position that can last be active in a subtree rooted at a *-node in the syntax tree, then all positions initially active in that subtree are in *follow*(i).

Let us compute the state Q representing the transition from the initial state $\{1, 2, 3\}$ on the input symbol a. The leaf corresponding to position 1 matches a, so we add *follow*$(1)=\{1, 2, 3\}$ to Q. Position 2 does not match a, but position 3 does, so we add *follow*$(3)=\{4\}$ to Q. Thus the transition from state $\{1, 2, 3\}$ on a is to state $\{1, 2, 3, 4\}$.

To create the complete DFA, we compute all transitions for each new state. Any state containing the position corresponding to the endmarker # is made an accepting state.

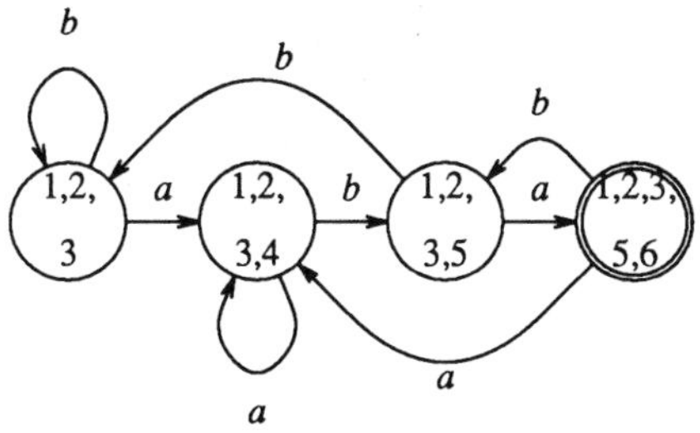

Fig. 19. DFA for $(a|b)^*aba$.

We do not compute transitions for the symbol #. Figure 19 contains the complete DFA for the regular expression $(a|b)^*aba$.

The transition function of a DFA can be simply represented as a two-dimensional array. A DFA can then be simulated efficiently by the table look-up program in Fig. 20.

```
begin
  Q := InitialState
  if Q is accepting then return "yes"
  for j := 1 to n do
  begin
    Q := goto[Q, s_j]
    if Q is accepting then return "yes"
  end
  return "no"
end
```

Fig. 20. Algorithm to simulate a DFA on an input string $s_1 \ldots s_n$.

Thus, once a DFA has been constructed from a regular expression, we can determine whether it matches the input string in O(n) time. Note that neither the size of the input alphabet nor the length of the regular expression affect the time to simulate the DFA. Berry and Sethi [26] relate this technique to the derivatives method for constructing a DFA from a regular expression proposed by Brzozowski [33].

If the transition function is stored as a two-dimensional array, then the storage requirement for a DFA is the product of the input alphabet size times the number of states. For some regular expressions even a minimum-state DFA recognizer may have 2^n states, where n is the length of the regular expression. In these situations, the complete DFA is time-consuming to construct, and the transition function requires lots of storage. Consider, for example, the regular expression

$$(a|b)^*a(a|b)(a|b)\ldots(a|b)$$

where there are k copies of $(a|b)$ at the end. This expression denotes all sequences of a's and b's in which the $(k+1)$st symbol from the end is an a. The smallest DFA recognizing this expression must have at least 2^k states to remember the 2^k possible sequences of the last k characters it has seen. Thus, to summarize, we have the following theorem.

5.3. THEOREM. *Problem* 5.1 *can be solved with a DFA in* $O(2^m+n)$ *time and* $O(2^m)$ *space.*

Several techniques are available to reduce the space requirements of the transition table [9]. An effective storage-reduction technique that has been used by the author in the UNIX system command *egrep* is "lazy transition evaluation." The transition function is only computed when the DFA is run. Computed transitions are kept in a cache. Before a transition is made, the cache is examined. If the required transition is not in the cache, it is computed and stored for subsequent use. If the cache becomes full, some or all of the previously computed transitions are removed to make room for the new transition. This method allows the pattern matcher to use a fixed-size storage area for the transitions with only a small run-time penalty. The observed performance in practice of this approach for solving Problem 5.1 is $O(m+n)$ time and $O(m)$ space, combining the best features of both the nondeterministic and deterministic approaches. It would be interesting to know whether it is possible to construct a regular-expression pattern-matching algorithm with this as its worst-case behavior.

The problem of storing a transition table compactly and yet providing constant-time access to its elements has stimulated recent research in perfect hashing [51, 126] and dynamic perfect hashing [8, 45]. Some additional issues in the efficient representation of transition tables are discussed in [12]. An alternative to creating a transition table is to generate machine code directly to simulate the automaton [109, 127].

An *extended* regular expression is one that also has operators for intersection and complement. Hopcroft and Ullman describe a dynamic programming algorithm to match extended regular expressions in time $O((n+m)^4)$ [70, Exercise 3.23].

6. Related problems

In the three previous sections we presented pattern-matching algorithms for keywords, sets of keywords, and regular expressions. There are many other string-matching problems that are of interest in computer science and in this section we will mention a few that are related to the ones that we have studied.

6.1. *Matching regular expressions with back referencing*

Let us briefly consider the problem of matching regular expressions with back referencing because it dramatically brings out the point that as we generalize the class of patterns, we can make the pattern-matching process much more difficult computationally.

6.1. PROBLEM. Given a pattern consisting of a rewbr r and an input string s, answer "yes" if s contains a substring matched by r; "no" otherwise.

6.2. THEOREM. *Problem* 6.1 *is* NP-*complete.*

PROOF. We will use a reduction from the vertex-cover problem. Let $E_1, E_2, \ldots, E_m$ be subsets of cardinality 2 of some finite set of vertices V. The vertex-cover problem is to determine, given a positive integer k, whether there exists a subset V' of V of cardinality at most k such that V' contains at least one element in each E_i. We can think of the E_i's as being edges of a graph and V' as being a set of vertices such that each edge contains at least one vertex in V'. The vertex-cover problem is a well-known NP-complete problem [59].

We can transform this problem into a pattern-matching problem for rewbrs as follows. Let N be the parenthesized string $(n_1|n_2|\ldots|n_f)$ where $V=\{n_1,n_2,\ldots,n_f\}$. Let $\#$ be a distinct marker symbol. For $1 \leqslant i \leqslant k$, let

$$x_i = N^*N\%v_i\,N^*\#$$

where the v_i's are distinct variable names. Likewise, for $1 \leqslant i \leqslant m$, let

$$y_i = N^*N\%w_i\,N^*\#$$

where the w_i's are distinct variable names. For $1 \leqslant i \leqslant m$, let

$$z_i = w_i^* v_1^* v_2^* \ldots v_k^* w_i^* \#$$

Let r be the rewbr $x_1 \ldots x_k y_1 \ldots y_m z_1 \ldots z_m$.

We shall now construct an input string s such that if the rewbr r matches s, then the vertex-cover problem has a solution of size k. Let u be the string $n_1 n_2 \ldots n_f \#$ repeated $k+m$ times. Let e_i be the string $ab\#$ where $E_i=\{a,b\}$ for $1 \leqslant i \leqslant m$. Finally, let s be the input string $ue_1 \ldots e_m$.

Now, notice that r matches s if and only if the set of vertices assigned to the variables $v_1, \ldots, v_k$ forms a vertex cover for the set of edges $\{E_1, E_2, \ldots, E_m\}$. Thus, r matches s if and only if the vertex-cover problem has a solution of cardinality at most k.

It is easy to match a rewbr in nondeterministic polynomial time. The most straightforward approach to matching a rewbr pattern r deterministically is to use backtracking to keep track of the possible substrings of the input string s that can be assigned to the variables in r. There are $O(n^2)$ possible substrings that can be assigned to any one variable in r, where n is the length of s. If there are k variables in r, then there are $O(n^{2k})$ possible assignments in all. Once an assignment of substrings to variables is fixed, the problem reduces to ordinary regular expression matching. Thus, rewbr matching can be done in at worst $O(n^{2k})$ time. □

This NP-completeness result implies that if $P \neq NP$, then there is no polynomial-time algorithm for matching rewbrs.

6.2. *Finding repeated patterns and palindromes*

Rewbrs allow one to find repeated patterns in an input string, but at a considerable cost. There are much more efficient techniques for finding certain classes of repeated patterns. Weiner [137] devised an efficient way to construct a compact index to all the distinct substrings of a set of strings in linear time. Variants of this index are called *position trees* [6, 91], *subword trees* [13], *complete inverted files* [28–31], *PATRICIA trees* [103], and *suffix trees* [41, 95, 115]. The position tree is useful for solving in linear time problems such as finding the longest repeated substring of an input string. Apostolico [13] discusses other innovative uses for this index. Chen and Seiferas [34] have a particularly clean construction for the index. Crochemore [42] and Perrin [110] relate the failure function of Section 5.1 to the construction of a minimal suffix automaton.

A related problem, more of mathematical interest, is to find all the squares in a string, where a *square* is a substring of the form xx with x nonempty. For example, the Fibonacci string F_n contains at least $(|F_n| \log |F_n|)/12$ different squares. Shortly after the turn of the century, Thue [128, 129] had asked how long a square-free string could be and showed that with an alphabet of three characters square-free strings of any length could be constructed. Several authors have devised $\Theta(n \log n)$ algorithms for finding all squares in a string of length n [20, 40, 89, 90]. One can determine in linear time whether a string s has a prefix that is a square, that is, whether $s = xxy$ for nonempty x and some y. It is an open problem as to whether the set of strings of the form xxy (*prefixsquares*) can be recognized by a 2DPDA. Rabin [112] gives a simple "fingerprinting" algorithm similar to the one in Section 3.2 for finding in $O(n \log n)$ expected time the earliest repetition in a string s, that is, the shortest w and x such that $s = wxxy$.

Palindromes, strings that read the same forwards and backwards, have provided amusement for centuries. Strings that begin with an even-length palindrome, that is, strings of the form xx^Ry with x nonempty, can be recognized by a 2DPDA, and hence in linear time on a RAM. (x^R is the reversal of x.) Strings of nontrivial palindromes, that is, strings of the form $x_1x_2 \ldots x_n$ (*palstars*), where each x_i is a palindrome of length greater than 1, can be recognized in linear time on a RAM but we do not know whether they can be recognized by a 2DPDA. See [52, 54, 56–58, 92, 121, 123] for more details and for algorithms that give real-time performance for finding palindromes in strings.

6.3. *Approximate string matching*

We now consider several important variants of pattern-matching problems that arise in areas such as file comparison, molecular biology, and speech recognition. Perhaps the simplest is the *file-difference* problem: Given two strings x and y, determine how "close" x is to y. A useful way to compare the two files is to print a minimal sequence of editing changes that will convert the first file into the second. Let us treat the files as two strings of symbols, $x = a_1 \ldots a_m$ and $y = b_1 \ldots b_n$, where a_i represents the ith line of the first file and b_j represents the jth line of the second. Define two editing transformations: the insertion of a line and the deletion of a line. The *edit distance* between the two files is the smallest number of editing transformations required to change the first file into the second. For example, if the first file is represented by the string *abcabba* and the second by *cbabac*, the edit distance is 5.

A simple dynamic programming algorithm to compute this edit distance has been discovered independently by many authors [118]. Let d_{ij} be the edit distance between the prefix of length i of the first string and the prefix of length j of the second string. Let d_{00} be 0, d_{i0} be i for $1 \leqslant m$, and d_{0j} be j for $1 \leqslant j \leqslant n$. Then, for $1 \leqslant i \leqslant m$ and $1 \leqslant j \leqslant n$, compute d_{ij} by taking the minimum of the three quantities:

$$(1)\ d_{i-1,j}+1, \qquad (2)\ d_{i,j-1}+1, \qquad (3)\ d_{i-1,j-1} \text{ if } a_i=b_j$$

The first quantity represents the deletion of the jth character from the first string, the second represents the insertion of a character after the $(j-1)$st position in the first string, and the third says $d_{ij} \leqslant d_{i-1,j-1}$ if the ith character in the first string agrees with the jth character in the second string. After completing this computation, we can easily show that d_{mn} gives the minimum number of edit changes required to transform the first file into the second. From these distance calculations we can construct the corresponding sequence of editing transformations. Figure 21 shows the matrix of edit distances for the strings *abcabba* and *cbabac*.

c	6	6	5	4	3	4	5	6	5
a	5	5	4	3	4	3	4	5	4
b	4	4	3	2	3	4	3	4	5
a	3	3	2	3	4	3	4	5	4
b	2	2	3	2	3	4	3	4	5
c	1	1	2	3	2	3	4	5	6
	0	0	1	2	3	4	5	6	7
		0	1	2	3	4	5	6	7
			a	b	c	a	b	b	a

Fig. 21. Edit distances d_{ij}.

The running time of this algorithm is always proportional to mn, the product of the sizes of the two files. In one sense, this algorithm minimizes the number of character comparisons: with equal-unequal comparisons, $\Omega(mn)$ character comparisons are necessary in the worst case [5].

Hunt and McIlroy [74] implemented Hirschberg's linear-space version of this algorithm [66] which worked very well for short inputs, but when it was applied to longer and longer files its performance became significantly slower. This behavior, of course, is entirely consistent with the quadratic nature of this algorithm's time complexity.

Shortly thereafter, Hunt and Szymanski [75] proposed another approach to the file-difference problem. They suggested extracting a longest common subsequence from the two strings and producing the editing changes from the subsequence.

DEFINITION. A *subsequence* of a string x is a sequence of characters obtained by deleting zero or more characters from x. A *common subsequence* of two strings x and y is a string that is a subsequence of both x and y, and a *longest common subsequence* is a common subsequence that is of greatest length. For example, *baba* and *cbba* are both longest common subsequences of *abcabba* and *cbabac*.

The problem of finding a longest common subsequence of two strings is closely related to the file-difference problem. The following formula shows the relationship between the edit distance and the length of a longest common subsequence of two strings of symbols, $x = a_1 \ldots a_m$ and $y = b_1 \ldots b_n$:

$$EditDistance(x, y) = m + n - 2length(lcs(x, y)).$$

To solve the longest-common-subsequence problem for x and y, Hunt and Szymanski constructed a matrix M, where M_{ij} is 1 if $a_i = b_j$ and 0 otherwise. They produced a longest common subsequence for x and y by drawing a longest strictly monotonically increasing line through the points of the M matrix as shown in Fig. 22. If there are r points, such a line can be found in time proportional to $(r+n)\log n$, assuming $n \geqslant m$. In the file-difference problem, r is usually on the order of n, so in practice the running time of the Hunt–Szymanski technique is $O(n \log n)$, significantly faster than the dynamic programming solution.

McIlroy put this new algorithm into his program and the inputs that used to take

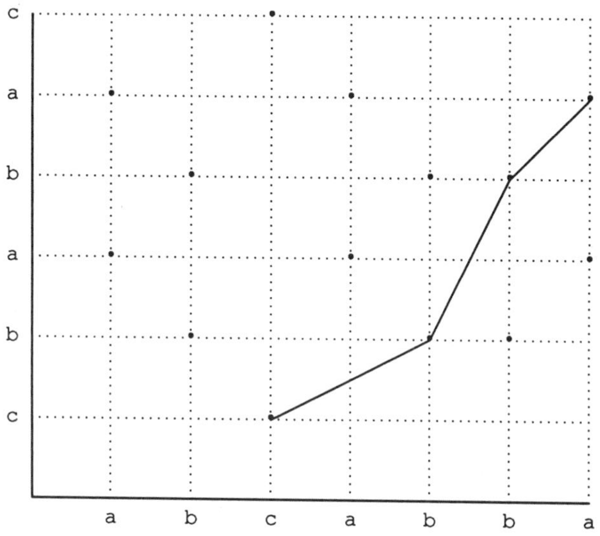

Fig. 22. A longest monotonically increasing line represents a longest common subsequence.

many minutes to compare now ran in a few seconds. The functionality of the program had not changed but its average performance had improved noticeably.

More recently, Myers [104] and Ukkonen [131] independently suggested a third algorithm for this problem: treat the problem as one of constructing a cheapest-cost path in a graph. The nodes of the graph consist of the intersection points on an $(m+1)\times(n+1)$ grid. In addition to the horizontal and vertical edges, there is a diagonal edge from node $(i-1, j-1)$ to (i, j) if $a_i = b_j$. Each horizontal and vertical edge has a cost of 1, and each diagonal edge a cost of 0. A cheapest path from the origin to node (m, n) uses as many diagonal edges as possible as illustrated in Fig. 23. The tails of the diagonal edges on the cheapest path define a longest common subsequence between the two strings. Dijkstra's algorithm [6] can be used to construct the cheapest path in time proportional to dn, where d is the edit distance between the two strings. In situations where d is small, this approach is the method of choice.

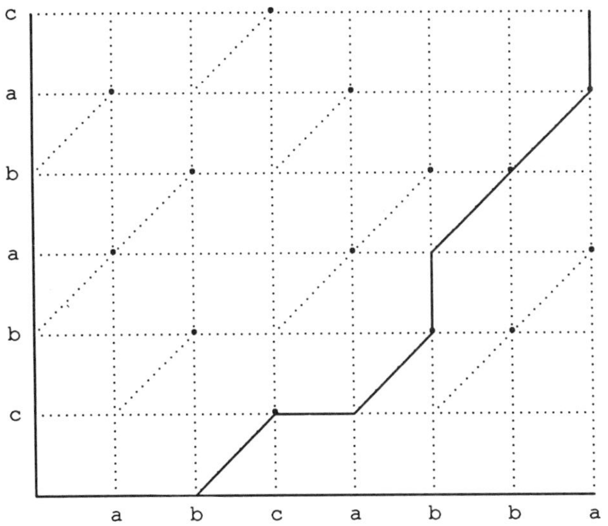

Fig. 23. A cheapest path represents a longest common subsequence.

Myers [103] describes a refinement of the algorithm above that runs in $O(n \log n + d^2)$ time using suffix trees. Masek and Paterson [93] present an $O(n^2/\log n)$ time algorithm using the "Four Russians" trick. So far, no single algorithm is known for the longest-common-subsequence problem that dominates all applications. More algorithms, programs, and other aspects of the problem are discussed in [10, 14, 15, 19, 36, 50, 64, 66, 67, 68, 101, 107, 130, 138].

Many generalizations of the file-difference problem arise in molecular biology, speech recognition, and related applications. We can define the *approximate string-matching* problem as follows: Given a pattern p, an input string s, an integer k, and a distance metric d, find all substrings x of s such that $d(p,x) \leqslant k$, that is, the distance between the pattern p and the substring x is at most k.

When p is a single string, several distance measures have been proposed. The two most common are the Hamming distance and Levenshtein distance. The *Hamming distance* between two strings of equal length is the number of positions with mismatching characters. For example, the Hamming distance between *abcabb* and *cbacba* is 4. The approximate string-matching problem with d being Hamming distance is called *string matching with k mismatches.* Algorithms for the k-mismatches problem are presented in [55, 76, 82].

The *Levenshtein distance* between two strings of not necessarily equal length is the minimum number of character changes, insertions, and deletions required to transform one string into the other. For example, the Levenshtein distance between *abcabba* and *cbabac* is 4. Algorithms for computing the Levenshtein distance between a pair of strings are contained in [91, 118, 131, 134]. The approximate string-matching problem with d being the Levenshtein distance is called *string matching with k differences.* Algorithms for the k-differences problem are described in [81, 83, 118, 132, 135].

Galil and Giancarlo [55] present an excellent survey of algorithms and data structures that have been devised to solve approximate string-matching problems efficiently. More general distance measures are considered in [1, 47, 87, 100]. The approximate string-matching problem where p is a regular expression has been considered by Wagner [133], Wagner and Seiferas [135], and Myers and Miller [106].

6.4. *String matching with "don't cares"*

We conclude by mentioning a variant of the string-matching problem in which there is a "don't care" symbol that matches any single character (including a don't care symbol). We say two strings $a_1 \ldots a_m$ and $b_1 \ldots b_m$ match if for $1 \leqslant i \leqslant m$ whenever a_i and b_i are not both don't cares, they are equal. The pattern-matching problem with don't cares comes in two flavors:

(1) only the pattern p contains don't cares, and

(2) both the pattern p and the input string s contain don't cares.

Pinter [111] uses an extension of the Aho–Corasick algorithm to solve the first problem. Fischer and Paterson [49] have shown that the second problem reduces to integer multiplication. If $M(m,n)$ is the amount of time needed to multiply an m-bit number by an n-bit number, then the time needed to solve the second version of the pattern-matching problem with don't cares is

$$O(M(|p|,|s|)\log|p|\log A)$$

where A is the size of the input alphabet. Using the Schönhage–Strassen integer-multiplication algorithm [6], the time bound becomes

$$O(|s|\log^2|p|\log\log|p|\log A).$$

7. Concluding remarks

In this chapter we have discussed algorithms for solving string-matching problems that have proven useful for text-editing and text-processing applications. As the reader can see, even in this restricted domain there is a rich literature with interesting ideas and deep analyses.

We should mention a few important generalizations of string pattern matching that we have not had the space to consider. There is an abundant literature on term-rewriting systems [72], and on efficient parsing methods for context-free grammars [70] and special cases of context-free grammars, such as the LL(k) and LR(k) grammars [6]. There are also methods for matching patterns in trees, graphs, and higher-dimensional structures [6, 27, 69, 77, 117].

In this paper, we have restricted ourselves to algorithms for the RAM of computation. In recent years, there has been accelerating interest in developing algorithms for various parallel models of computation, but practical experience with these algorithms has been much more limited than with those for the RAM. In the coming years, we hope to see for these new models of computation the same interplay between theory and practice that has enriched the field of pattern-matching algorithms for the random-access machine.

Acknowledgment

The author is grateful to David Lee for helpful comments on the manuscript.

References

[1] ABRAHAMSON, K., Generalized string matching, *SIAM J. Comput.* **16**(6) (1987) 1039–1051.

[2] AHO, A.V., Pattern matching in strings, in: R. Book, ed., *Formal Language Theory, Perspectives and Open Problems* (Academic Press, New York, 1980) 325–347.

[3] AHO, A.V., and M.J. CORASICK, Efficient string matching: an aid to bibliographic search, *Comm. ACM* **18**(6) (1975) 333–340.

[4] AHO, A.V., M. GANAPATHI and S.W.K. TJIANG, Code generation using tree matching and dynamic programming, *ACM Trans. Programming Languages and Systems* **11**(4) (1989).

[5] AHO, A.V., D.S. HIRSCHBERG and J.D. ULLMAN, Bounds on the complexity of the maximal common subsequence problem, *J. ACM* **23**(1) (1976) 1–12.

[6] AHO, A.V., J.E. HOPCROFT and J.D. ULLMAN, *The Design and Analysis of Computer Algorithms* (Addison-Wesley, Reading, MA, 1974).

[7] AHO, A.V., B.W. KERNIGHAN and P.J. WEINBERGER, *The AWK Programming Language* (Addison-Wesley, Reading, MA, 1988).

[8] AHO, A.V. and D. LEE, Storing a dynamic sparse table, in: *Proc. 27th IEEE Symp. on Foundations of Computer Science* (1986) 56–60.

[9] AHO, A.V., R. SETHI, and J.D. ULLMAN, *Compilers: Principles, Techniques, and Tools* (Addison-Wesley, Reading, MA, 1986).

[10] ALLISON, L. and T.I. DIX, A bit-string longest-common-subsequence algorithm, *Inform. Process. Lett.* **23** (1986) 305–310.

[11] ANGLUIN, D., Finding patterns common to a set of strings, in: *Proc. 11th Ann. ACM Symp. on Theory of Computing* (1979) 130–141.

[12] AOE, J., Y. YAMAMOTO and R. SHIMADA, A method for improving string pattern matching machines, *IEEE Trans. Software Engrg.* **10**(1) (1984) 116–120.

[13] APOSTOLICO, A., The myriad virtues of subword trees, in: A. Apostolico and Z. Galil, eds., *Combinatorial Algorithms on Words* (Springer, Berlin, 1985).

[14] APOSTOLICO, A., Improving the worst-case performance of the Hunt–Szymanski strategy for the longest common subsequence of two strings, *Inform. Process. Lett.* **23** (1986) 63–69.

[15] APOSTOLICO, A., Remark on the Hsu–Du new algorithm for the longest common subsequence problem, *Inform. Process. Lett.* **25** (1987) 235–236.

[16] APOSTOLICO, A. and Z. GALIL, eds., *Combinatorial Algorithms on Words* (Springer, Berlin, 1985).

[17] APOSTOLICO, A. and R. GIANCARLO, Pattern matching machine implementation of a fast test for unique decipherability, *Inform. Process. Lett.* **18** (1984) 155–158.

[18] APOSTOLICO, A. and R. GIANCARLO, The Boyer–Moore–Galil string searching strategies revisited, *SIAM J. Comput.* **15**(1) (1986) 98–105.

[19] APOSTOLICO, A. and G. GUERRA, The longest common subsequence problem revisited, *Algorithmica* **2** (1987) 315–336.

[20] APOSTOLICO, A. and F.P. PREPARATA, Optimal off-line detection of repetitions in a string, *Theoret. Comput. Sci.* **22** (1983) 297–315.

[21] BAEZA-YATES, R.A., Improved string searching, *Software—Practice and Experience* **19**(3) (1984) 257–271.

[22] BARTH, G., An alternative for the implementation of Knuth–Morris–Pratt algorithm, *Inform. Process. Lett.* **13** (1981) 134–137.

[23] BARTH, G., Relating the average-case costs of the brute-force and the Knuth–Morris–Pratt string matching algorithm, in: A. Apostolico and Z. Galil, eds., *Combinatorial Algorithms on Words* (Springer, Berlin, 1985).

[24] BEAN, D.R., A. EHRENFEUCHT and G.F. MCNULTY, Avoidable patterns in strings of symbols, *Pacific J. Math.* **85** (1985) 31–56.

[25] BEN-YEHUDA, S. and R.Y. PINTER, Symbolic layout improvement using string matching based local transformations, in: *Proc. Decennial Caltech Conf. on VLSI* (MIT Press, Cambridge, MA, 1989) 227–239.

[26] BERRY, G. and R. SETHI, From regular expressions to deterministic automata, *Theoret. Comput. Sci.* **48**(1) (1986) 117–126.

[27] BIRD, R.S., Two dimensional pattern matching, *Inform. Process. Lett.* **6**(5) (1977) 168–170.

[28] BLUMER, A., J. BLUMER, A. EHRENFEUCHT, D. HAUSSLER, M.T. CHEN and J. SEIFERAS, The smallest automaton recognizing the subwords of a text, *Theoret. Comput. Sci.* **40**(1) (1985) 31–56.

[29] BLUMER, A., J. BLUMER, A. EHRENFEUCHT, D. HAUSSLER and R. MCCONNELL, Building a complete inverted file for a set of text files in linear time, in: *Proc. 16th ACM Symp. on Theory of Computing* (1984) 349–358.

[30] BLUMER, A., J. BLUMER, A. EHRENFEUCHT, D. HAUSSLER and R. MCCONNELL, Building the minimal DFA for the set of all subwords of a word on-line in linear time in: J. Paredaens, ed., *Proc. 11th Internat. Coll. on Automata, Languages and Programming*, Lecture Notes in Computer Science, Vol. 172 (Springer, Berlin, 1984) 109–118.

[31] BLUMER, A., J. BLUMER, D. HAUSSLER, R. MCCONNELL and A. EHRENFEUCHT, Complete inverted files for efficient text retrieval and analysis, *J. ACM* **34**(3) (1987) 578–595.

[32] BOYER, R.S. and J.S. MOORE, A fast string searching algorithm, *Comm. ACM* **20**(10) (1977) 62–72.

[33] BRZOZOWSKI, J.A., Derivatives of regular expressions, *J. ACM* **11**(4) (1964) 481–494.

[34] CHEN, M.T. and J.I. SEIFERAS, Efficient and elegant subword tree construction, in: A. Apostolico and Z. Galil, eds., *Combinatorial Algorithms on Words* (Springer, Berlin, 1985).

[35] CHERRY, L.L., Writing tools, *IEEE Trans. Comm.* **30**(1) (1982) 100–105.

[36] CHVATAL, V. and D. SANKOFF, Longest common subsequence of two random sequences, *J. Appl. Probab.* **12** (1975) 306–315.

[37] COMMENTZ-WALTER, B., A string matching algorithm fast on the average, in: H.A. Maurer, ed., *Proc. 6th Internat. Coll. on Automata, Languages and Programming* (Springer, Berlin, 1979) 118–132.

[38] COOK, S.A., Linear time simulation of deterministic two-way pushdown automata, in: *Proc. IFIP Congress, 71, TA-2* (North-Holland, Amsterdam, 1971) 172–179.

[39] CHROBAK, M. and W. RYTTER, Remarks on string-matching and one-way multihead automata, *Inform. Process. Lett.* **24** (1987) 325–329.

[40] CROCHEMORE, M., An optimal algorithm for computing the repetitions in a word, *Inform. Process. Lett.* **12**(5) (1981) 244–250.

[41] CROCHEMORE, M., Transducers and repetitions, *Theoret. Comput. Sci.* **45** (1986) 63–86.

[42] CROCHEMORE, M., String matching with constraints, Tech. Report 88-5, Dept. de Mathématique et Informatique, Univ. de Paris-Nord, 1988.

[43] CROCHEMORE, M. and D. PERRIN, Two-way pattern matching, Tech. Report, Univ. de Paris, 1989.

[44] DAVIES, G. and S. BOWSHER, Algorithms for pattern matching, *Software—Practice and Experience* **16**(6) (1986) 575–601.

[45] DIETZFELBINGER, M., A. KARLIN, K. MEHLHORN, F. MEYER AUF DER HEIDE, H. ROHNERT and R.E. TARJAN, Dynamic perfect hashing: upper and lower bounds, in: *Proc. 29th Ann. IEEE Symp. on Foundations of Computer Science* (1988) 524–531.

[46] DUVAL, J.P., Factorizing words over an ordered alphabet, *J. Algorithms* **4** (1983) 363–381.

[47] EILAN-TZOREFF, T. and U. VISHKIN, Matching patterns in strings subject to multi-linear transformations, *Theoret. Comput. Sci.* **60**(3) (1988) 231–254.

[48] FARBER, D.J., R.E. GRISWOLD and I.P. POLONSKY, SNOBOL, a string manipulation language, *J. ACM* **11**(1) (1964) 21–30.

[49] FISCHER, M.J. and M.S. PATERSON, String matching and other products, in: R.M. Karp, ed., *Complexity of Computation*, SIAM–AMS Proceedings, Vol. 7 (Amer. Mathematical Soc., Providence, RI, 1974) 113–125.

[50] FREDMAN, M.L., On computing the length of longest increasing subsequences, *Discrete Math.* **11**(1) (1975) 29–35.

[51] FREDMAN, M.L., J. KOMLOS and E. SZEMEREDI, Storing a sparse table with O(1) worst case access time, *J. ACM* **31**(3) (1984) 538–544.

[52] GALIL, Z., Two fast simulations which imply some fast string matching and palindrome recognition algorithms, *Inform. Process. Lett.* **4** (1976) 85–87.

[53] GALIL, Z., On improving the worst case running time of the Boyer–Moore string matching algorithm, *Comm. ACM* **22**(9) (1979) 505–508.

[54] GALIL, Z., String matching in real time, *J. ACM* **28**(1) (1981) 134–149.

[55] GALIL, Z. and R. GIANCARLO, Data structures and algorithms for approximate string matching, *J. Complexity* **4**(1) (1988) 33–72.

[56] GALIL, Z. and J.I. SEIFERAS, A linear-time on-line recognition algorithm for 'Palstar', *J. ACM* **25** (1978) 102–111.

[57] GALIL, Z. and J.I. SEIFERAS, Saving space in fast string matching, *SIAM J. Comput.* **9**(2) (1980) 417–438.

[58] GALIL, Z. and J. SEIFERAS, Time-space-optimal string matching, *J. Comput. System Sci.* **26** (1983) 280–294.

[59] GAREY, M.R. and D.S. JOHNSON, *Computers and Intractability: A Guide to the Theory of NP-Completeness* (Freeman, San Francisco, CA, 1979).

[60] GRISWOLD, R.E., J.F. POAGE and I.P. POLONSKY, *The SNOBOL4 Programming Language* (Prentice Hall, Englewood Cliffs, NJ, 2nd ed., 1971).

[61] GUIBAS, L.J. and A.M. ODLYZKO, A new proof of the linearity of the Boyer–Moore string searching algorithm, *SIAM J. Comput.* **9**(4) (1980) 672–682.

[62] GUIBAS, L.J. and A.M. ODLYZKO, Periods in strings, *J. Combin. Theory Ser. A* **30** (1981) 19–42.

[63] GUIBAS, L.J. and A.M. ODLYZKO, String overlaps, pattern matching, and nontransitive games, *J. Combin. Theory Ser. A* **30**(2) (1981) 183–208.

[64] HALL, P.A.V. and G.R. DOWLING, Approximate string matching, *ACM Comput. Surveys* **12**(4) (1980) 381–402.

[65] HARTMANIS, J. On the succinctness of different representations of languages, *SIAM J. Comput.* **9**(1) (1980) 114–120.

[66] HIRSCHBERG, D.S., A linear space algorithm for computing maximal common subsequences, *Comm. ACM* **18**(6) (1975) 341–343.

[67] HIRSCHBERG, D.S., Algorithms for the longest common subsequence problem, *J. ACM* **24**(4) (1977) 664–675.

[68] HIRSCHBERG, D.S., An information-theoretic lower bound for the longest common subsequence problem, *Inform. Process. Lett.* **7** (1978) 40–41.

[69] HOFFMAN, C.W. and M.J. O'DONNELL, Pattern matching in trees, *J. ACM* **29**(1) (1982) 68–95.

[70] HOPCROFT, J.E. and J.D. ULLMAN, *Introduction to Automata Theory, Languages, and Computation* (Addison-Wesley, Reading, MA, 1979).

[71] HORSPOOL, R.N., Practical fast searching in strings, *Software—Practice and Experience* **10**(6) (1980) 501–506.

[72] HUET, G. and D.C. OPPEN, Equations and rewrite rules, in: R. Book, ed., *Formal Language Theory, Perspectives and Open Problems* (Academic Press, New York, 1980) 349–393.

[73] HUME, A.G., A tale of two greps, *Software—Practice and Experience* **18**(11) (1988) 1063–1072.

[74] HUNT, J.W. and M.D. MCILROY, An algorithm for differential file comparison, Computing Science Tech. Report 41, AT&T Bell Laboratories, Murray Hill, NJ, 1976.

[75] HUNT, J.W. and T.G. SZYMANSKI, A fast algorithm for computing longest common subsequences, *Comm. ACM* **20**(5) (1977) 350–353.

[76] IVANOV, A.G., Recognition of an approximate occurrence of words on a Turing machine in real time, *Math. USSR-Izv.* **24**(3) (1985) 479–522.

[77] KARP, R.M., R.E. MILLER, and A.L. ROSENBERG, Rapid identification of repeated patterns in strings, trees, and arrays, in: *Proc. 4th ACM Symp. on Theory of Computing* (1972) 125–136.

[78] KARP, R.M. and M.O. RABIN, Efficient randomized pattern-matching algorithms, *IBM J. Res. Develop.* **31**(2) (1987) 249–260.

[79] KLEENE, S.C., Representation of events in nerve nets and finite automata, in: C.E. Shannon and J. McCarthy, eds., *Automata Studies* (Princeton Univ. Press, Princeton, NJ, 1956) 3–42.

[80] KNUTH, D.E., J.H. MORRIS and V.R. PRATT, Fast pattern matching in strings, *SIAM J. Comput.* **6**(2) (1977) 323–350.

[81] LANDAU, G.M. and U. VISHKIN, Introducing efficient parallelism into approximate string matching and a new serial algorithm, in: *Proc. 18th ACM Symp. on Theory of Computing* (1986) 220–230.

[82] LANDAU, G.M. and U. VISHKIN, Efficient string matching with k mismatches, *Theoret. Comput. Sci.* **43** (1986) 239–249.

[83] LANDAU, G.M. and U. VISHKIN, Fast string matching with k differences, *J. Comput. System Sci.* **37**(1) (1988) 63–78.

[84] LESK, M.E., LEX—a lexical analyzer generator, Computing Science Tech. Report 39, Bell Laboratories, Murray Hill, NJ, 1975.

[85] LI, M. and Y. YESHA, String-matching cannot be done by a two-head one-way deterministic finite automata, *Inform. Process. Lett.* **22** (1986) 231–236.

[86] LIU, K.-C., On string pattern matching: a new model with a polynomial time algorithm, *SIAM J. Comput.* **10**(1) (1981) 118–140.

[87] LOWRANCE, R. and R.A. WAGNER, An extension of the string-to-string correction problem, *J. ACM* **22** (1975) 177–183.

[88] LYNDON, R.C. and M.P. SCHÜTZENBERGER, The equation $a^M = b^N c^P$ in a free group, *Michigan Math. J.* **9** (1962) 289–298.

[89] MAIN, M.G. and R.J. LORENTZ, An $O(n \log n)$ algorithm for finding all repetitions in a string, *J. Algorithms* **5**(3) (1984) 422–432.

[90] MAIN, M.G. and R.J. LORENTZ, Linear time recognition of square-free strings, in: A. Apostolico and Z. Galil, eds., *Combinatorial Algorithms on Words* (Springer, Berlin, 1985).

[91] MAJSTER, M.E. and A. REISNER, Efficient on-line construction and correction of position trees, *SIAM J. Comput.* **9**(4) (1980) 785–807.

[92] MANACHER, G., A new linear-time on-line algorithm for finding the smallest initial palindrome of a string, *J. ACM* **22** (1975) 346–351.

[93] MASEK, W.J. and M.S. PATERSON, A faster algorithm for computing string-edit distances, *J. Comput. System Sci.* **20**(1) (1980) 18–31.

[94] McCULLOCH, W.S. and W. PITTS, A logical calculus of the ideas immanent in nervous activity, *Bull. Math. Biophysics* **5** (1943) 115–133.

[95] McCREIGHT, E.M., A space-economical suffix tree construction algorithm, *J. ACM* **23**(2) (1976) 262–272.

[96] McILROY, M.D., ed., *UNIX Time-Sharing System Programmer's Manual, Vol. I* (AT&T Bell Laboratories, Murray Hill, NJ, 9th ed., 1986).

[97] McNAUGHTON, R. and H. YAMADA, Regular expressions and state graphs for automata, *IRE Trans. Electron. Comput.* **9**(1) (1960) 39–47.

[98] MEHLHORN, K., *Data Structures and Algorithms 1: Sorting and Searching* (Springer, Berlin, 1984).

[99] MEYER, A.R., and M.J. FISCHER, Economy of description by automata, grammars, and formal systems, in: *Proc. IEEE Symp. on Switching and Automata Theory* (1971) 188–190.

[100] MILLER, W., *A Software Tools Sampler* (Prentice Hall, Englewood Cliffs, NJ, 1987).

[101] MILLER, W., and E.W. MYERS, A file comparison program, *Software—Practice and Experience* **15**(11) (1985) 1025–1040.

[102] MILLER, W., and E.W. MYERS, Sequence comparison with concave weighting functions, *Bull. Math. Biol.* **50**(2) (1988) 97–120.

[103] MORRISON, D.R., PATRICIA—Practical algorithm to retrieve information coded in alphanumeric, *J. ACM* **15**(4) (1968) 514–534.

[104] MYERS, E.W., An O(*ND*) difference algorithm and its variations, *Algorithmica* **1** (1986) 251–266.

[105] MYERS, E.W., A four Russians algorithm for regular expression pattern matching, Tech. Report 88-34, Dept. of Computer Science, Univ. of Arizona, Tucson, AZ, 1988.

[106] MYERS, E.W. and W. MILLER, Approximate matching of regular expressions, *Bull. Math. Biol.* **51**(1) (1989) 5–37.

[107] NAKATSU, N., Y. KAMBAYASHI and S. YAJIMA, A longest common subsequence algorithm suitable for similar text strings, *Acta Inform.* **18** (1982) 171–179.

[108] NEEDLEMAN, S.B. and C.D. WUNSCH, A general method applicable to the search for similarities in the amino acid sequences of two proteins, *J. Molecular Biol.* **48** (1970) 443–453.

[109] PENNELLO, T.J., Very fast LR parsing, *SIGPLAN Notices* **21**(7) (1986) 145–150.

[110] PERRIN, D., Finite automata, in: J. van Leeuwen, ed., *Handbook of Theoretical Computer Science, Vol. B* (North-Holland, Amsterdam, 1990) 1–57.

[111] PINTER, R.Y., Efficient string matching with don't-care patterns, in: A. Apostolico and Z. Galil, eds., *Combinatorial Algorithms on Words* (Springer, Berlin, 1985).

[112] RABIN, M.-O., Discovering repetitions in strings, in: A. Apostolico and Z. Galil, eds., *Combinatorial Algorithms on Words* (Springer, Berlin, 1985).

[113] RABIN, M.O. and D. SCOTT, Finite automata and their decision problems, *IBM J. Res. Develop.* **3**(2) (1959) 114–125.

[114] RIVEST, R.L., On the worst-case behavior of string-searching algorithms, *SIAM J. Comput.* **6**(4) (1977) 669–674.

[115] RODEH, M., V.R. PRATT and S. EVEN, Linear algorithms for data compression via string matching, *J. ACM* **28**(1) (1981) 16–24.

[116] RYTTER, W., A correct preprocessing algorithm for Boyer–Moore string searching, *SIAM J. Comput.* **9**(3) (1980) 509–512.

[117] SALOMAA, K., Deterministic tree pushdown automata and monadic tree rewriting systems, *J. Comput. System Sci.* **37**(3) (1988) 367–394.

[118] SANKOFF, D. and J.B. KRUSKAL, *Time Warps, String Edits, and Macromolecules: The Theory and Practice of Sequence Comparison* (Addison-Wesley, Reading, MA, 1983).

[119] SCHABACK, R., On the expected sublinearity of the Boyer–Moore algorithm, *SIAM J. Comput.* **17**(4) (1988) 648–658.

[120] SEDGEWICK, R., *Algorithms* (Addison-Wesley, Reading, MA, 1983).

[121] SEIFERAS, J. and Z. GALIL, Real-time recognition of substring repetition and reversal, *Math. Systems Theory* **11** (1977) 111–146.

[122] SELLERS, P.H., The theory and computation of evolutionary distances: pattern recognition, *J. Algorithms* **1** (1980) 359–373.

[123] SLISENKO, A.O., Detection of periodicities and string-matching in real time *J. Soviet Math.* **22**(3) (1983) 1316–1386 (originally published in 1980).

[124] SMIT, G. DE V., A comparison of three string matching algorithms, *Software—Practice and Experience* **12** (1982) 57–66.

[125] TAKAOKA, T., An on-line pattern matching algorithm, *Inform. Process. Lett.* **22** (1986) 329–330.

[126] TARJAN, R.E. and A.C. YAO, Storing a sparse table, *Comm. ACM* **22**(11) (1979) 606–611.

[127] THOMPSON, K., Regular expression search algorithm, *Comm. ACM* **11**(6) (1968) 419–422.

[128] THUE, A., Über unendliche Zeichenreihen, *Norske Videnskabers Selskabs Skrifter Mat.-Nat. Kl.* (*Kristiania*) **1** (1906) 1–22.

[129] THUE, A., Über die gegenseitige Lage gleicher Teile gewisser Zeichenreihen, *Norske Videnskabers Selskabs Skrifter Mat.-Nat. Kl.* (*Kristiania*) **7** (1912) 1–67.

[130] TICHY, W., The string-to-string correction problem with block moves, *ACM Trans. Comput. Systems* **2** (1984) 309–321.

[131] UKKONEN, E., Algorithms for approximate string matching, *Inform. and Control* **64** (1985) 100–118.

[132] UKKONEN, E., Finding approximate patterns in strings, *J. Algorithms* **6** (1985) 132–137.

[133] WAGNER, R., Order-n correction of regular languages, *Comm. ACM* **11**(6) (1974) 265–268.

[134] WAGNER, R. and M. FISCHER, The string-to-string correction problem, *J. ACM* **21**(1) (1974) 168–178.

[135] WAGNER, R. and J.I. SEIFERAS, Correcting counter-automaton-recognizable languages, *SIAM J. Comput.* **7**(3) (1978) 357–375.

[136] WATERMAN, M.S., General methods for sequence comparison, *Bull. Math. Biol.* **46**(4) (1984) 473–500.

[137] WEINER, P., Linear pattern matching algorithms, in: *Proc. 14th IEEE Symp. on Switching and Automata Theory* (1973) 1–11.

[138] WONG, C.K. and A.K. CHANDRA, Bounds for the string editing problem, *J. ACM* **23**(1) (1976) 13–16.

[139] YAO, A.C., The complexity of pattern matching for a random string, *SIAM J. Comput.* **8** (1979) 368–387.

CHAPTER 6

Data Structures

K. MEHLHORN

Fachbereich 14, Informatik, Universität des Saarlandes, D-6600 Saarbrücken, FRG

A. TSAKALIDIS

Department of Computer Engineering and Informatics, University of Patras, 26500 Patras, Greece, and Computer Technology Institute, P.O. Box 1122, 26110 Patras, Greece

Contents

HANDBOOK OF THEORETICAL COMPUTER SCIENCE
Edited by J. van Leeuwen

1. Introduction

Data structuring is the study of *concrete implementations* of frequently occurring *abstract data types*. An abstract data type is a set together with a collection of operations on the elements of the set. We give two examples for the sake of concreteness.

In the data type *dictionary* the set is the powerset of a universe U, and the operations are *insertion* and *deletion* of elements and the *test of membership*. A typical application of dictionaries are symbol tables in compilers. In the data type *priority queues* the set is the powerset of an ordered universe U and the operations are insertion of elements and finding and deleting the minimal element of a set. Priority queues arise frequently in network optimization problems, see, e.g. Chapter 10 on graph algorithms (this Handbook).

This chapter is organized as follows. In Sections 2 and 3 we treat unweighted and weighted dictionaries respectively. The data structures discussed in these sections are ephemeral in the sense that a change to the structure destroys the old version, leaving only the new version available for use. In contrast, persistent structures, which are covered in Section 4, allow access to any version, old or new, at any time. In Section 5 we deal with the *Union-Split-Find* problem and in Section 6 with priority queues. Section 7 is devoted to operations on trees and Sections 8 and 9 cover selection and merging. Finally, Section 10 is devoted to dynamization techniques. Of course, we can only give an overview; we refer the reader to the textbooks of Knuth [112], Aho, Hopcroft, Ullman [2, 4], Wirth [219], Horowitz and Sahni [98], Standish [187], Tarjan [195], Gonnet [83], Mehlhorn [145], and Sedgewick [181] for a more detailed treatment. In this section we build the basis for the latter sections. In particular, we review the types of complexity analysis and the machine models used in the data structure area.

The machine models used in this chapter are the pointer machine (pm-machine) [112, 179, 193] and the random-access machine (RAM-machine) [2]. A thorough discussion of both models can be found in Chapter 1 of this Handbook. In a pointer machine memory consists of a collection of *records*. Each record consists of a fixed number of *cells*. The cells have associated *types*, such as *pointer*, *integer*, *real*, and access to memory is only possible by "pointers". In other words, the memory is structured as a directed graph with bounded out-degree. The edges of this graph can be changed during execution. Pointer machines correspond roughly to high-level programming languages without arrays. In contrast, the memory of a RAM consists of an array of cells. A cell is accessed through its address and hence address arithmetic is available. We assume in both models that storage cells can hold arbitrary numbers and that the basic arithmetic and pointer operations take constant time. This is called the uniform cost assumption. All our machines are assumed to be deterministic except when explicitly stated otherwise.

Three types of complexity analysis are customary in the data structure area: worst-case analysis, average-case analysis and amortized analysis. In a *worst-case analysis* one derives worst-case time bounds for each single operation. This is the most frequent type of analysis. A typical result is that dictionaries can be realized with

$O(\log n)$ worst-case cost for the operations *Insert*, *Delete*, and *Access*; here n is the current size of the set manipulated.

In an *average-case analysis* one postulates a probability distribution on the operations of the abstract data type and computes the expected cost of the operations under this probability assumption. A typical result is that static dictionaries (only operation *Access*) can be realized with $O(H(\beta_1, \ldots, \beta_n))$ expected cost per access operation; here n is the size of the dictionary, β_i, $1 \leqslant i \leqslant n$, the probability of access to the ith element and $H(\beta_1, \ldots, \beta_n)$ is the entropy of the probability distribution. (This type of analysis is extensively discussed in Chapter 9 on analysis of algorithms and data structures.)

In an *amortized analysis* we study the worst-case cost of a sequence of operations. Since amortized analysis is not very frequently used outside the data structure area, we discuss it in more detail. An amortized analysis is relevant whenever the cost of an operation can fluctuate widely and only averaging the costs of a sequence of operations over the sequence leads to good results. The technical instrument used to do the averaging is an account (banker's view of amortization) or potential (physicist's view of amortization) associated with the data structure.

In the banker's view of amortization, which was implicitly used by Brown and Tarjan [33] and then more fully developed by Huddleston and Mehlhorn [101], we view a computer as coin- or token-operated. Each token pays for a constant amount of computer time. Whenever a user of a data structure calls a certain operation of the data type, we charge him a certain amount of tokens; we call this amount the amortized cost of the operation. The amortized cost does in general not agree with the actual cost of the operation (that is the whole point of amortization). The difference between the amortized and the actual cost is either deposited into or withdrawn from the account associated with the data structure. Clearly, if we start with no tokens initially, then the total actual cost of a sequence of operations is just the total amortized cost minus the final balance of the account.

More formally, let *bal* be a function which maps the possible configurations of the data structure into the real numbers. For an operation *op* which transforms a configuration D into a configuration D', define the amortized cost [184] by

$$\mathit{amortized\_cost}(op) = \mathit{actual\_cost}(op) + \mathit{bal}(D') - \mathit{bal}(D).$$

Then for a sequence $op_1, \ldots, op_m$ of operations we have

$$\sum_i \mathit{amortized\_cost}(op_i) = \sum_i \mathit{actual\_cost} + \mathit{bal}(D_m) - \mathit{bal}(D_0)$$

where D_0 is the initial data structure and D_m is the final data structure. In particular, if $\mathit{bal}(D_0) = 0$ and $\mathit{bal}(D) \geqslant 0$ for all data structures D, i.e., we never borrow tokens from the bank, the actual cost of the sequence of operations is bounded by its amortized cost. In the physicist's view of amortization which was developed by Sleator and Tarjan [184], we write $\mathit{pot}(D)$ instead of $\mathit{bal}(D)$, call it the potential of the data structure and view deposits and withdrawals as increases and decreases of potential. Of course, the two viewpoints are equivalent. A more thorough discussion of amortized analysis can be found in [198] and [145, Vol. I, Section I.6.1.1].

Let us illustrate these concepts by an example, the data type *counter*. A counter takes nonnegative integer values, the two operations are *Set-to-zero* and *Increment*. We realize counters as sequences of binary digits. Then *Set-to-zero* returns a string of zeros and *Increment* has to add 1 to a number in binary representation. We implement *Increment* by first increasing the least significant digit by 1 and then calling a (hidden) procedure *Propagate-carry* if the increased digit is 2. This procedure changes the digit 2 into a zero, increases the digit to the left by 1 and then calls itself recursively. The worst-case cost of an increment operation is, of course, unbounded. For the amortized analysis we define the potential of a string $\ldots \alpha_1 \alpha_1 \alpha_0$ as the sum $\sum_i \alpha_i$ of its digits. Then *Set-to-zero* creates a string of potential zero. Also, a call of procedure *Propagate-carry* has actual cost 1 and amortized cost 0 since it decreases the potential of a string by 1. Finally, changing the last digit has actual cost 1 and amortized cost 2 since it increases the potential by 1. Thus the total cost of a sequence of one *Set-to-zero* operation followed by n *Increments* is bounded by $1+2\cdot n$, although some increments may cost as much as $\log n$.

There is one other view of amortization which we now briefly discuss: the buyer's and seller's view. A buyer (user) of a data structure wants to be charged in a predictable way for his uses of the data structure, e.g. the same amount for each call of *Increment*. A seller (implementer) wants to account for his actual uses of computing time. Amortization is the trick which makes both of them happy. The seller charges the buyer two tokens for each call of increment; the analysis given above ensures the seller that his actual costs are covered and the buyer pays a uniform price for each call of *Increment*.

2. The dictionary problem

Let $S \subseteq U$ be any subset of the universe U. We assume that an item of information $inf(x)$ is associated with every element $x \in S$. The *dictionary problem* asks for a data structure which supports the following three operations:

(1) *Access*(x): return the pair (**true**, $inf(x)$) if $x \in S$, and return **false** otherwise;
(2) *Insert*(x, inf): replace S by $S \cup \{x\}$ and associate the information inf with x;
(3) *Delete*(x): replace S by $S - \{x\}$.

Note that operations *Insert* and *Delete* are destructive, i.e., the old version of set S is destroyed by the operations. The nondestructive version of the problem is discussed in Section 4. In the *static dictionary problem* only operation *Access* has to be supported.

Since the mid-fifties many sophisticated solutions were developed for the dictionary problem. They can be classified into two large groups, the comparison-based and the representation-based data structures. Comparison-based data structures only use comparisons between elements of U in order to gain information, whilst the less restrictive representation-based structures mainly use the representation, say as a string over some alphabet, of the elements of U for that purpose. Typical representatives of the two groups are respectively search trees and hashing.

2.1. Comparison-based data structures

We distinguish two types of data structures: explicit and implicit. An implicit data structure for a set S, $|S| = n$, uses a single array of size n to store the elements of S and

```
(1) low←1; high←n;
(2) next←an integer in [low ... high];
(3) while x≠S[next] and high>low
(4) do if x<S[next]
(5)    then high←next−1
(6)    else low←next+1 fi;
(7)    next←an integer in [low ... high]
(8) od;
(9) if x=S[next] then "successful" else "unsuccessful".
```

Fig. 1. Program 1.

only O(1) additional storage. The explicit data structures use more storage; mostly for explicit pointers between elements.

The basis for all comparison-based methods is the following algorithm for searching ordered arrays. Let $S=\{x_1<x_2<\cdots<x_n\}$ be stored in array $S[1 \ldots n]$, i.e. $S[i]=x_i$, and let $x \in U$. In order to decide $x \in S$, we compare x with some table element and then proceed with either the lower or the upper part of the table (see Program 1 in Fig. 1). Various algorithms can be obtained from this scheme by replacing lines (2) and (7) by specific strategies for choosing next. Linear search is obtained by *next*←*low*, binary search by *next*←$[(low+high)/2]$. The worst-case access time is O(n) for linear search and O(log n) for binary search. It is clear that log n is also a lower bound for the access time for any comparison-based data structure.

2.1.1. Implicit data structures

There are a number of ways to implement dictionaries implicitly. If we store the elements in an unordered list, then update can be done in constant time, but searching requires linear time. On the other hand, if the dictionary is maintained as an array sorted in increasing order, then searching can be done in logarithmic time, but updates may require that all the elements in the array be moved. Rotated lists are arrays that can be sorted into increasing order by performing a cyclic shift (rotation) of the elements. They are not much more difficult to search, but only half the elements have to be moved in the worst case.

Munro and Suwanda [156] were the first to explicitly consider the dynamic implicit dictionary problem. They showed that if the elements are stored partially sorted in a triangular grid, then search and update can be performed in $O(\sqrt{n})$ steps. Using blocks of rotated lists (sorted relative to one another), they were able to improve the search time to O(log n), keep the number of moves per update at $O(\sqrt{n})$, and only increase the number of comparisons per update to $O(\sqrt{n}\log n)$. Combining these ideas, they also produced an implicit dictionary that can be searched or updated in $O(n^{1/3}\log n)$ time. By using rotated lists in a recursive manner, Frederickson [67] was able to achieve O(log n) search time and $O(n^{\sqrt{2/\log n}}(\log n)^{3/2})$ update time. Munro [152, 153] created implicit dictionaries that use $O((\log n)^2)$ time for both search and update. His basic approach is to have the order of elements within blocks of the array implicitly represent pointers and counters.

Since the $O((\log n)^2)$ upper bound falls short of the $O(\log n)$ upper bound for explicit structures, the interesting question of lower bounds arises. Munro and Suwanda [156] proved that their triangular grid scheme is optimal within a restricted class of implicit structures. Borodin et al. [28] prove a trade-off between search and update time which is valid for *all* implicit data structures. In particular, their result implies that if the update time is constant then the search time is $\Omega(n^\varepsilon)$ for some constant $\varepsilon > 0$.

A related problem is the searching problem for semisorted tables. Assume that the elements of the set S can be arranged to any one of p different permutations. Then Alt and Mehlhorn [9] showed that $\Omega(p^{1/n})$ comparisons are necessary for an access operation in the worst case, under the restrictive assumption that all comparisons must involve the element sought for. Their lower bound is even true in the average case and for nondeterministic algorithms. If the p permutations are the set of extensions of a single partial order, then a more precise bound was given by Linial and Saks [127]; they show that $O(\log N)$ stops are necessary and sufficient where N is the number of ideals in the partial order; we remind the reader that a subset A of a partially ordered set $(P, \geqslant)$ is an *ideal* if $x \in A$ and $y < x$ implies $y \in A$.

Partial orders naturally arise as the result of preprocessing. Borodin et al. [29] considered the problem of determining the trade-off between preprocessing and search time. If $P(n)$ and $S(n)$ denote the number of comparisons performed to respectively preprocess and search an initially unsorted array of length n, then, in the worst case, $P(n) + n \log S(n)$ is $\Omega(n \log n)$. Mairson [132] proved that this result also holds in the average case. Many of the lower-bound results mentioned were recently unified and extended by McDiarmid [137].

2.1.2. *Explicit data structures or search trees*

A search tree for a set $S = \{x_1 < \cdots < x_n\}$ is a full binary tree, i.e., each node has either two or no son, with n leaves (=nodes with no son). The n leaves are labelled by the elements from S from left to right. The internal nodes (=nodes with two sons) are labelled by elements of U such that elements in the left subtree of a node v are labelled no larger than v and elements in the right subtree are labelled larger than v. Figure 2 shows a search tree for $\{1, 7, 19, 23\} \subseteq \mathbb{Z}$. The root could be labelled by any integer between 7 and 18 inclusive. Our convention for storing the set S is frequently called *leaf-oriented*. The alternative *node-oriented* organization, where the elements are stored in the internal nodes will not be discussed; most results carry over with only small modifications.

An access operation takes time proportional to the height of the tree (=length of the

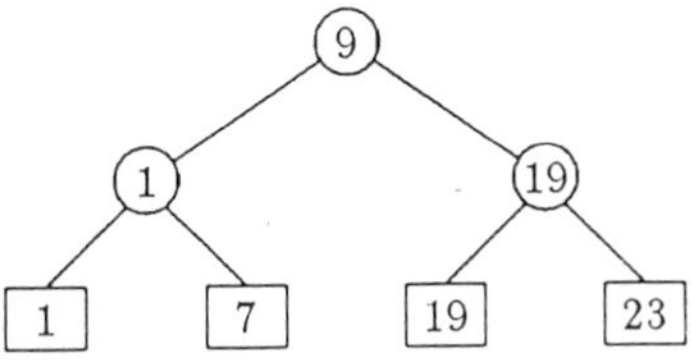

Fig. 2. A search tree for $\{1, 7, 19, 23\}$.

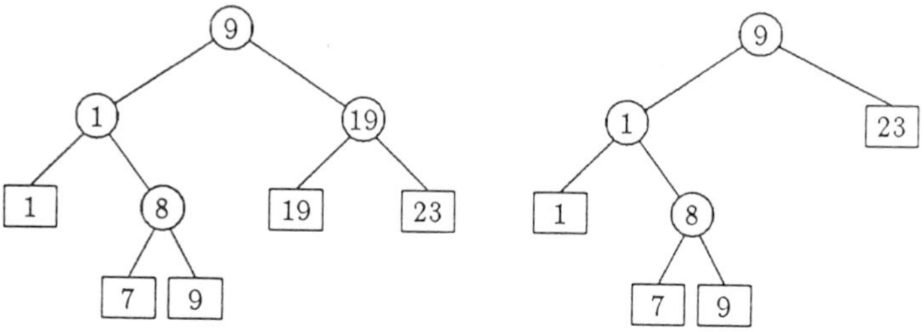

Fig. 3. Insertion of 9 and deletion of 19.

longest path from the root to a leaf) in the worst case: compare the element x to be sought for with the label of the root, proceed to the right subtree if it is larger and to the left subtree otherwise. Insertions and deletions are also easily accomplished. In order to insert x, we search for x and then replace the leaf where the search ended by a tree with two leaves. In order to delete x, we only have to delete the leaf labelled x and its parent from the tree. Figure 3 shows the tree of Fig. 2 after the insertion of 9 and the deletion of 19.

The worst-case time for an *Access*, *Insert* or *Delete* is $O(n)$ since trees may degenerate to linear lists. On the average, we can expect to do much better.

Consider the following scenario. We start with a tree consisting of a single leaf. Before the nth step we have a tree with n leaves. One of the leaves is chosen at random and replaced by a tree with two leaves. Let $D(n)$ be the expected total depth of the n leaves before the nth step, i.e.,

$$D(n)=\sum_T \left[p(T) \sum_i d_i(T) \right]$$

where $p(T)$ is the probability that a certain tree T with n leaves arises and $d_i(T)$ is the depth of the ith leaf of T. Then $D(1)=1$ and

$$D(n+1)-D(n)=\sum_T p(T)\left(2+\sum_i d_i(T)/n \right)$$

since the increase in depth is $d_i(T)+2$ if the ith leaf of T is replaced and the probability for this event is $p(T)/n$. Thus $D(n+1)-D(n)=2+1/n\cdot D(n)$ or $D(n+1)=2+(1+1/n)\cdot D(n)$. This recurrence relation has solution

$$D(n)=2n\cdot \sum_{i=1}^{n} 1/i-n=O(n \log n).$$

This shows that under random insertions we can expect an expected access time of $O(\log n)$. An excellent survey of the behavior of random trees is [83]. Recent results can be found in [44, 45].

We have seen that depth is a crucial parameter for the behavior of search trees. This led to the development of balanced trees where the depth is guaranteed to be $O(\log n)$. We distinguish two major types of balanced trees, the height- and the weight-balanced trees. In the first class the height of subtrees is balanced; it includes AVL-trees [1], (2–3)-trees [2], B-trees [15], HB-trees [162], red-black trees [89], half-balanced trees [160], (a, b)-trees [101] and balanced-binary-trees [196]. The behavior of AVL-trees was studied by Foster [65, 66], Knuth [112], Brown [31], Mehlhorn [142], Mehlhorn and Tsakalidis [151], and Tsakalidis [205]. In the second class the weight of the subtrees is balanced; it includes the BB[α]-trees [159] and the internal-path trees [82].

We treat only (2, 4)-trees and refer the reader to the textbooks mentioned above for the other classes. In a (2, 4)-tree all leaves have the same depth and every internal node has between 2 and 4 children. We use $\rho(v)$ to denote the number of children (= arity) of a node v. In a node v we store $\rho(v)-1$ labels in order to guide searches. We will describe the insertion and deletion algorithms and their *amortized analysis* next. For a tree T, let the potential of T be twice the number of 4-nodes (= nodes of arity 4) plus the number of 2-nodes.

Let us consider an insertion first. We start by adding a new leaf. This may convert the parent from a 4-node to a 5-node, which is not allowed in a (2, 4)-tree. We split such a 5-node into a 2-node and a 3-node. This may create a new 5-node, which we split in turn. We continue splitting newly created 5-nodes, moving up the tree, until either the root splits or no new 5-node is created, see Fig. 4. If the root splits, we create a new root, a 2-node, causing the tree to grow in height by 1. The time needed for the insertion is proportional to 1 plus the number of splits.

Let us define the actual time of an insertion to be 1 plus the number of splits. Then the amortized time of an insertion is at most 3: each split costs 1 but converts a node that was originally a 4-node into a 2-node and a 3-node, for a net potential drop of 1; in addition, the insertion can create one new 2-node or 4-node.

Deletion is an inverse process, only slightly more complicated. To delete a given item, we destroy the leaf containing it. This may make the parent a 1-node. If this 1-node has a neighboring sibling that is a 3-node or a 4-node, we move a child of this neighbor to the 1-node and the deletion stops. (This is called *borrowing*.) If the 1-node has a neighboring sibling that is a 2-node, we combine the 1-node and the 2-node. (This is called *fusing*.) Fusing may produce a new 1-node, which we eliminate in the same way. We move up the tree eliminating 1-nodes until either a borrowing occurs or the root

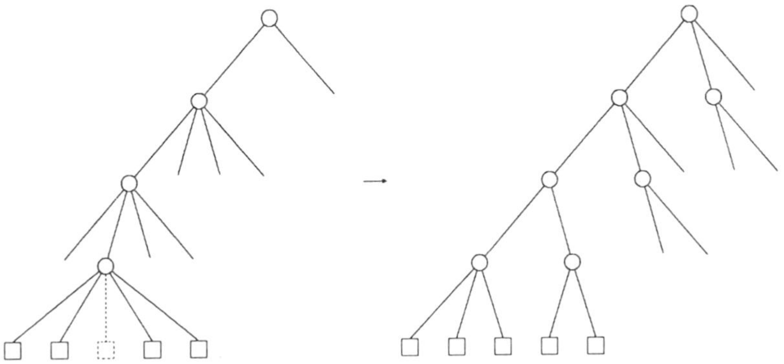

Fig. 4.

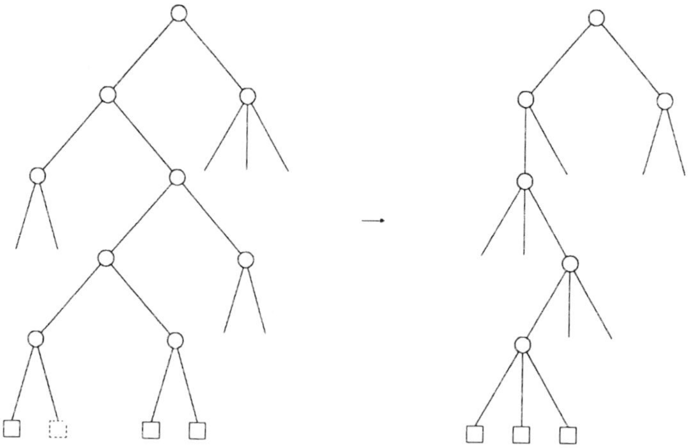

Fig. 5.

becomes a 1-node, which we destroy. (See Fig. 5). The time needed for the deletion is proportional to 1 plus number of fusings.

Let us define the actual time of a deletion to be 1 plus the number of fusings. Then the amortized cost is at most 2: each fusing costs 1 but converts two nodes that were originally 2-nodes into a 3-node, for a net potential drop of 2; in addition, the deletion can create one new 2-node.

We summarize. The amortized cost of an insertion or deletion is O(1) although the actual cost may be as large as O(log n). This result is due to Huddleston and Mehlhorn [101] and Maier and Salveter [131]. The particularly simple proof given above is due to Hoffmann et al. [95]. Levcopoulus and Overmars [126] have recently shown that constant insertion and deletion time can also be obtained in the worst case and not only in the amortized sense. It is open whether their scheme can also support the various extensions of (2, 4)-trees to be discussed next.

How frequently do expensive insertions and deletions arise? Huddleston and Mehlhorn [101] showed that the number of insertions and deletions which cause t fusings or splittings decreases as an exponential function of t. (2, 4)-Trees are strongly related to red-black trees [89], symmetric binary B-trees [14] and half-balanced trees [160]. All of the results about amortized rebalancing cost mentioned above carry over to these trees. Moreover, red-black trees and half-balanced trees have the additional property that only O(1) structural changes of the tree are required after an insertion or deletion; all other changes concern only bookkeeping information [160, 196]. These properties and their simplicity make red-black trees a good choice for practical implementations of balanced trees; they make them also the ideal basis for *finger search trees*. Finger search trees represent ordered lists into which one can maintain pointers, called fingers, from which searches can start; the time for a search, insertion or deletion is O(log d), where d is the number of items between the search starting point and the accessed item. The O(log d) bound can either be achieved in the amortized sense [33, 101] or in the worst-case sense [88, 117, 100, 194, 203, 204]. Finger trees lead to optimal algorithms for the basic set operations union, intersection, difference, . . . ; i.e., these operations take time

$$O\left(\log\binom{n+m}{\min(n,m)}\right)$$

when applied to sets of size n and m respectively [101]. Another application of finger trees is efficient list splitting. Suppose that we have a list of n items which we want to split into sublists of length d and $n-d$ respectively. The position of the split is determined by a finger search starting from both ends of the list. The finger search takes time $O(\log \min(d, n-d))$ and this is also a bound on the amortized cost of the split [95]. Consider now a sequence of splits, i.e., we start with a single list of length n which we split repeatedly until we obtain n lists of length 1 each. The worst-case cost $T(n)$ of this process is given by

$$T(1)=0 \quad \text{and} \quad T(n)=\max_{1\leqslant d\leqslant n-1}[T(d)+T(n-d)+O(\log \min(d,n-d))]$$
$$\text{for } n\geqslant 1.$$

This recurrence has solution $T(n)=O(n)$, i.e., the amortized cost of each split is $O(1)$.

Frequently, the same search has to be performed on many lists. Several researchers [209, 128, 216, 55, 41, 104] observed that the naive strategy of locating the key separately in each list by binary search is far from optimal and that more efficient techniques frequently exist. Chazelle and Guibas [39] distilled from these special-case solutions a general data structuring technique and called it *fractional cascading* which supports under very general assumptions the search for an item in d lists out of n lists in time $O(d+\log N)$, where N is the total size of the n lists. This has numerous applications to computational geometry. Mehlhorn and Näher [146] studied dynamic fractional cascading and showed that the amortized analysis of balanced trees carries over to the more general situation of fractional cascading; more precisely, they showed that insertions and deletions have amortized cost $O(\log\log N)$ and that a search for an item in d lists out of n lists takes time $O((d+\log N)\log\log N)$.

Weight-balanced trees [159] share many of the properties of height-balanced trees; their implementation is more cumbersome, however. On the other hand, they have the following *weight-property* [217, 25]: A node of weight w (i.e., w descendants) participates in only $O(n/w)$ structural changes of the tree when a sequence of n insertions and deletions is processed. The weight-property makes weight-balanced trees superior to height-balanced trees whenever structural changes of the tree are very costly, e.g., when their cost depends linearly on the size of the subtree changed. This is frequently the case in applications to multidimensional search or computational geometry where trees are augmented with substantial additional information. In these applications weight-balanced trees guarantee good amortized behavior which cannot be obtained with height-balanced trees.

2.2. *Representation-based data structures*

In these data structures the representation, say as a string of digits, of the elements stored is used to compute their position in memory. We consider interpolation search (Section 2.2.1), hybrid data structures (Section 2.2.2) and hashing (Section 2.2.3).

2.2.1. Interpolation search

We first consider static interpolation search, which is an implicit data structure. Interpolation search was suggested by Peterson [170] as a method for searching in sorted arrays. It is obtained from Program 1 (Fig. 1), by replacing lines (2) and (7) by

$$next \leftarrow (low-1) + \left\lceil \frac{x - S[low-1]}{S[high+1] - S[low-1]} \cdot (high - low + 1) \right\rceil.$$

It is assumed that positions $S[0]$ and $S[n+1]$ are added and filled with artificial elements. The worst-case complexity of interpolation search is clearly $O(n)$; consider the case that $S[0]=0$, $S[n+1]=1$, $x=1/(n+1)$ and $0<S[i]<x$ for $1 \leqslant i \leqslant n$. Then $next = low$ always and interpolation search deteriorates to linear search. The average-case behavior is much better. Average access time is $O(\log\log n)$ under the assumption that the keys $x_1, \ldots, x_n$ are drawn independently from a uniform distribution over the open interval (x_0, x_{n+1}). This was shown by Yao and Yao [222], Pearl et al. [168] and Gonnet et al. [86].

A very intuitive explanation of the behavior of interpolation search can be found in [169], where a variant called *binary interpolation search* is presented. We discuss this variant: Binary search has access time $O(\log n)$ because it consists of a single scanning of a path in a complete binary tree of depth $\log n$. If we could do binary search on the paths of the tree then we could obtain $\log\log n$ access time. So let us consider the question whether there is a fast (at least on the average) way to find the node on the search path which is halfway down the tree, i.e., the node on the path of search which has depth $\frac{1}{2}\log n$. There are $2^{1/2 \log n} = \sqrt{n}$ of these nodes and they are $\sqrt{n}$ apart in the array representation of the tree. Let us make an initial guess by interpolating and then search through these nodes by linear search. Note that each step of the linear search jumps over $\sqrt{n}$ elements of S and hence as we will see shortly only $O(1)$ steps are required on the average. Thus an expected cost of $O(1)$ has reduced the size of the set from n to $\sqrt{n}$ (or, in other words, determined the first half of the path of search) and hence total expected search time is $O(\log\log n)$.

The precise algorithm for *Access*(x, S) is as follows. Let $low=1$, $high=n$ and let $next=\lceil p \cdot n \rceil$ be defined as above; here $p=(x-x_0)/(x_{n+1}-x_0)$. If $x>S[next]$ then compare x with $S[next+\sqrt{n}]$, $S[next+2\sqrt{n}], \ldots$ until an i is found with $x \leqslant S[next+(i-1)\cdot\sqrt{n}]$. This will use up to i comparisons. If $x<S[next]$ then we proceed analogously. In any case, the subtable of size $\sqrt{n}$ thus found is then searched by applying the same method recursively. Let p_i be the probability that i or more steps are required in the search for the subtable. Then the expected cost of the search for the subtable is $O(\sum_i p_i)$. Also, if i or more steps are required then the actual position of element x in the array differs by more than $i\sqrt{n}$ from the expected position (which is pn) and hence $p_i = O(1/i^2)$ by an application of the Chebyshev inequality. This shows that time $O(1)$ is spent at each level of the recursion and hence the expected access time is $O(\log\log n)$.

Santoro and Sidney [176] proposed another variant called *interpolation-binary* search which has $O(\log\log n)$ average-case access time for the uniform distribution and $O(\log n)$ worst-case time.

Willard [215] has shown that the log log n asymptotic retrieval time of interpolation search does not hold for most nonuniform probability distributions. However, he was able to modify interpolation search such that its expected running time is O(log log n) on static μ-random files where μ is any *regular* probability density. A density μ is regular if there are constants b_1, b_2, b_3, b_4 such that $\mu(x)=0$ for $x<b_1$ or $x>b_2$, $\mu(x)\geqslant b_3\geqslant 0$ and $|\mu'(x)|\leqslant b_4$ for $b_1\leqslant x\leqslant b_2$. A file is μ-random if its elements are drawn independently according to density μ. It is important to observe that Willard's algorithm does not have to know the density μ. Rather, its running time is O(log log n) on μ-random files provided that μ is *regular*. Thus his algorithm is fairly robust.

Dynamic interpolation search, i.e., data structures which support insertions and deletions as well as interpolation search, was discussed by Frederickson [67] and Itai, Konheim, Rodeh [105]. Frederickson presents an implicit data structure which supports insertions and deletions in time $O(n^{\varepsilon})$, $\varepsilon>0$, and accesses with expected time O(log log n). The structure of Itai, Konheim and Rodeh has expected insertion time O(log n) and worst-case insertion time $O((\log n)^2)$. It is claimed to support interpolation search, although no formal analysis of its expected behavior is given. Both papers assume that the files are generated according to the uniform distribution.

Mehlhorn and Tsakalidis [150] extend dynamic interpolation search to nonuniform distributions. They achieve an amortized insertion and deletion cost of O(log n) and an expected amortized insertion and deletion cost of O(log log n). The expected search time on files generated by μ-random insertions and random deletions is O(log log n) provided that μ is a *smooth* density. An insertion is μ-random if the key to be inserted is drawn from density μ. A deletion is random if every key present in the current file is equally likely to be deleted. These notions of randomness are called I_r and D_r respectively in [113]. A density μ is *smooth* if there are constants d and $\alpha<1$ such that, for all a, b, c with $a<c<b$ and all integers m and n, $m=[n^{\alpha}]$,

$$\int_{c-(b-a)/m}^{c} \hat{\mu}(x)\,\mathrm{d}x\leqslant dn^{-1/2}$$

where $\hat{\mu}(x)=0$ for $x<a$ or $x>b$ and $\hat{\mu}(x)=\mu(x)/p$ for $a\leqslant x\leqslant b$ where $p=\int_a^b\mu(x)\,\mathrm{d}x$. Every regular density of Willard is smooth in this sense.

2.2.2. *Hybrid data structures*

This group of data structures combines arrays and pointers. A main representative is the digital search tree or trie. The trie was proposed by Fredkin [71] as a simple way to structure a file by using the digital representation of its elements; e.g. we may represent $S=\{121, 102, 211, 120, 210, 212\}$ by the trie in Fig. 6.

The general situation is as follows: the universe U consists of all strings of length l over some alphabet of say k elements, i.e. $U=\{0,\ldots,k-1\}^l$. A set $S\subseteq U$ is represented as the k-ary tree consisting of all prefixes of elements of S. An implementation which immediately comes to mind is to use an array of length k for every internal node of the tree. Then operations *Access*, *Insert* and *Delete* are very fast and are very simple to program; in particular, if the reverse of all elements of S are stored (in our example this would be the set {121, 201, 112, 021, 012, 212}), then the program in Fig. 7 will realize operation *Access*(x). This program takes time $O(l)=O(\log_k N)$ where $N=|U|$. Unfor-

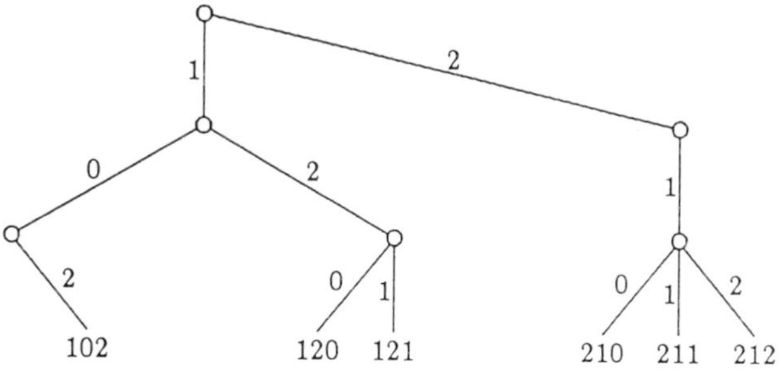

Fig. 6. A trie.

```
v←root;
y←x;
do l TIMES (i, y)←(y MOD k, y DIV k);
   v←ith son of v
od;
if x=CONTENT[v] then "yes" else "no".
```

Fig. 7. Program 2.

tunately, the space requirement of a trie as described above can be horrendous: $O(n \cdot l \cdot k)$. For each element of set S, $|S|=n$, we might have to store an entire path of l nodes, all of which have degree 1 and use up to space $O(k)$.

There is a very simple method to reduce the storage requirement to $O(n \cdot k)$. We only store internal nodes which are at least binary. Since a trie for a set S of size n has n leaves there will be at most $n-1$ internal nodes of degree 2 or more. Chains of internal nodes of degree 1 are replaced by a single number, the number of nodes in the chain. In our example we obtain Fig. 8. Here internal nodes are drawn as arrays of length 3. On the pointers from fathers to sons the numbers indicate the increase in depth, i.e., 1 plus the

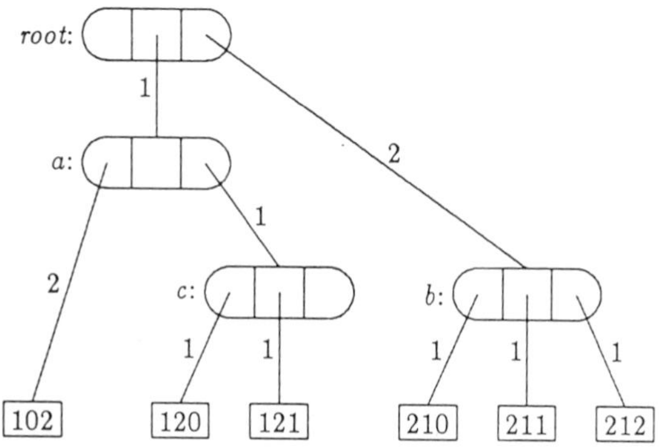

Fig. 8. The compressed trie.

length of the eliminated chain of nodes of degree 1. In our example the 2 on the pointer from the root to the son with name b indicates that after branching on the first digit in the root we have to branch on the third $(1+2)$ digit in the son. The algorithms for *Access, Insert* and *Delete* become slightly more complicated for compressed tries but still run in time $O(l)$.

Thus a compressed trie supports operations *Access, Insert* and *Delete* with time bound $O(\log_k N)$, where N is the size of the universe and k is the branching factor of the trie. A set S of n elements requires space $O(k \cdot n)$.

Note that tries exhibit an interesting time–space trade-off. Choosing k large will make tries faster but more space consuming, choosing k small will make tries slower but less space consuming. For static sets, i.e. only operation *Access* being supported, there is a technique developed by Tarjan and Yao [200] which can store a set S ($S \subseteq U, |S| = n$ and $|U| = N$) in an n-ary trie using $O(n)$ storage locations (of $O(\log n)$ bits each) such that it can support operation *Access* in time $O(\log_n N)$ worst case and $O(1)$ expected case. Their technique combines a trie structure with a method for compressing tables by using double displacements. This double displacement method is an elaboration of a single displacement method suggested by Aho and Ullman [6] and Ziegler [225] for compressing parsing tables.

Willard [214] introduces two dynamic trie structures, the P-fast trie and the Q-fast trie. A P-fast trie uses space $O(n\sqrt{\log N}\,2^{\sqrt{\log N}})$ for representing a set S ($S \subseteq U, |S| = n$, $|U| = N$) and can support access, insertion and deletion in worst-case time $O(\sqrt{\log N})$. The Q-fast trie is derived from the P-fast trie by using a pruning technique; it has the same time bounds but needs only $O(n)$ space.

Tries also support the following three additional operations:

(1) *Successor*(x): find the least element in the set S with key value greater than x;

(2) *Predecessor*(x): find the greatest element in the set S with key value less than x;

(3) *Subset*(x_1, x_2): find (and produce) the list of those elements of S whose key value lies between x_1 and x_2.

The P-fast and Q-fast tries can support the operations *Successor*(x) and *Predecessor*(x) in time $O(\sqrt{\log N})$ and *Subset*(x_1, x_2) in time $O(\sqrt{\log N} + k)$, where k is the size of the output.

Willard [213] also proposed another static trie structure, called Y-fast trie, which uses $O(n)$ space and supports *Access*(x), *Successor*(x) and *Predecessor*(x) in worst-case time $O(\log\log N)$ and *Subset*(x_1, x_2) in $O(\log\log N + k)$. This method uses perfect thashing (see Subsection 2.2.3.3.).

2.2.3. *Hashing*

The ingredients are very simple: an array $T[0 \ldots m-1]$, the hash table, and a function $h: U \to [0 \ldots m-1]$, the hash function. U is the universe; we will assume $U = [0 \ldots N-1]$ throughout this section. The basic idea is to store a set S as follows: $x \in S$ is stored in $T[h(x)]$. Then an access is an extremely simple operation: compute $h(x)$ and look up $T[h(x)]$. There is one immediate problem with this basic idea: what to do if $h(x) = h(y)$ for some $x, y \in S$, $x \neq y$. Such an event is called a *collision*. There are two main methods for dealing with collisions: chaining and open addressing. We briefly present both methods and then discuss two techniques for choosing the hash function h: perfect and universal hashing. We do not treat the average behavior of hashing in detail but rather refer the reader to Chapter 14 on the complexity of finite functions.

2.2.3.1. Hashing with chaining. The hash table T is an array of linear lists. A set $S \subseteq U$ is represented as m linear lists. The ith list contains all elements $x \in S$ with $h(x)=i$. Operation *Access*(x, S) is realized by the following program:

(1) compute $h(x)$;
(2) search for x in list $T[h(x)]$.

Operations *Insert*(x, S) and *Delete*(x, S) are implemented similarly. We only have to add x to or delete x from the list $T[h(x)]$. For the analysis of hashing we assume that h can be evaluated in constant time and therefore define the cost of an operation referring to key x as $O(1+\delta_h(x, S))$ where S is the set of stored elements and

$$\delta_h(x, S)=\sum_{y \in S} \delta_h(x, y), \quad \text{and}$$

$$\delta_h(x, y)=\begin{cases}1 & \text{if } h(x)=h(y) \text{ and } x \neq y, \\ 0 & \text{otherwise.}\end{cases}$$

The worst-case complexity of hashing is now easily determined. The worst case occurs when the hash function h restricted to set S is a constant, i.e. $h(x)=i_0$ for all $x \in S$. Then hashing deteriorates to searching through a linear list and any one of the three operations costs $O(|S|)$ time units.

The average-case behavior of hashing is much better. The expected cost of a sequence of n insertions, deletions and accesses is $O((1+\beta/2)n)$, where $\beta=n/m$ is the maximal load factor of the table. This is true under the following probability assumption:

(1) The hash function $h: U \rightarrow [0 \ldots m-1]$ distributes the universe uniformly over the interval $[0 \ldots m-1]$, i.e. for all $i, i' \in [0 \ldots m-1]$, $|h^{-1}(i)|=|h^{-1}(i')|$.

(2) All elements of U are equally likely as argument of any one of the operations in the sequence, i.e., the argument of the kth operation of the sequence is equal to a fixed $x \in U$ with probability $1/|U|$.

These assumptions imply that the value $h(x_k)$ of the hash function on the argument of the kth operation is uniformly distributed in $[0 \ldots m-1]$, i.e. $\Pr(h(x_k)=i)=1/m$ for all $k \in [1 \ldots n]$ and $i \in [0 \ldots m-1]$. Thus the expected length of any one of the lists is at most i/m after the ith operation and therefore the expected cost of the $(i+1)$st operation is $O(1+i/m)$. Hence the total expected cost of all n operations is $O((1+\beta/2)n)$. More on the expected analysis of hashing with chaining can be found in Chapter 14.

2.2.3.2. Hashing with open addressing. Each element $x \in U$ defines a sequence $h(x, i)$, $i=0, 1, 2, \ldots$ of table positions. This sequence of positions is searched through whenever an operation referring to key x is performed. A very popular method for defining the function $h(x, i)$ is to use the linear combination of two hash functions h_1 and h_2:

$$h(x, i)=[h_1(x)+i \cdot h_2(x)] \bmod m.$$

Hashing with open addressing does not require any additional space. However, its performance becomes poor when the load factor is nearly 1.

The average-case behavior of hashing with open addressing is easy to determine under the unrealistic assumption that the sequence $h(x, i)$ is a random permutation of the table positions and that the cost of an *Insert* is $1+\min\{i: T[h(x, i)]$ is not occupied$\}$. Let $C(n, m)$ be the expected cost of an *Insert* when n items are already stored in a table of size m. Then $C(n, m)=1+n/m \cdot C(n-1, m-1)$ if $n>0$, since one probe is always required, this probe inspects an occupied position with probability n/m and since in this case an insertion into a table of size $m-1$ holding already $n-1$ items is still required, and $C(0, m)=1$. Thus

$$C(n, m)=(m+1)/(m-n+1) \approx 1/(1-\beta),$$

where $\beta=n/m$ is the load factor. The expected cost of an *Access* operation is $O((1/\beta) \cdot \ln 1/(1-\beta))$ by a similar analysis. More on the expected-case behavior of hashing with open addressing can be found in Chapter 14.

2.2.3.3. Universal hashing. Universal hashing was first described by Carter and Wegman [37]. It is a method to deal with the basic problem of hashing: its linear worst-case behavior. We saw in Subsection 2.2.3.1 that hashing provides us with $O(1)$ expected access time and $O(n)$ worst-case access time. Thus it is always very risky to use hashing when the actual distribution of the inputs is not known to the designer of the hash function. It is always conceivable that the actual distribution favors worst-case inputs and hence will lead to large average access times.

Universal hashing is a way out of this dilemma. We work with an entire class H of hash functions instead of a single hash function; the specific hash function in use is selected randomly from the collection H. If H is chosen properly, i.e., for every subset $S \subseteq U$ almost all $h \in H$ distribute S fairly evenly over the hash table, then this will lead to small expected access time for every set S. Note that the average is now taken over the functions in the class H, i.e., the randomization is done by the algorithm itself, not by the user: the algorithm controls the dices.

Let us reconsider the symbol table example. At the beginning of each compiler run the compiler chooses a random element $h \in H$. It will use hash function h for the next compilation. In this way the time needed to compile any fixed program will vary over different runs of the compiler, but the time spent on manipulating the symbol table will have small mean.

What properties should the collection H of hash functions have? For any pair $x, y \in U, x \neq y$, a random element $h \in H$ should lead to collision, i.e. $h(x)=h(y)$, with fairly small probability. More precisely, let $c \in \mathbb{R}$ and $N, m \in \mathbb{N}$. A collection $H \subseteq \{h: h:[0 \ldots N-1] \rightarrow [0 \ldots m-1]\}$ is c-universal if for all $x, y \in [0 \ldots N-1]$, $x \neq y$,

$$|\{h: h \in H \text{ and } h(x)=h(y)\}| \leqslant c \cdot |H|/m.$$

Carter and Wegman [37] analyzed the expected behavior of universal hashing under the following assumptions:

(1) The hash function h is chosen at random from some c-universal class H, i.e., each $h \in H$ is chosen with probability $1/|H|$.

(2) Hashing with chaining is used.

They showed that the expected cost of an *Access*, *Insert* or *Delete* operation is $O(1+c\beta)$, where $\beta=n/m$ is the load factor, if a c-universal class of hash functions is used.

They also gave examples of universal classes. Let $m, N \in \mathbb{N}$ and let N be a prime. For $a, b \in [0 \ldots N-1]$ let $h_{a,b}(x) = ((ax+b) \bmod N) \bmod m$ for all x. Then the class $H = \{h_{a,b}\colon a, b \in [0 \ldots N-1]\}$ is 4-universal. A member of this class requires $O(\log N)$ random bits for its selection. Mehlhorn [143] shows that there is a 8-universal class whose members can be selected by $O(\log m + \log\log N)$ bits and also shows that this is optimal. Other universal classes are described in [134].

2.2.3.4. Perfect hashing. Perfect hashing is the simplest solution to the collision avoidance problem. We just postulate that there are no collisions, i.e., the hash function operates injectively on the set to be stored. A function $h: [0 \ldots N-1] \rightarrow [0 \ldots m-1]$ is a *perfect hash function* for $S \subseteq [0 \ldots N-1]$ if $h(x) \neq (y)$ for all $x, y \in S$, $x \neq y$.

Perfect hashing was first discussed by Sprugnoli [186]. He describes heuristic methods for constructing perfect hash functions. Fredman, Komlós and Szemerédi [73] show that very simple perfect hash functions exist whenever $m \geqslant 3n$, $n = |S|$. The hash functions can be found in deterministic time $O(nN)$ and probabilistic time $O(n)$, can be evaluated in time $O(1)$ and are given as programs of length $O(n \log N)$ bits. The program size was improved to $O(n \log n + \log\log N)$ by Mehlhorn [145, p. 138], and to $O(n \log\log n + \log\log N)$ by Jacobs and van Emde Boas [106]. A lower bound for program size is $\Omega(n^2/m + \log\log N)$ [143].

Perfect hashing was first developed for static sets S. Aho and Lee [5] showed how to deal with insertions and deletions. They describe a hashing scheme which supports accesses and deletions in worst-case time $O(1)$ and insertions in expected time $O(1)$. The expectation is computed with respect to random inputs; cf. assumption (2) in Subsection 2.2.3.1.

A further improvement was made by Dietzfelbinger et al. [46]. This scheme combines the advantages of universal and perfect hashing. Accesses and deletions take worst-case time $O(1)$ and insertions take time $O(1)$ on the average. However, the average is now computed with respect to the random choices made by the algorithm and is worst-case with respect to inputs.

2.2.3.5. Extendible hashing. Our treatment of hashing in the previous subsections was based on the assumption that main memory is large enough to completely contain the hash table. Let us assume now that this assumption is no longer warranted. Let us also assume that the transfer of information between main and secondary memory is in pages of size b. In this situation extendible hashing as described by Fagin et al. [59] can be used. Similar schemes are known under the names dynamic hashing [122], virtual hashing [129] and expandable hashing [110].

Let $h: U \rightarrow \{0, 1\}^*$ be an injective function. For an integer d and $x \in U$ let $h_d(x)$ be the prefix of $h(x)$ of length d and for a set $S \subseteq U$ let $d(S)$ be the minimal d such that $|\{x \in S\colon h_d(x) = a\}| \leqslant b$ for all $a \in \{0, 1\}^d$, i.e. h_d partitions the set S into subsets of size at most b. Extendible hashing uses a table $T[0 \ldots 2^{d(S)} - 1]$, called the directory, and some number of buckets of capacity b to store set S. The directory entries are pointers to buckets. An element $x \in S$ is accessed by computing $i = h_d(x)$, looking up $T[i]$ and then accessing the bucket pointed to by $T[i]$. In this way at most two accesses to secondary memory are ever needed. The first access is to the page containing $T[i]$ and the second access is to the bucket containing x. We allow different directory entries to point to the same bucket as long as the following buddy principle holds: If $r<d(S)$, $a\in\{0,1\}^r$, $a_1,a_2\in\{0,1\}^{d(S)-r}$, $a_1\neq a_2$ and $T[aa_1]=T[aa_2]$ then necessarily $T[aa_1]=T[aa_3]$ for all $a_3\in\{0,1\}^{d(S)-r}$.

Insertions are easily processed except when a bucket overflows. In this case $d(S)$ increases by 1, the size of the directory doubles, and the overflowing bucket is split into two.

The expected behavior of extendible hashing is treated in the papers mentioned above and more completely in Flajolet [62]; cf. also Chapter 14. He shows that the expected number of buckets is $n/(b \ln 2)$ and that the expected size of the directory is about $e/(b \ln 2)n^{1+1/b}$. A variant of extendible hashing [191] achieves linear expected directory size and O(1) expected access time.

3. The weighted dictionary problem and self-organizing data structures

In the weighted dictionary problem a weight (=access frequency) is associated with every element of the set S. The basic goal is to make accesses to high-frequency elements faster than to low-frequency elements without sacrificing the efficiency of the other set operations.

The first problem studied was the construction of an optimal search tree for given access frequencies. More precisely, let $S=\{x_1,\ldots,x_n\}$ and let w_i be the weight of item x_i. For a tree T let d_i be the depth of item i. Then $P=\sum_i w_i d_i$ is called the *weighted path length* of the tree T. It is not too hard to construct an optimum tree, i.e. a tree which minimizes the weighted path length, in time $O(n^3)$ by dynamic programming. This was improved to $O(n^2)$ by Knuth [111] (see also [224]). In the case of leaf-oriented storage an $O(n \log n)$ algorithm was found by Hu and Tucker [99] and Garsia and Wachs [81]. The latter algorithms resemble Huffman's algorithm [102] for the construction of optimal prefix codes. The original correctness proofs for both algorithms were fairly involved. A simple proof was recently found by Kingston [108].

A search tree for a set S yields directly a prefix code for the symbols $x_1,\ldots,x_n$ over the binary alphabet {"*go left*", "*go right*"}. In view of the noiseless coding theorem it is therefore not surprising that the normalized weighted path length $\bar{P}=\sum_i(w_i/W)d_i$, where $W=\sum_i w_i$, is strongly related to the entropy $H=\sum_i(w_i/W)\log(W/w_i)$ of the access frequency distribution. In particular, $\bar{P}\geqslant H-\log H$ for node-oriented storage and $\bar{P}\geqslant H$ for leaf-oriented storage [13]. Also, there are trees achieving $\bar{P}\leqslant H+2$ for both storage organizations [139]. Nearly optimal trees, i.e., trees which come within a constant factor of optimality and can be constructed in linear time $O(n)$ were described by Fredman [72], Mehlhorn [138], Bayer [13] and Korsch [116]. An implicit data structure achieving the same goal was found by Frederickson, [69].

We will now turn to the dynamic case. It has two facets: the underlying set S may change and the access frequencies of the elements of S may change.

Faller [60] and Gallager [78] proposed a scheme for dynamic Huffman coding. Knuth [114] gives a real-time implementation of this scheme. More precisely, suppose that the frequency w of a leaf at depth l changes by 1. Then an optimal Huffman tree for the new distribution can be derived from the old Huffman tree in time $O(l)$. Vitter [210] refined this result. His scheme does not only work on-line in real time, it also uses at most one more bit per letter than the standard off-line Huffman algorithm.

Let us consider search trees next. The relevant operations are as follows:

(1) *Access*(x, S): if item x is in set S then return a pointer to its location, otherwise return **nil**;

(2) *Insert*(x, S): insert x into set S;

(3) *Delete*(x, S): delete x from set S and return the resulting set;

(4) *Join*(S_1, S_2): return a set representing the items in S_1 followed by the items in S_2, destroying S_1 and S_2 (this assumes that all elements of S_1 are smaller than all elements of S_2);

(5) *Split*(x, S): returns two sets S_1 and S_2; S_1 contains all items of S smaller than x and S_2 contains all items of S larger than x (this assumes that x is in S); set S is destroyed;

(6) *ChangeWeight*(x, S, δ): changes the weight of element x by δ.

What time bounds can we hope for? We mentioned above that the normalized weighted path length is related to the entropy of the frequency distribution and therefore the best we can hope for operation *Access* is an $O(\log(W/w_i))$ time bound. We call this bound the *ideal access time* of item i. Similarly, inserting an element of weight w can be no cheaper than accessing the new element or one of its neighbors in the new tree and we should therefore content ourself with an

$$O(\log(W+w)/\min(w^+, w, w^{-1}))$$

time bound, where w^- and w^+ are the weights of the two neighbors of the new element.

Bayer [13] gave a heuristics for the weighted dynamic dictionary problem, but he gave no theoretical results, and indeed his trees do not have ideal access time in the worst case. Unterauer [207] achieves ideal access time, but does not analyze the worst-case time required for updates. Mehlhorn [140, 141] introduced *D-trees*; D-trees extend BB[α]-trees and achieve ideal worst-case *Access, Change, Insert* and *Delete* time. He did not consider *Join* and *Split*. D-trees have the weight-property of BB[α]-trees mentioned in Section 2. Operations *Join* and *Split* were then considered by Güting and Kriegel [87], Kriegel and Vaishnavi [119], Bent, Sleator and Tarjan [18], achieving good worst-case bounds for all five operations.

A very elegant alternative which achieves good amortized time bounds for all operations was found by Sleator and Tarjan [183]. Their splay trees show the following behavior:

- The amortized cost of *Access*(x, S) is $O(\log(W/w))$.
- The amortized cost of *Delete*(x, S) is $O(\log(W/\min(w, w^-)))$, where w^- is the weight of the predecessor of x in S.
- The amortized cost of *Join*(S_1, S_2) is $O(\log((w_1+w_2)/w))$, where w is the weight of the largest element in S_1 and w_i is the weight of S_i.
- The amortized cost of *Insert*(x, S) is $O(\log(W+w/\min(w^-, w^+, w)))$, where w^- is the weight of the predecessor, w^+ is the weight of the successor of x.
- The amortized cost of *Split*(x, S) is $O(\log(W/w))$.
- The amortized cost of *ChangeWeight*(x, S, δ) is $O(\log((W+\delta)/w))$.

Splay trees achieve this behavior *without* keeping any explicit information about the weights of the elements; they are a so-called *self-organizing data structure.* In self-organizing data structures an item x is moved closer to the entry point of the data structure whenever it is accessed. This will make subsequent accesses to x cheaper. In this way the elements of S compete for the good places in the data structure and high-frequency elements are more likely to be there. Note however, that we do *not* maintain any explicit frequency counts or weights; rather, we hope that the data structure self-organizes to a good data structure.

Since no global information is kept in a self-organizing data structure, the worst-case behavior of a single operation can always be horrible. However, the average and the amortized behavior may be good. For an average case analysis we need to have probabilities for the various operations. The data structure then leads to a Markov chain whose states are the different incarnations of the data structure. We can then use probability theory to compute the stationary probabilities of the various states and use these probabilities to derive bounds on the expected behavior of the data structure.

For the amortized behavior, we take a combinatorial point of view and analyze the cost of sequences of operations. Experiments [20] suggest that the behavior of the heuristics on real data is more closely described by the amortized analysis than by the probabilistic analysis. We will now describe self-organizing linear search in some detail and then briefly return to search trees.

Self-organizing linear search is quite simple. Let $S=\{x_1,\ldots,x_n\}$ and let us assume that S is organized as a linear list. For simplicity, we consider only the operation *Access*(x) and postulate that the cost of *Access*(x) is *pos*(x) where *pos*(x) denotes the position of x in the current list. Two popular strategies for self-organizing linear search are the Move-to-Front and the Transposition Rule.

Move-to-Front Rule (*MFR*): Operations *Access*(x) and *Insert*(x) make x the first element of the list and leave the order of the remaining elements unchanged.

Transposition Rule (*TR*): Operation *Access*(x) interchanges x with the element preceding x in the list.

For the average-case analysis we postulate a probability distribution $p_1,\ldots,p_n$ and assume that the accesses are independent random variables and that an access to item x_i has probability p_i. We may assume $p_1\geqslant p_2\geqslant\cdots\geqslant p_n$ w.l.o.g. Assuming complete knowledge of the distribution it is clearly optimal to arrange the items according to *decreasing frequency* and to never change this arrangement. The expected access cost is $E_{\mathrm{FD}}=\sum_i p_i\cdot i$ in this case. Let us denote the expected access cost of the two rules by E_{MF} and E_{T} respectively. The expected behavior of the Move-to-Front Rule was investigated by McCabe [136], Burville and Kingman [34], Knuth [112], Hendricks [93], Rivest [175] and Bitner [22]. In particular,

$$E_{\mathrm{MF}}=\sum_{i=1}^{n} p_i\left(1+\sum_{j\neq i}\frac{p_j}{p_i+p_j}\right)\leqslant 2\cdot E_{\mathrm{FD}}.$$

This can be seen as follows. The expected position of element x_i under the Move-to-Front Rule is 1 plus the expected number of x_j's which are in front of x_i. Next observe that x_j is ahead of x_i if there is an integer k such that the last k accesses were an access to x_j followed by $k-1$ accesses to items different from x_i and x_j. Hence the probability that x_j is ahead of x_i is given by

$$p_j \sum_{k\geq 0} (1-(p_i+p_j))^k = p_j/(p_i+p_j)$$

and the expression for $E_{\rm MF}$ follows. Next observe that $p_j/(p_i+p_j) \leq 1$ and hence

$$E_{\rm MF} = 1 + 2\cdot \sum_{i=1}^{n} p_i \sum_{j=1}^{i-1} \frac{p_j}{p_i+p_j} \leq 1 + 2\cdot \sum_i p_i(i-1) \leq 2\cdot E_{\rm FD}.$$

This shows, that the expected access cost under the Move-to-Front Rule is at most twice the optimum. The Transposition Rule is never worse. More precisely, Rivest [175] showed that $E_{\rm T} \leq E_{\rm MF}$, with strict inequality unless $n=2$ or $p_i = 1/n$ for all i. He further conjectured that *Transpose* minimizes the expected access time for any p, but Anderson et al. [10] found a counterexample. Chung et al. [40] showed $E_{\rm MF} \leq \frac{1}{2}\pi \cdot E_{\rm FD}$ and Gonnet et al. [85] gave an example where this bound is obtained. Thus the worst-case ratio $E_{\rm MF}/E_{\rm FD}$ is $\frac{1}{2}\pi$. The convergence speed of the two rules was compared by Bitner [22] and it was found that the Move-to-Front Rule converges more quickly. A combination of the Move-to-Front Rule with the more traditional method of keeping frequency counts is discussed in [121].

We will next present an analysis of the amortized behavior of the Move-to-Front Rule due to Bentley and McGeoch [20]. Assume that we perform a sequence of m accesses to this list. Let h_i, $1 \leq i \leq n$, be the number of accesses to x_i. We may assume w.l.o.g. that $h_1 \geq h_2 \geq \cdots \geq h_n$. If we start with the list $x_1, \ldots, x_n$ and never change it then the total cost of the m accesses is $C_{\rm FD} = \sum_i i \cdot h_i$. Suppose now that we start with the list $x_1, \ldots, x_n$ and use the Move-to-Front Rule. Let $C_{\rm MF}$ be the total cost of m accesses under the Move-to-Front Rule. Let t_j^i, $1 \leq j \leq h_i$, be the cost of the jth access to item x_i. Then

$$C_{\rm MF} - C_{\rm FD} \leq \sum_{i=1}^{n} \sum_{j=1}^{h_i} t_j^i - \sum_{i=1}^{n} ih_i = \sum_{i=1}^{n} \sum_{j=1}^{h_i} (t_j^i - i).$$

Consider an arbitrary term $t_j^i - i$ in this sum. If $t_j^i > i$ then there were at least $t_j^i - i$ accesses to items x_h, $h > i$, between the $(j-1)$st access and the jth access to x_i (the 0th access is the initial configuration). Thus

$$\sum_{j=1}^{h_i} (t_j^i - i) \leq h_{i+1} + \cdots + h_n$$

and hence

$$C_{\rm MF} - C_{\rm FD} \leq \sum_{i=1}^{n} (h_{i+1} + \cdots + h_n) = \sum_{i=1}^{n} (i-1)h_i = C_{\rm FD} - m.$$

In particular, $C_{\rm MF} \leq 2\cdot C_{\rm FD}$. A much stronger result was shown by Sleator and Tarjan [184]. They proved that no rule whatsoever can beat the Move-to-Front Rule by more than a factor of 2. This is even true for off-line rules which know the complete sequence of accesses in advance.

Allen and Munro [8] were the first to consider self-organizing binary search. They show that the expected search time under the Move-to-Root heuristics, where an accessed item is moved to the root by a sequence of rotations, is only a constant factor above the expected search time in an optimum search tree. They also showed that the transposition heuristics does not have this property. Bitner [22] investigates the expected behavior of several other heuristics.

Splay trees [183] refine the Move-to-Root heuristics by combining it with the concept of path compression, cf. Section 5 on the *Union-Split-Find* problem. The access time in splay trees is within a constant factor of the access time of static optimum search trees; this is the analogue of the relationship $C_{MF} \leqslant 2 \cdot C_{FD}$ derived above. It is open whether this result can be extended as it was in the case of the Move-to-Front Rule for linear search. A first step was taken by Tarjan [197] who showed that accessing the nodes of a splay tree in sequential order takes time $O(n)$.

4. Persistence

Ordinary data structures are *ephemeral* in the sense that a change to the structure destroys the old version, leaving only the new version available for use. In contrast, a *persistent* structure allows access to any version, old or new, at any time. We distinguish *partial* and *full* persistence. A data structure is partially persistent if all versions can be accessed but only the newest version can be modified, and fully persistent if every version can be both accessed and modified.

A number of researchers have developed partially or fully persistent forms of various data structures, including stacks [157], queues [96], search trees [158, 163, 174, 190, 177] and related structures [38, 41, 49, 53]. In the articles of Dobkin and Munro [49], Overmars [163, 165], Chazelle [38], the problem of persistence is called *searching in the past*.

Dobkin and Munro [49] consider the following problem. Let S be a universe of n objects subject over time to m deletions and insertions in arbitrary order. Assuming that there exists a total order among the objects, they describe a method for computing in time $O(\log n \log m)$, the *rank* of any object at any given time, i.e., the number of objects that precede it at that time. The space used is $O(n+m \log n)$. Overmars [164] improves the query time to $O(\log(n+m))$. Chazelle [38] presents a data structure that requires $O(n+m)$ storage and allows the computation of any neighbor of a new object at time θ, in $O(\log n \log m)$ time. The method also allows the set S to be only partially ordered.

Overmars [165] studies three simple but general ways to obtain partial persistence. One method is to explicitly store every version, copying the entire ephemeral structure after each update operation. This costs $\Omega(n)$ time and space per update. An alternative method is to store no versions but instead to store the entire sequence of update operations, rebuilding the current version from scratch each time an access is performed. A hybrid method is to store the entire sequence of update operations and in addition every kth version for some suitably chosen value of k.

Cole [41] devises a partial persistent representation of sorted sets that occupies $O(m)$ space and has $O(\log m)$ access time, where m is the total number of updates (insertions and deletions) starting from an empty set. His solution is off-line, i.e., the entire sequence of updates must be known in advance. Sarnak and Tarjan [177] propose a simple data structure that overcomes this drawback. They present a persistent form of binary search tree with an $O(\log m)$ worst-case access/insert/delete time and an amortized space requirement of $O(1)$ per update.

Swart [190], using the idea of *path copying*, presents a fully persistent representation of sorted sets and lists with an $O(\log m)$ time bound per operation and an $O(\log m)$ space bound per update.

Driscoll et al. [53] finally propose a general solution for the persistence problem. They develop simple, systematic and efficient techniques for making different linked data structures persistent. They show first that if an ephemeral structure has nodes of bounded in-degree, then the structure can be made partially persistent at an amortized space cost of $O(1)$ per update step and a constant-factor increase in the amortized cost of access and update operations. Second, they present a method which can make a linked structure of bounded in-degree fully persistent at an amortized time and space cost of $O(1)$ per update and a worst-case time of $O(1)$ per access step. At last they present a partial persistent implementation of balanced search tree with a worst-case time per operation of $O(\log n)$ and an amortized space cost of $O(1)$ per insertion or deletion. Combining this result with a *delayed updating* technique of Tsakalidis [204] they obtain a fully persistent form of balanced search trees with the same time and space bounds as in the partially persistent case. Using another technique they can make the $O(1)$ space bound for insertion and deletion worst-case instead of amortized. The technique employed by Driscoll et al. is strongly related to fractional cascading, cf. Subsection 2.1.2. This relationship can be used to support a forget operation which permits to explicitly delete versions and thus improves the space requirement [148].

5. The *Union-Split-Find* problem

We consider the problem of maintaining a collection of disjoint sets under the operations of *Union* and *Split*. More precisely, the problem is to carry out three kinds of operations on a partition of the universe $U=\{1,\ldots,n\}$: *Find*, which determines the set containing a given element; *Union*, which combines two sets into one; and *Split*, which splits a set at a given element into two. In order to identify the sets, we assume that the algorithms maintain with each set a unique name. The precise formulation of the three operations is as follows:

(1) *Find*(x): return the name of the set containing x;

(2) *Union*(A, B): combine the two sets with names A and B, destroying the two old sets, and return a name for the new set;

(3) *Split*(A, x): split the set with name A into the sets $A_1=\{y\in A: y\leqslant x\}$ and $A_2=A-A_1$, destroying the old set, and return names for the two sets formed.

5.1. *The Union-Find problem*

Here, we start with the partition of U into singleton sets and allow only the operations *Find* and *Union*. A popular solution, proposed by Galler and Fischer [79], represents each set by a rooted tree. The nodes of the tree are the elements of the set and the name of a set is the root of the tree. With each element x we store a pointer $p(x)$ to its parent; the parent pointer of a root is **nil**. Initially, all parent pointers are **nil**. To carry out *Find*(x), we follow parent pointers starting in x until we reach a root and return the root. To carry out *Union*(A, B), we define $p(A)$ to be B and return B.

The naive algorithm is not very efficient, requiring $O(n)$ time per find in the worst case. However, there are two simple heuristics which improve the efficiency dramatically: the *weighted union* [79] and the *path compression* heuristics, the latter of which was suggested by McIllroy and Morris. We store with each set A its cardinality *size*(A) and carry out *Union*(A, B) by defining $p(A)$ to be B if $size(A) \leqslant size(B)$, and $p(B)$ to be A otherwise. In the algorithm for *Find*(x) we not only traverse the path from x to the root but also redirect all parent pointers of the nodes traversed to the root.

Suppose now that we carry out $m \geqslant n$ *Finds* and $n-1$ *Unions*. Let $t(m, n)$ be the maximum time required by any such sequence of instructions. Fischer [61] showed that $t(m, n) = O(n + m \cdot \log \log m)$, Hopcroft and Ullman [97] improved the bound to $t(m, n) = O(n + m \cdot \log^* n)$ and finally Tarjan [192] and Banachowski [12] improved the bound to $t(m, n) = O(n + m \cdot \alpha(m, n))$, where α is the inverse of Ackermann's function, i.e.,

$$\alpha(m, n) = \min\{z \geqslant 1 : A(z, 4\lceil m/n \rceil) > \log n\}$$

and

$$\begin{aligned} &A(i, 0) = 1 && \text{for all } i \geqslant 1, \\ &A(0, x) = 2x && \text{for all } x \geqslant 0, \\ &A(i+1,\ x+1) = A(i, A(i+1, x)) && \text{for all } i, x \geqslant 0. \end{aligned}$$

Tarjan and van Leeuwen [199] show that the same time bound applies to several variants of the *Union-Find* algorithm described above.

How good is this algorithm? Tarjan [193] has shown that any pointer-machine algorithm which obeys the *separation assumption* requires time $\Omega(m \cdot \alpha(m, n))$ to solve the *Union-Find* problem. The separation assumption states that sets correspond to connected components in the pointer structure.

All time bounds quoted above are amortized. The worst-case behavior was considered by N. Blum [24] and he proves a $\Theta(\log n/\log \log n)$ bound; again, the lower bound used the separation assumption.

The average-case behavior was considered by Doyle and Rivest [51], Yao [220], Knuth and Schönhage [115], Yao [221] and Bollobás and Simon [26]. They show that under various probability assumptions the expected running time of the *Union-Find* algorithm is linear, even if only one of the two heuristics is used.

Manilla and Ukkonen [133] consider a variant of the general *Union-Find* problem where backtracking over the *last* union operations is possible by means of a *Deunion* operation. They present two methods which allow to process a sequence of m *Finds*, k *Unions* and k *Deunions* in time $O((m+k) \cdot \log n/\log \log n)$ and $O(k + m \log n)$ respectively [212]. A further generalization where weights are associated with the unions is considered by Gambosi et al. [80].

All of the algorithms mentioned above run on pointer machines. A linear-time RAM algorithm for the special case where the unions are known in advance was recently described by Gabow and Tarjan [77].

5.2. *The interval Split-Find problem*

Here, we start with the partiton of U into a single block and allow only the operations *Find* and *Split*. Clearly, all blocks arising during the execution are intervals. Hopcroft and Ullmann [97] gave a solution which processes n *Splits* in total time $O(n \log^* n)$ and each *Find* in worst-case time $O(\log^* n)$. Gabow [76] gave an improved algorithm which processes the n *Splits* in total time $O(n \cdot \alpha(n, n))$ and each *Find* in amortized time $O(\alpha(n, m))$. All of this runs on a pointer machine.

Again, one can do better on a RAM. Gabow and Tarjan [77] gave a linear-time solution, which was extended by Imai and Asano [104], to the incremental set-splitting problem. In this version the underlying universe can be expanded by adding (operation *Add*) a new element next to a given element. Of course, the *Access* operation, which is implicit in a *Find*, becomes nontrivial now and is not accounted for in the linear time bound. However, in the applications discussed by Imai and Asano this causes no problems.

5.3. *The interval Union-Split-Find problem*

We finally turn to the full problem. Here, we start with the partition of U into singletons and allow the operations *Find*, *Split* and *Union*. We postulate however, that the partition is always a partition into intervals, i.e., a union operation can only join adjacent intervals. It is customary, to use the right endpoint of an interval as the name of the interval in this problem.

Van Emde Boas, Kaas and Zijlstra [58] describe a pointer-machine solution which executes all three operations in worst-case time $O(\log \log n)$. The corresponding lower bound was shown by Mehlhorn, Näher and Alt [147]. The lower bound does not use the separation assumption and is even valid for the amortized complexity of the problem. Mehlhorn, Näher and Alt [147] also show that the lower bound increases to $\Omega(\log n)$ with the separation assumption.

Finally, Mehlhorn and Näher [146] show that the additional operations *Add* and *Erase*, which modify the underlying universe by adding an element next to a given element or removing a given element, can also be supported in amortized time $O(\log \log n)$.

6. Priority queues

A *priority queue* is an abstract data structure consisting of a set of *items*, each with a real-valued *key*, subject to the following operations:

(1) *Makequeue*: return a new, empty priority queue;

(2) *Insert*(i, h): insert a new item i with predefined key into priority queue h;

(3) *Findmin*(h): return an item of minimum key in priority queue h; this operation does not change h;

(4) *Deletemin*(h): delete an item of minimum key from h and return it.

In addition, the following operations on priority queues are often useful:

(5) *Meld*(h_1, h_2): return the priority queue formed by taking the union of the item-disjoint priority queues h_1 and h_2; this operation destroys h_1 ad h_2;
(6) *Decreasekey*(Δ, i, h): decrease the key of item i in priority queue h by subtracting the nonnegative real number Δ; this operation assumes that the position of i in h is known;
(7) *Delete*(i, h): delete arbitrary item i from priority queue h. This operation assumes that the position of i in h is known.

In our discussion of priority queues we shall assume that a given item is in only one priority queue at a time and that a pointer to its priority queue position is maintained. It is important to remember that priority queues do *not* support efficient searching for an item.

Priority queues have many applications, most noteworthy sorting and network optimization problems. A set of n items may be sorted by one *Makequeue*, n *Insert*, n *Findmin* and n *Deletemin* operations. The single-source shortest path problem (cf. Chapter 10) on a graph with n nodes and m edges of nonnegative weight can be solved by one *Makequeue*, n *Insert*, *Findmin* and *Deletemin* and m *Decreasekey* operations. We infer from these applications that at least one of the operations *Insert*, *Findmin* and *Deletemin* must have cost $\Omega(\log n)$ because of the $\Omega(n \log n)$ lower bound for sorting, and that the operations can occur with drastically different frequencies in some applications.

Some solutions, to be described in the sequel, do not support the *Meld* operation. We will refer to a priority queue without this operation as a simple priority queue. Also note, that a *Decreasekey* operation can always be simulated by a *Delete* followed by an *Insert*.

For small universes, say the keys are drawn from the range $[1 \ldots N]$, any solution for the *Union-Split-Find* problem (cf. Section 5) gives a simple priority queue: the set of items in the queue corresponds to the marked items of that section, *Find*(1) is *Findmin*, *Union*(i) is *Delete*(i), *Union*(*Find*(1)) is *Deletemin*, and *Split*(i) is *Insert*(i). In particular, the simple queue operations can all be realized in time $O(\log \log N)$ per operation. Orlin and Ahuja [161] give a solution with an O(1) bound for *Makequeue*, *Delete*, *Findmin*, *Decreasekey* and an $O(\log N)$ bound for *Insert* and *Deletemin*. Ahuja et al. [7] improve this to O(1) for *Delete*, *Findmin* and *Decreasekey*, $O(n)$ for *Makequeue* and $O(\sqrt{\log N})$ for *Insert* and *Deletemin*; here n is the number of *Insert* operations. This gives rise to an $O(m+n\sqrt{\log N})$ shortest path algorithm.

Most solutions for arbitrary universes are based on the concept of a *heap-ordered* tree. A heap-ordered tree is a rooted tree containing a set of items, one item in each node, with the items arranged in *heap order*: if v is any node, then the key of the item in v is no less than the key of the item in its parent $p(v)$, provided v has a parent. Thus the tree root contains an item of minimum key.

A first realization of simple priority queues was given by Williams [218]. He uses complete binary heap-ordered trees and shows that the simple queue operations all take time $O(\log n)$ and that complete binary trees can be stored implicitly in an array without the use of pointers. Floyd [63] shows that a *Makequeue* followed by n inserts can be made to run in time $O(n)$ instead of $O(n \log n)$.

Doberkat [47] gives an average-case analysis of Floyd's algorithm. Porter and Simon [171] analyzed the average cost of inserting a random element into a random heap in terms of exchanges. They proved that this average is bounded by the constant 1.61. Their proof does not generalize to sequences of insertions since random insertions into random heaps do not create random heaps. The repeated insertion problem was solved by Bollobás and Simon [27]; they show that the expected number of exchanges is bounded by 1.7645. The worst-case cost of *Inserts* and *Deletemins* was studied by Gonnet and Munro [84]; they give $\log\log n+O(1)$ and $\log n+\log^* n+O(1)$ bounds for the number of comparisons respectively.

In the 70s and 80s many different realizations of the full repertoire of priority queue operations were found: heap-ordered (2–3)-trees [2], leftist trees [42], binomial queues [211, 30], self-adjusting heaps [185] and pairing heaps [74]. They achieve $O(\log n)$ bounds for all operations.

Johnson [107] introduces a-ary heap-ordered trees, where a is a parameter, and shows how to implement *Makequeue* and *Findmin* in time $O(1)$, *Decreasekey* in time $O(\log n/\log a)$ and the other operations in time $O(a \log n/\log a)$. With $a=\max(2, \log(m/n))$ this gives rise to an

$$O(m \log n/\max(2, \log(m/n)))$$

shortest path algorithm.

Fredman and Tarjan [75] improve Johnson's result and obtain $O(1)$ amortized time bounds for *Makequeue, Findmin, Insert, Meld* and *Decreasekey* and $O(\log n)$ amortized time bounds for *Delete* and *Deletemin*. They call their data structure Fibonacci-heaps. F-heaps give rise to an $O(n \log n+m)$ shortest path algorithm. An alternative to F-heaps are the relaxed heaps of Driscoll et al. [52]. In contrast to F-heaps the time bound for *Delete* and *Deletemin* is worst-case instead of amortized.

A generalization of priority queues was considered by Atkinson et al. [11]. They add the operations *Findmax* and *Deletemax* and show that the $O(\log n)$ time bound can be maintained.

7. Nearest common ancestors

In this section we consider the following problem: given a dynamically changing collection of rooted trees, answer queries of the form: "What is the nearest common ancestor of two given nodes x and y, denoted by $nca(x, y)$?" Of course, this problem comes in many different flavors according to which update operations are permitted. We use n to denote the total number of nodes, m to denote the number of dynamic operations and k to denote the number of queries.

In the *off-line* version of the nearest common ancestor problem, the collection of trees is static and the sequence of queries is known in advance. Aho, Hopcroft and Ullman [3] describe an $O(n+k\cdot\alpha(k+n, n))$ pm-algorithm using $O(n)$ space. Here α is the functional inverse of Ackerman's function (cf. Section 5). Gabow and Tarjan [77] give an $O(n+k)$ time RAM algorithm using $O(n)$ space.

In the *static, on-line* version the collection of trees is static but the queries are given *on-line*, i.e., each query must be answered before the next one is given. Aho, Hopcroft and Ullman [3] propose a RAM algorithm requiring $O(n \log\log n)$ preprocessing time, $O(n \log\log n)$ space and $O(\log\log n)$ time per query. Harel and Tarjan [92] present a RAM algorithm running in $O(n)$ preprocessing time, $O(n)$ space and answering a query in $O(1)$ time; a simpler solution with the same performance is given by Schieber and Vishkin [178]. Van Leeuwen [123] gives a pm-algorithm requiring $O(n \log\log n)$ preprocessing time, $O(n \log\log n)$ space and $O(\log\log n)$ *optimal* query time. Harel and Tarjan [92] prove the optimality of this query time and claim that van Leeuwen's algorithm can be modified to run on a pointer machine in linear time and space. Another optimal pm-algorithm with $O(n)$ preprocessing time, $O(n)$ space and $O(\log\log n)$ query time is described in [125].

We now come to the dynamic versions of the problem. In the *linking roots version*, the queries are given on-line and two trees can be joined by linking their roots. Van Leeuwen [123] gives an $O(n+m \log\log n)$ time pm-algorithm answering each query in $O(\log\log n)$ time. Harel and Tarjan [92] present an $O(n+m\cdot\alpha(m+n, n))$ time RAM algorithm answering each query in $O(\alpha(m+n, n))$ amortized time.

In the *linking and cutting version* the queries are on-line and the collection of trees can be modified by two operations. *Link*(x, y), where y is a root, makes y a child of the vertex x (not necessarily a root) and *Cut*(x) deletes the edge connecting x to its parent and thus makes x an additional root. Aho, Hopcroft and Ullman [3] consider only linkings. Their algorithm runs on a RAM in time $O((m+n)\log n)$ and space $O(n \log n)$ and the time required for a query is $O(\log n)$. Maier [130] considers linkings and cuttings, achieves the same time bound and improves upon the space bound. His solution requires less than $O(n \log n)$ but still more than linear space; it also runs on a RAM. Sleator and Tarjan [182] give the fastest solution which in addition runs on a pointer machine. At first they propose a solution with total running time $O(n+m \log n)$ for m *Link* and *Cut* operations between n nodes using space $O(n)$; the nearest common ancestor can also be determined in time $O(\log n)$. By using a more complicated data structure they can improve the above result so that each individual *Link* and *Cut* operation takes time $O(\log n)$.

In the *dynamic tree version* the queries are on-line, there is only one tree, and this tree can be updated by the insertion of leaves or deletion of nodes. Tsakalidis [206] presents a pm-algorithm which needs $O(n)$ space, performs m arbitrary insertions and deletions on an initially empty tree in time $O(m)$ and allows to determine the nearest common ancestor of nodes x and y in time

$$O(\log(\min\{\mathit{depth}(x), \mathit{depth}(y)\})+\alpha(k, k)),$$

where the second term is amortized over the k queries and $\mathit{depth}(x)$ is the distance from node x to the root.

8. Selection

The selection problem can be stated as follows: we are given a sequence $x_1, \ldots, x_n$ of pairwise distinct elements and an integer k, $1 \leqslant k \leqslant n$, and want to find the kth smallest element of the sequence, i.e. an x_j such that there are $k-1$ keys x_l with $x_l < x_j$ and $n-k$ keys x_l with $x_l > x_j$. For $k = n/2$ such a key is called *median*. Of course, selection can be reduced to sorting. We might first sort sequence $x_1, \ldots, x_n$ and then find the kth smallest element by a single scan of the sorted sequence. This results in an $O(n \log n)$ algorithm. A linear expected time algorithm was found by Hoare [94]. One simply splits the set S into subsets $S_1 = \{x \in S: x \leqslant x_1\}$ and $S_2 = \{x \in s: x > x_1\}$ and then applies the algorithm recursively to the appropriate subset. If $T(i, n)$ denotes the expected time to determine the ith largest element in a collection of n elements then

$$T(i,n) = n + \frac{1}{n}\left(\sum_{k=1}^{i-1} T(i-k, n-k) + \sum_{k=i}^{n} T(i,k)\right)$$

since x_1 is the kth largest element with probability $1/n$. Then $T(i, n) = O(n)$ as a simple induction shows.

A worst-case running time $O(n)$ can be obtained by choosing the splitting element more carefully [23]. The set S is divided into subsets of five elements each and the median of each subset is computed. The algorithm is then used recursively to determine the median of the $n/5$ medians. The median of the medians is used as the splitting element. This choice guarantees that $\max(|S_1|, |S_2|) \leqslant 8n/11$ and therefore the worst-case running time is governed by the recurrence

$$T(n) \leqslant O(n) + T(n/5) + T(8n/11)$$

which has an $O(n)$ solution.

Knuth [112, Section 5.3.3], traces the history of the selection problem to the writings of Dodgson [50], and provides an excellent overview of the early developments of this problem. The original formulation of the problem by Dodgson [50] and Steinhaus [188] was in terms of lawn tennis tournaments. Let $V_k(n)$ $(\bar{V}_k(n))$ be the number (average number) of pairwise comparisons needed to find the kth smallest of n numbers $(k \geqslant 2)$ (assuming that all $n!$ orderings are equally likely). Of particular interest is the asymptotic behavior of $V_k(n)$ $(\bar{V}_k(n))$ when n goes to infinity either with k fixed or with $k = \alpha \cdot n$ for a fixed constant α in the range $0 \leqslant \alpha \leqslant \frac{1}{2}$.

Hadian and Sobel [91] described an algorithm, called *replacement selection* by Knuth, that proves the general bound

$$V_k(n) \leqslant n - k + (k-1)\lceil \log(n-k+2) \rceil.$$

For large k (in particular for $k = \alpha \cdot n$), this was improved by M. Blum et al. [23] and further improved by Schönhage et al. [180] to $V_k(n) \leqslant 3n + o(n)$. The first general lower bound, $V_k(n) \geqslant n + k - 2$, was proved by M. Blum et al. [23]. For a detailed survey of all the lower bounds see [109]. Bent and John [17] improved Kirkpatrick's lower bound in [109]. Especially, they show

$$V_{\alpha n}(n) \geqslant (1 + H_\alpha)n + O(\sqrt{n}),$$

where $H_\alpha = -\alpha \log \alpha - (1-\alpha)\log(1-\alpha)$ is the discrete entropy of α. In particular, the bound for the *median* is

$$V_m(n) \geqslant 2n + O(\sqrt{n}), \quad \text{where } m = \lceil \tfrac{1}{2} n \rceil.$$

A considerable amount of work has also been done on estimating $\bar{V}_k(n)$. Matula [135] proved that, for some absolute constant c, $\bar{V}_k(n) \leqslant n + ck \ln \ln n$ as $n \to \infty$. A.C. Yao and F.F. Yao [223] showed that there exists an absolute constant $c' > 0$, such that $\bar{V}_k(n) \geqslant n + c'k \ln \ln n$ as $n \to \infty$, proving a conjecture of Matula. An upper bound of $\bar{V}_k(n) \leqslant n + k + o(n)$ for all k and n is due to Floyd and Rivest [64]. Cunto and Munro [43] prove that $\bar{V}_k(n) \geqslant n + k - O(1)$ in general and that for the *max-min-median* problem

$$2n - o(n) \leqslant \bar{V}_t(n) \leqslant \tfrac{9}{4}n + o(n), \quad \text{where } t \in \{1, n, \lceil \tfrac{1}{2}n \rceil\}.$$

Dobkin and Munro [48] show that a linear-time algorithm is possible for the median problem which requires an additional work space of only $\lceil \frac{1}{2}n \rceil + 1$ data cells. Ramanan and Hyafil [172] present some algorithms which solve efficiently the problem of selecting the kth smallest elements, and give their respective order of a totally ordered set of n elements, when k is small compared to n.

Munro and Paterson [154] consider time–space trade-offs for selection in a *tape input model*, i.e., inputs are stored on a read-only input tape and may be accessed only in a sequential scan of the entire tape. Output is to a write-only tape and workspace registers may hold input values. Let T be the time, S the workspace and n the size of the set; then when time is measured in comparisons, their algorithm realizes a trade-off of $T \log S = O(n \log n)$ over the range $\Omega(\log^2 n) = S = O(2^{\sqrt{\log n}})$. Frederickson [70] modifies and extends their approach to realize the same trade-off over the broader range

$$\Omega(\log^2 n) = S = O(2^{(\log n / \log^* n)}).$$

9. Merging

The *two-way merging* problem consists of combining two sorted lists A and B to make one larger sorted list. The algorithms developed for this problem can be classified according to the following properties:

(i) Minimizing the number of comparisons and assignments.

(ii) Minimizing the workspace needed in addition to the space used to store the two files. An algorithm using O(1) additional workspace is said to be *in-place*.

(iii) Maintaining stability. A stable merging algorithm is one which preserves the relative orderings of equal elements from each of the sequences.

(iv) Maintaining searchability. An in-place merging algorithm is said to support searchability if at any stage in the process a search for an arbitrary element can be performed with a small number of comparisons.

If the lists A and B have m and n elements respectively, then there are $\binom{m+n}{n}$ possible placements of the elements of B in the combined list; it follows that $\lceil \log\binom{m+n}{n} \rceil$ comparisons are necessary to distinguish these possible orderings. If we take $m \leqslant n$ then

$$\left\lceil \log\binom{m+n}{n} \right\rceil = \Theta\left(m \log \frac{n}{m}\right).$$

The "binary merging" algorithm of Hwang and Lin [103] (see also [112, Section 5.3.2]) requires fewer than $\lceil \log\binom{m+n}{n} \rceil + \min(m, n)$ comparisons to combine sets of size m and n and uses $O(\log n)$ additional workspace. Kronrad [118] showed that merging may be performed in place in linear time. His algorithm does not have the stability property.

Brown and Tarjan [32] note that there is a problem in implementing the *binary merging* algorithm so that it runs in time proportional to the number of comparisons it uses. They present a fast-merging algorithm, based on AVL-trees (Adel'son-Vel'skii and Landis [1]), which runs in optimal time $O(m \log(n/m))$ and uses additional workspace $O(m+n)$. This was extended by Huddleston and Mehlhorn [101] to other set operations than merging.

Trabb Pardo [202] gives a *stable merging* algorithm performing $O(m+n)$ assignments and comparisons in $O(1)$ workspace It yields a stable in-place sorting algorithm running in optimal time. The stable merging algorithm of Dudziński and Dydek [54] performs $O(m \log(n/m))$ comparisons and $O((m+n)\log m)$ assignments in $O(\log m)$ workspace. Carlsson [35] presented a fast stable merging algorithm, called "split-merge", which performs $O(m \log(n/m))$ comparisons and $O(m+n)$ assignments in $O(m \log n)$ space.

Let $M(m, n)$ be the minimum number of pairwise comparisons which will always suffice to merge two ordered lists of lengths m and n. Stockmeyer and F.F. Yao [189] prove that $M(m, m+d) = 2m+d-1$ whenever $m \geqslant 2d-2$. (This shows that the standard linear merging algorithm is optimal whenever $m \leqslant n \leqslant \lfloor \frac{3}{2}m \rfloor + 1$.) Thanh et al. [201] present optimal expected-time algorithms for $(2, n)$ and $(3, n)$ merge problems.

Searchability is introduced by Munro and Poblete [155], where they show that a pair of sorted arrays can be merged in-place in linear time so that a logarithmic-time search may be performed at any point during the process.

10. Dynamization techniques

Data structures for static sets are always easier to discover than dynamic data structures which also support updates. We now discuss some general methods for turning static data structures into dynamic data structures.

Suppose that we know a *static* data structure, which can be constructed in time $P_S(n)$, requires space $S_S(n)$ and supports queries in time $Q_S(n)$ for a set of n items. We assume throughout that $Q_S(n)$, $P_S(n)/n$ and $S_S(n)/n$ are nondecreasing. We also assume that the queries are *decomposable* in the following sense [19]: Consider a set A and any partitition of A into disjoint sets B and C. It is then assumed that the answer to a query about A can be obtained from the answers to the queries about B and C in time $O(1)$. A membership query is decomposable in this sense.

We now show how to support insertions. A brute-force solution reconstructs the static data structure from scratch after every insertion. A more efficient strategy is to partition a set S of n items into blocks S_i, to answer queries by combining the answers to the queries about the blocks and to insert an item by forming a new block consisting of a single item. In order to avoid the proliferation of blocks, several blocks are merged into a single block from time to time. The details are as follows. Let $n = \sum_{i \geqslant 0} a_i 2^i$, $a_i \in \{0, 1\}$, be the binary representation of n. Let $S_0, S_1, \ldots$ be any partition of S with

$|S_i|=a_i2^i$, $0\leqslant i\leqslant \log n$. Then the dynamic structure D is just a collection of static data structures, one for each nonempty S_i. The space requirement of D is easily computed as

$$S_D(n)=\sum_i S_S(a_i2^i)=\sum_i (S_S(a_i2^i)/a_i2^i)a_i2^i$$

$$\leqslant \sum_i (S_S(n)/n)a_i2^i=S_S(n).$$

The inequality follows from our basic assumption that $S_S(n)/n$ is nondecreasing.

Next note that a query about S can be answered by combining the answers about the S_i's and that there are never more than log n nonempty S_i's. Hence a query can be answered in time

$$\log n+\sum_i Q_S(a_i2^i)\leqslant \log n(1+Q_S(n))=\mathrm{O}(\log nQ_S(n)).$$

Finally consider operation *Insert*(x, S). Let $n+1=\sum \beta_i 2^i$ and let j be such that $\alpha_j=0$, $\alpha_{j-1}=\alpha_{j-2}=\cdots=\alpha_0=1$. Then $\beta_j=1$, $\beta_{j-1}=\cdots=\beta_0=0$. We process the $(n+1)$st insertion by taking the new point x and the $2^j-1=\sum_{i=0}^{j-1}2^i$ points stored in structures $S_0, S_1,\ldots,S_{j-1}$ and constructing a new static data structure for $\{x\}\cup S_0\cup S_1\cup\cdots\cup S_{j-1}$. Thus the cost of the $(n+1)$st insertion is $P_S(2^j)$. Next note that a cost of $P_S(2^j)$ has to be paid after insertions $2^j(2l+1)$, $l=0,1,2,\ldots$, and hence at most $n/2^j$ times during the first n insertions. Thus the total cost of the first n insertions is bounded by

$$\sum_{j=0}^{\lfloor\log n\rfloor} P_S(2^j)n/2^j\leqslant n\sum_{j=0}^{\lfloor\log n\rfloor} P_S(n)/n\leqslant P_S(n)(\lfloor \log n\rfloor+1).$$

We may summarize the argument above as $Q_D(n)=\mathrm{O}(Q_S(n)\log n)$, $S_D(n)=\mathrm{O}(S_S(n))$ and $\bar{I}_D(n)=\mathrm{O}((P_S(n)/n)\log n)$ where $Q_D(n)$ is the worst-case query time, $S_D(n)$ is the worst-case space requirement and $\bar{I}_D(n)$ is the amortized insertion time for a set of n elements [21].

A static data structure for the membership problem is a sorted array. Its performance is $S_S(n)=n$, $Q_S(n)=\log n$ and $P_S(n)=n\log n$. The construction above yields a dynamic data structure with $S_D(n)=\mathrm{O}(n)$ and $\bar{I}_D(n)=Q_D(n)=\mathrm{O}((\log n)^2)$.

Overmars and van Leeuwen [167] turned the amortized insertion time into a worst-case bound by spreading work over time. Of course, there is nothing unique about the use of the binary representation in the construction above. Using b-ary notation for some $b>1$, i.e. expressing n as $\sum a_ib^i$, $0\leqslant a_i< b$, and using a_i sets of size b^i in the partition of S, a query time $Q_D(n)=\mathrm{O}(b\cdot Q_S(n)\log n/\log b)$ and an amortized insertion time of $\bar{I}_D(n)=\mathrm{O}((P_S(n)/n)\log n/\log b)$ results [21]. Mehlhorn and Overmars [149] describe a family of dynamization schemes which allow to trade query time for insertion time and vice versa. The corresponding lower bounds were first shown in [21] and later generalized in [144].

Deletions are somewhat harder to cope with. We need the additional assumption that the static data structure supports deletions (but not insertions) without increasing query time and storage requirement. With this additional assumption the results mentioned above can be extended to also include deletions [124, 166, 167]. Extensions and variations of the basic dynamization scheme are discussed in [164, 56, 173].

References

[1] ADEL'SON-VEL'SKII, G.M. and E.M. LANDIS, An algorithm for the organisation of information, *Dokl. Akad. Nauk SSSR* **146** (1962) 263–266 (in Russian); English translation in *Soviet. Math.* **3** (1962) 1259–1262.

[2] AHO, A.V., J.E. HOPCROFT and J.D. ULLMANN, *The Design and Analysis of Computer Algorithms* (Addison-Wesley, Reading, MA, 1974).

[3] AHO, A.V., J.E. HOPCROFT and J.D. ULLMAN, On finding lowest common ancestors in trees, *SIAM J. Comput.* **1** (1) (1976) 115–132.

[4] AHO, A.V., J.E. HOPCROFT and J.D. ULLMAN, *Data Structures and Algorithms* (Addison-Wesley, Reading, MA, 1983).

[5] AHO, A.V. and D. LEE, Storing a dynamic sparse table, in: *Proc 27th Ann. IEEE Symp. on Foundations of Computer Science* (1986) 55–60.

[6] AHO, A.V. and J.D. ULLMAN, *Principles of Compiler Design* (Addison-Wesley, Reading, MA, 1977).

[7] AHUJA, R.K., K. MEHLHORN, J.B. ORLIN and R.E. TARJAN, Faster algorithms for the shortest path problem, Tech. Report A04/88, Univ. of Saarland, Saarbrücken, FRG.

[8] ALLEN, B. and I. MUNRO, Self-organizing search trees, *J. ACM* **25**(4) (1978) 526–535.

[9] ALT H. and K. MEHLHORN, Searching semi-sorted tables, *SIAM.J. Comput.* **14**(4) (1985) 840–848.

[10] ANDERSON, E.J., P. NASH and R.R. WEBER, A counter example to a conjecture on optimal list ordering, *J. Appl. Probab.* **19**(3) (1982) 730–732.

[11] ATKINSON, M.D., J.R. SACK, N. SANTORO and T. STROTHOTTE, Min-max heaps and generalized priority queues, *Comm. ACM* **19**(10) (1986) 996–1000.

[12] BANACHOWSKI, L., A complement to Tarjan's result about the lower bound on the complexity of the set union problem, *Inform. Process. Lett.* **11**(2) (1980) 59–65.

[13] BAYER, P.J., Improved bounds on the cost of optimal and balanced binary search trees, Tech. Report, Dept. of Computer Science, Massachusetts Institute of Technology, Cambridge, MA, 1975.

[14] BAYER, R., Symmetric binary *B*-trees: data structure and maintenance algorithms, *Acta Inform.* **1** (1972) 290–306.

[15] BAYER, R. and E.M. MCCREIGHT, Organisation and maintenance of large ordered indices, *Acta Inform.* **1** (1972) 173–189.

[16] BEN-AMRAM, A.M. and Z. GALIL, On pointers versus adresses, in: *Proc. 29th Ann. IEEE Symp. on Foundations of Computer Science* (1988) 532–538.

[17] BENT, S.W. and J. JOHN, Finding the median requires $2n$ comparisons, in: *Proc. 17th Ann. ACM Symp. on Theory of Computing* (1985) 213–216.

[18] BENT, S.W., D.D. SLEATOR and R.E. TARJAN, Biased search trees, *SIAM J. Comput.* **14**(3) (1985) 545–568.

[19] BENTLEY, J., Decomposable searching problems, *Inform. Process. Lett.* **8**(5) (1979) 244–251.

[20] BENTLEY, J. and C.C. MCGEOCH, Amortized analysis of self-organizing sequential search heuristics, *Comm. ACM* **28**(4) (1985) 405–411.

[21] BENTLEY, J. and J.B. SAXE, Decomposable searching problems I: static-to-dynamic transformations, *J. Algorithms* **1** (1980) 301–358.

[22] BITNER, J.R., Heuristics that dynamically organize data structures, *SIAM J. Comput.* **8**(1) (1979) 82–110.

[23] BLUM, M., R.W. FLOYD, V. PRATT, R.L. LEWIS and R.E. TARJAN, Time bounds for selection, *J. Comput. System Sci.* **7** (1973) 448–461.

[24] BLUM, N., On the single-operation worst-case time complexity of the disjoint set union problem, *SIAM J. Comput.* **15**(4) (1986) 1021–1924.

[25] BLUM, N. and K. MEHLHORN, On the average number of rebalancing operations in weight-balanced trees, *Theoret. Comput. Sci.* **11** (1980) 303–320.

[26] BOLLOBÁS, B. and I. SIMON, On the expected behavior of disjoint set union algorithms, in: *Proc. 17th Ann. ACM Symp. on Theory of Computing* (1985) 224–231.

[27] BOLLOBÁS, B. and I. SIMON, Repeated random insertion into a priority queue, *J. Algorithms* **6** (1985) 466–477.

[28] BORODIN, A., F. FICH, F. MEYER AUF DER HEIDE, E. UPFAL and A. WIGDERSON, A tradeoff between search and update time for the implicit dictionary problem, in: *Proc. 13th Internat. Coll. on Automata, Languages an Programming*, Lecture Notes in Computer Science, Vol. 226 (Springer, Berlin, 1986) 50–59.

[29] BORODIN, A., L.J. GUIBAS, N.A. LYNCH and A.C. YAO, Efficient searching using partial ordering, *Inform. Process. Lett.* **12** (1981) 71 – 75.

[30] BROWN, M.R., Implementation and analysis of binomial queue algorithms, *SIAM J. Comput.* **7** (1978) 298–319.

[31] BROWN, M.R., A partial analysis of random height-balanced trees, *SIAM J. Comput.* **8**(1) (1979) 33–41.

[32] BROWN, M.R. and R.E. TARJAN, A fast merging algorithm, *J. ACM* **26**(2) (1979) 211–226.

[33] BROWN, M.R. and R.E. TARJAN, Design and analysis of a data structure for representing sorted lists, *SIAM J. Comput.* **9** (1980) 594–614.

[34] BURVILLE, P. and J. KINGMAN, On a model for storage and search, *J. Appl. Probab.* **10**(3) (1973) 697–701.

[35] CARLSSON, S., Splitmerge—a fast stable merging algorithm, *Inform. Process. Lett.* **22** (1986) 189–192.

[36] CARLSSON, S., J.F. MUNRO and P.V. POBLETE, An implicit binomial queue with constant insertion time, in: *Proc. 1st Scandinavian Workshop on Algorithm Theory (SWAT 88)*, Halmstad, Sweden, Lecture Notes in Computer Science, Vol. 318 (Springer, Berlin, 1988) 1–13.

[37] CARTER, J.L. and M.N. WEGMAN, Universal classes of hash functions, in: *Proc. 9th Ann. ACM Symp. on Theory of Computing* (1977) 106–112.

[38] CHAZELLE, B., How to search in history, *Inform. and Control* **63** (1985) 77–99.

[39] CHAZELLE, B. and L.J. GUIBAS, Fractional cascading I: a data structuring technique, *Algorithmica* **1**(2) (1986) 133–162.

[40] CHUNG, F.R., and D.J. HAJELA and P.D. SEYMOUR, Self-organizing sequential search and Hilbert's inequalities, *J. Comput. System Sci.* **36** (1988) 148–157.

[41] COLE, R., Searching and storing similar lists, *J. Algorithms* **7** (1986) 202–220.

[42] CRANE, C.A., Linear lists and priority queues as balanced binary trees, Tech. Report, STAN-CS-72-259, Computer Science Dept., Stanford Univ., Stanford, CA, 1972.

[43] CUNTO, W. and J.I. MUNRO, Average case selection, in: *Proc. 16th Ann. ACM Symp. on Theory of Computing* (1984) 369–375.

[44] DEVROYE, L., A note on the height of binary search trees, *J. ACM* **33**(3) (1986) 489–498.

[45] DEVROYE, L., Branching processes in the analysis of the heights of trees, *Acta Inform.* **24** (1987) 277–298.

[46] DIETZFELBINGER, M., A. KARLIN, K.MEHLHORN, F. MEYER AUF DER HEIDE, H. ROHNERT and R.E. TARJAN, Dynamic perfect hashing: upper and lower bounds, in: *Proc. 29th Ann IEEE Symp. on Foundations of Computer Science* (1988) 524–531.

[47] DOBERKAT, E.E., An average case analysis of Floyd's algorithm to construct heaps, *Inform. and Control* **61** (1984) 114–131.

[48] DOBKIN, D. and J.I. MUNRO, Optimal minimal space selection algorithms, *J. ACM* **28**(3) (1981) 454–461.

[49] DOBKIN, D. and J.I. MUNRO, Efficient uses of the past, *J. Algorithms* **6** (1985) 455–465.

[50] DODGSON, C.L., *St. James Gazette* **5–6** (1 August 1883).

[51] DOYLE, I. and R. RIVEST, Linear expected time of a simple union-find algorithm, *Inform. Process. Lett.* **5** (1976) 146–148.

[52] DRISCOLL, J.R., H.N. GABOW, R. SHRAIRMAN and R.E. TARJAN, Relaxed heaps: an alternative to Fibonacci heaps with applications to parallel computation, *Comm. ACM* **31**(11) (1988) 1343–1354.

[53] DRISCOLL, J.R., N. SARNAK, D.D. SLEATOR and R.E. TARJAN, Making data structures persistent, *J. Comput. System Sci.* **28** (1989) 86–124.

[54] DUDZIŃSKI, K. and A. DYDELL, On a stable minimum space merging algorithm, *Inform. Process. Lett.* **12**(1) (1981) 5–8.

[55] EDELSBRUNNER, H., L.J. GUIBAS and J. STOLFI, Optimal point location in a monotone subdivision, *SIAM J. Comput.* **15**(2) (1986) 317–340.
[56] EDELSBRUNNER, H. and M.H. OVERMARS, Batched dynamic solutions to decomposable searching problems, *J. Algorithms* **6** (1985) 515–542.
[57] EMDE BOAS, P. VAN, Preserving order in a forest in less than logarithmic time and linear space, *Inform. Process. Lett.* **6** (1977) 80–82.
[58] EMDE BOAS, P. VAN, R. KAAS and E. ZIJLSTRA, Design and implementation of an efficient priority queue, *Math. Systems Theory* **10** (1977) 99–127.
[59] FAGIN, R., J. NIEVERGELT, N. PIPPENGER and H.R. STRONG, Extendible hashing—a fast access method for dynamic files, *ACM Trans. Database Systems* **4** (1979) 315–344.
[60] FALLER, N., An adaptive system for data compression, in: *Record 7th Asilomar Conf. on Circuits, Systems, and Computers* (1973) 593–597.
[61] FISCHER, M.J., Efficiency of equivalence algorithms, in: R.E. Miller and J.W. Thatcher, eds., *Complexity of Computer Computation* (Plenum Press, New York, 1972) 153–168.
[62] FLAJOLET, PH., On the performance evaluation of extendible hashing and trie searching, *Acta Inform.* **20** (1983) 345–369.
[63] FLOYD, R.W., Algorithm 245: Treesort 3, *Comm. ACM* **7** (1964) 701.
[64] FLOYD, R.W. and R.L. RIVEST, Expected time bounds for selection, *Comm. ACM* **18** (1975) 165–172.
[65] FOSTER, C.C., Information storage and retrieval using AVL-trees, in: *Proc. ACM 20th Nat. Conf.* (1965) 192–205.
[66] FOSTER, C.C., A generalization of AVL-trees, *Comm. ACM* **16**(8) (1973) 513–517.
[67] FREDERICKSON, G.N., Implicit data structures for the dictionary problem, *J. ACM* **30**(1) (1983) 80–94.
[68] FREDERICKSON, G.N., Self-organizing heuristic for implicit data structures, *SIAM J. Comput.* **13**(2) (1984) 277–291.
[69] FREDERICKSON, G.N., Implicit data structures for weighted elements, *Inform. and Control* **66** (1985) 61–82.
[70] FREDERICKSON, G.N., Upper bounds for time-space tradeoffs in sorting and selection, *J. Comput. System Sci.* **34** (1987) 19–26.
[71] FREDKIN, E., Trie memory, *Comm. ACM* **3** (1962) 490–499.
[72] FREDMAN, M.L., Two applications of a probabilistic search technique: sorting $X+Y$ and building balanced search trees, in: *Proc. 7th Ann. ACM Symp. on Theory of Computing* (1975) 240–244.
[73] FREDMAN, M.L., J. KOMLÓS and E. SZEMERÉDI, Storing a sparse table with O(1) worst case access time, *J. ACM* **31**(3) (1984) 538–544.
[74] FREDMAN, M.L., R. SEDGEWICK, D.D. SLEATOR and R.E. TARJAN, The pairing heap: a new form of self-adjusting heap, *Algorithmica* **1**(1) (1986) 111–129.
[75] FREDMAN, M.L. and R.E. TARJAN, Fibonacci heaps and their uses in improved network optimization algorithms, *J. ACM* **34**(3) (1987) 596–615.
[76] GABOW, H.N., A scaling algorithm for weighted matching in general graphs, in: *Proc. 26th Ann. IEEE Symp. on Foundations of Computer Science* (1985) 90–100.
[77] GABOW, H.N. and R.E. TARJAN, A linear time algorithm for a special case of disjoint set union, *J. Comput. System Sci.* **30** (1985) 209–221.
[78] GALLAGER, R.G., Variations on a theme by Huffman, *IEEE Trans. Inform. Theory* **24** (1978) 668–674.
[79] GALLER, B.A. and M.J. FISCHER, An improved equivalence algorithm, *Comm. ACM* **7** (1964) 301–303.
[80] GAMBOSI, G., G.F. ITALIANO and M. TALAMO, Getting back to the past in the union-find problem, in: *Proc. STACS 88, 5th Ann. Symp.*, Lecture Notes in Computer Science, Vol. 294 (Springer, Berlin, 1988) 8–17.
[81] GARSIA, A.M. and M.L. WACHS, A new algorithm for minimal binary search trees, *SIAM J. Comput.* **6** (1977) 622–642.
[82] GONNET, G.H., Balancing binary trees by internal path reduction, *Comm. ACM* **26**(12) (1983) 1074–1081.
[83] GONNET, G.H., *Handbook of Algorithms and Data Structures*, International Computer Science Series (Addison-Wesley, Reading, MA, 1984).
[84] GONNET, G.H. and I. MUNRO, Heaps on heaps, *SIAM J. Comput.* **15**(4) (1986) 964–971.
[85] GONNET, G.H., I. MUNRO and H. SUWANDA, Exegesis of self-organizing linear search, *SIAM J. Comput.* **10**(3) (1981) 613–637.

[86] GONNET, G.H., L. ROGERS and I. GEORGE, An algorithmic and complexity analysis of interpolation search, *Acta Inform.* **3**(1) (1980) 39–52.

[87] GÜTING, H. and H.P. KRIEGEL, Multidimensional B-tree: an efficient dynamic file structure for exact match queries, *Informatik Fachberichte* **33** (1980) 375–388.

[88] GUIBAS, L.J., E.M. MCCREIGHT, M.F. PLASS and J.R. ROBERTS, A new representation for linear lists, in: *Proc. 9th Ann. ACM Symp. on Theory of Computing* (1977) 49–60.

[89] GUIBAS, L.J. and R. SEDGEWICK, A dichromatic framework for balanced trees, in: *Proc. 19th Ann. IEEE Symp. on Foundations of Computer Science* (1978) 8–21.

[90] GULBERSON, J., The effect of updates in binary search trees, in: *Proc. 17th Ann. ACM Symp. on Theory of Computing* (1985) 205–210.

[91] HADIAN, A. and M. SOBEL, Selecting the *t*th largest using binary errorless comparisons, in: P. Erdös, A. Renyi and V.T. Sós, eds., *Combinatorial Theory and its Applications II*, Colloquia Mathematica Societatis Janos Bolyai, Vol. 4 (North-Holland, Amsterdam, 1969) 585–599.

[92] HAREL, D. and R.E. TARJAN, Fast algorithms for finding nearest common ancestors, *SIAM J. Comput.* **13** (1984) 338–355.

[93] HENDRICKS, W.J., The stationary distribution of an interesting Markov chain, *J. Appl. Probab.* **9**(1) (1972) 231–233.

[94] HOARE, C.A.R., Quicksort, *Comput. J.* **5** (1962) 10–15.

[95] HOFFMANN, K., K. MEHLHORN, P. ROSENSTIEHL and R.E. TARJAN, Sorting Jordan sequences in linear time using level-linked search trees, *Inform. and Control* **68** (1986) 170–184.

[96] HOOD, R. and R. MELVILLE, Real-time queue operations in pure LISP, *Inform. Process. Lett.* **13** (1981) 50–54.

[97] HOPCROFT, J.E. and J.D. ULLMAN, Set merging algorithms, *SIAM J. Comput.* **2** (1973) 294–303.

[98] HOROWITZ, E. and S. SAHNI, *Fundamentals of Data Structures* (Computer Science Press, Potomac MD, 1976).

[99] HU, T.C. and A.C. TUCKER, Optimal computer-search trees and variable-length alphabetic codes, *SIAM J. Appl. Math.* **21** (1971) 514–532.

[100] HUDDLESTON, S., An efficient scheme for fast local updates in linear lists, Tech. Report, Dept. of Information and Computer Science, Univ. of California, Irvine, 1981.

[101] HUDDLESTON, S. and K. MEHLHORN, A new data structure for representing sorted lists, *Acta Inform.* **17** (1982) 157–184.

[102] HUFFMAN, D.A., A method for the construction of minimum redundancy codes, *Proc. IRE* **40** (1952) 1098–1101.

[103] HWANG, F.K. and S. LIN, A simple algorithm for merging two disjoint linearly ordered sets, *SIAM J. Comput.* **1**(1) (1972) 31–39.

[104] IMAI, H. and T. ASANO, Dynamic orthogonal segment intersection search, *J. Algorithms* **8** (1987) 1–18.

[105] ITAI, A., A.G. KONHEIM and M. RODEH, A sparse table implementation of priority queues, in: *Proc. 8th Internat. Coll. on Automata, Languages and Programming*, Lecture Notes in Computer Science, Vol. 115 (Springer, Berlin, 1981) 417–431.

[106] JACOBS, T.M. and P. VAN EMDE BOAS, Two results on tables, *Inform. Process. Lett.* **22** (1986) 43–48.

[107] JOHNSON, D., Efficient algorithms for shortest paths in sparse networks, *J. ACM* **24** (1977) 1–3.

[108] KINGSTON, J.H., A new proof of the Garsia–Wachs algorithm, *J. Algorithms* **9** (1988) 129–136.

[109] KIRKPATRICK, D.G., A unified lower bound for selection and set partitioning problem, *J. ACM* **28** (1981) 150–165.

[110] KNOTT, G.D., Expandable open addressing hash table storage and retrieval, in: *Proc. ACM SIGFIDET Workshop on Data Description, Access and Control* (1971) 186–206.

[111] KNUTH, D.E., Optimum binary search tree, *Acta Inform.* **1** (1971) 14–25.

[112] KNUTH, D.E., *The Art of Computer Programming, Vol. 3: Sorting and Searching* (Addison-Wesley, Reading, MA, 1973).

[113] KNUTH, D.E., Deletions that preserve randomness, *IEEE Trans. Software Engrg.* **3**(5) (1977) 351–359.

[114] KNUTH, D.E., Dynamic Huffman coding, *J. Algorithms* **6** (1985) 163–180.

[115] KNUTH, D.E. and A. SCHÖNHAGE, The expected linearity of a simple equivalence algorithm, *Theoret. Comput. Sci.* **6** (1978) 281–315.

[116] KORSCH, J.F., Greedy binary search trees are nearly optimal, *Inform. Process. Lett.* **13** (1981) 16–19.

[117] KOSARAJU, S.R., Localized search in sorted list, in: *Proc. 13th Ann. ACM Symp. on Theory of Computing* (1981) 62–69.

[118] KRONROD, M.A., Optimal ordering algorithm without operational field, *Soviet. Math. Dokl.* **10** (1969) 744–746.

[119] KRIEGEL, H.P. and V.K. VAISHNAVI, Weighted multidimensional B-trees used as nearly optimal dynamic dictionaries, in: *Proc. Mathematical Foundations of Computer Science*, Štrbské Pleso, Czechoslovakia, Lecture Notes in Computer Science, Vol. 118 (Springer, Berlin, 1981) 410–417.

[120] LAI, T.W. and D. WOOD, Implicit selection, in: *Proc. 1st Scandinavian Workshop on Algorithm Theory (SWAT 88)*, Halmstad, Sweden, Lecture Notes in Computer Science, Vol. 318 (Springer, Berlin, 1988) 14–23.

[121] LAM, K., M.K. SIU and C.T. YU, A generalized counter scheme, *Theoret. Comput. Sci.* **16** (1981) 271–278.

[122] LARSON, P.A., Dynamic hashing, *BIT* **18** (1978) 184–201.

[123] LEEUWEN, J. VAN, Finding lowest common ancestors in less than logarithmic time, Unpublished report, Amherst, NY, 1976.

[124] LEEUWEN, J. VAN and H.A. MAURER, Dynamic systems of static data structures, Bericht 42, Institut für Informationsverarbeitung, TU Graz, Austria, 1980.

[125] LEEUWEN, J. VAN and A.K. TSAKALIDIS, An optimal pointer machine algorithm for nearest common ancestors, Tech. Report, UU-CS-88-17, Dept. of Computer Science, Univ. of Utrecht, Utrecht, 1988.

[126] LEVCOPOULUS, C. and M.H. OVERMARS, A balanced search tree with O(1) worst-case update time, *Acta Inform.* **26** (1988) 269–277.

[127] LINIAL, N. and M.E. SAKS, Information bounds are good for search problems on ordered data structures, in: *Proc. 24th Ann. IEEE Symp. on Foundations of Computer Science* (1983) 473–475.

[128] LIPSKI, W., An $O(n \log n)$ Manhattan path algorithm, *Inform. Process. Lett.* **19** (1984) 99–102.

[129] LITWIN, W., Virtual hashing: a dynamically changing hashing, in: *Proc. Very Large Data Bases Conf.*, Berlin (1978) 517–523.

[130] MAIER, D., An efficient method for sorting ancestor information in trees, *SIAM J. Comput.* **8**(4) (1979) 599–618.

[131] MAIER, D. and S.C. SALVETER, Hysterical B-trees, *Inform. Process. Lett.* **12** (1981) 199–202.

[132] MAIRSON, H., Average case lower bounds on the construction and searching of partial orders, in: *Proc. 26th Ann. IEEE Symp. Foundation of Computer Science* (1985) 303–311.

[133] MANNILA., H. and E. UKKONEN, The set union problem with backtracking, in: *Proc. 13th Internat. Coll. on Automata, Languages and Programming*, Lecture Notes in Computer Science, Vol. 226 (Springer, Berlin, 1986) 236–243.

[134] MARKOWSKY, G., J.L. CARTER and M.N. WEGMAN, Analysis of a universal class of hash functions, in: *Proc. Mathematical Foundations of Computer Science*, Zakapone, Poland, Lecture Notes in Computer Science, Vol. 64 (Springer, Berlin, 1978) 345–354.

[135] MATULA, D.W., Selecting the t-th best in average $n+O(\log \log n)$ comparisons, Tech. Report 73-9, Dept. of Mathematics, Washington Univ., St. Louis, 1973.

[136] MCCABE, J., On serial file with relocatable records, *Oper. Res.* **12** (1965) 609–618.

[137] MCDIARMID, C., Average-case lower bounds for searching, *SIAM J. Comput.* **17**(1) (1988) 1044–1060.

[138] MEHLHORN, K., Nearly optimal binary search trees, *Acta Inform.* **5** (1975) 287–295.

[139] MEHLHORN, K., A best possible bound for the weighted path length of binary search trees, *SIAM J. Comput.* **6**(2) (1977) 235–239.

[140] MEHLHORN, K., Dynamic binary search, *SIAM J. Comput.* **8** (1979) 175–198.

[141] MEHLHORN, K., Arbitrary weight changes in dynamic trees, *RAIRO Inform. Théor.* **15**(3) (1981) 183–211.

[142] MEHLHORN, K., A partial analysis of height-balanced trees under random insertions and deletions, *SIAM J. Comput.* **11** (1982) 748–760.

[143] MEHLHORN, K., On the program size of perfect and universal hash functions, in: *Proc. 23rd Ann. IEEE Symp. on Foundations of Computer Science* (1982) 170–175.

[144] MEHLHORN, K., Lower bounds on the efficiency of transforming static data structures into dynamic data structures, *Math. Systems Theory* **15** (1982) 1–16.

[145] MEHLHORN, K., *Data Structures and Algorithms 1: Sorting and Searching*, EATCS Monographs on Theoretical Computer Science (Springer, Berlin, 1984).

[146] MEHLHORN, K. and ST. NÄHER, Dynamic fractional cascading, Tech. Report A 06/86, Univ. of Saarland, Saarbrücken, FRG; also: *Algorithmica*, to appear.

[147] MEHLHORN, K., ST. NÄHER and H. ALT, A lower bound for the complexity of the union-split-find problem, in: *Proc. 14th Internat. Coll. on Automata, Languages and Programming*, Lecture Notes in Computer Science, Vol. 267 (Springer, Berlin, 1987); also: *SIAM J. Comput.*, to appear.

[148] MEHLHORN, K., ST. NÄHER and CH. UHRIG, Deleting versions in persistent data structures, Tech. Report, Univ. of Saarland, Saarbrücken, 1989.

[149] MEHLHORN, K. and M.H. OVERMARS, Optimal dynamization of decomposable searching problems, *Inform. Process. Lett.* **12**(2) (1981) 93–98.

[150] MEHLHORN, K. and A.K. TSAKALIDIS, Dynamic interpolation search, in: *Proc. 12th Internat. Coll. on Automata, Languages and Programming*, Lecture Notes in Computer Science, Vol. 194 (Springer, Berlin, 1985) 424–434.

[151] MEHLHORN, K. and A.K. TSAKALIDIS, An amortized analysis of insertions into AVL-trees, *SIAM J. Comput.* **15**(1) (1986) 22–33.

[152] MUNRO, I., An implicit data structure supporting insertion, deletion and search in $O(\log^2 n)$ time, *J. Comput. System Sci.* **33** (1986) 66–74.

[153] MUNRO, I., Developing implicit data structures, in: *Proc. Mathematical Foundations of Computer Science*, Bratislava, Czechoslovakia, Lecture Notes in Computer Science, Vol. 233 (Springer, Berlin, 1986) 168–176.

[154] MUNRO, I. and M.S. PATERSON, Selecting and sorting with limited space, *Theoret. Comput. Sci.* **12** (1980) 315–323.

[155] MUNRO, I. and P. POBLETE, Searchability in merging and implicit data structures, in: *Proc. 10th Internat. Coll. on Automata, Languages and Programming*, Lecture Notes in Computer Science, Vol. 154 (Springer, Berlin, 1983) 527–535.

[156] MUNRO, I. and H. SUWANDA, Implicit data structures, *J. Comput. System Sci.* **21** (1980) 236–250.

[157] MYERS, E.W., An applicative random-access stack, *Inform. Process. Lett.* **17** (1983) 241–248.

[158] MYERS, E.W., Efficient applicative data types, in: *Conf. Record 11th Ann. ACM Symp. on Principles of Programming Languages* (1984) 66–75.

[159] NIEVERGELT, I. and E.M. REINGOLD, Binary search trees of bounded balance, *SIAM J. Comput.* **2** (1973) 33–43.

[160] OLIVIE, H.J., A new class of balanced search trees: half balanced binary search tree, *RAIRO Inform. Théor.* **16**(1) (1982) 51–71.

[161] ORLIN, J.B. and R.K. AHUJA, A fast and simple shortest path algorithm, in: *Cornell Workshop on Combinatorial Optimization*, Ithaca, NY, 1987.

[162] OTTMANN, TH. and H.W. SIX, Eine neue Klasse von ausgeglichenen Bäumen, *Angewandte Informatik* **18** (1976) 395–400.

[163] OVERMARS, M.H., Searching in the past I, Report RUU-CS-81-7, Dept. of Computer Science, Univ. of Utrecht, Utrecht, 1981.

[164] OVERMARS, M.H., Dynamization of order decomposable set problems, *J. Algorithms* **2** (1981) 245–260.

[165] OVERMARS, M.H., *The Design of Dynamic Data Structures*, Lecture Notes in Computer Science, Vol. 156 (Springer, Berlin, 1983).

[166] OVERMARS, M.H. and J. VAN LEEUWEN, Two general methods for dynamizing decomposable searching problems, *Comput.* **26** (1981) 155–166.

[167] OVERMARS, M.H. and J. VAN LEEUWEN, Worst-case optimal insertion and deletion methods for decomposable searching problems, *Inform. Process. Lett.* **12**(4) (1981) 168–173.

[168] PEARL, Y., A. ITAI and H. AVNI, Interpolation search—a log log N search, *Comm. ACM* **21**(7) (1978) 550–554.

[169] PEARL, Y. and E.M. REINGOLD, Understanding the complexity of interpolation search, *Inform. Process. Lett.* **6** (1977) 219–222.

[170] PETERSON, W.W., Addressing for random storage, *IBM J. Res. Develop.* **1** (1957) 131–132.

[171] PORTER, T. and I. SIMON, Random insertion into a priority queue structure, *IEEE Trans. Software Engrg.* **1** (1975) 292–298.
[172] RAMANAN, P.V. and L. HYAFIL, New algorithms for selection, *J. Algorithms* **5** (1984) 557–578.
[173] RAO, N.S.V., V.K. VAISHNAVI and S.S. IYENGAR, On the dynamization of data structures, *BIT* **28** (1988) 37–53.
[174] REPS, T., T. TEITELBAUM and A. DEMERS, Incremental context-dependent analysis for language-based editors, *ACM Trans. Programming Systems and Languages* **5** (1983) 449–477.
[175] RIVEST, R., On self-organizing sequential search heuristics, *Comm. ACM* **2** (1976) 63–67.
[176] SANTORO, N. and I. SIDNEY, Interpolation-binary search, *Inform. Process. Lett.* **20** (1985) 179–181.
[177] SARNAK, N. and R.E. TARJAN, Planar point location using persistent search trees, *Comm. ACM* **29** (1986) 669–679.
[178] SCHIEBER, B. and U. VISHKIN, On finding lowest common ancestors: simplification and parallelization, in: *Proc. 3rd Aegean Workshop on VLSI Algorithms and Architecture*, Lecture Notes in Computer Science, Vol. 319 (Springer, Berlin, 1988) 111–123.
[179] SCHÖNHAGE, A., Storage modification machines, *SIAM J. Comput.* **9**(3) (1980) 490–508.
[180] SCHÖNHAGE, A., M. PATERSON and N. PIPPENGER, Finding the median, *J. Comput. System Sci.* **13** (1976) 184–199.
[181] SEDGEWICK, R., *Algorithms* (Addison-Wesley, Reading, MA, 2nd ed., 1988).
[182] SLEATOR, D.D. and R.E. TARJAN, A data structure for dynamic trees, *J. Comput. System Sci.* **26** (1983) 362–391.
[183] SLEATOR, D.D. and R.E. TARJAN, Self-adjusting binary search trees, *J. ACM* **32**(3) (1985) 652–686.
[184] SLEATOR, D.D. and R.E. TARJAN, Amortized efficiency of list update and paging rules, *Comm. ACM* **28** (2) (1985) 202–208.
[185] SLEATOR, D.D. and R.E. TARJAN, Self adjusting heaps, *SIAM J. Comput.* **15**(1) (1986) 52–68.
[186] SPRUGNOLI, R., Perfect hash functions: a single probe retrieval method for static sets, *Comm. ACM* **20** (1977) 841–850.
[187] STANDISH, T.A., *Data Structure Techniques* (Addison-Wesley, Reading, MA, 1980).
[188] STEINHAUS, H., Some remarks about tournaments, in: *Calcutta Math. Soc. Golden Commemoration* **2** (1958).
[189] STOCKMEYER, P. and F.F. YAO, On the optimality of linear merge, *SIAM J. Comput.* **9**(1) (1980) 85–90.
[190] SWART, G.F., Efficient algorithms for computing geometric intersections, Tech. Report 85-01-02, Dept. of Computer Science, Univ. of Washington, Seattle, WA, 1985.
[191] TAMMINEN, M., Extendible hashing with overflow, *Inform. Process. Lett.* **15**(5) (1982) 227–233.
[192] TARJAN, R.E., Efficiency of a good but not linear set union algorithm, *J. ACM* **22** (1975) 215–225.
[193] TARJAN, R.E., A class of algorithms which require non-linear time to maintain disjoint sets, *J. Comput. System Sci.* **18** (1979) 110–127.
[194] TARJAN, R.E., Private communication, 1982.
[195] TARJAN, R.E., *Data Structures and Network Algorithms* (Society for Industrial and Applied Mathematics, Philadelphia, PA, 1983).
[196] TARJAN, R.E., Updating a balanced search tree in O(1) rotations, *Inform. Process. Lett.* **16** (1983) 253–257.
[197] TARJAN, R.E., Sequential access in splay trees takes linear time, *Combinatorica* **5**(4) (1985) 367–378.
[198] TARJAN, R.E., Amortized computational complexity, *SIAM J. Alg. Discrete Meth.* **2**(6) (1985) 306–318.
[199] TARJAN, R.E. and J. VAN LEEUWEN, Worst-case analysis of set union algorithms, *J. ACM* **31** (1984) 245–281.
[200] TARJAN, R.E. and A.C. YAO, Storing a sparse table, *Comm. ACM* **22**(11) (1979) 606–611.
[201] THANH, M., V.S. ALAGAR and T.O. BUI, Optimal expected-time algorithms for merging, *J. Algorithms* **7** (1986) 341–357.
[202] TRABB PARDO, L., Stable sorting and merging with optimal space and time, *SIAM J. Comput.* **6**(2) (1977) 351–372.
[203] TSAKALIDIS, A.K., AVL-trees for localized search, *Inform. and Control* **67**(1–3) (1985) 173–194.
[204] TSAKALIDIS, A.K., A simple implementation for localized search, in: *Proc. WG'85, Internat. Workshop on Graph-theoretical Concepts in Computer Science*, Würzburg, FRG (Trauner-Verlag, Linz, 1985) 363–374; also: *SIAM J. Comput.*, to appear.

[205] TSAKALIDIS, A.K., Rebalancing operations for deletions in AVL-trees, *RAIRO Inform. Théor.* **19**(4) (1985) 323–329.

[206] TSAKALIDIS, A.K., The nearest common ancestor in a dynamic tree, *Acta Inform.* **25** (1988) 37–54.

[207] UNTERAUER, K., Dynamic weighted binary search trees, *Acta Inform.* **11** (1979) 341–362.

[208] VAISHNAVI, V.K., Weighted leaf AVL-trees, *SIAM J. Comput.* **16**(3) (1987) 503–537.

[209] VAISHNAVI, V.K. and D. WOOD, Rectilinear segment intersection layered segment trees and dynamization, *J. Algorithms* **3** (1982) 160–176.

[210] VITTER, J., Design and analysis of dynamic Huffman coding, *J. ACM* **34**(4) (1987) 825–845.

[211] VUILLEMIN, J., A data structure for manipulating priority queues, *Comm. ACM* **21** (1978) 309–314.

[212] WESTBROOK, J., and R.E. TARJAN, Amortized analysis of algorithms for set union with backtracking, *SIAM J. Comput.* **18** (1989) 1–11.

[213] WILLARD, D.E., Log-logarithmic worst-case range queries are possible in space $\Theta(N)$, *Inform. Process. Lett.* **17** (1983) 81–84.

[214] WILLARD, D.E., New trie data structures which support very fast search operations, *J. Comput. System Sci.* **28** (1984) 379–394.

[215] WILLARD, D.E., Searching unindexed and nonuniformly generated files in $\log\log N$ time, *SIAM J. Comput.* **14**(4) (1985) 1013–1029.

[216] WILLARD, D.E., New data structures for orthogonal queries, *SIAM J. Comput.* **14**(1) (1985) 232–235.

[217] WILLARD, D.E. and G.S. LUECKER, Adding range restriction capability to dynamic data structures, *J. ACM* **32**(3) (1985) 597–617.

[218] WILLIAMS, J.W.J., Algorithm 232: Heapsort, *Comm. ACM* **7** (1964) 347–348.

[219] WIRTH, N., *Algorithms + Data Structures = Programs* (Prentice Hall, Englewood Cliffs, NJ, 1976).

[220] YAO, A.C., On the average behavior of set merging algorithms, in: *Proc. 8th Ann. ACM Symp. Theory of Computing* (1976) 192–195.

[221] YAO, A.C., On the expected performance of path compression algorithms, *SIAM J. Comput.* **14**(1) (1985) 129–133.

[222] YAO, A.C. and F.F. YAO, The complexity of searching on ordered random table, in: *Proc. 17th Ann. IEEE Symp. Foundations of Computer Science* (1976) 173–177.

[223] YAO, A.C. and F.F. YAO, On the average case complexity of selecting the k-th best, *SIAM J. Comput.* **11**(3) (1982) 428–447.

[224] YAO, F.F., Efficient dynamic programming using quadrangle inequalities, in: *Proc. 12th Ann. ACM Symp. Theory of Computing* (1980) 429–435.

[225] ZIEGLER, S.F., Smaller faster table driven parser, Unpublished manuscript, Madison Academic Computing Center, Univ. of Wisconsin, Madison, WI, 1977.

CHAPTER 7

Computational Geometry

F. Frances YAO

Xerox Palo Alto Research Center, 3333 Coyote Hill Road, Palo Alto, CA 94304, USA

Contents

HANDBOOK OF THEORETICAL COMPUTER SCIENCE
Edited by J. van Leeuwen

1. Introduction

Computational geometry is a relatively young field in theoretical computer science. Historically, geometry has been an important and venerable subject, but the study of efficient algorithms for solving geometric problems has been scattered until fairly recently. Around 1975 a systematic investigation of the computational complexity of some basic geometric problems was initiated in several papers by M.I. Shamos. In little over a decade, the subject has attracted enormous interest and accumulated an impressive body of results.

The emergence of computational geometry as an active area of research stems from its inherent interest as well as its close relationship with several applied areas of computer science. Many applications have presented geometric problems for which efficient algorithms are needed. For example, VLSI design aids require fast algorithms for solving problems on rectangles; computer graphics needs efficient methods for manipulating geometric figures; and speedy database retrieval depends on suitable multidimensional data structures and search algorithms. Moreover, the large input size typical in such applications makes it worthwhile to investigate algorithms with superior performance in the asymptotic sense. Robotics is another applied subject for which computational geometry provides an appropriate framework; see Chapter 8 on algorithmic motion planning in robotics (this Handbook) for results in that area.

This chapter can only attempt to cover some of the major topics in computational geometry, and provide a few references on each topic. Over two hundred references are cited in all, which represent only a small percentage of the published work in this area. Still, it is hoped that through these references additional sources of information can be located without much difficulty. The reader may also consult the textbooks by Edelsbrunner [80], Mehlhorn [163], Preparata and Shamos [184], and the lecture notes by Aggarwal and Wein [6].

2. Techniques and paradigms

In the study of computational geometry, we use concepts and results from classical geometry, topology, combinatorics, as well as standard algorithmic techniques such as sorting and searching, graph manipulations, dynamic programming, etc. In addition, certain special techniques and paradigms have emerged and found repeated application in computational geometry. We will briefly describe these techniques in the following and present two problems whose solutions use a combination of these techniques. Many more examples where these techniques are applied will be found throughout this chapter.

2.1. Divide-and-conquer

The basic idea is to divide the problem into two or more subproblems, solve them recursively and then merge the results to obtain a solution to the original problem. Although this technique is common in algorithm design, it seems particularly suited for solving geometric problems: the dividing step can naturally be accomplished by splitting along a line (or a hyperplane in d dimensions), and the merging step can take advantage of reduced problem complexity across the dividing hyperplane.

2.2. *Sweep technique*

This is a technique designed to reduce the dimension of a (static) geometric problem at the cost of changing its mode from static to dynamic. In the two-dimensional case, for example, we can regard the y-coordinate as time and move a horizontal *sweep-line* across the plane from $y=-\infty$ to $y=+\infty$. A cross-section induced by the sweep-line is a one-dimensional object, which will be updated as the sweep-line moves from one (discrete) position to the next. The sweep technique can also be used in higher dimensions.

2.3. *Geometric transformations*

Through a geometric transformation, one can often state a given problem in an alternative form which, though equivalent, may shed new light on the problem. Among the transformations used most often are the *dual transform* between points and hyperplanes, the *inversion* transform [33], and the *projective transform* through a paraboloid [87]. Other examples include the use of *Plücker coordinates* to represent a line in three dimensions as a point in five dimensions [232], and the *skewed projection* in three dimensions determined by a pair of lines [127].

2.4. *Locus approach*

This approach applies to geometric query problems and it works as follows. We interpret the query as a point in some space (through an initial transformation, if necessary), then partition that space into a set of nonoverlapping regions such that the answer is invariant for all query points that fall in the same region. Several ingredients are involved in this approach: transformation of the query into a point, definition and construction of a suitable space partition, and an algorithm for point-location. Using the Voronoi diagram to solve the nearest-neighbor query is a basic example of this approach (Section 4).

2.5. *Divide-and-query*

A large class of geometric problems is concerned with computing some binary predicate defined on a set $A\times B$, where A and B contain m and n objects respectively. For example, given a set of tetrahedra in space, we may want to determine if any pair of them intersect, or to find the closest pair among them. For an $m\times n$ "all-pairs" problem of this type, an $O(mn^{1-\delta})$ time algorithm can always be obtained provided that, with polynomial-time preprocessing of the set B, the $1\times n$ "query" version of the same problem for any query object $a\in A$ can be answered in $O(n^c)$ time for some $c<1$. The algorithm first divides the set B into subsets of size $O(n^{\varepsilon})$ for some suitably small ε, and then solves the query problem for all $a \in A$ within each subset (see [229]).

2.6. *Multidimensional search*

In a multidimensional space, we need to organize the data in a variety of different ways to suit different needs. Many data structure schemes have been invented, including *segment trees* [22], *k-d trees* [21], *polygon trees* [225], *octant trees* [234], *ε-nets* [119], etc.

2.7. *Random sampling*

The idea is to use random sampling of the input objects to carry out *divide-and-conquer* efficiently; that is, to split up the original problem into subproblems each guaranteed to be of small size [57, 58]. The expected performance of the resulting algorithm is with respect to the built-in randomization process, and does not depend on any assumption about the input distribution. This technique often yields simple algorithms with good expected performance.

2.8. *Lower bounds*

Lower bounds in computational geometry are usually proved for the *algebraic decision tree model* [205] (see Section 3.2). Quite often an $\Omega(n \log n)$ lower bound can be obtained through a simple reduction of *sorting* or some related problem. Sometimes direct counting is necessary to yield an information-theoretic bound. For example, in [205] lower bounds were derived by mapping polynomial inequalities to a suitable algebraic variety and then associating the number of computational steps with some characteristics (e.g., the Betti numbers) of the algebraic variety.

2.1. EXAMPLE. Suppose we are given a set S of n line segments in the plane, and asked to find out if any two of them intersect. A straightforward solution will have to check all $\binom{n}{2}$ pairs in the worst case. However, using the sweep approach, we can solve the problem in $O(n \log n)$ time as follows. First notice that in any cross-section of S, that is, the intersection of S with a horizontal line L: $Y = y_0$, the segments of S will appear as points, ordered linearly from left to right. If we move the line L up or down, the ordering will change only when L crosses an endpoint (or an intersection point if there exists any); even then the change is very local. On the other hand, if S has any intersections at all, there must be a *smallest* value y_0 such that say, of segments a and b, a and b appear adjacent in the cross-section $Y = y_0$. Thus, by sweeping the line L from $Y = -\infty$ to $Y = +\infty$, we can reduce the two-dimensional line intersection problem to a one-dimensional dynamic sorting problem. Specifically, we shall maintain a sorted list subject to n insertions and n deletions, and perform an intersection test for any new instance of adjacent segments. With a balanced search tree, this can be done in $O(n \log n)$ time. (See Section 11.1 for more discussions of this problem.)

2.2. EXAMPLE. How about the intersection problem for lines in three dimensions? (For simplicity, let us consider only the case when the lines are infinite.) This problem seems much harder than the two-dimensional case; nevertheless, it can still be solved in subquadratic time. We resort to a representation of lines by the *Plücker coordinates*, due to Julius Plücker dated 1865. Let L be an infinite line spanned by two distinct points $\boldsymbol{a}=(a_1,a_2,a_3,a_4)$ and $\boldsymbol{b}=(b_1,b_2,b_3,b_4)$ in homogeneous coordinates. The Plücker coordinates of L are

$$p(L)=(p_{12},p_{13},p_{14},p_{23},p_{24},p_{34})$$

where

$$p_{ij}=-p_{ji}=\begin{vmatrix} a_i & b_i \\ a_j & b_j \end{vmatrix} \quad \text{for } 1\leqslant i<j\leqslant 4.$$

Two infinite lines L and L' intersect if and only if

$$p_{12}p'_{34}+p_{13}p'_{42}+p_{14}p'_{23}+p_{34}p'_{12}+p_{42}p'_{13}+p_{23}p'_{14}=0$$

(This follows from the fact that L and L' intersect if and only if the 4×4 matrix given by four points spanning L and L' has rank less than 4.) If we view $p(L)$ as the (homogeneous) coordinates of a point in R^5, and, dually, $p(L')$ as the coefficients of a hyperplane in R^5, then L and L' intersect if and only if the point $p(L)$ lies on the hyperplane $p(L')$. By using the locus approach, it can be shown (see Section 10.2) that the query problem "Given a set of hyperplanes, does a query point lie on any one of them?" is solvable in $O(n^c)$ time for some $c<1$. Therefore, by the divide-and-query technique described above, we can solve in $O(n^{2-\delta})$ time the all-pairs point-hyperplane incidence problem, and hence the line intersection problem in three dimensions.

REMARK (*about models of computation*). The model of computation usually adopted for geometric algorithms is the *real Random-Access Machine*. This is similar to the ordinary RAM model, but with unit-cost assumption for infinite-precision real arithmetic and storage. Issues related to finite- vs. infinite-precision geometric computations will be discussed in Section 13.1.

3. Convex hulls

A set in R^d is *convex* if for any two points p,q in the set, the segment $\overline{pq}$ is entirely contained in the set. The *convex hull* of a set S of points in R^d is the boundary of the smallest convex set that contains S. Being a basic and natural concept, the convex hull has many applications as well as a rich mathematical theory behind it. The construction of the convex hull of a set has been a central topic in computational geometry. We will survey the construction algorithms, first in the plane and then in higher dimensions. We also examine lower bounds to the complexity of this problem, and discuss some other topics related to convex hulls. For theory on convex polytopes, the reader is referred to textbooks such as Grünbaum [114] or Bronsted [31].

3.1. Convex hull of a planar set

The first convex hull algorithm with $O(n \log n)$ running time for n points in R^2 is due to Graham [111]. The algorithm first sorts the points by polar angle about an interior point and then, in a linear scan of the sorted list, eliminates any point where a reflex angle occurs. The convex hull can also be constructed by a divide-and-conquer algorithm [80, 183, 198]. We partition the points into two equal subsets P_L and P_R by a vertical line, find their convex hulls recursively, and merge them to form the desired convex hull. The merging step is accomplished by finding the *upper bridge* b_1 and the *lower bridge* b_2 between P_L and P_R (see Fig. 1). To find the upper bridge, we take the line segment ℓ connecting the rightmost point p_k of P_L and the leftmost point p_{k+1} of P_R, move its endpoints upward iteratively along the boundaries of P_L and P_R, until ℓ reaches the highest position which defines b_1. The lower bridge b_2 is found analogously. The time required is $T(n) \leqslant 2T(n/2) + O(n) \leqslant O(n \log n)$.

Another method for computing convex hulls was proposed by Jarvis [128] (similar in spirit to the "gift wrapping" method of Chand and Kapur [36] designed for higher dimensions). It finds the convex hull edges one by one and runs in time $O(hn)$ where h is the number of edges on the convex hull. This is an example of an *output-sensitive* algorithm, i.e., an algorithm whose running time is dependent on the size of the output as well as the input. Kirkpatrick and Seidel [138] gave an improved algorithm with running time $O(n \log h)$ by employing the techniques of two-dimensional linear programming (Section 6.1). If the input points are sorted, or more generally, form the (ordered) vertices of a simple polygon (i.e., a polygon without self-intersections), the convex hull can be found in linear time [112, 145].

3.2. Lower bounds

The construction of convex hulls was one of the first geometric problems for which nontrivial lower bounds were established. The techniques developed for this problem

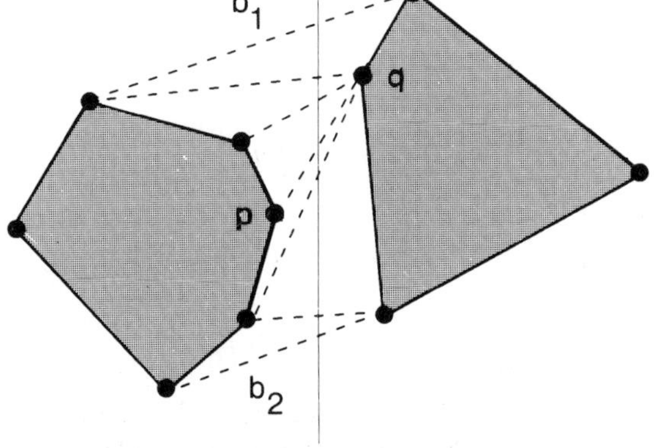

Fig. 1. Forming the union of two convex polygons.

have since been applied to many other situations. The only other topic in computational geometry that has been a similar focus of complexity studies is the range search problem (see Section 10.3).

Yao [228] showed that $\Omega(n \log n)$ is a lower bound just for determining the points belonging to the convex hull, that is, not necessarily producing them in cyclic order. This lower bound was proved for a decision tree model with quadratic tests, which accommodates all the known planar convex hull algorithms. In a more general *algebraic decision tree* model, polynomial tests of arbitrary degrees may be used. (Notice that in d dimensions, simply detecting whether a point lies on a hyperplane determined by d points involves evaluating a polynomial of degree d.) Steel and Yao [205] considered a *bounded degree* algebraic decision tree model, and devised a topological approach for deriving lower bounds in this model by using Milnor's theorem on the Betti numbers of algebraic varieties. Ben-Or [20] strengthened this approach by making more effective use of Milnor's theorem, and derived an $\Omega(n \log n)$ lower bound for the convex hull problem in this model. This lower bound is also valid when the degree is arbitrary, provided that multiplications/divisions as well as comparisons are counted. It remains an open question whether decision trees of height $o(n \log n)$ exist when polynomial tests of arbitrary degrees are allowed and all arithmetic operations are considered free.

3.3. *Convex hulls in higher dimensions*

The $\Omega(n \log n)$ lower bound mentioned above carries over for finding the convex hull of n points in space. In contrast to the planar case, only one algorithm is known that achieves this bound in three dimensions. The $O(n \log n)$ convex hull algorithm by Preparata and Hong [183] is a generalization of the divide-and-conquer algorithm in two dimensions. In the crucial merging step of the algorithm, the hull of the union of two nonintersecting polyhedra is formed by "gift wrapping" the two objects. The merging cost is proportional to the total size of the incidence graphs of the two polyhedra (i.e., the graphs induced by the vertices and edges of the polyhedra), and is $O(n)$ since these incidence graphs are planar.

For $d \geqslant 4$, the situation becomes more complex since a d-polytope with n vertices can have up to $O(n^{\lfloor d/2 \rfloor})$ faces by the *upper bound theorem* (see [114]). There are two types of convex hull algorithms, depending on whether the algorithm only enumerates the *facets*, i.e., the $(d-1)$-faces, of the convex hull or produces the complete *facial lattice*, i.e., a description of all faces and incidence relationships of the convex polytope. Two approaches for computing convex hulls have been proposed. One approach is the "gift-wrapping method" proposed by Chand and Kapur [36], and later analyzed by Seidel [197] and Swart [207]. For fixed d, it has worst-case time complexity either $O(nF)$ for enumerating F facets, or $O(\min(nL^2, n^{\lfloor d/2 \rfloor + 1}))$ for producing a facial lattice of size L. A different approach, called the "beneath-beyond" method, was proposed by Kallay [130], and independently by Seidel [195]. This is an incremental approach that constructs the convex hull by adding one point at a time. It has worst-case running time $O(n^{\lfloor (d+1)/2 \rfloor})$, and hence is optimal in terms of input size for even dimensions $d \geqslant 4$ [195]. By employing a "shelling" approach [35] related to the gift-wrapping method, Seidel

[197] gave an algorithm which has running time $O(n^2 + F \log n)$ for facet enumeration and $O(n^2 + L \log n)$ for computing the facial lattice.

Through duality, the problem of enumerating the facets of the convex hull of a point set is equivalent to the problem of enumerating the vertices of a polytope defined by a set of linear inequalities; in the latter form this problem has also been widely explored in the optimization literature (see [75] for a survey).

3.4. *Related problems*

3.4.1. *Dynamization*

In the *on-line* version of the problem, input points $p_1, p_2, \ldots$ are received in sequence, and the algorithm must find the convex hull of $p_1, p_2, \ldots, p_n$ after each p_n is processed. Shamos [199] gave an on-line algorithm with running time $O(n \log n)$, although the update time between two input points can be $O(\log^2 n)$ in the worst case. Preparata [182] improved the update time to $O(\log n)$. In the case when points may be deleted as well as inserted, Overmars and van Leeuwen [174] showed that convex hulls can be maintained with $O(\log^2 n)$ time per insertion/deletion. Not much appears to be known about dynamic maintenance of convex hulls in higher dimensions.

3.4.2. *Diameter*

For a set S of n points in R^d, the greatest distance between two points in S is called the *diameter* of S. There is an $\Omega(n \log n)$ lower bound for finding the diameter when $d \geqslant 2$. Shamos [199] gave an algorithm which finds the diameter of a planar set in $O(n)$ time once its convex hull has been computed. In d dimensions, an $O(n^{2-\varepsilon})$ algorithm based on the divide-and-query strategy was given by A.C. Yao [229]. For $d=3$, there is a randomized algorithm with $O(n \log n)$ expected time by Clarkson and Shor [61].

3.4.3. *Convex layers*

An interesting structure can be obtained for a set through repeated computation of convex hulls. Given a set S, we remove the convex hull of S, then compute and remove the convex hull of the remainder, and continue until no points are left. This procedure, known as "shelling" or "peeling", is due to Tukey (see [126]). In two dimensions, the resulting convex polygons are called the *convex layers* of S. Chazelle [37] showed that the convex layers can be computed in $O(n \log n)$ time. (See Section 10.2.1 for an application of this structure.)

3.4.4. *Intersection of convex polygons and polyhedra*

Consider the problem of forming the intersection of two convex polygons P and Q with a total of n vertices. An algorithm using the "slab method" was given by Shamos [198]. It first merges the vertex lists of P and Q, then draws vertical lines through each vertex to form "slabs". Within a slab, P and Q are reduced to two trapezoids whose intersection can be computed in constant time. A different algorithm, which works by directly traversing the boundaries of P and Q in a synchronized fashion, was given by O'Rourke, Chien, Olson and Naddor [171]. Both algorithms run in $O(n)$ time. It follows that the intersection of k convex polygons with a total of n vertices can be found in $O(n \log k)$ time.

How fast can we compute the intersection of two convex polyhedra in three dimensions? Muller and Preparata [167] gave the first efficient algorithm for this problem, using a combination of intersection detection and convex hull computation. The algorithm has running time $O(n \log n)$, where n is the total number of vertices of the two polyhedra. A different algorithm achieving the same time was given by Hertel, Mehlhorn, Mäntylä and Nievergelt [121]. It is one of the few examples where a plane sweep is used in three dimensions (cf. also [219]). For the simpler problem of detecting whether two polyhedra intersect, a linear-time algorithm was given by Dobkin and Kirkpatrick [72]. (It also follows from the linear-time algorithm for solving linear programs in three dimensions, see Section 6.) Recently, Chazelle [44] found an optimal linear-time algorithm for computing the intersection of two convex polyhedra in three dimensions. It also implies an $O(n \log k)$ algorithm for intersecting k convex polyhedra.

Chazelle and Dobkin [45], and also Dobkin and Kirkpatrick [72], considered the intersection problem with preprocessing. For example, if convex polygons P and Q are preprocessed into suitable hierarchical representations, then their intersection can be found in $O(\log n)$ time.

4. Voronoi diagrams

The Voronoi diagram is a classical object named after the French mathematician G. Voronoi ([220, 221], dated 1907 and 1908), although the concept can be traced further back to Dirichlet ([71], dated 1850) and thus is also called the *Dirichlet tessellation*. Its earlier mathematical applications were mainly in packing and covering problems [190]. However, the diagram has since been recognized as a basic tool for spatial data analysis and used widely in diverse fields of science including biology [30], physics [32] and archaeology [122].

Since Shamos and Hoey [200] introduced the Voronoi diagram to computer science, efforts have been made to find efficient algorithms for its construction and for solving problems with this structure. We will see that techniques such as the locus approach, planar subdivisions, point location, geometric transform, sweep-line approach, etc. will all be encountered in algorithms related to Voronoi diagrams. Indeed, the Voronoi diagram serves as a paradigm for most basic techniques in computational geometry. A survey of Voronoi diagrams and related structures can be found in [15].

4.1. *Preliminaries*

Consider the *post-office problem*: Given a set S of n points in the plane (called *post-offices* or *sites*), determine for an arbitrary point (x, y) which post-office is closest to (x, y). For each post-office $p \in S$, the locus of points (x, y) that are closer to p than to any other points of S is a convex polygon $V(p)$ obtained as the intersection of $n-1$ half-planes, $V(p) = \bigcap_{q \neq p} H(p, q)$, where $H(p, q)$ is the half-plane which contains p and is bounded by the perpendicular bisecting line of p and q. We call $V(p)$ the *Voronoi*

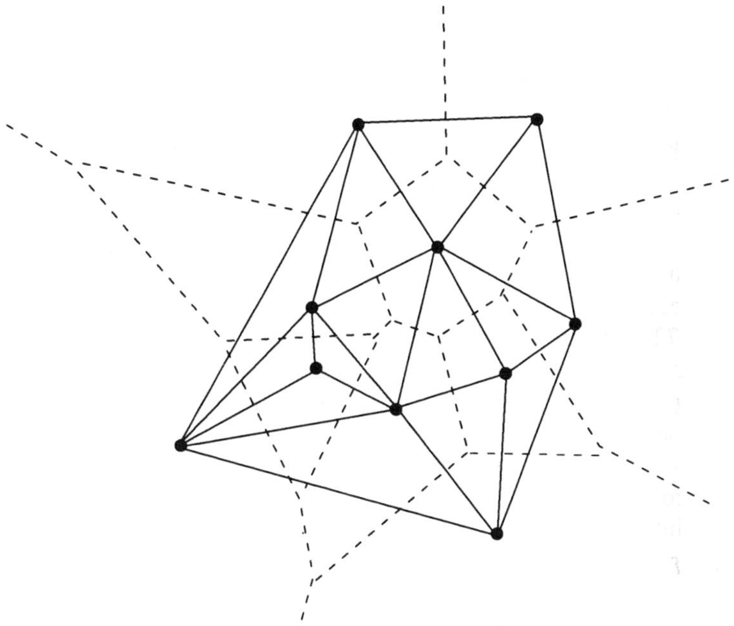

Fig. 2. Voronoi diagram and Delaunay triangulation.

polygon or *Voronoi region* associated with site p. The n polygons $V(p)$ form a partition of the plane, called the *Voronoi diagram* of S, denoted by $Vor(S)$ (see Fig. 2).

We list below some basic properties of $Vor(S)$. These properties are exploited in the construction and applications of the Voronoi diagram. For convenience, we avoid degenerate cases by assuming that no four sites in S are cocircular.

(P1) Each vertex v of $Vor(S)$ is equidistant from three sites that are closest to v. That is, v is the center of a circle going through three sites and containing no sites in its interior. The converse is also true: if a circle determined by three sites of S contains no sites in its interior, then the center of the circle is a vertex of $Vor(S)$.

(P2) If p is a nearest neighbor of q, then polygons $V(p)$ and $V(q)$ are adjacent (i.e., they share an edge) in the Voronoi diagram.

(P3) The straight-line dual of the Voronoi diagram is a triangulation of S, called the *Delaunay triangulation* [68].

4.2. *Construction*

Shamos and Hoey [200] gave an $O(n \log n)$ time divide-and-conquer algorithm for constructing the Voronoi diagram for a set S of n points. The points of S are partitioned into two equal subsets A and B by a vertical line; the Voronoi diagrams $Vor(A)$ and $Vor(B)$ are computed recursively, and then merged together along their common boundary $\partial(A, B)$ in $Vor(S)$, where $\partial(A,B) = \bigcup_{p\in A, q\in B} V(p) \cap V(q)$. The boundary $\partial(A, B)$ is a path monotonically increasing in y, and can be computed in $O(n)$ time by making a "zigzag" walk through the polygons of $Vor(A)$ and $Vor(B)$.

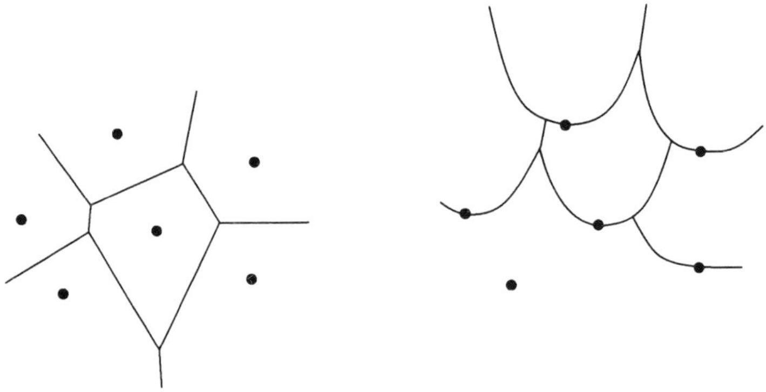

Fig. 3. A Voronoi diagram and its transform.

More recently, Fortune [94] presented an $O(n \log n)$ *sweep-line* algorithm for constructing a Voronoi diagram in one pass from bottom to top. The sweep-line approach is made possible through a geometric transformation of the Voronoi diagram. The transformed diagram has the property that the lowest point of the Voronoi region of a site appears at the site itself (Fig. 3). By sweeping a horizontal line upward, the algorithm needs to build a Voronoi region only when the site is reached by the sweep-line; furthermore, the boundary of a region can be computed through properly maintained information between adjacent regions along the sweep line. Finally, the real Voronoi diagram can be easily reconstructed from its transform. The geometric transformation used by Fortune maps a point r in the region $V(p)$ to $r+(0, y)$ where $y=d(r, p)$ is the distance from r to p. This can be intuitively understood as follows. Think of the normal Voronoi diagram as generated by a physical process where obstacles are placed at the sites, and an expandable disk is moved in between them such that at any moment the disk has a maximal radius as allowed by the obstacles. The locus of the center of the disk, then, defines the Voronoi diagram. If, instead of tracing the center, we trace out the locus of the *topmost* point of the disk at all times, the result is precisely the transformed Voronoi diagram of Fortune.

If the points form the ordered vertices of a convex polygon, Aggarwal, Guibas, Saxe and Shor [3] showed that the Voronoi diagram is computable in linear time. The technique can also be used to update a (general) Voronoi diagram in linear time when a site is deleted. The algorithm in [3] makes use of a linear-time procedure for merging two Voronoi diagrams by Kirkpatrick [136].

4.3. *Generalized Voronoi diagrams*

The elegance and utility of the Voronoi diagram inspired researchers to try to adapt the concept to more general situations, resulting in many variations of generalized Voronoi diagrams. Among them are Voronoi diagrams in R^d, higher-order Voronoi diagrams, Voronoi diagrams with weighted points [16, 94], Voronoi diagrams for line segments and circular arcs [94, 146, 235], etc. We will discuss the first two kinds of generalizations in the following, beginning with some useful insights into the connection between Voronoi diagrams and convex hulls.

4.3.1. Voronoi diagrams and convex hulls

Brown [33] was the first to show that, through an inversion transform f, the Voronoi diagram of a set S in R^d corresponds to the convex hull of $f(S)$ in R^{d+1}. We will describe a different correspondence due to Edelsbrunner and Seidel [87]. It is based on the paraboloid $\mathscr{R}$ defined by $x_{d+1}=x_1^2+x_2^2+\cdots+x_d^2$. We transform a point $p=(a_1,a_2,\ldots,a_d)$ of S to the hyperplane P tangent to the paraboloid $\mathscr{R}$ at the point $(a_1,a_2,\ldots,a_d,a_1^2+\cdots+a_d^2)$. Let $\mathscr{T}$ be the polytope defined by these hyperplanes and containing the paraboloid (see Fig. 4). We claim that the vertical projection of the boundary of $\mathscr{T}$ onto the hyperplane $x_{d+1}=0$ gives exactly the Voronoi diagram of S. This can be easily verified from the equation of P: $x_{d+1}=2a_1x_1+2a_2x_2+\cdots+2a_dx_d-(a_1^2+a_2^2+\cdots+a_d^2)$. Thus $Vor(S)$ can be obtained from $\mathscr{T}$, the intersection of a set of half-spaces in R^{d+1}, which by duality is equivalent to the convex hull of a point set in R^{d+1}.

4.3.2. Voronoi diagrams in higher dimensions

Preparata [181] observed that the Voronoi diagram for n points in R^3 can have $O(n^2)$ edges. Klee [139] showed that the maximum number of faces of a Voronoi diagram in R^d grows exponentially in d, and tight bounds have been obtained by

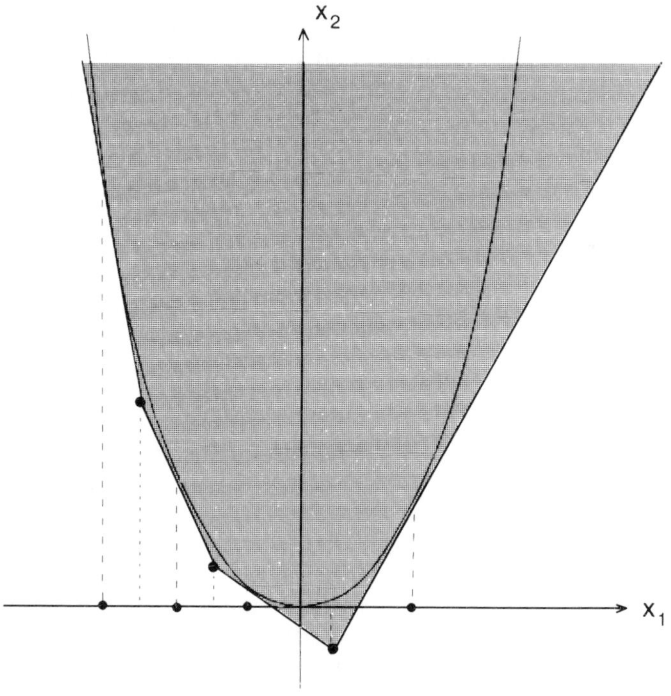

Fig. 4. Correspondence between Voronoi diagrams and convex hulls.

Paschinger [178] and Seidel [196]. Because of the correspondence between Voronoi diagrams in R^d and convex hulls in R^{d+1}, one can resort to the convex hull algorithms described in Section 3.3 to obtain Voronoi diagrams in higher dimensions. Thus, for $d \geqslant 3$, the Voronoi diagram of n points in R^d can be constructed in $O(n^{\lfloor (d+1)/2 \rfloor})$ time and $O(n^{\lfloor d/2 \rfloor})$ space.

4.3.3. *Higher-order Voronoi diagrams*

The *order-k Voronoi diagram* of a set S in R^d is an extension of the nearest-point (order-1) Voronoi diagram for the purpose of answering "What are the k sites closest to a query point?". In the one-dimensional case, the diagram for a set $S = \langle p_1 \leqslant p_2 \leqslant \cdots \leqslant p_n \rangle$ consists simply of the midpoints between p_i and p_{i+k} for $1 \leqslant i \leqslant n-k$. For $d=2$, it is a planar subdivision obtained by forming, for any k-subset T of S, the polygon $V_k(T)$ which is the locus of points that are closer to T than to any points of $S-T$. ($V_k(T)$ may be empty for some T.) The order-$(n-1)$ Voronoi diagram is also called the *furthest point Voronoi diagram*, since each polygon $V_{n-1}(T)$ is characterized by having $\{p\} = S-T$ as the furthest site from $V_{n-1}(T)$; clearly $V_{n-1}(T)$ is nonempty only if p is a point on the convex hull of S. (Some applications of the furthest-point Voronoi diagram are discussed in Section 5.7.)

Lee showed that the order-k Voronoi diagram of n planar points can be obtained in $O(k^2 n \log n)$ time [143]. Edelsbrunner, O'Rourke and Seidel [84] gave a uniform method for constructing all higher-order Voronoi diagrams in R^d based on the correspondence described before. As noted earlier, the vertical projection of the polytope $\mathscr{T}$ onto the hyperplane $x_{d+1}=0$ gives the order-1 Voronoi diagram of S. If the hyperplanes of $\mathscr{T}$ are extended to their second intersection by stellation, the projection is essentially the order-2 Voronoi diagram (see Fig. 5). The order-3 Voronoi diagrams can be obtained by further extending the hyperplanes, and so on. These observations lead to an algorithm, based on the construction of the *arrangement* of a set of hyperplanes (see Section 12.1), that computes the family of all higher-order Voronoi diagrams for a set of n points in R^d in optimal time $O(n^{d+1})$.

5. Proximity problems

Questions concerning various proximity relations among a set of points account for a large class of problems studied in computational geometry. We will consider several basic problems of this type; for most of them, the Voronoi diagram plays a central role in their solutions.

5.1. *Nearest neighbors and closest pair*

The *all-nearest-neighbors* problem is to find, for a given set of n points in R^d, a nearest neighbor for each point. For $d=2$, Shamos and Hoey [200] gave an $O(n \log n)$ solution which first builds the Voronoi diagram for the points and then, by property **P2**, finds a nearest neighbor for each p_i by searching the sites adjacent to p_i. The total searching time is proportional to the size of the diagram, hence $O(n)$. In d dimensions, Vaidya [214] gave an $O(d^d n \log n)$ algorithm for the all-nearest-neighbors problem which is

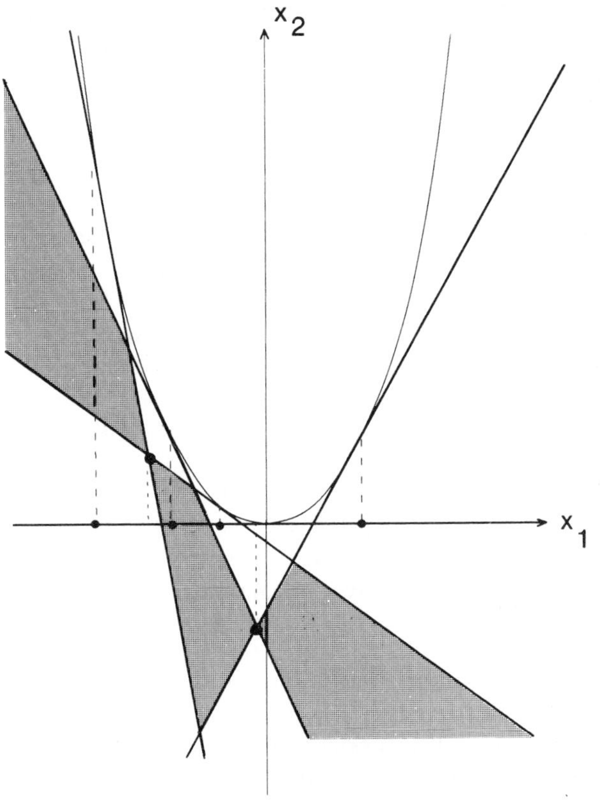

Fig. 5. The order-2 Voronoi diagram for four points on a line.

optimal for any fixed d. The algorithm uses a *box decomposition* scheme, which keeps a collection of nonoverlapping boxes containing all the points and maintains potential nearest-neighbor relationships among them. The largest box is split in each iteration until finally each box contains a single point.

Once all nearest neighbors are known, the *closest pair* for a set of points can be computed in additional O(n) time. For the closest pair problem, there is also a divide-and-conquer algorithm by Bentley and Shamos [25] that achieves optimal O($n \log n$) time in any dimension d. In two dimensions, the set is divided into two subsets by a median vertical line L. If δ_1 and δ_2 are recursively the closest distances in the two subsets, then the merging step only examines pairs of points across L which are potentially within distance $\delta = \min\{\delta_1, \delta_2\}$ of each other. It can be proved that there are at most O(n) such pairs. In d dimensions, they showed that a similar but more intricate "sparsity" condition still holds if the cut plane L is chosen suitably.

In the related *nearest-neighbor query* problem, a given set S of n points in R^d is to be preprocessed into a suitable data structure so that for any query point q, its nearest neighbor in S can be found quickly. This is the *post-office* problem raised at the beginning of Section 4.1. In the plane, this can be solved by storing the Voronoi diagram $Vor(S)$ and applying an O($\log n$) time *point-location* algorithm (Section 8) to find the region in $Vor(S)$ that contains q. In d dimensions, $O(\log n)$ query time can be achieved with $O(n^{2^{d+1}})$ space and preprocessing time (see [73, 229]). For $d=3$, the space and preprocessing time can be reduced to $O(n^2)$ while allowing $O(\log^2 n)$ query time [39]. The post-office problem is a basic subroutine used in the solution of many geometric problems (cf. Sections 5.2 and 5.3). No efficient *dynamic* scheme is currently available for this problem. (See Section 10 for discussions of *range query* problems in general.)

5.2. *Euclidean minimum spanning tree*

The *minimum spanning tree* (MST) is a well-studied subject in graph theory. Given a graph $G=(V,E)$ with n vertices and a weight associated with each edge, an MST for G is a spanning tree (i.e., a connected, cycle-free subgraph on V) whose total edge weight is as small as possible. A given set S of n points in R^d induces a complete graph on n vertices where the edge weights correspond to the interpoint distances. An MST for this complete graph is called a *Euclidean* MST for S.

An efficient algorithm for constructing a Euclidean MST must avoid explicitly computing all $\binom{n}{2}$ interpoint distances. As it turns out, it is sufficient to consider the much smaller edge set of the Delaunay triangulation of S. The following property is an extension of **P2**:

(**P2**$'$) For every subset S' of S, any shortest edge between S' and $S-S'$ must belong to the Delaunay triangulation of S.

This property, in conjunction with a well-known characterization of MST by Prim (see [184]), implies that a Euclidean MST is contained in the Delaunay triangulation of S. Therefore, once the latter has been computed, we can apply a linear-time MST algorithm for planar graphs (due to Yao, see [53]) to find a Euclidean MST for S. The total time $O(n \log n)$ is optimal, since sorting can be easily reduced to finding an MST.

Subquadratic MST algorithms in higher dimensions were developed by A.C. Yao [229], including an $O((n \log n)^{1.8})$ algorithm for $d=3$ which was later reduced to $O((n \log n)^{1.5})$ by Chazelle [39] as a result of an improved solution to the post-office query in three dimensions (cf. Section 5.1).

5.3. *Euclidean minimum matching*

Given a weighted graph on $2n$ vertices, a *matching* is a set of n edges such that each vertex has exactly one edge incident on it; a matching with the least weight possible is called a *minimum matching* (MM). When the $2n$ vertices are points in the plane with edge weights induced by distances, we refer to the MM problem as the *Euclidean* MM problem. In the bipartite version, half the points are painted with one color and the other half with another color, and an edge of the matching must be incident with two points of different colors. The fastest known MM algorithm for general graphs requires $O(n^3)$ time (see [104, 141]). In the Euclidean case, Vaidya [216] reduced the time to $O(n^{2.5}(\log n)^4)$ and, for the bipartite version, to $O(n^{2.5}\log n)$. The saving was obtained as a result of using dynamic nearest-neighbor query (in a weighted Voronoi diagram) to implement the augmenting path computation in the standard MM algorithm.

Further improvement to the dynamic query problem can lead to a corresponding improvement for the present problem (cf. Section 5.1).

5.4. *Euclidean travelling salesman problem*

In the Euclidean version of the *travelling salesman problem* (TSP), we are to find a shortest tour (closed path) through n given points in the plane. TSP for weighted graphs is a well-known NP-complete problem [132], and its Euclidean version is also NP-complete [177]. The following heuristics by Christofides [54] will find a tour whose length is guaranteed to be within $\frac{3}{2}$ of the optimal, which is currently the best ratio achievable by any polynomial-time algorithm. The heuristics makes use of the Euclidean MST and MM algorithms mentioned above. One first builds a minimum spanning tree on the points, then finds a minimum matching among the tree nodes of odd degrees. Let $|MST|$ and $|MM|$ denote their weights. Then the weight $|TSP|$ of the optimal tour satisfies $|TSP| \geqslant |MST|$ and $|TSP| \geqslant 2|MM|$, hence $|MST| + |MM| \leqslant \frac{3}{2}|TSP|$. But the union of the MST and the MM, being without odd-degree vertices, admits an Eulerian tour which can be reduced to a shorter simple tour T by using the triangle inequality. This results in a tour T whose length is at worst $\frac{3}{2}$ times the optimum. The time required to compute T is dominated by that for finding a Euclidean minimum matching.

5.5. *Largest empty circle*

A *largest empty circle* for a set S of n points in the plane is a circle with maximum radius which does not contain any points in its interior and whose center lies within $CH(S)$, the convex hull of S. It can be shown that the center must be either a vertex of $Vor(S)$, or the intersection of an edge of $Vor(S)$ and an edge of $CH(S)$. This fact, together with the observation that the portion of $Vor(S)$ lying outside of $CH(S)$ forms a tree, can be used to design an optimal $O(n \log n)$ algorithm for this problem (see [212]). We note that a related *largest empty rectangle* problem (with two points of S forming its opposite corners) has an $O(n \log^2 n)$ time solution by Aggarwal and Suri [5].

5.6. *Clustering problems*

In a clustering problem, one wishes to find a partition of a given set of n points in R^d into k groups, called clusters, so that the *cluster size* is small according to some measure. Clustering problems arise in applications such as data compression and pattern recognition. Optimal general solutions are often NP-hard (see [96]), but in practice efficient suboptimal solutions are useful. Recent results by Feder and Greene [91] provided tight bounds and efficient algorithms for many versions of clustering problems. For example, they showed that approximate clustering in $d \geqslant 2$ dimensions within a factor close to 2 is NP-hard, and presented an algorithm that achieves a factor of 2 in time $O(n \log k)$ using an extension of the box decomposition scheme of Vaidya [214]. This running time is optimal in the algebraic decision tree model. Clustering problems with small k can sometimes be solved exactly. For example, the two-cluster problem, where a planar set is to be partitioned into two subsets so that their larger diameter is minimized, can be solved in $O(n \log n)$ time by first computing a *Euclidean maximum spanning tree* of the set and then two-coloring the spanning tree [11, 165]. Other applications of geometric algorithms to pattern recognition can be found in [210, 211].

5.7. *Maximum distances*

Euclidean maximum spanning trees and furthest neighbors are examples where some distance function on a point set is to be *maximized* rather than *minimized*. In general, the two versions (maximization vs. minimization) are not symmetric via any known geometric transform, although they are symmetric by inverting weights in the underlying complete graph.

Several problems of this type can be solved in $O(n \log n)$ time by using the furthest-point Voronoi diagram, including finding all-furthest-neighbors [29, 142], Euclidean maximum spanning trees [164], and the smallest enclosing circle (see Section 6.3).

REMARK. For the problems discussed in this section, more efficient solutions are often possible in special cases, such as when the set of points form the ordered vertices of a convex polygon (see [3, 4, 154, 164]).

6. Linear programming

We have seen that convex hulls and Voronoi diagrams hold the key to the solution of many geometric problems. On the other hand, these structures require $O(n \log n)$ time to construct even in the two-dimensional case. It turns out that many geometric problems can be formulated as *linear programs*, which are problems of the form

$$\text{minimize } c^{\mathrm{T}}x \quad \text{subject to } Ax \leqslant b$$

where $A \in R^{n \times d}$, $b \in R^n$ and $c \in R^d$. We will present an algorithm by Megiddo [159, 160] for solving linear programs and consider its application to problems in computational geometry. The algorithm runs in time linear in n, the number of constraints, when the dimension d is fixed. (For $d=2$ and 3, linear-time algorithms were developed by Dyer [76] independently.) As the time complexity grows exponentially in d, this algorithm is mainly suited for use in low dimensions.

6.1. *Linear programs in two dimensions*

We sketch Megiddo's algorithm for solving a two-variable linear program with n constraints. The algorithm performs an iterative search for the optimal vertex; after each iteration a fixed fraction of the constraints are discarded until (in the case of a bounded, feasible optimum, say) only two constraints determining the optimal vertex are left. Each iteration requires time proportional to the number of constraints currently remaining, hence the entire procedure uses linear time. To illustrate how the constraints can be pruned, we assume that, through a linear transform, the problem is to maximize the value of Y subject to $a_iX+b_iY+c_i\leqslant 0$ for $1\leqslant i\leqslant n$. We separate the constraints with positive b_i from those with negative b_i since they impose, respectively, *upper* and *lower* bounds to the objective function Y. Now, by evaluating the minimum value of the upper bounds and the maximum value of the lower bounds at any particular value x_0 of X, one can decide whether an optimal vertex lies to the left or to the right of $X=x_0$ (or conclude that the problem is infeasible or unbounded). For example, if the min upper bound $(-a_ix_0-c_i)/b_i$ is smaller than the max lower bound $(-a_jx_0-c_j)/b_j$, and their slopes satisfy $-a_i/b_i<-a_j/b_j$, then the feasible region (if nonempty) must lie further to the right of $X=x_0$. We will choose the "evaluation point" x_0 so that no matter whether the search proceeds to the left or to the right, a fraction of the constraints can be discarded. A constraint can be discarded if it is *redundant*, i.e., it does not go through the optimum vertex we are converging to. Note that for any pair of upper-bounding constraints which intersect at $X=\tilde{x}$, it is always the case that one of them is redundant to the left of $\tilde{x}$ while the other is redundant to the right of $\tilde{x}$. This suggests the following choice for the evaluation point x_0. We arbitrarily pair up the lines corresponding to the upper-bounding constraints and compute the x-projection of their pairwise intersections; likewise for the lower-bounding constraints. Let x_0 be the median of all the x-projections $\{\tilde{x}\}$ so obtained. Now, if it is decided that the search will proceed to the right of x_0, for example, then for each $\tilde{x}$ lying to the left of x_0, the corresponding constraint redundant to the left of $\tilde{x}$ can be discarded. Hence approximately one quarter of the constraints are eliminated after the evaluation at x_0. The cost of each phase is clearly linear, resulting in a linear algorithm.

6.2. *Generalization to three and higher dimensions*

We sketch the generalization of the preceding algorithm to three dimensions; further generalization to dimensions $d>3$ is straightforward.

We assume that the objective function is to maximize the value of Z subject to n linear constraints in X, Y and Z. As before, we separate the set S^+ of constraints bounding Z from above and the set S^- bounding Z from below. We pair up the planes corresponding to the constraints in S^+ arbitrarily, and likewise for S^-. The line of intersection of each pair $\{P,P'\}$ is then projected vertically onto a line ℓ in the XY-plane. Consider the *vertical extension* $V(\ell)$ of ℓ, that is, the plane going through ℓ and parallel to the Z-axis. Suppose the location of an optimal solution is known relative to the plane $V(\ell)$ (we refer to this oracle as the *line query for* ℓ); then it will also be known which one of $\{P,P'\}$ is a redundant constraint for that solution. It can be shown that a line query for any given line in the XY-plane can be answered by solving three 2-dimensional linear programs of n constraints (i.e., the original linear program restricted to $V(\ell)$ and to two additional planes parallel to and on different sides of $V(\ell)$). However, instead of solving this query for every projected line ℓ, we claim that, by performing only *two* tests, the line query can be decided for about one-eighth of those $n/2$ projected lines $\mathscr{L}=\{\ell\}$, thus resulting in the elimination of approximately $\frac{1}{16}$ of the constraints. These two tests are themselves line queries for two special lines L_1 and L_2 in the XY-plane that are chosen as follows. We let α be the median slope of lines in $\mathscr{L}$, and pair up lines steeper than α with lines less steep than α, resulting in pairwise intersection points P. The line L_1 will be vertical, while L_2 will have slope α. Specifically, we will choose L_1 to be a vertical bisector for the point set P and, after solving the line query for L_1, choose L_2 to be a bisector with slope α for the subset of P on the side containing an optimal solution. After solving the line query for L_2, we arrive at a "quadrant" Q defined by L_1 and L_2 which contains an optimal solution. Then, for each pair of lines intersecting in the quadrant opposite to Q, the answer to the line query is decided for one of the two lines. This proves the claim that approximately $\frac{1}{16}$ of the constraints can be discarded after a linear amount of work, thus a solution can be found in linear time (see Fig. 6).

In d dimensions the recursion is analogous and can be summarized as follows: by solving at most $3 \cdot 2^{d-1}$ problems of order $n \times (d-1)$, we reduce a problem of order $n \times d$ to a problem of order $n(1-2^{1-2^d}) \times d$. The solution of this recursion yields a complexity function $C_1(n,d) \leqslant n \cdot 2^{2^{d+2}}$. Using a similar approach, Megiddo [160] also gave an algorithm with complexity $C_2(n,d) = O(n(\log n)^{d^2}))$. By carefully balancing these two recursive schemes, Clarkson [56] (also Dyer [77] independently) was able to derive a better bound $C'(n,d) = O(n \cdot 3^{d^2})$.

As for the practicality of the algorithm described above, we notice that the recursion unfolds into repeated median-finding in the one-dimensional case. In practice one may use random sampling to find approximate medians instead. Recently, Clarkson [59] gave a randomized algorithm for linear programming with expected time $O(d^2 n)$ for small d.

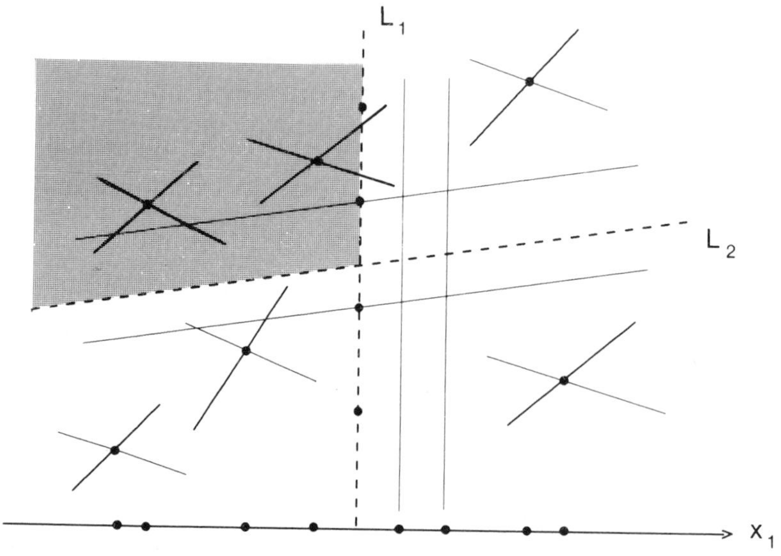

Fig. 6. Elimination of constraints.

6.3. *Applications*

We give some examples of geometric problems that can be solved by viewing them as linear programming problems. In some instances, such as Examples 6.3.2 and 6.3.3, the problems are not linear programs per se; nevertheless, the method described above can be adapted to yield linear-time solutions.

6.3.1. *Linear separability*

Given two finite sets of points S_1 and S_2 in R^d, find a hyperplane (if there is one) that separates the two sets. This problem is useful in statistics and pattern recognition. We are seeking a hyperplane $p^T x=0$ which satisfies the conditions $p^T a_i \leqslant 0$ for each $a_i \in S_1$, and $p^T b_j \geqslant 0$ for each $b_j \in S_2$. It can thus be solved as a d-dimensional linear program in time proportional to $|S_1|+|S_2|$.

6.3.2. *Quartering of points sets—the pancake problem*

Let P and Q be two sets of m and n points in the plane that are separated by a line ℓ_1. We are asked to find a line ℓ_2 that bisects P and Q simultaneously. This problem is dubbed the "pancake problem" when P and Q are density functions (pancakes); in that case the existence of ℓ_2 can be proven by a continuity argument. A quadratic-time constructive algorithm for the discrete case was given by Willard [225]. He used it to build a data structure where a planar set of t points is partitioned with two straight lines such that each open quadrant contains at most $t/4$ points (cf. Section 10.2). A linear-time algorithm for this problem was subsequently found by Megiddo [161]. Since the problem statement involves counting, there is no apparent way to formulate it directly as a linear program with $O(n+m)$ constraints. However, it can be solved by using a search strategy similar to that for linear programming in three dimensions. We first put the problem into the followng equivalent form: given m increasing linear functions P and n decreasing linear functions Q, find out where the pointwise median of P intersects the pointwise median of Q. We then apply the algorithm described in Section 6.2 to the set of lines in $P \cup Q$; in this case a line query involves median computations rather than two-dimensional linear programs.

6.3.3. *Smallest enclosing circle*

Given a set of n points in the plane, find a smallest circle that encloses them. This is the *minimax facility location problem* in operations research, as we are looking for a point (x, y) such that its maximum distance to any of n given sites (a_i, b_i) is minimized. Since the problem was first proposed by Sylvester [208] in 1857, many $O(n^3)$ solutions have been suggested. Shamos and Hoey [200] exhibited the first $O(n \log n)$ solution by observing that the smallest enclosing circle is either determined by the diameter of the set, or centered at a vertex of the furthest-point Voronoi diagram (cf. Section 5.7). Again, although the problem cannot be transformed into a linear program, the same three-dimensional search procedure can be used to find a solution [159]. The constraints $z \geqslant f_i(x, y)=(x-a_i)^2+(y-b_i)^2$ define paraboloids in the present case, but pairwise intersections still have linear projections in the XY-plane (namely, the equidistant lines between two sites). The solution of the line query here is more complicated than in the linear programming algorithm, and relies on the fact that the functions $f_i(x, y)$ are convex.

6.4. Remark

Linear programming is a fundamental problem studied in mathematical optimization (see [67, 55]). In most cases the number of variables d is considered to be of the same order of magnitude as the number of constraints n. The most widely used method is the simplex algorithm introduced by Dantzig in 1947. Klee and Minty [140] showed that the worst-case complexity of the simplex algorithm is exponential in the input size nd. However, the algorithm performs well in practice, and recent theoretical analyses by Smale [204] and Megiddo [162] lend support to its favorable average-case performance.

For linear programs with finite-precision input, Khachiyan [135] proposed an ellipsoid algorithm with complexity $O(dn^3L)$, where L is the maximum number of bits used to specify a coefficient. The bound was subsequently improved by Karmarkar [131], Renegar [189] and Vaidya [215]. In all these algorithms, one finds, through a sequence of iterations, a feasible point sufficiently close to an optimal vertex of the polytope, and then jumps to that vertex. In contrast, the simplex method and the algorithms by Megiddo are measured by the number of arithmetic operations, independent of the precision of the input.

7. Triangulation and decomposition

A common strategy in practice for solving problems on general polygons is to decompose the polygon first into simpler pieces, apply specialized algorithms to each individual piece, and then combine the partial solutions. There are two types of decompositions: *partition*, where the component parts must be disjoint, and *covering*, where the components are allowed to overlap. Sometimes additional points, called *Steiner points*, may be introduced in a decomposition. We choose some measure as the objective function to be minimized in a decomposition — usually the number of component parts, and occasionally the total "size" of component parts. Typical component types are triangles, trapezoids and convex polygons for decomposing a general polygon, and rectangles for decomposing a rectilinear polygon. A detailed survey of decomposition problems can be found in [134].

We consider the triangulation of simple polygons in Section 7.1, the polygon partition problem in Section 7.2, and the polygon covering problem in Section 7.3.

7.1. Triangulation of simple polygons

Given a simple polygon P of n vertices, a *triangulation* of P is a collection of $n-3$ nonintersecting diagonals that partition the interior of P into $n-2$ triangles. Garey, Johnson, Preparata, and Tarjan [106] gave an $O(n \log n)$ time triangulation algorithm for simple polygons. Their algorithm also works in $O(n)$ time in the special case of *monotone* polygons (a monotone polygon is a simple polygon such that any line in a certain direction intersects the polygon in at most two points). Tarjan and van Wyk [209] obtained an $O(n \log\log n)$ time algorithm for triangulating simple polygons. Their algorithm works by way of solving a related *visibility* problem which is known to be linear-time equivalent to triangulation (see [95, 52]). The visibility problem for a simple polygon is to compute all vertex-edge pairs that are *visible* in the sense that there exists a horizontal line segment inside the polygon connecting the vertex to the edge. The visibility algorithm by Tarjan and van Wyk makes use of *Jordan-sorting*, which is the problem of sorting the intersections of a simple polygon P with a horizontal line by x-coordinate, given as input only the intersections in the order in which they occur clockwise around P. For this latter problem there is a linear-time algorithm by Hoffman, Mehlhorn, Rosenstiehl, and Tarjan [123].

The above visibility algorithm can also be modified to produce an $O(n \log\log n)$ time algorithm for testing whether a given polygon is simple (see [210]). Note that this is faster than applying a general $O(n \log n)$ algorithm for detecting line-intersections (Section 11.1).

7.2. *Partition of polygons*

Generally speaking, the partitioning problem has polynomial-time solutions for polygons without holes (i.e., simply-connected polygons), and becomes NP-hard when holes are present (see [134] for more details). We will mention some results for the former case only.

First consider the problem of partitioning a simple polygon into a minimum number of convex polygons. When Steiner points are allowed, Chazelle and Dobkin [46] gave an $O(n+c^3)$ algorithm where c is the number of cusps, i.e., vertices with reflex angles. Te. Asano, Ta. Asano and Imai [10] gave an $O(n^3)$ algorithm for partitioning a simple polygon into a minimum number of trapezoids with two horizontal sides. For partitions without Steiner points, Keil [133] gave a dynamic programming algorithm with $O(n^2 c \log c)$ running time.

Ferrari, Sankar and Sklansky [92] gave an $O(n^{2.5})$ time algorithm for partitioning a rectilinear polygon into a minimum number of rectangles (see also [150]). For some applications, such as VLSI routing, it is important to minimize the total edge length of the rectangles used in the partition. Lingas, Pinter, Rivest and Shamir [149] gave an $O(n^4)$ dynamic programming algorithm for this problem.

For partitions without Steiner points, quadrilaterals rather than rectangles become the natural component type. Kahn, Klawe and Kleitman [129] proved that it is always possible to partition a rectilinear polygon (with rectilinear holes) into convex quadrilaterals. Sack [191] gave an $O(n \log n)$ algorithm for partitioning a rectilinear polygon without holes into convex quadrilaterals.

7.3. *Covering of polygons*

Two basic versions of the covering problem are to cover a simple polygon with convex polygons, and to cover a rectilinear polygon with rectangles. Culberson and

Reckhow [66] showed in 1988 that the minimum cover problem is NP-hard for both versions (even when the polygons do not have holes). We mention some algorithms which either produce approximate covers, or find optimal covers for special cases.

The rectangular covering problem has applications to processes where a rectilinear mask is to be produced by a rectangular pattern generator (see [158, pp. 93–98]). Aggarwal [2] gave an approximation algorithm for this problem which is an $O(\log n)$ factor away from the optimum. In the special case of *vertically convex* polygons (i.e., every column of squares in the polygon is connected), Franzblau and Kleitman [97] gave an $O(n^2)$ algorithm for finding an optimal rectangular cover.

Consider the *art gallery problem*: station a minimum number of guards in a polygon such that every point of the polygon can be seen by a guard. This problem can be restated as the problem of covering a polygon with a minimum number of star polygons (a star polygon is a polygon which contains a point, or guard, that "sees" every point of the polygon). This covering problem is also NP-hard [1, 172]. It is known that $\lfloor n/4 \rfloor$ guards are sufficient for a rectilinear, simply connected art gallery [117, 129]. Motwani, Raghunathan and Saran [166] recently showed that a rectilinear polygon can be optimally covered by rectilinear star polygons in polynomial time. The art gallery problem was discussed in detail in [1, 170].

8. Planar point location

One of the basic problems in computational geometry is to locate a point in a *planar subdivision*. That is, given a partition $\mathscr{P}$ of the plane into polygonal regions by n line segments, we want to determine which region contains any query point q. Note that $O(\log n)$ query time is a lower bound in a decision-tree model.

Many solutions have been proposed for the point location problem over the last decade. We will review these solutions and briefly describe the approaches used. Dobkin and Lipton [73] gave a solution that achieves $O(\log n)$ query time at a cost of $O(n^2)$ space and preprocessing time. Their method divides the plane into *slabs* by drawing vertical lines through each vertex of the subdivision. Point location is done with two binary searches, first with respect to x to locate the slab, and then with respect to y in the slab to determine the line segments enclosing q. Lipton and Tarjan showed that, as an application of their planar separator theorem [151], optimal space $O(n)$ and query time $O(\log n)$ can be achieved. Subsequent developments have been aimed at providing practical algorithms or alternative approaches for solving the problem within the same asymptotic bounds.

Kirkpatrick [137] described a conceptually simple algorithm which works for a triangulated subdivision $\mathscr{T}$. Starting with $\mathscr{T}$, he constructs a sequence of increasingly coarser subdivisions $\mathscr{T}_0\ (=\mathscr{T}), \mathscr{T}_1, \ldots, \mathscr{T}_k$, where $k = O(\log n)$, by repeatedly deleting a large independent set of vertices with small degree from the current subdivision. Point location in $\mathscr{T}$ begins with the coarsest subdivision $\mathscr{T}_k$ (consisting of a single triangle), and steps back through successive refinements $\mathscr{T}_{k-1}, \mathscr{T}_{k-2}, \ldots, \mathscr{T}_0$ where each step has constant cost.

Lee and Preparata [147] proposed the notion of *separating chains* as a replacement for the vertical slabs of Dobkin and Lipton, and gave a point-location algorithm for *monotone subdivisions* with O(n) space and O((log n)^2) time. A subdivision is *monotone* if every region is a monotone polygon (with respect to y, say), i.e., any horizontal line intersects it at no more than two points. Given a monotone subdivision $\mathcal{M}$, one can find a decomposition of $\mathcal{M}$ into a set of *separating chains* $\{\mathcal{C}_i\}$ which are monotone paths (with shared edges possibly) from the lowest point to the highest point of $\mathcal{M}$, such that any horizontal line will intersect $\mathcal{C}_1, \mathcal{C}_2, \ldots$ at points ordered from left to right. The significance of a chain decomposition is that it reduces point location to a twofold binary search: with respect to the chains in the outer loop, and with respect to y in the inner loop. Although the total size of the chains may be O(n^2), a reduced O(n) structure where an edge shared by chains $\mathcal{C}_i$ through $\mathcal{C}_j$ is stored only at $\mathcal{C}_k$, the *lowest common ancestor* of $\mathcal{C}_i$ and $\mathcal{C}_j$, is sufficient for carrying out the twofold binary search (see Fig. 7). The O((log n)^2) search time was later improved to O(log n) by Edelsbrunner, Guibas and Stolfi [81]. In their search structure, the chains stored at the two children are merged with the chain of the parent, so that once a binary search is performed at the root, by following pointers subsequent searches have cost O(1) per node of (cf. [226,

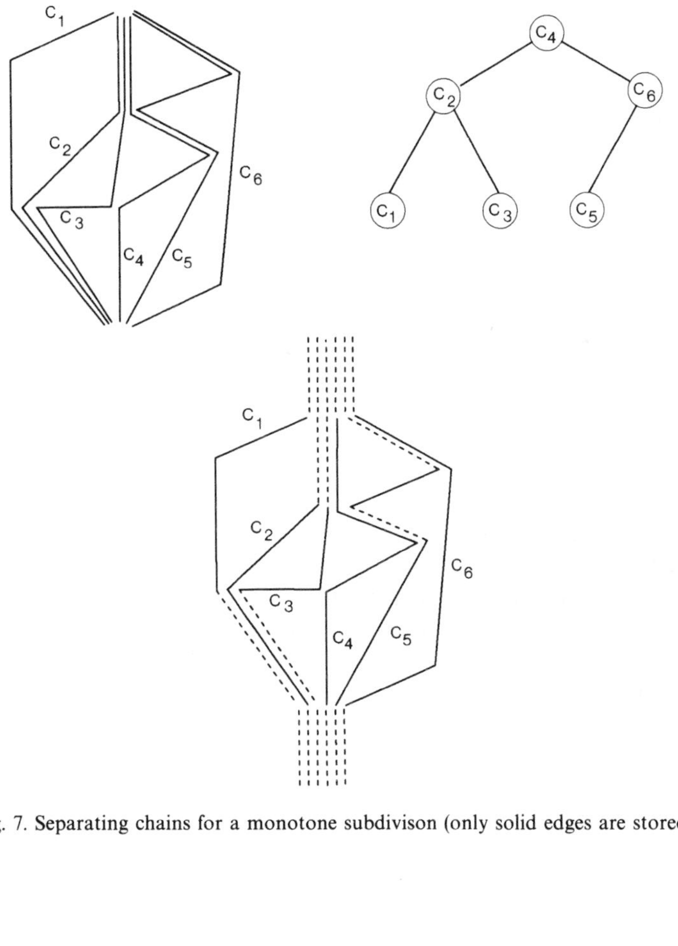

Fig. 7. Separating chains for a monotone subdivison (only solid edges are stored).

Section 10.1]). To keep the storage linear, only *every other* point on a chain is merged with its parent node, resulting in an expansion factor of only 2 for storage space.

Further developments in the point-location problem came from Cole [64] and Sarnak and Tarjan [192]. Cole observed that in the Dobkin–Lipton approach, the lists of line segments intersecting two contiguous slabs are "similar". In other words, one can regard the x-coordinate as time, and maintain the contents of a slab by insertions and deletions of line segments. Thus point location can be reduced to a data structure problem of storing a sorted set subject to insertions and deletions so that all past versions of the set, as well as the current version, can be accessed efficiently. Cole gave a solution to this "similar lists" problem that uses $O(n)$ space and has $O(\log n)$ access time for n updates. Sarnak and Tarjan coined the term of a "persistent" data structure to refer to a data structure where all past versions can be accessed and the current version can be modified. They gave a persistent form of binary search tree with $O(\log n)$ access/update time and $O(1)$ amortized space per update. (Cole's solution assumes all updates are known in advance.)

The preconditions of the methods by Kirkpatrick [137], Lee and Preparata [147], and Edelsbrunner, Guibas and Stolfi [81] can be met by applying an $O(n \log\log n)$ time triangulation algorithm (Section 7.1). It remains an open problem whether there is an optimal solution for point location with $O(n)$ preprocessing time, $O(n)$ space and $O(\log n)$ query time.

Techniques for dynamic point location have been studied by Preparata and Tamassia [185]. They showed that, for a monotone planar subdivison with n vertices, there is an $O(n)$ space dynamic point location data structure with query time $O(\log^2 n)$, which allows for insertion/deletion of a vertex in $O(\log n)$ time, and insertion/deletion of a chain of k edges in $O(\log^2 n + k)$ time.

9. Multidimensional trees

The solutions to geometric search problems in d dimensions depend on suitable data structures for representing a set $S \in R^d$. In this section we consider *orthogonal* data structures, where a set is organized by its projections onto individual coordinates; nonorthogonal data structures will be discussed in Section 10.2.

In contrast to the one-dimensional case, where balanced trees solve most searching problems efficiently, in higher dimensions we need to organize a set by coordinates in a variety of ways to solve different problems. We will present *segment trees* and *priority search trees* which are two-dimensional data structures, and *k-d trees* and *range trees* which are d-dimensional data structures. Some sample applications will be given; further applications to *orthogonal queries* are presented in Section 10.1.

9.1. *Segment tree*

The *segment tree* by Bentley [22] is a method for storing intervals whose endpoints belong to a set $S = \{x_1 < x_2 < \cdots < x_N\}$. Without loss of generality, assume S to be the integers in the range $[1, N]$. The segment tree is a binary tree where the leaves correspond to the segments $[1,2]$, $[2,3], \ldots, [N-1, N]$, and each internal node corresponds to a segment which is the union of the segments represented by its leaf descendants. Any interval $I=[\ell, r]$ with $1 \leqslant \ell < r \leqslant N$ can be stored as the disjoint union of up to $2\lceil \log N \rceil$ segments. Thus it takes $O(N+k \log N)$ space to store k intervals, and an interval can be inserted or deleted in $O(\log N)$ time. Each node of the tree points to a list of all intervals stored there, plus any additional information useful for a specific application.

Bentley used the segment tree to solve the *measure problem* posed by Klee, that is, to compute the measure (area) of the union of n given rectangles. For this application, we store a quantity $m(v)$ at each node v which is the measure of all intervals stored in the subtree rooted at v. The measure of n given rectangles can be computed by a sweep-line algorithm. Let S be the set of x coordinates of the corner of the rectangles. The intersection of a horizontal sweep-line with the rectangles gives a set of intervals, which we store in a segment tree over S. The y-coordinates of the corners of the rectangles represent events, which trigger updates of the segment tree, as well as accumulation of $m(root)\cdot(y_i - y_{i-1})$, the area covered between two recent events. The algorithm runs in $O(n \log n)$ time in the plane, which generalizes to $O(n^{d-1} \log n)$ time in d dimensions. Van Leeuwen and Wood [219] improved the bound to $O(n^{d-1})$. In 1988, Overmars and Yap [175] obtained an $O(n^{d/2} \log n)$ time algorithm for the measure problem by using the inclusion/exclusion principle and new data structures for d-space.

9.2. *Priority search tree*

Consider the problem of enumerating all intersecting pairs among a set of n rectangles with sides parallel to the x- and y-axes. Bentley and Wood [26] gave an optimal $O(n \log n + k)$ algorithm for reporting k intersecting pairs. Note that with the sweep approach, each cross-section induced by the sweep-line consists of a set of intervals. Thus the rectangle intersection problem can be reduced to *on-line intersection for a dynamic set of intervals.* We will describe an $O(\log n + k)$ solution to the latter problem with the *priority search tree* of McCreight [156]. (An alternative data structure, the *interval tree* by Edelsbrunner [78], also supports on-line intersection queries of intervals in logarithmic time.)

Let $[u_i, v_i]$, $1 \leqslant i \leqslant n$, be n input intervals to be processed. To simplify the exposition, we assume that the sorted sequence U of all the u_i's is known in advance. (The data structure works without this assumption so long as it is maintained by a suitable balancing scheme.) We will view each interval $[u_i, v_i]$ as a point (u_i, v_i) in two dimensions. After successive intervals have been processed, the root of the priority search tree T contains an interval (u_i, v_i) with maximum v_i, while the remaining intervals $\{(u_j, v_j)\}$ are split into two equal subsets by their u value, and stored recursively in the left and right subtrees of the root. (Thus the u values are stored as a binary search tree, while the v values are organized as a priority queue.) Each subtree of T corresponds to a vertical strip in the plane bounded on top by $v=v_i$, if (u_i, v_i) is the root node of the subtree. It is easy to see that each insertion or deletion of an interval can be handled in $O(\log n)$ time, essentially as a priority queue update. To report all intersections with a query interval (u^*, v^*), we need to identify those (u_i, v_i) with $u_i \leqslant v^*$ and $v_i \geqslant u^*$. These two constraints define a *northwest quadrant* Q in the plane which is bounded below by $v=u^*$. We first search T to obtain a decomposition of Q into $O(\log n)$ vertical strips, then traverse downward in each vertical strip, i.e., visit the nodes in preorder, until we cross the lower boundary $v=u^*$. The total number of points visited in each vertical strip is at most $2k+1$ if k of them are reported. Thus we have the claimed time bound.

The extension of the rectangle intersection problem to d dimensions has also been studied. Six and Wood [203] gave an $O(n(\log n)^{d-1})$ space, $O(n(\log n)^{d-1})$ time solution. Chazelle and Incerpi [52], also Edelsbrunner and Overmars [85], reduced the space to $O(n)$, and the time to $O(n(\log n)^{d-2})$.

9.3. *k-d Tree*

The *k-d tree*, which stands for *k-dimensional binary search tree*, was proposed by Bentley [21] as an extension of the one-dimensional binary search tree. Let S be a set of n points in R^d. At every level of a *k-d* tree for S, the set is divided into two halves according to one of the coordinates, chosen cyclically from $\{x_1, x_2, \ldots, x_d\}$. A *k-d* tree occupies $O(dn)$ space and can be constructed in $O(dn \log n)$ time. We will see in Section 10.1 that *k-d* trees can be used to solve *orthogonal queries*.

9.4. *Range tree*

The *range tree* represents yet another natural extension of binary search trees to higher dimensions (see [153, 226]). We will describe the structure for the 2-dimensional case, generalization to d dimensions is straightforward.

Given a set S of points in R^2, we build a binary search tree based on the x-value of the points. In addition, each node v of the tree points to an auxiliary y-tree, which is a binary search tree organized by y-value for all the descendants of v. A range tree for $d=2$ requires $O(n \log n)$ space, and can be constructed in $O(n \log n)$ time. For d dimensions, the space and construction time are $O(n(\log n)^{d-1})$. Insertions and deletions in a range tree can be handled by using a suitable balanced tree scheme [227]. Like the *k-d* tree, the range tree provides a solution to orthogonal queries (cf. Section 10.1).

REMARK. Dynamization of multidimensional data structures appears to be much harder than in the one-dimensional case. There are basically two approaches to dynamization: one approach is based on *balancing* techniques for trees, the other uses *decomposition* techniques to maintain a collection of data structures. For discussion of these techniques and their applications, see [24, 85, 173, 174, 218, 227].

10. Range search

A *range query* is a problem that fits the following general description: "Given a set of objects X in R^d, store X in a suitable data structure so that for any query object y (regarded as a "range"), those $x \in X$ such that $x \cap y \neq \emptyset$ can be identified quickly."

Range queries not only represent a large class of problems that occur naturally in computational geometry, they also arise as subproblems in solving other non-query type problems. We consider two major types of range queries: orthogonal queries and half-space queries. In these queries the set of objects X is a finite point set in R^d, and the range y is an orthogonal box and a half-space in R^d respectively. As one expects, orthogonal queries can be solved much more efficiently than general half-space queries.

There is usually a trade-off between the space required by a data structure and the time needed to answer a query. Of particular interest are solutions at the two extremes: best query time achievable by a *linear-space* data structure, and minimum storage space to enable *polylog* (i.e., a fixed power of the logarithm) query time. Distinction can also be made between problems where all objects in the query range are to be *reported* (*enumerated*), or merely to be *counted*. It is traditional to express the query time in the first case as $O(f(n)+k)$, where k is the number of objects reported, and as $O(f(n))$ in the second case. Another parameter of a solution is the preprocessing time of the data structure, which of course is just a one-time cost.

10.1 Orthogonal queries

For a set of n points in R^d, we would like to retrieve quickly all points in any specified orthogonal box. Range trees allow us to solve orthogonal range queries in time $O(\log^{d-1} n)$ and space $O(n \log^{d-1} n)$. Consider the case $d=2$. For a query rectangle $[a,b]\times[c,d]$, we decompose the x-range $[a,b]$ into $\log n$ intervals, and perform in each interval a one-dimensional search in the auxiliary y-tree. The query time is $O((\log n)^2+k)$ if k points are retrieved. The following observation due to Willard [226] enables the query time to be reduced to $O(\log n+k)$. As the y-tree of an internal node is the disjoint union of the y-trees of its two children, a y-search at the root already determines the outcome of a y-search at any descendant node. Thus we store only a master y-tree at the root, replacing the other y-trees with y-lists and using two pointers per entry to indicate its position in the two y-sublists.

Chazelle [41] reduced the storage space of the above solution by a factor of $\log\log n$. For orthogonal queries in three dimensions, a solution with $O(\log n+k)$ query time and $O(n(\log n)^3)$ space was given by Chazelle and Guibas [50].

We can also use k-d trees to solve orthogonal queries by recursively visiting each node v whose region $R(v)$ is partially contained in the query rectangle Q, until we find a set of regions contained entirely in Q. Lee and Wong [148] showed that the search algorithm has running time $O(dn^{1-1/d}+k)$. Consider the two-dimensional case: $T(n)$, the maximum number of regions $R(v)$ that can be intersected by a vertical (or horizontal) line, satisfies $T(n)\leqslant 2T(n/4)+O(1)$, which yields $T(n)=O(\sqrt{n})$.

There is also a related *point enclosure* problem, that is, given n rectangles, find those that enclose a query point. Solutions were provided by Vaishnavi [217] and McCreight [156]. An optimal $O(n)$ space, $O(\log n+k)$ time algorithm has been found by Chazelle [40].

As a footnote, the technique of introducing pointers between a list and its sublists to assist searching without expansion in space, as we saw above and also in planar point location, has found wide use in other data structures. This technique was put into a general framework under the name of "fractional cascading" by Chazelle and Guibas [49, 50].

10.2. *Half-space queries*

We will survey solutions to half-space queries in two, three and higher dimensions, with a focus on the *space-partition* approach which leads to linear-space and sublinear-query-time solutions. This approach generalizes to query problems where the range is the intersection of half-spaces, i.e., triangles and polygons in two dimensions and polytopes in higher dimensions; these problems are also referred to as *simplex queries* in the literature.

10.2.1. *Half-plane query in R^2*

As an example of a half-plane query, consider a database where the first two coordinates represent the income and the number of children respectively of a household, and a query asks for all households whose income exceeds $1000 plus $200 for each child. Note that a *k-d* tree is inadequate for handling half-plane queries, since it cannot guarantee a reduction in the cardinality of S after each level of recursion.

Willard [225] was the first to propose a linear-space data structure, called a *polygon tree*, that can achieve sublinear query time. A set S of n points is partitioned by two straight lines ℓ_1, ℓ_2 into four quadrants such that at most $n/4$ points are contained in the interior of each quadrant; the points in each quadrant are then partitioned recursively (see Example 6.3.2). For a query half-plane Q, since the line defining Q can intersect at most three of the four quadrants, one of the quadrants is either entirely inside or outside of Q and hence can be exempted from further search. The search time $T(n)$ thus satisfies $T(n)=3T(n/4)+O(\log n)$, which yields $T(n)=O(n^\alpha)$ with $\alpha=\log_4 3\cong 0.774$.

Edelsbrunner and Welzl [88] improved α to 0.695 by taking advantage of the fact that the first bisector ℓ_1 can be arbitrary. Thus a common bisector ℓ_1 may be used for two sibling quadrants in their own subdivisions. The resulting structure is called a *conjugate tree*; its search time satisfies the recurrence $T(n)=T(n/2)+T(n/4)+O(\log n)$ whose solution is as stated. The construction time for both the polygon tree and the conjugate tree is $O(n \log n)$ with the algorithm given in Section 4.3. Also, the query time holds for both reporting and counting. Recently, Welzl [223] improved the bound to $O(\sqrt{n}\log^3 n)$ using a very different approach based on the concept of Vapnik–Chernovenkis dimension.

With the above schemes, one can retrieve points in the intersection of r half-planes in the same order of time, but the constant increases with r. Paterson and F.F. Yao [179] gave a solution to the polygon query that achieves $O(r \log n)$ query time with $O(n^2)$ space. By using convex layers to represent a point set (cf. Section 3.4), Chazelle, Guibas and Lee [51] gave a solution which reports all points in a query half-plane in $O(\log n+k)$ time with $O(n)$ space. However, this solution does not generalize to the counting problem or the polygon query.

10.2.2. *Half-space query in R^3*

Can we extend *four-partitions* in R^2 to *eight-partitions* in R^3? The answer turns out to be yes. Given a set S of n points in R^3, one can always find three planes that form an *eight-partition* in the sense that at most $n/8$ points of S lie in each of the eight open regions (see [234]). This theorem is proved by using the *Borsuk–Ulam* theorem in algebraic topology. Thus, we can represent a point set in R^3 by recursive eight-partitions, resulting in an *octant tree*. An arbitrary plane in R^3 can intersect at most seven of the eight open regions defined by an eight-partition. This leads to a retrieval time of $O(n^\alpha)$ with $\alpha = \log_8 7 \cong 0.936$ for half-space queries.

As in the two-dimensional case, an octant tree can be refined by taking advantage of the fact that one of the three planes forming an eight-partition may be an arbitrary bisector. This reduces the query time to $\alpha \cong 0.899$. A *polyhedron query* defined by the intersection of r half-spaces can be solved in the same time, with the constant increasing with r. An octant tree for n points can be constructed in $O(n^6)$ time.

In the *circular query* problem for a planar set, one must retrieve all points lying in an arbitrary query circle. Using the correspondence between circles in R^2 and planes in R^3 through the mapping $(x, y) \leftrightarrow (x, y, x^2 + y^2)$, we can obtain a *circular* octant tree where an eight-partition is formed by three circles. Since a query circle can intersect at most six regions of the partition, this leads to a retrieval time of $O(n^\alpha)$ where $\alpha = \log_8 6 \cong 0.86$ [234].

10.2.3. Higher dimensions

Avis [17] showed that 2^d-partitions are not always possible in dimensions $d \geqslant 5$. A different proof was given in [234]. Both proofs are based on the idea that for $d \geqslant 5$, the number of constraints $2^d - 1$ exceeds the degree of freedom d^2. The case of $d = 4$ remains an interesting open question, that is, given any finite point set in R^4, whether one can always partition it with four hyperplanes such that each orthant contains at most $\frac{1}{16}$ of the points.

Generalizations of eight-partitions to arbitrary dimensions can be carried out along a different line, by relaxing the number of hyperplanes used in the partition; see [232, 63]. The resulting partitions have linear size and $O(n^{1-\varepsilon})$ query time for half-space queries, and they can be constructed in $O(n \log^f n)$ time where ε and f depend on d. It was also shown in [232] that a variety of range queries expressible in higher-order polynomial inequalities with $\{\cup, \cap, \min\}$ can be reduced to solving half-space queries. Haussler and Welzl [119] used random sampling to demonstrate the existence of partitions in R^d, called ε-nets, which achieve the best query time currently known for half-space queries in $d \geqslant 3$. In particular, their scheme gives query time with $\alpha = 0.857$ for $d = 3$. As their algorithms are probabilistic, it is an interesting open question to find deterministic algorithms for constructing partitions to realize similar or better query time in R^d. Some progress has been made in the two-dimensional case by Matoušek [155].

10.3. Lower bounds to range search

There appears to be some limit to how efficiently one can process range queries; certain types of range queries also seem harder to solve than others. We survey below a number of results which support such intuition. For example, if only linear space is used, then the query time per simplex query for $d=2$ must be of the order $\Omega(\sqrt{n})$, which compares well with the best upper bound $O(\sqrt{n}\log^3 n)$.

Fredman [98, 99, 100] introduced a model for studying the complexity of *dynamic queries*. In this model a range query problem is specified by a *key space* K, a family Γ of *regions*, i.e. subsets of keys, and an associated *value* in some semigroup S for each key in K. The answer to a query *Query*(R), where $R \in \Gamma$, is the semigroup sum of all the keys in region R. (The notion of a semigroup unifies the *reporting* and the *counting* problems.) A data structure for the dynamic range query problem must process an on-line sequence of n instructions of the form *Insert*(k), *Delete*(k), and *Query*(R). Fredman showed that, for orthogonal queries, any data structure must use $\Omega(n(\log n)^d)$ time in the worst case; for half-plane queries, polygonal queries and circular queries the lower bound is $\Omega(n^{4/3})$. A related model of partial sum query was discussed in [101].

A.C. Yao [230, 231] investigated the complexity of *static* range queries in Fredman's model, with *Query*(R) instructions only. It was shown that, for the static interval query problem with n records, the storage space S and query response time T must satisfy a trade-off relation

$$T=\Theta(\alpha(S,n)+n/(S-n+1)),$$

where α is the inverse of Ackermann's function. For the circular query, a trade-off $TS=\Omega(n^{1+\varepsilon})$ was given for some fixed $\varepsilon>0$; and for the two-dimensional orthogonal query, a trade-off

$$T\cdot\log((S\log n)/n)=\Omega(\log n)$$

was given. Further results on orthogonal range query were also obtained by Vaidya [213].

Chazelle [42] showed a trade-off of

$$T=\Omega((\log n)/(\log 2S/n)^{d-1})$$

for the static d-dimensional partial sum problem. Orthogonal query in a pointer machine model was also investigated. In [43], trade-offs were derived for the static d-dimensional simplex query problem:

$$T=\Omega(n/\sqrt{S})\quad\text{for } d=2,\quad\text{and}\quad T=\Omega((n/\log n)/S^{1/d})\quad\text{for } d\geqslant 3.$$

11. Visibility computations

The visibility problems originating in computer graphics have served as motivation for much research in computational geometry. We start by revisiting the basic problem of computing line intersections which was mentioned in Section 2. We then discuss algorithms for eliminating hidden surfaces and for computing visibility graphs.

11.1. Intersection of line segments

The $O(n\log n)$ sweep-line algorithm given in Section 2 for deciding whether any intersection exists among a set of n segments in the plane was due to Shamos and Hoey

[201]. Bentley and Ottmann [23] extended this algorithm to *report* all k intersections in $O((n+k)\log n)$ time and $O(n+k)$ space. The modification involves inserting each intersection point, once it is discovered, into the set of unprocessed endpoints which are now stored in a priority queue. The space requirement of this algorithm was later reduced to $O(n)$ by Brown [34]. An algorithm with running time

$$O(n \log^2 n/\log \log n + k)$$

was obtained by Chazelle [38] based on a hierarchical approach. In 1988, an optimal $O(n \log n + k)$ algorithm was found by Chazelle and Edelsbrunner [47]. Clarkson [58] and Mulmuley [168] gave simple randomized algorithms with $O(n \log n + k)$ expected time. The three-dimensional line intersection algorithm given in Section 2 was from A.C. Yao and F.F. Yao [232] (see also [48]).

The query version of the line intersection problem, namely, given a set S of n segments in the plane and a query segment q, report all segments of S that intersect q, can be solved in $O(\log n)$ time with an $O(n^2)$ data structure [41]. When q is an infinite line, the problem of deciding whether q intersects every segment of S (i.e., q is a *transversal* of S) or whether there exists any transversal for S has been studied by Edelsbrunner, Maurer, Preparata, Rosenberg, Welzl and Wood [82] who gave an $O(n \log n)$ solution. Algorithms for computing transversals of circles, rectangles, etc. are given in [79, 13], and algorithms for transversals of polyhedra are given in [18, 127].

11.2. Hidden-surface removal

To render a realistic picture of a scene on a display, one must decide, for any given viewpoint, which parts of the scene are visible and which parts are hidden. The problem of *hidden-line* and *hidden-surface* removal is one of the fundamental problems in computer graphics, and a variety of algorithms have been invented to suit different applications and image complexities. A description of representative hidden-surface algorithms can be found in [206].

Depending on the scene model, hidden-surface removal can be formulated in several different ways as a computational geometry problem. A general formulation is to consider the input scene as a collection of disjoint opaque polyhedra in E^3, and the output to be the image visible from a given viewpoint. Let n be the number of edges in the scene. Most of the algorithms published in the computer graphics literature (such as those mentioned in [206]) have worst-case complexity $O(n^3)$. Schmitt [194] and Devai [69] gave $O(n^2 \log n)$ time algorithms. Subsequently $O(n^2)$ time algorithms were found by Devai [70] for hidden-line removal, and independently by McKenna [157] for hidden-surface removal. Both of these algorithms approached the problem by extending the edges into infinite lines and then traversing the resulting *line arrangement* (see Section 12.1). Goodrich [110] gave an $O(n \log n + i)$ algorithm based on the new line-intersection algorithm by Chazelle and Edelsbrunner [47], where i is the number of intersecting pairs of edges in the projection plane.

The $O(n^2)$ running time of the algorithms mentioned above is unavoidable in the worst case if $\Omega(n^2)$ segments are visible [70, 193]. However, one may expect to find efficient output-sensitive algorithms which can take advantage of situations when the output size is small.

An interesting special case of the hidden-line removal problem is one when the scene consists of a set of n rectangles, each of which has sides parallel to the x- and y-axes and a constant z-coordinate. This situation arises, for example, in overlapping windowed systems. Guting and Ottmann [116] gave an $O(n \log^2 n + k \log^2 n)$ algorithm for static scenes with k visible windows. This algorithm was improved to $O((n+k)\log n)$ by Bern [27] (see also [186, 14]). The dynamic version of the problem when windows can be inserted and deleted was also studied in [27]. Other hidden-line removal algorithms for special scenes, such as polyhedral terrains, can be found in [65, 188, 28].

For some applications such as flight simulation where the viewpoint changes frequently, a *priority approach* proposed by Schumacker et al. (see [206]) for hidden-surface computation is particularly suitable. Efficient algorithms using this method have been investigated by Fuchs, Kedem and Naylor [103], F.F. Yao [233], Rappaport [187], Paterson and F.F. Yao [180]. The approach taken in [103, 180] addresses a general scheme for partitioning objects in d dimensions with hyperplanes, and it has applications to problems other than hidden-surface removal.

11.3. Visibility graphs

Another type of visibility problems arises in connection with *collision avoidance* in robotics. Given a set of polygonal obstacles in the plane, we would like to find all polygon vertices that are visible from a query point, or compute a shortest path that connects two points and avoids the obstacles (see [152]). For this purpose, we define the *visibility graph* for a set S of n nonintersecting line segments in the plane as an undirected graph in which the vertices are the endpoints of the segments, and two vertices are adjacent if they can "see" each other (i.e., the open line segment connecting them does not intersect any of the segments).

The straightforward method for constructing a visibility graph taks $O(n^3)$ time. Asano et al. [9] and Welzl [222] found $O(n^2)$ time algorithms which are best possible if the graph contains $O(n^2)$ edges. Ghosh and Mount [107] gave an optimal output-sensitive algorithm with $O(E+n \log n)$ time and $O(E+n)$ space, where E is the number of edges in the graph. Once the visibility graph has been constructed, the shortest path problem mentioned above can be solved in $O(E+n \log n)$ time by applying an algorithm for weighted graphs by Fredman and Tarjan [102]. In the special case that the segments form a simple polygon, visibility algorithms have been studied by [89, 115, 120, 144].

12. Combinatorial geometry

The analysis of geometric algorithms often requires detailed combinatorial knowledge of the geometric structures involved. The book by Edelsbrunner [80] is a good reference for problems which overlap both computational and combinatorial geometry. We will mention a few representative topics in the following.

12.1. *Arrangements*

A collection H of n lines in R^2 partitions the plane into a collection of $O(n^2)$ regions, edges and vertices. Such a partition is known as the *arrangement* $\mathscr{A}(H)$ of H [114]. Similarly, a set H of n hyperplanes in R^d dissects R^d into a collection of convex regions, or *d-faces*; the boundaries of these d-faces are $(d-1)$-dimensional convex polyhedra, called the $(d-1)$-faces, and so on. The *arrangement* $\mathscr{A}(H)$ of H is the partition of R^d into these $O(n^d)$ faces of dimension k, $0 \leqslant k \leqslant d$. An arrangement $\mathscr{A}(H)$ can be represented by an *incidence graph* $\mathscr{I}(H)$ which stores all incidence relations between k-faces and $(k+1)$-faces for $0 \leqslant k \leqslant d-1$. The incidence graph for an arrangement of n hyperplanes can be built incrementally by adding the hyperplanes one at a time to an existing arrangement. Initially, we pick out d hyperplanes whose normal vectors span R^d (if this is not the case, then the arrangement will be constructed in a subspace of R^d). The incidence graph of these d hyperplanes is isomorphic to a d-dimensional cube and can be set up straightforwardly. Each remaining hyperplane h of H is then added in turn to the existing arrangement $\mathscr{A}(H)$ in several steps: first an edge e of $\mathscr{A}(H)$ that intersects h is identified; then, starting with e, all faces that intersect h are marked and updated appropriately. The time needed for inserting h into $\mathscr{A}(H)$ turns out to be proportional to the cardinality of faces bounding the *cells* (i.e., d-faces) intersected by h, which can be shown to be $O(n^{d-1})$ if $|H|=n$. Thus the arrangement $\mathscr{A}(H)$ of n hyperplanes can be constructed in total time $O(n^d)$, which matches the size of $\mathscr{A}(H)$ and hence is optimal. This result was established by Edelsbrunner, O'Rourke and Seidel [84]. The incremental algorithm in the two-dimensional case was independently developed in [51].

Applications of the structure $\mathscr{A}(H)$ include computing order-k Voronoi diagrams (Section 4.3), testing for degeneracy, finding the minimum volume simplex spanned by a set of points, etc. (see [80, 84] for more details). Various combinatorial bounds for arrangements of curves and surfaces can be found in [60].

12.2. *Davenport–Schinzel sequences*

Many algorithms in computational geometry and motion planning involve calculation of the lower envelope (i.e., pointwise minimum) of a collection of functions. For example, given a polyhedral surface and a viewpoint a, the "upper rim" (also called *silhouette*) of the surface as seen from a can be expressed as the lower envelope of a set of line segments in two dimensions (see [65]).

Lower envelopes are closely related to *Davenport–Schinzel sequences.* (See Chapter 8 for definition and references on Davenport–Schinzel sequences.) In the above example, the lower envelope of a set of n segments $\{\ell_1, \ldots, \ell_n\}$ in the plane is a Davenport–Schinzel sequence of order 3, since no alternation of the form $\ell_i \ldots \ell_j \ldots \ell_i \ldots \ell_j \ldots \ell_i$ can occur in the lower envelope. The maximum length of such a sequence is $\lambda_3(n) = \Theta(n\alpha(n))$ where $\alpha(n)$ is the inverse of Ackermann's function [118]; and the worst case can be realized by line segments [224]. Thus the silhouette of a polyhedral surface has complexity $\Theta(n\alpha(n))$, and is computable in $O(n\alpha(n)\log n)$ time (see [65]). Other geometric applications of Davenport–Schinzel sequences can be found in [12, 86, 202].

12.3. Bisections and k-sets

Let S be a set in E^d containing n points. A set $S' \subseteq S$ with $|S'|=k$ is called a k-set if there is a hyperplane that separates S' from the rest of S. A k-set for $k=\lfloor n/2 \rfloor$ is called a bisection. The combinatorial function that counts the maximum number of k-sets has applications to higher-order Voronoi diagrams and partitions of space [80].

Erdös, Lovász, Simmons and Strauss [90] showed that the number of k-sets in the plane has an upper bound of $O(n\sqrt{k})$ and a lower bound of $\Omega(n \log k)$. The lower bound generalizes to d dimensions. For example, there are at least $\Omega(n^{d-1} \log n)$ bisections (and a more complicated expression holds for k-sets, see [80]). In 1989, Pach, Steiger and Szemerédi [176] improved the upper bound for k-sets to $O(n\sqrt{k}/(\log^* n))$ in two dimensions; and Barany, Furedi and Lovász [19] obtained an upper bound of $O(n^{3-\varepsilon})$ for the number of bisections in three dimensions. For general k and d, sometimes tight bounds can be obtained for the total number of j-sets with $j \leqslant k$, rather than for k-sets only (except when k is very small). Let $g_{k,d}(n)$ denote this total number. Alon and Györi [8] proved a tight bound $g_{k,2}(n)=nk$ for $k<n/2$. In three and higher dimensions, Clarkson [58] used a probabilistic argument to show that $g_{k,d}(n)=\Theta(n^{\lfloor d/2 \rfloor} k^{\lceil d/2 \rceil})$ as $n/k \to \infty$, for fixed d. Since a convex hull facet in E^d can be viewed as a d-set, this formula is related to the upper bound theorem.

12.4. λ-Matrix and order type

Given a set S of n labelled points $\{p_1, p_2, \ldots, p_n\}$ in the plane, the λ-matrix of S is the n by n matrix whose (i,j)th entry is the number of points to the left of the directed line determined by (p_i, p_j). It is a somewhat surprising fact that the λ-matrix actually determines *which* points of S lie to the left of each directed line (p_i, p_j). That is, it determines the *order type* of S, i.e. the orientation of every triple of points of S. The λ-matrix for n points can be constructed in optimal time $O(n^2)$ [84]. The preceding results all generalize to dimension d. The number of realizable order types in d dimensions has been shown by Goodman and Pollack [108] to have upper and lower bounds that agree in a leading term $O(n^{d^2 n})$ (see also [7]). Goodman, Pollack and Sturmfels [109] showed in 1989 that, to realize all order types in the plane by lattice points, the grid size must be exponential. This result has implications for representing configurations on a finite-resolution display (see Section 13.1).

13. Other topics

We will mention two additional topics in computational geometry that have received increasing attention in the eighties. Both of these topics are motivated by practical considerations, yet they have demonstrated to have significant mathematical content as well.

13.1. Finite-precision geometry

In designing geometric algorithms, one usually makes use of properties of the continuous Euclidean space. In reality, however, the algorithms are executed under finite-precision constraints due to the limitations of the internal computation and the display resolution. The naive approach of rounding the coordinates can create many problems including topological inversions. Greene and Yao [112] explored the question of transforming an algorithm from the continuous to the discrete domain, and proposed a solution for the line-intersection problem by allowing lines to be "bent" with a continued fraction representation.

A related, though somewhat different issue is the handling of degenerate cases in geometric algorithms. Yap described a symbolic perturbation scheme and proved its consistency in [236]. A similar technique and many case studies were given by Edelsbrunner and Mücke [83]. Another related issue is the question of numerical accuracy and robustness of geometric computations; see [74, 124, 125]. In comparison with the algorithmic issues mentioned above, the question of digitizing geometric figures (e.g., lines and curves) has been studied more extensively in the literature; see for example [169]. In order for computational geometry to have more practical impact, the adaptation of geometric computations to a finite-precision model is an important direction to investigate.

13.2. Random sampling

In recent years, randomization techniques have seen increasing use in computational geometry. The basic idea is to use random sampling to split up a given problem into subproblems of small size. Random sampling has been applied to solve many geometric problems, including computing convex hulls [58 61], diameters [61], line intersections [58, 168], triangulations of simple polygons [62], etc. We will describe a basic framework of random sampling due to Clarkson [58]. For detailed discussions of the framework and more applications, see [57, 58].

Let the input to a problem be a set S of n subsets of R^d, referred to as *objects*. Assume that $\mathcal{F}_S$ is a set of *regions* defined by the objects in some way, and that the desired computation is to find the subset $\mathcal{F}'_S \subseteq \mathcal{F}_S$ consisting of those regions that are not intersected by any object in S. For example, in computing planar convex hulls, the objects are points and the regions are half-planes defined by pairs of points; in computing Delaunay triangulations, the regions correspond to disks defined by triples of points, etc. We make the assumption that each region is defined by only a constant number of objects. (In the preceding two examples, the constant is equal to 2 and 3 respectively.)

Let $R \subseteq S$ be a random sample of size r, and let $\mathcal{F}_R$ and $\mathcal{F}'_R$ be defined as before, but with respect to the subset R. Then a theorem in [58] states that the average number of objects of S meeting any region $F \in \mathcal{F}'_R$ is $O(n/r)$. Intuitively, the fact that none of the sampled objects in R intersects F is evidence that most of the original objects of S miss F as well. Thus, a way to construct the desired $\mathcal{F}'_S$ is to take a random sample R, recursively compute $\mathcal{F}'_R$, and then modify the result by adding the remaining objects from $S-R$. The theorem mentioned above implies that the amount of work required for the modification is expected to be reasonable, thus leading to an efficient randomized algorithm for computing $\mathcal{F}'_S$.

14. Conclusion

We discussed in this chapter a wide range of topics in computational geometry. The space limitation forced us to leave out many topics, such as average-case analyses, parallel geometric algorithms, etc. Some of the results described here may also soon become outdated due to the fast-moving nature of the area. Nevertheless, it is hoped that the survey provides a broad view of computational geometry and its applications. We conclude by listing a few major open problems.

(1) Linear programming is used frequently as a tool for solving geometrical optimization problems. It remains a long-standing open problem whether there is a *strongly polynomial* algorithm for linear programming, that is, an algorithm that solves a linear program in a number of arithmetic operations that is polynomial in the input size nd. As an intermediate step, one may first try to find a *randomized* strongly polynomial algorithm for this problem.

(2) Two geometrical optimization problems stand out for their special interest: the Euclidean Minimum Spanning Tree problem (EMST) and the Euclidean Travelling Salesman problem (ETSP). The best time bound for constructing EMST in three dimensions is currently $O((n \log n)^{3/2})$. Can it be improved to $O(n^{1+\varepsilon})$? For ETSP in the plane, is there a polynomial-time algorithm that can achieve an approximation bound better than $\frac{3}{2}$?

(3) The fastest algorithm known for triangulating a simple polygon requires $O(n \log \log n)$ time, while no nonlinear lower bound to this problem is known. A faster triangulation algorithm would immediately lead to improved algorithms for many problems on simple polygons.

(4) Given n points in the plane, does there exist an $O(n)$ space, $O(\log n)$ time solution to the orthogonal range query problem? Again, such a result would lead to improved algorithms for many problems concerning orthogonal objects in two and higher dimensions.

(5) Does there exist an efficient output-sensitive algorithm for hidden-line removal, that is, an algorithm with running time $o(n^2)$ when the complexity of the scene is $o(n^2)$?

(6) The problem of testing whether n points in the plane are in general position has an upper bound of $O(n^2)$ and a lower bound of $\Omega(n \log n)$. It would be interesting to determine the exact complexity of this very basic problem.

(7) The lower bounds that can be proven for geometric problems are mostly of the form $\Omega(n \log n)$ and, occasionally, $\Omega(n^{1+\varepsilon})$ (as in range queries). Can one establish higher-order polynomial lower bounds for geometric problems in the algebraic decision tree model? One possible candidate is the measure problem, i.e., the problem of computing the volume of the union of n orthogonal boxes in d dimensions, for which the best upper bound at present is $O(n^{d/2} \log n)$.

References

[1] AGGARWAL, A., The art gallery theorem: its variations, applications and algorithmic aspects, Ph.D. Thesis, Computer Science Dept., The Johns Hopkins Univ., Baltimore, MD, 1984.

[2] AGGARWAL, A., Unpublished note, 1987.
[3] AGGARWAL, A., L.J. GUIBAS, J. SAXE and P. SHOR, A linear time algorithm for computing the Voronoi diagram of a convex polygon, in: *Proc. 19th Ann. ACM Symp. Theory Comput.* (1987) 39–45.
[4] AGGARWAL, A., M.M. KLAWE, S. MORAN, P.W. SHOR and R. WILBER, Geometric applications of a matrix-searching algorithm, *Algorithmica* **2** (1987) 195–208.
[5] AGGARWAL, A. and S. SURI, Fast algorithms for computing the largest empty rectangle, in: *Proc. 3rd Ann. ACM Symp. Comput. Geom.* (1987) 278–290.
[6] AGGARWAL, A. and J. WEIN, Computational geometry, Research Seminar Series MIT/LCS/RSS-3, Lab. for Computer Science, Massachusetts Institute of Technology, Cambridge, MA, 1988.
[7] ALON, N., The number of polytopes, configurations, and real matroids, *Mathematika* **33** (1986) 62–71.
[8] ALON, N. and E. GYŐRI, The number of small semispaces of a finite set of points in the plane, *J. Combin. Theory Ser. A* **41** (1986) 154–157.
[9] ASANO, TE., TA. ASANO, L.J. GUIBAS, J. HERSHBERGER and H. IMAI, Visibility of disjoint polygons, *Algorithmica* **1** (1986) 49–63.
[10] ASANO, TE., TA. ASANO and H. IMAI, Partitioning a polygonal region into trapezoids, *J. ACM* **33** (1986) 290–312.
[11] ASANO, T., B. BHATTACHARYA, M. KEIL and F.F. YAO, Clustering algorithms based on minimum and maximum spanning trees, in: *Proc. 4th Ann. ACM Symp. Comput. Geom.* (1988) 252–257.
[12] ATALLAH, M., Dynamic computational geometry, in: *Proc. 24th Ann. Symp. Found. Comput. Sci.* (1983) 92–99; Also: *Comp. Maths. with Appls.* **11** (1985) 1171–1181.
[13] ATALLAH, M.J. and C. BAJAJ, Efficient algorithms for common transversals, *Inform. Process. Lett.* **25** (1987) 87–91.
[14] ATTALAH, M.J. and M.T. GOODRICH, Output-sensitive hidden surface elimination for rectangles, Tech. Report 88-13, Computer Science Dept., The Johns Hopkins Univ., Baltimore, MD, 1988.
[15] AURENHAMMER, F., Voronoi diagrams — a survey, Manuscript, 1988.
[16] AURENHAMMER, F. and H. EDELSBRUNNER, An optimal algorithm for constructing the weighted Voronoi diagram in the plane, *Pattern Recognition* **17** (1984) 251–257.
[17] AVIS, D., Non-partitionable point sets, *Inform. Process. Lett.* **19** (1984) 125–129.
[18] AVIS, D. and R. WENGER, Polyhedral line transversals in space, *Discrete & Comput. Geometry* **3** (1988) 257–266.
[19] BARANY, I., Z. FUREDI and L. LOVÁSZ, On the number of halving planes, in: *Proc. 5th Ann. ACM Symp. Comput. Geom.* (1989) 140–144.
[20] BEN-OR, M., Lower bounds for algebraic computation trees, in: *Proc. 15th Ann. ACM Symp. Theory Comput.* (1983) 80–86.
[21] BENTLEY, J.L., Multidimensional binary search trees used for associative searching, *Comm. ACM* **18** (1975) 509–517.
[22] BENTLEY, J.L., Solution to Klee's rectangle problems, Unpublished notes, Dept. of Computer Science, Carnegie-Mellon Univ., Pittsburgh, PA, 1977.
[23] BENTLEY, J.L. and T.A. OTTMANN, Algorithms for reporting and counting geometric intersections, *IEEE Trans. Comput.* **28** (1979) 643–647.
[24] BENTLEY, J.L. and J.B. SAXE, Decomposable searching problems I: static-to-dynamic transformation, *J. Algorithms* **1** (1980) 301–358.
[25] BENTLEY, J.L. and M.I. SHAMOS, Divide and conquer in multidimensional space, in: *Proc. 8th Ann. ACM Symp. Theory Comput.* (1976) 220–230.
[26] BENTLEY, J.L. and D. WOOD, An optimal worst case algorithm for reporting intersection of rectangles, *IEEE Trans. Comput.* **29** (1980) 571–577.
[27] BERN, M., Hidden surface removal for rectangles, in: *Proc. 4th Ann. ACM Symp. Comput. Geom.* (1988) 183–192.
[28] BERN, M.W., D.P. DOBKIN and R. GROSSMAN, Visibility with a moving point of view, in: *Proc. 1st Ann. ACM–SIAM Symp. Discrete Algorithms* (1989) 107–117.
[29] BHATTACHARYA, B. and G.T. TOUSSAINT, On geometric algorithms that use the furthest-point Voronoi diagram, in: G.T. Toussaint, ed., *Computational Geometry* (North-Holland, Amsterdam 1985) 43–62.
[30] BOWYER, A., Computational Dirichlet tessellations, *Comput. J.* **24** (1981) 162–166.

[31] BRONSTED, A., *An Introduction to Convex Polytopes*, Graduate Texts in Mathematics (Springer, New York, 1983).
[32] BROSTOW, W., J.P. DUSSAULT, and B.L. FOX, Construction of Voronoi polyhedra, *J. Comput. Phys.* **29** (1978) 81–92.
[33] BROWN, K.Q., Voronoi diagrams from convex hulls, *Inform. Process. Lett.* **9** (1979) 223–228.
[34] BROWN, K.Q., Comments on "Algorithms for reporting and counting geometric intersections", *IEEE Trans. Comput.* **30** (1981) 147–148.
[35] BRUGGESSER, H. and P. MANI, Shellable decompositions of cells and spheres, *Math. Scand.* **29** (1971) 197–205.
[36] CHAND, D.R. and S.S. KAPUR, An algorithm for convex polytopes, *J. ACM* **17** (1970) 78–86.
[37] CHAZELLE, B.M., Optimal algorithms for computing depths and layers, in: *Proc. 21st Allerton Conf. on Commun., Control and Comput.* (1983) 427–436.
[38] CHAZELLE, B.M., Intersection is easier than sorting, in: *Proc. 16th Ann. ACM Symp. Theory Comput.* (1984) 125–134.
[39] CHAZELLE, B.M., How to search in history, *Inform. and Control* **64** (1985) 77–99.
[40] CHAZELLE, B.M., Slimming down search structures: a functional approach to algorithm design, in: *Proc. 26th Ann. IEEE Symp. Found. Comput. Sci.* (1985) 165–174.
[41] CHAZELLE, B.M., Filtering search: a new approach to query answering, *SIAM J. Comput.* **15** (1986) 703–724.
[42] CHAZELLE, B.M., Lower bounds on the complexity of multidimensional searching, in: *Proc. 27th Ann. IEEE Symp. Found. Comput. Sci.* (1986) 87–96.
[43] CHAZELLE, B.M., Polytope range searching and integral geometry, in: *Proc. 28th Ann. IEEE Symp. Found. Comput. Sci.* (1987) 1–10.
[44] CHAZELLE, B.M., An optimal algorithm for intersecting three-dimensional convex polyhedra, Tech. Report CS-TR-205-89, Dept. of Computer Science, Princeton Univ., Princeton, NJ, 1989.
[45] CHAZELLE, B.M. and D.P. DOBKIN, Detection is easier than computation, *J. ACM* **27** (1980) 146–153.
[46] CHAZELLE, B.M. and D.P. DOBKIN, Optimal convex decompositions, in: G.T. Toussaint, ed., *Computational Geometry* (North-Holland, Amsterdam, 1985) 63–134.
[47] CHAZELLE, B.M. and H. EDELSBRUNNER, An optimal algorithm for intersecting line segments in the plane, in: *Proc. 28th Ann. IEEE Symp. Found. Comput. Sci.* (1988) 590–600.
[48] CHAZELLE, B.M., H. EDELSBRUNNER, L.J. GUIBAS and M. SHARIR, Lines in space—combinatorics, algorithms and applications, in: *Proc. 13th Ann. ACM Symp. Theory Comput.* (1989) 382–393.
[49] CHAZELLE, B.M. and L.J. GUIBAS, Fractional cascading I: a data structuring technique, *Algorithmica* **1** (1986) 133–162.
[50] CHAZELLE, B.M. and L.J. GUIBAS, Fractional cascading II: applications, *Algorithmica* **1** (1986) 163–191.
[51] CHAZELLE, B.M., L.J. GUIBAS and D.T. LEE, The power of geometric duality, *BIT* **25** (1985) 76–90.
[52] CHAZELLE, B.M. and J. INCERPI, Triangulation and shape-complexity, *ACM Trans. Graphics* **3** (1984) 135–152.
[53] CHERITON, D. and R.E. TARJAN, Finding minimum spanning trees, *SIAM J. Comput.* **5** (1976) 724–742.
[54] CHRISTOFIDES, N., Worst-case analysis of a new heuristic for the travelling salesman problem, in: *Proc. Symp. on New Directions and Recent Results in Algorithms and Complexity*, Carnegie-Mellon Univ., Pittsburgh, (1976).
[55] CHVÁTAL, V., *Linear Programming* (Freeman, New York, 1983).
[56] CLARKSON, K.L., Linear programming in $O(n \cdot 3^{d^2})$ time, *Inform. Process. Lett.* **22** (1986) 21–24.
[57] CLARKSON, K.L., New applications of random sampling in computational geometry, *Discrete & Comput. Geometry* **2** (1987) 195–222.
[58] CLARKSON, K.L., Applications of random sampling in computational geometry II, in: *Proc. 4th Ann. ACM Symp. Comput. Geom.* (1988) 1–11.
[59] CLARKSON, K.L., A Las Vegas algorithm for linear programming when the dimension is small, in: *Proc. 29th Ann. IEEE Symp. Found. Comput. Sci.* (1988) 452–457.
[60] CLARKSON, K.L., H. EDELSBRUNNER, L.J. GUIBAS, M. SHARIR and E. WELZL, Combinatorial complexity bounds for arrangements of curves and surfaces, in: *Proc. 29th Ann. IEEE Symp. Found. Comput. Sci.* (1988) 568–579.

[61] CLARKSON, K.L. and P.W. SHOR, Algorithms for diametral pairs and convex hulls that are optimal, randomized and incremental, in: *Proc. 4th Ann. ACM Symp. Comput. Geom.* (1988) 12–17.
[62] CLARKSON, K.L., R.E. TARJAN and C.J. VAN WYK, A fast Las Vegas algorithm for triangulating a simple polygon, in: *Proc. 4th Ann. ACM Symp. Comput. Geom.* (1988) 18–22.
[63] COLE, R., Partitioning point sets in arbitrary dimensions, *Theoret. Comput. Sci.* **49** (1987) 239–266.
[64] COLE, R., Searching and storing similar sets, *J. Algorithms* **7** (1986) 202–220.
[65] COLE, R. and M. SHARIR, Visibility problems for polyhedral terrains, *J. Symbolic Comput.* **7** (1989) 11–30.
[66] CULBERSON, J.C. and R.A. RECKHOW, Covering polygons is hard, in: *Proc. 29th Ann. IEEE Symp. Found. Comput. Sci.* (1988) 601–611.
[67] DANTZIG, G.B., *Linear Programming and Extensions* (Princeton Univ. Press, Princeton, NJ, 1963).
[68] DELAUNAY, B., Sur la sphère vide, *Bull. Acad. Sci. USSR (VII), Classe Sci. Mat. Nat.* (1934) 793–800.
[69] DEVAI, F., Complexity of visibility computations, Dissertation for the Candidate of Sciences Degree, Hungarian Academy of Sciences, Budapest, Hungary, 1981.
[70] DEVAI, F., Quadratic bounds for hidden-line elimination, in: *Proc. 2nd Ann. ACM Symp. Comput. Geom.* (1986) 269–275.
[71] DIRICHLET, P.G.L., Über die Reduktion der positiven quadratischen Formen mit drei unbestimmten ganzen Zahlen, *J. Reine Angew. Math.* **40** (1850) 209–227.
[72] DOBKIN, D.P. and D.G. KIRKPATRICK, Fast detection of polyhedral intersection, *Theoret. Comput. Sci.* **27** (1983) 241–253.
[73] DOBKIN, D.P. and R.J. LIPTON, Multidimensional searching problems, *SIAM J. Comput.* **5** (1976) 181–186.
[74] DOBKIN, D.P. and D. SILVER, Recipes for geometry and numerical analysis—Part I: an empirical study, in: *Proc. 4th Ann. ACM Symp. Comput. Geom.* (1988) 93–105.
[75] DYER, M.E., The complexity of vertex enumeration methods, *Math. Oper. Res.* **8** (1983) 381–402.
[76] DYER, M.E., Linear time algorithms for two- and three-variable linear programs, *SIAM J. Comput.* **13** (1984) 31–45.
[77] DYER, M.E., On a multidimensional search technique and its application to the Euclidean one-center problem, *SIAM J. Comput.* **15** (1986) 725–738.
[78] EDELSBRUNNER, H., Dynamic rectangle intersection searching, Report F47, Institut für Informationsverarbeitung, Technical Univ. Graz, 1980.
[79] EDELSBRUNNER, H., Finding transversals for sets of simple geometric figures, *Theoret. Comput. Sci.* **35** (1985) 55–69.
[80] EDELSBRUNNER, H., *Algorithms in Combinatorial Geometry* (Springer, Heidelberg, 1987).
[81] EDELSBRUNNER, H., L.J. GUIBAS and J. STOLFI, Optimal point location in a monotone subdivision, *SIAM J. Comput.* **15** (1986) 317–340.
[82] EDELSBRUNNER, H., A. MAURER, F.P. PREPARATA, A.L. ROSENBERG, E. WELZL, D. WOOD, Stabbing line segments, *BIT* **22** (1982) 274–281.
[83] EDELSBRUNNER, H. and E.P. MÜCKE, Simulation of simplicity: a technique to cope with degenerate cases in geometric algorithms, in: *Proc. 4th Ann. ACM Symp. Comput. Geom.* (1988) 118–133.
[84] EDELSBRUNNER, H., J. O'ROURKE and R. SEIDEL, Constructing arrangements of lines and hyperplanes with applications, *SIAM J. Comput.* **15** (1986) 341–363.
[85] EDELSBRUNNER, H. and M.H. OVERMARS, Batched dynamic solutions to decomposable searching problems, *J. Algorithms* **6** (1985) 515–542.
[86] EDELSBRUNNER, H., J. PACH, J. SCHWARTZ and M. SHARIR, On the lower envelope of bivariate functions and its applications, in: *Proc. 28th Ann. IEEE Symp. Found Comput. Sci.* (1987) 27–37.
[87] EDELSBRUNNER, H. and R. SEIDEL, Voronoi diagrams and arrangements, *Discrete & Comput. Geom.* **1** (1986) 25–44.
[88] EDELSBRUNNER, H. and E. WELZL, Halfplanar range search in linear space and $O(n^{0.695})$ query time, *Inform. Process. Lett.* **23** (1986) 289–293.
[89] EL GINDY, H.A. and D. AVIS, A linear algorithm for computing the visibility polygon from a point, *J. Algorithms* **2** (1981) 186–197.
[90] ERDŐS, P., L. LOVÁSZ, A. SIMMONS and E.G. STRAUSS, Dissection graphs of planar point sets, in: J.N. Srivastava et al., eds., *A Survey of Combinatorial Theory* (North-Holland, Amsterdam, 1973).

[91] FEDER, T. and D.H. GREENE, Optimal algorithms for approximate clustering, in: *Proc. 20th Ann. ACM Symp. Theory Comput.* (1988) 434–444.

[92] FERRARI, L., P.V. SANKAR and J. SKLANSKY, Minimal rectangular partitions of digitized blobs, in: *Proc. 5th Internat. IEEE Conf. on Pattern Recognition* (1981) 1040–1043.

[93] FINKEL, R.A. and J.L. BENTLEY, Quad-trees: a data structure for retrieval on composite keys, *Acta Inform.* **4** (1974) 1–9.

[94] FORTUNE, S.J., A sweepline algorithm for Voronoi diagrams, in: *Proc. 2nd Ann. ACM Symp. Comput. Geom.* (1986) 313–322.

[95] FOURNIER, A. and D.Y. MONTUNO, Triangulating simple polygons and equivalent problems, *ACM Trans. Graphics* **3** (1984) 153–174.

[96] FOWLER, R.J., M.S. PATERSON and S.L. TANIMOTO, Optimal packing and covering in the plane are NP-complete, *Inform. Process. Lett.* **12** (1981) 133–137.

[97] FRANZBLAU, D. and D. KLEITMAN, An algorithm for constructing regions with rectangles: independence and minimum generating sets for collections of intervals, in: *Proc. 16th ACM Symp. Theory Comput.* (1984) 167–174.

[98] FREDMAN, M.L., The inherent complexity of dynamic data structures which accommodate range queries, in: *Proc. 21st Ann. IEEE Symp. Found. Comput. Sci.* (1980) 191–199.

[99] FREDMAN, M.L., Lower bounds on the complexity of some optimal data structures, *SIAM J. Comput.* **10** (1981) 1–10.

[100] FREDMAN, M.L., A lower bound of the complexity of orthogonal range queries, *J. ACM* **28** (1981) 696–705.

[101] FREDMAN, M.L., The complexity of maintaining an array and computing its partial sums, *J. ACM* **29** (1982) 250–260.

[102] FREDMAN, M.L. and R.E. TARJAN, Fibonacci heaps and their uses in improved network optimization algorithms, in: *Proc. 25th Ann. IEEE Symp. Found. Comput. Sci.* (1984) 338–346.

[103] FUCHS, H., Z. KEDEM and B. NAYLOR, On visible surface generation by a priori tree structures, in: *Computer Graphics (SIGGRAPH '80 Conf. Proc.)* 124–133.

[104] GABOW, H.N., An efficient implementation of Edmond's algorithm for maximum matching on graphs, *J. ACM* **23** (1976) 221–234.

[105] GAREY, M.R. and D.S. JOHNSON, *Computers and Intractability: A Guide to the Theory of NP-Completeness* (Freeman, San Francisco, CA, 1979).

[106] GAREY, M.R., D.S. JOHNSON, F.P. PREPARATA and R.E. TARJAN, Triangulating a simple polygon, *Inform. Process. Lett.* **7** (1978) 175–180.

[107] GHOSH, S.K. and D.M. MOUNT, An output-sensitive algorithm for computing visibility graphs, in: *Proc. 28th Ann. IEEE Symp. Found. Comput. Sci.* (1987) 11–19.

[108] GOODMAN, J.E. and R. POLLACK, Upper bounds for configurations and polytopes in R^d, *Discrete & Comput. Geometry* **1** (1976) 219–228.

[109] GOODMAN, J.E., R. POLLACK and B. STURMFELS, Coordinate representation of order types requires exponential storage, in: *Proc. 21th ACM Ann. Symp. Theory Comput.* (1989) 405–410.

[110] GOODRICH, M.T., A polygonal approach to hidden-line elimination, Tech. Report 87-18, Dept. of Computer Science, The Johns Hopkins Univ., Baltimore, MD, 1987.

[111] GRAHAM, R.L., An efficient algorithm for determining the convex hull of a finite planar set, *Inform. Process. Lett.* **1** (1972) 132–133.

[112] GRAHAM, R.L. and F.F. YAO, Finding the convex hull of a simple polygon, *J. Algorithms* **4** (1983) 324–331.

[113] GREENE, D.H. and F.F. YAO, Finite-resolution computational geometry, in: *Proc. 27th Ann. IEEE Symp. Found. Comput. Sci.* (1986) 143–152.

[114] GRÜNBAUM, B., *Convex Polytopes* (Wiley-Interscience, New York, 1967).

[115] GUIBAS, L.G., J. HERSHBERGER, D. LEVEN, M. SHARIR and R.E. TARJAN, Linear time algorithms for visibility and shortest path problems inside triangulated simple polygons, *Algorithmica*, to appear.

[116] GÜTING, R.H. and T. OTTMANN, New algorithms for special cases of the hidden line elimination problem, Tech. Report 184, Fachbereich Informatik, Univ. Dortmund, 1984.

[117] GYÖRI, E., A short proof of the rectilinear art gallery theorem, *SIAM J. Alg. Discrete Meth.* **7** (1986) 452–454.

[118] HART, H. and M. SHARIR, Nonlineararity of Davenport–Schinzel sequences and of generalized path compression schemes, *Combinatorica* **6** (1986) 151–177.

[119] HAUSSLER, D. and E. WELZL, ε-Nets and simplex range queries, *Discrete & Comput. Geometry* **2** (1987) 127–151.

[120] HERSHBERGER, J., Finding the visibility graph of a simple polygon in time proportional to its size, in: *Proc. 3rd Ann. ACM Symp. Comput. Geom.* (1987) 11–20.

[121] HERTEL, S., M. MÄNTYLÄ, K. MEHLHORN and J. NIEVERGELT, Space sweep solves intersection of two convex polyhedra, *Acta Inform.* **21** (1984) 501–519.

[122] HODDER, I. and C. ORTON, *Spatial Analysis in Archaeology* (Cambridge Univ. Press, Cambridge, UK, 1976).

[123] HOFFMAN, K., K. MEHLHORN, P. ROSENSTIEHL and R.E. TARJAN, Using level-linear search trees, *Inform. and Control* **2** (1986) 170–184.

[124] HOFFMANN, C.M., The problem of accuracy and robustness in geometric computation, Tech. Report CSD-TR-771, Dept. of Computer Science, Purdue Univ., W. Lafayette, IN, 1988.

[125] HOFFMANN, C.M., J.E. HOPCROFT and M.S. KARASICK, Towards implementing robust geometric computations, in: *Proc. 4th Ann. ACM Symp. Comput. Geom.* (1988) 106–117.

[126] HUBER, P.J., Robust statistics: a review, *Ann. Math. Stat.* **43** (1972) 1041–1067.

[127] JAROMCZYK, J.W. and M. KOWALUK, Skewed projections with an application to line stabbing in R^3, *Proc. 4th Ann. ACM Symp. Comput. Geom.* (1988) 362–370.

[128] JARVIS, R.A., On the identification of the convex hull of a finite set of points in the plane, *Inform. Process. Lett.* **2** (1973) 18–21.

[129] KAHN, J., M. KLAWE and D. KLEITMAN, Traditional galleries require fewer watchmen, *SIAM J. Alg. Discrete Math.* **4** (1983) 194–206.

[130] KALLAY, M., Convex hull algorithms in higher dimensions, Unpublished manuscript, Dept. Math., Univ. of Oklahoma, Norman, OK, 1981.

[131] KARMARKAR, N., A new polynomial time algorithm for linear programming, *Combinatorica* **4** (1984) 373–395.

[132] KARP, R.M., Reducibility among combinatorial problems, in: Miller and Thatcher, eds., *Complexity of Computer Computations* (Plenum Press, New York, 1972) 85–104.

[133] KEIL, J.M., Decomposing a polygon into simpler components, *SIAM J. Comput.* **14** (1985) 799–817.

[134] KEIL, J.M. and J.R. SACK, Minimum decomposition of polygonal objects, in: G.T. Toussaint, ed., *Computational Geometry* (North-Holland, Amsterdam, 1985).

[135] KHACHIYAN, L.G., Polynomial algorithms for linear programming, *Dokl. Akad. Nauk SSSR* **244** (1979) 1093–1096.

[136] KIRKPATRICK, D.G., Efficient computation of continuous skeletons, in: *Proc. 20th Ann. IEEE Symp. Found. Comput. Sci.* (1979) 18–27.

[137] KIRKPATRICK, D.G., Optimal search in planar subdivisions, *SIAM J. Comput.* **12** (1983) 28–35.

[138] KIRKPATRICK, D.G. and R. SEIDEL, The ultimate planar convex hull algorithm?, *SIAM J. Comput.* **15** (1986) 287–299.

[139] KLEE, V., On the complexity of d-dimensional Voronoi diagrams, *Arch. Math.* **34** (1980) 75–80.

[140] KLEE, V. and G.L. MINTY, How good is the simplex algorithm?, in: O. Shisha, ed., *Inequalities III*, (Academic Press, New York, 1972) 159–179.

[141] LAWLER, E., *Combinatorial Optimization: Networks and Matroids* (Holt Rinehart & Winston, New York, 1976).

[142] LEE, D.T., Furthest neighbor Voronoi diagrams and applications, Tech. Report 80-11-FC-04, Dept. of Electrical Engineering and Computer Science, Northwestern Univ., Evanston, IL, 1980.

[143] LEE, D.T., On k-nearest neighbor Voronoi diagrams in the plane, *IEEE Trans. Comput.* **31** (1982) 478–487.

[144] LEE, D.T., Visibility of a simple polygon, *Computer Vision, Graphics and Image Process.* **22** (1983) 207–221.

[145] LEE, D.T., On finding the convex hull of a simple polygon, *Internat. J. Comput. and Inform. Sci.* **12** (1983) 87–98.

[146] LEE, D.T. and R.L. DRYSDALE III, Generalized Voronoi diagrams in the plane, *SIAM J. Comput.* **10** (1981) 73–87.

[147] LEE, D.T. and F.P. PREPARATA, Location of a point in a planar subdivision and its applications, *SIAM J. Comput.* **6** (1977) 594–606.
[148] LEE, D.T., and C.K. WONG, Worst-case analysis for region and partial region searches in multidimensional binary search trees and balanced quad trees, *Acta Inform.* **9** (1977) 23–29.
[149] A. LINGAS, R. PINTER, R. RIVEST and A. SHAMIR, Minimum edge length partitioning of rectilinear polygons, in: *Proc. 20th Ann. Allerton Conf. on Commun., Control and Comput.* (1982) 53–63.
[150] LIPSKI, W., Finding a Manhattan path and related problems, Tech. Rept. ACT-17, Coordinated Science Lab., Univ. of Illinois, Urbana, IL, 1979.
[151] LIPTON, R.J. and R.E. TARJAN, Applications of a planar separator theorem, *SIAM J. Comput.* **9** (1980) 615–627.
[152] LOZANO-PEREZ, T. and M.A. WESLEY, An algorithm for planning collision-free paths among polyhedral obstacles, *Comm. ACM* **22** (1979) 560–570.
[153] LUEKER, G.S., A data structure for orthogonal range queries, in: *Proc. 19th Ann. IEEE Symp. Found. Comput. Sci.* (1978) 28–34.
[154] MARCOTTE, O. and S. SURI, Fast algorithms for finding weighted matching in simple polygons, in: *Proc. 5th Ann. ACM Symp. Comput. Geom.* (1989) 302–314.
[155] MATOUŠEK, J., Construction of epsilon nets, in: *Proc. 5th Ann. ACM Symp. Comput. Geom.* (1989) 1–10.
[156] MCCREIGHT, E.M., Priority search trees, *SIAM J. Comput.* **14** (1985) 257–276.
[157] MCKENNA, M., Worst-case optimal hidden-surface removal, *ACM Trans. Graphics* **6** (1987) 19–28.
[158] MEAD, C. and L. CONWAY, *Introduction to VLSI Systems* (Addison-Wesley, Reading, MA, 1980).
[159] MEGIDDO, N., Linear time algorithm for linear programming in R^3 and related problems, *SIAM J. Comput.* **12** (1983) 759–776.
[160] MEGIDDO, N., Linear programming in linear time when the dimension is fixed, *J. ACM* **31** (1984) 114–127.
[161] MEGIDDO, N., Partitioning with two lines in the plane, *J. Algorithms* **6** (1985) 430–433.
[162] MEGIDDO, N., Improved asymptotic analysis of the average number of steps performed by the selfdual simplex algorithm, *Math. Programming* **35** (1986) 140–172.
[163] MEHLHORN, K., *Data Structures and Algorithms: 3 Multi-dimensional Searching and Computational Geometry* (Springer, Berlin, 1984).
[164] MONMA, C., M.S. PATERSON, S. SURI, F.F. YAO, Computational Euclidean maximum spanning trees, in: *Proc. 4th Ann. ACM Symp. Comput. Geom.* (1988) 241–251.
[165] MONMA, C. and S. SURI, Partitioning points and graphs to minimize the maximum or the sum of diameters, in: *Proc. 6th Internat. Conf. Theory and Appl. of Graphs* (Wiley, New York, 1989).
[166] MOTWANI, R., A. RAGHUNATHAN and H. SARAN, Covering orthogonal polygons with star polygons: the perfect graph approach, in: *Proc. 4th Ann. ACM Symp. Comput. Geom.* (1988) 211–223.
[167] MULLER, D.E. and F.P. PREPARATA, Finding the intersection of two convex polyhedra, *Theoret. Comput. Sci.* **7** (1978) 217–236.
[168] MULMULEY, K., A fast planar partition algorithm I, in: *Proc. 4th Ann. Symp. Found. Comput. Sci.* (1988) 580–589.
[169] NEWMAN, W.M. and R.F. SPROULL, *Principles of Interactive Computer Graphics* (McGraw-Hill, New York, 1973).
[170] O'ROURKE, J., *Art Gallery Theorems and Algorithms* (Oxford Univ. Press, Oxford, 1987).
[171] O'ROURKE, J., C.B. CHIEN, T. OLSON and D. NADDOR, A new linear algorithm for intersecting convex polygons, *Computer Graphics and Image Process.* **19** (1982) 384–391.
[172] O'ROURKE, J. and K. SUPOWIT, Some NP-hard polygon decomposition problems, *IEEE Trans. Inform. Theory* **29** (1983) 181–190.
[173] OVERMARS, M.H., The design of dynamic data structures, in: Lecture Notes in Computer Science, Vol. 156 (Springer, Berlin, 1983).
[174] OVERMARS, M.H. and J. VAN LEEUWEN, Maintenance of configurations in the plane, *J. Comput. System Sci.* **23** (1981) 166–204.
[175] OVERMARS, M.H. and C.K. YAP, New upper bounds in Klee's measure problem, in: *Proc. 29th Ann. IEEE Symp. Found. Comput. Sci.* (1988) 550–556.

[176] PACH, J., W. STEIGER and E. SZEMEREDI, in: *Proc. 30th Ann. Symp. Found. Comput. Sci.* (1989) 72–81.

[177] PAPADIMITRIOU, C.H., The Euclidean travelling salesman problem is NP-complete, *Theoret. Comput. Sci.* **4** (1977) 237–244.

[178] PASCHINGER, I., Konvexe Polytope und Dirichletsche Zellenkomplexe, Ph.D. Thesis, Institut für Mathematik, Univ. Salzburg, Austria, 1982.

[179] PATERSON, M.S. and F.F. YAO, Point retrieval for polygons, *J. Algorithms* **7** (1986) 441–447.

[180] PATERSON, M.S. and F.F. YAO, Binary partitions with applications to hidden-surface removal and solid modelling, in: *Proc. 5th Ann. ACM Symp. Comput. Geom.* (1989).

[181] PREPARATA, F.P., Steps into computational geometry, Tech. Report, Coordinated Science Lab., Univ. of Illinois, Urbana, IL, 1977.

[182] PREPARATA, F.P., An optimal real time algorithm for planar convex hulls, *Comm. ACM* **22** (1979) 402–405.

[183] PREPARATA, F.P. and S.J. HONG, Convex hulls of finite sets of points in two and three dimensions, *Comm. ACM* **2** (1977) 87–93.

[184] PREPARATA, F.P. and M.I. SHAMOS, *Computational Geometry* (Springer, Berlin, 1985).

[185] PREPARATA, F.P. and R. TAMASSIA, Dynamic techniques for point location and transitive closure in planar structures, in: *Proc. 29th Ann. IEEE Symp. Found. Comput. Sci.* (1988) 558–567.

[186] PREPARATA, F.P., J.S. VITTER and M. YVINEC, Computation of the axial view of a set of isothetic parallelpipeds, *ACM Trans. Graphics*, to appear.

[187] RAPPAPORT, D., A linear algorithm for eliminating hidden-lines from a polygonal cylinder, *Visual Comput.* **2** (1986) 44–53.

[188] REIF, J. and S. SEN, An efficient output-sensitive hidden-surface removal algorithm and its parallelization, in: *Proc. 4th Ann. ACM Symp. Comput. Geom.* (1988) 193–200.

[189] RENEGAR, J., A polynomial-time algorithm based on Newton's method, for linear programming, Tech. Report, MSRI 07118-86, Mathematical Science Research Institute, Berkeley, CA, 1986.

[190] ROGERS, C.A., *Packing and Covering* (Cambridge Univ. Press, Cambridge, UK, 1964).

[191] SACK, J., An $O(n \log n)$ algorithm for decomposing a simple rectilinear polygon into convex quadrilaterals, in: *Proc. 20th Ann. Allerton Conf. Commun. Control and Comput.* (1982) 64–75.

[192] SARNAK, N. and R.E. TARJAN, Planar point location using persistent search trees, *Comm. ACM* **29** (1986) 669–679.

[193] SCHMITT, A., Time and space bounds for hidden line and hidden surface algorithms, in: *Eurographics '81* (North-Holland, Amsterdam, 1981) 43–56.

[194] SCHMITT, A., On the time and space complexity of certain exact hidden line algorithms, Tech. Report, 24/81, Fakultät für Informatik, Univ. Karlsruhe, 1981.

[195] SEIDEL, R., A convex hull algorithm optimal for point sets in even dimensions, Computer Science Tech. Report 81-14, Univ. of British Columbia, Vancouver, 1981.

[196] SEIDEL, R., The complexity of Voronoi diagrams in higher dimensions, in: *Proc. 20th Ann. Allerton Conf. Commun. Control and Comput.* (1982) 94–95.

[197] SEIDEL, R., Constructing higher-dimensional convex hulls at logarithmic cost per face, in: *Proc. 18th Ann. ACM Symp. Theory Comput.* (1986) 404–413.

[198] SHAMOS, M.I., Geometric complexity, in: *Proc. 7th Ann. ACM Symp. Theory Comput.* (1975) 224–233.

[199] SHAMOS, M.I., Computational Geometry, Ph.D. Thesis, Dept. Computer Science, Yale Univ., New Haven, CT, 1978.

[200] SHAMOS, M.I. and D. HOEY, Closest-point problems, in: *Proc. 16th IEEE Ann. Symp. Found. Comput. Sci.* (1975) 151–162.

[201] SHAMOS, M.I. and D. HOEY, Geometric intersection problems, in: *Proc. 17th Ann. IEEE Symp. Found. Comput. Sci.* (1976) 208–215.

[202] SHARIR, M., R. COLE, K. KEDEM, D. LEVEN, R. POLLACK and S. SIFRONY, Geometric applications of Davenport–Schinzel sequences, in: *Proc. 27th Ann. IEEE Symp. Found. Comput. Sci.* (1986) 77–86.

[203] SIX, H.W. and D. WOOD, Counting and reporting intersections of d-ranges, *IEEE Trans. Comput.* **31** (1982) 181–187.

[204] SMALE, S., On the average speed of the simplex method of linear programming, *Math. Programming* **27** (1983) 241–262.

[205] STEEL, J.M. and A.C. YAO, Lower bounds for algebraic decision trees, *J. Algorithms* **3** (1982) 1–8.

[206] SUTHERLAND, I.E., R.F. SPROULL and R.A. SCHUMACKER, A characterization of ten hidden-surface algorithms, *Comput. Serv.* **6** (1974) 1–55.

[207] SWART, G., Finding the convex hull facet by facet, *J. Algorithms* **6** (1985) 17–48.

[208] SYLVESTER, J.J., A question in the geometry of situation, *Quart. J. Math.* **1** (1857) 79.

[209] TARJAN, R.E. and C.J. VAN WYK, An $O(n \log n \log n)$-time algorithm for triangulating simple polygons, *SIAM J. Comput.* **17** (1988) 143–178.

[210] TOUSSAINT, G.T., Pattern recognition and geometrical complexity, in: *Proc. 5th Internat. Conf. Pattern Recognition* (1980) 1324–1347.

[211] TOUSSAINT, G.T., Computational geometric problems in pattern recognition, in: J. Kittler Fu, ed., *Pattern Recognition, Theory and Application* (Reidel, Dordrecht, 1982) 73–91.

[212] TOUSSAINT, G.T., Computational largest empty circles with location constraints, *Internat. J. Comput. and Inform. Sci.* **12** (1983) 347–358.

[213] VAIDYA, P.M., Space-time tradeoffs for orthogonal range queries, in: *Proc. 17th Ann. ACM Symp. Theory Comput.* (1985) 169–174.

[214] VAIDYA, P.M., An optimal algorithm for the all-nearest-neighbors problem, in: *Proc. 27th Ann. IEEE Symp. Found. Comput. Sci.* (1986) 117–122.

[215] VAIDYA, P.M., An algorithm for linear programming which requires $O(((m+n)n^2+(m+n)^{1.5}n)L)$ arithmetic operations, in: *Proc. 19th Ann. ACM Symp. Theory Comput.* (1987) 29–38.

[216] VAIDYA, P.M., Geometry helps in matching, in: *Proc. 20th Ann. ACM Symp. Theory Comput.* (1988) 422–425.

[217] VAISHNAVI, V., Computational point enclosures, *Pattern Recognition* **15** (1982) 22–29.

[218] VAN LEEUWEN, J., and D. WOOD, Dynamization of decomposable searching problems, *Inform. Process. Lett.* **10** (1980) 51–56.

[219] VAN LEEUWEN, J., and D. WOOD, The measure problem for rectangular ranges in d-space, *J. Algorithms* **2** (1981) 282–300.

[220] VORONOI, G., Nouvelles applications des paramètres continus à la théorie des formes quadratiques; premier mémoire: sur quelques propriétés des formes quadratiques positives parfaites, *J. Reine Angew. Math.* **133** (1907) 97–178.

[221] VORONOI, G., Nouvelles applications des paramètres continus à la théorie des formes quadratiques; deuxième mémoire: recherches sur les parallélloèdres primitifs, *J. Reine Angew. Math.* **134** (1908) 198–287.

[222] WELZL, E., Constructing the visibility graph for n line segments in $O(n^2)$ time, *Inform. Process. Lett.* **20** (1985) 167–171.

[223] WELZL, E., Partition trees for triangle counting and other range searching problems, in: *Proc. 20th Ann. ACM Symp. Comput. Geom.* (1988) 23–33.

[224] WIERNIK, A., and M. SHARIR, Planar realizations of nonlinear Davenport–Schinzel sequences, *Discrete & Comput. Geometry* **3** (1988) 15–48.

[225] WILLARD, D.E., Polygon retrieval, *SIAM J. Comput.* **11** (1982) 149–165.

[226] WILLARD, D.E., New data structures for orthogonal range queries, *SIAM J. Comput.* **14** (1985) 232–253.

[227] WILLARD, D.E. and G.S. LUEKER, Adding range restriction capability to dynamic data structures, *J. ACM* **32** (1985) 597–617.

[228] YAO, A.C., A lower bound to finding convex hulls, *J. ACM* **28** (1981) 780–787.

[229] YAO, A.C., On constructing minimum spanning trees in k-dimensional space and related problems, *SIAM J. Comput.* **11** (1982) 721–736.

[230] YAO, A.C., Space-time tradeoff for answering range queries, in: *Proc. 14th Ann. ACM Symp. Theory Comput.* (1982) 128–136.

[231] YAO, A.C., On the complexity of maintaining partial sums, *SIAM J. Comput.* **14** (1985) 277–288.

[232] YAO, A.C. and F.F. YAO, A general approach to s-dimensional geometric queries, in: *Proc. 17th Ann. ACM Symp. Theory Comput.* (1985) 163–168.

[233] YAO, F.F., On the priority approach to hidden-surface algorithms, in: *Proc. 21st Ann. Symp. Found. Comput. Sci.* (1980) 301–307.

[234] YAO, Y.Y., D.P. DOBKIN, H. EDELSBRUNNER and M.S. PATERSON, Partitioning space for range queries, *SIAM J. Comput.* **18** (1989) 371–384.
[235] YAP, C.K., An $(n \log n)$ algorithm for the Voronoi diagram of a set of simple curve segments, *Discrete & Comput. Geometry* **2** (1987) 365–394.
[236] YAP, C.K., A geometric consistency theorem for a symbolic perturbation scheme, in: *Proc. 4th Ann. ACM Symp. Comput. Geom.* (1988) 134–142.

CHAPTER 8

Algorithmic Motion Planning in Robotics

Jacob T. SCHWARTZ

Courant Institute of Mathematical Sciences, New York University, New York, NY 10012, USA

Micha SHARIR

Courant Institute of Mathematical Sciences, New York University, New York, NY 10012, USA, and School of Mathematical Sciences, Tel Aviv University, Tel Aviv, Israel

Contents

HANDBOOK OF THEORETICAL COMPUTER SCIENCE
Edited by J. van Leeuwen

1. Introduction

This chapter surveys recent progress in algorithmic motion planning in robotics. Research on theoretical problems in robotics looks ahead to a future generation of robots that will be considerably more autonomous than present robotic systems. The main objective is to endow robotic systems with various basic capabilities that they will need to possess in order to operate in an intelligent and independent manner. These improved capabilities can be grouped into three broad categories: *sensing, planning, and control.* That is, the system should be able to gather information about its workspace through a variety of sensing devices (vision, tactile, or proximity sensing, etc.), analyze and transform the raw sensory data into a world model of the environment, use this model to plan tasks that it is commanded to execute (navigation, assembly, inspection, machining, etc.), where planning amounts to breaking up the complex task into a sequence of simple subgoals, whose combined execution will accomplish the desired task, and finally obtain a low-level control loop which monitors the actual execution of each planned substep of the task.

Of these three categories, the planning stage aims to allow the robot's user to specify a desired activity in very high-level, general terms, and then have the system fill in the missing low-level details. For example, the user might specify the end product of some assembly process, and ask the system to construct a sequence of assembly substeps; or, at a less demanding level, to plan collision-free motions which pick up individual subparts of an object to be assembled, transport them to their assembly position, and insert them into their proper places.

Studies in this area have shown it to have significant mathematical content; tools drawn from various subfields of mathematics, such as classical geometry, topology, algebraic geometry, algebra, and combinatorics have all proved relevant. The algorithmic approach to robot planning, which is the one emphasized in this survey, uses a variety of tools drawn from *computational geometry* and other areas of computer science, and involves the design and analysis of geometric algorithms, which are required to be nonheuristic, exact (if the input data is accurate), and asymptotically efficient.

We will mainly concentrate in this survey on the (collision-avoiding) *motion planning* problem, and give only scant attention to the substantially more difficult area of general task planning; the motion planning problem was studied intensively during the 1980s, and has turned out to possess a substantial amount of mathematical depth and sophistication, as the reader will hopefully realize from this survey.

In addition, the interaction between motion planning and computational and discrete geometry has produced rich cross-fertilization between the fields, leading to many deep new results in geometry on the one hand (some of which will be reviewed below), and to better understanding of the motion planning problem on the other hand, reaching the stage of a few initial successful implementations of some of the efficient algorithms that have been devised.

The approach to motion planning that is surveyed here, which emphasizes object-oriented, exact and discrete (or combinatorial) algorithmic techniques, in which asymptotically worst-case efficient solutions are being sought, is relatively new. It has been preceded by (and currently "competes" with) several alternative, usually more engineering or AI-based, approaches with emphasize heuristic, approximating, rule-based, or best-case-tailored solutions. Several interesting solutions of this kind have been developed by Lozano-Perez, Brooks, Mason, Taylor and others; see [9, 10, 11, 26, 30, 60, 61, 67] for a sample. Other approaches are nondiscrete, in the sense that they view the motion planning problem as an integral part of the low-level control of the robot. A typical such approach regards the obstacles as sources of a repelling potential field while the target placement of the robot is regarded as a strong attractor. The hope is that the moving body, as it attempts to follow the potential gradient vector field, will be able to reach the destination while avoiding collision with the obstacles (see e.g. [44, 53]). Yet another approach, which is nonheuristic but not quite discrete, and applies in situations where the environment is not known a priori, has been developed by Lumelski and his collaborators (see [63, 64, 65]); it will be reviewed below.

We will not attempt to settle the issue of which method is superior, since each has its advantages and disadvantages. For example, many of the heuristic approaches have been implemented for several years now, and have working prototypical systems capable of planning motions in a variety of situations (usually involving only very simple configurations). However, our contention is that because the problem has such a rich geometric and combinatorial structure, it is imperative to at least understand this structure in detail from a mathematically rigorous point of view, and explore it for potential algorithmic improvements, before beginning to negotiate approximating or heuristic shortcuts, which, if done naively, can completely miss this structure, often resulting in considerable loss of efficiency at best, and in complete misunderstanding of the issues at the heart of the problem at worst. As we will see below, exact solutions of the problem for complex moving systems may be (provably) computationally intractable in the worst case, so in practice heuristic shortcuts may have to be made, but again we believe that they can be made most effectively once the exact solution and its limitations are well understood. A final way out of this issue is simply to ignore it; adopting a theoretical computer science point of view, we can regard the pragmatic context only as a source of motivation for the abstract geometric problems that arise, and enjoy their study just for the elegant and rich structure that they possess, and the deep mathematical and algorithmic challenges that they pose.

2. Statement of the problem

In its simplest form, the motion planning problem can be defined as follows. Let B be a robot system consisting of a collection of rigid subparts (some of which may be attached to each other at certain joints while others may move independently) having a total of k degrees of freedom, and suppose that B is free to move in a two- or three-dimensional space V amidst a collection of obstacles whose geometry is known to the robot system. Typical values of k range from two (e.g. for a rigid object translating on a planar floor (without rotating)) to six (which is the typical number of joints for a manipulator arm), but can also be much larger (e.g. when we need to coordinate the motion of several independent systems in the same workspace).

DEFINITION. The *motion planning problem* for B is: given an initial placement Z_1 and a desired target placement Z_2 of B, determine whether there exists a continuous obstacle-avoiding motion of B from Z_1 to Z_2, and if so plan such a motion. See Fig. 1 for an illustration of this problem.

Since we are interested in an exact, nonheuristic, and nonapproximating algorithmic approach, we want any algorithm for solving the motion planning problem to always be able to find a motion if one exists. It is intuitively clear that in favorable layouts of the obstacles (e.g. relatively few obstacles well separated from each other) even simple-minded techniques will generally yield a solution. However, if the obstacles are very densely cluttered, it might be quite difficult, often beyond normal "human intelligence" capabilities, to find a solution. It is in such circumstances that the exact algorithmic approach has the highest pay-off.

Before plunging into discussion of the problem as posed above, we note that there are many useful extensions and variants of the problem that can arise in practice. For example, if the environment is not fully known to the system, then it has to use its sensors to continually update the environment model and its planned motion. Another situation is that the objects might not be stationary, but rather move along predictable and known trajectories, in which case the system needs to "dodge" the obstacles while moving towards its goal. Yet another case to be considered is when the geometric model of the environment is not accurately known; the system will then need to allow for such errors and to come up with a robust planning strategy that would adjust itself automatically when unanticipated contacts are being made. This is especially crucial in cases of compliant motion, when the system moves in contact with some obstacle whose shape is known only approximately. Other variants require the planned motion to be *optimal*, in the sense that it minimizes the time needed for its execution, or the total distance travelled, or the total energy consumed, etc. Also, in the way in which the problem has been posed, it ignores the dynamics of the moving system. For example,

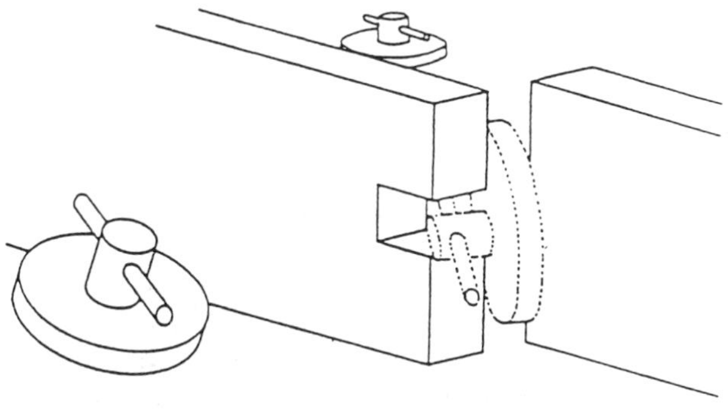

Fig. 1. Illustration of the motion planning problem.

the system might be constrained so that its velocity and/or acceleration cannot exceed certain thresholds, in which case even simple motions such as turning around a corner may be nontrivial to plan.

We will have more to say about these extensions of the problem later on, but for now let us focus on the simpler, static, and purely geometric version of the problem as posed above. As we shall see, it is difficult enough to solve the problem as it is, and each of the extensions just mentioned tends to make the solution much more complicated.

An alternative formulation of the problem, which will help us understand its geometric context, can be obtained as follows. Let FP denote the "free configuration space" of the moving system B; i.e. FP is the k-dimensional space of all free placements of B (i.e. the set of placements of B in which B does not intersect any obstacle). Each point Z in FP is a k-tuple giving the values of the parameters controlling the k degrees of freedom of B at the corresponding placement. For example, if B is a planar rigid body moving in a planar region, then it has three degrees of freedom, and each placement of B can be specified by three parameters (x, y, θ), where (x, y) are the coordinates of some fixed reference point on B, and θ is the angle between some fixed reference axis attached to B and the x-axis. As another example, if B is a line segment moving in three-dimensional space, B has five degrees of freedom and each placement of B can be specified by a 5-tuple giving the x-, y- and z-coordinates of one designated endpoint p of B and the spherical coordinates of the direction along B from p to the other endpoint.

Obviously, not every placement of a moving system B is free. For a placement Z to be free, it needs to satisfy a collection of *collision constraints*, each of which forbids a specific feature of B (a corner, straight link, etc.) to contact or to intersect a specific feature of some obstacle (a vertex, edge, face, etc.). Each such collision constraint corresponds to a "forbidden region" in the configuration space, and the free configuration space is simply the complement of the union of these regions.

In this setting, the motion planning problem becomes a problem in "computational topology", that is, the problem of calculating and obtaining a discrete representation of FP. More precisely, what one needs is to decompose FP into its *path-connected components*, because a collision-free path exists between the two given placements Z_1, Z_2 of B if and only if they both lie in the same connected component of FP. An appropriate discrete representation of the connected components of FP usually assumes the form of a *connectivity graph*, whose nodes represent certain local portions of FP and whose edges connect portions that are adjacent to each other in FP. Once such a representation has been obtained, actual motion planning reduces to searching for a path in the connectivity graph which connects between the initial and final placements of B. Having found such a path, the system finally transforms it into a sequence of actual "elementary" motions, each of which can be executed by a single command to the motors controlling the robot motion.

Section 3 of this survey will discuss the inherent computational complexity of the basic motion planning problem, detail several general techniques developed for treating it, and review various efficient algorithms for planning the motions of certain special robot systems having only few degrees of freedom.

In more detail, we will show that when the number k of degrees of freedom becomes arbitrarily large, then the problem complexity substantially increases, and, for certain types of systems with an arbitrarily large value of k, such as a collection of many independent rectangular objects moving in coordinated fashion inside a rectangular region, the problem becomes PSPACE-hard. Intuitively, this is because for large k the space FP is a high-dimensional space with irregular boundaries, and is thus hard to calculate efficiently. General techniques have been developed for solving the motion planning problem in rather full generality, and we will describe some of them, but, in accordance with what has just been said, these general algorithms are hopelessly inefficient in the worst case, and although their development is significant from a theoretical point of view, they are completely useless in practice. However, when k is small the problem becomes more tractable, and efficient algorithms have been developed for a variety of simple systems; a prototypical example has been that of a line segment (a "ladder", "rod" or "pipe") moving in the plane amidst polygonal obstacles (cf. [56, 78, 79, 80, 95, 112]), but many other systems have also been handled successfully. We will describe these specialized algorithms in Section 3.

In Section 4 we will survey some of the extensions to the motion planning problem mentioned above, such as the case of moving obstacles, the case of optimal motion, the case of unknown or only partially known environment, etc. As can be expected, these extensions are substantially more difficult to handle than the original problem version, which makes efficient solutions thereof much harder to obtain, except in certain very special cases.

To illustrate this, consider the issue of producing a collision-free path which satisfies some criterion of optimality. For example, if a mobile robot is approximated as a single moving point, one might want to find the shortest Euclidean path between the initial and final system placements, which avoids (the interiors of) the given obstacles. In more complex situations the notion of optimal motion is less clearly defined, and has as yet been little studied. However, even in the simplified case of the Euclidean shortest path problem, the situation becomes considerably complicated. While in the planar case the shortest path can be found in time that is worst-case quadratic in the number of obstacle corners and edges, finding the shortest path between two points in three dimensions, which avoids a collection of polyhedral obstacles, is already NP-hard (whereas the coarser original problem of just determining whether a path exists at all can be solved by a straightforward polynomial-time algorithm).

As already noted, studies of the motion planning problem tend to make heavy use of many algorithmic techniques in computational geometry, and to motivate the development of new and deep results in this area. Various motion-planning related problems in computational geometry will be reviewed in Section 5.

3. Motion planning in static and known environments

As above, let B be a moving robot system, k be its number of degrees of freedom, V denote the two- or three-dimensional space in which B is free to move, and FP denote the free configuration space of all free placements of B, as defined above. The space FP is determined by the collection of equalities and inequalities which express the fact that at a free placement the system B avoids collision with any of the obstacles present in its workspace. We will denote by n the number of equalities and inequalities needed to define FP, and call it the "geometric (or combinatorial) complexity" of the given instance of the motion planning problem. We will make the reasonable assumption that the parameters describing the degrees of freedom of B can be chosen in such a way that each of these inequalities is algebraic. Indeed, the group of motions (involving various combinations of translations and rotations) available to a given robot can ordinarily be given an algebraic representation, and the system B and its environment V can typically be modelled as objects bounded by a collection of algebraic surfaces (e.g., polyhedral, quadratic, or spline-based).

To illustrate these concepts and to get a somewhat better feeling for the structure of FP, consider the following simple example. Let B be a line segment pq translating in the plane amidst a collection of polygonal obstacles having a total of n corners (or edges). Since B is not allowed to rotate, its position is fully specified by the (x, y) coordinates of its endpoint p. The forbidden regions of collision are bounded by a collection of *constraint curves*, each expressing a possible contact between some fixed feature of B and some fixed obstacle feature. In this case, each such contact is either between an endpoint p or q of B and some obstacle edge, or between an obstacle corner and the line segment B. It is a fairly simple exercise in geometry to show that each of these curves (namely the locus of the endpoint p of B as it makes such a contact) is a line segment either at the orientation of B (in case of contact between B and an obstacle corner) or otherwise parallel to the corresponding obstacle edge. Thus each forbidden region is polygonal, so that their union (and thus its complement FP) is also a polygonal region bounded by portions of these constraint curves (see Fig. 2 for an illustration). Note that the number of constraint curves is proportional to n. Another way of thinking about FP is by considering the *arrangement* of the $O(n)$ constraint segments, i.e. the planar subdivision obtained by drawing all these segments. Each face of this arrangement (i.e.,

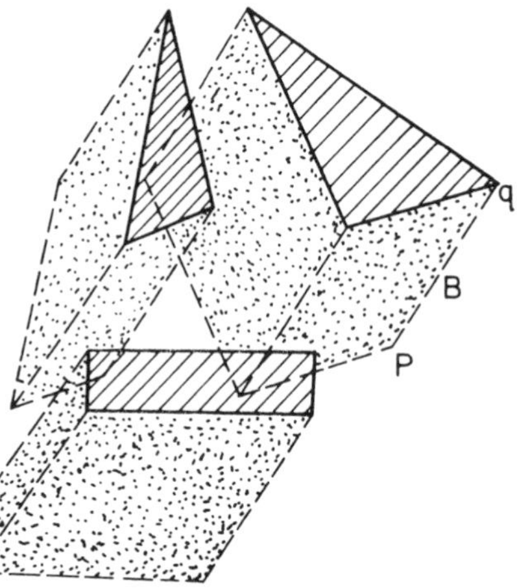

Fig. 2. Constraint curves and the space FP for a translating segment.

a connected component of the complement of the union of the segments) is either fully contained in FP or fully contained in a forbidden region. This follows easily from the fact that any transition between a free region and a forbidden one must occur at a point lying on some constraint segment. Thus FP is simply a collection of faces of the arrangement of the constraint curves. These arguments lead to a crucial observation which simplifies the problem still further. That is, if the initial placement Z_1 of B is known in advance, then it suffices to calculate only the connected component (face) of FP which contains Z_1, because no other free placement can ever be reached from Z_1 by a continuous obstacle-avoiding motion.

In this setting, the special case of the motion planning problem just discussed has been reduced to the problem of calculating a single connected component in an arrangement of $O(n)$ segments in the plane. Essentially the same arguments show that, to solve the motion planning problem for an arbitrary system with two degrees of freedom, it suffices to calculate a single face in an arrangement of arcs in the plane. When the system has more degrees of freedom, the problem becomes that of calculating a single cell in an arrangement of (algebraic) surface patches in a higher-dimensional space.

To recap, a typical instance of the motion planning problem will involve a collection of n constraint surfaces in the k-dimensional parametric space of all system placements, where each surface is a patch of some algebraic surface, and our goal is to analyze and calculate a single cell in the arrangement of these surfaces. We note here that in most practical instances the maximal degree of these algebraic surfaces is fixed and small (because these surfaces describe only a few types of possible contacts between B and the obstacles), while the number of surfaces is generally large. Thus in what follows we will tend to ignore the *algebraic complexity* of handling these surfaces (e.g. finding the intersection points of any k of these surfaces), and concentrate on the combinatorial complexity of their arrangement, i.e. the size of a discrete representation of the arrangement as a function of the number of surfaces n. Nevertheless, the algebraic complexity of the motion planning problem is also an important topic; see e.g. [7].

To get some intuition about the structure of such an arrangement, think of a simple special case involving flat triangular surfaces in 3-space (this special case actually arises in the motion planning problem for a polyhedral object B translating in 3-space amidst polyhedral obstacles; see [3] and below). These triangles decompose 3-space into connected cells, the boundary of which consists of 2-D portions of these triangles which meet one another along 1-D edges, each being the intersection segment of a pair of triangles, which terminate at endpoints, each being either an intersection point of three of the triangles or an intersection of an edge of one triangle with the interior of another. If we think of each triangle as the locus of placements of B in which it makes some specific contact with some obstacle, then edges of the arrangement (that appear along the boundary of FP) correspond to loci where B makes simultaneously two different obstacle contacts (while otherwise remaining free of collisions), and vertices correspond to placements of B at which it makes three simultaneous contacts with the obstacles. Abstracting to a general instance involving k degrees of freedom, we see that each face on the boundary of FP consists of placements of B in which it makes some $j \leqslant k$ distinct contacts with the obstacles simultaneously (while otherwise remaining free). A simple combinatorial argument implies that the number of faces comprising the boundary of FP is at most proportional to n^k. Specifically, consider the vertices of FP. Each vertex is an intersection point of k of the n surfaces. There are $\binom{n}{k}$ distinct k-tuples of the given surfaces, and, by the classical Bezout Theorem in algebraic geometry, each such k-tuple of surfaces intersects in at most d^k points, where d is the maximum degree of the surfaces. Assuming d is small, thus ignoring the factor d^k, the claim follows.

To see that the complexity of FP can be that high, consider a simple spider-like planar system B consisting of k rigid straight links all joined together at a common endpoint, which is constrained to lie at a fixed point O in the plane (so each link can rotate independently around O). This system has k (rotational) degrees of freedom. Suppose the obstacles consist of n small disks arranged along some circular arc about O (see Fig. 3). Thinking of FP as a semialgebraic set in E^k, a vertex of its boundary corresponds to a placement of B where k obstacle contacts are being made simultaneously, i.e. where each link touches some obstacle. It is thus easy to obtain an example where the boundary of FP has about n^k vertices, as claimed.

We thus summarize with the following theorem.

3.1. THEOREM. *The maximum combinatorial complexity of the free configuration space for a robot system with k degrees of freedom, whose motion is constrained by n geometric constraints of fixed maximum algebraic degree, is* $\Theta(n^k)$.

At this point it is prudent to add the following parenthetical disclaimer. Even though the combinatorial complexity of the entire FP may be high, it is conceivable that to plan a motion of B one does not need to calculate FP in full. Indeed, we have already remarked that it suffices to calculate just a single connected component of FP. But even that might be too much when all we really want to produce is a collision-free path, or, in a further restricted version of the problem, just to determine whether such a path exists. To bring this argument to the extreme, suppose FP is known to be a convex region. Then we do not need to compute it at all, because any two placements in FP can be connected by a straight segment joining them, which necessarily lies within FP. We will return to this issue later on, but to start, we first consider techniques that do aim to calculate the entire FP.

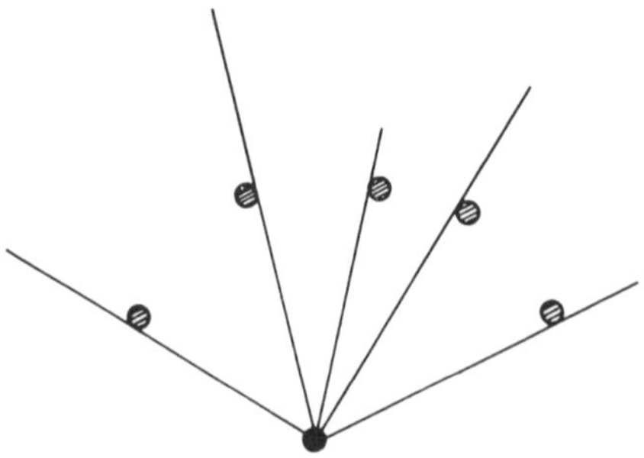

Fig. 3. An FP-vertex for a spider-like system.

3.1. *General solutions of the motion planning problem*

Assuming then that FP is an algebraic or semialgebraic set in E^k, it was shown in [96] that the motion planning problem can be solved in time polynomial in the number n of (algebraic) geometric constraints defining FP and in their maximal degree, but doubly exponential in k. This initial result has since been extended and improved in several papers [54, 90, 122], culminating in the recent fundamental work of Canny [12], which provides a general technique for solving the motion planning problem for a system with k degrees of freedom under n geometric constraints of fixed maximal degree, in time $O(n^k \log n)$, where unfortunately the constant of proportionality, which depends on the maximal degree of the collision constraints, is prohibitively large. Still, from a theoretical point of view this has been a significant progress, because, as we have noted above, the combinatorial complexity of FP can indeed be as high as $\Omega(n^k)$ in the worst case, which makes Canny's algorithm nearly worst-case optimal. Canny [13] and Renegar [92] have later improved the technique slightly, by showing that it can be implemented in PSPACE (i.e. using only a polynomial amount of storage).

Both techniques of [96] and of [12] are fairly complicated, and apply to general semialgebraic sets to produce some discrete representation of the topological structure of such a set. We will give here only a sketchy description of both techniques.

We begin with the technique of [96]. Even though this method gives inferior performance, it is significant because of its (relatively) simple structure, which makes variants of the method quite attractive for solving the motion planning problem for several special instances. This technique is a variant of Collins' *cylindrical algebraic decomposition* technique [21], which was originally applied to Tarski's theory of real closed fields. (In this theory one is given a quantified Boolean formula whose variables can assume real values, and whose atomic subformulae are equalities and inequalities involving polynomials in these variables, and the problem is to decide whether the formula is true.)

Collins' decomposition proceeds recursively through the dimension k (the number of degrees of freedom in our case). At each level j it maintains a polynomial $S_j(x_1, \ldots, x_j)$ in the j remaining variables (starting at $j=k$, in which case S_j is just the product of all the given constraint polynomials, and ending at $j=1$), and seeks a decomposition of E^j into cells (of various dimensions between 0 and j) such that S_j is sign-invariant over each of these cells (i.e. is everywhere positive, everywhere negative, or everywhere zero on each cell). When $j=1$ this amounts to finding all real roots of S_1, which can be done symbolically, using precise rational arithmetic, by various "root isolation" techniques. For $j>1$, the method first eliminates the variable x_j by computing from S_j certain polynomials in the first $j-1$ variables, using *resultants* and *subresultants* of S_j and some other related polynomials (see [96] for details). The collection of resulting polynomials has the property that if c is a connected cell of E^{j-1} in which all these polynomials are sign-invariant, then as $(x_1, \ldots, x_{j-1})$ varies over c, the polynomial $S_j(x_1, \ldots, x_{j-1}, x_j)$, viewed as a polynomial in x_j only, has the same number of real roots, and each root varies continuously with $x_1, \ldots, x_{j-1}$. We then multiply all the produced polynomials to obtain S_{j-1}, recursively obtain a cylindrical algebraic decomposition of E^{j-1} for S_{j-1}, and then, for each cell c in the decomposition, over which S_j has q distinct real roots, we obtain the following collection of cells in E^j (where $\boldsymbol{x}$ represents $(x_1, \ldots, x_{j-1})$):

$$\{[\boldsymbol{x}, x_j] : \boldsymbol{x} \in c,\ x_j < f_1(\boldsymbol{x})\} \quad \text{(lower semi-infinite "segment" of } c \times E^1\text{)},$$
$$\{[\boldsymbol{x}, f_i(\boldsymbol{x})] : \boldsymbol{x} \in c\} \quad \text{("section" of } c \times E^1\text{)},$$
$$\{[\boldsymbol{x}, x_j] : \boldsymbol{x} \in c, f_i(\boldsymbol{x}) < x_j < f_{i+1}(\boldsymbol{x})\} \quad \text{("segment" of } c \times E^1\text{)},$$
$$\{[\boldsymbol{x}, x_j] : \boldsymbol{x} \in c, f_q(\boldsymbol{x}) < x_j\} \quad \text{(upper semi-infinite "segment" of } c \times E^1\text{)},$$

where $f_1, \ldots, f_q$ denote the q distinct roots of S_j over c. Together, all these cells constitute the Collins decomposition of E^j.

It follows easily by induction that each of the sets constituting such a cylindrical algebraic decomposition K of E^k is topologically equivalent to an open cell of some dimension $j \leqslant k$. We can therefore refer to the elements $c \in K$ as the (open) *Collins cells* of the decomposition K.

3.2. THEOREM (Collins [21]). *The cylindrical algebraic decomposition defined above can be computed effectively, using only purely rational arithmetic, by representing each cell by an algebraic point inside it (given implicitly so that its coordinates are roots of certain specified polynomials), plus some other tricks. The number of cells, as well as the time needed to produce them all, is doubly exponential in the number of dimensions k, but depends only polynomially on the number of polynomials and on their maximal degree.*

In terms of the motion planning application, if we apply the Collins decomposition technique to the collection of collision constraints, each cell c in the resulting decomposition has the property that each constraint has a fixed sign over c. Consequently, either c is fully contained within FP or lies fully in the forbidden region. Thus by scaning each of the Collins cells we can collect all cells whose union forms FP.

This is still not quite sufficient to solve the motion planning problem, because we also need to know how these cells are connected ("glued") to each other in FP, so as to be able to deduce the topological structure of FP. This extra ingredient is supplied in [96], where a (somewhat complicated) method for determining all pairs of adjacent Collins cells in FP is given; subsequent improvements to this method have been provided in [54, 86] and others.

The motion planning problem is then solved by constructing a *connectivity graph* CG, whose nodes are the Collins cells constituting FP and whose edges connect pairs of adjacent such cells. Then, for a given pair of initial and final placements Z_1, Z_2 of B, we locate the two nodes c_1, c_2 of CG that contain these placements, and then search for a path in CG connecting c_1 and c_2. If such a path exists then B can be moved continuously from Z_1 to Z_2 without collision along a path which is easily obtained from the graph path; otherwise no such motion is possible.

Many ingredients of the above technique (decomposing FP into simple cells, establishing adjacency between the cells, transforming the problem to a discrete path searching in a graph) have proved useful in many related approaches, where efficient solutions for specific systems have been obtained (see below). However, the general method as sketched above is very inefficient. To get some intuitive feeling as to the main weakness of this decomposition technique, we illustrate it in the simple case where the constraint surfaces are all planes in three dimensions. It is easy to check that n such planes decompose 3-space into convex polyhedral cells whose total complexity is at most $O(n^3)$ (see e.g. [28]). However, the Collins decomposition will produce many more cells. Specifically, one can show that in this case the Collins technique first finds all $O(n^2)$ intersection lines between the given planes, and then projects all these lines into the xy-plane. This gives an arrangement of $\Theta(n^2)$ lines in the plane, which decomposes the plane into a $\Theta(n^4)$ cells (the actual Collins technique decomposes these cells still further, into a total of $\Theta(n^6)$ subcells). For each of these cells we then construct a vertical prism with that cell as its base, and then cut this prism into pieces by the n original planes. Altogether, we obtain $\Theta(n^5)$ (or even $\Theta(n^7)$) cells, much more than necessary to represent the way in which the given planes decompose 3-space (the actual complexity is only $\Theta(n^3)$; see [28] for a technique of obtaining such an efficient representation).

Canny's technique is quite different. Instead of decomposing FP into cells, his idea is to construct a 1-dimensional skeleton S (which he calls a "roadmap") of curves within FP, having two crucial properties:

(i) *reachability*: each placement Z in FP can be (effectively) moved continuously to a placement in S;

(ii) *connectivity*: the intersection of S with each connected component of FP is (nonempty and) connected.

Clearly, if such a roadmap can be found, the motion planning problem is easily solved as follows. Apply the reachability procedure to the initial and final placements Z_1, Z_2 of B to obtain two respective placements W_1, W_2 on S. Then search for a path in S (which is easy to do since S is 1-dimensional) between these two placements. If no path exists, then the connectivity property of S guarantees that Z_1 and Z_2 cannot be reached from one another by a collision-free continuous motion. Otherwise, the required motion is simply the concatenation of the paths connecting Z_1 to W_1 and Z_2 to W_2 and the path connecting W_1 to W_2 within S.

To obtain a roadmap, Canny uses a recursive procedure on the dimension k of FP. If $k=1$ then FP is a curve, and is taken to be its own roadmap. Otherwise, FP is projected onto some two-dimensional plane, and the *silhouette* (i.e. boundary curve) Σ of the projected image is being formed. Next one finds all critical levels of Σ, i.e. the y-coordinates (in the projected plane) of points at which Σ is either singular or has a horizontal tangent. Then FP is sliced at each of these critical levels, and the algorithm is applied recursively to each cross-section. Finally, all roadmaps obtained for each cross-section are glued to the silhouette Σ, and the union of all these curves and of Σ is the desired roadmap for FP. It is shown in [12] that this roadmap satisfies conditions (i) and (ii) above.

This basic algorithm is not as efficient as possible, because the slices of FP through successive critical levels of the silhouette are essentially very similar in shape, except for a "local" change that occurs near the points of criticality of Σ. Using this observation, Canny derives an improved algorithm, summarized in the following theorem.

3.3. THEOREM (Canny [12]). *Canny's roadmap technique, for n k-variate polynomials* *of maximum degree d, can be implemented in time*

$$O(n^k(\log n \cdot d^{O(k)} + (dk)^{O(k^2)}))$$

where d is the maximum algebraic degree of the constraint polynomials (and where we have omitted factors that depend on the magnitude of the coefficients of the constraint polynomials). In other words, if d and k are both small and fixed, the time complexity of Canny's algorithm is $O(n^k \log n)$.

3.2. Lower bounds

The results just cited suggest that motion planning becomes harder rapidly as the number k of degrees of freedom increases; this conjecture has in fact been proved for various model "robot" systems. Specifically, the following results have been obtained.

3.4. THEOREM. *Motion planning is* PSPACE*-hard for*
(a) *a certain* 3*-D system involving arbitrarily many links and moving through a complex system of narrow tunnels* [89];
(b) *certain* 2*-D systems of mechanical linkages* [38];
(c) *a system of* 2*-D independent rectangular blocks sliding inside a rectangular box* [41];
(d) *and a single* 2*-D arm with many links moving through a* 2*-D polygonal space* [47].

Several weaker results establishing NP-hardness for still simpler systems have also been obtained.

The Hopcroft–Joseph–Whitesides result [38] is established by showing that, given an arbitrary Turing machine T with a fixed bounded tape memory, one can construct a planar linkage L whose motions simulate the actions of T, so that L can move from a specified initial to a specified final configuration only if the Turing machine T eventually halts. The size of the linkage L constructed is polynomially bounded by the size of the state table and memory tape of T. (This suffices because it is well known that the halting problem for Turing machines of bounded tape size is PSPACE-hard [33].) One proceeds by noting that the actions of an arbitrary T can easily be characterized by a set of polynomial constraints, and then by using the century-old result of Kempe [52] which shows how to construct a mechanical linkage capable of representing any specified multivariate polynomial $P(x_1, \ldots, x_n)$.

In more detail, a *planar linkage* is a mechanism consisting of finitely many rigid rods, of prespecified lengths, joined together at some of their endpoints by hinge-pins about which they are free to rotate. Any number of rod-ends are allowed to share a common hinge-pin; and particular pins can be held at specified points by being "fastened to the plane". Aside from this, the hinge-pins and rods are free to move in the plane, and it is assumed that the motion of one rod does not impede the motion of any other (i.e. the rods are allowed to "pass over" each other). Such a linkage is said to represent the multivariate polynomial $P(x_1, \ldots, x_n)$ if there exist n hinge-pins $p_1, \ldots, p_n$ which the linkage constrains to move along the real axis, and an $(n+1)$st hinge-pin p_0 which the linkage constrains to lie at the real point $P(x_1, \ldots, x_n)$ whenever $p_1, \ldots, p_n$ are placed at the real points $x_1, \ldots, x_n$. (It is assumed that the linkage leaves $p_1, \ldots, p_n$ free to move independently over some large interval of the real axis.)

The existence of a linkage representing an arbitrary polynomial P in the case just explained is established by exhibiting linkages which realize the basic operations of addition, multiplication, etc. and then showing how to represent arbitrary combinations of these operations by fastening sublinkages together appropriately.

The Hopcroft–Schwartz–Sharir result [41] on PSPACE-hardness of the coordinated motion planning problem for an arbitrary set of rectangular blocks moving inside a rectangular frame is proved similarly. It is relatively easy to show that the actions of an arbitrary tape-bounded Turing machine can be imitated by the motions of a collection of similarly-sized nearly rectangular "keys" whose edges bear protrusions and indentations which constrain the manner in which these "keys" can be juxtaposed, and hence the manner in which they can move within a confined space. A somewhat more technical discussion then shows that these keys can be cut appropriately into rectangles without introducing any significant possibilities for motion of their independent parts that do not correspond to motions of entire keys.

The results of [38] and [41] are interesting because they provide two relatively simple physical devices for simulating bounded-tape Turing machines. In addition, these lower bounds are fairly strong, because they apply to the decision problem in motion planning (i.e. only determining whether two given placements are reachable from one another), and do not presume that the entire configuration space, or even the path itself, need to be constructed.

In concluding this subsection, we mention a few lower bounds for simple systems with just a few degrees of freedom. O'Rourke [80] has considered the problem of a line segment ("rod") moving in a 2-D polygonal space with n corners, and has shown that in the worst case there exist two placements of the rod that are reachable from one another, but any motion between these placements requires at least $\Omega(n^2)$ submotions (namely the rod has to be pushed forwards and backwards at least that many times to reach its destination). This has been extended by Ke and O'Rourke [48] to obtain a similar lower bound for a rod moving in 3-space amidst polyhedral obstacles with n vertices, where the minimum number of submotions needed to realize a motion of the rod between two placements can be in the worst case as high as $\Omega(n^4)$. We will return to these bounds later on.

3.3. *The "projection method"*

The negative worst-case results mentioned above suggest that when the number k of degrees of freedom is large, the problem becomes intractable. (This does not mean, of course, that there is no hope of attacking such general problems. As is the case with many other intractable problems, there might still be algorithms that work very well in the average case, or approximating algorithms that can guarantee, e.g. that they will find a path efficiently provided the "clearance" between obstacles is not too small, etc. However, we will not address these issues in this survey.) However, when k is small, the general algorithms of Canny and of Schwartz and Sharir show that the problem can be solved efficiently (in time polynomial in the number n of constraints). Still even in such special cases, the efficiency of these general techniques is very far from being satisfactory. Consequently, a lot of recent research has focussed on obtaining improved efficient algorithms for a variety of special systems, having a small number of degrees of freedom.

These algorithms can be broadly classified according to several general approaches that they exemplify. The first such approach, known as the *projection method*, uses ideas similar to those appearing in the Collins decomposition procedure described above. One fixes some of the problem's degrees of freedom (for the sake of exposition, suppose just one parameter y is fixed, and let x be the remaining parameters); then one solves the resulting restricted $(k-1)$-dimensional motion planning problem. This subproblem solution must be such as to yield a discrete combinatorial representation of the restricted free configuration space (essentially, a cross-section of the entire space FP) that does not depend continuously on the final parameter y, and changes only at a finite collection of "critical" values of y. These critical values are then calculated; they partition the entire space FP into connected cells, and by calculating relationships of adjacency between these cells one can describe the connectivity of FP by a discrete *connectivity graph* CG, in much the same way as in the general algorithm of [96].

This relatively straightforward technique was applied in a series of papers by Schwartz and Sharir on the "piano movers" problem, to yield polynomial-time motion planning algorithms for various specific systems, including a rigid polygonal object moving in 2-D polygonal space [95], two or three independent disks moving in coordinated fashion in 2-D polygonal space [97], certain types of multiarm linkages moving in 2-D polygonal space [108], and a rod moving in 3-D polyhedral space [98]. These initial solutions were coarse and not very efficient; subsequent refinements have improved their efficiency substantially.

A typical example that has been studied extensively is the case of a line segment B (a "rod") moving in two-dimensional polygonal space whose boundary consists of n segments ("walls"). Here the configuration space FP is three-dimensional, and it can be decomposed into cells efficiently using a modified projection technique developed by Leven and Sharir [56]. Their result is summarized in the following theorem.

3.5. THEOREM (Leven and Sharir [56]). *Motion planning for a rod in a 2-D polygonal space bounded by n edges can be done in time* $O(n^2 \log n)$.

In this approach one starts by restricting the motion of B to a single degree of freedom of translation along its length. For this trivial subproblem the restricted FP simply consists of an interval which can be represented by a discrete label $[w_1, w_2]$ consisting of the two walls against which B stops when moving backwards or forwards from its given placement.

Next one admits a second degree of freedom by allowing arbitrary translational motion of B. Assuming that B points in the direction of the positive y-axis, the second degree of freedom is parameterized by the x-coordinate of B. As x varies, the label $[w_1, w_2]$ of the placements of B remains unchanged until one reaches either an endpoint of w_1 or of w_2, or a wall corner lying between w_1 and w_2, or a placement in which B simultaneously touches both w_1 and w_2. Hence, given a placement Z of B with label $[w_1, w_2]$, we can define a 2-D "noncritical" region R consisting of all placements of B which are reachable from Z by a translational motion during which the label $[w_1, w_2]$ does not change. R itself can be uniquely represented by another discrete label of the form $\lambda(R)=[w_1, w_2, LEFT, RIGHT]$, where $LEFT$, $RIGHT$ describe the type of criticality (an endpoint of w_1 or of w_2, a new corner between w_1 and w_2, or a "dead-end" at which B gets stuck) defining respectively the left and right boundaries of R.

Finally one introduces the last rotational degree of freedom θ. Again one can show that the label $\lambda(R)$ of the noncritical region R containing a placement Z of B remains constant as θ varies, unless θ crosses a *critical orientation* at which the left or right boundary of R either

(i) comes to contain two wall corners; or

(ii) contains an endpoint of w_1 or of w_2 and the distance from that endpoint to the other wall (in the direction of θ or $\theta+\pi$) is equal to the length of B; or

(iii) contains another corner and the distance between w_1 and w_2 along this boundary is equal to the length of B

(see Fig. 4 for an illustration of these critical conditions).

One can therefore define a 3-D noncritical cell C of FP containing Z to consist of all placements of B reachable from Z by a motion during which the label of the 2-D region enclosing the placement of B remains constant, and represent each such cell by a triple $[\lambda, \theta_1, \theta_2]$, where λ is the label of the 2-D region enclosing Z and where θ_1, θ_2 are two critical orientations at which the label λ changes discontinuously. The collection of these cells yields the desired partitioning of FP.

Leven and Sharir show that the number of critical orientations is at most $O(n^2)$, and that, assuming B and the walls are in "general position", each critical orientation delimits only a small constant number of cells. Thus the total number of cells in FP is also $O(n^2)$. In [56] a fairly straightforward algorithm is presented, which runs in time $O(n^2 \log n)$, for constructing these cells and for establishing their adjacency in FP, a very substantial improvement of the $O(n^5)$ algorithm originally presented in [95]. Moreover, in view of O'Rourke's lower bound mentioned above [80], the Leven–Sharir algorithm is nearly optimal in the worst case.

3.4. *The "retraction method" and related techniques*

Several other important algorithmic motion planning techniques were developed subsequent to the simple projection technique originally considered. The so-called

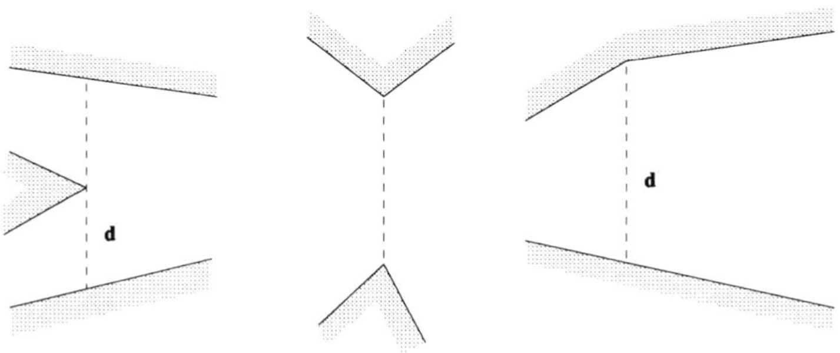

Fig. 4. Critical orientations for a moving rod in the plane.

retraction method is very similar to (and has in fact preceded) the general technique of Canny described above. Specifically, it proceeds by retracting the configuration space *FP* onto a lower-dimensional (usually a one-dimensional) subspace *N*, which satisfies both the reachability and the connectivity properties mentioned above, thus ensuring that two system placements in *FP* lie in the same connected component of *FP* if and only if their retractions to *N* lie in the same connected component of *N*. This reduces the dimension of the problem, and if *N* is one-dimensional, the problem again becomes one of searching a graph. (Rigorously speaking, a retraction is a continuous map from some space *E* onto a subspace *F* which becomes the identity when restricted to *F*. The retractions that are used in this approach are not always everywhere continuous, but their discontinuities are easy to characterize and to "patch-up".)

O'Dunlaing and Yap [77] have used this retraction technique in the simple case of a disk moving in 2-D polygonal space. In this case, each placement of the disk *D* can be fully specified by the position of its center in the plane. The retraction used attempts to repel *D* away from its nearest obstacle, until its center becomes equally nearest to at least two distinct obstacle features. The locus *N* of such points is known as the *Voronoi diagram* associated with the set of given polygonal obstacles (see also [123]). It is shown in [123] that the Voronoi diagram of *n* line segments in the plane can be calculated in time $O(n \log n)$. After computing the diagram, and pruning away portions of it where the nearest obstacle is too close for *D* to fit in, we obtain the desired network *N*, in total time $O(n \log n)$.

After this first paper O'Dunlaing, Sharir and Yap [78, 79] generalized the retraction approach to the case of a rod moving in 2-D polygonal space by defining a variant Voronoi diagram in the 3-D configuration space *FP* of the rod, and by retracting onto this diagram. The diagram consists of all placements of the rod at which it is simultaneously nearest to at least two obstacles. The Voronoi diagram defined by a set of obstacles in general position can readily be divided into 2-D Voronoi sheets (placements in which the rod is simultaneously nearest to two obstacles), which are bounded by 1-D Voronoi edges (placements in which the rod is nearest to three obstacles), which in turn are delimited by Voronoi vertices (placements in which the rod is nearest to four obstacles; cf. Fig. 5). The algorithm described in [78, 79] actually constructs a 1-D subcomplex within the Voronoi diagram; this complex consists of the Voronoi edges and vertices plus some additional connecting arcs. It is shown in [78]

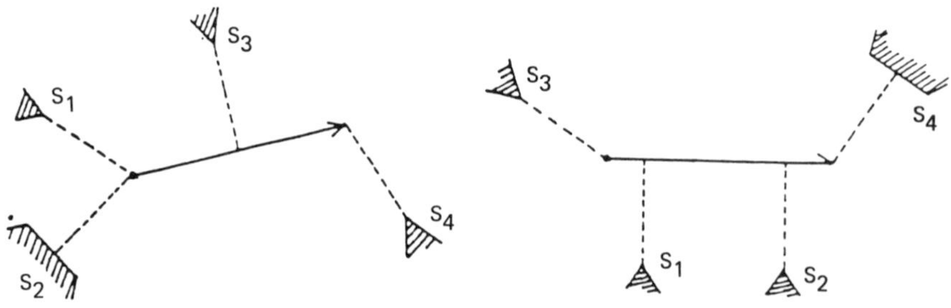

Fig. 5. Voronoi vertices for a ladder.

that this Voronoi "skeleton" characterizes the connectivity of FP, in the sense that each connected component of FP contains exactly one connected component of the skeleton. A fairly involved geometric analysis given in [79] shows that the total number of Voronoi vertices is $O(n^2 \log^* n)$, and that the entire "skeleton" can be calculated in time $O(n^2 \log n \log^* n)$, where $\log^* n$ is a very slowly growing function giving the number of repeated applications of the log function to n needed to bring it down to 1; this was a substantial improvement of the original projection technique, but nevertheless a result shortly afterward superceded by [56].

A similar retraction approach was used by Leven and Sharir [57] to obtain an $O(n \log n)$ algorithm for planning the purely translational motion of a simple convex object B amidst polygonal barriers. The idea is similar to that of [77]. Indeed, any placement of B is fully specified by the position of some fixed reference point O within B. As before, we would like to push B away from the nearest obstacle until O becomes equally nearest to more than one obstacle. However, for the method to work correctly here, we need to modify the notion of distance from the usual Euclidean distance (as used in [77]) to one which depends on B. To define this distance $d_B(p, q)$ from a point p to a point q, place B with O lying at p, and expand or shrink B in such a way that O remains at p and B remains similar to its original shape. The expansion (or shrinkage) factor λ which makes q lie on the boundary of B is the distance $d_B(p, q)$ (the reader is invited to check that this coincides with the usual Euclidean distance when B is a unit disk).

With this definition of distance, one can define a generalized Voronoi diagram, known as the B-Voronoi diagram, in much the same way as the standard diagram is defined for the Euclidean distance. See Fig. 6 for an illustration of such a diagram.

It is shown in [57] that, though of a more complex structure than standard Voronoi diagrams, the B-diagram retains most of the useful properties of standard Voronoi diagrams. In particular, its size is linear in the number of obstacles in S, and, if B has sufficiently simple shape, it can be calculated in time $O(n \log n)$, using a variant of the technique described in [123].

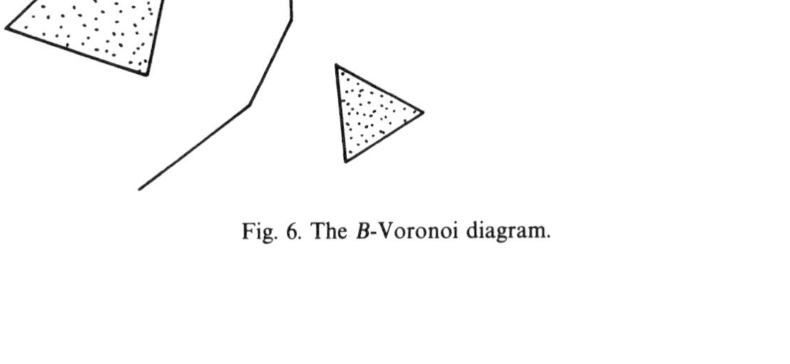

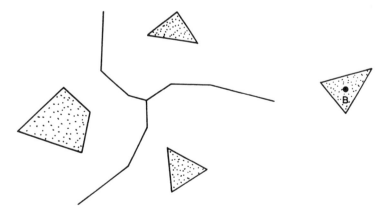

Fig. 6. The B-Voronoi diagram.

Next, as in [77], let N be the portion of the B-diagram consisting of points whose B-distance from the nearest obstacle(s) is greater than 1. Then any translate of B in which the reference point O on B is placed at a point in N is a free placement of B. It is proved in [57] that N characterizes the connectivity of the free configuration space of B, in the sense defined above, so that, for purpose of planning, motion of B can be restricted to have the reference point O move only along N (apart from an initial and final "pushes" towards N). This yields an $O(n \log n)$ motion planning algorithm for this case. See below for an alternative technique for this special case of motion planning.

We summarize all these results in the following theorem.

3.6. THEOREM. *Using various generalizations of Voronoi diagrams, one can*
(a) *plan the motion of a disk in a 2-D polygonal space in time* $O(n \log n)$ [77];
(b) *plan the purely translational motion of a convex k-gon in a 2-D polygonal space in time* $O(n \log n)$ (*here k is assumed to be small and fixed*) [57];
(c) *plan the motion of a rod in a 2-D polygonal space in time* $O(n^2 \log n \log^* n)$ [78, 79].

Recently Sifrony and Sharir [112] have devised another retraction-based algorithm for the motion of a rod in 2-D polygonal space. Their approach is to construct the 1-D network of arcs within FP consisting of all the 1-D edges on the boundary of FP (each such edge consists of semifree placements in which the rod simultaneously makes two specific contacts with the obstacles), plus several additional arcs which connect different components of the boundary of FP that bound the same connected component of FP. Again, this network characterizes the connectivity of FP, so that a motion planning algorithm need only consider motions proceeding within this network. The Sifrony–Sharir approach generates motions in which the rod is in contact with the obstacles, and is thus conceptually somewhat inferior to the Voronoi-diagram-based techniques, which aim to keep the moving system between obstacles, not letting it get too close to any single obstacle. However the network in [112] is much simpler to analyze and to construct. Specifically, it is shown in [112] that this network has $O(n^2)$ vertices and edges in the worst case, and can be constructed in time $O(n^2 \log n)$. (Actually, the network size is bounded by the number K of pairs of obstacles lying at distance less than or equal to the length of the moving rod, and the complexity of the algorithm is bounded by $O(K \log n)$. Thus if the obstacles are not too badly cluttered together, the Sifrony–Sharir algorithm will run quite efficiently; this makes the approach in [112] more attractive than the previous solutions in [56, 79].) In summary, we have the next theorem.

3.7. THEOREM (Sifrony and Sharir [112]). *Motion planning for a rod in a 2-D polygonal space can be done in time* $O(K \log n)$, *where K is the number of pairs of obstacle corners at distance at most the length of the rod apart.*

3.5. *Other techniques*

3.5.1. *The expanded-obstacles approach*

A rather early paper by Lozano-Perez and Wesley [62] has introduced a general approach to a variety of motion planning problems, known as the *expanded obstacles* approach. We illustrate this idea in the simple case of a convex polygon B, with k sides, translating in a 2-D polygonal space (an alternative retraction-based technique for this case has been mentioned above). Suppose the obstacles are decomposed into m convex pieces (with pairwise disjoint interiors) that have an overall of n corners. Let B_0 be a standard placement of B, in which some fixed reference point O within B lies at the origin. For each convex obstacle A_i, consider the *expanded obstacle* A_i^* which is defined as the *Minkowski* (*or vector*) *difference* of A_i and B_0; that is

$$A_i^* = \{x - y: x \in A_i, y \in B_0\}$$

where $x - y$ denotes vector difference. If B is moved so that O lies at some point z of A_i^*, then by definition $z = x - y$ for some x in A_i and y in B_0, so $z + y$, which by definition is a point occupied by the current placement of B, is equal to x, which is a point in A_i. In other words, A_i^* is precisely the set of all placements of the reference point O in which B intersects A_i. Thus we can shrink B to the single point O and replace the obstacles by their expanded versions, obtaining a much easier problem to solve. As a matter of fact, the free configuration space is simply the complement of the union of the expanded obstacles (the expanded obstacles are in fact concrete exemplifications of the somewhat vaguer notion of forbidden regions mentioned earlier).

While the conceptual idea is quite simple, and can be applied to more general instances (e.g. three-dimensional translational motion, translational motion of a non-convex object, etc.), it raises some interesting problems. One problem is that the expanded obstacles are not necessarily disjoint any more, so we need to analyze the pattern in which they intersect. Fortunately, when B is convex, it was shown by Kedem, Livne, Pach and Sharir [49] that the boundaries of any pair of expanded obstacles intersect in at most two points (assuming "general position" of the obstacles). Furthermore, in [49] it is also shown that, for any collection of n closed Jordan curves in the plane having the property that any pair of them intersect at most twice, the union of their interiors has only a linear number of intersections on its boundary (more precisely, at most $6n - 12$ such points). This implies the following theorem.

3.8. THEOREM (Kedem et al. [49]). *The complexity of the free configuration space FP, for a translating convex polygon, is only linear; it can be calculated using a simple divide-and-conquer algorithm, in time* $O(n \log^2 n)$.

3.5.2. The single-component approach

If, in the preceding discussion, the translating object B is not convex, it is easy to give examples where FP can have a quadratic number of corners (see Fig. 7). However, as has been pointed out earlier, we do not have to compute the entire FP; it suffices to compute just the connected component of FP that contains the initial placement of B. The hope is that this component will have a much smaller complexity than the entire FP, and that consequently it could be calculated by a more efficient algorithm.

While for general motion planning problems this hope is still far from being substantiated (see below), in the planar case (i.e. for systems having only two degrees of freedom) the situation is quite favorable. Following a series of papers [36, 87, 111, 118], it has been shown that the following theorem holds.

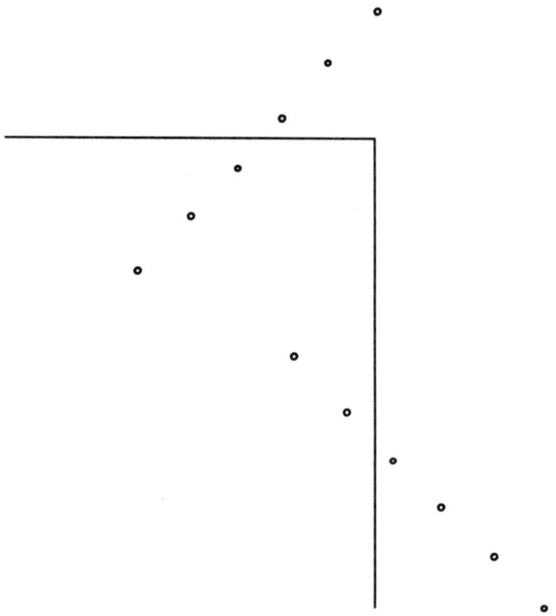

Fig. 7. Quadratic-size *FP* for a translating nonconvex planar object.

3.9. THEOREM (Guibas, Sharir and Sifrony [35]). *A single connected component (or face) in an arrangement of n Jordan arcs, each pair of which intersect in at most some fixed number s of points, has small combinatorial complexity. That is, the number of connected portions of these arcs which appear along the boundary of such a face (as illustrated in Fig. 8) is at most* $O(\lambda_{s+2}(n))$, *where* $\lambda_s(q)$ *is the maximum length of* (q, s) Davenport-Schinzel sequences.

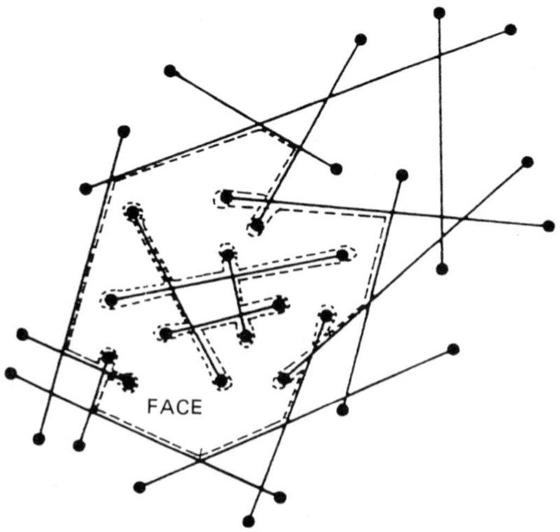

Fig. 8. A face in an arrangement of segments.

In Section 5 we will describe these sequences and their significance in many geometric problems. For now it will suffice to say that for any fixed s, $\lambda_s(q)$ is nearly linear in q, although, for $s \geqslant 3$, it is (ever so slightly) superlinear in q.

In addition, in [35] an algorithm is presented for calculating the face in such an arrangement which contains a given point x, in time $O(\lambda_{s+2}(n)\log^2 n)$, which is thus close to linear in n. Applying this algorithm to the collection of constraint curves and to the initial placement of the moving system (with two degrees of freedom) in question, we obtain a solution to the motion planning problem which has the same time complexity. This constitutes a significant improvement over any algorithm that attempts to calculate the entire space FP. The algorithm is also relatively simple to implement, and thus provides a general-purpose routine for solving efficiently practically any motion planning problem with two degrees of freedom. This class of systems is fairly rich; for example, it includes a translating rigid 3-D object on some 2-D surface, a two-link arm moving in the plane with one endpoint fixed, a spatial arm rotating around a fixed universal joint, etc.

For systems having three or more degrees of freedom, much less is known about the complexity of a single component of FP. Let us consider the case of three degrees of freedom. Using general algebraic and geometric arguments, as in the beginning of this section, we can show that the total complexity of FP for such systems, which is a manifold in 3-dimensional space, is at most $O(n^3)$, and that this complexity can be attained in the worst case for many specific systems. However, there exist systems with three degrees of freedom, such as a line segment (or even an arbitrary convex object) moving in the plane amidst polygonal obstacles, for which one can show that the entire FP has only quadratic, or near-quadratic complexity. When this is not the case, the question still remains whether this bound holds for just a single connected component of FP, so that one might be able to solve motion planning problems for systems with three degrees of freedom in close to quadratic time.

Unfortunately, this question is still largely unsettled. However, we have the following theorem.

3.10. THEOREM (Aronov and Sharir [3]). *If all the n constraint surfaces are flat triangles in 3-space, the complexity of a single connected component (cell) in their arrangement is at most* $O(n^{7/3}\alpha(n)^{2/3}\log^{4/3}n)$, *and the cell which contains a given point can be calculated by a probabilistic randomized algorithm, whose expected time complexity is also close to* $O(n^{7/3})$ *(here* $\alpha(n)$ *denotes the extremely slowly growing inverse of Ackermann's function—see Section 5 for more detail).*

As a corollary, it follows that planning a purely translational motion of a polyhedral object with k faces amidst polyhedral obstacles having a total of n faces (a system obviously having three degrees of freedom) can be accomplished in expected time close to $O((kn)^{7/3})$. However, for other nontrivial systems with three degrees of freedom, not to mention systems with more degrees of freedom, we still do not know whether a single component of FP always has combinatorial complexity that is significantly smaller than that of the entire FP.

3.5.3. The case of a mobile convex object moving in a 2-D polygonal space

Let B be a convex k-gon moving (translating and rotating) in the plane amidst polygonal obstacles having a total of n corners and edges. This system has three degrees of freedom, and is one of the simplest systems which arises in practical applications (think of a navigating vehicle, a household or factory mobile server, etc.). This particular system has been studied recently by Leven and Sharir [58] and by Kedem and Sharir [51]. They have shown the following theorem to hold.

3.11. THEOREM (cf. [51, 58]). *The entire free configuration space FP for a convex k-gon has combinatorial complexity* $O(kn\lambda_6(kn))$, *which is to say, nearly quadratic in kn.* (*Note that the number of collision constraints is* $O(kn)$, *where each constraint represents placements of B in which either a specific corner of B makes contact with some specific obstacle edge, or a specific side of B touches a specific obstacle corner.*) *Moreover, FP can be calculated in close to quadratic time, specifically in* $O(kn\lambda_6(kn)\log kn)$ *time.*

The algorithm of [51] for calculating FP is fairly elaborate and sophisticated. It has been recently implemented and tested successfully on commercial equipment. It is beyond the scope of this survey to go into details of this algorithm, and we merely note that it uses a mixture of the previous techniques, which involves slicing FP into 2-D cross sections FP_θ in each of which the orientation θ of B is fixed, followed by calculation of these restricted FPs using the techniques of [49] reviewed above. The overall representation of FP is obtained by extending the vertices of the 2-D cross-sections FP_θ of FP into edges of FP, by letting θ vary. Critical values of θ arise when the combinatorial structure of the cross-section changes discontinuously, and, as it turns out, the main (though not exclusive) source of criticalities are placements of B at which it makes simultaneously three contacts with the obstacles (while otherwise remaining free). A previous combinatorial result of Leven and Sharir [58] shows that the maximum number of such free triple contacts is at most $O(kn\lambda_6(kn))$. Using this result, [51] shows that the total complexity of the connectivity "edge graph" that represents FP is of the same order of magnitude, and develops the afore-mentioned algorithm for calculating that graph.

We mention this algorithm also because it is one of the few implemented motion planning algorithms that conform to the maxim of using exact, nonheuristic and worst-case efficient techniques. A related recent implementation, due to Avnaim, Boissonnat and Faberjon [6], solves the motion planning problem for an arbitrary mobile polygonal robot, but has worst-case cubic time complexity (and was thus simpler to implement).

3.5.4. Miscellaneous other results

We conclude this section with a somewhat arbitrary sample of a miscellany of other recent motion planning algorithms (it has been impossible to cover all recent developments in this area).

Yap [121] and later Madilla and Yap [66] have introduced the variant in which the environment is fixed and has simple structure, but the moving system is arbitrarily taken from a certain large class of systems. For example, Yap [121] considers the problem of moving a "chair", which is an arbitrary simple polygon B with n sides, through a "door", which is an interval opening along the y-axis, say, which is otherwise a solid wall; Maddila and Yap [66] consider a related problem of moving such a chair around a right-angle corner in a corridor. Their result has recently been improved by Agarwal and Sharir [1].

Another special case is that of coordinated motion planning, i.e. the case of several independent systems that move in the same workspace and are to be controlled by a common algorithm, which has to make sure that no two systems collide. This case can of course be viewed as a special case of the general motion planning problem, but the independence between the moving systems can be exploited to obtain improved algorithms. This has been recently done by Sharir and Sifrony [110], following previous works [32, 88, 120].

Pushing the number of degrees of freedom still higher, Ke and O'Rourke have studied the motion planning problem for a rod in 3-space, amidst a collection of polyhedral obstacles with a total of n faces. This problem has five degrees of freedom. Ke and O'Rourke [48] have first obtained a lower bound on the number of submotions needed to move the rod between any two placements, showing that in the worst case $\Omega(n^4)$ submotions might be necessary. As an upper bound, they have obtained an $O(n^6 \log n)$ algorithm, which is simpler than the general technique of Canny [12] but is nevertheless asymptotically inferior, because Canny's technique requires only $O(n^5 \log n)$ time. What is further intriguing here is that the number of vertices of FP can be shown to be only $O(n^4)$, but in spite of this fact, it is still not known how to calculate and represent FP in a connectivity-preserving manner within a similar time bound.

The problem of planning the motion of a general convex polyhedral object B in 3-space, including translation and rotation, has been studied by Canny and Donald [14], using the notion of "simplified Voronoi diagrams". However, in spite of the name, calculation and manipulation of that diagram in the general case (involving all six degrees of freedom of B) has been quite complex, although when considering restricted problem versions, such as translation only, the diagram becomes much easier to manipulate.

Looking ahead into the future, there are still many deep challenges down the road. Perhaps a major milestone will be reached when an exact and yet efficient algorithm for a manipulator arm with six degrees of freedom will be developed. Many simpler problems, some of which mentioned above, still defy a solution.

4. Variants of the motion planning problem

4.1. Optimal motion planning

The only optimal motion planning problem which has been studied extensively thus far is that in which the moving system is represented as a single point, in which case one aims to calculate the Euclidean shortest path connecting the initial and final system placements, given that specified obstacles must be avoided. Most existing work on this

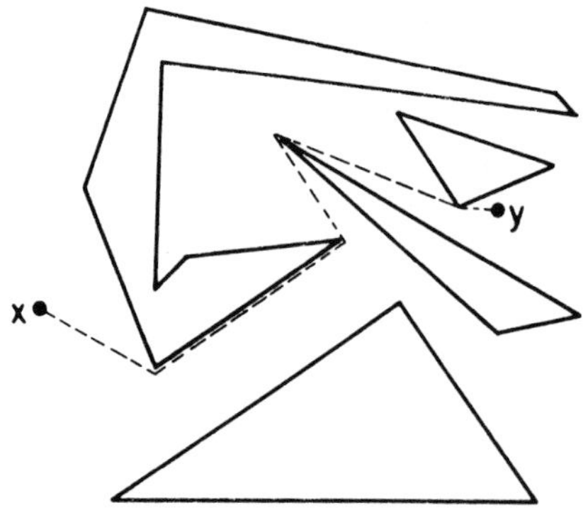

Fig. 9. An instance of the 2-D Euclidean shortest path problem.

problem assumes that the obstacles are either polygonal (in 2-space) or polyhedral (in 3-space).

The 2-D case is considerably simpler than the 3-D case (see Fig. 9). When the free space V in 2-D is bounded by n straight edges, it is easy to calculate the desired shortest path in time $O(n^2 \log n)$. This is done by constructing a *visibility graph VG* whose edges connect all pairs of boundary corners of V which are visible from each other through V, and then by searching for a shortest path (in the graph-theoretic sense) through VG (see [109] for a sketch of this idea). This procedure was improved to $O(n^2)$ by Asano et al. [4], by Welzl [117], and by Reif and Storer [91], using a cleverer method for constructing VG. This quadratic time bound has been improved in certain special cases (see [94, 91]). However, it is not known whether shortest paths for a general polygonal space V can be calculated in subquadratic time. Among the special cases allowing more efficient treatment the most important is that of calculating shortest paths inside a simple polygon P. Lee and Preparata [55] gave a linear-time algorithm for this case, assuming that a triangulation of P is given in advance. The Lee–Preparata result was recently extended by Guibas et al. [34], who gave a linear-time algorithm which calculates all shortest paths from a fixed source point to all vertices of (a triangulated polygon) P. We summarize these results in the following theorem.

4.1. THEOREM. *The Euclidean shortest path problem in a 2-D polygonal space can be solved in* $O(n^2)$ *time in the worst case, and in* $O(n)$ *time inside a triangulated simple polygon.*

Other results on 2-D shortest paths include an $O(n \log n)$ algorithm for finding *rectilinear* shortest paths which avoid n rectilinear disjoint rectangles [93]; an $O(n^2)$ algorithm for finding Euclidean shortest motion of a circular disk or a translating convex object in a 2-D polygonal region [37] (cf. also [18]); algorithms for cases in which the obstacles consist of a small number of disjoint convex regions [91, 94]; algorithms for the "weighted region" case (in which the plane is partitioned into polygonal regions in each of which the path has a different multiplicative cost factor) [72]; and various other special cases.

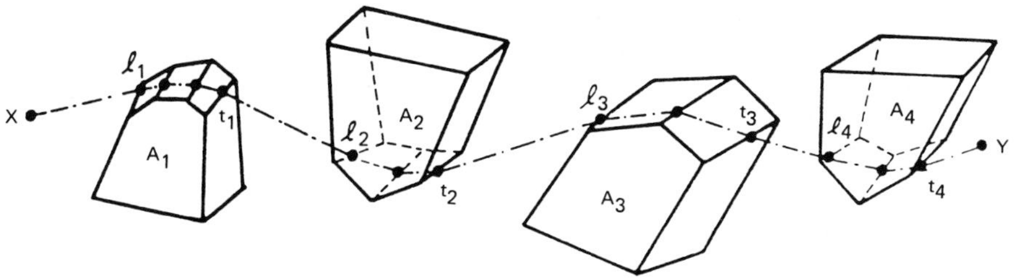

Fig. 10. A shortest path in 3-D polyhedral space.

The 3-D polyhedral shortest path problem is substantially more difficult. To calculate shortest paths in 3-space amidst polyhedral obstacles bounded by n edges, we can begin by noting that any such path must be piecewise linear, with corners lying on the obstacle edges, and that it must subtend equal incoming and outgoing angles at each such corner (see Fig. 10).

These remarks allow shortest path calculation to be split into two subproblems:

(i) find the sequence of edges through which the desired path must pass;

(ii) find the contact points of the path with these edges.

However, even when the sequence of edges crossed by the path is known, calculation of the contact points is still nontrivial, because it involves solution of a large system of quadratic equations, expressing the conditions of equal incoming and outgoing angles at each crossed edge, whose precise solution may require use of one of the algebraic techniques mentioned above (e.g. [13, 21]) thus requiring time exponential in n (see [91] for such an exponential-time solution). Subproblem (ii) can also be solved using numerical iterative procedures, but even if this approximating approach is taken, there still remains subproblem (i), whose only known solution to date is exhaustive enumeration of all possible edge sequences. A recent result of Canny and Reif [16] has finally settled the problem by showing it to be computationally intractable (more specifically, NP-hard) in three dimensions.

One of the reasons the problem is difficult in the general polyhedral case is that consecutive edges crossed by the shortest path can be skewed to one another. There are, however, some special cases in which this difficulty does not arise, and they admit efficient solutions. One such case is that in which we aim to calculate shortest paths lying along the surface of a single convex polyhedron having n edges. In this case subproblem (ii) can easily be solved by "unfolding" the polyhedron surface at the edges crossed by the path, thereby transforming the path to a straight segment connecting the unfolded source and destination points (cf. [109]). Extending this observation, Mount [74] has devised an $O(n^2 \log n)$ algorithm, which proceeds by sophisticating an algorithmic technique originally introduced by Dijkstra to find shortest paths in graphs, specifically by maintaining and updating a combinatorial structure characterizing shortest paths from a fixed initial point to each of the edges of the polyhedron (cf. also [109] for an initial version of this approach).

This result has recently been extended in several ways. A similar $O(n^2 \log n)$ algorithm for shortest paths along a (not necessarily convex) polyhedral surface is given in [71]. Baltsan and Sharir [8] consider the problem of finding the shortest path connecting two points lying on two disjoint convex polyhedral obstacles, and report a nearly cubic algorithm, which makes use of the Davenport–Schinzel sequences technique (to be described below). The case of shortest paths which avoid a fixed number k of disjoint convex polyhedral obstacles is analyzed in [104], which describes an algorithm that is polynomial in the total number of obstacle edges, but is exponential in k. Finally, an approximating pseudopolynomial scheme for the general polyhedral case is reported in [84]; this involves splitting each obstacle edge by adding sufficiently many new vertices and by searching for the shortest piecewise linear path bending only at those vertices.

In summary, we have the following theorem.

4.2. THEOREM. *The Euclidean shortest path problem amidst polyhedral obstacles in 3-space*

(a) *is* NP-*hard and can be solved in singly-exponential time in the general case;*

(b) *can be solved in* $O(n^2 \log n)$ *time, if motion is restricted to the surface of a convex (or simply-connected) polyhedron;*

(c) *can be solved in polynomial time for any fixed number of convex polyhedral obstacles;*

(d) *can be solved approximately by a pseudopolynomial algorithm.*

Finally, we mention a few initial works on optimal motion planning for non-point systems. Papadimitriou and Silverberg [85] consider the case of a rod in a 2-D polygonal space, in which one endpoint of the rod moves along a prespecified polygonal path and the aim is to minimize the total distance travelled by that endpoint (or, in another variant, the sum of this distance and the total rotation of the rod). O'Rourke [81] also considers shortest paths for a moving rod, but restricts the kinds of motions allowed and also the shape of the obstacles.

4.2. *Adaptive and exploratory motion planning*

If the environment of obstacles through which motion must be planned is not known to a robot system a priori, but the system is equipped with sensory devices, motion planning assumes a more "exploratory" character. If only tactile (or proximity) sensing is available, then a plausible strategy might be to move along a straight line (in physical or configuration space) directly to the target position, and when an obstacle is reached, to follow its boundary until the original straight line of motion is reached again. This technique has been developed and refined by Lumelski and his collaborators [63, 64, 65], and has been successful in handling systems with two degrees of freedom, in the sense that this strategy provably reaches the goal, if this is at all possible, with a reasonable bound on the length of the motion. This technique has been implemented on several real and simulated systems, and has applications to maze-searching problems.

Attempts to extend this technique to systems with three (or more) degrees of freedom are currently under way. One such extension, with a somewhat different flavor, was obtained in 1988 by Cox and Yap [22]. They apply the retraction technique to the free configuration space *FP*, and compute within it a certain one-dimensional skeleton *N* which captures the connectivity of *FP*. The twist here is that *FP* is not known in advance, so the construction of *N* has to be done in an incremental, exploratory manner, adding more features to *N* as the system explores its environment. Cox and Yap show how to implement this exploration in a controlled manner that does not require too many "probing" steps, and which enables the system to recognize when the construction of *N* has been completed (if the goal has not been reached beforehand).

If vision is also available, then other possibilities need to be considered, e.g. the system can obtain partial information about its environment by viewing it from the present placement, and then "explore" it to gain progressively more information until the desired motion can be fully planned. Some of these issues are considered in Lumelski's and others' works.

Even when the environment is fully known to the system, other interesting issues arise if the environment is changing. For example, when some of the objects in the robot's environment are picked up by the robot and moved to a different placement, one wants fast techniques for incremental updating of the environment model and the data structures used for motion planning. Moreover, whenever the robot grasps an object to move it, robot plus grasped object become a new moving system and may require a different motion planning algorithm, but one whose relationship to motions of the robot alone needs to be investigated. Adaptive motion planning problems of this kind still remain to be studied.

4.3. *Motion planning in the presence of moving obstacles*

Interesting generalizations of the motion planning problem arise when some of the obstacles in the robot's environment are assumed to be moving along known trajectories. In this case the robot's goal will be to "dodge" the moving obstacles while moving to its target placement. In this "dynamic" motion planning problem, it is reasonable to assume some limit on the robot's velocity and/or acceleration. Two initial studies of this problem by Reif and Sharir [90] and by Sutner and Maass [113] indicate that the problem of avoiding moving obstacles is substantially harder than the corresponding static problem. By using time-related configuration changes to encode Turing machine states, they show that the problem is PSPACE-hard even for systems with a small and fixed number of degrees of freedom (more negative related results have been obtained in [16]). However, polynomial-time algorithms are available in a few particularly simple special cases. Another variant of this problem has been recently considered by Wilfong [119], where the obstacles do not move by themselves, but some of them are movable by the system *B* (e.g. *B* can "push aside" such obstacles to clear its passage). Again, it is shown that the general problem of this kind is PSPACE-hard, but that polynomial-time algorithms are available in certain special cases.

4.4. *Miscellaneous other variants*

Many other variants of the motion planning problem have been considered. For lack of space we cannot cover in detail all of them, but we would like to mention a few more.

One variant deals with *constrained motion planning*, in which the motion of the object B is constrained not to exceed certain velocity or acceleration thresholds. For example, what is the fastest way for an object to go from an initial placement at some initial velocity to a final placement at some specified final velocity, assuming there is some bound on the maximum acceleration that B can exert. Even without any obstacles such problems are far from being trivial, and the presence of (stationary or moving) obstacles makes them extremely complicated to solve. Several recent works have addressed these issues (in addition to [90, 113] mentioned above); O'Dunlaing [76] discusses the bounded-acceleration case mentioned above, Fortune and Wilfong [31] study a variant in which the system has to avoid stationary obstacles in the plane, and the curvature of the motion path cannot exceed a certain threshold, and Canny, Donald and Reif [15] analyze similar near-optimal-time planning problems under various "kinodynamic" constraints.

Another class of problems deals with motion planning with uncertainty. Here the moving system has only approximate knowledge of the geometry of the obstacles and/or of its current position and velocity. Usually, in these cases the main "accurate" form of data that the system can acquire is a sensory feedback that tells it when it contacts an obstacle. The objective is to devise a strategy that will guarantee that the system reaches its goal, where such a strategy usually proceeds through a sequence of free motions (until an obstacle is hit) intermixed with compliant motions (sliding along surfaces of contacted obstacles) until it can be ascertained that the goal has been reached. This problem is studied, among other papers, in [67], where a few initial observations about the problem are made, in [27], where motion strategies for some special cases are obtained, and in [16], which shows that the general problem is computationally infeasible in the worst case.

4.5. *General task planning*

Finally we consider the general problem of task planning. Here the system is given a complex task to perform, for example, it is required to assemble some part from several components, or to restructure its workcell into some new layout, but the precise sequence of substeps needed to attain the final goal is not specified and must be figured out by the system.

Not much has been done about an exact and efficient algorithmic approach to this general problem, and it is clear that the problem complexity significantly increases as compared to the simpler motion planning problem (some instances of which, as we have seen, are already bad enough in terms of complexity). Still, several interesting basic problems can be distilled as special cases of task planning, and have recently been investigated. A typical example is the problem of "movable separability of sets", as studied by Toussaint and his collaborators (see e.g. [115]). In this problem, one reverses the assembly process by taking the final assembled object and attempting to take it apart by moving subcollections of its parts away from one another, and continue this process until all parts have been separated. Several algorithms have been obtained for various special cases (see [75, 87, 115]), and many interesting open problems have emerged. We mention one of them. Given a collection of *convex* polyhedra in 3-space (with pairwise disjoint interiors), can they always be partitioned into two disjoint and nonempty subcollections, so that the polyhedra in one collection can be translated rigidly together in some direction so as to be separated from those in the second collection (which remain stationary and should not collide with the moving polyhedra)? Some counterexamples (in [25, 75]) show that this conjecture is false if we are only allowed to move one polyhedron at a time away from the others.

Another research area related to task planning is that of *grasp planning*, where a multifinger system has to plan a grasping of a given object B, so as to minimize slippage, or to ensure that the fingers avoid interfering with subsequent tasks to be executed by the gripped object (e.g. if B is to be placed on a table, no finger should touch that face of B that is to lie on the table). See [70, 100] for two examples of recent work in this area.

5. Results in computational geometry relevant to motion planning

The various studies of motion planning described above make extensive use of efficient algorithms for the geometric subproblems which they involve, for which reason motion planning has been one of the major motivating forces that have encouraged research in computational geometry, and have led to rich cross-fertilization between these two areas. We have mentioned in the previous sections a variety of problems in computational geometry which arise in robotic motion planning, and will mention a few more below. Generally these problems fall into several subareas. We mention here a few, but note that many other areas are also involved, e.g. probabilistic algorithms in geometry, and the study of arrangements of curves or surfaces; unfortunately, for lack of space we cannot cover them all in this survey.

5.1. *Intersection detection*

The problem here is to detect intersections and to compute shortest distances, e.g. between moving subparts of a robot system and stationary or moving obstacles. Simplifications which have been studied include those in which all objects involved are circular disks (in the 2-D case) or spheres (in the 3-D case). In a study of the 2-D case of this problem, Sharir [102] developed a generalization of Voronoi diagrams for a set of (possibly intersecting) circles, and used this diagram to detect intersections and computing shortest distances between disks in time $O(n \log^2 n)$ (an alternative and more efficient approach to this appears in [45]). Hopcroft, Schwartz and Sharir [40] present an algorithm for detecting intersections among n 3-D spheres which also runs in time $O(n \log^2 n)$. However, this algorithm does not adapt in any obvious way to allow proximity calculation or other significant problem variants.

Other intersection detection algorithms appearing in the computational geometry literature involve rectilinear objects and use multidimensional searching techniques for achieving high efficiency (see [69] for a survey of these techniques).

A recent result in [1] gives efficient algorithms for detecting an intersection between any of m red Jordan arcs and any of n blue arcs in the plane (where red-red and blue-blue intersections are allowed and need not be reported). This result (see also Theorem 5.6) has applications to collision detection and certain instances of motion planning, as mentioned above.

5.2. *Generalized Voronoi diagrams*

The notion of Voronoi diagram has proven to be a useful tool in the solution of many motion planning problems. The discussion given previously has mentioned the use of various variants of Voronoi diagram in the retraction-based algorithms for planning the motion of a disk [77], or of a rod [78, 79], or the translational motion of a convex object [57], and in the intersection detection algorithm for disks mentioned above [102]. The papers just cited, and some related works [59, 123] describe the analysis of these diagrams and the design of efficient algorithms for their calculation.

There are many interesting further extensions of Voronoi diagrams that arise in connection with motion planning problems but which have not been studied yet. For example, one obvious approach to planning the motion of a ball in 3-space amidst polyhedral obstacles is to extend the 2-D approach of [77], that is, to calculate the standard (i.e. Euclidean) Voronoi diagram of the obstacles, and constrain the motion of the center of the ball to lie on the diagram. Nobody has yet analyzed the complexity of such diagrams (except in the special case where all obstacles are singleton points, in which case we obtain the standard 3-D Voronoi diagram of point sites). Can this complexity be shown to be subcubic, and perhaps close to quadratic, in the number of faces of the polyhedra? No comparably efficient (subcubic) algorithms for the calculation of such diagrams are known either. Even the simplest case of Voronoi diagrams of n lines in space seems to be completely open.

Various other types of 3-D generalized Voronoi diagrams arise in connection with motion planning and deserve to be studied. For example, planning the translational motion of a convex polyhedron B, even amidst point obstacles, calls for calculation of the Voronoi diagram of the obstacles under the "pseudometric" in which B serves as the unit ball (extending the 2-D approach of [57]). Nothing is yet known about the structure and complexity of such diagrams.

5.3. *Davenport–Schinzel sequences*

Davenport–Schinzel sequences are combinatorial sequences of n symbols which do not contain certain forbidden subsequences of alternating symbols. Sequences of this sort appear in studies of the combinatorial complexity and of efficient techniques for calculating the lower envelope of a set of n continuous functions, if it is assumed that the graphs of any two functions in the set can intersect in some fixed number of points at most. These sequences, whose study was initiated in [23, 24], have proved to be powerful tools for analysis (and design) of a variety of geometric algorithms, many of which are useful for motion planning. More specifically we have the following definition and subsequent theorem.

DEFINITION. An (n, s) *Davenport–Schinzel sequence* is defined to be a sequence U composed of n symbols, such that

(i) no two adjacent elements of U are equal, and

(ii) there do not exist $s+2$ indices $i_1 < i_2 < \cdots < i_{s+2}$ such that $u_{i_1} = u_{i_3} = u_{i_5} = \cdots = a$, $u_{i_2} = u_{i_4} = u_{i_6} = \cdots = b$, with $a \neq b$. Let $\lambda_s(n)$ denote the maximal length of an (n, s) Davenport–Schinzel sequence.

5.1. THEOREM (Szemeredi [114]). *$\lambda_s(n) \leqslant C_s n \log^* n$, where C_s is a constant depending on s, and $\log^* n$ is the "iterated logarithm" function defined above.*

Improving on this result, Hart and Sharir [36] proved that $\lambda_3(n)$ is proportional to $n\alpha(n)$ where $\alpha(n)$ is the very slowly growing inverse of the Ackermann function. (The inverse Ackermann function arises in many applications in logic, combinatorics and computer science. It approaches infinity as n grows, but does this extremely slowly; for example, it does not exceed 5 for all n up to an exponential tower $2^{2^{2^{\cdots}}}$ having 65536 2s. See [36] for more details concerning this function.) Following further improvements in [103, 106], the currently best known bounds on $\lambda_s(n)$ are due to Agarwal, Sharir and Shor [2], and are summarized in the next theorem.

5.2. THEOREM

$\lambda_1(n) = n; \qquad \lambda_2(n) = 2n-1$	*(easy)*.
$\lambda_3(n) = \Theta(n\alpha(n))$	[36].
$\lambda_4(n) = \Theta(n \cdot 2^{\alpha(n)})$	[2].
$\lambda_{2s}(n) = O(n \cdot 2^{O(\alpha(n)^{s-1})})$	*for* $s > 2$ [2].
$\lambda_{2s+1}(n) = O(n \cdot \alpha(n)^{O(\alpha(n)^{s-1})})$	*for* $s \geqslant 2$ [2].
$\lambda_{2s}(n) \geqslant cn \cdot 2^{c'\alpha(n)^{s-1}}$,	*for* $s > 2$ *and for appropriate constants depending on* s [2].

These results show that, in practical terms, $\lambda_s(n)$ is an almost linear function of n for any fixed s, and is slightly superlinear for all $s \geqslant 3$.

In the 80s, numerous applications of these sequences to motion planning have been found. These include the following (most of which have already been mentioned earlier, but are re-examined here in the context of Davenport–Schinzel sequences).

(i) Let B be a convex k-gon translating and rotating in a closed 2-D polygonal space V bounded by n edges. The *polygon containment* problem calls for determining whether there exists any free placement of B, i.e. a placement in which B lies completely within V. Some variants of this problem have been studied by Chazelle [17], who showed that if such a free placement of B exists, then there also exists a *stable* free placement of B, namely a placement in which B lies completely within V and makes three simultaneous contacts with the boundary of V (see Fig. 11).

5.3. THEOREM (Leven and Sharir [58]). *The number of free stable placements of a convex k-gon is at most $O(kn\lambda_6(kn))$, and they can all be calculated in time*

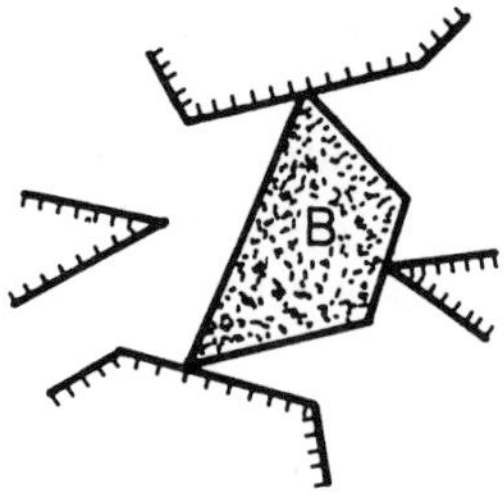

Fig. 11. Critical free contact of a convex polygon.

$O(kn\lambda_6(kn)\log kn)$. *Thus, within the same time bound, one can determine whether P can be placed inside Q.*

(ii) As already mentioned, Kedem and Sharir [51] have used this result (and further Davenport–Schinzel arguments) to produce an $O(kn\lambda_6(kn)\log kn)$ algorithm for planning the motion of a convex k-gon B in a 2-D polygonal space bounded by n edges.

(iii) We have also mentioned the result of Guibas, Sharir and Sifrony [35] that the complexity of a single face in an arrangement of n Jordan arcs in the plane, each pair of which intersect in at most s points, is $O(\lambda_{s+2}(n))$, and that such a face can be calculated in time $O(\lambda_{s+2}(n)\log^2 n)$. This result is presently the last step in a series of related results. The first, noted in [5] but going back as far as [23, 24], is that the lower envelope (i.e. pointwise minimum) of n continuous functions, each pair of which intersect in at most s points, has at most $\lambda_s(n)$ "breakpoints" (i.e. intersection points of the given functions which appear along the graph of that envelope). Moreover, if the given functions are only partially defined, each in some interval over which it is continuous, then the number of breakpoints in their lower envelope is at most $\lambda_{s+2}(n)$, and this bound is tight in the worst case for certain collections of such (rather irregularly shaped) functions. For example, when each function has a straight line segment for a graph (in which case $s=1$), the total complexity (i.e. number of breakpoints) of the lower envelope is at most $O(n\alpha(n))$. Wiernik and Sharir [118] have shown that this bound can actually be attained by certain collections of segments. Later, Pollack, Sharir and Sifrony [87] have extended this result to the complexity of a single face in an arrangement of segments, and paper [35] generalized the proof for arrangements of more general arcs as above.

(iv) The result concerning lower envelopes of segments in the plane has been generalized in [29, 83] to lower envelopes of simplices in any dimension. For example, we have this following theorem.

5.4. THEOREM. (a) *The lower envelope of n triangles in three dimensions has combinatorial complexity* $O(n^2\alpha(n))$, *which is tight in the worst case, and it can be calculated in* $O(n^2\alpha(n))$ *time.*

(b) *Hence, the boundary of the free configuration space FP of a polyhedral object with k faces flying (without rotating) over a polyhedral terrain with n faces, has combinatorial complexity* $O((kn)^2\alpha(kn))$.

(This should be contrasted with the weaker results of Aronov and Sharir [3] mentioned above, where a single cell in such an arrangement of triangles, rather than their easier-to-analyze lower envelope, is being considered.)

(v) Another application of Davenport–Schinzel sequences is given in this theorem.

5.5. Theorem (Baltsan and Sharir [8]). *Finding the shortest Euclidean path between two points in 3-space avoiding the interior of two disjoint convex polyhedra having n faces altogether, can be done in time* $O(n^2\lambda_{10}(n)\log n)$ *(that is, nearly cubic).*

(vi) Another recent application is the following one.

5.6. Theorem (Agarwal and Sharir [1]). *Detecting an intersection between a simple polygon B having m sides, and any of a collection of n arcs, each pair of which intersect in at most s points, can be done in time* $O(\lambda_{s+2}(n)\log^2 n + (\lambda_{s+2}(n)+m)\log(m+n))$.

The algorithm is based on the results of [35], and has applications in efficient motion planning for moving B around a right-angle corner in a corridor, as mentioned above.

(vii) The analysis in [79] of the complexity of the generalized Voronoi diagram, introduced in [78] for motion planning for a rod in 2-space, also involves Davenport–Schinzel sequences.

Many additional applications of Davenport–Schinzel sequences to other basic and applied geometric problems have been found recently. Among them we mention papers [5, 19, 20, 36, 46, 68, 82]; a survey on these sequences and their applications can be found in [107].

Acknowledgment

Work on this chapter has been supported by Office of Naval Research Grants N00014-87-K-0129, N00014-89-J-3042, and N00014-90-J-1284, by National Science Foundation Grants DCR-83-20085 and CCR-89-01484, and by grants from the U.S.–Israeli Binational Science Foundation and the Fund for Basic Research administered by the Israeli Academy of Sciences.

References

[1] Agarwal, P. and M. Sharir, Red-blue intersection detection algorithms with applications to motion planning and collision detection, in: *Proc. 4th Ann. ACM Symp. on Computational Geometry* (1988) 70–80.

[2] Agarwal, P., M. Sharir and P. Shor, Sharp upper and lower bounds for the length of general Davenport-Schinzel sequences, *J. Combin. Theory Ser. A* **52** (1989) 228–274.

[3] Aronov, B. and M. Sharir, Triangles in space, or: building (and analyzing) castles in the air, in: *Proc. 4th Ann. ACM Symp. on Computational Geometry* (1988) 381–391.

[4] Asano, T., T. Asano, L. Guibas, J. Hershberger and H. Imai, Visibility polygon search and

Euclidean shortest paths, in: *Proc. 26th Ann. IEEE Symp. on Foundations of Computer Science* (1985) 155–164.

[5] ATALLAH, M., Some dynamic computational geometry problems, *Comput. Math. Appl.* **11** (1985) 1171–1181.

[6] AVNAIM, F., J.D. BOISSONNAT and B. FABERJON, A practical exact motion planning algorithm for polygonal objects amidst polygonal obstacles, in: *Proc. IEEE Symp. on Robotics and Automation* (1988) 1656–1661.

[7] BAJAJ, C. and M.S. KIM, Compliant motion planning with geometric models, in: *Proc. 3rd Ann. ACM Symp. on Computational Geometry* (1987) 171–180.

[8] BALTSAN A. and M. SHARIR, On shortest paths between two convex polyhedra, *J. ACM* **35** (1988) 267–287.

[9] BROOKS, R.A., Solving the find-path problem by good representation of free space, *IEEE Trans. Systems Man & Cyber.* **13**(3) (1983) 190–197.

[10] BROOKS, R.A., Planning collision-free motions for pick-and-place operations, *Internat. J. Robotics Res.* **2**(4) (1983) 19–40.

[11] BROOKS, R.A. and T. LOZANO-PEREZ, A subdivision algorithm in configuration space for findpath with rotation, in: *Proc. IJCAI-83* (1983).

[12] CANNY, J., Complexity of robot motion planning, Ph.D. Dissertation, Computer Science Dept., Massachusetts Institute of Technology, Cambridge, MA, 1987.

[13] CANNY, J., Some algebraic and geometric computations in PSPACE, in: *Proc. 20th Ann. ACM Symp. on Theory of Computing* (1988) 460–467.

[14] CANNY, J. and B. DONALD, Simplified Voronoi diagrams, *Discrete Comput. Geom.* **3** (1988) 219–236.

[15] CANNY, J., B. DONALD, J. REIF and P. XAVIER, On the complexity of kinodynamic planning, in: *Proc. 29th Ann. IEEE Symp. on Foundations of Computer Science* (1988) 306–316.

[16] CANNY, J. and J. REIF, New lower bound techniques for robot motion planning problems, in: *Proc. 28th Ann. IEEE Symp. on Foundations of Computer Science* (1987) 49–60.

[17] CHAZELLE, B., On the polygon containment problem, in: F. Preparata, ed., *Advances in Computing Research, Vol. II: Computational Geometry* (JAI Press, Greenwich, CT, 1983) 1–33.

[18] CHEW, L.P., Planning the shortest path for a disc in $O(n^2 \log n)$ time, in: *Proc. ACM Symp. on Computational Geometry* (1985) 214–223.

[19] CLARKSON, K., Approximation algorithms for shortest path motion planning, in: *Proc. 19th Ann. ACM Symp. on Theory of Computing* (1987) 56–65.

[20] COLE, R. and M. SHARIR, Visibility problems for polyhedral terrains, *J. Symbolic Comput.* **7** (1989) 11–30.

[21] COLLINS, G.E., Quantifier elimination for real closed fields by cylindrical algebraic decomposition, in: *Proc. 2nd GI Conf. on Automata Theory and Formal Languages*, Lecture Notes in Computer Science, Vol. 33 (Springer, Berlin, 1975) 134–183.

[22] COX, J. and C.K. YAP, On-line motion planning: moving a planar arm by probing an unknown environment, Tech. Report, Courant Institute, New York, 1988.

[23] DAVENPORT, H., A combinatorial problem connected with differential equations, II, *Acta Arithmetica* **17** (1971) 363–372.

[24] DAVENPORT, H. and A. SCHINZEL, A combinatorial problem connected with differential equations, *Amer. J. Math.* **87** (1965) 684–694.

[25] DAWSON, R., On removing a ball without disturbing others, *Mathematics Magazine* **57** (1984).

[26] DONALD, B., Motion planning with six degrees of freedom, Tech. Report 791, AI Lab, Massachusetts Institute of Technology, Cambridge, MA, 1984.

[27] DONALD, B., The complexity of planar compliant motion planning under uncertainty, in: *Proc. 4th Ann. ACM Symp. on Computational Geometry* (1988) 309–318.

[28] EDELSBRUNNER, H., *Algorithms in Combinatorial Geometry* (Springer, Heidelberg, 1987).

[29] EDELSBRUNNER, H., L. GUIBAS and M. SHARIR, The upper envelope of piecewise linear functions: algorithms and applications, *Discrete Comput. Geom.*, **4** (1989) 311–336.

[30] ERDMANN, M., On motion planning with uncertainty, Tech. Report 810, AI Lab, Massachusetts Institute of Technology, Cambridge, MA, 1984.

[31] FORTUNE, S. and G. WILFONG, Planning constrained motion, in: *Proc. 20th Ann. ACM Symp. on Theory of Computing* (1988) 445–459.

[32] FORTUNE, S., G. WILFONG and C. YAP, Coordinated motion of two robot arms, in: *Proc. IEEE Conf. on Robotics and Automation* (1986) 1216–1223.

[33] GAREY, M.R. and D.S. JOHNSON, *Computers and Intractability—A Guide to the Theory of NP-Completeness* (Freeman, San Francisco, CA, 1979).

[34] GUIBAS, L., J. HERSHBERGER, D. LEVEN, M. SHARIR and R.E. TARJAN, Linear time algorithms for shortest path and visibility problems inside triangulated simple polygons, *Algorithmica* **2** (1987) 209–233.

[35] GUIBAS, L., M. SHARIR and S. SIFRONY, On the general motion planning problem with two degrees of freedom, *Discrete Comput. Geom.* **4** (1989) 491–521.

[36] HART, S. and M. SHARIR, Nonlinearity of Davenport–Schinzel sequences and of generalized path compression schemes, *Combinatorica* **6** (1986) 151–177.

[37] HERSHBERGER, J. and L. GUIBAS, An $O(n^2)$ shortest path algorithm for a non-rotating convex body, *J. Algorithms* **9** (1988) 18–46.

[38] HOPCROFT, J.E., D.A. JOSEPH and S.H. WHITESIDES, Movement problems for 2-dimensional linkages, *SIAM J. Comput.* **13** (1984) 610–629.

[39] HOPCROFT, J.E., D.A. JOSEPH and S.H. WHITESIDES, On the movement of robot arms in 2-dimensional bounded regions, *SIAM J. Comput.* **14** (1985) 315–333.

[40] HOPCROFT, J.E., J.T. SCHWARTZ and M. SHARIR, Efficient detection of intersections among spheres, *Internat. J. Robotics Res.* **2**(4) (1983) 77–80.

[41] HOPCROFT, J.E., J.T. SCHWARTZ and M. SHARIR, On the complexity of motion planning for multiple independent objects; PSPACE hardness of the "warehouseman's problem", *Internat. J. Robotics Res.* **3**(4) (1984) 76–88.

[42] HOPCROFT, J.E., J.T. SCHWARTZ and M. SHARIR, eds., *Planning, Geometry, and Complexity of Robot Motion* (Ablex, Norwood, NJ, 1987).

[43] HOPCROFT, J.E. and G. WILFONG, Motion of objects in contact, *Internat. J. Robotics Res.* **4**(4) (1986) 32–46.

[44] HWANG, Y.K. and N. AHUJA, Path planning using a potential field representation, Tech. Report UILU-ENG-88-2251, Dept. of Electrical Engineering, Univ. of Illinois at Urbana-Champaign, 1988.

[45] IMAI, H., M. IRI M. and K. MUROTA, Voronoi diagram in the Laguerre geometry and its applications, *SIAM J. Comput.* **14** (1985) 93–105.

[46] JAROMCZYK, J. and K. KOWALUK, Skewed projections with applications to line stabbing in R^3, in: *Proc. 4th Ann. ACM Symp. on Computational Geometry* (1988) 362–370.

[47] JOSEPH, D.A. and W.H. PLANTINGS, On the complexity of reachability and motion planning questions, in: *Proc. ACM Symp. on Computational Geometry* (1985) 62–66.

[48] KE, Y. and J. O'ROURKE, Moving a ladder in three dimensions: upper and lower bounds, in: *Proc. 3rd Ann. ACM Symp. on Computational Geometry* (1987) 136–146.

[49] KEDEM, K., R. LIVNE, J. PACH and M. SHARIR, On the union of Jordan regions and collision-free translational motion amidst polygonal obstacles, *Discrete and Comput. Geom.* **1** (1986) 59–71.

[50] KEDEM, K. and M. SHARIR, An efficient algorithm for planning translational collision-free motion of a convex polygonal object in 2-dimensional space amidst polygonal obstacles, in: *Proc. ACM Symp. on Computational Geometry* (1985) 75–80.

[51] KEDEM, K. and M. SHARIR, An efficient motion planning algorithm for a convex rigid polygonal object in 2-dimensional polygonal space, *Discrete Comput. Geom.*, to appear.

[52] KEMPE, A.B., On a general method of describing plane curves of the n-th degree by linkwork, *Proc. London Math. Soc.* **7** (1876) 213–216.

[53] KHATIB, O., Real-time obstacle avoidance for manipulators and mobile robots, *Internat. J. Robotics Res.* **5**(1) (1986) 90–98.

[54] KOZEN, D. and C.K. YAP, Algebraic cell decomposition in NC, in: *Proc. 26th Ann. IEEE Symp. on Foundations of Computer Science* (1985) 515–521.

[55] LEE, D.T. and F.P. PREPARATA, Euclidean shortest paths in the presence of rectilinear barriers, *Networks* **14**(3) (1984) 393–410.

[56] LEVEN, D. and M. SHARIR, An efficient and simple motion-planning algorithm for a ladder moving in two-dimensional space amidst polygonal barriers, *J. Algorithms* **8** (1987) 192–215.

[57] LEVEN, D. and M. SHARIR, Planning a purely translational motion of a convex object in two-dimensional space using generalized Voronoi diagrams, *Discrete Comp. Geom.* **2** (1987) 9–31.

[58] LEVEN, D. and M. SHARIR, On the number of critical free contacts of a convex polygonal object moving in 2-D polygonal space, *Discrete Comp. Geom.* **2** (1987) 255–270.

[59] LEVEN, D. and M. SHARIR, Intersection and proximity problems and Voronoi diagrams, in: J. Schwartz and C. Yap, eds. *Advances in Robotics, Vol. I* (Lawrence Erlbaum, Hillsdale, NJ, 1987) 187–228.

[60] LOZANO-PEREZ, T., Spatial planning: a configuration space approach, *IEEE Trans. Computers* **32**(2) (1983) 108–119.

[61] LOZANO-PEREZ, T., A simple motion planning algorithm for general robot manipulators, Tech. Report, AI Lab, Massachusetts Institute of Technology, Cambridge, MA, 1986.

[62] LOZANO-PEREZ, T. and M. WESLEY, An algorithm for planning collision-free paths among polyhedral obstacles, *Comm. ACM* **22** (1979) 560–570.

[63] LUMELSKI, V.J., Dynamic path planning for planar articulated robot arm moving amidst unknown obstacles, *Automatica* **23** (1987) 551–570.

[64] LUMELSKI, V.J., Algorithmic and complexity issues of robot motion in uncertain environment, *J. Complexity* **3** (1987) 146–182.

[65] LUMELSKI, V. and A. STEPANOV, Path planning strategies for a point mobile automaton moving amidst unknown obstacles of arbitrary shape, *Algorithmica* **2** (1987) 403–430.

[66] MADDILA, S. and C.K. YAP, Moving a polygon around the corner in a corridor, in: *Proc. 2nd Ann. ACM Symp. on Computational Geometry* (1986) 187–192.

[67] MASON, M., Automatic planning of fine motions: correctness and completeness, Tech. Report CMU-RI No. 83-18, Carnegie-Mellon Univ., Pittsburgh, PA, 1983.

[68] MCKENNA, M. and J. O'ROURKE, Arrangements of lines in 3-space: a data structure with applications, in: *Proc. 4th Ann. ACM Symp. on Computational Geometry* (1988) 371–380.

[69] MEHLHORN, K., *Data Structures and Algorithms 3: Multidimensional Searching and Computational Geometry* (Springer, Heidelberg, 1984).

[70] MISHRA, B., J.T. SCHWARTZ and M. SHARIR, On the existence and synthesis of multifinger positive grips, *Algorithmica* **2** (1987) 541–558.

[71] MITCHELL, J., D. MOUNT and C. PAPADIMITRIOU, The discrete geodesic problem, *SIAM J. Comput.* **16** (1987) 647–668.

[72] MITCHELL, J. and C. PAPADIMITRIOU, The weighted region problem, in: *Proc. 3rd Ann. ACM Symp. on Computational Geometry* (1987) 30–38.

[73] MORAVEC, H.P., Robot rover visual navigation, Ph.D. Dissertation, Stanford University (UMI Research Press, 1981).

[74] MOUNT, D.M., On finding shortest paths on convex polyhedra, Tech. Report, Computer Science Dept., Univ. of Maryland, 1984.

[75] NATARAJAN, B.K., On planning assemblies, in: *Proc. 4th Ann. ACM Symp. on Computational Geometry* (1988) 299–308.

[76] O'DUNLAING, C., Motion planning with inertial constraints, *Algorithmica* **2** (1987) 431–475.

[77] O'DUNLAING, C. and C. YAP, A "retraction" method for planning the motion of a disk, *J. Algorithms* **6** (1985) 104–111.

[78] O'DUNLAING, C., M. SHARIR and C. YAP, Generalized Voronoi diagrams for a ladder I: topological considerations, *Comm. Pure Appl. Math.* **39** (1986) 423–483.

[79] O'DUNLAING, C., M. SHARIR and C. YAP, Generalized Voronoi diagrams for a ladder II: efficient construction of the diagram, *Algorithmica* **2** (1987) 27–59.

[80] O'ROURKE, J., A lower bound for moving a ladder, Tech. Report JHU/EECS-85/20, The Johns Hopkins University, Baltimore, MD, 1985.

[81] O'ROURKE, J., Finding a shortest ladder path: a special case, Tech. Report 353, IMA Preprint Series, Univ. of Minnesota, 1987.

[82] OVERMARS, M. and E. WELZL, New methods for computing visibility graphs, in: *Proc. 4th Ann. ACM Symp. on Computational Geometry* (1988) 164–171.

[83] PACH, J. and M. SHARIR, The upper envelope of piecewise linear functions and the boundary of a region enclosed by convex plates: combinatorial analysis, *Discrete Comput. Geom.* **4** (1989) 291–309.

[84] PAPADIMITRIOU, C., An algorithm for shortest path motion in three dimensions, *Inform. Process. Lett.* **20** (1985) 259–263.

[85] PAPADIMITRIOU, C. and E. SILVERBERG, Optimal piecewise linear motion of an object among obstacles, *Algorithmica* **2** (1987) 523–539.

[86] PRILL, D., On approximations and incidence in cylindrical algebraic decompositions, preprint, Dept. of Mathematics, Univ. of Rochester, Rochester, NY, 1988.

[87] POLLACK, R., M. SHARIR and S. SIFRONY, Separating two simple polygons by a sequence of translations, *Discrete Comput. Geom.* **3** (1988) 123–136.

[88] RAMANATHAN, G. and V.S. ALAGAR, Algorithmic motion planning in robotics: coordinated motion of several discs amidst polygonal obstacles, in: *Proc. IEEE Symp. on Robotics and Automation* (1985) 514–522.

[89] REIF, J., Complexity of the mover's problem and generalizations, in: *Proc. 20th Ann. IEEE Symp. on Foundations of Computer Science* (1979) 421–427.

[90] REIF, J. and M. SHARIR, Motion planning in the presence of moving obstacles, in: *Proc. 26th Ann. IEEE Symp. on Foundations of Computer Science* (1985) 144–154.

[91] REIF, J.H. and J.A. STORER, Shortest paths in Euclidean space with polyhedral obstacles, Tech. Report CS-85-121, Computer Science Dept., Brandeis Univ., Waltham, MA, 1985.

[92] RENEGAR, J., A Faster, PSPACE algorithm for deciding the existential theory of the reals, Tech. Report 792, School of ORIE, Cornell Univ., Ithaca, NY, 1988.

[93] DEREZENDE, P.J., D.T. LEE and Y.F. WU, Rectilinear shortest paths with rectangular barriers, in: *Proc. ACM Symp. on Computational Geometry* (1985) 204–213.

[94] ROHNERT, H., A new algorithm for shortest paths avoiding convex polygonal obstacles, Tech. Report A86/02, Univ. des Saarlandes, Saarbrücken, 1986.

[95] SCHWARTZ, J.T. and M. SHARIR, On the piano movers' problem I: the case of a rigid polygonal body moving amidst polygonal barriers, *Comm. Pure and Appl. Math.* **36** (1983) 345–398.

[96] SCHWARTZ, J.T. and M. SHARIR, On the piano movers' problem II: general techniques for computing topological properties of real algebraic manifolds, *Advances in Appl. Math.* **4** (1983) 298–351.

[97] SCHWARTZ, J.T. and M. SHARIR, On the piano movers' problem III: coordinating the motion of several independent bodies: the special case of circular bodies moving amidst polygonal barriers, *Internat. J. Robotics Res.* **2**(3) (1983) 46–75.

[98] SCHWARTZ, J.T. and M. SHARIR, On the piano movers' problem V: the case of a rod moving in three-dimensional space amidst polyhedral obstacles, *Comm. Pure Appl. Math.* **37** (1984) 815–848.

[99] SCHWARTZ, J.T. and M. SHARIR, Efficient motion planning algorithms in environments of bounded local complexity, Tech. Report 164, Computer Science Dept., Courant Institute, New York, 1985.

[100] SCHWARTZ, J.T. and M. SHARIR, Finding effective "force targets" for two-dimensional multifinger frictional grips, in: *Proc. 25th Allerton Conf. on Communication, Control and Computing* (1987) 843–848.

[101] SCHWARTZ, J.T. and M. SHARIR, Motion planning and related geometric algorithms in robotics, in: *Proc. Internat. Congress of Mathematicians Vol. 2.* (1986) 1594–1611.

[102] SHARIR, M., Intersection and closest pair problem for a set of planar discs, *SIAM J. Comput.* **14** (1985) 448–468.

[103] SHARIR, M., Almost linear upper bounds for the length of general Davenport–Schinzel sequences, *Combinatorica* **7** (1987) 131–143.

[104] SHARIR, M., On shortest paths amidst convex polyhedra, *SIAM J. Comput.* **16** (1987) 561–572.

[105] SHARIR, M., Efficient algorithms for planning purely translational collision-free motion in two and three dimensions, in: *Proc. IEEE Symp. on Robotics and Automation* (1987) 1326–1331.

[106] SHARIR, M., Improved lower bounds on the length of Davenport–Schinzel sequences, *Combinatorica* **8** (1988) 117–124.

[107] SHARIR, M., Davenport–Schinzel sequences and their geometric applications, in: R.A.D. Earnshaw, ed., *Theoretical Foundations of Computer Graphics and CAD*, NATO ASI Series, Vol. F40 (Springer, Berlin, 1988) 253–278.

[108] SHARIR, M. and E. ARIEL-SHEFFI, On the piano movers' problem IV: various decomposable two-dimensional motion planning problems, *Comm. Pure Appl. Math.* **37** (1984) 479–493.

[109] SHARIR, M. and A. SCHORR, On shortest paths in polyhedral spaces, *SIAM J. Comput.* **15** (1986) 193–215.

[110] SHARIR, M. and S. SIFRONY, Coordinated motion planning for two independent robots, in: *Proc. 4th Ann. ACM Symp. on Computational Geometry* (1988) 319–328.
[111] SHOR, P., Geometric realizations of superlinear Davenport–Schinzel sequences, In preparation.
[112] SIFRONY, S. and M. SHARIR, An efficient motion planning algorithm for a rod moving in two-dimensional polygonal space, *Algorithmica* **2** (1987) 367–402.
[113] SUTNER, K. and W. MAASS, Motion planning among time dependent obstacles, *Acta Inform.* **26** (1988) 93–122.
[114] SZEMERÉDI, E., On a problem by Davenport and Schinzel. *Acta Arithmetica* **25** (1974) 213–224.
[115] TOUSSAINT, G., Movable separability of sets, in: G. Toussaint, ed., *Computational Geometry* (North-Holland, Amsterdam, 1985) 335–375.
[116] UDUPA, S., Collision detection and avoidance in computer-controlled manipulators, Ph.D. Dissertation, California Institute of Technology, Pasadena, CA, 1977.
[117] WELZL, E., Constructing the visibility graph for n line segments in $O(n^2)$ time, *Inform. Process. Lett.* **20** (1985) 167–172.
[118] WIERNIK, A. and M. SHARIR, Planar realization and nonlinear Davenport–Schinzel sequences by segments, *Discrete Comput. Geom.* **3** (1988) 15–47.
[119] WILFONG, G., Motion planning in the presence of movable obstacles, in: *Proc. 4th Ann. ACM Symp. on Computational Geometry* (1988) 279–288.
[120] YAP, C.K., Coordinating the motion of several disks, Tech. Report, Robotics Lab, Courant Institute, New York, 1984.
[121] YAP, C.K., How to move a chair through a door, Preprint, Robotics Lab, Courant Institute, New York, 1984.
[122] YAP, C.K., Algorithmic motion planning, in: J.T. Schwartz and C.K. Yap, eds., *Advances in Robotics, Vol. 1* (Lawrence Erlbaum, Hillsdale, NJ, 1987) 95–143.
[123] YAP, C.K., An $O(n \log n)$ algorithm for the Voronoi diagram of a set of simple curve segments, *Discrete Comput. Geom.* **2** (1987) 365–393.

CHAPTER 9

Average-Case Analysis of Algorithms and Data Structures

Jeffrey Scott VITTER

Department of Computer Science, Brown University, Providence, RI 02912, USA

Philippe FLAJOLET

INRIA, Domaine de Voluceau-Rocquencourt, 78153 Le Chesnay, France

Contents

HANDBOOK OF THEORETICAL COMPUTER SCIENCE
Edited by J. van Leeuwen

0. Introduction

Analyzing an algorithm means, in its broadest sense, characterizing the amount of computational resources that an execution of the algorithm will require when applied to data of a certain type. Many algorithms in classical mathematics, primarily in number theory and analysis, were analyzed by eighteenth and nineteenth century mathematicians. For instance, Lamé in 1845 showed that Euclid's GCD algorithm requires at most $\approx \log_\varphi n$ division steps (where φ is the "golden ratio" $(1+\sqrt{5})/2$) when applied to numbers bounded by n. Similarly, the well-known quadratic convergence of Newton's method is a way of describing its complexity/accuracy tradeoff.

This chapter presents analytic methods for average-case analysis of algorithms, with special emphasis on the main algorithms and data structures used for processing nonnumerical data. We characterize algorithmic solutions to a number of essential problems, such as sorting, searching, pattern matching, register allocation, tree compaction, retrieval of multidimensional data, and efficient access to large files stored on secondary memory.

The first step required to analyze an algorithm $\mathscr{A}$ is to define an *input data model* and a *complexity measure*:

(1) Assume that the input to $\mathscr{A}$ is data of a certain type. Each commonly used data type carries a natural notion of size: the size of an array is the number of its elements; the size of a file is the number of its records; the size of a character string is its length; and so on. An input model is specified by the subset $\boldsymbol{I}_n$ of inputs of size n and by a probability distribution over $\boldsymbol{I}_n$, for each n. For example, a classical input model for comparison-based sorting algorithms is to assume that the n inputs are real numbers independently and uniformly distributed in the unit interval $[0, 1]$. An equivalent model is to assume that the n inputs form a random permutation of $\{1, 2, \ldots, n\}$.

(2) The main complexity measures for algorithms executed on sequential machines are *time utilization* and *space utilization*. These may be either "raw" measures (such as the time in nanoseconds on a particular machine or the number of bits necessary for storing temporary variables) or "abstract" measures (such as the number of comparison operations or the number of disk pages accessed).

Let us consider an algorithm $\mathscr{A}$ with complexity measure μ. The worst-case and best-case complexities of algorithm $\mathscr{A}$ over $\boldsymbol{I}_n$ are defined in an obvious way. Determining the worst-case complexity requires constructing extremal configurations that force μ_n, the restriction of μ to $\boldsymbol{I}_n$, to be as large as possible.

The *average-case* complexity is defined in terms of the probabilistic input model:

$$\overline{\mu_n}[\mathscr{A}] = \mathbf{E}\{\mu_n[\mathscr{A}]\} = \sum_k k \Pr\{\mu_n[\mathscr{A}] = k\},$$

where $\mathbf{E}\{\cdot\}$ denotes expected value and $\Pr\{\cdot\}$ denotes probability with respect to the probability distribution over $\boldsymbol{I}_n$. Frequently, $\boldsymbol{I}_n$ is a finite set and the probabilistic model assumes a uniform probability distribution over $\boldsymbol{I}_n$. In that case, $\overline{\mu_n}[\mathscr{A}]$ takes the form

$$\overline{\mu_n}[\mathscr{A}] = \frac{1}{I_n} \sum_k k J_{nk},$$

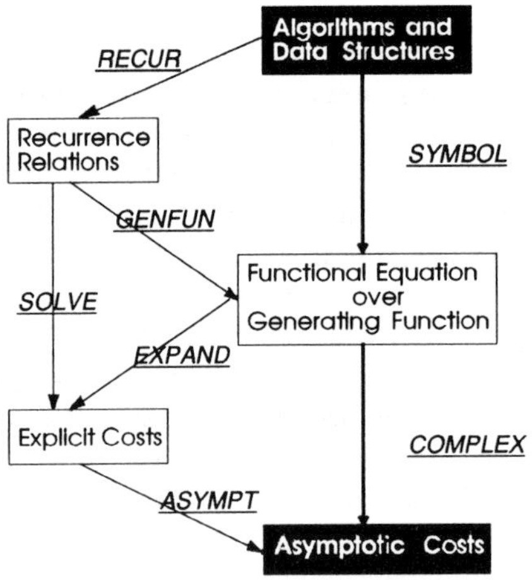

Fig. 1. Methods used in the average-case analysis of algorithms.

where $I_n = |\boldsymbol{I}_n|$ and J_{nk} is the number of inputs of size n with complexity k for algorithm $\mathscr{A}$. Average-case analysis then reduces to *combinatorial enumeration.*

The next step in the analysis is to express the complexity of the algorithm in terms of standard functions like $n^\alpha(\log n)^\beta(\log\log n)^\gamma$, where α, β, and γ are constants, so that the analytic results can be easily interpreted. This involves getting *asymptotic estimates.*

The following steps give the route followed by many of the average-case analyses that appear in the literature (see Fig. 1):

(1) *RECUR*: To determine the probabilities or the expectations in exact form, start by setting up recurrences that relate the behavior of algorithm $\mathscr{A}$ on inputs of size n to its behavior on smaller (and similar) inputs.

(2) *SOLVE*: Solve previous recurrences explicitly using classical algebra.

(3) *ASYMPT*: Use standard real asymptotics to estimate those explicit expressions. An important way to solve recurrences is via the use of *generating functions*:

(4) *GENFUN*: Translate the recurrences into equations involving generating functions. The coefficient of the nth term of each generating function represents a particular probability or expectation. In general we obtain a set of *functional equations.*

(5) *EXPAND*: Solve those functional equations using classical tools from algebra and analysis, then expand the solutions to get the coefficients in explicit form.

The above methods can often be bypassed by the following more powerful methods, which we emphasize in this chapter:

(6) *SYMBOL*: Bypass the use of recurrences and translate the set-theoretic definitions of the data structures or underlying combinatorial structures directly into functional equations involving generating functions.

(7) *COMPLEX*: Use complex analysis to translate the information available from

the functional equations directly into asymptotic expressions of the coefficients of the generating functions.

The symbolic method (*SYMBOL*) is often direct and has the advantage of characterizing the *special functions* that arise from the analysis of a natural class of related algorithms. The *COMPLEX* method provides powerful tools for direct asymptotics from generating functions. It has the intrinsic advantage in many cases of providing asymptotic estimates of the coefficients of functions known *only implicitly* from their functional equations.

In Sections 1 and 2 we develop general techniques for the mathematical analysis of algorithms, with emphasis on the *SYMBOL* and *COMPLEX* methods. Section 1 is devoted to exact analysis and combinatorial enumeration. We present the primary methods used to obtain counting results for the analysis of algorithms, with emphasis on symbolic methods (*SYMBOL*). The main mathematical tool we use is the generating function associated with the particular class of structures. A rich set of combinatorial constructions translates directly into functional relations involving the generating functions. In Section 2 we discuss asymptotic analysis. We briefly review methods for elementary real analysis and then concentrate on complex analysis techniques (*COMPLEX*). There we use analytic properties of generating functions to recover information about their coefficients. The methods are often applicable to functions known only indirectly via functional equations, a situation that presents itself naturally when counting recursively defined structures.

In Sections 3–6, we apply general methods for analysis of algorithms, especially those developed in Sections 1 and 2, to the analysis of a large variety of algorithms and data structures. In Section 3, we describe several important sorting algorithms and apply statistics of inversion tables to the analysis of the iteration-based sorting algorithms. In Section 4, we extend our approach of Section 1 to consider valuations on combinatorial structures, which we use to analyze trees and structures with a tree-like recursive decomposition; this includes plane trees, binary and multidimensional search trees, digital search trees, quicksort, radix-exchange sort, and algorithms for register allocation, pattern matching, and tree compaction. In Section 5, we present a unified approach to hashing, address calculation techniques, and occupancy problems. Section 6 is devoted to performance measures that span a period of time, such as the expected amortized time and expected maximum data structure space used by an algorithm.

General references

Background sources on combinatorial enumerations and symbolic methods include [52, 17]. General coverage of complex analysis appears in [105, 58, 59] and applications to asymptotics are discussed in [6, 87, 7, 19]. Mellin transforms are covered in [23], and limit probability distributions are studied in [29, 96, 9].

For additional coverage of average-case analysis of algorithms and data structures, the reder is referred to Knuth's seminal multivolume work *The Art of Computer Programming* [70, 71, 74] and to [31, 54, 100, 93, 34]. Descriptions of most of the algorithms analyzed in this chapter can be found in [70, 71, 74, 49, 100].

1. Combinatorial enumerations

Our main objective in this section is to introduce useful combinatorial constructions that translate directly into generating functions. Such constructions are called *admissible*. In Section 1.2 we examine admissible constructions for ordinary generating functions, and in Section 1.3 we consider admissible constructions for exponential generating functions, which are related to the enumeration of labeled structures.

1.1. Overview

The most elementary structures may be enumerated using sum/product rules[1].

1.0. THEOREM (Sum-product rule). *Let $\mathcal{A}$, $\mathcal{B}$, $\mathcal{C}$ be sets with cardinalities a, b, c. Then*

$$\mathcal{C}=\mathcal{A}\cup\mathcal{B} \text{ with } \mathcal{A}\cap\mathcal{B}=\emptyset \Rightarrow c=a+b;$$

$$\mathcal{C}=\mathcal{A}\times\mathcal{B} \Rightarrow c=a\cdot b.$$

Thus, the number of binary strings of length n is 2^n, and the number of permutations of $\{1, 2, \ldots, n\}$ is $n!$.

In the next order of difficulty, explicit forms are replaced by recurrences when structures are defined in terms of themselves. For example, let F_n be the number of coverings of the interval $[1, n]$ by disjoint segments of length 1 and 2. By considering the two possibilities for the last segment used, we get the recurrence

$$F_n=F_{n-1}+F_{n-2} \quad \text{for } n\geqslant 2, \tag{1.1a}$$

with initial conditions $F_0=0$, $F_1=1$. Thus, from the classical theory of linear recurrences, we find the Fibonacci numbers expressed in terms of the golden ratio φ:

$$F_n=\frac{1}{\sqrt{5}}(\varphi^n-\hat{\varphi}^n) \quad \text{with } \varphi, \hat{\varphi}=\frac{1\pm\sqrt{5}}{2}. \tag{1.1b}$$

This example illustrates recurrence methods (*RECUR*) in (1.1a) and derivation of explicit solutions (*SOLVE*) in (1.1b).

Another example, which we shall discuss in more detail in Section 4.1, is the number B_n of plane binary trees with n internal nodes [70]. By considering all possibilities for left and right subtrees, we get the recurrence

$$B_n=\sum_{k=0}^{n-1} B_k B_{n-k-1} \quad \text{for } n\geqslant 1, \tag{1.2a}$$

with the initial condition $B_0=1$. To solve (1.2a), we introduce a *generating function* (GF): Let $B(z)=\sum_{n\geqslant 0} B_n z^n$. From (1.2a) we get

$$B(z)=1+zB^2(z), \tag{1.2b}$$

[1] We also use the sum notation $\mathcal{C}=\mathcal{A}+\mathcal{B}$ to represent the union of $\mathcal{A}$ and $\mathcal{B}$ when $\mathcal{A}\cap\mathcal{B}=\emptyset$.

and solving the quadratic equation for $B(z)$, we get

$$B(z)=\frac{1-\sqrt{1-4z}}{2z}. \tag{1.2c}$$

Finally, the Taylor expansion of $(1+x)^{1/2}$ gives us

$$B_n=\frac{1}{n+1}\binom{2n}{n}=\frac{(2n)!}{n!(n+1)!}. \tag{1.2d}$$

In this case, we started with recurrences (*RECUR*) in (1.2a) and introduced generating functions (*GENFUN*), leading to (1.2b); solving and expanding (*EXPAND*) gave the explicit solutions (1.2c) and (1.2d). (This example dates back to Euler; the B_n are called *Catalan numbers*.)

The symbolic method (*SYMBOL*) that we are going to present can be applied to this last example as follows: The class $\mathscr{B}$ of binary trees is defined recursively by the equation

$$\mathscr{B}=\{\blacksquare\}\cup(\{\bigcirc\}\times\mathscr{B}\times\mathscr{B}), \tag{1.3a}$$

where ■ and ○ represent external nodes and internal nodes respectively. A standard lemma asserts that disjoint unions and Cartesian products of structures correspond respectively to sums and products of corresponding generating functions. Therefore, specification (1.3a) translates term by term directly into the generating function equation

$$B(z)=1+z\cdot B(z)\cdot B(z), \tag{1.3b}$$

which agrees with (1.2b).

DEFINITION. A *class of combinatorial structures* $\mathscr{C}$ is a finite or countable set together with an integer valued function $|\cdot|_{\mathscr{C}}$, called the *size* function, such that for each $n\geqslant 0$ the number C_n of structures in $\mathscr{C}$ of size n is finite. The *counting sequence* for class $\mathscr{C}$ is the integer sequence $\{C_n\}_{n\geqslant 0}$. The *ordinary generating function* (OGF) $C(z)$ and the *exponential generating function* (EGF) $\hat{C}(z)$ of a class $\mathscr{C}$ are defined respectively by

$$C(z)=\sum_{n\geqslant 0} C_n z^n \quad\text{and}\quad \hat{C}(z)=\sum_{n\geqslant 0} C_n\frac{z^n}{n!}. \tag{1.4}$$

The coefficient of z^n in the expansion of a function $f(z)$ (or simply, the nth coefficient of $f(z)$) is written $[z^n]f(z)$; we have $C_n=[z^n]C(z)=n![z^n]\hat{C}(z)$.

The generating functions $C(z)$ and $\hat{C}(z)$ can also be expressed as

$$C(z)=\sum_{\gamma\in\mathscr{C}} z^{|\gamma|} \quad\text{and}\quad \hat{C}(z)=\sum_{\gamma\in\mathscr{C}}\frac{z^{|\gamma|}}{|\gamma|!}, \tag{1.5}$$

which can be checked by counting the number of occurrences of z^n in the sums.

We shall adopt the notational convention that a class (say, $\mathscr{C}$), its counting sequence (say, C_n or c_n), its associated ordinary generating function (say, $C(z)$ or $c(z)$), and its associated exponential generating function (say, $\hat{C}(z)$ or $\hat{c}(z)$) are named by the same group of letters.

The basic notion for the symbolic method is that of an *admissible construction* in which the counting sequence of the construction depends only upon the counting sequences of its components (see [52, 31, 53]); such a construction thus "translates" over generating functions. It induces an operator of a more or less simple form over formal power series. For instance, let $\mathcal{U}$ and $\mathcal{V}$ be two classes of structures, and let

$$\mathcal{W} = \mathcal{U} \times \mathcal{V} \tag{1.6a}$$

be their Cartesian product. If the size of an ordered pair $w = (u, v) \in \mathcal{W}$ is defined as $|w| = |u| + |v|$, then by counting possibilities, we get

$$W_n = \sum_{k \geqslant 0} U_k V_{n-k}, \tag{1.6b}$$

so that (1.6a) has the corresponding (ordinary) generating function equation

$$W(z) = U(z)V(z). \tag{1.6c}$$

Such a combinatorial (set-theoretic) construction that translates in the manner of (1.6a)–(1.6c) is called *admissible.*

1.2. *Ordinary generating functions*

In this section we present a catalog of admissible constructions for ordinary generating functions (OGFs). We assume that the size of an element of a disjoint union $\mathcal{W} = \mathcal{U} \cup \mathcal{V}$ is inherited from its size in its original domain; the size of a composite object (product, sequence, subset, etc.) is the sum of the sizes of its components.

1.1. THEOREM (Fundamental sum/product theorem). *The disjoint union and Cartesian product constructions are admissible for OGFs:*

$$\mathcal{W} = \mathcal{U} \cup \mathcal{V} \text{ with } \mathcal{U} \cap \mathcal{V} = \emptyset \;\Rightarrow\; W(z) = U(z) + V(z);$$
$$\mathcal{W} = \mathcal{U} \times \mathcal{V} \;\Rightarrow\; W(z) = U(z)V(z).$$

PROOF. Use recurrences $W_n = U_n + V_n$ and $W_n = \sum_{0 \leqslant k \leqslant n} U_k V_{n-k}$. Alternatively,use (1.5) for OGFs, which yields for Cartesian products

$$\sum_{w \in \mathcal{W}} z^{|w|} = \sum_{(u,v) \in \mathcal{U} \times \mathcal{V}} z^{|u|+|v|} = \sum_{u \in \mathcal{U}} z^{|u|} \cdot \sum_{v \in \mathcal{V}} z^{|v|}. \qquad \square$$

Let $\mathcal{U}$ be a class of structures that have positive size. Class $\mathcal{W}$ is called the *sequence class* of class $\mathcal{U}$, denoted $\mathcal{W} = \mathcal{U}^*$, if $\mathcal{W}$ is composed of all sequences $(u_1, u_2, \ldots, u_k)$ with $u_j \in \mathcal{U}$. Class $\mathcal{W}$ is the (finite) *powerset* of class $\mathcal{U}$, denoted $\mathcal{W} = 2^{\mathcal{U}}$, if $\mathcal{W}$ consists of all finite subsets $\{u_1, u_2, \ldots, u_k\}$ of $\mathcal{U}$ (the u_j are distinct) for $k \geqslant 0$.

1.2. THEOREM. *The sequence and powerset constructs are admissible for OGFs:*

$$\mathcal{W}=\mathcal{U}^* \Rightarrow W(z)=\frac{1}{1-U(z)};$$

$$\mathcal{W}=2^{\mathcal{U}} \Rightarrow W(z)=e^{\Phi(U)(z)}, \quad \textit{where } \Phi(f)=\frac{f(z)}{1}-\frac{f(z^2)}{2}+\frac{f(z^3)}{3}-\cdots$$

PROOF. Let ε denote the empty sequence. Then, for the sequence class of $\mathcal{U}$, we have

$$\begin{aligned}&\mathcal{W}=\mathcal{U}^*\equiv\{\varepsilon\}+\mathcal{U}+(\mathcal{U}\times\mathcal{U})+(\mathcal{U}\times\mathcal{U}\times\mathcal{U})+\cdots;\\&W(z)=1+U(z)+U(z)^2+U(z)^3+\cdots=(1-U(z))^{-1}.\end{aligned}\tag{1.7}$$

The powerset class $\mathcal{W}=2^{\mathcal{U}}$ is equivalent to an infinite product:

$$\mathcal{W}=2^{\mathcal{U}}=\prod_{u\in\mathcal{U}}(\{\varepsilon\}+\{u\}); \qquad W(z)=\prod_{u\in\mathcal{U}}(1+z^{|u|})=\prod_n(1+z^n)^{U_n}.\tag{1.8}$$

Computing logarithms and expanding, we get

$$\log W(z)=\sum_{n\geq 1}U_n\log(1+z^n)=\sum_{n\geq 1}U_nz^n-\frac{1}{2}\sum_{n\geq 1}U_nz^{2n}+\cdots \qquad \square$$

Other constructions can be shown to be admissible:

(1) Diagonals and subsets with repetitions. The *diagonal* $\mathcal{W}=\{(u,u)\mid u\in\mathcal{U}\}$ of $\mathcal{U}\times\mathcal{U}$, written $\mathcal{W}=\Delta(\mathcal{U}\times\mathcal{U})$, satisfies $W(z)=U(z^2)$. The class of *multisets* (or *subsets with repetitions*) of class $\mathcal{U}$ is denoted $\mathcal{W}=\mathbf{R}\{\mathcal{U}\}$. It is isomorphic to $\Pi_{u\in\mathcal{U}}\{u\}^*$, so that its OGF satisfies

$$W(z)=e^{\Psi(U)(z)}, \quad \text{where } \Psi(f)=\frac{f(z)}{1}+\frac{f(z^2)}{2}+\frac{f(z^3)}{3}+\cdots\tag{1.9}$$

(2) Marking and composition. If $\mathcal{U}$ is formed with "atomic" elements (nodes, letters, etc.) that determine its size, then we define the *marking* of $\mathcal{U}$, denoted $\mathcal{W}=\mu\{\mathcal{U}\}$, to consist of elements of $\mathcal{U}$ with one individual atom marked. Since $W_n=nU_n$, it follows that $W(z)=z\mathrm{d}U(z)/\mathrm{d}z$. Similarly, the *composition* of $\mathcal{U}$ and $\mathcal{V}$, denoted $\mathcal{W}=\mathcal{U}[\mathcal{V}]$, is defined as the class of all structures resulting from substitutions of atoms of $\mathcal{U}$ by elements of $\mathcal{V}$, and we have $W(z)=U(V(z))$.

EXAMPLES. (1) Combinations. Let m be a fixed integer and $\mathcal{J}^{(m)}=\{1,2,\ldots,m\}$, each element of $\mathcal{J}^{(m)}$ having size 1. The generating function of $\mathcal{J}^{(m)}$ is $J^{(m)}(z)=mz$. The class $\mathcal{C}^{(m)}=2^{\mathcal{J}^{(m)}}$ is the set of all *combinations of* $\mathcal{J}^{(m)}$. By Theorem 1.2, the generating function of the number $C_n^{(m)}$ of n-combinations of a set with m elements is

$$C^{(m)}(z)=e^{\Phi(J^{(m)})(z)}=\exp\left(\frac{mz}{1}-\frac{mz^2}{2}+\frac{mz^3}{3}-\cdots\right)=\exp(m\log(1+z))=(1+z)^m,$$

and by extracting coefficients we find as expected

$$C_n^{(m)}=\binom{m}{n}=\frac{m!}{n!(m-n)!}.$$

Similarly, for $\mathscr{R}^{(m)} = \mathbf{R}\{\mathscr{J}^{(m)}\}$, the class of *combinations with repetitions*, we have from (1.9)

$$R^{(m)}(z) = (1-z)^{-m} \Rightarrow R_n^{(m)} = \binom{m+n-1}{m-1}.$$

(2) Compositions and partitions. Let $\mathscr{N} = \{1, 2, 3, \ldots\}$, each $i \in \mathscr{N}$ having size i. The sequence class $\mathscr{C} = \mathscr{N}^*$ is called the set of *integer compositions*. Since $N(z) = z/(1-z)$ and $C(z) = (1 - N(z))^{-1}$, we have

$$C(z) = \frac{1-z}{1-2z} \Rightarrow C_n = 2^{n-1} \quad \text{for } n \geqslant 1.$$

The class $\mathscr{P} = \mathbf{R}\{\mathscr{N}\}$ is the set of *integer partitions*, and we have

$$\mathscr{P} = \prod_{\alpha \in \mathscr{N}} \{\alpha\}^* \Rightarrow P(z) = \prod_{n \geqslant 1} \frac{1}{1-z^n}. \tag{1.10}$$

(3) Formal languages. Combinatorial processes can often be encoded naturally as strings over some finite alphabet $\mathscr{A}$. Regular languages are defined by regular expressions or equivalently by deterministic or nondeterministic finite automata. This is illustrated by the following two theorems, based upon the work of Chomsky and Schützenberger [118]. Further applications appear in [8].

1.3A. THEOREM (Regular languages and rational functions). *If $\mathscr{L}$ is a regular language, then its OGF is a rational function $L(z) = P(z)/Q(z)$, where $P(z)$ and $Q(z)$ are polynomials. The counting sequence L_n satisfies a linear recurrence with constant coefficients, and we have, when $n \geqslant n_0$, $L_n = \sum_j \pi_j(n)\omega_j^n$, for a finite set of constants ω_j and polynomials $\pi_j(z)$.*

PROOF (*Sketch*). Let D be a deterministic automaton that recognizes $\mathscr{L}$, and let $\mathscr{S}_j$ be the set of words accepted by D when D is started in state j. The $\mathscr{S}_j$ satisfy a set of linear equations (involving unions and concatenation with letters) constructed from the transition table of the automaton. For generating functions, this translates into a set of linear equations with polynomial coefficients that can be solved by Cramer's rule. □

1.3B. THEOREM (Context-free languages and algebraic functions). *If $\mathscr{L}$ is an unambiguous context-free language, then its OGF is an algebraic function. The counting sequence L_n satisfies a linear recurrence with polynomial coefficients: for a family $q_j(z)$ of polynomials and $n \geqslant n_0$, we have $L_n = \sum_{1 \leqslant j \leqslant m} q_j(n) L_{n-j}$.*

PROOF (*Sketch*). Since the language is unambiguous, its counting problem is equivalent to counting derivation trees. A production in the grammer $S \rightarrow aTbU + bUUa + abba$ translates into $S(z) = z^2 T(z)U(z) + z^2 U^2(z) + z^4$, where $S(z)$ is the generating function associated with nonterminal S. We obtain a set of polynomial equations that reduces to a single equation $P(z, L(z)) = 0$ through elimination. To obtain the recurrence, we use Comtet's theorem [16] (see also [33] for corresponding asymptotic estimates). □

EXAMPLES (*continued*). (4) Trees. We shall study trees in great detail in Section 4.1. All trees here are *rooted.* In *plane trees*, subtrees under a node are ordered; in *nonplane trees*, they are unordered. If $\mathcal{G}$ is the class of general plane trees with all node degrees allowed, then $\mathcal{G}$ satisfies an equation $\mathcal{G}=\{\bigcirc\}\times\mathcal{G}^*$, signifying that a tree is composed of a root followed by a sequence of subtrees. Thus, we have

$$G(z)=\frac{z}{1-G(z)} \Rightarrow G(z)=\frac{1-\sqrt{1-4z}}{2} \text{ and } G_n=\frac{1}{n}\binom{2n-2}{n-1}.$$

If $\mathcal{H}$ is the class of general nonplane trees, then $\mathcal{H}=\{\bigcirc\}\times\mathbf{R}\{\mathcal{H}\}$, so that $H(z)$ satisfies the functional equation

$$H(z)=z\,\mathrm{e}^{H(z)+H(z^2)/2+H(z^3)/3+\cdots}. \tag{1.11}$$

There are no closed form expressions for $H_n=[z^n]H(z)$. However, complex analysis methods make it possible to determine H_n asymptotically [92].

1.3. *Exponential generating functions*

Exponential generating functions are essentially used for counting *well-labeled structures.* Such structures are composed of "atoms" (the size of a structure being the number of its atoms), and each atom is labeled by a distinct integer. For instance, a labeled graph of size n is just a graph over the set of nodes $\{1,2,\ldots,n\}$. A permutation (respectively circular permutation) can be viewed as a linear (respectively cyclic) directed graph whose nodes are labeled by distinct integers.

The basic operation over labeled structures is the *partitional product* [45; 52; 31, Chapter I; 53]. The partitional product of $\mathcal{U}$ and $\mathcal{V}$ consists of forming ordered pairs (u, v) from $\mathcal{U}\times\mathcal{V}$ and relabeling them in all possible ways that preserve the order of the labels in u and v. More precisely, let $w\in\mathcal{W}$ be a labeled structure of size q. A 1-1 function θ from $\{1,2,\ldots,q\}$ to $\{1,2,\ldots,r\}$, where $r\geqslant q$, defines a relabeling, denoted $w'=\theta(w)$, where label j in w is replaced by $\theta(j)$. Let u and w be two labeled structures of respective sizes m and n. The partitional product of u and v is denoted by $u*v$, and it consists of the set of all possible relabelings (u',v') of (u,v) so that $(u',v')=(\theta_1(u),\theta_2(v))$, where $\theta_1:\{1,2,\ldots,m\}\to\{1,2,\ldots,m+n\}$, $\theta_2:\{1,2,\ldots,n\}\to\{1,2,\ldots,m+n\}$ satisfy the following:

(1) θ_1 and θ_2 are monotone increasing functions. (This preserves the order structure of u and v.)

(2) The ranges of θ_1 and θ_2 are disjoint and cover the set $\{1,2,\ldots,m+n\}$.

The partitional product of two classes $\mathcal{U}$ and $\mathcal{V}$ is denoted $\mathcal{W}=\mathcal{U}*\mathcal{V}$ and is the union of all $u*v$, for $u\in\mathcal{U}$ and $v\in\mathcal{V}$.

1.4. THEOREM (Sum/product theorem for labeled structures). *The disjoint union and partitional product over labeled structures are admissible for EGFs:*

$$\mathcal{W}=\mathcal{U}\cup\mathcal{V} \text{ with } \mathcal{U}\cap\mathcal{V}=\emptyset \Rightarrow \hat{W}(z)=\hat{U}(z)+\hat{V}(z);$$
$$\mathcal{W}=\mathcal{U}*\mathcal{V} \Rightarrow \hat{W}(z)=\hat{U}(z)\hat{V}(z).$$

PROOF. Obvious for unions. For products, observe that

$$W_q = \sum_{0 \leq m \leq q} \binom{q}{m} U_m V_{q-m}, \tag{1.12}$$

since the binomial coefficient counts the number of partitions of $\{1, 2, \ldots, q\}$ into two sets of cardinalities m and $q-m$. Dividing by $q!$ we get

$$\frac{W_q}{q!} = \sum_{0 \leq m \leq q} \frac{U_m}{m!} \frac{V_{q-m}}{(q-m)!}. \qquad \square$$

The *partitional complex* of $\mathcal{U}$ is denoted $\mathcal{U}^{\langle * \rangle}$. It is analogous to the sequence class construction and is defined by

$$\mathcal{U}^{\langle * \rangle} = \{\varepsilon\} + \mathcal{U} + (\mathcal{U} * \mathcal{U}) + (\mathcal{U} * \mathcal{U} * \mathcal{U}) + \cdots,$$

and its EGF is $(1 - \hat{U}(z))^{-1}$. The kth partitional power of $\mathcal{U}$ is denoted $\mathcal{U}^{\langle k \rangle}$. The *abelian partional power*, denoted $\mathcal{U}^{[k]}$, is the collection of all sets $\{v_1, v_2, \ldots, v_k\}$ such that $(v_1, v_2, \ldots, v_k) \in \mathcal{U}^{\langle k \rangle}$. In other words, the order of components is not taken into account. We can write symbolically $\mathcal{U}^{[k]} = \mathcal{U}^{\langle k \rangle}/k!$ so that the EGF of $\mathcal{U}^{[k]}$ is $\hat{U}^{\langle k \rangle}(z)/k!$ The *abelian partitional complex* of $\mathcal{U}$ is defined analogously to the powerset construction:

$$\mathcal{U}^{[*]} = \{\varepsilon\} + \mathcal{U} + \mathcal{U}^{[2]} + \mathcal{U}^{[3]} + \cdots$$

1.5. THEOREM. *The partitional complex and abelian partitional complex are admissible for EGFs*:

$$\begin{aligned} &\mathcal{W} = \mathcal{U}^{\langle * \rangle} \Rightarrow \hat{W}(z) = \frac{1}{1 - \hat{U}(z)}; \\ &\mathcal{W} = \mathcal{U}^{[*]} \Rightarrow \hat{W}(z) = \mathrm{e}^{\hat{U}(z)}. \end{aligned} \tag{1.13}$$

EXAMPLES. (1) Permutations and cycles. Let $\mathcal{P}$ be the class of all *permutations*, and let $\mathcal{C}$ be the class of *circular permutations* (or *cycles*). By Theorem 1.5, we have $\hat{P}(z) = (1-z)^{-1}$ and $P_n = n!$. Since any permutation decomposes into an ordered set of cycles, we have $\mathcal{P} = \mathcal{C}^{[*]}$, so that $\hat{C}(z) = \log(1/(1-z))$ and $C_n = (n-1)!$. This construction also shows that the EGF for permutations having k cycles is $\log^k(1/(1-z))$, whose nth coefficient is $s_{n,k}/n!$ where $s_{n,k}$ is a Stirling number of the first kind.

Let $\mathcal{Q}$ be the class of permutations without cycles of size 1 (that is, without fixed points). Let $\mathcal{D}$ be the class of cycles of size at least 2. We have $\mathcal{D} \cup \{(1)\} = \mathcal{C}$, and hence $\hat{D}(z) + z = \hat{C}(z)$, $\hat{D}(z) = \log(1-z)^{-1} - z$. Thus, we have

$$\hat{Q}(z) = \mathrm{e}^{\hat{D}(z)} = \frac{\mathrm{e}^{-z}}{1-z}. \tag{1.14}$$

Similarly, the generating function for the class $\mathcal{I}$ of *involutions* (permutations with cycles of lengths 1 and 2 only) is

$$\hat{I}(z) = \mathrm{e}^{z + z^2/2}. \tag{1.15}$$

(2) Labeled graphs. Let $\mathcal{G}$ be the class of all *labeled graphs*, and let $\mathcal{K}$ be the class of

connected labeled graphs. Then $G_n = 2^{n(n-1)/2}$, and $\hat{K}(z) = \log \hat{G}(z)$, from which we can prove that $K_n/G_n \to 1$, as $n \to \infty$.

(3) Occupancies and set partitions. We define the *urn* of size n, for $n \geqslant 1$, to be the structure formed from the unordered collection of the integers $\{1, 2, \ldots, n\}$; the urn of size 0 is defined to be the empty set. Let $\mathscr{U}$ denote the class of all urns; we have $\hat{U}(z) = e^z$. The class $\mathscr{U}^{\langle k \rangle}$ represents all possible ways of throwing distinguishable balls into k distinguishable urns, and its EGF is e^{kz}, so that as anticipated we have $U_n^{\langle k \rangle} = k^n$. Similarly, the generating function for the number of ways of throwing n balls into k urns, no urn being empty, is $(e^z - 1)^k$, and thus the number of ways is $n! [z^n](e^z - 1)^k$, which is equal to $k! S_{n,k}$, where $S_{n,k}$ is a Stirling number of the second kind.

If $\mathscr{S} = \mathscr{V}^{[*]}$, where $\mathscr{V}$ is the class of nonempty urns, then an element of $\mathscr{S}$ of size n corresponds to a *partition* of the set $\{1, 2, \ldots, n\}$ into equivalence classes. The number of such partitions is a Bell number

$$\beta_n = n! [z^n] \exp(e^z - 1). \tag{1.16}$$

In the same vein, the EGF of *surjections* $\mathscr{S} = \mathscr{V}^{\langle * \rangle}$ (surjective mappings from $\{1, 2, \ldots, n\}$ onto an initial segment $\{1, 2, \ldots, m\}$ of the integers, for some $1 \leqslant m \leqslant n$) is

$$\hat{S}(z) = \frac{1}{1 - (e^z - 1)} = \frac{1}{2 - e^z}. \tag{1.17}$$

For labeled structures, we can also define marking and composition constructions that translate into EGFs. Greene [53] has defined a useful *boxing operator*: $\mathscr{C} = \mathscr{A}^{\square} * \mathscr{B}$ denotes the subset of $\mathscr{A} * \mathscr{B}$ obtained by retaining only pairs $(u, v) \in \mathscr{A} * \mathscr{B}$ such that label 1 is in u. This construction translates into the EGF

$$\hat{C}(z) = \int_0^z \hat{A}'(t)\ B(t)\, dt.$$

1.4. From generating functions to counting

In the previous section we saw how generating function equations can be written directly from structural definitions of combinatorial objects. We discuss here how to go from the functional equations to exact counting results, and then indicate some extensions of the symbolic method to multivariate generating functions.

Direct expansions from generating functions

When a GF is given explicitly as the product or composition of known GFs, we often get an *explicit form* for the coefficients of the GF by using classical rules for Taylor expansions and sums. Examples related to previous calculations are the Catalan numbers (1.2), derangement numbers (1.14), and Bell numbers (1.16):

$$[z^n] \frac{1}{\sqrt{1-4z}} = \binom{2n}{n}; \qquad [z^n] \frac{e^{-z}}{1-z} = \sum_{0 \leqslant k \leqslant n} \frac{(-1)^k}{k!};$$

$$[z^n] \exp(e^z - 1) = e^{-1} \sum_{k \geqslant 0} \frac{k^n}{k!}.$$

Another method for obtaining coefficients of implicitly defined GFs is the method of indeterminate coefficients. If the coefficients of $f(z)$ are sought, we translate over coefficients the functional relation for $f(z)$. An important subcase is that of a first-order linear recurrence $f_n = a_n + b_n f_{n-1}$, whose solution can be found by iteration or summation factors:

$$f_n = a_n + b_n a_{n-1} + b_n b_{n-1} a_{n-2} + b_n b_{n-1} b_{n-2} a_{n-3} + \cdots \tag{1.18}$$

Solution methods for functional equations

Algebraic equations over GFs may be solved explicitly if of low degree, and the solutions can then be expanded (see the Catalan numbers (1.2d) in Section 1.1). For equations of higher degrees and some transcendental equations, the *Lagrange–Bürmann* inversion formula is useful.

1.6. THEOREM (Lagrange–Bürmann inversion formula). *Let $f(z)$ be defined implicitly by the equation $f(z) = z\varphi(f(z))$, where $\varphi(u)$ is a series with $\varphi(0) \neq 0$. Then the coefficients of $f(z)$, its powers $f(z)^k$, and an arbitrary composition $g(f(z))$ are related to the coefficients of the powers of $\varphi(u)$ as follows:*

$$[z^n]f(z) = \frac{1}{n}[u^{n-1}]\varphi(u)^n; \tag{1.19a}$$

$$[z^n]f(z)^k = \frac{k}{n}[u^{n-k}]\varphi(u)^n; \tag{1.19b}$$

$$[z^n]g(f(z)) = \frac{1}{n}[u^{n-1}]\varphi(u)^n g'(u). \tag{1.19c}$$

EXAMPLES. (1) Abel identities. By (1.19a), $f(z) = \sum_{n \geq 1} n^{n-1} z^n/n!$ is the expansion of $f(z) = ze^{f(z)}$. By taking coefficients of $e^{\alpha f(z)} e^{\beta f(z)} = e^{(\alpha+\beta) f(z)}$ we get the *Abel identity*

$$(\alpha+\beta)(n+\alpha+\beta)^{n-1} = \alpha\beta \sum_k \binom{n}{k} (k+\alpha)^{k-1} (n-k+\beta)^{n-k-1}.$$

(2) Ballot numbers. Letting $b(z) = z + zb^2(z)$ (which is related to $B(z)$ defined in (1.2b) by $b(z) = zB(z^2)$, see also Section 4.1) and $\varphi(u) = 1 + u^2$, we find that

$$[z^n]B^k(z) = \frac{k}{2n+k}\binom{2n+k}{n}$$

(these are the *ballot numbers*).

Differential equations occur especially in relation to binary search trees, as we shall see in Section 4.2. For the first-order linear *differential equation* $\mathrm{d}f(z)/\mathrm{d}z = a(z) + b(z)f(z)$, the *variation of parameter* (or *integration factor*) method gives us the solution

$$f(z) = e^{B(z)} \int_{z_0}^{z} a(t) e^{-B(t)}\, \mathrm{d}t, \quad \text{where } B(z) = \int_0^z b(u)\, \mathrm{d}u. \tag{1.20}$$

The lower bound z_0 is chosen to satisfy the initial conditions on $f(z)$.

For other functional equations, *iteration* (or *bootstrapping*) may be useful. For example, under suitable (formal or analytic) convergence conditions, the solution to $f(z)=a(z)+b(z)f(\gamma(z))$ is

$$f(z)=\sum_{k\geq 0}\left(a(\gamma^{((k))}(z))\prod_{0\leq j\leq k-1}b(\gamma^{((j))}(z)\right), \tag{1.21}$$

where $\gamma^{((k))}(z)$ denotes the kth iterate $\gamma(\gamma(\ldots(\gamma(z))\ldots))$ of $\gamma(z)$ (cf. (1.18)).

In general, the whole arsenal of algebra can be used on generating functions; the methods above represent only the most commonly used techniques. Many equations still escape exact solution, but asymptotic methods based upon complex analysis can often be used to extract asymptotic information about the GF coefficients.

Multivariate generating functions

If we need to count structures of size n with a certain combinatorial characteristic having value k, we can try to treat k as a parameter (see the examples above with Stirling numbers). Let $g_{n,k}$ be the corresponding counting sequence. We may also consider bivariate generating functions, such as

$$G(u,z)=\sum_{n,k\geq 0} g_{n,k}u^k z^n \quad \text{or} \quad G(u,z)=\sum_{n,k\geq 0} g_{n,k}u^k\frac{z^n}{n!}.$$

Extensions of previous translation schemes exist (see [52]). For instance, for the Stirling numbers $s_{n,k}$ and $S_{n,k}$, we have

$$\sum_{n,k\geq 0} s_{n,k}u^k\frac{z^n}{n!}=\exp(u\log(1-z)^{-1})=(1-z)^{-u}; \tag{1.22a}$$

$$\sum_{n,k\geq 0} S_{n,k}k!u^k\frac{z^n}{n!}=\frac{1}{1-u(e^z-1)}. \tag{1.22b}$$

Multisets

An extension of the symbolic method to multisets is carried out in [31]. Consider a class $\mathscr{S}$ of structures and for each $\sigma\in\mathscr{S}$ a "multiplicity" $\mu(\sigma)$. The pair $(\mathscr{S},\mu)$ is called a *multiset*, and its generating function is by definition $S(z)=\sum_{\sigma\in\mathscr{S}}\mu(\sigma)z^{|\sigma|}$ so that $S_n=[z^n]S(z)$ is the cumulated value of μ over all structures of size n. This extension is useful for obtaining generating functions of expected (or cumulated) values of parameters over combinatorial structures, since translation schemes based upon admissible constructions also exist for multisets. We shall encounter such extensions when analyzing Shellsort (Section 3.3), trees (Section 4), and hashing (Section 5.1).

2. Asymptotic methods

In this section, we start with elementary asymptotic methods. Next we present complex asymptotic methods, based upon singularity analysis and saddle point integrals, which allow in most cases a direct derivation of asymptotic results for coefficients of generating functions. Then we introduce Mellin transform techniques that permit asymptotic estimations of a large class of combinatorial sums, especially those involving certain arithmetic and number-theoretic functions. We conclude by a discussion of (asymptotic) limit theorems for probability distributions.

2.1. *Generalities*

We briefly recall in this subsection standard *real analysis* techniques and then discuss *complex analysis* methods.

Real analysis

Asymptotic evaluation of the most elementary counting expressions may be done directly, and a useful formula in this regard is *Stirling's formula*:

$$n! \sim \sqrt{2\pi n}\left(\frac{n}{\mathrm{e}}\right)^n\left(1+\frac{1}{12n}+\frac{1}{288n^2}-\frac{139}{51840n^3}-\cdots\right). \tag{2.1}$$

For instance, the central binomial coefficient satisfies $\binom{2n}{n}=(2n)!/n!^2 \sim 4^n/\sqrt{\pi n}$.

The *Euler-Maclaurin* summation formula applies when an expression involves a sum at regularly spaced points (a Riemann sum) of a continuous function: such a sum is approximated by the corresponding integral, and the formula provides a full expansion. The basic form is the following:

2.1. THEOREM (Euler–Maclaurin summation formula). *If $g(x)$ is C^∞ over $[0, 1]$, then for any integer m, we have*

$$\frac{g(0)+g(1)}{2} - \int_0^1 g(x)\,\mathrm{d}x$$
$$= \sum_{1 \leqslant j \leqslant m-1} \frac{B_{2j}}{(2j)!}(g^{(2j-1)}(1)-g^{(2j-1)}(0)) - \int_0^1 g^{(2m)}(x)\frac{B_{2m}(x)}{(2m)!}\,\mathrm{d}x, \tag{2.2a}$$

where $B_j(x) \equiv j![z^j]z\mathrm{e}^{xz}/(\mathrm{e}^z-1)$ is a Bernoulli polynomial, and $B_j=B_j(1)$ is a Bernoulli number.

We can derive several formulae by summing (2.2a). If $\{x\}$ denotes the fractional part of x, we have

$$\sum_{0\leqslant j\leqslant n} g(j) - \int_0^n g(x)\,\mathrm{d}x = \tfrac{1}{2}g(0)+\tfrac{1}{2}g(n) + \sum_{1\leqslant j\leqslant m-1}\frac{B_{2j}}{(2j)!}(g^{(2j-1)}(n)-g^{(2j-1)}(0))$$
$$- \int_0^n g^{(2m)}(x)\frac{B_{2m}(\{x\})}{(2m)!}\,\mathrm{d}x, \tag{2.2b}$$

which expresses the difference between a discrete sum and its corresponding integral. By a change of scale, for h small, setting $g(x)=f(hx)$, we obtain the asymptotic expansion of a Riemann sum, when the step size h tends to 0:

$$\sum_{0\leqslant jh\leqslant 1} f(jh) \sim \frac{1}{h}\int_0^1 f(x)\,\mathrm{d}x + \frac{f(0)+f(1)}{2} + \sum_{j\geqslant 1} \frac{B_{2j}h^{2j-1}}{(2j)!}(f^{(2j-1)}(1) - f^{(2j-1)}(0)). \tag{2.2c}$$

EXAMPLES. (1) The *harmonic numbers* are defined by $H_n = 1 + \frac{1}{2} + \frac{1}{3} + \cdots + \frac{1}{n}$, and they satisfy $H_n = \log n + \gamma + \frac{1}{2n} + \cdots$.

(2) The binomial coefficient $\binom{2n}{n-k}$, for $k < n^{2/3}$, is asymptotically equal to the central coefficient $\binom{2n}{n}$ times $\exp(-k^2/n)$, which follows from estimating its logarithm. This *Gaussian approximation* is a special case of the central limit theorem of probability theory.

Laplace's method for sums is a classical approach for evaluating sums $S_n = \sum_k f(k, n)$ that have a dominant term. First we determine the rank k_0 of the dominant term. We can often show for "smooth" functions $f(k, n)$ that $f(k, n) \approx f(k_0, n)\Phi((k-k_0)h)$, with $h = h(n)$ small (like $1/\sqrt{n}$ or $1/n$). We conclude by applying the Euler–Maclaurin summation to $\Phi(x)$. An example is the asymptotics of the Bell numbers defined in (1.16) [19, page 108] or the number of involutions (1.15) [71, page 65]. There are extensions to multiple sums involving multivariate Euler–Maclaurin summations.

Complex analysis

A powerful method (and one that is often computationally simple) is to use *complex analysis* to go directly from a generating function to the asymptotic form of its coefficients. For instance, the EGF for the number of 2-regular graphs [17, page 273] is

$$f(z) = \frac{\mathrm{e}^{-z/2 - z^2/4}}{\sqrt{1-z}}, \tag{2.3}$$

and $[z^n]f(z)$ is sought. A bivariate Laplace method is feasible. However, it is simpler to notice that $f(z)$ is analytic for complex z, except when $z = 1$. There a "singular expansion" holds:

$$f(z) \sim \frac{\mathrm{e}^{-3/4}}{\sqrt{1-z}}, \text{ as } z \to 1. \tag{2.4a}$$

General theorems that we are going to discuss in the next section allow us to "transfer" an approximation (2.4a) of the function to an approximation of the coefficients:

$$[z^n]f(z) \sim [z^n]\frac{\mathrm{e}^{-3/4}}{\sqrt{1-z}}. \tag{2.4b}$$

Thus, $[z^n]f(z) \sim \mathrm{e}^{-3/4}(-1)^n\binom{-1/2}{n} \sim \mathrm{e}^{-3/4}/\sqrt{\pi n}$.

2.2. *Singularity analysis*

A *singularity* is a point at which a function ceases to be analytic. A dominant singularity is one of smallest modulus. It is known that a function with positive coefficients that is not entire always has a dominant positive real singularity. In most cases, the asymptotic behavior of the coefficients of the function is determined by that singularity.

Location of singularities

The classical exponential-order formula relates the location of singularities of a function to the exponential growth of its coefficients.

2.2. THEOREM (Exponential growth formula). *If $f(z)$ is analytic at the origin and has nonnegative coefficients, and if ρ is its smallest positive real singularity, then its coefficients $f_n = [z^n]f(z)$ satisfy*

$$(1-\varepsilon)^n \rho^{-n} <_{\text{i.o.}} f_n <_{\text{a.e.}} (1+\varepsilon)^n \rho^{-n}, \tag{2.5}$$

for any $\varepsilon > 0$. Here "i.o." means infinitely often (for infinitely many values) and "a.e." means "almost everywhere" (except for finitely many values).

EXAMPLES. (1) Let $f(z) = 1/\cos z$ (EGF for "alternating permutations") and $g(z) = 1/(2-e^z)$ (EGF for "surjections"). Then bounds (2.5) apply with $\rho = \pi/2$ and $\rho = \log 2$ respectively.

(2) The solution $f(z)$ of the functional equation $f(z) = z + f(z^2 + z^3)$ is the OGF of 2–3 trees [85]. Setting $\sigma(z) = z^2 + z^3$, the functional equation has the following formal solution, obtained by iteration (see (1.21)):

$$f(z) = \sum_{m \geq 0} \sigma^{((m))}(z), \tag{2.6a}$$

where $\sigma^{((m))}(z)$ is the mth iterate of $\sigma(z)$. The sum in (2.6a) converges geometrically when $|z|$ is less than the smallest positive root ρ of the equation $\rho = \sigma(\rho)$, and it becomes infinite at $z = \rho$. The smallest possible root is $\rho = 1/\varphi$, where φ is the golden ratio $(1+\sqrt{5})/2$. Hence we have

$$\left(\frac{1+\sqrt{5}}{2}\right)^n (1-\varepsilon)^n <_{\text{i.o.}} [z^n]f(z) < \text{a.e.} \left(\frac{1+\sqrt{5}}{2}\right)^n (1+\varepsilon)^n. \tag{2.6b}$$

The bound (2.6b) and even an asymptotic expansion [85] are obtainable without an explicit expression for the coefficients (see Theorem 4.7).

Nature of singularities

Another way of expressing Theorem 2.2 is as follows: we have $f_n \sim \theta(n)\rho^{-n}$, where the *subexponential factor* $\theta(n)$ is i.o. larger than any decreasing exponential and a.e. smaller than any increasing exponential. Common forms for $\theta(n)$ are $n^\alpha \log^\beta n$, for some constants α and β. The subexponential factors are usually related to the growth of the function around its singularity. (The singularity may be taken equal to 1 by normalization.)

METHOD (Singularity analysis). *Assume that $f(z)$ has around its dominant singularity*

1 *an asymptotic expansion of the form*

$$f(z)=\sigma(z)+R(z), \quad \textit{with } R(z)\ll\sigma(z), \textit{ as } z\to 1, \tag{2.7a}$$

where $\sigma(z)$ *is in a standard set of functions that include* $(1-z)^a\log^b(1-z)$ *for constants* a *and* b. *Then, under general conditions,* (2.7a) *leads to*

$$[z^n]f(z)=[z^n]\sigma(z)+[z^n]R(z), \quad \textit{with } [z^n]R(z)\ll[z^n]\sigma(z), \textit{ as } n\to\infty. \tag{2.7b}$$

Applications of this principle are based upon a variety of conditions on function $f(z)$ or $R(z)$, giving rise to several methods:

(1) *Transfer methods* require only growth information on the remainder term $R(z)$, but the approximation has to be established for $z\to 1$ in some region of the complex plane. Transfer methods largely originate in [85] and are developed systematically in [37].

(2) *Tauberian theorems* assume only that (2.7a) holds when z is real and less than 1 (that is, as $z\to 1^-$), but they require a priori Tauberian side conditions (positivity, monotonicity) to be satisfied by the coefficients f_n and are restricted to less general types of growth for $R(z)$. (See [29, page 447] and [54, page 52] for a combinatorial application.)

(3) *Darboux's method* assumes smoothness conditions (differentiability) on the remainder term $R(z)$ [59, page 447].

Our transfer method approach is the one that is easiest to apply and the most flexible for combinatorial enumerations. First, we need the asymptotic growth of coefficients of standard singular functions. For $\sigma(z)=(1-z)^{-s}$, where $s>0$, by Newton's expansion the nth coefficient in its Taylor expansion is $\binom{n+s-1}{n}$, which is $\sim n^{s-1}/\Gamma(s)$. For many standard singular functions, like $(1-z)^{-1/2}\log^2(1-z)^{-1}$, we may use either Euler–Maclaurin summation on the explicit form of the coefficients or contour integration to find $\sigma_n\sim(\pi n)^{-1/2}\log^2 n$. Next we need to "transfer" coefficients of remainder terms.

2.3. THEOREM (Transfer lemma). *If* $R(z)$ *is analytic for* $|z|<1+\delta$ *for some* $\delta>0$ *(with the possible exception of a sector around* $z=1$, *where* $|\mathrm{Arg}(z-1)|<\varepsilon$ *for some* $\varepsilon<\pi/2$) *and if* $R(z)=\mathrm{O}((1-z)^r)$ *as* $z\to 1$ *for some real* r, *then*

$$[z^n]R(z)=\mathrm{O}(n^{-r-1}). \tag{2.8}$$

The proof proceeds by choosing a contour of integration made of part of the circle $|z|=1+\delta$ and the boundary of the sector, except for a small notch of diameter $1/n$ around $z=1$. Furthermore, when $r\leqslant -1$, we need only assume that the function is analytic for $|z|\leqslant 1$, $z\neq 1$.

EXAMPLES. (1) The EGF $f(z)$ of 2-regular graphs is given in (2.3). We can expand the exponential around $z=1$ and get

$$f(z)=\frac{\mathrm{e}^{-z/2-z^2/4}}{\sqrt{1-z}}=\mathrm{e}^{-3/4}(1-z)^{-1/2}+\mathrm{O}((1-z)^{1/2}), \quad \text{as } z\to 1. \tag{2.9a}$$

The function $f(z)$ is analytic in the complex plane slit along $z\geqslant 1$, and (2.9a) holds there in the vicinity of $z=1$. Thus, by the transfer lemma with $r=\frac{1}{2}$, we have

$$[z^n]f(z)=\mathrm{e}^{-3/4}\binom{n-\frac{1}{2}}{n}+\mathrm{O}(n^{-3/2})=\frac{\mathrm{e}^{-3/4}}{\sqrt{\pi}}n^{-1/2}+\mathrm{O}(n^{-3/2}). \tag{2.9b}$$

(2) The EGF of surjections was shown in (1.17) to be $f(z)=(2-\mathrm{e}^z)^{-1}$. It is analytic for $|z|\leqslant 3$, except for a simple pole at $z=\log 2$, where local expressions show that

$$f(z)=\frac{1}{2\log 2}\cdot\frac{1}{1-z/\log 2}+\mathrm{O}(1), \quad \text{as } z\to\log 2, \tag{2.10a}$$

so that

$$[z^n]f(z)=\frac{1}{2}\left(\frac{1}{\log 2}\right)^{n+1}\left(1+\mathrm{O}\left(\frac{1}{n}\right)\right). \tag{2.10b}$$

(3) A functional equation. The OGF of certain trees [92] $f(z)=1+z+z^2+2z^3+\cdots$ is known only via the functional equation $f(z)=(1-zf(z^2))^{-1}$. It can be checked that $f(z)$ is analytic at the origin. Its dominant singularity is a simple pole $\rho<1$ determined by cancellation of the denominator, $\rho f(\rho^2)=1$. Around $z=\rho=0.59475\ldots$, we have

$$f(z)=\frac{1}{\rho f(\rho^2)-zf(z^2)}=\frac{1}{c(\rho-z)}+\mathrm{O}(1), \quad \text{with } c=\left.\frac{\mathrm{d}}{\mathrm{d}z}zf(z^2)\right|_{z=\rho}. \tag{2.11a}$$

Thus, with $K=(c\rho)^{-1}=0.36071$, we find that

$$[z^n]f(z)=K\rho^{-n}\left(1+\mathrm{O}\left(\frac{1}{n}\right)\right). \tag{2.11b}$$

More precise expansions exist for coefficients of meromorphic functions (functions with poles only), like the ones in the last two examples (for example, see [71, 5.3.1–3, 4], and [59, 37]). For instance, the error of approximation (2.11b) is less than 10^{-13} when $n=100$. Finally, the OGF of 2–3 trees (2.6a) is amenable to transfer methods, though extraction of singular expansions is appreciably more difficult [85].

We conclude this subsection by citing the lemma at the heart of Darboux's method [59, page 447] and a classical Tauberian Theorem [29, page 447].

2.4. THEOREM (Darboux's method). *If $R(z)$ is analytic for $|z|<1$, continuous for $|z|\leqslant 1$, and d times continuously differentiable over $|z|=1$, then*

$$[z^n]R(z)=\mathrm{o}\left(\frac{1}{n^d}\right). \tag{2.12}$$

For instance, if $R(z)=(1-z)^{5/2}H(z)$, where $H(z)$ is analytic for $|z|<1+\delta$, then we can use $d=2$ for Theorem 2.4 and obtain $[z^n]R(z)=\mathrm{o}(1/n^2)$. The theorem is usually applied to derive expansions of coefficients of functions of the form $f(z)=(1-z)^rH(z)$, with $H(z)$ analytic in a larger domain than $f(z)$. Such functions can however be treated directly by transfer methods (Theorem 2.3).

2.5. THEOREM (Tauberian theorem of Hardy–Littlewood–Karamata). *Assume that the function* $f(z)=\sum_{n\geq 0} f_n z^n$ *has radius of convergence* 1 *and satisfies for real* z, $0\leq z<1$,

$$f(z)\sim\frac{1}{(1-z)^s}L\left(\frac{1}{1-z}\right), \quad \text{as } z\to 1^-, \tag{2.13a}$$

where $s>0$ *and* $L(u)$ *is a function varying slowly at infinity, like* $\log^b(u)$. *If* $\{f_n\}_{n\geq 0}$ *is monotonic, then*

$$f_n\sim\frac{n^{s-1}}{\Gamma(s)}L(n). \tag{2.13b}$$

An application to the function $f(z)=\Pi_k(1+z^k/k)$ is given in [54, page 52]; the function represents the EGF of permutations with distinct cycle lengths. That function has a natural boundary at $|z|=1$ and hence is not amenable to Darboux's or transfer methods.

Singularity analysis is used extensively in Sections 3–5 for asymptotics related to sorting methods, plane trees, search trees, partial match queries, and hashing with linear probing.

2.3. *Saddle point methods*

Saddle point methods are used for extracting coefficients of entire functions (which are analytic in the entire complex plane) and functions that "grow fast" around their dominant singularities, like $\exp(1/1-z))$. They also play an important rôle in obtaining limit distribution results and exponential tails for discrete probability distributions.

A simple bound

Assume that $f(z)=\sum_n f_n z^n$ is entire and has positive coefficients. Then by Cauchy's formula, we have

$$f_n=\frac{1}{2\pi \mathrm{i}}\int_\Gamma \frac{f(z)}{z^{n+1}}\,\mathrm{d}z. \tag{2.14}$$

We refer to (2.14) as a Cauchy coefficient integral. If we take as contour Γ the circle $|z|=R$, we get an easy upper bound

$$f_n\leq\frac{f(R)}{R^n}, \tag{2.15}$$

since the maximum value of $|f(z)|$, for $|z|=R$, is $f(R)$. The bound (2.15) is valid for *any* $R>0$. In particular, we have $f_n\leq\min_{R>0}\{f(R)R^{-n}\}$. We can find the minimum value by setting $\mathrm{d}(f(R)R^{-n})/\mathrm{d}R=(f'(R)-f(R)(n/R))R^{-n}=0$, which gives us the following bound:

2.6. THEOREM (Saddle point bound). *If $f(z)$ is entire and has positive coefficients, then for all n, we have*

$$[z^n]f(z) \leqslant \frac{f(R)}{R^n}, \tag{2.16}$$

where $R=R(n)$ is the smallest positive real number such that

$$R\frac{f'(R)}{f(R)}=n. \tag{2.17}$$

Complete saddle point analysis

The saddle point method is a refinement of the technique we used to derive (2.15). It applies in general to integrals depending upon a large parameter, of the form

$$I=\frac{1}{2\pi i}\int_{\Gamma} e^{h(z)}dz. \tag{2.18a}$$

A point $z=\sigma$ such that $h'(z)=0$ is called a *saddle point* owing to the topography of $|e^{h(z)}|$ around $z=\sigma$: There are two perpendicular directions at $z=\sigma$, one along which the integrand $|e^{h(z)}|$ has a local minimum at $z=\sigma$, and the other (called the *axis* of the saddle point) along which the integrand has a local maximum at $z=\sigma$. The principal steps of the saddle point method are as follows:

(1) Show that the contribution of the integral is asymptotically localized to a fraction Γ_ε of the contour around $z=\sigma$ traversed along its axis. (This forces ε to be not too small.)

(2) Show that, over this subcontour, $h(z)$ can be suitably approximated by $h(\sigma)+\frac{1}{2}(z-\sigma)^2h''(\sigma)$. (This imposes a conflicting constraint that ε should not be too large.)

If points (1) and (2) can be established, then I can be approximated by

$$I\approx\frac{1}{2\pi i}\int_{\Gamma_\varepsilon}\exp\left(h(\sigma)+\frac{(z-\sigma)^2}{2}h''(\sigma)\right)dz\approx\frac{e^{h(\sigma)}}{\sqrt{2\pi h''(\sigma)}}. \tag{2.18b}$$

Classes of functions such that the saddle point estimate (2.18b) applies to Cauchy coefficient integrals (2.14) are called *admissible* and have been described by several authors [57, 56, 86]. Cauchy coefficient integrals (2.14) can be put into the form (2.18a), where $h(z)=h_n(z)=\log f(z)-(n+1)\log z$, and a saddle point $z=R$ is a root of the equation

$$h'(z)=\frac{d}{dz}(\log f(z)-(n+1)\log z)=\frac{f'(z)}{f(z)}-\frac{(n+1)}{z}=0.$$

By the method of (2.18) we get the following estimate:

2.7. THEOREM (Saddle point method for Cauchy coefficient integrals). *If $f(z)$ has positive coefficients and is in a class of admissible functions, then*

$$f_n\sim\frac{f(R)}{\sqrt{2\pi C(n)}R^{n+1}},\quad \textit{with } C(n)=\frac{d^2}{dz^2}\log f(z)\bigg|_{z=R}+(n+1)R^{-2}, \tag{2.19}$$

where the saddle point R is the smallest positive real number such that

$$R\frac{f'(R)}{f(R)}=n+1. \tag{2.20}$$

EXAMPLES. (1) We get Stirling's formula (2.1) by letting $f(z)=\mathrm{e}^z$. The saddle point is $R=(n+1)$, and by Theorem 2.7 we have

$$\frac{1}{n!}=[z^n]\mathrm{e}^z\sim\frac{\mathrm{e}^{n+1}}{\sqrt{2\pi/(n+1)}\,(n+1)^{n+1}}\sim\frac{1}{\sqrt{2\pi n}}\left(\frac{\mathrm{e}}{n}\right)^n.$$

(2) By (1.15), the number of involutions is given by

$$I_n=n!\,[z^n]\mathrm{e}^{z+z^2/2}=\frac{n!}{2\pi\mathrm{i}}\int_\Gamma\frac{\mathrm{e}^{z+z^2/2}}{z^{n+1}}\,\mathrm{d}z,$$

and the saddle point is $R=\sqrt{n}+\frac{1}{2}+5/(8\sqrt{n})+\cdots$ We choose $\varepsilon=n^{-2/5}$, so that for $z=R\mathrm{e}^{\mathrm{i}\varepsilon}$, we have $(z-R)^2h''(R)\to\infty$ while $(z-R)^3h'''(R)\to 0$. Thus,

$$\frac{I_n}{n!}\sim\frac{\mathrm{e}^{3/4}}{2\sqrt{\pi}}n^{-n/2}\mathrm{e}^{n/8}.$$

The asymptotics of the Bell numbers can be done in the same way [19, page 104].

(3) A function with a finite singularity. For $f(z)=\mathrm{e}^{z/(1-z)}$, Theorem 2.7 gives us

$$f_n\equiv[z^n]\exp\left(\frac{z}{1-z}\right)\sim\frac{C\mathrm{e}^{d\sqrt{n}}}{n^\alpha}. \tag{2.21}$$

A similar method can be applied to the integer partition function $p(z)=\Pi_{n\geqslant 1}(1-z^n)^{-1}$ though it has a natural boundary, and estimates (2.21) are characteristic of functions whose logarithm has a pole-like singularity.

Specializing some of Hayman's results, we can define inductively a class $\mathscr{H}$ of admissible functions as follows: (i) If $p(z)$ denotes an arbitrary polynomial with positive coefficients, then $\mathrm{e}^{p(z)}\in\mathscr{H}$. (ii) If $f(z)$ and $g(z)$ are arbitrary functions of $\mathscr{H}$, then $\mathrm{e}^{f(z)}$, $f(z)\cdot g(z)$, $f(z)+p(z)$, and $p(f(z))$ are also in $\mathscr{H}$.

Several applications of saddle point methods appear in Section 5.1 in the analysis of maximum bucket occupancy, extendible hashing, and coalesced hashing.

2.4. Mellin transforms

The Mellin transform, a tool originally developed for analytic number theory, is useful for analyzing sums where arithmetic functions appear or nontrivial periodicity phenomena occur. Such sums often present themselves as expectations of combinatorial parameters or generating functions.

Basic properties

Let $f(x)$ be a function defined for real $x\geqslant 0$. Then its *Mellin transform* is a function

$f^*(s)$ of the complex variable s defined by

$$f^*(s)=\int_0^\infty f(x)x^{s-1}\,dx. \tag{2.22}$$

If $f(x)$ is continuous and is $O(x^\alpha)$ as $x\to 0$ and $O(x^\beta)$ as $x\to\infty$, then its Mellin transform is defined in the "fundamental strip" $-\alpha<\Re(s)<-\beta$, which we denote by $\langle-\alpha;-\beta\rangle$. For instance the Mellin transform of e^{-x} is the well-known *Gamma function* $\Gamma(s)$, with fundamental strip $\langle 0;+\infty\rangle$, and the transform of $\sum_{n\geqslant k}(-x)^n/n!$, for $k>0$, is $\Gamma(s)$ with fundamental strip $\langle -k;-k+1\rangle$. There is also an inversion theorem à la Fourier:

$$f(x)=\frac{1}{2\pi i}\int_{c-i\infty}^{c+i\infty} f^*(s)x^{-s}\,ds, \tag{2.23}$$

where c is taken arbitrarily in the fundamental strip.

The important principle for asymptotic analysis is that *under the Mellin transform, there is a correspondence between terms of asymptotic expansions of $f(x)$ at 0 (respectively $+\infty$) and singularities of $f^*(s)$ in a left (respectively right) half-plane.* To see why this is so, assume that $f^*(s)$ is small at $\pm i\infty$ and has only polar singularities. Then we can close the contour of integration in (2.23) to the left (for $x\to 0$) or to the right for ($x\to\infty$) and derive by Cauchy's residue formula

$$\begin{aligned} f(x)&=+\sum_\sigma \mathrm{Res}(f^*(s)x^{-s};\ s=\sigma)+O(x^{-d}), \quad \text{as } x\to 0;\\ f(x)&=-\sum_\sigma \mathrm{Res}(f^*(s)x^{-s};\ s=\sigma)+O(x^{-d}), \quad \text{as } x\to\infty. \end{aligned} \tag{2.24}$$

The sum in the first equation is extended to all poles σ where $d\leqslant\Re(\sigma)\leqslant-\alpha$; the sum in the second equation is extended to all poles σ with $-\beta\leqslant\Re(\sigma)\leqslant d$. Those relations have the character of asymptotic expansions of $f(x)$ at 0 and $+\infty$: We observe that if $f^*(s)$ has a kth-order pole at σ, then a residue in (2.24) is of the form $Q_{k-1}(\log x)x^{-\sigma}$, where $Q_{k-1}(u)$ is a polynomial of degree $k-1$.

There is finally an important functional property of the Mellin transform: If $g(x)=f(\mu x)$, then $g^*(s)=\mu^{-s}f^*(s)$. Hence, transforms of sums called "harmonic sums" decompose into the product of a generalized Dirichlet series $\sum\lambda_k\mu_k^s$ and the transform $f^*(s)$ of the basis function:

$$F(x)=\sum_k \lambda_k f(\mu_k x) \quad\Rightarrow\quad F^*(s)=\left(\sum_k \lambda_k\mu_k^{-s}\right)f^*(s). \tag{2.25}$$

Asymptotics of sums

The standard usage of Mellin transforms devolves from a combination of (2.24) and (2.25):

2.8. THEOREM (Mellin asymptotic summation formula). *Assume that in* (2.25) *the transform $f^*(s)$ of $f(x)$ is exponentially small towards $\pm i\infty$ with only polar singularities and that the Dirichlet series is meromorphic of finite order. Then the asymptotic behavior* *of a harmonic sum* $F(x)=\sum_k \lambda_k f(\mu_k x)$, *as* $x\to 0$ *(respectively* $x\to\infty$*), is given by*

$$\sum_k \lambda_k f(\mu_k x) \sim \pm\sum_\sigma \operatorname{Res}\left(\left(\sum_k \lambda_k \mu_k^{-s}\right) f^*(s) x^{-s};\ s=\sigma\right). \tag{2.26}$$

For an asymptotic expansion of the sum as $x\to 0$ *(respectively as* $x\to\infty$*), the sign in (2.26) is "+" (respectively "−"), and the sum is taken over poles to the left (respectively to the right) of the fundamental strip.*

EXAMPLES. (1) An arithmetical sum. Let $F(x)$ be the harmonic sum $\sum_{k\geqslant 1} d(k)e^{-k^2x^2}$, where $d(k)$ is the number of divisors of k. Making use of (2.25) and the fact that the transform of e^{-x} is $\Gamma(s)$, we have

$$F^*(s)=\tfrac{1}{2}\Gamma(s/2)\sum_{k\geqslant 1} d(k)k^{-s}=\tfrac{1}{2}\Gamma(s/2)\,\zeta^2(s), \tag{2.27a}$$

where $\zeta(s)=\sum_{n\geqslant 1} n^{-s}$. Here $F^*(s)$ is defined in the fundamental strip $\langle 1; +\infty\rangle$. To the left of this strip, it has a simple pole at $s=0$ and a double pole at $s=1$. By expanding $\Gamma(s)$ and $\zeta(s)$ around $s=0$ and $s=1$, we get for any $d>0$

$$F(x)=-\frac{\sqrt{\pi}}{2}\frac{\log x}{x}+\left(\frac{3\gamma}{4}-\frac{\log 2}{2}\right)\frac{\sqrt{\pi}}{x}+\frac{1}{4}+O(x^d), \quad \text{as } x\to 0. \tag{2.27b}$$

(2) A sum with hidden periodicities. Let $F(x)$ be the harmonic sum $\sum_{k\geqslant 0} 1-e^{-x/2^k})$. The transform $F^*(s)$ is defined in the fundamental strip $\langle -1; 0\rangle$, and by (2.25) we find

$$F^*(s)=-\Gamma(s)\sum_{k\geqslant 0} 2^{ks}=-\frac{\Gamma(s)}{1-2^s}. \tag{2.28a}$$

The expansion of $F(x)$ as $x\to\infty$ is determined by the poles of $F^*(s)$ to the right of the fundamental strip. There is a double pole at $s=0$ and the denominator of (2.28a) gives simple poles at $s=\chi_k=2k\pi i/\log 2$ for $k\neq 0$. Each simple pole χ_k contributes a fluctuating term $x^{-\chi_k}=\exp(2k\pi i \log_2 x)$ to the asymptotic expansion of $F(x)$. Collecting fluctuations, we have

$$F(x)=\log_2 x+P(\log_2 x)+O(x^{-d}), \quad \text{as } x\to\infty, \tag{2.28b}$$

where $P(u)$ is a periodic function with period 1 and a convergent Fourier expansion.

Mellin transforms are the primary tool to study tries and radix-exchange sort (Section 4.3). They are also useful in the study of certain plane tree algorithms (Section 4.1), bubble sort (Section 3.4), and interpolation search and extendible hashing (Section 5.1).

2.5. *Limit probability distributions*

General references for this section are [29, 9]. We recall that if X is a real-valued *random variable* (RV), then its *distribution function* is $F(x)=\Pr\{X\leqslant x\}$, and its *mean* and *variance* are $\mu_X=\bar{X}=\mathbf{E}\{X\}$ and $\sigma_X^2=\operatorname{var}(X)=\mathbf{E}\{X^2\}-(\mathbf{E}\{X\})^2$ respectively. The kth moment of X is $M_k=\mathbf{E}\{X^k\}$. We have $\mu_{X+Y}=\mu_X+\mu_Y$, and when X and Y are independent, we have $\sigma^2_{X+Y}=\sigma^2_X+\sigma^2_Y$. For a nonnegative integer-valued RV, its *probability generating function* (PGF) is defined by $p(z)=\sum_{k\geqslant 0}p_k z^k$, where $p_k=\Pr\{X=k\}$; the mean and variance are respectively $\mu=p'(1)$ and $\sigma^2=p''(1)+p'(1)-(p'(1))^2$. It is well known that the PGF of a sum of independent RVs is the product of their PGFs, and conversely, a product of PGFs corresponds to a sum of independent RVs.

A problem that naturally presents itself in the analysis of algorithms is as follows: Given a class $\mathscr{C}$ of combinatorial structures (such as trees, permutations, etc.), with X_n^* a "parameter" over structures of size n (path length, number of inversions, etc.), determine the *limit* (asymptotic) distribution of the normalized variable $X_n=(X_n^*-\mu_{X_n^*})/\sigma_{X_n^*}$. Simulations suggest that such a limit does exist in typical cases. The limit distribution usually provides more information than does a plain average-case analysis. The following two transforms are important tools in the study of limit distributions:

(1) *Characteristic functions* (or *Fourier transforms*), defined for RV X by

$$\varphi(t)=\mathbf{E}\{e^{itX}\}=\int_{-\infty}^{+\infty} e^{itx}\,dF(x). \tag{2.29}$$

For a nonnegative integer-valued RV X, we have $\varphi(t)=p(e^{it})$.

(2) *Laplace transforms*, defined for a nonnegative RV X by

$$g(t)=\mathbf{E}\{e^{-tX}\}=\int_{0}^{+\infty} e^{-tx}\,dF(x). \tag{2.30}$$

The transform $g(-t)$ is sometimes called the "moment generating function" of X since it is essentially the EGF of X's moments. For a nonnegative integer-valued RV X, we have $g(t)=p(e^{-t})$.

Limit theorems

Under appropriate conditions the distribution functions $F_n(x)$ of a sequence of RVs X_n converge pointwise to a limit $F(x)$ at each point of continuity of $F(x)$. Such a convergence is known as "weak convergence" or "convergence in distribution," and we denote it by $F=\lim F_n$ [9].

2.9. THEOREM (Continuity theorem for characteristic functions). *Let X_n be a sequence of RVs with characteristic functions $\varphi_n(t)$ and distribution functions $F_n(x)$. If there is a function $\varphi(t)$ continuous at the origin such that* $\lim \varphi_n(t)=\varphi(t)$, *then there is a distribution function $F(x)$ such that* $F=\lim F_n$. *Function $F(x)$ is the distribution function of the RV with characteristic function $\varphi(t)$.*

2.10. THEOREM (Continuity theorem for Laplace transforms). *Let X_n be a sequence of RVs with Laplace transforms $g_n(t)$ and distribution functions $F_n(x)$. If there is some a such that for all $|t|\leqslant a$ there is a limit* $g(t)=\lim g_n(t)$, *then there is a distribution function $F(x)$ such that* $F=\lim F_n$. *Function $F(x)$ is the distribution function of the RV with Laplace transform $g(t)$.*

Similar limit conditions exist when the moments of the X_n converge to the moments of an RV X, provided that the "moment problem" for X has a unique solution. A sufficient condition for this is $\sum_{j\geqslant 0}\mathbf{E}\{X^{2j}\}^{-1/(2j)}=+\infty$.

Generating functions

If the X_n are nonnegative integer-valued RVs, then they define a sequence $p_{n,k}=\Pr\{X_n=k\}$, and the problem is to determine the asymptotic behavior of the distributions $\pi_n=\{p_{n,k}\}_{k\geqslant 0}$ (or the associated cumulative distribution functions F_n), as $n\to\infty$. In simple cases, such as binomial distributions, explicit expressions are available and can be treated using the real analysis techniques of Section 2.1.

In several cases, either the "horizontal" GFs $p_n(u)$ or "vertical" GFs $q_k(z)$

$$p_n(u)=\sum_{k=0}^{\infty} p_{n,k}u^k, \qquad q_k(z)=\sum_{n=0}^{\infty} p_{n,k}z^n \tag{2.31}$$

have explicit expressions, and complex analysis methods can be used to extract their coefficients asymptotically.

Sometimes, only the bivariate generating function

$$P(u,z)=\sum_{n,k\geqslant 0} p_{n,k}u^k z^n \tag{2.32}$$

has an explicit form, and a two-stage method must be employed.

Univariate problems

The most well known application of univariate techniques is the *central limit theorem*. If $X_n=A_1+\cdots+A_n$ is the sum of independent identically distributed RVs with mean 0 and variance 1, then $X_n/\sqrt{n}$ tends to a normal distribution with unit variance. The classical proof [29, page 515] uses characteristic functions: the characteristic function of $X_n/\sqrt{n}$ is $\varphi_n(t)=\varphi^n(t/\sqrt{n})$, where $\varphi(t)$ is the characteristic function of each A_j, and it converges to $\mathrm{e}^{-t^2/2}$, the characteristic function of the normal distribution.

Another proof that provides information on the rate of convergence and on densities when the A_j are nonnegative integer-valued uses the saddle point method applied to

$$\Pr\{X_n=k\}=\frac{1}{2\pi\mathrm{i}}\oint\frac{p(z)^n}{z^{k+1}}\,\mathrm{d}z, \tag{2.33}$$

where $p(z)$ is the PGF of each A_j [54].

A general principle is that univariate problems can be often solved using either continuity theorems or complex asymptotics (singularity analysis or saddle point) applied to vertical or horizontal generating functions.

EXAMPLES. (1) A horizontal generating function. The probability that a random permutation of n elements has k cycles is $[u^k]p_n(u)$, where

$$p_n(u)=\frac{1}{n!}u(u+1)(u+2)\cdots(u+n-1).$$

Like for the basic central limit theorem above, either characteristic functions or saddle point methods can be used to establish normality of the limiting distribution as $n \to \infty$ (Goncharov's theorem). The same normality result holds for the distribution of inversions in permutations, for which

$$p_n(u) = \frac{1}{n!} \prod_{1 \leqslant j \leqslant n} \frac{1-u^j}{1-u}.$$

Inversions will be studied in Section 3.1 in connection with sorting.

(2) A vertical generating function. The probability that a random binary string with length n has no "1-run" of length k is $[z^n]q_k(z)$, where

$$q_k(2z) = \frac{1-z^k}{1-2z+z^{k+1}}.$$

A singularity analysis [73] can be used: the dominant singularity of $q_k(z)$ is at $z = \zeta_k \approx 1 + 2^{-k-1}$, and we have $[z^n]q_k(z) \approx e^{-n/2k+1}$.

Bivariate problems

For bivariate problems with explicit bivariate GFs (2.32), the following two-stage approach may be useful. First, we can often get a good approximation to the Cauchy coefficient integral

$$p_n(u) = \frac{1}{2\pi i} \oint \frac{P(z,u)}{z^{n+1}} \, dz$$

by treating u as a parameter and applying singularity analysis or the saddle point method.

Second, if u is real and close to 1 (for example, $u = e^{-t}$, with t close to 0), we may be able to conclude the analysis using the continuity theorem for Laplace transforms. If u is complex, $|u| = 1$ (that is, $u = e^{it}$), we try to use instead the continuity theorem for characteristic functions. For instance, Bender [5] and Canfield [14] have obtained general normality results for distributions corresponding to bivariate generating functions of the form

$$\frac{1}{1-ug(z)} \quad \text{and} \quad e^{ug(z)}.$$

These results are useful since they correspond to the distribution of the number of components in a sequence (or partitional complex) construct and an abelian partitional complex construct respectively.

Little is known about bivariate GFs defined only implicitly via nonlinear functional equations, a notable exception being [63, 64]. Finally, other multivariate (but less analytical) techniques are used in the analysis of random graph models [10].

3. Sorting algorithms

In this section we describe several important sorting algorithms, including insertion sort, Shellsort, bubble sort, quicksort, radix-exchange sort, selection sort, heapsort, and merge sort. Some sorting algorithms are more naturally described in an iterative fashion, while the others are more naturally described recursively. In this section we analyze the performance of the "iterative" sorting algorithms in a unified way by using some basic notions of inversion tables and lattice paths; we apply the combinatorial tools of Section 1 to the study of inversions and left-to-right maxima, and we use the techniques of Section 2 to derive asymptotic bounds.

We defer the analyses of the sorting algorithms that are more "recursive" in nature until Section 4, where we exploit the close connections between their recursive structures and the tree models studied in Section 4. Yet another class of sorting algorithms, those based upon distribution sorting, will be described and analyzed in Section 5.

For purposes of average-case analysis of sorting, we assume that the input array (or input file) $x[1]x[2]\ldots x[n]$ forms a random permutation of the n elements. A permutation of a set of n elements is a 1-1 mapping from the set onto itself. Typically we represent the set of n elements by $\{1, 2, \ldots, n\}$. We use the notation $\sigma = \sigma_1 \sigma_2 \ldots \sigma_n$ to denote the permutation that maps i to σ_i, for $1 \leqslant i \leqslant n$. Our input data model is often justified in practice, such as when the key values are generated independently from a common continuous distribution.

3.1. *Inversions*

The common thread running through the analyses of many sorting algorithms is the connection between the running time of the algorithm and the number of inversions in the input. An *inversion* in permutation σ is an "out of order" pair (σ_k, σ_j) of elements, in which $k<j$ but $\sigma_k > \sigma_j$. The number of inversions is thus a measure of the amount of disorder in a permutation. Let us define the RV I_n to be the number of inversions; the number of inversions in a particular permutation σ is denoted $I_n[\sigma]$. This concept was introduced two centuries ago as a means of computing the determinant of a matrix $A = (A_{ij})$:

$$\det A = \sum_{\sigma \in S_n} (-1)^{I_n[\sigma]} A_{1,\sigma_1} A_{2,\sigma_2} \cdots A_{n,\sigma_n}, \tag{3.1}$$

where S_n denotes the set of $n!$ possible permutations.

3.0. DEFINITION. The *inversion table* of the permutation $\sigma = \sigma_1 \sigma_2 \ldots \sigma_n$ is the ordered sequence

$$b_1, b_2, \ldots, b_n, \quad \text{where } b_k = |\{1 \leqslant j < \sigma_k^{-1} \mid \sigma_j > k\}|. \tag{3.2a}$$

(Here σ^{-1} denotes the inverse permutation to σ; that is, σ_k^{-1} denotes the index of k in σ.) In other words, b_k is equal to the number of elements in the permutation σ that precede k but have value $>k$.

The number of inversions can be expressed in terms of inversion tables as follows:

$$I_n[\sigma] = \sum_{1 \leqslant k \leqslant n} b_k.$$

It is easy to see that there is a 1-1 correspondence between permutations and inversion tables. An inversion table has the property that

$$0 \leqslant b_k \leqslant n-k \quad \text{for each } 1 \leqslant k \leqslant n, \tag{3.2b}$$

and moreover all such combinations are possible. We can thus view inversion tables as a cross product

$$\prod_{1 \leqslant k \leqslant n} \{0, 1, \ldots, n-k\}. \tag{3.2c}$$

We associate each inversion table $b_1, b_2, \ldots, b_n$ with the monomial $x_{b_1} x_{b_2} \ldots x_{b_n}$, and we define the generating function

$$F(x_0, x_1, \ldots, x_{n-1}) = \sum_{\sigma \in S_n} x_{b_1} x_{b_2} \ldots x_{b_n}, \tag{3.3}$$

which is the sum of the monomials over all $n!$ permutations. By (3.2b) the possibilities for b_k correspond to the term $(x_0 + x_1 + \cdots + x_{n-k})$, and we get the following fundamental formula, which will play a central rôle in our analyses.

3.1. THEOREM. *The generating function defined in* (3.3) *satisfies*

$$F(x_0, x_1, \ldots, x_{n-1}) = x_0(x_0 + x_1) \cdots (x_0 + x_1 + \cdots + x_{n-1}).$$

Theorem 3.1 is a powerful tool for obtaining statistics related to inversion tables. For example, let us define $I_{n,k}$ to be the number of permutations of n elements having k inversions. By Theorem 3.1, the OGF $I(z) = \Sigma_k I_{n,k} z^k$ is given by

$$I_n(z) = z^0(z^0 + z^1)(z^0 + z^1 + z^2) \cdots (z^0 + z^1 + \cdots + z^{n-1}), \tag{3.4a}$$

since each monomial $x_{b_1} x_{b_2} \ldots x_{b_n}$ in (3.3) contributes $z^{b_1 + b_2 + \cdots + b_n}$ to $I_n(z)$. We can convert (3.4a) into the PGF $\Phi_n(z) = \Sigma_k \Pr\{I_n = k\} z^k$ by dividing by $|S_n| = n!$:

$$\Phi_n(z) = \sum_k \frac{I_{n,k}}{n!} z^k = \prod_{1 \leqslant k \leqslant n} b_k(z), \quad \text{where } b_k(z) = \frac{z^0 + z^1 + \cdots + z^{n-k}}{n-k+1}. \tag{3.4b}$$

The expected number of inversions $\overline{I_n}$ and the variance $\operatorname{var}(I_n)$ are thus equal to

$$\overline{I_n} = \Phi_n'(1) = \tfrac{1}{4} n(n-1); \tag{3.4c}$$

$$\operatorname{var}(I_n) = \Phi_n''(1) + \Phi_n'(1) - (\Phi_n'(1))^2 = n(2n+5)(n-1)/72. \tag{3.4d}$$

The mean $\overline{I_n}$ is equal to half the worst-caste value of I_n. Note from (3.4b) that $\Phi_n(z)$ is a product of individual PGFs $b_k(z)$, which indicates by a remark at the beginning of Section 2.5 that I_n can be expressed as the sum of independent RVs. This suggests another way of looking at the derivation: The decomposition of I_n in question is the obvious one based upon the inversion table (3.2a); we have $I_n = b_1 + b_2 + \cdots + b_n$, and the PGF of b_k is $b_k(z)$ given above in (3.4b). Equations (3.4c) and (3.4d) follow from summing $\overline{b_k}$ and $\operatorname{var}(b_k)$, for $1 \leqslant k \leqslant n$. By a generalization of the central limit theorem to sums of independent but nonidentical RVs, it follows that $(I_n - \overline{I_n})/\sigma_{I_n}$ converges to the normal distribution, as $n \to \infty$.

Another RV important to our sorting analyses is the number of *left-to-right minima*, denoted by L_n. For a permutation $\sigma \in S_n$, $L_n[\sigma]$ is the number of elements in σ that are less than all the preceding elements in σ. In terms of (3.2a), we have

$$L_n[\sigma] = |\{1 \leqslant k \leqslant n \mid b_k = \sigma_k^{-1} - 1\}|.$$

Let us define $L_{n,k}$ to be the number of permutations of n elements having k left-to-right minima. By Theorem 3.1, the OGF $L_n(z) = \Sigma_k L_{n,k} z^k$ is given by

$$L_n(z) = z(z+1)(z+2)\cdots(z+n-1), \tag{3.5a}$$

since the contribution to $L_n(z)$ from the x_j term in $(x_0 + x_1 + \cdots + x_{k-1})$ in Theorem 3.1 is z if $j = k-1$ and 1 otherwise. The PGF $\Lambda_n(z) = \Sigma_k \Pr\{L_n = k\} z^k$ is thus

$$\Lambda_n(z) = \sum_k \frac{L_{n,k}}{n!} z^k = \prod_{1 \leqslant k \leqslant n} l_k(z), \quad \text{where } l_k(z) = \frac{z+k-1}{k}. \tag{3.5b}$$

Taking derivatives as above, we get

$$\overline{L_n} = \Lambda_n'(1) = H_n; \tag{3.5c}$$

$$\operatorname{var}(L_n) = \Lambda_n''(1) + \Lambda_n'(1) - (\Lambda_n'(1))^2 = H_n - H_n^{(2)}, \tag{3.5d}$$

where H_n is the nth harmonic number $\Sigma_{1 \leqslant k \leqslant n} 1/k$, and $H_n^{(2)} = \Sigma_{1 \leqslant k \leqslant n} 1/k^2$. The mean $\overline{L_n}$ is much less than the worst-case value of L_n, which is n. As above, we can look at this derivation in the way suggested by the product decomposition of $\Lambda_n(z)$ in (3.5b): We can decompose L_n into a sum of independent RVs $l_1 + l_2 + \cdots + l_n$, where $l_k[\sigma]$ is 1 if σ_k is a left-to-right minimum, and 0 otherwise. The PGF for l_k is $l_k(z)$ given in (3.5b), and summing $\overline{l_k}$ and $\operatorname{var}(l_k)$ for $1 \leqslant k \leqslant n$ gives (3.5c) and (3.5d). The central limit theorem shows that L_n, when normalized, converges to the normal distribution.

The above information about I_n and L_n suffices for our purposes of analyzing sorting algorithms, but it is interesting to point out that Theorem 3.1 has further applications. For example, let $T_{n,i,j,k}$ be the number of permutations $\sigma \in S_n$ such that $I_n[\sigma] = i$, $L_n[\sigma] = j$, and there are k *left-to-right maxima*. By Theorem 3.1, the OGF of $T_{n,i,j,k}$ is

$$\begin{aligned} T_n(x, y, z) &= \sum_{i,j,k} T_{n,i,j,k} x^i y^j z^k \\ &= yz(y + xz)(y + x + x^2 z)\cdots(y + x + x^2 + \cdots + x^{n-1} z). \end{aligned} \tag{3.6}$$

3.2. *Insertion sort*

Insertion sort is the method card players typically use to sort card hands. In the kth loop, for $1 \leqslant k \leqslant n-1$, the first k elements $x[1], \ldots, x[k]$ are already in sorted order, and the $(k+1)$st element $x[k+1]$ is inserted into its proper place with respect to the preceding elements.

In the simplest variant, called *straight insertion*, the correct position for $x[k+1]$ is found by successively comparing $x[k+1]$ with $x[k], x[k-1], \ldots$ until an element $\leqslant x[k+1]$ is found. The intervening elements are simultaneously bumped one position to the right to make room. (For simplicity, we assume that there is a dummy element

$x[0]$ with value $-\infty$ so that an element $\leqslant x[k+1]$ is always found.) When the values in the input file are distinct, the number of comparisons in the kth loop is equal to 1 plus the number of elements $>x[k+1]$ that precede $x[k+1]$ in the input. In terms of the inversion table (3.2a), this is equal to $1+b_{x[k+1]}$. By summing on k, we find that the total number of comparisons used by straight insertion to sort a permutation σ is $I_n[\sigma]+n-1$. The following theorem follows directly from (3.4c) and (3.4d):

3.2. THEOREM. *The mean and the variance of the number of comparisons performed by straight insertion when sorting a random permutation are* $n^2/4+3n/4-1$ *and* $n(2n+5)\times(n-1)/72$ *respectively.*

An alternative to straight insertion is to store the already-sorted elements in a binary search tree; the kth loop consists of inserting element $x[k+1]$ into the tree. After all n elements are inserted, the sorted order can be obtained via an in-order traversal. A balanced binary search tree can be used to ensure $O(n \log n)$ worst-case time performance, but the overhead of the balancing operations slows down the algorithm in practice. When the tree is not required to be balanced, there is little overhead, and the average running time is faster. We defer the analysis until our discussion of binary search trees in Section 4.2.

3.3. Shellsort

The main reason why straight insertion is relatively slow is that the items are inserted sequentially; each comparison reduces the number of inversions (which is $\Theta(n^2)$ on the average) by at most 1. Thus, the average running time is $\Theta(n^2)$. Shell [102] proposed an efficient variant (now appropriately called *Shellsort*) in which the insertion process is done in several passes of successive refinements. For a given input size n, the passes are determined by an "increment sequence" $h_t, h_{t-1}, \ldots, h_1$, where $h_1=1$. The h_i pass consists of straight insertion sorts of each of the h_i subfiles

$$\begin{array}{ll} \text{subfile 1:} & x[1]\,x[1+h_i]\,x[1+2h_i]\ldots \\ \text{subfile 2:} & x[2]\,x[2+h_i]\,x[2+2h_i]\ldots \\ \vdots & \vdots \\ \text{subfile } h_i\text{:} & x[h_i]\,x[2h_i]\,x[3h_i]\ldots \end{array} \tag{3.7}$$

In the early passes (when the increments are typically large), elements can be displaced far from their previous positions with only a few comparisons; the later passes "fine tune" the placement of elements. The last pass, when $h_1=1$, consists of a single straight insertion sort of the entire array; we know from Section 3.2 that this is fast when the number of remaining inversions is small.

Two-ordered permutations

A good introduction to the average-case analysis of Shellsort is the two-pass version with increment sequence (2, 1). We assume that the input is a random permutation. Our measure of complexity is the total number of inversions encountered in the subfiles (3.7) during the course of the algorithm. For simplicity, we restrict ourselves to the case when n is even.

The first pass is easy to analyze, since it consists of two independent straight insertion sorts, each of size $n/2$. We call a permutation *k-ordered* if $x[i] < x[i+k]$ for all $1 \leq i \leq n-k$. At the end of the first pass, the permutation is 2-ordered, and by our randomness assumption it is easy to see that each of the $\binom{n}{n/2}$ possible 2-ordered permutations is equally likely. The analysis of the last pass consists in determining the average number of inversions in a random 2-ordered permutation.

3.3. THEOREM. *The mean and the variance of the number of inversions I_{2n} in a random 2-ordered permutation of size $2n$ are*

$$\overline{I_{2n}} = \frac{n4^{n-1}}{\binom{2n}{n}} \sim \frac{\sqrt{\pi}}{4} n^{3/2}, \qquad \operatorname{var}(I_{2n}) \sim \left(\frac{7}{30} - \frac{\pi}{16}\right) n^3.$$

PROOF. The starting point of the proof is the important 1-1 correspondence between the set of 2-ordered permutations of $2n$ elements and the set of monotone paths from the upper-left corner $(0, 0)$ to the bottom-right corner (n, n) of the n-by-n lattice. The kth step of the path is $\downarrow$ if k appears in an odd position in the permutation, and it is $\rightarrow$ if k appears in an even position. The path for a typical permutation σ is given in Fig. 2. The sorted permutation has the "staircase path" shown by dotted lines. The important property of this representation is that *the number $I_{2n}[\sigma]$ of inversions in σ is equal to the area between the staircase path and σ's path.*

There is an easy heuristic argument to show why $\overline{I_{2n}}$ should be $\Theta(n^{3/2})$: Intuitively, the first n steps of a random path from $(0, 0)$ to (n, n) are like a random walk, and

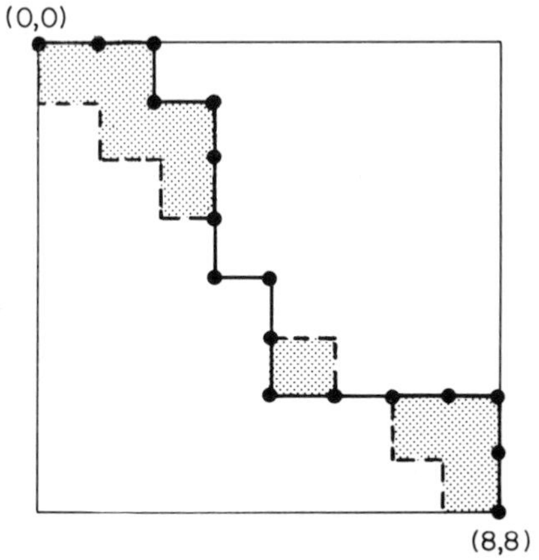

Fig. 2. Correspondence between 2-ordered permutations of $2n$ elements and monotone paths from $(0, 0)$ to (n, n), for $n=8$. The dotted path represents the sorted permutation. The dark path corresponds to the permutation $\sigma = 3\ 1\ 5\ 2\ 6\ 4\ 7\ 8\ 9\ 11\ 10\ 12\ 15\ 13\ 16\ 14$. The number of inversions in σ (namely, 9) is equal to the shaded area between σ's path and the staircase path.

similarly for the last n steps. (The transition probabilities are slightly different from those for a random walk since the complete path is constrained to have exactly $n \downarrow$ moves and $n \rightarrow$ moves.) It is well known that random walks tend to be $\Theta(\sqrt{n})$ units away from the diagonal after n steps, thus suggesting that the area between the walk and the staircase path is $\Theta(n^{3/2})$.

An extension to the notions of admissibility in Section 1.2 provides an elegant and precise way to count the area, cumulated among all possible 2-ordered permutations. Let us define $\mathscr{P}$ to be the set of all possible paths of length $\geqslant 0$ from $(0,0)$ to another point on the diagonal, and $\hat{\mathscr{P}}$ to be the subset of all "arches" (paths that meet the diagonal only at the endpoints) of positive length. Each path in $\mathscr{P}$ can be decomposed uniquely into a concatenation of zero or more arches in $\hat{\mathscr{P}}$. In the language of Theorem 1.2, $\mathscr{P}$ is the sequence class of $\hat{\mathscr{P}}$:

$$\mathscr{P} = \hat{\mathscr{P}}^*. \tag{3.8a}$$

For example, the path $p \in \mathscr{P}$ in Fig. 2 consists of four paths in $\hat{\mathscr{P}}$: from (0, 0) to (3, 3), from (3, 3) to (4, 4), from (4, 4) to (6, 6), and from (6, 6) to (8, 8). For reasons of symmetry, it is useful to look at paths that stay to one side (say, the right side) of the diagonal. The restrictions of $\mathscr{P}$ and $\hat{\mathscr{P}}$ to the right side of the diagonal are denoted $\mathscr{S}$ and $\hat{\mathscr{S}}$ respectively. As above, we have

$$\mathscr{S} = \hat{\mathscr{S}}^*. \tag{3.8b}$$

Each path has the same number of $\downarrow$ moves as $\rightarrow$ moves. We define the *size* of path p, denoted $|p|$, to be the number of $\downarrow$ moves in p. The other parameter of interest is the area between p and the staircase path; we call this area the *weight* of p and denote it by $\|p\|$. The size and weight functions are linear; that is, $|pq| = |p| + |q|$ and $\|pq\| = \|p\| + \|q\|$, where pq denotes the concatenation of paths p and q.

Let $P_{n,k}$ (respectively $\hat{P}_{n,k}$, $S_{n,k}$, $\hat{S}_{n,k}$) be the number of paths $p \in \mathscr{P}$ (respectively $\hat{\mathscr{P}}, \mathscr{S}, \hat{\mathscr{S}}$) such that $|p| = n$ and $\|p\| = k$. We define the OGF

$$P(u,z) = \sum_{k,n} P_{n,k} u^k z^n, \tag{3.9}$$

and we define the OGFs $\hat{P}(u,z)$, $S(u,z)$, and $\hat{S}(u,z)$ similarly. The mean and variance of I_{2n} can be expressed readily in terms of $P(u,z)$:

$$\overline{I_{2n}} = \frac{1}{\binom{2n}{n}} [z^n] \left.\frac{\partial P(u,z)}{\partial u}\right|_{u=1}; \tag{3.10a}$$

$$\overline{I_{2n}(I_{2n}-1)} = \frac{1}{\binom{2n}{n}} [z^n] \left.\frac{\partial^2 P(u,z)}{\partial u^2}\right|_{u=1}; \tag{3.10b}$$

$$\operatorname{var}(I_{2n}) = \overline{I_{2n}(I_{2n}-1)} + \overline{I_{2n}} - (\overline{I_{2n}})^2. \tag{3.10c}$$

We can generalize our admissibility approach in Theorem 1.2 to handle OGFs with two variables as follows.

3.4. LEMMA. *The sequence construct is admissible with respect to the size and weight functions*:

$$\mathscr{P}=\hat{\mathscr{P}}^* \;\Rightarrow\; P(u,z)=\frac{1}{1-\hat{P}(u,z)}; \tag{3.11}$$

$$\mathscr{S}=\hat{\mathscr{S}}^* \;\Rightarrow\; S(u,z)=\frac{1}{1-\hat{S}(u,z)}. \tag{3.12}$$

PROOF. An equivalent definition of the OGF $P(u, z)$ is

$$P(u,z)=\sum_{p\in\mathscr{P}} u^{\|p\|}z^{|p|}.$$

Each nontrivial path $p\in\mathscr{P}$ can be decomposed uniquely into a concatenation $\hat{p}_1\hat{p}_2\ldots\hat{p}_l$ of nontrivial paths in $\hat{\mathscr{P}}$. By the linearity of the size and weight functions, we have

$$P(u,z)=\sum_{\substack{\hat{p}_1,\hat{p}_2,\ldots,\hat{p}_l\in\hat{\mathscr{P}}\\ l\geqslant 0}} u^{\|\hat{p}_1\|+\|\hat{p}_2\|+\cdots+\|\hat{p}_l\|}z^{|\hat{p}_1|+|\hat{p}_2|+\cdots+|\hat{p}_l|}$$

$$=\sum_{l\geqslant 0}\left(\sum_{\hat{p}\in\hat{\mathscr{P}}} u^{\|\hat{p}\|}z^{|\hat{p}|}\right)^l=\sum_{l\geqslant 0}\left(\hat{P}(u,z)\right)^l=\frac{1}{1-\hat{P}(u,z)}.$$

The proof of (3.12) is identical. □

PROOF OF THEOREM 3.3 (*continued*). Lemma 3.4 gives us two formulae relating the four OGFs $P(u, z)$, $\hat{P}(u, z)$, $S(u, z)$, and $\hat{S}(u, z)$. The following decomposition gives us another two relations, which closes the cycle and lets us solve for the OGFs: Every path $\hat{s}\in\hat{\mathscr{S}}$ can be decomposed uniquely into the path $\rightarrow s\downarrow$, for some $s\in\mathscr{S}$, and the size and weight functions of $\hat{s}$ and s are related by $|\hat{s}|=|s|+1$ and $\|\hat{s}\|=\|s\|+|s|+1$. Hence, we have

$$\hat{S}(u,z)=\sum_{s\in\mathscr{S}} u^{\|s\|+|s|+1}z^{|s|+1}=uz\sum_{s\in\mathscr{S}} u^{\|s\|}(uz)^{|s|}=uz\,S(u,uz). \tag{3.13}$$

Each path in $\hat{\mathscr{P}}$ is either in $\hat{\mathscr{S}}$ or in the reflection of $\hat{\mathscr{S}}$ about the diagonal, which we call refl($\hat{\mathscr{S}}$). For $\hat{s}\in\hat{\mathscr{S}}$, we have $|\text{refl}(\hat{s})|=|\hat{s}|$ and $\|\text{refl}(\hat{s})\|=\|\hat{s}\|-|\hat{s}|$, which gives us

$$\hat{P}(u,z)=\sum_{\hat{s}\in\hat{\mathscr{S}}\cup\text{refl}(\hat{\mathscr{S}})} u^{\|\hat{s}\|}z^{|\hat{s}|}=\sum_{\hat{s}\in\hat{\mathscr{S}}}(u^{\|\hat{s}\|}z^{|\hat{s}|}+u^{\|\hat{s}\|-|\hat{s}|}z^{|\hat{s}|})=\hat{S}(u,z)+\hat{S}(u,z/u). \tag{3.14}$$

Equations (3.13) and (3.14) can be viewed as types of admissibility reductions, similar to those in Lemma 3.4, except that in this case the weight functions are not linear (the relations between the weight functions involve the size function), thus introducing the uz and z/u arguments in the right-hand sides of (3.13) and (3.14).

The four relations (3.11)–(3.14) allow us to solve for $P(u, z)$. Substituting (3.13) into (3.12) gives

$$S(u,z)=uzS(u,z)S(u,uz)+1, \tag{3.15}$$

and substituting (3.13) into (3.14) and the result into (3.11), we get

$$P(u,z)=(uzS(u,uz)+zS(u,z))P(u,z)+1. \tag{3.16}$$

Using (3.16), we can then express $(\partial/\partial u)P(u,z)|_{u=1}$ and $(\partial^2/\partial u^2)P(u,z)|_{u=1}$, which we need for (3.10), in terms of derivatives of $S(u,z)$ evaluated at $u=1$, which in turn can be calculated from (3.15). These calculations are straightforward, but are best done with a symbolic algebra system. □

ALTERNATE PROOF OF THEOREM 3.3. We can prove the first part of Theorem 3.3 in a less elegant way by studying how the file is decomposed after the first pass into the two sorted subfiles

$$X_1=x[1]x[3]x[5]\ldots \quad \text{and} \quad X_2=x[2]x[4]x[6]\ldots$$

We can express $\overline{I_{2n}}$ as

$$\overline{I_{2n}}=\frac{1}{\binom{2n}{n}}\sum_{\substack{1\leqslant i\leqslant n\\ 0\leqslant j\leqslant n-1}} A_{i,j}, \tag{3.17a}$$

where $A_{i,j}$ is the total number of inversions involving the ith element of X_1 (namely, $x[2i-1]$) among all 2-ordered permutations in which there are j elements in X_2 less than $x[2i-1]$. The total number of such 2-ordered permutations is

$$\binom{i+j+1}{i-1}\binom{2n-i-j}{n-j},$$

and a simple calculation shows that each contributes $|i-j|$ inversions to $A_{i,j}$. Substituting this into (3.17a), we get

$$\overline{I_{2n}}=\frac{1}{\binom{2n}{n}}\sum_{\substack{1\leqslant i\leqslant n\\ 0\leqslant j\leqslant n-1}} |i-j|\binom{i+j-1}{i-1}\binom{2n-i-j}{n-j}, \tag{3.17b}$$

and the rest of the derivation consists of manipulation of binomial coefficients. The derivation of the variance is similar. □

The increment sequence (2, 1) is not very interesting in practice, because the first pass still takes quadratic time, on the average. We can generalize the above derivation to show that the average number $\overline{I_n}$ of inversions in a random h-ordered permutation is

$$\overline{I_n}=\frac{2^{2q-1}q!q!}{(2q+1)!}\left(\binom{h}{2}q(q+1)+\binom{r}{2}(q+1)-\binom{h-r}{2}\frac{q}{2}\right), \tag{3.18}$$

where $q=\lfloor n/h\rfloor$ and $r=n \bmod h$ [61]. This allows us to determine (for large h and larger n) the best two-pass increment sequence $(h, 1)$. In the first pass, there are h insertion sorts of size $\sim n/h$; by (3.4c), the total number of inversions in the subfiles is $\sim n^2/(4h)$, on the average. By (3.18), we can approximate the average number of inversions encountered in the second pass by $\overline{I_n}\sim\frac{1}{8}\sqrt{\pi h}n^{3/2}$. The total number of inversions in both passes is thus $\sim n^2/(4h)+\frac{1}{8}\sqrt{\pi h n^{3/2}}$ on the average, which is minimized when $h\sim(16n/\pi)^{1/3}$. The resulting expected running time is $O(n^{5/3})$.

When there are more than two passes, an interesting phenomenon occurs: *if an h-sorted file is k-sorted, the file remains h-sorted.* A.C.-C. Yao [113] shows how to combine this fact with an extension of the approach used in (3.17a) and (3.17b) to analyze increments of the form $(h, k, 1)$, for constant values h and k. Not much else is known in the average case, except when each increment is a multiple of the next. In that case, the running time can be reduced to $O(n^{1.5+\varepsilon/2})$, where $\varepsilon=1/(2^t-1)$ and t is the number of increments.

Pratt discovered that we get an $O(n\log^2 n)$ time algorithm in the worst case if we use be in decreasing order, but $2^{p+1}3^q$ and 2^p3^{q+1} should precede 2^p3^q. This particular approach is typically not used in practice, since the number of increments (and hence the number of passes) is $O(\log^2 n)$. Sequences with only $O(\log n)$ increments that result in $O(n^{1+\varepsilon})$ running time are reported in [62]. Lower bounds on the worst-case sorting time for various types of increment sequences with $O(\log n)$ increments are given in [110, 18]. For example, Shellsort has $\Omega(n\log^2 n/\log\log n)$ worst-case time when the increment sequence is monotonically decreasing [18].

A possible application of Shellsort is to the construction of efficient networks for sorting. Sorting networks operate via a sequence of pairwise comparison/exchanges, where the choice of which pair of elements to compare next is made independently of the outcomes of the previous comparisons. Comparison/exchanges that involve different elements can be done in parallel by the network, so up to $n/2$ operations can be done simultaneously in one parallel step. Sorting networks thus require $\Omega(\log n)$ parallel steps (or depth). Batcher developed practical sorting networks of depth $\frac{1}{2}k^2$ for $n=2^k$, based upon his odd-even merge and bitonic sort networks [4, 71]. Later, Ajtai, Komlós, and Szemerédi [1] solved a longstanding open problem by constructing a sorting network of depth $O(\log n)$; a complete coverage is given in [88]. The AKS sorting network is a theoretical breakthrough, but in terms of practicality the network is not very useful since the coefficient implicit in the big-oh term is huge. If an $O(n\log n)$ time Shellsort is found, it might be possible to modify it to yield a sorting network of depth $O(\log n)$ that is practical. However, the lower bound result quoted above [18] shows that, in order to find an $O(n\log n)$ time Shellsort, the increment sequence will have to be fundamentally different from those used in practice.

3.4. *Bubble sort*

The bubble sort algorithm works by a series of passes. In each pass, some final portion $x[l+1]x[l+2]\ldots x[n]$ of the array is already known to be in sorted order, and the largest element in the initial part of the array $x[1]x[2]\ldots x[l]$ is "bubbled" to the right by a sequence of $l-1$ pairwise comparisons

$$x[1]:x[2],\; x[2]:x[3],\ldots, x[l-1]:x[l].$$

In each comparison, the elements exchange place if they are out of order. The value of l is reset to the largest t such that the comparison $x[t]:x[t+1]$ required an exchange, and then the next pass starts.

The bubble sort algorithm itself is not very useful in practice, since it runs more slowly than insertion sort and selection sort, yet is more complicated to program. However, its analysis provides an interesting use of inversion statistics and asymptotic techniques. The running time of bubble sort depends upon three quantities: the number of inversions I_n, the number of passes A_n, and the number of comparisons C_n. The analysis of I_n has already been given in Section 3.1.

3.5. THEOREM (Knuth [71]). *The average number of passes and comparisons done in bubble sort, on a random permutation of size n, is*

$$\overline{A_n}=n-\sqrt{\tfrac{1}{2}\pi n}+O(1), \qquad \overline{C_n}=\tfrac{1}{2}(n^2-n\log n-(\gamma+\log 2-1)n)+O(\sqrt{n}).$$

PROOF. Each pass in bubble sort reduces all the non-zero entries in the inversion table by 1. There are at most k passes in the algorithm when each entry in the original inversion table is $\leqslant k$. The number of such inversion tables can be obtained via Theorem 3.1 by substituting $x_i=\delta_{i\leqslant k}$ into $F(x_0,x_1,\ldots,x_n)$, which gives $k!\,k^{n-k}$. We use the notation δ_R to denote 1 if relation R is true and 0 otherwise. Plugging this into the definition of expected value, we get

$$\overline{A_n}=n+1-\frac{1}{n!}\sum_{0\leqslant k\leqslant n} k!k^{n-k}. \tag{3.19}$$

The summation can be shown to be equal to $\sqrt{\pi n/2}-\frac{2}{3}+O(1/\sqrt{n})$ by an application of the Euler–Maclaurin summation formula (Theorem 2.1).

The average number of comparisons $\overline{C_n}$ can be determined in a similar way. For the moment, let us restrict our attention to the jth pass. Let c_j be the number of comparisons done in the jth pass. We have $c_j=l-1$, where l is the upper index of the subarray processed in pass j. We can characterize l as the last position in the array at the beginning of pass j that contains an element which moved to the left one slot during the previous pass. We denote the value of the element in position l by i. It follows that all the elements in positions $l+1$, $l+2,\ldots,$ n have value $>i$. We noted earlier that each non-zero entry in the inversion table decreases by 1 in each pass. Therefore, the number of inversions for element i at the beginning of the jth pass is b_i-j+1, and element i is located in array position $l=i+b_i-j+1$. This gives us $c_j=i+b_i-j$.

Without a priori knowledge of l or i, we can calculate c_j by using the formula

$$c_j=\max_{1\leqslant i\leqslant n}\{i+b_i-j \mid b_i\geqslant j-1\}. \tag{3.20}$$

The condition $b_i\geqslant j-1$ restricts attention to those elements that move left one place in pass $j-1$; it is easy to see that the term in (3.20) is maximized at the correct i. To make use of (3.20), let us define $f_j(k)$ to be the number of inversion tables (3.2a) such that either $b_i<j-1$ or $i+b_i-j\leqslant k$. We can evaluate $f_j(k)$ from Theorem 3.1 by substituting $x_i=1$ if $b_i<j-1$ or $i+b_i-j\leqslant k$, and $x_i=0$ otherwise. A simple calculation gives

$$f_j(k)=(j+k)!\,(j-1)^{n-j+k} \quad \text{for } 0\leqslant k\leqslant n-j.$$

By the definition of expected value, we have

$$\overline{C_n}=\frac{1}{n!}\sum_{\substack{1\leq j\leq n\\0\leq k\leq n-j}} k(f_j(k)-f_j(k-1))=\binom{n+1}{2}-\frac{1}{n!}\sum_{\substack{1\leq j\leq n\\0\leq k\leq n-j}} f_j(k)$$

$$=\binom{n+1}{2}-\frac{1}{n!}\sum_{0\leq r<s\leq n} s!\,r^{n-s}. \tag{3.21a}$$

The intermediate step follows by summation by parts. The summation in (3.21a) can be simplified using the Euler–Maclaurin summation formula into series of sums of the form

$$\frac{1}{m!}\sum_{1\leq t<m}(m-t)!\,(m-t)^t t^q \quad \text{for } q\geq -1, \tag{3.21b}$$

where $m=n+1$. By applying Stirling's approximation, we find that the summand in (3.21b) decreases exponentially when $t>m^{1/2+\varepsilon}$, and we can reduce the problem further to that of approximating $F_q(1/m)$, where

$$F_q(x)=\sum_{k\geq 1} \mathrm{e}^{-k^2x/2}k^q \quad \text{for } q\geq -1, \tag{3.22a}$$

as $m\to\infty$. The derivation can be carried out using Laplace's method for sums and the Euler–Maclaurin summation formula (see Section 2.1), but things get complicated for the case $q=-1$. A more elegant derivation is to use Mellin transforms. The function $F_q(x)$ is a harmonic sum, and its Mellin transform $F_q^*(s)$ is

$$F_q^*(s)=\Gamma(s)\zeta(2s-q)2^s, \tag{3.22b}$$

defined in the fundamental strip $\langle 0;\ +\infty\rangle$. The asymptotic expansion of $F_q(1/m)$ follows by computing the residues in the left half-plane $\Re(s)\leq 0$. There are simple poles at $s=-1,-2,\ldots$ because of the term $\Gamma(s)$. When $q=-1$, the $\Gamma(s)$ and $\zeta(2s-q)$ terms combine to contribute a double pole at $s=0$. When $q\geq 0$, $\Gamma(s)$ contributes a simple pole at $s=0$, and $\zeta(2s-q)$ has a simple pole only at $s=(q+1)/2>0$. Putting everything together, we find that

$$\frac{1}{n!}\sum_{0\leq r<s\leq n} s!r^{n-s}=\tfrac{1}{2}n\log n+\tfrac{1}{2}(\gamma+\log 2)n+O(\sqrt{n}). \tag{3.23}$$

The formula for $\overline{C_n}$ follows immediately from (3.21a). □

After k passes in the bubble sort algorithm, the number of elements known to be in their final place is typically larger than k; the variable l is always set to be as small as possible so as to minimize the size of the array that must be considered in a pass. The fact that (3.23) is $O(n\log n)$ implies that the number of comparisons for large n is not significantly less than that of the more naïve algorithm in which the bubbling process in the kth pass is done on the subarray $x[1]x[2]\dots x[n-k+1]$.

3.5. *Quicksort*

We can get a more efficient exchange-based sorting algorithm by using a divide-and-conquer approach. In the *quicksort* method, due to Hoare, an element s is chosen (say, the first element in the file), the file is partitioned into the part $\leqslant s$ and the part $>s$, and then each half is sorted recursively. The recursive decomposition can be analyzed using the sophisticated tools we develop in Section 4.2 for binary search trees, which have a similar decomposition, and so we defer the analysis until Section 4.2. The expected number of comparisons is $\sim 2n \log n$. Analysis techniques and results for quicksort can be found in [71, 98].

Quicksort is an extremely good general-purpose sorting routine. A big drawback of the version described above is its worst-case performance: it requires $\Theta(n^2)$ time to sort a file that is already sorted or nearly sorted! A good way to guard against guaranteed bad behavior is to choose the partitioning element s to be a random element in the current subfile, in which case our above analysis applies. Another good method is to choose s to be the median of three elements from the subfile (say, the first, middle, and last elements). This also has the effect of reducing the average number of comparisons to $\sim \frac{12}{7} n \log n$.

If the smaller of the two subfiles is always processed first after each partition, then the recursion stack contains at most $\log n$ entries. But by clever programming, we can simulate the stack with only a constant amount of space, at a very slight increase in computing time [25, 60]. The analysis of quicksort in the presence of some equal keys is given in [97].

3.6. *Radix-exchange sort*

The *radix-exchange sort* algorithm is an exchange-based algorithm that uses divide-and-conquer in a different way than quicksort does. The recursive decomposition is based upon the individual bits of the keys. In pass k, the keys are partitioned into two groups: the group whose kth least significant bit is 0, and the group whose kth least significant bit is 1. The partitioning is done in a "stable" manner so that the relative order of the keys within each group is the same as before the partitioning. Then the 1-group is appended to the 0-group, and the next pass begins. After t passes, where t is the number of bits in the keys, the algorithm terminates.

In this case, the recursive decomposition is identical to radix-exchange tries, which we shall study in a general context in Section 4.3, and the statistics of interest for radix-exchange sorting can be expressed directly in terms of corresponding parameters of tries. We defer the details to Section 4.3.

3.7. *Selection sort and heapsort*

Selection sort in some respects is the inverse of insertion sort, because the order in which the elements are processed is based upon the output order rather than upon the input order. In the kth pass, for $1 \leqslant k \leqslant n-1$, the kth smallest element is selected and is put into its final place $x[i]$.

Straight selection

In the simplest variant, called *straight selection sort*, the kth smallest element is found by a sequential scan of $x[k]x[k+1]\ldots x[n]$, and it changes places with the current $x[k]$. Unlike insertion sort, the algorithm is not stable; that is, two elements with the same value might be output in the reverse of the order that they appear in the input.

The number of comparisons performed is always

$$C_n = (n-1)+(n-2)+\cdots+1 = \tfrac{1}{2}n(n-1).$$

The number of times a new minimum is found (the number of data movements) in the kth pass is the number of left-to-right minima L_{n-k+1} encountered in $x[k]x[k+1]\ldots x[n]$, minus 1. All permutations of $\{x[k], x[k+1], \ldots, x[n]\}$ are equally likely. By (3.5c), the average number of updates of the minimum over the course of the algorithm is

$$\sum_{1 \leqslant k \leqslant n-1} (\overline{L_{n-k+1}} - 1) = (n+1)H_n - 2n = n \log n + (\gamma - 2)n + \log n + O(1).$$

The variance is much more difficult to compute, since the contributions L_{n-k+1} from the individual passes are not independent. If the contributions were independent, then by (3.5d) the variance would be $\sim n \log n$. A.C.-C. Yao [114] shows by relating the variance to a geometric stochastic process that the variance is $\sim \alpha n^{3/2}$, and he gives the constant $\alpha = 0.91\ldots$ explicitly in summation form.

Heapsort

A more sophisticated way of selecting the minimum, called *heapsort*, due to Williams, is based upon the notion of tournament elimination. The $n-k+1$ elements to consider in the kth pass are stored in a heap priority queue. A heap is a tree in which the value of the root of each subtree is less than or equal to the values of the other elements in the subtree. In particular, the smallest element is always at the root. The heaps we use for heapsort have the nice property that the tree is always perfectly balanced, except possibly for the rightmost part of the last level. This allows us to represent the heap as an array h without need of pointers: the root is element $h[1]$, and the children of element $h[i]$ are stored in $h[2i]$ and $h[2i+1]$.

The kth pass of heapsort consists of outputing $h[1]$ and deleting it from the array; the element stored in $h[n-k+1]$ is then moved to $h[1]$, and it is "filtered" down to its correct place in $O(\log n)$ time. The creation of the initial heap can be done in linear time. The worst-case time to sort the n elements is thus $O(n \log n)$. In the average case, the analysis is complicated by a lack of randomness: the heap at the start of the kth pass, for $k \geqslant 2$, is not a random heap of $n-k+1$ elements. Heaps and other types of priority queues are discussed in Section 6.1 and [83].

3.8. *Merge sort*

The first sorting program ever executed on an electronic computer used the following divide-and-conquer approach, known as merge sort: The file is split into two subfiles of equal size (or nearly equal size), each subfile is sorted recursively, and then the two sorted subfiles are merged together in the obvious linear fashion. When done from bottom-up, the algorithm consists of several passes; in each pass, the sorted subfiles are paired off, and each pair is merged together.

The linear nature of the merging process makes it ideal for input files in the form of a linked list and for external sorting applications, in which the file does not fit entirely in internal memory and must instead be stored in external memory (like disk or tape), which is best accessed in a sequential manner. Typically, the merge is of a higher order than 2; for example, four subfiles at a time might be merged together, rather than just two. Considerations other than the number of comparisons, such as the rewind time on tapes and the seek time on disks, affect the running time. An encyclopedic collection of variants of merge sort and their analyses appears in [71]. Merge sort algorithms that are optimal for multiple disks are discussed in [117, 120].

For simplicity, we restrict our attention to the number of comparisons performed during a binary (order-2) merge sort, when $n=2^j$ for some $j\geqslant 0$. All the comparisons take place during the merges. For each $0\leqslant k\leqslant j-1$, there are 2^k merges of pairs of sorted subfiles, each subfile of size $n/2^{k+1}=2^{j-k-1}$. If we assume that all $n!$ permutations are equally likely, it is easy to see that, as far as relative order is concerned, the two subfiles in each merge form a random 2-ordered permutation, independent of the other merges. The number of comparisons to merge two random sorted subfiles of length p and q is $p+q-\mu$, where μ is the number of elements remaining to be output in one subfile when the other subfile becomes exhausted. The probability that $\mu\geqslant s$, for $s\geqslant 1$, is the probability that the s largest elements are in the same subfile, namely,

$$\frac{p^{\underline{s}}}{(p+q)^{\underline{s}}}+\frac{q^{\underline{s}}}{(p+q)^{\underline{s}}},$$

where $a^{\underline{b}}$ denotes the "falling power" $a(a-1)\cdots(a-b+1)$. Hence, we have

$$\bar{\mu}=\sum_{s\geqslant 1}\left(\frac{p^{\underline{s}}}{(p+q)^{\underline{s}}}+\frac{q^{\underline{s}}}{(p+q)^{\underline{s}}}\right)=\frac{p}{q+1}+\frac{q}{p+1}\,. \tag{3.24}$$

By (3.24), the total number of comparisons used by merge sort, on the average, is

$$\overline{C_n}=j2^j-\sum_{1\leqslant k\leqslant j-1}2^k\frac{2^{j-k}}{2^{j-k-1}+1}=n\log_2 n-\alpha n+\mathrm{O}(1),$$

where $\alpha=\sum_{n\geqslant 0}(2^n+1)^{-1}=1.2645\ldots$

The analysis when n is not a power of 2 involves arithmetic functions based upon the binary representation of n and can be found in [71]. Batcher's odd-even merge and bitonic sort networks, which can be used to construct sorting networks of depth $\frac{1}{2}(\log_2 n)^2$, are analyzed in [4, 71, 99]. Other merging algorithms are covered in [83].

4. Recursive decompositions and algorithms on trees

In this section we develop a uniform framework for obtaining average-case statistics on four classes of trees—binary and plane trees, binary search trees, radix-exchange tries, and digital search trees. Our statistics, which include number of trees, number of nodes, height, and path length, have numerous applications to the analysis of tree-based searching and symbolic processing algorithms, as well as to the sorting algorithms whose analysis we deferred from Section 2, such as quicksort, binary tree sort, and radix-exchange sort. Our approach has two parts:

(1) Each of the four classes of trees has a recursive formulation that lends itself naturally to the symbolic generating function method described in Section 1. The statistic of interest for each tree t corresponds naturally to a *valuation function* (*VF*) $v[t]$. The key idea which unifies our analyses is an extension of the admissibility concept of Section 1: A recursive definition of the VF translates directly into a functional equation involving the generating function. The type of generating function used (either OGF or EGF) and the type of functional equation that results depend upon the particular nature of the recursion.

(2) We determine the coefficients of the GF based upon the functional equation resulting from step (1). Sometimes an explicit closed form is obtained, but typically we apply the asymptotic methods of Section 2, our particular approach depending upon the nature of the functional equation.

4.1. *Binary trees and plane trees*

Binary trees and plane trees provide a natural representation for many types of symbolic expressions and recursive structures. This section studies statistical models under which all trees of a given size are equally likely. Such models are not applicable to the study of binary search trees, radix-exchange tries, and digital search trees, which we cover in Sections 4.2, 4.3, and 4.4, but when enriched slightly, they provide good models for algorithms operating on expression trees, term trees, and Lisp structures [15].

We begin by considering the class $\mathcal{B}$ of *binary trees* defined in Section 1.1:

$$\mathcal{B}=\{\blacksquare\}+(\{\bigcirc\}\times\mathcal{B}\times\mathcal{B}), \tag{4.1}$$

where ■ represents an external node and ○ an internal node. The size of a binary tree is the number of internal nodes in the tree.

The Cartesian product decomposition in (4.1) suggests that we represent our statistic of interest via an OGF. Further motivation for this choice is given in (1.2) and (1.3). We use $v[t]$ to represent the valuation function v applied to tree t. We define v_n to be its cumulated value $\sum_{|t|=n} v[t]$ among all trees of size n, and $v(z)$ to be the OGF $\sum_{n\geqslant 0} v_n z^n$. The recursive decomposition of $\mathcal{B}$ leads directly to the following fundamental relations:

4.1. THEOREM. *The sum and recursive product valuation functions are admissible for the* *class $\mathscr{B}$ of binary trees:*

$$v[t]=u[t]+w[t] \quad \Rightarrow \quad v(z)=u(z)+w(z);$$
$$v[t]=u[t_{\text{left}}]\cdot w[t_{\text{right}}] \quad \Rightarrow \quad v(z)=z\cdot u(z)\cdot w(z),$$

where t_{left} and t_{right} denote the left and right subtrees of t.

The proof is similar to that of Theorem 1.1. The importance of Theorem 4.1 is due to the fact that it provides an automatic translation from VF to OGF, for many VFs of interest.

EXAMPLES. (1) Enumeration. The standard trick we shall use throughout this section for counting the number of trees of size n in a certain class is to use the unit VF $I[t]\equiv 1$. For example, let B_n, for $n\geqslant 0$, be the number of binary trees with n internal nodes. By our definitions above, B_n is simply equal to I_n, and thus the OGF $B(z)$ is equal to $I(z)$. We can solve for $B(z)$ via Theorem 4.1 if we use the following recursive definition of $I[t]$,

$$I[t]=\delta_{|t|=0}+I[t_{\text{left}}]\cdot I[t_{\text{right}}], \tag{4.2a}$$

which is a composition of the two types of VF expressions handled by Theorem 4.1. Again, we use δ_R to denote 1 if relation R is true and 0 otherwise. Since $B_0=1$, the OGF for $\delta_{|t|=0}$ is 1. Theorem 4.1 translates (4.2a) into

$$B(z)=1+zB(z)^2. \tag{4.2b}$$

The solution $B(z)=(1-\sqrt{1-4z})/2z$ follows by the quadratic formula. By expanding coefficients, we get $B_n=\binom{2n}{n}/(n+1)$, as in Section 1.1.

(2) Internal path length. The standard recursive tree traversal algorithm uses a stack to keep track of the ancestors of the current node in the traversal. The average stack size, amortized over the course of the traversal, is related to the *internal path length* of the tree, divided by n. The VF corresponding to the cumulated internal path lengths among all binary trees with n nodes can be expressed in the following form suitable for Theorem 4.1:

$$\begin{aligned} p[t]&=|t|-1+\delta_{|t|=0}+p[t_{\text{left}}]+p[t_{\text{right}}]\\ &=|t|-1+\delta_{|t|=0}+p[t_{\text{left}}]\cdot I[t_{\text{right}}]+I[t_{\text{left}}]\cdot p[t_{\text{right}}]. \end{aligned} \tag{4.3a}$$

We computed the OGFs for $I[t]\equiv 1$ and $\delta_{|t|=0}$ in the last example, and the OGF for the size VF $S[t]=t$ is easily seen to be $S(z)=\sum_{n\geqslant 0}nB_nz^n=zB'(z)$. Applying Theorem 4.1, we have

$$p(z)=zB'(z)-B(z)+1+2z\,p(z)B(z),$$

which gives us

$$p(z)=\frac{zB'(z)-B(z)+1}{1-2zB(z)}=\frac{1}{1-4z}-\frac{1}{z}\left(\frac{1-z}{\sqrt{1-4z}}-1\right). \tag{4.3b}$$

We get p_n by expanding (4.3b). The result is given below; the asymptotics follow from Stirling's formula.

4.2. THEOREM. *The cumulated internal path length over all binary trees of n nodes is*

$$p_n = 4^n - \frac{3n+1}{n+1}\binom{2n}{n},$$

and the expected internal path length p_n/B_n is asymptotically $n\sqrt{\pi n} - 3n + O(\sqrt{n})$.

Theorem 4.2 implies that the time for a traversal from leaf to root in a random binary tree is $O(\sqrt{n})$, on the average.

In a similar derivation, Knuth in [70] considers the bivariate OGF $B(u,z) = \sum_{n,k\geqslant 0} b_{n,k} u^k z^n$, where $b_{n,k}$ is the number of binary trees with size n and internal path length k. It satisfies the functional equation $B(u,z) = 1 + zB(u,uz)^2$ (cf. (4.2b)). The expected internal path length and the variance can be formed in the usual way in terms of the coefficients of z^n in the derivatives of $B(u,z)$, evaluated at $u=1$.

The two examples above illustrate our general philosophy that it is useful to compute the OGFs for a standard catalog of valuation functions, so as to handle a large variety of statistics. The most important VFs are clearly $I[t]$ and $S[t]$.

Another important class of trees is the class $\mathscr{G}$ of *plane trees* (also known as *ordered trees*). Each tree in $\mathscr{G}$ consists of a root node and an arbitrary number of ordered subtrees. This suggests the recursive definition

$$\mathscr{G} = \{\bigcirc\} \times \mathscr{G}^*, \tag{4.4}$$

where $\bigcirc$ represents a node, and $\mathscr{G}^* = \{\bigcirc\} \times \sum_{k\geqslant 0} \mathscr{G}^k$ is the sequence class of $\mathscr{G}$, defined in Section 1.2. The size of a tree is defined to be the number of its nodes. An interesting subclass of $\mathscr{G}$ is the class $\mathscr{T} = \mathscr{T}^{\Omega}$ of plane trees in which the degrees of the nodes are constrained to be in some subset Ω of the nonnegative integers. We require that $0 \in \Omega$ or else the trees will be infinite. The class $\mathscr{G}$ is the special case of $\mathscr{T}^{\Omega}$ when Ω is the set of nonnegative integers. It is possible to mimic (4.4) and get the corresponding representation for $\mathscr{T}$

$$\mathscr{T} = \{\bigcirc\} \times \sum_{k\in\Omega} \mathscr{T}^k, \tag{4.5}$$

but we shall see that it is just as simple to deal directly with (4.4) by using the appropriate VF to restrict the degrees. There are two important correspondences between $\mathscr{B}$, $\mathscr{G}$, and $\mathscr{T}^{\{0,2\}}$:

(1) The set of binary trees of size n is isomorphic to the set of plane trees of size $n+1$. A standard technique in data structures illustrates the correspondence: We represent a general tree of $n+1$ nodes as a binary tree with no right subtree and with a binary tree of n internal nodes as its left subtree; the left link denotes "first child" and the right link denotes "next sibling."

(2) If we think of the bottommost nodes of trees in $\mathscr{T}^{\{0,2\}}$ as "external nodes", we get a 1-1 correspondence between binary trees of size n and plane trees with degree constraint $\{0,2\}$ of size $2n+1$.

Theorem 4.1 generalizes in a straightforward manner for $\mathscr{G}$ in the following way.

4.3. THEOREM. *The sum and recursive product valuation functions are admissible for the class $\mathscr{G}$ of plane trees:*

$$v[t]=u[t]+w[t] \quad\Rightarrow\quad v(z)=u(z)+w(z);$$

$$v[t]=\delta_{\deg t\in\Omega}\prod_{1\leqslant i\leqslant \deg t} u_{i,\deg t}[t_i] \quad\Rightarrow\quad v(z)=z\sum_{k\in\Omega}\prod_{1\leqslant i\leqslant k} u_{i,k}(z),$$

where $t_1, t_2, \ldots, t_{\deg t}$ represent the subtrees attached to the root of t.

EXAMPLES (*Enumerations*). (1) The number G_n of plane trees with n nodes is obtained via the unit VF $I[t]\equiv 1=\prod_{0\leqslant i\leqslant \deg t} I[t_i]$. (Plane trees are always nonempty, so $|t|\geqslant 1$.) By Theorem 4.3, we get $G(z)=I(z)=z\sum_{k\leqslant 0} I(z)^k=z/(1-I(z))$. This implies $G(z)=I(z)=\frac{1}{2}(1-\sqrt{1-4z})=zB(z)$, and thus we have $G_{n+1}=B_n$, which illustrates correspondence (1) mentioned above.

(2) For the number T_n of trees of size n with degree constraint Ω, we apply to $\mathscr{G}$ the constrained unit VF $I^{\Omega}[t]=\delta_{t\in\mathscr{T}}=\delta_{\deg t\in\Omega}\prod_{0\leqslant i\leqslant \deg t} I^{\Omega}[t_i]$. For the special case $\Omega=\{0,2\}$, Theorem 4.3 gives us $T(z)=I^{\Omega}(z)=z+zI^{\Omega}(z)^2$. The solution to this quadratic equation is $T(z)=I^{\Omega}(z)=(1-\sqrt{1-4z^2})/2z=zB(z^2)$, and thus we have $T_{2n+1}=B_n$, illustrating correspondence (2).

(3) For $d\geqslant 2$, let us define the class $\mathscr{D}=\mathscr{D}_d$ of d-ary trees to be $\mathscr{D}=\{\blacksquare\}+(\{\bigcirc\}\times\mathscr{D}^d)$. Binary trees are the special case $d=2$. The number D_n of d-ary trees can be obtained by generalizing our derivation of B_n at the beginning of this section. The derivation we present, though, comes from generalizing correspondence (2) and staying within the framework of plane trees: Each d-ary tree corresponds to a plane tree of $dn+1$ nodes with degree constraint $\Omega=\{0,d\}$. The same derivation used in the preceding example gives $T(z)=I(z)=z+zI(z)^d$. By Lagrange–Bürmann inversion, with $f(z)=T(z)$, $\varphi(u)=1+u^d$, we get

$$D_n=T_{dn+1}=[z^{dn+1}]T(z)=\frac{1}{dn+1}[u^{dn}]((1+u^d)^{dn+1})=\frac{1}{dn+1}\binom{dn+1}{n}.$$

In each of the examples above, the functional equation involving the OGF was simple enough so that either the OGF could be solved in explicit closed form or else the Lagrange–Bürmann inversion theorem could be applied easily (that is, the coefficients of powers of $\varphi(u)$ were easy to determine). More advanced asymptotic methods are needed, for example, to determine the number T_n of plane trees with arbitrary degree constraint Ω. Let us assume for simplicity that Ω is *aperiodic*, that is, Ω consists of 0 and any sequence of positive integers with a greatest common divisor of 1.

To count T_n, we start out as in the second example above. By applying Theorem 4.3, we get

$$T(z)=z\omega(T(z)), \tag{4.6}$$

where $\omega(u)=\sum_{k\in\Omega}u^k$. Lagrange–Bürmann inversion is of little help when $\omega(u)$ has several terms, so we take another approach. The singularities of $T(z)$ are of an algebraic nature. We know from Section 2.2 that the asymptotic behavior of the coefficients T_n are related to the dominant singularities of $T(z)$, that is, the ones with smallest modulus.

To find the singularities of $T(z)$, let us regard $T(z)$ as the solution y of the equation

$$F(z,y)=0, \quad \text{where } F(z,y)=y-z\omega(y). \tag{4.7}$$

The function $y=T(z)$ is defined implicitly as a function of z. By the Implicit Function Theorem, the solution y with $y(z_0)=y_0$ is analytically continuable at z_0 if $F_y(z_0,y_0)\neq 0$, where $F_y(z,y)$ denotes the partial derivative with respect to y. Hence, the singularities of y (the values of z where y is not analytic) are the values ρ where

$$F(\rho,\tau)=\tau-\rho\omega(\tau)=0, \qquad F_y(\rho,\tau)=1-\rho\omega'(\tau)=0. \tag{4.8}$$

This gives $\rho=\tau/\omega(\tau)=1/\omega'(\tau)$, where τ is a root of the equation

$$\omega(\tau)-\tau\omega'(\tau)=0. \tag{4.9}$$

We denote the dominant singularity of $T(z)$ on the positive real line by ρ^*, and we let τ^* be the (unique) corresponding value of τ from (4.9). Since $T(\rho^*)=\tau^*$, it follows that τ^* is real. If Ω is aperiodic, then by examining the power series equation corresponding to (4.9), we see that τ^* is the unique real solution to (4.9), and any other solution τ must have larger modulus.

Around the point (ρ^*,τ^*), the dependence between y and z is locally of the form

$$\begin{aligned}0=F(z,y)=F_z(\rho^*,\tau^*)(z-\rho^*)+0\cdot(y-\tau^*)+\tfrac{1}{2}F_{yy}(\rho^*,\tau^*)(y-\tau^*)^2\\ +\text{smaller-order terms.}\end{aligned} \tag{4.10}$$

By iteration and bounding the coefficients, we can show that $y(z)$ has the form

$$y(z)=f(z)+g(z)\sqrt{1-\frac{z}{\rho^*}}, \tag{4.11a}$$

where $f(z)$ and $g(z)$ are analytic at $z=\rho^*$, and $g(\rho^*)=-\sqrt{2\omega(\tau^*)/\omega''(\tau^*)}$. Hence, we have

$$y(z)=f(z)+g(\rho^*)\sqrt{1-\frac{z}{\rho^*}}+O\left(\left(1-\frac{z}{\rho^*}\right)^{3/2}\right). \tag{4.11b}$$

Theorem 2.2 shows that the contribution of $f(z)$ to T_n is insignificant. By applying the transfer lemma (Theorem 2.3), we get our following final result.

4.4. THEOREM (Meir and Moon [84]). *If Ω is aperiodic, we have $T_n\sim c\rho^{-n}n^{-3/2}$, where the constants c and ρ are given by $c=\sqrt{\omega(\tau)/(2\pi\omega''(\tau))}$ and $0<\rho=\tau/\omega(\tau)<1$, and τ is the smallest positive root of the equation $\omega(\tau)-\tau\omega'(\tau)=0$.*

For brevity, we expanded only the first terms of $y(z)$ in (4.11b), but we could easily have expanded $y(z)$ further to get the full asymptotic expansion of T_n. In the periodic case, which is also considered in [84], the generating function $T(z)$ has more than one dominant singularity, and the contributions of these dominant singularities must be added together.

In the rest of this section we show how several other parameters of trees can be analyzed by making partial use of this approach. The asymptotics are often determined via the techniques described in Sections 2.2–2.4.

Height of plane trees

For example, let us consider the expected maximum stack size during a recursive tree traversal. (We earlier considered the expected stack size amortized over the course of the traversal.) The maximum stack size is simply the height of the tree, namely, the length of the longest path from the root to a node.

4.5. THEOREM (De Bruijn, Knuth, and Rice [20]). *The expected height $\overline{H_n}$ of a plane tree with n nodes, where each tree in $\mathscr{G}$ is equally likely, is $\overline{H_n}=\sqrt{\pi n}+O(1)$.*

PROOF (*Sketch*). The number $G_n^{[h]}$ of plane trees with height $\leqslant h$ corresponds to using the 0-1 VF $G^{[h]}$, which is defined to be 1 if the height of t is $\leqslant h$, and 0 otherwise. It has a recursive formulation

$$G^{[h+1]}[t]=\prod_{1\leqslant i\leqslant \deg t} G^{[h]}[t_i], \tag{4.12a}$$

which is in the right form to apply Theorem 4.3. From it we get

$$G^{[h+1]}(z)=\sum_{k\geqslant 0} z(G^{[h]}(z))^k=\frac{z}{1-G^{[h]}(z)}, \tag{4.12b}$$

where $G^{[0]}(z)=z$. Note the similarity to the generating function for plane trees, which satisfies $G(z)=z/(1-G(z))$. It is easy to transform (4.12b) into

$$G^{[h+1]}(z)=z\frac{F_h(z)}{F_{h+1}(z)}, \quad \text{where } F_{h+2}(z)=F_{h+1}(z)-zF_h(z). \tag{4.13}$$

The polynomials $F_h(z)$ are Chebyshev polynomials. From the linear recurrence that $F_h(z)$ satisfies, we can express $F_h(z)$ as a rational function of $G^{[h]}(z)$. Then applying Lagrange–Bürmann inversion, we get the following expression:

$$G_{n+1}-G_{n+1}^{[h]}=\sum_{j\geqslant 0}\left(\binom{2n}{n+1-j(h+2)}-2\binom{2n}{n-j(h+2)}+\binom{2n}{n-1-j(h+2)}\right). \tag{4.14}$$

The expected tree height $\overline{H_{n+1}}$ is given by

$$\overline{H_{n+1}}=\frac{1}{G_{n+1}}\sum_{h\geqslant 1} h(G_{n+1}^{[h]}-G_{n+1}^{[h-1]})=\frac{1}{G_{n+1}}\sum_{h\geqslant 0}(G_{n+1}-G_{n+1}^{[h]}). \tag{4.15}$$

By substituting (4.14), we see that the evaluation of (4.15) is related to sums of the form

$$S_n=\sum_{k\geqslant 0} d(k)\binom{2n}{n-k}\Big/\binom{2n}{n}, \tag{4.16}$$

where $d(k)$ is the number of divisors of k. By Stirling's approximation, we can approximate S_n by $T(1/\sqrt{n})$, where

$$T(x)=\sum_{k\geqslant 1} d(k)e^{-k^2x^2}. \tag{4.17}$$

The problem is thus to evaluate $T(x)$ asymptotically as $x \to 0$. This is one of the expansions we did in Section 2.4 using Mellin transforms, where we found that

$$T(x) \sim -\frac{\sqrt{\pi}}{2}\frac{\log x}{x} + \left(\frac{3\gamma}{4} - \frac{\log 2}{2}\right)\frac{\sqrt{\pi}}{x} + \frac{1}{4} + O(x^m), \tag{4.18}$$

for any $m > 0$, as $x \to 0$. The theorem follows from the appropriate combination of terms of the form (4.18). □

4.6. THEOREM (Flajolet and Odlyzko [36]). *The expected height $\overline{H_n}$ of a plane tree with degree constraint Ω, where Ω is aperiodic and each tree in $\mathcal{T}$ is equally likely, is $\sim \lambda\sqrt{n}$, where λ is a constant that depends upon Ω.*

PROOF (*Sketch*). By analogy with (4.12), we use the VF $T^{[h]}[t] = \delta_{\text{height}\, t \leqslant h}$. This VF can be expressed recursively as

$$T^{[h+1]}[t] = \delta_{\deg t \in \Omega} \prod_{1 \leqslant i \leqslant \deg t} T^{[h]}[t_i]. \tag{4.19a}$$

Theorem 4.3 gives us

$$T^{[h+1]}(z) = z\omega(T^{[h]}(z)), \tag{4.19b}$$

where $T^{[0]}(z) = z$ and $\omega(u) = \sum_{k \in \Omega} u^k$. The generating function $H(z)$ of the cumulated height of trees is equal to

$$H(z) = \sum_{h \geqslant 0} (T(z) - T^{[h]}(z)). \tag{4.20}$$

One way to regard (4.19b) is simply as the iterative approximation scheme to the fixed point equation (4.6) that determines $T(z)$. A delicate singularity analysis leads to the result. To do the analysis, we need to examine the behavior of the iterative scheme near the singularity $z = \rho$, which is an example of a *singular iteration problem*. We find in the neighborhood of $z = \rho$ that

$$H(z) \sim \frac{d}{1 - z/\rho} \log \frac{1}{1 - z/\rho},$$

where d is the appropriate constant. The theorem follows directly. □

Methods similar to those used in the proof of this theorem had been used by Odlyzko to prove the following theorem.

4.7. THEOREM (Odlyzko [85]). *The number E_n of balanced 2-3 plane trees with n "external" nodes is $\sim \varphi^n W(\log n)/n$, where φ is the golden ratio $(1+\sqrt{5})/2$, and $W(x)$ is a continuous and periodic function.*

This result actually extends to several families of balanced trees, which are used as search structures with guaranteed $O(\log n)$ access time. The occurrence of the golden ratio in Theorem 4.7 is not surprising, given our discussion in Section 2.2 of the equation $f(z)=z+f(z^2+z^3)$, which is satisfied by the OGF of E_n.

Pattern matching

Another important class of algorithms on trees deals with *pattern matching*, the problem of detecting all occurrences of a given pattern tree inside a larger text tree, which occurs often in symbolic manipulation systems. Unlike the simpler case of string matching, where linear-time worst-case algorithms are known, it is conjectured that no linear-time algorithm exists for tree pattern matching.

The following straightforward algorithm, called *sequential matching*, has quadratic running time in the worst case, but can be shown to run in linear time on the average. For each node of the tree, we compare the subtree rooted at that node with the pattern tree by doing simultaneous preorder traversals. Whenever a mismatch is found, the preorder traversal is aborted, and the next node in the tree is considered. If a preorder traversal successfully finishes, then a match is found.

4.8. THEOREM (Steyaert and Flajolet [103]). *The expected running time of the sequential matching algorithm, when applied to a fixed pattern P and all trees in $\mathscr{T}$ of size n, is $\sim cn$, where c is a function of the degree constraint Ω and the structure of pattern P, and is uniformly bounded by an absolute constant, for all P.*

PROOF (*Sketch*). The proof depends upon a lemma that the probability that P occurs at a random node in the tree is asymptotically $\tau^{e-1}\rho^i$, where i and e are the numbers of internal and external nodes in P. The algebraic part of the proof of the lemma is a direct application of the method of Theorems 4.1 and 4.3 applied to multisets of trees. Generating functions for the number of pattern matches have simple expressions in terms of $T(z)$; a singularity analysis finishes the proof. □

The same type of analysis can be applied to a large variety of tree algorithms in a semiautomatic way. One illustration is the following theorem.

4.9. THEOREM (Flajolet and Steyaert [44]). *For any set Θ of operators and Δ of differentiation rules with at least one "expanding rule", the average-case complexity of the symbolic differentiation algorithm is asymptotically $cn^{3/2}+\mathrm{O}(n)$, where the constant c depends upon Θ and Δ.*

Tree compaction

A different kind of singular behavior occurs in the problem known as *common subexpression elimination* or *tree compaction*, where a tree is compacted into a directed acyclic graph by avoiding duplication of identical substructures. This has applications to the compaction of Lisp programs and to code optimization.

4.10. THEOREM (Flajolet, Sipala, and Steyaert [43]). *The expected size of the maximally compacted dag representation of a random tree in $\mathscr{T}$ of size n is $cn/\sqrt{\log n}+\mathrm{O}(n/\log n)$, where the constant c depends upon Ω.*

The dominant singularity in this case is of the form $1/\sqrt{(1-z)}\ \log(1-z)^{-1}$. The theorem shows that the space savings to be expected when compacting trees approaches 100 percent as $n \to \infty$, though convergence is slow.

Register allocation

The *register allocation* problem consists of finding an optimal strategy for evaluating expressions that can be represented by a tree. The optimal pebbling strategy, due to Ershov, requires only $O(\log n)$ registers to evaluate a tree of size n. The following theorem determines the coefficient in the average case for evaluating expressions involving binary operators.

4.11. THEOREM (Flajolet, Raoult, and Vuillemin [39] and Kemp [65]). *The expected optimum number of registers to evaluate a random binary tree of size n is* $\log_4 n + P(\log_4 n) + o(1)$, *where* $P(x)$ *is a periodic function with period* 1 *and small amplitude.*

PROOF (*Sketch*). The analysis involves the combinatorial sum

$$V_n = \sum_{k \geqslant 1} v_2(k) \binom{2n}{n-k},$$

where $v_2(k)$ is the number of 2s in the decomposition of n into prime factors. If we normalize and approximate the binomial coefficient by an exponential term, as in (4.17), we can compute the approximation's Mellin transform

$$\frac{1}{2} \frac{\zeta(s)}{2^s - 1} \Gamma(\tfrac{1}{2}).$$

The set of regularly spaced poles $s = 2k\pi i/\log 2$ corresponds to periodic fluctuations in the form of a Fourier series. □

4.2. Binary search trees

We denote by $\mathscr{BST}(S)$ the *binary search tree* formed by inserting a sequence S of elements. It has the recursive decomposition

$$\mathscr{BST}(S) = \begin{cases} \langle \mathscr{BST}(S_{\leqslant}),\ s_1,\ \mathscr{BST}(S_{>}) \rangle & \text{if } |S| \geqslant 1; \\ \langle \blacksquare \rangle & \text{if } |S| = 0, \end{cases} \tag{4.21}$$

where s_1 is the first element in S, $S_{\leqslant}$ is the subsequence of the other elements that are $\leqslant s_1$, and $S_{>}$ is the subsequence of elements $> s_1$. An empty binary search tree is represented by the external node ■.

The search for an element x proceeds as follows, starting with the root s_1 as the current node y: We compare x with y, and if $x < y$ we set y to be the left child of y, and if $x > y$ we set y to be the right child of y. The process is repeated until either $x = y$ (successful search) or else an external node is reached (unsuccessful search). (Note that this process finds only the *first* element with value x. If the elements' values are all distinct, this is no problem; otherwise, the left path should be searched until a leaf or an element of smaller value is reached.) Insertion is done by inserting the new element into the tree at the point where the unsuccessful search ended. The importance of binary search trees to sorting and range queries is that a linear-time in-order traversal will output the elements in sorted order.

Well-known data structures, such as 2-3 trees, AVL trees, red-black trees, and self-adjusting search trees, do some extra work to ensure that the insert, delete, and query operations can be done in $O(\log n)$ time, where n is the size of the tree. (In the first three cases, the times are logarithmic in the worst case, and in the latter case they are logarithmic in the amortized sense.) Balanced trees are discussed further in [83].

In this section we show that the same logarithmic bounds hold in the average case without need for any balancing overhead. Our probability model assumes that the sequence S of n elements $s_1, s_2, \ldots, s_n$ is picked by random sampling from a real interval, or equivalently, as far as relative ordering is concerned, the elements form a random permutation of size n. The dynamic version of the problem, which corresponds to an average-case amortized analysis, appears in Section 6.

We define $\mathcal{BST}$ to be the class of all binary search trees corresponding to permutations, $\mathcal{BST}=\{\mathcal{BST}(\sigma)\mid \sigma\in S_n\}$. We use K to denote the random variable describing the size of the left subtree; that is, $|S_\leqslant|=K$ and $|S_>|=n-1-K$. By our probability model, the splitting probabilities become

$$\Pr\{K=k\}=\frac{1}{n}, \quad \text{for } 0\leqslant k\leqslant n-1. \tag{4.22}$$

One consequence of this is that not all trees in $\mathcal{BST}$ are equally likely to occur. For example, the perfectly balanced tree of three nodes (which occurs for $\sigma=2\,1\,3$ and $\sigma=2\,3\,1$) is twice as likely to occur as the tree for $\sigma=1\,2\,3$.

The powerful valuation function method that we introduced in the last section applies equally well to binary search trees. In this case, however, the nature of recurrence (4.21) suggests that we use EGFs of cumulative values (or equivalently OGFs of expected values). For VF $v[t]$, we let v_n be its expected value for trees of size n, and we define $v(z)$ to be the OGF $\sum_{n\geqslant 0} v_n z^n$.

4.12. THEOREM. *The sum and subtree product valuation functions are admissible for the class $\mathcal{BST}$ of binary search trees:*

$$v[t]=u[t]+w[t] \quad \Rightarrow \quad v(z)=u(z)+w(z);$$

$$v[t]=u[t_\leqslant]\cdot w[t_>] \quad \Rightarrow \quad v(z)=\int_0^z u(t)w(t)\,\mathrm{d}t,$$

where $t_\leqslant$ and $t_>$ denote the left and right subtrees of t.

The subtree product VF typically results in an integral equation over the OGFs, which by differentiation can be put into the form of a differential equation. This differs from the equations that resulted from Theorems 4.1 and 4.3, which we used in the last section for binary and plane trees.

A good illustration of these techniques is to compute the expected number of probes $\overline{C}_n$ per successful search on a random binary search tree of size n. We assume that each of the n elements is equally likely to be the object of the search. It is easy to see that $\overline{C_n}$ is equal to the expected internal path length p_n, divided by n, plus 1, so it suffices to compute p_n. The recursive definition of the corresponding VF $p[t]$ is

$$\begin{aligned} p[t] &= |t| - 1 + \delta_{|t|=0} + p[t_{\leqslant}] + p[t_{>}] \\ &= |t| - 1 + \delta_{|t|=0} + p[t_{\leqslant}] \cdot I[t_{>}] + I[t_{\leqslant}] \cdot p[t_{>}], \end{aligned} \tag{4.23a}$$

where $I[t] \equiv 1$ is the unit VF, whose OGF is $I(z) = \sum_{n \geqslant 0} z^n = 1/(1-z)$. The size VF $S[t] = |t|$ has OGF $\sum_{n \geqslant 0} nz^n = z/(1-z)^2$. Theorem 4.12 translates (4.23a) into

$$p(z) = \frac{z^2}{(1-z)^2} + 2\int_0^z \frac{p(t)}{1-t}\,dt. \tag{4.23b}$$

Differentiating (4.23b), we get a linear first-order differential equation

$$p'(z) - \frac{2p(z)}{1-z} - \frac{2z}{(1-z)^3} = 0, \tag{4.23c}$$

which can be solved using the variation-of-parameter method (1.20) to get

$$p(z) = -2\frac{\log(1-z)+z}{(1-z)^2} = 2H'(z) - \frac{2(1+z)}{(1-z)^2},$$

where $H(z) = -\log(1-z)/(1-z)$ is the OGF of the harmonic numbers. The following theorem results by extracting $[z^n]p(z)$.

4.13. THEOREM. *The expected internal path length of a random binary search tree with n internal nodes is*

$$p_n = 2(n+1)H_n - 4n \sim 2n\log n + (2\gamma - 4)n + O(\log n).$$

Theorem 4.13 shows that the average search time in a random binary search tree is about 39 percent longer than in a perfectly balanced binary search tree.

There is also a short ad hoc derivation of p_n: In a random binary search tree, s_i is an ancestor of s_j when s_i is the first element of $\{s_{\min\{i,j\}}, s_{\min\{i,j\}+1}, \ldots, s_{\max\{i,j\}}\}$ inserted into the tree, which happens with probability $1/(|i-j|+1)$. Thus we have $p_n = \sum_{1 \leqslant i,j \leqslant n} 1/(|i-j|+1)$, which readily yields the desired formula.

The expected internal path length p_n has direct application to other statistics of interest. For example, p_n is the expected number of comparisons used to sort a sequence of n elements by building a binary search tree and then performing an in-order traversal. The expected number $\overline{U_n}$ of probes per unsuccessful search (which is also the average number of probes per insertion, since insertions are preceded by an unsuccessful search) is the average *external path length* $\overline{EP_n}$, divided by $n+1$. The well-known correspondence

$$EP_n = IP_n + 2n \tag{4.24a}$$

between the external path length EP_n and the internal path length IP_n of a binary tree with n internal nodes leads to

$$\overline{U_n}=\frac{n}{n+1}(\overline{C_n}+1), \tag{4.24b}$$

which readily yields an expression for $\overline{U_n}$ via Theorem 4.13. We can also derive $\overline{U_n}$ directly as we did for $\overline{C_n}$ or via the use of PGFs. Yet another alternative is an ad hoc proof that combines (4.24b) with a different linear relation between $\overline{U_n}$ and $\overline{C_n}$, namely,

$$\overline{C_n}=1+\frac{1}{n}\sum_{0\leqslant i\leqslant n-1}\overline{U_i}. \tag{4.25}$$

Equation (4.25) follows from the observation that the n possible sucessful searches on a tree of size n retrace the steps taken during the n unsuccessful searches that were done when the elements were originally inserted.

Quicksort

We can apply our valuation function machinery to the analysis of quicksort, as mentioned in Section 3.5. Let q_n be the average number of comparisons used by quicksort to sort n elements. Quicksort works by choosing a partitioning element s (say, the first element), dividing the file into the part $\leqslant s$ and the part $>s$, and recursively sorting each subfile. The process is remarkably similar to the recursive decomposition of binary search trees. The version of quicksort in [71, 98] uses $n+1$ comparisons to split the file into two parts. (Only $n-1$ comparisons are needed, but the extra two comparisons help speed up the rest of the algorithm in actual implementations.) The initial conditions are $q_0=q_1=0$. The corresponding VF $q[t]$ is

$$\begin{aligned}q[t]&=|t|+1-\delta_{|t|=0}-2\delta_{|t|=1}+q[t_{\leqslant}]+q[t_{>}]\\&=|t|+1-\delta_{|t|=0}-2\delta_{|t|=1}+q[t_{\leqslant}]\cdot I[t_{>}]+I[t_{\leqslant}]\cdot q[t_{>}].\end{aligned} \tag{4.26a}$$

As before, the OGFs for $I[t]\equiv 1$ and $S[t]=|t|$ are $1/(1-z)$ and $z/(1-z)^2$. By the translation of Theorem 4.12, we get

$$q(z)=\frac{z}{(1-z)^2}+\frac{1}{1-z}-1-2z+2\int_0^z\frac{q(t)}{1-t}\,dt; \tag{4.26b}$$

$$q'(z)=\frac{2}{(1-z)^3}-2+\frac{2q(z)}{1-z}. \tag{4.26c}$$

The linear differential equation (4.26c) can be solved via the variation-of-parameter method to get

$$q(z)=-2\frac{\log(1-z)}{(1-z)^2}-\frac{2}{3(1-z)^2}+\tfrac{2}{3}(1-z)=2H'(z)-\frac{8}{3(1-z)^2}+\tfrac{2}{3}(1-z). \tag{4.27a}$$

We can then extract $q_n=[z^n]q(z)$ to get

$$q_n=2(n+1)(H_{n+1}-\tfrac{4}{3})\sim 2n\log n+2n(\gamma-\tfrac{4}{3})+O(\log n). \tag{4.27b}$$

In practice, quicksort can be optimized by stopping the recursion when the size of the subfile is $\leqslant m$, for some parameter m. When the algorithm terminates, a final insertion sort is done on the file. (We know from Section 3.2 that insertion sort is very efficient when the number of inversions is small.) The analysis of quicksort can be modified to give the average running time as a function of n and m. The optimum m can then be determined, as a function of n. This is done in [71, 98], where it is shown that $m=9$ is optimum in typical implementations once n gets large enough. The average number of comparisons can be derived using the truncated VF

$$q_m[t]=\delta_{|t|>m}(|t|+1)+q_m[t_\leqslant]\cdot I[t_>]+I[t_\leqslant]\cdot q_m[t_>] \tag{4.28}$$

(cf. (4.26a)). The truncated unit VF $I_m[t]=\delta_{|t|>m}$ and the truncated size VF $S_m[t]=\delta_{|t|>m}|t|$ have the OGFs $\sum_{n>m}z^n=z^{m+1}/(1-z)$ and $\sum_{n>m}nz^n=((m+1)z^{m+1}-mz^{m+2})/(1-z)^2$ respectively. The rest of the derivation proceedes as before (and should be done with a symbolic algebra system); the result (cf. (4.27b)) is

$$q_{m,n}=2(n+1)(H_{n+1}-H_{m+2}+\tfrac{1}{2}). \tag{4.29}$$

Height of binary search trees

The analysis of the height of binary search trees involves interesting equations over generating functions. By analogy with (4.12), let $G_n^{[h]}$ denote the probability that a random binary search tree of size n has height $\leqslant h$. The corresponding VF $G^{[h]}[t]=\delta_{\text{height}\,t\leqslant h}$ is of the form

$$G^{[h+1]}[t]=\delta_{|t|=0}+G^{[h]}[t_\leqslant]\cdot G^{[h]}[t_>]. \tag{4.30a}$$

Theorem 4.12 translates this into

$$G^{[h+1]}(z)=1+\int_0^z (G^{[h]}(t))^2\,dt, \tag{4.30b}$$

where $G^{[0]}(z)=1$ and $G(z)=G^{[\infty]}(z)=1/(1-z)$. The sequence $\{G^{[h]}(z)\}_{h\geqslant 0}$ forms a sequence of Picard approximants to $G(z)$. The OGF for the expected height is

$$H(z)=\sum_{h\geqslant 0}(G(z)-G^{[h]}(z)). \tag{4.31}$$

It is natural to conjecture that $H(z)$ has the singular expansion

$$H(z)\sim\frac{c}{1-z}\log\frac{1}{1-z}, \tag{4.32}$$

as $z\to 1$, for some constant c, but no one has succeeded so far in establishing it directly. Devroye [21] has determined the asymptotic form of H_n using the theory of branching processes:

4.14. THEOREM (Devroye [21]). *The expected height H_n of a binary search tree of size n is $\sim c\log n$, where $c=4.311070\ldots$ is the root $\geqslant 2$ of the equation* $(2e/c)^c=e$.

Theorems 4.13 and 4.14 point out clearly that a random binary search tree is fairly balanced, in contrast to the random binary trees $\mathscr{B}$ we studied in Section 4.1. The expected height and path lengths of binary search trees are $O(\log n)$ and $O(n \log n)$, whereas by Theorem 4.2 the corresponding quantities for binary trees are $O(\sqrt{n})$ and $O(n\sqrt{n})$.

Interesting problems in average-case analysis also arise in connection with balanced search trees, but interest is usually focused on storage space rather than running time. For example, a fringe analysis is used in [112, 12] to derive upper and lower bounds on the expected storage utilization and number of balanced nodes in random 2-3 trees and B-trees. These techniques can be extended to get better bounds, but the computations become prohibitive.

Multidimensional search trees

The binary search tree structure can be generalized in various ways to two dimensions. The most obvious generalization, called *quad trees*, uses internal nodes of degree 4. The quad tree for a sequence $S = s_1, s_2, \ldots, s_n$ of n inserted elements is defined by

$$\mathscr{Q}(S) = \begin{cases} \langle s_1, \mathscr{Q}(S_{>,>}),\ \mathscr{Q}(S_{\leqslant,>}),\ \mathscr{Q}(S_{\leqslant,\leqslant}),\ \mathscr{Q}(S_{>,\leqslant})\rangle & \text{if } |S| \geqslant 1; \\ \langle\blacksquare\rangle & \text{if } |S| = 0. \end{cases} \tag{4.33}$$

Here each element s in S is a two-dimensional number, and the four quadrants determined by s are denoted $S_{>,>}$, $S_{\leqslant,>}$, $S_{\leqslant,\leqslant}$, and $S_{>,\leqslant}$. Quad trees support general range searching, and in particular partially specified queries of the form "Find all elements $s = (s_x, s_y)$ with $s_x = c$". The search proceeds recursively to all subtrees whose range overlaps the query range.

4.15. THEOREM (Flajolet et al. [35]). *The expected number of comparisons $\overline{C_n}$ for a partially specified query in a quad tree of size n is $\overline{C_n} = bn^{(\sqrt{17}-3)/2} + O(1)$, where b is a positive real number.*

PROOF (*Sketch*). The splitting probabilities for quad trees are not in as simple a form as in (4.22), but they can be determined readily. By use of the appropriate VF $c[t]$, we get

$$c_n = 1 + \frac{4}{n(n+1)} \sum_{0 \leqslant l \leqslant n-1} \sum_{0 \leqslant k \leqslant l} c_k. \tag{4.34a}$$

In terms of the OGF $d(z) = zc(z)$, this becomes a second-order differential equation

$$d''(z) - \frac{4}{z(1-z)^2} d(z) = \frac{2}{(1-z)^3}. \tag{4.34b}$$

It is not clear how to solve explicitly for $d(z)$, but we can get asymptotic estimates for d_n based upon the fact that $d(z) \sim a(1-z)^{-\alpha}$, as $z \to 1$, for some positive real a and α. We cannot determine a in closed form in general for this type of problem, but α can be determined by substituting $a(1-z)^{-\alpha}$ into (4.34b) to get the "indicial equation"

$$\alpha(\alpha+1) - 4 = 0, \tag{4.35}$$

whose positive solution is $\alpha = (\sqrt{17} - 1)/2$. The transfer lemma (Theorem 2.3) gives us our final result. □

Quad trees can be generalized to k dimensions, $k \geqslant 2$, but the degrees of the nodes become 2^k, which is too large to be practical. A better alternative, called *k-d trees*, is a binary search tree in which the splitting at each node on level i, $i \geqslant 0$, is based upon ordinate $i \bmod k+1$ of the element stored there.

4.16. THEOREM (Flajolet and Puech [38]). *The expected number of elementary comparisons needed for a partially specified query in a k-d tree of size n, in which s of the k fields are specified, is* $\sim a n^{1-s/k+\theta(s/k)}$, *where* $\theta(u)$ *is the root* $\theta \in [0,1]$ *of the equation* $(\theta+3-u)^u(\theta+2-u)^{1-u}-2=0$.

PROOF (*Sketch*). The proof proceeds by first developing a system of integral equations for the OGFs of expected costs using the appropriate VFs and applying Theorem 4.12. This reduces to a differential system of order $2k-s$. It cannot be solved explictly in terms of standard transcendental functions, but a singularity analysis can be done to get the result, based upon a generalization of the approach for quad trees. □

Data structures of multidimensional search and applications in computational geometry are given in [116].

Heap-ordered trees

We conclude this section by considering *heap-ordered trees*, in which the value of the root node of each subtree is $\leqslant$ the values of the other elements in the subtree. We discussed the classical array representation of a perfectly balanced heap in connection with the heapsort algorithms in Section 3.5. Heap-ordered trees provide efficient implementations of priority queues, which support the operations *insert*, *find_min*, and *delete_min*. Additional operations sometimes include *merge* and *decrease_key*. Pagodas [47] are a direct implementation of heap-ordered trees that also support the merge operation.

For the sequence S of n elements $s_1, s_2, \ldots, s_n$, we define $\mathcal{HOT}(S)$ to be the (canonical) heap-ordered tree formed by S. It has the recursive definition

$$\mathcal{HOT}(S)=\begin{cases}\langle \mathcal{HOT}(S_{\text{left}}),\ s_{\min(S)},\ \mathcal{HOT}(S_{\text{right}})\rangle & \text{if } |S|\geqslant 1;\\ \langle\blacksquare\rangle & \text{if } |S|=0.\end{cases} \tag{4.36}$$

where $\min(S)$ is the index of the rightmost smallest element in S, S_{left} is the initial subsequence $s_1, \ldots, s_{\min(S)-1}$, and S_{right} is the final subsequence $s_{\min(S)+1}, \ldots, s_n$. We assume in our probability model that S is a random permutation of n elements. Analysis of parameters of heap-ordered trees and pagodas is similar to the analysis of binary search trees, because of the following equivalence principle due to Burge [13, 109]:

4.17. THEOREM. *For each pair of inverse permutations* σ *and* σ^{-1}, *we have*

$$\mathcal{BST}(\sigma) \equiv_{\text{shape}} \mathcal{HOT}(\sigma^{-1}),$$

where $t \equiv_{\text{shape}} u$ *means that the unlabeled trees associated with trees t and u are identical.*

For purposes of analysis, any parameter of permutations defined inductively over the associated heap-ordered tree can thus be analyzed using the admissibility rules of Theorem 4.12 for binary search trees. Heap-ordered trees in the form of Cartesian trees can also be used to handle a variety of two-dimensional search problems [109].

4.3. *Radix-exchange tries*

Radix-exchange tries are binary search trees in which the elements are stored in the external nodes, and navigation through the trie at level i is based upon the ith bit of the search argument. Bit 0 means "go left", and bit 1 means "go right". We assume for simplicity that each element is a real number in $[0, 1]$ of infinite precision. The trie $\mathcal{TR}(S)$ for a set S of elements is defined recursively by

$$\mathcal{TR}(S)=\begin{cases}\langle \mathcal{TR}(S_0),\ \bigcirc,\ \mathcal{TR}(S_1)\rangle & \text{if } |S|>1;\\ \langle \blacksquare\rangle & \text{if } |S|=1;\\ \langle \emptyset\rangle & \text{if } |S|=0,\end{cases} \tag{4.37}$$

where S_0 and S_1 are defined as follows: If we take the elements in S that have 0 as their first bit and then throw away that bit, we get S_0, the elements in the left subtrie. The set S_1 of elements in the right subtrie is defined similarly for the elements starting with a 1 bit. The elements are stored in the external nodes of the trie. When S has a single element s, the trie consists of the external node $\blacksquare$ with value s; an empty trie is represented by the null external node $\emptyset$. The size of the trie is the number of $\blacksquare$ external nodes.

The trie $\mathcal{TR}(S)$ does not depend upon the order in which the elements in S are inserted; this is quite different from the case of binary search trees, where order can make big difference upon the shape of the tree. In tries, the shape of the trie is based upon the distribution of the elements' values.

We use the probability model that the values of the elements are independent and uniform in the real interval [0, 1]. We define the class of all tries to be $\mathcal{TR}$. The probability that a trie of n elements has a left subtrie of size k and a right subtrie of size $n-k$ is the Bernoulli probability

$$p_{n,k}=\frac{1}{2^n}\binom{n}{k}. \tag{4.38}$$

This suggests that we use EGFs of expected values to represent trie statistics. We denote the expected value of VF $v[t]$ among trees of size n by v_n, and the EGF $\sum_{n\geqslant 0} v_n z^n/n!$ by $v(z)$. The admissibility theorem takes yet another form as shown below.

4.18. THEOREM. *The sum and subtree product valuation functions are admissible for the class $\mathcal{TR}$ of tries*:

$$v[t]=u[t]+w[t] \quad\Rightarrow\quad v(z)=u(z)+w(z);$$
$$v[t]=u[t_0]\cdot w[t_1] \quad\Rightarrow\quad v(z)=u(\tfrac{1}{2}z)w(\tfrac{1}{2}z),$$

where t_0 and t_1 represent the left and right subtries of t.

A framework for the analysis of tries via valuation functions is given in [41]. In typical cases, the EGFs for the VFs that we encounter are in the form of difference equations that we can iterate.

The expected number of bit inspections per successful search in a trie with n external nodes is equal to the expected external path length p_n, divided by n. The following theorem shows that the search times are logarithmic, on the average, when no balancing is done.

4.19. THEOREM (Knuth [71]). *The average external path length p_n of a random trie of size n is $p_n = n\log_2 n + (\gamma/\log 2 + \frac{1}{2} + R(\log_2 n))n + O(\sqrt{n})$, where $R(u)$ is a periodic function of small amplitude with period 1 and mean value 0.*

PROOF. The VF corresponding to external path length is

$$p[t] = [t] - \delta_{|t|=1} + p[t_0]\cdot I[t_1] + I[t_0]\cdot p[t_1]. \tag{4.39a}$$

The unit VF $I[t] \equiv 1$ has EGF $\sum_{n\geq 0} z^n/n! = e^z$, and the size VF $S[t] = |t|$ has EGF $\sum_{n\geq 0} nz^n/n! = ze^z$. By Theorem 4.18, (4.39a) translates to

$$p(z) = ze^z - z + 2e^{z/2}p(\tfrac{1}{2}z). \tag{4.39b}$$

By iterating the recurrence and then extracting coefficients, we get

$$p(z) = z\sum_{k\geq 0}(e^z - e^{z(1-1/2^k)}); \tag{4.39c}$$

$$p_n = n\sum_{k\geq 0}\left(1 - \left(1 - \frac{1}{2^k}\right)^{n-1}\right). \tag{4.39d}$$

It is easy to verify the natural approximation $p_n \sim nP(n)$, where $P(x)$ is the harmonic sum

$$P(x) = \sum_{k\geq 0}(1 - e^{-x/2^k}). \tag{4.40}$$

We have already derived the asymptotic expansion of $P(x)$, as $x\to\infty$, by use of Mellin transforms in (2.28). The result follows immediately. □

Theorem 4.19 generalizes to the biased case, where the bits of each element are independently 0 with probability p, and 1 with probability $q = 1 - p$. The average external path length is asymptotically $(n\log n)/H$, where H is the entropy function $H = -p\log p - q\log q$. In the case of unsuccessful searches, a similar approach shows that the average number of bit inspections is $\sim(\log n)/H$. The variance is $O(1)$ with fluctuation in the unbiased case [63, 68]. Variance estimates in this range of problems involve interesting connections with modular functions [69]. The variance increases to $\sim c\log n + O(1)$, for some constant c, in the biased case [63]. The limiting distributions are studied in [63, 89]. The height of a trie has mean $\sim 2\log_2 n$ and variance $O(1)$ [32]. Limiting distributions of the height are studied in [32, 89].

Another important statistics on tries, besides search time, is storage space. Unlike binary search trees, the amount of auxiliary space used by tries, measured in terms of the number of internal nodes, is variable. The following theorem shows that the average number of internal nodes in a trie is about 44 percent more than the number of elements stored in the trie.

4.20. THEOREM (Knuth [71]). *The expected number i_n of internal nodes in a random unbiased trie with n external nodes is $(n/\log 2)(1+Q(\log_2 n))+O(\sqrt{n})$, where $Q(u)$ is a periodic function of small amplitude with period 1 and mean value 0.*

PROOF. The VF corresponding to the number of internal nodes is

$$i[t]=\delta_{|t|>1}+i[t_0]\cdot I[t_1]+I[t_0]\cdot i[t_1]. \tag{4.41a}$$

Theorem 4.18 translates this to

$$i(z)=e^z-1-z+2e^{z/2}i(\tfrac{1}{2}z). \tag{4.41b}$$

By iterating the recurrence and then extracting coefficients, we get

$$i(z)=\sum_{k\geqslant 0} 2^k\left(e^z-\left(1+\frac{z}{2^k}\right)e^{(1-1/2^k)z}\right); \tag{4.41c}$$

$$i_n=\sum_{k\geqslant 0} 2^k\left(1-\left(1-\frac{1}{2^k}\right)^n-\frac{n}{2^k}\left(1-\frac{1}{2^k}\right)^{n-1}\right). \tag{4.41d}$$

We can approximate i_n to within $O(\sqrt{n})$ in a natural way by $S(n)$, where

$$S(x)=\sum_{k\geqslant 0} 2^k\left(1-e^{-x/2^k}\left(1+\frac{x}{2^k}\right)\right). \tag{4.42a}$$

Equation (4.42a) is a harmonic sum, and its Mellin transform $S^*(s)$ can be computed readily:

$$S^*(s)=\frac{(s+1)\Gamma(s)}{1-2^{s+1}}, \tag{4.42b}$$

where the fundamental strip of $S^*(s)$ is $\langle -2; -1\rangle$. The result follows by computing the residues in the right half-plane $\Re(s)\geqslant -1$. There is a simple pole at $s=0$ due to $\Gamma(s)$ and poles at $-1+2k\pi i/\log 2$ due to the denominator of (4.42b). □

In the biased case, the expected number of internal nodes is $\approx n/H$. The variance for both the unbiased and biased case is $O(n)$, which includes a fluctuating term [64]; the distribution of the number of internal nodes is normal [63].

Theorem 4.20, as a number of the results about tries do, generalizes to the case in which each external node in the trie represents a page of secondary storage capable of storing $b\geqslant 1$ elements. Such tries are generally called *b-tries*. The analysis uses truncated VFs, as in the second quicksort example in Section 4.2, to stop the recursion when the subtrie has $\leqslant b$ elements. The result applies equally well to the extendible hashing scheme of [27], where the trie is built upon the hashed values of the elements, rather than upon the elements themselves. Extendible hashing will be considered further in Section 5.1.

4.21. THEOREM (Knuth [71]). *The expected number of pages of capacity b needed to store a file of n records using b-tries or extendible hashing is*

$$(n/(b\log 2))(1+R(\log_2 n))+O(\sqrt{n}),$$

where $R(u)$ is periodic with period 1 *and mean value* 0.

Patricia tries

Every external node in a trie of size $\geqslant 2$ has a sibling, but that is not generally the case for internal nodes. A more compact form of tries, called *Patricia tries*, can be obtained by collapsing the internal nodes with no sibling. Statistics on Patricia tries are analyzed in [71, 68].

Radix-exchange sorting

It is no accident that radix-exchange tries and the radix-exchange sorting algorithm have a common name. Radix-exchange sorting is related to tries in a way very similar to how quicksort is related to binary search trees, except that the relationship is even closer. All the average-case analyses in this section carry over to the analysis of radix-exchange sorting: The distribution of the number of partitioning stages used by radix-exchange sorting to sort n numbers is the same as the distribution of the number of internal nodes in a trie, and the distribution of the number of bit inspections done by radix-exchange sorting is the same as the distribution of the external path length of a trie.

4.4. Digital search trees

Digital search trees are like tries except that the elements are stored in the internal nodes, or equivalently, they are like binary search trees except that the branching at level i is determined by the $(i+1)$st bit rather than by a full element-to-element comparison. The digital search tree $\mathscr{DST}(S)$ for a sequence S of inserted elements is defined recursively by

$$\mathscr{DST}(S)=\begin{cases}\langle\mathscr{DST}(S_0),s_1,\mathscr{DST}(S_1)\rangle & \text{if } |S|\geqslant 1;\\ \langle\blacksquare\rangle & \text{if } |S|=0,\end{cases} \tag{4.43}$$

where s_1 is the first element of S, and the sequence S_0 of elements in the left subtree is formed by taking the elements in $S-\{s_1\}$ that have 0 as the first bit and then throwing away the first bit. The sequence S_1 for the right subtree is defined symmetrically for the elements with 1 as their first bit. Like binary search trees, the size of the tree is its number of internal nodes, and its shape is sensitive to the order in which the elements are inserted. The empty digital search tree is denoted by the external node ■. Our probability model is the same as for tries, except that the probability that a tree of n elements has a left subtree of size k and a right subtree of size $n-k-1$ is $\binom{n-1}{k}/2^{n-1}$. The class of all digital search trees is denoted $\mathscr{DST}$.

The nature of the decomposition in (4.43) suggests that we use EGFs of expectations in our analysis, as in the last section, but the admissibility theorem takes a different form as shown below.

4.22. THEOREM. *The sum and subtree product valuation functions are admissible for the class $\mathscr{DST}$ of digital search trees:*

$$v[t]=u[t]+w[t] \quad \Rightarrow \quad v(z)=u(z)+w(z);$$

$$v[t]=u[t_0]\cdot w[t_1] \Rightarrow v(z)=\int_0^z u(\tfrac{1}{2}t)w(\tfrac{1}{2}t)\,\mathrm{d}t,$$

where t_0 and t_1 denote the left and right subtrees of t.

Tries are preferred in practice over digital search trees, since the element comparison done at each node in a digital search tree takes longer than the bit comparison done in a trie, and the elements in a trie are kept in sorted order. We do not have the space in this chapter to include the relevant analysis; instead we refer the reader to [71, 77, 42, 68]. The key difference between the analysis of digital search trees and the analysis of tries in the last section is that the equations over the EGFs that result from Theorem 4.22 are typically difference-differential equations, to which the Mellin techniques that worked so well for tries cannot be applied directly. Instead the asymptotics come by an application due to Rice of the following classical formula from the calculus of finite differences; the proof of the formula is an easy application of Cauchy's formula.

4.23. THEOREM. *Let C be a closed curve encircling the points $0, 1, \ldots, n$, and let $f(z)$ be analytic inside C. Then we have*

$$\sum_k \binom{n}{k}(-1)^k f(k)=\frac{1}{2\pi\mathrm{i}}\int_C B(n+1,-z)f(z)\,\mathrm{d}z,$$

where $B(x, y)=\Gamma(x)\Gamma(y)/\Gamma(x+y)$ is the classical Beta function.

4.24. THEOREM (Knuth [71] and Konheim and Newman [77]). *The expected internal path length of a random digital search tree is*

$$(n+1)\log_2 n+\left(\frac{\gamma-1}{\log\,2}+\tfrac{1}{2}-\alpha+P(\log_2 n)\right)n+\mathrm{O}(\sqrt{n}),$$

where $\gamma=0.57721\ldots$ is Euler's constant, $\alpha=1+\frac{1}{3}+\frac{1}{7}+\frac{1}{15}+\cdots=1.606695\ldots$, and $P(u)$ is a periodic function with period 1 and very small amplitude.

5. Hashing and address computation techniques

In this section we consider several well-known hashing algorithms, including separate chaining, coalesced hashing, uniform probing, double hashing, secondary clustering, and linear probing, and we also discuss the related methods of interpolation search and distribution sorting. Our machine-independent model of search performance for hashing is the number of probes made into the hash table during the search. We are primarily interested in the expected number of probes per search, but in some cases we also consider the distribution of the number of probes and the expected maximum number of probes among all the searches in the table.

With hashing, searches can be performed in constant time, on the average, regardless of the number of elements in the hash table. All hashing algorithms use a predefined *hash function*

$$hash\colon \{\text{all possible elements}\} \to \{1, 2, \ldots, m\} \tag{5.1}$$

that assigns a *hash address* to each of the n elements. Hashing algorithms differ from one another in how they resolve the *collision* that results when an element's hash address is already occupied. The two main techniques for resolving collisions are chaining (in which links are used to explicitly link together elements with the same hash address) and open addressing (where the search path through the table is defined implicitly). We study these two classes of hashing algorithms in the next two sections.

We use the *Bernoulli probability model* for our average-case analysis: We assume that all m^n possible sequences of n hash addresses (or *hash sequences*) are equally likely. Simulation studies confirm that this is a reasonable assumption for well-designed hash functions. Further discussion of hash functions, including universal hash functions, appears in [83]. We assume that an unsuccessful search can begin at any of the m slots in the hash table with equal probability, and the object of a successful search is equally likely to be any of the n elements in the table. Each insertion is typically preceded by an unsuccessful search to verify that the element is not already in the table, and so for simplicity we shall identify the insertion time with the time for the unsuccessful search. We denote the expected number of probes per unsuccessful search (or insertion) in a hash table with n elements by $\overline{U_n}$, and the expected number of probes per successful search by $\overline{C_n}$.

5.1. *Bucket algorithms and hashing by chaining*

Separate chaining

One of the most obvious techniques for resolving collisions is to link together all the elements with the same hash address into a list or chain. The generic name for this technique is *separate chaining.* The first variant we shall study stores the chains in auxiliary memory; the ith slot in the hash table contains a link to the start of the chain of elements with hash address i. This particular variant is typicaly called *indirect chaining*, because the hash table stores only pointers, not the elements themselves.

Search time clearly depends upon the number of elements in the chain searched. For each $1 \leqslant i \leqslant m$, we refer to the set of elements with hash address i as the ith bucket. We define ${}^m_n X_i$ (or simply X_i) in the Bernoulli model to be the RV describing the number of elements in bucket i. This model is sometimes called the *urn model*, and the distribution of X_i is called the *occupancy distribution.* Distributions of this sort appear in the analyses of each of the chaining algorithms we consider in this section, and they serve to unify our analyses. Urn models were discussed in Section 1.3.

An unsuccessful search on a chain of length k makes one probe per element, plus one probe to find the link to the beginning of the chain. This allows us to express the expected number of probes per unsuccessful search as

$$\overline{U_n}=\mathbf{E}\left\{\frac{1}{m}\sum_{1\leqslant i\leqslant m}(1+X_i)\right\}. \tag{5.2a}$$

By symmetry, the expected values $\mathbf{E}\{X_i\}$ are the same for each $1\leqslant i\leqslant m$, so we can restrict our attention to one particular bucket, say, bucket 1. (For simplicity, we shall abbreviate X_1 by X.) Equation (5.2a) simplifies to

$$\overline{U_n}=1+\mathbf{E}\{X\}. \tag{5.2b}$$

For successful searches, each chain of length k contributes $2+3+\cdots+(k+1)=3k/2+k^2/2$ probes. The expected number of probes per successful search is thus

$$\overline{C_n}=\mathbf{E}\left\{\frac{1}{n}\sum_{1\leqslant i\leqslant m}(\tfrac{3}{2}X_i+\tfrac{1}{2}X_i^2)\right\}=\frac{m}{n}(\tfrac{3}{2}\mathbf{E}\{X\}+\tfrac{1}{2}\mathbf{E}\{X^2\}). \tag{5.3}$$

We can compute (5.2b) and (5.3) in a unified way via the PGF

$$X(u)=\sum_{k\geqslant 0}\Pr\{X=k\}u^k. \tag{5.4}$$

Equations (5.2b) and (5.3) for $\overline{U_n}$ and $\overline{C_n}$ are expressible simply in terms of derivatives of $X(u)$:

$$\overline{U_n}=1+X'(1);\qquad \overline{C_n}=\frac{m}{n}(2X'(1)+\tfrac{1}{2}X''(1)). \tag{5.5}$$

We shall determine $X(u)$ by extending the admissible constructions we developed in Section 1.3 for the urn model. This approach will be especially useful for our analysis later in this section of the maximum bucket occupancy, extendible hashing, and coalesced hashing. We consider the hash table as the m-ary partitional product of the individual buckets

$$\mathcal{H}=\mathcal{B}*\mathcal{B}*\cdots*\mathcal{B}. \tag{5.6a}$$

The new twist here is that some of the elements in the table are "marked" according to some rule. (We shall explain shortly how this relates to separate chaining.) We let $H_{k,n,m}$ be the number of m^n hash sequences for which k of the elements are marked, and we denote its EGF by

$$\hat{H}(u,z)=\sum_{k,n\geqslant 0}H_{k,n,m}u^k\frac{z^n}{n!}. \tag{5.6b}$$

By analogy with Theorem 1.4 for EGFs, the following theorem shows that the partitional product in (5.6a) translates into a product of EGFs; the proof is similar to that of Theorem 1.4 and is omitted for brevity.

5.1. THEOREM. *If the number of marked elements in bucket i is a function of only the* *number of elements in bucket i, then the EGF* $\hat{H}(u,z)=\sum_{k,n\geq 0} H_{k,n,m} u^k z^n/n!$ *can be decomposed into*

$$\hat{H}(u,z)=\hat{B}_1(u,z)\cdot\hat{B}_2(u,z)\cdots\hat{B}_m(u,z), \tag{5.6c}$$

where

$$\hat{B}_i(u,z)=\sum_{t\geq 0} u^{f_i(t)}\frac{z^t}{t!}, \tag{5.6d}$$

and $f_i(t)$ *is the number of marked elements in bucket i when there are t elements in bucket i.*

We are interested in the number of elements in bucket 1, so we adopt the marking rule that all elements in bucket 1 are marked, and the other elements are left unmarked. In terms of Theorem 5.1, we have $f_1(t)=t$ and $\hat{B}_1(u,z)=\sum_{t\geq 0} u^t z^t/t!=e^{uz}$ for bucket 1, and $f_i(t)=0$ and $\hat{B}_i(u,z)=\sum_{t\geq 0} z^t/t!=e^z$ for the other buckets $2\leq i\leq m$. Substituting this into Theorem 5.1, we have

$$\hat{H}(u,z)=e^{(m-1+u)z}. \tag{5.7a}$$

We can obtain the EGF of $X(u)$ by dividing each $H_{k,n,m}$ term in (5.6b) by m^n, or equivalently by replacing z by z/m. Combining this with (5.7a) gives

$$\hat{H}\left(u,\frac{z}{m}\right)=\sum_{n\geq 0} X(u)\frac{z^n}{n!}=e^{(m-1+u)z/m}. \tag{5.7b}$$

Hence, we have

$$X(u)=\left(\frac{m-1+u}{m}\right)^n; \qquad X'(1)=\frac{n}{m}; \qquad X''(1)=\frac{n(n-1)}{m^2}. \tag{5.7c}$$

We get the following theorem by substituting the expressions for $X'(1)$ and $X''(1)$ into (5.5); the term $\alpha=n/m$ is called the *load factor.*

5.2. THEOREM. *The expected number of probes per unsuccessful and successful search for indirect chaining, when there are n elements in a hash table of m slots, is*

$$\overline{U_n}=1+\frac{n}{m}=1+\alpha; \qquad \overline{C_n}=2+\frac{n-1}{2m}\sim 2+\frac{\alpha}{2}. \tag{5.8}$$

We can also derive (5.8) in an ad hoc way by decomposing X into the sum of n independent 0-1 RVs

$$X=x_1+x_2+\cdots+x_n, \tag{5.9}$$

where $x_i=1$ if the ith element goes into bucket i, and $x_i=0$ otherwise. Each x_i has the same distribution as ${}^m_1X_1$, and its PGF is clearly $(m-1+u)/m$. Since the PGF of a sum of independent RVs is equal to the product of the PGFs, we get (5.7c) for $X(u)$. We can also derive the formula for $\overline{C_n}$ from the one for $\overline{U_n}$ by noting that (4.25) for binary search trees holds for separate chaining as well.

EXAMPLES. (1) Direct chaining. A more efficient version of separate chaining, called *direct chaining*, stores the first element of each chain directly in the hash table itself. This shortens each successful search time by one probe, and the expected unsuccessful search time is reduced by $\Pr\{X>0\}=1-X(0)=1-(1-1/m)^n \sim 1-e^{-\alpha}$ probes. We get

$$\overline{U_n}=\left(1-\frac{1}{m}\right)^n+\frac{n}{m}\sim e^{-\alpha}+\alpha; \qquad \overline{C_n}=1+\frac{n-1}{2m}\sim 1+\alpha. \tag{5.10}$$

(2) Direct chaining with relocation. The above variant is wasteful of hash table slots, because auxiliary space is used to store colliders even though there might be empty slots available in the hash table. (And, for that reason, the load factor $\alpha=n/m$ defined above is not a true indication of space usage.) The method of *direct chaining with relocation* stores all elements directly in the hash table. A special situation arises when an element x with hash address $hash(x)$ collides during insertion with another element x'. If x' is the first element in its chain, then x is inserted into an empty slot in the table and linked to the end of the chain. Otherwise, x' is not at the start of the chain, so x' is relocated to an empty slot in order to make room for x; the link to x' from its predecessor in the chain must be updated. The successful search time is the same as before. Unsuccessful searches can start in the middle of a chain; each chain of length $k>0$ contributes $k(k+1)/2$ probes. A search starting at one of the $m-n$ empty slots takes one probe. This gives us

$$\begin{aligned}\overline{U_n}&=X'(1)+\tfrac{1}{2}X''(1)+\Pr\{\text{slot } hash(x) \text{ is empty}\}\\ &=1+\frac{n(n-1)}{2m^2}\sim 1+\frac{\alpha^2}{2}.\end{aligned} \tag{5.11}$$

The main difficulty with this algorithm is the overhead of moving elements, which can be expensive for large record elements and might not be allowed if there are pointers to the elements from outside the hash table. Updating the previous link requires either the use of bidirectional or circular chains or recomputing the hash address of x' and following links until x' is reached. None of these alternatives is attractive, and we shall soon consider a better alternative called *coalesced hashing*, which has nearly the same number of probes per search, but without the overhead.

Distribution sorting

Bucketing can also be used to sort efficiently in linear time when the values of the elements are smoothly distributed. Suppose for simplicity that the n values are real numbers in the unit interval $[0, 1)$. The *distribution sort* algorithm works by breaking up the range of values into m buckets $[0, 1/m)$, $[1/m, 2/m), \ldots, [(m-1)/m, 1)$; the elements are partitioned into the buckets based upon their values. Each bucket is sorted using selection sort (cf. Section 3.5) or some other simple quadratic sorting method. The sorted buckets are then appended together to get the final sorted output.

Selection sort uses $\binom{k}{2}$ comparisons to sort k elements. The average number of comparisons $\overline{C_n}$ used by distribution sort is thus

$$\overline{C_n}=\mathbf{E}\left\{\sum_{1\leqslant i\leqslant m}\binom{X_i}{2}\right\}=\sum_{1\leqslant i\leqslant m}\mathbf{E}\left\{\frac{X_i(X_i-1)}{2}\right\}=\frac{1}{2}\sum_{1\leqslant i\leqslant m}X_i''(1). \tag{5.12a}$$

By (5.7c), we have $\overline{C_n}=n(n-1)/(2m)$ when the values of the elements are independently and uniformly distributed. The other work done by the algorithm takes $O(n+m)$ time, so this gives a linear-time sorting algorithm when we choose $m=\Theta(n)$. (Note that this does not contradict the well-known $\Omega(n \log n)$ lower bound on the average sorting time in the comparison model, since this is not a comparison-based algorithm.)

The assumption that the values are independently and uniformly distributed is not always easy to justify, unlike for the case of hashing, because there is no hash function to scramble the values; the partitioning is based upon the elements' raw values. Suppose that elements are independently distributed according to density function $f(x)$. In the following analysis, suggested by Karp [71, 5.2.1–38], [22], we assume that $\int_0^1 f(x)^2\,dx<\infty$, which assures that $f(x)$ is well-behaved. For each n we choose m so that $n/m\to\alpha$, for some positive constant α, as $n\to\infty$. We define $p_i=\int_{A_i} f(x)\,dx$ to be the probability that an element falls into the ith bucket $A_i=[(i-1)/m, i/m)$. For general $f(x)$, (5.7c) for $X_i(u)$ becomes

$$X_i(u)=((1-p_i)+p_i u)^n. \tag{5.12b}$$

By (5.12a) and (5.12b) we have

$$\overline{C_n}=\binom{n}{2}\sum_{i\leqslant i\leqslant m} p_i^2=\binom{n}{2}\sum_{1\leqslant i\leqslant m}\left(\int_{A_i} f(x)\,dx\right)^2. \tag{5.12c}$$

The last summation in (5.12c) can be bounded by an application of Jensen's inequality treating $f(x)$ as a RV with x uniformly distributed:

$$\sum_{1\leqslant i\leqslant m}\left(\int_{A_i} f(x)\,dx\right)^2=\frac{1}{m^2}\sum_{1\leqslant i\leqslant m}\left(\int_{A_i} f(x)\,d(mx)\right)^2$$

$$\leqslant\frac{1}{m^2}\sum_{1\leqslant i\leqslant m}\int_{A_i} f(x)^2\,d(mx)=\frac{1}{m}\int_0^1 f(x)^2\,dx. \tag{5.12d}$$

We can show that the upper bound (5.12d) is asymptotically tight by computing a corresponding lower bound. We have

$$n\sum_{1\leqslant i\leqslant m}\left(\int_{A_i} f(x)\,dx\right)^2=\frac{n}{m}\int_0^1 f_n(x)^2\,dx,$$

where $f_n(x)=mp_i$ for $x\in A_i$ is the histogram approximation of $f(x)$, which converges to $f(x)$ almost everywhere. By Fatou's Lemma, we get the lower bound

$$\liminf_{n\to\infty}\frac{n}{m}\int_0^1 f_n(x)^2\,dx=\alpha\liminf_{n\to\infty}\int_0^1 f_n(x)^2\,dx\geqslant\alpha\int_0^1\liminf_{n\to\infty} f_n(x)^2\,dx$$

$$=\alpha\int_0^1 f(x)^2\,dx.$$

Substituting this approximation and (5.12d) into (5.12c), we find that the average number of comparisons is

$$\overline{C_n} \sim \frac{\alpha n}{2} \int_0^1 f(x)^2 \, dx. \tag{5.12e}$$

The coefficient of the linear term in (5.12e) is proportional to $\int_0^1 f(x)^2 \, dx$, which can be very large. The erratic behavior due to nonuniform $f(x)$ can be alleviated by one level of recursion, in which the above algorithm is used to sort the individual buckets: Let us assume that $m = n$. If the number X_i of elements in bucket i is more than 1, we sort the bucket by breaking up the range $[(i-1)/n, i/n)$ into X_i subbuckets and proceed as before. The surprising fact, which can be shown by techniques similar to those above, is that $\overline{C_n}$ is bounded by $n/2$ in the limit, regardless of $f(x)$ (assuming of course that our assumption $\int_0^1 f(x)^2 \, dx < \infty$ is satisfied).

5.3. THEOREM (Devroye [22]). *The expected number of comparisons $\overline{C_n}$ done by two-level bucketing to sort n elements that are independently distributed with density function $f(x)$, which satisfies $\int_0^1 f(x)^2 \, dx < \infty$, is*

$$\overline{C_n} \sim \tfrac{1}{2} n \int_0^1 e^{-f(x)} \, dx \leqslant \tfrac{1}{2} n.$$

The variance and higher moments of the number of probes are also small. If the unit interval assumption is not valid and the values of the elements are not bounded, we can redefine the interval to be $[x_{\min}, x_{\max}]$ and apply the same basic idea given above. The analysis becomes a little more complicated; details appear in [22].

For the actual implementation, a hashing scheme other than separate chaining can be used to store the elements in the table. Sorting with linear probing is discussed at the end of the section. An application of bucketing to fast sorting on associative secondary storage devices appears in [79]. A randomized algorithm that is optimal for sorting with multiple disks is given in [121].

Interpolation search

Newton's method and the secant method are well-known schemes for determining the leading k bits of a zero of a continuous function $f(x)$ in $O(\log k)$ iterations. (By "zero", we mean a solution x to the equation $f(x) = 0$.) Starting with an initial approximation x_0, the methods produce a sequence of refined approximations $x_1, x_2, \ldots$ that converge to a zero x^*, assuming that $f(x)$ is well-behaved. The discrete analog is called *interpolation search*, in which the n elements are in a sorted array, and the goal is to find the element x^* with a particular value c. Variants of this method are discussed in [83].

5.4. THEOREM (Yao and Yao [115] and Gonnet, Rogers, and George [51]). *The average number of comparisons per successful search using interpolation search on a sorted array is $\sim \log_2 \log_2 n$, assuming that the elements' values are independently and identically distributed.*

This bound is similar to that of the continuous case; we can think of the accuracy k as being $\log_2 n$, the number of bits needed to specify the array position of x^*. A more detailed probabilistic analysis connecting interpolation search with Brownian motion appears in [80].

PROOF (*Sketch*). We restrict ourselves to considering the upper bound. Gonnet, Rogers, and George [51] show by some probabilistic arguments that the probability that at least k probes are needed to find x^* is bounded by

$$p_k(t) = \prod_{1 \leqslant i \leqslant k} (1 - \tfrac{1}{2}e^{-t2^{-i}}), \tag{5.13a}$$

where $t = \log(\pi n/8)$. Hence, the expected number of probes is bounded by

$$F(t) = \sum_{k \geqslant 0} p_k(t), \tag{5.13b}$$

which can be expressed in terms of the harmonic sum

$$F(t) = \frac{1}{Q(t)} \sum_{k \geqslant 0} Q(t2^k), \tag{5.13c}$$

where

$$Q(t) = \prod_{i \geqslant 1} (1 - \tfrac{1}{2}e^{-t2^{-i}})^{-1}. \tag{5.13d}$$

The sum in (5.13c) is a harmonic sum to which Mellin transforms can be applied to get

$$F(t) \sim \log_2 t + \alpha + P(\log_2 t) + o(1) \quad \text{as } t \to \infty, \tag{5.13e}$$

where α is a constant and $P(u)$ is a periodic function associated with the poles at $\chi_k = 2k\pi i/\log 2$. The $\log_2 \log_2 n$ bound follows by substituting $t = \log(\pi n/8)$ into (5.13e). □

Maximum bucket occupancy

An interesting statistic that lies between average-case and worst-case search times is the expected number of elements in the *largest* bucket (also known as the *maximum bucket occupancy*). It has special significance in parallel processing applications in which elements are partitioned randomly into buckets and then the buckets are processed in parallel, each in linear time; in this case, the expected maximum bucket occupancy determines the average running time.

We can make use of the product decomposition (5.6a) and Theorem 5.1 to count the number of hash sequences that give a hash table with $\leqslant b$ elements in each bucket. We mark all the elements in a bucket if the bucket has $>b$ elements; otherwise, the elements are left unmarked. In this terminology, our quantity of interest is simply the number $H_{0,n,m}$ of hash sequences with no marked elements:

$$\begin{aligned} H_{0,n,m} &= n!\,[u^0 z^n]\hat{H}(u,z) = n!\,[z^n]\hat{H}(0,z) \\ &= n!\,[z^n](\hat{B}_1(0,z)\cdot\hat{B}_2(0,z)\cdots\hat{B}_m(0,z)), \end{aligned} \tag{5.14a}$$

where

$$\hat{B}_i(0, z) = \sum_{0 \leqslant n \leqslant b} \frac{z^n}{n!}, \tag{5.14b}$$

for $1 \leqslant i \leqslant m$. The sum in (5.14b) is the truncated exponential which we denote by $e_b(z)$. Hence, the number of hash sequences with $\leqslant b$ elements per bucket is

$$H_{0,n,m} = n![z^n](e_b(z))^m. \tag{5.14c}$$

We use $q_n^{[b]}$ to denote the probability that a random hash sequence puts at most b elements into each bucket. As in (5.7b), we can convert from the enumeration $H_{0,n,m}$ to the probability $q_n^{[b]}$ by replacing z in (5.14c) by z/m:

$$q_n^{[b]} = n![z^n]\left(e_b\left(\frac{z}{m}\right)\right)^m. \tag{5.14d}$$

There is a close relation between the EGF of Bernoulli statistics and the corresponding Poisson statistics:

5.5. THEOREM. *If $\hat{g}(z) = \sum_{n \geqslant 0} g_n z^n/n!$ is the EGF for a measure g_n (for example, probability, expectation, moment) in the Bernoulli model, then $e^{-\alpha}\hat{g}(\alpha)$ is the corresponding measure if the total number of elements $X_1 + \cdots + X_m$ is Poisson with mean α.*

PROOF. The measure in the Poisson model is the conditional expectation of the measure in the Bernoulli model, namely,

$$\sum_{n \geqslant 0} g_n \Pr\{X_1 + \cdots + X_m = n\} = \sum_{n \geqslant 0} g_n \frac{e^{-\alpha}\alpha^n}{n!} = e^{-\alpha}\hat{g}(\alpha). \qquad \square$$

We shall use Theorem 5.5 and direct our attention to the Poisson model, where the number of elements in each bucket is Poisson distributed with mean α, and hence the total number of elements is a Poisson RV with mean $m\alpha$. (The analysis of the Bernoulli case can be handled in much the same way as we shall handle the analysis of extendible hashing later in this section, so covering the Poisson case will present us with a different perspective.)

We let M_α denote the maximum number of elements per bucket in the Poisson model, and we use $p_\alpha^{[b]}$ to denote the probability that $M_\alpha \leqslant b$. By Theorem 5.5, we have

$$p_\alpha^{[b]} = (e^{-\alpha} e_b(\alpha))^m.$$

(This can also be derived directly by noting that the m buckets are independent and that the Poisson probability that a given bucket has $\leqslant b$ elements is $e^{-\alpha}e_b(\alpha)$.) What we want is to compute the expected maximum bucket occupancy

$$\overline{M_\alpha} = \sum_{b \geqslant 1} b(p_\alpha^{[b]} - p_\alpha^{[b-1]}) = \sum_{b \geqslant 0} (1 - p_\alpha^{[b]}). \tag{5.15}$$

We shall consider the case $\alpha = o(\log m)$ (although the basic principles of our analysis will apply for any α). A very common occurrence in occupancy RVs is a sharp rise in the distribution $p_\alpha^{[b]}$ from ≈ 0 to ≈ 1 in the "central region" near the mean value. Intuitively, a good approximation for the mean is the value $\tilde{b}$ such that $p_\alpha^{[\tilde{b}]}$ is sufficiently away from 0 and 1. We choose the value $\tilde{b} > \alpha$ such that

$$\frac{e^{-\alpha}\alpha^{\tilde{b}+1}}{(\tilde{b}+1)!} \leqslant \frac{1}{m} < \frac{e^{-\alpha}\alpha^{\tilde{b}}}{\tilde{b}!}. \tag{5.16a}$$

When $\alpha = \Theta(1)$, it is easy to see from (5.16) that $\tilde{b} \sim \Gamma^{-1}(m) \sim (\log m)/\log\log m$. (Here $\Gamma^{-1}(y)$ denotes the inverse of the Gamma function $\Gamma(x)$.) We define λ using the left-hand side of (5.16a) so that

$$\frac{e^{-\alpha}\alpha^{\tilde{b}+1}}{(\tilde{b}+1)!} = \frac{\lambda}{m}. \tag{5.16b}$$

In particular we have $\alpha/(\tilde{b}+1) < \lambda \leqslant 1$. The following bound illustrates the sharp increase in $p_\alpha^{[b]}$ as a function of b in the vicinity $b \approx \tilde{b}$:

$$p_\alpha^{[\tilde{b}+k]} = (e^{-\alpha} e_{\tilde{b}+k}(\alpha))^m = \left(1 - e^{-\alpha} \sum_{b > \tilde{b}+k} \frac{\alpha^b}{b!}\right)^m$$

$$\sim \left(1 - \frac{e^{-\alpha}\alpha^{\tilde{b}+k+1}}{(\tilde{b}+k+1)!}\right)^m \sim \left(1 - \frac{\lambda\alpha^k}{m\tilde{b}^k}\right)^m \sim e^{-\lambda\alpha^k/\tilde{b}^k}. \tag{5.17}$$

The approximation is valid uniformly for $k = o(\sqrt{\tilde{b}})$. The expression $1 - p_\alpha^{[\tilde{b}+k]}$ continues to decrease exponentially as $k \to \infty$. The maximum bucket size is equal to $\tilde{b}$ with probability $\sim e^{-\lambda}$ and to $\tilde{b}+1$ with probability $\sim 1 - e^{-\lambda}$. The net effect is that we can get an asymptotic approximation for $\overline{M_\alpha}$ by approximating $1 - p_\alpha^{[b]}$ in (5.15) by a 0-1 step function with the step at $b = \tilde{b}$:

$$\overline{M_\alpha} \sim \sum_{0 \leqslant b < \tilde{b}} (1) + \sum_{b \geqslant \tilde{b}} (0) \sim \tilde{b}. \tag{5.18}$$

The same techniques can be applied for general α. The asymptotic behavior of $\overline{M_\alpha}$ for the Bernoulli and Poisson models is the same.

5.6. THEOREM (Kolchin et al. [76]). *In the Bernoulli model with n elements inserted in m buckets ($\alpha = n/m$) or in the Poisson model in which the occupancy of each bucket is an independent Poisson RV with mean α, the expected maximum bucket occupancy is*

$$\overline{M_\alpha} \sim \begin{cases} \dfrac{\log m}{\log\log m} & \text{if } \alpha = \Theta(1); \\ \tilde{b} & \text{if } \alpha = o(\log m); \\ \alpha & \text{if } \alpha = \omega(\log m), \end{cases}$$

where $\tilde{b}$ is given by (5.16a).

When α gets large, $\alpha = \omega(\log m)$, the bucket occupancies become fairly uniform; the difference $\overline{M_\alpha} - \alpha$ converges in probability to $\sim\sqrt{2\alpha\log m}$, provided that $\alpha = m^{O(1)}$.

Extendible hashing

A quantity related to maximum bucket occupancy is the expected directory size used in *extendible hashing*, in which the hash table is allowed to grow and shrink dynamically [27, 78]. Each slot in the hash table models a page of secondary storage capable of storing up to b elements. If a bucket overflows, the number of buckets in the table is successively doubled until each bucket has at most b elements. The directory acts as a b-trie, based upon the infinite precision hash addresses of the elements (cf. Section 4.3). For this reason, the analyses of directory size and trie height are very closely related.

At any given time, the directory size is equal to the number of buckets in the table, which is always a power of 2. The probability $\pi_n^{[h]}$ that the directory size is $\leqslant 2^h$ is

$$\pi_n^{[h]} = n![z^n]\left(e_b\left(\frac{z}{2^h}\right)\right)^{2^h}. \tag{5.19a}$$

This is identical to (5.14d) with $m = 2^h$, except that in this case m is the parameter that varies, and the bucket capacity b stays fixed. We can also derive (5.19a) via the admissibility theorem for tries (Theorem 4.18): $\pi_n^{[h]}$ is the probability that the height of a random b-trie is $\leqslant h$. The EGF $\pi^{[h]}(z) = \sum_{n\geqslant 0} \pi_n^{[h]} z^n/n!$ satisfies

$$\pi^{[h]}(z) = (\pi^{[h-1]}(\tfrac{1}{2}z))^2; \qquad \pi^{[0]}(z) = e_b(z). \tag{5.19b}$$

Hence, (5.19a) follows.

5.7. THEOREM (Flajolet [32]). *In the Bernoulli model, the expected directory size in extendible hashing for bucket size $b > 1$ when there are n elements present is*

$$\overline{S_n} \sim \left(\frac{\Gamma(1-1/b)}{(\log 2)(b+1)!^{1/b}} + Q\left(\left(1+\frac{1}{b}\right)\log_2 n\right)\right) n^{1+1/b},$$

where $Q(u)$ is a periodic function with period 1 and mean value 0.

PROOF (*Sketch*). We can express the average directory size $\overline{S_n}$ in terms of $\pi_n^{[h]}$ in a way similar to (5.15):

$$\overline{S_n} = \sum_{h\geqslant 1} 2^h(\pi_n^{[h]} - \pi_n^{[h-1]}) = \sum_{h\geqslant 0} 2^h(1 - \pi_n^{[h]}). \tag{5.20}$$

The first step in the derivation is to apply the saddle point method of Section 2.3. We omit the details for brevity. As for the maximum bucket occupancy, the probabilities $\pi_n^{[h]}$ change quickly from ≈ 0 to ≈ 1 in a "central region," which in this case is where $h = \tilde{h} = (1+1/b)\log_2 n$. By saddle point, we get the uniform approximation

$$\pi_n^{[h]} \sim \exp\left(\frac{-n^{b+1}}{(b+1)!2^{bh}}\right), \tag{5.21}$$

for $|h - \tilde{h}| < \log_2 \log n$. When $h \geqslant \tilde{h} + \log_2 \log n$, approximation (5.21) is no longer valid, but the terms $1 - \pi_n^{[h]}$ continue to decrease exponentially with respect to h. Hence, we can substitute approximation (5.21) into (5.20) without affecting the leading terms of

$\overline{S_n}$. We get $\overline{S_n} \sim T(n^{b+1}/(b+1)!)$, where $T(x)$ is the harmonic sum

$$T(x) = \sum_{h \geqslant 0} 2^h(1 - e^{-x/2^{bh}}), \tag{5.22a}$$

whose Mellin transform is

$$T^*(s) = \frac{-\Gamma(s)}{1 - 2^{bs+1}}, \tag{5.22b}$$

in the fundamental strip $\langle -1; -1/b \rangle$. The asymptotic expansion of $T(x)$, as $x \to \infty$, is given by the poles of $T^*(s)$ to the right of the strip. There is a simple pole at $s = 0$ due to $\Gamma(s)$ and simple poles at $s = -1/b + 2k\pi i/\log 2$ due to the denominator. The result follows immediately from (2.24). □

Theorem 5.7 shows that the leading term of $\overline{S_n}$ oscillates with $(1 + 1/b)\log_2 n$. An intuition as to why there is oscillation can be found in the last summation in (5.20). The sum samples the terms $1 - \pi_n^{[h]}$ at each nonnegative integer h using the exponential weight function 2^h. The value of h where the approximation (5.21) changes quickly from ≈ 0 to ≈ 1 is close to an integer value only when $(1 + 1/b)\log_2 n$ is close to an integer, thus providing the periodic effect.

It is also interesting to note from Theorem 5.7 that the directory size is asymptotically superlinear, that is, the directory becomes larger than the file itself when n is large! Fortunately, convergence is slow, and the nonlinear growth of $\overline{S_n}$ is not noticeable in practice when b is large enough, say $b \geqslant 40$. Similar results for the Poisson model appear in [94].

The same techniques apply to the analysis of the expected height $\overline{H_n}$ of tries:

$$\overline{H_n} = \sum_{h \geqslant 1} h(\pi_n^{[h]} - \pi_n^{[h-1]}) = \sum_{h \geqslant 0} (1 - \pi_n^{[h]}). \tag{5.23}$$

This is the same as (5.20), but without the weight factor 2^h. (When trie height grows by 1, directory size doubles.)

5.8. THEOREM (Flajolet [32]). *The expected height in the Bernoulli model of a random b-trie of n elements is*

$$\overline{H_n} = \left(1 + \frac{1}{b}\right)\log_2 n + \tfrac{1}{2} + \frac{\gamma - \log((b+1)!)}{b \log 2} + P\left(\left(1 + \frac{1}{b}\right)\log_2 n\right) + o(1),$$

where $P(u)$ is periodic with period 1, *small amplitude, and mean value* 0.

In the biased case, where 0 occurs with probability p and 1 occurs with probability $q = 1 - p$, we have

$$\pi^{[h]}(z) = \pi^{[h-1]}(pz) \cdot \pi^{[h-1]}(qz),$$

which gives us

$$\pi^{[h]}(z) = \prod_{1 \leqslant k \leqslant h} (e_b(p^k q^{n-k} z))^{\binom{h}{k}}$$

(cf. (5.19a) and (5.19b)). Multidimensional versions of extendible hashing have been studied in [95].

Coalesced hashing

We can bypass the problems of direct chaining with relocation by using *hashing with coalescing chains* (or simply *coalesced hashing*). Part of the hash table is dedicated to storing elements that collide when inserted. We redefine m' to be the number of slots in the hash table. The range of the hash function is restricted to $\{1, 2, \ldots, m\}$. We call the first m slots the *address region*; the bottom $m'-m$ slots, which are used to store colliders, comprise the *cellar*.

When a collision occurs during insertion (because the element's hash address is already occupied), the element is inserted instead into the largest-numbered empty slot in the hash table and is linked to the end of the chain it collided with. This means that the colliding record is stored in the cellar if the cellar is not full. But if there are no empty slots in the cellar, the element ends up in the address region. In the latter case, elements inserted later could collide with this element, and thus their chains would "coalesce".

If the cellar size is chosen so that it can accommodate all the colliders, then coalesced hashing reduces to separate chaining. It is somewhat surprising that average-case performance can be improved by choosing a smaller cellar size so that coalescing usually occurs. The intuition is that by making the address region larger, the hash addresses of the elements are spread out over a larger area, which helps reduce collisions. This offsets the disadvantages of coalescing, which typically occurs much later. We might be tempted to go to the extreme and eliminate the cellar completely (this variant is called *standard coalesced hashing*), but performance deteriorates. The theorem below gives the expected search times as a function of the load factor $\alpha = n/m'$ and the address factor $\beta = m/m'$. We can use the theorem to determine the optimum β. It turns out that β_{opt} is a function of α and of the type of search, but the compromise value $\beta = 0.86$ gives near-optimum search performance for a large range of α and is recommended for general use.

5.9. THEOREM (Vitter [107]). *The expected number of probes per search for coalesced hashing in an m'-slot table with address size $m = \beta m'$ and with $n = \alpha m'$ elements is*

$$\overline{U_n} \sim \begin{cases} e^{-\alpha/\beta} + \dfrac{\alpha}{\beta} & \text{if } \alpha \leqslant \lambda\beta; \\[2ex] \dfrac{1}{\beta} + \dfrac{1}{4}\left(e^{2(\alpha/\beta - \lambda)} - 1\right)\left(3 - \dfrac{2}{\beta} + 2\lambda\right) - \dfrac{1}{2}\left(\dfrac{\alpha}{\beta} - \lambda\right) & \text{if } \alpha \geqslant \lambda\beta; \end{cases}$$

$$\overline{C_n} \sim \begin{cases} 1 + \dfrac{\alpha}{2\beta} & \text{if } \alpha \leqslant \lambda\beta; \\[2ex] 1 + \dfrac{\beta}{8\alpha}\left(e^{2(\alpha/\beta - \lambda)} - 1 - 2\left(\dfrac{\alpha}{\beta} - \lambda\right)\right)\left(3 - \dfrac{2}{\beta} + 2\lambda\right) & \\[2ex] \quad + \dfrac{1}{4}\left(\dfrac{\alpha}{\beta} + \lambda\right) + \dfrac{\lambda}{4}\left(1 - \dfrac{\lambda\beta}{\alpha}\right) & \text{if } \alpha \geqslant \lambda\beta, \end{cases}$$

where λ is the unique nonnegative solution to the equation $e^{-\lambda} + \lambda = 1/\beta$.

The method described above is formally known as *late-insertion coalesced hashing* (LICH). Vitter and Chen [108] also analyze two other methods called *early-insertion coalesced hashing* (EICH) and *varied-insertion coalesced hashing* (VICH). In EICH, a colliding element is inserted immediately after its hash address in the chain, by rerouting pointers. VICH uses the same rule, except when there is a cellar slot in the chain following the element's hash address; in that case, the element is linked into the chain immediately after the last cellar slot in the chain. VICH requires slightly fewer probes per search than the other variants and appears to be optimum among all possible linking methods. Deletion algorithms and implementation issues are also covered in [108].

PROOF (*Sketch*). We first consider the unsuccessful search case. We count the number $m^{n+1}\overline{U_n}$ of probes needed to perform all possible unsuccessful searches in all possible hash tables of n elements. Each chain of length l that has t elements in the cellar (which we call an (l, t)-chain) contributes

$$\delta_{l=0}+l+(l-t-1)+(l-t-2)+\cdots+1=\delta_{l=0}+l+\binom{l-t}{2} \tag{5.24a}$$

probes, and we have

$$m^{n+1}\overline{U_n}=m^{n+1}p_n+\sum_{l,t} lc_n(l,t)+\sum_{l,t}\binom{l-t}{2}c_n(l,t), \tag{5.24b}$$

where p_n is the probability that the hash address of the element is unoccupied, and $c_n(l, t)$ is the number of (l, t)-chains. The second term is easily seen to be nm^n, since there are n elements in each hash table. The evaluations of the first term $m^{n+1}p_n$ and the third term

$$S_n=\sum_{l,t}\binom{l-t}{2}c_n(l,t) \tag{5.24c}$$

are similar; for brevity we restrict our attention to the latter. There does not seem to be a closed form expression for $c_n(l, t)$, but we can develop a recurrence for $c_n(l, t)$ which we can substitute into (5.24c) to get

$$S_n=(m+2)^{n-1}\sum_{0\leqslant j\leqslant n-1}\left(\frac{m}{m+2}\right)^j(j-m'+m)\frac{F_j}{m^j}, \tag{5.25}$$

where F_j is the number of hash sequences of j elements that yield full cellars. In terms of probabilities, F_j/m^j is the probability that the cellar is full after j insertions.

This provides a link to the occupancy distributions studied earlier. We define the RV N_j to be the number of elements that collide when inserted, in a hash table of j elements. The cellar is full after k insertions if and only if $N_j \geqslant m'-m$; by taking probabilities we get

$$\frac{F_j}{m^j}=\Pr\{N_j\geqslant m'-m\}. \tag{5.26}$$

This expresses F_j/m^j as a tail of the probability distribution for N_j. We can determine the distribution by applying Theorem 5.1. We let $H_{k,j,m}$ be the number of hash sequences of n elements for which k elements are marked. Our marking rule is that all elements that collide when inserted are marked; that is, for each bucket (hash address) i with s elements, there are $f_i(s)=s-1+\delta_{s=0}$ marked elements. By Theorem 5.1, we have

$$\hat{H}(u,z)=\sum_{k,j\geqslant 0} H_{k,j,m}u^k\frac{z^j}{j!}=\prod_{1\leqslant i\leqslant m}\hat{B}_i(u,z), \tag{5.27a}$$

where

$$\hat{B}_i(u,z)=\sum_{t\geqslant 0} u^{t-1+\delta_{t=0}}\frac{z^t}{t!}=\frac{e^{uz}-1+u}{u}, \tag{5.27b}$$

for each $1\leqslant i\leqslant m$. Substituting (5.27b) into (5.27a), we get

$$\hat{H}(u,z)=\left(\frac{e^{uz}-1+u}{u}\right)^m. \tag{5.27c}$$

This allows us to solve for $F_j/m^j=\Pr\{N_j\geqslant m'-m\}$:

$$\Pr\{N_j=k\}=\frac{H_{k,j,m}}{m^j}=j![u^kz^j]\hat{H}\left(u,\frac{z}{m}\right); \tag{5.28a}$$

$$F_j/m^j=1-j![u^{m'-m-1}z^j]\frac{1}{1-u}\hat{H}\left(u,\frac{z}{m}\right). \tag{5.28b}$$

We can get asymptotic estimates for (5.28b) by use of the saddle point method as in our analysis of extendible hashing or by Chernoff bounds as in [107, 108]; the details are omitted for brevity. The distribution F_j/m^j increases sharply from ≈ 0 to ≈ 1 in the "central region" where

$$\overline{N_j}\approx m'-m. \tag{5.29a}$$

The expected number of collisions $\overline{N_j}$ is

$$\overline{N_j}=j![z^j]\frac{\partial}{\partial u}\hat{H}\left(u,\frac{z}{m}\right)\bigg|_{u=1}=j-m+m\left(1-\frac{1}{m}\right)^j. \tag{5.29b}$$

We can solve for the value of $j=\tilde{\jmath}$ that satisfies (5.29a) by combining (5.29a) and (5.29b) and using the ratios $\tilde{\alpha}=\tilde{\jmath}/m'$ and $\beta=m/m'$. We have

$$e^{-\tilde{\alpha}}+\tilde{\alpha}\approx\frac{1}{\beta}. \tag{5.29c}$$

In a way similar to (5.18), we approximate F_j/m^j by a 0-1 step function with the step at $j=\tilde{\jmath}$, and we get

$$S_n\sim(m+2)^{n-1}\sum_{\tilde{\jmath}\leqslant j\leqslant n-1}\left(\frac{m}{m+2}\right)^j(j-m'+m), \tag{5.30}$$

which can be summed easily. The error of approximation can be shown to be negligible, as in the analysis of maximum bucket occupancy and extendible hashing. The analysis of the term $m^{n+1}p_n$ in (5.24b) is based upon similar ideas and is omitted for brevity.

For the case of successful searches, the formula for $\overline{C_n}$ follows from a formula similar to (4.25) for binary search trees and separate chaining:

$$\overline{C_n}=1+\frac{1}{n}\sum_{0\leqslant i\leqslant n-1}(\overline{U_i}-p_i), \tag{5.31}$$

where p_i is the term in (5.24b). □

A unified analysis of all three variants of coalesced hashing—LICH, EICH, and VICH—is given in [108]. The maximum number of probes per search among all the searches in the same table, for the special case of standard LICH (when there is no cellar and $m=m'$), is shown in [91] to be $\log_c n-2\log_c\log n+O(1)$ with probability ~ 1 for successful searches and $\log_c n-\log_c\log n+O(1)$ with probability ~ 1 for unsuccessful searches, where $c=1/(1-e^{-\alpha})$.

5.2. *Hashing by open addressing*

An alternative to chaining is to probe an *implicitly* defined sequence of locations while looking for an element. If the element is found, the search is successful; if an "open" (empty) slot is encountered, the search is unsuccessful, and the new element is inserted into the empty slot that terminated the search.

Uniform probing

A simple scheme to analyze, which serves as a good approximation to more practical methods, is *uniform probing*. The probing sequence for each element x is a random permutation $h_1(x)h_2(x)\ldots h_m(x)$ of $\{1,2,\ldots,m\}$. For an unsuccessful search, the probability that k probes are needed when there are n elements in the table is

$$p_k=\frac{n^{\underline{k-1}}(m-n)}{m^{\underline{k}}}=\binom{m-k}{m-n-1}\Big/\binom{m}{n}. \tag{5.32a}$$

Hence, we have

$$\overline{U_n}=\sum_{k\geqslant 0}kp_k=\frac{1}{\binom{m}{n}}\sum_{k\geqslant 0}k\binom{m-k}{m-n-1}. \tag{5.32b}$$

We split k into two parts $k=(m+1)-(m-k+1)$, because we can handle each separately; the $m-k+1$ term gets "absorbed" into the binomial coefficient.

$$\begin{aligned}\overline{U_n}&=\frac{1}{\binom{m}{n}}\sum_{k\geqslant 0}(m+1)\binom{m-k}{m-n-1}-\frac{1}{\binom{m}{n}}\sum_{k\geqslant 0}(m-k+1)\binom{m-k}{m-n-1}\\&=m+1-\frac{m-n}{\binom{m}{n}}\sum_{k\geqslant 0}\binom{m-k+1}{m-n}\\&=m+1-\frac{m-n}{\binom{m}{n}}\binom{m+1}{m-n+1}=\frac{m+1}{m-n+1}.\end{aligned} \tag{5.32c}$$

Successful search time for open addressing algorithms satisfy a relation similar to (4.25):

$$\overline{C_n}=\frac{1}{n}\sum_{0\leqslant i\leqslant n-1}\overline{U_i}\,. \tag{5.33}$$

Putting this all together, we get the following theorem.

5.10. THEOREM. *The expected number of probes per unsuccessful and successful search for uniform probing, when there are $n=m\alpha$ elements in a hash table of m slots, is*

$$\overline{U_n}=\frac{m+1}{m-n+1}\sim\frac{1}{1-\alpha};\qquad \overline{C_n}=\frac{m+1}{n}(H_{m+1}-H_{m-n+1})\sim\frac{1}{\alpha}\log\frac{1}{1-\alpha}\,.$$

The asymptotic formula $\overline{U_n}\sim 1/(1-\alpha)=1+\alpha+\alpha^2+\cdots$ has the following intuitive interpretation: With probability α we need more than one probe, with probability α^2 we need more than two probes, and so on. The expected maximum search time per hash table is studied in [48].

Double hashing and secondary clustering

The practical limitation on uniform probing is that computing several hash functions is very expensive. However, the performance of uniform probing can be approximated well by use of just two hash functions. In the *double hashing* method, the ith probe is made at slot

$$h_1(x)+ih_2(x)\bmod m\quad\text{for } 0\leqslant i\leqslant m-1; \tag{5.34}$$

that is, the probe sequence starts at slot $h_1(x)$ and steps cyclically through the table with step size $h_2(x)$. (For simplicity, we have renumbered the m table slots to be 0, 1, ..., $m-1$.) The value of the second hash function must be relatively prime to m so that the probe sequence (5.34) gives a full permutation. Guibas and Szemerédi [55] show using interesting probabilistic techniques that when $\alpha<0.319$, the number of probes per search is asymptotically equal to that of uniform probing, and this has been extended to all fixed $\alpha<1$ [81].

It is often faster in practice to use only one hash function and to define $h_2(x)$ implicitly in terms of $h_1(x)$. For example, if m is prime, we could set

$$h_2(x)=\begin{cases}1 & \text{if } h_1(x)=0;\\ m-h_1(x) & \text{if } h_1(x)>0.\end{cases} \tag{5.35}$$

A useful approximation to this variant is the *secondary clustering* model, in which the initial probe location $h_1(x)$ uniquely determines the remaining probe locations $h_2h_3\ldots h_m$, which form a random permutation of the other $m-1$ slots.

5.11. THEOREM (Knuth [71]). *The expected number of probes per unsuccessful and successful search for hashing with secondary clustering, when there are $n=m\alpha$ elements in a hash table of m slots, is*

$$\overline{U_n}\sim\frac{1}{1-\alpha}-\alpha+\log\frac{1}{1-\alpha};\qquad \overline{C_n}\sim 1+\log\frac{1}{1-\alpha}-\frac{\alpha}{2}\,.$$

The proof is a generalization of the method we use below to analyze linear probing. The number of probes per search for secondary clustering is slightly more than for double hashing and uniform probing, but slightly less than for linear probing.

Linear probing

Perhaps the simplest implementation of open addressing is a further extension of (5.34) and (5.35), called *linear probing*, in which the unit step size $h_2(x) \equiv 1$ is used. This causes *primary clustering* in the table, because all the elements with the same hash address follow the same probing sequence.

5.12. THEOREM (Knuth [71]). *The expected number of probes per unsuccessful and successful search for linear probing, when there are* $n = m\alpha$ *elements in a hash table of* m *slots, is*

$$\overline{U_n} = \tfrac{1}{2}(1 + Q_1(m, n)); \qquad \overline{C_n} = \tfrac{1}{2}(1 + Q_0(m, n-1)), \tag{5.36}$$

where

$$Q_r(m, n) = \sum_{k \geqslant 0} \binom{r+k}{k} \frac{n^{\underline{k}}}{m^k}.$$

If $\alpha = n/m$ *is a constant bounded away from* 1, *then we have*

$$\overline{U_n} \sim \frac{1}{2}\left(1 + \frac{1}{(1-\alpha)^2}\right); \qquad \overline{C_n} \sim \frac{1}{2}\left(1 + \frac{1}{1-\alpha}\right).$$

For full tables, we have

$$\overline{U_{m-1}} = \frac{m+1}{2}; \qquad \overline{C_m} \sim \frac{1}{2}\sqrt{\frac{\pi m}{2}} + \frac{1}{3} + \frac{1}{24}\sqrt{\frac{\pi}{2m}} + O(1/m).$$

PROOF. The derivation of the exact formulæ for $\overline{U_n}$ and $\overline{C_n}$ is an exercise in combinatorial manipulation. The number of the m^n hash sequences such that slot 0 is empty is

$$f(m, n) = \left(1 - \frac{n}{m}\right) m^n. \tag{5.37}$$

By decomposing the hash table into two separate parts, we find that the number of hash sequences such that position 0 is empty, positions 1 through k are occupied, and postion $k+1$ is empty is

$$g(m, n, k) = \binom{n}{k} f(k+1, k) f(m-k-1, n-k). \tag{5.38}$$

The probability that $k+1$ probes are needed for a unsuccessful search is thus

$$p_k = \frac{1}{m^n} \sum_{k \leqslant j \leqslant n} g(m, n, j). \tag{5.39}$$

The formulae (5.36) for $\overline{U_n}=\sum_{0\leqslant k\leqslant n}(k+1)p_k$ and $\overline{C_n}=(1/n)\sum_{0\leqslant i\leqslant n-1}\overline{U_i}$ in terms of $Q_0(m,n-1)$ and $Q_1(m,n)$ follow by applications of Abel's identity (cf. Section 1.4).

When α is bounded below 1, then we can evaluate $Q_r(m,n)$ asymptotically by approximating $n^{\underline{k}}/m^k$ in the summations by $n^k/m^k=\alpha^k$. The interesting case as far as analysis is concerned is for the full table when $\alpha\approx 1$. It is convenient to define the new notation

$$Q_{\{b_k\}}=\sum_{k\geqslant 0} b_k\frac{(m-1)^{\underline{k}}}{m^k}. \tag{5.40a}$$

The link between our new and old notation is

$$Q_{\left\{\binom{r+k}{k}\right\}}(m)=Q_r(m,m-1). \tag{5.40b}$$

Note that the Q functions each have only a finite number of non-zero terms. The following powerful theorem provides asymptotic expansions for several choices of $\{b_k\}$:

5.13. THEOREM (Knuth and Schönhage [75]). *We have*

$$Q_{\{b_k\}}(m)=\frac{m!}{m^{m-1}}[z^m]A(y(z)), \tag{5.41a}$$

where $y(z)$ is defined implicitly by the equation

$$y(z)=ze^{y(z)}, \tag{5.41b}$$

and $A(u)=\sum_{k\geqslant 0}b_ku^{k+1}/(k+1)$ is the antiderivative of $B(u)=\sum_{k\geqslant 0}b_ku^k$.

PROOF. The proof uses an application of Lagrange–Bürmann inversion (Theorem 1.6) applied to $y(z)$. The motivation is based upon the fact that sums of the form (5.40a) are associated with the implicitly defined function $y(z)$ in a natural way; the number of labeled oriented trees on m vertices is m^{m-1}, and its EGF is $y(z)$. Also, $y(z)$ was used in Section 1.4 to derive Abel's identity, and Abel's identity was used above to derive (5.36). By applying Lagrange–Bürmann inversion to the right-hand side of (5.41a), with $\varphi(u)=e^u$, $g(u)=A(u)$, and $f(z)=y(z)$, we get

$$\begin{aligned}\frac{m!}{m^{m-1}}[z^m]A(y(z))&=\frac{m!}{m^{m-1}}\frac{1}{m}[u^{m-1}]e^{um}A'(u)=\frac{(m-1)!}{m^{m-1}}[u^{m-1}]e^{um}B(u)\\&=\frac{(m-1)!}{m^{m-1}}\sum_{0\leqslant k\leqslant m-1}b_k\frac{m^{m-k-1}}{(m-k-1)!}\\&=\sum_{0\leqslant k\leqslant m-1}b_k\frac{(m-1)^{\underline{k}}}{m^k}=Q_{\{b_k\}}(m).\qquad\square\end{aligned}$$

PROOF OF THEOREM 5.12 (*continued*). By (5.40b) the sequences $\{b_k\}$ corresponding to $Q_0(m,m-1)$ and $Q_1(m,m-1)$ have generating functions $B_0(u)=1/(1-u)$ and $B_1(u)=$

$1/(1-u)^2$, and the corresponding antiderivatives are $A_0(u)=\log(1/(1-u))$ and $A_1(u)=1/(1-u)$. By applying Theorem 5.13, we get

$$Q_0(m, m-1)=\frac{m!}{m^{m-1}}[z^m]\log\frac{1}{1-y(z)}; \tag{5.42a}$$

$$Q_1(m, m-1)=\frac{m!}{m^{m-1}}[z^m]\frac{1}{1-y(z)}. \tag{5.42b}$$

It follows from the method used in Theorem 4.4 to count the number T_n of plane trees with degree constraint Ω that the dominant singularity of $y(z)$ implicitly defined by (5.41b) is $z=1/\mathrm{e}$, where we have the expansion $y(z)=1-\sqrt{2}\sqrt{1-\mathrm{e}z}+\mathrm{O}(1-\mathrm{e}z)$. Hence, $z=1/\mathrm{e}$ is also the dominant singularity of $\log(1/(1-y(z)))$, and we get directly

$$\log\frac{1}{1-y(z)}=\tfrac{1}{2}\log\frac{1}{1-\mathrm{e}z}-\tfrac{1}{2}\log 2+\mathrm{O}(\sqrt{1-\mathrm{e}z}). \tag{5.43}$$

The approximation for $\overline{C_m}$ follows by extracting coefficients from (5.43). The formula $\overline{U_{m-1}}=(m+1)/2$ can be obtained by an analysis of the right-hand side of (5.42b), or more directly by counting the number of probes needed for the m unsuccessful searches in a hash table with only one empty slot. □

The length of the maximum search per table for fixed $\alpha<1$ is shown in [90] to be $\Theta(\log n)$, on the average. Amble and Knuth [2] consider a variation of linear probing, called *ordered hashing*, in which each cluster of elements in the table is kept in the order that would occur if the elements were inserted in increasing order by element value. Elements are relocated within a cluster, if necessary, during an insertion. The average number of probes per unsuccessful search decreases to $\overline{C_{n-1}}$.

Linear probing can also be used for sorting, in a way similar to distribution sort described earlier, except that the elements are stored in a linear probing hash table rather than in buckets. The n elements in the range $[a, b]$ are inserted into the table using the hash function $h_1(x)=\lfloor(x-a)/(x-b)\rfloor$. The elements are then compacted and sorted via insertion sort. The hash table size m should be chosen to be somewhat larger than n (say, $n/m\approx 0.8$), so that insertion times for linear probing are fast. A related sorting method described in [50] uses ordered hashing to keep the elements in sorted order, and thus the final insertion sort is not needed.

6. Dynamic algorithms

In this section we study performance measures that reflect the dynamic performance of a data structure over an interval of time, during which several operations may occur. In Section 6.1, we consider priority queue algorithms and analyze their performance integrated over a sequence of operations. The techniques apply to other data structures as well, including dictionaries, lists, stacks, and symbol tables. In Section 6.2 we analyze the maximum size attained by a data structure over time, which has important applications to preallocating resources. We conclude in Section 6.3 with the probabilistic analysis of union-find algorithms.

6.1. Integrated cost and the history model

Dynamic data structures such as lists, search trees, and heap-ordered trees can be analyzed in a dynamic context under the effect of *sequences of operations.* Knuth [72] considers various models of what it means for deletions to "preserve randomness", and this has been applied to the study of various data structures (for example, see [108]). Françon [46] proposed a model called the "history model" which amounts to analyzing dynamic structures under all possible evolutions up to order-isomorphism. Using combinatorial interpretations of continued fractions and orthogonal polynomials [30], several data structures, including dictionaries, priority queues, lists, stacks, and symbol tables, can be analyzed under this model [119]. In this section, we shall present an overview of the theory, with special emphasis on priority queues.

A priority queue (see Section 4.2) supports the operations of inserting an element x into the queue (which we denote by $I(x)$) and deletion of the minimum element (which we denote by D). An example of a particular sequence of operations is

$$s = I(3.1)\, I(1.7)\, I(2.9)\, I(3.7)\, D\, D\, I(3.4). \tag{6.1}$$

Such a sequence s consists of a *schema IIIIDDI*, from which we see that s causes the queue size to increase by 3. If we restrict attention to structures that operate by comparison between elements, the effect of s on an initially empty queue is fully characterized by the following information: The second operation $I(1.7)$ inserts an element smaller than the first, the third operation $I(2.9)$ inserts an element that falls in between the two previous ones, and so on. We define the *history* associated with a sequence s to consist of the schema of s in which each operation is labeled by the *rank* of the element on which it operates (relative to the current state of the structure). If we make the convention that ranks are numbered starting at 0, then all delete-minimum operations must be labeled by 0, and each insert is labeled between 0 and k, where k is the size of the priority queue at the time of the insert. The history associated with (6.1) is

$$I_0\, I_0\, I_1\, I_3\, D_0\, D_0\, I_1\,. \tag{6.2}$$

We let $\mathscr{H}$ denote the set of all histories containing as many inserts as delete-minimum operations, and we let $\mathscr{H}_n$ be the subset of those that have length n. We define $h_n = |\mathscr{H}_n|$ and $h(z) = \sum_{n \geq 0} h_n z^n$.

6.1. THEOREM. *The OGF of priority queue histories has the continued fraction expansion*

$$h(z) = \cfrac{1}{1 - \cfrac{1 \cdot z^2}{1 - \cfrac{2 \cdot z^2}{1 - \cfrac{3 \cdot z^2}{1 - \cfrac{4 \cdot z^2}{\cdots}}}}}\,. \tag{6.3}$$

Theorem 6.1 is a special case of a general theorem of [30] that expresses generating functions of labeled schemas in terms of continued fractions. Another special case is the enumeration of plane trees, given in Section 4.1.

Returning to priority queue histories, from a classical theorem of Gauss (concerning the continued fraction expansion of quotients of hypergeometric functions) applied to continued fraction (6.3), we find

$$h_{2n} = 1 \times 3 \times 5 \times \cdots \times (2n-1),$$

with $h_{2n+1} = 0$. Thus the set of histories has an explicit and simple counting expression. Let us define the *height* of a history to be the maximum size that the priority queue attains over the course of the operations. From the same theory, it follows that the OGF $h^{[k]}(z)$ for histories with height bounded by an integer k is the kth convergent of (6.3):

$$h^{[k]}(z) = \frac{P_k(z)}{Q_k(z)}. \tag{6.4}$$

where P_k and Q_k are closely related to Hermite polynomials. From (6.3) and (6.4), it is possible to determine generating functions for extended sets of histories (such as the set of histories $\mathscr{H}^{\langle k \rangle}$ that start at size 0 and end at size k) and then to find the number of times a given operation is performed on a priority queue structure of size k in the course of all histories of $\mathscr{H}_n$. The expressions involve the continued fraction $h(z)$ in (6.3) and its convergents given in (6.4). From there, for a given priority queue structure, we can compute the *integrated cost* $\overline{K}_n$ defined as the expected cost of a random history in $\mathscr{H}_n$: If $\overline{CI}_k$ (respectively, $\overline{CD}_k$) is the individual expected cost of an insert (respectively, delete-minimum) operation on the priority queue structure when it has size k, then we have

$$K_n = \frac{1}{h_n} \sum_k (\overline{CI}_k \cdot NI_{n,k} + \overline{CD}_k \cdot ND_{n,k}), \tag{6.5}$$

where $NI_{n,k}$ (respectively $ND_{n,k}$) is the number of times operation insert (respectively delete-minimum) occurs at size k inside all histories in $\mathscr{H}_n$. Manipulations with EGFs make it possible to express the final result in simple form. For instance, we have the following two EGFs of histories $\mathscr{H}$ with a simple expression:

$$\sum_{n \geqslant 0} h_n \frac{z^n}{n!} = e^{z^2/2} \quad \text{and} \quad \sum_{n \geqslant 0} h_{2n} \frac{z^n}{n!} = \frac{1}{\sqrt{1-2z}}.$$

The main theorem 6.2 (below) is that the following two GFs

$$C(x) = \sum_{k \geqslant 0} (\overline{CI}_k + \overline{CD}_{k+1}) x^k \quad \text{and} \quad K(z) = \sum_{n \geqslant 0} \overline{K}_{2n}\, h_{2n} \frac{z^n}{n!}, \tag{6.6}$$

where $C(x)$ is an OGF of individual costs and $K(z)$ is a modified EGF of integrated costs (after normalization by h_n), are closely related.

6.2. THEOREM. *The GFs $C(x)$ and $K(z)$ defined above satisfy*

$$K(z)=\frac{1}{\sqrt{1-2z}}C\left(\frac{z}{1-z}\right). \tag{6.7}$$

If we plug into (6.7) the OGF $C(x)$ corresponding to a particular implementation of priority queues, and then extract coefficients, we get the integrated cost for that implementation. For instance, for histories of length $2n$ for sorted lists (SL) and binary search trees (BST), we have

$$\overline{K_{2n}^{\mathrm{SL}}}=n(n+5)/6 \quad \text{and} \quad \overline{K_{2n}^{\mathrm{BST}}}=n\log n+\mathrm{O}(n). \tag{6.8}$$

A variety of other dynamic data structures, including dictionaries, lists, stacks, and symbol tables, can be analyzed under the history model with these techniques. Each data type is associated with a continued fraction of the form (6.3), a class of orthogonal polynomials (such as Laguerre, Meixner, Chebyshev, and Poisson–Charlier) related to (6.4), and finally a transformation analogous to (6.6) that describes the transition from a GF of individual costs to the corresponding GF of integrated costs and that is usually expressed as an integral transform.

6.2. *Size of dynamic data structures*

We can model the effect of insertions and deletions upon the size of a dynamic data structure by regarding the ith element as being a subinterval $[s_i, t_i]$ of the unit interval; the ith element is "born" at time s_i, "dies" at time t_i, and is "living" when $t \in [s_i, t_i]$. At time t, the data structure must store the elements that are "living" at time t.

It is natural to think of the data structure as a statistical queue, as far as size is concerned. Let us denote the number of living elements at time t by *Size*(t). If we think of the elements as horizontal intervals, then *Size*(t) is just the number of intervals "cut" by the vertical line at position t. In many applications, such as in VLSI artwork analysis, for example, the number of living elements at any given time tends to be the square root of the total number of elements; thus for purposes of storage efficiency the data structure should expunge dead elements.

In the *hashing with lazy deletion* (HwLD) data structure, we assume that each element has a unique key. The data structure supports dynamic searching of elements by key value, which is useful in several applications. The elements are stored in a hash table of H buckets, based upon the hash addresses of their keys. Typically separate chaining is used. The distinguishing feature of HwLD is that an element is not deleted as soon as it dies; the "lazy deletion" strategy deletes a dead element only when a later insertion accesses the same bucket. The number H of buckets is chosen so that the expected number of elements per bucket is small. HwLD is thus more time-efficient than doing "vigilant-deletion", at a cost of storing some dead elements.

Expected queue sizes

We define *Use*(t) to be the number of elements in the HwLD data structure at time t; that is, *Use*(t) = *Size*(t) + *Waste*(t), where *Waste*(t) is the number of dead elements stored in the data structure at time t. Let us consider the M/M/∞ queueing model, in which the births form a Poisson process, and the lifespans of the intervals are independently and exponentially distributed.

6.3. THEOREM (Feller [28] and Van Wyk and Vitter [106]). *In the stationary M/M/∞ model, both Size and Use $-$ H are identically Poisson distributed with mean λ/μ, where λ is the birth rate of the intervals and $1/\mu$ is the average lifetime per element.*

PROOF. We define the notation $p_{m,n}(t) = \Pr\{Size(t)=m,\ Waste(t)=n\}$ for $m, n \geqslant 0$, and $p_{m,n}(t)=0$ otherwise. In the M/M/∞ model, we have

$$\begin{aligned} p_{m,n}(t+\Delta t) = {} & ((1-\lambda\Delta t)(\mathrm{e}^{-\mu\Delta t})^m + \mathrm{o}(\Delta t))p_{m,n}(t) \\ & + ((1-\lambda\Delta t)(m+1)(1-\mathrm{e}^{-\mu\Delta t})(\mathrm{e}^{-\mu\Delta t})^m + \mathrm{o}(\Delta t))p_{m+1,n-1}(t) \\ & + \delta_{n=0}(\lambda\Delta t + \mathrm{o}(\Delta t)) \sum_{j\geqslant 0} p_{m-1,j}(t) + \mathrm{o}(\Delta t). \end{aligned} \tag{6.9a}$$

By expanding the exponential terms in (6.9a) and rearranging, and letting $\Delta t \to 0$, we get

$$p'_{m,n}(t) = (-\lambda - m\mu)p_{m,n}(t) + (m+1)\mu p_{m+1,n-1}(t) + \delta_{n=0}\lambda \sum_{j\geqslant 0} p_{m-1,j}(t). \tag{6.9b}$$

In the stationary model, the probabilities $p_{m,n}(t)$ are independent of t, and thus the left-hand side of (6.9b) is 0. For notational simplicity we shall drop the dependence upon t. The rest of the derivation proceeds by considering the multidimensional OGF $P(z,w) = \sum_{m,n} p_{m,n} z^m w^n$. Equation (6.9b) becomes

$$\mu(z-w)\frac{\partial P(z,w)}{\partial z} = -\lambda P(z,w) + \lambda z P(z,1). \tag{6.9c}$$

This provides us with the distribution of *Size*:

$$\Pr\{Size=m\} = [z^m]P(z,1) = [z^m]\mathrm{e}^{(z-1)\lambda/\mu} = \frac{(\lambda/\mu)^m}{m!}\mathrm{e}^{\lambda/\mu}. \tag{6.10}$$

To find the distribution of *Use*, we replace w by z in (6.9c), which causes the left-hand side of (6.9c) to become 0. We get

$$\Pr\{Use=k\} = [z^k]P(z,z) = [z^k]zP(z,1). \tag{6.11}$$

The rest follows from (6.10). $\square$

Maximum queue size

A more interesting statistic, which has direct application to matters of storage preallocation, is the maximum values of $Size(t)$ and $Use(t)$ as the time t varies over the entire unit interval.

Orthogonal polynomials arise in an interesting way when considering the more general model of a stationary birth-and-death process, which is a Markov process in which transitions from level k are allowed only to levels $k+1$ and $k-1$. The infinitesimal birth and death rates at level k are denoted by λ_k and μ_k:

$$\Pr\{Size(t+\Delta t)=j \mid Size(t)=k\} = \begin{cases} \lambda_k \Delta t + o(\Delta t) & \text{if } j=k+1; \\ \mu_k \Delta t + o(\Delta t) & \text{if } j=k-1; \\ o(\Delta t) & \text{otherwise.} \end{cases}$$

For the special case of the M/M/∞ model, we have $\lambda_0 = \lambda_1 = \cdots = \lambda$ and $\mu_k = k\mu$; for the M/M/1 model, we have $\lambda_0 = \lambda_1 = \cdots = \lambda$ and $\mu_0 = \mu_1 = \cdots = \mu$.

6.4. THEOREM (Mathieu and Vitter [82]). *The distribution of* $\max_{0 \leqslant t \leqslant 1}\{Size(t)\}$ *can be expressed simply in terms of Chebyshev polynomials (for the M/M/1 process) and Poisson–Charlier polynomials (for the M/M/∞ process). For several types of linear birth-and-death processes, of the form* $\lambda_k = \alpha k + \beta$, $\mu_k = \gamma k + \delta$, $Q_j(x)$ *can be expressed in terms of either Laguerre polynomials or Meixner polynomials of the second kind.*

It is interesting to note that orthogonal polynomials arose in a similar way in Section 6.1. The formulae referred to in Theorem 6.4 can be used to calculate the distributions of $\max_{0 \leqslant t \leqslant 1}\{Size(t)\}$ numerically, but they do not seem to yield asymptotics directly. Instead we rely on a different approach:

6.5. THEOREM (Kenyon–Mathieu and Vitter [67]). *In the stationary M/M/∞ probabilistic model, assuming either that* $\mu \to 0$ *or that* $\mu = \Omega(1)$ *and* $\lambda \to \infty$, *we have*

$$\mathbf{E}\left\{\max_{t\in[0,1]}\{Size(t)\}\right\} \sim \begin{cases} \dfrac{\lambda}{\mu} & \text{if } f(\lambda,\mu)\to 0; \\ d\dfrac{\lambda}{\mu} & \text{if } f(\lambda,\mu)\to c; \\ \dfrac{f(\lambda)}{\ln f(\lambda)}\dfrac{\lambda}{\mu} & \text{if } f(\lambda,\mu)\to\infty, \end{cases} \tag{6.12}$$

where $f(\lambda,\mu) = (\ln\lambda)/(\lambda/\mu)$ *and the constant d is defined implicitly from the constant c by* $d\ln d - d = c-1$. *In the first case* $\ln\lambda = o(\lambda/\mu)$, *we also have*

$$\mathbf{E}\left\{\max_{t\in[0,1]}\{Use(t)\}\right\} \sim \frac{\lambda}{\mu} + H. \tag{6.13}$$

When $\ln\lambda = o(\lambda/\mu)$, Theorem 6.5 says that the expected maximum value of $Size(t)$ (respectively $Use(t)$) is asymptotically equal to the maximum of its expected value. For example, in VLSI artwork applications, we might have $\lambda = n$, $\mu = \sqrt{n}$, so that the average number of living elements at any given time is $\lambda/\mu = \sqrt{n}$; by Theorem 6.5, the expected maximum data structure size is asymptotically the same. Kenyon-Mathieu and Vitter [67] also study the expected maximum under history models as in Section 6.1 and under other probabilistic models.

It is no accident that the theorem is structurally similar to Theorem 5.6 for maximum bucket occupancy. The quantity $\max_{t\in[0,1]}\{Size(t)\}$ can be regarded as the continuous counterpart of the maximum bucket occupancy. The proof below makes use of that relation.

PROOF (*Sketch*). For brevity, we consider only the analysis of $\mathbf{E}\{\max_{t\in[0,1]}\{Size(t)\}\}$. We shall concentrate primarily on case 1, in which $\ln\lambda = o(\lambda/\mu)$. The lower bound follows immediately by Theorem 6.3:

$$\mathbf{E}\{\max_{t\in[0,1]}\{Size(t)\}\} \geqslant \mathbf{E}\{Size(0)\} = \frac{\lambda}{\mu}.$$

We get the desired upper bound by looking at the following discretized version of the problem: Let us consider a hash table with $m = g\mu$ slots, where $g\mu$ is an integer and $g\to\infty$ very slowly, as $\lambda\to\infty$. The jth slot, for $1\leqslant j\leqslant g\mu$, represents the time interval $((j-1)/(g\mu), j/(g\mu)]$. For each element we place an entry into each slot whose associated time interval intersects the element's lifetime. If we define $N(j)$ to be the number of elements in slot j, we get the following upper bound:

$$\max_{0\leqslant t\leqslant 1}\{Size(t)\} \leqslant \max_{1\leqslant j\leqslant g\mu}\{N(j)\}.$$

The slot occupancies $N(j)$ are Poisson distributed with mean $(\lambda/\mu)(1+1/g)$. However they are not independent, so our analysis of maximum bucket occupancy in Theorem 5.6 does not apply to this case. The main point of the proof is showing that the lack of independence does not significantly alter the expected maximum:

$$\mathbf{E}\left\{\max_{1\leqslant j\leqslant g\mu}\{N(j)\}\right\} \sim \frac{\lambda}{\mu}.$$

This gives us the desired upper bound, which completes the proof for case 1. The formula for $\mathbf{E}\{\max_{t\in[0,1]}\{Use(t)\}\}$ can be derived in the same way.

This approach when applied to cases 2 and 3 of Theorem 6.5 gives upper bounds on $\mathbf{E}\{\max_{t\in[0,1]}\{Size(t)\}\}$ that are off by a factor of 2. To get asymptotic bounds, different techniques are used, involving probabilistic arguments that the distribution of $\max_{t\in[0,1]}\{Size(t)\}$ is peaked in some "central region" about the mean. The technique is similar in spirit to those used in Section 5.1 for the analyses of extendible hashing, maximum bucket occupancy, and coalesced hashing. □

The probabilistic analysis of maximum size has also been successfully carried out for a variety of combinatorial data structures, such as dictionaries, linear lists, priority queues, and symbol tables, using the history model discussed in Section 6.1.

6.3. *Set-Union-Find algorithms*

The *Set-Union-Find* data type is useful in several computer applications, such as computing minimum spanning trees, testing equivalence of finite-state machines, performing unification in logic programming and theorem proving, and handling

COMMON blocks in FORTRAN compilers. The operation *Union*(x, y) merges the equivalence classes (or simply *components*) containing x and y and chooses a unique "representative" element for the combined component. Operation *Find*(x) returns the representative of x's component, and *Make_set* (x) creates a singleton component $\{x\}$ with representative x.

Union-find algorithms have been studied extensively in terms of worst-case and amortized performance. Tarjan and van Leeuwen [104] give matching upper and lower amortized bounds of $\Theta(n+m\alpha(m+n, n))$ for the problem, where n is the number of make-set operations, m is the number of finds, and $\alpha(a, b)$ denotes a functional inverse of Ackermann's function. The lower bound holds in a separable pointer machine model, and the upper bound is achieved by the well-known tree data structure that does weighted merges and path compression. A more extensive discussion appears in [83].

In this section we study the average-case running time of more simple-minded algorithms, called "quick find" (QF) and "quick find weighted" (QFW). The data structure consists of an array called *rep*, with one slot per element; *rep*$[x]$ is the representative for element x. Each find can thus be done in constant time. In addition, the elements in each component are linked together in a list. In the QF algorithm, *Union*(x, y) is implemented by setting *rep*$[z]$:= *rep*$[x]$, for all z in y's component. The QFW algorithm is the same, except that when x's component is smaller than y's, we take the quicker route and set *rep*$[z]$:=*rep*$[y]$ for all z in x's component. An auxiliary array is used to keep track of the size of each component.

Since all finds take constant time, we shall confine our attention to the union operations. We consider $n-1$ unions performed on a set of n elements, so that we end up with a single component of size n. Our performance measure, which we denote by T_n^{QF} and T_n^{QFW}, is the total number of updates to slots of *rep* made during the unions. We consider three models of "random" input.

Random graph model

Probably the most realistic model was proposed in [111], based upon the random graph model of [26]. Each of the $\binom{n}{2}$ undirected edges between n vertices "fires" independently, governed by a Poisson process. Each order of firings is thus equally likely. When an edge $\{x,y\}$ fires, we execute *Union*(x,y) if x and y are in different components.

6.6. THEOREM (Knuth and Schönhage [75], and Bollobás and Simon [11]). *The average number of updates done by QF and QFW in the random graph model is*

$$\overline{T_n^{\mathrm{QF}}} = (n^2/8) + o(n(\log n)^2);$$
$$\overline{T_n^{\mathrm{QFW}}} = cn + o(n/\log n), \quad \textit{where } c = 2.0847\ldots.$$

We shall restrict ourselves to showing that $\overline{T_n^{\mathrm{QF}}} \sim n^2/8$ and $\overline{T_n^{\mathrm{QFW}}} = O(n)$ using the derivation from [75]. The techniques in [11] are needed to determine the coefficient c and to get better bounds on the second-order terms. In addition, in [11] sequences of fewer than n unions are considered. They show that, on the average, QF performs $(\frac{1}{2}-\varepsilon)n$ unions in $O(n \log n)$ time, and QFW does k unions in $O(k)$ time, for any $k \leqslant n$.

PROOF (*Sketch*). The proof is based upon the intuition from [26] that with probability $1-O(1/\log n)$ a random graph on n vertices with $\frac{1}{2}n\log n+\frac{1}{2}cn$ edges, where c is a constant, consists of one giant connected component of size $\geqslant n-\log\log n$ and a set of isolated vertices. The graph is connected with probability $e^{-e^{-c}}$. In terms of union operations, it is very likely that the last few unions joined the giant component to singleton components; the cost for each such union would be $O(n)$ for QF and $O(1)$ for QFW. The proof of Theorem 6.6 consists in showing that this behavior extends over the entire sequence of unions.

For QFW, we find by recurrences and asymptotic approximations that

$$E_{n,k,m}=O\left(\frac{n}{k^{3/2}m^{3/2}(k+m)^{3/2}}\right) \quad \text{for } k,m<n^{2/3} \text{ and } k,m>n^{2/3}, \tag{6.14}$$

where $E_{n,k,m}$ is the expected number of times a component of size k is merged with one of size m. Hence,

$$\overline{T_n^{\mathrm{QFW}}}=\sum_{1\leqslant k,m<n}\min\{k,m\}E_{n,k,m}\leqslant\sum_{1\leqslant k\leqslant m<n}k(E_{n,k,m}+E_{n,m,k}). \tag{6.15}$$

For the portion of the sum to which (6.14) applies, we can bound (6.15) by $O(n)$. For the rest of the range, in which $1\leqslant k\leqslant n^{2/3}\leqslant m<n$, the sum is bounded by n, since each element can be merged at most once from a component of size $<n^{2/3}$ into one of size $\geqslant n^{2/3}$. The analysis of QF is similar. □

Several combinatorial algorithms have been designed and analyzed using the random graph model. For example, Babai, Erdös, and Selkow [3] give an algorithm for testing graph isomorphism that runs in $O(n^2)$ average time, though all known algorithms require exponential time in the worst case.

Random spanning tree model

Each sequence of union operations corresponds to a "union tree", in which the directed edge $\langle x, y\rangle$ means that the component with representative y is merged into the component with representative x. In the random spanning tree model, all possible union trees are equally likely; there are n^{n-2} possible unoriented trees and $(n-1)!$ firing orders of the edges in each tree.

6.7. THEOREM (A.C.-C. Yao [111], and Knuth and Schönhage [75]). *The average number of updates done by QF and QFW in the random spanning tree model is*

$$\overline{T_n^{\mathrm{QF}}}=\sqrt{\frac{\pi}{8}}n^{3/2}+O(n\log n);\qquad \overline{T_n^{\mathrm{QFW}}}=\frac{1}{\pi}n\log n+O(n).$$

PROOF (*Sketch*). An admissibility argument similar to those in Section 4 allows us to compute the probability $p_{n,k}$ that the last union does a merge of components of sizes k and $n-k$:

$$p_{n,k}=\frac{1}{2(n-1)}\binom{n}{k}\left(\frac{k}{n}\right)^{k-1}\left(\frac{n-k}{n}\right)^{n-k-1}. \tag{6.16}$$

And it is easy to show that

$$\overline{T_n} = c_n + 2 \sum_{0<k<n} p_{n,k} \overline{T_k}, \tag{6.17}$$

where $c_n = \sum_{0<k<n} k p_{n,k}$ for QF and $c_n = \sum_{0<k<n} \min\{k, n-k\} p_{n,k}$ for QFW. By symmetry we have $\sum_{0<k<n} k p_{n,k} = n/2$, and arguments similar to those used for Theorem 6.6 show that $\sum_{0<k<n} \min\{k, n-k\} p_{n,k} = (2n/\pi)^{1/2} + O(1)$. Recurrence (6.17) is in a special linear form that allows us to solve it "by repertoire": the solution of (6.17) for $c_n = a_n + b_n$ is the sum of the solutions for $c_n = a_n$ and for $c_n = b_n$. Hence, if we can find a "basis" of different c_n for which (6.17) can be solved easily, then we can solve (6.17) for QF and QFW by linear combinations of the basis functions. It turns out that the basis in our case is the set of Q-functions $Q_{\{1/k^r\}}(n)$, for $r = -1, 0, 1, 2, \ldots$, which we studied in connection with linear probing in Section 5.2. □

Random components model

In the simplest model, and also the least realistic, we assume that at any given time each pair of existing components is equally likely to be merged next. The union tree in this framework is nothing more than a random binary search tree, which we studied extensively in Section 4.2. Admissibility arguments lead directly to the following result:

6.8. THEOREM (Doyle and Rivest [24]). *The average number of updates done by QF and QFW in the random components model is*

$$\overline{T_n^{QF}} = n(H_n - 1) = n \log n + O(n);$$
$$\overline{T_n^{QFW}} = nH_n - \tfrac{1}{2} n H_{\lfloor n/2 \rfloor} - \lceil n/2 \rceil = \tfrac{1}{2} n \log n + O(n).$$

Acknowledgment

We thank Don Knuth for several helpful comments and suggestions.

References

[1] AJTAI, M., J. KOMLÓS and E. SZEMERÉDI, An O($n \log n$) sorting network, in: *Proc. 15th Ann. Symp. on Theory of Computing* (1983) 1–9.
[2] AMBLE, O. and D.E. KNUTH, Ordered hash tables, *Comput. J.* **17**(2) (1974) 135–142.
[3] BABAI, L., P. ERDŐS and S.M. SELKOW, Random graph isomorphisms, *SIAM J. Comput.* **9** (1980) 628–635.
[4] BATCHER, K.E., Sorting networks and their applications, in: *Proc. AFIPS Spring Joint Computer Conf.* **32** (1968) 307–314.
[5] BENDER, E., Central and local limit theorems applied to asymptotic enumeration, *J. Combin. Theory Ser. A* **15** (1973) 91–111.
[6] BENDER, E., Asymptotic methods in enumeration, *SIAM Rev.* **16** (1974) 485–515.
[7] BENDER, C. and S. ORSZAG, *Advanced Mathematical Methods for Scientists and Engineers* (McGraw-Hill, New York, 1978).

[8] BERSTEL, J. and L. BOASSON, Context-free languages, in: J. van Leeuwen, ed., *Handbook of Theoretical Computer Science, Vol. B* (North-Holland, Amsterdam, 1990) Chapter 2.
[9] BILLINGSLEY, P., *Probability and Measure* (Academic Press, New York, 1986).
[10] BOLLOBÁS, B., *Random Graphs* (Academic Press, New York, 1985).
[11] BOLLOBÁS, B. and I. SIMON, On the expected behavior of disjoint set union algorithms, in: *Proc. 17th Ann. Symp. on Theory of Computing* (1985) 224–231.
[12] BROWN, M.R., A partial analysis of random height-balanced trees, *SIAM J. Comput.* **8**(1) (1979) 33–41.
[13] BURGE, W.H., An analysis of a tree sorting method and some properties of a set of trees, in: *Proc. 1st USA–Japan Computer Conf.* (1972) 372–378.
[14] CANFIELD, E.R., Central and local limit theorems for coefficients of binomial type, *J. Combin. Theory Ser. A* **23** (1977) 275–290.
[15] CLARK, D.W., Measurements of dynamic list structure use in Lisp, *IEEE Trans. Software Eng.* **5**(1) (1979) 51–59.
[16] COMTET, L., Calcul pratique des coefficients de Taylor d'une fonction algébrique, *Enseign. Math.* **10** (1969) 267–270.
[17] COMTET, L., *Advanced Combinatorics* (Reidel, Dordrecht, 1974).
[18] CYPHER, R., A lower bound on the size of Shellsort sorting networks, in: *Proc. 1st Ann. ACM Symp. on Parallel Algorithms and Architectures* (1989) 58–63.
[19] DE BRUIJN, N.G., *Asymptotic Methods in Analysis* (Dover, New York, 1981).
[20] DE BRUIJN, N.G., D.E. KNUTH and S.O. RICE, The average height of planted plane trees, in: R.-C. Read, ed., *Graph Theory and Computing* (Academic Press, New York, 1972) 15–22.
[21] DEVROYE, L., A note on the height of binary search trees, *J. ACM* **33** (1986) 489–498.
[22] DEVROYE, L., *Lecture Notes on Bucket Algorithms* (Birkhäuser, Boston, 1986).
[23] DOETSCH, G., *Handbuch der Laplace Transformation Volumes 1–3* (Birkhäuser, Basel, 1955).
[24] DOYLE, J. and R.L. RIVEST, Linear expected time of a simple union-find algorithm, *Inform. Process. Lett.* **5** (1976) 146–148.
[25] ĎURIAN, B., Quicksort without a stack, in: *Proc. 12th Ann. Symp. on Mathematical Foundations of Computer Science*, Lecture Notes in Computer Science, Vol. 233 (Springer, Berlin, 1986) 283–289.
[26] ERDŐS, P. and A. RÉNYI, On the evolution of random graphs, *Publ. Math. Inst. Hungarian Acad. Sci.* **5** (1960) 17–61.
[27] FAGIN, R., J. NIEVERGELT, N. PIPPENGER and H.R. STRONG, Extendible hashing—a fast access method for dynamic files, *ACM Trans. Database Systems* **4**(3) (1979) 315–344.
[28] FELLER, W., *An Introduction to Probability Theory and Its Applications, Vol. 1* (Wiley, New York, 3rd ed., 1968).
[29] FELLER, W., *An Introduction to Probability Theory and its Applications, Vol. 2* (Wiley, New York, 2nd ed., 1971).
[30] FLAJOLET, PH., Combinatorial aspects of continued fractions, *Discrete Math.* **32** (1980) 125–161.
[31] FLAJOLET, PH., Analyse d'algorithmes de manipulation d'arbres et de fichiers, in: *Cahiers du BURO* **34–35** (1981) 1–209.
[32] FLAJOLET, PH., On the performance evaluation of extendible hashing and trie searching, *Acta Inform.* **20** (1983) 345–369.
[33] FLAJOLET, PH., Analytic models and ambiguity of context-free languages, *Theoret. Comput. Sci.* **49** (1987) 283–309.
[34] FLAJOLET, PH., Mathematical methods in the analysis of algorithms and data structures, in: E. Börger, ed., *Trends in Theoretical Computer Science* (Computer Science Press, Rockville, MD, 1988) 225–304.
[35] FLAJOLET, PH., G. GONNET, C. PUECH and M. ROBSON, Analytic variations on quad trees, Draft, 1989.
[36] FLAJOLET, PH. and A.M. ODLYZKO, The average height of binary trees and other simple trees, *J. Comput. System Sci.* **25** (1982) 171–213.
[37] FLAJOLET, PH. and A.M. ODLYZKO, Singularity analysis of generating functions, *SIAM J. Discrete Math.* **3**(1) (1990).
[38] FLAJOLET, PH. and C. PUECH, Partial match retrieval of multidimensional data, *J. ACM* **33**(2) (1986) 371–407.
[39] FLAJOLET, PH., J.-C. RAOULT and J. VUILLEMIN, The number of registers required to evaluate arithmetic expressions, *Theoret. Comput. Sci.* **9** (1979) 99–125.

[40] FLAJOLET, PH., M. RÉGNIER and R. SEDGEWICK, Some uses of the Mellin integral transform in the analysis of algorithms, *Combinatorics on Words, NATO ASI Series F, Vol. 12* (Springer, Berlin, 1985).
[41] FLAJOLET, PH., M. RÉGNIER and D. SOTTEAU, Algebraic methods for trie statistics, *Ann. Discrete Math.* **25** (1985) 145–188.
[42] FLAJOLET, PH. and R. SEDGEWICK, Digital search trees revisited, *SIAM J. Comput.* **15**(3) (1986) 748–767.
[43] FLAJOLET, PH., P. SIPALA and J.M. STEYAERT, Analytic variations on the common subexpression problem, in: *Proc. 17th Ann. Internat. Coll. on Automata, Languages and Programming* (*ICALP*), Warwick, Lecture Notes in Computer Science (Springer, Berlin, 1990).
[44] FLAJOLET, PH. and J.M. STEYAERT, A complexity calculus for recursive tree algorithms, *Math. Systems Theory* **19** (1987) 301–331.
[45] FOATA, D., *La Série Génératrice Exponentielle dans les Problèmes d'Énumération*, Series S.M.S. (Montreal University Press, 1974).
[46] FRANÇON, J., Histoire de fichiers, *RAIRO Inform. Theor.* **12** (1978) 49–67.
[47] FRANÇON, J., G. VIENNOT and J. VUILLEMIN, Description and analysis of an efficient priority queue representation, in: *Proc. 19th Ann. Symp. on Foundations of Computer Science* (1978) 1–7.
[48] GONNET, G.H., Expected length of the longest probe sequence in hash code searching, *J. ACM* **28**(2) (1981) 289–304.
[49] GONNET, G.H., *Handbook of Algorithms and Data Structures* (Addison-Wesley, Reading, MA., 1984).
[50] GONNET, G.H. and J.I. MUNRO, A linear probing sort and its analysis, in: *Proc. 13th Ann. Symp. on Theory of Computing* (1981) 90–95.
[51] GONNET, G.H., L.D. ROGERS and A. GEORGE, An algorithmic and complexity analysis of interpolation search, *Acta Inform.* **13**(1) (1980) 39–46.
[52] GOULDEN, I. and D. JACKSON, *Combinatorial Enumerations* (Wiley, New York, 1983).
[53] GREENE, D.H., Labelled formal languages and their uses, Tech. Report STAN-CS-83-982, Stanford Univ., 1983.
[54] GREENE, D.H. and D.E. KNUTH, *Mathematics for the Analysis of Algorithms* (Birkhäuser, Boston, 2nd ed., 1982).
[55] GUIBAS, L.J. and E. SZEMERÉDI, The analysis of double hashing, *J. Comput. System Sci.* **16**(2) (1978) 226–274.
[56] HARRIS, B. and L. SCHOENFELD, Asymptotic expansions for the coefficients of analytic functions, *Illinois J. Math.* **12** (1968) 264–277.
[57] HAYMAN, W.K., A generalization of Stirling's formula, *J. Reine Angew. Math.* **196** (1956) 67–95.
[58] HENRICI, P., *Applied and Computational Complex Analysis, Vol. 1* (Wiley, New York, 1974).
[59] HENRICI, P., *Applied and Computational Complex Analysis, Vol. 2* (Wiley, New York, 1977).
[60] HUANG, B.-C. and D.E. KNUTH, A one-way, stackless quicksort algorithm, *BIT* **26** (1986) 127–130.
[61] HUNT, D.H., Bachelor's thesis, Princeton Univ., 1967.
[62] INCERPI, J.M. and R. SEDGEWICK, Improved upper bounds on Shellsort, *J. Comput. System Sci.* **31** (1985) 210–224.
[63] JACQUET, PH. and M. RÉGNIER, Trie partitioning process: limiting distributions, in: *Proc. CAAP '86*, Lecture Notes in Computer Science, Vol. 214 (Springer, Berlin, 1986) 196–210.
[64] JACQUET, PH. and M. RÉGNIER, Normal limiting distribution of the size of tries, in: *Performance '87, Proc. 13th Internat. Symp. on Computer Performance* (1987).
[65] KEMP, R., The average number of registers needed to evaluate a binary tree optimally, *Acta Inform.* **11**(4) (1979) 363–372.
[66] KEMP, R., *Fundamentals of the Average Case Analysis of Particular Algorithms* (Teubner-Wiley, Stuttgart, 1984).
[67] KENYON-MATHIEU, C.M. and J.S. VITTER, General methods for the analysis of the maximum size of data structures, in: *Proc. 16th Ann. Internat. Coll. on Automata, Languages, and Programming* (*ICALP*), Stresa, Lecture Notes in Computer Science, Vol. 372 (Springer, Berlin, 1989) 473–487.
[68] KIRSCHENHOFER, P. and H. PRODINGER, Some further results on digital search trees in: *Proc. 13th Ann. Internat. Coll. on Automata, Languages, and Programming* (*ICALP*), Rennes, Lecture Notes in Computer Science, Vol. 226 (Springer, Berlin, 1986) 177–185.
[69] KIRSCHENHOFER, P. and H. PRODINGER, On some applications of formulæ of Ramanujan in the analysis of algorithms, Preprint, 1988.

[70] KNUTH, D.E., *The Art of Computer Programming, Vol. 1: Fundamental Algorithms* (Addison-Wesley, Reading, MA, 2nd ed., 1973).
[71] KNUTH, D.E., *The Art of Computer Programming, Vol. 3: Sorting and Searching* (Addison-Wesley, Reading, MA, 1973).
[72] KNUTH, D.E., Deletions that preserve randomness, *IEEE Trans. Software Eng.* **3** (1977) 351–359.
[73] KNUTH, D.E., The average time for carry propagation, *Indag. Math.* **40** (1978) 238–242.
[74] KNUTH, D.E., *The Art of Computer Programming, Vol. 2: Semi-Numerical Algorithms* (Addison-Wesley, Reading, MA, 1981).
[75] KNUTH, D.E. and A. SCHÖNHAGE, The expected linearity of a simple equivalence algorithm, *Theoret. Comput. Sci.* **6** (1978) 281–315.
[76] KOLCHIN, V.F., B.A. SEVAST'YANOV and V.P. CHISTYAKOV, *Random Allocations* (Winston & Sons, Washington, 1978).
[77] KONHEIM, A.G. and D.J. NEWMAN, A note on growing binary trees, *Discrete Math.* **4** (1973) 57–63.
[78] LARSON, P.A., Dynamic hashing, *BIT* **20**(2) (1978) 184–201.
[79] LINDSTROM, E.E. and J.S. VITTER, The design and analysis of bucketsort for bubble memory secondary storage, *IEEE Trans. Comput.* **34**(3) (1985) 218–233.
[80] LOUCHARD, G., The Brownian motion: a neglected tool for the complexity analysis of sorted table manipulation, *RAIRO Inform. Théor.* **17** (1983) 365–385.
[81] LUEKER, G.S. and M. MOLODOWITCH, More analysis of double hashing, in: *Proc. 20th Ann. ACM Symp. on Theory of Computing* (1988) 354–359.
[82] MATHIEU, C.M. and J.S. VITTER, Maximum queue size and hashing with lazy deletion, in: *Proc. 20th Ann. Symp. on the Interface of Computing Science and Statistics* (1988) 743–748.
[83] MEHLHORN, K. and A. TSAKALIDIS, Data structures, in: J. van Leeuwen, ed., *Handbook of Theoretical Computer Science, Vol. A* (North-Holland, Amsterdam, 1990) 301–341.
[84] MEIR, A. and J.W. MOON, On the altitude of nodes in random trees, *Canad. J. Math.* **30** (1978) 997–1015.
[85] ODLYZKO, A.M., Periodic oscillations of coefficients of power series that satisfy functional equations, *Adv. Math.* **44** (1982) 180–205.
[86] ODLYZKO, A.M. and L.B. RICHMOND, Asymptotic expansions for the coefficients of analytic generating functions, *Aequationes Math.* **28** (1985) 50–63.
[87] OLVER, F.W.J., *Asymptotics and Special Functions* (Academic Press, New York, 1974).
[88] PIPPENGER, N.J., Communication networks, in: J. van Leeuwen, ed., *Handbook of Theoretical Computer Science, Vol. A.* (North-Holland, Amsterdam, 1990) 805–833.
[89] PITTEL, B., Paths in a random digital tree: limiting distributions, *Adv. Appl. Prob.* **18** (1986) 139–155.
[90] PITTEL, B., Linear probing: the probable largest search time grows logarithmically with the number of records, *J. Algorithms* **8**(2) (1987) 236–249.
[91] PITTEL, B., On probabilistic analysis of a coalesced hashing algorithm, *Ann. Probab.* **15** (1987).
[92] PÓLYA, G., Kombinatorische Anzahlbestimmungen für Gruppen, Graphen und chemische Verbindungen, *Acta Math.* **68** (1937) 145–254; translated in: G. Pólya and R.C. Read, *Combinatorial Enumeration of Groups, Graphs, and Chemical Compounds* (Springer, New York, 1987).
[93] PURDOM, P.W. and C.A. BROWN, *The Analysis of Algorithms* (Holt, Rinehart and Winston, New York, 1985).
[94] RÉGNIER, M., On the average height of trees in digital search and dynamic hashing, *Inform. Process. Lett.* **13** (1981) 64–66.
[95] RÉGNIER, M., Analysis of grid file algorithms, *BIT* **25** (1985) 335–357.
[96] SACHKOV, V.N., *Verojatnosnie Metody v Kombinatornom Analize* (Nauka, Moscow, 1978).
[97] SEDGEWICK, R., Quicksort with equal keys, *SIAM J. Comput.* **6**(2) (1977) 240–267.
[98] SEDGEWICK, R., The analysis of quicksort programs, *Acta Inform.* **7** (1977) 327–355.
[99] SEDGEWICK, R., Data movement in odd-even merging, *SIAM J. Comput.* **7**(3) (1978) 239–272.
[100] SEDGEWICK, R., Mathematical analysis of combinatorial algorithms, in: G. Louchard and G. Latouche, eds., *Probability Theory and Computer Science* (Academic Press, London, 1983).
[101] SEDGEWICK, R., *Algorithms* (Addison-Wesley, Reading, MA, 2nd ed., 1988).
[102] SHELL, D.L., A high-speed sorting procedure, *Comm. ACM* **2**(7) (1959) 30–32.
[103] STEYAERT, J.M. and PH. FLAJOLET, Patterns and pattern-matching in trees: an analysis, *Inform. and Control* **58** (1983) 19–58.

[104] TARJAN, R.E. and J. VAN LEEUWEN, Worst-case analysis of set union algorithms, *J. ACM* **31**(2) (1984) 245–281.

[105] TITCHMARSH, E.C., *The Theory of Functions* (Oxford Univ. Press, London, 2nd ed., 1939).

[106] VAN WYK, C.J. and J.S. VITTER, The complexity of hashing with lazy deletion, *Algorithmica* **1**(1) (1986) 17–29.

[107] VITTER, J.S., Analysis of the search performance of coalesced hashing, *J. ACM* **30**(2) (1983) 231–258.

[108] VITTER, J.S. and W.-C. CHEN, *Design and Analysis of Coalesced Hashing* (Oxford Univ. Press, New York, 1987).

[109] VUILLEMIN, J., A unifying look at data structures, *Comm. ACM* **23**(4) (1980) 229–239.

[110] WEISS, M.A. and R. SEDGEWICK, Tight lower bounds for Shellsort, Tech. Report CS-TR-137-88, Princeton Univ., 1988.

[111] YAO, A.C.-C., On the average behavior of set merging algorithms, in: *Proc. 8th Ann. ACM Symp. on Theory of Computing* (1976) 192–195.

[112] YAO, A.C.-C., On random 2–3 trees, *Acta Inform.* **9**(2) (1978) 159–170.

[113] YAO, A.C.-C., An analysis of $(h, k, 1)$ Shellsort, *J. Algorithms* **1**(1) (1980) 14–50.

[114] YAO, A.C.-C., On straight selection sort, Tech. Report CS-TR-185-88, Princeton Univ., 1988.

[115] YAO, A.C.-C. and F.F. YAO, The complexity of searching an ordered random table, in: *Proc. 17th Ann. Symp. on Foundations of Computer Science* (1976) 173–177.

[116] YAO, F.F., Computational geometry, in: J. van Leeuwen, ed., *Handbook of Theoretical Computer Science, Vol. A* (North-Holland, Amsterdam, 1990) 343–389.

[117] AGGARWAL, A. and J.S. VITTER, The input-output complexity of sorting and related problems, *Comm. ACM* **31**(9) (1988) 1116–1127.

[118] CHOMSKY, N. and M.P. SCHÜTZENBERGER, The algebraic theory of context-free languages, in: *P. Braffort and D. Hirschberg*, eds., *Computer Programming and Formal Languages* (North-Holland, Amsterdam, 1963) 118–161.

[119] FLAJOLET, PH., J. FRANÇON and J. VUILLEMIN, Sequence of operations analysis for dynamic data structures, *J. Algorithms* **1**(2) (1980) 111–141.

[120] NODINE, M.H. and J.S.VITTER, Greed Sort: optimal sorting with multiple disks, Tech. Report CS-90-04, Brown Univ., 1990.

[121] VITTER, J.S. and E.A.M. SHRIVER, Optimal I/O with parallel block transfer, in: *Proc. 22nd Ann. ACM Symp. on Theory of Computing* (1990).

CHAPTER 10

Graph Algorithms

J. van LEEUWEN

Department of Computer Science, University of Utrecht,
P.O. Box 80.089, 3508 TB Utrecht, Netherlands

Contents

HANDBOOK OF THEORETICAL COMPUTER SCIENCE
Edited by J. van Leeuwen

Prelude

Graphs are the most common "abstract" structure encountered in computer science. Any system that consists of discrete states or sites (called nodes or vertices) and connections between it can be modeled by a graph. The connections between nodes of a graph are called edges (in case the connections are between unordered pairs of nodes), directed edges (in case the connections are between ordered pairs of nodes), or hyperedges (in case the connections are arbitrary nonempty sets of nodes). Connections may also carry additional information as labels or weights, related to the interpretation of the graph. Consequently there are many types of graphs and many basic notions that capture aspects of the structure of graphs. Also, many applications require efficient algorithms that essentially operate on graphs.

In this chapter we give an overview of the common paradigms and results in graph algorithms with an emphasis on the more recent developments in the field. We outline the underlying principles of many essential graph algorithms, but usually omit the details of the carefully tuned datastructures that may be needed to achieve the ultimate bounds on time or space complexity for the algorithms. We will only deal with "ordinary" sequential algorithms, and consider a discussion of parallel and distributed algorithms on graphs as beyond the scope of this chapter.

The chapter deals with three main areas of concern in graph algorithms: the representation and exploration of graphs, basic structure algorithms for graphs, and algorithms for some common optimalization problems on graphs. This by no means exhausts the subject of graph algorithms, but covers many of the techniques and results that are commonly needed. In particular we do not cover the existing algorithms for NP-complete graph problems and for enumerations in graphs. Some familiarity with elementary graph theory and the theory of NP-completeness is assumed. References are given for each (sub)section separately but are listed at the end of the chapter, thus facilitating the search for source material on graph algorithms by "subject". Within the subsections, results are usually credited by the name of the author(s) of the corresponding source paper rather than by a numbered reference. The references for this section include the major textbooks and monographs on graphs and graph algorithms.

1. Representation of graphs

When a computational problem is modeled in terms of graphs, the resulting graphs must be generated and represented in some form in order to facilitate the operations of an algorithm. The desired representation may be an "abstract" datastructure that supports certain operations very efficiently, or it may be a "concrete" display (or drawing) of the graph that allows visual inspection and interactive manipulations. We will discuss the typical problems that arise in drawing graphs "neatly" first, in order to show the interrelations between graph theory and the study of graph algorithms through a rather direct (and aesthetically pleasing) example. In the remainder of this section we discuss the different approaches to the compact representation of graphs by datastructures, the basic algorithms for exploring graphs, the generation of random graphs, and the representation of classes of graphs in relation to efficient algorithm design.

1.1. Drawing a graph

In this section we give a first introduction to the "representation" of graphs and to the challenge of finding efficient graph algorithms, without being very specific about the necessary details of implementation on a computer.

DEFINITION. A *graph* is a structure $G=\langle V, E\rangle$ in which V is a finite set of nodes and $E \subseteq V \oplus V$ is a finite set of edges (unordered pairs). A *directed graph* is a structure $G=\langle V, E\rangle$ in which V is again a finite set of nodes and $E \subseteq V \times V$ is a finite set of directed edges (ordered pairs). Given a graph G, we will denote $|V|$ by n and $|E|$ by e or m.

It is common to "draw" graphs in the (two-dimensional) plane, with nodes represented by dots and edges by lines between the dots. Directed edges (arcs) are drawn as arrows (see Fig. 1). If graphs are to be drawn "automatically" (from a specification of the node adjacencies), then some standard must be adopted for the format and desired aesthetics of the drawings, and an algorithm is needed to generate drawings according to the set standard. For example, one might want to draw a graph in the plane allowing curved lines for edges and minimize the number of edge-intersections ("crossings"). Whereas this may be desired in a standard, no efficient (i.e., polynomial-time bounded) algorithm is presently known for actually generating such drawings. In fact, the problem of deciding for graphs G and integers K whether the "crossing number" of G is bounded by K is known to be NP-complete (Garey & Johnson).

Many approaches have been suggested to the problem of automatic graph drawing (see Tamassia et al.) which immediately lead to interesting graph-theoretic and algorithmic problems.

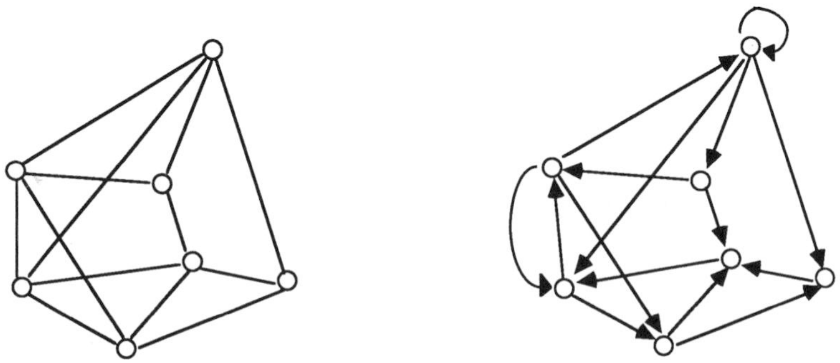

(a) a graph (b) a directed graph

Fig. 1. Drawing a graph.

DEFINITION. A graph G is called *planar* if it can be drawn in the plane without crossing edges. A plane graph, or a "planar embedding", is any drawing of a (planar) graph in the plane without crossing edges. A *face* of a plane graph is any topologically connected region surrounded by edges of the plane graph. The one unbounded face of a plane graph is sometimes called the "exterior face".

DEFINITION. An (undirected) graph $G=\langle V, E\rangle$ is called *connected* if for every two nodes x, $y \in V$ there exists a path of edges from E joining x and y. A graph is called disconnected if it is not connected. A graph G is called *k-connected* ($k \geqslant 1$) if there does not exists a set of $k-1$ or fewer nodes $V' \subseteq V$ such that the removal of all nodes of V' and their incident edges from G results in a disconnected graph. A *k-connected component* of a graph G ($k \geqslant 1$) is any maximal k-connected subgraph of G. A 1-connected component is usually referred to simply as a *connected component*, and a 2-connected component as a *biconnected component* or a "block".

Both the study of planar graphs and the study of connectivity properties of graphs are major themes in the analysis of graphs. Consequently there is a substantial interest in graph algorithms that are especially suitable for planar or k-connected graphs for some $k \geqslant 1$. k-Connected graphs are especially interesting for networking applications in computer science because of the following version of Menger's theorem due to Whitney (see Whitney, or Harary): a graph G is k-connected if and only if every two distinct nodes x and y of G are connected by at least k node-disjoint paths. In this section we continue with the problem of automatic graph drawing, and consider the important and practical case of planar graphs.

Before we consider drawing a planar embedding of a graph, the question arises how one can actually determine whether a given graph is planar or not. This classical problem in graph theory has a fundamental answer in the form of Kuratowski's theorem: a graph G is planar if and only if it has no subgraph "homeomorphic" to $K_{3,3}$ or K_5. ($K_{3,3}$ is the complete bipartite graph on 2 sets of 3 nodes and K_5 is the complete graph on 5 nodes.) This characterization seems far from a feasible computational recipe for testing planarity, and a different approach is called for. A simple observation shows that we can actually restrict the planarity test and, later, the design of a suitable drawing algorithm to the biconnected components of graphs: a graph is planar if and only if its biconnected components are. (This follows because the biconnected components of a graph can intersect in at most one node.) Tarjan proved that the connected and the biconnected components of a graph can be determined by an algorithm that operates in $O(n+e)$ time on a graph (see also Sections 1.3 and 2.1). Hopcroft and Tarjan succeeded in proving the following fact, which is essential for all algorithms that deal with planar graphs.

1.1. THEOREM. *There exists an algorithm that tests whether a graph is planar or not in* $O(n+e)$ *time.*

There are several other "linear-time" algorithms for planarity testing now (see e.g. Booth & Lueker, De Fraysseix & Rosenstiehl, and Williamson), and the algorithms usually give a concrete plane representation if a graph is found to be planar (see also Chiba et al.). Williamson's algorithm even finds an embedded $K_{3,3}$ or K_5 within the same time bound if a graph is found to be nonplanar, thus showing that Kuratowski's theorem can be turned into a feasible planarity test after all. (See Tarjan for an account of the history of the subject.)

Efficient planarity testing algorithms do not really settle the automatic graph-drawing problem for (planar) graphs, and the plane representation they yield as output should be subjected to further algorithms that actually yield "nice" drawings according to whatever criteria one likes to adopt. We pursue this a little further, because it shows once again how simple questions can lead to challenging problems in graph theory on the one hand and the design of efficient graph algorithms on the other. Observe that any algorithm for generating "nice" drawings of planar graphs may also be applied to embedded nonplanar graphs, provided an auxiliary node is introduced for every edge-crossing in the embedding.

In 1948, Fáry proved that every planar graph admits a planar embedding in which all edges are straight-line segments. (Planar embeddings of this kind will be called Fáry embeddings.) Stein showed that every plane graph in which the common boundary (lines and/or points) of any two touching faces is connected even admits a Fáry embedding in which every face, except possibly the exterior face, is a convex polygon. If we adopt "straight-line segments and convexity" as a desired standard for planar graph drawings, then the question arises again whether the graph-theoretic result has an efficient algorithmic counterpart.

It appears that the generation of a Fáry embedding with convex (bounded) faces when possible requires that the given (planar) graph is decomposed into its 3-connected components first. A result of Hopcroft & Tarjan shows how this can be achieved algorithmically for any graph in $O(n+e)$ time. The reason for being interested in 3-connected graphs in this context derives from a result of Steinitz which asserts, in our terminology, that a graph is isomorphic to the "nodes and edges of a three-dimensional convex polytope" (and thus has a Fáry embedding in which *all* faces are bounded by a convex polygon) if and only if the graph is planar and 3-connected. (See also Grünbaum's account of this result.) We also mention Maclane's result that a graph is planar if and only if its 3-connected components are. Tutte showed that 3-connected planar graphs admit a convex drawing in the plane in which the node positions are determined by a simple set of linear conditions. Using a further analysis of convex representations of 2-connected graphs by Thomassen, Chiba et al. succeeded in proving the following result.

1.2. THEOREM. *There exists a linear-time, linear-space algorithm for generating a Fáry embedding of a planar graph, with the additional property that the (bounded) faces of each 3-connected component are drawn as convex polygons.*

This is not necessarily the last word on planar graph drawing. For example, one may want further criteria to be satisfied in order for a convex drawing to look nice (and have faces that are e.g. balanced in size or have an otherwise regular form). For concrete CAD applications it is desirable to consider graph drawings on a grid, in which each node is mapped to a grid-point. De Fraysseix et al. have shown that every planar graph has a Fáry embedding in a $2n-4$ by $n-2$ grid which, moreover, can be computed by an $O(n \log n)$ time and linear-space bounded algorithm. It can be improved to an $O(n)$ time algorithm. For planar graphs with node degrees bounded by 4, one might also want embeddings in a grid in which the edges are represented as disjoint paths along grid-lines. By transforming it to a special case of the "minimum cost flow problem" (discussed in Section 3.5.2 of this chapter), Tamassia showed that an embedding of this kind that requires the smallest possible number of "bends" can be computed in $O(n^2 \log n)$ time. And finally, for some classes of graphs one may want to preserve some overall structure aspects in the actual drawings. Especially well-known is the case of trees, for which various criteria and good drawing algorithms have been developed (see e.g. Wetherell & Shannon, Reingold & Tilford, and Vaucher). Various desired optimizations in e.g. grid-based embeddings can lead to computationally intractable problems, even for trees (see e.g. Supovit & Reingold, Johnson, and Brandenburg).

The drawing problem for general graphs tends to be very much harder, because of the possibly large number of edges that may be involved and the lack of adequate structural decomposition results. One possible approach for a given graph $G=\langle V, E\rangle$ could be to partition the set E into subsets $E_1, \ldots, E_k$ for some $k \geqslant 1$ such that each graph $G_i=\langle V, E_i\rangle$ $(1 \leqslant i \leqslant k)$ is planar (which would imply a k-layered planar drawing for G). The smallest value of k for which this is possible is known as the "thickness" of a graph (cf. Harary), denoted by $\Theta(G)$. Unfortunately, as we saw in other cases before, no efficient (i.e., polynomial-time bounded) algorithm is presently known to actually decompose an arbitrary graph G into $\Theta(G)$ planar subgraphs. It has, in fact, been shown that the problem of deciding for graphs G and integers K whether $\Theta(G)$ is less than or equal to K or not, is NP-complete (Mansfield).

An interesting heuristics for achieving "a small k" in the decomposition above could be to try and extract a "maximal planar subgraph" from G, and iterate the procedure on the resulting graph. The approach is generally referred to as the *planarization* of a graph and algorithms for graph planarization, not surprisingly, tend to derive from efficient planarity testers. Both Chiba, Nishioka & Shirakawa and Ozawa & Takahashi have presented $O(ne)$ algorithms for the general construction of a maximal planar subgraph of a graph along this line, with several optimizations to achieve good practical running times. Jayakumar et al. showed that, given a biconnected spanning planar subgraph H, a maximal planar subgraph of G which contains H can be constructed in only $O(n^2)$ time.

There are many other approaches to the graphical representation of graphs, and many lead to interesting algorithmic questions which are not immediately solved by current graph theory. For example, one might want to draw a graph with all nodes on a line and edges drawn as horizontal strips on the grid such that no two edges in one strip overlap (see Fig. 2). Define the cutwidth of a layout as the maximum number of edges that pass over any point of the line, and define the *cutwidth* of the graph G as the minimum of these numbers over all linear arrangements of the nodes on a line. It is a simple result of graph theory that the cutwidth of a graph G equals the minimum number of horizontal grid lines required to layout the graph as desired (the MIN-CUT LINEAR ARRANGEMENT problem). On the other hand, determining the cutwidth of a graph is an example of a problem in the theory of graph algorithms.

Aside from the review of some important results for the problem of automatic graph

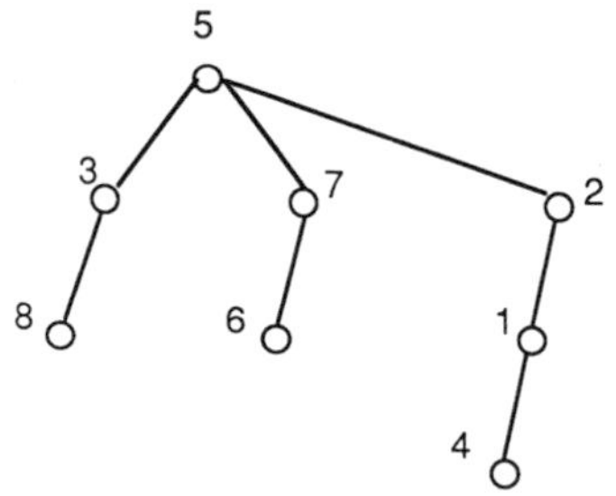

(a) a graph G

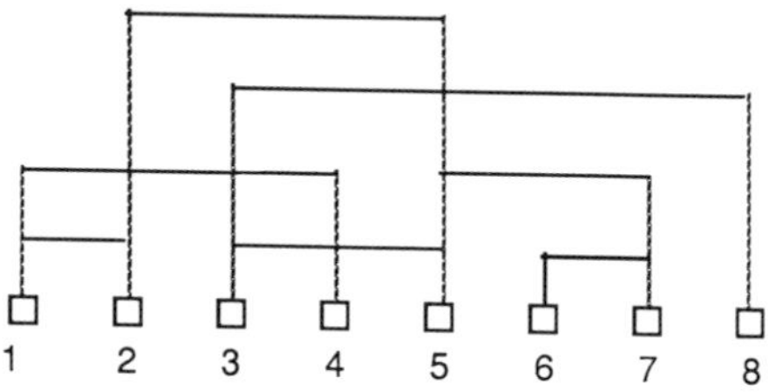

(b) a possible layout of G

Fig. 2.

drawing, it also was the purpose of this section to show that there often is a very close connection between graph theory and the design of efficient graph algorithms. It is a connection that can be observed throughout this chapter. While graph theory is concerned with general results about the structure of graphs, the design and analysis of effective methods to compute on graphs often seems to require complementary insights and new results. Very often the designer of an algorithm must delve deeply into the structure of a graph before a really practical algorithm of low complexity can be concluded. The quest for algorithms of low complexity has lead to many intriguing problems in the study of graph algorithms. For example, no efficient, i.e., polynomial-time algorithm is presently known for the graph layout problem discussed above. In fact, the MIN-CUT LINEAR ARRANGEMENT problem is known to be NP-complete. It is common in the theory of graph algorithms to study problems for special classes of graphs also, and to show that the special graphs are easier to deal with algorithmically. For example, the MIN-CUT LINEAR ARRANGEMENT problem is solvable by an $O(n \log n)$ algorithm for the case of n-node trees. Also, it has been shown that for any fixed k the problem of deciding whether the cutwidth of a graph is less than or equal to k is solvable by an $O(n^2)$ algorithm (Fellows & Langston, Bodlaender). This chapter should provide many of the tools needed for the design of efficient graph algorithms.

1.2. *Computer representations of graphs*

In order to compute with a graph G, it must be represented in computer storage somehow. The particular representation chosen for G can have a profound effect on the complexity of an algorithm. Thus graph representation is a form of datastructuring. We discuss some representations and their impact.

1.2.1. *Edge query representation*

The most obvious representations for a graph G use a Boolean subroutine $S(x, y)$ with the property that $S(i,j)=1$ if and only if $(i,j) \in E$. There generally is a trade-off between the time it takes S to respond to a query $S(i,j)$ ("is there an edge between i and j") and the data storage it uses. One possibility for S is to use the *adjacency matrix* A of

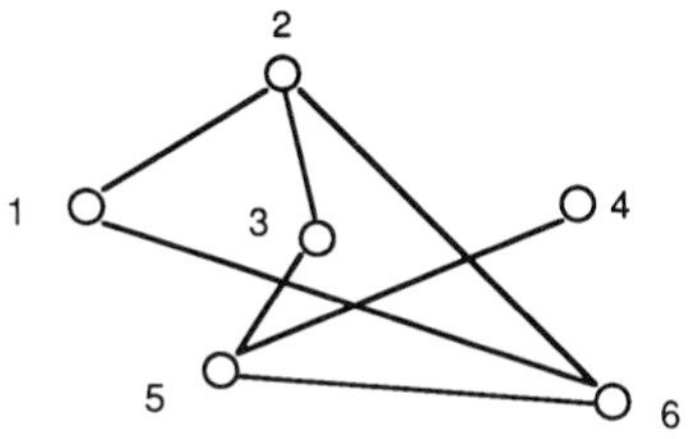

(a) a graph G

	1	2	3	4	5	6
1	0	1	0	0	0	1
2	1	0	1	0	0	1
3	0	1	0	0	1	0
4	0	0	0	0	1	0
5	0	0	1	1	0	1
6	1	1	0	0	1	0

(b) the adjacency matrix of G

Fig. 3.

G, with A the two-dimensional 0-1 matrix defined by $A(i,j)=1$ (0) if and only if $(i,j)\in(\notin)E$ (see Fig. 3). It enables query answering in O(1) time but requires $O(n^2)$ storage. For undirected graphs the adjacency matrix is symmetric, and space can be saved by storing only the $\frac{1}{2}n(n-1)$ upperdiagonal elements. In so-called loop-free (directed) graphs we have $A(i,j)\cdot A(j,i)=0$ for all i,j and all information can be comprised in the $\frac{1}{2}n(n-1)$ upperdiagonal elements, e.g. by setting $A(i,j)$ to -1 when $A(j,i)=1$ $(i<j)$.

1.3. Lemma. *Any $n\times n$ adjacency matrix can be represented in $O(n^2/\log n)$ storage locations while retaining O(1) time access to its elements.*

Proof. Choose $k=\sqrt{\log n}$ and observe that A can have at most 2^{k^2} different $k\times k$ submatrices with 0-1 elements. Write A as an $(n/k)\times(n/k)$ matrix of pointers to $k\times k$ submatrices. This requires

$$\frac{n^2}{k^2}+2^{k^2}\cdot k^2=O\left(\frac{n^2}{\log n}\right)$$

storage locations, while the access time for every element is still O(1). (Observe that this does not save on the total number of bits needed to represent A because of the pointers that are needed.) □

Adjacency matrices are very handy when dealing with path problems in graphs.

1.4. Lemma. *Nodes i,j are connected by a path of length k if and only if $A^k(i,j)=1$.*

Proof. By induction on k. The induction basis $(k=1)$ is trivial. For $k>1$, observe that

$$A^k(i,j)=(A^{k-1}\cdot A)(i,j)=\sum_x A^{k-1}(i,x)\cdot A(x,j)=1$$

if and only if there is an $x\in V$ with $A^{k-1}(i,x)=1$ and $A(x,j)=1$. □

Consider undirected or loop-free directed graphs G with n nodes. Extremely interesting questions arise if we wish to know the *minimum number of probes* $c(P)$ of the adjacency matrix of a graph required in the worst case, in order to determine whether the graph possesses a given *nontrivial property* P. A probe is simply the inspection of one entry of the adjacency matrix (and we assume that an algorithm will forever "remember" what the entry was after it probed it), and a property is a subclass of the class of all n-node graphs that is closed under isomorphism. A property P is called "nontrivial" if P holds for some graphs but not for all. At first sight it may seem that $c(P)=\frac{1}{2}n(n-1)$ for all nontrivial graph properties, but a classical example due to Aanderaa (see Rosenberg) shows that this is not the case. The following, slightly better result is due to King and Smith-Thomas.

1.5. LEMMA. *If P is the property of having a sink node (for loop-free directed graphs), then* $c(P)=3n-[\log n]-3$.

PROOF. (A sink is defined to be a node with in-degree $n-1$ and out-degree 0.) We only show that $c(P)\leqslant 3n-[\log n]-3$. Arrange the n nodes at the leaves of a tournament tree (or heap). Run tournaments, with node i defeating node j when $A(i,j)=0$ and node j defeating node i when $A(i,j)=1$. Observe that losers cannot be sinks. Clearly it takes $n-1$ probes to determine the winner v of the tournament, which is the only possible candidate for being a sink. By probing the $A(v,j)$- and $A(j,v)$-values for all $n-1$ nodes $j\neq v$ it is easily decided whether v is indeed a sink. We do not have to probe the $\geqslant[\log n]$ entries $A(v,j)$ or $A(j,v)$ that were already queried during the tournament. Thus $(n-1)+2(n-1)-[\log n]$ probes suffice. □

Bollobás and Eldridge have established that for all nontrivial graph properties P, $c(P)\geqslant 2n-4$.

On the other hand, there are many graph properties for which $c(P)$ is not linear. Define a graph property P to be *elusive* (*evasive*) if all essential entries of the adjacency matrix must be probed (in the worst case) in order to establish P, i.e., when $c(P)=\frac{1}{2}n(n-1)$ in the undirected case and $c(P)=n(n-1)$ in the directed case for general graphs. The following theorem combines results of Holt & Reingold, of Best, van Emde Boas & Lenstra, of Milner & Welsh, of Bollobás, of Kirkpatrick, and of Yap.

1.6. THEOREM. *The following graph properties are elusive:* (i) *having* $\leqslant k$ *edges* $(0\leqslant k\leqslant\frac{1}{2}n(n-1))$, (ii) *having a vertex of degree* $\geqslant 2$ $(n\geqslant 3)$, (iii) *being a tree*, (iv) *having a triangle*, (v) *having a cycle*, (vi) *being connected*, (vii) *being Eulerian*, (viii) *being Hamiltonian* (*n prime*), (ix) *being biconnected*, (x) *being planar* $(n\geqslant 5)$, (xi) *containing a complete graph of order* k $(2\leqslant k\leqslant n)$, *and* (xii) *being k-chromatic.*

Bollobás and Eldridge have shown that every nonelusive property of graphs must be satisfied by at least three, nonisomorphic graphs. Finally, define a graph property P to be *monotone* when every n-node supergraph of an n-node graph G which satisfies P also satisfies P. In 1973 Rosenberg formulated *the Aanderaa–Rosenberg conjecture* asserting that for every nontrivial, monotone property of graphs P, $c(P)=\Omega(n^2)$. Much of the results stated earlier actually arose from attempts to resolve the conjecture. The Aanderaa–Rosenberg conjecture was settled by Rivest and Vuillemin, who proved the following result.

1.7. THEOREM. *If P is a nontrivial, monotone property of graphs, then* $c(P) \geqslant n^2/16$.

Kleitman and Kwiatkowski improved the bound to $c(P) \geqslant n^2/9$, Kahn et al. improved it to $c(P) \geqslant n^2/4 + O(n^2)$ and King to $c(P) \geqslant n^2/2 + O(n^2)$ for directed graph properties. Kahn et al. also proved that every nontrivial, monotone property for n-node graphs with n a prime power is elusive. Yao proved that every nontrivial, monotone bipartite graph property is elusive.

Similar to $c(P)$ one can define $R(P)$, the minimum number of probes required by any *randomized* algorithm in the worst case in order to establish a property P on an n-node graph. It is known that $R(P) > n^{5/4}/44$ for all n sufficiently large, for any nontrivial monotone property P (King).

In a different approach to obtain a compact edge query representation one might want to represent S by a combinational circuit with two ($\log n$)-bit input lines for node names and a 3-valued output line such that

$$P(x, y) = \begin{cases} ? & \text{if } x \notin V \text{ or } y \notin V, \\ 0 & \text{if } x, y \in V \text{ and } (x, y) \notin E, \\ 1 & \text{if } x, y \in V \text{ and } (x, y) \in E. \end{cases}$$

Define a graph to have a *small circuit representation* if it can be represented by a combinational circuit of size $O(\log^k n)$ for some k. Many simple graph problems are surprisingly complex for graphs that are given by small circuit representations, i.e., the succinctness of the graph encoding saves space but not time (Galperin & Wigderson).

In a related, more concrete approach one can try to assign $k \log n$ bit names (some fixed k) to the nodes of an n-node graph in such a way that adjacency of nodes can be decided quickly from an inspection of their names. This eliminates the need for an adjacency matrix and is (thus) referred to as an *implicit representation* of the graph (Kannan et al.). We say that a class of graphs C can be implicitly represented if there exists a k such that for every graph $G \in C$ there is a labeling with ($k \log n$)-bit names with the desired property. To show a simple but rather far-reaching result, we observe the following (cf. the discussion of graphs of bounded treewidth in Section 1.6).

1.8. THEOREM. *Every class of graphs C of bounded treewidth can be implicitly represented.*

PROOF. Let treewidth$(C) \leqslant b$. Consider any n-node graph $G = \langle V, E \rangle \in C$ with $n \geqslant b$, and let treewidth$(G) = k \leqslant b$. In Section 1.6 it is shown that G is a partial k-tree, i.e., it is a subgraph of a k-tree on V. For any partial k-tree there is a way to number the nodes from 1 to n such that for every i $(1 \leqslant i \leqslant n)$ there are at most k nodes $j < i$ that are adjacent to i. (This follows from the definition of partial k-trees.) Label any node i with the list consisting of i followed by the names of its neighbors j in G with $j<i$. Names are bounded to $(k+1)\log n$, and hence to $(b+1)\log n$ bits, and the adjacency of two nodes is easily decided from their names. For graphs G with $n<b$ a similar scheme works with names of no more than $(b+1)\log n$ bits. Thus C can be implicitly represented. □

Many familiar classes of graphs are of bounded treewidth (cf. Section 1.6) and thus can be implicitly represented. The same technique can be used to show the more general result that any class of graphs C in which the n-node members have at most cn edges (for some fixed constant c) and that is closed under taking subgraphs can be implicitly represented. This not only includes the class of partial k-trees (for any fixed k) but also the classes of graphs of bounded genus, and thus e.g. the class of planar graphs.

1.2.2. Node query representation

Another common representation for a graph G uses a subroutine $S(x)$ which generates the edges incident to x for any $x \in V$. A typical example is the *adjacency list* representation of G in which for each $i \in V$, $S(i)$ produces a pointer to the "list" of edges incident to i. An edge (i, j) is represented by a simple record (see Fig. 4).

The adjacency lists of a graph use $O(n+e)$ storage and thus are usually a more compact representation than the adjacency matrix. The fact that the adjacency list produces the edges incident to a node in $O(1)$ time per (next) edge makes it the primary representation of graphs in many algorithms. Note that for undirected graphs one can save space by numbering the nodes from 1 to n, and including in the adjacency list of a node i only those edges that lead to nodes $j \geqslant i$. Itai and Rodeh show by an information-theoretic argument that in many cases this is about the tightest possible representation of a graph (in terms of the number of bits used). Sometimes it is already advantageous (from the algorithmic point of view) to have the edges (i, j) in the adjacency list of i ordered by increasing value of j. We call this *the ordered list representation* of G.

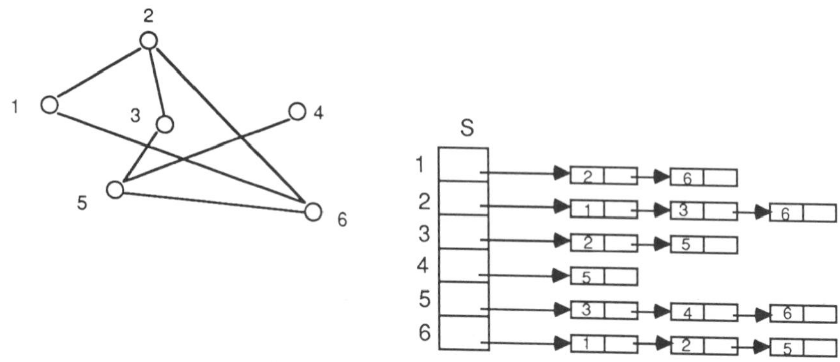

(a) a graph G (b) an adjacency list representation of G

Fig. 4.

1.9. LEMMA. *Given any representation of G by adjacency lists, the ordered list representation of G can be constructed in* $O(n+e)$ *time.*

PROOF. Use buckets B_1 through B_n. Go through the given adjacency lists edge by edge, and throw edge (i,j) into bucket B_j. Now purge the adjacency lists and build new ones by emptying the buckets B_1 through B_n in this order edge-wise, appending any edge $(i,j)\in B_j$ to the end of the (current) adjacency list for i. □

1.10. LEMMA. *The representation of G by adjacency lists can be kept within* $O(n^2/\log n)$ *storage locations while retaining the essential properties of the representation.*

PROOF. Use the ordered version of the representation. Choose $k=\frac{1}{2}\log n$, and observe that there can be at most 2^k different sublists on the adjacency lists involving the nodes $\alpha k+1$ through $(\alpha+1)k$ for $\alpha=0,\ldots,n/k-1$. Represent the different lists separately, and include a reference to the proper sublist for every nonempty interval in the adjacency lists. This requires

$$n\cdot\frac{n}{k}+\frac{n}{k}\cdot k2^k=\frac{n^2}{k}+n2^k=O\left(\frac{n^2}{\log n}\right)$$

storage locations, and still allows for a simple traversal of the lists in $O(1)$ time per edge. □

Given the adjacency list representations of G, one can effectively obtain an adjacency matrix for G in $O(e)$ time despite the fact that $O(n^2)$ storage is required (cf. Aho, Hopcroft & Ullman, p. 71, Exercise 2.12).

1.2.3. Structural representation

When a graph G has a particular structure, it may well be exploited for a suitable representation in computer storage. For example, an interval graph is defined such that every node x corresponds to an interval $I(x)$ on the real line and $(i,j)\in E$ if and only if $I(i)\cap I(j)\neq\emptyset$. The representation could simply consist of the intervals $I(x)$ for $x\in V$. It is an example of an "edge query representation" that uses only $O(n)$ storage.

When G is planar it is useful to record the edges in the adjacency list of every node i in counterclockwise order and to include in the record of every edge (i,j) a pointer to the record corresponding to (j,i) in the adjacency list of j. This representation has the added advantage that one can easily traverse the consecutive edges of any face of G in clockwise order in $O(1)$ time per edge. Itai and Rodeh have devised an ingenious scheme to represent every n-node planar graph using only $1.5n\log n+O(n)$ bits.

Sometimes graphs can be *hierarchically defined*. In this case some nodes of G can be "expanded" and replaced by other graphs (again hierarchically defined), with suitable rules that specify the connection of the edges incident to an expansion node and the nodes of the replacement graph. Observe that the complete expansion could lead to a graph of exponential size. Many basic graph problems like planarity testing and minimum spanning tree contruction for hierarchically defined graphs can be solved in time and space linear in the size of the hierarchical description (Lengauer, Wagner).

Acyclic directed graphs can be represented with the nodes linearly arranged in such a way that all (directed) edges run strictly from left to right. The representation is often used and known as the *topological ordering* or topological sort of the graph.

1.11. THEOREM. *A topological ordering of an acyclic directed graph can be computed in* $O(n+e)$ *time.*

PROOF. By enumerating the edges one by one, one can compute the in-degree $c(i)$ of all nodes i in linear time. Make one sweep over the list of $c(i)$ values to compile the "batch" L of nodes with $c(i)$-value 0 (the current sinks). Now execute the following code, using an auxiliary list L' that is initially empty.

```
repeat
   output the nodes of L (in any order);
   for each node i ∈ L do
      for each edge (i, j) do
      begin
         c(j):=c(j)−1;
         if c(j)=0 then append j to L'
      end;
      L:=L';
      L':=empty
until L=empty.
```

Note that in the **for**-loop the computational effort can be accounted for by charging a cost of $O(1)$ to each node $i \in L$ and $O(1)$ to each edge (i, j). Because every node (and hence, every edge) appears precisely once in the process, the computation takes only $O(n+e)$ time total. □

Computing the lexicographically first topological order is NLOG-complete (Shoudai). For general directed graphs, Shiloach has shown that the nodes can be linearly arranged as 1, ... n such that for every $i<j$ the minimum number of edges blocking every path from i to j is greater than or equal to the minimum number of edges blocking every path from j to i. The ordering can be computed in $O((n+e)^2)$ time.

1.3. Graph exploration by traversal

Consider the following traversal of a (connected) graph G. Initially all nodes are marked unvisited and all edges are colorless. Start at a fixed but arbitrary node i_0 and proceed in the following way whenever a node i is reached. When i was not reached before, mark it as visited. If there are no unexplored edges left in the edge list of i, then backtrack to the node from which i was reached. Otherwise traverse the first unexplored edge (i, j) in the edge list of i (and mark it explored implicitly by moving some pointer to the next element of the list). When j was visited before, color the edge red and backtrack to i. When j was not visited before, color the edge green. Now "visit" the node reached in a similar manner. The algorithm is known as *depth-first search*

(DFS) and is easily implemented using a stack and suitable pointer moves over the adjacency lists. The algorithm terminates when all edges incident to i_0 have been explored and the search has returned to i_0.

1.12. THEOREM. *Let G be a connected graph.*

(i) *The edges colored green by DFS form a spanning tree of G, called a* DFS tree.
(ii) *The edges colored red by DFS always connect two nodes of which one is an ancestor of the other in the DFS tree.*
(iii) *DFS takes* $O(n+e)$ *time.*

DFS trees are very special because of (ii), and normally reveal quite a bit of structure of a graph. For example, DFS leads to an algorithm for finding the biconnected components of a graph in $O(n+e)$ time. Depth-first search can also be applied to directed graphs. An important application is Tarjan's algorithm for determining the strongly connected components of a directed graph in $O(n+e)$ time. For undirected graphs G it can be decided in $O(\max\{n,e\})$ time whether a given spanning tree is a DFS tree (Korach & Ostfeld).

Another popular technique of traversing a (connected) graph G proceeds as follows and assigns a number $L(i)$ to every node i. Initially all nodes are marked unvisited and all edges are colorless. Start with $\{i_0\}$, mark i_0 as visited, set $L(i_0)=0$, and do the following whenever a set of nodes I is reached. For every $i\in I$ and edge (i,j) on the edge list of i, color the edge red when j was visited before. Otherwise, color the edge green, mark j as visited, set $L(j)$ to $L(i)+1$, and add j to a set I'. Repeat the step with I equal to I' and I' initially empty, and continue until the "new" I' is empty. The algorithm is known as *breadth-first search* (BFS) and is easily implemented y maintaining the set I in a queue. All nodes in a set I have the same L-value.

1.13. THEOREM. *Let G be a connected graph.*

(i) *The edges colored green by BFS form a spanning tree of G, called a* BFS tree.
(ii) *For every* $i\in V$, $L(i)$ *is the shortest distance (in edges) from* i_0 *to* i.
(iii) *For every green edge* (i,j) *one has* $|L(i)-L(j)|=1$. *For all edges* (i,j) *one has* $|L(i)-L(j)|\leqslant 1$.
(iv) *BFS takes* $O(n+e)$ *time.*

$L(i)$ is called the "level" of i. Observe from (ii) that edges always connect nodes that are at most one level apart. Define *truncated* BFS to be the BFS procedure up to and including the moment that the first red edge is encountered. Truncated BFS finishes in $O(n)$ time because it traces a tree-like environment of i_0. (The red edge closes the first cycle, if there is one). Clearly other forms of truncated BFS can be defined.

1.4. Transitive reduction and transitive closure

When one is only interested in representing the path information of a (directed) graph G, two extreme approaches can be followed:

(i) (*minimum storage representation*) determine a "minimal" graph $G^-=\langle V,E^-\rangle$

with the property that there is a (directed) path from i to j in G if and only if there is a (directed) path from i to j in G^-.

(ii) (*minimum query time representation*) determine the (unique) graph $G^* = \langle V, E^* \rangle$ with the property that there is a (directed) path from i to j in G if and only if there is a (directed) edge $(i, j) \in E^*$ in G^*. The graph G^* is traditionally known as the *transitive closure* of G.

We outline the known results for both types of representation. A simple idea for obtaining a minimum storage representation of G is to delete as many edges as possible without destroying the existing connections by directed paths. Any minimal subgraph of G that can be obtained through edge deletions in this way is called an *irreducible kernel* of G. An irreducible kernel can be computed in $O(n^3)$ time for general graphs, in $O(ne^-)$ time for acyclic graphs (where e^- is the number of edges in the resulting irreducible kernel) and in $O(n^2)$ time when G is strongly connected (Noltemeier). Any graph with the smallest number of remaining edges that can be obtained in this way is called a *minimum equivalent graph* (or MEG). Moyles and Thompson (see also Hsu) have shown that a MEG for G can be constructed out of the MEGs of the individual strongly connected components of G and a MEG of the "condensed" graph (i.e., the acyclic graph with the strongly connected components as nodes). Sahni proved that the problem of deciding whether G has a MEG with at most k edges (for specified k) is NP-complete. If we drop the requirement that the "reduced" graph be a subgraph of G, then one can sometimes save more edges. Define a *transitive reduction* of G to be any graph G^- with the smallest possible number of edges, with the property that there is a directed path from i to j in G if and only if there is a directed path from i to j in G^-. For acyclic graphs the transitive reductions are MEGs (and vice versa). Aho, Garey and Ullman have shown that each graph has a transitive reduction that can be computed in polynomial time and that, algorithmically, computing transitive reductions is of the same time complexity as computing transitive closures and, hence, of multiplying Boolean matrices (see below).

The problem of computing the *transitive closure* G^* of a (directed) graph was first considered in 1959 (Roy) and a variety of algorithms have been proposed for it since. One popular approach is based on the use of the adjacency matrix A of G, considered as a Boolean matrix. Assume that ordinary matrix multiplication takes $O(n^\alpha)$ time and that the elementary arithmetic operations (no division) on n-bit numbers can be performed in $m(n)$ "steps". It is known that one can choose $\alpha < 2.4$, $m(n) = 1$ in the uniform model and $m(n) = O(n \log n \log\log n)$ if we count bit operations. Furman observed that the adjacency matrix A^* of G^* can be written as $A^* = I \vee A \vee A^2 \vee \cdots \vee A^{n-1} = (I \vee A)^{n-1}$. It suggests that A^* can be computed in $O(n^\alpha m(\log n) \log n)$ time using repeated squaring, by computing each necessary matrix product over the ring of integers modulo $n+1$ and recovering the true Boolean matrix product by changing non-zero entries to ones. Munro and Fischer & Meyer proved the following key result. Let $M(n)$ be a function satisfying $M(2n) \geqslant 4M(n)$ and $M(3n) \leqslant 27M(n)$.

1.14. THEOREM. *A^* can be computed in $O(M(n))$ time if and only if the product of two arbitrary $n \times n$ Boolean matrices can be computed in $O(M(n))$ time.*

It follows that transitive closures can be computed in $O(n^\alpha m(\log n))$ time, because two $n \times n$ Boolean matrices can be multiplied within this time bound (by the same technique as before). Adleman et al. have improved this bound to $O(n^\alpha \log n)$ bit operations and even to

$$O\left(n^\alpha \log n \left(\frac{\log\log n}{\log n}\right)^{(\alpha/2)-1}\right)$$

bit operations, by means of modular arithmetic and table look-up techniques using only $O(n^2 \log n)$ bits of storage. The well-known "four Russians" algorithm (apparently due to Kronrod) is much less sophisticated and gives a bound of only $O(n^3/\log n)$ steps for Boolean matrix multiplication both in the worst case and in an average sense (van Leeuwen). O'Neill and O'Neill show that the simple technique of computing every row $\times$ column product by stepping through the ones in the row (only) and stopping when the corresponding column position has a one (in which case the product will be 1), yields an algorithm for computing the product of two arbitrary Boolean matrices in $O(n^2)$ expected time.

Other approaches use the adjacency matrix in more direct terms as a problem representation. The classical algorithm of Warshall computes A^* as the "limit" of a series of matrices $A(0) = I \vee A$, $A(1)$, $A(2), \ldots$ with $a_{ij}(s) = 1$ if and only if there is a directed path from i to j passing through intermediate nodes from $\{1, \ldots, s\}$ only. $A(s)$ can be computed in $O(n^2)$ time from $A(s-1)$ by observing that

$$a_{ij}(s) = a_{ij}(s-1) \vee a_{is}(s-1)a_{sj}(s-1).$$

As $A^* = A(n)$, Warshall's algorithm computes transitive closures in $O(n^3)$ time. Several improvements of Warshall's algorithm have been proposed that are more efficient when one analyzes the number of rows that must be "paged in", assuming that matrices are stored row-wise on disk (Warren, Martynyuk). An interesting, space-efficient variant was proposed by Thorelli. Put all directed edges in a queue Q, introduce two pointers α and β, and execute the following code. (Suppose that the successive queue elements are numbered $1, 2, \ldots$ for purposes of presentation and let the edge pointed at by α be (i_α, j_α).)

for $\alpha := 1$ **to** end(Q) **do**
  for $\beta := 1$ **to** end(Q) **do**
    if $j_\alpha = i_\beta$ and $i_\alpha \neq j_\beta$
    then if $(i_\alpha, j_\beta) \notin Q$ **then** append (i_α, j_β) to Q.

Note that Q keeps expanding as the algorithm proceeds, until it contains the full $e^* = |E^*|$ edges of the transitive closure. Another interesting variant was considered by Ibaraki and Katoh. Suppose we have the A^* of G and wish to update it when, say, k edges are added. Clearly no one can change to a zero, and we can confine ourselves to considering the question which zeros of A^* must be turned into ones. In case an edge (i, j) is added and $a^*(u, v) = 0$ then $a^*(u, v)$ must only be changed to 1 when $a^*(u, j) = 0$, $a^*(u, i) = 1$ and $a^*(j, v) = 1$. Process the tth edge as follows. Determine the sets $U_t = \{u \mid a^*(u, j) = 0,\ a^*(u, i) = 1\}$ and $V_t = \{v \mid a^*(j, v) = 1\}$ (always $j \in V_t$) by straightforward inspection in $O(n)$ time and, when $U_t \neq \emptyset$, set $a^*(u, v)$ to 1 for all $u \in U_t$ and $v \in V_t$.

Note that the latter costs $O(u_t v_t)$ time and creates at least u_t more ones in A^* (because all $a^*(u,j)$ were 0 for $u \in U_t$), with $u_t = |V_t|$. Thus the addition of k edges requires a total of $O(kn + \sum_1^k u_t v_t) = O(ne^*)$ time, using that $k \leqslant e^*$, $v_t \leqslant n$ for all t and $\sum_1^k u_t \leqslant e^*$. The algorithm suggests an interesting, incremental approach to the construction of transitive closures. Ibaraki and Katoh also prove that the transitive closure can be updated in $O(n^3 + n^2 e)$ time when any number of edges is deleted from G.

A considerable number of transitive closure algorithms is based on the following lemma implicit in the work of Munro and Purdom. Let $l(n)$ denote the time for computing the transitive closure of an acyclic directed graph of n nodes, and let the condensed graph of G (i.e., the induced graph with the strongly connected components of G as nodes) have n_c nodes and e_c edges.

1.15. THEOREM. *The transitive closure of a directed graph can be computed in* $O(n + e^* + l(n_c))$ *time.*

PROOF. Compute the strongly connected components of G and the condensed graph G_c in $O(n+e)$ time, using DFS. The transitive closure of G can be computed in $O(n+e^*)$ time from the strongly connected components and the transitive closure of G_c by enumerating all implied pairs, i.e., all pairs (i,j) with i and j in the same strongly connected component or in components that are directly connected in G_c^* by an edge from the i-component to the j-component. Because G_c is acyclic, its transitive closure can be computed in $l(n_c)$ time. □

Eve & Kurki-Suonio, Ebert and Schmitz all show that the DFS algorithm for computing the strongly connected components of a graph gives enough structural information to generate the transitive closure directly, typically in $O(n^2 + ne)$ time. The theorem shows that this bound may be overly pessimistic, especially when n_c is small. Consider any acyclic directed graph G of n nodes, let $G^- = \langle V, E^- \rangle$ be its minimum equivalent graph (or transitive reduction) and $e^- = |E^-|$ (hence $e^- \leqslant e$). One can compute a topological ordering of G in $O(n+e)$ time (cf. Section 1.2), with no additional overhead for having each adjacency list $L(i)$ sorted (in increasing order). The following result is due to Goralcikova and Koubek.

1.16. THEOREM. $l(n) = O(n + e + ne^-)$.

PROOF. We have spent $O(n+e)$ time to represent G in topological order. Define $R(i) = \{j \mid \text{there is a directed path from } i \text{ to } j\}$. It is clear that the sets $R(i)$ represent G^*. (Observe that $j \in R(i)$ implies $j > i$.) Compute the sets $R(i)$ inductively as follows for i from n down to 1. Suppose $R(j)$ has been computed for all $j > i$. Compute $R(i) = \{i\} \cup \{R(j) \mid (i,j) \in E\}$ as follows, using an auxiliary bit vector γ that is initially empty. Begin by putting i into $R(i)$ and setting the ith bit of γ to 1. In fact we will use γ as the characteristic vector of $R(i)$ at all times. Now process all edges $(i,j) \in E$ in order of increasing j as follows: when j already appears in $R(i)$ (which we can tell in $O(1)$ time using γ), ignore this edge and proceed; otherwise, append the elements of $R(j)$ to $R(i)$ (while updating γ and, in fact, using γ to avoid adding nodes to $R(i)$ that already appear in it). Note that the edges (i,j) with $j \in R(i)$ are correctly ignored because, when j was added to $R(i)$, necessarily all nodes reachable from j must have been added to $R(i)$ as well. The edges (i,j) with $j \notin R(i)$ necessarily belong to G^-. For suppose $(i,j) \notin G^-$. Then there would be a path $i \to u \to \cdots \to j$ with $(i,u) \in E$ and (necessarily) $u < j$, and j must have been added to $R(i)$ by the time the edge (i,u) was processed. Contradiction. It follows that the time complexity of the algorithm is $O(n+e)$ plus $(\sum_{(i,j) \in E^-} |R(j)|) = O(ne^-)$. □

It can be argued that for random acyclic directed graphs the expected value of e^- is $O(n^{1.5})$. Jaumard and Minoux have given a slightly different implementation of the same algorithm, showing that $l(n) = O(n + de^*)$ for acyclic directed graphs with in-degrees bounded by d. Mehlhorn and Simon have developed another algorithm showing that $l(n) = O(n + e + pe^-)$, where p is the number of chains (paths) in a chain decomposition of the acyclic directed graph. The expected time complexity of their algorithm is $O(n^2 \log \log n)$.

Following yet another approach, several algorithms try to construct the transitive closure of G by progressively searching for the "successors" of every node. Bloniarz et al. considered the following simple algorithm for determining all nodes j that are reachable by a directed path from i, for each i. We use the adjacency lists $L(j)$, an auxiliary list $R(i)$ to accumulate the j's that are reachable from i, an auxiliary bit vector γ that will serve as the characteristic vector for $R(i)$ (as before), and an auxiliary pushdown list P to record the nodes which we need to explore further in DFS-like order. For every node i, the following code is executed. Let "record(k)" denote the combined operation of appending k to $R(i)$, setting the kth bit of γ to 1, and pushing k onto P.

```
record(i);
count:=1;
while P≠∅ & count<n  do
begin
  pop the top element of P and assign it to j;
  for each node k∈L(j) do
    if k has not been recorded yet (kth bit of γ is 0) then
    begin
      record(k);
      count:=count+1
    end
end;
output R(i).
```

For each i the algorithm may need up to $O(n+e)$ time, and its total routine will be $O(n^2+ne)$ in worst case. Bloniarz et al. prove that over wide classes of "random" graphs the algorithm has an average runtime of $O(n^2 \log n)$. A transitive closure algorithm with an even better expected time complexity has been devised by Schnorr. The algorithm differs from the preceding one in two ways: the DFS-like search is replaced by a BFS-like search, and advantage is taken of the fact that the sets $R(i)$ are computed for all i and thus allow for a combination of effort. More precisely, using a BFS-like search procedure on G and G^r (the graph G with all edges reversed) respectively, the algorithm determines sets $R_+(i)$ and $R_-(i)$ for every i such that $R_+(i)$ contains the maximum number $\leqslant [\frac{1}{2}n]+1$ of successors of i and $R_-(i)$ contains the maximum number $\leqslant [\frac{1}{2}n]+1$ of predecessors of i. The following lemma shows that these sets are sufficient to obtain G^* in another $O(n+e^*)$ time.

1.17. LEMMA. *For each i,*

$$R(i) = R_+(i) \cup \{j \mid i \in R_-(j)\} \cup \{j \mid |R_+(i)| = |R_-(j)| = [\tfrac{1}{2}n]+1\}.$$

PROOF. Observe that $R_+(i)$ and $R_-(j)$ contain all successors of i and predecessors of j respectively, whenever their cardinality is $< [\frac{1}{2}n]+1$. When they both have cardinality $= [\frac{1}{2}n]+1$, $R_+(i) \cap R_-(j) \neq \emptyset$ by the pigeon hole principle and there must be a node u such that $i \to^* u \to^* j$ and (hence) j is reachable from i. □

Schnorr proves that the transitive closure algorithm has an expected time complexity of $O(n+e^*)$ over large classes of random graphs and that the probability that the algorithm will run for more than cn^2 steps (some constant c) is exponentially small in n.

1.5. Generating an arbitrary graph

One of the most immediate problems in dealing with graphs may be the question of generating an "arbitrary" graph with a given number of nodes (n). More specifically, the problem is to generate (or select) a graph G uniformly at random from the set of unlabeled graphs with n nodes. The problem is not entirely straightforward because graphs may be written down in (usually) many isomorphic forms and different graphs can have different numbers of isomorphisms. Let G_n be the set of undirected graphs on the node set $1, \ldots, n$ and let $E_n = \{(i,j) \mid 1 \leqslant i, j \leqslant n \text{ and } (i,j) \text{ unordered}\}$. Clearly any $G \in G_n$ has an edge set E with $E \subseteq E_n$. Consider the symmetric group S_n acting on G_n by permuting the node labels. The orbits under S_n correspond precisely to the isomorphism classes of graphs on n nodes. Dixon and Wilf have given a simple algorithm for selecting an orbit uniformly at random for the general case of a group acting on a set, and obtained the following result by specializing the algorithm to the present situation. Assume that the value $g_n = |G_n|$ has been precomputed.

1.18. THEOREM. *There is an algorithm for selecting an n-node graph G uniformly at random in* $O(n^2)$ *average time.*

PROOF. We only sketch the algorithm. For each $\pi \in S_n$, define the permutation π^* on E_n by $\pi^*((i,j)) = (\pi(i), \pi(j))$. Considering the action of S_n on G_n, one verifies that a graph G is left fixed by π if and only if $\pi^*(E) = E$. Thus $G \in Fix(\pi)$ if and only if for every cycle C of π^* (as a permutation) either all pairs in C occur as edges of G or none of them does. Hence $|Fix(\pi)| = 2^{c(\pi)}$, where $c(\pi)$ is the number of cycles of π^*. The value of $c(\pi)$ can be computed "by formula" from the cycle structure of π. Let $[k_1, \ldots, k_n]$ denote the conjugacy class of S_n consisting of the permutations having k_i cycles of length i ($1 \leqslant i \leqslant n$), and let

$$W([k_1,\ldots,k_n]) = |[k_1,\ldots,k_n]| \cdot 2^{c(\pi)} = \frac{n!2^{c(\pi)}}{\pi_i(i^{k_i}k_i!)}$$

be its "weight". The following algorithm will select an n-node graph G uniformly at random:

(1) Choose nonnegative integers $k_1,\ldots,k_n$ with $\Sigma_1^n\, i \cdot k_i = n$ such that the probability of choosing an n-tuple $[k_1,\ldots,k_n]$ is $W([k_1,\ldots,k_n])/(n!g_n)$.

(2) Take (any) representative π from the class $[k_1,\ldots,k_n]$ chosen in step 1, and choose a graph G uniformly at random from $Fix(\pi)$.

(3) "Ignore" the node labels of G and output it.

Steps 2 and 3 are easily implemented in $O(n^2)$ time, which is essentially the time needed for writing down G. Dixon and Wilf show that step 1 can be implemented in $O(n^2)$ average time, provided g_n is known. □

The Dixon–Wilf algorithm is particularly useful for selecting graphs "fairly", without the need for extra storage (i.e., beyond what is needed to represent the selected graph itself). The integers involved in the computation get very large for $n \geqslant 10$ and the procedure prohibitively complex in practical terms, but this seems inherent to the problem. Dixon and Wilf also prove that the $[g_n \log(g_n/\varepsilon)]$-fold iteration of the algorithm, with all choices independently and uniformly at random, leads to a sequence which, with probability $> 1-\varepsilon$, includes each n-node graph at least once. The algorithm seems particularly useful for testing hypotheses for n-node graphs, without having to list all graphs and go through expensive isomorphism rejection routines.

The approach of Dixon and Wilf can be used, at least in principle, for generating arbitrary graphs of a more restrictive type as well, but much work remains to be done here. For selecting a connected n-node graph uniformly at random one could simply iterate the Dixon–Wilf algorithm until a connected graph is output. (Note that by e.g. a DFS-based routine, the connectedness of a graph can be checked in $O(n+e)$ time.) On the average no more than two iterations will suffice for this, by known results on the population of connected graphs among all unlabeled n-node graphs. Simpler techniques can be used for generating random regular graphs of low degree (Wormald).

1.6. Classes of graphs

Many different types of graphs have been distinguished. There are usually good theoretical and practical reasons for studying special classes of graphs, and it can be very worthwhile to explore a graph-algorithmic problem for special types of graphs first. This leads to the important recognition problem for every type X of interest: given a graph, is it of type X? Related questions are: given a graph G and an integer k, can one turn G into a graph of type X by adding/deleting k nodes/edges? There are numerous versions of this question.

Interestingly enough, the recognition problem for classes of graphs can be quite tricky. It should probably be required of any class of graphs that is distinguished that its recognition problem is of polynomial-time bounded complexity. We have listed the known bounds for a number of classes of graphs that are commonly encountered, in Table 1 (cf. Johnson).

There are many related questions that can be asked: given a graph of type X, is it of type Y? There is also the following range of questions: given a graph of type X, does it have property P? (Here a "property" can be any graph-theoretic property that one may wish to test like planarity, having a Hamiltonian cycle, etc.). Also, many graph-theo-

Table 1
Some classes of graphs

Category	Class of graphs	Complexity of recognition
trees and related graphs	trees/forests	linear
	almost trees (k)	linear
	treewidth-k = partial k-trees	$O(n^2)$ for fixed k (Robertson & Seymour, Bodlaender), $O(n^{k+2})$ (Arnborg et al.)
	bandwidth-k	polynomial in n, for fixed k (Saxe)
	degree-k	linear
planar graph	planar	linear (Hopcroft & Tarjan, Booth & Lueker)
	series parallel	linear (Valdes et al.)
	outerplanar	linear (Mitchell)
	Halin	linear
	k-outerplanar	polynomial (Bienstock & Monma)
	grid	linear
	$K_{3,3}$-free	linear (Asano, Williamson)
	thickness-k	NP-complete for $k \geq 2$
	genus-k	NP-complete for k arbitrary (Thomassen), $O(n^{O(k)})$ (Filotti et al.), $O(n^2)$ for fixed k (Robertson & Seymour)
perfect graphs	perfect	in co-NP
	chordal	linear (Gavril, Tarjan & Yannakakis)
	split	linear (Golumbic)
	strongly chordal	polynomial (Farber)
	planar perfect	$O(n^3)$ (Hsu)
	comparability	$O(\delta \cdot e)$ with δ = max degree (Golumbic)
	bipartite	linear
	permutation	$O(n^3)$ (Golumbic)
	cographs	linear
intersection graphs	undirected path	$O(n^4)$ (Gavril)
	directed path	$O(n^4)$ (Gavril)
	interval	linear (Booth & Lueker)
	circular arc	$O(n^3)$ (Tucker)
	circle	polynomial (Gabor et al.)
	proper circular arc	polynomial (Tucker)
	edge (or line)	linear (Lehot, Syslo)
	claw-free	linear

retic constructions can be specialized to graphs of some type X and take advantage of its characteristic properties to achieve a lower or otherwise interesting complexity bound.

The given list of graphs is a natural hierarchy for investigating the complexity of graph problems. There is a vast body of literature resulting from this "research program", especially dealing with the study of problems that are NP-complete for general graphs (see Johnson).

An intriguing, new perspective on the design of algorithms for certain classes of graphs derives from the work by Robertson & Seymour. In order to explain the far-reaching consequences for our study of graph algorithms, we need a number of important concepts. Recall (cf. Harary) that in an "elementary contraction" of a graph, two adjacent nodes u and v are contracted to form one new node w that is adjacent to all nodes that were adjacent to u or to v (thus "collapsing" the edge between u and v to "length zero").

DEFINITION. Let G and H be graphs, and let C be a class of graphs.

(i) H is a *contraction* of G if H can be obtained from G by a sequence of elementary contractions.

(ii) H is a *minor* of G, denoted by $H \leqslant G$, if H is the contraction of a subgraph of G.

(iii) H is a *minor-minimal* in C if $H \in C$ and no graph $G \in C$ exists such that $G \leqslant H$ and $G \neq H$.

(iv) C is *minor-closed* if the minors of every graph in C again belong to C.

It is well-known that for arbitrary graphs G and H, the problem of deciding whether H is a minor of G is NP-complete (cf. Johnson). A first important result due to Robertson & Seymour is that for fixed graphs H the minor-containment problem is always decidable in polynomial time.

1.19. THEOREM. *For every fixed graph H, the problem of deciding for arbitrary graphs G whether $H \leqslant G$ is solvable by an* $O(n^3)$ *time bounded algorithm.*

A second, important result claimed by Robertson & Seymour solves a long-standing open problem in graph theory known as Wagner's Conjecture, and is crucial in all applications that we will discuss.

1.20. THEOREM. *Every class of graphs C contains only finitely many distinct, minor-minimal elements.*

The two theorems combined lead to the conclusion that for every class of graphs C that is minor-closed there exists a polynomial-time bounded algorithm for the recognition problem. In order to see the simple reason for this, we introduce yet another auxiliary notion. Let $\bar{C}$ denote the complement of C.

DEFINITION. Let C be a class of graphs. The *obstruction set* for C is the set $Obstr(C) = \{H \mid H \text{ is minor-minimal in } \bar{C}\}$.

By the theorem of Robertson & Seymour, every obstruction set will be finite, and we observe that for every minor-closed class of graphs C and every graph G we have: $G \in C$ if and only if there exists no graph $H \in Obstr(C)$ such that $H \leqslant G$. Thus the recognition problem for a minor-closed class of graphs is reduced to a finite number of minor-containment tests with the (fixed) elements of the obstruction set. This yields a polynomial-time decision algorithm. As a simple example we can take for C the class of planar graphs, which is minor-closed. By Kuratowski's theorem we know that $Obstr(C) = \{K_5, K_{3,3}\}$ (cf. Section 1.1).

A severe difficulty in applying the theory is the fact that the proof of Robertson & Seymour's theorem is not constructive. Thus, unlike the case for planar graphs, it may well be that it is very hard to explicitly determine the obstruction set for some class of graphs that interests us. In fact, Fellows & Langston have shown that there cannot exist a uniform procedure for constructing obstruction sets (in the sense of computability theory).

1.21. THEOREM. *The problem of determining obstruction sets from machine descriptions of minor-closed classes of graphs is recursively unsolvable.*

PROOF. (The proof assumes some familiarity with the theory of computation and the framework of abstract complexity theory.) Let $\gamma_0, \gamma_1, \ldots$ be an acceptable numbering of the computable functions, and let Φ be a complexity measure. Also let $G_0, G_1, \ldots$ be an effective enumeration of all graphs such that, for all G, H, if $H \leqslant G$, then H precedes G in the enumeration. (This condition is satisfied e.g. when the graphs are enumerated by increasing number of nodes.) Let Dec be the effective procedure with $Dec(j) = G_j$ for all $j \geqslant 0$. Consider the following computable function:

$$\varphi(i, g) = \begin{cases} 1 & \text{if } \Phi_i(i) > g \text{ or} \\ & \quad (\Phi_i(i) < g \text{ and } Dec(\Phi_i(i)) \text{ is not a minor of } Dec(g)), \\ 0 & \text{otherwise, i.e. if } \Phi_i(i) = g \text{ or} \\ & \quad (\Phi_i(i) < g \text{ and } Dec(\Phi_i(i)) \text{ is a minor of } Dec(g)). \end{cases}$$

The classes $C_i = \{Dec(g) \mid \varphi(i, g) = 1\}$ $(i \geqslant 0)$ are minor-closed and can be efficiently described by a machine for their recognition problem. Note that $Obstr(C_i) = \emptyset$ when $\Phi_i(i)\uparrow$, and $Obstr(C_i) = \{Dec(\Phi_i(i))\}$ otherwise. It follows that $Obstr(C_i) \neq \emptyset$ if and only if $i \in K$, with $K = \{i \mid \varphi_i(i)\downarrow\}$. As K is not recursive, there can be no effective procedure to construct the obstruction sets of the C_i. □

Despite this result, the theorem of Robertson & Seymour can be extremely useful for concluding that a polynomial-time recognition or decision algorithm must exist for some graph property P. All it requires is a verification that the class $C = \{G \mid G$ satisfies $P\}$ or its complement is minor-closed. We state the result, together with two further intriguing claims by Robertson & Seymour and Fellows & Langston respectively.

1.22. THEOREM. *Let C be any class of graphs.*

(i) *If C is minor-closed, then its recognition problem can be solved by an* $O(n^3)$ *time bounded algorithm.*

(ii) *If C is a minor-closed and $\bar{C}$ contains at least one planar graph, then the recognition problem for C can be solved by an* $O(n^2)$ *time bounded algorithm.*

(iii) *If C is minor-closed and $\bar{C}$ contains at least one cycle, then the recognition problem from C can be solved by an* $O(n)$ *time bounded algorithm.*

It appears that in many practical cases the theorem actually leads to a constructive result, but only after additional considerations (cf. Bodlaender, Fellows & Langston). There is a further development which has important algorithmic consequences as well.

DEFINITION. Let $G=\langle V, E\rangle$ be a graph.

(i) *A tree decomposition* of G is any pair $\langle\{X_i | i\in I\}, T\rangle$ with $\{X_i | i\in I\}$ a family of subsets of V and $T=\langle I, F\rangle$ a tree with the elements of I as nodes, such that the following properties hold:

- $\bigcup_{i\in I} X_i = V$,
- for every edge $e=(u, v)\in E$ there is an $i\in I$ such that $u\in X_i$ and $v\in X_i$, and
- for all $i, j, k\in I$, if j lies on the path between i and k in T, then $X_i\cap X_k\subseteq X_j$.

The *width* of a tree decomposition $\langle\{X_i | i\in I\}, T\rangle$ is equal to $\max_{i\in I}\{|X_i|-1\}$.

(ii) The *treewidth* of G, denoted by treewidth(G), is the minimum value of k such that G has a tree decomposition of width k.

DEFINITION. A class of graphs C is said to be of *bounded treewidth* if there exists a $k\geqslant 0$ such that all graphs $G\in C$ have treewidth$(G)\leqslant k$.

The notion of (bounded) treewidth again derives from the work of Robertson & Seymour, and is interesting for a number of reasons. First of all it ties in with the results for minor-closed classes outlined before. The following theorem combines results of Robertson & Seymour and of Fellows & Langston.

1.23. THEOREM. *Let C be any class of graphs.*

(i) *If C is minor-closed and $\bar{C}$ contains at least one planar graph H, then C has bounded treewidth (and the treewidth of all elements of C is bounded by a constant depending on H).*

(ii) *If C is minor-closed and $\bar{C}$ contains at least one cycle of length $k\geqslant 3$, then C is of bounded treewidth (and the treewidth of all elements of C is bounded by $k-1$).*

In addition to this result, it appears that many well-known classes of graphs that have interested researchers in the past actually are of bounded treewidth (cf. Bodlaender, see Table 2).

Bodlaender has shown that there is a concrete $O(n^2)$ algorithm to decide for arbitrary graphs G whether treewidth$(G)\leqslant k$ for any fixed k. The algorithm constructs a tree decomposition of width $\leqslant k$ for G when treewidth$(G)\leqslant k$.

In the context of this chapter the notion of treewidth is important, because it appears that many "hard" (e.g. NP-complete) problems can be solved efficiently for any class of graphs of bounded treewidth. A similar phenomenon was first observed for classes of graphs that are *partial k-trees* (cf. Arnborg) which arose from different, algorithmic considerations.

Table 2
Some classes of graphs of bounded treewidth

Class of graphs	Bound on treewidth
trees/forests	1
almost trees (k)	$k+1$
partial k-trees	k
bandwidth-k	k
cutwidth-k	k
planar with radius k	$3k$
series parallel	2
outerplanar	2
Halin	3
k-outerplanar	$3k-1$
chordal with maximal clique size k	$k-1$
undirected path with maximal clique size k	$k-1$
directed path with maximal clique size k	$k-1$
interval with maximal clique size k	$k-1$
proper interval with maximal clique size k	$k-1$
circular arc with maximal clique size k	$2k-1$
proper circular arc with maximal clique size k	$2k-2$

DEFINITION. Let $G=\langle V, E\rangle$ be a graph, and $k \geqslant 1$.

(i) G is called a *k-tree* if either G is a complete graph on k nodes or G has a node $u \in V$ of degree k such that $E(u)=\{v \in V \mid v \text{ is adjacent to } u\}$ is a size-k clique of G and the graph obtained by deleting u and all its incident edges from G is again a k-tree. (The node u is called a simplicial node of G.)

(ii) G is a *partial* k-tree if G is a subgraph of a k-tree.

1.24. LEMMA. *Let $G=\langle V, E\rangle$ be a graph with $|V| \geqslant k$. Then G is a partial k-tree if and only if* treewidth$(G) \leqslant k$.

PROOF. Let G be a partial k-tree. Without loss of generality we may assume that G is a k-tree (because treewidth can only decrease when we take subgraphs). We prove by induction that G has a tree decomposition $\langle\{X_i \mid i \in I\}, T\rangle$ in which $|X_i| \leqslant k+1$ for every $i \in I$ and every size-k clique occurs as a subset of some X_i. The hypothesis holds trivially when G is a complete graph on k nodes (take $I=\{0\}$ and $X_0=V$). Thus let $|V| \geqslant k+1$ and let u be a simplicial node of G. Let $E(u)=\{v \in V \mid v \text{ is adjacent to } u\}$. By induction, there exists a tree decomposition $\langle\{X_i \mid i \in I\}, T\rangle$ with the desired property for $G-\{u\}$. As $E(u)$ is a size-k clique, there exists a node $i \in T$ such that $E(u) \subseteq X_i$. Allocate a "new" node i' and attach it to i in T, and let $X_{i'}=E(u) \cup \{u\}$. The result is a tree decomposition for G with the desired properties.

Conversely, let G be a graph with treewidth$(G) \leqslant k$. Without loss of generality we may assume that treewidth$(G)=k$. Thus let $\langle\{X_i \mid i \in I\}, T\rangle$ be a tree decomposition of G with $\max_{i \in I}|X_i|=k+1$. We prove by induction (on T) that G is the subgraph of a k-tree on V in which every subset of at most k elements of an X_i $(i \in I)$ is part of a size-k clique. Starting with the fullest node of T, this follows trivially for the part of G that it contains (take the complete graph on $k+1$ nodes, which is a k-tree). Let i be a leaf of T.

By induction, the graph G' that corresponds to the tree decomposition with i (and X_i) removed from T is the subgraph of a k-tree S with the desired properties. Let i be attached to j in T, and let $X_i \cap X_j = X'_i$ and $X_i - X'_i = X''_i$. If $X''_i = \emptyset$ then $X_i = X_j$ and $G = G'$. (Note that the edges of G are specified implicitly with the decomposition.) If $X''_i \neq \emptyset$ then we note that its elements cannot occur in any other node of T. Also X'_i is a proper subset of X_i and hence a subset of at most k elements of X_j, and by induction it follows that X'_i is part of a size-k clique Y of S. Add X''_i to S one node at a time according to the following method:

while $X''_i \neq \emptyset$ **do**
begin
  select a node $u \in X''_i$;
  $X''_i := X''_i - \{u\}$;
  add u to S and include edges between u and all nodes of Y;
  select a node $v \in Y$ such that $v \notin X_i$ (if it exists);
  if v is defined **then** $Y := (Y - \{v\}) \cup \{u\}$
end.

The resulting graph S' is again a k-tree, and one easily verifies that the entire set X''_i is added together with all possible edges that can exist between nodes in $X_i = X'_i \cup X''_i$. Thus G is a subgraph of S' and S' has the desired properties w.r.t. the tree decomposition. □

The notion of (bounded) treewidth thus seems to be very natural in capturing a general characteristic that makes "hard" problems feasible for special classes of graphs. The underlying intuition is that many computational problems for graphs are easily solved for trees, and that graphs of bounded treewidth sufficiently resemble trees when given by their tree decomposition, so a similar solution strategy will work. Both Bodlaender and Arnborg, Lagergren & Seese, among others, provide general frameworks for efficiently (i.e., polynomial-time) solvable problems on graphs of bounded treewidth and give long lists of problems for which concrete polynomial-time and often even O(n) algorithms result for any class of graphs of bounded treewidth or, alternatively, for any class of partial k-trees for fixed k. The problems include e.g. VERTEX COVER, DOMINATING SET, CHROMATIC NUMBER, ACHROMATIC NUMBER, FEEDBACK VERTEX SET, MINIMUM MAXIMAL MATCHING, PARTITION INTO FORESTS, COVERING BY COMPLETE BIPARTITE SUBGRAPHS, CLIQUE, INDEPENDENT SET, HAMILTONIAN COMPLETION, HAMILTONIAN CIRCUIT, PLANAR SUBGRAPH, LONGEST PATH, DISJOINT CONNECTING PATHS for fixed k, and THICKNESS $\leqslant k$ for fixed k. (The problems are named as in Garey & Johnson.) An interesting, foundational view of the notion of treewidth is given by Courcelle.

2. Basic structure algorithms

Graph theory provides us with a wealth of results about the structure of graphs. In graph algorithms the aim is to identify substructures or properties algorithmically, by a program that can be run on every admissible input graph. Thus, the theory of graph algorithms will answer questions like "does a graph G have property P?" by providing an algorithm that tests graphs for property P, instead of by theorems about property P alone. The theory is sophisticated because of the concern for algorithms of a provably low worst-case or average (expected) complexity. To reach low complexity, graph algorithms usually exploit theorems from graph theory (or finds them, when they are lacking). In this section we review the essential structure algorithms for general, weighted and unweighted graphs.

2.1. *Connectivity*

Two nodes i, j of G are said to be k-connected if k is the largest integer such that there exist k node-disjoint paths from i to j in G. We denote the connectivity of i and j by $cn(i,j)$. By Menger's theorem $cn(i,j)$ is equal to the minimum size of any set of nodes S whose removal disconnects i and j. (If i and j are adjacent, we take $cn(i,j)=n-1$.) Define the *connectivity* $cn(G)$ of G by $cn(G)=\min_{i,j} cn(i,j)$. It is common to study the question: given some k, is $cn(G)\geqslant k$? The question is relevant in networks where one wants to know for each pair of nodes a guaranteed bound on the number of alternate routes between the nodes.

For $k=1$ the problem reduces to the question of whether G is connected. By DFS this is easily answered in $O(n+e)$ steps. One of the classical results in graph algorithms is the fact that for $k=2$ and $k=3$ the problem can be solved in $O(n+e)$ steps also. We only look at the case $k=2$ and assume without loss of generality that G is connected. Do DFS starting at some i_0 and number nodes consecutively the first time they are visited. Let $v(i)$ be the number assigned to node i, $v(i_0)=1$. Let T be the DFS tree, with all green edges (see Section 1.3) For any j, define $LOW(j)=\min\{v(i) \mid i$ is reachable from j by a downward path of green edges followed by at most one red edge$\}$. Clearly $LOW(i_0)=1$ and $LOW(j)\leqslant v(j)$ for all j.

2.1. LEMMA. (i) *i_0 is a cutpoint of G if and only if the degree of i_0 in T is at least* 2.
(ii) *For all $j\neq i_0$, j is a cutpoint of G if and only if it has a son j' in T with $LOW(j')\geqslant v(j)$.*

PROOF. (i) Let i_0 be a cutpoint and let x, y be two nodes such that every path from x to y passes through i_0. Then x and y cannot be in the same subtree under i_0, and i_0 has degree $\geqslant 2$. Conversely, take any two nodes x, y in separate subtrees under i_0. Because red edges cannot reach across subtrees (cf. Section 1.3), every path from x to y must pass through i_0. Hence i_0 is a cutpoint.

(ii) Let $j\neq i_0$ be a cutpoint, and let x and y be two nodes such that every path from x to y passes through j. At least one of x, y must belong to a subtree under j, say y. If for every subtree under j there is at least one red edge connecting it to some ancestor of j, then every subtree can be reached from x by a path that avoids j. As this is impossible, there must be one subtree with no red edges leading to an ancestor of j. Its root j' is a son of j and satisfies $LOW(j')\geqslant v(j)$. The converse is easy, as any path connecting i_0 and j' must pass through j. □

Observe that the numbers $v(i)$ are easily assigned while DFS proceeds, and $LOW(j)$

can be computed by the time the last subtree under j has been fully explored and DFS backtracks to the father of j in T. Using the lemma, all cutpoints can be identified during DFS with only little extra work and (hence) $cn(G) \geq 2$ if in fact no cutpoint is found. The algorithm can be modified to find all biconnected components in $O(n+e)$ steps as well: stack edges when they are traversed for the first time, and output and pop whatever edges still are on the stack and were put on it after j was visited as a component, whenever j is identified as a cutpoint. A more complex modification of DFS generates all triconnected components in $O(n+e)$ time as well. Kanevsky & Ramachandran proved that 4-connectivity can be tested in $O(n^2)$ time. An interesting (and practical) problem is to determine the smallest number of edges that must be added to a graph in order to make it k-connected. The problem can be solved by an $O(n+e)$ algorithm for $k \leq 2$ (see e.g. Eswaran & Tarjan and Rosenthal & Goldner), but is largely open for other values of k (see e.g. Watanabe & Nakamura).

Testing whether $cn(G) \geq k$ for $k \geq 5$ is considerably harder. Some algorithms compute $cn(G)$ exactly using the following observation.

2.2. LEMMA. *For every i, j the value of $cn(i, j)$ can be computed by solving an integer $\{0, 1\}$ maximum flow problem on a graph of $O(n)$ nodes and $O(e)$ edges.*

Suppose we have an $O(n^{\alpha} e^{\beta})$ time maxflow algorithm for $\{0, 1\}$ networks (take e.g. $\alpha = \frac{1}{2}$ and $\beta = 1$, see Section 3.4). Write $c = cn(G)$. Assume G is not a complete graph, hence $c \leq n-2$. Let i_0 be a node that is not connected to every other node and let $i_0, i_1, \ldots$ be some fixed listing of the nodes. Consider the following algorithm:

$\gamma_{-1} := n-2$;
$t := -1$;
repeat
  $t := t+1$;
  compute $cn(i_t, i)$ for every $i \in \{i_{t+1}, \ldots, i_{n-1}\}$ and
  set γ_t equal to the minimum of the computed values
  and γ_{t-1}
until $t \geq \gamma_t$.

Clearly, $c \leq \gamma_t \leq \gamma_{t-1} \leq n-2$ for all $t \geq 0$ and, because t increases in every iteration, the algorithm is guaranteed to terminate. Suppose it terminates when t has value k, i.e., right after node i_k was considered. Then $c \leq \gamma_k \leq k$. Let x, y be two nonadjacent nodes with $cn(x, y) = c$ and S be a set of c nodes whose removal disconnects x and y. Because $k \geq c$ there must be an i_j with $0 \leq j \leq k$ such that $i_j \notin S$ and, because S "separates" the graph, there must be a node i such that the removal of the nodes in S in fact disconnects i_j and i as well. It means that $cn(i_j, i) \leq c$ (hence $cn(i_j, i) = c$) and γ_j must have been set to c during the iteration with $t = j$. Consequently $\gamma_t = c$ when the algorithm terminates and termination occurs exactly for t with $t = c$. The jth iteration costs (at most) n maxflow computations and thus takes $O(n^{\alpha+1} e^{\beta})$ time. The complete algorithm computes $c = cn(G)$ in $O(c \cdot n^{\alpha+1} e^{\beta})$ time. (Note that $c \leq 2e/n$, because every node must have degree $\geq c$.) By a different algorithm the question "is $cn(G) \geq k$?" can be solved in $O(k^3 e + kne)$ time for every fixed k.

An interesting probabilistic approach to determining $c = cn(G)$ was suggested by Becker et al. It is based on the following observation.

2.3. LEMMA. *Let* $c = cn(G) \leqslant n-2$, $\varepsilon > 0$ *and* $r = \log(1/\varepsilon)/(\log n - \log c)$. *Draw r random nodes* $i_1, \ldots, i_r$ *of G and determine* $c' = \min_{1 \leqslant j \leqslant r} \min_v cn(i_j, v)$, *where the inner "min" is taken over all v not adjacent to* i_j. *Then* $\Pr(c' > c) \leqslant \varepsilon$.

PROOF. Let S be any node separator with $|S| = c$. For all nodes $i \in V - S$ one has $\min_v cn(i, v) = c$. (Note that S separates the graph in at least two components. For any node v in a component different from the one containing i one has $cn(i, v) \leqslant |S| = c$, hence $cn(i, v) = c$.) Thus for c' to be strictly greater than c, all random nodes that were drawn must belong to S. It follows that $\Pr(c' > c) \leqslant (c/n)^r \leqslant \varepsilon$. □

The lemma shows that c can be computed with error probability $\leqslant \varepsilon$ by minimizing over the values $\min_v cn(i, v)$ for a sufficiently large random sample of nodes i. By solving $O(n)$ maximum flow problems (Lemma 2.2), each $\min_v cn(i, v)$ value can be computed in $O(n^{\alpha+1} e^{\beta})$ time. Yet the minimum sample size r we need is hard to determine, because we need c itself to compute it. A (poor) upper bound can be obtained by setting c' to $n-2$ and substituting c' for c in the formula for r. Now observe that by increasing the sample size, the value for c' can only decrease (approaching c with high probability for a sample of size approaching r) and the corresponding estimate for r can only decrease as well, approaching the correct value of r with high probability. One can show that the expected number of random drawings needed for the sample to reach the size estimated as the algorithm goes on is $O(r + \varepsilon n)$. Thus for $\varepsilon \leqslant 1/n$ the algorithm computes $cn(G)$ with error probability $\leqslant \varepsilon$ in only

$$O(rn^{\alpha+1} e^{\beta}) = O\left(\frac{\log(1/\varepsilon)}{\log n - \log c}\, n^{\alpha+1} e^{\beta}\right)$$

time, which can be significantly faster than the $O(cn^{\alpha+1} e^{\beta})$ algorithm discussed above. In concrete terms it runs in $O(\log(1/\varepsilon) \cdot n^{3/2} e)$ expected time whenever $e \leqslant \frac{1}{2}\delta n^2$ for a fixed $\delta < 1$, as in that case $c \leqslant 2e/n \leqslant \delta n$. A different approach by Linial et al. yields a "Monte Carlo type" probabilistic algorithm to determine c in $O(c^{2.5} n + n^{2.5})$ expected time with error probability $< 1/n$ and a "Las Vegas type" probabilistic algorithm to determine c in $O(c^{3.5} n + cn^{2.5})$ expected time. (A "Las Vegas type" probabilistic algorithm always gives a value with no probability of error.)

Two nodes i, j of G are said to be *k-edge-connected* if k is the largest integer such that there exist k edge-disjoint paths from i to j in G. We denote the edge connectivity of i and j by $ecn(i, j)$. By Menger's theorem $ecn(i, j)$ is equal to the minimum size of any set of edges whose removal disconnects i and j. Define the edge connectivity $ecn(G)$ by $ecn(G) = \min_{i,j} ecn(i, j)$. Using network flow techniques the edge connectivity $\lambda = ecn(G)$ of a directed graph G can be determined in $O(ne \min\{n^{2/3} e^{1/2}\})$ time (Even & Tarjan) and in (λne) time (Schnorr). Ma & Chen have improved the Even–Tarjan bound to

$$O(ne \min\{d_{\min}, n^{2/3} e^{1/2}\})$$

by a suitable modification of the algorithm, with $d_{\min}$ the smallest degree of any node in the graph. Matula has shown that there exists an $O(kn^2)$ algorithm to decide the question "is $ecn(G) \geqslant k$?" for undirected graphs G, which leads to an $O(\lambda n^2)$ algorithm to determine the edge connectivity of an undirected graph. Extending the algorithm, Mansour and Schieber have shown that the edge connectivity of a directed graph can be determined in $O(\lambda^2 n^2)$ time. Watanabe and Nakamura have given a polynomial-time algorithm to determine the smallest number of edges (and, in fact, a concrete minimum set of edges) that must be added to a graph in order to make it k-edge-connected for any specified $k \geqslant 2$.

2.2. *Minimum spanning trees*

Let $G = \langle V, E, w \rangle$ be a weighted graph, with w a function that assigns a weight to every edge. A *minimum spanning tree* (MST) is a spanning tree of G of which the sum of all edge weights is minimum, over all spanning trees. The problem of computing a MST has an interesting history that goes back to 1926. The problem is interesting in many networking contexts. Weights might be distances, traffic densities, costs, etc.

DEFINITION. (i) An *MST family* in G is a collection of disjoint subtrees $F = \{T_1, \ldots, T_k\}$ for which there exists an MST T with $T_i \subseteq T$ for every i, $1 \leqslant i \leqslant k$, and $\bigcup_1^k T_i$ spans G.

(ii) Let $F = \{T_1, \ldots, T_k\}$ be an MST family in G. The *shrinking* of G modulo F, denoted by G mod F, is the graph $\langle V', E', w' \rangle$ with $V' = \{1, \ldots, k\}$, $E' = \{(i, j) \mid$ there is an edge connecting T_i and T_j $(i \neq j)\}$ and for every edge $(i, j) \in E'$, $w'(i, j)$ is the weight of the least weighted edge connecting T_i and T_j.

For every MST family F, G mod F can be constructed in $O(n+e)$ time. All MST algorithms start from the trivial MST family consisting of all single nodes and try to reduce it to a one-element MST family by a suitable process of combining trees and shrinking, i.e., reducing the problem to G mod F. The known algorithms differ by the organization of the combining and shrinking steps, and by the datastructures employed to keep the complexity low. The following two lemmas are essential to all MST algorithms.

COMBINING LEMMA. *Let $F = \{T_1, \ldots, T_k\}$ be a MSTfamily in G and $1 \leqslant i \leqslant k$, $k \geqslant 2$. Let the least weighted edge from T_i to another member of F be (x, y) with $x \in T_i$ and $y \in T_j$ $(i \neq j)$. Let T_{ij} be the tree obtained by connecting T_i and T_j by the edge (x, y). Then the collection F' obtained from F by deleting T_i and T_j and adding T_{ij} is another MST family in G (with one element less than F).*

SHRINKING LEMMA. *Let $F = \{T_1, \ldots, T_k\}$ be a MST family in G, and T' a MST of G mod F. Expand T' by replacing every node i by the subtree T_i and every edge $(i, j) \in T'$ by the least weighted edge connecting T_i and T_j. The resulting tree T is a MST of G.*

Several algorithms only use the Combining Lemma:

(A) Boruvka (1926): apply the lemma for every $1 \leqslant i \leqslant k$ simultaneously (assuming all edge weights are different),

(B) Kruskal (1956): take the least weighted edge connecting two distinct subtrees and let i, j be accordingly,

(C) Jarnik (1930), Prim (1957), Dijkstra (1959): take $i=1$ always.

Variant A is of interest for distributed computing. Variant B runs very smoothly after all edges are sorted by increasing weight, in $O(e \log n)$ time. Note that every tree T_i can (also) be represented as a set, and deciding whether (x, y) connects two different subtrees amounts to comparing the outcome of a *Find*(x) and a *Find*(y). Combining the two trees and joining the corresponding sets is a *Union* operation. Thus the nonsorting phase of variant B can be implemented by a sequence of $O(e)$ *Finds* and $n-1$ *Unions*, which can be achieved in $O(e\alpha(n, e))$ time where $\alpha(n, e)$ is a very slowly growing function (practically a constant). Variant C is usually implemented by maintaining a datastructure Q which contains for every node $x \notin T_1$ the least weighted edge connecting x to T_1. Say x is joined to T_1, in the combining step. Then the edge for x must be deleted from Q, and for every neighbor y of x with $y \notin T_1$ we must consider whether its entry in Q must be updated to (y, x) (which leads to at most $\deg(x)$ update operations on Q, hence, to e updates alltogether). Implementing Q by any priority queue, leads to an $O(e \log n)$ time bound for variant C. A more clever priority queue leads to an

$$O(e \log n/\max\{1, \log(e/n)\})$$

time bounded algorithm, which is linear in e whenever $e=\Omega(n^{\alpha})$.

Now consider the following variant. Keep the subtrees of a MST family in a queue with a marker # at the end. Apply the Combining Lemma to whatever T_i appears up front and pull the desired T_j from the row, put the resulting T_{ij} at the end of the queue and proceed until the # appears up front. Now move the marker to the end of the queue and continue with another phase, unless there is only one tree left (which must be a MST). Since every node occurs precisely once before the # at the beginning of a phase, the process of finding the least cost edge out of T_i (for every T_i that appears up front in this phase) inspects every node once. Let the edges incident to every node x be divided into $\lceil \deg(x)/f \rceil$ groups of size f and sort every group, at a total cost of about

$$O\left(\sum_x (1+\deg(x)/f) f \log f\right) = O(nf \log f + e \log f).$$

Clearly this is done before phase 1. The least weight edge incident to x can now be found by inspecting every group, throwing away the edges that do not lead out of T_i, and minimizing over $\leqslant 1+\deg(x)/f$ edges that do. This leads to a cost of $O(n+e/f)$ per phase, and a total term of $O(e)$ for eliminating edges. By induction one easily proves that in phase l, every T_i in the queue has $\geqslant 2^l$ nodes. Thus there are (at most) $\log n$ phases, and the complexity of the algorithm after s phases is bounded by $O(nf \log f + ns + e \log f + e \cdot s/f)$. Here is how the lemma comes in. After $\log\log n$ phases we have spent $O(e \log\log n)$ time (take $f=1$) and the MST family F has trees of size $\geqslant 2^{\log\log n} = \log n$ each. $G \bmod F$ is a graph of $\leqslant n/\log n$ nodes and $\leqslant e$ edges, and by running the same algorithm on it for the full $O(\log n)$ phases we obtain a MST for $G \bmod F$ in $O(e \log\log n)$ time as well (take $f=\log n$). Using the Shrinking Lemma, it follows that a MST is obtained in $O(e \log\log n)$ time.

For special graphs a better bound may be obtained. For example, by shrinking after every phase a MST of a planar graph can be obtained in O(n) steps. Also, there is still a lot of freedom left in the algorithm for the general case. By using another organization of the steps and refined data structures, the complexity of the MST problem can be reduced to O($e\beta(n,e)$) with $\beta(n,e)=\min\{i \mid \log^i n \leqslant e/n\}$ (a very slowly growing function) (Fredman and Tarjan) and even to O($e \log \beta(n,e)$) (Gabow et al.). Note that $\beta(n,e) \leqslant \log^* n$ for $e \geqslant n$.

An interesting problem concerns the "maintenance" of minimum spanning trees. Spira & Pan showed that a minimum spanning tree can be updated in O(n) time when a new node is inserted in the graph. Chin & Houck developed an O(n^2) algorithm for handling deletions. Frederickson improved the bound to O($\sqrt{e}$) for inserting or deleting an edge or changing its weight, and to O($\log^2 e$) for planar graphs.

Sometimes it is of interest to have a number of pairwise edge-disjoint spanning trees of a graph (if they exist), typically with the condition that the total costs of the spanning trees is minimized. Roskind and Tarjan presented an O($e \log e + k^2 n^2$) algorithm to find k edge-disjoint spanning trees of minimum total cost. They show that the maximum k for which there exist k edge-disjoint spanning trees in the graph can be determined in O(e^2) time.

2.3. *Shortest paths*

Let $G=\langle V, E, w\rangle$ be a weighted, directed graph. For $s, t \in V$ we let $l(s,t)$ be the weight of a least weight path from s to t and π_{st} a path of weight $l(s,t)$ from s to t, provided such a path exists. (The weight of a path is the sum of the weights of the edges of the path.) If $w(x,y)=1$ for every $(x,y) \in E$, then a least weight path is a "shortest" path in the traditional sense, but we will use the phrase for the general case as well. For $A, B \subseteq V$ the *(A, B)-shortest path problem* asks for the shortest path (and its length) from s to t, for every $s \in A$ and $t \in B$. The problem is well-defined if the following assumptions are in effect:

(∗) for every $s \in A$ and $t \in B$ there is a path from s to t, and
(∗∗) no path from s to t contains a cycle of negative weight (a "negative" cycle).

Traditionally the $(\{s\}, V)$- and (V, V)-shortest path problems have received most attention, and we will restrict ourselves to it here. The problems are known as the "single source" and the "all-pairs" shortest path problem, respectively.

2.3.1. *Single-source shortest paths*

Let s be a fixed source node and let (∗), (∗∗) be in effect. Write $l(t)$ for $l(s,t)$, $l(s)=0$. Define a shortest path tree (*SP-tree*) to be any spanning tree rooted at s such that the tree path from s to t is a shortest path from s to t in G for every $t \in V$.

2.4. THEOREM. *With (∗), (∗∗) in effect, there exist SP-trees for every s. If the values $l(t)$ are known ($t \in V$), then a SP-tree can be constructed in* O(e) *time.*

The theorem shows that we can restrict the problem to computing the weights $l(t)$. In order to do so, it is common to view the $l(t)$ as solutions to a set of equations, known as

Bellman's equations:

$$u_s = 0,$$
$$u_t = \min_{x \neq t}\{u_x + w(x,t)\} \quad \text{for } t \neq s.$$

(the u_t for $t \neq s$ are the unknowns). It can be shown that when G has no cycles of weight $\leqslant 0$, then the solution $l(t)$ to Bellman's equations is unique. Otherwise it is unique under the additional assumption that $\sum_t u_t$ is maximal, over all solutions. Thus the shortest path lengths are the solution to the following linear programming problem (set $w(x,y) = \infty$ for $(x,y) \notin E$):

$$\text{maximize: } \sum_t u_t$$
$$\text{subject to: } u_s = 0,$$
$$u_t \leqslant u_x + w(x,t) \quad \text{for } t \neq s, x.$$

We will not explore this connection to linear programming, but several solution methods of Bellman's equation directly fit in with this theory. Nevertheless, the characterization suggests the following general strategy for computing a solution and a SP-tree (with $f(t)$ pointing to the father of t):

initialize: set $u_s = 0$, $u_t = \infty$ for $t \neq s$, $f(t) = \textbf{nil}$ for all t.

algorithm: **while** not all inequalities are satisfied **do**
begin
  scan: determine a $t \neq s$ for which there is an $x \neq t$ with $u_t > u_x + w(x,t)$;
  label: set u_t to $u_x + w(x,t)$ and $f(t)$ to x
end.

The algorithm is due to Ford and known as the "labeling (and scanning) method". It can be shown that the algorithm converges regardless of the choice made in the scanning step, provided (*) and (**) are in effect. Some scanning orders may be better than others however, and this distinguishes the various algorithms known for the problem. (Some scanning orders may lead to exponential-time computations.) A simple example is provided by the acyclic graphs. Let an acyclic graph be given in topologically sorted order. Considering the nodes t in this order in Ford's algorithm must produce the (final) value of u_t when its turn is there. The algorithm will finish in $O(e)$ steps.

Another efficient scanning order exists when $w(x,y) \geqslant 0$, although it takes more work per step to pull out the right node t (i.e., out of a suitable datastructure). There will be a set F of nodes t for which u_t already has a final value, and a set I of nodes t for which u_t is not known to be final. Initially, $F = \{s\}$ and $I = V - \{s\}$. The algorithm is slightly changed as follows:

initialize: set $u_s = 0, u_t = w(s,t)$ for $t \neq s$, $f(t) = s$ for all t;

algorithm: **while** I is not $\emptyset$ **do**
begin
scan: choose $t \in I$ for which u_t is minimal;
$F := F \cup \{t\}$; $I := I - \{t\}$;
update: for every neighbor x of $t, x \in I$, set u_x to $\min\{u_x, u_t + w(x,t)\}$ and set $f(x)$ to t when $u_t + w(x,t)$ was smaller than u_x
end.

The algorithm is due to Dijkstra, and operates on the observation that every node closest to s by a path through the current fragment of the final SP-tree (F) must be exposed in the scanning step. The algorithm is easily seen to require $O(e)$ steps, except for the operations for maintaining a priority queue of the u_x-values (selection and deletion of an element in the scanning step, and $\deg(t)$ priority updates in the update step). A simple list will lead to $O(n)$ steps per iteration, thus $O(n^2)$ total time, and a standard priority queue to $O(\deg(t) \cdot \log n)$ steps per iteration, hence $O(e \log n)$ steps total. By using a refined datastructure this can be improved to

$$O(e \log n / \log \max\{2, e/n\})$$

steps, and even to $O(e + n \log n)$ steps (Fredman and Tarjan). When the edge weights all belong to a fixed range $0, \ldots, K$, the algorithm can be implemented to run in

$$O(\min\{(n+e)\log\log K,\ nK + e\}).$$

Assuming (*) and (**), there is a good scanning order for general graphs as well: assume a fixed ordering of the nodes $\neq s$, and make consecutive passes over the list. Observe that it takes $O(\deg(t))$ time to see if a node t must be selected and its u_t updated accordingly, and one pass takes $O(e)$ time. By induction one sees that in the kth pass all nodes t for which a shortest path from s exists with k edges, must get their final u_t-value. Thus in $O(n)$ passes all u-values must converge, and the algorithm finishes in $O(ne)$ time. Observe that for planar graphs this gives an $O(n^2)$ algorithm, because $e \leqslant 3n - 6$ (for $n \geqslant 3$). By a partitioning technique based on the planar separator theorem, this can be improved to $O(n^{1.5} \log n)$ time for planar graphs (Mehlhorn & Schmidt). Frederickson even obtained an $O(n\sqrt{\log n})$ algorithm for the single-source shortest path problem for planar graphs with nonnegative edge weights. He also proved that a planar graph can be preprocessed in $O(n \log n)$ time such that each subsequent single-source shortest path computation can be performed in only $O(n)$ time. For general graphs an $O(hn^2)$ algorithm has been shown, where h is the minimum of n and the number of edges with negative weight (Yap).

Another interesting problem concerns finding shortest paths of even or odd length between two specified vertices s and t. Derigs shows that this can be solved in $O(n^2)$ time, or in $O(e \log n)$ time when suitable datastructures are used.

2.3.2. All-pairs shortest paths

The all-pairs shortest path problem has been subjected to a wide variety of solution methods, and illustrates better than any other graph problem the available techniques in the field. We assume throughout that the conditions (*) and (**) are in effect, in particular we assume that there are no negative cycles in the graph. One approach to the all-pairs shortest path problem is to solve n single-source shortest path problems, one problem for every possible source. The resulting complexity bounds are $O(ne+n^2\log n)$ for graphs with nonnegative edges and $O(n^2e)$ for general graphs, using the results from the previous section. For planar graphs Frederickson proved an $O(n^2)$ bound. By Theorem 2.4 it takes another $O(ne)$ time to construct all SP-trees.

By an ingenious device, Edmonds & Karp observed that the all-pairs shortest path problem for general graphs can be reduced to the problem for graphs with nonnegative edge weights. Let $f: V \to \mathbb{R}$ be a function such that for all $x, y \in V$, $f(x)+w(x, y) \geqslant f(y)$ and consider the graph $G' = \langle V, E, w' \rangle$ with $w'(x, y) = f(x)+w(x, y)-f(y)$. One easily verifies that $l'(s, t) = l(s, t)+f(s)-f(t)$ for $s, t \in V$, where $l'(s, t)$ is the length of a shortest path from s to t in G'. As G' has only nonnegative weights, the all-pairs shortest path problem is solved in $O(ne+n^2\log n)$ steps for G. A suitable function f can be obtained by considering the graph G'' obtained from G by adding a (new) source z and edges (z, x) for all $x \in V$, and taking $f(x) = l''(x)$ (the length of the shortest path from z to x in G''). Computing f takes one single-source shortest path problem computation on a general graph, and (thus) no more than $O(ne)$ time. The total computation still takes no more than $O(ne+n^2\log n)$ time.

A variety of methods for the all-pairs shortest path problem is based on the use of a variant of the adjacency matrix A_G defined by $A_G(i, j) = w(i, j)$ for $i, j \in V$, with $w(i, i) = 0$. (We assume that the nodes are numbered $1, \ldots, n$ here.) Consider matrices $A = (a_{ij})$, $B = (b_{ij})$ over the reals and define $A \times B = C = (c_{ij})$ to be the matrix with $c_{ij} = \min_{1 \leqslant k \leqslant n} (a_{ik}+b_{kj})$, a term reminiscent of matrix product with operations min and $+$ for $+$ and $\cdot$. It can be shown that all algebraic laws for matrix multiplication hold for the (min, $+$) product. Observe that $A_G^k = A_G \times \cdots \times A_G$ (k times) is the matrix with $A_G^k(i, j)$ equal to the length of the shortest path from i to j with $\leqslant k$ edges. Define the transitive closure of any matrix A as $A^* = \min\{I, A, A, A^2, A^3, \ldots\}$ with I the matrix with 0 on the main diagonal and ∞ elsewhere, and min taken componentwise. Clearly, A_G^* is the matrix with $A_G^*(i, j) = l(i, j)$ and, in the absence of negative cycles, $A_G^*(i, j) = A_G^{n-1}(i, j)$.

The conclusions in the preceding paragraph give rise to several interesting algorithms and complexity considerations. For example, the Floyd–Warshall algorithm for computing transitive closures can be adapted to compute the all-pairs shortest paths in $O(n^3)$ steps. A beautiful result of Romani shows that every $O(n^\alpha)$ matrix multiplication algorithm (over rings) can be adapted to compute the transitive closure of any A_G in $O(n^\alpha)$ steps, provided A_G has elements from $\mathbb{Z} \cup \{\infty\}$. Watanabe (with corrections) has shown that in this case a suitable representation of the actual shortest paths can be computed in $O(n^{\alpha})$ time as well. The representation is a matrix $\tilde{A}_G$ with $\tilde{A}_G(i,j)$ equal to a node $x \neq i,j$ on a shortest path from i to j.

Several algorithms have been proposed that compute the all-pairs shortest path information probabilistically fast, over a large class of random graphs. The best complexity bound so far is $O(n^2 \log n)$ expected time (Moffat and Takaoka).

2.4. *Paths and cycles*

For most of this section we assume that G is an (unweighted) undirected graph, although for most problems discussed there also is a "directed" version. Whenever we speak of a path and a cycle, we shall mean a simple path and a simple cycle (or circuit) respectively. Paths and cycles have been a predominant issue in the analysis of graphs for ages. Consequently we can touch only on a number of topics under this category.

2.4.1. *Paths of length k*

Probably the simplest question is, given k, whether there exists a (simple) path of length k from i and j. Algorithmically the problem can be solved in $O(k(n-2)^{k-1})$ time by considering all possible choices of $k-1$ intermediate nodes, and verifying that a sequence is a path. An efficient solution is surprisingly hard to obtain. In fact, with k arbitrary the problem is NP-complete and (thus) probably not polynomial-time computable. Monien has proved the following result.

2.5. THEOREM. *Let G be a (directed or undirected) graph, $k \geqslant 0$. Then a path of length k can be computed for all pairs i, j (with some default indication for the pairs for which no path of length k exists) in time* $O(k!ne)$.

By applying the theorem, one can find a longest path in G in $O((\mu+1)!ne)$ time with μ the length of the longest path: search for paths of length k for $k=1, 2, 3, \ldots$ until the last value for which a path is found. The longest path problem can be solved in polynomial time for the class of outerplanar graphs (Ellis et al.).

For any fixed k there exists an $O(n)$ time algorithm for deciding whether a graph has a (simple) path of length $\geqslant k$ or not. This follows from a result of Fellows & Langston (cf. Section 1.6), by observing that $C = \{G \mid G \text{ has no simple path of length } \geqslant k\}$ is minor-closed and excludes a length-$(k+1)$ cycle. Bodlaender gives an explicit $O(2^k k!n)$ algorithm.

2.4.2. *Disjoint paths*

We shall only consider vertex-disjoint paths, although the case of edge-disjoint paths is interesting as well. The problem of finding a maximum set of vertex-disjoint paths between two nodes i and j is of obvious interest in many network problems. The (few) algorithms for the problem make use of the connection to a 0–1 maximum flow problem, which makes it polynomial in n and e.

2.6. LEMMA. *The number of vertex-disjoint paths between i and j may be found as the maximum flow in an augmented (directed) graph G^+ with $2n$ nodes, $2e+n$ edges, and edge capacities* 1.

PROOF. Let $G^+ = \langle V^+, E^+, c\rangle$ be the (directed) graph defined as follows. For every $x \in V$ let there be two nodes $x', x'' \in V^+$ and an edge $(x'', x') \in E^+$. For every (undirected) edge $(x, y) \in E$ let there be two edges $(x', y''), (y', x'') \in E^+$. All edges $\alpha \in E^+$ have capacity $c(\alpha) = 1$. If there are k vertex-disjoint paths from i to j in G, then it is seen that there can be a flow of k from i to j in G^+. Consider a maximum flow in G^+. By the integer maxflow theorem there is a maximum flow that is integral on all edges, hence 0 or 1 in this case. By tracing the 1s that flow out of i, one necessarily obtains maxflow vertex-disjoint traces in G^+, which translate to vertex-disjoint paths in G. □

Frisch has exploited the lemma for the construction of a maximum set of vertex-disjoint paths, see also Steiglitz & Bruno. Clearly the algorithm is too complex, e.g. if one only wants to find k vertex-disjoint paths between i and j for some k (provided they exist), but by exploiting the connection to integer flow it should be clear that in this case one only needs to find (and trace) k augmenting paths to get a flow of $\geqslant k$. Suurballe extended the connection to a minimal cost flow problem (see Section 3.5.2) and derived an algorithm to find a set of k vertex-disjoint paths with minimal total length. No complexity analysis was given. Menger's theorem gives a necessary and sufficient condition for the existence of k vertex-disjoint paths.

A simple and practical generalization of the problem is the following: let l pairs $(i_1, j_1), \ldots, (i_l, j_l)$ be given; does G contain l vertex-disjoint paths, one connecting i_k and j_k for every $1 \leqslant k \leqslant l$? (There are many variants, e.g. one may wish to have more than one connecting vertex-disjoint path for every pair.) With G and l arbitrary, the problem is known as the DISJOINT CONNECTING PATHS problem and is NP-complete. (It is a very simple instance of the multicommodity flow problem.) For fixed $l \geqslant 2$ little is known about the complexity of the problem. For $l = 2$ Shiloach obtained an $O(ne)$ algorithm, by a tedious analysis. For planar graphs (and $l = 2$) Krishan et al. gave an $O(n)$ algorithm. The problem is an interesting example where graph theory is advanced because of the need of an efficient algorithm. Shiloach was able to conclude that in every 4-connected nonplanar graph there are always two connecting vertex-disjoint paths for every choice of i_1 and j_1 and i_2 and j_2. Seymour observed that the general problem is equivalent to the following: define a set of edges $E' \subseteq E$ to be G-separable if for each $\alpha \in E'$ there is a (simple) cycle C_α containing α and $C_\alpha \cap C_\beta = \emptyset$ for $\alpha \neq \beta$; under what conditions is a set G-separable? Cypher proved that for $l \leqslant 5$ the problem is solvable in polynomial time for graphs that are $(l+2)$-connected. Robertson & Seymour claim that for any fixed l the problem is indeed solvable by an $O(n^2e)$ algorithm.

2.4.3. *Cycles*

Throughout this section we only consider connected, undirected graphs. The cycle structure of graphs has inspired many studies and analyses, and a variety of types of cycles have been considered. The set of cycles can be viewed as a vector space over the integers mod 2 where "addition" ($\oplus$) is defined as follows: for $G_1 = \langle V_1, E_1\rangle$ and $G_2 = \langle V_2, E_2\rangle$, $G_1 \oplus G_2 = \langle V_1 \cup V_2, (E_1 \cup E_2) - (E_1 \cap E_2)\rangle$. The cycle space of a graph has dimension $e - n + 1$, and every basis for it is called a *fundamental set of cycles* or a "cycle basis". Thus every cycle is the sum of fundamental cycles. It is known that for every spanning tree T the cycles that can be obtained by adding an arbitrary non-tree edge to T are a fundamental set. A fundamental set of cycles can be generated easily during a DFS of a graph. Each time a red edge (i, j) is encountered while exploring node i, another fundamental circuit is closed and can easily be output (because j is an ancestor of i) in $O(n)$ steps. The method is due to Paton and runs in $O(n(e-n+1))$ time. Horton has shown that a fundamental set of cycles never needs to contain more than $\frac{3}{2}(n-1)(n-2)$ edges and gives an $O(ne^3)$ algorithm for computing a fundamental set with the least possible total number of edges. Finding specific cycles by enumerating the $2^{\gamma}-1$ nontrivial combinations of fundamental cycles ($\gamma = e-n+1$) is usually inefficient. Dixon & Goodman use a branch-and-bound search in the cycle space to find a longest cycle in the graph.

The most common problem is that of finding cycles with a length constraint: specified length (k), shortest, longest, odd length, even length. Consider first the simplest case of all, namely, finding a cycle of length 3 (a triangle). An $O(ne)$ algorithm is obtained by inspecting every combination of an edge (i, j) and a node x, and deciding in $O(1)$ time using the adjacency matrix A_G whether (i, x) and (j, x) are edges. A possibly better worst-case bound results by observing that triangles of G correspond to off-diagonal ones in the (Boolean) matrix $A_G^2 \wedge A_G$, where $\wedge$ denotes the elementwise "and". The complexity is bounded by $O(n^{\alpha})$, with α the exponent of a fast matrix multiplication algorithm. More intriguing is the following observation. Let T be any rooted spanning tree of G (or, a spanning forest in case G is not connected).

2.7. LEMMA. *G has a triangle which contains a tree edge if and only if G has a non-tree edge (x, y) for which* $(\text{father}(x), y) \in E$.

The lemma suggests the following algorithm: find a spanning tree (forest) of G, test every non-tree edge in $O(1)$ time per edge, and delete the tree edges and repeat the same procedure on the resulting graph if no triangle was found. Itai & Rodeh show that at most $2\sqrt{e}$ iterations will lead to a graph of isolated vertices, and (thus) the algorithm is $O(e^{3/2})$ time bounded. Observing that $e \leqslant 3n-6$ ($n \geqslant 3$) for planar graphs, each iteration removes at least $\frac{1}{3}$ of the edges of G in the planar case and the total algorithm finishes in $O(n)$ time (in this case). For planar graphs Richards gives an $O(n)$ algorithm for finding a 4-cycle, as well as $O(n \log n)$ algorithms for finding a 5-cycle and a 6-cycle. Chiba and Nishizeki prove that 4-cycles in general graphs G can be found in $O(n \cdot \Gamma(G))$ time, with $\Gamma(G)$ the arboricity of G. As $\Gamma(G) < \min\{\frac{1}{2}n, \frac{1}{2}\sqrt{n+2e}\} + 1$, it follows that the algorithm is $O(ne^{1/2})$ time bounded for connected graphs. For planar graphs $\Gamma(G) \leqslant 3$ and the algorithm is linear.

Algorithms to find a shortest cycle in general usually rely on a form of "truncated BFS". Let i be the first node during BFS from x where a red edge is created. Let the edge be (i, j) and define $L_x = L(i)$. Observe that necessarily $L(j) = L(i)$ or $L(j) = L(i)+1$. Let $l_0 = \min_x L_x$ and x_0 be such that $l_0 = L_{x_0}$. Let k be the length of a shortest cycle in G, with $k = \infty$ if no cycle exist. l_0 follows in $O(n^2)$ time (when defined).

2.8. LEMMA. *Let $k < \infty$. The red edge in the truncated* BFS *started at x_0 closes a cycle of length k' with $k \leqslant k' \leqslant k+1$ (and $2l_0+1 \leqslant k' \leqslant 2l_0+2$).*

Note that a cycle of length $2l_0+1$ containing x_0 can exist only if level l_0 of the BFS tree contains two nodes connected by a (red) edge. Thus it is easily decided within $O(e)$ steps what the smallest cycle through x_0 is. If G is bipartite and (thus) has cycles of even length only, this test can be skipped. Considering all possible choices for x_0, the smallest cycle of G is found in $O(ne)$ steps. When G is bipartite, $O(n^2)$ steps suffice. Monien has shown that the technique can be modified to find the smallest cycle of odd length in $O(ne)$ time and the smallest cycle of even length in $O(n^2 \min\{\alpha(n,n), \lambda\})$ where $\alpha(m,n)$ is an "almost constant" function (cf. *Union-Find* programs) and λ the length of the shortest even cycle. Vazirani and Yannakakis give interesting perspectives on the complexity of determining even cycles in directed graphs.

Note that deciding whether level l_0 of the truncated BFS tree of x_0 contains two nodes i and j connected by an edge can be formulated as the problem of finding a triangle $\triangle(x_0, i, j)$ in a collapsed version of the tree. Itai & Rodeh show that in $O(n^2)$ time a graph G' of at most $2n$ nodes can be built with the property that G contains a cycle of length $2l_0+1$ (and thus has a shortest cycle of this length) if and only if G' contains a triangle. Triangle finding was discussed above.

Little is known about the complexity of finding cycles of a specified length. Monien's techniques (see Section 2.4.1) lead to an algorithm for finding a longest cycle in $O((2\mu)!ne)$ time, where μ is the length of the longest cycle of G. For any fixed k, there is an $O(2^k k!n)$ algorithm for deciding whether a graph has a cycle of length $\geqslant k$ (Bodlaender). Note that the longest cycle problem in graphs is NP-complete. Another problem asks for the existence of a cycle through k specified vertices. The problem is polynomial for fixed k: this was proved for $k=2$ and $k=3$ by LaPaugh, and in general it follows from Robertson & Seymour's result for the DISJOINT CONNECTING PATHS problem for fixed k (see Section 2.4.2). In directed graphs the problem is NP-complete already for $k=2$. Yet another problem asks for determining a set F of at most k nodes such that F contains at least one vertex from every cycle in G. (F is called a *feedback vertex set*.) The problem is NP-complete, but for fixed k it can be solved in $O(n^{k-1}e)$ time. (Note that an algorithm of $O(n^k e)$ time results if one considers every subset of k vertices and tests whether removing these vertices yields an acyclic graph.) Fellows & Langston have even shown the existence of an $O(n^2)$ algorithm for the k-feedback vertex set problem, for any fixed k.

2.4.4. *Weighted cycles*

Now we let $G=\langle V, E, w\rangle$ be a weighted graph. Assuming that all weights are nonnegative, Horton gives an $O(ne^3)$ algorithm for computing a fundamental set of cycles of smallest total weight. Consider the problem of finding cycles with a length and/or a weight constraint. If all edge weights are $\geqslant 0$, then it is not hard to modify any of the existing shortest path algorithms to find a cycle of least (total) weight in polynomial time. Monien shows that the least cost cycle of odd length can be computed in $O(n^3)$ time. If edge weights <0 are allowed, the problem of finding a least weight cycle is NP-complete. However, in this case there are polynomial-time algorithms like Karp's for finding just a negative cycle (if one exists). Define the mean weight of a cycle as the total weight divided by the number of edges in the cycle and define a *minimum-mean cycle* to be any cycle whose mean weight is as small as possible. Karp showed that a minimum-mean cycle can always be determined in $O(ne)$ time. (See Karp and Orlin for an $O(ne \log n)$ algorithm.) Clearly, negative cycles exist if and only if the minimum cycle mean is negative. The problem of finding a negative cycle in a graph is interesting e.g. in the context of shortest path finding or applications in optimization theory. An early heuristics is due to Florian & Robert and based on the following observation. (We now assume that G is directed.)

2.9. LEMMA. *Every negative cycle C contains a vertex i such that the partial sums of the edge weights along the cycle starting and ending at i are all negative.*

By applying a branch-and-bound strategy at every node i, one hopes to be able to "guide" the search quickly to negative cycles (if there are). The method seems to perform well in practice, but Yen proved that on certain graphs the algorithm does very badly.

More interesting techniques are based on shortest path algorithms for general graphs (see Section 2.3). Recall that these algorithms assume that no negative cycles exist only for the sake of valid termination. Thus, if, e.g. in Ford's algorithm applied to a source s, no termination has occurred after $O(ne)$ steps, then there must be a negative cycle accessible from s (and vice versa). In fact, a negative cycle is detected as soon as some node x is added into the (current fragment of the) SP-tree that appears to be its own ancestor (i.e., x already occurs earlier on the path towards the root). A direct and (hence) more efficient approach was developed by Maier, by building an SP-tree in breadth-first fashion as follows. Suppose we have a tree T which is an SP-tree of paths from s of length $\leqslant k-1$. In stage k we look for paths of length k to nodes that were visited in earlier stages that are less costly than the paths of length $\leqslant k-1$ discovered so far. To this end, consider the "frontier" nodes i at depth $k-1$: if $(i,j)\in E$ and $u_j > u_i + w(i,j)$ then we clearly have a "better" path via i to j. If j is an ancestor of i, then we have a negative cycle. Otherwise we purge whatever subtree was attached to j, and make j a son of i. (The nodes of the subtree will automatically come in again with better paths at later stages.) Repeat this for all i, except those for which a better path was found (they are moved to the next stage). The algorithm requires "fast" tree-grafting and ancestor-query techniques. Tsakalidis has shown that this can all be handled in $O(n+e)$ time and $O(n+e)$ space.

2.4.5. Eulerian (and other) cycles

Define an *Eulerian cycle* to be a (nonsimple) cycle that contains each edge of G exactly once. An *Eulerian path* (or *covering trail*) is defined to be a (nonsimple) path with the same property. An excellent survey of results was given by Fleischner. The characterization of Eulerian graphs is classical.

2.10. THEOREM. *Let G be connected. The following conditions are equivalent:*
(i) *G contains an Eulerian cycle,*
(ii) *each node of G has even degree,*
(iii) *G can be covered by (edge-disjoint) simple cycles.*

Testing G for the existence of an Eulerian cycle is simple by (ii), but efficient algorithms to find Eulerian cycles all seem to be based on (iii) and the following lemma. Define a (maximal) random trail of untraversed edges from x to be any trail that starts at x and takes a previously untraversed edge each time a node is reached, unless no such edge exists (in which case the trail ends at this node).

2.11. LEMMA. *Let G be connected. The following conditions are equivalent:*

(i) *G contains an Eulerian cycle,*

(ii) *for each node x, every random trail from x ends at x (even if at some nodes an even number of edges is eliminated at the beginning).*

PROOF. (i) ⇒ (ii). Because all degrees are even, every node $\neq x$ that is reached over an edge can be left again. Thus, eventually, the trail must return to x. If there are untraversed edges incident to x, the trail is continued. Eventually the trail must end, necessarily at x.

(ii) ⇒ (i). It is not hard to argue (by induction on $|G|$) that (ii) implies that G is covered by edge-disjoint cycles. Apply Theorem 2.10(iii). □

The simplest approach to finding an Eulerian cycle is to pick a node x and follow a random trail. With some luck the trail will be Eulerian, but this need not be so. Ore has shown that every random trail from x is an Eulerian cycle if and only if x is a feedback vertex (i.e., x lies on every cycle). He also noted that graphs with this property are necessarily planar. If the random trail is not Eulerian, it must contain a node y which still has an (even) number of untraversed edges. One can now enlarge the random trail as follows: insert a random trail of untraversed edges from y at some point in the random trail from x where y is reached. Repeating this will eventually yield an Eulerian cycle (or a proof that none exists). The algorithm is due to Hierholzer, and is easily implemented in $O(n+e)$ steps by linked-list techniques. The algorithm still has the (minor) defect that it does not produce an Eulerian cycle in traversal order immediately. But a simple modification of DFS will. Start from a node i_0 and follow previously untraversed edges (i.e., a random trail) until a node (i_0) is reached where one cannot proceed. Backtrack to the last node visited on the path which still has untraversed edges and proceed again. Repeat it until one backtracks to the very beginning of the path. One can observe that, when G is Eulerian, outputting edges in the order in which they are traversed the second time (i.e., when backtracking) produces an Eulerian cycle in traversal order. Rather similar results are known for Eulerian paths, and for the case of directed graphs.

In case a graph is not Eulerian, one might ask for a shortest nonsimple cycle containing every edge (necessarily, some edges more than once). Any cycle of this sort is called a *postman's walk* and the problem is known as the *Chinese postman problem*, after Kwan Mei-Ko.

2.12. THEOREM. *Let G be connected. The following conditions are equivalent:*

(i) *P is a postman's walk,*

(ii) *the set of edges that appear more than once in P contains precisely a minimum number of edges that need to be doubled in order to obtain an Eulerian graph,*

(iii) *no edge occurs more than twice in P and for every cycle C, the number of edges of C that occur twice in P is at most* $\frac{1}{2}|C|$.

More generally one can consider weighted graphs $G=\langle V,E,w\rangle$ and ask for a postman's walk of minimum weight (or "length"). The following result is due to Edmonds & Johnson. (It is assumed that weights are nonnegative.)

2.13. LEMMA. *The Chinese postman problem can be solved by means of an all-pairs shortest path computation, solving a minimum weight perfect matching problem, and tracing an Eulerian cycle in an (Eulerian) graph.*

PROOF. In analogy to Theorem 2.12(ii) the minimum cost set of edges that must be doubled in G to make G "Chinese postman optimal" must be the union of minimum cost paths connecting nodes of odd degree, and (conversely) any union of this kind added to G will yield an Eulerian graph. Thus, let A be the set of nodes of odd degree and solve the (A, A)-shortest path problem in G. To select a minimum cost set of paths, design a complete graph G' on node set A with $w'(i,j)$ equal to $d(i,j)$ for $i,j \in A$ (the shortest distance between i and j in G), and determine a minimum weight perfect matching in G'. (Note that, necessarily, $|A|$ is even.) For every edge (i,j) of the matching, double the edges of the shortest path from i to j in G. Tracing an Eulerian cycle in the resulting graph will yield an (optimal) postman walk. □

It follows (see Section 3) that the Chinese postman problem is solvable in polynomial time. A similar result holds in the directed case, but quite surprisingly the problem is NP-complete for mixed graphs.

Considering Theorem 2.12(iii) one might wish to approach non-Eulerian graphs as follows. One problem is to determine a maximum set of edge-disjoint cycles in G (a "cycle packing"). In the weighted case one asks for a cycle packing of maximum total weight. Meigu proves that this problem is equivalent to the Chinese postman problem (hence polynomially solvable). Another problem asks for a minimum set of cycles such that each edge is contained in at least one cycle from the set (a "cycle covering"). In the weighted case one asks for a cycle covering of minimum total weight. Itai et al. have shown that every 2-connected graph has a cycle covering of total length at most $\min\{3e, 6n+e\}$, and that this covering can be found in $O(n^2)$ time. (No results seem to be known for the weighted case.)

2.4.6. *Hamiltonian (and other) cycles*

Define a *Hamiltonian cycle* to be a (simple!) cycle that contains each node of G exactly once. Define a Hamiltonian path to be any (simple) path with the same property. Unlike the "Eulerian" case, there appears to be no easily recognized characterization of Hamiltonian graphs. Thus all traditional algorithms to find a Hamiltonian cycle in a graph are based on exhaustive search from a (random) starting point. When a partial cycle $i_0, i_1, \ldots, i_{j-1}$ has been formed and i_j (a neighbor of i_{j-1}) is the next node tried, then the path is extended by i_j and one of its neighbors is considered next if i_j did not already occur on the partial cycle. If this neighbor happens to be i_0 and all nodes have been visited, then a Hamiltonian cycle is formed. Otherwise we just consider the neighbor like we did i_j. If i_j did occur on the partial cycle, then we back up to i_{j-1} and consider another untried neighbor. If there are no more untried neighbors, then we back up to i_{j-2} and repeat. The algorithm ends with a Hamiltonian cycle if one exists, otherwise it ends with all neighbors of i_0 "tried". Rubin discusses a variety of techniques to design a more efficient branch-and-bound algorithm for the problem.

2.14. LEMMA. *For every graph G there is a graph G′ with n+1 nodes and e+n edges such that G has a Hamiltonian path if and only if G′ has a Hamiltonian cycle.*

PROOF. Let s be a new node, and let $G' = \langle V \cup \{s\}, E \cup \{(s, i) \mid i \in V\} \rangle$. The lemma is now easily verified. □

There is an interesting analogy between the lack of an easily tested criterion for the existence of a Hamiltonian cycle and the lack of an efficient (viz. polynomial-time) algorithm for finding a Hamiltonian cycle in a graph. In fact, even knowing that a Hamiltonian cycle exists does not seem to be of much help in finding one. Of course the Hamiltonian cycle problem is a "classical" example of an NP-complete problem. The problem is even NP-complete for rather restricted classes of graphs and demonstrates the "thin" dividing line that sometimes exists between tractable and untractable problems. We mention only two examples.

(i) It is not hard to see that the Hamiltonian cycle problem is polynomial for directed graphs with all out-degrees or all in-degrees at most 1. Plesnik has shown that the problem is NP-complete already for planar directed graphs with in-degrees and out-degrees at most 2.

(ii) Garey, Johnson & Tarjan have shown that the Hamiltonian cycle problem is NP-complete for 3-connected planar graphs. On the other hand a beautiful theorem of Tutte asserts that every 4-connected planar graph must be Hamiltonian (although this does not imply in itself that the cycle is easy to find!). Gouyou-Beauchamps has shown that in the latter case a Hamiltonian cycle can indeed be found in polynomial time. For 4-connected planar graphs there even exists a linear-time algorithm for the problem (Asano et al. and Chiba & Nishizeki).

As in the "Eulerian" case (see Section 2.4.5), several approaches have been considered for non-Hamiltonian graphs. Define the Hamiltonian completion number $hc(g)$ to be the smallest number of edges that must be added to G to make it Hamiltonian. Goodman, Hedetniemi, & Slater prove that the Hamiltonian completion number can be computed in $O(n)$ time for graphs with (at most) one cycle. Arnborg et al. prove that for any fixed k it can be decided whether $hc(G) \leqslant k$ in $O(n)$ time for all graphs of bounded treewidth. (Of course the problem is NP-complete in general.) Rather more interesting perhaps is the approach in which one tries to minimize the number of "double" visits. Define a *Hamiltonian walk* (or shortest closed spanning walk) to be any minimum length cycle that contains each node of G at least once. If G admits a closed spanning walk of length $n+h$, then we say that G has Hamiltonian excess bounded by h. (Clearly,

h is equal to the number of nodes visited twice in the walk.) One can show that

$$h \leqslant 2n - \lfloor \tfrac{1}{2}k \rfloor (2d-2) - 2$$

for k-connected graphs of diameter d.

2.15. LEMMA. *Let W be a Hamiltonian walk. Then*

(i) *no edge of G appears more than twice in W, and*

(ii) *for every cycle C, the number of edges that appear twice in W is at most $\frac{1}{2}|C|$.*

PROOF. (i) Suppose (i, j) occurs $\geqslant 3$ times in W. Then there are (possibly empty) sub-walks W_1, W_2, W_3 and W_4 such that W can be written in one of the following forms:

(a) $W_1 ij W_2 ij W_3 ij W_4$,

(b) $W_1 ij W_2 ji W_3 ij W_4$, and

(c) $W_1 ij W_2 ij W_3 ji W_4$.

In case (a) $W_1 i \bar{W}_2 j W_3 ij W_4$ would be a shorter walk, contradicting the minimality of W. ($\bar{W}'_2$ denotes the reverse of W_2.) Similar contradictions are obtained in the other cases.

(ii) Let $G_W = \langle V, E_W \rangle$ be the (multi)graph obtained from G by only including the edges traversed by W, taking an edge "twice" if it is doubled by W. G_W is Eulerian. Suppose C is a cycle such that more than $\frac{1}{2}|C|$ edges of C appear twice in W (hence in G_W). Delete one copy of every "doubled" edge of C from G_W, but add every non-traversed edge of C to it. G_W is still connected and has all even degrees, and hence is again connected. The Eulerian cycle in the (modified) graph G_W traces a closed spanning walk of G that is shorter than W. Contradiction. □

The following result due to Takamizawa, Nishizeki & Saito gives a sufficient condition for the Hamiltonian excess to be bounded by $n-c$ for given c. (By traversing e.g. a spanning tree one observes that the Hamiltonian excess is always bounded by n.) The result is remarkable, because for $c=n$ (excess 0, i.e., the case of a Hamiltonian graph) it subsumes several known sufficient conditions for the existence of Hamiltonian cycles.

2.16. THEOREM. *Let G be connected, $n \geqslant 3$ and $0 \leqslant c \leqslant n$. Suppose there exists a labeling $i_1, \ldots, i_n$ of the nodes such that for all j, k, $j<k$, $(i_j, i_k) \notin E$, $\deg(i_j) \leqslant j$, and $\deg(i_k) \leqslant k-1$ implies $\deg(i_j) + \deg(i_k) \geqslant c$. Then G contains a Hamiltonian walk of length $\leqslant 2n-c$ (i.e., excess $\leqslant n-c$).*

Takamizawa et al. prove that if the conditions of the theorem are satisfied, then a closed spanning walk of length $\leqslant 2n-c$ can be found in $O(n^2 \log n)$ time. (For $c=n$ this would necessarily be a Hamiltonian cycle.)

While the Hamiltonian cycle problem is NP-complete (thus "hard") in general, there may be algorithms for it that do quite well in practice. There are two ways to make this more precise:

(i) by considering the problem for random graphs (with a certain edge probability) or,

(ii) by averaging over all graphs with N edges (taking each graph with equal probability).

The starting point are results of the following form: if $N \geqslant cn \log n$, then the probability that G is Hamiltonian tends to 1 for $n \to \infty$. (The sharpest bound for which this holds is

$$N \geqslant \tfrac{1}{2}n \log n + \tfrac{1}{2}n \log \log n + w(n) \quad \text{for } w(n) \to \infty,$$

cf. Bollobás.) Angluin & Valiant proposed the following randomized algorithm for finding a Hamiltonian cycle with high probability (averaged over all graphs with $N \geqslant cn \log n$). In the course of the algorithm explored edges will be deleted ("blocked") from G. Let s be a specified starting node and suppose we succeeded building a partial cycle C from s to x for some x. If the partial cycle contains all nodes of G and if we previously deleted (x, s) from G, then add the edge (x, s) to C and report success (C is now a Hamiltonian cycle). Otherwise, select (and delete) a random edge $(x, y) \in E$. If no edges (x, y) exist, then stop and report failure. If $y \neq s$ and y does not already occur on C, then add (x, y) to C and continue by exploring y by the same algorithm. If $y \neq s$ but y does occur on C (see Fig. 5), then locate the neighbor z of y on C in the direction of x, delete the edge (y, z) from C and add (y, x), modify the traversal order of the edges so z becomes the "head" of C, and continue exploring z. (The case $y = s$ is handled as in the opening clause of the algorithm.) By using suitable datastructures each "step" of the algorithm can be executed in $O(\log n)$ time. It can be shown that for $N \geqslant cn \log n$, the probability that the algorithm reports success within $O(n \log n)$ steps tends to 1 for $n \to \infty$. The result implies that the randomized algorithm "almost certainly" finds a Hamiltonian cycle within $O(n \log^2 n)$ time, for $N \geqslant cn \log n$ (and averaged over all graphs with N edges). More recent results have lowered the threshold for N to $\tfrac{1}{2}n \log n + \tfrac{1}{2}n \log \log n + w(n)$ while still guaranteeing a polynomial-time algorithm that is successful almost always. Similar results with rather more complicated algorithms hold for the directed case.

2.5. *Decomposition of graphs*

"Decomposition" is a broad term, and is generally used to indicate a technique for unraveling a graph in terms of simpler (smaller) structures. Traditionally graphs are decomposed into components with a certain degree of connectivity, or in a (small) number of subgraphs of a specified kind. For example, a connected graph can be decomposed as a "tree" of biconnected components, and this decomposition can be computed in $O(n+e)$ time and space. The biconnected components are the (unique)

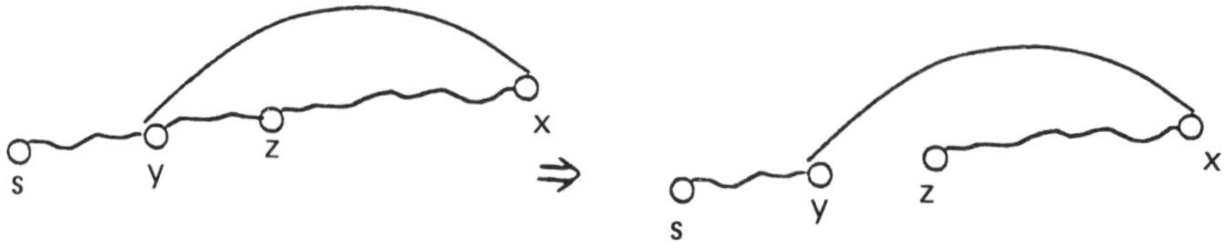

Fig. 5.

maximal connected subgraphs without cutpoints. The biconnected components can be decomposed further into triconnected components (roughly speaking, the components without nontrivial 2-element separating sets) which can also be computed in $O(n+e)$ time and space by a rather more complicated DFS-based algorithm. The decomposition of a graph into its k-connected components can be defined for any $k \geqslant 1$ (Cunningham and Edmonds, Hohberg and Reischuk) and is computable by an $O(k^6n^2+k^4ne)$ algorithm for $k \geqslant 4$. Decompositions can be very helpful in reducing problems for general graphs. Many problems that are "hard" for general graphs can be solved efficiently for graphs that have a decomposition into "small" k-connected components.

Directed graphs can be decomposed as an "acyclic graph" of strongly connected components, again in $O(n+e)$ time by a suitable version of DFS. One can group strongly connected components into "weak components" W, which have the property that for any x, y either x and y belong to the same strongly connected component, or there is no (directed) path from x to y nor from y to x. The weak components of a graph can again be identified in $O(n+e)$ time in a rather elegant fashion from the topologically sorted arrangement of the strongly connected components. Yet another notion is the following. Define a unilaterally connected component as a maximal subgraph U with the property that for any $x, y \in U$ there is a path from x to y or from y to x (or both). It is easily seen that the unilaterally connected components are maximal chains of strongly connected components. These chains can be identified in $O(n+e+nc)$ time when c is the number of chains to be output, provided one works with the transitive reduction of the underlying acyclic graph (otherwise nonmaximal chains must be explicitly removed at extra cost). Finally, strongly connected components can be decomposed into a hierarchy of (sub)components as follows. In its general form the process assumes that edges have distinct, nonnegative weights. Begin with the isolated nodes of C, and add the edges in order of increasing weight one-at-a-time. Gradually, strongly connected clusters are formed and merged into larger strongly connected components, until one strongly connected component (namely C itself) results. The process is represented by a tree in which every internal node v corresponds with a strongly connected cluster that is formed, its sons are the strongly connected clusters of which it is the immediate merge (after a number of edge insertions), and a label is assigned that is equal to the weight of the "last" edge that formed the component associated with v. Tarjan showed that strong component decomposition trees can be constructed in $O(e \log n)$ time.

2.17. LEMMA. *Any algorithm for constructing strong component decomposition trees can be used to construct minimum spanning trees of undirected graphs within the same time bound.*

PROOF. Let $G=\langle V, E, w\rangle$ be an undirected graph with distinct edge weights. Let G' be the directed graph obtained by taking a directed version in both directions of every edge of G. (G' is strongly connected.) The internal labels of any strong component decomposition tree correspond precisely to the edges of a minimum spanning tree of G. □

Another classical approach to decomposition involves the notion of a factor of a graph. A *factor F* is any nontrivial (i.e., not totally disconnected) spanning subgraph of G. If F is connected, then it is called a *connected* factor. A decomposition (or, factorization) of G is any expression of G as the union of a number of factors no two of which have an edge in common. Tutte proved that G can be decomposed into c connected factors if and only if

$$c\cdot(q(V,E-E')-1)\leqslant|E'| \quad \text{for every } E'\subseteq E,$$

where $q(V,E)$ denotes the number of components of the graph $\langle V,E\rangle$. Computational results on factors are few. Define an H-factor to be any factor that consists of a number of disjoint copies of H. The question whether G contains an H-factor is NP-complete for any fixed H that is not a disjoint union of K_1's (single nodes) and K_2's (complete graphs on two nodes). Note that for the existence of an H-factor it is required that $|H|$ divides $|G|$. The special case that $H=K_2$ can be recognized as the perfect matching problem, which is solvable in polynomial time. Define a k-factor ($k\geqslant 1$) to be any factor that is regular of degree k. The problem of computing a 1-factor is again another formulation of the perfect matching problem. 2-Factors can also be identified by means of the perfect matching algorithm as follows.

2.18. LEMMA. *For every graph G there exists a (bipartite) graph G′ with 2n nodes and e edges such that G has a 2-factor if and only if G′ has a perfect matching.*

PROOF. Define $G'=\langle V\cup V',E'\rangle$ with $(i,j')\in E'$ iff $(i,j)\in E$ and no other edges in E'. (V' consists of nodes i' for $i\in V$.) Clearly, any 2-factor of G translates into a perfect matching of G' and vice versa. □

Note that the problem of finding a connected 2-factor is just the Hamiltonian cycle problem and (thus) NP-complete. Next, define an odd (even) factor to be any factor in which all nodes have odd (even) degree. Ebert showed that a maximal odd (even) factor of a graph (when it exists) can be found in $O(n+e)$ time by an application of DFS. Finally, a result due to Lovász asserts that for any k-connected graph G and partition of n into $a_1,\ldots,a_k\geqslant 1$, G admits a factor with k connected components $C_1,\ldots,C_k$ and $|C_i|=a_i$ $(1\leqslant i\leqslant k)$. For $k=2$ an $O(ne)$ algorithm is known to compute a factor of this form.

In a related approach to decomposition one tries to partition the nodes rather than the edges of G. Define a (node-disjoint) covering of G by graphs of type X to be any collection of subgraphs $G_1,\ldots,G_k$ that are node-disjoint and of type X and together contain all nodes of G. (k is called the size of the covering.) In many cases the problem of determining whether a graph has a covering of size $\leqslant k$ is NP-complete, e.g. when G is a general undirected graph and X is the collection of triangles, circuits or cliques. Some interesting results are known when X is the collection of (directed) paths. The minimum size of a covering with node-disjoint paths is known as the *path-covering number* of G. The following result is quite straightforward, except part (ii) which is due to Noorvash.

2.19. THEOREM. (i) *Unless G is Hamiltonian, the path-covering number of G is equal to the Hamiltonian completion number of G.*

(ii) *A path covering of G has minimum size if and only if it contains the maximum number of edges among all path coverings of G.*

(iii) *If $g(n, K)$ is the smallest number such that every n-node graph with $\geqslant g(n, K)$ edges has a path-covering number $\leqslant K$, then*

$$\tfrac{1}{2}(n-K)(n-K-1)+1 \leqslant g(n, K) \leqslant \tfrac{1}{2}(n-1)(n-K-1)+1.$$

One concludes that the path-covering number of a graph is computationally hard to determine in general. Misra and Tarjan prove that it can be computed in $O(n \log n)$ time for rooted trees (with all edges directed away from the root), even in a weighted version. The following, more general result is due to Boesch and Gimpel.

2.20. THEOREM. *The path-covering number of a directed acyclic graph can be computed by means of maximum matching algorithm in a suitable bipartite graph.*

PROOF. Construct a graph G' by replacing each node i by a pair of nodes i' and i'' such that all edges directed into i are directed into i' and all edges directed out of i are directed out of i''. One easily verifies that a minimum path covering of G corresponds to a maximum matching of G', and vice versa. □

(In Section 3.2 we will see that the maximum matching problem in bipartite graphs is efficiently solvable in polynomial time.)

The last approach we discuss is based on pulling a graph apart or "separating" it by cutting a small number of edges (or, deleting a few nodes). The approach is crucial in order to apply a divide-and-conquer strategy on graphs that admit this kind of decomposition.

One approach is the following. Let us say that $G = \langle V, E \rangle$ "decomposes" if there is a nontrivial partition $V_1 \cup V_2 \cup V_3 \cup V_4$ of V such that for every $\alpha \in E$ there is an i such that α connects two nodes in V_i or two nodes in V_i and V_{i+1}, and furthermore $V_2 \times V_3 \subseteq E$ ("all nodes in V_2 are connected to all nodes in V_3"). When G "decomposes", it is completely determined by its "components" $G/(V_1 \cup V_2 \cup \{i_3\})$ and $G/(\{i_2\} \cup V_3 \cup V_4)$ for some $i_3 \in V_3$ and some $i_2 \in V_2$. Clearly one may try to decompose the components again. When a graph does not decompose this way, it is called "prime". A decomposition of a graph into prime components can be computed in $O(ne)$ time. Prime graphs with $\geqslant 5$ nodes have several useful properties, e.g. they have no cutpoints and every two nodes i and j in it are nonsimilar (i.e., there is a node x that is adjacent to one but not to the other). Another approach is explicitly based on separation. Let G be connected, and $1 \leqslant k \leqslant n$.

DEFINITION. (i) A *k-separator* is a set $U \subseteq V$ such that $G-U$ partitions into two sets A and B of at most k nodes each and with no edge between A and B. (Sometimes a k-separator is defined to be a set $U \subseteq V$ such that each {strongly} connected component of $G-U$ has at most k nodes.)

(ii) A *k-edge separator* is a set $F \subseteq E$ such $G-F$ partitions into two sets A and B of at most k nodes each and with no edge between A and B. (It means that every edge between A and B must belong to F.)

(A separator is called "planar" if it applies to a planar graph.)

The celebrated planar separator theorem asserts that planar graphs have "small" separators. Ungar proved that for every k there exists a planar k-separator of size $O((n/\sqrt{k})\log^{3/2}k)$, Lipton & Tarjan proved that there always exists a planar $\frac{2}{3}n$-separator of size $\leqslant \sqrt{8n}$ (which, moreover, can be determined in linear time). Djidjev showed that there always exists a planar $\frac{2}{3}n$-separator of size $\leqslant \sqrt{6n}$, and Gazit even claims a bound of $\frac{7}{3}\sqrt{n}$. Djidjev showed several further results. For example for every $\varepsilon\,(0<\varepsilon<1)$ there is a planar εn-separator of size $4\sqrt{n/\varepsilon}$ (which can be determined in linear time). Planar graphs do not necessarily always have small edge separators as well. For example, any $\frac{2}{3}n$-edge separator of the star graph of degree $n-1$ must contain at least $\frac{1}{3}n$ edges. Diks et al. proved that every planar graph in which the degrees are bounded by k, must have a $\frac{2}{3}n$-edge separator of size $O(\sqrt{kn})$ which, moreover, can be determined again in linear time.

Miller has shown that every 2-connected planar graph of which every face has at most d edges, has a $\frac{2}{3}n$-separator that forms a simple cycle of size $O(\sqrt{dn})$. Moreover, this separating cycle can be found in $O(n)$ time. There are several other classes of graphs besides the planar ones that can be separated evenly by deleting only a "small" number of nodes or edges. We note that trees have $\frac{2}{3}n$-separators of size 1. Chordal graphs admit a $\frac{1}{2}n$-separator of size $O(\sqrt{e})$ that can be determined in linear time (Gilbert et al.). Graphs of genus g admit separators of no more than $O(\sqrt{gn})$ nodes, which can be computed in $O(n+g)$ time (Gilbert et al., Djidjev). All directed graphs have $\frac{1}{2}n$-separators that form a simple path or a simple cycle (Kao). Similar separation results exists for node-weighted graphs, in which the "weight" of the components after separation is bounded. Separator theorems find many uses in the design of divide-and-conquer algorithms for graphs. There also are applications to the layout problem for graphs (Leiserson).

2.6. *Isomorphism testing*

In many applications of graphs one or both of the following two problems arise: (i) given graphs G and H, determine whether G is isomorphic to a subgraph of H (the "*subgraph isomorphism problem*") and (ii) given graphs G and H, determine whether G is isomorphic to H (the "*graph isomorphism problem*"). Both problems are notoriously hard in general, and no polynomial-time bounded algorithm is presently known for either of them. We will outline some of the more recent results.

2.6.1. *Subgraph isomorphism testing*

The subgraph isomorphism problem is well-known to be NP-complete. Even in the restricted cases that G is an n-node circuit and H an n-node planar graph of degree $\leqslant 3$ or G is a forest and H is a tree, the problem is NP-complete. Matula has shown that the subgraph isomorphism problem can be solved in polynomial time when G is a tree and

H is a forest, by the following simple technique. Assume that G is a tree with n nodes, H is a tree with m nodes ($n \leqslant m$), and both G and H are rooted (all without loss of generality). Let the root of G have degree p and the root of H have degree q. Now execute the following (recursive) procedure **TEST**.

PROCEDURE **TEST**(G, H):
begin
(1) delete the roots of G and H, and isolate the (rooted) subtrees $G_1, \ldots, G_p$ of G and $H_1, \ldots, H_q$ of H;
(2) form a bipartite graph with nodes corresponding to G_1 through G_p and H_1 through H_q, and draw an edge between "G_i" and "H_j" if and only if **TEST**(G_i, H_j) returns true (meaning that G_i is a rooted subtree of H_j);
(3) compute a maximum matching in the bipartite graph to determine whether the rooted subtrees of G can be matched (mapped) to distinct, rooted subtrees of H;
(4) if the outcome of step 3 is successful then return true, otherwise return false.
end.

The procedure is easily modified to actually give a concrete isomorphism embedding of G into H when it returns the value true. Suppose we have an $O(nm^\alpha)$ algorithm for computing a maximum matching in an (n, m)-bipartite graph ($n \leqslant m$, $\alpha > 1$). (Such algorithms exist e.g. for $\alpha = 1.5$, see Section 3.)

2.21. THEOREM. *The subgraph isomorphism problem for rooted trees can be solved in* $O(nm^\alpha)$ *time.*

PROOF. Assume inductively that the problem can be solved in $\leqslant cnm^\alpha$ time for small n and m (c sufficiently large). The running time of **TEST** can be estimated by

$$cpq^\alpha + \sum_{i=1}^{p} \sum_{j=1}^{q} cn_i m_j^\alpha, \quad \text{where } n_i = |G_i| \text{ and } m_j = |H_j|.$$

(We assume that the linear amount of time for step 1 of **TEST** is subsumed by the first term.) Using that $\sum_{i=1}^{p} n_i = n-1$ and $\sum_{j=1}^{q} m_j = m-1$, this is easily bounded by cnm^α. □

By using a faster maximum matching algorithm, the subtree isomorphism test can be improved as well (Reyner, Chung). Lingas has extended the result by showing that for k-trees G and H with n and m nodes respectively, the subgraph isomorphism problem can be solved in $O(k \cdot k! n^{1.5} m)$ time.

For the subgraph isomorphism problem for more general classes of graphs, very little is known. Lingas has studied the problem for classes of graphs that are $s(N)$-separable for some function s (see Section 2.5). A graph is $s(N)$-separable if it consists of one node or has a $\frac{2}{3}n$-separator of size $\leqslant s(n)$ whose removal disconnects the graph into two parts of (say) n_1 and n_2 nodes that are $s(n_1)$-separable and $s(n_2)$-separable respectively. Even for 1-separable graphs the subgraph isomorphism problem remains NP-complete.

2.22. THEOREM. *If G and H are n-node graphs that are s(n)-separable and have degrees $\leqslant d(n)$, then the subgraph isomorphism problem can be solved in $2^{O(\gamma(\log n+d(n)))}$ time for*

$$\gamma=\sum_{i=0}^{\lceil\log_{3/2}n\rceil} s((\tfrac{2}{3})^i n).$$

By using the planar separator theorem (see Section 2.5) it follows, for example, that for planar graphs that are (log n)-degree bounded, the subgraph isomorphism problem can be solved in $2^{O(\sqrt{n}\log^2 n)}$ time. Lingas has shown that for biconnected outerplanar graphs G and H with n and m nodes respectively, the subgraph isomorphism problem can be solved in $O(nm^2)$ time. Lingas and Syslo proved that for biconnected series-parallel graphs the problem can be solved in $O(n^{4.5}m^2+m^3)$ time.

In general there is hardly any alternative to the brute-force approach of enumerating all n-node subgraphs of H and testing for isomorphism with G or enumerating and testing all feasible embeddings of G into H in some suitable representation (see e.g. Bertziss). Ullman describes an interesting technique for pruning the systematic enumeration process.

2.6.2. Graph isomorphism testing

The problem of finding an efficient (i.e., polynomial-time) algorithm for testing whether two n-node graphs G and H are isomorphic is of fundamental importance in the theory of graphs, but has withstood all attempts at a solution to date. The graph isomorphism problem is especially tantalizing because it is neither known to be NP-complete nor known to be polynomially solvable. There is at least some theoretical evidence that the graph isomorphism problem is not NP-complete because, if it were NP-complete, the Meyer–Stockmeyer polynomial hierarchy would collapse to its second level, which seems an unlikely event (Schöning). In this section we will sketch the main developments concerning the graph isomorphism problem only. For more background, see Read and Corneil and the book of Hoffmann.

Let us look first at some "easy" cases. It is well-known that for rooted n-node trees the isomorphism problem can be solved in $O(n)$ time (see Aho, Hopcroft and Ullman, Theorem 3.3). Weinberg first studied the isomorphism problem for planar graphs and obtained an $O(n^3)$ algorithm for the case of triconnected planar graphs, heavily relying on a theorem of Whitney asserting that a triconnected graph has a unique embedding on the sphere. It was eventually shown that for any two planar n-node graphs the isomorphism problem can be solved in $O(n)$ time (Hopcroft & Wong, Fontet). Lueker and Booth give an $O(n+e)$ algorithm for the isomorphism problem for interval graphs.

For most other classes of graphs the graph isomorphism problem quickly becomes very hard. For example, for bipartite graphs the isomorphic problem is as hard as it is for general graphs. It is easy to see that two graphs G and H are isomorphic if and only if the bipartite graphs G' and H' are isomorphic, with G' and H' obtained from G and H by inserting an auxiliary node in every edge.

It has been shown that the graph isomorphism problem is polynomially equivalent to the problem of deciding for any graph G and a node v of G whether G admits an automorphism that "moves" v (Lubiw). Interestingly, the problem of deciding whether

G admits an automorphism that moves every node is NP-complete. To solve the general graph isomorphism problem, several approaches have been followed. The older techniques are often based on a type of brute-force backtracking procedure (see e.g. Schmidt and Druffel, McKay). An interesting technique has been proposed by Deo et al., using a very natural incremental approach. Suppose the nodes of G are numbered from 1 to n (based on any reasonable scheme). At the kth level of the algorithm we always consider the subgraph $G(k)$ of G induced by the nodes 1 through k and a subgraph H' of H isomorphic to $G(k)$. Essentially the algorithm now proceeds as follows, exploiting a call of the (recursive) procedure **TEST** which we describe very informally.

```
PROCEDURE TEST(k, H'):
begin
  if k=n then return true
  else
  begin
    b:=false;
    {consider all possible extensions of H'...}
    while ¬b and there is an unexplored node v ∉ H' left do
    begin
      extend H' by v to obtain an (induced) graph H";
      if the isomorphism between G(k) and H' can be extended to an isomorphism
         between G(k+1) and H"
      then assign to b the value returned by TEST(k+1, H")
      else "try the next v"
    end;
    return b
  end
end.
```

Deo et al. argue that for random graphs the algorithm has an expected computation time of $O(n \log n)$. Apparently the best worst-case algorithm for the graph isomorphism problem runs in $O(c^{n^{1/2+o(1)}})$ time (Babai and Luks). For graphs of genus $\leqslant g$ the graph isomorphism problem can be solved in $O(n^{O(g)})$ time (Filotti & Mayer, Miller). For partial k-trees the problem can be solved in $O(n^{k+4.5})$ time (Bodlaender). From the latter result it follows that the isomorphism problem is polynomial for every class of graphs that is of bounded treewidth.

A second approach that has considerably advanced the understanding of the graph isomorphism problem has been based on the study of various types of automorphism groups associated with graphs. The connection has considerably spurred interest in the aspects of polynomial computability in groups (see e.g. Hoffmann). In fact, the graph isomorphism problem can be reduced entirely to computational problems for suitable groups. Let $\mathrm{Aut}(G)$ denote the automorphism group of G. Consider the following series of problems.

(1) GRAPH ISOMORPHISM: given two n-node graphs G and H, decide whether they are isomorphic.
(2) REGULAR GRAPH ISOMORPHISM: given two regular n-node graphs G and H, decide whether they are isomorphic.
(3) LABELED GRAPH ISOMORPHISM: given two labeled n-node graphs G and H, decide whether they are isomorphic by means of a "label-preserving" isomorphism.
(4) COMPLEMENT ISOMORPHISM: given an n-node graph G, is G isomorphic to its complement?
(5) GRAPH ISOMORPHISM CONSTRUCTION: given two n-node graphs G and H, decide whether they are isomorphic and if so, construct an isomorphism from G to H.
(6) GRAPH ISOMORPHISM COUNTING: given two n-node graphs G and H, determine the number of isomorphisms from G to H.
(7) GRAPH AUTOMORPHISM WITH RESTRICTION: given a n-node graph G and a node v, decide whether G has an automorphism that moves v.
(8) GRAPH AUTOMORPHISM GROUP: given an n-node graph G, determine a generating set for $\mathrm{Aut}(G)$.
(9) GRAPH AUTOMORPHISM GROUP ORDER: given an n-node graph G, determine the order of $\mathrm{Aut}(G)$.
(10) GRAPH AUTOMORPHISM PARTITION: given an n-node graph G, determine the orbits of $\mathrm{Aut}(G)$ on G.

The intriguing fact is that when any of these ten problems is solvable in polynomial time, then so are all the others. Combining results of especially Mathon, Babai, Booth, Lubiw and Colbourn & Colbourn one has the following fact.

2.23. THEOREM. GRAPH ISOMORPHISM *through* GRAPH AUTOMORPHISM PARTITION *are all polynomially equivalent.*

While the theorem has set the scene for many studies in group-theoretic complexity theory, it does not in itself lead to all the benefits of the study of automorphism groups. A crucial type of result is the following. Let $\mathrm{Aut}_e(G)$ be the group of automorphisms of G that leave a particular edge e fixed.

2.24. THEOREM. *The isomorphism problem for trivalent graphs is polynomially reducible to the problem of determining a set of generators for* $\mathrm{Aut}_e(X)$, *where X is a trivalent connected graph and e a distinguished edge.*

Luks succeeded in finding a polynomial-time algorithm for the group-theoretic problem mentioned in the theorem, resulting in an $O(n^6)$ algorithm for the isomorphism problem for trivalent graphs. More generally he proved that for every fixed d, the isomorphism problem for graphs of degree $\leqslant d$ can be solved in polynomial time. Currently an $O(n^3)$ probabilistic algorithm and $O(n^3 \log n)$ deterministic algorithm are known for the isomorphism problem for trivalent graphs (Galil et al.). Babai et al. have shown that for n-node graphs with eigenvalues of multiplicities $\leqslant k$, the isomorphism problem can be solved by an $O(n^{4k+O(1)})$ deterministic algorithm and by an $O(n^{2k+O(1)})$ probabilistic algorithm.

A third, and computationally very intriguing approach to the graph isomorphism problem is based on the use of "signatures". A *signature* is a (partial) mapping s defined on the set of n-node graphs such that for all graphs G,

(i) if $s(G)$ is defined then so is $s(H)$ for all graphs H isomorphic to G and $s(G)=s(H)$, and

(ii) if $s(H)$ is defined for some graph H and $s(G)=s(H)$ then G and H are isomorphic.

Assume that all n-node graphs are defined on a standard node set $\{v_1,\ldots,v_n\}$.

2.25. LEMMA. *Signature functions which are defined for all graphs exist.*

PROOF. A classical example is due to Heap. For each graph G, let $s(G)$ be the lexicographically largest 0-1 vector that can be obtained by listing the nodes in some permuted order and concatenating the rows of the adjacency matrix of G corresponding to this order. □

Clearly Heap's signature function can be exceedingly hard to compute in general, although it may be reasonable for small graphs (Proskurowski). An interesting problem is to find efficiently computable, possibly partial signature functions that apply to large classes of graphs.

A number of studies have shown considerable success with signature functions that merely label the nodes of a given graph in some "canonical" way. Babai, Erdös and Selkow define an $O(n^2)$ time computable canonical labeling scheme s with the property that for random graphs G, $s(G)$ is defined with probability $\geqslant 1-O(n^{-1/7})$. Karp has given an $O(n^2 \log n)$ time computable canonical labeling scheme s such that for random graphs G, $s(G)$ is even defined with a probability of at least $1-O(n^{3/2}2^{-n/2})$. Babai and Kučera have shown that there are $O(n)$ time computable labeling schemes s with exponentially small failure probability. They also show the following remarkable result.

2.26. THEOREM. *There exists a canonical labeling scheme (a signature) defined for all n-node graphs that is computable in $O(n^2)$ expected time for random graphs.*

Again group-theoretic considerations have been applied to the study of signatures. For any fixed d, the class of graphs of degree $\leqslant d$ has a signature function that is polynomial-time computable. For general graphs the best signature function known to date is computable in $O(c^{n^{2/3+o(1)}})$ time (Babai & Luks).

3. Combinatorial optimization on graphs

The richest source of computational problems on graphs probably is the theory of combinatorial optimization, where the underlying structures usually are *networks*. Roughly speaking, a network is a graph in which the edges are labeled by (positive) edge weights or capacities. The labels have a natural interpretation when certain transports are carried out over the edges of the graph. Traditionally two main areas have manifested themselves here:

(a) *matching problems*: determine a (maximum cardinality, maximum weight) set of edges $E' \subseteq E$ such that no two edges of E' are incident.

(b) *flow problems*: one or more commodities must be transported through the network, under constraints of maximum throughput and (perhaps) minimum cost.

Both matching and network flow are important graph-theoretic principles: many other problems can often be solved by restating ("reducing") these problems in terms of matching or flow problems. In all cases the task consists of maximizing a goal function, under suitable constraints.

3.1. *Maximum matching*

An edge belonging to a given matching M is called matched, the other edges are called free. If (x, y) is matched, then x and y are called "mates". A node incident to a matched edge is called matched as well, the other nodes are free or "exposed". There are two types of maximum matching:

(i) *maximum cardinality matching*, which asks for a matching of maximum size (and called *perfect* when all nodes are matched),

(ii) *maximum weight matching*, which asks for a matching of maximum total weight (assuming that fixed weights $\geqslant 0$ are assigned to the edges).

A useful generalization is the notion of a "b-matching". Suppose (fixed) integers b_i are assigned to the nodes i such that $0 \leqslant b_i \leqslant \deg(i)$. A b-matching is a set of edges $E' \subseteq E$ such that at most b_i edges of E' are incident to node i for all i. (Taking $b_i = 1$ results in an "ordinary" matching.) One can now ask for maximum cardinality and maximum weight b-matchings.

3.1. LEMMA. *For every graph G there is a (polynomial-sized) graph G' such that b-matchings in G correspond exactly to certain matchings in G'.*

PROOF. Construct G' as follows. For each node $i \in V$, include $\deg(i)$ nodes $\alpha_i^{(j_1)}, \ldots$ corresponding to the edges $(i, j), \ldots$ incident to i and $\deg(i) - b_i$ nodes $\beta_i^{(1)}, \ldots$ Connect all $\alpha_i^{(j)}$ nodes to all $\beta_i^{(k)}$ nodes for every i, and connect $\alpha_i^{(j)}$ and $\alpha_j^{(i)}$ for all $(i, j) \in E$. We claim that b-matchings in G correspond precisely to matchings in G' that leave no β-node exposed. $\square$

For computational purposes we like to have a "direct" reduction to maximum matching. Construct the following graph G''. For each node $i \in V$ include b_i nodes $i^{(1)}, \ldots, i^{(b_i)}$ and for each edge $e \in E$ include two nodes x_e and y_e. For each $e = (i, j) \in E$ connect the nodes according to the schema $S(e)$ in Fig. 6.

3.2. THEOREM. *Let E' be a maximum cardinality b-matching in G, M a maximum cardinality matching in G''. Then*

(a) $|M| = |E'| + e$,

(b) *a maximum cardinality b-matching E_0 can be obtained from M by including every edge e for which $|S(e) \cap M| = 2$.*

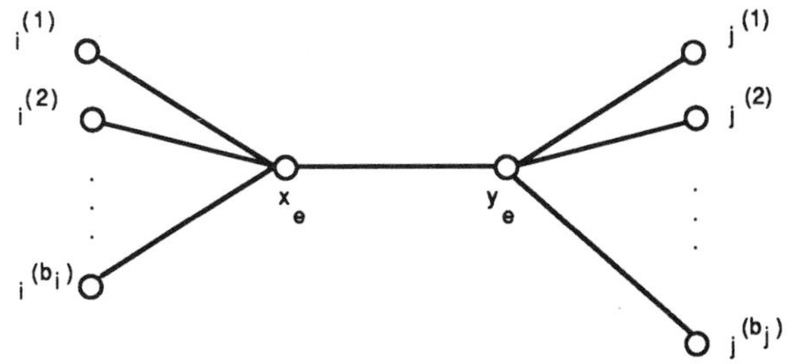

Fig. 6.

PROOF. Observe that a maximum matching M of G'' contains one or two edges from every $S(e)$. Let $E_0=\{e\in E \,||S(e)\cap M|=2\}$. It follows that $|M|=|E_0|+e$. On the other hand, if we "choose" the edges from E_0, then we obtain a b-matching, hence $|E_0|\leqslant|E'|$ and $|M|\leqslant|E'|+e$. Conversely, given E', we can construct a matching M_0 of G'' as follows: If $e\notin E'$, then $(x_e, y_e)\in M_0$. If $e=(i,j)\in E'$ then include $(i^{(k)}, x_e)$ and $(y_e, j^{(l)})$ in M_0 for suitable k, l. Because E' is a b-matching, one can always choose k, l such that $i^{(k)}$ and $j^{(l)}$ are still exposed. It follows that $|M_0|=|E'|+e$. As $|M_0|\leqslant|M|$, we have $|M|=|E'|+e$, and the b-matching E_0 was maximum. □

A similar reduction applies to the weighted case. (Let all edges in $S(e)$ have weight $w(e)$.) The theorem shows that the techniques for maximum matching can be used to solve the more general maximum b-matching problem as well. We will see later that there are algorithms of low polynomial time complexity for it. (Note that G'' has $e+\sum_1^n b_i \deg(i)\leqslant 2ne$ edges and $2e+\sum_1^n b_i\leqslant 4e$ nodes.) b-Matchings can be computed in linear time when G is a tree.

An interesting variant of b-matching requires to find a (maximum) set of edges E' such that at least a_i and at most b_i edges are incident to node i for a given assignment of values a_i and b_i with $a_i\leqslant b_i$ to the nodes. This is probably the most general version of what is usualy called a "degree-constrained subgraph construction problem". (Note that matching is a very special case of this problem.) Again, this general version can be solved in polynomial time.

b-Matching is a useful intermediate step towards the theory of coverings.

DEFINITION. A *node* (*edge*) *cover* of G is any set of nodes V' (edges E') such that each edge (node) of G is incident to at least one of the nodes in V' (or edges in E' respectively).

Observe that every edge covering is the complement (in E) of a b-matching with $b_i=\deg(i)-1$ for all $i\in V$. Thus the problem of computing a minimum cardinality or minimum weight edge cover is equivalent to computing a maximum cardinality or maximum weight b-matching respectively, and hence both problems are polynomial-time computable. For the case of bipartite graphs, the correspondence takes a particularly attractive form, known as the König–Egervary theorem (the "duality theorem" of matching).

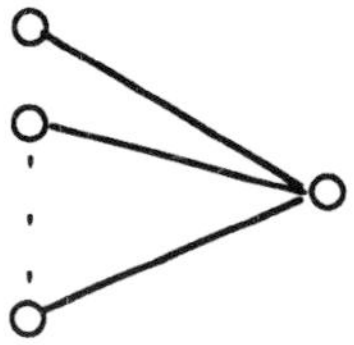

Fig. 7.

3.3. THEOREM. *For any bipartite graph G, the number of edges in a maximum matching is equal to the number of edges in a minimum edge-covering.*

There is an analogous result for general graphs due to Edmonds. We note that the problem of computing minimum node covers is NP-complete.

Of related interest is the question what fraction of the nodes is actually covered by a maximum matching. In general, no particular bound can be given, and Fig. 7 shows an example that the fraction can be arbitrarily close to 0.

Papadimitriou & Yannakakis proved that in any planar graph with minimum node degree $\geqslant 3$, any maximum matching will contain $\geqslant \frac{1}{3}n$ edges. (For minimum node degree $\geqslant 4$ and $\geqslant 5$, they prove a bound of $\geqslant \frac{2}{5}n$ and $\geqslant \frac{5}{11}n$ edges respectively.)

Finally we need the following crucial concept (apparently due to Berge).

DEFINITION. An *alternating path* or *cycle* is a simple path or cycle whose edges are alternately matched and free. An alternating path is called *augmenting* if both its end nodes are exposed. The weight of an alternating path or cycle is equal to the weight of the free edges minus the weight of the matched edges.

If M admits an augmenting path, then it cannot be maximum: reversing the roles of the matched and free edges in the path results in a new matching of size $|M|+1$. A central result of matching theory states that repeated augmentation must result in a maximum matching.

3.4. THEOREM. (i) *A matching is of maximum cardinality if and only if it has no augmenting path.*

(ii) *A matching is of maximum weight if and only if it has no alternating path or cycle of weight* >0.

PROOF. (i) $\Rightarrow$ is trivial. To prove $\Leftarrow$, let M be not of maximum cardinality. Let M' be a matching with $|M'|>|M|$. Consider the graph G' on V with edge set $E'=M \triangle M'$ (the symmetric difference of M and M'). Clearly, each node of G' is incident to at most one edge of M and at most one edge of M'. It easily follows that the connected components of G' will be paths or circuits of even length. In all circuits we have the same number of edges from M as from M'. Because $|M'|>|M|$ there must be a path with more edges from M' than from M. This path necessarily is an augmenting path for M.

(ii) $\Rightarrow$ is trivial. (Reversing the roles of matched and free edges in an alternating path or cycle of weight >0 would result in a "heavier" matching.) To prove $\Leftarrow$, we proceed as before, by a very similar argument. $\square$

Theorem 3.4(ii) does not have the same constructive flavor as (i). In 1967, White proved the following more useful result.

3.5. THEOREM. *Let M be a maximum weight matching of size q (i.e., maximum among all matchings of size q), and let L be an augmenting path of maximum weight. Then the matching M' obtained from M by reversing the roles of the edges along L is of maximum weight among all matchings of size $q+1$.*

PROOF. Let M'' be a maximum weight matching among all matchings of weight $q+1$. Consider the graph $\tilde{G}$ on V with edge set $M \triangle M''$. As before, the components of $\tilde{G}$ are paths or cycles of even length. Any even-length path or (even-length) cycle must have weight 0 with respect to M: if not, then reversing the roles of the edges would lead to a matching of larger weight than M or M''. Consider the paths of odd length. Because $M \triangle M''$ has precisely one more edge in M'' than in M, the number of odd-length paths is odd and we can combine all but one of the paths in pairs such that each pair has an equal number of edges in M and in M''. Each pair of paths must have total weight 0 with respect to M (otherwise the choice of M or M'' as maximum weight matchings is contradicted). The one remaining path must have precisely one more edge in M' than in M, and is augmenting with respect to M. Augmenting along this path gives a matching of $q+1$ edges necessarily of the same weight of M''! The theorem now follows. □

Theorem 3.5 is intriguing because it shows that in the weighted case the "augmenting path method" can be used to construct maximum weight matchings of all possible sizes $(0, 1, 2, \ldots)$ if we always augment using maximum weight augmenting paths. While the theory is strongly suggestive of the types of algorithms we need to use, more details are needed to show that maximum matchings can be computed in low polynomial time.

In one case maximum matchings can be characterized in another way, namely when G is a tree.

DEFINITION. Let G be a tree rooted at r, and let $f(i)$ denote the father of a node i. A matching $M \subseteq E$ is termed *proper* if for every exposed node i there is a brother j of i such that $(j, f(i)) \in M$.

3.6. LEMMA. *Let G be a tree. If M is a proper matching, then it is a maximum matching.*

(The easy proof proceeds by showing that no proper matching can have an augmenting path.) One can show that a proper matching in a tree can be constructed in $O(n)$ time.

3.2. Computing maximum matchings

We consider the simplest case first, namely maximum cardinality matching. All algorithms are based on the augmenting path idea, with special techniques to find augmenting paths efficiently. We outline some of the underlying ideas.

Consider the case that G is bipartite. One can search for an augmenting path as follows. (We assume that a matching M is given as a start.) Build a BFS tree starting from an exposed node i_0. (If there is no exposed node, M is a perfect matching.) Because G is bipartite, there can be no edges connecting nodes within the same level. If an exposed node j appears for the first time (necessarily in an odd level), then the "green" path from i_0 to j is an augmenting path. If no odd level contains exposed nodes, then the BFS-tree is called "Hungarian" and we can forget i_0. If the trees from all exposed nodes are Hungarian, the matching we have built clearly is maximum. Clearly there can be at most $\frac{1}{2}n$ augmentations (as the number of matched nodes increases by 2 in every round) and every round takes $O(e)$ time. Thus the algorithm computes a maximum matching in $O(ne)$ time. A better result can be obtained by doing more augmentations "in one step". In particular, the idea is to compute a maximal set of vertex-disjoint augmenting paths $L_1, \ldots, L_r$ that are all of equal, shortest length (under the current matching M), and to augment along all paths simultaneously. Hopcropt & Karp showed that this must result in a maximum matching after $O(\sqrt{n})$ steps for all graphs, and that for bipartite graphs each step can be implemented in $O(e)$ time. Thus maximum cardinality matching on bipartite graphs needs only $O(\sqrt{n} \cdot e)$ time.

There is another way to see this, by observing a simple connection between maximum matching and flow theory (even in the weighted case).

3.7. Theorem. *For every bipartite graph G there exists a network G' with integer edge capacities such that maximum matchings in G correspond to maximum flows in G'.*

Proof. Let $V = X \cup Y$ $(X \cap Y = \emptyset)$ such that $E \subseteq X \times Y$. Design G' by adding a source node s, a target node t, and edges (s, x) and (y, t) for all $x \in X$, $y \in Y$. Let the edges (s, x) and (y, t) have capacity 1 (cost 0), and the other edges $(x, y) \in E$ have capacity ∞ (cost $-w(x, y)$). By the integer flow theorem, maximum flows in G' are integral. Clearly every flow of size f must identify f matching edges for G and vice versa. Cost $-c$ of a maximum flow corresponds to weight $-c$ of the corresponding matchings. □

As maximum flows in 0-1 networks are computable in $O(\sqrt{n} \cdot e)$ time, so can maximum cardinality matchings.

Maximum cardinality matching for general (connected, nonbipartite) graphs is considerably more difficult and requires an additional technique due to Edmonds. Once again we try to find an augmenting path from an exposed node i_0, by building a "tree" of alternating paths starting at i_0 very much like in the bipartite case. Note that we can assume that odd-level nodes are always extended by a matched edge (otherwise an augmenting path is found) and, hence, that the "tree" is grown by building on the even-level nodes. The "Hungarian" case (no augmenting path from i_0 exists) is special, and implies that we can effectively delete the "tree" from the graph.

3.8. Lemma. *If there is no augmenting path from node i_0 (at some stage), then there never will be an augmenting path from i_0.*

Proof. We show that the lemma must hold after every augmentation of a matching M for which there is no augmenting path from i_0. (The lemma then follows by induction.) Let L be an augmenting path from x to y, necessarily with x and y exposed

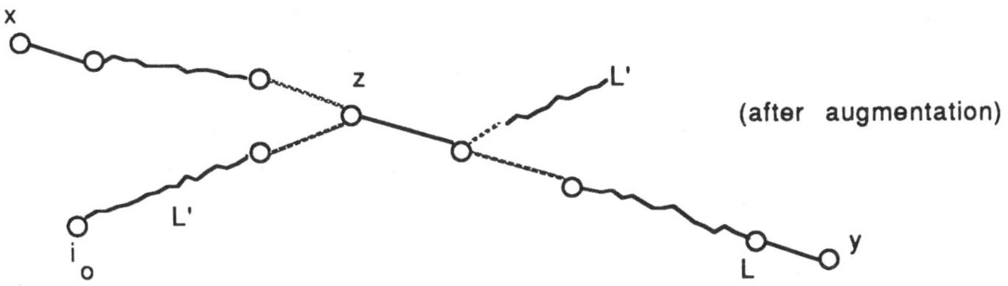

Fig. 8.

and $x, y \neq i_0$. Suppose there is an augmenting path L' from i_0 with respect to the matching $M \triangle L$. If L and L' are node-disjoint, then L' would already have been augmenting in M. Contradiction (cf. Fig. 8). Thus let $L \cap L' \neq \emptyset$, and let z be the first node on L' that is also on L. Consider the situation before augmenting along L. The path from i_0 to x via z (along a section of L' and of L) must have been an augmenting path in M, contrary to our assumption. □

Consider the process of building a "tree" of alternating paths from i_0. We are especially interested in what happens when we examine even-level nodes x. (It is not necessary to build the tree from one node, one can just work on matched or exposed nodes, orienting the incident matched edges such that the nodes are even.) If $(x, y) \in E$ and y is not matched, then there is an augmenting path from i_0 to y. If y is matched, there are two cases. If y is odd, there is nothing special happening (like in the bipartite case). If y is even, there is a special situation. After all, there is an alternating path of even length from i_0 to x and one from i_0 to y, and the edge (x, y) closes a cycle of odd length. If we let b be the last node common to both paths (the base node), then the resulting structure is as shown in Fig. 9.

The cycle is a special case of a substructure known as a blossom, which can be formed if we do the building process in a distributed manner over many clusters simultaneously.

DEFINITION. Let M be a matching, B set of edges connecting nodes in $V_B \subseteq V$, $|B| = 2r+1$ $(r \geqslant 1)$. B is called a *blossom* of M if the following conditions hold:

(i) $|M \cap B| = r$ (the unique node of B left exposed in $M \cap B$ is called the base b of the blossom),

(ii) there is an alternating path L of even length with $L \cap B = \emptyset$ from an exposed node (the "root") to the base of the blossom, and

(iii) for each node $i \in V_B$ there is an alternating path of even length from the base of the blossom to i.

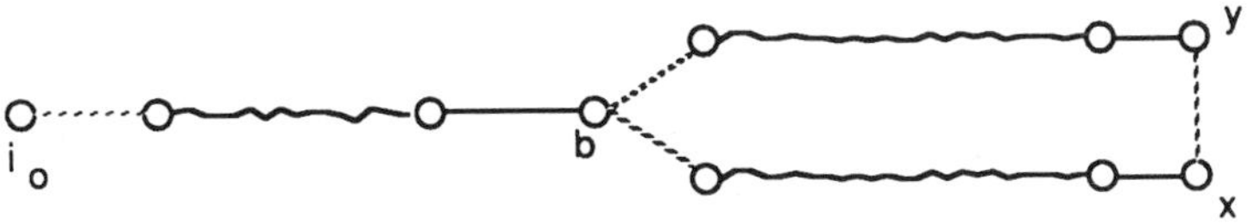

Fig. 9.

3.9. Lemma. *Let B be a blossom. There exists an augmenting path in G* mod *B with respect to $M-B$ if and only if there exists an augmenting path in G with respect to M.*

The construction of G mod B is known as "blossom shrinking". Algorithmically, the lemma is of crucial importance for finding augmenting paths. The different algorithms that exist are usually based on (very) efficient techniques of finding and manipulating blossoms. The algorithm of Micali & Vazirani follows the Hopcroft–Karp strategy, but succeeds in handling blossoms, so a maximal set of augmenting paths is found in each phase in $O(e)$ time. See Table 3 for the known complexity results.

The construction of maximum weight matchings roughly follows similar techniques and is based on White's theorem (cf. Section 3.1).

Once again the bipartite case is the easiest to handle. In this case the problem is also known as the assignment problem, because it can be interpreted as the problem of finding an optimal cost assignment of tasks to agents (machines, workers). For an explanation of the so-called Hungarian method to solve the assignment problem, see Papadimitriou & Steiglitz. (It runs in $O(n^3)$ time.) The problem can be formulated as a minimum cost maximum flow problem, which can be solved in

$$O(ne \log n/\log(2+m/n))$$

time. The general case is solved by a technique that exploits the formulation as a linear programming problem (due to Edmonds). Define the variables x_{ij} to mean $x_{ij}=1$ if $(i,j)\in M$, and 0 otherwise. Let B_k be any set of $2r_k+1$ nodes (i.e., an odd set). The fact that we want the x-variables to represent a matching leads to the constraint $\sum_{i,j\in B_k} x_{ij} = r_k$ for all k and sets B_k. Now maximum weight matching can be formulated as the following linear programming problem:

$$(*)\qquad \begin{array}{lll} \text{maximize:} & \sum_{i,j} w_{ij}x_{ij} & \\ \text{subject to:} & \sum_{\langle i,j\rangle\in E} x_{ij}\leqslant 1 & \text{for all } i\in V, \\ & \sum_{i,j\in B_k} x_{ij}\leqslant r_k & \text{for all sets } B_k,\ |B_k|=2r_k+1, \\ & x_{ij}\geqslant 0 & \text{for all } i,j. \end{array}$$

Table 3
Maximum cardinality matching algorithms for general (nonbipartite) graphs

Approximate year	Author(s)	Complexity
1965	Edmonds	$O(n^4)$
1967	Balinski	$O(n^3)$
1974	Kameda & Munro	$O(ne)$
1976	Lawler	$O(n^3)$
1976	Gabow	$O(n^3)$
1983	Gabow & Tarjan	$O(ne)$
1975	Even & Kariv	$O(\min\{n^{5/2}, \sqrt{n}\cdot e\log\log n\})$
1980	Micali & Vazirani	$O(\sqrt{n}\cdot e)$

3.10. THEOREM. *Every solution to the linear programming problem* (*) *has* $x_{ij} \in \{0, 1\}$ *for all* i, j *and hence can be interpreted as a matching.*

All that is "left" is to solve (*)! It turns out that the dual problem holds important clues to the efficient solution of (*). The dual problem has a variable u_i for every vertex i and a variable z_k for every set B_k, and reads as follows (by applying the duality theorem of linear programming):

$$\begin{aligned} \text{minimize:} \quad & \sum_i u_i + \sum_k r_k z_k \\ \text{subject to:} \quad & u_i + u_j + \sum_{k:i,j \in B_k} z_k \geqslant w_{ij} \quad \text{for all } i, j, \\ & u_i \geqslant 0 \quad \text{for all } i, \\ & z_k \geqslant 0 \quad \text{for all } k. \end{aligned}$$

One can redefine the computational problem as follows. Let $\pi_{ij} = u_i + u_j + \sum_{k:i,j \in B_k} z_k - w_{ij}$.

3.11. LEMMA. *A matching M has maximum weight if and only if the following conditions are satisfied for all nodes i, edges* $(i, j) \in E$, *and odd sets* $B_k \subseteq V$ *with* $|B_k| = 2r_k + 1$:

(a) $u_i, \pi_{ij}, z_k \geqslant 0$,
(b) *if* $\langle i, j \rangle$ *is matched, then* $\pi_{ij} = 0$,
(c) *if i is exposed, then* $u_i = 0$,
(d) *if* $z_k > 0$, *then* B_k *is maximally matched* (*i.e.*, $|\{\langle i, j \rangle \mid i, j \in B_k, (i, j) \in M\}| = r_k$ *and* B_k *is a blossom*).

PROOF. "Only if" follows from duality. The "if" part is seen as follows. Let M satisfy (a) through (d) and let M' be another matching. Observe

$$\sum_{(i,j) \in M'} w_{ij} \leqslant \sum_{(i,j) \in M'} \left(u_i + u_j + \sum_{k:i,j \in B_k} z_k \right) \leqslant \sum_i u_i + \sum_k r_k z_k = \sum_{(i,j) \in M} w_{ij},$$

by applying known facts for M (and the constraints for M'). □

The algorithms for maximum weight matching start with an empty matching M, $z_k = 0$ for all k, and $u_i = G$ for all i and a suitably large G (e.g. $G = \frac{1}{2} \max w_{ij}$). This satisfies (a), (b) and (d) but not (c). One now tries to make changes such that (a), (b) and (d) are preserved but the number of violations of (c) is reduced (this is done by finding augmenting paths between nodes i, j with $u_i = u_j > 0$). Again one uses blossom shrinking to find the augmenting paths. The further details are explained in Lawler. The most efficient implementations of maximum weight matching due to Gabow, Galil and Spencer run in time

$$O(ne \log n) \quad \text{and} \quad O(ne \log \log \log_d n + n^2 \log n) \quad \text{for } d = \max\{2, e/n\},$$

and an approach due to Gabow even achieves a bound of $O(ne + n^2 \log n)$ time. For planar graphs, a different algorithm based on the planar separator theorem yields a maximum weight matching in $O(n^{1.5} \log n)$ time.

The maximum matching algorithms are still fairly complex and invite to alternative approaches. One approach is to drop the strict requirement that a matching is maximum, and to construct a matching that is "approximately" maximum. A simple but rather crude bound is obtained as follows: construct a DFS tree T of G (in linear time), and construct a maximum matching in T (in linear time, cf. Section 3.1). Let the matching be M'.

3.12. THEOREM. $|M'| \geq \frac{1}{2}|M|$ *for any maximum matching* M *of* G.

A different approach aims at finding probabilistic algorithms for maximum matchings. The basic starting point is a result of Erdös & Renyi that asserts that the probability that an n-node graph (n even) with N edges has a perfect matching tends to 1 for $n \to \infty$, provided $N \geq \frac{1}{2}n \log n + w(n)n$ (for some $w(n)$ with $w(n) \to \infty$ for $n \to \infty$). Angluin & Valiant proposed the following randomized algorithm for finding a perfect matching with high probability (averaged over all graphs with $N \geq cn \log n$ for c sufficiently large). Suppose we succeeded building a (partial) matching M. If M has $\frac{1}{2}n$ edges, we can stop and report success (M is perfect). Otherwise, let x be the last-numbered exposed node in M and select (and delete) a random edge $(x, y) \in E$. If no edge exists, then stop and report failure. If y is exposed in M, then (x, y) is added to M and we repeat the entire procedure for the new matching. If y is matched in M with mate z, then delete (y, z) from M but add (x, y) to it (cf. Fig. 10) and repeat the procedure at node z. By using a suitable datastructure, the overhead in each step of the algorithm takes O(1) time. It can be shown that for $N \geq cn \log n$ the probability that the algorithm reports success within $O(n \log n)$ steps tends to 1 for $n \to \infty$. It follows that the algorithm "almost certainly" finds a perfect matching in $O(n \log n)$ time.

3.3. Maximum flow

Let G be a directed graph with a nonnegative capacity $c(e)$ assigned to every edge e. (We shall henceforth refer to G as a "network".) We are interested in solving the s-t maximum flow problem, normally referred to as *the maximum flow problem*, defined as follows: determine a maximum (legal) flow f from s to t, where s and t are specified nodes of G called the source and the target of the flow respectively. It is normally assumed that G is "loop-free" (i.e., when $(i, j) \in E$ then $(j, i) \notin E$), that s has in-degree 0 and that t has out-degree 0. The maximum flow problem is probably one of the most classical problems in combinatorial optimization on graphs. We will outline the basic theory and some very efficient, polynomial-time algorithms for it. The results are usually refined for special subcases of the problem, such as

(i) the network is of a special type (e.g. planar),
(ii) the edge capacities are integral, or
(iii) the edge capacities are all 0 or 1.

Fig. 10.

We will usually mention the results for these cases, but omit the details. Let $\overleftarrow{E} = \{(j,i)|(i,j)\in E\}$, and $\overleftarrow{e}=(j,i)$ for $e=(i,j)$.

DEFINITION. An *s-t flow* on G (or simply, a flow, when s and t are understood) is a function $f: E\cup\overleftarrow{E}\to\mathbb{R}$ satisfying the following constraints:

(i) for every $e\in E$, $0\leqslant f(e)\leqslant c(e)$ and $f(\overleftarrow{e})=-f(e)$, and

(ii) for every $i\neq s,t$, $\sum_{j:(j,i)\in E} f(j,i)=\sum_{j:(i,j)\in E} f(i,j)$.

The latter constraint is the basic conservation law, which requires that what flows in flows out for every node i different from s and t. The value $|f|$ of a flow is defined as $\sum_{i:(s,i)\in E} f(s,i)$. It is seen to be equal to $\sum_{i:(i,t)\in E} f(i,t)$: what flows out of s is precisely what flows into t.

DEFINITION. An *s-t cut* is a partition $X;\bar{X}$ of V such that $s\in X$ and $t\in\bar{X}$. The capacity of an s-t cut $X;\bar{X}$ is defined to be $c(X,\bar{X})=\sum_{(i,j)\in E: i\in X,\, j\in\bar{X}} c(i,j)$ and the flow across the cut is defined to be $f(X,\bar{X})=\sum_{(i,j)\in E: i\in X,\, j\in\bar{X}} f(i,j)$.

The following lemma can be shown by induction on $|X|$.

3.13. LEMMA. $|f|=f(X,\bar{X})-f(\bar{X},X)\leqslant c(X,\bar{X})$ *for any s-t cut* $X;\bar{X}$.

We can now establish the following basic fact, with a proof that is typical for many optimization problems on networks. Consider an arbitrary network G with source s and target t.

3.14. THEOREM. *Maximum flows exist.*

PROOF. By Lemma 3.13, all flow values are uniformly bounded (by the minimum capacity of any s-t cut). Let α be the least upper bound of the set of flow values. Then $\alpha=\lim_{i\to\infty}|f_i|$ with $\{f_i\}_{i\geqslant 0}$ a sequence of flows such that $|f_0|\leqslant|f_1|\leqslant\cdots$. Every flow f_i can be viewed as a tuple of flow values $f_i(e)$ $(e\in E)$ in finite-dimensional space. By induction on $|E|$ one proves that there must be an infinite subsequence $\{f'_i\}_{i\geqslant 0}$ of $\{f_i\}_{i\geqslant 0}$ such that $\lim_{i\to\infty}|f'_i|=\alpha$ and $\{f'_i(e)\}_{i\geqslant 0}$ is monotone (increasing or decreasing) for every $e\in E$. Define f by setting $f(e)=\lim_{i\to\infty} f_i(e)$ and $f(\overleftarrow{e})=-f(e)$ for every $e\in E$. Clearly f is well-defined and satisfies the flow conditions, hence f is a flow. As $|f|=\alpha$, it follows that f is a maximum flow. □

Other existence proofs follow e.g. from the formulation of the maximum flow problem as a linear programming problem or from the later algorithms that we shall present for computing maximum flows. The methods of finding a maximum flow mostly use the basic technique of successively augmenting a given flow. An *augmenting path* from s to t is any "path" π of edges $\in E\cup\overleftarrow{E}$ with the following properties:

(i) for any edge $e\in\pi(e\in E)$, $f(e)<c(e)$,

(ii) for any edge $\overleftarrow{e}\in\pi(e\in E)$, $f(e)>0$.

The edges $e\in\pi$ are sometimes called "forward edges", and the edges $\overleftarrow{e}\in\pi$ "backward edges". It should be clear why π is called an augmenting path. Define $\varepsilon_1=\min\{c(e)-$

$f(e)|e \in \pi(e \in E)\}, \varepsilon_2 = \min\{f(e)|\overleftarrow{e} \in \pi(e \in E)\}$ and $\varepsilon = \min\{\varepsilon_1, \varepsilon_2\}$. Adding ε flow to every "edge" in π results in a flow f' with $|f'| = |f| + \varepsilon$, i.e., a flow of larger value! (Note that the process of augmentation effectively increases the flow by ε on every forward edge, and decreases the flow by ε on every backward edge.) The following result is crucial and due to Ford & Fulkerson.

3.15. THEOREM. (i) (the augmenting path theorem) *f is a maximum flow if and only if there is no augmenting path for f.*

(ii) (the max-flow min-cut theorem) *The maximum value of an s-t flow is equal to the minimum capacity of an s-t cut.*

PROOF. (i) Let f be a maximum flow. It is clear that there can be no augmenting path, otherwise a flow augmentation along the path would yield a larger flow (contradiction). Next suppose there is no augmenting path from s to t. Any attempt to build a path of edges $\in E \cup \overleftarrow{E}$ from s to t must end at a node where all next edges $e \in E$ are saturated (i.e., $f(e) = c(e)$) and all next edges $\overleftarrow{e}$ ($e \in E$) have $f(e) = 0$. Define X to be the set of nodes v (including s) such that there is an augmenting path from s to v. It follows that $X;\bar{X}$ is an s-t cut and $|f| = f(X, \bar{X}) = c(X, \bar{X})$. As this is the largest value a flow can have, f must be maximum.

(ii) Let f be a maximum s-t flow. By the preceding argument there is an s-t cut $X;\bar{X}$ such that $|f| = c(X, \bar{X})$. As $|f| \leqslant c(Y, \bar{Y})$ for any s-t cut $Y;\bar{Y}$, it follows that $X;\bar{X}$ is an s-t cut of minimum capacity. □

From the proof it follows that for any maximum s-t flow and minimum capacity s-t cut $X;\bar{X}$ the edges leading "across the cut" from X to $\bar{X}$ are saturated and the edges leading from $\bar{X}$ to X carry flow 0. The max-flow min-cut theorem can also be derived from the duality theorem of linear programming.

COROLLARY (the integral flow theorem). *If all capacities are integers, there is a maximum flow which is integral (i.e., with an integral flow through every edge).*

PROOF. Consider any feasible integral flow f. If f is not maximum, there must be an augmenting path. Augmenting the flow along this path yields another feasible integral flow f' with $|f'| \geqslant |f| + 1$. By repeating this we must arrive at an integral flow that is maximum. □

While these are the main theorems on which the construction of maximum flows is based, the theory can be ramified further. For example, one may want to impose limits on the capacity of the various nodes and wish to determine a maximum flow f subject to the (additional) requirement that $\sum_{j:(j,i) \in E} f(j, i) \leqslant c_i$, i.e., the total flow into every node i does not exceed its capacity c_i ($i \neq s, t$). This problem can be reduced to an ordinary maximum flow problem as follows: split each node $i \neq s, t$ into two nodes i' and i'' connected by an edge (i', i''), let the edge (i', i'') have capacity c_i, and lead all incoming edges of i to i' and all outgoing edges from i''. By virtue of this transformation the previous theorems go through virtually unchanged.

Another common constraint is to impose a lower bound on the flow through each edge, in addition to the upper bound (or: capacity) for each edge. The results here require that we change our view from flows to circulations. Let $l(e)$ and $c(e)$ denote the lower and upper bound respectively, of the flow through edge $e \in E$.

DEFINITION. A *circulation* in G is a function ("flow") $f: E \cup \overleftarrow{E} \to \mathbb{R}$ satisfying the following constraints:

(i) for every edge $e \in E$, $0 \leqslant f(e) \leqslant c(e)$ and $f(\overleftarrow{e}) = -f(e)$, and

(ii) for every i, $\sum_{j:(j,i)\in E} f(j,i) = \sum_{j:(i,j)\in E} f(i,j)$.

Thus circulations simply are flows that observe the basic conservation law at all nodes. The maximum flow problem can easily be converted to circulation form as follows: add an edge (t, s) to the flow network with $l(t, s) = 0$ and $c(t, s) = \infty$, and ask for a circulation f for which $f(t, s)$ is maximum. A typical scenario for determining a maximum flow "with lower bounds" is the following: first determine a feasible circulation, next augment the implicit (feasible) s-t flow to a maximum flow. The following result is due to Hoffman. A circulation is called feasible if $l(e) \leqslant f(e) \leqslant c(e)$.

3.16. THEOREM (the circulation theorem). *In a network with lower bounds and capacities, a feasible circulation exists if and only if* $\sum_{i\in Y,\, j\in \bar{Y}} l(j,i) \leqslant \sum_{i\in Y,\, j\in \bar{Y}} c(i,j)$ *for every cutset* $Y;\bar{Y}$.

PROOF. Let f be a feasible circulation. One can prove by induction that the effective flow across any cut must be zero, i.e., $f(Y, \bar{Y}) - f(\bar{Y}, Y) = 0$. Now

$$\sum_{i\in Y,\, j\in \bar{Y}} l(j,i) \leqslant f(\bar{Y}, Y) = f(Y, \bar{Y}) \leqslant \sum_{i\in Y,\, j\in \bar{Y}} c(i,j)$$

as claimed, for any cut $Y;\bar{Y}$.

Conversely, assume that $\sum_{i\in Y,\, j\in \bar{Y}} l(j,i) \leqslant \sum_{i\in Y,\, j\in \bar{Y}} c(i,j)$ for every cut $Y;\ \bar{Y}$. Suppose that the network admits no feasible circulation. Consider any circulation f and let u, v be two nodes with $(u, v) \in E$ and $f(u, v) < l(u, v)$. We claim that there must be an augmenting path from v to u, i.e., a path in which each forward edge e has $f(e) > l(e)$ and each backward edge $\overleftarrow{e}$ has $f(e) < c(e)$. For suppose this is not the case. Let Y be the set of nodes which can be reached from v by an augmenting path. Clearly $Y;\bar{Y}$ is a v-u cut, every edge from Y and $\bar{Y}$ must be saturated and every edge e from $\bar{Y}$ to Y must have $f(e) \leqslant l(e)$. Assuming (as we may) that f does satisfy the basic conservation law, we have

$$\sum_{i\in Y,\, j\in \bar{Y}} c(i,j) = f(Y, \bar{Y}) = f(\bar{Y}, Y) < \sum_{i\in Y,\, j\in \bar{Y}} l(j,i)$$

(with strict inequality because of the edge (u, v)). This contradicts the assumption. Thus consider any augmenting path from v to u, and let δ be the smallest surplus on any edge (i.e., $f(e) - l(e)$ on a forward edge and $c(e) - f(e)$ on a backward edge). We can now "augment" f by redistributing flow as follows: add a flow of δ to edge (u, v) and subtract a flow of δ from the edges in the augmenting path (i.e., subtract δ from every forward edge and add δ to every backward edge in it). Apply this procedure to any network with a circulation f for which $\min_{e\in E}|f(e)-l(e)|$ is smallest, and it follows that we obtain a circulation for which this minimum is smaller yet unless it is zero. Contradiction. □

In a flow network with lower and upper bounds, it may be of interest to determine both a maximum and a minimum flow that satisfy the contraints. The following result is again due to Ford and Fulkerson.

3.17. THEOREM. *Let G be a flow network with lower bounds and capacities, which admits a feasible s-t flow. Define the capacity of a cutset* $Y;\bar{Y}$ *as*

$$\sum_{i\in Y,\, j\in \bar{Y}} c(i,j) - \sum_{i\in Y,\, j\in \bar{Y}} l(j,i) = cap(Y;\bar{Y})$$

(i) (the generalized max-flow min-cut theorem) *The maximum value of an s-t flow is equal to the minimum capacity of an s-t cut.*

(ii) (the min-flow max-cut theorem) *The minimum value of an s-t flow is equal to the maximum of* $\sum_{i\in Y, j\in \bar{Y}} l(j,i) - \sum_{i\in Y, j\in \bar{Y}} c(i,j)$ *over all s-t cuts* $Y;\bar{Y}$, *i.e., minus the minimum of the capacity of an t-s cut.*

PROOF. (i) Convert the flow network into a circulation network by adding an edge (t,s) with $l(t,s)=l$ and $c(t,s)=\infty$, for a "parametric" value of l. Because G admits a feasible flow f, the modified network admits a feasible circulation for any $l\leqslant|f|$. Clearly, the maximum s-t flow in G has a value equal to the maximum value of l for which the modified network admits a feasible circulation. By the circulation theorem a feasible circulation exists if and only if $l\leqslant cap(Y;\bar{Y})$. The result follows.

(ii) Similarly, by considering the circulation network obtained by adding an edge (t,s) with $l(t,s)=0$ and $c(t,s)=l$. □

An important type of flow problem arises when a cost (or weight) $b(i,j)\geqslant 0$ is involved with each unit of flow through an edge (i,j). The cost of a flow f is defined as $\sum_{e\in E} b(e)f(e)$. The *minimum cost flow problem* asks for the s-t flow of some value v and (among all flows of this value) of least cost. Usually, the minimum cost flow problem is understood to be the problem of finding a maximum flow of least cost. The methods for finding a minimum cost flow typically start with a feasible flow (i.e., a flow of value v or a maximum flow) and transform it into another flow of some value and e.g. lesser cost by working on "cost-reducing augmenting paths". The cost of an augmenting path or cycle is defined as the sum of the costs of the flow through the forward edges minus the sum of the costs of the flow through the backward edges.

3.18. THEOREM (the minimum cost flow theorem). (i) *A flow of value v has minimum cost if and only if it admits no augmenting cycle of negative cost.*

(ii) *Given a minimum cost flow of value v, the augmentation by δ along an s-t augmenting path of minimum cost yields a minimum cost flow of value* $v+\delta$.

PROOF. (i) Suppose f has value v and minimum cost. Clearly no negative cost augmenting cycle can exist, otherwise we could introduce a (small) amount of extra flow around the cycle without changing the value of the flow but reducing its cost. Conversely, let f have value v and no negative cost augmenting cycle. Suppose f does not have minimum cost, but let f^* be a flow of value v that has minimum cost. Now $f^* - f$ is a "flow" of value 0 and negative cost. The "flow" can be decomposed into cycles of flow (by reasoning backwards from every edge carrying non-zero flow) at least one of which apparently must have negative cost. Contradiction.

(ii) Let f be a minimum cost flow of value v, and suppose the augmentation by δ along the minimum cost s-t augmenting path π introduced an augmenting cycle C of negative cost. Thus C includes one or more edges affected by the augmentation along π in reversed direction, to account for the negative cost of C. Define $\pi \oplus C$ to be the set of edges of $\pi \cup C$ except those which occur in π and reversed on C. The set $\pi \oplus C$ partitions into an s-t path π' and a number of cycles, and the total flow cost in the edges of $\pi \oplus C$ is less than the cost of π. Clearly π' must be an s-t augmenting path of cost less than the cost of π, as the cycles cannot have nonnegative cost by (i) and the assumption of f. Contradiction. □

The minimum cost flow theorem is implicit in the work of Jewell and of Busacker and Gowen.

COROLLARY (the integrality theorem for minimum cost flows). *If all capacities are integer, there is a maximum flow which is integral and has minimum cost (over all maximum flows).*

PROOF. Similar to the proof of the integrality theorem, by using (ii) of the minimum cost flow theorem. □

In the subsequent sections we will see that many versions of the maximum flow problem can be solved by polynomial-time algorithms. Goldberg et al. give a detailed account of later developments and efficient algorithms for many network flow problems.

It has been shown that the maximum flow problem (i.e., the problem to determine whether there exists a flow of value $\geqslant K$ for specified K) is logspace-complete for the class P of all polynomially computable problems.

3.4. *Computing maximum flows*

3.4.1. *Augmenting path methods*

Many algorithms to determine a maximum flow in a network start from Ford and Fulkerson's augmenting path theorem. Given a flow f and an augmenting path π, let the maximum amount by which we can augment f along π be called the *residual capacity* $res(\pi)$ of π. (Thus $res(\pi) = \min\{res(e) | e \in \pi\}$ with $res(e) = c(e) - f(e)$ if $e \in E$ and $res(e) = -f(e)$ if $e \in \overleftarrow{E}$). Augmenting path methods are based on the iteration of one of the following types of steps, starting from any feasible flow (e.g. $f \equiv 0$):

(1) *clairvoyant augmentation*: augment the flow along some s-t augmenting path π by an amount $\leqslant res(\pi)$,
(2) *Ford–Fulkerson augmentation*: find an augmenting path π and augment the flow by $res(\pi)$ along π,
(3) *maximum capacity augmentation*: find an augmenting path π of maximum residual capacity and augment the flow by $res(\pi)$ along π,
(4) *Edmonds–Karp augmentation*: find a shortest augmenting path π and augment the flow by $res(\pi)$ along π.

Clairvoyant augmentation is only of theoretical interest, but the following observation will be useful.

PROPOSITION. *A maximum flow can be constructed using at most e clairvoyant augmentation steps (along augmenting paths without backward edges).*

PROOF. Let f be a (maximum) non-zero flow in G, and let $e \in E$ be an edge with the smallest non-zero amount of flow $f(e)$. One easily verifies that there must be an s-t path π containing only forward edges including e. Decrease the flow in every edge along π by $f(e)$, obtaining a flow f' with $|f'|<|f|$. (Note that f is obtained from f' by a clairvoyant augmentation along π.) Repeat this process as long as a non-zero flow is obtained. Because in each step at least one more edge is given a zero flow, the process goes on for at most e steps (and thus leads to at most e paths) and ends with the zero flow. By reversing the process and augmenting the flow by the right amount along the paths (taken in reverse order) one obtains f again. □

The other augmenting path methods are more interesting and lead to a variety of polynomial-time algorithms for computing maximum flows. The main problem in a Ford–Fulkerson augmentation step is finding an augmenting path. The usual procedure for this is reminiscent of a breadth-first search algorithm and inductively generates for every node u (beginning with the neighbors of s) a "label" that indicates whether there exists an augmenting path from s to u. The label will contain a pointer to the predecessor on the augmenting path, if one exists, in order to be able to trace an augmenting path when the labeling procedure terminates at t. If the labeling procedure gets stuck before t is reached, then no augmenting path from s to t exists and the flow must be maximum. (Observe that in this case $X;\bar{X}$ with X consisting of s and the labeled vertices must be an s-t cut of minimum capacity.) When suitably implemented a Ford–Fulkerson augmentation step takes only $O(e)$ time. It follows that in a network with all integer capacities, a maximum flow can be constructed by Ford–Fulkerson augmentation in $O(ef^*)$ time, where f^* is the value of the maximum flow. On the other hand, Ford and Fulkerson have shown that Ford–Fulkerson augmentation on an arbitrary network does not necessarily converge to the maximum flow. (Their example shows that the algorithm may get bogged down on a never-ending sequence of augmenting paths with residual capacity converging to zero, without ever coming close to the maximum flow.) Thus at each step of an augmenting path method, the augmenting path to work on must be properly chosen in order to obtain a viable maximum flow algorithm on general networks. We will first look at maximum capacity augmentation, suggested by Edmonds and Karp.

3.19. THEOREM. (i) *Maximum capacity augmentation produces a sequence of flows that always converges to a maximum flow.*

(ii) *If the capacities are all integers, maximum capacity augmentation finds a maximum flow in*

$$O(\min\{e \log c^*, 1+\log_{M/M-1} f^*\})$$

iterations where c^* is the maximum edge capacity, f^* the value of a maximum flow and $M>1$ is the maximum number of edges across any s-t cut.

PROOF. (i) Let f be a flow and f^* a maximum flow, chosen such that $f^* \geqslant f$ (by the augmenting path theorem). One easily verifies that f^*-f is a (maximum) flow of value $|f^*|-|f|$ in the residual graph, i.e., the graph G in which each edge is given its current residual capacity. By the preceding proposition there must be an augmenting path over which one can augment the flow by at least $(|f^*|-|f|)/e$ which is thus a lower bound on the residual capacity of the current maximum capacity augmenting path. On the other hand, over the next $2e$ augmentations (as long as the maximum flow is not reached) the residual capacity of a maximum augmenting path must go down to $(|f^*|-|f|)/2e$ or less. Thus after $2e$ or fewer augmentations, the residual capacity of the maximum augmenting path is reduced by at least a factor of 2.

(ii) In the special case that all capacities are integers, the residual capacity of a maximum augmenting path is initially at most c^* and will never be less than 1. Thus a maximum flow must be obtained after at most $O(e \log c^*)$ maximum capacity augmentations. To obtain another bound, let $f_0, f_1, \ldots, f_k = f^*$ be the sequence of flows obtained by iterating maximum capacity augmentation and let $\varepsilon_i = |f_{i+1}|-|f_i|$ $(1 \leqslant i < k)$ be the ith flow increment. Let X_i consist of s and all points reachable from s by an augmenting path of residual capacity $>\varepsilon_i$. Clearly, $X_i;\bar{X}_i$ is an s-t cut and all edges across the cut have either a residual capacity $\leqslant \varepsilon_i$ (for edges from X_i to $\bar{X}_i$) or capacity $\leqslant \varepsilon_i$ (for edges from $\bar{X}_i$ to X_i). Thus

$$|f^*|-|f_i| \leqslant c(X,\bar{X}) - f(X,\bar{X}) + f(X,\bar{X}) \leqslant \varepsilon_i M = (|f_{i+1}|-|f_i|)M$$

or, equivalently $|f^*|-|f_{i+1}| \leqslant (|f^*|-|f_i|)(1-1/M)$. By induction we have

$$|f^*|-|f_i| \leqslant |f^*|(1-1/M)^i$$

and, hence, $1 \leqslant |f^*|(1-1/M)^{k-1}$. The bound on k follows. □

Finding an augmenting path of maximum residual capacity can be done by a procedure that is reminiscent of Dijkstra's shortest path algorithm, within $O(e \log n)$ or even less time. Thus, if all capacities are integers, maximum capacity flow augmentation yields a maximum flow in $O(m^2 \log n \log c^*)$ or less time.

Polynomial-time algorithms for general networks (with a complexity independent of c^* or f^*) require a different approach. A classical starting point is Edmonds–Karp augmentation, in view of the following result.

3.20. THEOREM. *Edmonds–Karp augmentation yields a maximum flow after not more than $\frac{1}{2}ne$ iterations.*

As a shortest augmenting path is easily found in O(e) time (by a suitable version of breadth-first search), Edmonds–Karp augmentation yields an O(ne^2) time algorithm for determining a maximum flow in any network. By extending the idea, Dinic observed that a more efficient algorithm results if one can construct all shortest augmenting paths first in one step (or phase) before performing the necessary augmentations on the subgraph. Define the distance $d(u)$ of a node u as the length of the shortest augmenting path from s to u. (Set $d(u)$ to ∞ if no such path exists.) Define the critical graph N as the subgraph of G containing only the edges $e=(u,v)\in E\cup \overleftarrow{E}$ with $d(v)=d(u)+1$, with all nodes not on a path from s to t omitted. Every edge is assigned the residual capacity with respect to f. Clearly $s\in N$ and, provided f is not maximum, $t\in N$ as well. Observe that N contains precisely the shortest augmenting paths from s to t. N is also called the level graph or the layered network of f. Finally, let us call g a *blocking* flow if every path from s to t contains an edge of residual capacity zero.

(5) *Dinic augmentation*: construct the critical graph N, find a blocking flow g in N, and "augment" f by replacing it by the flow $f+g$.

3.21. THEOREM. *Dinic augmentation yields a maximum flow after at most $n-1$ iterations.*

PROOF. The idea of the proof is that $d(t)$, i.e. the length of the shortest augmenting path from s to t, increases with every Dinic augmentation. Let f be the current flow, d the distance function, and N the critical graph. Let f', d', and N' be the corresponding entities after one Dinic augmentation (assuming that f is not maximum). If $t\notin N'$, the flow f' is maximum, $d(t)=\infty$ and we are done. Thus assume f' is not maximum (hence $t\in N'$). Consider a path $s=u_0, u_1, \ldots, u_k=t$ from s to t in N' with $d'(u_i)=i$. We distinguish two cases:

Case I: $u_i\in N$ *for* $0\leqslant i\leqslant k$. We argue by induction that $d'(u_i)\geqslant d(u_i)$. For $i=0$ this is obvious. Suppose the claim holds for i, and consider u_{i+1}. If $d(u_{i+1})\leqslant d(u_i)+1$, the induction step follows immediately. Thus let $d(u_{i+1})>d(u_i)+1$. Then the edge (u_i, u_{i+1}) is not in N despite the fact that it has a non-zero residual capacity (because it was not affected by the augmentation and is an edge in N'). This contradicts that $d(u_{i+1})> d(u_i)+1$, thus completing the induction. It follows in particular that $d'(t)\geqslant d(t)$. Suppose that $d(t')=d(t)$. Then $d'(u_i)=d(u_i)$ for $0\leqslant i\leqslant k$, and the entire path must have been a path in N. This contradicts the fact that all paths in N were "blocked" after the Dinic augmentation. Hence $d'(t)>d(t)$.

Case II: $u_j\notin N$ *for some* j. Choose u_j to be the first node on the path with $u_j\notin N$. Clearly $j>0$ and, by the preceding argument, $d'(u_i)\geqslant d(u_i)$ for $0\leqslant i<j$. Consider the edge $(u_{j-1}, u_j)\in N'$. Because it has non-zero residual capacity now, it must have had so before the augmentation. The fact that $u_{j-1}\in N$ but $u_j\notin N$ implies that necessarily $d(u_j)=d(t)$ and (hence) $d(t)\leqslant d(u_{j-1})+1$. We conclude that $d'(t)>d'(u_{j-1})+1\geqslant d(u_{j-1})+1\geqslant d(t)$.

As the distance of t is at least 1 and at most $n-1$, and increases by at least 1 with each Dinic augmentation, the number of Dinic augmentations required to obtain a maximum flow is at most $n-1$. □

Given a flow f, its critical graph can be determined in O(e) time by means of breadth-first search. Thus implementations of Dinic augmentation for computing maximum flows require $O(ne+nF)$ time, where $F=F(n,e)$ is a bound on the time to determine a blocking flow in a network of (at most) n nodes and e edges. Dinic's original algorithm determines a blocking flow by saturating some edges in a path from s to t (in N) path after path, and achieves $F=O(ne)$ and (hence) a total running time of $O(n^2e)$. Later algorithms greatly improved the efficiency of a Dinic augmentation, by constructing a blocking flow in a different manner and/or using special datastructures. The most efficient implementation to date is due to Sleator and Tarjan and achieves $F=O(e \log n)$, which thus yields an $O(ne \log n)$ time algorithm to determine maximum flows. Table 4 shows the interesting sequence of maximum flow algorithms that developed over the years. (Goldberg's algorithm will be outlined below.) If all capacities are integers, an approach due to Gabow yields an $O(ne \log U)$ algorithm, where U is the maximum edge capacity. Ahuja and Orlin developed a general $O(ne+n^2\log U)$ algorithm based on Goldberg's techniques, which was improved by Ahuja, Orlin and Tarjan to an $O(ne \log(n\sqrt{\log U+2}/e))$ algorithm.

There are a number of results for more specialized networks that deserve mentioning at this point. For a network with all edge capacities equal to 1, Even and Tarjan proved that a maximum flow is obtained after no more than $O(\min\{n^{2/3}, e^{1/2}\})$ Dinic augmentations. In a network in which all edge capacities are integers and each node $u \neq s, t$ has either one single incoming edge, of capacity 1, or one single outgoing edge, of capacity 1, at most $2\lceil\sqrt{n-2}\,\rceil$ Dinic augmentations are needed.

Next observe that for planar (directed) networks the best maximum flow algorithms we have seen so far still require $O(n^2\log n)$ time or more (take $e=O(n)$). There is at least one case in which the maximum flow in a planar network can be obtained much easier, by an intuitively appealing technique. Define a flow network to be *s-t planar* if it is planar and s and t are nodes on the same face. (The notion is due to Ford and Fulkerson, on a suggestion of Dantzig.) Let G be an s-t planar network, and assume without loss of generality that s and t lie on the boundary of the exterior face. Embed

Table 4
Some maximum flow algorithms and their complexity. (U denotes the maximum edge capacity.)

Approximate year	Author(s)	Complexity
1956	Ford and Fulkerson	–
1969	Edmonds and Karp	$O(ne^2)$
1970	Dinic	$O(n^2e)$
1974	Karzanov	$O(n^3)$
1977	Cherkasky	$O(n^2\sqrt{e})$
1978	Malhotra, Pramodh Kumar and Maheshwari	$O(n^3)$
1978	Galil	$O(n^{5/3}e^{2/3})$
1980	Sleator and Tarjan	$O(ne \log n)$
1984	Tarjan	$O(n^3)$
1985	Goldberg	$O(n^3)$
1986	Goldberg and Tarjan	$O(ne \log(n^2/e))$
1988	Ahuja, Orlin and Tarjan	$O(ne \log(n\sqrt{\log U+2}/e))$

G such that it lies in a vertical strip with s located on the left-bounding line of the strip and t on the right. Let π be the topmost path from s to t, obtained by always choosing the leftmost edge in a node on the way from s to t and backtracking when one gets stuck. The following result is due to Ford and Fulkerson.

3.22. LEMMA. *The path π contains precisely one edge "across the cut" from X to $\bar{X}$, for every s-t cut $X;\bar{X}$ of minimum capacity.*

PROOF. Consider any minimum capacity s-t cut $X;\bar{X}$. Clearly π must contain at least one edge across the cut from X to $\bar{X}$. Suppose it contains more than one edge across the cut, say $e_1=(u_1,v_1)$ and $e_2=(u_2,v_2)$, with e_1 occurring before e_2 on the path. Because the s-t cut has minimum capacity there must be directed paths π_1 and π_2 from s to t which contain e_1 and e_2 respectively, as the only edge across the cut from X to $\bar{X}$. Let π'_1 be the part of π_1 from v_1 to t, π'_2 the part of π_2 from s to u_2. Because π runs "above" π_1 and π_2, π'_1 and π'_2 must intersect a node u. But now the path from s to t obtained by following π'_2 from s to u and π'_1 from u to t has no edge across the cut from X to $\bar{X}$! Contradiction. □

The lemma suggests a simple way of computing a maximum flow due to Dantzig (but sometimes referred to as Berge's algorithm).

(6) *Dantzig augmentation* (for s-t planar networks): find the topmost path π (forward edges only) with no saturated edges, and augment the flow by $res(\pi)$ along π.

3.23. THEOREM. *Given an s-t planar network, Dantzig augmentation yields a maximum flow after at most $e=O(n)$ iterations.*

PROOF. Each Dantzig augmentation saturates at least one more edge, and hence there can be no more than e iterations. By the lemma it follows that whenever no further Dantzig augmentation can be carried out, the edges across the cut from X to $\bar{X}$ for any minimum capacity s-t cut $X;\bar{X}$ must all be saturated. By the max-flow min-cut theorem, the flow must be maximum. □

A straightforward implementation of Dantzig augmentation requires a total of $O(e)$ edge inspections, and $O(n)$ time to augment the flow per iteration. Thus one can construct maximum flows in s-t planar networks in $O(n^2)$ time. Itai and Shiloach have developed an implementation of the algorithm that requires only $O(n \log n)$ time. As an interesting generalization Johnson and Venkatesan proved that a maximum flow in a planar network can be determined in $O(pn \log n)$ time, where p is the fewest number of faces that needs to be traversed to get from s to t. Kučera proved that maximum flows in s-t planar networks can be computed in linear expected time.

There have been various approaches to the maximum flow problem for general (undirected) planar graphs. Using a divide-and-conquer approach Johnson & Venkatesan obtained an $O(n^{3/2}\log n)$ algorithm for the problem. Further progress came by utilizing suitable algorithms for the determination of minimum s-t cuts in planar graphs. Itai and Shiloach gave an $O(n^2\log n)$ algorithm for this problem, but Reif obtained a much faster $O(n \log^2 n)$ algorithm for it. Using additional techniques Hassin and Johnson obtained an $O(n \log^2 n)$ algorithm for the complete maximum flow problem. Frederickson has shown how both minimum s-t cuts and maximum flows can be computed in $O(n \log n)$ time for general planar networks, using further techniques for shortest path computations.

3.4.2. Maximum flow algorithms (*continued*)

In 1985 Goldberg presented an entirely new approach to the construction of maximum s-t flows in directed networks. The approach uses several insights from the preceding algorithms, but abandons the idea of flow augmentations along entire paths and of maintaining legal flows at all times. The generic version of Goldberg's algorithm consists of two phases. In phase I an infinite amount of flow is put on s, and an iterative scheme is called upon to let each node push as much flow as it can according to the residual capacities of the links "in the direction of the shortest path (in the residual graph)" to t. The result will be that usually more flow is entering a node than is leaving it, but that eventually all edges across any minimum cut get saturated. In phase II of the algorithm all excess flow that is entering the nodes is removed without altering the overal (maximum) value of the flow, thus resulting in a legal flow that is maximum. We describe the algorithm in more detail to appreciate some of the underlying mechanisms. (See also the survey by Goldberg et al. for an account of the algorithm and its ramifications.)

DEFINITION. A preflow on G is a function $g: E \cup \overleftarrow{E} \to \mathbb{R}$ satisfying the following constraints:

(i) for every $e \in E$, $0 \leqslant g(e) \leqslant c(e)$ and $g(\overleftarrow{e}) = -g(e)$, and

(ii) for every $i \neq s, t$, $\sum_{j:(j,i) \in E} g(j,i) \geqslant \sum_{j:(i,j) \in E} g(i,j)$.

Throughout phase I of Goldberg's algorithm, every node v maintains two items of information:

(i) the current flow excess $\Delta(v)$ defined by $\Delta(s) = \infty$ and $\Delta(i) = \sum_{j:(j,i) \in E} g(j,i) - \sum_{j:(i,j) \in E} g(i,j)$ for all i, and

(ii) a lower bound $d(v)$ on the length of the shortest path from v to t in the residual graph.

For the estimates $d(v)$ we only require that $d(t) = 0$ and $d(w) \geqslant d(v) - 1$ for every residual edge (v, w). At the start of phase I we set out with the zero flow for g and the ordinary distance $d(v, t)$ from v to t for $d(v)$. The initialization only requires $O(n + e)$ time. In the subsequent algorithm a node v will be called active if $\Delta(v) > 0$ and $0 < d(v) < n$. The idea of phase I is to let active nodes release as much flow as possible towards t, according to some iterative scheme, until no more active nodes remain. In the generic version of the algorithm, one of the following actions can be carried out in every step:

(A) *Push*: Select an active node v and a residual edge (v, w) with $d(w) = d(v) - 1$. Let the edge have residual capacity r. Now send a flow of $\delta = \min\{\Delta(v), r\}$ from v to w, updating g and the flow excesses in v and w accordingly. (The push is called saturating if $\delta = r$ and nonsaturating otherwise.)

(B) *Relabel*: Select a node v with $0 < d(v) < n$, and replace $d(v)$ by the minimum value of $d(w) + 1$ over all nodes w for which (v, w) is a residual edge (∞ if no such nodes exist).

Of course *Relabel* steps are only useful if they actually change (i.e., increase) a $d(v)$-value. The following lemma is easily verified.

3.24. Lemma. *Every step of the generic algorithm maintains the following invariants:*
(i) *g is a preflow,*
(ii) *d is a valid labeling, and*
(iii) *if $\Delta(v)>0$ then there is a directed path from s to v in the residual graph.*
Furthermore, the $d(v)$-values never decrease.

We now show that the generic algorithm always terminates within polynomially many steps regardless of the regime of *Push* and *Relabel* instructions, provided we ignore *Relabel* instructions that do not actually change a label ("void relabelings").

3.25. Theorem. *Phase* I *terminates within* $O(n^2e)$ *steps, provided no void relabelings are counted.*

Proof. Every counted *Relabel* step increments a $d(v)$-value with $v\neq t$ and $d(v)<n$ by at least 1. Thus there can be at most $n-1$ *Relabel* steps that affect a given $d(v)$-value for $v\neq t$, thus $O(n^2)$ *Relabel* steps in all.

Next we estimate the number of saturating pushes involving a given edge $(u, v)\in E$. If $v=t$, the edge can be saturated at most once (because $d(t)$ remains 0 forever). For $v\neq t$, a saturating *Push* from u to v can only be followed (eventually) by another saturating *Push* over the same edge, necessarily from v to u, if $d(v)$ has increased by at least 2 in the meantime. A similar observation holds for $d(u)$ if we start with a saturating *Push* from v to u. (Note that $d(u)$ and $d(v)$ can only increase.) Thus the edge can be involved in saturating *Push* steps at most $n-2$ times, and the total number of saturating *Push* steps is bounded by $O(ne)$.

To estimate the number of nonsaturating *Push* steps, we use the entropy function $\Phi=\sum_{v\,\text{active}} d(v,t)$ with distances being measured in the residual graph of the current preflow. A saturating *Push* causes Φ to increase by at most $n-2$. A nonsaturating *Push* causes Φ to decrease by at least 1 (because the node sending flow pushes all its excess flow and thus turns nonactive). As the total increase of Φ is thus bounded by $O(ne)\cdot(n-2)=O(n^2e)$ and the initial value is bounded by $O(n^2)$, the number of nonsaturating *Push* steps is bounded by $O(n^2e)$. It follows that phase I terminates within $O(n^2e)$ steps, provided no void relabelings are counted as steps. □

One can implement phase I to run in $O(n^2e)$ time, using a queue A to maintain the set of active nodes. It is clear from the theorem that we can do better only if we change the regime of instructions so as to get by with fewer nonsaturating *Push* steps. Note that *Relabel* steps are best executed when we visit an active node, and thus add only $O(1)$ overhead to a *Push* step. Assume that at every node the outgoing and incoming edges are stored in a circular list, with a pointer to the "next" edge that must be considered. (Initially it is a random first edge.) Consider the following type of step:

(C) *Discharge*: Select the next active node v from the front of A (and delete it from A). Consider the edges incident to v in circular order starting from the "current" edge, and apply *Push*/*Relabel* steps until $\Delta(v)$ becomes 0 or $d(v)$ increases. (Note that at least one of the two must arise before a full cycle around the list is completed.) If a *Push* from v to some node u causes $\Delta(u)>0$ (necessarily implying that u turns active), then add u to the rear of the queue A. After handling v, add it to the rear of A also, provided it is still active.

Define a "pulse" as follows. The first pulse consists of applying *Discharge* to the initial queue $A=\{s\}$. For $i>1$, the ith pulse consists of applying the *Discharge* steps to all nodes that were put into the queue during the $(i-1)$st pulse. Consider applying *Discharge* steps until $A=\emptyset$.

3.26. LEMMA. *Phase* I *terminates after at most* $O(n^2)$ *pulses.*

PROOF. Consider the entropy function $\Phi=\max\{d(v)\mid v \text{ active}\}$. If Φ does not decrease during a pulse, then the $d(v)$-value of some node must increase by at least 1. Thus there can be at most $(n-1)^2$ pulses in which Φ does not decrease. As $0<\Phi<n$ until there are no more active nodes, the total number of pulses in which Φ decreases can be no more than $(n-1)^2+(n-1)$. Thus phase I terminates after $2(n-1)^2+(n-1)=O(n^2)$ pulses. □

Observe that by the way a *Discharge* step is defined, there can be at most one nonsaturating *Push* step for each node v (namely the one that makes $\Delta(v)=0$) in every pulse. Thus the number of nonsaturating *Push* steps in this implementation of phase I is bounded by $O(n^2)\cdot(n-1)=O(n^3)$. Given that there are at most $O(n^2)$ *Relabel* steps and $O(ne)$ saturating *Push* steps as proved in the theorem, it easily follows that this implementation of phase I has a total runtime bounded by $O(n^3)$. A more detailed analysis leads to a bound of $O(n^2\sqrt{e})$ (Cherigan and Maheshwari). Goldberg and Tarjan have shown that by exploiting very special datastructures and yet another regime to cut down on the number of nonsaturating *Push* steps, the complexity of phase I can be further reduced to $O(ne\log n^2/e)$.

When phase I terminates, it does not necessarily terminate with a maximum flow. (Indeed g need not be a legal flow yet.) But the following lemma confirms the intuition that the current preflow saturates at least one s-t cut. Let $X=\{u\in G\mid d(u)\geqslant n\}$ (the set of nodes from which t is not reachable in the residual graph) and $\bar{X}=\{v\in G\mid 0<d(v)<n\}$ (the set of nodes from which t is reachable in the residual graph). Because phase I has terminated and (thus) no more active nodes exist, it should be clear that $X;\bar{X}$ indeed is an s-t cut.

3.27. LEMMA. *The g-flow across the cut $X;\bar{X}$ equals the capacity of the cut.*

PROOF. Consider any edge (u,v) or (v,u) with $u\in X$ and $v\in\bar{X}$. Because d is a valid labeling w.r.t. the residual graph at all times, we can have $d(u)\geqslant n$ only if (u,v) and (v,u) have no residual capacity left. This means that the flow through (u,v) is maximum and the flow through (v,u) is 0. □

Phase II of Goldberg's algorithm consists of eliminating the excess flow in all nodes except s and t, without affecting the value of the flow across the cut $X;\bar{X}$ just defined. It follows that the resulting legal flow must be maximum and ipso facto that $X;\bar{X}$ must have been a minimum capacity s-t cut.

Algorithmically, phase II can be implemented in several ways. In one approach all circulations are eliminated first by means of an algorithm of Sleator and Tarjan in O($e \log n$) time. (Note that no circulation or "cycle of flow" can go across the $X;\bar{X}$ cut and that their elimination does not change the value of the flow.) Next the nodes of G are processed in reverse topological order, and for every node v with $\Delta(v)>0$ the inflow is reduced by a total of $\Delta(v)$. This increases the flow excess in the predecessors, but they are handled next anyway. (Note that the nodes in $\bar{X}$ have flow excess 0 and thus remain unaffected, as does the flow across the $X;\bar{X}$ cut.) The entire procedure remains bounded by O($e \log n$) time. Originally Goldberg formulated a very similar algorithm to phase I for taking away excess flow. Its complexity was O(n^3).

Goldberg's approach has led to the most efficient maximum flow algorithm to date for the general flow problem, with a running time of O($ne \log n^2/e$) time when a suitable implementation is made (Goldberg & Tarjan). At the same time his algorithm is ideally suited for parallel or distributed computation models.

3.5. *Related flow problems*

There is a large variety of optimization problems on graphs that can be formulated in terms of flows in networks. We briefly review the main results in this direction.

3.5.1. *Networks with lower bounds*

It is common to also impose a lower bound on the amount of flow through each edge, in addition to the given upper bound ("capacity") of each edge. The main problem now is to find a feasible flow, i.e., a flow that satisfies the given constraints in each edge. Taking a feasible flow as the starting point, any of the maximum flow algorithms based on flow augmentation can be called upon to construct a maximum feasible flow. In a similar way the minimum feasible flow can be constructed. For both types of flow there are results relating the value of the flow to the capacity of cuts (see Section 3.2).

Lawler describes the following procedure for finding a feasible flow (if one exists). Add the edge (t, s) with lower bound 0 and capacity ∞, and start with a suitable initial circulation f that satisfies the capacity constraints (e.g. $f \equiv 0$). If f is not feasible, there will be an edge (u, v) for which $f(u, v) < l(u, v)$. Find a flow augmenting path from v to u in which each backward edge carries an amount of flow strictly larger than its lower bound. Now augment the flow from v to u along the path (and, of course, along the edge (u, v)) by the maximum amount δ that keeps the flow within the constraints. Repeat this until $f(u, v) \geqslant l(u, v)$ and likewise for every other edge. The procedure will construct a feasible flow provided the conditions of Hoffman's circulation theorem are satisfied (see Section 3.2). When suitably implemented, the algorithm finds a feasible flow in O(n^3) time or less.

Simon considered the computation of minimum flows in networks that have lower bounds but no specific upper bounds on the edge capacities. Given an edge $e=(u, v)$ with lower bound $l(e)$ and a directed path π from u to v not involving e, one can delete e and implicitly force its minimum flow requirement by adding $l(e)$ to the capacity lower bounds along π. It follows that the minimum flow problem on a network G with lower bounds only is equivalent to the minimum flow problem on any irreducible kernel G' of G (cf. Section 1.4) with modified lower bounds, and a minimum flow on G' can be transformed back into a minimum flow on G in only $O(n+e)$ time. Simon proves that this leads to an $O(n^2)$ minimum flow algorithm for strongly connected graphs, and to an $O(ne^- \log n)$ algorithm for acyclic graphs (where e^- is the number of edges in the transitive reduction). Note that for cyclic directed graphs, e^- is usually much less than e, with an expected value of $O(n \log n)$ for random graphs.

Curiously the problem of finding a feasible flow seems to be much harder for undirected networks. In this case a (feasible) flow is defined to be any function $f: V \times V \to \mathbb{R}$ with the following properties:

(i) $f(i,j) \neq 0$ only if $(i,j) \in E$ and $f(i,j) = -f(j,i)$,

(ii) $l(i,j) \leqslant |f(i,j)| \leqslant c(i,j)$ and

(iii) $\sum_{u \in V} f(u,j) = 0$ for all $j \neq s,t$.

(Interpret $f(i,j) > 0$ as the amount of flow from i to j.) It is assumed that $l(i,j) = l(j,i)$ and $c(i,j) = c(j,i)$. Itai has shown that the problem of determining whether there exists a feasible flow is an undirected network with lower and upper bounds is NP-complete.

3.5.2. Minimum cost flow

Next suppose that each unit of flow through an edge (i,j) takes a cost $b(i,j) \geqslant 0$, and consider the problem of constructing a minimum cost flow of value v (if it exists). The problem is discussed at length in Ford and Fulkerson. A classical result of Wagner shows that the minimum cost flow problem is polynomially equivalent to the so-called transportation problem (or "Hitchcock problem") in linear programming, which has been extensively studied. Assuming integer capacities, Edmonds and Karp developed an algorithm for finding a minimum cost maximum flow exploiting this connection, achieving a running time polynomial in n, e and $\log|f^*|$ (with f^* the value of the maximum flow).

The minimum cost flow Theorem 3.18 suggests the following algorithm for finding a minimum cost flow of value v. Set f^0 equal to the zero flow and construct a sequence of (minimum cost) flows $f^1, f^2, \ldots$ as follows, for as long as $|f^k| < v$ and f^k is not maximal: find an s-t augmenting path π of minimum cost for f^k, and construct a new flow (f^{k+1}) by augmenting the flow by $\min\{res(\pi), v - |f^k|\}$ along π. The algorithm is known as the minimum cost augmentation algorithm. Zadeh has shown that in general the algorithm is not polynomially bounded in the parameters of the network. If all capacities are integers, the algorithm clearly is polynomial, as there can be at most $v+1$ stages in the algorithm (each augmentation increments the flow value by an integer $\geqslant 1$) and each stage requires the computation of a minimum cost path from s to t in the residual graph (which takes $O(ne)$ time when a general single-source shortest path algorithm is used, where we note that the residual graph of any minimum cost flow is known to have no negative cost cycles). The shortest path computation in every stage can be considerably simplified by the following observation due to Edmonds and Karp.

3.28. LEMMA. *The minimum cost augmenting path computation required in each stage can be carried out in a network with all weights (costs) nonnegative.*

Using this fact one can show that for integer capacity networks the minimum cost augmentation algorithm produces a minimum cost flow of value v in about $O(vS(n,e))$

time, with $S(n,e)$ the time needed for solving a single-source shortest path problem in a network with all weights nonnegative. The best bound known to date is $S(n,e) = e + n \log n$ (see Section 2.3).

For general networks the minimum cost augmentation algorithm may behave very poorly. Zadeh showed that even convergence of the algorithm is not guaranteed. Consider the following variant:

(7) *Edmonds–Karp minimum cost augmentation*: find a shortest minimum cost augmenting path π and augment the flow by $\min\{res(\pi), v - |f^k|\}$ along π.

(Here f^k denotes the flow in the kth iteration of the algorithm.) Assume that Edmonds–Karp minimum cost augmentation is used in every stage of the minimum cost flow algorithm.

3.29. THEOREM. *Edmonds–Karp minimum cost augmentation produces a minimum cost flow of value v in finitely many steps.*

PROOF. We need the following property of the general minimum cost augmentation algorithm. Consider the kth flow (f^k) and let $\pi^k(v)$ denote the minimum cost of a path from s to v in the residual graph R^k. For $(u,v) \in R^k$ we have $\pi^k(u) + b \geqslant \pi^k(v)$, where b is the cost of the edge ($b(u,v)$ when $(u,v) \in E$ and $-b(u,v)$ when $(u,v) \in \overleftarrow{E}$) and equality holds if (u,v) is on a minimum cost path from s to v. Consider the flow f^{k+1} obtained after augmenting the flow along a minimum cost s-t path (in R^k) and its corresponding residual graph R^{k+1}. R^{k+1} consists of a subset of the edges of R^k and, possibly, some new edges. One verifies that the only edges $(v,u) \in R^{k+1} - R^k$ are edges such that $(u,v) \in R^k$ and $\pi^k(u) + b = \pi^k(v)$ with b as before (using that the augmentation was done along a minimum cost path). It easily follows that $\pi^{k+1}(v) \geqslant \pi^k(v)$ for all nodes v. Taking $v = t$ this shows that the cost of the successive augmenting paths used in the minimum cost augmentation algorithm is nondecreasing.

Now consider the Edmonds–Karp minimum cost augmentation algorithm. We know that $\pi^k(t)$ is nondecreasing in k. During any period in which $\pi^k(t)$ remains constant, Edmonds–Karp minimum cost augmentation behaves exactly like Edmonds–Karp augmentation in a maximum flow algorithm. It follows (cf. Section 3.4.1) that $\pi^k(t)$ remains constant for at most $\frac{1}{2}ne$ iterations in a row before it necessarily must increase. But $\pi^k(t)$ can only increase a finite number of times because, for each k, $\pi^k(t)$ is the cost of some loop-free path from s to t and the number of such paths is finite. □

In 1985 Tardos obtained a polynomial-time minimum cost flow algorithm for the general case that is "strongly polynomial", i.e., that has the additional property that the numbers occurring during the computation remain polynomially bounded in the size of the problem as well. There have been various improvements to this result, by a variety of techniques and using very efficient datastructures (see Table 5).

The fastest polynomial-time bounded algorithm known to date is due to Ahuja et al. and runs in $O(ne \log\log U \log nC)$, where U is the maximum capacity and C the maximum cost of an edge. (See the survey by Goldberg et al. for a further account of minimum cost flow algorithms.)

Table 5
Some polynomial-time algorithms for the minimum cost flow problem. ($S(n, e)$ is the complexity of any shortest path algorithm e.g. $S(n, e) = O(e + n \log n)$, U is the maximum capacity, C is the maximum cost of an edge.)

Approximate year	Author(s)	Complexity
1972	Edmonds & Karp	$O(e \log U\, S(n, e))$
1980	Röck	$O(e \log U\, S(n, e))$
1980	Röck	$O(e \log C\, ne \log n^2/e)$
1985	Tardos	$O(e^4)$
1984	Orlin	$O(e^2 \log n\, S(n, e))$
1985	Bland & Jensen	$O(e \log C\, ne \log n^2/e)$
1986	Fujishige	$O(e^2 \log n\, S(n, e))$
1986	Galil & Tardos	$O(n^2 \log n\, S(n, e))$
1987	Goldberg & Tarjan	$O(\min\{ne \log n, n^{5/3}e^{2/3}, n^3\} \log nC)$
1988	Goldberg & Tarjan	$O(ne \log(n^2/e) \min\{e \log n, \log nC\})$
1988	Orlin	$O(e \log n\, S(n, e))$
1988	Ahuja, Goldberg, Orlin & Tarjan	$O(ne \log\log U \log nC)$

3.5.3. *Multiterminal flow*

The *multiterminal flow problem* requires that we compute the maximum flow between every two of p given nodes in a network G for some $p > 1$. Clearly this can be accomplished in polynomial time, by solving $\frac{1}{2}p(p-1)$ maximum flow problems (one for each pair of distinct nodes). Sometimes a much more efficient algorithm can be found. A classical case is provided by the "symmetric" networks, which have $(v, u) \in E$ whenever $(u, v) \in E$ (i.e., the network is essentially undirected) and $c(v, u) = c(u, v)$. We outline the results of Gomory and Hu that show that for symmetric networks the multiterminal flow problem can be answered by solving only $p-1$ maximum flow problems. Several ramifications of the problem will be discussed afterwards.

Let G be symmetric, and let v_{ij} be the value of the maximum flow from i to j $(1 \leqslant i, j \leqslant n)$. Let $v_{ii} = \infty$ for all i, for consistency.

3.30. LEMMA. (i) *For all i, j $(1 \leqslant i, j \leqslant n)$, $v_{ij} = v_{ji}$.*

(ii) *For all i, j, k $(1 \leqslant i, j, k \leqslant n)$, $v_{ik} \geqslant \min\{v_{ij}, v_{jk}\}$.*

(iii) *There exists a tree T on the nodes 1 through n such that for all i, j $(1 \leqslant i, j \leqslant n)$, $v_{ij} = \min\{v_{ij_1}, v_{j_1 j_2}, \ldots, v_{j_k j}\}$, where $i - j_1 - \cdots - j_k - j$ is the (unique) path from i to j in T.*

PROOF. (i) By symmetry of the network.

(ii) Let $X;\bar{X}$ be a minimum capacity i-k cut saturated by a maximum flow from i to k, i.e., $v_{ik} = c(X, \bar{X})$. If $j \in \bar{X}$ then necessarily $v_{ij} \leqslant c(X, \bar{X}) = v_{ik}$. If $j \in X$ then $v_{jk} \leqslant c(X, \bar{X}) = v_{ik}$.

(iii) Consider the complete graph on n nodes with the edges (i, j) labeled by the "weight" $v_{ij} (= v_{ji})$. Let T be a maximum weight spanning tree. By an inductive argument using (ii) one verifies that for each pair of nodes i, j $(1 \leqslant i, j \leqslant n)$, $v_{ij} \geqslant \min\{v_{ij_1}, v_{j_1 j_2}, \ldots, v_{j_k j}\}$, where $i - j_1 - \cdots - j_k - j$ is the path from i to j in T. Suppose for some i, j there is strict inequality. Then there is an edge (i', j') on the path between

i and j with $v_{ij} > v_{i'j'}$. This means that the tree T' obtained by replacing the edge (i',j') of T by the edge (i,j) will have a larger weight than T. Contradiction. □

Part (iii) of the lemma has the interesting consequence that among the $\frac{1}{2}n(n-1)$ maximum flow values v_{ij} there exist at most $n-1$ different values (namely, the values associated with the edges of T). It also leads to the intuition that perhaps $n-1$ maximum flow computations suffice to build a tree T with the desired property, in order to determine all v_{ij}-values. An algorithm due to Gomory and Hu indeed achieves this.

The crucial observation of Gomory and Hu is expressed in the following lemma. Given two nodes i,j and a minimum capacity i-j cut $X;\bar{X}$, define the "condensed" network G^c as the network obtained from G by contracting the nodes in $\bar{X}$ to a single (special) node $u_{\bar{X}}$ and replacing all edges leading from a node $v \in X$ across the cut by a single edge between v and $u_{\bar{X}}$ having the total capacity (for all $v \in X$).

3.31. LEMMA. *For every two ordinary nodes i',j' of G^c (i.e., i' and $j' \in X$), the maximum flow from i' to j' in G^c has the same value as the maximum flow from i' to j' in the original network.*

(Note that for nodes $i',j' \in \bar{X}$ a similar statement holds after contracting X in G.) An important consequence of the lemma is that for every two nodes $i',j' \in X$ (or $\bar{X}$) there is a minimum capacity i'-j' cut such that all nodes of $\bar{X}$ (or X respectively) are on one side of the cut. Given two nodes i,j and a minimum capacity i-j cut $X;\bar{X}$, we represent the current situation by a tree with two nodes, one representing X and the other $\bar{X}$, and one edge with label v_{ij}. (Thus the edge "represents" the cut $X;\bar{X}$.) Of course we begin by choosing i and j from among the p nodes between which the maximum flow value must be computed. We will now show how to expand the tree. For convenience we denote the nodes of the tree by capital letters. Let A be the set of designated nodes, $|A|=p$.

DEFINITION. A tree T is called a semicut tree if it satisfies the following properties:

(i) every node U of T corresponds to a subset of the nodes in G and contains at least one node of A,

(ii) every edge (U, V) carries one label v, and there are nodes $i,j \in A$ with $i \in U$ and $j \in V$ such that the maximum flow between i and j has value v,

(iii) every edge (U, V) represents a minimum capacity i-j cut with $i,j \in A$ and i contained in one of the nodes of T in the subtree headed by U and j contained in one of the nodes of T in the subtree headed by V (and the cut consisting of the two collective sets of nodes).

If every node of a semicut tree T contains precisely one node of A, then T is called a cut tree for A. (If A consists of all nodes of G, then a cut tree for A is simply called a cut tree.)

3.32. THEOREM. *Let T be a cut tree for A. Then for every $i,j \in A$ one has $v_{ij} = \min\{v_1, \ldots, v_{k+1}\}$, where v_1 through v_{k+1} are the edge labels in T on the path from the (unique) node containing i to the (unique) node containing j.*

PROOF. Let i—j_1— ⋯ —j_k—j be the "path" of A-nodes leading from i to j in T. (Each A-node listed is in fact the unique A-node contained in the corresponding node on the path in T.) By the properties of the cut tree the labels on the edges are simply $v_{ij_1}, \ldots, v_{j_kj}$. By the lemma we have $v_{ij} \geqslant \min\{v_{ij_1}, \ldots, v_{j_kj}\}$. On the other hand each edge corresponds to a cutset separating i and j, with a capacity equal to the corresponding v-label. Thus $v_{ij} \leqslant \min\{v_{ij_1}, \ldots, v_{j_kj}\}$ and the desired equality follows. □

The algorithm of Gomory and Hu simply begins with the 2-node semicut tree for A that we outlined above and shows how it can be transformed in a step-by-step fashion into a cut tree for A. By the theorem a cut tree for A has precisely the property we want.

3.33. THEOREM. *A cut tree for A exists and can be constructed by means of only* $p-1$ *maximum flow computations.*

PROOF. Suppose we have a semicut tree T for A. (This is certainly true at the beginning of the algorithm.) Suppose T still has a node U which contains two nodes $i, j \in A$. By the lemma the maximum flow between i and j in G has the same value as the maximum flow between i and j in the condensed graph G^c obtained by contracting the sets of nodes in each subtree attached to U to a single (special) node, and for each $v \in U$ replacing the edges that lead from v to a node in a particular subtree by one edge with the combined total capacity between v and the corresponding special node. Let $X;\bar{X}$ be a minimum capacity i-j cut in G^c (necessarily with capacity v_{ij}). Construct a tree T' as follows: split U into the nodes X and $\bar{X}$, connect X and $\bar{X}$ by an edge labeled v_{ij}, and (re)connect every neighbor V of U to either X or $\bar{X}$ depending on which side of the cut the special node corresponding to V's subtree was in G^c (with original edge label).

We claim that T' is again a semicut tree for A. The only nontrivial property to verify is part (ii) of the definition. Thus let V be any neighbor of U in T and let the label on the edge (U, V) correspond to the maximum flow value between $r \in U$ and $s \in V$ $(r, s \in A)$. Assume without loss of generality that $j, r \in \bar{X}$. Now the following two cases can arise in the construction:

Case I: *V gets connected to $\bar{X}$*. This trivially preserves the properties of the semicut tree.

Case II: *V gets connected to X*. This situation is more subtle (see Fig. 11). Clearly the label v_{ij} on the edge $(X, \bar{X})$ is appropriate to satisfy the definition for the new edge. Now

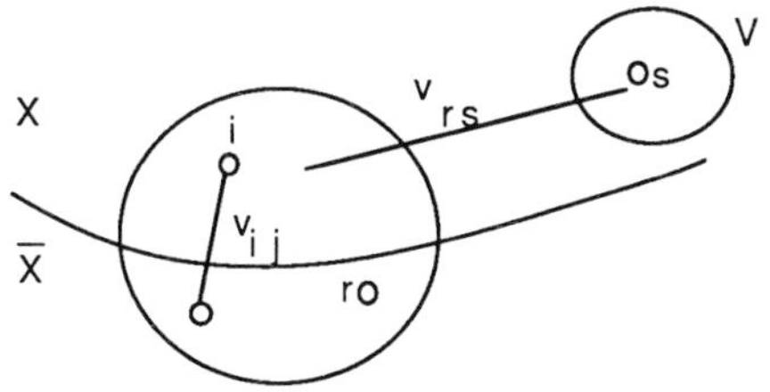

Fig. 11.

consider the label (v_{rs}) on the edge (X, V). We show that, in fact, $v_{is} = v_{rs}$. The argument is as follows. By the lemma we know $v_{is} \geqslant \min\{v_{ij}, v_{jr}, v_{rs}\}$. Because i and s are on one side of the cut, we know that the flow values between nodes in $\bar{X}$ do not affect v_{is} and we have $v_{is} \geqslant \min\{v_{ij}, v_{rs}\}$. On the other hand v_{is} must be bounded by the capacity of the minimum cut corresponding to the edge $(U, V) \in T$, i.e., $v_{is} \leqslant v_{rs}$. By the lemma $v_{ij} \geqslant \min\{v_{is}, v_{rs}\}$, hence $v_{ij} \geqslant v_{rs}$. It follows that $v_{is} = v_{rs}$, and the edge (X, V) corresponds to the maximum flow between i and s (with the same value and the same cutset corresponding to it as in the original tree T).

By repeating the construction it is clear that a cut tree for A is obtained, at the expense of a total of $p-1$ maximum flow (i.e., minimum cut) computations. □

The Gomory–Hu construction of a cut tree clearly solves the multiterminal flow problem in an elegant fashion and saves a large number of computations over the naive algorithm. Using the maximum flow algorithm of Goldberg and Tarjan, the construction needs essentially $O(pne \log n^2/e)$ time. Hu and Sing have shown that a cut tree for a network with t triconnected components can be computed in $O(n^4/t^3 + n^2 + t)$ time. For outerplanar graphs a cut tree can be constructed in linear time, for planar graphs in $O(n^2 \log^* n / \log n)$ time.

Several ramifications of the multiterminal flow problem can be considered. Gupta has shown that the results of Gomory and Hu hold for the larger class of so-called pseudosymmetric networks, which are defined as the networks in which there exists a circulation that saturates all edges. The multiterminal flow problem can also be studied for e.g. minimum cost flows, but few results along these lines have been reported. Some attention has been given to the "synthesis" version of the multiterminal flow problem. In its simplest form the question is to determine a network (i.e., a graph and suitable capacities for the edges) that realizes a given set of maximum flow values. For the restriction to symmetric networks again some results are known. The following result is due to Gomory and Hu.

3.34. THEOREM. *The values v_{ij} $(1 \leqslant i, j \leqslant n)$ are the maximum flow values of an n-node symmetric network if and only if the values are nonnegative, for all i, j $(1 \leqslant i, j \leqslant n)$ $v_{ij} = v_{ji}$ and for all i, j, k $(1 \leqslant i, j, k \leqslant n)$ $v_{ik} \geqslant \min\{v_{ij}, v_{jk}\}$.*

PROOF. The necessity follows from Lemma 3.30. The sufficiency follows by considering the tree T constructed in part (iii) of the same lemma. Let the edge labels in T serve as edge capacities. One verifies that the v_{ij}-values actually arise as maximum flow values in the network so obtained. □

In its more interesting form, the synthesis version of the multiterminal flow problem asks for a network that realizes a given set of maximum flow values (or a set of flows with these values as lower bounds) with minimum total edge capacity. For the symmetric case Gomory and Hu have again devised an efficient algorithm for this problem. Gusfield has shown that one can construct an optimal symmetric network G in $O(\max\{e, n \log n\})$ time, where the network G that is obtained has the additional properties that it is planar, has degrees bounded by 4 and the smallest possible number of edges (in a well-defined sense).

3.5.4. Multicommodity flow

In the discussion of (maximum) flow problems we have hitherto assumed that the flow concerns a single "commodity" that must be transported from a source node s to a target node t. The *m-commodity (maximum) flow problem* is the natural generalization in which there are m commodities ($m \geqslant 2$) and m source–target pairs (s_i, t_i). The problem is to simultaneously transport any required number of units of the ith commodity from s_i to t_i for $1 \leqslant i \leqslant m$ through the same network, i.e., under the constraint that the total flow of commodities 1 to m through each edge is bounded by the capacity of the edge. Let f_i denote the flow function for the ith commodity ($1 \leqslant i \leqslant m$). Traditionally one seeks to maximize the total flow of commodities $\Sigma_1^m |f_i|$, or even to maximize the flow and at the same time minimize the total cost of all flows (assuming that cost functions are defined for the edges as in the minimum cost flow problem). Both problems are easily formulated as linear programming problems, and early studies by Ford and Fulkerson and by Tomlin have attemped to exploit the special structure of these linear programs to obtain practical algorithms. Because of the discovery of several polynomial-time bounded linear programming algorithms (e.g. Khachian's ellipsoid method), we know that both the maximum multicommodity flow problem and the minimum cost multicommodity flow problem can be solved in polynomial time when all network parameters are rational. Kapoor and Vaidya have adapted Karmarkar's linear programming algorithm to show that both multicommodity flow problems can be solved in $O(m^{3.5}n^{2.5}eL)$ steps, where each step involves an arithmetic operation to a precision of $O(L)$ bits. (L is the size of the network specification in bits.) Assad has given an excellent account of the linear programming aspects of the many variants of the multicommodity flow problem. Itai has shown that the maximum m-commodity flow problem is in fact polynomially equivalent to linear programming for every $m \geqslant 2$.

In this section we restrict ourselves to the (maximum) m-commodity flow problem in undirected networks. The ith commodity flow ($1 \leqslant i \leqslant m$) is to be a function $f_i: V \times V \to \mathbb{R}$ that satisfies the following properties:

(i) $f_i(u,v) \neq 0$ only if $(u,v) \in E$ and $f_i(u,v) = -f_i(v,u)$,

(ii) $0 \leqslant |f_i(u,v)| \leqslant c(u,v)$ and

(iii) $\Sigma_{u\in V} f_i(u,v) = 0$ for all $v \neq s_i, t_i$.

It is assumed that $c(u,v) > 0$ only if $(u,v) \in E$ and that $c(u,v) = c(v,u)$ for all u, v. (Interpret $f_i(u,v) > 0$ as the flow of the ith commodity from u to v.) The value of f_i is denoted as $|f_i|$. An m-commodity flow is called feasible if all f_i ($1 \leqslant i \leqslant m$) are legal flows and for every $(u,v) \in E$, $|f_1(u,v)| + \cdots + |f_m(u,v)| \leqslant c(u,v)$. An m-commodity flow is called maximum if it is feasible and the total flow $|f_1| + \cdots + |f_m|$ is maximum (over all feasible flows).

The m-commodity flow problem in undirected networks is especially interesting for $m=2$, because in this case there is an analog to the max-flow min-cut theorem that leads to a fairly efficient algorithm. Let $c(s_1, s_2;\ t_1, t_2)$ be the minimum capacity of any cut separating s_i and t_i ($1 \leqslant i \leqslant 2$). Let $\tau(u_1, u_2;\ v_1, v_2)$ denote the minimum capacity of any cut $X;\bar{X}$ such that $u_1, u_2 \in X$ and $v_1, v_2 \in \bar{X}$ (i.e., u_1 and u_2 are on the same side of the cut, as v_1 and v_2 are). The following fact is due to Hu.

3.35. LEMMA. *In undirected networks* $c(s_1, s_2;\ t_1, t_2) = \min\{\tau(s_1, s_2;\ t_1, t_2),\ \tau(s_1, t_2;\ s_2, t_1)\}$.

Next define $c(s_i; t_i)$ as the minimum capacity of any s_i-t_i cut (for $1 \leqslant i \leqslant 2$). Observe that for every feasibly flow, $|f_i| \leqslant c(s_i; t_i)$ $(1 \leqslant i \leqslant 2)$, by virtue of the max-flow min-cut theorem. The theory of the 2-commodity flow problem is governed by the following results due to Hu and complemented by Itai.

3.36. THEOREM. (i) (the 2-commodity feasible flow theorem) *There exists a feasible 2-commodity flow with* $|f_i| = v_i$ $(1 \leqslant i \leqslant 2)$ *if and only if* $v_1 \leqslant c(s_1; t_1)$, $v_2 \leqslant c(s_2; t_2)$ *and* $v_1 + v_2 \leqslant c(s_1, s_2; t_1, t_2)$.

(ii) (the 2-commodity max-flow min-cut theorem) *A 2-commodity flow is maximum if and only if it is feasible and* $|f_1| + |f_2| = c(s_1, s_2; t_1, t_2)$.

3.37. THEOREM. (i) (the 2-commodity maximum component flow theorem) *For each i,* $1 \leqslant i \leqslant 2$, *there exists a maximum 2-commodity flow in which* $|f_i|$ *is maximum as well (as a single commodity flow).*

(ii) (the even-integer theorem) *If all capacities are even integers, there is a maximum 2-commodity flow* f_1, f_2 *which is integral (i.e., with all flow values integer).*

(iii) (the half-units theorem) *If all capacities are integers, there is a maximum 2-commodity flow* f_1, f_2 *in which all flow values are integer multiples of* $\frac{1}{2}$.

Hu has shown that most of the results do not generalize to maximum m-commodity flows for $m > 2$. Rothschild and Whinston have shown that an integral maximum 2-commodity flow exists also when all capacities are integers and for each node v, $\sum_{u \in V} c(u, v)$ is even.

The proofs of the (maximum) 2-commodity flow theorems essentially all follow from an algorithm due to Hu for constructing a maximum 2-commodity flow. It will be useful to keep the analogy with the ordinary maximum flow problem in mind. The idea of Hu's algorithm is to start with a maximum flow for the first commodity and a zero flow for the second, and then to use a suitable augmentation device to increase the flow of the second commodity while recirculating some flow of the first (without changing its value). Define a bipath to be any pair of loop-free paths (π_α, π_β) such that π_α leads from s_2 to t_2 and π_β from t_2 to s_2. Given a feasible 2-commodity flow f_1, f_2 define for each (directed) edge $(u, v) \in E$,

$$A(u, v) = \tfrac{1}{2}\{c(u, v) - f_1(u, v) - f_2(u, v)\}$$

and

$$B(u, v) = \tfrac{1}{2}\{c(u, v) - f_1(u, v) - f_2(v, u)\}.$$

(Note that $A(u, v)$ is half the "residual capacity" of the flow from u to v, and that $B(u, v)$ is half the "residual capacity" for augmenting the first flow from u to v and the second from v to u.) For paths π_α and π_β as above define $A(\pi_\alpha) = \min\{A(u, v) \mid (u, v) \in \pi_\alpha\}$ and $B(\pi_\beta) = \min\{B(u, v) \mid (u, v) \in \pi_\beta\}$. Define the residual capacity of a bipath $\pi = (\pi_\alpha, \pi_\beta)$ as $res(\pi) = \min\{A(\pi_\alpha), B(\pi_\beta)\}$. A bipath π is called an augmenting bipath if $res(\pi) > 0$. Given an augmenting bipath $\pi = (\pi_\alpha, \pi_\beta)$ we can augment the current 2-commodity flow by "biaugmentation" as follows: increase the flow of both commodities by $res(\pi)$ along π_α

("going from s_2 to t_2"), and increase the flow of the first commodity and decrease the flow of the second commodity by $res(\pi)$ along π_β ("going from t_2 to s_2").

3.38. LEMMA. *Biaugmentation preserves the feasibility of 2-commodity flows.*

3.39. THEOREM (the augmenting bipath theorem). *A 2-commodity flow with f_1 a maximum flow is maximum if and only if it is feasible and admits no augmenting bipath.*

PROOF. Let f_1, f_2 be a maximum 2-commodity flow. Suppose there was an augmenting bipath π. Biaugmentation along π leaves the value of f_1 unchanged but increments the value of f_2 by $2res(\pi)$. This contradicts that f_1, f_2 is maximum. Next, let f_1, f_2 be feasible with f_1 maximum, and suppose there exists no augmenting bipath. By the nature of the problem we must have $|f_1| \leqslant c(s_1;t_1)$, $|f_2| \leqslant c(s_2;t_2)$ and $|f_1|+|f_2| \leqslant c(s_1,s_2;\ t_1,t_2)$. We show that $|f_1|+|f_2|=c(s_1,s_2;\ t_1,t_2)$, which means that the 2-commodity flow f_1, f_2 must be maximum. We distinguish two cases:

Case I: *there exists no "forward" path π_α with $A(\pi_\alpha)>0$.* Let X be the set of all nodes reachable from s_2 by a loop-free path of edges (u,v) with $A(u,v)>0$. Clearly $X;\bar{X}$ is an s_2-t_2 cut and for all $(u,v) \in E$ with $u \in X$ and $v \in \bar{X}$ we must have $A(u,v)=0$, i.e., $f_1(u,v)+f_2(u,v)=c(u,v)$. Because $|f_1(u,v)|+|f_2(u,v)| \leqslant c(u,v)$ by feasibility, it follows that $f_1(u,v)$ and $f_2(u,v)$ are both nonnegative and (also) that (u,v) is "saturated".

If $f_1(u,v)$ is zero for all edges across the cut, then the cut is entirely saturated by f_2 and $|f_2|=c(s_2;t_2)$ (by the max-flow min-cut theorem). It follows that f_1, f_2 must be a maximum 2-commodity flow in this case, with $|f_1|+|f_2|=c(s_1;t_2) \geqslant \tau(s_1,s_2;\ t_1,t_2)$ and (hence) $|f_1|+|f_2|=c(s_1,s_2;t_1,t_2)$. Next, consider the case that $f_1(u,v)>0$ for some edge (u,v) across the cut. Now $s_1 \in X$ and $t_1 \in \bar{X}$, otherwise there would be an edge across the cut with negative f_1-flow (to conserve flow) and this is impossible. Thus $X;\bar{X}$ is also an s_1-t_1 cut, and we have $|f_1|+|f_2|=c(X,\bar{X}) \geqslant \tau(s_1,s_2;t_1,t_2)$. Thus $|f_1|+|f_2|=c(s_1,s_2;\ t_1,t_2)$ and the flow is again maximum.

Case II: *there is no "backward" path π_β with $B(\pi_\beta)>0$.* Similar to Case I, now proving that $|f_1|+|f_2| \geqslant \tau(s_1,t_2;\ s_2,t_1)$. Using Hu's lemma, the maximality of f_1, f_2 again follows. □

The augmenting bipath theorem provides exactly the tool we need to construct maximum 2-commodity flows, in perfect analogy to the single commodity case. Augmenting bipath methods are based on the iteration of one of the following types of steps, starting with a maximum flow for f_1 and any flow f_2 such that f_1, f_2 is a feasible 2-commodity flow (e.g. $f_2 \equiv 0$). (To obtain a feasible 2-commodity flow with preassigned values for $|f_1|$ and $|f_2|$ the steps should be suitably modified.)

(8) *Hu biaugmentation*: find a biaugmenting path π and augment the 2-commodity flow by $res(\pi)$ along π.
(9) *Itai biaugmentation*: find a biaugmenting path $\pi(\pi_\alpha, \pi_\beta)$ with π_α and π_β of shortest possible length, and augment the 2-commodity flow by $res(\pi)$ along π.

Both types of biaugmentation can be implemented in $O(e)$ time per step.

Clearly Hu biaugmentation yields a maximum 2-commodity flow with all the desired properties in finitely many steps in case the capacities are integers. For arbitrary (real) capacities Hu biaugmentation may not converge to a maximum 2-commodity flow, very much like Ford–Fulkerson augmentation may not converge to a maximum flow in the single commodity flow problem. Itai biaugmentation is reminiscent of Edmonds–Karp augmentation in the ordinary maximum flow case, and can be shown to lead to a maximum 2-commodity flow in $O(n^2e)$ time for general capacities. Itai has shown that an $O(n^3)$ algorithm can be obtained by suitably adapting Karzanov's maximum flow algorithm to the 2-commodity case. Note that the fact that Itai biaugmentation terminates proves the 2-commodity max-flow min-cut theorem for general undirected networks. (Seymour has given a proof that does not rely on this fact, solely based on the circulation theorem.) One easily verifies that the maximum 2-commodity flow obtained satisfies the additional properties of the theorem. The half-units theorem is perhaps most interesting of all. Even, Itai and Shamir have shown that the undirected 2-commodity flow problem becomes NP-complete once we insist on both flows being integral!

Hu's theory for the 2-commodity flow problem does not seem to generalize for $m > 2$. In particular there seems to be no suitable version of an m-commodity max-flow min-cut theorem. (Okamura and Seymour have shown that such a theorem does exist for planar graphs with all source and target nodes on the boundary of the infinite face.) Even the maximum m-commodity integral flow problem with all capacities equal to 1 is NP-complete: it can be recognized as the problem of finding m pairwise disjoint paths, with the ith path connecting s_i and t_i $(1 \leqslant i \leqslant m)$.

References

Textbooks and monographs on graphs and graph algorithms

[1] BERGE, C., *The Theory of Graphs and its Applications* (Wiley, New York, 1962).
[2] BERGE, C., *Graphs and Hypergraphs* (North-Holland, Amsterdam, 1973).
[3] BERGE, G., *Graphs* (North-Holland, Amsterdam, 2nd rev. ed., 1985).
[4] BOLLOBÁS, B., *Extremal Graph Theory* (Academic Press, New York, 1978).
[5] BOLLOBÁS, B., *Graph Theory: an Introductory Course*, Graduate Texts in Mathematics, Vol. 63 (Springer, New York, 1979).
[6] BOLLOBÁS, B., *Random Graphs* (Academic Press, London, 1985).
[7] BONDY, J.A. and U.S.R. MURTY, *Graph Theory with Applications* (American Elsevier, New York, MacMillan, London, 1976).
[8] BUSACKER, R.G. and T.L. SAATY, *Finite Graphs and Networks: an Introduction with Applications* (McGraw-Hill, New York, 1965).
[9] CHRISTOFIDES, N., *Graph Theory: an Algorithmic Approach* (Academic Press, New York, 1975).
[10] DEO, N., *Graph Theory with Applications to Engineering and Computer Science* (Prentice Hall, Englewood Cliffs, NJ, 1974).
[11] EBERT, J., *Effiziente Graphenalgorithmen* (Akademische Verlagsgesellschaft, Wiesbaden, 1981).
[12] EVEN, S., *Graph Algorithms* (Computer Science Press, Potomac, MD, 1979).
[13] FRANK, H. and I.T. FRISCH, *Communication, Transmission, and Transportation Networks* (Addison-Wesley, Reading, MA, 1971).
[14] GIBBONS, A., *Algorithmic Graph Theory* (Cambridge Univ. Press, Cambridge, UK, 1985).
[15] GOLUMBIC, M.C., *Algorithmic Graph Theory and Perfect Graphs* (Academic Press, New York, 1980).
[16] GONDRAN, M. and M. MINOUX, *Graphs and Algorithms* (Wiley, Chichester, UK, 1984).

[17] HARARY, F., *Graph Theory* (Addison-Wesley, Reading, MA, 1969).
[18] KÖNIG, D., *Theorie der Endlichen und Unendlichen Graphen* (Akademie Verlagsgesellschaft, Leipzig, 1936; reprinted by Chelsea, New York, 1950).
[19] LAWLER, E., *Combinatorial Optimization: Networks and Matroids* (Holt, Rinehart and Winston, New York, 1976).
[20] MAYEDA, W., *Graph Theory* (Wiley, New York, 1972).
[21] MEHLHORN, K., *Graph Algorithms and NP-Completeness*, Data Structures and Algorithms Vol. 2, (Springer, Berlin, 1984).
[22] MINIEKA, E., *Optimization Algorithms for Network and Graphs* (Dekker, New York, NY, 1978.
[23] NISHIZEKI, T. and N. CHIBA, *Planar Graphs: Theory and Algorithms* (North-Holland, Amsterdam, 1989).
[24] ORE, O., *The Four-Color Problem* (Academic Press, New York, 1967).
[25] PAPADIMITRIOU, C.H. and K. STEIGLITZ, *Combinatorial Optimization: Algorithms and Complexity* (Prentice Hall, Englewood Cliffs, NJ, 1982).
[26] ROBERTS, F.S., *Graph Theory and its Applications to the Problems of Society* (SIAM, Philadelphia, PA, 1978.
[27] ROCKAFELLAR, R.T., *Network Flows and Monotropic Optimization* (Wiley, New York, 1984).
[28] TARJAN, R.E., *Data Structures and Network Algorithms* (SIAM, Philadelphia, PA, 1983).
[29] TUTTE, W.T., *Connectivity of graphs*, Mathematicae Expositiones, Vol. 15 (Univ. of Toronto Press, Toronto and Oxford Univ. Press, London, 1966.
[30] TUTTE, W.T., *Graph Theory*, Encyclopedia of Mathematics and Its Applications, Vol. 21 (Addison-Wesley, Menlo Park, CA, 1984).
[31] YAP, H.P., *Some Topics in Graph Theory*, London Mathematical Society Lecture Note Series, Vol. 108 (Cambridge Univ. Press, Cambridge, UK, 1986).

Section 1.1. Drawing a graph

[1] BODLAENDER, H.L., Improved self-reduction algorithms for graphs with bounded tree-width, Tech. Report RUU-CS-88-29, Dept. of Computer Science, Univ. of Utrecht, Utrecht, 1988.
[2] BOOTH, K.S. and G.S. LUEKER, Testing for the consecutive ones property, interval graphs, and graph planarity using *PQ*-trees, *J. Comput. System Sci.* **13** (1976) 335–379.
[3] BRANDENBURG, F.J., Nice drawing of graphs and of trees are computationally hard, Bericht MIP-8820, Fak. Mathematik und Informatik, Universität Passau, Passau, 1988.
[4] CHIBA, N., T. NISHIZEKI, S. ABE and T. OZAWA, A linear algorithm for embedding planar graphs using *PQ*-trees. *J. Comput. System Sci.* **30** (1985) 54–76.
[5] CHIBA, N., K. ONOGUCHI and T. NISHIZEKI, Drawing plane graphs nicely, *Acta Inform.* **22** (1985) 187–201.
[6] CHIBA, N., T. YAMANOUCHI and T. NISHIZEKI, Linear algorithms for convex drawings of planar graphs, in: J.A. Bondy and U.S.R. Murty, ed., *Progress in Graph Theory* (Academic Press, Toronto, 1984) 153–173.
[7] CHIBA, T., I. NISHIOKA and I. SHIRAKAWA, An algorithm for maximum planarization of graphs, in: *Proc. IEEE Int. Symp. Circuits and Systems* (1979) 649–652.
[8] DE FRAYSSEIX, H., J. PACH and R. POLLACK, Small sets supporting Fáry embeddings of planar graphs, in: *Proc. 20th Ann. ACM Symp. Theory of Computing*, Chicago (1988) 426–433.
[9] DE FRAYSSEIX, H. and P. ROSENSTIEHL, A depth-first search characterisation of planarity, *Ann. Discrete Math.* **13** (1982) 75–80.
[10] FÁRY, I., On straight line representations of planar graphs, *Acta Sci. Math. Szeged* **11** (1948) 229–233.
[11] FELLOWS, M.R. and M.A. LANGSTON, Layout permutation problems and well-partially-ordered sets, in: *Proc. 5th MIT Conf. Adv. Research in VLSI*, Cambridge, MA (1988) 315–327.
[12] GAREY, M.R. and D.S. JOHNSON, *Computers and Intractability: A Guide to the Theory of NP-Completeness* (Freeman, San Francisco, CA, 1979).
[13] GAREY, M.R. and D.S. JOHNSON, Crossing number is NP-complete, *SIAM J. Algebraic Discrete Meth.* **4** (1983) 312–316.

[14] GRÜNBAUM, B., *Convex Polytopes* (Wiley, London, 1967).
[15] HARARY, F., *Graph Theory* (Addison-Wesley, Reading, MA, 1969).
[16] HOPCROFT, J.E. and R.E. TARJAN, Dividing a graph into triconnected components, *SIAM J. Comput.* **2** (1973) 135–158.
[17] HOPCROFT, J.E. and R.E. TARJAN, Efficient planarity testing, *J. ACM* **21** (1974) 549–568.
[18] JAYAKUMAR, R., K. THULASIRAMAN and M.N.S. SWAMY, On maximal planarization of non-planar graphs, *IEEE Trans. Circuits and Systems* **33** (1986) 843–844.
[19] JAYAKUMAR, R., K. THULASIRAMAN and M.N.S. SWAMY, Planar embedding: linear-time algorithms for vertex placements and edge ordering, *IEEE Trans. Circuits and Systems* **35** (1988) 334–344.
[20] JAYAKUMAR, R., K. THULASIRAMAN and M.N.S. SWAMY, O(n^2) algorithms for graph planarization, in: J. van Leeuwen, ed., *Graph-Theoretic Concepts in Computer Science, Proc. Int. Workshop WG88*, Lecture Notes in Computer Science, Vol. 344 (Springer, Berlin, 1989) 352–377.
[21] JOHNSON, D.S., The NP-completeness column: an ongoing guide, *J. of Algorithms* **3** (1982) 89–99.
[22] JOHNSON, D.S., The NP-completeness column: an ongoing guide, *J. of Algorithms* **3** (1982) 381–395.
[23] KURATOWSKI, K., Sur le problème des courbes gauches en topologie, *Fund. Math.* **15** (1930) 271–283
[24] MACLANE, S., A structural characterization of planar combinatorial graphs, *Duke Math. J.* **3** (1937) 460–472.
[25] MANSFIELD, A., Determining the thickness of graphs is NP-hard, *Math. Proc. Cambridge Philos. Soc.* **93** (1983) 9–23.
[26] OZAWA, T. and H. TAKAHASHI, A graph-planarization algorithm and its application to random graphs, in: *Graph Theory and Algorithms*, Lecture Notes in Computer Science, Vol. 108 (Springer, Berlin, 1981) 95–107.
[27] REINGOLD, E.M. and J.S. TILFORD, Tidier drawings of trees, *IEEE Trans. Softw. Eng.* **7** (1981) 223–228.
[28] STEIN, S.K., Convex maps, *Proc. Amer. Math. Soc.* **2** (1951) 464–466.
[29] STEINITZ, E., Polyeder und Raumeinteilungen, in: *Enzyklopädie der Mathematischen Wissenschaften, Dritter Band (Geometrie), Erster Teil, Zweite Hälfte* (Teubner, Leipzig, 1916) 1–139.
[30] SUPOWIT, K.J. and E.M. REINGOLD, The complexity of drawing trees nicely, *Acta Inform.* **18** (1983) 377–392.
[31] TAMASSIA, R., G. DI BATTISTA and C. BATINI, Automatic graph drawing and readability of diagrams, *IEEE Trans. Systems Man Cybernet.* **18** (1988) 61–79.
[32] TARJAN, R.E., Depth-first search and linear graph algorithms, *SIAM J. Comput.* **1** (1972) 146–160.
[33] TARJAN, R.E., Algorithm design, *Comm. ACM* **30** (1987) 205–212.
[34] THOMASSEN, C., Planarity and duality of finite and infinite planar graphs, *J. Combin. Theory Ser. B* **29** (1980) 244–271.
[35] THOMASSEN, C., Plane representations of graphs, in: J.A. Bondy and U.S.R. Murty, eds., *Progress in Graph Theory* (Academic Press, Toronto, 1984) 43–69.
[36] TUTTE, W.T., Convex representations of graphs, *Proc. London Math. Soc.* (3) **10** (1960) 304–320.
[37] TUTTE, W.T., How to draw a graph, *Proc. London Math. Soc.* (3) **12** (1963) 743–768.
[38] VAUCHER, J.C., Pretty-printing of trees, *Software—Pract. & Exper.* **10** (1980) 553–561.
[39] WETHEREL, C. and A. SHANNON, Tidy drawings of trees, *IEEE Trans Softw. Eng.* **5** (1979) 514–520.
[40] WHITNEY, H., Congruent graphs and the connectivity of graphs, *Amer. J. Math.* **54** (1932) 150–168.
[41] WILLIAMSON, S.G., Embedding graphs in the plane—algorithmic aspects, *Ann. Discrete Math.* **6** (1980) 349–384.
[42] WILLIAMSON, S.G., Depth-first search and Kuratowski subgraphs, *J. ACM* **31** (1984) 681–693.
[43] YANNAKAKIS, M., A polynomial time algorithm for the min-cut linear arrangement of trees, *J. ACM* **32** (1985) 950–988.

Section 1.2. Computer representations of graphs

[1] AHO, A.V., J.E. HOPCROFT and J.D. ULLMAN, *The Design and Analysis of Computer Algorithms* (Addison-Wesley, Reading, MA, 1974).
[2] BEST, M.R., P. VAN EMDE BOAS and H.W. LENSTRA JR, A sharpened version of the Aanderaa–Rosenberg conjecture, Tech. Report ZW 30/74, Mathematical Centre, Amsterdam, 1974.

[3] BOLLOBÁS, B., Complete subgraphs are elusive, *J. Combin. Theory Ser. B* **21** (1976) 1–7.
[4] BOLLOBÁS, B. and S.E. ELDRIDGE, Packings of graphs and applications to computational complexity, *J. Combin. Theory Ser. B* **25** (1978) 105–124.
[5] GALPERIN, H. and A. WIGDERSON, Succinct representations of graphs, *Inform. and Control* **56** (1983) 183–198.
[6] HOLT, R.C. and E.M. REINGOLD, On the time required to detect cycles and connectivity in graphs, *Math. Systems Theory* **6** (1972) 103–106.
[7] ILLIES, N., A counterexample to the generalized Aanderaa–Rosenberg conjecture, *Inform. Process. Lett.* **7** (1978) 154–155.
[8] ITAI, A. and M. RODEH, Representation of graphs, *Acta Inform.* **17** (1982) 215–219.
[9] KAHN, A.B., Topological sorting of large networks, *Comm. ACM* **5** (1962) 558–562.
[10] KAHN, J., M. SAKS and D. STURTIVANT, A topological approach to evasiveness, *Combinatorica* **4** (1984) 297–306.
[11] KANNAN, S., M. NAOR and S. RUDLICH, Implicit representation of graphs, in: *Proc. 20th Ann. ACM Symp. Theory of Computing*, Chicago (1988) 334–343.
[12] KING, K.N. and B. SMITH-THOMAS, An optimal algorithm for sink-finding, *Inform. Process. Lett.* **14** (1982) 109–111.
[13] KING, V., Lower bounds on the complexity of graph properties, in: *Proc. 20th Ann. ACM Symp. Theory of Computing*, Chicago (1988) 468–476.
[14] KIRKPATRICK, D., Determining graph properties from matrix representations, in: *Proc. 6th Ann. ACM Symp. Theory of Computing*, Seattle (1974) 84–90.
[15] KIRKPATRICK, D., Optimal search in planar subdivisions, *SIAM J. Comput.* **12** (1983) 28–35.
[16] KLEITMANN, D. and D.J. KWIATKOWSKI, Further results on the Aanderaa–Rosenberg conjecture, *J. Combin. Theory Ser. B* **28** (1980) 85–95.
[17] KNUTH, D.E., *The Art of Computer Programming, Vol. 1: Fundamental Algorithms* (Addison-Wesley, Reading, MA, 1968).
[18] LENGAUER, T., Hierarchical graph algorithms, Tech. Report SFB 124, Fachbereich 10, Angewandte Mathematik und Informatik, Univ. des Saarlandes, Saarbrücken, 1984.
[19] LENGAUER, T., Efficient solution of connectivity problems on hierarchically defined graphs, Bericht 24, Fachbereich Informatik, Gesamthochshule Paderborn, Paderborn, 1985.
[20] LENGAUER, T., Efficient algorithms for finding minimum spanning forests of hierarchically defined graphs, *J. Algorithms* **8** (1987) 260–284.
[21] LENGAUER, T., Hierarchical planarity testing algorithms, J. *ACM* **36** (1989) 474–509.
[22] LENGAUER, T. and K. WAGNER, The correlation between the complexities of the non-hierarchical and hierarchical versions of graph problems, in: F.J. Brandenburg et al., eds., *Proc. STACS 87, 4th Ann. Symp.*, Lecture Notes in Computer Science, Vol. 247 (Springer, Berlin, 1987) 100–113.
[23] MILNER, E.C. and D.J.A. WELSH, On the computational complexity of graph theoretical properties, in: C.St.J.A. Nash-Williams and J. Sheenan, eds., *Proc. 5th British Combin. Conf.* (1976) 471–487.
[24] PAPADIMITRIOU, C.H. and M. YANNAKAKIS, A note on succinct representations of graphs, *Inform. and Control* **71** (1986) 181–185.
[25] RIVEST, R.L. and J. VUILLEMIN, On recognizing graph properties from adjacency matrices, *Theoret. Comput. Sci.* **3** (1976) 371–384.
[26] ROSENBERG, A.L., On the time required to recognize properties of graphs, *SIGACT News* **5** (4) (1973) 15–16.
[27] SHILOACH, Y., Strong linear orderings of a directed network, *Inform. Process. Lett.* **8** (1979) 146–148.
[28] SHOUDAI, T., The lexicographically first topological order problem is NLOG-complete, *Inform. Process. Lett.* **33** (1989/1990) 121–124.
[29] SLISENKO, A., Context-free grammars as a tool for describing polynomial time subclasses of hard problems, *Inform. Process. Lett.* **14** (1982) 52–56.
[30] WAGNER, K., The complexity of problems concerning graphs with regularities, in: *Mathematical Foundations of Computer Science, Proc. 11th Symp.*, Lecture Notes in Computer Science, Vol. 176 (Springer, Heidelberg, 1984) 544–552.
[31] WAGNER, K., The complexity of combinatorial problems with succinct input representation, *Acta Inform.* **23** (1986) 325–356.

[32] YAO, A.C-C., Lower bounds to randomized algorithms for graph properties, in: *Proc. 28th Ann. IEEE Symp. Foundations of Computing*, Los Angeles (1987) 393–400.
[33] YAO, A.C-C., Monotone bipartite graph properties are evasive, *SIAM J. Comput.* **17** (1988) 517–520.
[34] YAP, H.P., Computational complexity of graph properties, in: K.M. Koh and H.P. Yap, eds., *Graph Theory, Singapore 1983*, Lecture Notes in Mathematics, Vol. 1073, (Springer, Heidelberg, 1984) 35–54.

Section 1.3. Graph exploration by traversal

[1] KORACH E. and Z. OSTFELD, DFS tree construction: algorithms and characterizations, in: J. van Leeuwen, ed., *Graph-Theoretic Concepts in Computer Science, Int. Workshop*, Lecture Notes in Computer Science, Vol. 344 (Springer, Berlin, 1989) 87–106.
[2] MOORE, E.F., The shortest path through a maze, in: *Proc. Int. Symp. Switching Theory 1957, Part II* (1959) 285–292.
[3] REIF, J.H. and W.L. SHERLIS, Deriving efficient graph algorithms, Tech. Report TR-30-82, Aiken Computation Laboratory, Harvard Univ., Cambridge, MA, 1982.
[4] TARJAN, R.E., Depth-first search and linear graph algorithms, *SIAM J. Comput.* **1** (1972) 146–160.
[5] TARRY, G., Le problème des labyrinthes, *Nouv. Ann. Math.* **14** (1895) 187.

Section 1.4. Transitive reduction and transitive closure

[1] ADLEMAN, L., K.S. BOOTH, F.P. PREPARATA and W.L. RUZZO, Improved time and space bounds for Boolean matrix multiplication, *Acta Inform.* **11** (1978) 61–75.
[2] AHO, A.V., M.R. GAREY and J.D. ULLMAN, The transitive reduction of a directed graph, *SIAM J. Comput.* **1** (1972) 131–137.
[3] ANGLUIN, D., The four Russians' algorithm for Boolean matrix multiplication is optimal in its class, *SIGACT News* **8** (1976) 29–33.
[4] ARLAZAROV, V.L., E.A. DINIC, M.A. KRONROD and I.A. FARADŽEV, On economical construction of the transitive closure of an oriented graph, *Soviet Math. Dokl.* **11** (1970) 1209–1210.
[5] BAKER, J., A note on multiplying Boolean matrices, *Comm. ACM* **5** (1962) 102.
[6] BLONIARZ, P.A., M.J. FISCHER and A.R. MEYER, A note on the average time to compute transitive closures, in: S. Michaelson and R. Milner, eds., *Proc. 3rd Int. Coll. Automata, Languages and Programming* (Edinburgh Univ. Press, Edinburgh, 1976) 425–434.
[7] EBERT, J., A sensitive closure algorithm, *Inform. Process Lett.* **12** (1981) 255–258.
[8] EVE, J. and R. KURKI-SUONIO, On computing the transitive closure of a relation, *Acta Inform.* **8** (1977) 303–314.
[9] FISCHER, M.J. and A.R. MEYER, Boolean matrix multiplication and transitive closure, in: *Conf. Rec. 12th Ann. IEEE Symp. Switching and Automata Theory* (1971) 129–131.
[10] FURMAN, M.E., Application of a method of fast multiplication of matrices in the problem of finding the transitive closure of a graph, *Soviet Math. Dokl.* **11** (1970) 1252.
[11] GORALCIKOVA, A. and V. KOUBEK, A reduct and closure algorithm for graphs, in: J. Bečvář, ed., *Proc. Conf. Mathematical Foundations of Computer Science*, Lecture Notes in Computer Science, Vol. 74 (Springer, Heidelberg, 1979) 301–307.
[12] HSU, H.T., An algorithm for finding a minimal equivalent graph of a digraph, *J. ACM* **22** (1975) 11–16.
[13] IBARAKI, I. and N. KATOH, On-line computation of transitive closures of graphs, *Inform. Process Lett.* **16** (1983) 95–97.
[14] JAUMARD, B. and M. MINOUX, An efficient algorithm for the transitive closure and a linear worst-case complexity result for a class of sparse graphs, *Inform. Process. Lett.* **22** (1986) 163–169.
[15] MARTELLO, S., An algorithm for finding a minimal equivalent graph of a strongly connected graph, *Comput.* **15** (1979) 183–194.
[16] MARTELLO, S. and P. TOTH, Finding a minimum equivalent graph of a digraph, *Networks* **12** (1982) 89–100.

[17] MARTYNYUK, V.V., On economical construction of the transitive closure of binary relations, *Ž. Vyčisl. Mat. i. Mat. Fiz* **2** (1962) 723–725.
[18] MOYLES, D.M. and G.L. THOMPSON, An algorithm for finding a minimum equivalent graph of a digraph, *J. ACM* **16** (1969) 455–460.
[19] MUNRO, I., Efficient determination of the transitive closure of a directed graph, *Inform. Process. Lett.* **1** (1971) 56–58.
[20] NOLTEMEIER, H., Reduction of directed graphs to irreducible kernels, Discussion paper 7505, Lehrstuhl Mathematische Verfahrenforschung (Operations Research) und Datenverarbeitung, Univ. Göttingen, Göttingen, 1975.
[21] O'NEILL, P.E. and E.J. O'NEILL, A fast expected time algorithm for Boolean matrix multiplication and transitive closure, *Inform. and Control* **22** (1973) 132–138.
[22] PURDOM JR., P., A transitive closure algorithm, *BIT* **10** (1970) 76–94.
[23] ROY, B., Transitivité et connexité, *C.R. Acad. Sci. Paris* **249** (1959) 216–218.
[24] SAHNI, S., Computationally related problems, *SIAM J. Comput.* **3** (1974) 262–279.
[25] SCHMITZ, L., An improved transitive closure algorithm, *Comput.* **30** (1983) 359–371.
[26] SCHNORR, C.P., An algorithm for transitive closure with linear expected time, *SIAM J. Comput.* **7** (1987) 127–133.
[27] SIMON, K., An improved algorithm for transitive closure on acyclic digraphs, in: L. Kott, ed., *Proc 13th Int. Coll. Automata, Languages and Programming*, Lecture Notes in Computer Science, Vol. 226 (Springer, Berlin, 1988) 376–386.
[28] SIMON, K., On minimum flow and transitive reduction, in: T. Lepistö and A. Salomaa, eds., *Proc. 15th Int. Coll. Automata, Languages and Programming*, Lecture Notes in Computer Science, Vol. 317 (Springer, Berlin, 1988) 535–546.
[29] SYSLO, M.M. and J. DZIKIEWICZ, Computational experiences with some transitive closure algorithms, *Comput.* **15** (1975) 33–39.
[30] THORELLI, L.E., An algorithm for computing all paths in a graph, *BIT* **6** (1966) 347–349.
[31] VAN LEEUWEN, J., Efficiently computing the product of two binary relations, *Int. J. Comput. Math.* **5** (1976) 193–201.
[32] WARREN JR., H.S., A modification of Warshall's algorithm for the transitive closure of binary relations, *Comm. ACM* **18** (1975) 218–220.
[33] WARSHALL, S., A theorem on Boolean matrices, *J. ACM* **9** (1962) 11–12.

Section 1.5. Generating an arbitrary graph

[1] DIXON, J.D. and H.S. WILF, The random selection of unlabeled graphs, *J. Algorithms* **4** (1983) 205–213.
[2] NIJENHUIS, A. and H.S. WILF, The enumeration of connected graphs and linked diagrams, *J. Combin. Theory Ser. A.* **27** (1979) 356–359.
[3] SINCLAIR, A. and M. JERRUM, Approximate counting, uniform generation and rapidly mixing Markov chains, *Inform. and Comput.* **82** (1989) 93–133.
[4] WILF, H.S., The uniform selection of free trees *J. Algorithms* **2** (1981) 204–207.
[5] WORMALD, N.C., Generating random regular graphs, *J. Algorithms* **5** (1984) 247–280.
[6] WORMALD, N.C., Generating random unlabeled graphs, *SIAM J. Comput.* **16** (1987) 717–727.

Section 1.6. Classes of graphs

[1] ARNBORG, S., Efficient algorithms for combinatorial problems on graphs with bounded decomposability — a survey, *BIT* **25** (1985) 2–23.
[2] ARNBORG, S., D.G. CORNEIL and A. PROSKUROWSKI, Complexity of finding embeddings in a k-tree, Report TRITA-NA-8407, Dept. of Numerical Analysis and Computer Science, Royal Institute of Technology, Stockholm, 1984; revised version: *SIAM J. Algebraic Discrete Meth.* **8** (1987) 277–284.
[3] ARNBORG, S., J. LAGERGREN and D. SEESE, Problems easy for tree-decomposable graphs (extended

abstract), in: T. Lepistö and A. Salomaa, ed., *Proc. 15th Int. Coll. Automata, Languages and Programming*, Lecture Notes in Computer Science, Vol. 317 (Springer, Berlin, 1988) 38–51.

[4] ASANO, T., A linear time algorithm for the subgraph homeomorphism problem for the fixed graph $K_{3,3}$, Report AL 83-20, Dept. of Mathematical Engineering and Instrumental Physics, Univ. of Tokyo, Tokyo, 1983.

[5] BODLAENDER, H.L., Classes of graphs with bounded tree-width, Tech. Report RUU-CS- 86-22, Dept. of Computer Science, Univ. of Utrecht, Utrecht, 1986.

[6] BODLAENDER, H.L., Dynamic programming on graphs with bounded tree-width, in: T. Lepistö and A. Salomaa, ed., *Proc. 15th Int. Coll. Automata, Languages and Programming*, Lecture Notes in Computer Science, Vol. 317 (Springer, Berlin, 1988) 103–118.

[7] BODLAENDER H.L., Improved self-reduction algorithms for graphs with bounded tree-width, Tech. Report RUU-CS-88-29, Dept. of Computer Science, Univ. of Utrecht, Utrecht, 1988.

[8] BODLAENDER, H.L., Planar graphs with bounded tree-width, Tech. Report RUU-CS-88-14, Dept. of Computer Science, Univ. of Utrecht, Utrecht, 1988; also: *EATCS Bull.* **36** (1988) 116–126.

[9] BOOTH, K.S. and G.S. LUEKER, Testing for the consecutive ones property, interval graphs, and graph planarity using *PQ*-trees. *J. Comput. System Sci.* **13** (1976) 335–379.

[10] BROWN, D.J., M.R. FELLOWS and M.A. LANGSTON, Polynomial time of selfreducibility: theoretical motivations and practical results, Tech. Report CS-87-171, Computer Science Dept., Washington State Univ., Pullman, WA, 1988.

[11] COURCELLE, B., Graph rewriting: an algebraic and logic approach, in: J. van Leeuwen, ed., *Handbook of Theoretical Computer Science, Vol. B* (North-Holland, Amsterdam, 1990) Chapter 5.

[12] FARBER, M., Characterizations of strongly chordal graphs, *Discrete Math.* **43** (1983) 173–189.

[13] FELLOWS, M.R. and M.A. LANGSTON, Nonconstructive tools for proving polynomial time decidability, *J. ACM* **35** (1988) 727–739.

[14] FELLOWS, M.R. and M.A. LANGSTON, On search, decision and the efficiency of polynomial time algorithms, Tech. Report CS-88-190. Computer Science Dept., Washington State University, Pullman, WA, 1988.

[15] FILOTTI, I.S., An efficient algorithm for determining whether a cubic graph is toroidal, in: *Proc. 10th Ann. ACM Symp. Theory of Computing*, San Diego (1978) 133–142.

[16] FILOTTI, I.S., G.L. MILLER and J. REIF, On determining the genus of a graph in $O(v^{O(g)})$ steps, in: *Proc. 11th Ann. ACM Symp. Theory of Computing*, Atlanta, GA (1979) 27–37.

[17] GABOR, C.P., W.L. HSU and K.J. SUPOWIT, Recognizing circle graphs in polynomial time, in: *Proc. 26th Ann. IEEE Symp. Foundations of Computer Science*, Portland, OR (1985) 106–116; also: *J. ACM* **36** (1989) 435–473.

[18] GAREY, M.R. and D.S. JOHNSON, *Computers and Intractability: A Guide to the Theory of NP-Completeness* (Freeman, San Francisco, CA, 1979).

[19] GAVRIL, F., An algorithm for testing chordality of graphs, *Inform. Process. Lett.* **3** (1974) 110–112.

[20] GAVRIL, F., A recognition algorithm for the intersection graphs of directed paths in directed trees, *Discrete Math.* **13** (1975) 327–249.

[21] GAVRIL, F., A recognition algorithm for the intersection graphs of paths in trees, *Discrete Math.* **23** (1978) 211–227.

[22] GOLUMBIC, M.C., *Algorithmic Graph Theory and Perfect Graphs* (Academic Press, New York, 1980).

[23] HARARY, F., *Graph Theory* (Addison-Wesley, Reading, MA, 1969).

[24] HOPCROFT, J.E. and R.E. TARJAN, Efficient planarity testing *J. ACM* **21** (1974) 549–568.

[25] HSU, W.L., Recognizing planar perfect graphs, *J. ACM* **34** (1987) 255–288.

[26] JOHNSON, D.S., The NP-completeness column: an ongoing guide, *J. Algorithms* **6** (1985) 431–451.

[27] LEHOT, P.G.H., An optimal algorithm to detect a line graph and output its root graph, *J. ACM* **21** (1974) 569–575.

[28] MANSFIELD, A., Determining the thickness of graphs is NP-hard, *Math. Proc. Cambridge Philos. Soc.* **93** (1983) 9–23.

[29] MITCHELL, S.L., Linear algorithms to recognize outerplanar and maximal outerplanar graphs, *Inform. Process. Lett.* **9** (1979) 229–232.

[30] ROBERTSON, N. and P.D. SEYMOUR, Graph minors: I excluding a forest, *J. Combin. Theory Ser. B.* **35** (1983) 39–61.

[31] ROBERTSON, N. and P.D. SEYMOUR, Graph minors III: planar treewidth, *J. Combin Theory Ser. B.* **36** (1984) 49–64.
[32] ROBERTSON, N. and P.D. SEYMOUR, Graph minors II: algorithmic aspects of tree-width, *J. Algorithms* **7** (1986) 309–322.
[33] ROBERTSON, N. and P.D. SEYMOUR, Graph minors V: excluding a planar graph, *J. Combin. Theory Ser. B.* **41** (1986) 92–114.
[34] ROBERTSON, N. and P.D. SEYMOUR, Graph minors IV: tree-width and well quasi-ordering, *J. Combin. Theory Ser. B.*, to appear.
[35] SAXE, J.B., Dynamic programming algorithms for recognizing small-bandwidth graphs in polynomial time, *SIAM J. Algebraic Discrete Meth.* **1** (1980) 363–369.
[36] SYSLO, M.M., A labeling algorithm to recognize a line graph and output its root graph, *Inform. Process. Lett.* **15** (1982) 28–30.
[37] TARJAN, R.E. and M. YANNAKAKIS, Simple linear time algorithms to test chordality of graphs, test acyclicity of hypergraphs, and selectively reduce acyclic hypergraphs, *SIAM J. Comput.* **13** (1984) 566–579.
[38] THOMASSEN, C., The graph genus problem is NP-complete, *J. Algorithms* **10** (1989) 568–576.
[39] TUCKER, A., Matrix characterization of circular-arc graphs, *Pacific J. Math.* **39** (1971) 535–545.
[40] TUCKER, A., An efficient test for circular-arc graphs, *SIAM J. Comput.* **9** (1980) 1–24.
[41] VALDES, J., R.E. TARJAN and E.L. LAWLER, The recognition of series parallel graphs, *SIAM J. Comput.* **11** (1982) 298–313.
[42] WILLIAMSON, S.G., Depth-first search and Kuratowski subgraphs, *J. ACM* **31** (1984) 681–693.

Section 2.1. Connectivity

[1] BECKER, M., W. DEGENHARDT, J. DOENHART, S. HERTEL, G. KANINKE, W. KEBER, K. MEHLHORN, S. NÄHER, H. ROHNERT and T. WINTER, A probabilistic algorithm for vertex connectivity of graphs, *Inform. Process. Lett.* **15** (1982) 135–136.
[2] ESWARAN, K.P. and R.E. TARJAN, Augmentation problems, *SIAM J. Comput.* **5** (1976) 653–665.
[3] EVEN, S., An algorithm for determining whether the connectivity of a graph is at least *k*, *SIAM J. Comput.* **4** (1975) 393–396.
[4] EVEN, S. and R.E. TARJAN, Network flow and testing graph connectivity, *SIAM J. Comput.* **4** (1975) 507–518.
[5] GALIL, Z., Finding the vertex connectivity of a graph, *SIAM J. Comput.* **9** (1980) 197–199.
[6] HOPCROFT, J.E. and R.E. TARJAN, Dividing a graph into triconnected components, *SIAM J. Comput.* **2** (1973) 135–158.
[7] KANEVSKY, A. and V. RAMACHANDRAN, Improved algorithms for graph four-connectivity, in: *Proc. 28th Ann. IEEE Symp. Foundations of Computer Science*, Los Angeles (1987) 252–259.
[8] LINIAL, N., L. LOVÁSZ and A. WIGDERSON, A physical interpretation of graph connectivity and its algorithmic implications, in: *Proc. 27th Ann. IEEE Symp. Foundations of Computer Science*, Toronto (1986) 39–48.
[9] MA, Y-W.E. and C.-M. CHEN, Connectivity analysis of dynamic computer networks, Tech. Report MS-CIS-86-30, Dept. of Computer and Information Science, Univ. of Pennsylvania, Philadelphia, PA, 1986.
[10] MANSOUR, Y. and B. SCHIEBER, Finding the edge-connectivity of directed graphs, *J. Algorithms* **10** (1989) 76–85.
[11] MATULA, D., Determining edge connectivity in O(*nm*), in: *Proc. 28th Ann. IEEE Symp. Foundations of Computer Science*, Los Angeles (1987) 249–251.
[12] ROSENTHAL, A. and A. GOLDNER, Smallest augmentations to biconnect a graph, *SIAM J. Comput.* **6** (1977) 55–66.
[13] SCHNORR, C.P., Bottlenecks and edge connectivity in unsymmetrical networks, *SIAM J. Comput.* **8** (1979) 265–274.
[14] TARJAN, R.E., Depth-first search and linear graph algorithms, *SIAM J. Comput.* **1** (1972) 146–160.

[15] WATANABE, T. and A. NAKAMURA, Edge-connectivity augmentation problems, *J. Comput. System Sci.* **35** (1987) 96–144.

Section 2.2. Minimum spanning trees

[1] BORŮVKA, O., O jistém problému minimálním, *Práca Moravske Přírodovědecké Společnosti* **3** (1926) 37–58.
[2] CHERITON, D. and R.E. TARJAN, Finding minimum spanning trees, *SIAM J. Comput.* **5** (1976) 724–742.
[3] CHIN, F. and D. HOUCK, Algorithms for updating minimal spanning trees, *J. Comput. System Sci.* **16** (1978) 333–344.
[4] DIJKSTRA, E.W., A note on two problems in connexion with graphs, *Numer. Math.* **1** (1959) 269–271.
[5] FREDERICKSON, G.N., Data structures for on-line updating of minimum spanning trees, with applications, *SIAM J. Comput.* **14** (1985) 781–798.
[6] FREDMAN, M.L. and R.E. TARJAN, Fibonacci heaps and their uses in improved network optimalization problems, in: *Proc. 25th Ann. IEEE Symp. Foundations of Computer Science*, Singer Island (1984) 338–346; revised version: *J. ACM* **34** (1987) 596–615.
[7] GABOR, C.P., W.L. HSU and K.J. SUPOWIT, Recognizing circle graphs in polynomial time, in: *Proc. 26th Ann. IEEE Symp. Foundations of Computer Science*, Portland, OR (1985) 106–116.
[8] GABOW, H.N., Z. GALIL and T.H. SPENCER, Efficient implementation of graph algorithms using contraction, in: *Proc. 25th Ann. IEEE Symp. Foundations of Computer Science*, Singer Island (1984) 347–357.
[9] GABOW, H.N., Z. GALIL, T. SPENCER and R.E. TARJAN, Efficient algorithms for finding minimum spanning trees in undirected and directed graphs, *Combinatorica* **6** (1986) 109–122.
[10] GRAHAM, R.L. and P. HELL, On the history of the minimum spanning tree problem, *Ann. Hist. Comput.* **7** (1985) 43–57.
[11] JARNIK, V., O jistém problému minimálním, *Práca Moravské Přírodovědecké Společnosti* **6** (1930) 57–63.
[12] JOHNSON, D.B., Priority queues with update and finding minimum spanning trees, *Inform. Process. Lett.* **4** (1975) 53–57.
[13] KÓMLOS, J., Linear verification for spanning trees, in: *Proc. 25th Ann. IEEE Symp. Foundations of Computer Science*, Singer Island (1984) 201–206.
[14] PRIM, R.C., Shortest connection networks and some generalizations, *Bell Syst. Tech. J.* **36** (1957) 1389–1401.
[15] ROSKIND, J. and R.E. TARJAN, A note on finding minimum cost edge-disjoint spanning trees, *Math. Oper. Res.* **10** (1985) 701–708.
[16] SPIRA, P.M. and A. PAN, On finding and updating spanning trees and shortest paths, *SIAM J. Comput.* **4** (1975) 375–380.
[17] YAO, A., An $O(|E|\log\log|V|)$ algorithm for finding minimum spanning trees, *Inform. Process. Lett.* **4** (1975) 21–23.

Section 2.3. Shortest paths

Section 2.3.1. Single-source shortest paths

[1] DERIGS, U., An efficient Dijkstra-like labeling method for computing shortest odd/even paths, *Inform. Process. Lett.* **21** (1985) 253–258.
[2] DIJKSTRA, E.W., A note on two problems in connexion with graphs, *Numer. Math.* **1** (1959) 269–271.
[3] FORD, L.R. and D.R. FULKERSON, *Flows in Networks* (Princeton Univ. Press, Princeton, NJ, 1962).
[4] FREDERICKSON, G.N., Fast algorithms for shortest paths in planar graphs, with applications, *SIAM J. Comput.* **16** (1987) 1004–1022.
[5] FREDMAN, M.L., New bounds on the complexity of the shortest path problem, *SIAM J. Comput.* **5** (1976) 83–89.

[6] FREDMAN, M.L. and R.E. TARJAN, Fibonacci heaps and their uses in improved network optimalization problems, in: *Proc. 25th Ann. IEEE Symp. Foundations of Computer Science*, Singer Island (1984) 338–346; revised version: *J. ACM* **34** (1987) 596–615.
[7] JOHNSON, D.B., Effect algorithms for shortest paths in sparse networks, *J. ACM* **24** (1977) 1–13.
[8] MEHLHORN, K. and B.H. SCHMIDT, A single source shortest path algorithm for graphs with separators, in: *Proc. FCT 83*, Lecture Notes in Computer Science, Vol. 158 (Springer, Berlin, 1983) 302–309.
[9] YAP, C.K., A hybrid algorithm for the shortest path between two nodes in the presence of few negative arcs, *Inform. Process. Lett.* **16** (1983) 181–182.

Section 2.3.2. All pairs shortest paths

[1] BLONIARZ, P., A shortest-path algorithm with expected time $O(n^2 \log n \log^* n)$, *SIAM J. Comput.* **12** (1983) 588–600.
[2] EDMONDS, J. and R.M. KARP, Theoretical improvements in algorithmic efficiency for network flow problems, *J. ACM* **19** (1972) 248–264.
[3] FREDERICKSON, G.N., Fast algorithms for shortest paths in planar graphs, with applications, *SIAM J. Comput.* **16** (1987) 1004–1022.
[4] MOFFAT, and T. TAKAOKA, An all pairs shortest path algorithm with expected running time $O(n^2 \log n)$, *SIAM J. Comput.* **16** (1987) 1023–1031.
[5] ROMANI, F., Shortest path problem is not harder than matrix multiplication, *Inform. Process. Lett.* **11** (1980) 134–136.
[6] WATANABE, A fast algorithm for finding all shortest paths, *Inform. Process. Lett.* **13** (1981) 1–3.

Section 2.4. Paths and cycles

Section 2.4.1. Paths of length k

[1] BODLAENDER, H.L., On linear time minor tests and depth-first search, in: F. Dehne et al., eds. *Proc. 1st Workshop on Algorithms and Data Structures*, Lecture Notes in Computer Science, Vol. 382 (Springer, Berlin, 1989) 577–590.
[2] ELLIS, J.A., M. MATA and G. MACGILLIVRAY, A linear time algorithm for longest (s, t)-paths in weighted outerplanar graphs, *Inform. Process. Lett.* **32** (1989) 199–204.
[3] MONIEN, B., The complexity of determining paths of length k, in: M. Nagl and J. Perl, eds., *Proc. Int. Workshop on Graph Theoretical Concepts in Computer Science* (Trauner Verlag, Linz, 1983) 241–251; revised version: How to find long paths efficiently, *Ann. Discrete Math.* **25** (1985) 239–254.

Section 2.4.2. Disjoint paths

[1] CYPHER, A., An approach to the k paths problem, in: *Proc. 12th Ann. ACM Symp. Theory of Computing*, Los Angeles (1980) 211–217.
[2] FORD, L.R. and D.R. FULKERSON, *Flows in Networks* (Princeton Univ. Press, Princeton, NJ, 1962).
[3] FRISCH, I.T., An algorithm for vertex-pair connectivity, *Int. J. Control* **6** (1967) 579–593.
[4] KRISHNAN, S.V., C. PANDU RANGAN and S. SESHADRI, A new linear algorithm for the two path problem on planar graphs, Preprint, Dept. of Computer Science and Engineering, Indian Institute of Technology, Madras, 1987.
[5] PERL, Y. and Y. SHILOACH, Finding two disjoint paths between two pairs of vertices in a graph, *J. ACM* **25** (1978) 1–9.
[6] ROBERTSON, N. and P.D. SEYMOUR, Graph minors XIII: The disjoint paths problem, to appear.
[7] ROBERTSON, N. and P.D. SEYMOUR, Disjoint paths—a survey, *SIAM J. Algebraic Discrete Meth.* **6** (1985) 300–305.
[8] SEYMOUR, P.D., Disjoint paths in graphs, *Discrete Math.* **29** (1980) 293–309.
[9] SHILOACH, Y., A polynomial solution to the undirected two paths problem, *J. ACM* **27** (1980) 445–456.
[10] STEIGLITZ, K. and J. BRUNO, A new derivation of Frisch's algorithm for calculating vertex-pair connectivity, *BIT* **11** (1971) 94–106.

[11] SUURBALLE, J.W., Disjoint paths in a network, *Networks* **4** (1974) 125–145.
[12] SUURBALLE, J.W. and R.E. TARJAN, A quick method for finding shortest pairs of disjoint paths, *Networks* **14** (1984) 325–336.

Section 2.4.3. Cycles

[1] BODLAENDER, H.L., On linear time minor tests and depth-first search, in: F. Dehne et al., eds., *Proc. 1st Workshop on Algorithms and Data Structures*, Lecture Notes in Computer Science, Vol. 382 (Springer, Berlin, 1989) 577–590.
[2] CHIBA, N. and T. NISHIZEKI, Arboricity and subgraph listing algorithms, *SIAM J. Comput.* **14** (1985) 210–223
[3] DEO, N., Minimum length fundamental cycle set, *IEEE Trans. Circuits and Systems* **26** (1979) 894–895.
[4] DEO, N., G.M. PRABHU and M.S. KRISHNAMOORTHY, Algorithms for generating fundamental cycles in a graph, *ACM Trans. Math. Software* **8** (1982) 26–42.
[5] DIXON, E.T. and S.E. GOODMAN, An algorithm for the longest cycle problem, *Networks* **6** (1976) 139–149.
[6] FELLOWS, M.R. and M.A. LANGSTON, Nonconstructive tools for proving polynomial time decidability, *J. ACM* **35** (1988) 727–739.
[7] GAREY, M.R. and R.E. TARJAN, A linear time algorithm for finding all feedback vertices, *Inform. Process. Lett.* **7** (1978) 274–276.
[8] HORTON, J.D., A polynomial time algorithm to find the shortest cycle basis of a graph, *SIAM J. Comput.* **16** (1987) 358–366.
[9] ITAI, A. and M. RODEH, Finding a minimum circuit in a graph, *SIAM J. Comput.* **7** (1978) 413–423.
[10] JOVANOVICH, A.D., Note on a modification of the fundamental cycles finding algorithm, *Inform. Process. Lett.* **3** (1974) 33.
[11] LAPAUGH, A.S. and R.L. RIVEST, The subgraph homeomorphism problem, in: *Proc. 10th Ann. ACM Symp. Theory of Computing*, San Diego (1978) 40–50.
[12] MONIEN, B., The complexity of determining a shortest cycle of even length, *Comput.* **31** (1983) 355–369.
[13] PATON, K., An algorithm for finding a fundamental set of cycles of a graph, *Comm. ACM* **12** (1969) 514–518.
[14] RICHARDS, D. and A.L. LIESTMAN, Finding cycles of a given length, *Ann. Discrete Math.* **27** (1985) 249–256.
[15] RICHARDS, D., Finding short cycles in planar graphs using separators, *J. Algorithms* **7** (1986) 382–394.
[16] RYAN, D.R. and S. CHEN, A comparison of three algorithms for finding fundamental cycles in a directed graph, *Networks* **11** (1981) 1–12.
[17] THOMASSEN, C., Even cycles in directed graphs, *European J. Combin.* **6** (1985) 85–89.
[18] VAZIRANI, V.V. and M. YANNAKAKIS, Pfaffian orientations, 0/1 permanents, and even cycles in directed graphs, in: T. Lepistö and A. Salomaa, eds., *Proc. 15th Int. Coll. Automata, Languages and Programming*, Lecture Notes in Computer Science, Vol. 317 (Springer, Verlag, Berlin, 1988) 667–681.

Section 2.4.4. Weighted cycles

[1] DOMSCHKE, W., Zwei Verfahren zur Suche negativer Zyklen in bewerteten Digraphen, *Comput.* **11** (1973) 125–136.
[2] FLORIAN, M. and P. ROBERT, A direct search method to locate negative cycles in a graph, *Manag. Sci.* **17** (1971) 307–310.
[3] HORTON, J.D., A polynomial time algorithm to find the shortest cycle basis of a graph, *SIAM J. Comput.* **16** (1987) 358–366.
[4] KARP, R.M., A characterization of the minimum cycle mean in a digraph, *Discrete Math.* **23** (1978) 309–311.
[5] KARP, R.M. and J.B. ORLIN, Parametric shortest path algorithms with an application to cycle staffing, *Discrete Appl. Math.* **3** (1981) 37–45.
[6] MAIER, D., A space efficient method for the lowest common ancestor problem and an application to

finding negative cycles, in: *Proc. 18th Ann. IEEE Symp. Foundations of Computer Science*, Providence, RI (1977) 132–141.
[7] TOBIN, R.L., Minimal complete matchings and negative cycles, *Networks* **5** (1975) 371–387.
[8] TSAKALIDIS, A.K., Finding a negative cycle in a directed graph, Tech. Report A85/05, Angewandte Mathematik u. Informatik, Fb-10, Univ. des Saarlandes, Saarbrücken, 1985.
[9] YEN, J.Y., On the efficiency of a direct search method to locate negative cycles in a network, *Manag. Sci.* **19** (1972) 335–336; rejoinder by M. Florian and P. Robert, *ibidem* 335–336.

Section 2.4.5. Eulerian (and other) cycles

[1] ALON, N. and M. TARSI, Covering multigraphs by simple circuits, *SIAM J. Algebraic Discrete Meth.* **6** (1985) 345–350.
[2] EBERT, J., Computing Eulerian trails, *Inform. Process. Lett.* **28** (1988) 93–97.
[3] EDMONDS, J. and E.L. JOHNSON, Matching, Euler tours and the Chinese postman, *Math. Progr.* **5** (1973) 88–124.
[4] FLEISCHER, H., Eulerian graphs, in: L.W. Beineke and R.J. Wilson, eds., *Selected Topics in Graph Theory* **2** (Academic Press, London, 1983) 17–53.
[5] GOODMAN, S. and S. HEDETNIEMI, Eulerian walks in graph, *SIAM J. Comput.* **2** (1973) 16–27.
[6] HIERHOLZER, C., Über die Möglichkeit, einen Linienzug ohne Wiederholung und ohne Unterbrechung zu umfahren, *Math. Ann.* **6** (1873) 30–32.
[7] ITAI, A., R.J. LIPTON, C.H. PAPADIMITRIOU, and M. RODEH, Covering graphs by simple circuits, *SIAM J. Comput.* **10** (1981) 746–750.
[8] KWAN MEI-KO, Graphic programming using odd or even points, *Acta Math. Sinica* **10** (1960) 263–266; *Chin. Math.* **1** (1962) 273–277.
[9] MEIGU GUAN, The maximum weighted cycle-packing problem and its relation to the Chinese postman problem, in: J.A. Bondy and U.S.R. Murty, eds., *Progress in Graph Theory* (Academic Press, Toronto, 1984) 323–326.
[10] ORE, O., A problem regarding the tracing of graphs, *Elem. Math.* **6** (1951) 49–53.
[11] PAPADIMITRIOU, C.H., On the complexity of edge traversing, *J. ACM* **23** (1976) 544–554.

Section 2.4.6. Hamiltonian (and other) cycles

[1] ANGLUIN, D. and L.G. VALIANT, Fast probabilistic algorithms for Hamiltonian circuits and matchings, *J. Comput. System Sci.* **18** (1979) 155–193.
[2] ARNBORG, S., J. LAGERGREN and D. SEESE, Problems easy for tree-decomposable graphs (extended abstract), in: T. Lepistö and A. Salomaa, eds., *Proc. 15th. Int. Coll. Automata, Languages and Programming*, Lecture Notes in Computer Science, Vol. 317 (Springer, Berlin, 1988) 38–51.
[3] ASANO, T., S. KIKUCHI and N. SAITO, An efficient algorithm to find a hamiltonian circuit in a 4-connected maximal planar graph, in: N. Saito and T. Nishizeki, eds., *Proc. Graph Theory and Algorithms*, Lecture Notes in Computer Science, Vol. 108 (Springer, Berlin, 1981) 182–195; also: *Discrete Appl. Math.* **7** (1984) 1–15.
[4] BOLLOBÁS, B., *Random Graphs* (Academic Press, London, 1985).
[5] CHIBA, N. and T. NISHIZEKI, The hamiltonian cycle problem is linear-time solvable for 4-connected planar graphs, *J. Algorithms* **10** (1989) 187–211.
[6] DANIELSON, G.H., On finding simple paths and circuits in a graph, *IEEE Trans. Circuit Theory* **15** (1968) 294–295.
[7] GAREY, M.R., D.S. JOHNSON and R.E. TARJAN, The planar Hamiltonian circuit problem is NP-complete, *SIAM J. Comput.* **5** (1976) 704–714.
[8] GOODMAN, S.E. and S.T. HEDETNIEMI, On Hamiltonian walks in graphs, *SIAM J. Comput.* **3** (1974) 214–221.
[9] GOODMAN, S.E., S.T. HEDETNIEMI and P.J. SLATER, Advances on the Hamiltonian completion problem, *J. ACM* **22** (1975) 352–360.

[10] GOUYOU-BEAUCHAMPS, D., The Hamiltonian circuit problem is polynomial for 4-connected planar graphs, *SIAM J. Comput.* **11** (1982) 529–539.
[11] ITAI, A., C.H. PAPADIMITRIOU and J.L. SZWARCFITER, Hamilton paths in grid graphs, *SIAM J. Comput.* **11** (1982) 676–686.
[12] KUNDU, S., A linear algorithm for the Hamiltonian completion number of a tree, *Inform. Process. Lett.* **5** (1976) 55–57.
[13] PLESŃIK, J., The NP-completeness of the Hamiltonian cycle problem in planar digraphs with degree bound two, *Inform. Process. Lett.* **8** (1979) 199–201.
[14] RUBIN, F., A search procedure for Hamilton paths and circuits, *J. ACM* **21** (1974) 576–580.
[15] SLATER, P.J., S.E. GOODMAN and S.T. HEDETNIEMI, On the optional Hamiltonian completion problem, *Networks* **6** (1976) 35–51.
[16] TAKAMIZAWA, K., T. NISHIZEKI and N. SAITO, An algorithm for finding a short closed spanning walk of a graph, *Networks* **10** (1980) 249–263.

Section 2.5. Decomposition of graphs

[1] AKIYAMA, J. and M. KANO, Factors and factorizations of graphs—a survey, *J. Graph Theory* **9** (1985) 1–42.
[2] ARJOMANDI, E., On finding all unilaterally connected components of a digraph, *Inform. Process. Lett.* **5** (1976) 8–10.
[3] BOESCH, F.T. and J.F. GIMPEL, Covering the points of a digraph with point-disjoint paths and its application to code optimalization, *J. ACM* **24** (1977) 192–198.
[4] CHESTON, G.A., A correction to a unilaterally connected components algorithm, *Inform. Process. Lett.* **7** (1978) 125.
[5] CUNNINGHAM, W. and J. EDMONDS, A combinatorial decomposition theory, *Canad. J. Math.* **32** (1980) 734–765.
[6] CUNNINGHAM, W.H., Decomposition of directed graphs, *SIAM J. Algebraic Discrete Meth.* **3** (1982) 214–228.
[7] DIKS, K., H.N. DJIDJEV, O. SÝKORA and I. VRŤO, Edge separators for planar graphs and their applications, in: M.P. Chytil et al., eds., *Mathematical Foundations of Computer Science 1988*, Lecture Notes in Computer Science, Vol. 324 (Springer, Berlin, 1988) 280–290.
[8] DJIDJEV, H.N., On the problem of partitioning planar graphs, *SIAM J. Discrete Appl. Meth.* **3** (1982) 229–240.
[9] DJIDJEV, H.N., Linear algorithms for graph separation problems, in: R. Karlsson and A. Lingas, eds., *Proc. SWAT '88, 1st Scandinavian Workshop*, Lecture Notes in Computer Science, Vol. 318 (Springer, Berlin, 1988) 216–222.
[10] EBERT, J., A note on odd and even factors of undirected graphs, *Inform. Process. Lett.* **11** (1980) 70–72.
[11] GAZIT, H. and G.L. MILLER, A parallel algorithm for finding a separator in planar graphs, in: *Proc. 28th Ann. IEEE Symp. Foundations of Computer Science*, Los Angeles (1987) 238–248.
[12] GILBERT, J.R., J.P. HUTCHINSON and R.E. TARJAN, A separator theorem for graphs of bounded genus, *J. Algorithms* **5** (1984) 391–407.
[13] GILBERT, J.R., D.J. ROSE and A. EDENBRANDT, A separator theorem for chordal graphs, *SIAM J. Discrete and Appl. Meth.* **5** (1984) 306–313.
[14] HOHBERG, W., and R. REISCHUK, Decomposition of graphs—an approach for the design of fast sequential and parallel algorithms on graphs, Tech. Report, Fb. Informatik, Technische Hochschule Darmstadt, Darmstadt, 1989.
[15] HOPCROFT, J.E. and R.E. TARJAN, Dividing a graph into triconnected components, *SIAM J. Comput.* **2** (1973) 135–158.
[16] KAO, M-Y., All graphs have cycle separators and planar directed depth-first search is in DNC, in: J.H. Reif, ed., *Proc. 3rd Aegean Workshop VLSI Algorithms and Architectures*, Lecture Notes in Computer Science, Vol. 319 (Springer, Berlin, 1988) 53–63.
[17] KIRKPATRICK, D.G. and P. HELL, On the complexity of general graph factor problems, *SIAM J. Comput.* **12** (1983) 601–609.

[18] LEISERSON, C.E., *Area Efficient Graph Algorithms (for VLSI)* (MIT Press, Cambridge, MA, 1983).
[19] LIPTON, R.J. and R.E. TARJAN, A separator theorem for planar graphs, *SIAM J. Appl. Math.* **36** (1979) 177–189.
[20] LIPTON, R.J. and R.E. TARJAN, Applications of a planar separator theorem, *SIAM J. Comput.* **9** (1980) 615–627.
[21] LOVÁSZ, L., A homology theory for spanning trees of a graph, *Acta Math. Acad. Sci. Hungary* **30** (1977) 241–251.
[22] MAURER, S.B., Vertex colorings without isolates, *J. Combin. Theory Ser. B* **27** (1979) 294–319.
[23] MILLER, G.L., Finding small simple cycle separators for 2-connected planar graphs, in: *Proc. 16th Ann. ACM Symp. Theory of Computing*, Washington DC (1984) 376–382; revised version in: *J. Comput. System Sci.* **32** (1986) 265–279.
[24] MISRA, J. and R.E. TARJAN, Optimal chain partitions of trees, *Inform. Process. Lett.* **4** (1975) 24–26.
[25] NOORVASH, S., Covering the vertices of a graph by vertex-disjoint paths, *Pacif. J. Math.* **58** (1975) 159–168.
[26] PACAULT, J.F., Computing the weak components of a directed graph, *SIAM J. Comput.* **3** (1974) 56–61.
[27] PHILIPP, R. and E.J. PRAUSS, Über Separatoren in planaren Graphen, *Acta Inform.* **14** (1980) 87–106.
[28] RAO, S., Finding near optimal separators in planar graphs, in: *Proc. 28th Ann. IEEE Symp. Found. of Computer Science*, Los Angeles (1987) 225–237.
[29] TARJAN, R.E., A new algorithm for finding weak components, *Inform. Process. Lett.* **3** (1974) 13–15.
[30] TARJAN, R.E., An improved algorithm for hierarchical clustering using strong components, *Inform. Process. Lett.* **17** (1983) 37–41.
[31] TARJAN, R.E., Decomposition by clique separators, *Discrete Math.* **55** (1985) 221–232.
[32] TUTTE, W.T., On the problem of decomposing a graph into n-connected factors, *J. London Math. Soc.* **36** (1961) 221–230.
[33] UNGAR, P., A theorem on planar graphs, *J. London Math. Soc.* **26** (1951) 256–262.
[34] VENKATESAN, S.M., Improved constants for some separator theorems, *J. Algorithms* **8** (1987) 572–578.

Section 2.6. Isomorphism testing

[1] AHO, A.V., J.E. HOPCROFT and J.D. ULLMAN, *The Design and Analysis of Computer Algorithms* (Addison-Wesley, Reading, MA, 1974).
[2] BABAI, L., Moderately exponential bound for graph isomorphism, in: F. Gécseg, ed., *Proc. Fundamentals of Computation Theory*, Lecture Notes in Computer Science, Vol. 117 (Springer, Berlin, 1981).
[3] BABAI, L., P. ERDŐS and S.M. SELKOW, Random graph isomorphism, *SIAM J. Comput.* **9** (1980) 628–635.
[4] BABAI, L., D.YU. GRIGORYEV and D.M. MOUNT, Isomorphism of graphs with bounded eigenvalue multiplicity, in: *Proc. 14th Ann. ACM Symp. Theory of Computing*, San Francisco (1982) 310–324.
[5] BABAI, L. and L. KUČERA, Canonical labelling of graphs in linear average time, in: *Proc. 20th Ann. IEEE Symp. Foundations of Computer Science*, San Juan (Puerto Rico) (1979) 39–46.
[6] BABAI, L. and E.M. LUKS, Canonical labeling of graphs, in: *Proc. 15th Ann. ACM Symp. Theory of Computing*, Boston (1983) 171–183.
[7] BERTZISS, A.T., A backtrack procedure for isomorphism of directed graphs, *J. ACM* **20** (1973) 365–377.
[8] BODLAENDER, H.L., Polynomial algorithms for graph isomorphism and chromatic index on partial k-trees, in: R. Karlsson and A. Lingas, eds., *Proc. SWAT '88, 1st Scandinavian Workshop*, Lecture Notes in Computer Science, Vol. 318 (Springer, Berlin, 1988) 223–232.
[9] BOOTH, K.S., Isomorphism testing for graphs, semigroups and finite automata are polynomially equivalent problems, *SIAM J. Comput.* **7** (1978) 273–279.
[10] CHUNG, M.J., $O(n^{2.5})$ time algorithms for the subgraph homeomorphism problem on trees, *J. Algorithms* **8** (1987) 106–122.
[11] COLBOURN, C.J. and K.S. BOOTH, Linear time automorphism algorithms for trees, interval graphs, and planar graphs, *SIAM J. Comput.* **10** (1981) 203–225.
[12] COLBOURN, M.J. and C.J. COLBOURN, Graph isomorphism and self-complementary graphs, *SIGACT News* **10**(1) (1978) 25–29.

[13] CORNEIL, D.G. and C.C. GOTLIEB, An efficient algorithm for graph isomorphism, *J. ACM* **17** (1970) 51–64.

[14] CORNEIL, D.G. and D.G. KIRKPATRICK, Theoretical analysis of various heuristics for the graph isomorphism problem, *SIAM J. Comput.* **9** (1980) 281–297.

[15] DEO, N., J.M. DAVIS and R.E. LORD, A new algorithm for digraph isomorphism, *BIT* **17** (1977) 16–30.

[16] FILOTTI, I.S. and J.N. MAYER, A polynomial time algorithm for determining the isomorphism of graphs of fixed genus, in: *Proc. 12th Ann. ACM Symp. Theory of Computing*, Los Angeles (1980) 236–243.

[17] FONTET, M., Linear algorithms for testing isomorphism of planar graphs, in: S. Michaelson and R. Milner, eds., *Proc. 3rd Coll. Automata, Languages and Programming* (Edinburgh Univ. Press, Edinburgh, 1976) 411–423.

[18] FÜRER, M., W. SCHNYDER and E. SPECKER, Normal forms for trivalent graphs and graphs of bounded valence, in: *Proc. 15th Ann. ACM Symp. Theory of Computing*, Boston (1983) 161–170.

[19] GALIL, Z., C.M. HOFFMANN, E.M. LUKS, C.P. SCHNORR and A. WEBER, An $O(n^3 \log n)$ deterministic and an $O(n^3)$ probabilistic isomorphism test for trivalent graphs, in: *Proc. 23rd Ann. IEEE Symp. Foundations of Computer Science*, Chicago (1982) 118–125; revised version: *J. ACM* **34** (1987) 513–531.

[20] HEAP, B.R., The production of graphs by computer, in: R.C. Read, ed., *Graph Theory and Computing* (Academic Press, New York, NY, 1972) 47–62.

[21] HOFFMAN, C.M., *Group-Theoretic Algorithms and Graph Isomorphism*, Lecture Notes in Computer Science, Vol. 136 (Springer, Berlin, 1982).

[22] HOPCROFT, J.E. and C.K. WONG, Linear time algorithms for isomorphism of planar graphs, in: *Proc. 6th Ann. ACM Symp. Theory of Computing*, Seattle (1974) 172–184.

[23] JOHNSON, D.S., The NP-completeness column: an ongoing guide, *J. Algorithms* **2** (1981) 393–405.

[24] KARP, R.M., Probabilistic analysis of a canonical numbering algorithm for graphs, in: D.K. Ray-Chaudhuri, ed., *Relations Between Combinatorics and Other Parts of Mathematics*, Proceedings Symposia in Pure Mathematics, Vol. 34 (Amer. Math. Soc., Providence, RI, 1979) 365–378.

[25] LINGAS, A., Subgraph isomorphism for easily separable graphs of bounded valence, in: H. Noltemeier, ed., *Proc. WG'85* (*Int. Workshop on Graph-theoretic Concepts in Computer Science*) (Trauner Verlag, Linz, 1985) 217–229.

[26] LINGAS, A., Subgraph isomorphism for biconnected outerplanar graphs in cubic time, in: B. Monien and G. Vidal-Naquet, eds., *Proc. STACS 86—3rd Ann. Symp. Theoretical Aspects of Computer Science*, Lecture Notes in Computer Science, Vol. 210 (Springer, Berlin, 1986) 98–103.

[27] LINGAS, A. and M.M. SYSLO, A polynomial time algorithm for subgraph isomorphism of two-connected series-parallel graphs, in: T. Lepistö and A. Salomaa, eds., *Proc. 15th Int. Coll. Automata, Languages and Programming*, Lecture Notes in Computer Science, Vol. 317 (Springer, Berlin, 1988) 394–409.

[28] LUBIW, A., Some NP-complete problems similar to graph isomorphism, *SIAM J. Comput.* **10** (1981) 11–21.

[29] LUEKER, G.S. and K.S. BOOTH, A linear time algorithm for deciding interval graph isomorphism, *J. ACM* **26** (1979) 183–195.

[30] LUKS, E.M., Isomorphism of graphs of bounded valence can be tested in polynomial time, in: *Proc. 21st Ann. IEEE Symp. Foundations of Computer Science* Syracuse (1980) 42–49; extended version in: *J. Comput. System Sci.* **25** (1982) 42–65.

[31] MACKAY, B.D., Practical graph isomorphism, in: *Proc. 10th Manitoba Conf. Numerical Mathematics and Computing, Vol. 1* (*1980*), *Congr. Numer.* **30** (1981) 45–87.

[32] MATHON, R., A note on the graph isomorphism counting problem, *Inform. Process. Lett.* **8** (1979) 131–132.

[33] MATULA, D.W., Subtree isomorphism in $O(n^{5/2})$, *Ann. Discrete Math.* **2** (1978) 91–106.

[34] MILLER, G., Isomorphism testing for graphs of bounded genus, in: *Proc. 12th Ann. ACM Symp. Theory of Computing*, Los Angeles (1980) 225–235.

[35] MILLER, G.L., Graph isomorphism, general remarks, *J. Comput. System Sci.* **18** (1979) 128–142.

[36] PROSKUROWSKI, A., Search for a unique incidence matrix of a graph, *BIT* **14** (1974) 209–226.

[37] READ, R.C. and D.G. CORNEIL, The graph isomorphism disease, *J. Graph Theory* **1** (1977) 239–363.

[38] REYNER, S.W., An analysis of a good algorithm for the subtree problem, *SIAM J. Comput.* **6** (1977) 730–732.

[39] SCHMIDT, D.C. and L.E. DRUFFEL, A fast backtracking algorithm to test directed graphs for isomorphism using distance matrices, *J. ACM* **23** (1976) 433–445.
[40] SCHÖNING, U., Graph isomorphism is in the low hierarchy, *J. Comput. System Sci.* **37** (1988) 312–323.
[41] ULLMANN, J.R., An algorithm for subgraph isomorphism, *J. ACM* **23** (1976) 31–42.
[42] WEINBERG, L., A simple and efficient algorithm for determining isomorphism of planar triply connected graphs, *IEEE Trans. Circuit Theory* **13** (1966) 142–148.

Section 3. Combinatorial optimization on graphs

[1] LAWLER, E., *Combinatorial Optimization: Networks and Matroids* (Holt, Rinehart and Winston, New York, 1976).
[2] LOVÁSZ, L. and M.D. PLUMMER, *Matching Theory*, Annals of Discrete Mathematics Vol. 29, North-Holland Mathematical Studies, Vol. 121 (North-Holland, Amsterdam, 1986).
[3] PAPADIMITRIOU, C.H. and K. STEIGLITZ, *Combinatorial Optimization: Algorithms and Complexity* (Prentice Hall, Englewood Cliffs, NJ, 1982).

Section 3.1. Maximum matching

[1] BERGE, C., Two theorems in graph theory, in: *Proc. Nat. Acad. Sci.* **43** (1957) 842–844.
[2] EDMONDS, J., Paths, trees, and flowers, *Canad. J. Math.* **17** (1965) 449–467.
[3] GOODMAN, S.E., S. HEDETNIEMI and R.E. TARJAN, *b*-Matchings in trees, *SIAM J. Comput.* **5** (1976) 104–108.
[4] NORMAN, R.Z. and M.O. RABIN, An algorithm for a minimum cover of a graph, *Proc. Amer. Math. Soc.* **10** (1959) 315–319.
[5] PAPADIMITRIOU, C.H. and M. YANNAKAKIS, Worst-case ratios for planar graphs and the method of induction on faces, in: *Proc. 22nd Ann. IEEE Symp. Foundations of Computer Science*, Nashville (1981) 358–363.
[6] SAVAGE, C., Maximum matchings and trees, *Inform. Process. Lett.* **10** (1980) 202–205.
[7] SHILOACH, Y., Another look at the degree constrained subgraph problem, *Inform. Process. Lett.* **12** (1981) 89–92.
[8] URQUHART, R.J., Degree constrained subgraphs of linear graphs, Ph.D. Thesis, Univ. of Michigan, Ann Arbor, 1967.
[9] WHITE, L.J., A parametric study of matchings and coverings in weighted graphs, Ph.D. Thesis, Dept. of Electrical Engineering, Univ. of Michigan, Ann Arbor, 1967.

Section 3.2. Computing maximum matchings

[1] ANGLUIN, D. and L.G. VALIANT, Fast probabilistic algorithms for Hamiltonian circuits and matchings, *J. Comput. System Sci.* **18** (1979) 155–193.
[2] BALINSKI, M.L., Labelling to obtain a maximum matching, in: *Proc. Combin. Math. and its Applic.*, Chapel Hill, NC (1967) 585–602.
[3] EDMONDS, J., Maximum matching and a polyhedron with 0,1 vertices, *J. Res. Nat. Bur. Standards* **69B** (1965) 125–130.
[4] EDMONDS, J., Paths, trees, and flowers, *Canad. J. Math.* **17** (1965) 449–467.
[5] ERDÖS, P. and A. RENYI, On the existence of a factor of degree one of connected random graphs, *Acta Math. Acad. Sci. Hungar.* **17** (1966) 359.
[6] EVEN, S. and O. KARIV, An O($n^{2.5}$) algorithm for maximum matching in general graphs, in: *Proc. 16th Ann. IEEE Symp. Foundations of Computer Science*, Berkeley (1975) 100–112.

[7] EVEN, S. and R.E. TARJAN, Network flow and testing graph connectivity, *SIAM J. Comput.* **4** (1975) 507–518.
[8] GABOW, H.N., An efficient implementation of Edmonds' maximum matching algorithm, *J. ACM* **23** (1976) 221–234.
[9] GABOW, H.N., Data structures for weighted matching and nearest common ancestors with linking, in: *Proc. 1st Ann. ACM-SIAM Symp. Discrete Algorithms*, San Francisco (1990) (to appear).
[10] GABOW, H.N. and R.E. TARJAN, A linear time algorithm for a special case of disjoint set union, in: *Proc. 15th Ann. ACM Symp. Theory of Computing*, Boston (1983) 246–251.
[11] GABOW, H. N., Z. GALIL and T. SPENCER, Efficient implementation of graph algorithms using contraction, *J. ACM* **36** (1989) 540–572.
[12] GALIL, Z., Efficient algorithms for finding maximum matchings in graphs, *Comput. Surveys* **18** (1986) 23–38.
[13] GALIL, Z., S. MICALI and H.N.GABOW, An $O(EV \log V)$ algorithm for finding a maximal weighted matching in general graphs, *SIAM J. Comput.* **15** (1986) 120–130.
[14] HOPCROFT, J.E. and R.M. KARP, An $n^{5/2}$ algorithm for maximum matchings in bipartite graphs, *SIAM J. Comput.* **4** (1973) 225–231.
[15] KAMEDA, T. and I. MUNRO, An $O(|V| \cdot |E|)$ algorithm for maximum matching of graphs, *Comput.* **12** (1974) 91–98.
[16] KARP, R.M. and M. SIPSER, Maximum matchings in sparse random graphs, in: *Proc. 22nd Ann. IEEE Symp. Found. of Computer Science*, Nashville (1981) 364–375.
[17] LAWLER, E., *Combinatorial Optimization: Networks and Matroids* (Holt, Rinehart and Winston, New York, 1976).
[18] LIPTON, R.J. and R.E. TARJAN, Applications of a planar separator theorem, *SIAM J. Comput.* **9** (1980) 615–627.
[19] MICALI, S. and V.V. VAZIRANI, An $O(\sqrt{V} \cdot E)$ algorithm for finding maximum matching in general graphs, in: *Proc. 21st Ann. IEEE Symp. Foundations of Computer Science*, Syracuse (1980) 17–27.
[20] SAVAGE, C., Maximum matchings and trees, *Inform. Process. Lett.* **10** (1980) 202–205.

Section 3.3. Maximum flow

[1] BUSACKER, R.G. and P.J. GOWEN, A procedure for determining a family of minimal-cost network flow patterns, O.R.O. Technical paper 15, 1961.
[2] FORD, L.R. and D.R. FULKERSON, *Flows in Networks* (Princeton Univ. Press, Princeton, NJ, 1962).
[3] GOLDBERG, A.V., E. TARDOS and R.E. TARJAN, Network flow algorithms, Tech. Report STAN-CS-89-1252, Dept. of Computer Science, Stanford Univ., Stanford, CA, 1989.
[4] GOLDSCHLAGER, L.M., R.A. SHAW and J. STAPLES, The maximum flow problem is log-space complete for P, *Theoret. Comput. Sci.* **21** (1982) 105–111.
[5] HOFFMAN, A.J., Some recent applications of the theory of linear inequalities to extremal combinatorial analysis, in: *Combinatorial Analysis*, Proceedings Symposia in Applied Mathematics, Vol. 10 (Amer. Math. Soc., Providence, RI, 1960) 113–127.
[6] JEWELL, W.S., Optimal flow through networks, Interim Tech. Report 8, MIT, Cambridge, MA, 1958.
[7] LAWLER, E., *Combinatorial Optimization: Networks and Matroids* (Holt, Rinehart and Winston, New York, 1976).
[8] SCHRIJVER, A., *Theory of Linear and Integer Programming* (Wiley, Chichester, UK, 1986).

Section 3.4. Computing maximum flows

Section 3.4.1. Augmenting path methods

[1] AHUJA, R.K. and J.B. ORLIN, A fast and simple algorithm for the maximum flow problem, Tech. Report 1905–87, Sloan School of Management, MIT, Cambridge, MA, 1987.
[2] AHUJA, R.K., J.B. ORLIN and R.E. TARJAN, Improved timebounds for the maximum flow problem, *SIAM J. Comput.* **18** (1989) 939–954.

[3] BARATZ, A.E., The complexity of the maximum network flow problem, Tech. Report MIT/LCS/TR-230, Lab. for Computer Science, MIT, Cambridge, MA, 1980.

[4] BERGE, C. and A. GHOUILA-HOURI, *Programmes, Jeux et Réseaux de Transport* (Dunod, Paris, 1962).

[5] CHERKASKY, R.V., Algorithm of construction of maximal flow in networks with complexity of $O(V^2\sqrt{E})$ operations, *Math. Methods of Solution of Econ. Problems* **7** (1977) 112–125 (in Russian).

[6] DINIC, E.A., Algorithm for solution of a problem of maximum flow in networks with power estimation, *Soviet Math. Dokl.* **11** (1970) 1277–1280.

[7] EDMONDS, J. and R.M. KARP, Theoretical improvements in algorithmic efficiency for network flow problems, *J. ACM* **19** (1972) 248–264.

[8] EVEN, S. and R.E. TARJAN, Network flow and testing graph connectivity, *SIAM J. Comput.* **4** (1975) 507–518.

[9] FORD JR, L.R. and D.R. FULKERSON, Maximal flow through a network, *Canad. J. Math.* **8** (1956) 399–404.

[10] FREDERICKSON, G.N., Fast algorithms for shortest paths in planar graphs, with applications, *SIAM J. Comput.* **16** (1987) 1004–1022.

[11] GABOW, H.N., Scaling algorithms for network problems, in: *Proc. 24th Ann. IEEE Symp. Foundations of Computer Science*, Tucson (1983) 248–258.

[12] GALIL, Z., An $O(V^{5/3} E^{2/3})$ algorithm for the maximal flow problem, *Acta Inform.* **14** (1980) 221–242.

[13] GALIL, Z., On the theoretical efficiency of various network flow algorithms, *Theoret. Comput. Sci.* **14** (1981) 103–111.

[14] GALIL, Z. and A. NAAMAD, An $O(EV \log^2 V)$ algorithm for the maximal flow problem, *J. Comput. System Sci.* **21** (1980) 203–217.

[15] GUSFIELD, D., C. MARTEL, and D. FERNANDEZ-BACA, Fast algorithms for bipartite network flow, *SIAM J. Comput.* **16** (1987) 237–251.

[16] HASSIN, R., Maximum flow in (s,t) planar networks, *Inform. Process. Lett.* **13** (1981) 107.

[17] HASSIN, R. and D.B. JOHNSON, An $O(n \log^2 n)$ algorithm for maximum flow in undirected planar networks, *SIAM J. Comput.* **14** (1985) 612–624.

[18] ITAI, A. and Y. SHILOACH, Maximum flow in planar networks, *SIAM J. Comput.* **8** (1979) 135–150.

[19] JOHNSON, D.B. and S.M. VENKATESAN, Partition of planar flow networks, in: *Proc. 24th Ann. IEEE Symp. Foundations of Computer Science*, Tucson (1983) 259–263.

[20] KARZANOV, A.V., Determining the maximal flow in a network by the method of preflows, *Soviet Math. Dokl.* **15** (1974) 434–437.

[21] KUČERA, L., Finding a maximum flow in *s-t* planar networks in linear expected time, in: M.P. Chytil and V. Koubek, eds., *Proc. Mathematical Foundations of Computer Science 1984*, Lecture Notes in Computer Science, Vol. 176 (Springer, Berlin, 1984) 370–377.

[22] MALHOTRA, V.M., M. PRAMODH KUMAR and S.N. MAHESWARI, An $O(|V|^3)$ algorithm for finding maximum flows in networks, *Inform. Process. Lett.* **7** (1978) 277–278.

[23] QUEYRANNE, M., Theoretical efficiency of the algorithm "capacity" for the maximum flow problem, *Math. Oper. Res.* **5** (1980) 258–266.

[24] REIF, J.H., Minimum *s-t* cut of a planar undirected network in $O(n \log^2 n)$ time, *SIAM J. Comput.* **12** (1983) 71–81.

[25] SLEATOR, D.D., An $O(nm \log n)$ algorithm for maximum network flow, Tech. Report STAN-CS-80-831, Dept. of Computer Science, Stanford Univ., Stanford, CA 1980.

[26] SLEATOR, D.D. and R.E. TARJAN, A data structure for dynamic trees, *J. Comput. System Sci.* **24** (1983) 362–391.

[27] TARJAN, R.E., A simple version of Karzanov's blocking flow algorithm, *Oper. Res. Lett.* **2** (1984) 265–268.

Section 3.4.2. Maximum flow algorithms (*continued*)

[1] CHERIYAN, J. and S.N. MAHESHWARI, Analysis of preflow push algorithms for maximum network flow, *SIAM J. Comput.* **18** (1989) 1057–1086.

[2] GOLDBERG, A.V., A new max-flow algorithm, Tech. Memorandum MIT/LCS/TM-291, Lab. for Computer Science, MIT, Cambridge, MA, 1985.

[3] GOLDBERG, A.V., E. TARDOS and R.E. TARJAN, Network flow algorithms, Tech. Report STAN-CS-89-1252, Dept. of Computer Science, Stanford Univ., Stanford, CA, 1989.
[4] GOLDBERG, A.V. and R.E. TARJAN, A new approach to the maximum flow problem, in: *Proc. 18th Ann. ACM Symp. Theory of Computing*, Berkeley (1986) 136–146; also: *J. ACM* **35** (1988) 921–940.
[5] SLEATOR, D.D. and R.E. TARJAN, Self-adjusting binary search trees, *J. ACM* **32** (1985) 652–686.

Section 3.5. Related flow problems

Section 3.5.1. Networks with lower bounds

[1] ARKIN, E.M. and C.H. PAPADIMITRIOU, On the complexity of circulations, *J. Algorithms* **7** (1986) 134–145.
[2] ITAI, A., Two-commodity flow, *J. ACM* **25** (1978) 596–611.
[3] LAWLER, E., *Combinatorial Optimization: Networks and Matroids* (Holt, Rinehart and Winston, New York, 1976).
[4] SIMON, K., On minimum flow and transitive reduction, in: T. Lepistö and A. Salomaa, eds., *Proc. 15th Int. Coll. Automata, Languages and Programming*, Lecture Notes in Computer Science, Vol. 317 (Springer, Berlin, 1988) 535–546.

Section 3.5.2. Minimum cost flow

[1] AHUJA, R.K., A.V. GOLDBERG, J.B. ORLIN and R.E. TARJAN, Finding minimum-cost flows by double scaling, Tech. Report STAN-CS-88-1227, Dept. of Computer Science, Stanford Univ., Stanford, CA, 1988.
[2] EDMONDS, J. and R.M. KARP, Theoretical improvements in algorithmic efficiency for network flow problems, *J. ACM* **19** (1972) 248–264.
[3] FORD, L.R. and D.R. FULKERSON, *Flows in Networks* (Princeton Univ. Press, Princeton, NJ, 1962).
[4] FUJISHIGE, S., A capacity-rounding algorithm for the minimum-cost circulation problem: a dual framework of the Tardos algorithm, *Math. Progr.* **35** (1986) 298–308.
[5] FULKERSON, D.R., An out-of-kilter method for minimal-cost flow problems, *J. SIAM* **9** (1961) 18–27.
[6] GABOW, H.N. and R.E. TARJAN, Faster scaling algorithms for network problems, *SIAM J. Comput.* **18** (1989) 1013–1036.
[7] GALIL, Z. and E. TARDOS, An $O(n^2(m+n \log n)\log n)$ min-cost flow algorithm, *J. ACM* **35** (1988) 374–386.
[8] GOLDBERG, A.V., E. TARDOS and R.E. TARJAN, Network flow algorithms, Tech. Report STAN-CS-89-1252, Dept. of Computer Science, Stanford Univ., Stanford. CA, 1989.
[9] GOLDBERG, A.V. and R.E. TARJAN, Solving minimum-cost flow problems by successive approximations, in: *Proc. 19th Ann. ACM Symp. Theory of Computing*, New York City (1987) 7–18.
[10] GOLDBERG, A.V. and R.E. TARJAN, Finding minimum-cost circulations by canceling negative cycles, in: *Proc. 20th Ann. ACM Symp. Theory of Computing*, Chicago (1988) 388–397; also: *J. ACM* **36** (1989) 873–886.
[11] ORLIN, J.B., Genuinely polynomial simplex and non-simplex algorithms for the minimum cost flow problem, Working paper 1615-84, Sloan School of Management, MIT, Cambridge, MA, 1984.
[12] ORLIN, J.B., A faster strongly polynomial minimum cost flow algorithm, in: *Proc. 20th Ann. ACM Symp. Theory of Computing*, Chicago (1988) 377–387.
[13] RÖCK, H., Scaling techniques for minimal cost network flows, in: U. Pape, ed., *Discrete Structures and Algorithms* (Hanser, Munich, 1980) 181–191.
[14] TARDOS, E., A strongly polynomial minimum cost circulation algorithm, *Combinatorica* **5** (1985) 247–255.
[15] WAGNER, H.M., On a class of capacitated transportation problems, *Manag. Sci.* **5** (1959) 304–318.
[16] ZADEH, N., A bad network for the simplex method and other minimum cost flow algorithms, *Math. Progr.* **5** (1973) 255–266.
[17] ZADEH, N., More pathological examples for network flow problems, *Math. Progr.* **5** (1973) 217–224.

Section 3.5.3. Multiterminal flow

[1] GOMORY, R.E. and T.C. HU, Multi-terminal network flows, *J. SIAM* **9** (1961) 551–570.
[2] GOMORY, R.E. and T.C. HU, An application of generalized linear programming to network flows, *J. SIAM* **10** (1962) 260–283.
[3] GUPTA, R.P., On flows in pseudosymmetric networks, *J. SIAM* **14** (1966) 215–225.
[4] GUSFIELD, D., Simple constructions for multi-terminal network flow synthesis, *SIAM J. Comput.* **12** (1983) 157–165.
[5] HU, T.C., *Integer Programming and Network Flows* (Addison-Wesley, Reading, MA, 1969).
[6] HU, T.C. and M.T. SHING, Multi-terminal flow in outerplanar networks, *J. Algorithms* **4** (1983) 241–261.
[7] HU, T.C. and M.T. SHING, A decomposition algorithm for multi-terminal network flows, Tech. Report TRCS 84-08, Dept. of Computer Science, Univ. of California, Santa Barbara, CA, 1984.
[8] SHILOACH, Y., Multi-terminal 0-1 flow; *SIAM J. Comput.* **8** (1979) 422–430.
[9] SHING, M.T. and P.K. AGARWAL, Multi-terminal flows in planar networks, Tech. Report TRCS 86-07, Dept. of Computer Science, Univ. of California, Santa Barbara, CA, 1986.

Section 3.5.4. Multicommodity flow

[1] ASSAD, A.A., Multicommodity network flows—a survey, *Networks* **8** (1978) 37–91.
[2] EVEN, S., A. ITAI and A. SHAMIR, On the complexity of timetable and multicommodity flow problems, *SIAM J. Comput.* **5** (1976) 691–703.
[3] FORD, L.R. and D.R. FULKERSON, A suggested computation for maximal multicommodity network flows, *Manag. Sci.* **5** (1958) 97–101.
[4] GAREY, M.R. and D.S. JOHNSON, *Computers and Intractability: A Guide to the Theory of NP-Completeness* (Freeman, San Francisco, CA, 1979).
[5] GOMORY, R.E. and T.C. HU, Synthesis of a communication network, *J. SIAM* **12** (1964) 348–369.
[6] HU, T.C., Multicommodity network flow, *Oper. Res.* **11** (1963) 344–360.
[7] HU, T.C., *Integer Programming and Network Flows* (Addison-Wesley, Reading, MA, 1969).
[8] ITAI, A., Two-commodity flow, *J. ACM* **25** (1978) 596–611.
[9] KAPOOR, S. and P.M. VAIDYA, Fast algorithms for convex quadratic programming and multicommodity flows, in: *Proc. 18th Ann. ACM Symp. Theory of Computing*, Berkeley (1986) 147–159.
[10] KENNINGTON, J.L., A survey of linear cost multicommodity network flows, *Oper. Res.* **26** (1978) 209–236.
[11] MATSUMOTO, K., T. NISHIZEKI and N. SAITO, Planar multicommodity flows, maximum matchings and negative cycles, *SIAM J. Comput.* **15** (1986) 495–510.
[12] OKAMURA, H. and P.D. SEYMOUR, Multicommodity flows in planar graphs, *J. Combin. Theory Ser. B* **31** (1981) 75–81.
[13] ROTHSCHILD, B. and A. WHINSTON, Feasibility of two commodity network flows, *Oper. Res.* **14** (1966) 1121–1129.
[14] ROTHSCHILD, B. and A. WHINSTON, On two commodity network flows, *Oper. Res.* **14** (1966) 377–387.
[15] SEYMOUR, P.D., A short proof of the two-commodity flow problem, *J. Combin. Theory Ser. B* **26** (1979) 370–371.
[16] SOUN, Y. and K. TRUEMPER, Single commodity representation of multicommodity networks, *SIAM J. Discrete Appl. Meth.* **1** (1980) 348–358.
[17] SUZUKI, H., T. NISHIZEKI and N. SAITO, Algorithms for multicommodity flows in planar graphs, *Algorithmica* **4** (1989) 471–501.
[18] TOMLIN, J.A., Minimum-cost multicommodity network flows, *Oper. Res.* **14** (1966) 45–51.

CHAPTER 11

Algebraic Complexity Theory

Volker STRASSEN

Universität Konstanz, Fakultät für Mathematik
Postfach 5560, D-7750 Konstanz 1, FRG

Contents

HANDBOOK OF THEORETICAL COMPUTER SCIENCE
Edited by J. van Leeuwen

1. Introduction

Complexity theory, as a project of "lower bounds" and optimality, unites two quite different traditions. The first comes from mathematical logic and the theory of recursive functions. Here the basic computational model is the Turing machine. One shows that certain decision problems (say) are inherently difficult in much the same way as one proves the undecidability of theories in mathematical logic: by diagonalization and by reduction of one problem to another.

The second tradition has developed from questions of numerical algebra. The problems here typically have a fixed finite size. Consequently, the computational model is based on something like an ordinary (finite, random access) computer, which however is supplied with the ability to perform any arithmetic operation with infinite precision and which in turn is required to deliver exact results. The formal model is that of *straight-line program* or *arithmetic circuit* or *computation sequence* (see Section 3 below), more generally that of *computation tree* (see Section 8 below). Since the origin is numerical algebra and the methods of investigation have turned out to be predominantly algebraic, this field of research is called *algebraic complexity theory.*

Its activities have so far been more influential in the area of symbolic or algebraic computation than in numerical mathematics. Apart from sociological grounds the reason may be the arithmetic model's disregard for the finitary representation of real or complex numbers and for the incomplete precision of their manipulation. While this disregard does no harm to the lower bounds proved for the complexity of a numerical problem (as long as one restricts oneself to direct methods, i.e. those that in principle give exact solutions), a fast algorithm quite acceptable in complexity theory may be rejected by a numerical analyst because of its lack of numerical stability. In applications to algebraic computation, on the other hand, one is most often concerned with finite groundfields. (Even if the original problem is formulated over the integers, finite fields appear by reduction modulo prime numbers.) Hence the question of numerical stability does not arise.

Although the models of Turing machine and of arithmetic circuit are not directly comparable, the former is clearly the more fundamental (even for numerical problems, compare Schönhage [179], Ritzmann [161]). On the other hand, the arithmetic model has many appealing properties: When it applies, it is usually much closer to actual programming and computing (also to hardware realizations). Since it respects the underlying algebraic structure, it behaves well under homomorphisms, forming direct products, changing groundfields etc. and often reveals a close relationship between algebraic and algorithmic concepts. Moreover, within the realm of computationally feasible problems, there seem to be few lower bound techniques known for Turing complexity, whereas there is a number of such techniques and results for the arithmetic model.

Algebraic complexity theory is one of the oldest branches of theoretical computer science, originating with two short papers of A. Scholz (1937) and A. Ostrowski (1954) and beginning to form a coherent body of results in the late sixties and early seventies. (See Iri [108] for a Japanese retrospective of this period.) The present survey is a compromise between a historical and a dogmatic exposition of the main developments in this field. The limitations of time, space and competence, which shape the subject matter of our chapter, have also resulted in a number of regrettable omissions, the most notable concerning parallel complexity. (We refer the interested reader to Chapter 17 of this Handbook for a survey of parallel algorithms for shared-memory machines.) On the other hand, we have included a list of unsolved research problems. They address themselves chiefly to the specialist, but are also intended to convey an impression of the liveliness of the field to the casual reader.

We have defined algebraic complexity theory by the underlying model of computation (the arithmetic model). There is a more general definition that encompasses the study of the complexity of *any* algebraic problem under *any* model of computation. Among the developments in this broader area are the following (not covered by the present chapter):

Univariate Polynomial Factorization: See Berlekamp [17, 18] (also Rabin [155], Ben-Or [14]) for theoretically as well as practically fast algorithms over finite fields, Zassenhaus [215] for a practically fast factorization over $\mathbb{Q}$, Lenstra–Lenstra–Lovàsz [126] for the theoretical breakthrough over $\mathbb{Q}$, A.K. Lenstra [127], Chistov–Grigoryev [44], Chistov [43], Kronecker [121], Trager [202], Landau [124] for algebraic number and function fields, Schönhage [179] for the classical numerical problem of (approximate) factorization over $\mathbb{C}$.

Multivariate Factorization: Out of a flood of activities we only mention the von zur Gathen–Kaltofen theory of manipulating polynomials coded by arithmetic circuits, in particular Kaltofen's polynomial factorization, [68, 112, 113, 114]. In this connection see also Grigoryev–Karpinski [81].

Representation Theory: See Friedl–Rónyai [62], Rónyai [163, 164, 165] for a general investigation of representations of finite-dimensional algebras. Much of the classical work on the symmetric and alternating groups and the classical groups is of course quite computational; see e.g. Weyl [209], James–Kerber [110]. The problem of evaluating a Wedderburn isomorphism is treated in Section 13 below.

Permutation Groups: See e.g. Sims [183, 184], Babai [6, 7], Furst–Hopcroft–Luks [65], Luks [137], Filotti–Mayer [60], Babai–Grigoryev–Mount [8], Hoffmann [102], Fürer–Schnyder–Specker [64] and the literature cited there.

First-Order Algebraic Theories: The highlights on lower bounds are the papers by Fischer–Rabin [61], who prove e.g. an exponential lower bound for the nondeterministic time complexity of the elementary theory of addition of real numbers (see also the pioneering paper by Meyer–Stockmeyer [140], further Ferrante–Rackoff [57], Berman [19]), and Mayr–Meyer [139], who show that membership in polynomial ideals over $\mathbb{Q}$ is exponential space hard, thereby complementing the classical result of Hermann [101]. In this connection see also the work on Gröbner bases, surveyed in Buchberger [36], and on membership in radical ideals by Brownawell [35] and Caniglia–Galligo–Heintz [40, 41]. Doubly exponential lower bounds for the complexity of quantifier elimination in the elementary theory of real numbers are given in Weispfenning [208] and Davenport–Heintz [53]. Decision procedures for the elementary theory of real numbers as well as that of algebraically closed fields of any characteristic are treated e.g. in Tarski [200], Collins [48], Monk–Solovay (see the article of Wüthrich in [186]), Heintz [94, 95], Grigoryev [78, 80], Ben-Or–Kozen–

Reif [16]. For the solution of systems of algebraic equations and inequalities see Chistov–Grigoryev [45], Grigoryev [79], Caniglia–Galligo–Heintz [40], Grigoryev–Vorobjov [82].

The subsequent sections allow a natural division into mutually almost independent parts: Problems with general coefficients in Sections 3–5, problems of high degree in Sections 2 and 6–9, problems of low degree in Sections 10–13, and complete problems in Section 14.

There are no special prerequisites. The reader is only supposed to have a general familiarity with algebra, and in some places with the language of classical algebraic geometry.

The greater part of the present chapter is a translation of [197]. I am indebted to S. De Martino, V. Schkölziger and V. Tobler for their help.

2. Addition chains

In 1937, a short note by A. Scholz [171] was published in the yearly report of the Deutsche Mathematiker-Vereinigung. There, he calls a sequence $1=a_0<\cdots<a_r=n$ of integers an addition chain for n, when each a_i with $i\geqslant 1$ is a sum of two preceding elements of the sequence: $a_i=a_j+a_k$ for some $j\leqslant k<i$. He designates with $l(n)$ the shortest length r of an addition chain for n. $l(n)$ is motivated as the optimal cost for the computation of the nth power of x, when only multiplications are permitted.

This work seems to be the first evidence for the notion of *computation sequence* (essentially equivalent to *straight-line program* or *arithmetic circuit*) as well as for a question concerning the complexity of a numerical problem. A. Scholz was a number theorist. His booklet "Einführung in die Zahlentheorie" reveals the unusual interest of the author in algorithmic questions. Of course, the notion of computation had been analyzed already before by Turing and others, in a much broader and more fundamental manner than by Scholz's work, yet without regard to complexity issues.

Let us return briefly to the function l. The decisive insight was soon obtained: $l(n)\sim\log n$ (Scholz [171], Brauer [30]; in this chapter "log" denotes the binary logarithm. The fine structure of l, however, still provides many exciting and difficult problems. (See Knuth [120], Schönhage [175].)

3. Computation sequences

Seventeen years after Scholz—the era of programmable computers had already begun—Ostrowski published his work [146] "On two problems in abstract algebra connected with Horner's rule", that led to numerous investigations on the complexity of algebraic problems.

It is well-known that, according to Horner, a polynomial $a_0x^n+a_1x^{n-1}+\cdots+a_n$ can be evaluated by means of n additions and n multiplications. Now Ostrowski inquires about the optimality of Horner's rule and, among other things, states the conjecture that there exists no general evaluation procedure which requires less than n multiplications/divisions. Additions and subtractions are not counted. Guided by good mathematical intuition, Ostrowski also allows for the costless multiplication with scalars. (One can think of scalars as constants in the computer program.)

A precise formulation of the conjecture naturally demands the specification of a model of computation. Following Ostrowski's train of thoughts, let us regard the intermediate results of a computation as functions of the input variables (which, in our case of the evaluation of a general polynomial, are the variables $a_0, \ldots, a_n, x$) and count only the results of the significant steps of the computation.

DEFINITION. Let k be an infinite field, and $x_1, \ldots, x_m$ indeterminates over k. A sequence $(g_1, \ldots, g_r)$ in $k(x_1, \ldots, x_m)$ is called a *computation sequence* if there exists for each $i \leqslant r$ a representation $g_i = u_i v_i$ or $g_i = u_i / v_i$ such that $u_i, v_i \in \mathit{lin}\{1, x_1, \ldots, x_m, g_1, \ldots, g_{i-1}\}$, where *lin* denotes the linear hull.

Thus, for Horner's rule, $r = n$, $g_i = a_0 x^i + \cdots + a_{i-1} x \in k(a_0, \ldots, a_n, x)$, $u_i = g_{i-1} + a_{i-1}$ and $v_i = x$.

DEFINITION. Let $f_1, \ldots, f_p \in k(x_1, \ldots, x_m)$. The (*nonscalar*) *complexity* $L(f_1, \ldots, f_p)$ is the smallest natural number r for which there exists a computation sequence $(g_1, \ldots, g_r)$ with $f_1, \ldots, f_p \in \mathit{lin}\{1, x_1, \ldots, x_m, g_1, \ldots, g_r\}$.

Horner's rule yields an upper bound for the complexity:

$$L(a_0 x^n + \cdots + a_n) \leqslant n.$$

Ostrowski's conjecture reads

$$L(a_0 x^n + \cdots + a_n) = n.$$

4. Substitution

Twelve years later, Pan [147] succeeded in proving Ostrowski's conjecture (and afterwards Borodin showed that Horner's rule is essentially the unique optimal algorithm for the problem). The substitution method used in the demonstration was further developed in [211, 190, 93, 125]. We shall illustrate it by applying it to the particular example of the sum of squares $f = x_1^2 + \cdots + x_m^2 \in \mathbb{R}[x_1, \ldots, x_m]$, in the calculation of which, for the sake of simplicity, division will not be allowed. If $(g_1, \ldots, g_r)$ is an "optimal" computation sequence then u_1 and v_1 are nonconstant linear polynomials. Thus, if for instance x_m appears in u_1, there exists an affine linear substitution $x_m \mapsto \Sigma_{j=1}^{m-1} \lambda_j x_j + \lambda$, which annihilates u_1 and therefore g_1. The substitution is a ring endomorphism of $\mathbb{R}[x_1, \ldots, x_m]$ which maps $(g_1, \ldots, g_r)$ to a computation sequence in which the first element is superfluous. f is thereby transformed into a polynomial with smaller complexity. Now it is easy to see that the sum of squares cannot be linearized with less than m such substitutions, which together with the trivial upper bound yields

$$L^{\mathbb{R}}(x_1^2 + \cdots + x_m^2) = m. \qquad (4.1)$$

This example can be generalized: If char(k)$\neq 2$ and $f \in k(x_1, \ldots, x_m)$ is a nondegenerate quadratic form, then $L(f)=m-q$, where q is the dimension of the maximal null spaces of f. Thus, for example,

$$L^{\mathbb{C}}(x_1^2+\cdots+x_m^2)=\lceil m/2 \rceil.$$

Comparison with (4.1) shows that the complexity may decrease by an extension of the groundfield k. (This effect is absent in the case of algebraically closed groundfields. However, it may appear even if the groundfield is algebraically closed in the considered extension.)

The use of the substitution method for the polynomial $a_0x^n+\cdots+a_n \in k(a_0, \ldots, a_n, x)$ encounters the further difficulty that a substitution of x would obviously spoil the entire process. Pan averts this problem by handling the intermediate field $k(x)$ in a particular way.

As an additional example, let us consider the evaluation of a "general" rational function of degree n

$$f_1=\frac{a_0x^n+\cdots+a_n}{x^n+b_1x^{n-1}+\cdots+b_n} \in k(a_0, \ldots, a_n, b_1, \ldots, b_n, x). \tag{4.2}$$

Here again we obtain the expected result $L(f_1)=2n$. Now any rational function can be expanded in a partial fraction or a continued fraction. If, for the sake of simplicity, we restrict ourselves to the type of expansions normally expected in the case of algebraically closed groundfields, namely

$$f_2=a_0+\frac{a_1}{x-b_1}+\cdots+\frac{a_n}{x-b_n}$$

and

$$f_3=a_0+\cfrac{a_1}{(x+b_1)+\cfrac{a_2}{(x+b_2)+\cfrac{a_3}{\ddots+\cfrac{a_n}{x+b_n}}}},$$

the substitution methods yields $L(f_2)=L(f_3)=n$. Thus, the evaluation of these expansions requires exactly half of the number of multiplications/divisions necessary for the natural representation (4.2).

At this point, the following questions remain unanswered: Do there exist representations of rational functions or polynomials for which the evaluation is more efficient than for the partial and continued fraction expansion or the representation of polynomials by their coefficient vector?

What is the situation with respect to the number of additions and subtractions in the cases we have considered?

PROBLEM. Determine the complexity of one Newton step for the approximation of a root of a general polynomial

$$L\left(x-\frac{a_0x^n+\cdots+a_n}{na_0x^{n-1}+\cdots+a_{n-1}}\right).$$

5. Degree of transcendency

Surprisingly, the questions above were clarified even before Ostrowski's conjecture was settled (Motzkin [145]; see also [13, 157]).

Let K be an infinite field and $f\in K(x)$. We denote by $L_+(f)$ the minimum number of additions/subtractions sufficient to evaluate f. Multiplications, divisions and recourse to arbitrary constants in K are regarded as costless. $L_*(f)$ is similarly defined: here multiplications/divisions including scalar multiplications are being counted. $L_+(f)$ and $L_*(f)$ are called respectively *additive* and *multiplicative* complexity of f.

For $u, v\in K[x]$, let $T(u, v)$ be the degree of transcendency of the set of coefficients of u and v over the primefield of K.

THEOREM. *If $u, v\in K[x]$ are relatively prime and the leading coefficient of v is* 1, *then*

$$L_+(u/v)\geqslant T(u,v)-1,\qquad L_*(u/v)\geqslant\tfrac{1}{2}(T(u,v)-1)).$$

Our questions are hereby readily answered. In order, for instance, to estimate from below the additive complexity of $a_0x^n+\cdots+a_n\in k(a_0,\ldots,a_n,x)$, we enlarge the groundfield k to (say) the algebraic closure K of $k(a_0,\ldots,a_n)$ (whereby the complexity can only decrease) and apply the theorem to $a_0x^n+\cdots+a_n\in K(x)$. Taking into account the trivial upper bound, we obtain

$$L_+(a_0x^n+\cdots+a_n)=n$$

and thus complete optimality for Horner's rule.

For the three representations of rational functions of degree n considered above, we get $L_+(f_i)=2n$. To prove this, we simply take for K the algebraic closure of $k(a_0,\ldots,a_n,b_1,\ldots,b_n)$ and note that the set of coefficients of f_i, written as a polynomial quotient, has at least a degree of transcendency $2n+1$. Similarly, one gets

$$L_*(f_1)\geqslant L_*(f_2)=L_*(f_3)=n,$$

which, in fact, already ensues from the results of the last section and the trivial fact that $L_*\geqslant L$. The present demonstrations are however more robust than the ones used there, since they allow for a "preparation of coefficients" (due to the extension of the groundfield). This means that arbitrary algebraic functions of the coefficients a_i, b_j may be precomputed without cost. Now any representation of polynomials or rational functions can be interpreted as such a preparation of coefficients. For this reason, the partial and continued fraction expansions indeed possess optimal efficiency. On the other hand, the natural representation is not optimal even for the evaluation of polynomials (see Belaga [13], Pan [147]).

The theorem of this section can be extended in a straightforward manner to the case of several indeterminates $x_1, \dots, x_m$. A less obvious generalization was given by Baur–Rabin [11]: If k is some subfield of K (for instance the primefield) then K and $k(x)$ are linearly disjoint over k. In the mentioned work, $K(x)$ is replaced by some field extension of k provided with two linearly disjoint intermediate fields.

6. Geometric degree

In the present section, we shall describe a method which will allow us to give meaningful lower bounds for the nonscalar complexity of some more complicated computational tasks.

Let us assume that k is algebraically closed, and start out with the rational functions $f_1, \dots, f_p \in k(x_1, \dots, x_m)$. Let W denote the graph of the corresponding rational map from k^m to k^p. W is a locally closed irreducible subset of k^{m+p} in the sense of the Zariski topology and, as such, has a degree deg W. (deg W is the maximal finite number of intersection points of W with an affine linear subspace of dimension p.)

THEOREM. $L(f_1, \dots, f_p) \geq \log \deg W$.

If, for instance, $f_1, \dots, f_m$ are polynomials in $x_1, \dots, x_m$ (so $p = m$), and the system of equations

$$f_1(\xi) = \cdots = f_m(\xi) = 0$$

has exactly N solutions $\xi \in k^m$, then $L(f_1, \dots, f_m) \geq \log N$.

To prove the theorem, one looks at a hypothetical computation of $f_1, \dots, f_p$ from a "graphical" point of view, interprets the arithmetic operations as intersections with hypersurfaces of degree 1 or 2 followed by projections and applies the theorem of Bezout ([192], see also [176]).

As it seems, the natural field of application of the above theorem is the algebraic manipulation of polynomials and rational functions. Since these can be represented in different ways, the transition from one representation to another also plays a role. Only the simplest of such tasks have a (nonscalar) complexity that may be estimated by a linear function of the degree of the considered polynomials: the multiplication (that is, the calculation of the coefficients of the product polynomial), the division with remainder (Sieveking [182], Strassen [192], Kung [122]) and the expansion in powers of $x - a$ (Shaw–Traub [181]).

Let us now consider the computation of the coefficients of a polynomial in one variable from its roots, in other words, the evaluation (up to the sign) of the elementary symmetric functions $\sigma_1, \dots, \sigma_m \in k(x_1, \dots, x_m)$. If $\lambda_1, \dots, \lambda_m \in k$, the system of equations

$$\sigma_1(\xi) - \lambda_1 = \cdots = \sigma_m(\xi) - \lambda_m = 0$$

is equivalent to the one equation

$$(t - \xi_1) \cdots (t - \xi_m) = t^m - \lambda_1 t^{m-1} + \cdots + (-1)^m \lambda_m$$

in $k[t]$ and therefore has exactly $m!$ solutions $\xi \in k^m$ for a proper choice of the λ_i. This yields $L(\sigma_1, \ldots, \sigma_m) = L(\sigma_1 - \lambda_1, \ldots, \sigma_m - \lambda_m) \geq \log m!$ and hence the lower bound in

$$m(\log m - 2) \leq L(\sigma_1, \ldots, \sigma_m) \leq m \log m. \tag{6.1}$$

(The upper bound can be obtained through inductive bisection of $(t - x_1)\cdots(t - x_m)$.)

Another frequently occurring problem is the evaluation of a polynomial $f = a_0 t^n + \cdots + a_n$ of degree n at many (say $n+1$) points, that is, the calculation of $f(x_0), \ldots, f(x_n) \in k(a_0, \ldots, a_n, x_0, \ldots, x_n)$. One might think that the separate evaluation by Horner's rule of the polynomial at every single x_i is optimal. In such a process the nonscalar cost amounts to $n(n+1)$. Surprisingly there exist much faster algorithms (Borodin–Munro [29], Fiduccia [58], Moenck–Borodin [142], Strassen [192]): Think of the input variables $a_0, \ldots, a_n, x_0, \ldots, x_n$ as concrete elements of k. Assuming that n be odd, say $n = 2q+1$, we try to evaluate f at first only at the points $x_0, \ldots, x_q$. Division with remainder

$$f(t) = g(t)(t - x_0)\cdots(t - x_q) + h(t)$$

yields a polynomial h of degree $q = \frac{1}{2}(n-1)$, which takes the same values as f for $x_0, \ldots, x_q$. Thus to begin with we can calculate the coefficients of $(t - x_0)\cdots(t - x_q)$ (elementary symmetric functions!), then execute the division with remainder, and finally evaluate the polynomial h at the points $x_0, \ldots, x_q$. If we proceed analogously for the remaining positions $x_{q+1}, \ldots, x_n$, we have reduced the initial task to two problems of the same type, but half the size. One can apply this method recursively and obtain for the overall nonscalar cost γ_n the recursion formula $\gamma_{2q+1} \leq \text{const} \cdot q \log q + 2\gamma_q$, hence $\gamma_n O(n(\log n)^2)$. A more careful proceeding yields the upper bound in

$$(n+1)\log n \leq L(f(x_0), \ldots, f(x_n)) = O(n \log n). \tag{6.2}$$

How does one show the lower bound? The system of equations

$$f(x_0) = \cdots = f(x_n) = a_0 - 1 = a_1 = \cdots = a_{n-2} = a_{n-1} - \alpha = a_n - \beta = 0$$

with suitable $\alpha, \beta \in k$ has n^{n+1} solutions in k^{2n+2}, so that

$$\begin{aligned} L(f(x_0), \ldots, f(x_n)) &= L(f(x_0), \ldots, f(x_n), a_0 - 1, a_1, \ldots, a_{n-2}, a_{n-1} - \alpha, a_n - \beta) \\ &\geq \log(n^{n+1}). \end{aligned}$$

Inverting the problem we just dealt with results in the following interpolation task. The base points $x_0, \ldots, x_n$ as well as the values $y_0, \ldots, y_n$ are now given. The objective is to determine the coefficients of the polynomial p of degree n with $p(x_i) = y_i$ for all i. Since this problem is inverse to the previous one, their graphs are essentially the same. Its complexity is therefore at least $(n+1)\log n$. On the other hand, the coefficients of the interpolation polynomial can be computed with nonscalar cost $O(n \log n)$ (Horowitz [105]). One obtains a similar result when one seeks to calculate the barycentric coordinates $w_0, \ldots, w_n$ of p in the sense of Rutishauser [166]. The question of a most efficient representation of the interpolation polynomial will be discussed at the end of the next section.

The hypothesis of algebraic closedness made in the formulation of the theorem of this section may be abandoned again in its applications since a field extension would only reduce the complexity. For the same reason we may drop our general assumption that the groundfield k be infinite, when discussing lower bounds. (It is not surprising that one is sometimes able to prove better lower bounds over finite fields than over infinite ones, see Brockett–Dobkin [33], Brown–Dobkin [34], Kaminski–Bshouty [115].) However, one has to be cautious with interpreting the results obtained. They are quite meaningful, when the algorithms under consideration are to be applied (say) to arbitrary finite extension fields of k. On the other hand, when one is interested only in computations in the finite groundfield itself, the notions of polynomial and rational function become inadequate: Two different polynomials may represent the same function, a rational function may be nowhere defined. Hence a different model has to be adopted. (See [194, 195] for a detailed discussion.)

Can the degree method be applied to additive complexity? The polynomial $x^n - 1 \in \mathbb{C}[x]$ seems to dash this hope. For $f \in \mathbb{R}[x]$ however, Borodin-Cook [27], Grigoryev [77] and Risler [159] have proved that

$$L_+^{\mathbb{R}}(f) \geqslant c\sqrt{\log N} \tag{6.3}$$

where N is the number of real roots of f and $c > 0$ a universal constant. Inequality (6.3) is based on a theorem of Hovanskii [106] concerning the roots of a system of real equations with few monomials.

Applying a Möbius transformation to $x^n - 1 \in \mathbb{C}[x]$, van de Wiele [210] constructed a real polynomial f of degree n that can be computed over $\mathbb{C}$ with three additions/subtractions, although it possesses n distinct roots in $\mathbb{R}$. According to (6.3), $L_+(f) \geqslant c\sqrt{\log n}$ over $\mathbb{R}$. Thus, a field extension of degree 2 may have disastrous consequences for the additive complexity.

It is somewhat surprising that the degree method may be applied to the problem of approximating the zeros of a complex polynomial by iteration procedures. (See Ritzmann [160]; also Baur [10] and the literature cited there for a general discussion of iteration from a complexity point of view.)

PROBLEM 1. Decide whether, using the above notation and a universal constant $c_1 > 0$,

$$L_+^{\mathbb{R}}(f) \geqslant c_1 \log N. \tag{6.4}$$

The Chebyshev polynomials T_{3^k} show that (6.4) is the best result of its kind which could be expected (Borodin–Cook [27], van de Wiele [210]). The interest in the still rather small lower bound (6.4) lies in the fact that its validity would suggest the possibility to prove $n \log n$ bounds for the additive complexity of systems of real polynomials in n variables in analogy to the degree method.

PROBLEM 2. Find the maximal number of integer roots which a polynomial $f \in \mathbb{Q}[x]$ with $L_+^{\mathbb{C}}(f) \leqslant r$ (or $L^{\mathbb{C}}(f) \leqslant r$) can possess.

7. Derivatives

The method of the last section provides us only with the trivial lower bound $L(f)\geqslant\log\deg f$ for the complexity of a single polynomial $f\in k(x_1,\ldots,x_m)$. Could the degree method be used here in a more efficient way? This question was first investigated in Schnorr [169]. The following inequality (Baur–Strassen [12]) together with the theorem of the previous section yields in many cases lower bounds of the correct order of magnitude.

THEOREM. *If* $f\in k(x_1,\ldots,x_m)$ *then*

$$L\left(f,\frac{\partial f}{\partial x_1},\ldots,\frac{\partial f}{\partial x_m}\right)\leqslant 3L(f).$$

The proof of this result is constructive. Note that the form of the inequality is independent of the number of variables and that no hypotheses have been made concerning the groundfield. Similar inequalities are valid when all multiplications/divisions or all arithmetic operations are counted.

As a first application we consider the power sum $f=x_1^n+\cdots+x_m^n$, assuming that n is not a multiple of char(k). Obviously, $L(f)\leqslant ml(n)$, where l is the Scholz function (Section 2). On the other hand, we have $\partial f/\partial x_i=nx_i^{n-1}$, so that

$$\begin{aligned} L(f)&\geqslant\tfrac13 L(x_1^{n-1},\ldots,x_m^{n-1})\\ &=\tfrac13 L(x_1^{n-1}+\alpha x_1+\beta,\ldots,x_m^{n-1}+\alpha x_m+\beta)\\ &\geqslant\tfrac13 m\log(n-1)\quad\text{(as in the demonstration of (6.2)).}\end{aligned}$$

Hence

$$L(x_1^n+\cdots+x_m^n)\asymp m\log n.$$

($\asymp$ denotes the equality of orders of magnitude.)

Somewhat more tedious is the proof of

$$L(\sigma_q)\asymp m\log\min\{q,m-q\},\tag{7.0}$$

where σ_q is the qth elementary symmetric function in m variables and $q<m-1$. The computation of a coefficient in the middle range from the roots of a polynomial thus requires approximately as many steps as the calculation of all the coefficients.

Resultant and discriminant as functions of the roots of a polynomial also fall in the field of application of the described methods:

$$L\left(\prod_{i,j=1}^{m}(x_i-y_j)\right)\asymp L\left(\prod_{i\neq j}(x_i-x_j)\right)\asymp m\log m.\tag{7.1}$$

In particular, the complexity of the Vandermonde determinant $\det[x_i^j]_{0\leqslant i,j\leqslant m-1}$ has an order of magnitude of at least $m\log m$, since its square is the discriminant. The best known algorithm, however, yields only $L(\det[x_i^j])=O(m(\log m)^2)$.

To conclude, here is a beautiful application of the above by Stoss [188]: Let char(k)=0. If $x_0,\dots,x_n, y_0,\dots,y_n$ are indeterminates over $k, p(x)$ is the interpolation polynomial for the (x_i, y_i) and $\xi \in k$, then

$$L(p(\xi)) \geqslant \tfrac{1}{3} n \log n - n.$$

Thus the complexity of the value of the interpolation polynomial given by the x_i, y_i at one single further point already has the order of magnitude $n \log n$. Hence it follows that every representation of p that can be evaluated in linear time requires a nonscalar cost of order $n \log n$ for its construction. Thus, in this sense, the coefficient as well as the barycentric representation of the interpolation polynomial have optimal efficiency.

PROBLEM **1.** Determine the order of magnitude of the complexity of resultant and discriminant as functions of the polynomial coefficients.

PROBLEM **2.** Find a useful generalization of the theorem in Section 7 to several functions or higher (say second) derivatives. These questions are equivalent: the second derivatives of f are the first derivatives of $\partial f/\partial x_1,\dots,\partial f/\partial x_m$, and the first derivatives of $f_1,\dots,f_p$ appear among the second derivatives of $a_1 f_1+\cdots+a_p f_p$. (The a_i are additional indeterminates.). Note that if $f=x_1,\dots,x_m$, then $L(f)=m-1$ (substitution), but

$$L\left(\left\{\frac{\partial^2 f}{\partial x_i \partial x_j}: 1 \leqslant i, j \leqslant m\right\}\right) \geqslant \binom{m}{2},$$

as can be shown with a simple dimension argument.

8. Branching

When transforming a computation sequence (in the sense of Section 3) to a computer program and applying the latter to a concrete input ξ, it is possible that the calculation is blocked by a required division by 0, although the functions $f_1,\dots,f_p$ to be calculated are defined at ξ. If the fact that this occurs only for a "thin" set of inputs is not considered satisfactory, it is necessary to complete the program with branching instructions in such a way that it yields the proper result for any input in the domain of the f_i (and otherwise outputs say $\emptyset$). To make the discussion independent of the groundfield, only null tests will be considered, that is, instructions of the form

$$\textbf{if } a=0 \textbf{ then goto } i \textbf{ else goto } j. \tag{8.1}$$

(Here a designates an already calculated intermediate result.)

It is reasonable to admit the possibility of such branching instructions in the abstract computation mode as well. A computation sequence thus becomes a *computation tree*; that is, a (finite) binary tree with arithmetic instructions attached at its simple nodes and instructions of the form (8.1) at its branch points. (For a more precise definition, see for instance [195].) The input ξ singles out a path beginning at the root of the tree and ending at a leaf where it is indicated which intermediate results form the output. We now demand, of course, that divisions by 0 be excluded even for particular values of the input. The cost of the computation tree for the input ξ is defined as the number of nonscalar operations and branching instructions that are to be executed along the path determined by ξ. The maximum of the cost function is called the cost of the tree. Thus we can introduce a new measure of complexity $C(f_1,\ldots,f_p)$ as the minimal cost of computation trees for the calculation of $f_1,\ldots,f_p$. We call C the *branching complexity*.

It is straightforward that

$$L(f_1,\ldots,f_p)\leqslant C(f_1,\ldots,f_p). \tag{8.2}$$

Indeed we may assume for a computation tree for $f_1,\ldots,f_p$ that no branching instruction is superfluous. If one chooses in such a tree the "thick" path, along which each branching instruction (8.1) is passed in the sense $a\neq 0$, one obtains a computation sequence for $f_1,\ldots,f_p$.

From (8.2) it follows that all lower bounds for the nonscalar complexity that we have given so far remain valid for the branching complexity. It seems that in concrete cases C can be considerably bigger than L (see for instance Keller-Gehrig [117]). This is due to the fact that the computation tree is expected to carry out the evaluation of the functions f_i for any given input, whereas the computation sequence is not. (Of course, L coincides with C if no divisions are permitted.) In other words, C measures the "worst-case" complexity and L the "almost-everywhere" complexity.

The greater flexibility of the computation tree model arises from the fact that it is not limited to the evaluation of finite sequences of rational functions, but can be applied to a larger class of computational tasks. Consider the following example.

In Section 4 we have determined the nonscalar complexity of the evaluation of a continued fraction in the case where it appeared in a form one could normally expect over an algebraically closed groundfield. (With the help of the substitution method used there, a corresponding result can be proved for the evaluation of an arbitrary continued fraction as well.) Now let us investigate the *expansion* of a polynomial quotient into a continued fraction. (Collins [48] and Loos [135] give information about the algorithmic significance of the continued fraction expansion and the closely related Euclidian expansion.) As is well-known, if $A_1, A_2\in k[x]$ and $A_2\neq 0$, one obtains the continued fraction expansion as the sequence $(Q_1,\ldots,Q_t)$ of quotients in the Euclidian algorithm

$$A_1=Q_1A_2+A_3,\ A_2=Q_2A_3+A_4,\ldots,A_t=Q_tA_{t+1}.$$

Calculation of the coefficient sequence of the polynomials $Q_1,\ldots,Q_t$ from the coefficients of A_1, A_2 by means of the Euclidian algorithm normally amounts to a nonscalar cost of order n^2. (Here it is assumed that deg $A_1=n$, deg $A_2\leqslant n$.) Knuth [119] and Schönhage [172] found a subtle recursive procedure for the computation of the continued fraction expansion, in which only suitable initial segments of the A_i are used and which reaches its goal with $O(n\log n)$ nonscalar operations. (Actually they work in the analogous number-theoretic situation; see [143, 195] for the polynomial setting.)

Note that the number of coefficients to be calculated is a nonconstant function of the input polynomials A_1, A_2. This means that in the present case the model of computation sequence falls short and should be replaced by that of computation tree. Formulating the Knuth-Schönhage algorithm as a computation tree results in an upper $O(n \log n)$ estimate for the branching complexity (defined above) of the continued fraction expansion. This is the optimal order of magnitude as can be proved by means of the degree method of Section 6. The information thereby obtained is even much more precise [195]: *The Knuth–Schönhage algorithm possesses a cost function of order* $n(H+1)$, *where* H *denotes the entropy of the "output format"* (deg $Q_1, \ldots,$ deg Q_t) *normalized to a probability vector. The cost function of an arbitrary computation tree for the continued fraction expansion is* $\geqslant n(H+1)$ *outside a "uniformly thin" set of inputs.*

In statistical terminology, the Knuth–Schönhage algorithm is thus "uniformly most powerful". Only slightly weaker lower bounds have been obtained by Schuster [180] for the problem of computing the output format alone of the continued fraction expansion. Lower bounds of order $n \log n$ for a number of interesting algebraic decision problems over algebraically closed fields appear in Lickteig [131].

In the following we shall suppose $k = \mathbb{R}$ and allow, besides the null tests $a = 0$, comparisons $a \geqslant 0$ as well. We shall again designate the corresponding branching complexity with C. Inequality (8.2) still holds, so the lower bounds for the nonscalar complexity proved in former sections remain valid if L is replaced by C.

The branching complexity of decision problems $W \subset \mathbb{R}^n$ (here, the indicator function of W is to be calculated) has been investigated in the past decade by a number of authors. Only recently, however, has it been possible to prove nonlinear lower bounds.

THEOREM (Ben-Or [15]). *Suppose that* $W \subset \mathbb{R}^n$ *possesses* N *connected components. Then*

$$C(W) \geqslant \frac{1}{1+\log 3}(\log N - n \log 3).$$

As an application, one obtains estimates of order $n \log n$ for the complexity of many natural decision problems. Here are a few examples.

(A) Given $(x_1, \ldots, x_n) \in \mathbb{R}^n$, decide whether the x_i are pairwise distinct.

(B) Given $(x_1, \ldots, x_n, y_1, \ldots, y_n) \in \mathbb{R}^{2n}$, decide whether $\{x_1, \ldots, x_n\} = \{y_1, \ldots, y_n\}$.

(C) Given $(x_1, \ldots, x_n, y_1, \ldots, y_n) \in \mathbb{R}^{2n}$, decide whether $\{x_1, \ldots, x_n\} \cap \{y_1, \ldots, y_n\} = \emptyset$.

In the case of the "knapsack problem":

(D) given $(y_1, \ldots, y_n) \in \mathbb{R}^n$, decide whether $\Sigma_{i \in I} y_i = 1$ for some $I \subset \{1, \ldots, n\}$,

the theorem even yields $C(D) \geqslant \text{const} \cdot n^2$. This result has been proved before in more restrictive computation models by Dobkin–Lipton [55] and Steele–Yao [187]. The latter authors have already used the essential ingredient of the proof of the above theorem, an estimate by Milnor [141] and Thom [201] for the number of connected components of a real algebraic variety.

In characteristic zero a weakened form of the degree bound of Section 6 may be derived from the above theorem by means of an elementary argument. Direct application of Ben-Or's theorem to resultant and discriminant (see (7.1)), as proposed in [15], appear to be sound only when divisions are excluded.

PROBLEM **1.** Decide whether the uniform optimality property of the Knuth–Schönhage algorithm still holds if comparisons are allowed as branching instructions.

PROBLEM **2.** Find out whether the lower bounds given for the branching complexity of problems (A)–(D) remain valid for groundfields of finite characteristic.

9. Complexity classes

The methods we have discussed until now yield only logarithmic lower bounds for the complexity of polynomials in one variable whose coefficients are algebraic over the prime field (e.g. of polynomials in $\mathbb{Q}[x]$). As a remedy to this deficiency, we take a closer look at the approach of Motzkin [145] and Belaga [13], and this in the form given to them by Paterson–Stockmeyer [152] for the nonscalar complexity. For simplicity, divisions will not be permitted (although the results remain correct without this condition) and the groundfield will be the field of complex numbers.

The idea is to investigate the dependence of the coefficients of the polynomials to be evaluated on the parameters appearing in a computation. For this, the suitable notion is that of a "generic computation": Let x, c_{ij}, d_{ij} $(1\leqslant i\leqslant r, 0\leqslant j\leqslant i)$ and e_j $(0\leqslant j\leqslant r+1)$ be indeterminates over $\mathbb{C}$. We define $U_i, V_i, G_i\in\mathbb{C}[c,d,e,x]$ for $1\leqslant i\leqslant r$ inductively by

$$U_i=c_{i0}+c_{i1}x+\sum_{j=1}^{i-1} c_{i,j+1}G_j,\qquad V_i=d_{i0}+d_{i1}+\sum_{j=1}^{i-1} d_{i,j+1}G_j,\qquad G_i=U_iV_i$$

(compare with Section 3). Then we set

$$F=e_0+e_1x+\sum_{j=1}^{r} e_{j+1}G_j\in\mathbb{C}[c,d,e,x].$$

For $f\in\mathbb{C}[x]$ it is now clear that

$$L(f)\leqslant r \Leftrightarrow \exists\gamma_{ij},\delta_{ij},\varepsilon_j\in\mathbb{C}\, f(x)=F(\gamma,\delta,\varepsilon,x). \tag{9.1}$$

Let us denote the linear space of all polynomials in $\mathbb{C}[x]$ with degree $\leqslant n$ by $\mathbb{C}[x]_n$. Then the *complexity class* $\{f\in\mathbb{C}[x]_n: L(f)\leqslant r\}$ is constructible and its Zariski closure X_{nr} is a subvariety of $\mathbb{C}[x]_n$ of a dimension $<(r+2)^2$ (since there are less than $(r+2)^2$ indeterminates c, d, e). If we take here $r=\lfloor\sqrt{n}-2\rfloor$, we see that almost all polynomials of degree n have a complexity $\geqslant\sqrt{n}-1$. (on the other hand, according to Paterson–Stockmeyer [152] any polynomial of degree n may be evaluated with a nonscalar cost $O(\sqrt{n})$.)

An "ideal" method to estimate from below the complexity of a concrete polynomial would consist in determining the vanishing ideal of the affine variety X_{nr}. Its elements are those complex polynomials in $n+1$ variables that vanish for the coefficient vectors of all $f \in X_{nr}$ (or equivalently for the lowest $n+1$ coefficients of $F \in \mathbb{C}[c, d, e][x]$). We shall call them *resultants*. Unfortunately, it seems difficult to exhibit interesting resultants. Nevertheless very fragmentary information such as the *existence* of integer resultants of small degree and small height turns out to be sufficient to obtain meaningful lower bounds for the complexity of concrete polynomials. (Strassen [193], Schnorr [168].) Here is an example:

$$L\left(\sum_{j=0}^{n} 2^{2^j} x^j\right) \geqslant \text{const} \sqrt{n/\log n}.$$

Note that the difficulty in the evaluation of $\Sigma_{j=0}^{n} 2^{2^j} x^j$ does not consist at all in producing the enormous coefficients, since arbitrary complex numbers are for free at our disposal.

If one is looking for lower bounds for the complexity of polynomials whose coefficients are algebraic but not necessarily rational, one can improve the method indicated above by using, instead of the existence of resultants of small degree, the fact that X_{nr} itself has a small degree (Heintz–Sieveking [99]). One thus obtains the following beautiful result.

THEOREM. *Let $\alpha_1, \ldots, \alpha_n \in \mathbb{C}$ be algebraic over a subfield k_0 of $\mathbb{C}$ and let N be the size of the orbit of $(\alpha_1, \ldots, \alpha_n)$ under the Galois group of $\mathbb{C}$ over k_0. If there exist polynomials $h_1, \ldots, h_n \in k_0[a_1, \ldots, a_n]$ of degree $\leqslant M$ whose common set of zeros in $\mathbb{C}^n$ is finite and contains $(\alpha_1, \ldots, \alpha_n)$, then*

$$L(\alpha_n x^n + \cdots + \alpha_1 x) \geqslant \left(\frac{\log N}{24 \log(nM)}\right)^{1/2}.$$

Although the proof once again involves the Theorem of Bezout, there is no perceptible connection with the method of Section 6.

As applications we have for example

$$L\left(\sum_{j=1}^{n} \alpha^{1/j} x^j\right) \geqslant \text{const} \sqrt{n/\log n}$$

for positive real $\alpha \neq 1$, and

$$L\left(\sum_{j=1}^{n} j^q x^j\right) \geqslant \text{const} \frac{\sqrt{n}}{\log n}$$

for $q \in \mathbb{Q} \setminus \mathbb{Z}$ (von zur Gathen–Strassen [72]). For $q \in \mathbb{N}$ one has $L(\Sigma_{j=1}^{n} j^q x^j) = O(\log n)$.

An interesting application is due to Lipton–Stockmeyer [133] (see also Schnorr [168]): Every polynomial $f \in \mathbb{C}[x]$ of degree n with only simple roots possesses divisors whose complexity have an order of magnitude $\sqrt{n/\log n}$ (although there exist such f's that are very easy to calculate, e.g. $f = x^n - 1$).

The methods of this section are applicable to other measures of complexity as well (counting of additions/subtractions or counting of all arithmetic operations). We refer to Schnorr–van de Wiele [170] and the literature cited there.

PROBLEM 1. For the *typical* nonscalar complexity

$$\lambda_n = \min\{r: X_{nr} = \mathbb{C}[x]_n\}$$

of polynomials of degree $\leqslant n$ Paterson–Stockmeyer [152] show that

$$\sqrt{n} - 1 \leqslant \lambda_n \leqslant \sqrt{2n} + O(\log n).$$

Improve this estimate (compare also with Section 12).

PROBLEM 2. Decide whether $L(\Sigma_{j=0}^{n} x^j/j!)$ and $L(\Sigma_{j=1}^{n} x^j/j)$ respectively have an order of magnitude $\log n$ (see also Brent [31]).

PROBLEM 3. Find a sequence (f_n) of concrete polynomials with coefficients in $\{0, 1\}$ such that $\deg f_n = n$ and the complexity of (f_n) in $\mathbb{C}[x]$ does not have the order $\log n$. According to Lipton [132] there exist such sequences. (See also Schnorr–van de Wiele [170] and Heintz–Schnorr [98].)

10. Matrix multiplication and bilinear complexity

The exponent ω of matrix multiplication over a field k is the infimum of all real numbers τ such that multiplication of square matrices of order h may be achieved with $O(h^\tau)$ arithmetical operations (or equivalently, that the *arithmetic* complexity of matrix multiplication of order h is $O(h^\tau)$). Obviously, $\omega \leqslant 3$.

The significance of this notion lies, above all, in the key role of matrix multiplication for numerical linear algebra. Thus the following problems all have "exponent" ω: matrix inversion, LR-decomposition, evaluation of the determinant or of all coefficients of the characteristic polynomial and, for $k = \mathbb{C}$, also QR-decomposition and unitary transformation to Hessenberg form (see Strassen [189, 191], Bunch-Hopcroft [37], Schönhage [174], Baur–Strassen [12], Keller-Gehrig [117], Kalorkoti [111]). Apart from this, such diverse computational problems as finding the transitive closure of a finite relation, parsing a context-free language or multiplying elements of the group algebra of a finite group are reducible to matrix multiplication (see the survey article of Paterson [151], and Section 13).

The exponent ω is quite robust with respect to the measure of complexity used in its definition. Instead of the arithmetic complexity we could take as well the nonscalar complexity L or even the so-called *rank* R, which allows the following simple algebraic definition: Let U, V, W be finite-dimensional k-vector spaces and let $f: U \times V \to W$ be a bilinear map (such as matrix multiplication of a particular size or the structure map of a finite-dimensional associative algebra or of a finite-dimensional module over such an algebra). The rank of f is the smallest integer r with the property that there are $x_\rho \in U^*$ (the dual of U), $y_\rho \in V^*$ and $z_\rho \in W$ $(\rho = 1, \ldots, r)$ such that

$$f(u, v) = \sum_{\rho=1}^{r} x_\rho(u) y_\rho(v) z_\rho \tag{10.0}$$

for all $u \in U$, $v \in V$. We denote the rank of f by $R(f)$. Given bases (u_i), (v_j), (w_l), for U, V, W, the map f may be represented by its coordinate tensor $[f_{ijl}]$ via $f(u_i, v_j) = \Sigma_l f_{ijl} w_l$. Then $R(f)$ is the smallest $r \in \mathbb{N}$ with the property that there are $x_{\rho i}, y_{\rho j}, z_{\rho l} \in k$ such that for all i, j, l

$$f_{ijl} = \sum_{\rho=1}^{r} x_{\rho i} y_{\rho j} z_{\rho l}. \tag{10.1}$$

($R(f)$ differs by at most a factor 2 from $L(f)$, which is defined as the nonscalar complexity of the task of computing the coefficients of $f(u, v)$ from the coefficients of u, v with respect to any bases.) Following Schönhage [178], we denote by $\langle e, h, l \rangle$ the rectangular matrix multiplication $k^{e \times h} \times k^{h \times l} \to k^{h \times l}$. Then according to our previous remarks

$$\omega = \inf\{\tau: R(\langle h, h, h \rangle) = O(h^\tau)\}. \tag{10.2}$$

The rank function has a number of pleasant algebraic properties, such as

$$R(f \oplus f') \leqslant R(f) + R(f'), \quad R(f \otimes f') \leqslant R(f) R(f'). \tag{10.3}$$

(In the special case of associative algebras $\oplus$ is the direct product, $\otimes$ the tensor product.) The so-called *additivity conjecture* [191, 214] asserts that the left-hand inequality is always an equality. Its correctness would for example reduce the rank of an arbitrary semisimple $\mathbb{C}$-algebra to the rank of matrix multiplication. It is probably wrong.

The right-hand inequality of (10.3) is of particular importance for estimating ω, due to the well-known fact that

$$\langle h, h, h \rangle \otimes \langle l, l, l \rangle \cong \langle hl, hl, hl \rangle. \tag{10.4}$$

For instance, it can be shown by a direct construction that $R(\langle 2, 2, 2 \rangle) \leqslant 7$. From this we may conclude

$$R(\langle 2^q, 2^q, 2^q \rangle) = R(\langle 2, 2, 2 \rangle \otimes \langle 2^{q-1}, 2^{q-1}, 2^{q-1} \rangle) \leqslant 7 R(\langle 2^{q-1}, 2^{q-1}, 2^{q-1} \rangle)$$

by (10.4). Hence $R(\langle 2^q, 2^q, 2^q \rangle) \leqslant 7^q$ and therefore $R(\langle h, h, h \rangle) = O(h^{\log_2 7})$ by imbedding. As a result [189],

$$\omega \leqslant \log_2 7 < 2.81.$$

It is clear that any upper bound for the rank of matrix multiplication of a particular size gives an upper estimate for ω in an analogous way. On account of a result of Hopcroft–Musinski [104] this is still true if one starts with multiplication of rectangular matrices:

$$R(\langle e, h, l \rangle) \leqslant r \;\Rightarrow\; (ehl)^{\omega/3} \leqslant r. \tag{10.5}$$

At this point one would expect further progress from more effective initial constructions. Indeed, Pan [148, 149] showed $\omega < 2.79$ in such a way. The next step, however was the introduction of a new and powerful concept by Bini–Capovani–Lotti–Romani [25] and Bini [22], namely that of the *border rank* $\underline{R}(f)$ of a bilinear map f. Using coordinates, $\underline{R}(f)$ may be defined as the smallest $r \in \mathbb{N}$ with the property that there exist

$g_{ijl} \in k[\varepsilon]$ and $x_{\rho i}, y_{\rho j}, z_{\rho l} \in k(\varepsilon)$ (where ε is an indeterminate over k) such that for all i, j, l

$$f_{ijl} + \varepsilon g_{ijl} = \sum_{\rho=1}^{r} x_{\rho i} y_{\rho j} z_{\rho l}$$

(compare with (10.1)). Clearly $\underline{R}(f) \leqslant R(f)$, and sometimes this inequality is strict. Nevertheless (10.5) generalizes:

$$\underline{R}(\langle e, h, l\rangle) \leqslant r \;\Rightarrow\; (ehl)^{\omega/3} \leqslant r. \tag{10.6}$$

As an immediate pay-off Bini–Capovani–Lotti–Romani [25] and Bini [22] only obtained a marginally better estimate of ω than Pan, namely $\omega < 2.78$. But already in Schönhage [178] the idea of border rank was put into effective use by combining it with an extension of (10.6), the celebrated τ-theorem:

$$\underline{R}\left(\bigoplus_{1}^{q} \langle e_i, h_i, l_i\rangle\right) \leqslant r \;\Rightarrow\; \sum_{1}^{q} (e_i h_i l_i)^{\omega/3} \leqslant r. \tag{10.7}$$

The τ-theorem is an immediate consequence of (10.6) if the additivity conjecture is assumed to hold for border rank. Schönhage [178] showed by a novel recursion technique that such an assumption is unnecessary. Viewing this from a different angle, one may interpret (10.7) as an asymptotic confirmation of the additivity conjecture for border rank. Ironically, Schönhage discovered in the same paper that this analogue of the additivity conjecture is in fact wrong. His counterexample is

$$\underline{R}(\langle e, 1, l\rangle \oplus \langle 1, (e-1)(l-1), 1\rangle) = el + 1, \tag{10.8}$$

while $\underline{R}(\langle e, 1, l\rangle) = el$ and $\underline{R}(\langle 1, (e-1)(l-1), 1\rangle) = (e-1)(l-1)$. Applying the τ-theorem to (10.8) with $e = l = 4$ he obtained $\omega < 2.55$.

After further improvements by Pan [150] ($\omega < 2.53$) and Romani [162] ($\omega < 2.52$) Coppersmith–Winograd [51] showed $\omega < 2.50$ by a dynamic generalization of (10.8). An interesting byproduct of their work is the discovery that in the conclusion of (10.5), (10.6) or of the τ-theorem one practically always has strict inequality. (This result is occasionally given the interpretation that "ω is only an infimum, not a minimum". Such an interpretation does not refer to (10.2), however. According to our present knowledge, $R(\langle h, h, h\rangle) = O(h^{\omega})$ is very well possible, even with $\omega = 2$ or with R replaced by the arithmetical complexity.)

Since our emphasis is on general concepts such as rank and border rank that apply to arbitrary bilinear maps, not just to matrix multiplication, let us also put the exponent of matrix multiplication into a general perspective (Gartenberg [67]): We define the *asymptotic rank* (or *asymptotic complexity*) $\tilde{R}(f)$ of a bilinear map f by

$$\tilde{R}(f) = \lim_{N \to \infty} R(f^{\otimes N})^{1/N}. \tag{10.9}$$

The limit always exists, since R is submultiplicative (10.3). It is easy to see (using (10.2)) that

$$\tilde{R}(\langle 2, 2, 2\rangle) = 2^{\omega}. \tag{10.10}$$

The robustness of ω generalizes to $\tilde{R}$: in (10.9) the rank may be replaced by the border rank or by the nonscalar or even arithmetic complexity of f.

PROBLEM 1. Let η be the exponent for the problem of solving linear systems of size n. Clearly, $\eta \leqslant \omega$. Decide whether $\eta = \omega$ (try to use Section 7).

PROBLEM 2. Prove or disprove the additivity conjecture. (The additivity conjecture implies that a scalar extension can reduce the rank of a bilinear map by at most a factor $\frac{1}{4}$, see [198, p. 427]. This may be helpful in disproving the conjecture over nonclosed groundfields.)

11. Degeneration and asymptotic spectrum

The exponent ω of matrix multiplication and the asymptotic rank $\tilde{R}$ depend on the groundfield only through its characteristic [178, 199]. Thus without loss of generality (for asymptotic considerations) we may assume that k is algebraically closed. For concreteness we take $k = \mathbb{C}$.

Let f, g be bilinear maps. f is called a *degeneration* of g (notation: $f \trianglelefteq g$) when, roughly speaking, f may be approximated with arbitrary precision by isomorphic (equivalent) copies of g.

For example, we may take for f the multiplication map of the algebra $\mathbb{C}[t]/(t^n)$ and for g the *diagonal map* $\langle n \rangle$ (componentwise multiplication of vectors in $\mathbb{C}^n$). Then $f \trianglelefteq g$, since for any non-zero $\lambda \in \mathbb{C}$ the maps $\mathbb{C}[t]/(t^n - \lambda)$ and $\langle n \rangle$ are isomorphic by the Chinese Remainder Theorem, and obviously multiplication mod $(t^n - \lambda)$ converges to multiplication mod t^n as λ tends to 0.

It is a fundamental fact that border rank may be expressed in terms of degeneration:

$$\underline{R}(f) = \min\{r : f \trianglelefteq \langle r \rangle\} \tag{11.1}$$

(see [198] and the literature cited there). (Thus $\underline{R}(\mathbb{C}[t]/(t^n)) \leqslant n$ by the previous remarks.) On the one hand, this relationship opens the door to applying techniques from deformation theory to problems of bilinear complexity; on the other hand it introduces into deformation theory the asymptotic point of view as a new principle.

For what follows let us adopt a convenient language: denote by $\mathscr{B}^+$ the "set" of equivalence classes $[f]$ of bilinear maps f. $\mathscr{B}^+$ becomes a commutative semiring, when addition and multiplication are defined by $[f]+[g]=[f \oplus g]$ and $[f][g]=[f \otimes g]$. Since $\mathscr{B}^+$ allows additive cancellation, it may be imbedded into a commutative ring $\mathscr{B}$ as a subcone ($\mathscr{B}^+ + \mathscr{B}^+ \subset \mathscr{B}^+, \mathscr{B}^+\mathscr{B}^+ \subset \mathscr{B}^+, 0, 1 \in \mathscr{B}^+, -1 \notin \mathscr{B}^+$) such that $\mathscr{B} = \mathscr{B}^+ - \mathscr{B}^+$. $\mathscr{B}$ is essentially unique. (The construction is entirely analogous to the imbedding of $\mathbb{N}$ into $\mathbb{Z}$. $\mathscr{B}$ is a variant of the Grothendieck ring of the category of bilinear maps.) The class of the diagonal map $\langle r \rangle$ is simply the element $r \in \mathscr{B}$. So $\mathscr{B}$ has characteristic zero.

Degeneration lifts to a partial order $\trianglelefteq$ on $\mathscr{B}^+$ compatible with addition and multiplication and $\underline{R}$, $\tilde{R}$ become functions on $\mathscr{B}^+$ which may be expressed in terms of $\trianglelefteq$ by the formulas

$$\underline{R}(b) = \min\{r \in \mathbb{N} : b \trianglelefteq r\}, \tag{11.2}$$

$$\tilde{R}(b) = \lim_{N \to \infty} \underline{R}(b^N)^{1/N} \tag{11.3}$$

for $b \in \mathscr{B}^+$. Also the exponent ω can be recovered with (10.10), where $\langle 2, 2, 2\rangle$ now means the *class* of matrix multiplication of order 2.

Equations (11.2) and (11.3) easily imply that $\underline{R}$ and $\tilde{R}$ are subadditive and submultiplicative. R has a curious additional property, a special case of which follows from the τ-theorem: formulating (10.7) for classes and as an inequality we get

$$\underline{R}\left(\sum_1^q \langle e_i, h_i, l_i\rangle\right) \geqslant \sum_1^q (e_i h_i l_i)^{\omega/3}.$$

Using (11.3) it is easy to see that $\underline{R}$ may here be replaced by $\tilde{R}$. Thus for $e_i = h_i = l_i = 2^{N_i}$ we get, in view of (10.4) and (10.10),

$$\tilde{R}\left(\sum_1^q \langle 2, 2, 2\rangle^{N_i}\right) = \tilde{R}\left(\sum_1^q \langle 2^{N_i}, 2^{N_i}, 2^{N_i}\rangle\right) \geqslant \sum_1^q 2^{\omega N_i} = \sum_1^q \tilde{R}(\langle 2, 2, 2\rangle)^{N_i},$$

hence

$$\tilde{R}\left(\sum_1^q \langle 2, 2, 2\rangle^{N_i}\right) = \sum_1^q \tilde{R}(\langle 2, 2, 2\rangle)^{N_i}$$

in view of the above-mentioned general properties of $\tilde{R}$.

It turns out that this equality holds for any $b \in \mathscr{B}^+$ in place of $\langle 2, 2, 2\rangle$. In other words, $\tilde{R}$ is subadditive and submultiplicative on all of $\mathscr{B}^+$ and becomes additive and multiplicative when restricted by any subcone of $\mathscr{B}^+$ generated by a single element b. This means that $\tilde{R}$ *behaves like the maximum functional* on the cone of nonnegative functions of some function ring.

Is there a mathematical explanation for this analogy? To see that this is indeed the case we have to introduce an asymptotic version of the degeneration order:

$$a \lesssim b \Leftrightarrow a^N \trianglelefteq b^{N+o(N)} \quad \text{for } a, b \in \mathscr{B}^+.$$

$\lesssim$ is a preorder on $\mathscr{B}^+$ compatible with addition and multiplication. Surprisingly, it may be extended to all of $\mathscr{B}$ maintaining these properties. ($\trianglelefteq$ does not allow such an extension, see [198, p. 419] and Bürgisser [39].) In this way, $\mathscr{B}$ becomes a preordered ring, which, as it turns out, possesses the crucial property that allows one to apply the structure theory of Stone, Kadison and Dubois. One obtains the following result.

THEOREM 1 (cf. [199]). *Take any $X \subset \mathscr{B}^+$. Then there is a compact space Δ and a ring homomorphism $\varphi: \mathbb{Z}[X] \to C(\Delta)$ (where $\mathbb{Z}[X]$ denotes the subring of $\mathscr{B}$ generated by X, and $C(\Delta)$ the ring of continuous real functions on Δ) such that $\varphi(X)$ separates the points of Δ and such that for any $a, b \in \mathbb{Z}[X]$*

$$a \lesssim b \Leftrightarrow (\varphi(a) \leqslant \varphi(b) \text{ pointwise}). \tag{11.4}$$

An easy corollary is

$$\tilde{R}(a) = \max \varphi(a) \tag{11.5}$$

for any $a \in \mathbb{N}[X] \subset \mathscr{B}^+$. Δ is called an *asymptotic spectrum* for X via φ. Actually (Δ, φ) is unique up to canonical isomorphism. However, the definite article is reserved for the following concrete version: Δ can always be realized by a unique compact subset $\Delta(X)$ of $\mathbb{R}^X$ (endowed with the product topology) in such a way that $\varphi(x)$ is the function x (the xth coordinate projection restricted to $\Delta(X)$) for any $x \in X$. $\Delta(X)$ is called *the* asymptotic spectrum of X.

In view of (11.4) and (11.5) it is clear that from an asymptotic standpoint the central problem in the deformation as well as the complexity theory of bilinear maps is to compute asymptotic spectra of interesting families of classes of bilinear maps.

Let $\Delta_c := \Delta(\langle 2,2,2\rangle) \subset \mathbb{R}$ be the asymptotic spectrum of matrix multiplication of order 2. Δ_c is actually an asymptotic spectrum for all $\langle h,h,h\rangle$ via

$$\varphi(\langle h,h,h\rangle) = x^{\log h} \mid \Delta_c.$$

This is an easy consequence of the fact that large matrix multiplications may be well approximated by tensor powers of $\langle 2,2,2\rangle$ (which follows from (10.4)). How does Δ_c look like? Its right endpoint may be determined immediately:

$$\max \Delta_c = \max_{\Delta_c} x = \max \varphi(\langle 2,2,2\rangle) = \tilde{R}(\langle 2,2,2\rangle) = 2^\omega$$

by (11.5) and (10.10). This already leads to a simple proof of the τ-theorem (for square matrix multiplication):

$$\begin{aligned} \underline{R}\left(\bigoplus_1^q \langle h_i,h_i,h_i\rangle\right) \leqslant r \Rightarrow & \sum_1^q \langle h_i,h_i,h_i\rangle \trianglelefteq r && \text{(by (11.1) and transition to classes)} \\ \Rightarrow & \sum_1^q x^{\log h_i} \leqslant r && \text{on } \Delta_c \text{ (by (11.4))} \\ \Rightarrow & \sum_1^q h_i^\omega \leqslant r && \text{(since } 2^\omega \in \Delta_c\text{).} \end{aligned}$$

THEOREM **2** (cf. [199]). $\Delta_c = [4, 2^\omega]$.

As a corollary one obtains the following definitive extension of the τ-theorem for square matrix multiplication:

$$\sum_1^p \langle h_i,h_i,h_i\rangle \lesssim \sum_1^p \langle l_j,l_j,l_j\rangle \Leftrightarrow \forall \sigma \in [2,\omega] \sum_1^q h_i^\sigma \leqslant \sum_1^q l_j^\sigma.$$

The proof of Theorem 2 consists in determining both endpoints of Δ_c and showing that Δ_c is convex. (Or rather logarithmically convex; in general, logarithmic convexity of the asymptotic spectrum of an $X \subset \mathscr{B}^+$ has a good characterization in terms of asymptotic degeneration.)

As we have seen, the assertion $\max \Delta_c = 2^\omega$ reduces via (11.5) and (10.10) to the definition of ω. The corresponding statement $\min \Delta_c = 4$ has a less tautological interpretation:

$$4^{N - o(N)} \trianglelefteq \langle 2^N, 2^N, 2^N\rangle. \tag{11.6}$$

Thus a matrix multiplication of order 2^N degenerates into a diagonal map of size approximately 4^N, i.e. into about 4^N separate multiplications in $\mathbb{C}$. This information may be used to derive an improved estimate for ω, namely $\omega < 2.48$, by means of the so-called *laser method* [198]. (Actually (11.6) is not quite sufficient. One has to use that there are degenerations achieving (11.6) that respect coordinates.) The laser method has been taken up by Coppersmith–Winograd [52] with great success. They show $\omega < 2.38$, which is the best estimate known today. Of course their reasoning also has a natural spectral interpretation.

What is known about general (rectangular) matrix multiplication?

$$\Delta_{\mathrm{m}} := \Delta(\langle 2,1,1\rangle, \langle 1,2,1\rangle, \langle 1,1,2\rangle) \subset \mathbb{R}^3$$

is an asymptotic spectrum for all $\langle e, h, l\rangle$ via

$$\varphi(\langle e,h,l\rangle) = x_1^{\log e} x_2^{\log h} x_3^{\log l} \mid \Delta_{\mathrm{m}}.$$

Δ_{m} lies in the strictly positive octant of $\mathbb{R}^3$ and

$$\log \Delta_{\mathrm{m}} \subset conv\{(1,1,0),(1,0,1),(0,1,1),(1,1,1)\}.$$

This is only a rough estimate. Indeed any degeneration relation

$$\bigoplus_1^q \langle e_i, h_i, l_i\rangle \trianglelefteq \bigoplus_{q+1}^r \langle e_i, h_i, l_i\rangle$$

leads by (11.4) to an inclusion relation for $\log \Delta_{\mathrm{m}}$:

$$\log \Delta_{\mathrm{m}} \subset \left\{\sigma \in \mathbb{R}^3 : \sum_1^q e_i^{\sigma_1} h_i^{\sigma_2} l_i^{\sigma_3} \leqslant \sum_{q+1}^r e_i^{\sigma_1} h_i^{\sigma_2} l_i^{\sigma_3}\right\}.$$

As to inner estimates for $\log \Delta_{\mathrm{m}}$ we have

$$\Gamma := conv\{(1,1,0),(1,0,1),(0,1,1)\} \subset \log \Delta_{\mathrm{m}}. \tag{11.7}$$

Moreover, $\log \Delta_{\mathrm{m}}$ is star-shaped relative to Γ. ($\Gamma = \log \Delta_{\mathrm{m}}$ if and only if $\omega = 2$.)

PROBLEM **1.** Determine ω.

PROBLEM **2.** Prove that Δ_{m} is logarithmically convex (compare [199, Corollary 6.6]).

12. Lower bounds for rank and border rank

Although no estimate for the exponent of matrix multiplication from below is known except the trivial $\omega \geqslant 2$, numerous lower bounds for the rank or border rank of particular classes of bilinear maps and over particular fields have been established. The majority of results is concerned either with multiplication maps of associative algebras, with bilinear maps of extreme format (the *format* of $f: U \times V \to W$ is the triple (m, n, p) of dimensions of U, V, W) or with the *typical* rank and border rank of bilinear maps of a given format.

Algebras

The natural question as to whether the rank estimate $R(\langle 2,2,2\rangle)\leqslant 7$ can be improved was answered by Hopcroft–Kerr [103] and Winograd [212]: $R(\langle 2,2,2\rangle)=7$. Somewhat more generally, Brockett–Dobkin [32] and Lafon–Winograd [123] have shown $R(\langle n,n,n\rangle)\geqslant 2n^2-1$. For the rank of some other classes of associative algebras there exist nontrivial lower bounds as well. Thus, one has for instance for division algebras A of dimension n

$$R(A)\geqslant 2n-1 \tag{12.1}$$

(Fiduccia–Zalcstein [59]). According to de Groote [86], equality holds in (12.1) iff A is a simple field extension. Applied to the algebra $\mathbb{H}$ of real quaternions, this yields one half of a result already known before

$$R(\mathbb{H})=8 \tag{12.2}$$

(Dobkin [54], de Groote [83], Howell–Lafon [107]). Fiduccia–Zalcstein [59] and Winograd [214] have investigated the multiplication of two polynomials modulo a given polynomial f, and have found

$$R(k[x]/(f))=2n-t,$$

where n is the degree of f, and t the number of its different prime factors. Here, the dependence of the complexity (in this case rank) on the groundfield appears clearly again. (For instance, the group algebra $k[x]/(x^n-1)$ of the cyclic group C_n has rank $2n-d(n)$ for $k=\mathbb{Q}$ where $d(n)$ is the number of divisors of n, but rank n only for $k=\mathbb{C}$.)

Except for (12.2), the lower bounds mentioned above are corollaries of the inequality

$$R(A)\geqslant 2n-t, \tag{12.3}$$

where $n=\dim A$ and t designates the number of maximal two-sided ideals of A. (Alder–Strassen [2]). This inequality proves to be suitable for the determination of the rank of certain noncommutative group algebras as well, for instance of the dihedral group.

As with (12.1), the question arises as to when (12.3) becomes an equality. One is led to the concept of *algebra of minimal rank*. De Groote–Heintz [87] have classified the commutative algebras of minimal rank: they are the finite products of simply generated algebras (that is, algebras $k[x]/(f)$) and so-called generalized null algebras (that is, algebras $k[\omega_1,\ldots,\omega_q]$ where all ω_i are nilpotent and annihilate each other). This result has been generalized to certain classes of noncommutative algebras by Büchi–Clausen [38] and Heintz–Morgenstern [97].

In a further work [88] de Groote and Heintz investigated the complexity and rank of simple Lie algebras (see also de Groote–Heintz–Möhler–Schmidt [89]). One reason for the interest in the rank of Lie algebras is the fact that the complexity of multiplication in a connected linear algebraic group is estimated from below, up to a constant factor, by the rank of the corresponding Lie algebra [191]. For the complexity of modules over associative algebras, we refer to Hartmann [92] and the literature cited there.

If the rank of an algebra A is known, it is natural to inquire about the structure of the set of optimal bilinear computations (i.e. representations (10.0)). The symmetry group

Γ of the algebra (a group of generalized automorphisms) operates on this set. The beautiful result that the operation is transitive for $A=k^{2\times 2}$ (the algebras of matrices of order 2) is due to de Groote [84]. So for this problem there is essentially a unique optimal algorithm. On the other hand, field extensions generally possess higher-dimensional orbit varieties of optimal algorithms (Winograd [214], see also de Groote [85]).

Very little is known about the problem of the inversion (of units) and division (by units) in finite-dimensional associative algebras. Alt–van Leeuwen [3] have shown that complex inversion has multiplicative complexity 4, and Lickteig [129, 130] has proved that the multiplicative complexity of complex division (as well as that of solving a linear system of size two) is 6.

Bilinear maps of extreme format

Let k be algebraically closed and let $f: U \times V \to W$ have format (m, n, p). For reasons of symmetry (see e.g. (10.1)) we may assume $m \geqslant n \geqslant p$. We also may take $W=k^p$.

For $p=1$, f is a bilinear form, a coordinate tensor of f is a matrix and rank and border rank coincide with matrix rank.

For $p=2$, f is a pair of bilinear forms and a coordinate tensor of f is given by a pair $A, B \in k^{m\times n}$ of matrices. When $m=n$ and A is invertible we have $\underline{R}(f)=n+l$, where l denotes the maximal number of nontrivial Jordan blocks for the same eigenvalue of the matrix $A^{-1}B$. This follows from a general result of Grigoryev [76] and Ja'Ja' [109], which expresses $R(f)$ by the invariants which appear in the Kronecker normal form of the matrix pair (A, B). (Bini [23] treats the border rank in an analogous way.)

For $p=3$ (three bilinear forms), a coordinate tensor for f is represented by a triple (A, B, C) of matrices in $k^{m\times n}$. Suppose $m=n$ is odd. Let X, Y, Z be matrices of order n with indeterminate coefficients and define a rational function F by

$$F(X, Y, Z)=(\det X)^2 \det(YX^{-1}Z-ZX^{-1}Y). \tag{12.4}$$

Then F turns out to be an irreducible polynomial and one has the following criterion ([196], see also Griesser [74]): $\underline{R}(f)=\frac{1}{2}(3n+1)$, unless $F(A, B, C)=0$, in which case $\underline{R}(f)<\frac{1}{2}(3n+1)$. In other words, the set of bilinear maps of border rank $<\frac{1}{2}(3n+1)$ is a hypersurface, whose vanishing ideal is generated by (12.4). As an application of this and an analogous statement for even n, the border rank of an arbitrary sl_2-module may be determined.

Typical rank and border rank

Let $k=\mathbb{C}$. (Any algebraically closed field would do as well.) Fix U, V, W of dimensions m, n, p and denote the vector space of bilinear maps $f: U\times V\to W$ by $Bil(U, V; W)$. The image of the polynomial map

$$\varphi_r: (U^* \times V^* \times W)^r \to Bil(U, V; W),$$

$$(x_\rho, y_\rho, z_\rho)_{\rho\leqslant r} \mapsto \Big((u, v)\mapsto \sum_{\rho=1}^{r} x_\rho(u) y_\rho(v) z_\rho\Big)$$

consists exactly of the bilinear maps of rank $\leqslant r$ (by (10.0)) and the closure of the image coincides with the set of bilinear maps of border rank $\leqslant r$. Thus $X_r := \{f \in Bil(U, V; W): \underline{R}(f) \leqslant r\}$ is a closed irreducible subvariety of $Bil(U, V; W)$. Taking into account trivial fibres of φ_r one obtains $\dim X_r \leqslant r(m+n+p-2)$ and therefore

$$R(m, n, p) := \max_{f \in Bil(U,V;W)} \underline{R}(f) \geqslant mnp/(m+n+p-2). \tag{12.5}$$

$R(m, n, p)$ is at the same time the rank and the border rank of almost all $f \in Bil(U, V; W)$. (But in general $R(m, n, p) < \max_{f \in Bil(U,V;W)} R(f)$.) (12.5) is a good estimate:

$$R(m, n, p) \sim mnp/(m+n+p-2)$$

as $m, n, p \to \infty$, and equality holds in (12.5) for a variety of formats, such as $(n, n, n+2)$ with $n \not\equiv 2 \bmod 3$ or $(n, n, n-1)$ with $n \equiv 0 \bmod 3$ [196]. Lickteig [128] proves, for the particularly interesting cubic format,

$$R(n, n, n) = \lceil n^3/(3n-2) \rceil$$

unless $n=3$. (For $n=3$ we have seen above that $\lceil n^3/(3n-2) \rceil = 4 < 5 = \frac{1}{2}(3n+1) = R(n, n, n)$.)

PROBLEM 1. Determine $\underline{R}(k^{2\times 2})$ for algebraically closed k. (Bini [24] has shown $\underline{L}(k^{2\times 2}) = 6$, where $\underline{L}$ is the border complexity [193, 73].)

PROBLEM 2. Decide whether the following analogue of (12.3) is correct:

$$\underline{R}(A) \geqslant 2 \dim A - \dim(\operatorname{rad} A) - t,$$

where t is again the number of maximal two-sided ideals of A and k is assumed to be algebraically closed (compare [197, p. 533]).

PROBLEM 3. Determine $R(K^{2\times 2})$, where K is a simple algebraic field extension of k and $K^{2\times 2}$ is understood as a k-algebra.

PROBLEM 4. Determine the algebras of minimal rank.

PROBLEM 5. Generalize the result of Feig [56] that a division algebra of "minimal complexity" has minimal rank (compare [2]).

PROBLEM 6. The inversion of units in $k[x]/(f)$ with $\deg f = n$ has a nonscalar complexity $O(n \log n)$, as follows from the Knuth–Schönhage algorithm for the Euclidean expansion [195]. Improve this estimate. Observe that e.g. for $f = x^n$ the complexity of inversion is only $O(n)$ [182, 122].

PROBLEM 7. Construct analogues of (12.4) for other formats.

PROBLEM 8. Determine all formats for which the inequality in (12.5) is an equality.

13. Fourier transform

In analogy to the case of bilinear maps (compare with (10.0)) it is possible for the evaluation of a system of linear forms

$$\sum_{j=1}^{n} \alpha_{ij}x_j, \ldots, \sum_{j=1}^{n} \alpha_{mj}x_j \tag{13.1}$$

to restrict oneself to the use of linear combinations $(g, h) \mapsto \beta g + \gamma h$ (with $\beta, \gamma \in k$) as computing operations without significant loss of efficiency. We shall call the complexity of (13.1) defined by means of such a linear computation model *complexity of the matrix* $a = [\alpha_{ij}] \in k^{m \times n}$ and we shall denote it by $L_s(a)$. It is the minimal number of linear combinations sufficient to compute the linear mapping $k^m \to k^n$ corresponding to the matrix a. (We can also speak of the complexity of linear maps between abstract vector spaces, but should then keep in mind that this notion presupposes the choice of bases.)

Of great practical significance is the algebra isomorphism

$$D_{(n)}: \mathbb{C}[x]/(x^n - 1) \to \mathbb{C}^n,$$

supplied by the Chinese Remainder Theorem, the Discrete Fourier Transform (DFT). Relative to the natural bases, $D_{(n)}$ (and essentially also $D_{(n)}^{-1}$) is represented by the matrix $[\zeta^{ij}] \in \mathbb{C}^{n \times n}$, where ζ is a primitive nth root of unity. The algorithm of the "Fast Fourier Transform" (FFT) (re)discovered in 1965 by Cooley–Tukey [49] yields, together with Bluestein [26] (or Rader [156] and Winograd [213]), the complexity estimate

$$L_s(D_{(n)}) = \mathrm{O}(n \log n). \tag{13.2}$$

Among the multiple applications of the FFT we only mention the multiplication of numbers or polynomials (Schönhage–Strassen [173], Schönhage [177], Cantor–Kaltofen [42]).

What is known about lower complexity bounds? Morgenstern [144] showed in a surprisingly simple way that an algorithm for the DFT using only linear combinations $\beta g + \gamma h$ with $|\beta| + |\gamma| \leqslant c$ for $c \geqslant 2$ costs at least $(n \log n)/(2 \log c)$.

The attempt to prove lower bounds for $L_s(D_{(n)})$ without such restrictions on the computational model has directed attention to a class of interesting graphs: the *n-superconcentrators*. These are graphs with n input and n output nodes such that for every $q \leqslant n$ one can connect any set of q input nodes with any set of q output nodes through q disjoint paths. An algorithm for the computation of the DFT with r linear combinations leads to an n-superconcentrator with $4r$ edges. The minimal number of edges of n-superconcentrators is thus a lower bound to $4L_s(D_{(n)})$. Pinsker [153] and Valiant [203] obtained the Janus-faced result that there exist n-superconcentrators with $\mathrm{O}(n)$ edges. An effective construction of such graphs is due to Margulis [138] (see also Gabber–Galil [66], Pippenger [154], Klawe [118]). The literature cited above also gives information about the numerous applications of the concept of superconcentrator.

The discrete Fourier transform is a particular case of the Wedderburn isomorphism

$$D: \mathbb{C}[G] \rightarrow \mathbb{C}^{n_1 \times n_1} \times \cdots \times \mathbb{C}^{n_t \times n_t}, \qquad a \mapsto (D_1(a), \ldots, D_t(a))$$

from the group algebra of a finite group G to a product of full matrix algebras. The D_i represent the equivalence classes of the irreducible matrix representations of G, and are consequently determined only up to conjugation. Thus it is natural to consider

$$L_s(G) := \min_{D_1, \ldots, D_t} L_s(D),$$

where $(D_1, \ldots, D_t)$ ranges over all transversals for the set of equivalence classes of irreducible matrix representations. In this language (13.2) says that

$$L_s(G) = O(n \log n)$$

for cyclic G of order n. Since any finite abelian group is a direct product of cyclic groups, the estimate extends to abelian groups [4, 116]. In [47] Clausen further extends this $O(n \log n)$ upper bound to all metabelian groups. (G is metabelian iff G has an abelian normal subgroup A such that G/A is abelian as well.) For the class of solvable groups Beth [21] proves that

$$L_s(G) = O(p^{-1/2} n^{3/2}), \tag{13.3}$$

where p is the greatest prime divisor of n. Clausen [46] shows that a slightly weakened version of (13.3) holds for arbitrary finite groups:

$$L_s(G) = O(\sqrt{q} n^{3/2})$$

when there exists a subgroup tower $1 = G_0 < \cdots < G_l = G$ with $(G_{i+1}:G_i) \leqslant q$ for all i. His method also yields the interesting result that the symmetric groups S_m admit a "Fast Wedderburn Transform":

$$L_s(S_m) \leqslant m^3 \cdot m! = O(n(\log n)^3),$$

where $n := \# S_m = m!$.

PROBLEM. Determine the order of magnitude of $L_s(D_{(n)})$.

14. Complete problems

The theory of NP-completeness, developed by Cook and Karp and centering around *Cook's Hypothesis* $P \neq NP$, has become the model of numerous investigations in theoretical computer science (see Johnson's Chapter 2 in this Handbook on complexity theory and complexity classes).

As one of a few exceptions, algebraic complexity had seemed for a number of years to remain unaffected by it. In 1979, however, Valiant [206, 207] developed a convincing analogue of this theory entirely in the algebraic framework. His starting point was the concept of #P-complete counting problem, which he had introduced in [204]. Consider the NP-complete problem of deciding whether a graph with the nodes

$1, \ldots, n$ and the incidence matrix $[\alpha_{ij}]$ ($\alpha_{ij}=1$ in case i and j are connected by an edge, otherwise $\alpha_{ij}=0$) possesses a Hamiltonian cycle. It is natural to make use of the enumerator

$$HC_{n\times n}([a_{ij}])=\sum_{\sigma\in S_n\text{ cycle of length } n} a_{1\sigma(1)}\cdots a_{n\sigma(n)}\in\mathbb{Q}[a_{ij}].$$

As can easily be seen, $HC_{n\times n}[\alpha_{ij}]$ is the number of Hamiltonian cycles of the graph. A fast procedure for the evaluation of the polynomial $HC_{n\times n}$ on a Turing machine would therefore yield a fast decision procedure for the existence of Hamiltonian cycles.

Many other NP-complete problems possess natural enumerators, and it is plausible that these do not differ essentially from each other in their computational complexity [204]. On the other hand, Valiant made the surprising discovery that the enumerator of the well-known marriage problem belongs to the same complexity class: among n boys and n girls, some have become friends; can every boy marry a friend of his? Let us set $\alpha_{ij}=1$ if the boy i is a friend of the girl j, and $\alpha_{ij}=0$ otherwise; then we can view the marriage of all boys as a permutation $\sigma\in S_n$ with the property $\alpha_{i\sigma(i)}=1$ for $1\leqslant i\leqslant n$. The question whether a matrix $[\alpha_{ij}]$ with coefficients 0 or 1 admits a marriage is, according to Hall [90], decidable in polynomial time (see also Specker's contribution V in [186] and the literature mentioned there). Valiant [205] now shows that the natural enumerator of the marriage problem, the permanent

$$Perm_{n\times n}([a_{ij}])=\sum_{\sigma\in S_n} a_{1\sigma(1)}\cdots a_{n\sigma(n)}$$

has the complexity of enumerators of NP-complete problems. Hence, under the assumption of Cook's Hypothesis, neither HC nor $Perm$ are computable in polynomial time.

One cannot simply take over these statements into the algebraic computational model of the previous sections. Valiant [206, 207] (see also von zur Gathen [69]) therefore developed a theory analogous to that of Cook and Karp in the framework of algebraic complexity.

The objects of his investigations are not decision problems, but sequences $(F_m)_{m\geqslant 1}$ of polynomials $F_m\in k[x_1,\ldots,x_m]$ such that $\deg F_m$ is P-bounded in m. (A function $t:\mathbb{N}\to\mathbb{N}$ is called P-bounded iff it is majorized by a polynomial.) The permanent leads to such a polynomial sequence, by setting for instance $F_m=0$, in case m is not a square, and $F_m=Perm_{n\times n}([a_{ij}])$ with $a_{ij}=x_{(i-1)n+j}$, in case $m=n^2$.

The sets decidable in polynomial time correspond to the "P-*computable*" polynomial sequences (F_m) for which $L(F_m)$ is P-bounded in m. (The question of a finite Turing machine program that controls the computation of all F_m is here irrelevant.)

Valiant calls a sequence (F_m) P-*definable* iff there exist a P-computable sequence (G_m) and a P-bounded t such that for all m

$$F_m(x_1,\ldots,x_m)=\sum_{\xi_{m+1},\ldots,\xi_{t(m)}\in\{0,1\}} G_{t(m)}(x_1,\ldots,x_m,\xi_{m+1},\ldots,\xi_{t(m)}).$$

The P-definable sequences correspond to the sets in NP. (The summation replaces the projection.) Most naturally occurring polynomial sequences turn out to be P-definable.

Lastly, the notion of reduction (called "projection") is very simple: $(F_m) \leqslant (G_m)$ iff there exists a P-bounded t such that for all m

$$F_m(x_1, \ldots, x_m) = G_{t(m)}(c_1, \ldots, c_{t(m)})$$

with $c_i \in \{x_1, \ldots, x_m\} \cup k$. $(F_m) \leqslant (G_m)$ thus means that any F_m can be derived from a not too remote (G_n) through substitution by indeterminates and constants. (Recall equation (9.1) where we have seen that computation may be interpreted as a substitution.)

THEOREM (Valiant [206]). *HC is complete with respect to* $\leqslant$ *in the class of* P-*definable polynomial sequences. The same is true for Perm unless* $\operatorname{char}(k) = 2$.

The analogue of Cook's Hypothesis, we call it *Valiant's hypothesis*, now reads: the sequence HC is not P-computable. In case $\operatorname{char}(k) \neq 2$ this is equivalent to saying that the sequence of permanents is not polynomially computable. (Over fields of characteristic 2, the permanent coincides with the determinant, and is thus P-computable.)

Assuming Valiant's hypothesis, it is possible to show that certain natural operations on polynomials, such as multiple differentiation, multiple integration, taking coefficients with respect to subsets of indeterminates, lead from easy polynomial sequences to hard ones [207]: Consider the sequence T whose non-zero terms are

$$T_{n^2+n} = \prod_{l=1}^{n} \sum_{i=1}^{n} x_{li} y_i.$$

(T_{n^2+n} is the product of n general linear forms in n variables.) Clearly T is P-computable. In contrast,

$$\frac{\partial}{\partial y_1} \cdots \frac{\partial}{\partial y_n} T_{n^2+n} = \text{coefficient of } y_1 \cdots y_n \text{ in } T_{n^2+n}$$

as well as

$$\left(\frac{3}{2}\right)^n \int_{-1}^{1} \cdots \int_{-1}^{1} (y_1 \cdots y_n T_{n^2+n}) \, \mathrm{d}y_1 \ldots \mathrm{d}y_n$$

coincide with the permanent of the matrix $[x_{li}]$. The corresponding polynomial sequences are therefore complete (unless $\operatorname{char}(k) = 2$), hence not P-computable by assumption.

Without using any hypothesis we can only prove a much weaker result [12]: for any n there is a polynomial $F(x_1, \ldots, x_{2n}, y)$ such that

$$L(F) = 2n - 1, \qquad L(\text{coefficient of } y^n \text{ in } F) \asymp n \log n.$$

(Take $F = (y - x_1) \cdots (y - x_{2n})$ and use (7.0).)

It is easy to see that substitution of polynomials into polynomials and the application of a fixed differential or integral operator does not lead out of the class of P-computable polynomial sequences. Valiant [207] shows that the class of P-definable sequences is closed under *all* operations considered above. Moreover, the analogue of the Meyer–Stockmeyer hierarchy (see Chapter 2 of this Handbook) collapses.

PROBLEM **1.** A function t: $\mathbb{N} \to \mathbb{N}$ is called QP-bounded iff $t(n) = 2^{(\log n)^{O(1)}}$. Let $\text{char}(k) \neq 2$. Decide whether there exists a QP-bounded t such that $Perm_{n \times n}[x_{ij}]$ can be derived from $Det_{t(n) \times t(n)}[y_{rs}]$ through substitution by indeterminates and constants. (If there exists no such t, Valiant [206] implies that $L(Perm_{n \times n}[x_{ij}])$ is not QP-bounded in n. In particular *Perm*, *HC* and the other complete polynomial sequences given in Valiant [206] are not P-computable then.) The problem of deriving the permanent from the determinant by substitution is classical. The best result known says that the $n \times n$ permanent is not a projection of the $m \times m$ determinant if $m \leqslant \sqrt{2}n - 6\sqrt{n}$ (see von zur Gathen [70]).

PROBLEM **2.** Hartmann [91] has determined the location of various immanents (see Littlewood [134]) between complete and P-computable. Prove or disprove the following conjecture: under Valiant's hypothesis the sequence of immanents belonging to partitions of the form (n^n) is not P-computable.

15. Conclusion

There has been continuous research in algebraic complexity since the sixties. Outstanding problems, such as matrix multiplication, the Discrete Fourier Transform and the evaluation of the permanent remain unsolved. Today (i.e. 1990) however, one has a much clearer conception of their significance and of the difficulties to be expected in their investigation. The prognosis that these problems are going to be solved in the context of algebraic geometry is certainly controversial—it would have been inconceivable twenty years ago.

Perhaps more than any other area, the one presented here is shaped by the tension between algebraic and algorithmic thinking. Perhaps more than any other algorithmic discipline it resists the conventional classification into computer science, pure or applied mathematics. While it draws its deeper methods of proof from algebra, logic and combinatorics, it owes to numerical analysis a treasure of beautiful and important problems, as well as many insights into their structure. And in its open-minded spirit it clearly belongs to computer science.

References

[1] AHO, A.V., J.E. HOPCROFT and J.D. ULLMANN, *The Design and Analysis of Computer Algorithms* (Addison Wesley, Reading, MA, 1974).

[2] ALDER, A. and V. STRASSEN, On the algorithmic complexity of associative algebras, *Theoret. Comput. Sci.* **15** (1981) 201–211.

[3] ALT, H. and J. VAN LEEUWEN, Complexity of basic complex operations, *Comput.* **27** (1981) 205–215.

[4] ATKINSON, M.D., The complexity of group algebra computations, *Theoret. Comput. Sci.* **1** (1977) 205–209.

[5] ATKINSON, M.D., and S. LLOYD, Bounds on the ranks of some 3-tensors, *Linear Algebra Appl.* **31** (1980) 19–31.

[6] BABAI, L., Monte-Carlo algorithms in graph isomorphism testing, Tech. Report 79-10, Dept. of Mathematics and Statistics, Univ. of Montreal, Montreal, 1979.

[7] BABAI, L., Moderately exponential bound for graph isomorphism, in: *Proc. Conf. Fundamentals of Computation Theory*, Lecture Notes in Computer Science, Vol. 117 (Springer, Berlin, 1981) 34–50.

[8] BABAI, L., D.-YU. GRIGORYEV and D.M. MOUNT, Isomorphism of graphs with bounded eigenvalue multiplicity, in: *Proc. 14th. Ann. ACM Symp. Theory of Computing* (1982) 310–324.

[9] BABAI, L., P.J. CAMERON and P.P. PÀLFY, On the order of primitive permutation groups with bounded nonabelian composition factors, to appear.

[10] BAUR, W., On the algebraic complexity of rational iteration procedures, Preprint, Univ. Konstanz, 1988.

[11] BAUR, W. and M.O. RABIN, Linear disjointness and algebraic complexity, in: *Logic and Algorithmic: an International Sympsosium held in honour of Ernst Specker*, Monographies de l'Enseignement Mathématique, Vol. 30 (Univ. Genève, Geneva, 1982) 35–46.

[12] BAUR, W. and V. STRASSEN, The complexity of partial derivatives, *Theoret. Comput. Sci.* **22** (1982) 317–330.

[13] BELAGA, E.G., Evaluation of polynomials of one variable with preliminary processing of the coefficients, *Problemy Kibernet.* **5** (1961) 7–15.

[14] BEN-OR, M., Probabilistic algorithms in finite fields, in: *Proc. 22nd IEEE Symp. Foundations of Computer Science* (1981) 394–398.

[15] BEN-OR, M., Lower bounds for algebraic computation trees, in: *Proc. 15th Ann. ACM Symp. Theory of Computing* (1983) 80–86.

[16] BEN-OR, M., D. KOZEN and J. REIF, The complexity of elementary algebra and geometry, *J. Comput. System Sci.* **32** (1986) 251–264.

[17] BERLEKAMP, E.R., Factoring polynomials over finite fields, *Bell Systems Tech. J.* **46** (1967) 1853–1859.

[18] BERLEKAMP, E.R., Factoring polynomials over large finite fields, *Math. Comp.* **24**(111) (1970) 713–735.

[19] BERMAN, L., The complexity of logical theories, *Theoret. Comput. Sci.* **11** (1980) 71–77.

[20] BETH, T., *Verfahren der schnellen Fourier-Transformation* (Teubner, Stuttgart, 1984).

[21] BETH, T., On the computational complexity of the general discrete Fourier transform, *Theoret. Comput. Sci.* **51** (1987) 331–339.

[22] BINI, D., Relations between EC-algorithms and APA-algorithms, applications, Nota Interna B79/8, I.E.I. Pisa, 1979.

[23] BINI, D., Border rank of a $p \times q \times 2$ tensor and the optimal approximation of a pair of bilinear forms, in: *Proc. Internat. Coll. Automata Languages and Programming*, Lecture Notes in Computer Science, Vol. 85 (Springer, Berlin, 1980) 98–108.

[24] BINI, D., A note on commutativity and approximation, Nota Interna B81 9, I.E.I. Pisa, 1981.

[25] BINI, D., M. CAPOVANI, G. LOTTI and F. ROMANI, $O(n^{2.7799})$ complexity for matrix multiplication, *Inform. Process. Lett.* **8** (1979) 234–235.

[26] BLUESTEIN, L.I., A linear filtering approach to the computation of the discrete Fourier transform, *IEEE Trans. Electroacoustics* **18** (1970) 451–455.

[27] BORODIN, A. and S. COOK, On the number of additions to compute specific polynomials, *SIAM J. Comput.* **5** (1976) 146–157.

[28] BORODIN, A. and I. MUNRO, Evaluating polynomials at many points, *Inform. Process. Lett.* **1** (1971) 66–68.

[29] BORODIN, A. and I. MUNRO, *Computational Complexity of Algebraic and Numeric Problems* (American Elsevier, New York, 1975).

[30] BRAUER, A., On addition chains, *Bull. Amer. Math. Soc.* **45** (1939) 736–739.

[31] BRENT, R.P., Multiple-precision zero-finding methods and the complexity of elementary function evaluation, in: J.F. Traub, ed., *Analytic Computational Complexity* (Academic Press, New York, 1976).

[32] BROCKETT, R.W. and D. DOBKIN, On the number of multiplications required for matrix multiplication, *SIAM J. Comput.* **5** (1976) 624–628.

[33] BROCKETT, R.W. and D. DOBKIN, On the optimal evaluation of a set of bilinear forms, *Linear Algebra Appl.* **19** (1978) 207–235.

[34] BROWN, M.R. and D. DOBKIN, An improved lower bound on polynomial multiplication, *IEEE Trans. Comput.* **29** (1980) 337–340.

[35] BROWNAWELL, W.D., Bounds for the degree in the Nullstellensatz, *Ann. of Math.* (2) **126**(3) (1987) 287–290.

[36] BUCHBERGER, B., Gröbner bases: an algorithmic method in polynomial ideal theory, in: N.K. Bose, ed., *Multidimensional Systems Theory* (Reidel, Dordrecht, 1985) Chapter 6.

[37] BUNCH, J. and J. HOPCROFT, Triangular factorization and inversion by fast matrix multiplication, *Math. Comp.* **28** (1974) 231–236.

[38] BÜCHI, W. and M. CLAUSEN, On a class of primary algebras of minimal rank, *Linear Algebra Appl.* **69** (1985) 249–268.

[39] BÜRGISSER, P., Doctoral Dissertation, Univ. Konstanz, in preparation.

[40] CANIGLIA, L., A. GALLIGO and J. HEINTZ, Some new effectivity bounds in computational geometry, in: *Applied Algebra, Algebraic Algorithms and Error Correcting Codes, Proc. 6th Internat. Conf.*, Lecture Notes in Computer Science, Vol. 357 (Springer, Berlin, 1989) 131–151.

[41] CANIGLIA, L., A. GALLIGO and J. HEINTZ, Borne simple exponentielle pour les degrés dans le théorème des zéros sur un corps de caractéristique quelconque, *C.R. Acad. Sci. Paris. Sér. I Math.* **309** (1988) 255–258.

[42] CANTOR, D.G. and E. KALTOFEN, Fast multiplication of polynomials over arbitrary rings, *Acta Inform.*, to appear.

[43] CHISTOV, A.L., Efficient factorization of polynomials over local fields, *Soviet. Math. Dokl.* **35**(2) (1984) 430–433.

[44] CHISTOV, A.L. and D.YU. GRIGORYEV, Polynomial time factoring of the multivariable polynomials over a global field, Steklov Mathematical Institute, USSR Academy of Sciences, Leningrad, 1982.

[45] CHISTOV, A.L. and D.YU. GRIGORYEV, Subexponential-time solving systems of algebraic equations I and II, LOMI Preprints, E-9-1983, E-10-1983, Leningrad, 1983.

[46] CLAUSEN, M., Fast generalized Fourier transforms, *Theoret. Comput. Sci.* **67** (1989) 55–63.

[47] CLAUSEN, M., Fast Fourier-transforms for metabelian groups, *SIAM J. Comput.*, to appear.

[48] COLLINS, G.E., Computer algebra of polynomials and rational functions, *Amer. Math. Monthly* **80**(7) (1973) 725–755.

[49] COOLEY, J.W. and J.W. TUKEY, An algorithm for the machine calculation of complex Fourier series, *Math. Comp.* **19**(90) (1965) 297–301.

[50] COPPERSMITH, D., Rapid multiplication of rectangular matrices, *SIAM J. Comput.* **11**(3) (1982) 467–471.

[51] COPPERSMITH, D. and S. WINOGRAD, On the asymptotic complexity of matrix multiplication, *SIAM J. Comput.* **11** (1982) 472–492.

[52] COPPERSMITH, D. and S. WINOGRAD, Matrix multiplication via arithmetic progression, in: *Proc. 19th Ann. ACM Symp. Theory of Computing* (1987) 1–6.

[53] DAVENPORT, J.H. and J. HEINTZ, Real quantifier elimination is doubly exponential, *J. Symbolic Comput.* **5** (1988) 29–35.

[54] DOBKIN, D., On the arithmetic complexity of a class of arithmetic computations, Thesis, Aiken Comput. Lab., Harvard Univ., 1973.

[55] DOBKIN, D. and R.J. LIPTON, A lower bound of n^2 on linear search programs for the Knapsack problem, *J. Comput. System Sci.* **16** (1978) 413–417.

[56] FEIG, E., On systems of bilinear forms whose minimal divisor-free algorithms are all bilinear, *J. Algorithms* **2** (1981) 261–281.

[57] FERRANTE, J. and C.W. RACKOFF, *The Computational Complexity of Logical Theories*, Lecture Notes in Mathematics, Vol. 718 (Springer, Berlin, 1979).

[58] FIDUCCIA, C., Polynomial evaluation via the division algorithm: the fast Fourier transform revisited, in: *Proc. 4th Ann. ACM Symp. Theory of Comping* (1972) 88–93.

[59] FIDUCCIA, C. and I. ZALCSTEIN, Algebras having linear multiplicative complexity, *J. ACM* **24** (1977) 311–331.

[60] FILOTTI, I.S. and J.N. MAYER, A polynomial time algorithm for determining isomorphism of graphs of fixed genus, in: *Proc. 12th Ann. ACM Symp. Theory of Computing* (1980) 236–243.

[61] FISCHER, M.J. and M. RABIN, Super-exponential complexity of Presburger arithmetic, in: R.M. Karp,

ed., *Complexity of Computation* (Amer. Mathematical Soc., Providence, RI, 1974) 27–41.
[62] FRIEDL, K. and L. RÓNYAI, Polynomial time solutions of some problems in computational algebra, in: *Proc. 17th Ann. ACM Symp. Theory of Computing* (1985) 153–162.
[63] FÜRER, M., The complexity of Presburger arithmetic with bounded quantifier alternation depth, *Theoret. Comput. Sci.* **18** (1982) 105–111.
[64] FÜRER, M., W. SCHNYDER and E. SPECKER, Normal forms for trivalent graphs and graphs of bounded valence, in: *Proc. 15th Ann. ACM Symp. Theory of Computing* (1983) 161–179.
[65] FURST, M., J. HOPCROFT and E. LUKS, Polynomial-time algorithms for permutation groups, in: *Proc. 21st IEEE Symp. Foundations of Computer Science* (1980) 36–41.
[66] GABBER, O. and Z. GALIL, Explicit constructions of linear superconcentrators, *J. Comput. System Sci.* **22** (1981) 407–420.
[67] GARTENBERG, P.A., Fast rectangular matrix multiplication, Ph.D. Thesis, Dept. of Mathematics, Univ. of California, Los Angeles, 1985.
[68] VON ZUR GATHEN, J., Irreducibility of multivariate polynomials, *J. Comput. System Sci.* **31** (1985) 225–264.
[69] VON ZUR GATHEN, J., Feasible arithmetic computations, *J. Symbolic Comput.* **4** (1987) 137–172.
[70] VON ZUR GATHEN, J., Permanent and determinant, *Linear Algebra Appl.* **96** (1987) 87–100.
[71] VON ZUR GATHEN, J. and E. KALTOFEN, Factoring sparse multivariate polynomials, *J. Comput. System Sci.* **31** (1985) 265–287.
[72] VON ZUR GATHEN, J. and V. STRASSEN, Some polynomials that are hard to compute, *Theoret. Comput. Sci.* **11**(3) (1980) 331–336.
[73] GRIESSER, B., Lower bounds for the approximative complexity, *Theoret. Comput. Sci.* **46** (1986) 329–338.
[74] GRIESSER, B., A lower bound for the border rank of a bilinear map, *Calcolo* **23**(2) (1986) 105–114.
[75] GRIGORYEV, D.YU., Some new bounds on tensor rank, LOMI preprint E-2-1978, Leningrad, 1978.
[76] GRIGORYEV, D.YU., Multiplicative complexity of a pair of bilinear forms and polynomial multiplication, in: *Proc. 7th Conf. Mathematical Foundations of Computer Science*, Lecture Notes in Computer Science, Vol. 64 (Springer, Berlin, 1978) 250–256.
[77] GRIGORYEV, D.YU., Notes of Scientific Seminars of LOMI, Vol. 118, Leningrad (1982) 25–82.
[78] GRIGORYEV, D.YU., The complexity of the decision for the first order theory of algebraically closed fields, *Math. USSR-Izv.* **29**(2) (1984) 459–475.
[79] GRIGORYEV, D.YU., Computational complexity in polynomial algebra, in: *Proc. Internat. Congress of Mathematics* (1987) 1452–1460.
[80] GRIGORYEV, D.YU., The complexity of deciding Tarski Algebra, *J. Symbolic Comput.* **5** (1988) 65–108.
[81] GRIGORYEV, D.YU. and M. KARPINSKI, A deterministic algorithm for rational function interpolation, *SIAM J. Comput.* (1988) (submitted).
[82] GRIGORYEV, D.YU. and N.N. VOROBYOV JR, Solving systems of polynomial inequalities in subexponential time, *J. Symbolic Comput.* **5** (1988) 37–64.
[83] DE GROOTE, H.F., On the complexity of quaternion multiplication, *Inform. Process. Lett.* **3** (1975) 177–179.
[84] DE GROOTE, H.F., On varieties of optimal algorithms for the computation of bilinear mappings II: optimal algorithms for 2×2-matrix multiplication, *Theoret. Comput. Sci.* **7** (1078) 127–148.
[85] DE GROOTE, H.F., *Lectures on the Complexity of Bilinear Problems*, Lecture Notes in Computer Science, Vol. 245 (Springer, Berlin, 1987).
[86] DE GROOTE, H.F., Characterization of division algebras of minimal rank and the structure of their algorithm varieties, *SIAM J. Comput.* **12**(1) (1983) 101–117.
[87] DE GROOTE, H.F. and J. HEINTZ, Commutative algebras of minimal rank, *Linear Algebra Appl.* **55** (1983) 37–68.
[88] DE GROOTE, H.F. and J. HEINTZ, A lower bound for the bilinear complexity of some semisimple Lie-algebras, in: *Proc. AAECC-3*, Lecture Notes in Computer Science, Vol. 229 (Springer, Berlin, 1986) 211–227.
[89] DE GROOTE, H.F., J. HEINTZ, S. MÖHLER and H. SCHMIDT, On the complexity of Lie algebras, in: *Proc. Fundamentals of Computation Theory*, Lecture Notes in Computer Science, Vol. 278 (Springer, Berlin, 1986) 172–179.

[90] HALL JR, M., An algorithm for distinct representatives, *Amer. Math. Monthly* **63** (1956) 716–717.
[91] HARTMANN, W., On the complexity of immanants, *Linear and Multilinear Algebra* **18** (1985) 127–140.
[92] HARTMANN, W., On the multiplicative complexity of modules over associative algebras, *SIAM J. Comput.* **14**(2) (1985) 383–395.
[93] HARTMANN, W. and P. SCHUSTER, Multiplicative complexity of some rational functions, *Theoret. Comput. Sci.* **10** (1980) 53–61.
[94] HEINTZ, J., Definability bounds of first order theories of algebraically closed fields (abstract), in: *Proc. Fundamentals of Computation Theory FCT '79* (1979) 160–166.
[95] HEINTZ, J., Definability and fast quantifier elimination in algebraically closed fields, *Theoret. Comput. Sci.* **24** (1983) 239–277.
[96] HEINTZ, J., A note on polynomials with symmetric Galois group which are easy to compute, *Theoret. Comput. Sci.* **47** (1986) 99–105.
[97] HEINTZ, J. and J. MORGENSTERN, On associative algebras of minimal rank, in: *Proc. AAECC-2*, Lecture Notes in Computer Science, Vol. 228 (Springer, Berlin, 1986) 1–24.
[98] HEINTZ, J. and C.P. SCHNORR, Testing polynomials which are easy to compute, *Logic and Algorithmic: an International Symposium held in honour of Ernst Specker*, Monographies de l'Enseignement Mathématique (Univ. Genève, Geneva, 1982) 237–254.
[99] HEINTZ, J. and M. SIEVEKING, Lower bounds for polynomials with algebraic coefficients, *Theoret. Comput. Sci.* **11** (1980) 321–330.
[100] HENRICI, P., *Applied and Computational Complex Analysis, Vols. I & II* (Wiley, New York, 1974/1977).
[101] HERMANN, G., Die Frage der endlich vielen Schritte in der Theorie der Polynomideale, *Math. Ann.* **95** (1926) 736–788.
[102] HOFFMANN, CH.M., *Group-theoretic Algorithms and Graph Isomorphism*, Lecture Notes in Computer Science, Vol. 136 (Springer, Berlin, 1982).
[103] HOPCROFT, J. and L. KERR, On minimizing the number of multiplications necessary for matrix multiplication, *SIAM J. Appl. Math.* **20** (1971) 30–36.
[104] HOPCROFT, J.E. and J. MUSINKSI, Duality applied to the complexity of matrix multiplications and other bilinear forms, *SIAM J. Comput.* **2** (1973) 159–173.
[105] HOROWITZ, E., A fast method for interpolation using preconditioning, *Inform. Process. Lett.* **1** (1972) 157–163.
[106] HOVANSKII, A.G., On a class of systems of transcendental equations, *Soviet Math. Dokl.* **22**(3) (1980) 762–765.
[107] HOWELL, T.D. and J.C. LAFON, The complexity of the quaternion product, Tech. Report 75-245, Dept. of Computer Science, Cornell Univ., Ithaca, NY, 1975.
[108] IRI, M., A very personal reminiscence of the problem of computational complexity, *Discrete Appl. Math.* **17** (1987) 17–27.
[109] JA'JA', J., Optimal evaluation of pairs of bilinear forms, in: *Proc. 10th Ann. ACM Symp. on Theory of Computing* (1978) 173–183.
[110] JAMES, G. and A. KERBER, *The Representation Theory of the Symmetric Group*, Encyclopedia of Mathematics and its Applications, Vol. 16 (Addison-Wesley, London, 1981).
[111] KALORKOTI, K., The trace invariant and matrix inversion, *Theoret. Comput. Sci.* **59** (1988) 277–286.
[112] KALTOFEN, E., Computing with polynomials given by straight-line programs II: sparse factorization, in: *Proc. 26th IEEE Symp. Foundations of Computer Science* (1985) 451–458.
[113] KALTOFEN, E., Uniform closure properties of *p*-computable functions, in: *Proc. 18th Ann. ACM Symp. Theory of Computing* (1986) 330–337.
[114] KALTOFEN, E., Factorization of polynomials given by straight-line programs, Preprint 02018-86 Mathematical Science Research Institute, Berkeley, CA, 1986; to appear in: S. Micali, ed., *Randomness in Computation, Advances in Computing Research* (JAI Press Inc., Greenwich).
[115] KAMINSKI, M. and N.H. BSHOUTY, Multiplicative complexity of polynomial multiplication over finite fields, Preprint, Technion, Haifa.
[116] KARPOVSKI, M.G., Fast Fourier transforms on finite non-abelian groups, *IEEE Trans. Comput.* **26** (10) (1977) 1082–1030.
[117] KELLER-GEHRIG, W., Fast algorithms for the characteristic polynomial, *Theoret. Comput. Sci.* **36** (1985) 309–317.

[118] KLAWE, M., Limitations on explicit constructions of expanding graphs, Preprint, Computer Science Dept., IBM Research, San Jose, USA, 1983.

[119] KNUTH, D.E., The analysis of algorithms, in: *Actes du Congrès International des Mathématiciens, Tome 3* (1970) 269–274.

[120] KNUTH, D.E., *The Art of Computer Programming, Vol. II: Seminumerical Algorithms* (Addison-Wesley, Reading, MA, 2nd ed., 1980).

[121] KRONECKER, L., *Grundzüge einer Arithmetischen Theorie der Algebraischen Grössen* (Riemer, Berlin, 1882).

[122] KUNG, H.T., On computing reciprocals of power series, *Numer. Math.* **22** (1974) 341–348.

[123] LAFON, J.C. and W. WINOGRAD, A lower bound for the multiplicative complexity of the product of two matrices, Unpublished manuscript, 1978.

[124] LANDAU, S., Factoring polynomials over algebraic number fields, *SIAM J. Comput.* **14**(1) (1985) 184–195.

[125] VAN LEEUWEN, J., and P. VAN EMDE BOAS, Some elementary proofs of lower bounds in complexity theory, *Linear Algebra Appl.* **19** (1978) 63–80.

[126] LENSTRA, A.K., H.W. LENSTRA and L. LOVÀSZ, Factoring polynomials with rational coefficients, *Math. Ann.* **261** (1982) 515–534.

[127] LENSTRA, A.K., Factoring polynomials over algebraic number fields, in: *Proc. Eurocal*, Lecture Notes Computer Science, Vol. 162 (Springer, Berlin, 1984) 458–465.

[128] LICKTEIG, T., Typical tensorial rank, *Linear Algebra Appl.* **69** (1985) 95–120.

[129] LICKTEIG, T., The computational complexity of the division in quadratic extension fields, *SIAM J. Comput.* **16** (1987) 278–311.

[130] LICKTEIG, T., Gaussian elimination is optimal for solving linear equations in dimension two, *Inform. Process. Lett.* **22** (1986) 277–279.

[131] LICKTEIG, T., Testing polynomials to zero, Preprint, Univ. Tübingen, 1989.

[132] LIPTON, R.J., Polynomials with 0-1 coefficients that are hard to evaluate, in: *Proc. 16th Ann. IEEE Symp. Foundations of Computer Science* (1975) 6–10.

[133] LIPTON, R.J. and L.J. STOCKMEYER, Evaluation of polynomials with superpreconditioning, in: *Proc. 8th Ann. ACM Symp. Theory of Computing* (1976) 174–180.

[134] LITTLEWOOD, D., *The Theory of Group Characters and Matrix Representations of Groups* (Clarendon Press, Oxford, 1950).

[135] LOOS, R., Generalized polynomial remainder sequences, *Comput. Suppl.* **4** (1982) 115–137.

[136] LOTTI, G. and F. ROMANI, On the asymptotic complexity of rectangular matrix multiplication, *Theoret. Comput. Sci.* **23** (1982) 1–15.

[137] LUKS, E.M., Isomorphism of graphs of bounded valence can be tested in polynomial time, *J. Comput. System Sci.* **25** (1982) 42–65.

[138] MARGULIS, G., Explicit constructions of concentrators, English translation in: *Problems of Information Transmission* (Plenum, New York, 1975).

[139] MAYR, E.W. and A.R. MEYER, The complexity of the word problem for commutative semigroups and polynomial ideals, *Adv. in Math.* **46** (1982) 305–329.

[140] MEYER, A.R. and L.J. STOCKMEYER, The equivalence problem for regular expressions with squaring requires exponential time, in: *Proc. 13th Ann. IEEE Symp. Switching and Automata Theory* (1972) 125–129.

[141] MILNOR, J., On the Betti numbers of real varieties, *Proc. Amer. Math. Soc.* **15** (1964) 275–280.

[142] MOENCK, R. and A. BORODIN, Fast modular transform via division, in: *Proc. 13th Ann. IEEE Symp. Switching and Automata Theory* (1972) 90–96.

[143] MOENCK, R., Fast computation of GCD's, in: *Proc. 5th Ann. ACM Symp. Theory of Computing* (1973) 142–151.

[144] MORGENSTERN, J., Note on a lower bound of the linear complexity of the fast Fourier transform, *J. ACM* **20** (2) (1973) 305–306.

[145] MOTZKIN, T.S., Evaluation of polynomials and evaluation of rational functions, *Bull. Amer. Math. Soc.* **61** (1955) 163.

[146] OSTROWSKI, A.M., On two problems in abstract algebra connected with Horner's rule, in: *Studies in Mathematics and Mechanics presented to Richard von Mises* (Academic Press, New York, 1954) 40–48.

[147] PAN, V.YA., Methods of computing values of polynomials, *Russian Math. Surveys* **21**(1) (1966) 105–136.

[148] PAN, V.YA., Strassen's algorithm is not optimal., in: *Proc. 19th Ann. IEEE Symp. Foundations of Computer Science* (1978) 166–176.

[149] PAN, V.YA., New fast algorithms for matrix operations *SIAM J. Comput.* **9** (1980) 321–342.

[150] PAN, V.YA., New combinations of methods for the acceleration of matrix multiplication, *Comput. Math. Appl.* **7** (1981) 73–125.

[151] PATERSON, M.S., Complexity of product and closure algorithms for matrices, in: *Proc. Internat. Congress of Mathematicians, Vol. 2* (1974) 483–489.

[152] PATERSON, M.S. and L. STOCKMEYER, Bounds of the evaluation time for rational polynomials, in: *IEEE Conf. Record 12th Ann. Symp. Switching and Automata Theory* (1971) 140–143.

[153] PINSKER, M., On the complexity of a concentrator, in: *Proc. 7th Internat. Teletraffic Conf.* (1973) 318/1–318/4.

[154] PIPPENGER, N., Superconcentrators, *SIAM J. Comput.* **6** (1978) 298–304.

[155] RABIN, M., Probabilistic algorithms in finite fields, *SIAM J. Comput.* **9** (1980) 273–280.

[156] RADER, C.M., Discrete Fourier transforms when the number of data samples is prime, *Proc. IEEE* **56** (1968) 1107-1108.

[157] REINGOLD, E.M. and I. STOCKS, Simple proofs of lower bounds for polynomial evaluation, in: R. Miller and J. Thatcher, eds., *Complexity of Computer Computations* (Plenum, New York, 1972) 21–30.

[158] RIFFELMACHER, D., Multiplicative complexity and algebraic structure, *J. Comput. System Sci.* **26** (1983) 92–106.

[159] RISLER, J.J., Additive complexity and zeros of real polynomials, *SIAM J. Comput.* **14** (1985) 178–183.

[160] RITZMANN, P., Ein numerischer Algorithmus zur Komposition von Potenzreihen und Komplexitätsschranken für die Nullstellenberechnung von Polynomen, Dissertation, Univ. Zürich, 1984.

[161] RITZMANN, P., A fast numerical algorithm for the composition of power series with complex coefficients, *Theoret. Comput. Sci.* **44** (1986) 1–16.

[162] ROMANI, F., Some properties of disjoint sums of tensors related to matrix multiplication, *SIAM J. Comput.* **11**(2) (1982) 263–267.

[163] RÓNYAI, L., Computing the structure of finite algebras, Tech. Report, Computer and Automation Institute, Hungarian Academy of Sciences, 1985.

[164] RÓNYAI, L., Zero divisors and invariant subspaces, Tech. Report IV/67, Computer and Automation Institute, Hungarian Academy of Sciences, 1985.

[165] RÓNYAI, L., Zero divisors in quaternion algebras, *J. of Algorithms* (submitted).

[166] RUTISHAUSER, H., *Vorlesungen über Numerische Mathematik, Band 1* (Birkhäuser, Basel, 1976).

[167] SAVAGE, J.E., *The Complexity of Computing* (Wiley, New York, 1976).

[168] SCHNORR, C.P., Improved lower bounds on the number of multiplications/divisions which are necessary to evaluate polynomials, *Theoret. Comput. Sci.* **7** (1978) 251–261.

[169] SCHNORR, C.P., An extension of Strassen's degree bound, *SIAM J. Comput.* **10** (1981) 371–382.

[170] SCHNORR, C.P. and J.P. VAN DE WIELE, On the additive complexity of polynomials, *Theoret. Comput. Sci.* **10** (1980) 1–18.

[171] SCHOLZ, A., *Jahresber. Deutsch. Math.-Verein* **47** (1937) 41–43.

[172] SCHÖNHAGE, A., Schnelle Berechnung von Kettenbruchentwicklungen, *Acta Inform.* **1**(1) (1971) 139–144.

[173] SCHÖNHAGE, A. and V. STRASSEN, Schnelle Multiplikation grosser Zahlen, *Comput.* **7** (1971) 281–292.

[174] SCHÖNHAGE, A., Unitäre Transformationen grosser Matrizen, *Numer. Math.* **20** (1973) 409–417.

[175] SCHÖNHAGE, A., A lower bound for the length of addition chains, *Theoret. Comput. Sci.* **1** (1975) 1–12.

[176] SCHÖNHAGE, A., An elementary proof for Strassen's degree bound, *Theoret. Comput. Sci.* **3** (1976) 267–272.

[177] SCHÖNHAGE, A., Schnelle Multiplikation von Polynomen über Körpern der Charakteristik 2, *Acta Inform.* **7** (1977) 395–398.

[178] SCHÖNHAGE, A., Partial and total matrix multiplication, *SIAM J. Comput.* **10** (1981) 343–455.

[179] SCHÖNHAGE, A., Equation solving in terms of computational complexity, in: *Proc. Internat. Congress of Mathematicians* (1987) 131–153.

[180] SCHUSTER, P., Interpolation und Kettenbruchentwicklung, Dissertation, Univ. Zürich, 1980.

[181] SHAW, M. and J.F. TRAUB, On the number of multiplications for the evaluation of a polynomial and some of its derivatives, *J. ACM* **21** (1974) 161–167.
[182] SIEVEKING, M., An algorithm for division of power series, *Comput.* **10** (1972) 153–156.
[183] SIMS, C.C., Computational methods in the study of permutation groups, in: J. Leech, ed., *Computational Problems in Abstract Algebra* (Pergamon, Oxford 1970) 169–183.
[184] SIMS, C.C., Some group-theoretic algorithms, in: Lecture Notes in Mathematics, Vol. 697 (Springer, Berlin, 1978) 108–124.
[185] SOLODOVNIKOV, V.I., Extension of Strassen's estimate to the solution of arbitrary systems of linear equations, *U.S.S.R. Comput. Maths. Math. Phys.* **19** (1979) 21–33.
[186] SPECKER, E. and V. STRASSEN, *Komplexität von Entscheidungsproblemen*, Lecture Notes in Computer Science, Vol. 43 (Springer, Berlin, 1976).
[187] STEELE, J. and A. YAO, Lower bounds for algebraic decision trees., *J. Algorithms* **3** (1982) 1–8.
[188] STOSS, J., The complexity of evaluating interpolation polynomials, *Theoret. Comput. Sci.* **41** (1985) 319–323.
[189] STRASSEN, V., Gaussian elimination is not optimal, *Numer. Math.* **13** (1969) 354–356.
[190] STRASSEN, V., Evaluation of rational functions, in: R. Miller and J. Thatcher, eds., *Complexity of Computer Computations* (Plenum, New York, 1972) 1–10.
[191] STRASSEN, V., Vermeidung von Divisionen, *Crelles J. Reine Angew. Math.* **264** (1973) 184–202.
[192] STRASSEN, V., Die Berechnungskomplexität von elementarsymmetrischen Funktionen und von Interpolationskoeffizienten, *Numer. Math.* **20**(3) (1973) 238–251.
[193] STRASSEN, V., Polynomials with rational coefficients which are hard to compute, *SIAM J. Comput.* **3**(2) (1974) 128–149.
[194] STRASSEN, V., Computational complexity over finite fields, *SIAM J. Comput.* **5**(2) (1976) 324–331.
[195] STRASSEN, V., The computational complexity of continued fractions, *SIAM J. Comput.* **12**(1) (1983) 1–27.
[196] STRASSEN, V., Rank and optimal computation of generic tensors, *Linear Algebra Appl.* **52** (1983) 645–685.
[197] STRASSEN, V., Algebraische Berechnungskomplexität, *Perspectives in Mathematics, Anniversary of Oberwolfach 1984* (Birkhäuser, Basel, 1984).
[198] STRASSEN, V., Relative bilinear complexity and matrix multiplication, *J. Reine Angew. Math.* **375/376** (1987) 406–443.
[199] STRASSEN, V., The asymptotic spectrum of tensors, *J. Reine Angew. Math.* **384** (1988) 102–152.
[200] TARSKI, A., *A Decision Method for Elementary Algebra and Geometry* (Univ. of California Press, Berkeley, CA, 1948, 2nd ed. 1951).
[201] THOM, R., Sur l'homologie des variétés algébriques réelles, in: S.S. Cairns, ed., *Differential and Combinatorial Topology. A Symposium in Honor of Marston Morse* (Princeton Univ. Press, Princeton, NJ, 1965) 255–265.
[202] TRAGER, B., Algebraic factoring and rational function integration, in: *Proc. 1976 ACM Symp. Symbolic and Algebraic Computation* (1976) 219–228.
[203] VALIANT, L.G., On non-linear lower bounds in computational complexity, in: *Proc. 7th Ann. ACM Symp. Theory of Computing* (1975) 45–53.
[204] VALIANT, L.G., The complexity of enumeration and reliability problems, *SIAM J. Comput.* **8**(3) (1979) 410–421.
[205] VALIANT, L.G., The complexity of computing the permanent, *Theoret. Comput. Sci.* **8** (1979) 189–201.
[206] VALIANT, L.G., Completeness classes in algebra, in: *Proc. 11th Ann. ACM Symp. Theory of Computing* (1979) 259–261.
[207] VALIANT, L.G., Reducibility by algebraic projections, in: *Logic and Algorithmic: an Internat. Symp. held in honour of Ernst Specker*, Monographies de l'Enseignement Mathématique, Vol. 30 (1982) 365–380.
[208] WEISPFENNING, V., The complexity of linear problems in fields, *J. Symbolic Comput.* **5** (1988) 3–28.
[209] WEYL, H., *The Classical Groups* (Princeton Univ. Press, Princeton, NJ, 1939/1946/1973).
[210] VAN DE WIELE, J.-P., Complexité additive et zéros des polynômes à coefficients réels et complexes, Rapport IRIA No. 292, INRIA, Rocquencourt, 1978.

[211] WINOGRAD, S., On the number of multiplications necessary to compute certain functions, *Comm. Pure Appl. Math.* **23** (1970) 164–179.

[212] WINOGRAD, S., On multiplication of 2×2 matrices, *Linear Algebra Appl.* **4** (1971) 381–388.

[213] WINOGRAD, S., On computing the discrete Fourier transform, *Proc. Nat. Acad. Sci. USA* **73** (1976) 1005–1006.

[214] WINOGRAD, S., Some bilinear forms whose multiplicative complexity depends on the field of constants, *Math. Systems Theory* **10** (1977) 169–180.

[215] ZASSENHAUS, H., On Hensel factorization I, *J. Number Theory* **1** (1969) 291–311.

CHAPTER 12

Algorithms in Number Theory

A.K. LENSTRA*

Department of Computer Science, The University of Chicago, Chicago, IL 60637, USA

H.W. LENSTRA, Jr

Department of Mathematics, University of California, Berkeley, CA 94720, USA

Contents

* Present affiliation: Bell Communications Research, 435 South Street, Morristown, NJ 07960, USA.

HANDBOOK OF THEORETICAL COMPUTER SCIENCE
Edited by J. van Leeuwen

1. Introduction

In this chapter we are concerned with algorithms that solve two basic problems in computational number theory: factoring integers into prime factors, and finding discrete logarithms.

In the factoring problem one is given an integer $n>1$, and one is asked to find the decomposition of n into prime factors. It is common to split this problem into two parts. The first is called *primality testing*: given n, determine whether n is prime or composite. The second is called *factorization*: if n is composite, find a nontrivial divisor of n.

In the discrete logarithm problem one is given a prime number p, and two elements h, y of the multiplicative group $\mathbf{F}_p^*$ of the field of integers modulo p. The question is to determine whether y is a power of h, and, if so, to find an integer m with $y=h^m$. The same problem can be posed for other explicitly given groups instead of $\mathbf{F}_p^*$.

We shall present a detailed survey of the best currently available algorithms to solve these problems, paying special attention to what is known, or believed to be true, about their time complexity. The algorithms and their analyses depend on many different parts of number theory, and we cannot hope to present a complete exposition from first principles. The necessary background is reviewed in the first few sections of the present chapter. The remaining sections are then devoted to the problems mentioned above. It will be seen that only the primality testing problem may be considered to be reasonably well solved. No satisfactory solution is known for the factorization problem and the discrete logarithm problem. It appears that these two problems are of roughly the same level of difficulty.

Number theory is traditionally believed to be the purest of all sciences, and within number theory the hunt for large primes and for factors of large numbers has always seemed particularly remote from applications, even to other questions of a number-theoretic nature. Most number theorists considered the small group of colleagues that occupied themselves with these problems as being inflicted with an incurable but harmless obsession. Initially, the introduction of electronic computers hardly changed this situation. The factoring community was provided with a new weapon in its eternal battle, and the fact that their exacting calculations could be used to test computing equipment hardly elevated their scientific status.

In the 1970s two developments took place that entirely altered this state of affairs. The first is the introduction of *complexity theory*, and the second is the discovery that computational number theory has applications in *cryptology*.

The formalism of complexity theory enabled workers in the field to phrase the fruits of their intellectual labors in terms of theorems that apply to more than a finite number of cases. For example, rather than saying that they proved certain specific numbers prime by means of a certain method, they could now say that the same method can be used to test *any* number n for primality within time $f(n)$, for some function f. Although this is doubtlessly a more respectable statement from a mathematical point of view, it turned out that such asymptotic assertions appealed mainly to theoretical computer scientists, and that many mathematicians had a tendency to regard these results as being of an exclusively theoretical nature, and of no interest for practical computations. It has since been interesting to observe that the practical validity of asymptotic time bounds increased with the speed of computers, and nowadays an algorithm is considered incomplete without a complexity analysis.

The area of number-theoretic complexity lost its exclusive function as a playground for theoretical computer scientists with the discovery, by Rivest, Shamir and Adleman [67], that the difficulty of factorization can be applied for cryptological purposes. We shall not describe this application, but we note that for the construction of the cryptographic scheme that they proposed it is important that primality testing is easy, and that for the unbreakability of the scheme it is essential that factorization is hard. Thus, as far as factorization is concerned, this is a *negative* application: a break-through might make the scheme invalid and, if not restore the purity of computational number theory, at least clear the way for applications that its devotees would find more gratifying.

It is important to point out that there is only *historical* evidence that factorization is an intrinsically hard problem. Generations of number theorists, a small army of computer scientists, and legions of cryptologists spent a considerable amount of energy on it, and the best they came up with are the relatively poor algorithms that Section 4 will be devoted to. Of course, as long as the widely believed $P \neq NP$-conjecture remains unproved, complexity theory will not have fulfilled its originally intended mission of proving certain algorithmic problems to be intrinsically hard; but with factorization the situation is worse, since even the celebrated conjecture just mentioned has no implications about its intractability. Factorization is considered easier than NP-complete and although the optimistic conjecture that it might be doable in polynomial time is only rarely publicly voiced, it is not an illegitimate hope to foster.

Proving *upper bounds* for the running time of number-theoretic algorithms also meets with substantial difficulties. We shall see that in many cases we have to be satisfied with results that depend on certain heuristic assumptions, of which the rigorous confirmation must perforce be left to posterity.

Several other applications of computational number theory in cryptology have been found, a prominent role being played by the discrete logarithm problem that we formulated above. For more information about these applications we refer to [12, 53]. Although the discrete logarithm problem has classically attracted less attention than the factoring problem, it does have a venerable history, see [27, Chapter VIII], [35, 81]. The methods that have been proposed for its solution are also important for factorization algorithms, and we discuss them in Section 3. What we have said above about the complexity of factorization applies to the discrete logarithm problem as well.

Many more problems than those that we deal with would fit under the heading *algorithms in number theory*, and we have preferred a thorough treatment of a few representative topics over a more superficial discussion of many. As guides for subjects that we left out we mention Knuth's book [37, Chapter 4] and the collection of articles published in [47]. Up-to-date information can often be traced through the current issues of *Mathematics of Computation*. An important subject that is much different in spirit is *computational geometry of numbers*, in particular the basis reduction algorithm of Lovász [43]. For a discussion of this area and its applications in linear programming and combinatorial optimization we refer to [31, 72].

Throughout this paper, *time* will mean *number of bit operations*. We employ the following notation. By $\mathbf{Z}$ we denote the ring of integers, and by $\mathbf{R}$ the set of real numbers. For a positive integer n we denote by $\mathbf{Z}/n\mathbf{Z}$ the ring of integers modulo n. For a prime power q, the finite field containing q elements is denoted by $\mathbf{F}_q$, and its multiplicative group by $\mathbf{F}_q^*$; notice that for a prime number p we have that $\mathbf{F}_p \cong \mathbf{Z}/p\mathbf{Z}$. The number of primes $\leqslant x$ is denoted by $\pi(x)$; the function π is called the prime counting function.

2. Preliminaries

Subsections 2.A–2.D contain some background for the material presented in the remainder of this chapter. We suggest that the reader only consults one of these first four subsections as the need arises.

2.A. Smoothness

In many of the algorithms that we will present, the notion of *smoothness* will play an important role. We say that an integer is *smooth with respect to y*, or *y-smooth*, if all its prime factors are $\leqslant y$. In what follows, we will often be interested in the probability that a random integer between 1 and x is smooth with respect to some y.

To derive an expression for this probability, we define $\psi(x, y)$ as the number of positive integers $\leqslant x$ that are smooth with respect to y. Lower and upper bounds for $\psi(x, y)$ are known from [15, 25]. Combination of these results yields the following. For a fixed arbitrary $\varepsilon > 0$, we have that for $x \geqslant 10$ and $u \leqslant (\log x)^{1-\varepsilon}$,

$$\psi(x, x^{1/u}) = x \cdot u^{-u + f(x,u)},$$

for a function f that satisfies $f(x, u)/u \to 0$ for $u \to \infty$ uniformly in x. For fixed $\alpha, \beta \in \mathbf{R}_{>0}$ we find that for $n \to \infty$

$$\psi(n^\alpha, n^{\beta\sqrt{(\log\log n)/\log n}}) = n^\alpha \cdot ((\alpha/\beta)\sqrt{\log n/\log\log n})^{-(1+o(1))(\alpha/\beta)\sqrt{\log n/\log\log n}},$$

which can conveniently be written as

$$\psi(n^\alpha, L(n)^\beta) = n^\alpha \cdot L(n)^{-\alpha/(2\beta)+o(1)}$$

where $L(n) = e^{\sqrt{\log n \log\log n}}$. It follows that a random positive integer $\leqslant n^\alpha$ is smooth with respect to $L(n)^\beta$ with probability $L(n)^{-\alpha/(2\beta)+o(1)}$, for $n \to \infty$.

For $\beta \in \mathbf{R}$ we will often write $L_n[\beta]$ for $L(n)^\beta$, and we will abbreviate $L_n[\beta + o(1)]$ to $L_n[\beta]$, for $n \to \infty$. Notice that in this notation $L_n[\alpha] + L_n[\beta] = L_n[\max(\alpha, \beta)]$, and that the prime counting function π satisfies $\pi(L_n[\beta]) = L_n[\beta]$.

2.B. Elliptic curves

We give an introduction to elliptic curves. For details and proofs we refer to [45, 75]. Our presentation is by no means conventional, but reflects the way in which we apply elliptic curves.

Let p be a prime number. The projective plane $\mathbf{P}^2(\mathbf{F}_p)$ over $\mathbf{F}_p$ consists of the equivalence classes of triples $(x, y, z) \in \mathbf{F}_p \times \mathbf{F}_p \times \mathbf{F}_p$, $(x, y, z) \neq 0$, where two triples (x, y, z) and (x', y', z') are equivalent if $cx = x'$, $cy = y'$, and $cz = z'$ for some $c \in \mathbf{F}_p^*$; the equivalence class containing (x, y, z) is denoted by $(x:y:z)$.

Now assume that p is unequal to 2 or 3. An *elliptic curve* over $\mathbf{F}_p$ is a pair $a, b \in \mathbf{F}_p$ for which $4a^3 + 27b^2 \neq 0$. These elements are to be thought of as the coefficients in the Weierstrass equation

(2.1) $$y^2 = x^3 + ax + b.$$

An elliptic curve a, b is denoted by $E_{a,b}$, or simply by E.

2.2. SET OF POINTS OF AN ELLIPTIC CURVE. Let E be an elliptic curve over $\mathbf{F}_p$. The *set of points* $E(\mathbf{F}_p)$ of E over $\mathbf{F}_p$ is defined by

$$E(\mathbf{F}_p) = \{(x:y:z) \in \mathbf{P}^2(\mathbf{F}_p) : y^2 z = x^3 + axz^2 + bz^3\}.$$

There is one point $(x:y:z) \in E(\mathbf{F}_p)$ for which $z = 0$, namely the *zero point* $(0:1:0)$, denoted by O. The other points of $E(\mathbf{F}_p)$ are the points $(x:y:1)$, where $x, y \in \mathbf{F}_p$ satisfy (2.1). The set $E(\mathbf{F}_p)$ has the structure of an *abelian group*. The group law, which we will write additively, is defined as follows.

2.3. THE GROUP LAW. For any $P \in E(\mathbf{F}_p)$ we define $P + O = O + P = P$. For non-zero $P = (x_1:y_1:1)$, $Q = (x_2:y_2:1) \in E(\mathbf{F}_p)$ we define $P + Q = O$ if $x_1 = x_2$ and $y_1 = -y_2$. Otherwise, the sum $P + Q$ is defined as the point $(x:-y:1) \in E(\mathbf{F}_p)$ for which (x, y) satisfies (2.1) and lies on the line through (x_1, y_1) and (x_2, y_2); if $x_1 = x_2$, we take the tangent line to the curve in (x_1, y_1) instead. With $\lambda = (y_1 - y_2)/(x_1 - x_2)$ if $x_1 \neq x_2$, and $\lambda = (3x_1^2 + a)/(2y_1)$ otherwise, we find that $x = \lambda^2 - x_1 - x_2$ and $y = \lambda(x - x_1) + y_1$. The proof that $E(\mathbf{F}_p)$ becomes an abelian group with this group law can be found in [75, Chapter 3].

2.4. THE ORDER OF $E(\mathbf{F}_p)$. The order $\# E(\mathbf{F}_p)$ of the abelian group $E(\mathbf{F}_p)$ equals $p + 1 - t$ for some integer t with $|t| \leqslant 2\sqrt{p}$, a theorem due to Hasse (1934). Conversely, a result of Deuring [26] can be used to obtain an expression for the number of times a given integer of the above form $p + 1 - t$ occurs as $\# E(\mathbf{F}_p)$, for a fixed p, where E ranges over all elliptic curves over $\mathbf{F}_p$. This result implies that for any integer t with $|t| < 2\sqrt{p}$ there is an elliptic curve E over $\mathbf{F}_p$ for which $\# E(\mathbf{F}_p) = p + 1 - t$. A consequence of this result that will prove to be important for our purposes is that $\# E(\mathbf{F}_p)$ is approximately uniformly distributed over the numbers near $p + 1$ if E is uniformly distributed over all elliptic curves over $\mathbf{F}_p$.

2.5. PROPOSITION (cf. [45, Proposition (1.16)]). *There are positive effectively computable constants c_1 and c_2 such that for any prime number $p \geqslant 5$ and any set S of integers s for which $|s - (p + 1)| < \sqrt{p}$ one has*

$$\frac{\# S - 2}{2[\sqrt{p}] + 1} \cdot c_1 (\log p)^{-1} \leqslant \frac{N}{p^2} \leqslant \frac{\# S}{2[\sqrt{p}] + 1} \cdot c_2 (\log p) \cdot (\log \log p)^2,$$

where N denotes the number of pairs $a, b \in \mathbf{F}_p$ that define an elliptic curve $E = E_{a,b}$ over $\mathbf{F}_p$ with $\#E(\mathbf{F}_p) \in S$.

Because N/p^2 is the probability that a random pair a, b defines an elliptic curve E over $\mathbf{F}_p$ for which $\#E(\mathbf{F}_p) \in S$, this proposition asserts that this probability is essentially equal to the probability that a random integer near p is in S.

2.6. COMPUTING THE ORDER OF $E(\mathbf{F}_p)$. For an elliptic curve E over $\mathbf{F}_p$ the number $\#E(\mathbf{F}_p)$ can be computed by means of the *division points method*, due to Schoof [71]. This method works by investigating the action of the Frobenius endomorphism on the l-division points of the curve, for various small prime numbers l. An l-division point is a point P over an extension of $\mathbf{F}_p$ for which $l \cdot P = O$, and the Frobenius endomorphism is the map sending $(x:y:z)$ to $(x^p:y^p:z^p)$. The division points method is completely deterministic, guaranteed to work if p is prime, and runs in $O((\log p)^8)$ bit operations (cf. [46]); with fast multiplication techniques this becomes $(\log p)^{5+o(1)}$. Its practical value is questionable, however.

Another method makes use of the *complex multiplication field*. The complex multiplication field L of an elliptic curve E with $\#E(\mathbf{F}_p) = p+1-t$ is defined as the imaginary quadratic field $\mathbf{Q}((t^2-4p)^{1/2})$ (cf. (2.4)). For certain special curves the field L is known; for instance for the curve $y^2 = x^3 + 4x$ and $p \equiv 1 \bmod 4$ we have $L = \mathbf{Q}(i)$, a fact that was already known to Gauss. Knowing L gives a fast way of computing $\#E(\mathbf{F}_p)$. Namely, suppose that L is known for some elliptic curve E; then the ring of integers A of L contains the zeros $\rho, \bar{\rho}$ of the polynomial $X^2 - tX + p$, and $\#E(\mathbf{F}_p) = (\rho - 1)(\bar{\rho} - 1)$. Although this polynomial is not known, a zero can be determined by looking for an element π in A for which $\pi\bar{\pi} = p$ (see (5.9)). This π can be shown to be unique up to complex conjugation and units in A. For a suitable unit u in A we then have that $\rho = u\pi$, so that $\#E(\mathbf{F}_p) = (u\pi - 1)(\bar{u}\bar{\pi} - 1)$. In most cases A will have only two units, namely 1 and -1; only if $L = \mathbf{Q}(i)$ (or $L = \mathbf{Q}(\sqrt{-3})$) we have four (or six) units in A. In the case that A has only the units 1 and -1, an immediate method to decide whether $\#E(\mathbf{F}_p)$ equals $(\pi - 1)(\bar{\pi} - 1) = m'$ or $(-\pi - 1)(-\bar{\pi} - 1) = m''$ does not yet exist, as far as we know; in practice one could select a random point $P \in E(\mathbf{F}_p)$ such that not both $m' \cdot P$ and $m'' \cdot P$ are equal to O, so that $\#E(\mathbf{F}_p) = m$ for the unique $m \in \{m', m''\}$ for which $m \cdot P = O$. If A contains four or six units there exists a more direct method [33, Chapter 18].

In (5.9) we will use this method in the situation where L, A, and p are known; the elliptic curve E will then be *constructed* as a function of L and p.

2.7. ELLIPTIC CURVES MODULO n. To motivate what follows, we briefly discuss elliptic curves modulo n, for a positive integer n. First we define what we mean by the projective plane $\mathbf{P}^2(\mathbf{Z}/n\mathbf{Z})$ over the ring $\mathbf{Z}/n\mathbf{Z}$. Consider the set of all triples $(x, y, z) \in (\mathbf{Z}/n\mathbf{Z})^3$ for which x, y, z generate the unit ideal of $\mathbf{Z}/n\mathbf{Z}$, i.e., the x, y, z for which $\gcd(x, y, z, n) = 1$. The group of units $(\mathbf{Z}/n\mathbf{Z})^*$ acts on this set by $u(x, y, z) = (ux, uy, uz)$. The orbit of (x, y, z) under this action is denoted by $(x:y:z)$, and $\mathbf{P}^2(\mathbf{Z}/n\mathbf{Z})$ is the set of all orbits.

We now restrict to the case that $\gcd(n, 6) = 1$. An elliptic curve $E = E_{a,b}$ modulo n is a pair $a, b \in \mathbf{Z}/n\mathbf{Z}$ for which $4a^3 + 27b^2 \in (\mathbf{Z}/n\mathbf{Z})^*$. It follows from Subsection 2.B that for any prime p dividing n, the pair $\bar{a}=a \bmod p$, $\bar{b}=b \bmod p$ defines an elliptic curve $E_{\bar{a},\bar{b}}$ over $\mathbf{F}_p$. The set of points of this latter curve will be denoted by $E(\mathbf{F}_p)$.

The set of points $E(\mathbf{Z}/n\mathbf{Z})$ of E modulo n is defined by

$$E(\mathbf{Z}/n\mathbf{Z})=\{(x:y:z)\in \mathbf{P}^2(\mathbf{Z}/n\mathbf{Z}): y^2z=x^3+axz^2+bz^3\}.$$

Clearly, for any $(x:y:z)\in E(\mathbf{Z}/n\mathbf{Z})$ and for any prime p dividing n, we have that $((x \bmod p):(y \bmod p):(z \bmod p))\in E(\mathbf{F}_p)$. It is possible to define a group law so that $E(\mathbf{Z}/n\mathbf{Z})$ becomes an abelian group, but we do not need this group structure for our purposes. Instead it suffices to define the following "pseudoaddition" on a subset of $E(\mathbf{Z}/n\mathbf{Z})$.

2.8. PARTIAL ADDITION ALGORITHM. Let $V_n \subset \mathbf{P}^2(\mathbf{Z}/n\mathbf{Z})$ consist of the elements $(x:y:1)$ of $\mathbf{P}^2(\mathbf{Z}/n\mathbf{Z})$ together with the zero element $(0:1:0)$, which will be denoted by O. For any $P\in V_n$ we define $P+O=O+P=P$. For non-zero $P=(x_1:y_1:1)$, $Q=(x_2:y_2:1)\in V_n$, and any $a\in \mathbf{Z}/n\mathbf{Z}$ we describe an addition algorithm that *either* finds a divisor d of n with $1<d<n$, *or* determines an element $R\in V_n$ that will be called the *sum* of P and Q:

(1) If $x_1=x_2$ and $y_1=-y_2$ put $R=O$ and stop.

(2) If $x_1\neq x_2$, perform step (2)(a), otherwise perform step (2)(b).

(2)(a) Use the Euclidean algorithm to compute $s,t\in \mathbf{Z}/n\mathbf{Z}$ such that $s(x_1-x_2)+tn=\gcd(x_1-x_2,n)$. If this gcd is not equal to 1, call it d and stop. Otherwise put $\lambda=s(y_1-y_2)$, and proceed to step (3). (It is not difficult to see that in this case $P=Q$.)

(2)(b) Use the Euclidean algorithm to compute $s,t\in \mathbf{Z}/n\mathbf{Z}$ such that $s(y_1+y_2)+tn=\gcd(y_1+y_2,n)$. If this gcd is not equal to 1, call it d and stop. Otherwise put $\lambda=s(3x_1^2+a)$, and proceed to step (3).

(3) Put $x=\lambda^2-x_1-x_2$, $y=\lambda(x-x_1)+y_1$, $R=(x:-y:1)$, and stop.

This finishes the description of the addition algorithm. Clearly the algorithm requires $O((\log n)^2)$ bit operations. Notice that this algorithm can be applied to any $P,Q\in V_n$, for any $a\in \mathbf{Z}/n\mathbf{Z}$, irrespective as to whether there exists $b\in \mathbf{Z}/n\mathbf{Z}$ such that a,b defines an elliptic curve modulo n with $P,Q\in E_{a,b}(\mathbf{Z}/n\mathbf{Z})$.

2.9. PARTIAL ADDITION WHEN TAKEN MODULO p. Let p be any prime dividing n, and let P_p denote the point of $\mathbf{P}^2(\mathbf{F}_p)$ obtained from $P\in V_n$ by reducing its coordinates modulo p.

Assume that, for some $a\in \mathbf{Z}/n\mathbf{Z}$ and $P,Q\in V_n$, the algorithm in (2.8) has been successful in computing the sum $R=P+Q\in V_n$. Let $\bar{a}$ denote $a \bmod p$, and suppose that there exists an element $b\in \mathbf{F}_p$ such that $4\bar{a}^3+27b^2\neq 0$ and such that $P_p, Q_p\in E_{\bar{a},b}(\mathbf{F}_p)$. It then follows from (2.3) and (2.8) that $R_p=P_p+Q_p$ in the group $E_{\bar{a},b}(\mathbf{F}_p)$.

Notice also that $P=O$ if and only if $P_p=O_p$, for $P\in V_n$.

2.10. MULTIPLICATION BY A CONSTANT. The algorithm in (2.8) allows us to multiply an element $P\in V_n$ by an integer $k\in \mathbf{Z}_{>0}$ in the following way. By repeated application of the addition algorithm in (2.8) for some $a\in \mathbf{Z}/n\mathbf{Z}$, we either find a divisor d of n with $1<d<n$, or determine an element $R=k\cdot P\in V_n$ such that according to (2.9) the following holds: for any prime p dividing n for which there exists an element $b\in \mathbf{F}_p$ such

that $4\bar{a}^3+27b^2\neq 0$ and $P_p\in E_{\bar{a},b}(\mathbf{F}_p)$, we have $R_p=k\cdot P_p$ in $E_{\bar{a},b}(\mathbf{F}_p)$ where $\bar{a}=a \bmod p$.

Notice that in the latter case $R_p=O_p$ if and only if the order of $P_p\in E_{\bar{a},b}(\mathbf{F}_p)$ divides k. But $R_p=O_p$ if and only if $R=O$, as we noted in (2.9), which is equivalent to $R_q=O_q$ for any prime q dividing n. We conclude that if $k\cdot P$ has been computed successfully, and if q is another prime satisfying the same conditions as p above, then k is a multiple of the order of P_p if and only if k is a multiple of the order of P_q.

By repeated duplications and additions, multiplication by k can be done in $O(\log k)$ applications of Algorithm (2.8), and therefore in $O((\log k)(\log n)^2)$ bit operations.

2.11. RANDOMLY SELECTING CURVES AND POINTS. In Subsection 5.C we will be in the situation where we suspect that n is prime and have to select elliptic curves E modulo n (in (5.7)) and points in $E(\mathbf{Z}/n\mathbf{Z})$ (in (5.6)) at random. This can be accomplished as follows. Assume that $\gcd(n,6)=1$. Randomly select $a,b\in\mathbf{Z}/n\mathbf{Z}$ until $4a^3+27b^2\neq 0$, and verify that $\gcd(n,4a^3+27b^2)=1$, as should be the case for prime n; per trial the probability of success is $(n-1)/n$, for n prime. The pair a,b now defines an elliptic curve modulo n, according to (2.7).

Given an elliptic curve $E=E_{a,b}$ modulo n, we randomly construct a point in $E(\mathbf{Z}/n\mathbf{Z})$. First, we randomly select an $x\in\mathbf{Z}/n\mathbf{Z}$ until x^3+ax+b is a square in $\mathbf{Z}/n\mathbf{Z}$. Because we suspect that n is prime, this can be done by checking whether $(x^3+ax+b)^{(n-1)/2}=1$. Next, we determine y as a zero of the polynomial $X^2-(x^3+ax+b)\in(\mathbf{Z}/n\mathbf{Z})[X]$ using for instance the probabilistic method for finding roots of polynomials over finite fields described in [37, Section 4.6.2]. The resulting point $(x{:}y{:}1)$ is in $E(\mathbf{Z}/n\mathbf{Z})$.

For these algorithms to work, we do not need a proof that n is prime, but if n is prime, they run in expected time polynomial in $\log n$.

2.C. Class groups

We review some results about class groups. For details and proofs we refer to [9, 70]. A polynomial $aX^2+bXY+cY^2\in\mathbf{Z}[X,Y]$ is called a *binary quadratic form*, and $\Delta=b^2-4ac$ is called its *discriminant*. We denote a binary quadratic form $aX^2+bXY+cY^2$ by (a,b,c). A form for which $a>0$ and $\Delta<0$ is called *positive*, and a form is *primitive* if $\gcd(a,b,c)=1$. Two forms (a,b,c) and (a',b',c') are *equivalent* if there exist $\alpha,\beta,\gamma,\delta\in\mathbf{Z}$ with $\alpha\delta-\beta\gamma=1$ such that $a'U^2+b'UV+c'V^2=aX^2+bXY+cY^2$, where $U=\alpha X+\gamma Y$, and $V=\beta X+\delta Y$. Notice that two equivalent forms have the same discriminant.

Now fix some negative integer Δ with $\Delta\equiv 0$ or $1 \bmod 4$. We will often denote a form (a,b,c) of discriminant Δ by (a,b), since c is determined by $\Delta=b^2-4ac$. The set of equivalence classes of positive, primitive, binary quadratic forms of discriminant Δ is denoted by C_Δ. The existence of the form $(1,\Delta)$ shows that C_Δ is nonempty.

2.12. REDUCTION ALGORITHM. It has been proved by Gauss that each equivalence class in C_Δ contains precisely one *reduced form*, where a form (a,b,c) is reduced if

$$\begin{cases}|b|\leqslant a\leqslant c,\\ b\geqslant 0 \quad\text{if } |b|=a \text{ or if } a=c.\end{cases}$$

These inequalities imply that $a\leqslant\sqrt{|\Delta|/3}$; it follows that C_Δ is finite. For any form (a,b,c)

of discriminant Δ we can easily find the reduced form equivalent to it by means of the following reduction algorithm:

(1) Replace (a, b) by $(a, b-2ka)$, where $k \in \mathbf{Z}$ is such that $-a<b-2ka \leqslant a$.
(2) If (a, b, c) is reduced, then stop; otherwise, replace (a, b, c) by $(c, -b, a)$ and go back to step (1).

It is easily verified that this is a polynomial-time algorithm. Including the observation made in [37, Exercise 4.5.2.30] in the analysis from [39], the reduction algorithm can be shown to take $O((\log a)^2+\log c)$ bit operations, where we assume that the initial b is already $O(a)$. It is not unlikely that with fast multiplication techniques one gets $O((\log a)^{1+\varepsilon}+\log c)$ by means of a method analogous to [69].

If the reduction algorithm applied to a form (a', b', c') yields the reduced form (a, b, c), then for any value $ax^2+bxy+cy^2$ a pair $u=\alpha x+\gamma y, v=\beta x+\delta y$ with $a'u^2+b'uv+c'v^2=ax^2+bxy+cy^2$ can be computed if we keep track of a (2×2)-transformation matrix in the algorithm. This does not affect the asymptotic running time of the reduction algorithm.

2.13. COMPOSITION ALGORITHM. The set C_Δ, which can now be identified with the set of reduced forms of discriminant Δ, is a finite abelian group, the *class group*. The group law, which we will write multiplicatively, is defined as follows. The inverse of (a, b) follows from an application of the reduction algorithm to $(a, -b)$, and the unit element 1_Δ is $(1, 1)$ for Δ odd, and $(1, 0)$ for Δ even. To compute $(a_1, b_1)\cdot(a_2, b_2)$, we use the Euclidean algorithm to determine $d=\gcd(a_1, a_2, (b_1+b_2)/2)$, and $r, s, t \in \mathbf{Z}$ such that $d=ra_1+sa_2+t(b_1+b_2)/2$. The product then follows from an application of the reduction algorithm to

$$(a_1a_2/d^2, b_2+2a_2(s(b_1-b_2)/2-tc_2)/d),$$

where $c_2=(b_2^2-\Delta)/(4a_2)$. It is again an easy matter to verify that this is a polynomial-time algorithm.

2.14. AMBIGUOUS FORMS. A reduced form is *ambiguous* if its square equals 1_Δ; for an ambiguous form we have $b=0$, or $a=b$, or $a=c$. From now on we assume that $\Delta\equiv 1 \bmod 4$. It was already known to Gauss that for these Δ's there is a bijective correspondence between ambiguous forms and factorizations of $|\Delta|$ into two relatively prime factors. For relatively prime p and q, the factorization $|\Delta|=pq$ corresponds to the ambiguous form (p, p) for $3p\leqslant q$, and to $((p+q)/4, (q-p)/2)$ for $p<q\leqslant 3p$. Notice that the ambiguous form $(1, 1)$ corresponds to the factorization $|\Delta|=1\cdot|\Delta|$.

2.15. THE CLASS NUMBER. The *class number* h_Δ of Δ is defined as the cardinality of the class group C_Δ. Efficient algorithms to compute the class number are not known. In [70] an algorithm is given that takes time $|\Delta|^{1/5+o(1)}$, for $\Delta\to-\infty$; both its running time and correctness depend on the assumption of the generalized Riemann hypothesis (GRH). It follows from the Brauer–Siegel theorem (cf. [41, Chapter XVI]) that $h_\Delta=|\Delta|^{1/2+o(1)}$ for $\Delta\to-\infty$. Furthermore, $h_\Delta<(\sqrt{|\Delta|}\log|\Delta|)/2$ for $\Delta<-3$. It follows from (2.14) that h_Δ is even if and only if $|\Delta|$ is not a prime power.

2.16. FINDING AMBIGUOUS FORMS. The ambiguous forms are obtained from forms whose order is a power of 2. Namely, if (a, b) has order 2^k with $k>0$, then $(a, b)^{2^{k-1}}$ is an ambiguous form. Because of the bound on h_Δ, we see that an ambiguous form can be computed in $O(\log|\Delta|)$ squarings, if a form (a, b) of 2-power order is given.

Such forms can be determined if we have an odd multiple u of the largest odd divisor of h_Δ, because for any form (c, d), the form $(c, d)^u$ is of 2-power order. Forms of 2-power order can therefore be determined by computing $(c, d)^u$ for randomly selected forms (c, d), or by letting (c, d) run through a set of generators for C_Δ; if in the latter case no (c, d) is found with $(c, d)^u \neq 1_\Delta$, then h_Δ is odd, so that Δ is a prime power according to (2.15).

2.17. PRIME FORMS. For a prime number p we define the Kronecker symbol $(\frac{\Delta}{p})$ by

$$\left(\frac{\Delta}{p}\right)=\begin{cases}1 & \text{if } \Delta \text{ is a quadratic residue modulo } 4p \text{ and } \gcd(\Delta, p)=1,\\ 0 & \text{if } \gcd(\Delta, p)\neq 1,\\ -1 & \text{otherwise.}\end{cases}$$

For a prime p for which $(\frac{\Delta}{p})=1$, we define the *prime form* I_p as the reduced form equivalent to (p, b_p), where $b_p=\min\{b\in \mathbf{Z}_{>0}: b^2\equiv \Delta \bmod 4p\}$. It follows from a result in [40] that, if the generalized Riemann hypothesis holds, then there is an effectively computable constant c, such that C_Δ is generated by the prime forms I_p with $p\leqslant c\cdot(\log|\Delta|)^2$, where we only consider primes p for which $(\frac{\Delta}{p})=1$ (cf. [70, Corollary 6.2]); according to [6] it suffices to take $c=48$.

2.18. SMOOTHNESS OF FORMS. A form (a, b, c) of discriminant Δ, with $\gcd(a, \Delta)=1$, for which the prime factorization of a is known, can be factored into prime forms in the following way. If $a=\prod_{p\,\text{prime}} p^{e_p}$ is the prime factorization of a, then $(a, b)=\prod_{p\,\text{prime}} I_p^{s_p e_p}$, where $s_p\in\{-1, +1\}$ satisfies $b\equiv s_p b_p \bmod 2p$, with b_p as in (2.17). Notice that the prime forms I_p are well-defined because the primes p divide a, $\gcd(a, \Delta)=1$, and $b^2\equiv \Delta \bmod 4a$.

We say that a form (a, b) is y-smooth if a is y-smooth. In [74] it has been proved that, under the assumption of the GRH, a random reduced form $(a, b)\in C_\Delta$ is $L_{|\Delta|}[\beta]$-smooth with probability at least $L_{|\Delta|}[-1/(4\beta)]$, for any $\beta\in \mathbf{R}_{>0}$. Since $a\leqslant\sqrt{|\Delta|/3}$, this is what can be expected on the basis of Subsection 2.A; the GRH is needed to guarantee that there are sufficiently many primes $\leqslant L_{|\Delta|}[\beta]$ for which $(\frac{\Delta}{p})=1$.

2.D. *Solving systems of linear equations*

Let A be an $(n\times n)$-matrix over a finite field, for some positive integer n, and let b be an n-dimensional vector over the same field. Suppose we want to solve the system $Ax=b$ over the field. It is well-known that this can be done by means of Gaussian elimination in $O(n^3)$ field operations. This number of operations can be improved to $O(n^{2.376})$ (cf. [23]).

A more important improvement can be obtained if the matrix A is *sparse*, i.e., if the number of non-zero entries in A is very small. This will be the case in the applications below. There are several methods that take advantage of sparseness. For two of those algorithms, we refer to [22, 53]. There it is shown that both the *conjugate gradient method* and the *Lanczos method*, methods that are known to be efficient for sparse systems over the real numbers, can be adapted to finite fields. These algorithms, which are due to Coppersmith, Karmarkar, and Odlyzko, achieve, for sparse systems, essentially the same running time as the method that we are going to present here.

2.19. THE COORDINATE RECURRENCE METHOD. This method is due to Wiedemann [82]. Assume that A is nonsingular. Let F be the minimal polynomial of A on the vector space spanned by b, Ab, $A^2b, \ldots$ Because F has degree $\leqslant n$ we have

$$F(A)b = \sum_{i=0}^{n} f_i A^i b = 0,$$

and for any $t \geqslant 0$,

$$\sum_{i=0}^{n} f_i A^{i+t} b = 0.$$

Let $v_{i,j}$ be the jth coordinate of the vector $A^i b$; then

$$\textbf{(2.20)} \qquad \sum_{i=0}^{n} f_i v_{i+t,j} = 0$$

for every $t \geqslant 0$ and $1 \leqslant j \leqslant n$. Fixing j, $1 \leqslant j \leqslant n$, we see that the sequence $(v_{i,j})_{i=0}^{\infty}$ satisfies the linear recurrence relation (2.20) in the yet unknown coefficients f_i of F. Suppose we have computed $v_{i,j}$ for $i = 0, 1, \ldots, 2n$ as the jth coordinate of $A^i b$. Given the first $2n+1$ terms $v_{0,j}, v_{1,j}, \ldots, v_{2n,j}$ of the sequence satisfying a recurrence relation like (2.20), the minimal polynomial of the recurrence can be computed in $O(n^2)$ field operations by means of the Berlekamp–Massey algorithm [48]; denote by F_j this minimal polynomial. Clearly F_j divides F.

If we compute F_j for several values of j, it is not unlikely that F is the least common multiple of the F_j's. We expect that a small number of F_j's, say 20, suffice for this purpose (cf. [53, 82]). Suppose we have computed F in this way. Because of the nonsingularity of A we have $f_0 \neq 0$, so that

$$\textbf{(2.21)} \qquad x = -f_0^{-1} \sum_{i=1}^{n} f_i A^{i-1} b$$

satisfies $Ax = b$.

To analyze the running time of this algorithm for a sparse matrix A, let $w(A)$ denote the number of field operations needed to multiply A by a vector. The vectors $A^i b$ for $i = 0, 1, \ldots, 2n$ can then be computed in $O(nw(A))$ field operations. The same estimate holds for the computation of x. Because we expect that we need only a few F_j's to compute F, the applications of the Berlekamp–Massey algorithm take $O(n^2)$ field operations. The method requires storage for $O(n^2)$ field elements. At the cost of recomputing the $A^i b$ in (2.21), this can be improved to $O(n) + w(A)$ field elements if we store only those coordinates of the $A^i b$ that we need to compute the F_j's. For a rigorous proof of these timings and a deterministic version of this probabilistic algorithm we refer to [82]. How the singular case should be handled can be found in [82, 53].

2.22. SOLVING EQUATIONS OVER THE RING $\mathbf{Z}/m\mathbf{Z}$. In the sequel we often have to solve a system of linear equations over the ring $\mathbf{Z}/m\mathbf{Z}$, where m is not necessarily prime. We briefly sketch how this can be done using Wiedemann's coordinate recurrence method. Instead of solving the system over $\mathbf{Z}/m\mathbf{Z}$, we solve the system over the fields $\mathbf{Z}/p\mathbf{Z}$ for the primes $p|m$, lift the solutions to the rings $\mathbf{Z}/p^k\mathbf{Z}$ for the prime powers $p^k|m$, and finally combine these solutions to the solution over $\mathbf{Z}/m\mathbf{Z}$ by means of the Chinese remainder algorithm. In practice we will not try to obtain a complete factorization of m, but we just start solving the system modulo m, and continue until we try to divide by a zero-divisor, in which case a factor of m is found.

Lifting a solution $Ax_0 = b$ modulo p to a solution modulo p^k can be done by writing $Ax_0 - b = py$ for some integer vector y, and solving $Ax_1 = y$ modulo p. It follows that $A(x_0 - px_1) = b$ modulo p^2. This process is repeated until the solution modulo p^k is determined. We conclude that a system over $\mathbf{Z}/m\mathbf{Z}$ can be solved by $O(\log m)$ applications of Algorithm (2.19).

3. Algorithms for finite abelian groups

3.A. Introduction

Let G be a finite abelian group whose elements can be represented in such a way that the group operations can be performed efficiently. In the next few sections we are interested in two computational problems concerning G: finding the order of G or of one of its elements, and computing discrete logarithms in G. For the latter problem we will often assume that the order n of G, or a small multiple of n, is known.

By computing discrete logarithms we mean the following. Let H be the subgroup of G generated by an element $h \in G$. For an element y of G, the problem of computing the *discrete logarithm* $\log_h y$ *of* y *with respect to* h, is the problem to decide whether $y \in H$, and if so, to compute an integer m such that $h^m = y$; in the latter case we write $\log_h y = m$. Evidently, $\log_h y$ is only defined modulo the order of h. Because the order of h is an unknown divisor of n, we will regard $\log_h y$ as a not necessarily well-defined integer modulo n, and represent it by an integer in $\{0, 1, \ldots, n-1\}$. Although $\log_h y$ is often referred to as the *index of* y *with respect to* h, we will only refer to it as the discrete logarithm, or logarithm, of y.

Examples of groups we are interested in are: multiplicative groups of finite fields, sets of points of elliptic curves modulo primes (cf. Subsection 2.B), class groups (cf. Subsection 2.C), and multiplicative groups modulo composite integers. In the first example n is known, and for the second example two methods to compute n were mentioned in (2.6).

In all examples above, the group elements can be represented in a unique way. Equality of two elements can therefore be tested efficiently, and membership of a sorted list of cardinality k can be decided in $\log k$ comparisons. Examples where unique representations do *not* exist are for instance multiplicative groups modulo an *unspecified* prime divisor of an integer n, or sets of points of an elliptic curve modulo n, when taken modulo an *unspecified* prime divisor of n (cf. (2.7)). In these examples *inequality* can be tested by means of a gcd-computation. If two nonidentically represented elements are equal, the gcd will be a nontrivial divisor of n. In Subsection 4.B we will see how this can be exploited.

In Subsection 3.B we present some algorithms for both of our problems that can be applied to any group G as above. By their general nature they are quite slow; the number of group operations required is an exponential function of $\log n$. Algorithms for groups with *smooth order* are given in Subsection 3.C (cf. Subsection 2.A). For groups containing many *smooth elements*, subexponential discrete logarithm algorithms are given in Subsection 3.D. Almost all of the algorithms in Subsection 3.D are only applicable to the case where G is the multiplicative group of a finite field, with the added restriction that h is a primitive root of the same field. In that case $G=H$, so that the decision problem becomes trivial. An application of these techniques to class groups is presented in Remark (3.13).

For practical consequences of the algorithms in Subsections 3.B through 3.D we refer to the original papers and to [53].

3.B. *Exponential algorithms*

Let G be a finite abelian group as in Subsection 3.A, let $h \in G$ be a generator of a subgroup H of G, and let $y \in G$. In this section we discuss three algorithms to compute $\log_h y$. The algorithms have in common that, with the proper choice for y, they can easily be adapted to compute the order n_h of h, or a small multiple of n_h.

Of course, $\log_h y$ can be computed deterministically in at most n_h multiplications and comparisons in G, by computing h^i for $i=1,2,\ldots$ until $h^i=y$ or $h^i=1$; here 1 denotes the unit element in G. Then $y \in H$ if and only if $h^i=y$ for some i, and if $y \notin H$ the algorithm terminates after $O(n_h)$ operations in G; in the latter case (and if $y=1$), the order of h has been computed. The method requires storage for only a constant number of group elements.

3.1. SHANK'S BABY-STEP-GIANT-STEP ALGORITHM (*cf.* [38, *Exercise* 5.17]). We can improve on the number of operations of the above algorithm if we allow for more storage being used, and if a unique representation of the group elements exists; we describe an algorithm that takes $O(\sqrt{n_h}\log n_h)$ multiplications and comparisons in G, and that requires storage for $O(\sqrt{n_h})$ group elements. The algorithm is based on the following observation. If $y \in H$ and $\log_h y < s^2$ for some $s \in \mathbf{Z}_{>0}$, then there exist integers i and j with $0 \leqslant i, j < s$ such that $y=h^{is+j}$. In this situation $\log_h y$ can be computed as follows. First, make a sorted list of the values h^j for $0 \leqslant j < s$ in $O(s \log s)$ operations in G. Next, compute yh^{-is} for $i=0,1,\ldots,s-1$ until yh^{-is} equals one of the values in the list; this search can be done in $O(\log s)$ comparisons per i because the list is sorted. If yh^{-is} is found to be equal to h^j, then $\log_h y = is+j$. Otherwise, if yh^{-is} is not found in the list for any of the values of i, then either $y \notin H$ or $\log_h y \geqslant s^2$.

This method can be turned into a method that can be guaranteed to use $O(\sqrt{n_h} \times$

$\log n_h$) operations in G, both to compute discrete logarithms and to compute n_h. For the latter problem, we put $y=1$, and apply the above method with $s=2^k$ for $k=1,2,\ldots$ in succession, excluding the case where both i and j are zero. After

$$\mathrm{O}\left(\sum_{k=1}^{\lceil \log_2 n_h \rceil/2} 2^k \log 2^k\right) = \mathrm{O}(\sqrt{n_h} \log n_h)$$

operations in G, we find i and j such that $h^{i2^k+j}=1$, and therefore a small multiple of n_h. To compute $\log_h y$ we proceed similarly, but to guarantee a timely termination of the algorithm in case $y \notin H$, we look for h^{-is} in the list as well; if some h^{-is} is in the list, but none of the yh^{-is} is, then $y \notin H$. We could also first determine n_h, and put $s=\lceil \sqrt{n_h} \rceil$.

We conclude that both the order of h and discrete logarithms with respect to h can be computed deterministically in $n_h^{1/2+o(1)}$ multiplications and comparisons in G, for $n_h \to \infty$. The method requires storage for $\mathrm{O}(\sqrt{n_h})$ group elements. In practice it can be recommended to use hashing (cf. [38, Section 6.4]) instead of sorting.

3.2. MULTIPLE DISCRETE LOGARITHMS TO THE SAME BASIS. If $e>1$ discrete logarithms with respect to the same h of order n_h have to be computed, we can do better than $\mathrm{O}(e\sqrt{n_h} \times \log n_h)$ group operations, if we allow for more than $\mathrm{O}(\sqrt{n_h})$ group elements being stored. Of course, if $e \geqslant n_h$, we simply make a sorted list of h^i for $i=0,1,\ldots,n_h-1$, and look up each element in the list; this takes $\mathrm{O}(e \log n_h)$ group operations and storage for n_h group elements. If $e<n_h$, we put $s=[\sqrt{e \cdot n_h}]$, make a sorted list of h^j for $0 \leqslant j < s$, and for each of the e elements y we compute yh^{-is} for $i=0,1,\ldots,\lceil n_h/s \rceil$ until yh^{-is} equals one of the values in the list. This takes

$$\mathrm{O}(\sqrt{e \cdot n_h} \log(e \cdot n_h))$$

group operations, and storage for $\mathrm{O}(\sqrt{e \cdot n_h})$ group elements.

3.3. POLLARD'S RHO METHOD (*cf.* [58]). The following randomized method needs only a constant amount of storage. It is randomized in the sense that we cannot give a worst-case upper bound for its running time. We can only say that the *expected* number of group operations to be performed is $\mathrm{O}(\sqrt{n})$ to compute discrete logarithms, and $\mathrm{O}(\sqrt{n_h})$ to compute the order n_h of h; here n is the order of G. Let us concentrate on computing discrete logarithms first.

Assume that a number n is known that equals the order of G, or a small multiple thereof. We randomly partition G into three subsets G_1, G_2, and G_3, of approximately the same size. By an *operation* in G we mean either a group operation, or a membership test $x \overset{?}{\in} G_i$. For $y \in G$ we define the sequence $y_0, y_1, y_2, \ldots$ in G by $y_0 = y$, and

(3.4)
$$y_i = \begin{cases} h \cdot y_{i-1} & \text{if } y_{i-1} \in G_1, \\ y_{i-1}^2 & \text{if } y_{i-1} \in G_2, \\ y \cdot y_{i-1} & \text{if } y_{i-1} \in G_3, \end{cases}$$

for $i>0$. If this sequence behaves as a random mapping from G to G, its expected cycle length is $\mathrm{O}(\sqrt{n})$ (see [37, Exercise 4.5.4.4]). Therefore, when comparing y_i and y_{2i} for

$i=1,2,\ldots$, we expect to find $y_k=y_{2k}$ for $k=O(\sqrt{n})$. The sequence has been defined in such a way that $y_k=y_{2k}$ easily yields $y^{e_k}=h^{m_k}$ for certain $e_k, m_k \in \{0,1,\ldots,n-1\}$. Using the extended Euclidean algorithm we compute s and t such that $s\cdot e_k+t\cdot n=d$ where $d=\gcd(e_k,n)$; if $d=1$, which is not unlikely to occur, we find $\log_h y=s\cdot m_k \bmod n$.

If $d>1$ then we do not immediately know the value of $\log_h y$, but we can exploit the fact that $y^{e_k}=h^{m_k}$ as follows. We introduce a number $l>0$, to be thought of as the smallest known multiple of n_h. Initially we put $l=n$. Every time that l is changed, we first check that $y^l=1$ (if $y^l\neq 1$ then clearly $y\notin H$), and next we compute new s, t, and d with $d=\gcd(e_k,l)=s\cdot e_k+t\cdot l$. Note that $h^{lm_k/d}=y^{le_k/d}=1$, so that $n_h|lm_k/d$. If d does not divide m_k, then change l to $\gcd(l,lm_k/d)$. Ultimately, d divides m_k. We have that $y^d=h^{sm_k}$, so we may stop if $d=1$. Otherwise, we determine the order d' of $h^{l/d}$ by means of any of the methods described in Subsections 3.B and 3.C. If this is difficult to do then d is large (which is unlikely), and it is probably best to generate another relation of the sort $y^{e_k}=h^{m_k}$. If $d'<d$ then change l to ld'/d. Finally, suppose that $d'=d$. Let $y'=yh^{-sm_k/d}$, then $y\in H$ if and only if $y'\in$ H, and since $(y')^d=1$, this is the case if and only if y' belongs to the subgroup generated by $h'=h^{l/d}$. The problem with y and h is now reduced to the same problem with y' and h', with the added knowledge that the order of h' equals d. The new problem can be solved by means of any of the methods described in Subsections 3.B and 3.C.

Of course, we could define the recurrence relation (3.4) in various other ways, as long as the resulting sequence satisfies our requirements.

Notice that, if $y\in H$, the recurrence relation (3.4) is defined over H. If also the $G_i\cap H$ are such that the sequence behaves as a random mapping from H to H, then we expect the discrete logarithm algorithm to run in $O(\sqrt{n_h})$ operations in G. In the case that n or some multiple of n_h is *not* known, a multiple of n_h can be computed in a similar way in about $O(\sqrt{n_h})$ operations in G. To do this, one partitions G into a somewhat larger number of subsets G_j, say 20, and one defines $y_0=1$, and $y_i=h^{t_j}\cdot y_{i-1}$ if $y_{i-1}\in G_j$; here the numbers t_j are randomly chosen from $\{2,3,\ldots,B-1\}$, where B is an estimate for n_h (cf. [68]).

We conclude this section by mentioning another randomized algorithm for computing discrete logarithms, the so-called *Lambda method for catching kangaroos*, also due to Pollard [58]. It can only be used when $\log_h y$ is *known* to exist, and lies in a specified interval of width w; it is not necessary that the order of G, or a small multiple thereof, is known. The method requires $O(\sqrt{w})$ operations in G, and a small amount of storage (depending on the implementation), but cannot be guaranteed to have success; the failure probability ε, however, can be made arbitrarily small, at the cost of increasing the running time which depends linearly on $\sqrt{\log(1/\varepsilon)}$. We will not pursue this approach further, but refer the interested reader to [58]. Notice that, with $w=n$, this method can be used instead of the rho method described above, if at least $y\in H$.

3.C. *Groups with smooth order*

In some cases one might suspect that the order of G, or of h, has only small prime factors, i.e., is s-smooth for some small $s\in \mathbf{Z}_{>0}$ (cf. Subsection 2.A). If one also knows an upper bound B on the order, this smoothness can easily be tested. Namely, in these circumstances the order should divide

$$(3.5)\qquad k=k(s,B)=\prod_{\substack{p\leqslant s\\ p\,\text{prime}}} p^{t_p},$$

where $t_p\in\mathbf{Z}_{\geqslant 0}$ is maximal such that $p^{t_p}\leqslant B$. Raising h to the kth power should yield the unit element in G; this takes $O(s\log_s B)$ multiplications in G to verify. If h^k indeed equals the unit element, the order of h can be deduced after some additional computations.

3.6. THE CHINESE REMAINDER THEOREM METHOD (*cf.* [56]). Also for the discrete logarithm problem a smooth order is helpful, as was first noticed by Silver, and later by Pohlig and Hellman [56]. Let $n_h=\prod_{p|n_h}p^{e_p}$ be the prime factorization of n_h. If $y\in H$, then it suffices to determine $\log_h y=m$ modulo each of the p^{e_p}, followed by an application of the Chinese remainder algorithm. This observation leads to an algorithm that takes

$$\sum_{\substack{p|n_h\\ p\text{ prime}}} O(\sqrt{e_p\,\max(e_p,p)}\log(p\cdot\min(e_p,p)))$$

group operations, and that needs storage for

$$O\Bigl(\max_{\substack{p|n_h\\ p\text{ prime}}}(\sqrt{p\cdot\min(e_p,p)})\Bigr)$$

group elements.

To compute m modulo p^e, where p is one of the primes dividing n_h and $e=e_p$, we proceed as follows. Write $m\equiv\sum_{i=0}^{e-1}m_ip^i$ modulo p^e, with $m_i\in\{0,1,\ldots,p-1\}$, and notice that

$$(m-(m\bmod p^i))n_h/p^{i+1}\equiv(n_h/p)m_i \bmod n_h$$

for $i=0,1,\ldots,\ e-1$. This implies that, if $y\in H$, then

$$(y\cdot h^{-(m\bmod p^i)})^{n_h/p^{i+1}}=(h^{n_h/p})^{m_i}.$$

Because $\bar{h}=h^{n_h/p}$ generates a cyclic subgroup $\bar{H}$ of G of order p, we can compute $m_0,m_1,\ldots,m_{e-1}$ in succession by computing the discrete logarithms of $\bar{y}_i=(y\cdot h^{-(m\bmod p^i)})^{n_h/p^{i+1}}$ with respect to $\bar{h}$, for $i=0,1,\ldots,e-1$. This can be done by means of any of the methods mentioned in Subsection 3.B. If $\bar{y}_i\notin\bar{H}$ for some i, then $y\notin H$, and the algorithm terminates. With (3.2) we now arrive at the estimates mentioned above.

3.D. Subexponential algorithms

In this subsection we will concentrate on algorithms to compute discrete logarithms with respect to a primitive root g of the multiplicative group G of a finite field. In this case the order of G is known. In principle the methods to be presented here can be applied to any group for which the concept of smoothness makes sense, and that contains sufficiently many smooth elements. This is the case for instance for class groups, as is shown in Remark (3.13).

We do not address the problem of finding a primitive root of G, or deciding whether a given element is a primitive root. Notice however that the latter can easily be accomplished if the factorization of the order of G is known. It would be interesting to analyze how the algorithms in this subsection behave in the case where it not known whether g is a primitive root or not.

A rigorous analysis of the expected running time has only been given for a slightly different version of the first algorithm below [61]. The timings of the other algorithms in this section are heuristic estimates.

3.7. REMARK. Any algorithm that computes discrete logarithms with respect to a primitive root of a finite field can be used to compute logarithms with respect to any non-zero element of the field. Let g be a primitive root of a finite field, G the multiplicative group of order n of the field, and h and y any two elements of G. To decide whether $y \in \langle h \rangle = H$ and, if so, to compute $\log_h y$, we proceed as follows. Compute $\log_g h = m_h$, $\log_g y = m_y$, and $\text{ind}(h) = \gcd(n, m_h)$. Then $y \in H$ if and only if $\text{ind}(h)$ divides m_y, and if $y \in H$ then

$$\log_h y = (m_y/\text{ind}(h))(m_h/\text{ind}(h))^{-1} \bmod n_h,$$

where $n_h = n/\text{ind}(h)$ is the order of h.

3.8. SMOOTHNESS IN $(\mathbf{Z}/p\,\mathbf{Z})^*$. If $G = (\mathbf{Z}/p\,\mathbf{Z})^*$ for some prime p, we identify G with the set $\{1, 2, \ldots, p-1\}$ of least positive residues modulo p; the order n of G equals $p-1$. It follows from Subsection 2.A that a randomly selected element of G that is $\leqslant n^\alpha$ is $L_n[\beta]$-smooth with probability $L_n[-\alpha/(2\beta)]$, for $\alpha, \beta \in \mathbf{R}_{>0}$ fixed with $\alpha \leqslant 1$, and $n \to \infty$. The number of primes $\leqslant L_n[\beta]$ is $\pi(L_n[\beta]) = L_n[\beta]$. In Subsection 4.B we will see that an element of G can be tested for $L_n[\beta]$-smoothness in expected time $L_n[0]$; in case of smoothness, the complete factorization is computed at the same time (cf. (4.3)).

3.9. SMOOTHNESS IN $\mathbf{F}^*_{2^m}$. If $G = \mathbf{F}^*_{2^m}$, for some positive integer m, we select an irreducible polynomial $f \in \mathbf{F}_2[X]$ of degree m, so that $\mathbf{F}_{2^m} \cong (\mathbf{F}_2[X])/(f)$. The elements of G are then identified with non-zero polynomials in $\mathbf{F}_2[X]$ of degree $< m$. We define the *norm* $\mathrm{N}(h)$ of an element $h \in G$ as $\mathrm{N}(h) = 2^{\text{degree}(h)}$. Remark that $\mathrm{N}(f) = \#\mathbf{F}_{2^m}$, and that the order n of G equals $2^m - 1$.

A polynomial in $\mathbf{F}_2[X]$ is *smooth with respect to* x for some $x \in \mathbf{R}_{>0}$, if it factors as a product of irreducible polynomials of norm $\leqslant x$. It follows from a theorem of Odlyzko [53] that a random element of G of norm $\leqslant n^\alpha$ is $L_n[\beta]$-smooth with probability $L_n[-\alpha/(2\beta)]$, for $\alpha, \beta \in \mathbf{R}_{>0}$ fixed with $\alpha < 1$, and $n \to \infty$. Furthermore, an element of G of degree k can be factored in time polynomial in k (cf. [37]). The number of irreducible polynomials of norm $\leqslant L_n[\beta]$ is about

$$L_n[\beta]/\log_2(L_n[\beta]) = L_n[\beta].$$

These results can all easily be generalized to finite fields of arbitrary, but fixed, characteristic.

3.10. THE INDEX-CALCULUS ALGORITHM. Let g be a generator of a group G of order n as in

(3.8) or (3.9); "prime element" will mean "prime number" (3.8) or "irreducible polynomial" (3.9), and for $G=(\mathbf{Z}/p\,\mathbf{Z})^*$ the "norm" of $x \in G$ will be x itself. Let $y \in G$, and let S be the set of prime elements of norm $\leqslant L_n[\beta]$ for some $\beta \in \mathbf{R}_{>0}$. We abbreviate $L_n[\beta]$ to $L[\beta]$. The algorithms to compute $\log_g y$ that we present in this subsection consist of two stages (cf. [81]):

(1) *precomputation*: compute $\log_g s$ for all $s \in S$;

(2) *computation of* $\log_g y$: find a multiplicative relation between y and the elements of S, and derive $\log_g y$ using the result from the precomputation stage.

This gives rise to an algorithm whose expected running time is bounded by a polynomial function of $L(n)$; notice that this is better than $O(n^{\varepsilon})$ for every $\varepsilon > 0$ (cf. [1]).

First, we will describe the second stage in more detail, and analyze its expected running time. Suppose that the discrete logarithms of the prime elements of norm $\leqslant L[\beta]$ all have been computed in the first stage. We determine an integer e such that $y \cdot g^e$ factors as a product of elements of S, by randomly selecting integers $e \in \{0, 1, \ldots, n-1\}$ until $y \cdot g^e \in G$ is smooth with respect to $L[\beta]$. For the resulting e we have

$$y \cdot g^e = \prod_{s \in S} s^{e_s},$$

so that

$$\log_g y = \left(\left(\sum_{s \in S} e_s \log_g s\right) - e\right) \bmod n,$$

where the $\log_g s$ are known from the precomputation stage. By the results cited in (3.8) and (3.9) we expect that $L[1/(2\beta)]$ trials suffice to find e. Because the time per trial is bounded by $L[0]$ for both types of groups, we expect to spend time $L[1/(2\beta)]$ for each discrete logarithm.

Now consider the precomputation stage, the computation of $\log_g s$ for all $s \in S$. We collect multiplicative relations between the elements of S, i.e., linear equations in the $\log_g s$. Once we have sufficiently many relations, we can compute the $\log_g s$ by solving a system of linear equations.

Collecting multiplicative relations can be done by randomly selecting integers $e \in \{0, 1, \ldots, n-1\}$ until $g^e \in G$ is smooth with respect to $L[\beta]$. For a successful e we have

$$g^e = \prod_{s \in S} s^{e_s}$$

which yields the linear equation

(3.11) $$e = \left(\sum_{s \in S} e_s \log_g s\right) \bmod n.$$

We need about $|S| \approx L[\beta]$ equations of the form (3.11) to be able to solve the resulting system of linear equations, so we repeat this step about $L[\beta]$ times.

It follows from the analysis of the second stage that collecting equations can be done in expected time $L[\beta + 1/(2\beta)]$. Because the system can be solved in time $L[3\beta]$ by ordinary Gaussian elimination (cf. Subsection 2.D and (2.22)), the precomputation stage takes expected time $L[\max(\beta+1/(2\beta), 3\beta)]$, which is $L[\frac{3}{2}]$ for the optimal choice $\beta=\frac{1}{2}$. This dominates the cost of the second stage which takes, for $\beta=\frac{1}{2}$, time $L[1]$ per logarithm. The storage requirements are $L[1]$ for the precomputation (to store the system of equations), and $L[\frac{1}{2}]$ for the second stage (to store the $\log_g s$ for $s \in S$).

An important improvement can be obtained by noticing that in the equations of the form (3.11) at most $\log_2 n$ of the $|S| \approx L[\beta]$ coefficients e_s can be non-zero. This implies that we can use the coordinate recurrence method described in (2.19), which has, combined with (2.22), the following consequence. Multiplying the matrix defining the system by a vector can be done in time $(\log_2 n)L[\beta]$, which is $L[\beta]$. The system can therefore be solved in time $L[2\beta]$, so that the expected time for the precomputation stage becomes $L[\max(\beta+1/(2\beta), 2\beta)]$. For $\beta=\sqrt{\frac{1}{2}}$, we get $L[\sqrt{2}]$ arithmetic operations in G or $\mathbf{Z}/n\mathbf{Z}$ for the precomputation, and $L[\sqrt{\frac{1}{2}}]$ operations per logarithm. The method requires storage for $L[\sqrt{\frac{1}{2}}]$ group elements both in the precomputation and in the second stage. We refer to [61] for a rigorous proof that a slightly modified version of the index-calculus algorithm runs in time $L[\sqrt{2}]$, for both of our choices of G.

3.12. REMARK. As suggested at the end of (3.9), the algorithm in (3.10), and the modifications presented below, can be adapted to finite fields of arbitrary, but fixed, characteristic. For $\mathbf{F}_{p^2}$ a modified version of the index-calculus algorithm is presented in [29]; according to Odlyzko [53] this method applies to $\mathbf{F}_{p^m}$, for fixed m, as well. It is an as yet unanswered question how to compute discrete logarithms when *both p and m* tend to infinity.

3.13. REMARK. The ideas from the index-calculus algorithm can be applied to other groups as well. Consider for instance the case that G is a class group as in Subsection 2.C, of unknown order n. Suppose we want to compute the discrete logarithm of y with respect to h, for $h, y \in G$. Let S be a set of prime forms that generates G (cf. (2.17)). The mapping φ from $\mathbf{Z}^S$ to G that maps $(e_s)_{s \in S} \in \mathbf{Z}^S$ to $\prod_{s \in S} s^{e_s} \in G$ is a surjection. The kernel of φ is a sublattice of the lattice $\mathbf{Z}^S$, and $\mathbf{Z}^S/\ker(\varphi) \cong G$. In particular the determinant of $\ker(\varphi)$ equals n.

To calculate $\ker(\varphi)$, we introduce a subgroup Λ of $\mathbf{Z}^S$, to be thought of as the largest subgroup of $\ker(\varphi)$ that is known. Initially one puts $\Lambda=\{0\}$. To enlarge Λ, one looks for relations between the elements of S. Such relations can be found in a way similar to the precomputation stage of (3.10)), as described in (4.12); the primitive root g is replaced by a product of random powers of elements of S, thus producing a random group element. Every relation gives rise to an element $r \in \ker(\varphi)$. One tests whether $r \in \Lambda$, and if not one replaces Λ by $\Lambda+\mathbf{Z}r$; if Λ is given by a basis in Hermite form, this can be done by means of the algorithm of [36]. Repeating this a number of times, one may expect to find a lattice Λ containing $|S|$ independent vectors. The determinant of Λ is then a non-zero multiple of n. After some additional steps it will happen that Λ does not change any more, so that one may hope that $\Lambda=\ker(\varphi)$. In that case, $\det(\Lambda)=n$, and $\mathbf{Z}^S/\Lambda \cong G$.

Supposing that $\Lambda=\ker(\varphi)$, we can write G as a direct sum of cyclic groups by bringing the matrix defining Λ to diagonal form [36]. This may change the set of generators of G. To solve the discrete logarithm problem one expresses both h and y as products of powers of the new generators, and applies (3.7) repeatedly. Notice that if the assumption $\Lambda=\ker(\varphi)$ is wrong (i.e., we did not find sufficiently many relations), we may incorrectly decide that $y\notin\langle h\rangle$.

3.14. A METHOD BASED ON THE RESIDUE-LIST SIEVE FROM [22]. We now discuss a variant of the index-calculus algorithm that yields a better heuristic running time, namely $L[1]$ for the precomputation and $L[\frac{1}{2}]$ per individual logarithm. Instead of looking for random smooth group elements that yield equations like (3.11), we look for smooth elements of much smaller norm that still produce the necessary equations. Because elements of smaller norm have a higher probability of being smooth, we expect that this will give a faster algorithm.

For ease of exposition we take $G=(\mathbf{Z}/p\,\mathbf{Z})^*$, as in (3.8), so that $n=p-1$. Let the notation be as in (3.10). Linear equations in the $\log_g s$ for $s\in S$ are collected as follows. Let $\alpha\in\mathbf{R}_{>0}$ and let u and v be two integers in $\{[\sqrt{p}]+1,\ldots,[\sqrt{p}+L[\alpha]]\}$, both smooth with respect to $L[\beta]$. If $uv-p$ is also smooth with respect to $L[\beta]$, then we have found an equation of the type we were looking for, because $\log_g u+\log_g v=\log_g(uv-p)$.

We analyze how much time it takes to collect $L[\beta]$ equations in this way. The probability of $uv-p=O(L[\alpha]\sqrt{p})$ being smooth with respect to $L[\beta]$ is $L[-1/(4\beta)]$, so we have to consider $L[\beta+1/(4\beta)]$ smooth pairs (u,v), and test the corresponding $uv-p$ for smoothness. This takes time $L[\beta+1/(4\beta)]$. It follows that we need $L[\beta/2+1/(8\beta)]$ integers $u\in\{[\sqrt{p}]+1,\ldots,[\sqrt{p}+L[\alpha]]\}$ that are smooth with respect to $L[\beta]$. For that purpose we take $L[\beta/2+1/(8\beta)+1/(4\beta)]$ integers in $\{[\sqrt{p}]+1,\ldots,[\sqrt{p}+L[\alpha]]\}$ and test them for smoothness, because the probability of smoothness is $L[-1/(4\beta)]$. Generating the u's therefore takes time $L[\beta/2+3/(8\beta)]$. Notice that we can take $\alpha=\beta/2+3/(8\beta)$. Notice also that u, v, and $uv-p$ are not generated randomly, but instead are selected in a deterministic way. Although we cannot justify it theoretically, we assume that these numbers have the same probability of smoothness as random numbers of about the same size. The running times we get are therefore only heuristic estimates.

Combined with the coordinate recurrence method (cf. (2.19), (2.22)), we find that the precomputation takes time $L[\max(\beta+1/(4\beta),\beta/2+3/(8\beta),2\beta)]$. This is minimized for $\beta=\frac{1}{2}$, so that the precomputation can be done in expected time $L[1]$ and storage $L[\frac{1}{2}]$. Notice that for $\beta=\frac{1}{2}$ we have $\alpha=1$.

The second stage as described in (3.10) also takes time $L[1]$. If we keep the $L[\frac{1}{2}]$ smooth u's from the precomputation stage, then the second stage can be modified as follows. We find e such that $y\cdot g^e \bmod p$ is smooth with respect to $L[2]$ in time $L[\frac{1}{4}]$. To calculate $\log_g y$, it suffices to calculate $\log_g x$ for each prime factor $x\leqslant L[2]$ of $y\cdot g^e \bmod p$. For fixed x this is done as follows. Find v in an interval of size $L[\frac{1}{2}]$ around $\sqrt{p/x}$ that is smooth with respect to $L[\frac{1}{2}]$ in time $L[\frac{1}{2}]$. Finally, find one of the $L[\frac{1}{2}]$ smooth u's such that $uvx-p=O(L[\frac{5}{2}]\sqrt{p})$ is smooth with respect to $L[\frac{1}{2}]$ in time $L[\frac{1}{2}]$. The value of $\log_g x$ now follows. Individual logarithms can therefore be computed in expected time and storage $L[\frac{1}{2}]$.

Generalization of this idea to $G=\mathbf{F}^*_{2^m}$, as in (3.9), follows immediately if we select some polynomial $g\in\mathbf{F}_2[X]$ of norm about $2^{m/2}$ (for instance $g=X^{[m/2]}$), and compute

$q, r \in \mathbf{F}_2[X]$ such that $f = qg + r$ (cf. (3.9)) with degree(r) < degree(g). In the precomputation we consider $u = g + \bar{u}$, $v = q + \bar{v}$ for polynomials $\bar{u}, \bar{v} \in \mathbf{F}_2[X]$ of norm $\leqslant L[\alpha]$, so that $\mathbf{N}(uv - f)$ is close to $L[\alpha]2^{m/2}$; here $L[\alpha] = L_{2^m-1}[\alpha]$. In the second stage we write $q = hx + \bar{x}$ for $h, \bar{x} \in \mathbf{F}_2[X]$ with degree($\bar{x}$) < degree(x), where x is as above, choose $v = h + \bar{v}$ with $\mathbf{N}(\bar{v}) < L[\frac{1}{2}]$, and consider $uvx - f$. The running time analysis remains unchanged. Instead of finding g, q, r as above, we could also choose f in (3.9) such that $f = X^m + f_1$ with degree(f_1) < $m/2$, so that we can take $g = q = X^{[(m+1)/2]}$.

3.15. A METHOD BASED ON THE LINEAR SIEVE ALGORITHM FROM [22]. Again we consider $G = (\mathbf{Z}/p\,\mathbf{Z})^*$. An improvement of (3.14) that is of practical importance, although it does not affect the timings when expressed in $L(n)$, can be obtained by including the numbers $u \in \{[\sqrt{p}] + 1, \ldots, [\sqrt{p} + L[\alpha]]\}$ in the set S as well. For such u and v we again have $uv - p = O(L[\alpha]\sqrt{p})$, but now we only require that $uv - p$ is smooth with respect to $L[\beta]$, without requiring smoothness for u or v. It follows in a similar way as above that the $L[\beta] + L[\alpha]$ equations can be collected in time $L[1]$ and storage $L[\frac{1}{2}]$ for $\beta = \frac{1}{2}$ and $\alpha = \beta/2 + 1/(8\beta) = \frac{1}{2}$. The reason that this version will run faster than the algorithm from (3.14) is that $uv - p$ is now only $O(L[\frac{1}{2}]\sqrt{p})$, whereas it is $O(L[1]\sqrt{p})$ in (3.14). In practice this will make a considerable difference in the probability of smoothness. The second stage can be adapted in a straightforward way. The running times we get are again only heuristic estimates.

In the methods for $G = (\mathbf{Z}/p\,\mathbf{Z})^*$ described in (3.14) and (3.15), the use of the smoothness test referred to in (3.8) can be replaced by sieving techniques. This does not change the asymptotic running times, but the resulting algorithms will probably be faster in practice [22].

3.16. A MORE GENERAL L FUNCTION. For the description of the last algorithm in this subsection, the bimodal polynomials method, it will be convenient to extend the definition of the function L from Subsection 2.A slightly. For $\alpha, r \in \mathbf{R}$ with $0 \leqslant r \leqslant 1$, we denote by $L_x[r;\alpha]$ any function of x that equals

$$e^{(\alpha + o(1))(\log x)^r (\log\log x)^{1-r}}, \quad \text{for } x \to \infty.$$

Notice that this is $(\log x)^\alpha$ for $r = 0$, and x^α for $r = 1$, up to the o(1) in the exponent. For $r = \frac{1}{2}$ we get the L from Subsection 2.A.

The smoothness probabilities from Subsection 2.A and (3.9) can now be formulated as follows. Let $\alpha, \beta, r, s \in \mathbf{R}$ be fixed with $\alpha, \beta > 0$, $0 < r \leqslant 1$, and $0 < s < r$. From Subsection 2.A we find that a random positive integer $\leqslant L_x[r;\alpha]$ is $L_x[s;\beta]$-smooth with probability $L_x[r - s; -\alpha(r - s)/\beta]$, for $x \to \infty$. From the same theorem of Odlyzko referred to in (3.9) we have that, for $r/100 < s < 99r/100$, a random polynomial in $\mathbf{F}_2[X]$ of norm $\leqslant L_x[r;\alpha]$ is smooth with respect to $L_x[s;\beta]$ with probability $L_x[r - s; -\alpha(r - s)/\beta]$, for $x \to \infty$.

3.17. COPPERSMITH'S BIMODAL POLYNOMIALS METHOD (*cf.* [21]). We conclude this subsection with an algorithm that was especially designed for $G = \mathbf{F}^*_{2^m}$, as in (3.9). This algorithm does not apply to fields with a large characteristic. It is again a variant of the index-calculus algorithm (3.10). Let f be a monic polynomial in $\mathbf{F}_2[X]$ of degree m as in (3.9), so that $\mathbf{F}_{2^m} \cong (\mathbf{F}_2[X])/(f)$. We assume that f can be written as $X^m + f_1$, for $f_1 \in \mathbf{F}_2[X]$ of degree $< m^{2/3}$. Because about one out of every m polynomials in $\mathbf{F}_2[X]$ of degree m is irreducible, we expect that such an f can be found.

We use the function L from (3.16), and we abbreviate $L_{2^m-1}[r;\alpha]$ to $L[r;\alpha]$. Notice that with this notation

$$L[r;\alpha] = 2^{\alpha(1+o(1))m^r(\log_2 m)^{1-r}}, \quad \text{for } \alpha > 0, \text{ and } m \to \infty.$$

We will see that the precomputation stage can be carried out in expected time $L[\frac{1}{3};2^{2/3}]$ and that individual logarithms can be computed in expected time $L[\frac{1}{3};\sqrt{\frac{1}{2}}]$. Notice that this is substantially faster than any of the other algorithms in this section.

Let S be the set of irreducible polynomials in $\mathbf{F}_2[X]$ of norm $\leqslant L[\frac{1}{3};\beta]$, for some $\beta \neq 0$. Furthermore, let k be a power of 2 such that $\mathbf{N}(X^{[m/k]})$ is as close as possible to $\mathbf{N}(v^k)$, for a polynomial $v \in \mathbf{F}_2[X]$ of norm $L[\frac{1}{3};\beta]$; this is achieved for a power of 2 close to $\beta^{-1/2}m^{1/3}(\log_2 m)^{-1/3}$. We find that

$$t = k/(\beta^{-1/2}m^{1/3}(\log_2 m)^{-1/3})$$

satisfies $\sqrt{\frac{1}{2}} < t \leqslant \sqrt{2}$ and that $\mathbf{N}(X^{[m/k]}) \leqslant L[\frac{2}{3};\sqrt{\beta}/t]$ and $\mathbf{N}(v^k) \leqslant L[\frac{2}{3};t\sqrt{\beta}]$. For polynomials $v_1, v_2 \in \mathbf{F}_2[X]$ of norm $\leqslant L[\frac{1}{3};\beta]$, we take $u_1 = X^{[m/k]+1}v_1 + v_2$, and $u_2 = u_1^k \bmod f$. Remark that the polynomial u_1 can be considered as a string of bits with two peaks; this explains the name of the method. Since

$$\log_g u_2 = (k \cdot \log_g u_1) \bmod (2^m - 1),$$

we find a linear equation in the $\log_g s$ for $s \in S$, if both u_i's are smooth with respect to $L[\frac{1}{3};\beta]$. Because the equations generated in this way are homogeneous, we assume that g is smooth with respect to $L[\frac{1}{3};\beta]$ as well. To analyze the probability that both u_i's are smooth, we compute their norms. By the choice of k we have that $\mathbf{N}(u_1) \leqslant L[\frac{2}{3};\sqrt{\beta}/t]$. Because k is a power of 2, we have

$$\begin{aligned} u_2 &= (X^{([m/k]+1)k}v_1^k + v_2^k) \bmod f \\ &= X^{([m/k]+1)k-m}f_1 v_1^k + v_2^k, \end{aligned}$$

so that $\mathbf{N}(u_2) \leqslant L[\frac{2}{3};t\sqrt{\beta}]$. The probability that both are smooth with respect to $L[\frac{1}{3};\beta]$ therefore is assumed to be

$$L[\tfrac{1}{3};-1/(3t\sqrt{\beta})] \cdot L[\tfrac{1}{3};-t/(3\sqrt{\beta})] = L[\tfrac{1}{3};-(t+t^{-1})/(3\sqrt{\beta})].$$

The $L[\frac{1}{3};\beta]^2$ pairs (v_1, v_2) must suffice to generate the $\approx L[\frac{1}{3};\beta]$ equations that we need (where we only consider polynomials v_1, v_2 that are relatively prime because the pairs (v_1, v_2) and $(w \cdot v_1, w \cdot v_2)$ yield the same equation). It follows that β must satisfy

$$L[\tfrac{1}{3};2\beta] \geqslant L[\tfrac{1}{3};\beta + (t+t^{-1})/(3\sqrt{\beta})].$$

The optimal choice for β is $((t+t^{-1})/3)^{2/3}$, and the value for t then follows by taking

t with $\sqrt{\frac{1}{2}}<t\leqslant\sqrt{2}$ such that

$$t((t+t^{-1})/3)^{-1/3}m^{1/3}(\log_2 m)^{-1/3}$$

is a power of 2. In the worst case $t=\sqrt{2}$ we find $\beta=(\frac{1}{2})^{1/3}\approx 0.794$, so that the precomputation can be done in time $2^{(1.588+o(1))m^{1/3}(\log_2 m)^{2/3}}$ (cf. (2.19), (2.22)). If we are so lucky that t can be chosen as 1, we find $\beta=(\frac{4}{9})^{1/3}\approx 0.764$, which makes the precomputation slightly faster.

To compute $\log_g y$ for $y\in\mathbf{F}_{2^m}^*$ we proceed as follows. We find e such that $y\cdot g^e \bmod f$ of norm $\leqslant L[1;1]$ is smooth with respect to $L[\frac{2}{3};1]$ in time $L[\frac{1}{3};\frac{1}{3}]$. Let $\bar{y}$ be one of the irreducible factors of $y\cdot g^e \bmod f$ with $\mathbf{N}(\bar{y})\leqslant L[\frac{2}{3};1]$. Let k be a power of 2 such that $\mathbf{N}(X^{[m/k]})\approx\mathbf{N}(v^k)$ for a polynomial $v\in\mathbf{F}_2[X]$ of norm $L[\frac{2}{3};1]$; in the worst case we get $\mathbf{N}(X^{[m/k]})=L[\frac{5}{6};\sqrt{\frac{1}{2}}]$ and $\mathbf{N}(v^k)=L[\frac{5}{6};\sqrt{2}]$. Find polynomials $v_1, v_2\in\mathbf{F}_2[X]$ of norm $\leqslant L[\frac{2}{3};1]$ such that $\bar{y}$ divides $u_1=X^{[m/k]+1}v_1+v_2$, and such that both $u_1/\bar{y}$ and $u_2=u_1^k \bmod f$ are smooth with respect to $L[\frac{1}{2};1]$. It follows from the choice for k that $u_1/\bar{y}$ and u_2 have norms bounded by $L[\frac{5}{6};\sqrt{\frac{1}{2}}]$ and $L[\frac{5}{6};\sqrt{2}]$, respectively, so that the probability that both are smooth with respect to $L[\frac{1}{2};1]$ is assumed to be

$$L[\tfrac{1}{3};-\sqrt{2}/6]\cdot L[\tfrac{1}{3};-\sqrt{2}/3]=L[\tfrac{1}{3};-\sqrt{\tfrac{1}{2}}].$$

Because $L[\frac{2}{3};1]^2/L[\frac{2}{3};1]$ of the pairs (v_1, v_2) satisfy the condition that $\bar{y}$ divides u_1, we must have that

$$L[\tfrac{2}{3};1]\geqslant L[\tfrac{1}{3};\sqrt{\tfrac{1}{2}}].$$

This condition is satisfied, and we find that the computation of the u_i's can be done in time

$$L[\tfrac{1}{3};\sqrt{\tfrac{1}{2}}]=2^{(\sqrt{1/2}+o(1))m^{1/3}(\log_2 m)^{2/3}}.$$

Because $\log_g u_2=(k\cdot(\log_g(u_1/\bar{y})+\log_g\bar{y})) \bmod (2^m-1)$, we have reduced the problem of computing the discrete logarithm of a polynomial of norm $L[\frac{2}{3};1]$ (the factor $\bar{y}$ of $y\cdot g^e \bmod f$) to the problem of computing the discrete logarithms of polynomials of norm $\leqslant L[\frac{1}{2};1]$ (the irreducible factors of $u_1/\bar{y}$ and u_2). To express $\log_g y$ in terms of $\log_g s$ for $s\in S$, we apply the above method recursively to each of the irreducible factors of $u_1/\bar{y}$ and u_2, thus creating a sequence of norms

$$L[\tfrac{1}{3}+\tfrac{1}{3};1],\quad L[\tfrac{1}{3}+\tfrac{1}{6};1],\quad L[\tfrac{1}{3}+\tfrac{1}{12};1],\ldots$$

that converges to $L[\frac{1}{3};1]$. The recursion is always applied to $<m$ polynomials per recursion step, and at recursion depth $O(\log m)$ all factors have norm $\leqslant L[\frac{1}{3};1]$, so that the total time to express $\log_g y$ in terms of $\log_g s$ for $s\in S$ is bounded by

$$m^{O(\log m)}L[\tfrac{1}{3};\sqrt{\tfrac{1}{2}}]=2^{(\sqrt{1/2}+o(1))m^{1/3}(\log_2 m)^{2/3}}.$$

We refer to [21] for some useful remarks concerning the implementation of this algorithm.

4. Factoring integers

4.A. Introduction

Finite abelian groups play an important role in several factoring algorithms. To illustrate this, we consider *Pollard's $p-1$ method*, which attempts to factor a composite number n using the following observation. For a prime p and any multiple k of the order $p-1$ of $(\mathbf{Z}/p\,\mathbf{Z})^*$, we have $a^k \equiv 1 \bmod p$, for any integer a that is not divisible by p. Therefore, if p divides n, then p divides $\gcd(a^k-1, n)$, and it is not unlikely that a nontrivial divisor of n is found by computing this gcd. This implies that prime factors p of n for which $p-1$ is s-smooth (cf. Subsection 2.A), for some $s \in \mathbf{Z}_{>0}$, can often be detected in $O(s \log_s n)$ operations in $\mathbf{Z}/n\,\mathbf{Z}$, if we take $k=k(s,n)$ as in (3.5). Notice that, in this method, we consider a multiplicative group modulo an unspecified prime divisor of n, and that we *hope* that the order of this group is smooth (cf. Subsections 3.A and 3.C).

Unfortunately, this method is only useful for composite numbers that have prime factors p for which $p-1$ is s-smooth for some *small* s. Among the generalizations of this method [7, 57, 84], one method, the *elliptic curve method* [45], stands out: instead of relying on fixed properties of a factor p, it depends on properties that can be randomized, independently of p. To be more precise, the multiplicative group $(\mathbf{Z}/p\,\mathbf{Z})^*$ of fixed order $p-1$ is replaced by the set of points of an elliptic curve modulo p (cf. (2.2)). This set of points is a group whose order is *close to p*; varying the curve will vary the order of the group and trying sufficiently many curves will almost certainly produce a group with a smooth order.

Another way of randomizing the group is by using class groups (cf. Subsection 2.C). For a small positive integer t with $t \equiv -n \bmod 4$, we have that $\Delta = -tn$ satisfies $\Delta \equiv 1 \bmod 4$ if n is odd. According to (2.14) and (2.16) a factorization of Δ can be obtained if we are able to compute an odd multiple of the largest odd divisor of the class number h_Δ. If h_Δ is s-smooth, such a multiple is given by the odd part of $k(s, B)$ as in (3.5), where $B = |\Delta|^{1/2+o(1)}$ (cf. (2.15)). By varying t, we expect to find a smooth class number after a while: with $s = L_n[\frac{1}{2}]$, we expect $L_n[\frac{1}{2}]$ trials (cf. Subsection 2.A, (2.15)), so that, with Subsection 3.C and (2.16), it takes expected time $L_n[1]$ to factor n. For details of this method, the *class group method*, we refer to [68].

In the next few subsections we will discuss the elliptic curve method (Subsection 4.B), its consequences for other methods (Subsection 4.C), and a very practical factoring algorithm that does not depend on the use of elliptic curves, the multiple polynomial variation of the quadratic sieve algorithm (Subsection 4.D). In Subsection 4.E we mention an open problem whose solution would lead to a substantially faster factoring algorithm.

Other methods and extensions of the ideas presented here can be found in [37, 47, 66]. The running times we derive are only informal upper bounds. For rigorous proofs of some of the results below, and for lower bounds, we refer to [59, 61].

4.B. Factoring integers with elliptic curves

Let n be a composite integer that we wish to factor. In this subsection we present an algorithm to factor n that is based on the theory of elliptic curves (cf. Subsection 2.B).

The running time analysis of this factoring algorithm depends upon an as yet unproved hypothesis, for which we refer to Remark (4.4).

4.1. THE ELLIPTIC CURVE METHOD (*cf.* [45]). We assume that $n>1$, that $\gcd(n, 6)=1$, and that n is not a power with exponent >1; these conditions can easily be checked. To factor n we proceed as follows:

> Randomly draw $a, x, y \in \mathbf{Z}/n\mathbf{Z}$, put $P=(x{:}y{:}1) \in V_n$ (cf. (2.8)), and select an integer $k=k(s, B)$ as in (3.5) (with s and B to be specified below). Attempt to compute $k\cdot P$ by means of the algorithm described in (2.10). If the attempt fails, a divisor d of n with $1<d<n$ is found, and we are done; otherwise, if we have computed $k\cdot P$, we start all over again.

This finishes the description of the algorithm.

4.2. EXPLANATION OF THE ELLIPTIC CURVE METHOD. We expect this algorithm to work, for a suitable choice of k, for the following reason. Let p and q be primes dividing n with $p<q$. In most iterations of the algorithm it will be the case that the pair a, y^2-x^3-ax when taken modulo p (modulo q) defines an elliptic curve over $\mathbf{F}_p$ (over $\mathbf{F}_q$). Now suppose that k is a multiple of the order of P_p; the value for k will be chosen such that a certain amount of luck is needed for this to happen. *If* it happens, it is unlikely that we are so lucky for q as well, so that k is *not* a multiple of the order of P_q. Then $k\cdot P$ cannot have been computed successfully (see (2.10)), and therefore a factorization of n has been found instead.

4.3. RUNNING TIME ANALYSIS. Let p be the smallest prime divisor of n, and let $\beta \in \mathbf{R}_{>0}$. We assume that the probability that the order of P_p is smooth with respect to $L_p[\beta]$ is approximately $L_p[-1/(2\beta)]$ (cf. Subsection 2.A and (2.4), and see Remark (4.4)). Therefore, if we take $k=k(L_p[\beta], p+2\sqrt{p}+1)$ as in (3.5) (cf. (2.4)), then about one out of every $L_p[1/(2\beta)]$ iterations will be successful in factoring n. According to Subsection 3.C and (2.10) each iteration takes $O(L_p[\beta]\cdot\log p)$ additions in V_n, which amounts to

$$O(L_p[\beta](\log p)(\log n)^2)$$

bit operations. The total expected running time therefore is

$$O((\log p)(\log n)^2 L_p[\beta+1/(2\beta)])$$

which becomes $O((\log n)^2 L_p[\sqrt{2}])$ for the optimal choice $\beta=\sqrt{\tfrac{1}{2}}$.

Of course the above choice for k depends on the divisor p of n that we do not know yet. This can be remedied by replacing p by a tentative upper bound v in the above analysis. If one starts with a small v that is suitably increased in the course of the algorithm, one finds that a nontrivial factor of n can be found in expected time $O((\log n)^2 L_p[\sqrt{2}])$ under the assumption made in (4.4). In the worst case $v=\sqrt{n}$ this becomes $L_n[1]$. The storage required is $O(\log n)$.

Another consequence is that for any fixed $\alpha \in \mathbf{R}_{>0}$, an integer n can be tested for smoothness with respect to $v=L_n[\alpha]$ in time $L_n[0]$; in case of smoothness the complete factorization of n can be computed in time $L_n[0]$ as well.

For useful remarks concerning the implementation of the elliptic curve method we refer to [13, 50, 44].

4.4. REMARK. A point that needs some further explanation is our assumption in (4.3) that the order of P_p is $L_p[\beta]$-smooth with probability approximately $L_p[-1/(2\beta)]$. Let $E_{\bar{a},\bar{b}}(\mathbf{F}_p)$ be the group under consideration. Regarding $\bar{a}$ and $\bar{b}$ as random integers modulo p, Proposition (2.5) asserts that the probability that $\# E_{\bar{a},\bar{b}}(\mathbf{F}_p)$ is smooth with respect to $L_p[\beta]$ and contained in the interval $(p-\sqrt{p}+1, p+\sqrt{p}+1)$ is essentially the same as the probability that a random integer in $(p-\sqrt{p}+1, p+\sqrt{p}+1)$ is $L_p[\beta]$-smooth.

From Subsection 2.A we know that a random integer $\leqslant p$ is $L_p[\beta]$-smooth with probability $L_p[-1/(2\beta)]$, and we *assume* here that the same holds for random integers in $(p-\sqrt{p}+1, p+\sqrt{p}+1)$. Because this has not been proved yet, the running times in (4.3) are conjectural.

Of course, if $\# E_{\bar{a},\bar{b}}(\mathbf{F}_p)$ is $L_p[\beta]$-smooth, then the order of P_p is $L_p[\beta]$-smooth as well.

4.5. A RIGOROUS SMOOTHNESS TEST. As explained in (4.4), the running times in (4.3) are conjectural. The result concerning the elliptic curve smoothness test can, however, be rigorously proved, in a slightly weaker and average sense. Briefly, the following has been shown in [61].

4.6. PROPOSITION. *There is a variant of the elliptic curve method for which the following statement is true. For each positive real number α there exists a function θ with $\theta(x)=o(1)$ for $x\to\infty$, such that the number $\psi'(x,y)$ of y-smooth integers $k\leqslant x$ that with probability at least $1-(\log k)/k$ are factored completely by the method in time at most $L_x[\theta(x)]$ satisfies*

$$\psi'(x, L_x[\alpha]) = \psi(x, L_x[\alpha])(1+O((\log\log x)^{11/2}(\log x)^{-1/2})),$$

with the O-constant depending on α.

In other words, apart from a small proportion, all smooth numbers behave as one would expect based on Subsection 2.A. The "variant" mentioned in Proposition (4.6) is very simple: first remove prime factors $\leqslant e^{64(\log\log x)^6}$ by trial division, and next apply the elliptic curve method to the remaining quotient, if it is not already equal to 1.

4.C. Methods depending on smoothness tests

The factoring algorithms presented so far are successful as soon as we find a certain abelian group with smooth order. In Subsections 4.C through E we will see a different application of smoothness. Instead of waiting for the occurrence of one lucky *group* with smooth order, the algorithms in this subsection combine many lucky instances of smooth *group elements*. For the algorithms in the present subsection the elliptic curve smoothness test that we have seen at the end of (4.3) will be very useful to recognize those smooth group elements. The algorithms in Subsections 4.D and 4.E do not need smoothness tests, but instead rely on sieving techniques. We abbreviate $L_n[\beta]$ to $L[\beta]$ (cf. Subsection 2.B).

4.7. DIXON'S RANDOM SQUARES ALGORITHM (*cf.* [28, 59]). Let n be a composite integer that we wish to factor, and let $\beta \in \mathbf{R}_{>0}$. In this algorithm one attempts to find integers x and y such that $x^2 \equiv y^2 \bmod n$ in the following way:

(1) Randomly select integers m until sufficiently many are found for which the least positive residue $r(m)$ of $m^2 \bmod n$ is $L[\beta]$-smooth.
(2) Find a subset of the m's such that the product of the corresponding $r(m)$'s is a square, say x^2.
(3) Put y equal to the product of the m's in this subset; then $x^2 \equiv y^2 \bmod n$.

Dixon has shown that, if n is composite, not a prime power, and free of factors $\leqslant L[\beta]$, then with probability at least $\frac{1}{2}$, a factor of n will be found by computing $\gcd(x+y, n)$, for x and y as above (cf. [28]). Therefore, we expect to factor n if we repeat the second and third step a small number of times. We will see that this leads to an algorithm that takes expected time $L[\sqrt{2}]$, and storage $L[\sqrt{\frac{1}{2}}]$.

Before analyzing the running time of this algorithm, let us briefly explain how the second step can be done. First notice that $\pi(L[\beta]) = L[\beta]$ (cf. Subsection 2.A). Therefore, each $r(m)$ can be represented by an $L[\beta]$-dimensional integer vector whose ith coordinate is the number of times the ith prime occurs in $r(m)$. A linear dependency modulo 2 among those vectors then yields a product of $r(m)$'s where all primes occur an even number of times, and therefore the desired x^2. This idea was first described in [52].

To analyze the running time of the random squares algorithm, notice that we need about $L[\beta]$ smooth m's in the first step to be able to find a linear dependency in the second step. According to Subsection 2.A a random integer $\leqslant n$ is $L[\beta]$-smooth with probability $L[-1/(2\beta)]$, and according to (4.3) such an integer can be tested for smoothness with respect to $L[\beta]$ in time $L[0]$. One $L[\beta]$-smooth $r(m)$ can therefore be found in expected time $L[1/(2\beta)]$, and $L[\beta]$ of them will take time $L[\beta + 1/(2\beta)]$. It is on this point that the random squares algorithm distinguishes itself from many other factoring algorithms that we discuss in these sections. Namely, it can be *proved* that, for random m's, the $r(m)$'s behave with respect to smoothness properties as random integers $< n$ (cf. [28]). This makes it possible to give a *rigorous* analysis of the expected running time of the random squares algorithm. For practical purposes, however, the algorithm cannot be recommended.

The linear dependencies in the second step can be found by means of Gaussian elimination in time $L[3\beta]$. The whole algorithm therefore runs in expected time $L[\max(\beta + 1/(2\beta), 3\beta)]$. This is minimized for $\beta = \frac{1}{2}$, so that we find that the random squares algorithm takes time $L[\frac{3}{2}]$ and storage $L[1]$.

As in Algorithm (3.10), however, we notice that at most $\log_2 n$ of the $L[\beta]$ coordinates of each vector can be non-zero. To multiply the matrix consisting of the vectors representing $r(m)$ by another vector takes therefore time at most $(\log_2 n)L[\beta] = L[\beta]$. Applying the coordinate recurrence method (cf. (2.19)) we conclude the dependencies can be found in expected time $L[2\beta]$, so that the random squares algorithm takes expected time $L[\max(\beta + 1/(2\beta), 2\beta)]$, which is $L[\sqrt{2}]$ for $\beta = \sqrt{\frac{1}{2}}$. The storage needed is $L[\sqrt{\frac{1}{2}}]$. For a rigorous proof using a version of the smoothness test from (4.5) that applies to this algorithm we refer to [61]. Notice that the random squares algorithm is in a way very similar to the index-calculus algorithm (3.10).

4.8. VALLÉE'S TWO-THIRDS ALGORITHM (*cf.* [79]). The fastest, fully proved factoring algorithm presently known is Vallée's two-thirds algorithm. The algorithm is only different from Dixon's random squares algorithm in the way the integers m in step (4.7)(1) are selected. Instead of selecting the integers m at random, as in (4.7), it is shown in [79] how those m can be selected in an almost uniform fashion in such a way that the least absolute remainder of $m^2 \bmod n$ is at most $4n^{2/3}$. According to Subsection 2.A the resulting factoring algorithm then takes expected time $L[\max(\beta+(\frac{2}{3})/(2\beta), 2\beta)]$, which is $L[\sqrt{\frac{4}{3}}]$ for $\beta=\sqrt{\frac{1}{3}}$. The storage needed is $L[\sqrt{\frac{1}{3}}]$. For a description of this algorithm and for a rigorous proof of these estimates we refer to [79].

4.9. THE CONTINUED FRACTION ALGORITHM (*cf.* [52]). If we could generate the m's in step (1) of the random squares algorithm in such a way that the $r(m)$'s are small, say $\leqslant\sqrt{n}$, then the $r(m)$'s would have a higher probability of being smooth, and that would probably speed up the factoring algorithm. This is precisely what is done in the continued fraction algorithm. We achieve an expected time $L[1]$ and storage $L[\frac{1}{2}]$.

Suppose that n is not a square, let a_i/b_i denote the ith continued fraction convergent to $\sqrt{n}$, and let $r(a_i)=a_i^2-nb_i^2$. It follows from the theory of continued fractions (cf. [32, Theorem 164]) that $|r(a_i)|<2\sqrt{n}$. Therefore we replace the first step of the random squares algorithm by the following:

> Compute $a_i \bmod n$ and $r(a_i)$ for $i=1,2,\ldots$ until sufficiently many $L[\beta]$-smooth $r(a_i)$'s are found.

The computation of the $a_i \bmod n$ and $r(a_i)$ can be done in $O((\log n)^2)$ bit operations (given the previous values) by means of an iteration that is given in [52]. The second step of the random squares algorithm can be adapted by including an extra coordinate in the vector representing $r(a_i)$ for the factor -1. The smoothness test is again done by means of the elliptic curve method. Assuming that the $|r(a_i)|$ behave like random numbers $<2\sqrt{n}$ *the probability of smoothness is* $L[1/(4\beta)]$, so that the total running time of the algorithm becomes $L[\max(\beta+1/(4\beta), 2\beta)]$. With the optimal choice $\beta=\frac{1}{2}$ we find that time and storage are bounded by $L[1]$ and $L[\frac{1}{2}]$, respectively.

We have assumed that the $|r(a_i)|$ have the same probability of smoothness as random numbers $<2\sqrt{n}$. The fact that all primes p dividing $r(a_i)$ and not dividing n satisfy $(\frac{n}{p})=1$, is not a serious objection against this assumption; this follows from [74, Theorem 5.2] under the assumption of the generalized Riemann hypothesis. More serious is that the $r(a_i)$ are generated in a deterministic way, and that the period of the continued fraction expansion for $\sqrt{n}$ might be short. In that case one may replace n by a small multiple.

The algorithm has proved to be quite practical, where we should note that in the implementations the smoothness of the $r(a_i)$ is usually tested by other methods. For a further discussion of the theoretical justification of this method we refer to [59].

4.10. SEYSEN'S CLASS GROUP ALGORITHM (*cf.* [74]). Another way of achieving time $L[1]$ and storage $L[\frac{1}{2}]$ is by using class groups (cf. Subsection 2.C). The advantage of the method to be presented here is that its expected running time can be proved rigorously, under the assumption of the generalized Riemann hypothesis (GRH). Let n be the composite integer to be factored. We assume that n is odd, and that $-n \equiv 1 \bmod 4$, which can be achieved by replacing n by $3n$ if necessary. Put $\Delta = -n$, and consider the class group C_Δ. We introduce some concepts that we need in order to describe the factorization algorithm.

4.11. RANDOMLY GENERATING REDUCED FORMS WITH KNOWN FACTORIZATION. Consider the prime forms I_p, with $p \leqslant c\cdot(\log|\Delta|)^2$, that generate C_Δ under the assumption of the GRH (cf. (2.17)). Let $e_p \in \{0, 1, \ldots, |\Delta|-1\}$ be randomly and independently selected, for every I_p. It follows from the bound on the class number h_Δ (cf. (2.15)) and from the fact that the I_p generate C_Δ that the reduced form $\prod I_p^{e_p}$ behaves approximately as a random reduced form in C_Δ; i.e., for any reduced form $f \in C_\Delta$ we have that $f = \prod I_p^{e_p}$ with probability $(1+o(1))/h_\Delta$, for $n \to \infty$ (cf. [74, Lemma 8.2]).

4.12. FINDING AN AMBIGUOUS FORM. Let $\beta \in \mathbf{R}_{>0}$; notice that $L[\beta] > c\cdot(\log|\Delta|)^2$. We attempt to find an ambiguous form (cf. (2.14)) in a way that is more or less similar to the random squares algorithm (4.7).

A randomly selected reduced form $(a, b) \in C_\Delta$ can be written as $\prod_{p \leqslant L[\beta]} I_p^{t_p}$ with probability $L[-1/(4\beta)]$ (cf. (2.18)), where at most $O(\log|\Delta|)$ of the exponents t_p are non-zero. According to (4.11) we get the same probability of smoothness if we generate the forms (a, b) as is done in (4.11). Therefore, if we use (4.11) to generate the random reduced forms, we find with probability $L[-1/(4\beta)]$ a relation

$$\prod_{\substack{p \leqslant c\cdot(\log|\Delta|)^2 \\ p \text{ prime}}} I_p^{e_p} = \prod_{\substack{p \leqslant L[\beta] \\ p \text{ prime}}} I_p^{t_p}.$$

With $r_p = e_p - t_p$, where $e_p = 0$ for $p > c\cdot(\log|\Delta|)^2$, we get

$$\textbf{(4.13)} \qquad \prod_{\substack{p \leqslant L[\beta] \\ p \text{ prime}}} I_p^{r_p} = 1_\Delta.$$

Notice that at most $c\cdot(\log|\Delta|)^2 + \log|\Delta|$ of the exponents r_p are non-zero. If all exponents are even, then the left-hand side of (4.13) with r_p replaced by $r_p/2$ is an ambiguous form. Therefore, if we have many equations like (4.13), and combine them in the proper way, we might be able to find an ambiguous form; as in the random squares algorithm (4.7) this is done by looking for a linear dependency modulo 2 among the vectors consisting of the exponents r_p.

There is no guarantee, however, that the thus constructed ambiguous form leads to a nontrivial factorization of $|\Delta|$. Fortunately, the probability that this happens is large enough, as shown in [74, Proposition 8.6] or [42, Section (4.6)]: if $L[\beta]$ equations as in (4.13) have been determined in the way described above, then a random linear dependency modulo 2 among the exponent vectors leads to a nontrivial factorization with probability at least $\frac{1}{2} - o(1)$.

4.14. RUNNING TIME ANALYSIS. The $(L[\beta]\times L[\beta])$-matrix containing the exponent vectors is sparse, as reasoned above, so that a linear dependency modulo 2 can be found in expected time $L[2\beta]$ by means of the coordinate recurrence method (cf. (2.19)). For a randomly selected reduced form (a, b), we assume that a can be tested for $L[\beta]$-smoothness in time $L[0]$ (cf. (4.3)). Generation of the $L[\beta]$ equations like (4.13) then takes time $L[\beta+1/(4\beta)]$, under the assumption of the GRH. The whole algorithm therefore takes expected time $L[\max(\beta+1/(4\beta), 2\beta)]$, which is $L[1]$ for $\beta=\frac{1}{2}$, under the assumption of the generalized Riemann hypothesis.

We can prove this expected running time rigorously under the assumption of the GRH, if we adapt the smoothness test from Proposition (4.6) to this situation. The argument given in [42] for this proof is not complete; the proof can however be repaired by incorporating [61, Theorem B′] in the proof of [74, Theorem 5.2].

4.D. *The quadratic sieve algorithm*

In this subsection we briefly describe practical factoring algorithms that run in expected time $L_n[1]$, and that existed before the elliptic curve method. As the methods from the previous subsection, but unlike the elliptic curve method, the running times of the algorithms to be presented here do not depend on the size of the factors. Nevertheless, the methods have proved to be very useful, especially in cases where the elliptic curve method performs poorly, i.e., if the number n to be factored is the product of two primes of about the same size. We abbreviate $L_n[\beta]$ to $L[\beta]$.

4.15. POMERANCE'S QUADRATIC SIEVE ALGORITHM (*cf.* [59]). The quadratic sieve algorithms only differ from the algorithms in (4.7), (4.8), and (4.9) in the way the $L[\beta]$-smooth quadratic residues modulo n are determined, for some $\beta\in\mathbf{R}_{>0}$. In the ordinary quadratic sieve algorithm that is done as follows. Let $r(X)=([\sqrt{n}]+X)^2-n$ be a quadratic polynomial in X. For any $m\in\mathbf{Z}$ we have that

$$r(m)\equiv([\sqrt{n}]+m)^2 \bmod n$$

is a square modulo n, so in order to solve $x^2\equiv y^2 \bmod n$ we look for $\approx L[\beta]$ integers m such that $r(m)$ is $L[\beta]$-smooth.

Let $\alpha\in\mathbf{R}_{>0}$ and let $|m|\leqslant L[\alpha]$. Then $|r(m)|=O(L[\alpha]\sqrt{n})$, so that $|r(m)|$ is $L[\beta]$-smooth with probability $L[-1/(4\beta)]$ according to Subsection 2.A, if $|r(m)|$ behaves as a random integer $\leqslant L[\alpha]\sqrt{n}$. Under this assumption we find that we must take $\alpha\geqslant\beta+1/(4\beta)$, in order to obtain sufficiently many smooth $r(m)$'s for $|m|\leqslant L[\alpha]$.

We have that $(\frac{n}{p})=1$ for primes $p\neq 2$ not dividing n, because if $p|r(m)$, then $([\sqrt{p}]+m)^2\equiv n \bmod p$. As in (4.9), this is not a serious objection against our assumption that the $r(m)$'s have the same probability of smoothness as random numbers of order $L[\alpha]\sqrt{n}$ (cf. [74, Theorem 5.3] under the GRH). The problem is to prove that at least a certain fraction of the $r(m)$'s with $|m|\leqslant L[\alpha]$ behave with respect to smoothness properties as random numbers of order $L[\alpha]\sqrt{n}$. For a further discussion of this point see [59].

Now consider how to test the $L[\alpha]$ numbers $r(m)$ for smoothness with respect to $L[\beta]$. Of course, this can be done by means of the elliptic curve smoothness test in time

$L[\alpha]$ (cf. (4.3)), thus giving a method that runs in time $L[\max(\beta+1/(4\beta), 2\beta)]=L[1]$ for $\beta=\frac{1}{2}$ (cf. (2.19)). The same time can, however, also be achieved without the elliptic curve method. Let p be a prime $\leqslant L[\beta]$ not dividing n such that $p\neq 2$ and $(\frac{n}{p})=1$. The equation $r(X)\equiv 0 \bmod p$ then has two solutions $m_1(p)$ and $m_2(p)$, which can be found by means of a probabilistic method in time polynomial in $\log p$ (cf. [37, Section 4.6.2]). But then $r(m_i(p)+kp)\equiv 0 \bmod p$ for any $k\in\mathbf{Z}$. Therefore, if we have a list of values of $r(m)$ for all consecutive values of m under consideration, we easily find the multiples of p among them at locations $m_i(p)+kp$ for any $k\in\mathbf{Z}$ such that $|m_i(p)+kp|\leqslant L[\alpha]$, and $i=1, 2$. For every p this takes twice time $L[\alpha]/p$, so that for all $p\leqslant L[\beta]$ with $(\frac{n}{p})=1$ together, this so-called *sieving* can be done in time $\Sigma_p L[\alpha]/p=L[\alpha]$. A similar procedure takes care of the powers of p and $p=2$. We conclude that we indeed get the same time $L[1]$ as with the elliptic curve smoothness test, but now we need to store all $L[1]$ values $r(m)$. We should note, however, that sieving is in practice much faster than applying the elliptic curve smoothness test, and that the sieving interval can easily be divided into smaller consecutive intervals, to reduce the storage requirements. (Actually, not the $r(m)$'s, but their logarithms are stored, and the $r(m)$'s are not divided by p but $\log p$ is subtracted from $\log r(m)$ during the sieving.) For other practical considerations we refer to [59].

4.16. THE MULTIPLE POLYNOMIAL VARIATION (*cf.* [60, 76]). Because there is only one polynomial in (4.15) that generates all smooth numbers that are needed, the size of the sieving interval must be quite large. Also, the quadratic residues $r(m)$ grow linearly with the size of the interval, which reduces the smoothness probability. If we could use *many* polynomials as in (4.15) and use a smaller interval for each of them, we might get a faster algorithm. This idea is due to Davis (cf. [24]); we follow the approach that was independently suggested by Montgomery (cf. [60, 76]). This algorithm still runs in expected time $L[1]$.

Let $r(X)=a^2X^2+bX+c$, for $a, b, c\in\mathbf{Z}$. In order for $r(m)$ to be a quadratic residue modulo n, we require that the discriminant $D=b^2-4a^2c$ is divisible by n, because then $r(m)\equiv(am+b/(2a))^2 \bmod n$. We show how to select a, b and c so that $|r(m)|=O(L[\alpha]\sqrt{n})$ for $|m|\leqslant L[\alpha]$. Let $D\equiv 1 \bmod 4$ be a small multiple of n, and let $a\equiv 3 \bmod 4$ be free of primes $\leqslant L[\beta]$ (if p divides a then $r(X)$ has at most one root modulo p), such that $a^2\approx\sqrt{D}/L[\alpha]$ and the Jacobi symbol $(\frac{a}{D})$ equals 1. For a we take a probable prime satisfying these conditions (cf. (5.1)). We need an integer b such that $b^2\equiv D \bmod 4a^2$; the value for c then follows. We put $b_1=D^{(a+1)/4} \bmod a$, so that $b_1^2\equiv D \bmod a$ because a is a quadratic residue modulo D and $D\equiv 1 \bmod 4$. Hensel's lemma now gives us

$$b=b_1+a\cdot((2b_1)^{-1}((D-b_1^2)/a) \bmod a);$$

if b is even, we replace b by $b-a^2$, so that the result satisfies $b^2\equiv D \bmod 4a^2$.

It follows from $a^2\approx\sqrt{D}/L[\alpha]$ that $b=O(\sqrt{D}/L[\alpha])$, so that $c=O(L[\alpha]\sqrt{D})$. We find that $r(m)=O(L[\alpha]\sqrt{D})$ for $|m|\leqslant L[\alpha]$. For any a as above, we can now generate a quadratic polynomial satisfying our needs. Doing this for many a's, we can sieve over many shorter intervals, with a higher probability of success. Remark that this can be done in parallel and independently on any number of machines, each machine working on its own sequence of a's; see [17, 44, 60, 76, 78] for a discussion of the practical problems involved. The multiple polynomial variation of the quadratic sieve algorithm is the only currently available method by which an arbitrary 100-digit number can be factored within one month [44].

4.E. The cubic sieve algorithm

In this final subsection on factorization we mention an open problem whose solution would lead to a factorization algorithm that runs in expected time $L_n[s]$ for some s with $\sqrt{\frac{2}{3}} \leqslant s < 1$. Instead of generating sufficiently many smooth quadratic residues modulo n close to $\sqrt{n}$ as in Subsection 4.D, one attempts to find identities modulo n that involve substantially smaller smooth numbers, and that still can be combined to yield solutions to $x^2 \equiv y^2 \bmod n$. The idea presented here was first described in [22]; it extends a method by Reyneri [65] to factor numbers that are close to perfect cubes. We again abbreviate $L_n[\beta]$ to $L[\beta]$.

Suppose that for some $\beta < \frac{1}{2}$ we have determined integers a, b, c such that

$$\textbf{(4.17)} \qquad \begin{cases} |a|, |b|, |c| \leqslant n^{2\beta^2}, \\ b^3 \equiv a^2 c \bmod n, \\ b^3 \neq a^2 c. \end{cases}$$

Notice that the last two conditions imply that at least one of $|a|$, $|b|$, and $|c|$ is $\geqslant (n/2)^{1/3}$, so that $\beta \geqslant \sqrt{\frac{1}{6} - (\log_n 2)/6}$.

Consider the cubic polynomial $(aU+b)(aV+b)(a(-U-V)+b)$. Fix some α with $\alpha > \beta$. There are $L[2\alpha]$ pairs (u, v) such that $|u|$, $|v|$, and $|u+v|$ are all $\leqslant L[\alpha]$. For each of these $L[2\alpha]$ pairs we have

$$\begin{aligned}(au+b)(av+b)(a(-u-v)+b) &= -a^3uv(u+v) - a^2b(u^2+v^2+uv) + b^3 \\ &\equiv a^2(-auv(u+v) - b(u^2+v^2+uv) + c) \bmod n,\end{aligned}$$

due to (4.17). Because $-auv(u+v) - b(u^2+v^2+uv) + c = O(L[3\alpha]n^{2\beta^2})$ (cf. (4.17)), we assume that each pair has a probability $L[-2\beta^2/(2\beta)] = L[-\beta]$ to produce a relation modulo n between integers $au+b$ with $|u| \leqslant L[\alpha]$ and primes $\leqslant L[\beta]$ (cf. Subsection 2.A). The $L[2\alpha]$ pairs taken together therefore should produce $L[2\alpha-\beta]$ of those relations. Since there are $L[\alpha]$ integers of the form $au+b$ with $|u| \leqslant L[\alpha]$ and $\pi(L[\beta])$ primes $\leqslant L[\beta]$, and since $L[\alpha] + \pi(L[\beta]) = L[\alpha]$, these $L[2\alpha-\beta]$ relations should suffice to generate a solution to $x^2 \equiv y^2 \bmod n$.

For each fixed u with $|u| \leqslant L[\alpha]$, the $L[\beta]$-smooth integers of the form $-auv(u+v) - b(u^2+v^2+uv) + c$ for $|v|$ and $|u+v|$ both $\leqslant L[\alpha]$ can be determined in time $L[\alpha]$ using a sieve, as described in (4.15). Thus, finding the relations takes time $L[2\alpha]$. Finding a dependency modulo 2 among the relations to solve $x^2 \equiv y^2 \bmod n$ can be done by means of the coordinate recurrence method in time $L[2\alpha]$(cf. (2.19)).

With the lower bound on β derived above, we see that this leads to a factoring algorithm that runs in expected time $L_n[s]$ for some s with $\sqrt{\frac{2}{3}} \leqslant s < 1$, at least if we can find a, b, c as in (4.17) within the same time bound. If a, b, and c run through the integers $\leqslant n^{1/3+o(1)}$ in absolute value, one gets $n^{1+o(1)}$ differences $b^3 - a^2c$, so we expect that a solution to (4.17) exists. The problem is of course that nobody knows how to find such a solution efficiently for general n.

The cubic sieve algorithm might be useful to factor composite n of the form b^3-c with c small, and for numbers of the form $y^n \pm 1$ (cf. [14]). For a discrete logarithm algorithm that is based on the same idea we refer to [22].

5. Primality testing

5.A. Introduction

As we will see in Subsection 5.B, it is usually easy to prove the compositeness of a composite number, without finding any of its factors. Given the fact that a number is composite, it is in general quite hard to find its factorization, but once a factorization is found it is an easy matter to verify its correctness. For prime numbers it is just the other way around. There it is easy to find the answer, i.e., prime or composite, but in case of primality it is not at all straightforward to verify the correctness of the answer. The latter problem, namely *proving* primality, is the subject of Subsections 5.B and 5.C. By *primality test* we will mean an algorithm to prove primality.

In Subsection 4.B we have seen that replacing the multiplicative group $(\mathbf{Z}/p\,\mathbf{Z})^*$ in Pollard's $p-1$ method by the group $E(\mathbf{Z}/p\,\mathbf{Z})$, for an elliptic curve E modulo p (cf. Subsection 2.B), resulted in a more general factoring algorithm. In Subsection 5.C we will see that a similar change in an older primality test that is based on the properties of $(\mathbf{Z}/p\,\mathbf{Z})^*$ leads to new primality tests.

This older algorithm is reviewed in Subsection 5.B, together with some well-known results concerning probabilistic compositeness algorithms. The primality tests that depend on the use of elliptic curves are described in Subsection 5.C.

More about primality tests and their implementations can be found in [83, 47].

5.B. Some classical methods

Let n be a positive integer to be tested for primality. In this subsection we review a method, based on a variant of *Fermat's theorem*, by which compositeness of n can easily be proved. If several attempts to prove the compositeness of n by means of this method have failed, then it is considered to be very likely that n is a prime; actually, such numbers are called *probable primes*. It remains to *prove* that such a number is prime. For this purpose, we will present a method that is based on a theorem of Pocklington, and that makes use of the factorization of $n-1$.

5.1. A PROBABILISTIC COMPOSITENESS TEST. Fermat's theorem states that, if n is prime, then $a^n \equiv a \bmod n$ for all integers a. Therefore, to prove that n is composite, it suffices to find an integer a for which $a^n \not\equiv a \bmod n$; such an a is called a *witness* to the compositeness of n. Unfortunately, there exist composite numbers, the so-called *Carmichael numbers*, for which no witnesses exist, so that a compositeness test based on Fermat's theorem cannot be guaranteed to work.

The following variant of Fermat's theorem does not have this disadvantage: if n is prime, then $a^u \equiv \pm 1 \bmod n$ or $a^{u \cdot 2^i} \equiv -1 \bmod n$ for an integer $i \in \{1, 2, \ldots, k-1\}$, where

$0<a<n$ and $n-1=u\cdot 2^k$ with u odd. Any a for which no such i exists is again called a witness to the compositeness of n; if a is not a witness, we say that n passes the test for this a. It has been proved [64] that for an odd composite n, there are at least $3(n-1)/4$ witnesses among $\{1,2,\ldots,n-1\}$. Therefore, if we randomly select some a's from this interval, and subject n to the test using these a's, it is rather unlikely that a composite n passes all tests. A number passing several tests, say 10, is called a *probable prime*.

In [49] Miller has shown that, if the generalized Riemann hypothesis holds, then there is for each composite n a witness in $\{2,3,\ldots,c\cdot(\log n)^2\}$ for some effectively computable constant c; according to [5] the value $c=2$ suffices. Notice that a proof of the generalized Riemann hypothesis therefore would lead to a primality test that runs in time polynomial in $\log n$. For a weaker probabilistic compositeness test, based on Jacobi symbols, see [77]; it is weaker in the sense that each witness for this test is also a witness for the above test, but not conversely.

Now that we can recognize composite numbers, let us consider how to prove the primality of a probable prime.

5.2. POCKLINGTON'S THEOREM (cf. [55]). *Let n be an integer >1, and let s be a positive divisor of $n-1$. Suppose there is an integer a satisfying*

$$a^{n-1}\equiv 1 \bmod n,$$

$$\gcd(a^{(n-1)/q}-1,n)=1 \quad \textit{for each prime } q \textit{ dividing } s.$$

Then every prime p dividing n is 1 mod *s, and if $s>\sqrt{n}-1$ then n is prime.*

We omit the proof of this theorem, as it can easily be deduced from the proof of a similar theorem below (cf. 5.4)), by replacing the role that is played by $E(\mathbf{Z}/p\,\mathbf{Z})$ in that proof by $(\mathbf{Z}/p\,\mathbf{Z})^*$ here. Instead, let us consider how this theorem can be employed to prove the primality of a probable prime n.

Apparently, to prove the primality of n by means of this theorem, we need a factor s of $n-1$, such that $s>\sqrt{n}-1$, and such that the complete factorization of s is known. Given such an s, we simply select non-zero integers $a\in\mathbf{Z}/n\,\mathbf{Z}$ at random until both conditions are satisfied. For such a, the first condition must be satisfied, unless n is composite. The second condition might cause more problems, but if n is prime then $q-1$ out of q choices for a will satisfy it, for a fixed q dividing s. Therefore, if an a satisfying both conditions has not been found after a reasonable number of trials, we begin to suspect that n is probably not prime, and we subject n to some probabilistic compositeness tests as in (5.1).

The main disadvantage of this method is that an s as above is not easy to find, because factoring $n-1$ is usually hard. If n is prime, then $n-1$ is the order of $(\mathbf{Z}/p\,\mathbf{Z})^*$ for the only prime p dividing n; in the next subsection we will randomize this order by replacing $(\mathbf{Z}/p\,\mathbf{Z})^*$ by $E(\mathbf{Z}/p\,\mathbf{Z})$. For other generalizations of this method we refer to the extensive literature on this subject [14, 47, 66, 73, 83].

5.3. THE JACOBI SUM TEST (*cf.* [3, 20]). The first primality test that could routinely handle numbers of a few hundred decimal digits was the Cohen–Lenstra version [20] of the primality test by Adleman, Pomerance, and Rumely [3]. It runs in time $(\log n)^{O(\log\log\log n)}$, which makes it the fastest deterministic primality test. Details concerning the implementation of this algorithm can be found in [19]. For a description of an improved version and its implementation we refer to [11].

5.C. *Primality testing using elliptic curves*

We assume that the reader is familiar with the material and the notation introduced in Subsection 2.B. In this subsection we discuss the consequences of the following analogue of Theorem (5.2).

5.4. THEOREM. *Let $n>1$ be an integer with* $\gcd(n,6)=1$. *Let $E=E_{a,b}$ be an elliptic curve modulo n (cf.* (2.7)), *and let m and s be positive integers with s dividing m. Suppose there is a point $P\in(V_n-\{O\})\cap E(\mathbf{Z}/n\mathbf{Z})$ (cf.* (2.8)) *satisfying*

$$m\cdot P=O \quad (cf.\ (2.10)),$$

$$(m/q)\cdot P \text{ is defined and different from } O, \text{ for each prime } q \text{ dividing } s,$$

where in (2.10) *we choose the a that is used in the definition of the elliptic curve $E_{a,b}$. Then $\#E(\mathbf{F}_p)\equiv 0 \bmod s$ for every prime p dividing n (cf.* (2.2), (2.7)), *and if $s>(n^{1/4}+1)^2$ then n is prime.*

PROOF. Let p be a prime dividing n, and let $Q=(m/s)\cdot P_p\in E(\mathbf{F}_p)$. By (2.10) we have $s\cdot Q=m\cdot P_p=(m\cdot P)_p=O_p$, so the order of Q divides s. If q is a prime dividing s then $(s/q)\cdot Q=(m/q)\cdot P_p=((m/q)\cdot P)_p\neq O_p$, because $(m/q)\cdot P\neq O$ (cf. (2.9)). The order of Q is therefore not a divisor of s/q, for any prime q dividing s, so this order equals s, and we find that $\#E(\mathbf{F}_p)\equiv 0 \bmod s$.

In (2.4) we have seen that $\#E(\mathbf{F}_p)=p+1-t$, for some integer t with $|t|\leqslant 2\sqrt{p}$ (Hasse's inequality). It follows that $(p^{1/2}+1)^2\geqslant \#E(\mathbf{F}_p)$. With $s>(n^{1/4}+1)^2$ and $\#E(\mathbf{F}_p)\equiv 0 \bmod s$ this implies that $p>\sqrt{n}$, for any prime p dividing n, so that n must be prime. This proves the theorem. □

5.5. REMARK. The proof of Theorem (5.2) follows the same lines, with $p-1$ replacing m.

Theorem (5.4) can be used to prove the primality of a probable prime n in the following way, an idea that is due to Goldwasser and Kilian (cf. [30]); for earlier applications of elliptic curves to primality tests see [10, 18].

5.6. OUTLINE OF THE PRIMALITY TEST. First, select an elliptic curve E over $\mathbf{Z}/n\mathbf{Z}$ and an integer m, such that $m=\#E(\mathbf{Z}/n\mathbf{Z})$ if n is prime, and such that m can be written as kq for a small integer $k>1$ and probable prime $q>(n^{1/4}+1)^2$; in (5.7) and (5.9) we will present two methods to select E and m. Next, find a point $P\in E(\mathbf{Z}/n\mathbf{Z})$ satisfying the requirements in Theorem (5.4) with $s=q$, on the assumption that q is prime. This is done as follows. First, use (2.11) to find a random point $P\in E(\mathbf{Z}/n\mathbf{Z})$. Next, compute $(m/q)\cdot P=k\cdot P$; if $k\cdot P$ is undefined, we find a nontrivial divisor of n, which is exceedingly unlikely. If $k \cdot P = O$, something that happens with probability $<\frac{1}{2}$ if n is prime, select a new P and try again. Otherwise, verify that $q \cdot (k \cdot P) = m \cdot P = O$, which must be the case if n is prime, because in that case $\# E(\mathbf{Z}/n\mathbf{Z}) = m$. The existence of P now proves that n is prime if q is prime, by (5.4). Finally, the primality of q is proved recursively.

We will discuss two methods to select the pair E, m.

5.7. THE RANDOM CURVE TEST (*cf.* [30]). Select a random elliptic curve E modulo n as described in (2.11), and attempt to apply the division points method mentioned in (2.6) to E. If this algorithm works, then it produces an integer m that is equal to $\# E(\mathbf{Z}/n\mathbf{Z})$ if n is prime. If the algorithm does not work, then n is not prime, because it is guaranteed to work for prime n.

This must be repeated until m satisfies the requirements in (5.6).

5.8. THE RUNNING TIME OF THE RANDOM CURVE TEST. First remark that the recursion depth is $O(\log n)$, because $k > 1$ so that $q \leqslant (\sqrt{n}+1)^2/2$ (cf. (2.4)). Now consider how often a random elliptic curve E modulo n has to be selected before a pair E, m as in (5.6) is found. Assuming that n is prime, $\# E(\mathbf{Z}/n\mathbf{Z})$ behaves approximately like a random integer near n, according to Proposition (2.5). Therefore, the probability that $m = kq$ with k and q as in (5.6) should be of the order $(\log n)^{-1}$, so that $O(\log n)$ random choices for E should suffice to find a pair E, m.

The problem is to *prove* that this probability is indeed of the order $(\log n)^{-c}$, for a positive constant c. This can be shown to be the case if we suppose that there is a positive constant c such that for all $x \in \mathbf{R}_{\geqslant 2}$ the number of primes between x and $x + \sqrt{2x}$ (cf. (2.4)) is of the order $\sqrt{x}(\log x)^{-c}$. Under this assumption, the random curve test proves the primality of n in expected time $O((\log n)^{9+c})$ (cf. [30]).

By a theorem of Heath-Brown, the assumption is *on the average* correct. In [30] it is shown that this implies that the fraction of primes n for which the algorithm runs in expected time polynomial in $\log n$, is at least $1 - O(2^{-l^{1/\log\log l}})$, where $l = [\log_2 n]$. In their original algorithm, however, Goldwasser and Killian only allow $k = 2$, i.e., they wait for an elliptic curve E such that $\# E(\mathbf{Z}/n\mathbf{Z}) = 2q$. By allowing more values for k, the fraction of primes for which the algorithm runs in polynomial time can be shown to be much higher [62] (cf. [2]). For a primality test that runs in expected polynomial time for all n, see (5.12) below.

Because the random curve test makes use of the division points method, it is not considered to be of much practical value. A practical version of (5.6) is the following test, due to Atkin [4]. Details concerning the implementation of this algorithm can be found in [51].

5.9. THE COMPLEX MULTIPLICATION TEST (*cf.* [4]). Here one does not start by selecting E, but by selecting the complex multiplication field L of E (cf. (2.6)). The field L can be used to calculate m, and only if m is of the required form kq (cf. (5.6)), one determines the pair a, b defining E.

This is done as follows. Let Δ be a negative fundamental discriminant $\leqslant -7$, i.e., $\Delta \equiv 0$ or $1 \bmod 4$ and there is no $s \in \mathbf{Z}_{>1}$ such that Δ/s^2 is a discriminant. Denote by

L the imaginary quadratic field $\mathbf{Q}(\sqrt{\Delta})$ and by $A=\mathbf{Z}[(\Delta+\sqrt{\Delta})/2]$ its ring of integers (cf. (2.6)). We try to find v with $v\bar{v}=n$ in A. It is known that $(\frac{\Delta}{n})=1$ and $(\frac{n}{p})=1$ for the odd prime divisors p of Δ are necessary conditions for the existence of v, where we assume that $\gcd(n, 2\Delta)=1$. If these conditions are not satisfied, select another Δ and try again. Otherwise, compute an integer $b \in \mathbf{Z}$ with $b^2 \equiv \Delta \bmod n$. This can for instance be done using a probabilistic method for finding the roots of a polynomial over a finite field [37, Section 4.6.2], where we assume that n is prime; for this algorithm to work, we do not need a proof that n is prime. If necessary add n to b to achieve that b and Δ have the same parity. We then have that $b^2 \equiv \Delta \bmod 4n$, and that $\mathbf{n}=\mathbf{Z}n+\mathbf{Z}((b+\sqrt{\Delta})/2)$ is an ideal in A with $\mathbf{n}\cdot\bar{\mathbf{n}}=A\cdot n$. Attempt to solve $\mathbf{n}=A\cdot v$ by looking for a shortest non-zero vector μ in the lattice $\mathbf{n}$. If $\mu\bar{\mu}=n$ then take $v=\mu$; otherwise $v\bar{v}=n$ is unsolvable.

Finding μ, and v if it exists, can for example be done by means of the reduction algorithm (2.12). With b as above, consider the form (a, b, c) with $a=n$ and $c=(b^2-\Delta)/(4n)$. For any two integers x and y the value $ax^2+bxy+cy^2$ of the form at x, y equals $|xn+y((b+\sqrt{\Delta})/2)|^2/n$, the square of the absolute value of the corresponding element of $\mathbf{n}$ divided by n. It follows that μ can be determined by computing integers x and y for which $ax^2+bxy+cy^2$ is minimal. More in particular, it follows that v with $v\bar{v}=n$ exists if and only if there exist integers x and y for which the form assumes the value 1.

Because $\gcd(n, 2\Delta)=1$, we have that $\gcd(n, b)=1$, so that the form (a, b, c) is primitive, which makes the theory of Subsection 2.C applicable. Apply the reduction algorithm (2.12) to (a, b, c); obviously, the set $\{ax^2+bxy+cy^2 : x, y \in \mathbf{Z}\}$ does not change in the course of the algorithm. Because a reduced form assumes its minimal value for $x=1$ and $y=0$, the x and y for which the original form (a, b, c) is minimized now follow, as mentioned in the last paragraph of (2.12). The shortest non-zero vector $\mu \in \mathbf{n}$ is then given by $xn+y((b+\sqrt{\Delta})/2)$. Now remark that $ax^2+bxy+cy^2=1$ if and only if the reduced form equivalent to (a, b, c) is the unit element 1_Δ. Therefore, if the reduced form equals 1_Δ, put $v=\mu$; otherwise select another Δ and try again because v with $v\bar{v}=n$ does not exist.

Assuming that v has been computed, consider $m=(v-1)(\bar{v}-1)$, and $m'=(-v-1)(-\bar{v}-1)$. If neither m nor m' is of the required form kq, select another Δ and try again. Supposing that $m=kq$, an elliptic curve E such that $\#E(\mathbf{Z}/n\,\mathbf{Z})=m$ if n is prime can be constructed as a function of a zero in $\mathbf{Z}/n\,\mathbf{Z}$ of a certain polynomial $F_\Delta \in \mathbf{Z}[X]$. To determine this polynomial F_Δ define, for a complex number z with $\operatorname{Im} z>0$,

$$j(z)=\frac{\left(1+240\sum_{k=1}^{\infty}\frac{k^3q^k}{1-q^k}\right)^3}{q\cdot\prod_{k=1}^{\infty}(1-q^k)^{24}}$$

where $q=e^{2\pi i z}$. Then

$$F_\Delta=\prod_{(a,b)}\left(X-j\left(\frac{b+\sqrt{\Delta}}{2a}\right)\right)$$

with (a, b) ranging over the set of reduced forms of discriminant Δ, see (2.12). The degree of F_Δ equals the class number of L, and is therefore $\approx\sqrt{|\Delta|}$. As these polynomials depend only on Δ, they should be tabulated. More about the computation of these polynomials can be found in [80, Sections 125–133].

Compute a zero $j\in \mathbf{Z}/n\mathbf{Z}$ of F_Δ over $\mathbf{Z}/n\mathbf{Z}$, and let c be a quadratic non-residue modulo n (assuming that n is prime). Put $k=j/(1728-j)$; then k is well-defined and non-zero because $\Delta\leqslant -7$. Finally, choose E as the elliptic curve $E_{3k,2k}$ or $E_{3kc^2,2kc^3}$ in such a way that $\#E(\mathbf{Z}/n\mathbf{Z})=m$ if n is prime; the right choice can be made as described at the end of (2.6).

We made the restriction $\Delta\leqslant -7$ only to simplify the exposition. If $n\equiv 1 \bmod 3$ (respectively $n\equiv 1 \bmod 4$), one should also consider $\Delta=-3$ (respectively $\Delta=-4$), as it gives rise to six (four) pairs E, m; the equations for the curves can in these cases be determined in a more straightforward manner, cf. [46].

5.10. THE RUNNING TIME OF THE COMPLEX MULTIPLICATION TEST. We present a heuristic analysis of the running time of the method just described. The computation of v is dominated by the computation of $\sqrt{\Delta} \bmod n$ and therefore takes expected time $O((\log n)^3)$ (cf. [37, Section 4.6.2]); with fast multiplication techniques this can be reduced to $O((\log n)^{2+\varepsilon})$. It is reasonable to expect that one has to try $O((\log n)^{2+\varepsilon})$ values of Δ before m (or m') has the required form, so that we may assume that the final Δ is $O((\log n)^{2+\varepsilon})$. For a reduced form (a, b) and $z=(b+\sqrt{\Delta})/(2a)$, $q=e^{2\pi i z}$, one can show that $|j(z)-q^{-1}|<2100$; and if, with the same notation, the summation in the definition of $j(z)$ is terminated after K terms and the product after K factors, then the error is $O(K^3q^K)$. To bound the coefficients of F_Δ we notice that $j(z)$ can only be large for small a. Since the number of reduced forms (a, b) with a fixed a is bounded by the number of divisors of a, there cannot be too many large $j(z)$'s. It follows that one polynomial F_Δ can be computed in time $|\Delta|^{2+o(1)}=O((\log n)^{4+\varepsilon})$; it is likely that it can be done in time $|\Delta|^{1+o(1)}=O((\log n)^{2+\varepsilon})$ using fast multiplication techniques. Assuming that n is prime, a zero of F_Δ can be computed in time

$$O((\deg F_\Delta)^2(\log n)^3)=O((\log n)^{5+\varepsilon})$$

(ordinary), or

$$O((\deg F_\Delta)(\log n)^{2+\varepsilon})=O((\log n)^{3+\varepsilon})$$

(fast). Heuristically, it follows that the whole primality proof takes time $O((\log n)^{6+\varepsilon})$, which includes the $O(\log n)$ factor for the recursion. The method has proved to be quite practical as shown in [51].

With fast multiplication techniques one gets $O((\log n)^{5+\varepsilon})$. As Shallit observed, the latter result can be improved to $O((\log n)^{4+\varepsilon})$, if we only use Δ's that can be written as the product of some small primes; to compute the square roots modulo n of the Δ's, it then suffices to compute the square roots of those small primes, which can be done at the beginning of the computation.

5.11. REMARK. It should be noted that both algorithms based on (5.6), if successful, yield a certificate of primality that can be checked in polynomial time.

5.12. THE ABELIAN VARIETY TEST (*cf.* [2]). A primality test that runs in expected polynomial time for all n can be obtained by using abelian varieties of higher dimensions, as proved by Adleman and Huang in [2]. We explain the basic idea underlying their algorithm, without attempting to give a complete description.

Abelian varieties are higher-dimensional analogues of elliptic curves. By definition, an abelian variety over a field K is a projective group variety A over K. The set of points $A(K)$ of an abelian variety over a field K has the structure of an abelian group. Moreover, if $K=\mathbf{F}_p$ then $\#A(\mathbf{F}_p)=p^g+O((4p)^{g-1/2})$, where g is the dimension of A. One-dimensional abelian varieties are the same as elliptic curves.

Examples of abelian varieties over $\mathbf{F}_p$, for an odd prime p, can be obtained as follows. Let f be a monic square-free polynomial of odd degree $2g+1$ over $\mathbf{F}_p$, and consider the *hyperelliptic curve* $y^2=f(x)$ over $\mathbf{F}_p$. Then the Jacobian A of this curve is an abelian variety of dimension g over $\mathbf{F}_p$. The elements of $A(\mathbf{F}_p)$ can in this case be regarded as pairs (a,b) with $a,b\in\mathbf{F}_p[T]$, a monic, $b^2\equiv f \bmod a$ and $\text{degree}(b)<\text{degree}(a)\leqslant g$. Note the analogy with the definition of reduced forms in Subsection 2.C and (2.12), with f playing the role of Δ. The composition in the abelian group $A(\mathbf{F}_p)$ can be done as in (2.13) (cf. [16]). The order of $A(\mathbf{F}_p)$ can be computed as described in [2], or by an extension of a method by Pila, who generalized the division points method (cf. (2.6)) to curves of higher genus and to abelian varieties of higher dimension [54].

The abelian variety test proceeds in a similar way as the random curve test, but with $g=1$ replaced by $g=2$. The order of $A(\mathbf{F}_p)$ is then in an interval of length $O(x^{3/4})$ around $x=p^2$. The main difference with the random curve test is that it can be *proved* that this interval contains sufficiently many primes [34]. The problem of proving the primality of a probable prime n is then reduced, in expected polynomial time, to proving the primality of a number of order of magnitude n^2. Although the recursion obviously goes in the wrong direction, it has been proved in [2] that, after a few iterations, we may expect to hit upon a number whose primality can be proved in polynomial time by means of the random curve test (5.7).

Acknowledgment

The second author is supported by the National Science Foundation under Grant No. DMS-8706176.

References

[1] ADLEMAN, L.M., A subexponential algorithm for the discrete logarithm problem with applications, in: *Proc. 20th Ann. IEEE Symp. on Foundations of Computer Science* (1979) 55–60.

[2] ADLEMAN, L.M. and M.A. HUANG, Recognizing primes in random polynomial time, Research report, Dept. of Computer Science, Univ. of Southern California, 1988; extended abstract in: *Proc. 19th Ann. ACM Symp. on Theory of Computing* (1987) 462–469.

[3] ADLEMAN, L.M., C. POMERANCE and R.S. RUMELY, On distinguishing prime numbers from composite numbers, *Ann. of Math.* **117** (1983) 173–206.

[4] ATKIN, A.O.L. Personal communication, 1985.

[5] BACH, E. *Analytic Methods in the Analysis and Design of Number-theoretic Algorithms* (MIT Press, Cambridge, MA, 1985).

[6] BACH, E., Explicit bounds for primality testing and related problems, *Math. Comp.*, to appear.

[7] BACH, E. and J. SHALLIT, Cyclotomic polynomials and factoring, in: *Proc. 26th Ann. IEEE Symp. on Foundations of Computer Science* (1985) 443–450; also: *Math. Comp.* **52** (1989) 201–219.

[8] BETH, T, N. COT and I. INGEMARSSON, eds., *Advances in Cryptology*, Lecture Notes in Computer Science, Vol. 209 (Springer, Berlin, 1985).

[9] BOREVIČ, Z.I. and I.R. ŠAFAREVIČ, *Teorija Čisel* (Moscow 1964; translated into German, English and French).

[10] BOSMA, W., Primality testing using elliptic curves, Report 85-12, Mathematisch Instituut, Univ. van Amsterdam, Amsterdam, 1985.

[11] BOSMA, W. and M.-P. VAN DER HULST, Fast primality testing, In preparation.

[12] BRASSARD, G., *Modern Cryptology*, Lecture Notes in Computer Science, Vol. 325 (Springer, Berlin, 1988).

[13] BRENT, R.P., Some integer factorization algorithms using elliptic curves, Research Report CMA-R32-85, The Australian National Univ., Canberra, 1985.

[14] BRILLHART, J., D.H. LEHMER, J.L. SELFRIDGE, B. TUCKERMAN and S.S. WAGSTAFF JR, *Factorizations of* $b^n \pm 1$, $b=2,3,5,6,7,10,11,12$ *up to High Powers*, Contemporary Mathematics, Vol. 22 (Amer. Mathematical Soc., Providence, RI, 2nd ed., 1988).

[15] CANFIELD, E.R., P. ERDŐS and C. POMERANCE, On a problem of Oppenheim concerning "Factorisatio Numerorum", *J. Number Theory* **17** (1983) 1–28.

[16] CANTOR, D.G., Computing in the Jacobian of a hyperelliptic curve, *Math. Comp.* **48** (1987) 95–101.

[17] CARON, T.R. and R.D. SILVERMAN, Parallel implementation of the quadratic sieve, *J. Supercomput.* **1** (1988) 273–290.

[18] CHUDNOVSKY, D.V. and G.V. CHUDNOVSKY, Sequences of numbers generated by addition in formal groups and new primality and factorization tests, *Adv. in Appl. Math.* **7** (1986) 187–237.

[19] COHEN, H. and A.K. LENSTRA, Implementation of a new primality test, *Math. Comp.* **48** (1987) 103–121.

[20] COHEN, H. and H.W. LENSTRA, JR, Primality testing and Jacobi sums, *Math. Comp.* **42** (1984) 297–330.

[21] COPPERSMITH, D., Fast evaluation of logarithms in fields of characteristic two, *IEEE Trans. Inform. Theory* **30** (1984) 587–594.

[22] COPPERSMITH, D., A.M. ODLYZKO and R. SCHROEPPEL, Discrete logarithms in GF(p), *Algorithmica* **1** (1986) 1–15.

[23] COPPERSMITH, D. and S. WINOGRAD, Matrix multiplication via arithmetic progressions, *J. Symbolic Comput.*, to appear; extended abstract in: *Proc. 19th ACM Symp. on Theory of Computing* (1987) 1–6.

[24] DAVIS, J.A. and D.B. HOLDRIDGE, Factorization using the quadratic sieve algorithm, Tech. Report SAND 83-1346, Sandia National Laboratories, Albuquerque, NM, 1983.

[25] DE BRUIJN, N.G., On the number of positive integers $\leqslant x$ and free of prime factors $> y$, II, *Indag. Math.* **38** (1966) 239–247.

[26] DEURING, M., Die Typen der Multiplikatorenringe elliptischer Funktionenkörper, *Abh. Math. Sem. Hansischen Univ.* **14** (1941) 197–272.

[27] DICKSON, L.E., *History of the Theory of Numbers, Vol. I* (Carnegie Institute of Washington, 1919; Chelsea, New York, 1971).

[28] DIXON, J.D., Asymptotically fast factorization of integers, *Math. Comp.* **36** (1981) 255–260.

[29] EL GAMAL, T., A subexponential-time algorithm for computing discrete logarithms over GF(p^2), *IEEE Trans. Inform. Theory* **31** (1985) 473–481.

[30] GOLDWASSER, S. and J. KILIAN, Almost all primes can be quickly certified, in: *Proc. 18th Ann. ACM Symp. on Theory of Computing* (1986) 316–329.

[31] GRÖTSCHEL, M., L. LOVÁSZ and A. SCHRIJVER, *Geometric Algorithms and Combinatorial Optimization* (Springer, Berlin, 1988).

[32] HARDY, G.H. and E.M. WRIGHT, *An Introduction to the Theory of Numbers* (Oxford Univ. Press, Oxford, 5th ed., 1979).

[33] IRELAND, K. and M. ROSEN, *A Classical Introduction to Modern Number Theory*, Graduate Texts in Mathematics, Vol. 84 (Springer, New York, 1982).

[34] IWANIEC, H. and M. JUTILA, Primes in short intervals, *Ark. Mat.* **17** (1979) 167–176.
[35] JACOBI, C.G.J., *Canon Arithmeticus* (Berlin, 1839).
[36] KANNAN, R. and A. BACHEM, Polynomial algorithms for computing the Smith and Hermite normal forms of an integer matrix, *SIAM J. Comput.* **8** (1979) 499–507.
[37] KNUTH, D.E., *The Art of Computer Programming, Vol 2, Seminumerical Algorithms* (Addison-Wesley, Reading, MA, 2nd ed., 1981).
[38] KNUTH, D.E., *The Art of Computer Programming, Vol 3, Sorting and Searching* (Addison-Wesley, Reading, MA, 1973).
[39] LAGARIAS, J.C., Worst-case complexity bounds for algorithms in the theory of integral quadratic forms, *J. Algorithms* **1** (1980) 142–186.
[40] LAGARIAS, J.C., H.L. MONTGOMERY and A.M. ODLYZKO, A bound for the least prime ideal in the Chebotarev density theorem, *Invent. Math.* **54** (1975) 137–144.
[41] LANG, S., *Algebraic Number Theory* (Addison-Wesley, Reading, MA, 1970).
[42] LENSTRA, A.K., Fast and rigorous factorization under the generalized Riemann hypothesis, *Proc. Kon. Ned. Akad. Wet. Ser. A* **91** (*Indag. Math.* **50**) (1988) 443–454.
[43] LENSTRA, A.K., H.W. LENSTRA, JR and L. LOVÁSZ, Factoring polynomials with rational coefficients, *Math. Ann.* **261** (1982) 515–534.
[44] LENSTRA, A.K. and M.S. MANASSE, Factoring by electronic mail, to appear.
[45] LENSTRA, JR, H.W., Factoring integers with elliptic curves, *Ann. of Math.* **126** (1987) 649–673.
[46] LENSTRA, JR, H.W., Elliptic curves and number-theoretic algorithms, in: *Proc. Internat. Congress of Mathematicians*, Berkeley, 1986 (Amer. Mathematical Soc., Providence, RI, 1988) 99–120.
[47] LENSTRA, JR, H.W. and R. TIJDEMAN, eds., *Computational Methods in Number Theory*, Mathematical Centre Tracts, Vol. 154/155 (Mathematisch Centrum, Amsterdam, 1982).
[48] MASSEY, J.L., Shift-register synthesis and BCH decoding, *IEEE Trans. Inform. Theory* **15** (1969) 122–127.
[49] MILLER, G.L., Riemann's hypothesis and tests for primality, *J. Comput. System. Sci.* **13** (1976) 300–317.
[50] MONTGOMERY, P.L., Speeding the Pollard and elliptic curve methods of factorization, *Math. Comp.* **48** (1987) 243–264.
[51] MORAIN, F., Implementation of the Goldwasser–Kilian–Atkin primality testing algorithm, INRIA Report 911, INRIA-Rocquencourt, 1988.
[52] MORRISON, M.A. and J. BRILLHART, A method of factoring and the factorization of F_7, *Math. Comp.* **29** (1975) 183–205.
[53] ODLYZKO, A.M., Discrete logarithms and their cryptographic significance, in: [8] 224–314.
[54] PILA, J., Frobenius maps of abelian varieties and finding roots of unity in finite fields, Tech. Report, Dept. of Mathematics, Stanford Univ., Standford, CA, 1988.
[55] POCKLINGTON, H.C., The determination of the prime and composite nature of large numbers by Fermat's theorem, *Proc. Cambridge Philos. Soc.* **18** (1914–16) 29–30.
[56] POHLIG, S.C. and M.E. HELLMAN, An improved algorithm for computing logarithms over GF(p) and its cryptographic significance, *IEEE Trans. Inform. Theory* **24** (1978) 106–110.
[57] POLLARD, J.M., Theorems on factorization and primality testing, *Proc. Cambridge Philos. Soc.* **76** (1974) 521–528.
[58] POLLARD, J.M., Monte Carlo methods for index computation (mod p), *Math. Comp.* **32** (1978) 918–924.
[59] POMERANCE, C., Analysis and comparison of some integer factoring algorithms, in: [47] 89–139.
[60] POMERANCE, C., The quadratic sieve factoring algorithm, in: [8] 169–182.
[61] POMERANCE, C., Fast, rigorous factorization and discrete logarithm algorithms, in: D.S. Johnson, T. Nishizeki, A. Nozaki and H.S. Wilf, eds., *Discrete Algorithms and Complexity* (Academic Press, Orlando, FL, 1987) 119–143.
[62] POMERANCE, C., Personal communication.
[63] POMERANCE, C., J.W. SMITH and R. TULER, A pipeline architecture for factoring large integers with the quadratic sieve algorithm, *SIAM J. Comput.* **17** (1988) 387–403.
[64] RABIN, M.O., Probabilistic algorithms for testing primality, *J. Number Theory* **12** (1980) 128–138.
[65] REYNERI, J.M., Unpublished manuscript.
[66] RIESEL, H., *Prime Numbers and Computer Methods for Factorization*, Progress in Mathematics, Vol. 57 (Birkhäuser, Boston, 1985).

[67] RIVEST, R.L., A. SHAMIR and L. ADLEMAN, A method for obtaining digital signatures and public-key cryptosystems, *Comm. ACM* **21** (1978) 120–126.
[68] SCHNORR, C.P. and H.W. LENSTRA, JR, A Monte Carlo factoring algorithm with linear storage, *Math. Comp.* **43** (1984) 289–311.
[69] SCHÖNHAGE, A., Schnelle Berechnung von Kettenbruchentwicklungen, *Acta Inform.* **1** (1971) 139–144.
[70] SCHOOF, R.J., Quadratic fields and factorization, in: [47] 235–286.
[71] SCHOOF, R.J., Elliptic curves over finite fields and the computation of square roots mod *p*, *Math. Comp.* **44** (1985) 483–494.
[72] SCHRIJVER, A., *Theory of Linear and Integer Programming* (Wiley, New York, 1986).
[73] SELFRIDGE, J.L. and M.C. WUNDERLICH, An efficient algorithm for testing large numbers for primality, in: *Proc. 4th Manitoba Conf. Numerical Math.*, University of Manitoba, Congressus Numerantium, Vol. XII (Utilitas Math., Winnipeg, Canada, 1975).
[74] SEYSEN, M., A probabilistic factorization algorithm with quadratic forms of negative discriminant, *Math. Comp.* **48** (1987) 757–780.
[75] SILVERMAN, J.H., *The Arithmetic of Elliptic Curves*, Graduate Texts in Mathematics, Vol. 106 (Springer, New York, 1986).
[76] SILVERMAN, R.D., The multiple polynomial quadratic sieve, *Math. Comp.* **48** (1987) 329–339.
[77] SOLOVAY, R., and V. STRASSEN, A fast Monte-Carlo test for primality, *SIAM J. Comput.* **6** (1977) 84–85; Erratum, *ibidem* **7** (1978) 118.
[78] TE RIELE, H.J.J., W.M. LIOEN and D.T. WINTER, Factoring with the quadratic sieve on large vector computers, Report NM-R8805, Centrum voor Wiskunde en Informatica, Amsterdam, 1988.
[79] VALLÉE, B., Provably fast integer factoring algorithm with quasi-uniform small quadratic residues, INRIA Report, INRIA-Rocquencourt, 1988.
[80] WEBER, H., *Lehrbuch der Algebra, Band 3* (Vieweg, Braunschweig, 1908).
[81] WESTERN, A.E. and J.C.P. MILLER, *Tables of Indices and Primitive Roots*, Royal Society Mathematical Tables, Vol. 9 (Cambridge Univ. Press, Cambridge, 1968).
[82] WIEDEMANN, D.H., Solving sparse linear equations over finite fields, *IEEE Trans. Inform. Theory* **32** (1986) 54–62.
[83] WILLIAMS, H.C., Primality testing on a computer, *Ars Combin.* **5** (1978) 127–185.
[84] WILLIAMS, H.C., A $p+1$ method of factoring, *Math. Comp.* **39** (1982) 225–234.

CHAPTER 13

Cryptography

Ronald L. RIVEST

MIT Laboratory for Computer Science, 545 Technology Square, Cambridge, MA 02139, USA

Contents

HANDBOOK OF THEORETICAL COMPUTER SCIENCE
Edited by J. van Leeuwen

1. Introduction

In 1976 Diffie and Hellman [52] proclaimed: "*We stand today on the brink of a revolution in cryptography.*" Today we are in the midst of that revolution. We have seen an explosion of research in cryptology. Many cryptosystems have been proposed, and many have been broken. Our understanding of the subtle notion of "cryptographic security" has steadily increased, and cryptosystems today are routinely *proven* to be secure (after making certain plausible assumptions). The fascinating relationships between cryptology, complexity theory, and computational number theory have gradually unfolded, enriching all three areas of research.

This chapter surveys the field of cryptology as it now exists, with an attempt to identify the key ideas and contributions. The reader who wishes to explore further will find available many excellent texts, collections, and survey articles [9, 13, 31, 46, 50, 49, 52, 54, 55, 67, 90 99, 99, 102, 117, 146, 148–151.], works of historical or political interest [12, 70, 92, 138, 157], relevant conference proceedings (CRYPTO, EUROSCRIPT, FOCS, STOC) [47, 23, 100] and bibliographies [14, 129].

2. Basics

Cryptography is about *communication in the presence of adversaries*. As an example, a classic goal of cryptography is *privacy*: two parties wish to communicate privately, so that an adversary knows nothing about what was communicated.

The invention of radio gave a tremendous impetus to cryptography, since an adversary can eavesdrop easily over great distances. The course of World War II was significantly affected by the use, misuse, and breaking of cryptographic systems used for radio traffic [92]. It is intriguing that the computational engines designed and built by the British to crack the German *Enigma* cipher are deemed by some to be the first real "computers"; one could argue that cryptography is the mother (or at least the midwife) of computer science.

A standard cryptographic solution to the privacy problem is a *secret-key cryptosystem*, which consists of the following:

- A *message space* $\mathcal{M}$: a set of strings (*plaintext messages*) over some alphabet.
- A *ciphertext space* $\mathcal{C}$: a set of strings (*ciphertexts*) over some alphabet.
- A *key space* $\mathcal{K}$: a set of strings (*keys*) over some alphabet.
- An *encryption algorithm* E mapping $\mathcal{K} \times \mathcal{M}$ into $\mathcal{C}$.
- A *decryption algorithm* D mapping $\mathcal{K} \times \mathcal{C}$ into $\mathcal{M}$. The algorithms E and D must have the property that $D(K, E(K, M)) = M$ for al $K \in \mathcal{K}$, $M \in \mathcal{M}$.

To use a secret-key cryptosystem, the parties wishing to communicate privately agree on a key K which they will keep secret (hence the name secret-key cryptosystem). They communicate a message M by transmitting the ciphertext $C = E(K, M)$. The recipient can decrypt the ciphertext to obtain the message M using K, since $M = D(K, C)$.

The cryptosystem is considered *secure* if it is infeasible in practice for an eavesdropper who learns $E(K, M)$, but who does not know K, to deduce M or any portion of M. This is an informal definition of security; we shall later see how this notion has been formalized, refined and improved.

As cryptography has matured, it has addressed many goals other than privacy, and considered adversaries considerably more devious than a mere passive eavesdropper. One significant new goal is that of *authentication*, where the recipient of a message wishes to verify that the message he has received has not been forged or modified by an adversary and that the alleged sender actually sent the message exactly as it was received. *Digital signatures* are a special technique for achieving authentication; they are to electronic communication what handwritten signatures are to paper-based communication.

But we are getting ahead of our story. In the next two subsections we review two "pre-revolutionary" (that is, pre-1976) cryptographic techniques which are "musts" for any survey of the field: the *one-time pad* and the *Data Encryption Standard*. After that we begin our survey of "modern" cryptology.

A note on terminology: the term *cryptosystem* refers to any scheme designed to work with a communication system in the presence of adversaries, for the purpose of defeating the adversaries' intentions. This is rather broad, but then so is the field. *Cryptography* refers to the art of *designing* cryptosystems, *cryptanalysis* refers to the art of *breaking* cryptosystems, and *cryptology* is the union of cryptography and cryptanalysis. It is not uncommon, however, even among professionals working in this area (including the author), to (mis)use the term "cryptography" to refer to the field of cryptology.

2.1. *The one-time pad*

The *one-time pad* is a nearly perfect cryptographic solution to the privacy problem. It was invented in 1917 by Gilbert Vernam [92] for use in telegraphy and has stimulated much subsequent work in cryptography. The one-time pad is a secret-key cryptosystem where the key is as long as the message being encrypted. Furthermore the key, once used, is discarded and never reused.

Suppose parties A and B wish to communicate privately using the one-time pad, and suppose further that they have previously agreed upon a secret key K which is a string of n randomly chosen bits. Then if A wishes to send an n-bit message M to B, she sends to B the ciphertext $C = M \oplus K$, where C is the bit-wise exclusive-or (mod-2 sum) of M and K. (For example, $0011 \oplus 0101 = 0110$.) The received ciphertext can be decrypted by B to obtain M, since $M = C \oplus K$. When another message is to be sent, another key K must be used—hence the name "one-time pad".

Russian spies have allegedly been captured with unused paper pads of printed key material for use in a one-time pad scheme. After a message was encrypted for transmission the spy would destroy the page of key material used to encrypt the message. The spy could then relax, since even if the ciphertext was intercepted it is *provably* impossible for the interceptor to decrypt the ciphertext and incriminate the spy. The proof is simple: the intercepted ciphertext C provides the interceptor no information whatsoever about M since *any* message M could have yielded C (if the key K is equal to $C \oplus M$).

This one-time pad is thus *provably* secure in an *information-theoretic* sense since the interceptor never has enough information to decrypt the ciphertext, and no amount of computational power could help him. The information-theoretic basis for cryptographic security was developed by Shannon [146] and later refined by Hellman [89]. Gilbert, MacWilliams and Sloane [72] have also extended information-theoretic approaches to handle the *authentication* problem.

While the one-time pad provides provable security, it is awkward to use since a large key must be generated, shared, and stored. As a consequence, the one-time pad is rarely used.

2.2. *The Data Encryption Standard* (*DES*)

This subsection describes the Data Encryption Standard, our second "pre-revolutionary" cryptosystem. Modern cryptosystems use relatively short keys (56 to 1000 bits), and are secure in a *computational* sense rather than in an information-theoretic sense. By this we mean that the adversary's task is *computationally infeasible* rather than information-theoretically impossible. The Data Encryption Standard (or DES) is a good example of a secret-key cryptosystem designed to be computationally secure. Researchers at IBM designed DES in the 1970s, and the U.S. National Bureau of Standards adopted DES as a standard for encrypting commercial and government unclassified information [120]. It is widely used, particularly in the banking industry. However its future as a standard is unclear since it is not certain for how much longer the NBS will support DES as a standard.

We now sketch the operation of DES. The DES algorithm takes as input a 64-bit message M and a 56-bit key K, and produces a 64-bit ciphertext C. DES first applies an initial fixed bit permutation IP to M to obtain M'. This permutation has no apparent cryptographic significance. Second, DES divides M' into a 32-bit left half L_0 and a 32-bit right half R_0. Third, DES executes the following operations for $i=1,\ldots,16$ (there are 16 "rounds"):

$$L_i = R_{i-1}, \tag{1}$$

$$R_i = L_{i-1} \oplus f(R_{i-1}, K_{i-1}). \tag{2}$$

Here f is a function that takes a 32-bit right half and a 48-bit *round key* and produces a 32-bit output. The function f is defined using eight substitution functions or *S-boxes*, each of which maps a 6-bit input into a 4-bit output. Each round key K_i contains a different subset of the 56 key bits. Finally the *pre-ciphertext* $C' = (R_{16}, L_{16})$ (note that the halves are swapped) is permuted according to IP^{-1} to obtain the final ciphertext C. It is easy to verify that DES is invertible from the above definition, independent of the definition of f.

In a typical application of DES, a message M is encrypted by breaking it into 64-bit blocks, and then encrypting each block *after XOR-ing it with the ciphertext for the previous block*. This is known as *cipher-block chaining;* it prevents repeated text in the message from yielding correspondingly repeated sections of ciphertext. Other such *modes of operation* for the use of DES, as well as proposed techniques for key management, have been published by the National Bureau of Standards.

Diffie and Hellman [53] argue that the choice of a 56-bit key makes DES vulnerable to a brute-force attack. For $20 million one might be able to build a machine consisting of 2^{20} chips, each of which can test 2^{20} keys/second, so that in 2^{16} seconds (18.2 hours) the entire key space can be searched for the key which maps a given plaintext into a given ciphertext.

Using a known-plaintext attack, Hellman shows how to break DES by performing a large pre-computation which essentially searches the entire key space, and which saves selected results, so that a *time–memory trade-off* results for the problem of later determining an unknown key used to encrypt the known plaintext [91].

3. The goals and tools of cryptology

As cryptology has developed, the number of goals addressed has expanded, as has the number of tools available for achieving those goals. In this section we survey some key goals and tools.

Cryptology provides methods that enable a communicating party to develop trust that his communications have the desired properties, in spite of the best efforts of an untrusted party (or adversary). The desired properties may include:

- *Privacy.* An adversary learns nothing useful about the message sent.
- *Authentication.* The recipient of a message can *convince himself* that the message as received originated with the alleged sender.
- *Signatures.* The recipient of a message can *convince a third party* that the message as received originated with the alleged signer.
- *Minimality.* Nothing is communicated to other parties except that which is specifically desired to be communicated.
- *Simultaneous exchange.* Something of value (e.g., a signature on a contract) is not released until something else of value (e.g., the other party's signature) is received.
- *Coordination.* In a multi-party communication, the parties are able to coordinate their activities toward a common goal even in the presence of adversaries.
- *Collaboration threshold.* In a multi-party situation, the desired properties hold as long as the number of adversaries does not exceed a given threshold.

At a high level, the tools available for the attainment of these goals include:

- *Randomness.* Each party may use a private natural source of randomness (such as a noise diode) to produce "truly random" bits in order to generate his own secret keys or to perform randomized computations [73].
- *Physical protection.* Each party must physically protect his secrets from the adversary. His most important secret is usually the secret key that he has randomly generated—this key will provide him with unique capabilities.

  By contrast, design information, such as equipment blueprints or cryptographic algorithm details, is usually assumed to be unprotectable, so security does not usually require the secrecy of such design information. (Kerckhoff's second requirement [92, p. 235] of a cryptosystem was that "compromise of the system should not inconvenience the correspondents.")

- *Channel properties.* Unusual properties of the communication channel can sometimes be exploited. For example, Alpern and Schneider [7] show how to communicate securely on channels for which an eavesdropper cannot tell *who* broadcasts each bit. Wyner [158] defeats eavesdroppers for whom reception is less reliable than for the intended receiver, or when the channel is analog rather than digital [159, 160]. Bennett et al. exploit the peculiarities of quantum effects in their channels [17]. And spread-spectrum channels are effectively unobservable to enemies who do not know the details of their use [71]. We do not pursue these variations further in this paper.
- *Information theory.* Some systems, such as the Vernam one-time pad [92] are secure in an information-theoretic sense: the adversary is never given enough information to work with to break the code; no amount of computational power can help him overcome this (see [146, 89]).
- *Computational complexity theory.* The adversary's task is more often *computationally infeasible*, rather than information-theoretically impossible. Modern cryptography uses computational complexity theory to design systems that one has reason to believe cannot be broken with any reasonable amount of computation in practice, even though they are breakable in principle (with extraordinary luck—by guessing a secret key—or by using inordinate amounts of computation).
- *Cryptographic operators.* These computational mappings—such as encryption and decryption functions, one-way functions, and pseudo-random sequence generators —are basic building blocks for constructing cryptographic systems. Note that these need not be *functions*, since they may use *randomization*, so that different computations may yield different outputs, even for the same input. Complex operators may be created by composing simpler ones.
- *Cryptographic protocols.* A *protocol* specifies how each party is to initiate and respond to messages, including erroneous or illegal messages. The protocol may also specify initialization requirements, such as setting up a directory of public keys. A party following the protocol will be protected against certain specified dangers, even if the other parties do not follow the protocol.

The design of protocols and the design of operators are rather independent, in the same sense that the implementation of an abstract data type may be independent of its use. The protocol designer creates protocols assuming the existence of operators with certain security properties. The operator designer proposes implementations of those operators, and tries to prove that the proposed operators have the desired properties.

4. Mathematical preliminaries

Many recently proposed cryptographic techniques depend heavily on number-theoretic concepts. In this section we review some basic number-theorectic and computational facts. For a more extensive review of elementary number theory see [122, 105, 8]. An excellent overview of the problems of factoring integers, testing primality, and computing discrete logarithms also appears in Chapter 12 of this Handbook [103].

It is apparently the case that it is dramatically easier to tell whether a given number is prime or composite than it is to factor a given composite number into its constituent prime factors; this difference in computational difficulty is the basis for many cryptosystems. Finding large prime numbers is useful for constructing cryptographic operators, while for many cryptosystems the adversary's task is provably as hard as factoring the product of two large prime numbers.

There are efficient algorithms for generating random k-bit prime numbers; these algorithms run in time polynomial in k [152, 132, 78, 5, 3] and come in two flavors: Monte Carlo probabilistic algorithms which may err with small probability, always terminate in polynomial time and are quite efficient in practice; and Las Vegas probabilistic algorithms which are always correct, generate as output a deterministic polynomial-time checkable proof of correctness, and run in expected polynomial time. Not only can random primes be generated, but Bach [11] has also shown how to create a random k-bit composite number in a factored form which uses primality testing as a subroutine.

On the other hand, to factor a number n seems to require time proportional to

$$e^{c\cdot\sqrt{\ln n\cdot\ln\ln n}}, \tag{3}$$

where the constant c is 1 for the fastest algorithms. Factoring numbers of more that 110 decimal digits is currently infeasible in general. Pomerance [127], Pomerance et al. [128], Riesel [135] and Dixon [56] discuss recent factoring methods.

Let Z_n denote the set of residue classes modulo n, and let Z_n^* denote the multiplicative subgroup of Z_n consisting of those residues which are relatively prime to n. We let $\phi(n)=|Z_n^*|$; this is called *Euler's totient function.* Let Q_n denote the set of all *quadratic residues* (or *squares*) modulo n; that is, $x\in Q_n$ iff there exists a y such that $x\equiv y^2 \pmod n$.

The Jacobi symbol $(\frac{x}{n})$ is defined for any $x\in Z_n^*$ and has a value in $\{-1, 1\}$; this value is easily computed by using the law of quadratic reciprocity, even if the factorization of n is unknown. If n is prime then $x\in Q_n \Leftrightarrow (\frac{x}{n})=1$; and if n is composite, $x\in Q_n \Rightarrow (\frac{x}{n})=1$. We let J_n denote the set $\{x \mid x\in Z_n^* \wedge (\frac{x}{n})=1\}$, and we let $\tilde{Q}_n$ denote the set of *pseudo-squares* modulo n: those elements of J_n which do *not* belong to Q_n. If n is the product of two primes, then $|Q_n|=|\tilde{Q}_n|$, and for any pseudo-square y the function $f_y(x)=y\cdot x$ maps Q_n one-to-one onto $\tilde{Q}_n$.

The *quadratic residuousity problem* is: given a composite n and $x\in J_n$, determine whether x is a square or a pseudo-square modulo n. This problem is believed to be computationally difficult, and is the basis for a number of cryptosystems.

Squaring and extracting square roots modulo n are frequently used operations in the design of cryptographic operators. We say that x is a *square root of* y, *modulo* n if $x^2\equiv y \pmod n$. If n has t prime factors, then x may have up to 2^t square roots. Rabin [131] proved that finding square roots modulo n is polynomial-time equivalent to factoring n; given an efficient algorithm for extracting square roots modulo n one can construct an efficient algorithm for factoring n, and vice versa. The following fact observed by Williams and Blum is also frequently useful: if n is the product of two primes each congruent to 3 mod 4, then squaring modulo n effects a permutation of Q_n.

It is frequently useful to use exponents larger than 2: the function $x^e \bmod n$ is called *modular exponentiation*; the modulus n may be either prime or composite. Unlike the squaring operator, modular exponentiation is one-to-one over Z_n if $\gcd(e, \phi(n)) = 1$ [137].

There are two ways to define an inverse operation to modular exponentiation, depending on whether e or x is to be solved for. In the first case, given x, y, and n, to compute an e (if any) such that $x^e \equiv y \pmod{n}$ is called computing the *discrete logarithm* of y, modulo n (with logarithm base x). We denote such an e as $index_{x,n}(y)$. We note that when n is prime there are many x such that $x^e \equiv y \pmod{n}$ has solutions for all $y \in Z_n^*$; such x's are called *generators*. Computing discrete logarithms seems to be difficult in general. However, Pohlig and Hellman [126] present effective techniques for this problem when n is prime and $n-1$ has only small prime factors. Adleman [1] shows how to compute discrete logarithms in the time as given in (3), so that computing discrete logarithms and factoring integers seem to have essentially the same difficulty, as a function of the size of the numbers involved. It is interesting to note that working over the finite field $GF(2^k)$ rather than working modulo n seems to make the problem substantially easier (see [43, 124]). See [45] for further improvements and general discussion.

The other way to invert modular exponentiation is to solve for x: given y, e, and n, to compute an x (if any) such that $x^e \equiv y \pmod{n}$ is called computing the *e-th root of y modulo n*. If n is prime, this computation can be performed in polynomial time [22, 4, 133], while if n is composite, this problem seems to be as hard as factoring n or computing discrete logarithms modulo n.

We say that a binary random variable X is *ε-biased* if the probability that $X=0$ is within ε of $\frac{1}{2}$: that is, if $\Pr(X=0) \in [\frac{1}{2}-\varepsilon, \frac{1}{2}+\varepsilon]$. This notion will be useful in our discussion of pseudo-random bit sequences.

5. Complexity-theoretic foundations of cryptography

Modern cryptography is founded on computational complexity theory. When we say that a system has been proven secure, we mean that a lower bound has been proven on the number of computational steps required to break the system. At this time, however, the young field of complexity theory has yet to prove a nonlinear lower bound for even one NP-complete problem. Thus the theory of cryptography today is based on certain unproved but seemingly plausible assumptions about the computational difficulty of solving certain problems, and the assumed existence of operators like one-way functions and trapdoor functions. In this section we review these operators.

5.1. Checksums and one-way functions

It is often useful to work with a function f which can take as input a long message M and produce as output a shorter value $f(M)$. Depending on the application, the function f we choose may need to have different properties. Typically, verifying the correspondence between M and $f(M)$ allows one to verify , with high confidence, that the message M has not been altered since $f(M)$ was computed.

The simplest such f are *checksums*. For example, a simple checksum of M is obtained by interpreting the bits of M as coefficients of a polynomial $M(x)$ over GF(2) and taking $f(M)$ as the remainder when $M(x)$ is divided by a fixed polynomial $p(x)$. Cyclic redundancy checksums are of this type. If the pair $(M, f(M))$ is transmitted over a noisy channel, transmission errors can be detected when the received pair (x, y) does not satisfy $y=f(x)$. Such a strategy works well against random errors, but not against malicious tampering by an adversary. Since anyone can compute f, this procedure provides the recipient no authentication regarding the identity of the sender. Checksums are therefore *not* suitable for typical cryptographic applications.

A more useful function for cryptographic applications is a *one-way function*. Such a function takes a message M and efficiently produces a value $f(M)$ such that it is computationally infeasible for an adversary, given $f(M)=z$, to find *any* message M' whatsoever (including $M'=M$) such that $f(M')=z$. A slightly stronger requirement is that it should be computationally infeasible for the adversary, given f, to come up with any pair of messages (x, y) such that $f(x)=f(y)$; such a function we call *claw-free*. (Because of the *birthday paradox*, for small message spaces it may be feasible in practice to find such a pair, even though it is infeasible in practice to invert f at a given point z [164].)

A publicly available one-way function has a number of useful applications.

(1) In a time-shared computer system, instead of storing a table of login passwords, one can store, for each password w, the value $f(w)$. Passwords can easily be checked for correctness at login, but even the system administrator cannot deduce any user's password by examining the stored table [60].

(2) In a public-key cryptosystem (see the following sections), it may be more efficient to sign $f(M)$ rather than signing M itself, since M may be relatively long whereas f can be designed to return a fixed-length result. Thus using $f(M)$ keeps the size of a signature bounded. In this application $f(M)$ is sometimes called a *message digest* or *fingerprint* for the message M. Since nobody can come up with two messages that have the same digest, the signer is protected against an adversary altering his signed messages.

5.2. *Trapdoor functions*

A *trapdoor function* f is like a one-way function except that there also exists a secret inverse function f^{-1} (the *trapdoor*) that allows its possessor to efficiently invert f at any point of his choosing. It should be easy to compute f on any point, but infeasible to invert f on any point without knowledge of the inverse function f^{-1}. Moreover, it should be easy to generate matched pairs of f's and corresponding f^{-1}'s. Once such a matched pair is generated, the publication of f should not reveal anything about how to compute f^{-1} on any point.

Trapdoor functions have many applications, as we shall see in the next sections. They are the basis for public-key cryptography. The ability to invert a particular trapdoor function f will uniquely identify a given party. That is, each party will publish his own trapdoor function f, and only the party who publishes a given trapdoor function f will be able to invert that function.

5.3. *One-way (and trapdoor) predicates*

A *one-way predicate*, which was first introduced in [79, 80], is a Boolean function $B: \{0, 1\}^* \rightarrow \{0, 1\}$ such that

(1) on input $v \in \{0, 1\}$ and 1^k, in expected polynomial time one can choose an x such that $B(x) = v$ and $|x| \leqslant k$, randomly and uniformly from the set of all such x;

(2) for all $c > 0$, for all k sufficiently large, no polynomial-time adversary given $x \in \{0, 1\}^*$ such that $|x| \leqslant k$ can compute $B(x)$ with probability greater than $\frac{1}{2} + 1/k^c$. (The probability is taken over the random choices made by the adversary and x such that $|x| \leqslant k$.)

A *trapdoor predicate* is a one-way predicate for which there exists, for every k, trapdoor information t_k the size of which is bounded by a polynomial in k and whose knowledge enables the polynomial-time computation of $B(x)$, for all x such that $|x| \leqslant k$.

These primitives are the basis for the probabilistic constructions for privacy and pseudo-random number generation discussed in later sections. Each party publishes his own trapdoor predicate B, and keeps private the associated trapdoor information which enables him alone to compute $B(x)$.

5.4. *Making appropriate complexity-theoretic assumptions*

Computational complexity theory works primarily with asymptotic complexity—what happens as the size of the problem becomes large. In order to apply notions of computational complexity theory to cryptography we must typically envisage not a single cryptosystem or cryptographic function but a family of them, parameterized by a *security parameter* k. That is, for each value of the security parameter k there is a specific cryptosystem or function. Or there may be a family of cryptosystems or functions for each value of k. We can imagine that a cryptosystem with security parameter k has inputs, outputs, and keys which are all strings of length k (or some suitable polynomial function of k). As the security parameter k becomes large, the complexity of the underlying mathematical problems embedded in the cryptosystem should become sufficiently great that one can hope that cryptographic security is obtained. Since we do not know for sure that $P \neq NP$, a "proof" of security is necessarily dependent on the assumption that certain computational problems are difficult as the inputs become large.

As an example, the assumption that factoring integers is hard might be formalized as: for any probabilistic polynomial-time (factoring) algorithm A, for all constants $c > 0$ and sufficiently large k, the chance that A can produce a nontrivial divisor of its input (where the input is the product of two randomly chosen k-bit primes) is at most $1/k^c$. Note that A may be a probabilistic algorithm, since any adversary worth his salt can also flip coins.

Then a careful proof of security would show that the ability of the adversary to defeat the cryptographic system a significant fraction of the time would contradict the assumed difficulty of the hard problem (e.g., factoring). This generally takes the form of a reduction, where it is shown how to solve the hard problem, given the ability to break the cryptographic system.

When formally defining the above primitives (one-way and trapdoor functions and predicates) one must carefully choose what computational power to give to the adversary attempting to break (i.e., invert) the primitive. The computational model should be strong enough to capture the computational power of real-life adversaries. To this end, we let the polynomial-time adversary be not only probabilistic but also nonuniform. The latter is appropriate since cryptosystems are often designed with a fixed size (that is, security parameter) in mind. Moreover, the most meaningful proofs of security are necessarily those proved with respect to the most powerful adversary. Thus, the adversary is modeled as an infinite family of probabilistic circuits (one for every security parameter k), the sizes of which grow as polynomials in k. However, the extent to which allowing an adversary to be nonuniform is really meaningful remains to be seen, since making this assumption normally just means that the assumption that (say) factoring is difficult, when formalized, has to be reformulated to say that factoring is even difficult for a nonuniform factoring procedure.

6. Privacy

The goal of privacy is to ensure that an adversary who overhears a given transmission learns nothing useful about the message transmitted. There are several ways of achieving this goal, which are sketched in this section.

6.1. Secret-key cryptosystems

A simple method to achieve privacy is to use a conventional secret-key cryptosystem such as DES. The parties wishing to communicate must initially arrange to share a common secret key K; this key is then used to encrypt all messages transmitted. An adversary who does not possess the shared secret key will not be able to decrypt any encrypted messages he happens to overhear. The difficulty with secret-key cryptosystems is that it is often awkward to establish the initial distribution of the secret shared key, a problem we now discuss.

A simple solution to the initialization problem is to use a courier to distribute the keys. This method requires that the courier visit each party who must be given the secret key. The courier must be trusted to ensure that the key is only given to the appropriate parties.

Sometimes it is known (or assumed) a priori that an adversary will be passive—that is, the adversary may eavesdrop on transmitted messages but will not transmit any messages himself. In such a case two parties can establish a shared secret key using *exponential key exchange*; an elegant technique proposed by Diffie and Hellman [52]. Let A and B be the two parties, and suppose that A and B agree (via a public dialogue that anyone can overhear) on a large prime p and a generator g of the multiplicative group Z_p^*. Then A and B choose respective large secret integers a and b, and they exchange with each other the values $g^a \bmod p$ and $g^b \bmod p$. Now A can compute $g^{ab} \bmod p$ (from a and g^b), and B can compute the same value (from b and g^a). Thus A and B can use the value $g^{ab} \bmod p$ as their shared secret key. An adversary who wants to determine this key is left with the problem of computing $g^{ab} \bmod p$ from $g^a \bmod p$ and $g^b \bmod p$; an apparently intractable problem.

6.2. *Deterministic public-key encryption*

The notion of a *public-key cryptosystem* was first published by Diffie and Hellman in 1976 [51, 52], although Merkle had earlier developed some of the conceptual framework [114]. The central idea is that of a trapdoor function, as defined earlier. According to Diffie and Hellman, a public-key cryptosystem should contain the following parts:

- A key-generation algorithm. This is a randomized polynomial-time algorithm $\mathscr{G}$, which, on input k (the security parameter), produces a public key E, and a corresponding private key D.
- An encryption algorithm. This algorithm takes as input a public key E and a message M, produces a ciphertext C, which we denote $E(M)$. This operation is one-to-one.
- A decryption algorithm. This algorithm takes as input a private key D and a ciphertext C, and produces a corresponding message $M = D(C)$.

These algorithms have the following properties:

- For every message M, $D(E(M)) = M$.
- For every message M, $E(D(M)) = M$.
- The key-generation, encryption, and decryption algorithms run in time polynomial in the length of their inputs. The key-generation algorithm is a randomized algorithm; the encryption and decryption algorithms are deterministic.
- Given the public key E, but not the private key D, the chance that a polynomial-time adversary can decrypt a random ciphertext $C = E(M)$ is less than any polynomial fraction. (Here M is chosen at random from the set of all messages, and we may take "polynomial fraction" to mean at least k^{-c} for some constant c and all sufficiently large k.) This implies that E is a trapdoor function.

We might also call the encryption algorithm a *trapdoor one-way permutation*, since it provides a permutation of the message space and since you can only go one way (from message to ciphertext but not vice versa) without knowledge of the secret trapdoor information D.

If user A of a communication network publishes her public key E_A in a directory of public keys, then anyone can send A private mail by encrypting a message M with A's public key. Only A possesses the decryption key D_A, so only A can decrypt such a message.

6.2.1. *RSA*

In 1977 Rivest, Shamir and Adleman [137] proposed a public-key cryptosystem satisfying the requirements proposed by Diffie and Hellman. In their scheme, each user's public key is a pair (e, n) of integers, such that n is the product of two large primes p and q and $\gcd(e, \phi(n)) = 1$. The encryption operation is then

$$C = M^e (\text{mod } n). \tag{4}$$

The corresponding private key is the pair (d, n), where $d \cdot e \equiv 1 \pmod{\phi(n)}$, and the decryption operation is

$$M = C^d (\text{mod } n). \tag{5}$$

There are several reasons to believe that RSA is secure. For example, it is provably as hard to derive the private key from the public key as it is to factor n. Furthermore, the RSA system is *multiplicative*: $E(X)\cdot E(Y)=E(X\cdot Y)$. For this reason, if an adversary could decrypt any polynomial fraction of the ciphertexts in polynomial time, then he could decrypt all ciphertexts in random polynomial time: to decrypt $E(X)$ it suffices to find (by random trial and error) a Y such that $E(X\cdot Y)$ (which is the same as $E(X)E(Y)$) can be decrypted (yielding XY), and then dividing the result by Y to obtain X. One might interpret this as saying that either RSA is uniformly secure or it is uniformly insecure.

Even stronger results have been proven. For example, it has been shown [83, 6, 18] that if a polynomial fraction of RSA ciphertexts cannot be decrypted in polynomial time, then neither can just the least significant bit of the message be guessed from the ciphertext with better than an ε bias.

Hastad [88] shows that it is unwise to use a low encryption exponent e, such as 3, if it is likely that a user may send the same message (or the same message with known variations) to a number of other users.

6.2.2. *Knapsacks*

A number of public-key cryptosystems have been proposed which are based on the *knapsack* (or—more properly—the *subset sum*) problem: given a vector $\boldsymbol{a}=(a_1, a_2, \ldots, a_n)$ of integers, and a target value C, determine if there is a length-n vector $\boldsymbol{x}$ of zeros and ones such that $\boldsymbol{a}\cdot\boldsymbol{x}=C$. This problem is NP-complete [69].

To use the knapsack problem as the basis for a public-key cryptosystem, you create a public key by creating a knapsack vector $\boldsymbol{a}$, and publish that as your public key. Someone else can send you the encryption of a message M (where M is a length-n bit vector), by sending you the value of the inner product $C=M\cdot\boldsymbol{a}$. Clearly, to decrypt this ciphertext is an instance of the knapsack problem. To make this problem easy for you, you need to build in a hidden structure (that is, a trapdoor) into the knapsack so that the encryption operation becomes one-to-one and so that you can decrypt a received ciphertext easily. It seems, however, that the problem of solving knapsacks containing a trapdoor is *not* NP-complete, so that the difficulty of breaking such a knapsack is no longer related to the P=NP question.

In fact, history has not been kind to knapsack schemes; most of them have been broken by extremely clever analysis and the use of the powerful L^3 algorithm [104] for working in lattices. See [116, 140, 142, 2, 144, 100, 33, 123].

Some knapsack or knapsack-like schemes are still unbroken. The Chor–Rivest scheme [40], and the multiplicative versions of the knapsack [116] are examples. McEliece has a knapsack-like public-key cryptosystem based on error-correcting codes [113]. This scheme has not been broken, and was the first scheme to use randomization in the encryption process.

6.3. *Probabilistic public-key encryption*

With the introduction of randomized or probabilistic cryptographic techniques, it becomes possible to propose satisfactory definitions of the mathematical security of a cryptographic system, and to prove that certain cryptosystems are secure under this definition (under suitable complexity-theoretic assumptions, as usual).

These probabilistic cryptosystems will make use of the one-way function and trapdoor functions primitives in a much more complex fashion than their earlier deterministic counterparts which (although quite useful and secure in practice) will not be able to satisfy the security definitions given here.

The pioneering work in this direction was performed by Goldwasser and Micali [79, 80]. Although the use of randomized techniques was itself not new (for example, [113]), using randomization to achieve a *provable* level of security was novel.

We begin by examining the rather subtle notion of cryptographic security.

6.3.1. *Attacks against a cryptosystem*

When proving that a cryptographic system is secure, it is important to carefully specify what sort of "attacks" the adversary may mount. It is not uncommon for a system to be secure against a weak attack (such as passive eavesdropping) but to be insecure against a more powerful attack (such as active eavesdropping and manipulation of the communication line).

In the simplest form of attack, the passive adversary merely observes legitimate users using the cryptographic system. He may see ciphertext only, or he may be able to see some plaintext/ciphertext pairs (a *known-plaintext* attack). In a public key set-up he can always generate a polynomial number of pairs of plaintext/ciphertext himself, using the public key.

A potentially more powerful attack, which has been considered and is probably more realistic in practice, is that of an adversary who is a legitimate user of the system himself. The adversary is then able to perform all of the actions permitted by a legitimate user before trying to break (decipher) ciphertexts sent between pairs of users.

More generally, we might assume that an adversary can manipulate the communications between any pair of legitimate users, and can even temporarily run their cryptographic equipment (but cannot take it apart to see its secret keys). For example, in a *chosen-ciphertext* attack the adversary is assumed to be able to see the plaintext corresponding to ciphertexts of his choice.

6.3.2. *The goals of the adversary*

"Success" for the adversary should be defined in the most generous manner, so that a proof of security rules out even the weakest form of success.

A modest goal to aim for in the design of a cryptosystem is that for most messages the adversary cannot derive the entire message from its ciphertext. This is too modest a goal, since it does not preclude the following problems:

- The cryptosystem is not secure for some probability distributions on the message space (e.g., the message space consists of only a polynomial number of messages which are known to the adversary).
- Partial information about messages may be easily computed from the ciphertext.
- It may be easy to detect simple but useful facts about traffic of messages, such as when the same message is sent more than once.

Note that deterministic public key cryptosystems as proposed by Diffie and Hellman achieve the modest goal above and yet suffer from the above problems:

- If m is drawn from a highly structured sparse message space, then $f(m)$ may be easy to invert (e.g., for the RSA function $f(0)$ and $f(1)$ are easily detectable).
- Some information about m is always easy to compute from $f(m)$ for any f, such as "the parity of $f(m)$" (or worse: for one-way candidate $f(x) = g^x \bmod p$ where p is prime and g is a generator, the least significant bit of x is easily computable from $f(x)$).

A much more ambitious goal is to require that for all probability distributions over the message spaces, the adversary cannot predict from the ciphertext with significantly increased accuracy any bit of information about the corresponding messages.

Essentially, this coincides with the existing formal security notions. Achieving it rules out all of the problems listed above.

6.3.3. *Definition of security: polynomial-time security*

Several definitions of security for probabilistic encryption schemes have been proposed and studied in [79, 161, 80]. All definitions proposed so far have been shown to be equivalent in [80, 118]. We provide one definition in detail, due to Goldwasser and Micali [80].

DEFINITION. We say that a probabilistic encryption scheme is *polynomial-time secure* if for all sufficiently large security parameters k, any probabilistic polynomial-time procedure that takes as input k (in unary) and a public key, and that produces as output two messages m_0 and m_1, cannot distinguish between a random encryption of m_0 and a random encryption of m_1 with probability greater than $\frac{1}{2} + 1/k^c$ for all c.

6.3.4. *Probabilistic encryption*

We begin by describing the probabilistic encryption technique proposed in [79, 80].

Alice creates a public key consisting of two parts: an integer n which is the product of two large primes p and q, and a pseudo-square $y \in \tilde{Q}_n$. We assume that the quadratic residuousity problem is hard, so that Alice (who knows the factorization of n) can distinguish squares from pseudo-squares modulo n, but Bob (who knows only n) cannot decide in probabilistic polynomial time whether x in J_n is a square or not mod n. The following theorem due to Goldwasser and Micali shows that if this decision problem is hard at all, then it is everywhere hard.

1. THEOREM (Goldwasser–Micali [79, 80]). *Let* $S \subset \{n \mid n = pq,\ |p| \approx |q|\}$. *If there exists a probabilistic polynomial-time algorithm A such that for $n \in S$,*

$$\Pr(A(n, x) \text{ decides correctly whether } x \in Q_n \mid x \in J_n) > \tfrac{1}{2} + \varepsilon, \tag{6}$$

where this probability is taken over the choice of $x \in J_n$ and A's random choices, then there exists a probabilistic algorithm B with running-time polynomial in ε^{-1}, δ^{-1} and $|n|$ such that for all $n \in S$, for all $x \in J_n$,

$$\Pr(B(x, n) \text{ decides correctly whether } x \in Q_n \mid x \in J_n) > 1 - \delta, \tag{7}$$

where this probability is taken over the random coin tosses of B.

Namely, a probabilistic polynomial-time bounded adversary cannot do better (except by a smaller than any polynomial advantage) than guess at random whether $x \in J_n$ is a square mod n, if the quadratic residuousity problem is not in BPP.

To send a message M to Alice using a probabilistic encryption scheme, Bob proceeds as follows. Let $M = m_1 m_2 \ldots m_k$ in binary notation. For $i = 1, \ldots, k$, Bob:

(1) randomly chooses an integer r from Z_n;

(2) if $m_i = 0$, sends $c_i = r^2 \bmod n$ to Alice; if $m_i = 1$, sends $c_i = y \cdot r^2 \bmod n$ to Alice.

When $m_i = 0$, Bob is sending a random square to Alice, whereas if $m_i = 1$, Bob is sending a random pseudo-square. (Alice needs to include y in her public key just so that Bob will be able to generate random pseudo-squares.) Since Alice can distinguish squares from pseudo-squares modulo n, she can decode the message.

Although decryption is easy for Alice, for an adversary the problem of distinguishing whether a given piece of ciphertext c_i represents a 0 or a 1 is *precisely* the problem of distinguishing squares from pseudo-squares that was assumed to be hard.

A natural generalization of the scheme based on any trapdoor predicate follows from [80]. Recall that a trapdoor predicate is a Boolean function $B: \{0,1\}^* \to \{0,1\}$ such that it is easy in expected polynomial time on input 1^k and bit v to choose an x randomly such that $B(x) = v$ and $|x| \leqslant k$; and no polynomial-time adversary given random $x \in X$ such that $|x| \leqslant k$ can compute $B(x)$ with probability greater than $\frac{1}{2} + 1/k^c$, for all $c > 0$ and for all sufficiently large k; if the trapdoor information is known however, it is easy to compute $B(x)$.

Fix a trapdoor predicate B. Let the security parameter of the system be k. Alice's public key contains a description of B, and her secret key contains the trapdoor information. Now Bob can send Alice a private message $M = m_1 m_2 \ldots m_k$ (this is M in binary notation) as follows. For $i = 1, \ldots, k$, Bob:

(1) randomly chooses a binary string x_i of length at most k such that $B(x) = m_i$;

(2) sends x_i to Alice.

To decrypt, Alice, who knows the trapdoor information, computes $m_i = B(x_i)$ for all $i = 1, \ldots, k$.

2. THEOREM (Goldwasser–Micali [80]). *If trapdoor predicates exist, then the above probabilistic public-key encryption scheme is polynomial-time secure.*

Implementation of trapdoor predicates based on the problem of factoring integers, and of inverting the RSA function can be found in [6]. We outline the RSA-based implementation. Let n be the public modulus, e the public exponent, and d the secret exponent. Let $B(x)$ be the least significant bit of $x^d \bmod n$ for $x \in Z_n^*$. Then, to select uniformly an $x \in Z_n^*$ such that $B(x) = v$ simply select a $y \in Z_n^*$ whose least significant bit is v and set $x = y^e \bmod n$.

A.C. Yao, in a pioneering paper [161], showed that the existence of any trapdoor length-preserving permutation implies the existence of a trapdoor predicate. Recently, Goldreich and Levin simplified Yao's construction as follows.

3. THEOREM (Goldreich–Levin [76]). *If trapdoor length-preserving permutations exist, then B is a trapdoor predicate, where B is defined as follows. Let $f: \{0,1\}^* \to \{0,1\}^*$ be* *a trapdoor length-preserving permutation. Let* $B(f(x), y) = xy \bmod 2$ (*the inner product of* x *and* y).

Now, to encrypt a single bit v, Alice simply selects at random a pair x, y of values from $\{0,1\}^k$ such that $xy \bmod 2 = v$ and obtains the ciphertext as $c = f(x)y$, the concatenation of $f(x)$ and y. How efficient are the probabilistic schemes? In the schemes described so far, the ciphertext is longer than the cleartext by a factor proportional to the security parameter. However, it has been shown [24, 28] using later ideas on pseudo-random number generation how to start with trapdoor functions and build a probabilistic encryption scheme that is polynomial-time secure for which the ciphertext is longer than the cleartext by only an additive factor. The most efficient probabilistic encryption scheme is due to Blum and Goldwasser [28] and is comparable with the RSA deterministic encryption scheme in speed and data expansion.

6.4. *Composition of cryptographic operators and multiple encryption*

Sometimes new cryptographic operators can be obtained by composing existing operators. The simplest example is that of multiple encryption.

Does multiple encryption increase security? Not always: consider the class of *simple substitution ciphers*, where the plaintext is turned into the ciphertext by replacing each plaintext letter with the corresponding ciphertext letter, determined according to some table. (Newspaper cryptograms are usually of this sort.) Since composing two simple substitution ciphers yields another simple substitution cipher, multiple encryption does not increase security.

On the other hand, multiple encryption using DES probably does increase security somewhat (see [95, 147]).

It is worth noting that the composition of two cryptosystems with n-bit keys can be broken in time $O(2^n)$ and space $O(2^n)$, using a *meet-in-the-middle* attack. Given a matching plaintext/ciphertext pair for the composed system, one can make a table of size 2^n of all possible intermediate values obtainable by encrypting the plaintext with the first system, and a second table of size 2^n of all possible intermediate values obtainable by decrypting the ciphertext with the second system. By sorting the two tables (which can be done in linear time using a bucket sort), and looking for overlaps, one can identify a pair of keys that take the given plaintext to the given ciphertext. Thus the composed system has difficulty proportional to 2^n, rather than proportional to 2^{2n}, as one would naively expect.

One very intuitive result, due to Even and Goldreich [62] (see also [10]), is that the composition of encryption schemes A and B is no weaker than A or B individually—security is not lost by composition. They assume that the two encryption keys are chosen *independently* and that the adversary can request the encryption of arbitrary text (that is, he can use a *chosen-plaintext* attack).

Luby and Rackoff [111, 112] prove a more powerful result, which shows that the composition generally *increases* security. Define an encryption scheme A to be $(1-\varepsilon)$-*secure* if no polynomial time procedure has a chance greater than $(1+\varepsilon)/2$ of distinguishing encryption functions from scheme A from truly random functions over the same domain (see [111] for a more precise definition and details). Then the composition of a $(1-\varepsilon)$-secure encryption scheme with a $(1-\delta)$-secure encryption scheme is $(\varepsilon\delta(2-\max(\varepsilon, \delta)))$-secure. This is an improvement whenever $\max(\varepsilon, \delta) < 1$.

In a similar vein, a number of researchers [80, 161, 107] have developed and refined proofs that the bit-wise XOR of several independent pseudo-random bit sequence generators is harder to predict (by a quantifiable amount) than any of the component generators.

7. Generating random or pseudo-random sequences and functions

We now examine in some detail the problem of generating random and pseudo-random sequences. One motivation for generating random or pseudo-random sequences is for use in the one-time pad, as described previously.

7.1. Generating random bit sequences

Generating a one-time pad (or, for that matter, any cryptographic key) requires the use of a "natural" source of random bits, such as a coin, a radioactive source or a noise diode. Such sources are absolutely essential for providing the initial secret keys for cryptographic systems.

However, many natural sources of random bits may be defective in that they produce *biased* output bits (so that the probability of a one is different from the probability of a zero), or bits which are *correlated* with each other. Fortunately one can remedy these defects by suitably processing the output sequences produced by the natural sources.

To turn a source which supplies biased but uncorrelated bits into one which supplies unbiased uncorrelated bits, von Neumann proposed grouping the bits into pairs, and then turning 01 pairs into 0s, 10 pairs into 1s, and discarding pairs of the form 00 and 11 [156]. The result is an unbiased uncorrelated source, since the 01 and 10 pairs will have an identical probability of occurring. Elias [59] generalizes this idea to achieve an output rate near the source entropy.

Handling a correlated bit source is more difficult. Blum [27] shows how to produce unbiased uncorrelated bits from a biased correlated source which produces output bits according to a known finite Markov chain.

For a source whose correlation is more complicated, Santha and Vazirani [139] propose modeling it *as a slightly random source*, where each output bit is produced by a coin flip, but where an adversary is allowed to choose *which* coin will be flipped, from among all coins whose probability of yielding "heads" is between δ and $1-\delta$. (Here δ is a small fixed positive quantity.) This is an extremely pessimistic view of the possible correlation; nonetheless U.V. Vazirani [153] shows that if one has *two, independent*, slightly random sources X and Y then one can produce "almost independent" ε-biased bits by breaking the outputs of X and Y into blocks $\boldsymbol{x},\boldsymbol{y}$ of length $k=\Omega(1/\delta^2 \log(1/\delta) \times \log(1/\varepsilon))$ bits each, and for each pair of blocks $\boldsymbol{x},\boldsymbol{y}$ producing as output the bit $\boldsymbol{x}\cdot\boldsymbol{y}$ (the inner product of $\boldsymbol{x}$ and $\boldsymbol{y}$ over GF(2)). This is a rather practical and elegant solution.

Chor and Goldreich [38] generalize these results, showing how to produce independent ε-biased bits from even worse sources, where some output bits can even be completely determined.

These results provide effective means for generating truly random sequences of bits—an essential requirement for cryptography—from somewhat defective natural sources of random bits.

7.2. *Generating pseudo-random bit or number sequences*

The one-time pad is generally impractical because of the large amount of key that must be stored. In practice, one prefers to store only a short random key, from which a long pad can be produced with a suitable cryptographic operator. Such an operator, which can take a short *random* sequence x and deterministically "expand" it into a *pseudo-random* sequence y, is called a *pseudo-random sequence generator*. Usually x is called the *seed* for the generator. The sequence y is called "pseudo-random" rather than random since not all sequences y are possible outputs; the number of possible y's is at most the number of possible seeds. Nonetheless, the intent is that for all practical purposes y should be indistinguishable from a truly random sequence of the same length.

It is important to note that the use of a pseudo-random sequence generator reduces *but does not eliminate* the need for a natural source of random bits; the pseudo-random sequence generator is a "randomness expander", but it must be given a truly random seed to begin with.

To achieve a satisfactory level of cryptographic security when used in a one-time pad scheme, the output of the pseudo-random sequence generator must have the property that an adversary who has seen a portion of the generator's output y must remain unable to efficiently predict other unseen bits of y. For example, note that an adversary who knows the ciphertext C can guess a portion of y by correctly guessing the corresponding portion of the message M, such as a standardized closing "Sincerely yours,". We would not like him thereby to be able to efficiently read other portions of M, which he could do if he could efficiently predict other bits of y. Most importantly the adversary should not be able to efficiently infer the seed x from the knowledge of some bits of y.

How can one construct secure pseudo-random sequence generators?

7.2.1. *Classical pseudo-random generators are unsuitable*

Classical techniques for pseudo-random number generation [98, Chapter 3] which are quite useful and effective for Monte Carlo simulations are typically unsuitable for cryptographic applications. For example, *linear* feedback shift registers [86] are well known to be cryptographically insecure; one can solve for the feedback pattern given a small number of output bits.

Linear congruential random number generators are also insecure. These generators use the recurrence

$$X_{i+1} = aX_i + b \pmod{m} \tag{8}$$

to generate an output sequence $\{X_0, X_1, \dots\}$ from secret parameters a, b, and m, and starting point X_0. It is possible to infer the secret parameters given just a few of the X_i [125]. Even if only a fraction of the bits of each X_i are revealed, but a, b, and m are known, Frieze, Hastad, Kannan, Lagarias and Shamir show how to determine the seed X_0 (and thus the entire sequence) using the marvelous *lattice basis reduction* (or "L^3") algorithm of Lenstra, Lenstra and Lovász [66, 104].

As a final example of the cryptographic unsuitability of classical methods, Kannan, Lenstra and Lovász [96] use the L^3 algorithm to show that the binary expansion of any algebraic number y (such as $\sqrt{5}=10.001111000110111\dots$) is insecure, since an adversary can identify y exactly from a sufficient number of bits, and then extrapolate y's expansion.

7.2.2. *Provably secure pseudo-random generators*

The first pseudo-random sequence generator proposed which was *provably secure* (assuming that it is infeasible to invert the RSA function (see Section 6.2.1)) is due to Shamir [143]. However, this scheme generates a sequence of *numbers* rather than a sequence of *bits*, and the security proof shows that an adversary is unable to predict the next *number*, given the previous numbers output. This is not strong enough to prove that, when used in a one-time pad scheme, each *bit* of the message will be well-protected.

M. Blum and Micali [29] introduced the first method for designing provably secure pseudo-random *bit* sequence generators, based on the use of one-way predicates. Let D denote a finite domain, let $f: D \to D$ denote a permutation of D, and let B denote a function from D to $\{0, 1\}$ such that (i) it is easy to compute $B(y)$, given $x=f^{-1}(y)$, and (ii) it is difficult to compute $B(y)$, given only y.

Given a seed $x_0 \in D$, we can create the sequence $x_0, x_1, \dots, x_n$ using the recurrence $x_{i+1}=f(x_i)$. To produce an output binary sequence $b_0, \dots, b_{n-1}$ of length n, define $b_i=B(x_{n-i})$. Note the reversal of order relative to the x sequence; we must compute $x_0, \dots, x_n$ first, and then compute b_0 from x_{n-1}, b_1 from $x_{n-2}, \dots$, and b_{n-1} from x_0.

If a pseudo-random bit sequence generator has the property that it is difficult to predict b_{i+1} from $b_0, \dots, b_i$ with accuracy greater than $(1+\varepsilon)/2$ in time polynomial in $1/\varepsilon$ and the size of the seed, then we say that the generator *passes the "next-bit" test*.

Blum and Micali prove that their generator passes the next-bit test as follows. Suppose otherwise. Then we derive a contradiction by showing how to compute $B(y)$ from y. Because f is a permutation, there is an x_0 such that $y=x_{n-i-1}$ in the sequence generated starting from x_0. Given $y=x_{n-i-1}$, we can compute $x_{n-i}, x_{n-i+1}, \dots, x_n$ as well, using f, and thus we can compute $b_0, \dots, b_i$ from y. If we can then predict $b_{i+1}=B(x_{n-i-1})=B(y)$ efficiently, we have contradicted our assumption that $B(y)$ is difficult to compute from y alone. (The above proof sketch is made rigorous in [29]; the phrases "easy" and "hard" are made precise, and the definition of computing $B(y)$ from y is generalized to include being able to predict $B(y)$ with an accuracy greater than $\frac{1}{2}$.)

Blum and Micali then proposed a particular generator based on the difficulty of computing discrete logarithms and the above method, as follows. Define $B(x)=B_{g,p}(x)$ to be 1 if $index_{g,p}(x) \leqslant (p-1)/2$, and 0 otherwise, where p is a prime, g is a generator of

Z_p^*, and $x \in Z_p^*$, and define $f(x) = f_{g,p}(x) = g^x \bmod p$. If computing discrete logarithms modulo p is indeed difficult, then the sequences produced will be unpredictable.

L. Blum, M. Blum and Shub [24] propose another generator, called the *$x^2 \bmod n$ generator*, which is simpler to implement and also provably secure (assuming that the quadratic residuousity problem is hard). This generator follows the Blum–Micali general method, with $B(x) = 1$ iff x is odd, and $f(x) = x^2 \bmod n$. Alexi, Chor, Goldreich and Schnorr [6] show that the assumption that the quadratic residuousity problem is hard can be replaced by the weaker assumption that factoring is hard. A related generator is obtained by using the RSA function $x^e \bmod n$ where $\gcd(e, \phi(n)) = 1$ [155, 6]. Kaliski shows how to extend these methods so that the security of the generator depends on the difficulty of computing elliptic logarithms; his techniques also generalize to other groups [93, 94].

To improve efficiency, it is desirable to obtain as many random bits as possible from each application of f. That is, $B(x)$ should return more than one bit of information. Long and Wigderson [109] show how to extract $c \log \log p$ pseudo-random bits from each x_i instead of just one bit as in the Blum–Micali generator. A similar result has been shown for the RSA generator [6].

A.C. Yao [161] shows that the pseudo-random generators defined above are *perfect* in the sense that no probabilistic polynomial-time algorithm can guess with probability greater than $\frac{1}{2} + \varepsilon$ whether an input string of length k was randomly selected from $\{0, 1\}^k$ or whether it was produced by one of the above generators. One can rephrase this to say that a generator that passes the next-bit test is perfect in the sense that it will *pass all polynomial-time statistical tests*. The Blum–Micali and Blum–Blum–Shub generators, together with the proof of Yao, represent a major achievement in the development of provably secure cryptosystems. Levin [107] shows that perfect pseudo-random bit generators exist if and only if there exists a one-way function f that cannot be inverted easily at points of the form $f^t(x)$, the tth iterate of f applied to a random point $x \in \{0, 1\}^k$.

7.2.3. *Pseudo-random functions and permutations*

More generally, one can imagine having a family $f_j(\cdot)$ of (pseudo-random) functions. The index j can be thought of as the *key* selecting which function is in use.

Such a family of functions can be used for authentication. If two users share a secret key j, then they can authenticate their messages to each other by appending the tag $f_j(M)$ to a message M. It should be infeasible for an adversary, seeing $f_j(M_1), \ldots, f_j(M_n)$, to produce a valid tag $f_j(M)$ for any other message M with a probability of success greater than random guessing.

Gilbert, MacWilliams and Sloane [72] present techniques which make it *information-theoretically impossible* for an adversary to forge a valid tag with probability greater than random guessing.

Goldreich, Goldwasser and Micali [75] show how to construct a family of pseudo-random functions from a cryptographically secure pseudo-random bit sequence generator, such that an adversary cannot distinguish between $f_j(M)$ and a randomly chosen string of the same length, even if the adversary is first allowed to examine $f_j(x_i)$

for many x_i's of his choice, and is allowed to even pick M (as long as it is different from every x_i he previously asked about). Knowledge of the index j allows efficient computation of $f_j(x)$ for any x. To the adversary (who does not know j), the function f_j is indistinguishable in polynomial time from a truly random function (that is, one picked at random from the set of all functions mapping $\{0,1\}^n$ into itself).

Since permutations are invertible, the Goldwasser–Goldreich–Micali construction provides a probabilistic private-key cryptosystem, one that is provably secure even against an adaptive chosen-ciphertext attack. To send a message M, the sender picks an r at random from the domain of the previously agreed-upon secret pseudo-random function f_j and then sends the pair $(r, M+f_j(r))$.

Luby and Rackoff [111] have extended the previous result by showing how to construct a family of pseudo-random *permutations* which is secure in the same sense and under the same assumptions. Curiously, their construction is based on the structure of DES. Since permutations are invertible, the Luby–Rackoff construction provides a provably secure deterministic private-key cryptosystem, one that is provably secure even against an adaptive chosen-ciphertext attack.

8. Digital signatures

The notion of a *digital signature* may prove to be one of the most fundamental and useful inventions of modern cryptography. A signature scheme provides a way for each user to *sign* messages so that the signatures can later be *verified* by anyone else. More specifically, each user can create a matched pair of private and public keys so that only he can create a signature for a message (using his private key), but anyone can verify the signature for the message (using the signer's public key). The verifier can convince himself that the message contents have not been altered since the message was signed. Also, the signer cannot later repudiate having signed the message, since no-one but the signer possesses his private key.

Diffie and Hellman [52] propose that with a public-key cryptosystem, a user A can sign any message M by appending his digital signature $D_A(M)$ to M. Anyone can check the validity of this signature using A's public key E_A from the public directory, since $E_A(D_A(M))=M$. Note also that this signature becomes invalid if the message is changed, so that A is protected against modifications after he has signed the message, and the person examining the signature can be sure that the message he has received is the one that was originally signed by A.

By analogy with the paper world, where one might sign a letter and seal it in an envelope, one can sign an electronic message using one's private key, and then *seal* the result by encrypting it with the recipient's public key. The recipient can perform the inverse operations of opening the letter and verifying the signature using his private key and the sender's public key, respectively. These applications of public-key technology to electronic mail are likely to be widespread in the near future.

The RSA public-key cryptosystem allows one to implement digital signatures in a straightforward manner. The private exponent d now becomes the *signing exponent*, and the signature of a message M is now the quantity $M^d \bmod n$. Anyone can verify that this signature is valid using the corresponding public *verification exponent* e by checking the identity $M=(M^d)^e \pmod{n}$. If this equation holds, then the signature M^d must have been created from M by the possessor of the corresponding signing exponent d. (Actually, it is possible that the reverse happened and that the "message" M was computed from the "signature" M^d using the verification equation and the public exponent e. However, such a message is likely to be unintelligible. In practice, this problem is easily avoided by always signing $f(M)$ instead of M, where f is a standard public one-way function.)

If the directory of public keys is accessed over the network, one needs to protect the users from being sent fraudulent messages purporting to be public keys from the directory. An elegant solution is the use of a *certificate*—a copy of a user's public key digitally signed by the public key directory manager or other trusted party. If user A keeps locally a copy of the public key of the directory manager, he can validate all the signed communications from the public-key directory and avoid being tricked into using fraudulent keys. Moreover, each user can transmit the certificate for his public key with any message he signs, thus removing the need for a central directory and allowing one to verify signed messages with no information other than the directory manager's public key. Needham and Schroeder [121] examine some of the protocol issues involved in such a network organization, and compare it to what might be accomplished using conventional cryptography.

Just as some cryptographic schemes are suited for encryption but not signatures, some proposals have been made for *signature-only* schemes. Some early suggestions were made that were based on the use of one-way functions or conventional cryptography [101, 130]. For example, if f is a one-way function, and Alice has published the two numbers $f(x_0)=y_0$ and $f(x_1)=y_1$, then she can sign the message 0 by releasing x_0 and she can similarly sign the message 1 by releasing the message x_1. Merkle [115] introduced some extensions of this basic idea, involving building a tree of authenticated values whose root is stored in the public key of the signer.

Rabin [131] proposed a method where the signature for a message M was essentially the square root of M, modulo n, the product of two large primes. Since the ability to take square roots is provably equivalent to the ability to factor n, an adversary should not be able to forge any signatures unless he can factor n. This argument assumes that the adversary only has access to the public key containing the modulus n of the signer. In practice an enemy may break this scheme with an *active* attack by asking the real signer to sign $M=x^2 \bmod n$, where x has been chosen randomly. If the signer agrees and produces a square root y of M, there is half a chance that $\gcd(n, x-y)$ will yield a nontrivial factor of n—the signer has thus betrayed his own secrets! Although Rabin proposed some practical techniques for circumventing this problem, they have the effect of eliminating the constructive reduction of factoring to forgery.

In general, knapsack-type schemes are not well-suited for use as signature schemes, since the mapping $M \cdot \boldsymbol{a}$ is not "onto", and no effective remedy has been proposed. A review of signature schemes can be found in [82].

8.1. *Proving security of signature schemes*

A theoretical treatment of digital signatures security was started by Goldwasser, Micali and A.C. Yao in [84] and continued in [82, 15], and more recently [119]. We first address what attacks are possible on digital signature schemes.

8.1.1. *Attacks against digital signatures*

We distinguish three basic kinds of attacks, listed below in the order of increasing severity.

- *Key-only attack:* In this attack the adversary knows only the public key of the signer and therefore only has the capability of checking the validity of signatures of messages given to him.
- *Known-signature attack*: The adversary knows the public key of the signer and has seen message/signature pairs chosen and produced by the legal signer. In reality, this is the minimum an adversary can do.
- *Chosen-message attack*: The adversary is allowed to ask the signer to sign a number of messages of the adversary's choice. The choice of these messages may depend on previously obtained signatures. For example, one may think of a notary public who signs documents on demand.

For a finer subdivision of the adversary's possible attacks, see [82].

8.1.2. *What does it mean to successfully forge a signature?*

We distinguish several levels of success for an adversary, listed below in the order of increasing success for the adversary.

- *Existential forgery*: The adversary succeeds in forging the signature of one message, not necessarily of his choice.
- *Selective forgery*: The adversary succeeds in forging the signature of some message of his choice.
- *Universal forgery*: The adversary, although unable to find the secret key of the signer, is able to forge the signature of any message.
- *Total break*: The adversary can compute the signer's secret key.

Clearly, different levels of security may be required for different applications. Sometimes, it may suffice to show that an adversary who is capable of a known-signature attack cannot succeed in selective forgery, while for other applications (for example when the signer is a notary public or a tax-return preparer) it may be required that an adversary capable of a chosen-message attack cannot succeed even at existential forgery with nonnegligible probability.

8.2. *Probabilistic signature schemes*

Probabilistic techniques have also been applied to the creation of digital signatures. This approach was pioneered by Goldwasser, Micali and A.C. Yao [84], who presented signature schemes based on the difficulty of factoring and on the difficulty of inverting the RSA function for which it is provably hard for the adversary to existentially forge using a known-signature attack.

Goldwasser, Micali and Rivest [82] have strengthened this result by proposing a signature scheme which is not existentially forgeable under a chosen-message attack. Their scheme is based on the difficulty of factoring, and more generally on the existence of claw-free trapdoor permutations (that is, pairs f_0, f_1 of trapdoor permutations defined on a common domain for which it is hard to find x, y such that $f_0(x)=f_1(y)$).

We briefly sketch their digital signature scheme. Let (f_0, f_1) and (g_0, g_1) be two pairs of claw-free permutations. Let $b=b_1 \dots b_k$ be a binary string, and define $F_b^{-1}(y)= f_{b_k}^{-1}(\dots(f_{b_2}^{-1}(f_{b_1}^{-1}(y))))$, and $G_b^{-1}(y)=g_{b_k}^{-1}(\dots(g_{b_2}^{-1}(g_{b_1}^{-1}(y))))$.

The signer makes public (x, f_0, f_1, g_0, g_1) where x is a randomly chosen element in the domain of f_0, f_1, and keeps secret the trapdoor information enabling him to compute F_b^{-1} and G_b^{-1}. The variable *history* is kept by the signer in memory as well, but need not be private. (For simplicity of exposition we assume a prefix-free message space.)

The signature of the ith message m_i is created as follows. The signer:

(1) picks r_i at random in the domain of g_0, g_1 and sets $history = history . r_i$ where . denotes concatenation;

(2) computes $l_i = F_{history}^{-1}(x)$ and $t_i = G_{m_i}^{-1}(r_i)$;

(3) produces the signature of m_i as the triplet $(history, l_i, t_i)$.

To check the validity of the signature (h, l, t) of message m, anyone with access to the signer's public key can check that:

(1) $F_h(l)=x$ where x is in the public file;

(2) $G_m(t)=r$ where r is a suffix of h.

If both conditions hold, the signature is valid.

The scheme as we describe it, although attractive in theory, is quite inefficient. However, it can be modified to allow more compact signatures, to make no use of memory between signatures other than for the public and secret keys, and even to remove the need of making random choices for every new signature. In particular, Goldreich [74] has made suggestions that make the factoring-based version of this scheme more practical while preserving its security properties.

Bellare and Micali [15] have shown a digital signature scheme whose security can be based on the existence of any trapdoor permutation (a weaker requirement than claw-freeness).

A major leap forward has been recently made by Naor and Yung [119] who have shown how, starting with any *one-way* permutation, to design a digital signature scheme which is not existentially forgeable by chosen-message attack.

9. Two-party protocols

In this section we sketch a number of cryptographic protocol problems that have been addressed in the literature; see [48] for additional examples.

9.1. Examples

9.1.1. User identification (friend-or-foe)

Suppose A and B share a secret key K. Later A is communicating with someone and he wishes to verify that it is B. A simple *challenge-response* protocol to achieve this identification goes as follows:

- A generates a random value r and transmits r to the other party.
- The other party (assuming it is B) encrypts r using the shared secret key K and transmits the resulting ciphertext back to A.
- A compares the received ciphertext with the result he obtains by encrypting r himself using the secret key K. If they agree, he knows that the other party is B; otherwise he assumes that the other party is an impostor.

This protocol is generally more useful than the transmission of an unencrypted shared password from B to A, since an eavesdropper could learn the password and then pretend to be B later. With the challenge-response protocol an eavesdropper presumably learns nothing about K by hearing many values of r encrypted with K as key.

9.1.2. Mental poker

How might one play a game of cards, such as poker, over the telephone? The difficulty, of course, is in dealing the cards. Shamir, Rivest and Adleman [145] proposed the following simple strategy (here Alice and Bob are the two players):

- The players jointly select 52 distinct messages $M_1, \ldots, M_{52}$ to represent the cards, and a large prime p.
- The players secretly choose exponents e_A and e_B respectively, so that Alice has a secret encryption function $E_A(M) = M^{e_A} \bmod p$ (similarly for Bob). Each player computes the corresponding decryption exponents d_A, d_B defining decryption functions $D_A(C) = C^{d_A} \bmod p$ (and similarly for Bob).
- Alice (the dealer) encrypts the cards $M_1, \ldots, M_{52}$, shuffles them (permutes their order), and sends the resulting list to Bob.
- Bob selects five of the cards and returns them to Alice; she decrypts these for her hand.
- Bob selects five of the remaining cards for his own hand, encrypts them with his encryption function, and sends the result to Alice. Note that each such card has the form $E_B(E_A(M_i)) = E_A(E_B(M_i))$ since E_A and E_B commute.
- Alice decrypts the five cards with D_A, and returns the result to Bob, who decrypts them with D_B to obtain his hand.

At the end of the play, the parties can reveal their secret keys so they can assure themselves that the other has not cheated.

This technique requires the use of *commutative* encryption functions. The particular scheme as proposed may be less satisfactory than desired (depending on the coding of the cards) since, as pointed out by Lipton [108], M will have Jacobi symbol 1 if and only if $E_A(M)$ does—the function E_A *leaks a bit*. Another attack has been proposed by Coppersmith [44]. A number of authors have proposed extensions and variations of this protocol.

9.1.3. Coin flipping

M. Blum [25] has proposed the problem of *coin flipping over the telephone.* If Alice and Bob do not trust each other, then if they wish to flip a coin, they need a procedure that will produce an outcome (head or tails) that cannot be biased by either party.

Using probabilistic encryption, Alice could send encrypted versions of the messages "Heads" and "Tails" to Bob. Bob picks one of the ciphertexts, and indicates his choice to Alice. Alice then reveals the secret encryption key to Bob. There are a number of interesting variations and subtleties to this problem (see, for example [79, 41]).

9.1.4. Oblivious transfer

An *oblivious transfer* is an unusual protocol wherein Bob transfers a message M to Alice in such a way that:

- with probability $\frac{1}{2}$, Alice receives the message, and with probability $\frac{1}{2}$, Alice receives garbage instead;
- at the end of protocol Bob does not know whether or not Alice received the message.

This strange-sounding protocol has a number of useful applications (see, for example [134, 21]). In fact, Kilian has shown in 1988 [97] that the ability to perform oblivious transfers is a sufficiently strong primitive to enable *any* two-party protocol to be performed.

9.1.5. Other examples

The problem of *contract signing* is that of simultaneously exchanging digital signatures to a contract. That is, we assume that an ordinary digital signature scheme is available, and want to arrange a two-party protocol so that the signatures are effectively simultaneous—neither party obtains the other's signature before giving up his own. Several interesting solutions to the problem have been proposed (see [63] or [19]).

The *certified electronic mail problem* is similar to the contract-signing problem above. The goal is to achieve a simultaneous exchange of an electronic letter M and a signed receipt for M from the recipient.

Another exchange problem is the simultaneous exchange of secrets. This has been studied in [26, 154, 110, 163].

9.2. Zero-knowledge protocols

The previous section listed a number of cryptographic protocol applications. In this section we review the theory that has been developed to prove that these protocols are secure, and to design protocols that are "provably secure by construction".

9.2.1. Zero-knowledge interactive proofs

An elegant way to prove a cryptographic protocol secure is to prove that it is a *zero-knowledge protocol.* Informally, a protocol is a zero-knowledge protocol if one party learns nothing (zero) from the protocol above and beyond what he is supposed to learn. This theory was originated by Goldwasser, Micali and Rackoff [81], and has been extended by many others, including Galil, Haber and Yung [68], Chaum [36], Goldwasser and Sipser [85], and Brassard and Crepeau [32].

Zero-knowledge protocols are two-party protocols; one party is called the *power* and the other the *verifier*. The prover knows some fact and wishes to convince the verifier of this fact. The verifier wants a protocol that will allow the prover to convince him of the validity of the fact, if and only if the fact is true. More precisely, the prover (if he follows the protocol) will be able (with extremely high probability) to convince the verifier of the validity of the fact if the fact is true, but the prover (even if he attempts to cheat) will not have any significant chance of convincing the verifier of the validity of the fact if the fact is false. However, the prover does not wish to disclose any information above and beyond the validity of the fact itself (for example, the reason why the fact holds). This condition can be very useful in cryptographic protocols, as we shall see. Examples of nontrivial NP-languages for which there exist zero-knowledge protocols include quadratic residuousity and graph-isomorphism. (Curiously, the complements of both of these languages also possess zero-knowledge proofs.)

More generally it has been shown by Goldreich, Micali and Wigderson [77] that *every* language in NP possesses a zero-knowledge protocol, on the assumption that there secure encryption is possible. (Probabilistic encryption schemes work fine here.)

As a concrete example of a zero-knowledge protocol, suppose I wish to convince you that a certain input graph is three-colorable, without revealing to you the coloring that I know. I can do so in a sequence of $|E|^2$ stages, each of which goes as follows.

- I switch the three colors at random (e.g., switching all red nodes to blue, all blue nodes to yellow, and all yellow nodes to red).
- I encrypt the color of each node, using a different probabilistic encryption scheme for each node, and show you all these encryptions, together with the correspondence indicating which ciphertext goes with which vertex.
- You select an edge of the graph.
- I reveal the decryptions of the colors of the two nodes that are incident to this edge by revealing the corresponding decryption keys.
- You confirm that the decryptions are proper, and that the two endpoints of the edge are colored with two different but legal colors.

If the graph is indeed three-colorable (and I know the coloring), then you will never detect any edge being incorrectly labeled. However, if the graph is not three-colorable, then there is a chance of at least $|E|^{-1}$ on each stage that I will be caught trying to fool you. The chance that I could fool you for $|E|^2$ stages without being caught is exponentially small.

Note that the history of our communications—in the case that the graph is three-colorable—consists of the concatenation of the messages sent during each stage. It is possible to prove (on the assumption that secure encryption is possible) that the probability distribution defined over these histories by our set of possible interactions is indistinguishable in polynomial time from a distribution that you can create on these histories by yourself, without my participation. This fact means that you gain zero (additional) knowledge from the protocol, other than the fact that the graph is three-colorable.

The proof that graph three-colorability has such a zero-knowledge interactive proof system can be used to prove that every language in NP has such a zero-knowledge proof system.

9.2.2. Applications to user identification

Zero-knowledge proofs provide a revolutionary new way to realize passwords [81, 65]. The idea is for every user to store a statement of a theorem in his publicly readable directory, the proof of which only he knows. Upon login, the user engages in a zero-knowledge proof of the correctness of the theorem. If the proof is convincing, access permission is granted. This guarantees that even an adversary who overhears the zero-knowledge proof cannot learn enough to gain unauthorized access. This is a novel property which cannot be achieved with traditional password mechanisms. Recently Fiat and Shamir [65] have developed variations on some of the previously proposed zero-knowledge protocols [80] which are quite efficient and particularly useful for user identification and passwords.

10. Multi-party protocols

In a typical multi-party protocol problem, a number of parties wish to coordinate their activities to achieve some goal, even though some (sufficiently small) subset of them may have been corrupted by an adversary. The protocol should guarantee that the "good" parties are able to achieve the goal even though the corrupted parties send misleading information or otherwise maliciously misbehave in an attempt to prevent the good parties from succeeding.

10.1. Examples

10.1.1. Secret sharing

In 1979 Shamir [141] considered the problem of *sharing a secret*, defined as follows. Suppose n people wish to share a secret (such as a secret cryptographic key) by dividing it into pieces in such a way that *any* subset of k people can recreate the secret from their pieces, but *no* subset of less than k people can do so. Here k is a given fixed positive integer less than n.

Shamir proposed the following elegant solution to this problem. Let the secret be represented as an integer s, which is less than some large convenient prime p. The secret can be "divided into pieces" by:

(1) generating k coefficients $a_0, a_1, \ldots, a_{k-1}$, where $a_0 = s$ but all the other coefficients are randomly chosen modulo p;

(2) defining the polynomial $f(x)$ to be $\Sigma_{0 \leqslant i < k} a_i x^i \bmod p$;

(3) giving party i the "piece" $f(i)$, for $1 \leqslant i \leqslant n$.

Now, given any k values for f, one can interpolate to find f's coefficients (including the secret $a_0 = s$). However, a subset of $k-1$ values for f provides absolutely no information about s, since for any possible s there is a polynomial of degree $k-1$ consistent with the given values and the possible value for s.

Shamir's scheme suffers from two problems. If the dealer of the secret is dishonest, he can give pieces which when put together do not uniquely define a secret. Secondly if some of the players are dishonest, at the reconstruction stage they may provide other players with different pieces than they received and again cause an incorrect secret to be reconstructed.

Chor, Goldwasser, Micali and Awerbuch [39] have observed the above problems and showed how to achieve secret sharing based on the intractability of factoring which does not suffer from the above problems. They call the new protocol *verifiable secret sharing* since now every party can verify that the piece of the secret he received is indeed a proper piece. Their protocol tolerated up to $O(\log n)$ colluders. Benaloh [16], and others [77, 64] showed how to achieve verifiable secret sharing if any one-way function exists which tolerates a minority of colluders. In 1988 [20] it was shown how to achieve verifiable secret sharing against a minority of colluders using error correcting codes, without making cryptographic assumptions.

10.1.2. Anonymous transactions

Chaum has advocated the use of *anonymous transactions* as a way of protecting individuals from the maintenance by "Big Brother" of a database listing all their transactions, and proposes using *digital pseudonyms* to do so. Using pseudonyms, individuals can enter into electronic transactions with assurance that the transactions cannot be later traced to the individual. However, since the individual is anonymous, the other party may wish assurance that the individual is authorized to enter into the transaction, or is able to pay [34, 35].

10.1.3. Voting

Cryptographic technology can be used to manage an election so that every voter's vote remains private, but yet every voter can be sure that the vote-counting was not manipulated. (See Cohen and Fischer [42].)

10.2. Multi-party ping-pong protocols

One way of demonstrating that a cryptographic protocol is secure is to show that the primitive operations that each party performs cannot be composed to reveal any secret information.

Consider a simple example due to Dolev and A.C. Yao [58] involving the use of public keys. Alice sends a message M to Bob, encrypting it with his public key, so that the ciphertext C is $E_B(M)$ where E_B is Bob's public encryption key. Then Bob "echos" the message back to Alice, encrypting it with Alice's public key, so that the ciphertext returned is $C' = E_A(M)$. This completes the description of the protocol.

Is this secure? Since the message M is encrypted on both trips, it is clearly infeasible for a *passive* eavesdropper to learn M. However, an *active* eavesdropper X can defeat this protocol. Here's how: the eavesdropper X overhears the previous conversation, and records the ciphertext $C = E_B(M)$. Later, X starts up a conversation with Bob using this protocol, and sends Bob the encrypted message $E_B(M)$ that he has recorded. Now

Bob dutifully returns to X the ciphertext $E_X(M)$, which gives X the message M he desires!

The moral is that an adversary may be able to "cut and paste" various pieces of the protocol together to break the system, where each "piece" is an elementary transaction performed by a legitimate party during the protocol, or a step that the adversary can perform himself.

It is sometimes possible to *prove* that a protocol is invulnerable to this style of attack. Dolev and Yao [58] pioneered this style of proof; additional work was performed by Dolev, Even and Karp [57], Yao [162], and Even and Goldreich [61]. In other cases a modification of the protocol can eliminate or alleviate the danger; see [136] as an example of this approach against the danger of an adversary "inserting himself into the middle" of a public-key exchange protocol.

10.3. Multi-party protocols when most parties are honest

Goldreich, Micali and Wigderson [77] have shown how to "compile" a protocol designed for honest parties into one which will still work correctly even if some number less than half of the players try to "cheat". While the protocol for the honest parties may involve the disclosure of secrets, at the end of the compiled protocol none of the parties know any more than what they knew originally, plus whatever information is disclosed as the "official output" of the protocol. Their compiler correctness and privacy is based on the existence of trapdoor functions.

Ben-Or, Goldwasser and Wigderson [20] and Chaum, Crepeau and Damgård [37] go one step further. They assume secret communication between pairs of users as a primitive. Making no intractability assumption, they show a "compiler" which, given a description (e.g., a polynomial-time algorithm or circuit) of any polynomial-time function f, produces a protocol which always computes the function correctly and guarantees that no additional information to the function value is leaked to dishonest players. The "compiler" withstands up to $\frac{1}{3}$ of the parties acting dishonestly in a manner directed by a worst-case unbounded-computation-time adversary.

These "master theorems" promise to be a very powerful tool in the future design of secure protocols.

11. Cryptography and complexity theory

Some cryptographic operators are secure in an information-theoretic sense; examples are the Vernam one-time pad, the secret-sharing scheme of Shamir, and the authentication scheme of Gilbert et al. In these cases the adversary never obtains enough information to enable him to set up a problem having a unique solution. As a consequence no amount of computational power can help him resolve this intrinsic uncertainty.

However, most practical cryptographic schemes are not information-theoretically secure. While the adversary is given enough information to determine a unique solution, the problem of actually computing this solution is deemed to be computationally intractable. We may term this *computational security*. Computational security depends critically on the existence and careful exploitation of "hard" computational problems.

While one goal of computational complexity theory is to be able to precisely characterize the computational difficulty of arbitrary problems, the state of this theory is such that we can currently only derive some tools and suggestive guidelines.

For example, we may observe that if it turns out to be the case that $P = NP$ (see [69]), then computational security would be unachievable, since the adversary's problem is easily seen to be in NP. (The adversary can just guess the secret keys, check their correctness against some known plaintext/ciphertext pairs, and thereby "break" the system.)

If we assume that $P \neq NP$, then it would seem natural to try to design cryptographic systems such that the problem of breaking them is NP-complete. However, this is difficult to arrange, since cryptographic problems usually have *unique* solutions, whereas NP-complete problems may have many solutions. It is not easy to reduce a problem with many solutions to a problem having a unique solution. Grollman and Selman [87] show that one-way functions exist if and only if $P \neq UP$, where UP is the class of languages accepted by a nondeterministic Turing machine which has at most one accepting computation for any input. They also show that secure public-key cryptosystems exist only if $P \neq UP$. Lempel, Even and Yacobi (see [102]) present a curious cryptographic system for which the cryptanalytic problem is NP-complete in general, even for chosen-plaintext attacks, but which is likely to be easily breakable in practice.

The Lempel/Yacobi example illustrates another difficulty with attempting to use the traditional notions of computational complexity in the design of cryptographic systems: the traditional notions relate to the *worst-case* complexity of the problem, whereas cryptographic security more realistically depends on the *average-case* complexity of the problem. There is much yet to be learned about the average-case complexity of problems. However, Levin [106] has introduced a formal notion of what it means for a problem to be "complete" in the sense of its average-case complexity.

Brassard [30] has shown that in certain relativized models of computation secure cryptography is possible.

Yao [161] develops the theory of trapdoor functions and pseudo-random bit sequence generators, and shows that if a strong one-way function exists, then

$$R \subseteq \bigcap_{\varepsilon > 0} \mathrm{DTIME}(2^{n^{\varepsilon}}), \tag{9}$$

since any algorithm in R can be adapted to use a pseudo-random bit sequence generator with a seed of size n^{ε} instead of a true random bit generator. (Note that all seeds have to be tried.)

Acknowledgment

I would like to thank Shafi Goldwasser for her extensive help in preparing this chapter, and also Benny Chor, Oded Goldreich, Silvio Micali, Phil Rogaway, and Alan Sherman for their comments and suggestions for improvements.

This chapter was prepared with support from NSF Grant DCR-8607494.

References

[1] ADLEMAN, L.M., A subexponential algorithm for the discrete logarithm problem with applications to cryptography, in: *Proc. 18th IEEE Symp. on Foundations of Computer Science* (1977) 55–60.

[2] ADLEMAN, L.M., On breaking generalized knapsack public key cryptosystems, in: *Proc. 15th ACM Symp. on Theory of Computing* (1983) 402–412.

[3] ADLEMAN, L.M. and M.A. HUANG, Recognizing primes in random polynomial time, in: *Proc. 19th ACM Symp. on Theory of Computing* (1987) 462–469.

[4] ADLEMAN, L.M., K. MANDERS and G. MILLER, On taking roots in finite fields, in: *Proc. 18th IEEE Symp. on Foundations of Computer Science* (1977) 175–177.

[5] ADLEMAN, L.M., C. POMERANCE and R.S. RUMELY, On distinguishing prime numbers from composite numbers, *Ann. Math.* **117** (1983) 173–206.

[6] ALEXI, W.B., B. CHOR, O. GOLDREICH and C.P. SCHNORR, RSA and Rabin functions: certain parts are as hard as the whole, *SIAM J. Comput.* **17**(2) (1988) 194–209.

[7] ALPERN, B. and F.B. SCHNEIDER, Key exchange using "keyless cryptography", *Inform. Process. Lett.* **16** (1983) 79–81.

[8] ANGLUIN, D., Lecture notes on the complexity of some problems in number theory, Tech. Report TR-243, Comput. Sci. Dept., Yale Univ., 1982.

[9] ANGLUIN, D. and D. LICHTENSTEIN, Provable security of cryptosystems: a survey, Tech. Report TR-288, Comput. Sci. Dept., Yale Univ., 1983.

[10] ASMUTH, C.A. and G.R. BLAKLEY, An efficient algorithm for constructing a cryptosystem which is harder to break than two other cryptosystems, *Comput. Math. Appl.* **7** (1981) 447–450.

[11] BACH, E., How to generate factored random numbers, *SIAM J. Comput.* **17**(2) (1988) 179–193.

[12] BAMFORD, J., *The Puzzle Palace–A Report on NSA, America's Most Secret Agency* (Houghton Mifflin, Boston, MA, 1982).

[13] BEKER, H. and F. PIPER, *Cipher Systems: The Protection of Communications* (Northwood, London, 1982).

[14] BELL, D.A. and S.E. OLDING, An annotated bibliography of cryptography, Tech. Report COM-100, National Physical Laboratory, 1978.

[15] BELLARE, M. and S. MICALI, How to sign given any trapdoor function, in: *Proc. 20th ACM Symp. on Theory of Computing* (1988) 32–42.

[16] BENALOH, J., Secret sharing homomorphisms: keeping shares of a secret secret, in: A.M. Odlyzko, ed., *Advances in Cryptology–CRYPTO 86*, Lecture Notes in Computer Science, Vol. 263 (Springer, Berlin, 1987) 251–260.

[17] BENNETT, C.H., G. BRASSARD, S. BREIDBARD and S. WIESNER, Quantum cryptography, in: R.L. Rivest, A. Sherman and D. Chaum, eds., *Advances in Cryptology, Proc. CRYPTO 82* (Plenum, New York, 1983) 267–275.

[18] BEN-OR, M., B. CHOR and A. SHAMIR, On the cryptographic security of single RSA bits, in: *Proc. 15th ACM Symp. on Theory of Computing* (1983) 421–430.

[19] BEN-OR, M., O. GOLDREICH, S. MICALI and R.L. RIVEST, A fair protocol for signing contracts, in: *Proc. 12th Internat. Coll. on Automata, Languages and Programming*, Lecture Notes in Computer Science, Vol. 194 (Springer, Berlin, 1985) 43–52.

[20] BEN-OR, M., S. GOLDWASSER and A. WIGDERSON, Completeness theorems for fault-tolerant distributed computing, in: *Proc. 20th ACM Symp. on Theory of Computing* (1988) 1–10.

[21] BERGER, R., R. PERALTA and T. TEDRICK, A provably secure oblivious transfer protocol. in: T. Beth, N. Cot and I. Ingemarsson, eds., *Advances in Cryptology, Proc. EUROCRYPTO 84*, Lecture Notes in Computer Science, Vol. 209 (Springer, Berlin, 1983) 379–386.

[22] BERLEKAMP, E., Factoring polynomials over large finite fields, *Math. Comp.* **24** (1970) 713–735.

[23] BETH, T., ed., *Cryptography, Proceedings*, Lecture Notes in Computer Science, Vol. 149 (Springer, Berlin, 1983).

[24] BLUM, L., M. BLUM and M. SHUB, A simple unpredictable pseudo-random number generator, *SIAM J. Comput.* **15**(2) (1986) 364–383.

[25] BLUM, M., Coin flipping by telephone: a protocol for solving impossible problems, in: *Proc. 24th IEEE Spring Computer Conf. COMPCOM* (1982) 133–137.

[26] BLUM, M., How to exchange (secret) keys, *ACM Trans. Comput. Systems* **1** (1983) 175–193.

[27] BLUM, M., Independent unbiased coin flips from a correlated biased source: a finite state Markov chain, in: *Proc. 25th IEEE Symp. on Foundations of Computer Science* (1984) 425–433.

[28] BLUM, M. and S. GOLDWASSER, An *efficient* probabilistic public-key encryption scheme which hides all partial information, in: G.R. Blakley and D.C. Chaum, eds., *Advances in Cryptology, Proc. CRYPTO 84*, Lecture Notes in Computer Science, Vol. 196 (Springer, Berlin, 1985) 289–299.

[29] BLUM, M. and S. MICALI, How to generate cryptographically strong sequences of pseudo-random bits, *SIAM J. Comput.* **13**(4) (1984) 850–863.

[30] BRASSARD, G., Relativized cryptography, in: *Proc. 20th Ann. IEEE Symp. on Foundations of Computer Science* (1979) 383–391.

[31] BRASSARD, G., *Modern Cryptology*, Lecture Notes in Computer Science, Vol. 325 (Springer, Berlin, 1988).

[32] BRASSARD, G. and C. CREPEAU, Nontransitive transfer of confidence: a *perfect* zero-knowledge interactive protocol for SAT and beyond, in: *Proc. 27th IEEE Symp. on Foundations of Computer Science* (1986) 188–195.

[33] BRICKELL, E.F., Breaking iterated knapsacks, in: G.R. Blakley and D.C. Chaum, eds., *Advances in Cryptology, Proc. CRYPTO 84*, Lecture Notes in Computer Science, Vol. 196 (Springer, Berlin, 1985) 342–358.

[34] CHAUM, D., Untraceable electronic mail, return addresses, and digital pseudonyms, *Comm. ACM* **24** (1981) 84–88.

[35] CHAUM, D., Blind signatures for untraceable payments, in: R.L. Rivest, A. Sherman and D. Chaum, eds., *Advances in Cryptology, Proc. CRYPTO 82* (Plenum, New York, 1983) 199–204.

[36] CHAUM, D., Demonstrating that a public predicate can be satisfied without revealing any information about how, in: A.M. Odlyzko, ed., *Advances in Cryptology–CRYPTO 86*, Lecture Notes in Computer Science, Vol. 263 (Springer, Berlin, 1987) 195–199.

[37] CHAUM, D., C. CREPEAU and I. DAMGÅRD, Multi-party unconditionally secure protocols, in: *Proc. 20th ACM Symp. on Theory of Computing* (1988) 11–19.

[38] CHOR, B. and O. GOLDREICH, Unbiased bits from sources of weak randomness and probabilistic communication complexity, *SIAM J. Comput.* **17**(2) (1988) 230–261.

[39] CHOR, B., S. GOLDWASSER, S. MICALI and B. AWERBUCH, Verifiable secret sharing and achieving simultaneity in the presence of faults, in: *Proc. 26th IEEE Symp. on Foundations of Computer Science* (1985) 383–395.

[40] CHOR, B. and R.L. RIVEST, A knapsack type public-key cryptosystem based on arithmetic in finite fields, *IEEE Trans. Inform. Theory* **34**(5) (1988) 901–909.

[41] CLEVE, R., Limits on the security of coin flips when half the processors are faulty, in: *Proc. 18th ACM Symp. on Theory of Computing* (1986) 364–369.

[42] COHEN, J.D. and M.J. FISCHER, A robust and verifiable cryptographically secure election scheme, in: *Proc. 26th IEEE Symp. on Foundations of Computer Science* (1985) 372–382.

[43] COPPERSMITH, D., Evaluating logarithms in GF(2^n), in: *Proc. 16th ACM Symp. on Theory of Computing* (1984) 201–207.

[44] COPPERSMITH, D., Cheating at mental poker, in: H.C. Williams, ed., *Advances in Cryptology—CRYPTO 85*, Lecture Notes in Computer Science, Vol. 218 (Springer, Berlin, 1986) 104–107.

[45] COPPERSMITH, D., A.M. ODLYZKO and R. SCHROEPPEL, Discrete logarithms in GF(p), *Algorithmica* **1**(1) (1986) 1–16.

[46] DAVIES, D.W., ed., *Tutorial: The Security of Data in Networks*, IEEE Computer Society Order #366 (IEEE Computer Soc. Press, Silver Spring, MD, 1981).

[47] DEMILLO, R.A., D.P. DOBKIN, A. JONES and R.J. LIPTON, eds., *Foundations of Secure Computation* (Academic Press, New York, 1978).

[48] DEMILLO, R.A., N. LYNCH and M.J. MERRITT, Cryptographic protocols, in: *Proc. 14th Ann. ACM Symp. on Theory of Computing* (1982) 383–400.

[49] DENNING, D.E., *Cryptography and Data Security* (Addison-Wesley, Reading, MA, 1982).

[50] DENNING, D.E. and P.J. DENNING, Data security, *ACM Comput. Surveys* **11** (1979) 227–249.

[51] DIFFIE, W. and M.E. HELLMAN, Multiuser cryptographic techniques, in: *Proc. AFIPS 1976 National Computer Conf.* (1976) 109–112.

[52] DIFFIE, W. and M.E. HELLMAN, New directions in cryptography, *IEEE Trans. Inform. Theory* **22** (1976) 644–654.

[53] DIFFIE, W. and M.E. HELLMAN, Exhaustive cryptanalysis of the NBS data encryption standard, *Computer* **10** (1977) 74–84.

[54] DIFFIE, W. and M.E. HELLMAN, Privacy and authentication: an introduction to cryptography, *Proc. IEEE* **67** (1979) 397–427.

[55] DIFFIE, W. and M.E. HELLMAN, An introduction to cryptography, in: Slonim, Unger and Fisher, eds., *Advances in Data Communication Management* (Wiley, New York, 1984) 44–134.

[56] DIXON, J.D., Factorization and primality tests, *Amer. Math. Monthly* **91**(3) (1984) 333–352.

[57] DOLEV, D., S. EVEN and R.M. KARP, On the security of ping-pong protocols, in: R.L. Rivest, A. Sherman and D. Chaum, eds., *Advances in Cryptology, Proc. CRYPTO 82* (Plenum, New York, 1983) 177–186.

[58] DOLEV, D. and A.C. YAO, On the security of public key protocols, in: *Proc. 22nd IEEE Symp. on Foundations of Computer Science* (1981) 350–357.

[59] ELIAS, P., The efficient construction of an unbiased random sequence, *Ann. Math. Statist.* **43**(3) (1972) 865–870.

[60] EVANS, A., W. KANTROWITZ, and E. WEISS, A user authentication scheme not requiring secrecy in the computer, *Comm. ACM* **17** (1974) 437–442.

[61] EVEN, S. and O. GOLDREICH, On the security of multi-party ping-pong protocols, in: *Proc. 24th IEEE Symp. on Foundations of Computer Science* (1983) 34–39.

[62] EVEN, S. and O. GOLDREICH, On the power of cascade ciphers, *ACM Trans. Comput. Systems* **3** (1985) 108–116.

[63] EVEN, S., O. GOLDREICH and A. LEMPEL, A randomized protocol for signing contracts, in: R.L. Rivest, A. Sherman and D. Chaum, eds., *Advances in Cryptology, Proc. CRYPTO 82* (Plenum, New York, 1983) 205–210.

[64] FELDMAN, P., A practical scheme for non-interactive verifiable secret sharing, in: *Proc. 28th IEEE Symp. on Foundations of Computer Science* (1985) 427–438.

[65] FIAT, A. and A. SHAMIR, How to prove yourself: practical solutions to identification and signature problems, in: A.M. Odlyzko, ed., *Advances in Cryptology, CRYPTO 86*, Lecture Notes in Computer Science, Vol. 263 (Springer, Berlin, 1987) 186–194.

[66] FRIEZE, A.M., J. HASTAD, R. KANNAN, J.C. LAGARIAS and A. SHAMIR, Reconstructing truncated integer variables satisfying linear congruences, *SIAM J. Comput.* **17**(2) (1988) 262–280.

[67] GAINES, H.F., *Cryptanalysis: A Study of Ciphers and Their Solutions* (Dover, New York, 1956).

[68] GALIL, Z., S. HABER and M. YUNG, A private interactive test of a boolean predicate and minimum-knowledge public-key cryptosystems, in: *Proc. 26th IEEE Symp. on Foundations of Computer Science* (1985) 360–371.

[69] GAREY, M. and D.S. JOHNSON, *Computers and Intractability: A Guide to the Theory of NP-Completeness* (Freeman, San Francisco, 1979).

[70] GARLIŃSKI, J., *Intercept: The Enigma War* (Dent, London, 1979).

[71] GERHARDT, L.A. and R.C. DIXON, eds., *Spread Spectrum Communications, IEEE Trans. Comm.* **25** (1977) (Special Issue).

[72] GILBERT, E.N., F.J. MACWILLIAMS and N.J.A. SLOANE, Codes which detect deception, *Bell System Tech. J.* **53** (1974) 405–424.

[73] GILL, J., Computational complexity of probabilistic Turing machines, *SIAM J. Comput.* **6** (1977) 675–695.

[74] GOLDREICH, O., Two remarks concerning the Goldwasser–Micali–Rivest signature scheme, Tech. Report MIT/LCS/TM-315, MIT Lab. Comput. Sci., 1986.

[75] GOLDREICH, O., S. GOLDWASSER and S. MICALI, How to construct random functions, in: *Proc. 25th IEEE Symp. on Foundations of Computer Science* (1984) 464–479.

[76] GOLDREICH, O. and L. LEVIN, A hard-core predicate for all one-way functions, in: *Proc. 21st Ann. ACM Symp. on Theory of Computing* (1989) 25–32.

[77] GOLDREICH, O., S. MICALI and A. WIGDERSON, Proofs that yield nothing but their validity and

a methodology of cryptographic protocol design, in: *Proc. 27th IEEE Symp. on Foundations of Computer Science* (1986) 174–187.
[78] GOLDWASSER, S. and J. KILIAN, Almost all primes can be quickly certified, in: *Proc. 18th Ann. ACM Symp. on Theory of Computing* (1986) 316–329.
[79] GOLDWASSER, S. and S. MICALI, Probabilistic encryption and how to play mental poker keeping secret all partial information, in: *Proc. 14th Ann. ACM Symp. on Theory of Computing* (1982) 365–377.
[80] GOLDWASSER, S. and S. MICALI, Probabilistic encryption, *J. Comput. System Sci.* **28**(2) (1984) 270–299.
[81] GOLDWASSER, S., S. MICALI and C. RACKOFF, The knowledge complexity of interactive proof-systems, *SIAM. J. Comput.* **18**(1) (1989) 186–208.
[82] GOLDWASSER, S., S. MICALI and R. RIVEST, A digital signature scheme secure against adaptive chosen-message attacks, *SIAM J. Comput.* **17**(2) (1988) 281–308.
[83] GOLDWASSER, S., S. MICALI and P. TONG, Why and how to establish a private code on a public network, in: *Proc. 23rd IEEE Symp. on Foundations of Computer Science* (1982) 134–144.
[84] GOLDWASSER, S., S. MICALI and A. YAO, Strong signature schemes, in: *Proc. 15th Ann. ACM Symp. on Theory of Computing* (1983) 431–439.
[85] GOLDWASSER, S. and M. SIPSER, Private coins versus public coins in interactive proof systems, in: *Proc. 18th Ann. ACM Symp. on Theory of Computing* (1986) 59–68.
[86] GOLOMB, S.W., *Shift Register Sequences* (Aegean Park, Laguna Hills, rev. ed., 1982).
[87] GROLLMAN, J. and A.L. SELMAN, Complexity measures for public-key cryptosystems, *SIAM J. Comput.* **17**(2) (1988) 309–335.
[88] HASTAD, J., Solving simultaneous modular equations of low degree, *SIAM J. Comput.* **17**(2) (1988) 336–341.
[89] HELLMAN, M.E., An extension of the Shannon theory approach to cryptography, *IEEE Trans. Inform. Theory* **23** (1977) 289–294.
[90] HELLMAN, M.E., The mathematics of public key cryptography, *Sci. Amer.* **241** (1979) 146–157.
[91] HELLMAN, M.E., A cryptanalytic time-memory trade off, *IEEE Trans. Inform. Theory* **26** (1980) 401–406.
[92] KAHN, D., *The Codebreakers* (Macmillan, New York, 1967).
[93] KALISKI, JR, B.S., A pseudo-random bit generator based on elliptic logarithms, in: A.M. Odlyzko, ed., *Advances in Cryptology–CRYPTO 86*, Lecture Notes in Computer Science, Vol. 263 (Springer, Berlin, 1987) 84–103.
[94] KALISKI, JR, B.S., Elliptic curves and cryptography: a pseudorandom bit generator and other tools, Ph.D. Thesis, MIT EECS Dept., 1988; published as MIT LCS Tech. Report MIT/LCS/TR-411, 1988.
[95] KALISKI, JR, B.S., R.L. RIVEST and A. SHERMAN, Is DES a pure cipher? (results of more cycling experiments on DES), in: H.C. Williams, ed., *Advances in Cryptology–CRYPTO 85*, Lecture Notes in Computer Science, Vol. 218 (Springer, Berlin, 1986) 212–222.
[96] KANNAN, R., A. LENSTRA and L. LOVÁSZ, Polynomial factorization and non-randomness of bits of algebraic and some transcendental numbers, in: *Proc. 16th ACM Symp. on Theory of Computing* (1984) 191–200.
[97] KILIAN, J., Founding cryptography on oblivious transfer, in: *Proc. 20th Ann. ACM Symp. on Theory of Computing* (1988) 20–31.
[98] KNUTH, D.E., *The Art of Computer Programming: Vol. 2, Seminumerical Algorithms* (Addison-Wesley, Reading, MA, 1969).
[99] KONHEIM, A.G., *Cryptography: A Primer* (Wiley, New York, 1981).
[100] LAGARIAS, J.C. and A.M. ODLYZKO, Solving low-density subset sum problems, in: *Proc. 24th IEEE Symp. on Foundations of Computer Science* (1983) 1–10.
[101] LAMPORT, L., Constructing digital signatures from a one-way function, Tech. Report CSL-98, SRI International, Palo Alto, 1979.
[102] LEMPEL, A., Cryptology in transition: a survey, *Comput. Surv.* **11** (1979) 285–304.
[103] LENSTRA, A.K. and H.W. LENSTRA, JR, Algorithms in number theory, in: J. van Leeuwen, ed., *Handbook of Theoretical Computer Science, Vol. A* (North-Holland, Amsterdam, 1990) 673–715.
[104] LENSTRA, A.K., H.W. LENSTRA, JR. and L. LOVÁSZ, Factoring polynomials with rational coefficients, *Math. Ann.* **261** (1982) 513–534.
[105] LEVEQUE, W.J., *Fundamentals of Number Theory* (Addison-Wesley, Reading, MA, 1977).

[106] LEVIN, L.A., Problems, complete in "average" instance, in: *Proc. 16th Ann. ACM Symp. on Theory of Computing* (1984) 465.

[107] LEVIN, L.A., One-way functions and pseudorandom generators, in: *Proc. 17th Ann. ACM Symp. on Theory of Computing* (1985) 363–365.

[108] LIPTON, R., How to cheat at mental poker, in: *Proc. AMS Short Course on Cryptography* (1981).

[109] LONG, D.L., and A. WIGDERSON, The discrete logarithm problem hides $O(\log n)$ bits, *SIAM J. Comput.* **17**(2) (1988) 363–372.

[110] LUBY, M., S. MICALI and C. RACKOFF, How to simultaneously exchange a secret bit by flipping a symmetrically biased coin, in: *Proc. 24th IEEE Symp. on Foundations of Computer Science* (1983) 11–22.

[111] LUBY, M. and C. RACKOFF, Pseudo-random permutation generators and cryptographic composition, in: *Proc. 18th Ann. ACM Symp. on Theory of Computing* (1986) 356–363.

[112] LUBY, M. and C. RACKOFF, How to construct pseudorandom permutations and pseudorandom functions, *SIAM J. Comput.* **17**(2) (1988) 373–386.

[113] MCELIECE, R.J., A public-key system based on algebraic coding theory, DSN Progress Report 44, Jet Propulsion Lab., 1978, 114–116.

[114] MERKLE, R.C., Secure communications over insecure channels, *Comm. ACM* **21** (1978) 294–299.

[115] MERKLE, R.C., Secrecy, authentication, and public key systems, Tech. Report, Stanford Univ., 1979.

[116] MERKLE, R.C. and M. HELLMAN, Hiding information and signatures in trapdoor knapsacks, *IEEE Trans. Inform. Theory* **24** (1978) 525–530.

[117] MEYER, C.H. and S.M. MATYAS, *Cryptography: A New Dimension in Computer Data Security* (Wiley, New York, 1982).

[118] MICALI, S., C. RACKOFF and R.H. SLOAN, The notion of security for probabilistic cryptosystems, *SIAM J. Comput.* **17**(2) (1988) 412–426.

[119] NAOR, M. and M. YUNG, Universal one-way hash functions and their cryptographic applications, in: *Proc. 21th Ann. ACM Symp. on Theory of Computing* (1989) 33–43.

[120] NATIONAL BUREAU OF STANDARDS, Announcing the data encryption standard, Tech. Report FIPS Publication 46, 1977.

[121] NEEDHAM, R.M. and M.D. SCHROEDER, Using encryption for authentication in large networks of computers, *Comm. ACM* **21**(12) (1978) 933–999.

[122] NIVEN, I and H.S. ZUCKERMAN, *An Introduction to the Theory of Numbers* (Wiley, New York, 1972).

[123] ODLYZKO, A.M., Cryptanalytic attacks on the multiplicative knapsack scheme and on Shamir's fast signature scheme, *IEEE Trans. Inform. Theory* **30** (1984) 594–601.

[124] ODLYZKO, A.M., Discrete logarithms in finite fields and their cryptographic significance, in: T. Beth, N. Cot and I. Ingemarsson, eds., *Advances in Cryptology, Proc. EUROCRYPTO 84*, Lecture Notes in Computer Science, Vol. 218 (Springer, Berlin, 1985) 516–522.

[125] PLUMSTEAD, J., Inferring a sequence generated by a linear congruence, in: *Proc. 23rd IEEE Symp. on Foundations of Computer Science* (1982) 153–159.

[126] POHLIG, S.C. and M.E. HELLMAN, An improved algorithm for computing logarithms over $GF(p)$ and its cryptographic significance, *IEEE Trans. Inform. Theory* **24** (1978) 106–110.

[127] POMERANCE, C., Analysis and comparison of some integer factoring algorithms, in: H.W. Lenstra, Jr. and R. Tijdeman, eds., *Computational Methods in Number Theory*, Math. Centrum Tract, Vol. 153 (CWI, Amsterdam, 1982), 89–139.

[128] POMERANCE, C., J.W. SMITH and R. TULER, A pipeline architecture for factoring large integers with the quadratic sieve algorithm, *SIAM J. Comput.* **17**(2) (1988) 387–403.

[129] PRICE, W.L., Annotated bibliographies of cryptography; published as National Physical Laboratories Tech. Reports since 1978.

[130] RABIN, M., Digitalized signatures, in: R.A. DeMillo, D.P. Dobkin, A.K. Jones and R.J. Lipton, eds., *Foundations of Secure Computation* (Academic Press, New York, 1978) 155–168.

[131] RABIN, M., Digitalized signatures as intractable as factorization, Tech. Report MIT/LCS/TR-212, MIT Lab. Comput. Sci., 1979.

[132] RABIN, M., Probabilistic algorithms for testing primality, *J. Number Theory* **12** (1980) 128–138.

[133] RABIN, M., Probabilistic algorithms in finite fields, *SIAM J. Comput.* **9** (1980) 273–280.

[134] RABIN, M., How to exchange secrets by oblivious transfer, Tech. Report TR-81, Harvard Univ., Aiken Comput. Lab., 1981.
[135] RIESEL, H., *Prime Numbers and Computer Methods for Factorization* (Birkhäuser, Boston, 1985).
[136] RIVEST, R.L. and A. SHAMIR, How to expose an eavesdropper, *Comm. ACM* **27** (1984) 393–395.
[137] RIVEST, R.L., A. SHAMIR and L.M. ADLEMAN, A method for obtaining digital signatures and public-key cryptosystems, *Comm. ACM* **21** (1978) 120–126.
[138] SANDERS, S., Data privacy: what Washington doesn't want you to know, *Reason* (1981) 24–37.
[139] SANTHA, M. and U.V. VAZIRANI, Generating quasi-random sequences from slightly-random sources, in: *Proc. 25th IEEE Symp. on Foundations of Computer Science* (1984) 434–440.
[140] SCHROEPPEL, R. and A. SHAMIR, A $TS^2 = O(2^n)$ time/space tradeoff for certain NP-complete problems, in: *Proc. 20th IEEE Symp. on Foundations of Computer Science* (1979) 328–336.
[141] SHAMIR, A., How to share a secret, *Comm. ACM* **22** (1979) 612–613.
[142] SHAMIR, A., On the cryptocomplexity of knapsack schemes, in: *Proc. 11th Ann. ACM Symp. on Theory of Computing* (1979) 118–129.
[143] SHAMIR, A., On the generation of cryptographically strong pseudo-random sequences, in: *Proc. 8th Internat. Coll. on Automata, Languages and Programming*, Lecture Notes in Computer Science, Vol. 115 (Springer, Berlin, 1981) 544–550.
[144] SHAMIR, A., A polynomial-time algorithm for breaking the basic Merkle–Hellman cryptosystem, in: *Proc. 23rd IEEE Symp. on Foundations of Computer Science* (1982) 145–152.
[145] SHAMIR, A., R.L. RIVEST and L.M. ADLEMAN, Mental poker, in: D. Klarner, ed., *The Mathematical Gardner* (Wadsworth, Belmont, CA, 1981) 37–43.
[146] SHANNON, C.E., Communication theory of secrecy systems, *Bell System Tech. J.* **28** (1949) 657–715.
[147] SHERMAN, A., Cryptology and VLSI (a two-part dissertation), Ph.D. thesis, MIT EECS Dept., 1986; published as MIT LCS Tech. Report MIT/LCS/TR-381, 1986.
[148] SIMMONS, G.J., Symmetric and asymmetric encryption, *ACM Comput. Surveys* **11** (1979) 305–330.
[149] SIMMONS, G.J., *Secure Communications and Asymmetric Cryptosystems*, Selected Symposia, Vol. 69, 1982.
[150] SIMMONS, G.J., Cryptology, in: *The New Encyclopaedia Brittanica*, Vol. 16 (1989) 860–873.
[151] SLOANE, N.J.A., Error-correcting codes and cryptography, in: D. Klarner, ed., *The Mathematical Gardner* (Wadsworth, Belmont, CA, 1981) 346–382).
[152] SOLOVAY, R. and V. STRASSEN, A fast Monte-Carlo test for primality, *SIAM J. Comput.* **6** (1977) 84–85.
[153] VAZIRANI, U.V., Towards a strong communication complexity theory, or generating quasi-random sequences from two communicating slightly-random sources, in: *Proc. 17th Ann. ACM Symp. on Theory of Computing* (1985) 336–378.
[154] VAZIRANI, U.V. and V.V. VAZIRANI, Trapdoor pseudo-random number generators, with applications to protocol design, in: *Proc. 24th IEEE Symp. on Foundations of Computer Science* (1983) 23–30.
[155] VAZIRANI, U.V. and V.V. VAZIRANI, Efficient and secure pseudo-random number generation, in: *Proc. 25th IEEE Symp. on Foundations of Computer Science* (1984) 485–463.
[156] VON NEUMANN, J., Various techniques for use in connection with random digits, in: *von Neumann's Collected Works* (Pergamon, New York, 1963) 768–770.
[157] WINTERBOTHAM, F.W., *The Ultra Secret* (Futura, London, 1975).
[158] WYNER, A.D., The wire-tap channel, *Bell System Tech. J.* **54** (1975) 1355–1387.
[159] WYNER, A.D., An analog scrambling scheme which does not expand bandwidth, part 1, *IEEE Trans. Inform. Theory* **25**(3) (1979) 261–274.
[160] WYNER, A.D., An analog scrambling scheme which does not expand bandwidth, part 2, *IEEE Trans. Inform. Theory* **25**(4) (1979) 415–425.
[161] YAO, A.C., Theory and application of trapdoor functions, in: *Proc. 23rd IEEE Symp. on Foundations of Computer Science* (1982) 80–91.
[162] YAO, A.C., Protocols for secure computations, in: *Proc. 23rd IEEE Symp. on Foundations of Computer Science* (1982) 160–164.
[163] YAO, A.C., How to generate and exchange secrets, in: *Proc. 27th IEEE Symp. on Foundations of Computer Science* (1986) 162–167.
[164] YUVAL, G., How to swindle Rabin, *Cryptologia* **3** (1979) 187–189.

CHAPTER 14

The Complexity of Finite Functions

Ravi B. BOPPANA

Department of Computer Science, New York University, New York, NY 10012, USA

Michael SIPSER

Mathematics Department, Massachusetts Institute of Technology, Cambridge, MA 02139, USA

Contents

HANDBOOK OF THEORETICAL COMPUTER SCIENCE
Edited by J. van Leeuwen

1. Introduction

The classification of problems according to computational difficulty comprises two subdisciplines of different character. One, the theory of algorithms, gives upper bounds on the amount of computational resource needed to solve particular problems. This endeavor has enjoyed much success in recent years with a large number of strong results and fruitful connections with other branches of mathematics and engineering. The other, called the theory of computational complexity, is an attempt to show that certain problems cannot be solved efficiently by establishing lower bounds on their inherent computational difficulty. This has not been as successful. Except in a few special circumstances, we have been unable to demonstrate that particular problems are computationally difficult, even though there are many which appear to be so. The most fundamental questions, such as the famous P-versus-NP question which simply asks whether it is harder to find a proof than to check one, remain far beyond our present abilities. This chapter surveys the present state of understanding on this and related questions in complexity theory.

The difficulty in proving that problems have high complexity seems to lie in the nature of the adversary: the algorithm. Fast algorithms may work in a counterintuitive fashion, using deep, devious, and fiendishly clever ideas. How does one prove that there is no clever way to quickly solve a problem? This is the issue confronting the complexity theorist. One way to make some progress on this is to limit the capabilities of the computational model, thereby limiting the class of potential algorithms. In this way it has been possible to achieve some interesting results. Perhaps these may lead the way to lower bounds for more powerful computational models.

There is by now an extensive literature on bounds for various models of computation. We have excluded some of these from this survey using the following criteria. First, we will concentrate on recent, mostly combinatorial bounds for Boolean circuits, formulas, and branching programs. We will not include the older work on lower bounds via the diagonalization method. Though important early progress was made in that way, the relativization results of Baker, Gill and Solovay [10] show that such methods from recursive function theory are inadequate for the remaining interesting questions. Second, we will mostly focus on bounds that differentiate polynomial from nonpolynomial rather than lower level bounds. This reflects our feeling that this type of result is closer to the interesting unsolved questions. Finally, in several cases we refrain from giving the tightest result known to keep the exposition as clear as possible.

1.1. *Brief history*

Circuit complexity theory dates from Shannon's seminal 1949 paper [79]. There he proposed the size of the smallest circuit computing a function as a measure of its complexity. His motivation was simply to minimize the hardware necessary for computation. He proved an upper bound on the complexity of all n input functions and used a counting argument to show that for most functions this bound is not too far off.

The 1960s saw the introduction of the algorithms as a way of measuring the complexity of functions. Edmonds [25] gave a polynomial-time algorithm for the matching problem and foresaw the issue of polynomial-versus-exponential complexity. Hartmanis and Stearns [33] formalized this measure as time on a Turing machine. Savage [77] established a close relationship between the time required to compute a function on a Turing machine and its circuit complexity. At this point the importance of proving good lower bounds on circuit size was apparent, but it was also becoming clear that this was going to be difficult to accomplish. See [32] for a discussion on this. By the end of the 1970s essentially the only results known were the linear lower bound of Paul [61] and the nearly quadratic lower bound of Nečiporuk [54] for the special case of formula size.

A new direction in the 1980s brought on a burst of activity. By placing sufficiently strong restrictions on the class of circuits it became possible to prove strong lower bounds. The first results of this kind were independently obtained by Furst, Saxe and Sipser [30] and Ajtai [2] for the bounded-depth circuit model. Razborov [72] and subsequently Andreev [7] gave strong lower bounds for the monotone circuit model. Numerous papers strengthening these results and giving others in the same vein have since appeared.

Our chapter emphasizes this latter work giving only a sketchy overview of the preceding period. For a more comprehensive discussion of the earlier work, see the survey paper of Paterson [59] and the book of Savage [78]. Much of the more recent work is covered in the books of Dunne [99] and Wegener [96].

2. General circuits

In an early paper Shannon [79] considered the size of Boolean circuits as a measure of computational difficulty. Circuits are an attractive model for proving lower bounds for several reasons. They are closely related in computational power to Turing machines so that a good lower bound on circuit size directly gives a lower bound on time complexity. Among computational models the circuit model has an especially simple definition and so may be more amenable to combinatorial analysis. Even so, we are currently able to prove only very weak lower bounds on circuit size.

A *Boolean circuit* is a directed acyclic graph. The nodes of in-degree 0 are called *inputs*, and are labeled with a variable x_i or with a constant 0 or 1. The nodes of in-degree $k>0$ are called *gates* and are labeled with a Boolean function on k inputs. We refer to the in-degree of a node as its *fan-in* and its out-degree as its *fan-out*. Unless otherwise specified we restrict to the Boolean functions AND, OR, and NOT. One of the nodes is designated the *output* node. The *size* is the number of gates, and the *depth* is the maximum distance from an input to the output. A *Boolean formula* is a special kind of circuit whose underlying graph is a tree.

A Boolean circuit represents a Boolean function in a natural way. Let N denote the natural numbers, $\{0,1\}^n$ the set of binary strings of length n, and $\{0,1\}^*$ the set of all finite binary strings. Let $f:\{0,1\}^n \to \{0,1\}$. Then $C(f)$ is the size of the smallest circuit representing f. Let $g:\{0,1\}^* \to \{0,1\}$ and $h: N \to N$. Say g has *circuit complexity* h if for all n, $C(g_n)=h(n)$ where g_n is g restricted to $\{0,1\}^n$. A *language* is a subset of $\{0,1\}^*$. The circuit complexity of a language is that of its characteristic function.

2.1. Boolean circuits and Turing machines

In this subsection we establish a relationship between the circuit complexity of a problem and the amount of time that is required to solve it. First we must select a model of computation on which to measure time. The *Turing machine* (*TM*) model was proposed by Alan Turing [89] as a means of formalizing the notion of effective procedure. This intuitively appealing model serves as a convenient foundation for many results in complexity theory. The choice is arbitrary among the many polynomially equivalent models. The complexity of Turing machine computations was first considered by Hartmanis and Stearns [33]. We briefly review here the variant of this model that we will use. For a more complete introduction see [36] or Chapter 1 on machine models by van Emde Boas [26].

A *deterministic Turing machine* consists of a finite control and a finite collection of tapes each with a head for reading and writing. The finite control is a finite collection of states. A tape is an infinite list of cells each containing a symbol. Initially, all tapes have blanks except for the first, which contains the input string. Once started, the machine goes from state to state, reading the symbols under the heads, writing new ones, and moving the heads. The exact action taken is governed by the current state, the symbols read, and the next-move function of the machine. This continues until a designated halt state is entered. The machine indicates its output by the halting condition of the tapes.

In a *nondeterministic Turing machine* the next-move function is multivalued. There may be several computations on a given input, and several output values. We say the machine accepts its input if at least one of these outputs signals acceptance. An *alternating Turing machine* is a nondeterministic Turing machine whose states are labeled $\wedge$ and $\vee$. Acceptance of an input is determined by evaluating the associated $\wedge$, $\vee$ tree of computations in the natural way. A Σ_i (Π_i) Turing machine is an alternating Turing machine which may have at most i runs of $\vee$ and $\wedge$ states and must begin with $\vee$ ($\wedge$) states. See [20] for more information about alternating Turing machines.

Say a Turing machine M *accepts* language A if M accepts exactly those strings in A. We now define Turing machine computations within a time bound T. Let $T: N \to N$. A deterministic, nondeterministic, or alternating Turing machine M *runs in time* $T(n)$ if for every input string w of length n, every computation of M on w halts within $T(n)$ steps. We define the time complexity classes.

$$\mathrm{TIME}(T(n))=\{A\text{: some deterministic TM accepts } A \text{ in time } \mathrm{O}(T(n))\},$$
$$\mathrm{NTIME}(T(n))=\{A\text{: some nondeterministic TM accepts } A \text{ in time } \mathrm{O}(T(n))\},$$
$$\mathrm{ATIME}(T(n))=\{A\text{: some alternating TM accepts } A \text{ in time } \mathrm{O}(T(n))\},$$
$$\Sigma_i\mathrm{TIME}(T(n))=\{A\text{: some } \Sigma_i \text{ TM accepts } A \text{ in time } \mathrm{O}(T(n))\},$$
$$\Pi_i\mathrm{TIME}(T(n))=\{A\text{: some } \Pi_i \text{ TM accepts } A \text{ in time } \mathrm{O}(T(n))\}.$$

We will use the following standard notation for the polynomial-time complexity classes and the polynomial-time hierarchy:

$$P = \bigcup_k \mathrm{TIME}(n^k),$$

$$NP = \bigcup_k \mathrm{NTIME}(n^k),$$

$$\Sigma_i P = \bigcup_k \Sigma_i \mathrm{TIME}(n^k),$$

$$\Pi_i P = \bigcup_k \Pi_i \mathrm{TIME}(n^k).$$

The connection between circuit complexity and Turing machine complexity was first shown by Savage [77]. The bound in the following theorem is due to Pippenger and Fischer [66].

2.1. THEOREM. *If language A is in* $\mathrm{TIME}(T(n))$, *then A has circuit complexity* $O(T(n)\log(T(n)))$.

PROOF (*sketch*). To see the main idea of the simulation we first prove the weaker bound $O(T^4(n))$. Let M accept A in time $T(n)$. Convert M to a 1-tape machine. This increases the time to $O(T^2(n))$. View a computation of M on some input as a table where row i is the tape at the ith step. At any point in time we will consider the cell currently scanned by the head to contain a symbol representing both the actual symbol and the state of the machine. Let cell(i, j) be the contents of the ith cell at time j. It is easy to see that cell(i,j) only depends upon its predecessors cell$(i-1,j-1)$, cell$(i,j-1)$, and cell$(i+1, j-1)$. We may encode the possible values of a cell in binary and build a small circuit for each cell which computes its value from its predecessors. Assuming the machine indicates its acceptance in the first tape cell upon halting, we designate the appropriate gate from cell$(1, l)$ to be the output where l is the index of the last row. Since the total number of cells is $O((T^2(n))^2)=O(T^4(n))$, the simulating circuit has size $O(T^4(n))$.

To obtain the tighter bound of the theorem, we first modify the original machine so that it is *oblivious*, i.e., the head motions are only dependent upon the length of the input but not the input itself. An $O(T(n)\log T(n))$ time simulation using two tapes accomplishes this. Once the motions of the heads are known the above construction may be repeated, except now it is only necessary to build circuitry for the cells which contain the head because the others do not change symbol. □

Thus every language in P has polynomial circuit complexity. The converse of this is false since there are nonrecursive languages with low circuit complexity. To obtain a converse we define the nonuniform extension of P. Alternatively we may impose a uniformity condition on the circuit families saying that the nth circuit is easy to find as a function of n. But, since lower bounds for nonuniform classes imply lower bounds for the corresponding uniform classes, we only consider the former here.

DEFINITIONS. Let $f: N \to N$. An $f(n)$ *advice sequence* is a sequence of binary strings

$A=(a_1, a_2, \ldots)$, where $|a_n| \leqslant f(n)$. For a language $B \subseteq \{0, 1, \#\}^*$ let $B@A=\{x: x\# a_{|x|} \in B\}$. Let $P/f(n)=\{B@A: B \in P$ and A is an $f(n)$ advice sequence$\}$. Let $P/\text{poly}=\bigcup_k P/n^k$. Apply this notation to other complexity classes, e.g., NP/poly. These classes are sometimes referred to as "nonuniform P" or "nonuniform NP".

2.2. THEOREM. *C is in* P/poly *iff C has polynomial circuit complexity.*

PROOF. If C is in P/poly, then $C=B@A$ for some B in P and polynomial advice sequence A. By the above theorem B has polynomial circuit complexity. Presetting the advice strings into the appropriate inputs of the circuits for B obtains the polynomial-size circuits for C.

For the other direction we encode the polynomial-size circuits for C as an advice sequence. □

Thus one may hope to show $P \neq NP$ by giving a superpolynomial lower bound on circuit size for an NP problem. Of course, it may be that all NP problems have polynomial circuit complexity even though $P \neq NP$. The following theorem of Karp and Lipton [39], and Sipser shows that if this were the case, then there would be other peculiar consequences.

2.3. THEOREM. *If* $NP \subseteq P/\text{poly}$, *then the polynomial-time hierarchy collapses to* $\Sigma_2 P$.

2.2. Nonexplicit lower bounds

Although we are mostly concerned with explicit problems (i.e., those in NP), it is worth understanding what can be said about nonexplicit problems. Muller [52], based on an argument of Shannon [79], proved an exponential lower bound on the circuit complexity of a nonexplicit problem. We prove this lower bound below.

Counting arguments can help establish lower bounds for nonexplicit problems. On the one hand, there are not too many small circuits; on the other hand, there are very many Boolean functions of n variables. Thus some Boolean functions must require circuits of exponential size. Muller obtains the following result.

2.4. THEOREM. *Almost every Boolean function of n variables requires circuits of size* $\Omega(2^n/n)$.

PROOF. We first show that the number of circuits with n variables and size s is bounded above by $(2\cdot(s+2n+2)^2)^s$. Each gate in a circuit is assigned an AND or OR operator that acts on two previous nodes. Each previous node can either be a previous gate (at most s choices), a literal, i.e., a variable or its negation ($2n$ choices), or a constant (2 choices). Thus each gate has at most $2\cdot(s+2n+2)^2$ choices. Compounding these choices for all s gates gives the claimed upper bound.

Notice that for $s=2^n/(10n)$, the above bound is approximately $2^{2^n/5} \ll 2^{2^n}$. Since there are 2^{2^n} Boolean function of n variables, almost every Boolean function requires circuits of size larger than $2^n/(10n)$. □

The above lower bound is optimal up to a constant factor. Expressed in disjunctive normal form (i.e., as an OR of ANDs of literals), every Boolean function has circuits of size $O(2^n \cdot n)$. By being more careful, Muller showed that every Boolean function has circuits of size $O(2^n/n)$.

Theorem 2.4 shows the existence of a Boolean function with exponential circuit complexity, but says nothing about the explicitness of the function. The theorem can be sharpened to yield a Boolean function computable in exponential space but requiring exponential-size circuits. Let f_n be the lexically-first function of n variables that requires circuits of size $2^n/(10n)$, and let f be the union of these functions over all n. By definition f requires circuits of exponential size. By enumerating in lexical order all functions of n variables, and also enumerating all small circuits, we can compute f in space $O(2^n)$. Unfortunately, this argument fails to provide a example of a problem in NP (or even in exponential time) that provably requires circuits of superpolynomial size.

As we did for circuit complexity above, one can ask about the complexity of most Boolean functions under other complexity measures. For example, the formula complexity (defined in Section 5) of almost every Boolean function is $\Theta(2^n/\log n)$. The branching program complexity (defined in Section 6) of almost every Boolean function is $\Theta(2^n/n)$.

2.3. *Explicit lower bounds*

Despite the importance of lower bounds on the circuit complexity of explicit problems, the best bounds known are only linear. N. Blum [15], improving a bound of Paul [61], proved a $3n-o(n)$ lower bound. In this subsection, we prove a weaker but easier lower bound of size $2n-O(1)$. These bounds apply to circuits with all binary gates allowed.

The proofs of these lower bounds use the following gate-elimination method. Given a circuit for the function in question, we first argue that some variable (or set of variables) must fan out to several gates. Setting this variable to a constant will eliminate several gates. By repeatedly applying this process, we conclude that the original circuit must have had many gates. As an example, we apply the gate-elimination method to threshold functions. Let $\mathrm{TH}_{k,n}$ be the function that outputs 1 iff at least k of its n variables are 1. We have the following lower bound.

2.5. Theorem. *For* $n \geq 2$, *the function* $\mathrm{TH}_{2,n}$ *requires circuits of size at least* $2n-4$.

Proof. The proof is by induction on n. For $n=2$ and $n=3$, the bound is trivial. Otherwise, let C be an optimal circuit for $\mathrm{TH}_{2,n}$, and suppose without loss of generality that the bottom-most gate of C acts on variables x_i and x_j (where $i \neq j$). Notice that under the four possible settings of x_i and x_j, the function $\mathrm{TH}_{2,n}$ has three possible subfunctions (namely $\mathrm{TH}_{0,n-2}$, $\mathrm{TH}_{1,n-2}$, and $\mathrm{TH}_{2,n-2}$). It follows that either x_i or x_j fans out to another gate in C, for otherwise C would have only two inequivalent subcircuits under the settings of x_i and x_j; suppose it is x_i that fans out to another gate. Setting x_i to 0 will eliminate the need for at least two gates from C. The resulting function is $\mathrm{TH}_{2,n-1}$, which by induction requires circuits of size $2(n-1)-4$. Adding the two eliminated gates to this bound shows that C has at least $2n-4$ gates, which completes the induction. □

3. Bounded-depth circuits

As we have seen in the previous section, our methods for proving lower bounds on the circuit complexity of explicit Boolean functions are presently very weak. We have only been able to obtain strong lower bounds by imposing sharp limitations on the types of computation performed by the circuit. In this and the next section we will survey the results on two types of limited circuit models, bounded-depth circuits and monotone circuits.

The first strong lower bounds for bounded-depth circuits were given by Furst, Saxe and Sipser [30] and Ajtai [2] who demonstrated a superpolynomial lower bound for constant-depth circuits computing the parity function. Subsequently Yao [97], by giving a deeper analysis of the method of Furst, Saxe and Sipser, was able to give a much sharper exponential lower bound. Hastad [34] further strengthened and simplified this argument, obtaining near optimal bounds. Ajtai used different but related probabilistic-combinatorial methods in his proof. We present Hastad's proof here.

3.1. Definitions

In this section we consider circuits whose depth is very much smaller than n, the number of inputs. We allow arbitrary fan-in so that the circuit may access the entire input. Equivalently one may allow only bounded fan-in and measure alternation depth.

For the following definitions we assume that the circuits are of the special form where all AND- and OR-gates are organized into alternating levels with edges only between adjacent levels and all negations appear only on the inputs. Any circuit may be converted to one of this form without increasing the depth and by at most squaring the size.

DEFINITIONS. A circuit C is called a Σ_i^S circuit if it has a total of at most S gates organized into at most i levels with an OR-gate at the output. Say C is a $\Sigma_i^{S,t}$ circuit if there are a total of at most S gates organized into at most $i+1$ levels with an OR-gate at the output and gates of fan-in at most t at the input level. Call C a *t-open* circuit if it is $\Sigma_1^{S,t}$ for some S, i.e., an OR of ANDs of fan-in at most t. Dually define Π_i^S, $\Pi_i^{S,t}$, and *t-closed* by exchanging AND and OR.

A peculiar feature of these definitions is that we define a $\Sigma_i^{S,t}$ circuit to be one of depth $i+1$. We defend this terminology by noting the analogy of unbounded fan-in gates to unbounded quantifiers and bounded fan-in gates to bounded quantifiers. Generally the subscript in the Σ_i symbol refers to the number of unbounded quantifiers. A further justification is that the statements of a number of the coming theorems are rendered more elegant by adopting this convention.

In the remainder of this section the statements made regarding Σ circuits have a natural dual form for Π circuits. To avoid continual repetition we omit this dual form where obvious.

We will speak of t-open, t-closed, Σ_i^S, and $\Sigma_i^{S,t}$ functions and subsets of $\{0,1\}^n$ as those which are defined by the respective type of circuits.

A *family* of circuits is a sequence $(C_1, C_2, \ldots)$ where C_n takes n input variables. A family may be used to define a language (subset of $\{0,1\}^*$) or a function from $\{0,1\}^*$ to $\{0,1\}$. We may use the notation $t(n)$-closed, $\Sigma_i^{f(n)}$, and $\Sigma_i^{f(n),t(n)}$ to describe the complexities of families and their associated languages and functions. Occasionally we will use $\Sigma_i^{f(n)}$, or $\Sigma_i^{f(n),t(n)}$ to mean the collection of such families, languages, or functions. A *uniform family* is one where the description of C_n may be easily computed from n. Since the lower bounds we present later in the chapter apply even in the stronger nonuniform case we will not occupy ourselves further with notions of uniformity.

Let Σ_i^{poly} and $\Sigma_i^{\text{poly,const}}$ mean $\bigcup_k \Sigma_i^{n^k}$ and $\bigcup_k \Sigma_i^{n^k,k}$, respectively.

DEFINITION. Let $AC^0 = \bigcup_i \Sigma_i^{\text{poly}}$. The classes NC^i and AC^i for $i \geqslant 0$ were proposed by Pippenger [64] and Cook [22] to be those functions computable by a uniform family of polynomial size, $O(\log^i n)$ depth circuits with constant and unbounded fan-in, respectively. For the remainder of this chapter we ignore the uniformity condition.

Thus AC^0 is the class of polynomial size, constant depth, unbounded fan-in circuits. This is equivalent to the above definition. NC^0 is the class of functions depending upon a number of input variables which is a constant, independent of n. As an aside to our main story, the following theorem, discovered independently by a number of people, is an interesting exercise.

3.1. THEOREM. $NC^0 = \Sigma_1^{\text{poly,const}} \cap \Pi_1^{\text{poly,const}}$.

In fact, one may prove the following stronger separation property.

3.2. THEOREM. *If A and B are disjoint $\Sigma_1^{\text{poly,const}}$ languages, then there is an* NC^0 *language C separating them, i.e., $A \subseteq C \subseteq \bar{B}$.*

This theorem is false for disjoint $\Pi_1^{\text{poly,const}}$ languages.

3.2. Restrictions

Furst, Saxe and Sipser introduced the method of probabilistic restrictions as a way of proving lower bounds on the size of bounded-depth circuits. Here a randomly selected subset of the variables is preset so that some of the gates in the circuit become determined, resulting in a simpler circuit on fewer variables.

DEFINITION. Let $X = \{x_1, \ldots, x_n\}$ be the input variables to a circuit C computing a function f. A *restriction* ρ is a mapping from X to $\{0, 1, *\}$.

We interpret ρ as presetting the variables assigned 0 or 1 and leaving variable those assigned star. Under ρ we may simplify C by eliminating gates whose values become determined. Call this the *induced circuit* $C|_\rho$ computing the *induced function* $f|_\rho$.

In the probabilistic arguments to follow we will be selecting restrictions from certain probability distributions. Fix $0<p<1$. Let $\mathscr{R}_p$ be the probability distribution on restrictions over X where each $x_i \in X$ is independently assigned a value in $\{0, 1, *\}$ so that $\Pr[\rho(x_i)=*]=p$ and $\Pr[\rho(x_i)=0]=\Pr[\rho(x_i)=1]=\frac{1}{2}(1-p)$.

3.3. *Hastad switching lemma*

The following lemma of Hastad [34] states that a t-closed function is very likely to become s-open under a restriction chosen at random from a suitable probability distribution.

3.3. LEMMA. *Let f be a t-closed function and ρ a random restriction from $\mathscr{R}_p$. Then*

$$\Pr[f|_\rho \textit{ is not s-open}] \leqslant \alpha^s$$

where $\alpha=\gamma pt$ and $\gamma=2/\ln\varphi\approx 4.16$ for $\varphi=(1+\sqrt{5})/2$, the golden ratio.

PROOF. We introduce some notation. A *minterm* of f is a minimal assignment to some subset of the variables which determines f to be 1. Say that $min(f)$ is the size of the largest minterm of f. By showing that $min(f|_\rho)\leqslant s$ we show that $f|_\rho$ is s-open because a function may be written as an OR of its minterms. Below we will prove that the slightly stronger inequality $min(f|_\rho)<s$ occurs with high probability.

To better understand the proof of this lemma it is helpful to first consider a special case. Say f is given by a t-closed circuit C where we assume that all of the ORs of C are on disjoint sets of variables. Then the following straightforward induction on l, the number of ORs in C, gives the lemma. The intuition behind this induction is easy. Consider the ORs one by one. If some variable in an OR gets a 1 under ρ, then the OR is determined to be 1 and it contributes nothing to any minterm. Only a starred variable in an OR without 1s may be part of a minterm. But such an OR is more likely to get all 0s. If this ever happens, then the function is determined to be 0 and has no minterms at all. Thus it is very unlikely that there will be large minterms. The details follow.

Basis ($l=0$): Obvious since f is the constant function.

Induction (Assume for $l-1$ and prove for l): Let C_1 be the first OR in C. By renaming variables we may assume that all of the variables in C_1 occur without negation. Let D be C without C_1. Then

$$\Pr[min(C|_\rho)\geqslant s]=\Pr[C_1|_\rho\equiv 1]\cdot(\mathrm{A})+\Pr[C_1|_\rho\not\equiv 1]\cdot(\mathrm{B})$$

where

$$(\mathrm{A})=\Pr[min(C|_\rho)\geqslant s \mid C_1|_\rho\equiv 1]$$

and

$$(\mathrm{B})=\Pr[min(C|_\rho)\geqslant s \mid C_1|_\rho\not\equiv 1].$$

Since the right-hand side is a weighted average of (A) and (B) it is sufficient to show that both (A) and (B) are at most α^s to conclude that the left-hand side is at most α^s. Here and in all expressions throughout the proofs of Lemmas 3.3 and 3.3′ we adopt the convention that $\Pr[A|B]=0$ if $\Pr[B]=0$.

Bounding (A): If $C_1|_\rho \equiv 1$, then C_1 will not contribute to the minterm and so $min(C|_\rho)=min(D|_\rho)$. Since we are assuming that the ORs are disjoint, the conditioning is irrelevant and so $(A)=\Pr[min(D|_\rho)\geqslant s]$. Now the bound follows from the induction hypothesis.

Bounding (B): We first divide up the probability, summing it according to the number of stars in C_1. Since the number of stars in C_1 is an upper bound on its contribution to any minterm we have $(B)\leqslant\sum_{k>0}^{t}(C)\cdot(D)$ where

$$(C)=\Pr[C_1|_\rho \text{ gets } k \text{ stars} \mid C_1|_\rho \not\equiv 1]$$

and

$$(D)=\Pr[min(D|_\rho)\geqslant s-k \mid C_1|_\rho \text{ gets } k \text{ stars and } C_1|_\rho \not\equiv 1].$$

Actually, 1 is an upper bound on the contribution of C_1, so we can say something a bit stronger here. We chose to estimate it this way to preserve the similarity with the following Lemma 3.3′. It is important to note that the term $k=0$ is excluded from the above sum since it implies that $C_1|_\rho$ and hence $C|_\rho$ are equivalent to 0.

To bound (C) observe that the condition $C_1|_\rho \not\equiv 1$ is equivalent to saying that all variables in C_1 receive 0 or star in ρ. The conditional probability that a variable receives a star is

$$\frac{p}{1-(1-p)/2}=\frac{p}{(1+p)/2}=\frac{2p}{1+p}\leqslant 2p.$$

So $(C)\leqslant\binom{t}{k}(2p)^k$.

To bound (D), again observe that the conditioning has no effect under the disjointness assumption, so by induction $(D)\leqslant\alpha^{s-k}$. Therefore,

$$\begin{aligned}(B)&\leqslant\sum_{k>0}^{t}\binom{t}{k}(2p)^k\alpha^{s-k}=\alpha^s\sum_{k>0}^{t}\binom{t}{k}\left(\frac{2p}{\alpha}\right)^k\\&=\alpha^s((1+2p/\alpha)^t-1)=\alpha^s((1+2p/(\gamma pt))^t-1)=\alpha^s((1+2/\gamma t)^t-1)\\&\leqslant\alpha^s(e^{2/\gamma}-1)<\alpha^s.\end{aligned}$$

To eliminate the disjointness assumption we may no longer ignore the conditioning which occurs in (A) and (B). To handle this we will prove a version of the lemma with a stronger induction hypothesis. The intuition behind this is that at every stage in the induction the only information about the probability distribution that is known is that it sets certain variables to star, others to 0 or 1, and that some ORs are determined to be 1. Conditioning on some variable being set to star does not hurt since these variables were already counted. The other conditioning does not hurt since it can only make it less likely that a variable is starred and contributes to a big minterm. In the following lemma, F is an arbitrary function defined on $X=\{x_1,\ldots,x_n\}$, the variables of C, as well as any other variables.

3.3′. STRONGER LEMMA. *Let f be a t-closed function, F an arbitrary function, and ρ a random restriction from $\mathcal{R}_p$. Then*

$$\Pr[f|_\rho \text{ not } s\text{-open} \mid F|_\rho \equiv 1] \leqslant \alpha^s$$

where $\alpha = \gamma pt$ and $\gamma = 2/\ln \varphi \approx 4.16$ for $\varphi = (1+\sqrt{5})/2$, the golden ratio.

PROOF. Let C be a t-closed circuit for f. We modify the first equation in the previous argument to include the new condition $F|_\rho \equiv 1$ in all probability statements. As before, we assume that the first OR C_1 has only positive occurrences of variables. Again we show that the modified (A) and (B) are at most α^s.

$$\Pr[\mathit{min}(C|_\rho) \geqslant s \mid F|_\rho \equiv 1]$$
$$= \Pr[C_1|_\rho \equiv 1 \mid F|_\rho \equiv 1] \cdot (\mathrm{A}) + \Pr[C_1|_\rho \not\equiv 1 \mid F|_\rho \equiv 1] \cdot (\mathrm{B})$$

where

$$(\mathrm{A}) = \Pr[\mathit{min}(C|_\rho) \geqslant s \mid C_1|_\rho \equiv 1 \text{ and } F|_\rho \equiv 1]$$

and

$$(\mathrm{B}) = \Pr[\mathit{min}(C|_\rho) \geqslant s \mid C_1|_\rho \not\equiv 1 \text{ and } F|_\rho \equiv 1].$$

Bounding (A): $\Pr[\mathit{min}(C|_\rho) \geqslant s \mid C_1|_\rho \equiv 1 \text{ and } F|_\rho \equiv 1]$ is clearly equivalent to $\Pr[\mathit{min}(D|_\rho) \geqslant s \mid (C_1 \wedge F)|_\rho \equiv 1]$. This is bounded by α^s using the induction hypothesis.

Bounding (B): Recalling (B) we have

$$(\mathrm{B}) = \Pr[\mathit{min}(C|_\rho) \geqslant s \mid C_1|_\rho \not\equiv 1 \text{ and } F|_\rho \equiv 1].$$

In the special case of the lemma we estimated (B) by breaking up the probability, dividing the minterms according to the number of variables in C_1 that they contain. Now we must use a finer scalpel and divide the minterms according to precisely which variables in C_1 they contain.

Let T be the collection of variables in C_1 and let $Y \subseteq T$. Let $\mathit{min}^Y(C)$ be the size of the largest minterm of C which sets all of the variable of Y and no others in T. Let $\sigma \in \{0,1\}^Y$ be an assignment of Y to 0 and 1. Let $\mathit{min}^{Y \leftarrow \sigma}(C)$ be the size of the largest minterm of C which sets all of the variables in Y to σ and sets no other variables in T. Both $\mathit{min}^Y(C)$ and $\mathit{min}^{Y \leftarrow \sigma}(C)$ are defined to be 0 if such minterms do not exist. Let $\rho(Y) = *$ denote the event that ρ assigns star to all variables in Y. Then we can write

$$(\mathrm{B}) = \Pr[\mathit{min}(C|_\rho) \geqslant s \mid C_1|_\rho \not\equiv 1 \text{ and } F|_\rho \equiv 1]$$

$$\leqslant \sum_{Y \subseteq T} \Pr[\mathit{min}^Y(C|_\rho) \geqslant s \mid C_1|_\rho \not\equiv 1 \text{ and } F|_\rho \equiv 1]$$

$$= \sum_{Y \subseteq T} \begin{pmatrix} \Pr[\rho(Y) = * \mid C_1|_\rho \not\equiv 1 \text{ and } F|_\rho \equiv 1] \\ \cdot \Pr[\mathit{min}^Y(C|_\rho) \geqslant s \mid \rho(Y) = * \text{ and } C_1|_\rho \not\equiv 1 \text{ and } F|_\rho \equiv 1] \end{pmatrix}.$$

We estimate the first term in this sum using an elementary fact from probability theory.

3.4. LEMMA. *For arbitrary events A and B,* $\Pr[A|B] \leqslant \Pr[A] \Leftrightarrow \Pr[B|A] \leqslant \Pr[B]$.

This is easily verified using the definition of conditional probability.

3.5. LEMMA. $\Pr[\rho(Y)=* \mid C_1|_\rho \not\equiv 1 \text{ and } F|_\rho \equiv 1] \leqslant (2p)^{|Y|}$.

PROOF. Without the condition $F|_\rho \equiv 1$ this lemma would follow immediately. We have already done essentially that calculation in the proof of Lemma 3.3. The present lemma also holds in the presence of the extra condition because the condition can only decrease the probability and so works in our favor. Intuitively this is true because knowing that a restriction forces some function to 1 cannot make it more likely that any given variable is assigned a star. To prove this formally we use the previous lemma. Let A be the event $\rho(Y)=*$ and B the event $F|_\rho \equiv 1$. To show

$$\Pr[\rho(Y)=* \mid F|_\rho \equiv 1 \text{ and } C_1|_\rho \not\equiv 1] \leqslant \Pr[\rho(Y)=* \mid C_1|_\rho \not\equiv 1]$$

it is enough to show that

$$\Pr[F|_\rho \equiv 1 \mid \rho(Y)=* \text{ and } C_1|_\rho \not\equiv 1] \leqslant \Pr[F|_\rho \equiv 1 \mid C_1|_\rho \not\equiv 1].$$

This last inequality holds because the condition $C_1|_\rho \not\equiv 1$ means that ρ assigns the variables in T either 0 or star and the condition $\rho(Y)=*$ means that the variables in the subset Y of T must all be star. Any restriction ρ satisfying both conditions corresponds to a set of $2^{|Y|}$ restrictions satisfying only the condition $C_1|_\rho \not\equiv 1$. If $F|_\rho \equiv 1$, then $F|_{\rho'} \equiv 1$ for every ρ' in that set, since any restriction forcing F to be 1 still does so even if some of the starred variables are assigned 0.

We now estimate the second term of the sum in (B). That is,

$$(\mathrm{D}) = \Pr[\mathit{min}^Y(C|_\rho) \geqslant s \mid \rho(Y)=* \text{ and } C_1|_\rho \not\equiv 1 \text{ and } F|_\rho \equiv 1].$$

First we further divide it according to how the minterm sets Y.

$$(\mathrm{D}) \leqslant \sum_{\substack{\sigma\in\{0,1\}^Y \\ \sigma\neq 0^Y}} \Pr[\mathit{min}^{Y\leftarrow\sigma}(C|_\rho) \geqslant s \mid \rho(Y)=* \text{ and } C_1|_\rho \not\equiv 1 \text{ and } F|_\rho \equiv 1].$$

The case $\sigma = 0^Y$ is excluded because a minterm must set some variable in Y to 1.

Estimating (D) is now greatly simplified. Break up the restriction ρ into ρ_1 which assigns values to T and ρ_2 which assigns values to the remaining variables. Then

$$(\mathrm{D}) \leqslant \sum_{\substack{\sigma\in\{0,1\}^Y \\ \sigma\neq 0^Y}} \max_{\rho_1} \Pr_{\rho_2}[\mathit{min}^{Y\leftarrow\sigma}(C|_\rho) \geqslant s \mid \rho(Y)=* \text{ and } C_1|_\rho \not\equiv 1 \text{ and } F|_\rho \equiv 1].$$

The maximum is taken over all ρ_1 assigning 0s and stars to the variables in T and only stars to the variables in Y. Having now fixed how ρ_1 sets the variables in C_1 we would like to estimate the above probability over ρ_2 using the induction hypothesis. To do this we must set all of the variables to which ρ_1 assigns a star to 0 or 1. For the variables in Y we take the assignments given by σ. For the variables in $T-Y$ we take the worst case setting τ of these variables to 0 and 1.

Fix σ and ρ_1. Let W be the variables in $T-Y$ that are assigned a star by ρ_1. Recall that D is C without C_1. We may obtain an upper bound on the above probability by writing

$$\max_{\tau \in \{0,1\}^W} \Pr_{\rho_2}[min((D|_{\sigma\tau\rho_1})|_{\rho_2}) \geq s - |Y| \mid \rho(Y) = * \text{ and } C_1|_\rho \not\equiv 1 \text{ and } F|_\rho \equiv 1].$$

Since the probability is only over ρ_2 and the first two conditions do not depend upon ρ_2, they may be dropped. Fixing the maximizing τ this is

$$\Pr_{\rho_2}[min((D|_{\sigma\tau\rho_1})|_{\rho_2}) \geq s - |Y| \mid (F|_{\rho_1})|_{\rho_2} \equiv 1].$$

By induction this is at most $\alpha^{s-|Y|}$. Pulling this together we get

$$(\mathrm{D}) \leq \sum_{\substack{\sigma \in \{0,1\}^Y \\ \sigma \neq 0^Y}} \max_{\rho_1} \alpha^{s-|Y|} = \sum_{\substack{\sigma \in \{0,1\}^Y \\ \sigma \neq 0^Y}} \alpha^{s-|Y|} = (2^{|Y|} - 1)\alpha^{s-|Y|}$$

and thus

$$(\mathrm{B}) \leq \sum_{Y \subseteq T} (2p)^{|Y|}(2^{|Y|} - 1)\alpha^{s-|Y|} = \alpha^s \sum_{k \geq 0} \binom{|T|}{k} (2p)^k \frac{2^k - 1}{\alpha^k}$$

$$\leq \alpha^s \left(\sum_{k \geq 0} \binom{t}{k} \left(\frac{4p}{\alpha}\right)^k - \sum_{k \geq 0} \binom{t}{k} \left(\frac{2p}{\alpha}\right)^k \right) = \alpha^S \left(\left(1 + \frac{4}{\gamma t}\right)^t - \left(1 + \frac{2}{\gamma t}\right)^t \right)$$

$$\leq \alpha^s(\mathrm{e}^{4/\gamma} - \mathrm{e}^{2/\gamma}) \leq \alpha^s \quad (\text{recalling } \gamma = 2/\ln \varphi). \quad \Box$$

3.4. Lower bound for the parity function

Now we use the Hastad Switching Lemma to derive a lower bound for small-depth circuits computing the parity function.

3.6. THEOREM. *For all p, d, $0 \leq k \leq d-1$, if f is $\Sigma_d^{S,t}$, then for a random ρ from $\mathscr{R}_{p^k}$*

$$\Pr[f|_\rho \text{ is not } \Sigma_{d-k}^{S,t}] < S(\gamma pt)^t$$

where $\gamma = 2/\ln \varphi \approx 4.16$.

PROOF. Consider the random restriction ρ as being composed from k restrictions $\rho = \rho_1 \rho_2 \ldots \rho_k$ each drawn from $\mathscr{R}_p$. Obtain the sequence of functions $f = f_1, f_2, \ldots, f_{k+1}$ where $f_{i+1} = f_i|_{\rho_i}$. At each step of this sequence there is a collection of t-open (or t-closed if $d-i$ is odd) bottom-level subcircuits in the circuit for f_i which may become t-closed (or t-open) under ρ_i and then merge with the gates above them. If this successfully occurs for each subcircuit in every f_i, then f_{k+1} is $\Sigma_{d-k}^{S,t}$. The probability that it fails at any particular subcircuit is at most $(\gamma pt)^t$ by the Hastad Switching Lemma. Hence the probability that it fails at any of the at most S subcircuits encountered is bounded above by $S(\gamma pt)^t$. $\Box$

The following corollary is independently interesting as a type of Ramsey theorem (see [31]).

3.7. COROLLARY. *If f is $\Sigma_d^{S,t}$ where $t \geq \log S$, then there is a restriction ρ assigning at least $n/3(10t)^{d-1} - t$ stars such that $f|_\rho$ is a constant function.*

PROOF. By the theorem above, if $p=1/10t$ and ρ is drawn from $\mathcal{R}_{p^{d-1}}$, then the probability that $f|_\rho$ is not $\Sigma_1^{S,t}$ is at most $S(\gamma pt)^t = S\beta^t \leq (2\beta)^t$ where $\beta=\gamma/10<0.42$. Hence $\Pr[f|_\rho$ is not $\Sigma_1^{S,t}]<0.84$. Furthermore ρ is expected to have np^{d-1} stars. An easy calculation shows that $\Pr[\rho$ has fewer than $np^{d-1}/3$ stars$]<0.15$. Since the sum of these probabilities is less than 1, there is a restriction ρ for which neither event occurs. Finally, since any nonconstant $\Sigma_1^{S,t}$ function may be forced to 1 by setting at most t inputs we may extend ρ by including these t additional settings and guarantee that $f|_\rho$ is constant. □

Using the preceding corollary we can now obtain the desired lower bounds for the parity function $\text{PARITY}(x_1, \ldots, x_n) = \Sigma x_i \bmod 2$.

3.8. THEOREM. *For all $n, d>0$, PARITY is not $\Sigma_d^{S,\log S}$ where $S<2^{(1/10)n^{1/d}}$.*

PROOF. If PARITY were $\Sigma_d^{S,\log S}$ for $S<2^{(1/10)n^{1/d}}$, then by the above corollary there would be a restriction ρ assigning at least one star such that $\text{PARITY}|_\rho$ is constant. This contradiction proves the theorem. □

3.9. COROLLARY. PARITY $\notin \mathrm{AC}^0$.

3.10. COROLLARY. *Polynomial-size parity circuits must have depth at least* $\log n/(c+\log\log n)$ *for some constant c.*

The bound in the above theorem cannot be significantly improved as it is quite close to the easily obtained upper bound.

3.11. THEOREM. *For all n and d, PARITY is $\Sigma_d^{S,\log S}$ where $S=n2^{n^{1/d}}$.*

We may also derive lower bounds for other simple Boolean functions as a corollary to the lower bound for the parity function, using the notion of AC^0-reducibility. This notion was proposed by Furst, Saxe and Sipser who gave a few examples of reductions, and further investigated by Chandra, Stockmeyer and Vishkin [21] who gave additional examples.

DEFINITION. Let $f, g: \{0,1\}^* \to \{0,1\}^*$. Say f is AC^0-*reducible* to g if there is a family of constant-depth, polynomial-size circuits computing f using AND-, OR-, and NOT-gates, and polynomial fan-in gates computing g.

As an example it is easy to see that PARITY is AC^0-reducible to the majority function $\text{MAJORITY}(x_1, \ldots, x_n)$ which is 1 iff at least half of the x_i are 1.

3.12. COROLLARY. MAJORITY $\notin \mathrm{AC}^0$.

3.5. *Depth hierarchy*

Thus far, restrictions have been used to establish the limitations of small, shallow circuits. In a similar way they may be used to show that the power of small circuits forms a hierarchy according to depth. Sipser [81] showed that $\Sigma_{d+1}^{\text{poly}} \neq \Sigma_d^{\text{poly}}$ for every d. It follows as an easy corollary that $\Sigma_d^{\text{poly, const}} \neq \Pi_d^{\text{poly, const}}$ for every d. Exponential lower bounds for the depth d to $d-1$ conversion were claimed without proof by A.C. Yao [97] and proved by Hastad [34] as a consequence of a variant of his switching lemma.

Let

$$f_d^m = \bigwedge_{i_1 \leqslant m_1} \bigvee_{i_2 \leqslant m_2} \bigwedge_{i_3 \leqslant m_3} \cdots \bigodot_{i_d \leqslant md} [x_{i_1 \ldots i_d} = 1],$$

where $\bigodot = \wedge$ or $\vee$ depending on the parity of d. The variables $x_1, \ldots, x_n$ appear as $x_{i_1 \ldots i_d}$ for $i_j \leqslant m_j$ where $m_1 = \sqrt{m/\log m}$, $m_2 = m_3 = \cdots m_{d-1} = m$, $m_d = \sqrt{dm \log m/2}$, and $m = (n\sqrt{2/d})^{1/(d-1)}$. Since $(\prod_{i \leqslant d} m_i) = n$ we see that f_d^m is Π_d^n.

3.13. THEOREM. *For every $m, d > 1$ the function f_d^m is not $\Sigma_d^{S, \log S}$ for $S < 2^{(1/20) n^{1/2d} (\log n)^{-1/2}}$.*

3.6. *Monotone bounded-depth circuits*

Monotone circuits have AND- and OR- but no NOT-gates. Boppana [16] and Klawe, Paul, Pippenger and Yannakakis [43] were the first to obtain strong lower bounds on the size of constant-depth monotone circuits. Boppana gave an exponential lower bound for the majority function and Klawe, Paul, Pippenger and Yannakakis showed that the depth d to $d-1$ conversion is exponential. These results were superseded by the previously described results of Yao and Hastad. Ajtai and Gurevich [4] and Okol'nishnikova [57] consider the relative power of monotone and general circuits for computing monotone functions. They give the following family of AC^0-functions not computable by monotone, constant-depth, polynomial-size circuits.

For each m we define the function $f_m(x_1, \ldots, x_n)$ where $n = m \log m$. Index the variables $x_1, \ldots, x_n$ as x_{ij} for $i \leqslant \log m$ and $j \leqslant m$. For each row i let s_i be the index of the last 1 such that there are no 0s preceding it in that row and set $s_i = 0$ if the row has all 0s. Then f_m is 1 if $\sum s_i \geqslant n/2$.

3.7. *Probabilistic bounded-depth circuits*

A probabilistic circuit is one which has in addition to its standard inputs some specially designated inputs called random inputs. When these random inputs are chosen from a uniform distribution the output of the circuit is a random variable. We say a probabilistic circuit *(a, b)-accepts* a language if it outputs 1 with probability at most a for strings not in the language and outputs 1 with probability at least b for strings in the language. The circuit accepts with *ε-error* if it $(\varepsilon, 1-\varepsilon)$-accepts and it accepts with *ε-advantage* if it $(\frac{1}{2}, \frac{1}{2}+\varepsilon)$-accepts.

A simple nonconstructive argument due to Adleman [1] and Bennett and Gill [13] shows that any language accepted by polynomial-size probabilistic circuits with $(1/n^k)$-advantage for any fixed k may also be accepted by polynomial-size deterministic circuits. Ajtai and Ben-Or [3] consider the analogous question for constant-depth circuits and give a similar answer provided the advantage is at least $1/\log^k n$.

3.14. THEOREM. *Every probabilistic circuit C of size s and depth d that accepts a language with* ($\log^{-k} n$)*-advantage (for fixed k) has an equivalent deterministic circuit of size* poly(n)$\cdot s$ *and depth* $d+2k+2$.

PROOF. Our plan is to first amplify the advantage until the error is exponentially small. Then there exists a setting of the random inputs which always gives the correct answer. In order to amplify the advantage we define two mappings on circuits. Let $and^l(C) = \bigwedge_{1 \leqslant i \leqslant l} C_i$ and $or^l(C) = \bigvee_{1 \leqslant i \leqslant l} C_i$ where each C_i is an independent copy of C sharing the same standard inputs but with its own set of random inputs. It is easy to see that if C is (a, b)-accepting, then $and^l(C)$ is (a^l, b^l)-accepting and $or^l(C)$ is $(1-(1-a)^l, 1-(1-b)^l)$-accepting. It follows that for $k>1$ if C accepts with ($\log^{-k} n$)-advantage, then $or^{l_2}(and^{l_1}(C))$ accepts with ($\log^{-(k-1)} n$)-advantage for $l_1 = 2\log n$ and $l_2 = n^2 \ln 2$. If C accepts with ($\log^{-1} n$)-advantage, then $or^{l_4}(and^{l_3}(or^{l_2}(and^{l_1}(C))))$ has error at most $e^{-n} < 2^{-n}$ for $l_1 = 2\log n$, $l_2 = 2n^2 \log n$, $l_3 = n^3$, and $l_4 = n$. □

3.8. *Razborov–Smolensky lower bound for circuits with* MOD$_p$ *gates*

A natural way to extend the results of the last few subsections is to obtain lower bounds for more powerful circuit models. Razborov [74] took an important step in this direction when he proved an exponential lower bound for computing the majority function with small-depth circuits having AND-, OR-, and PARITY-gates. His method is similar to that of his earlier paper on monotone circuits for the clique function, described in Section 4. Subsequently Smolensky [84] simplified and strengthened this theorem showing that for any p and q powers of distinct primes, the MOD$_p$ function cannot be computed with AND, OR, NOT, MOD$_q$ circuits of polynomial size and constant depth.

The majority function is the Boolean function MAJORITY$(x_1, \ldots, x_n) = 1$ iff $\Sigma x_i \geqslant n/2$. The MOD$_p$ function is the Boolean function which is 1 iff $\Sigma x_i \not\equiv 0 \pmod p$. We will prove the special case of the Razborov–Smolensky theorem where $p=2$ and $q=3$.

3.15. THEOREM. *The* MOD$_2$ *function cannot be computed with a size* $\frac{1}{50}2^{(1/2)n^{1/2d}}$, *depth d circuit of* AND-, OR-, NOT-, *and* MOD$_3$*-gates for sufficiently large n.*

PROOF. Let C be a depth d circuit computing MOD$_2$ with these types of gates. Think of the gates of C as operating on functions rather than merely on Boolean values. Our plan is to slightly adjust the results of each of the AND- and OR-operators in such a way that (1) each adjustment alters the output of C on few input settings, while (2) the end result is a function which differs from MOD$_2$ in many input settings. Hence many adjustments must occur. This gives a lower bound on the circuit size.

More precisely, we will assign to each subcircuit of C a new function called a *b-approximator*. A b-approximator is a polynomial on the input variables $x_1, \ldots, x_n$

of degree at most b over GF(3), the three-element field $\{-1, 0, 1\}$, where on inputs from $\{0, 1\}$ it takes values in $\{0, 1\}$. A subcircuit of height h is assigned an $n^{h/2d}$-approximator. The assignments are done inductively, first to the inputs, then working up to the output. Each assignment introduces some *error* which is the number of output deviations between it and the result of applying the true operator at that gate to the approximators of the subcircuits feeding into it, looking only at inputs drawn from $\{0, 1\}^n$.

Let D be a subcircuit of C all of whose proper subcircuits have been assigned b-approximators. If the top gate of D is a NOT-gate and the unique subcircuit feeding into it has approximator f, then we assign the b-approximator $1-f$ to D. Since f and $1-f$ are correct on the same input settings, this approximator introduces no new error. If the top gate of D is a MOD_3-gate and its inputs have b-approximators $f_1, \dots, f_k$, then we assign the $2b$-approximator $(\Sigma f_i)^2$ to D. Since $0^2=0$ and $-1^2=1^2=1$ this introduces no new error.

Before considering the case where the top gate is an AND or an OR let us examine how these operators affect degree. If $f_1, \dots, f_k$ are Boolean functions that are polynomials of degree at most b, then $\mathrm{AND}(f_1, \dots, f_k)$ is represented by the kb degree polynomial $f_1 \cdot f_2 \cdot \cdots \cdot f_k$. Since f_i and $\neg f_i = 1-f_i$ have the same degree, $\mathrm{OR}(f_1, \dots, f_k)$ also has degree at most kb. We cannot in general use this as an approximator because k may be polynomial in n and this bound is too large to use directly. To find a low degree approximation for the results of these operators we will allow some error to be introduced. Let us first fix a parameter l which determines the trade-off between error and degree.

If the top gate of D is an OR and its subcircuits have approximators $f_1, \dots, f_k$, then we find a $2lb$-approximator for D as follows. Select $F_1, \dots, F_l$ subsets of $\{f_i\}$, let $g_i = (\Sigma_{f \in F_i} f)^2$, and let $a = \mathrm{OR}(g_1, \dots, g_l)$. For an appropriate choice of Fs, the polynomial a is the desired $2lb$-approximator. To bound the error observe that if for a particular input setting $\mathrm{OR}(f_1, \dots, f_k) = 0$, then all f_i and hence all g_i are 0 and so a is 0. To analyze the case where the OR is 1 take the Fs to be independently selected random subsets of $\{f_i\}$. Since at least one f_i is 1, each g_i independently has at least a $\frac{1}{2}$ chance of being 1, so $\Pr[a=0] \leqslant 2^{-l}$. By an averaging argument there must be some collection of Fs so that the number of input settings on which $\mathrm{OR}(f_1, \dots, f_k) = 1$ and $a=0$ is at most 2^{n-l}. A dual argument applies if the top gate of D is an AND-gate using the identity $\neg f = 1 - f$. Thus, in either case, the approximator introduces at most 2^{n-l} error. This gives the following lemma.

3.16. LEMMA. *Every circuit C of depth d has a $(2l)^d$-approximator differing with C on at most size$(C) \cdot 2^{n-1}$ input settings.*

PROOF. The inputs to C are assigned the corresponding 1-approximators. Each level increases the degree of the approximators by a factor of at most $2l$. Each assignment of an approximator contributes at most 2^{n-l} error. □

Fixing $l = n^{1/2d}/2$ we obtain a $\sqrt{n}$-approximator for C.

3.17. LEMMA. *Any $\sqrt{n}$-approximator a differs from the* MOD_2 *function on at least* $\frac{1}{50}2^n$ *input settings for sufficiently large n.*

PROOF. Let U be the 2^n possible input settings and $G \subseteq U$ be the settings on which a agrees with MOD_2. For each input variable x_i let $y_i = x_i + 1$. Consider MOD_2 as a function from $\{-1,1\}^n \to \{-1,1\}$ under this change of variables. We may write $\mathrm{MOD}_2(y_1, \ldots, y_n)$ as the nth degree polynomial Πy_i. Since this change of variables does not alter the degrees of polynomials, a is a polynomial of degree at most $\sqrt{n}$ which agrees with Πy_i on G. Let $\mathscr{F}_G$ be the collection of $f: G \to \{-1,0,1\}$. Since $|\mathscr{F}_G| = 3^{|G|}$ we may bound the size of G by showing that $|\mathscr{F}_G|$ is small. We do this by assigning each $f \in \mathscr{F}_G$ a different polynomial of degree at most $\frac{1}{2}n + \sqrt{n}$ and then estimating the number of such polynomials.

Let $f \in \mathscr{F}_G$. Extend it in an arbitrary way to an $\bar{f}: U \to \{-1,0,1\}$. Let p be a polynomial in the y variables representing $\bar{f}$. Let $cy_{i_1} \cdots y_{i_l}$ be a term of p where $c \in \{-1,1\}$. Notice that it is multilinear, i.e. without powers higher than 1, since $y^2 = 1$ for $y \in \{-1,1\}$. Let $Y = \{y_1, \ldots, y_n\}$ and $T \subseteq Y$ be the set of y_{i_j} appearing in this term and $\bar{T} = Y - T$. Then we may rewrite this term $c\Pi T$ as $c\Pi Y \Pi \bar{T}$ again using $y^2 = 1$. Within G this is equivalent to the polynomial $c \cdot a \cdot \Pi \bar{T}$ of degree $\sqrt{n} + (n - |T|)$. If we rewrite in this way all terms in p of degree greater than $\frac{1}{2}n$ we obtain a polynomial that agrees with p in G and hence with f. It has degree at most $(n - \frac{1}{2}n) + \sqrt{n} = \frac{1}{2}n + \sqrt{n}$.

The number of multilinear monomials of degree at most $\frac{1}{2}n + \sqrt{n}$ is $\sum_{i=0}^{n/2+\sqrt{n}} \binom{n}{i}$ which for large n is approximately $0.9772 \cdot 2^n < \frac{49}{50}2^n$. The number of polynomials with monomials of this kind is thus less than $3^{(49/50)2^n}$, so this is an upper bound on the size of $\mathscr{F}_G$. Hence $|G| = \log_3 |\mathscr{F}_G| \leqslant \frac{49}{50}2^n$. Therefore the number of input settings where a differs from MOD_2 is at least $2^n - \frac{49}{50}2^n = \frac{1}{50}2^n$. □

Now we may conclude the proof of Theorem 3.15. We know that *size*(*C*) is at least the total error introduced divided by the error introduced at each assignment of an approximator. Thus

$$\mathit{size}(C) \geqslant \frac{\frac{1}{50}2^n}{2^{n-l}} \geqslant \frac{1}{50}2^{(1/2)n^{1/2d}}. \quad \square$$

3.9. *Applications*

In this subsection we give some consequences of the previous lower bounds for relativized computation, log-time computation, and first-order definability.

3.9.1. *Relativization of the polynomial-time hierarchy*

Baker, Gill and Solovay [10] proposed relativization as a means of showing that many familiar problems in complexity theory cannot be settled within pure recursive function theory. Among their results they give oracles under which P =? NP has an affirmative and a negative answer. They leave open the problem of finding an oracle under which the levels of the polynomial-time hierarchy are all distinct. Baker and Selman [11] extended this giving an oracle separating Σ_2P from Π_2P. Furst, Saxe and

Sipser [30] established a connection between exponential lower bounds for constant-depth circuits and the existence of the sought after oracle. The bounds given by A.C. Yao [97] and Hastad [34] settled this question.

3.18. THEOREM. *There is an oracle A such that for every i,* $\Sigma_i P^A \neq \Pi_i P^A$.

3.9.2. Log-time hierarchy

Chandra, Kozen and Stockmeyer [20] introduced the notion of alternating Turing machines operating in sublinear time. In their definition the sublinear-time alternating Turing machine may access the input in a random-access manner, by specifying the address of the bit to be read on a special address tape which by convention is reset to blank after each read. The alternating log-time hierarchy is defined by analogy with the polynomial-time hierarchy. Sipser [81] showed that as a consequence of the AC^0-hierarchy theorem, the levels of the log-time hierarchy are all distinct.

3.19. THEOREM. $\Sigma_i \mathrm{TIME}(\log n)/\mathrm{poly} = \Sigma_i^{\mathrm{poly,\,const}}$.

PROOF. Given a $\Sigma_i\mathrm{TIME}(\log n)$ machine M, we may convert it to a machine M' which reads its input at most once on any computation path and then only after all nondeterministic branches have been made. M' operates by simulating M except that it guesses the answer whenever M reads the input. The guesses are made with the same type of branching ($\wedge$ or $\vee$) in effect at the time. Then M' accepts iff M accepted on that branch and all guesses $g_1, \ldots, g_c$ are correct or if the first incorrect guess was made with $\wedge$-branching. Since M runs in time $O(\log n)$ and each input address has length $\log n$, we know that c is bounded above by a constant independent of n.

We convert M' to a $\Sigma_i^{\mathrm{poly,\,const}}$ circuit family by taking its $\vee$-$\wedge$ computation tree, collapsing adjacent $\vee$'s and adjacent $\wedge$'s on every branch and adding subcircuits to check the g's. This gives a level of gates for each nonalternating run of M' where the fan-in is given by the number of possible configurations (excluding the input tape) of M', except for the lowest level of the circuit where the g's are checked with fixed-size subcircuits. Then the inputs to the circuit corresponding to the advice may be preset accordingly.

For the other direction, we may represent each circuit in a $\Sigma_i^{\mathrm{poly,\,const}}$ circuit family as an advice string of polynomial length in such a way that a $\Sigma_i\mathrm{TIME}(\log n)$ may use its alternation to select a path of the circuit to one of the constant-size subcircuits and then evaluate it deterministically. □

3.20. COROLLARY. *For all i,* $\Sigma_i\mathrm{TIME}(\log n) \neq \Pi_i\mathrm{TIME}(\log n)$.

PROOF. The function f_i is in $\Pi_i\mathrm{TIME}(\log n)$ but by Theorem 3.13 not in $\Sigma_i^{\mathrm{poly,\,const}}$ and hence not in $\Sigma_i\mathrm{TIME}(\log n)$. □

3.9.3. First-order definability on finite structures

As observed by Ajtai [2] and Immerman [37], there is a close correspondence between the descriptive power of first-order sentences and AC^0. By a *first-order sentence* we mean an expression built up from relations, equality, quantifiers ranging over a specified universe, and variables taking values in the specified universe all appearing within the scope of some quantifier. A *structure* specifies the universe and the values of the relations. We say a structure $\mathcal{M}$ *satisfies* a sentence φ, or also φ *is true in* $\mathcal{M}$, written $\mathcal{M} \models \varphi$, if it specifies all of the relation symbols appearing in φ and φ is true in the familiar mathematical sense with these specifications.

To develop the relation with AC^0, let us focus on structures with universes U of finite cardinality. We fix a special relation symbol R to be used as the *input relation*. The other predicate symbols $S_1, \ldots, S_l$ are called the *built-in relations*. Given a sentence φ, if we specify U and the values of the S_i, then we may associate with φ the collection of R values which, together with the prespecified S_i values, make φ true. For example, if $\varphi = \forall x \forall y[(S(x) \wedge \neg S(y)) \rightarrow \neg R(x, y)]$ and the universe $U_n = \{1, \ldots, n\}$ with $S = \{1, \ldots, \lfloor n/2 \rfloor\}$, then φ defines the class of labeled directed graphs associated with R which contain no edges from the first half of the nodes to the second half. More precisely, if $\mathcal{S} = \{\mathcal{S}_1, \mathcal{S}_2, \ldots\}$ where each $\mathcal{S}_n = \langle \mathcal{S}_{n1}, \ldots, \mathcal{S}_{nl} \rangle$ and each $\mathcal{S}_{ni}$ is a value for the relation symbol S_i in the universe U_n and if $\mathcal{R} = \{\mathcal{R}_1, \mathcal{R}_2, \ldots\}$ where each $\mathcal{R}_n = \{R_{n1}, R_{n2}, \ldots\}$ and each R_{ni} is a value for R in the universe U_n, then we say φ *defines* $\mathcal{R}$ given $\mathcal{S}$ if for each n the structure $\langle U_n, S_{n1}, \ldots, S_{nl}, R \rangle \models \varphi$ iff $R \in \mathcal{R}_n$.

Say φ is Σ_d if it is in prenex normal form, i.e., all quantifiers are out in front, and it has d alternating blocks beginning with $\exists$. Such an $\mathcal{R}$ above is then called *nonuniformly* Σ_d *first-order definable*. Say that $\mathcal{R}$ is *nonuniformly first-order definable* if there exists a d such that it is nonuniformly Σ_d first-order definable. The built-in predicates S_{ni} play the same role in providing the nonuniform information as does the advice in nonuniform Turing machines. We identify collections of sets of relations $\mathcal{R}$ with the language $A_{\mathcal{R}}$ by encoding each relation with its characteristic binary string.

3.21. THEOREM. *For each* d, $\mathcal{R}$ *is nonuniformly* Σ_d *first-order definable iff* $A_{\mathcal{R}}$ *is in* $\Sigma_d^{\text{poly,const}}$.

PROOF. The easier direction ($\rightarrow$) of this equivalence is seen by constructing circuits simulating a given sentence φ. Each nonalternating block of quantifiers becomes a level of unbounded gates, the quantifier-free part of φ becomes fixed-size subcircuits, each atomic formula $S_i(x_1, \ldots, x_{j_i})$ becomes a preset input to the circuit and each $R(x_1, \ldots, x_k)$ becomes an input variable.

For the other direction we are given a $\Sigma_d^{n^l,c}$ circuit C to convert to a Σ_d sentence φ. The variables of φ are w_i and x_i for $1 \leqslant i \leqslant c$ and y_{jk} for $1 \leqslant j \leqslant d$, and $1 \leqslant k \leqslant l$. Let y_j denote the sequence $y_{j1}, \ldots, y_{jl}$, and y denote the sequence $y_1, \ldots, y_d$. For quantifier Q write Qy_j as a shorthand for $Qy_{j1} Qy_{j2} \ldots Qy_{jk}$. Then φ is $\exists y_1 \forall y_2 \ldots Qy_d[\psi]$ where ψ will be described shortly.

Expand all gates of C, except those at the bottom level, so that they have fan-in n^l, by adding redundant copies of subcircuits. Then each value for y gives a path through C from the output gate to one of the bottom gates of size at most c. For each y, let g_y denote the associated gate. Expand all bottom gates, by adding redundant variables if necessary, so that each contains exactly c exclusively positive variables, exactly c exclusively negative variables, or a combination of c positive and c negative variables.

Call these gates of type p, n, and np respectively. Now we define several relations. Let w denote the sequence $w_1,\ldots,w_c$ and x the sequence $x_1,\ldots,x_c$.

$$I^n=\{y: g_y \text{ is type } n \text{ or } np\},\qquad I^p=\{y: g_y \text{ is type } p \text{ or } np\},$$

$$P=\{(y,x)\colon x_1,\ldots,x_c \text{ are the positive variables in } g_y\},$$
$$N=\{(y,x)\colon x_1,\ldots,x_c \text{ are the negative variables in } g_y\}.$$

To complete the description of φ, if d is odd we let $\psi=\exists w\exists x[\zeta]$ where

$$\zeta=\begin{bmatrix} I^p(y)\rightarrow\left(P(y,w)\wedge\bigwedge_{b\leqslant c}R(w_b)\right) \\ \wedge\, I^n(y)\rightarrow\left(N(y,x)\wedge\bigwedge_{b\leqslant c}\neg R(x_b)\right)\end{bmatrix}$$

If d is even we let $\psi=\forall w\forall x[\zeta]$ where

$$\zeta=\begin{bmatrix} I^p(y)\rightarrow\left(P(y,w)\rightarrow\bigvee_{b\leqslant c}R(w_b)\right) \\ \wedge\, I^n(y)\rightarrow\left(N(y,x)\rightarrow\bigvee_{b\leqslant c}\neg R(x_b)\right)\end{bmatrix}. \qquad \square$$

As a corollary to this theorem and Theorem 3.8, we have the theorem of Ajtai [2] that parity is not nonuniformly first-order definable. Let $\mathrm{ODD}=\{\mathcal{R}_1,\mathcal{R}_2,\ldots\}$, where $\mathcal{R}_n=\{R: R$ is a unary relation on U_n defining a set of odd cardinality$\}$.

3.22. COROLLARY. ODD *is not nonuniformly first-order definable.*

We also obtain a hierarchy theorem as a corollary to the $\Sigma_d^{\mathrm{poly,const}}$ hierarchy theorem.

3.23. COROLLARY. *For all d there is a class of Σ_d first-order definable relations which is not nonuniformly Π_d first-order definable.*

It is interesting to contrast these results about nonuniform definability over finite structures with results about definability with built-in relations over infinite structures. For example let $\mathcal{G}=\{R: R$ is a binary relation representing a connected infinite graph$\}$. It is easy to show that $\mathcal{G}$ is not first-order definable. But if we build in the $+$ and $\times$ relations, then all recursive, in fact all arithmetic, relations are definable including $\mathcal{G}$. In essence, the sentence Gödel-encodes a path from a to b as a single integer. The finite analog of $\mathcal{G}$ is not first-order definable with any built-in relations since, by reduction from parity, we know that connectedness is not in AC^0. The same proof fails because the encoded path is an exponentially large integer not representable in the finite universe.

4. Monotone circuits

4.1. Background

This section surveys the lower bounds known for monotone circuits. A *monotone circuit* is a circuit with AND-gates and OR-gates, but with no NOT-gates; the gates may have fan-in 2 and unlimited fan-out. A Boolean function f is called *monotone* if $x \leqslant y$ implies that $f(x) \leqslant f(y)$, under the usual Boolean ordering. Notice that the only functions computable by monotone circuits are monotone functions. The *monotone circuit complexity* of a monotone function is the size of the smallest monotone circuit computing it.

Many important functions in complexity theory are monotone. As an example, consider the clique function from graph theory. The clique function (written $\text{CLIQUE}_{k,n}$) has $\binom{n}{2}$ variables, one for each potential edge in a graph on n vertices, and outputs 1 if the associated graph contains a clique (complete subgraph) on some k vertices. The clique function is monotone because setting more edges to 1 can only increase the size of the largest clique.

Strong lower bounds are known for monotone circuits. Razborov [72], in a major breakthrough, obtained a superpolynomial lower bound of size $n^{\Omega(\log n)}$ for the monotone circuit complexity of the clique function. (To appreciate the significance of this result, note that the best previous lower bound on the monotone circuit complexity of an explicit, single-output, monotone problem was only $4n$, due to Tiekenheinrich [88].) Shortly thereafter Andreev [7], using methods similar to Razborov, proved an exponential (not just superpolynomial) lower bound for a monotone problem in NP. This implies an exponential lower bound for the clique function since the clique function is complete (with respect to polynomial monotone projections) for "monotone NP" (see [91] and [83]). Alon and Boppana [6], by strengthening the combinatorial arguments of Razborov, proved a lower bound for $\text{CLIQUE}_{k,n}$ (where $k = n^{2/3}$) of size exponential in $\Omega((n/\log n)^{1/3})$.

If general circuits computing monotone functions could be converted into equivalent monotone circuits with only a polynomial blow-up in size, then the above lower bounds would extend to general circuit complexity. Razborov [73], using methods similar to his clique lower bound, dashed this possibility by proving that the perfect matching problem for bipartite graphs, known to be in P, requires monotone circuits of superpolynomial size. Tardos [87] improved the gap to truly exponential for another monotone problem in P.

In spite of the exponential gap between monotone and general circuit complexity, there is a special class of functions for which the two complexities are polynomially related. This is the class of slice functions introduced by Berkowitz [14]. A function f is called a *slice function* if for some integer k, the value of $f(x)$ is 0 when the number of 1s in x is fewer than k, and $f(x)$ is 1 when the number is more than k (but $f(x)$ may be arbitrary when the number is exactly k). Although slice functions may appear to be limited, there do exist NP-complete slice functions. Berkowitz showed that a general circuit computing a slice function can be converted into a monotone circuit by adding only a polynomial number of extra gates. Superpolynomial lower bounds on the *monotone* circuit complexity of explicit slice functions would thus imply that $P \neq NP$.

4.2. *Razborov lower bound for monotone circuit size*

To prove lower bounds on monotone circuit complexity, the behavior of small monotone circuits must be shown to be constrained. In Razborov's method to be described below, certain input settings will be designated "test" inputs that compare the circuit's behavior with the behavior of the clique function.

A *positive test graph* is a graph on n vertices that consists of a clique on some set of k vertices, and no other edges; these graphs are called "positive" because the function $\mathrm{CLIQUE}_{k,n}$ outputs 1 on them. Observe that there are $\binom{n}{k}$ such graphs. A *negative test graph* is formed by assigning each vertex a color from the set $\{1, 2, \ldots, k-1\}$, and then putting edges between those pairs of vertices with different colors; these graphs are called "negative" because the function $\mathrm{CLIQUE}_{k,n}$ outputs 0 on them. There are $(k-1)^n$ possible colorings, and although different colorings can lead to the same graph, negative test graphs formed from different colorings will be considered different for counting purposes.

Positive and negative test graphs are designed to measure how closely a circuit agrees with the function $\mathrm{CLIQUE}_{k,n}$. The main goal of Razborov's method is the following.

GOAL. Show that every small monotone circuit either outputs 0 on most positive test graphs or outputs 1 on most negative test graphs.

How can the goal be established? Monotone circuits can be amorphous, so to analyze their behavior directly is difficult. Instead, every small monotone circuit will be approximated by a special type of monotone circuit, called an *approximator circuit*. The behavior of approximator circuits will be much easier to analyze than the behavior of arbitrary monotone circuits.

The class of approximator circuits will now be defined. For a subset X of vertices, set the *clique indicator* of X (written $\lceil X \rceil$) to be the function of $\binom{n}{2}$ variables that is 1 if the associated graph contains a clique on the vertices X, and is 0 otherwise. An *approximator circuit* is an OR of at most m clique indicators, each of whose underlying vertex sets have cardinality at most l. Here $l \geqslant 2$ and $m \geqslant 2$ will have fixed values, depending only upon the values of k and n.

Approximator circuits will be important for establishing the goal. Every monotone circuit C will be assigned an approximator $\tilde{C}$. The goal will be proved by dividing it into the following two subgoals.

SUBGOAL 1. Show that if C is a small monotone circuit, then $C \leqslant \tilde{C}$ holds for most positive test graphs, and $C \geqslant \tilde{C}$ holds for most negative test graphs.

SUBGOAL 2. Show that every approximator either outputs 0 on most positive test graphs or outputs 1 on most negative test graphs.

How can arbitrary monotone circuits be approximated by such special approximator circuits? The approach to be taken is a "bottom-up" construction. Every subcircuit of the original circuit is assigned its own approximator, starting from the input variables and then working up. An input variable is of the form $x_{i,j}$, where i and j are two different vertices; it is equivalent to the clique indicator $\lceil\{i,j\}\rceil$. Hence an input variable is already an approximator.

Suppose that each proper subcircuit of a circuit C has been assigned an approximator circuit. What approximator should be assigned to the entire circuit? Assume, for argument's sake, that the top gate of the circuit C is an OR-gate. One natural idea to form the desired approximator is to OR together the approximators of the two subcircuits feeding into the top gate. Let the two approximators be denoted by $A=\bigvee_{i=1}^{r}\lceil X_i\rceil$ and $B=\bigvee_{i=1}^{s}\lceil Y_i\rceil$, where r and s are at most m. The OR of the two approximators is an OR of $r+s$ clique indicators. Unfortunately, $r+s$ can be as large as $2m$, so the OR of the two approximators need not be an approximator itself.

How can the number of clique indicators be reduced? The procedure used here is to replace several clique indicators with their "common" part. To implement this procedure, a combinatorial object called a sunflower is introduced. A *sunflower* is a collection of distinct sets $Z_1, Z_2, \ldots, Z_p$, called *petals*, such that the intersection $Z_i\cap Z_j$ is the same for every pair of distinct indices i and j; the common part $Z_i\cap Z_j$ is called the *center* of the sunflower. In the application to approximator circuits, each petal will be a subset of vertices.

Sunflowers can be used to reduce the number of clique indicators. Fix a value for $p\geqslant 2$, and look at the current collection of vertex sets $\{X_1,\ldots,X_r, Y_1,\ldots, Y_s\}$. If some p of these vertex sets form a sunflower, replace these p sets with their center. This operation is called a *plucking*. Repeatedly perform such pluckings until no more are possible. This entire procedure is called the *plucking procedure*. Since the number of vertex sets decreases with each plucking, at most $2m$ pluckings will occur. Regarding the number of vertex sets remaining after the plucking procedure is completed, the following combinatorial lemma on sunflowers is useful, due to Erdös and Rado [27].

4.1. LEMMA. *Let $\mathscr{Z}$ be a collection of sets each of cardinality at most l. If $|\mathscr{Z}|>(p-1)^l\cdot l!$, then the collection contains a sunflower with p petals.*

PROOF. The proof is by induction on l. The case $l=1$ is obvious. For $l\geqslant 2$, let $\mathscr{M}$ be a maximal subcollection of disjoint sets in $\mathscr{Z}$, and let S be the union of the sets in $\mathscr{M}$. If $|\mathscr{M}|\geqslant p$, then $\mathscr{M}$ itself forms the desired sunflower and we are done. Otherwise we have $|S|\leqslant(p-1)l$. Since $\mathscr{M}$ is maximally disjoint, the set S intersects every set in $\mathscr{Z}$. By averaging, some element i in S intersects a fraction at least $1/((p-1)l)$ of the sets in $\mathscr{Z}$. Consider the following collection of sets of cardinality at most $l-1$:

$$\mathscr{Z}'=\{Z-\{i\}: i\in Z \text{ and } Z\in\mathscr{Z}\}.$$

From the choice of i, we have

$$|\mathscr{Z}'|\geqslant\frac{|\mathscr{Z}|}{(p-1)l}>(p-1)^{l-1}\cdot(l-1)!.$$

Thus by induction, the collection $\mathscr{Z}'$ contains a sunflower with p petals. Adding i back to all these petals gives the desired sunflower in $\mathscr{Z}$. □

To apply the Erdös–Rado lemma to the present situation, set $m=(p-1)^l \cdot l!$. The lemma implies that after the plucking procedure is completed, at most m vertex sets remain. The clique indicators of the remaining vertex sets are then ORed together to form the approximator for the entire circuit. The resulting approximator is called the *approximate* OR of the two approximators A and B, written $A \sqcup B$.

The second case to consider is when the top gate is an AND-gate; again, let $A = \bigvee_{i=1}^{r} \lceil X_i \rceil$ and $B = \bigvee_{i=1}^{s} \lceil Y_i \rceil$ be the approximators of the two subcircuits feeding into the top gate. (For technical reasons, assume without loss of generality that none of the sets X_i or Y_i are singleton sets.) Forming the AND of the two approximators yields, by the distributive law, the expression $\bigvee_{i=1}^{r} \bigvee_{j=1}^{s} (\lceil X_i \rceil \wedge \lceil Y_j \rceil)$. Two reasons why this expression is not an approximator itself are that the term $\lceil X_i \rceil \wedge \lceil Y_j \rceil$ is not a clique indicator and that there can be as many as m^2 terms.

To overcome these difficulties, apply the following three steps. First, replace the term $\lceil X_i \rceil \wedge \lceil Y_j \rceil$ by the clique indicator $\lceil X_i \cup Y_j \rceil$. Second, erase those clique indicators $\lceil X_i \cup Y_j \rceil$ for which the cardinality of $X_i \cup Y_j$ is more than l. Finally, apply the plucking procedure (described above for OR-gates) to the remaining clique indicators; there will be at most m^2 pluckings. These three steps guarantee that a valid approximator is formed. The resulting approximator is called the *approximate* AND of the approximators A and B, written $A \sqcap B$.

The two operations described above, approximate OR and approximate AND, complete the bottom-up construction of the approximator $\tilde{C}$ from the monotone circuit C.

4.3. Lower bound for the clique function

The previous subsection observed that lower bounds on the monotone circuit complexity of the clique function follow from proving two subgoals. In this subsection, the two subgoals will be formally stated and proved. The proof given here will combine Razborov's original proof with some of the improvements due to Alon and Boppana [6]. The second subgoal is demonstrated first, since it is the easier of the two subgoals.

4.2. LEMMA. *Every approximator circuit either is identically* 0 *or outputs* 1 *on at least* $[1-\binom{l}{2}/(k-1)]\cdot(k-1)^n$ *of the negative test graphs.*

PROOF. Let $A = \bigvee_{i=1}^{r} \lceil X_i \rceil$ be an approximator circuit. If A is identically 0, then the first conclusion holds. If not, then $A \geqslant \lceil X_1 \rceil$. A negative test graph is rejected by the clique indicator $\lceil X_1 \rceil$ iff its associated coloring assigns some two vertices of X_1 the same color. Suppose a random coloring is chosen, with each of the $(k-1)^n$ possible colorings equally likely. The probability that some two vertices of X_1 are assigned the same color is bounded above by $\binom{|X_1|}{2}/(k-1) \leqslant \binom{l}{2}/(k-1)$. Hence the probability that $\lceil X_1 \rceil$ outputs 1 on the associated negative test graph is at least $1-\binom{l}{2}/(k-1)$. Rewriting this probabilistic statement as a counting statement yields the desired result. □

Subgoal 1 will be established by the following two lemmas on the relationship of a circuit C to its approximator $\tilde{C}$.

4.3. LEMMA. *For every monotone circuit C, the number of positive test graphs for which the inequality $C \leqslant \tilde{C}$ does not hold is at most size* $(C) \cdot m^2 \cdot \binom{n-l-1}{k-l-1}$.

PROOF. Let $A = \bigvee_{i=1}^{r} \lceil X_i \rceil$ and $B = \bigvee_{i=1}^{s} \lceil Y_i \rceil$ be two approximators. Both of the inequalities $A \vee B \leqslant A \sqcup B$ and $A \wedge B \leqslant A \sqcap B$ will be shown to fail for at most $m^2 \cdot \binom{n-l-1}{k-l-1}$ positive test graphs. This will imply the lemma because in the transformation from C to $\tilde{C}$ there are *size*(C) approximate AND- and OR-gates.

The inequality $A \vee B \leqslant A \sqcup B$ is always true, since $A \sqcup B$ is obtained from $A \vee B$ by the plucking procedure. Each plucking can only enlarge the class of accepted graphs.

Next, consider the inequality $A \wedge B \leqslant A \sqcap B$. The first step in the transformation from $A \wedge B$ to $A \sqcap B$ is to replace $\lceil X_i \rceil \wedge \lceil Y_j \rceil$ by $\lceil X_i \cup Y_j \rceil$. These two functions behave identically on positive test graphs. The second step is to erase those clique indicators $\lceil X_i \cup Y_j \rceil$ for which $|X_i \cup Y_j| \geqslant l+1$. For each such clique indicator, at most $\binom{n-l-1}{k-l-1}$ of the positive test graphs are lost. Since there are at most m^2 such clique indicators, at most $m^2 \cdot \binom{n-l-1}{k-l-1}$ positive test graphs are lost in the second step. The third and final step, applying the plucking procedure, only enlarges the class of graphs accepted, as noted in the previous paragraph. Summing up the three steps, at most $m^2 \cdot \binom{n-l-1}{k-l-1}$ positive test graphs fail to satisfy $A \wedge B \leqslant A \sqcap B$, completing the proof. □

4.4. LEMMA. *For every monotone circuit C, the number of negative test graphs for which $C \geqslant \tilde{C}$ does not hold is at most size*$(C) \cdot m^2 \cdot [\binom{l}{2}/(k-1)]^p \cdot (k-1)^n$.

PROOF. Let $A = \bigvee_{i=1}^{r} \lceil X_i \rceil$ and $B = \bigvee_{i=1}^{s} \lceil Y_i \rceil$ be two approximators. The inequalities $A \vee B \geqslant A \sqcup B$ and $A \wedge B \geqslant A \sqcap B$ will be shown to fail for at most $m^2 \cdot [\binom{l}{2}/(k-1)]^p \cdot (k-1)^n$ negative test graphs. As in the proof of Lemma 4.3, this will imply the desired result.

First, consider the inequality $A \vee B \geqslant A \sqcup B$. Recall that $A \sqcup B$ is obtained by performing at most $2m$ pluckings on $A \vee B$. Each plucking will be shown to accept only a few additonal negative test graphs. Color the vertices randomly, with all $(k-1)^n$ possible colorings equally likely, and let G be the associated negative test graph. Let $Z_1, Z_2, \ldots, Z_p$ be the petals of a sunflower with center Z. What is the probability that $\lceil Z \rceil$ accepts G, but none of the terms $\lceil Z_1 \rceil, \lceil Z_2 \rceil, \ldots, \lceil Z_p \rceil$ accept G? This event occurs iff the vertices of Z are assigned distinct colors (called a proper coloring, or PC), but every petal Z_i has two vertices colored the same. We have

$$\begin{aligned}\Pr[Z \text{ is PC and } Z_1, \ldots, Z_p \text{ are not PC}] &\leqslant \Pr[Z_1, \ldots, Z_p \text{ are not PC} \mid Z \text{ is PC}] \\ &= \prod_{i=1}^{p} \Pr[Z_i \text{ is not PC} \mid Z \text{ is PC}] \\ &\leqslant \prod_{i=1}^{p} \Pr[Z_i \text{ is not PC}].\end{aligned}$$

The first inequality holds by the definition of conditional probability; the second inequality holds by the mutual independence of the events $\{Z_i \text{ is not PC} \mid Z \text{ is PC}\}$; and the third inequality holds because the event "Z is PC" is negatively correlated with the other events.

As in the proof of Lemma 4.2, we have $\Pr[Z_i \text{ is not PC}] \leqslant \binom{l}{2}/(k-1)$. Substituting this inequality into the chain of inequalities in the previous paragraph shows that

$$\Pr[Z \text{ is PC and } Z_1, \ldots, Z_p \text{ are not PC}] \leqslant \left[\binom{l}{2}/(k-1)\right]^p.$$

Thus to the class of negative test graphs accepted each plucking adds at most $[\binom{l}{2}/(k-1)]^p \cdot (k-1)^n$ new graphs. There are at most $2m$ pluckings, so the number of negative test graphs violating the inequality $A \vee B \geqslant A \sqcup B$ is at most $2m \cdot [\binom{l}{2}/(k-1)]^p \cdot (k-1)^n$. This settles the case of approximate ORs.

Next, consider the inequality $A \wedge B \geqslant A \sqcap B$. In the transformation from $A \wedge B$ to $A \sqcap B$, the first step introduces no new violations, since $\lceil X_i \rceil \wedge \lceil Y_j \rceil \geqslant \lceil X_i \cup Y_j \rceil$. The second step of erasing large clique indicators also introduces no new violations. Only the third step, the plucking procedure, introduces new violations. This step was analyzed in the previous two paragraphs; the only difference now is that there can be m^2 pluckings instead of just $2m$. This settles the case of approximate ANDs, thus completing the proof. □

Subgoals 1 and 2 have thus been proved; combining them yields the following exponential lower bound on the monotone circuit complexity of the clique function.

4.5. THEOREM. *For* $k \leqslant n^{1/4}$, *the monotone circuit complexity of the function* $\text{CLIQUE}_{k,n}$ *is* $n^{\Omega(\sqrt{k})}$.

PROOF. Set $l = \lfloor \sqrt{k} \rfloor$ and $p = \lceil 10\sqrt{k} \log_2 n \rceil$, and recall that $m = (p-1)^l \cdot l!$. Let C be a monotone circuit that computes the function $\text{CLIQUE}_{k,n}$. By Lemma 4.2, the approximator $\tilde{C}$ either is identically 0 or outputs 1 on at least $\frac{1}{2} \cdot (k-1)^n$ of the negative test graphs. If the former case holds, then apply Lemma 4.3 to obtain

$$\mathit{size}(C) \cdot m^2 \cdot \binom{n-l-1}{k-l-1} \geqslant \binom{n}{k}.$$

A simple calculation shows that in this case $\mathit{size}(C)$ is $n^{\Omega(\sqrt{k})}$. Suppose instead that the latter case holds. Applying Lemma 4.4 shows that

$$\mathit{size}(C) \cdot m^2 \cdot 2^{-p} \cdot (k-1)^n \geqslant \tfrac{1}{2} \cdot (k-1)^n.$$

Another simple calculation shows that in this case $\mathit{size}(C)$ is $n^{\Omega(\sqrt{k})}$. □

4.4. Polynomial lower bounds

This subsection will describe the known results on functions with polynomial monotone circuit complexity. We present results for both single-output functions and multi-output functions.

Before Razborov's work, only linear lower bounds were known for the monotone circuit complexity of single-output monotone functions in NP. Tiekenheinrich [88] gave a $4n$ monotone lower bound for a simple explicit function. Dunne [24] proved a $3.5n$ lower bound on the monotone circuit complexity of the majority function. Majority is known to have monotone circuits of $O(n \log n)$ size by the work of Ajtai, Komlós and Szemerédi [5] discussed below.

Turning to multi-output functions, consider the Boolean sorting problem: given n Boolean variables, output their values in nondecreasing order. Boolean sorting is the same problem as simultaneously computing the threshold functions $TH_{k,n}$ (defined in Subsection 2.3) for all k between 1 and n. Ajtai, Komlós and Szemerédi [5] gave a very clever construction of monotone circuits of size $O(n \log n)$ for Boolean sorting. Lamagna and Savage [46] established an $\Omega(n \log n)$ lower bound on the monotone circuit complexity of Boolean sorting. Later, Pippenger and Valiant [67] and Lamagna [45] independently showed that Boolean merging, a special case of Boolean sorting, has monotone circuit complexity $\Omega(n \log n)$. Muller and Preparata [53] observed that Boolean sorting has nonmonotone circuits of linear size, exposing a small gap between monotone and general circuit complexity.

A larger gap was obtained for the problem of Boolean matrix multiplication. This problem takes two n-by-n Boolean matrices as input, and outputs their n-by-n Boolean matrix product. It is trivial to show that Boolean matrix multiplication has monotone circuits of size $2n^3 - n^2$. Pratt [68] was the first to demonstrate that its monotone circuit complexity is $\Omega(n^3)$. Later, Paterson [58] and Mehlhorn and Galil [50] independently showed that its monotone circuit complexity is exactly $2n^3 - n^2$. Interestingly enough, its general circuit complexity is known to be asymptotically smaller. For example, Coppersmith and Winograd [23] show that Boolean matrix multiplication has non-monotone circuits of size $O(n^{2.38})$.

Wegener [94] has proved monotone circuit lower bounds for a generalization of matrix multiplication, called Boolean direct product, which takes as input several n-by-n matrices. Wegener's results on Boolean direct product give an $\Omega(n^2/\log n)$ lower bound on the monotone circuit complexity of an explicit monotone problem with n inputs and n outputs.

Finally, lower bounds have been discovered for a class of multi-ouput functions called Boolean sums. A *Boolean sum* has n inputs and n outputs, each output being an OR of some subset of the inputs. Nečiporuk [55] constructed an explicit Boolean sum that has $\Omega(n^{3/2})$ monotone circuit complexity. Later, Mehlhorn [49] and Pippenger [65] independently obtained $\Omega(n^{5/3})$ monotone lower bounds for another explicit Boolean sum. Since then, Andreev [8] has explicitly contructed, for every fixed $\varepsilon > 0$, a Boolean sum with monotone complexity $\Omega(n^{2-\varepsilon})$.

5. Formulas

A formula is the special type of circuit whose gates have fan-out 1, i.e., a circuit whose underlying graph is a tree. The main motivation for studying formulas is their close relationship to circuit depth. Spira [85] showed that, over a complete basis, a Boolean function is computable by polynomial-size formulas iff it is computable by logarithmic-depth circuits. Wegener [95] showed the analogous result for the monotone basis, and Ugol'nikov [90] announced the analogous result for all fixed bases. Thus strong lower bounds on formula size may lead to lower bounds on circuit depth. In this section, we present the major results known on lower bounds for formula size.

The *size* of a formula is defined to be the number of occurrences of literals (variables or their negations) in the formula. Notice that the size of a formula with binary gates is precisely one more than the number of gates in the formula. Given a collection of Boolean functions Ω, called a *basis*, define the *formula complexity* $L_\Omega(f)$ to be the size of the smallest formula with gates from Ω computing the Boolean function f.

We will consider three bases: the full binary basis B consisting of all 16 binary gates, the *DeMorgan* basis $D=\{\text{AND, OR, NOT}\}$, and the monotone basis $M=\{\text{AND, OR}\}$. We refer to formulas over these bases as *binary formulas*, *DeMorgan formulas*, and *monotone formulas*, respectively. Trivially we have $L_B \leqslant L_D \leqslant L_M$. Pratt [69] showed that $L_D = O(L_B^c)$ for $c=\log_3 10 \approx 2.09$.

The best explicit lower bound known differs for the three bases. For monotone formulas, Karchmer and Wigderson [38] showed a superpolynomial lower bound on a problem with polynomial monotone circuits. For DeMorgan formulas, Andreev [9] showed a lower bound of size $\Omega(n^{5/2-\varepsilon})$ for every fixed $\varepsilon>0$. For binary formulas, Nečiporuk [54] showed an $\Omega(n^2/\log n)$ lower bound. All of these results will be proved in this section.

5.1. *Karchmer–Wigderson's lower bound for monotone formula size*

This subsection deals with lower bounds on the size of monotone formulas. Of course, the lower bounds for monotone circuits discussed in Section 4 imply as a special case lower bounds for monotone formulas. In fact, Razborov [76] gives a simpler proof that some monotone NP problems require monotone formulas of superpolynomial size[1]. Nevertheless, these results leave open the question of whether some problem with monotone circuits of polynomial size requires monotone formulas of superpolynomial size. Karchmer and Wigderson [38] answered this question affirmatively.

The graph *s-t* connectivity function (written CONNECT_n) takes as input an undirected graph on n ordinary vertices and two distinguished vertices s and t; the function outputs 1 iff the graph has a path from s to t. This function is well known to be computable in polynomial time, to have monotone circuits of polynomial size and $O((\log n)^2)$ depth, and to have monotone formulas of size $n^{O(\log n)}$. Karchmer and Wigderson originally proved that this function requires monotone circuits of $\Omega((\log n)^2/\log\log n)$ depth and monotone formulas of $n^{\Omega(\log n/\log\log n)}$ size. Later Boppana (unpublished) and Hastad (unpublished) independently simplified the proof and improved the bounds to the optimal $\Omega((\log n)^2)$ depth and $n^{\Omega(\log n)}$ formula size. By the result of Wegener [95], the depth bound and size bound are actually equivalent, so the proof

[1] In 1990, Raz and Wigderson [101] proved that the perfect matching problem for n-by-n bipartite graphs requires monotone formulas of $2^{\Omega(n)}$ size.

just focuses on circuit depth. Below we give an overview of the proof, and in the next subsection we give the complete proof.

The Karchmer–Wigderson method uses certain graphs to test a circuit's behavior. An *l-path graph* consists of a path from s to t with a sequence of l ordinary vertices (not necessarily distinct) in between, and no other edges. (If the same vertex appears in two consecutive positions of the sequence, then ignore that edge.) There are n^l possible sequences, and though two of them can lead to the same graph, graphs formed from different sequences will be considered different for counting purposes. A *cut graph* is formed by partitioning the vertices into two components, with s and t in different components, and placing edges between those vertex pairs in the same component. There are 2^n possible cut graphs.

These test graphs will measure how closely a circuit agrees with the function CONNECT$_n$. Say that a monotone circuit is an (α, β) separator if it outputs 1 for a fraction at least α of the l-path graphs, and outputs 0 for a fraction at least β of the cut graphs. The main goal of the proof is to show that no shallow monotone circuit can be a good separator; the goal will be established by a "top-down" argument on the structure of a circuit.

Assume C is a shallow monotone circuit computing the function CONNECT$_n$; clearly C is a $(1, 1)$ separator. We will explore C from top to bottom. If the top gate of C is an OR-gate, then it is easy to see that one of the two subcircuits feeding into this gate must be a $(\frac{1}{2}, 1)$ separator. Similarly, if the top gate of C is an AND-gate, one of the two subcircuits must be a $(1, \frac{1}{2})$ separator. In either case, this subcircuit is certainly a $(\frac{1}{2}, \frac{1}{2})$ separator. By repeating this argument j times, we see that one of the subcircuits j levels from the top is a $(2^{-j}, 2^{-j})$ separator.

Unfortunately, repeating the above argument too many times makes the path-accepting and cut-rejecting densities too low to be of interest. Is there a way to increase the densities? Yes, as follows. Select $\sqrt{n}$ vertices at random and collapse them into vertex s, identifying their edges with the corresponding edges of s. Similarly, select $\sqrt{n}$ other vertices at random and collapse them into vertex t. At the same time, halve the value of the path length l. The crucial fact, and the hardest one to prove, is that performing this vertex-collapsing step will likely convert an (α, β) separator into a $(\sqrt{\alpha}/2, \beta/n)$ separator. In other words, the path-accepting density increases greatly, while the cut-rejecting density decreases only slightly.

With the above tools, the proof outline becomes clearer. We constantly strive to maintain a high path-accepting density. When the path-accepting density becomes too low, we apply a vertex-collapsing step to make it high again. The cut-rejecting density constantly, but controllably, decreases. When we finally reach the bottom of the circuit, we will have a depth-0 monotone circuit that is a good separator; since we assumed the original circuit was shallow, the path length l will still be large. But a depth-0 monotone circuit is too limited to be a good separator, yielding a contradiction. Thus the original circuit must have been deep to begin with.

A similar top-down method was used by Klawe, Paul, Pippenger and Yannakakis [43] to prove a lower bound for monotone constant-depth circuits. Both they and Karchmer–Wigderson point out a connection between circuit depth and communication complexity.

5.2. Lower bound for the connectivity function

This subsection will present a formal proof of the Karchmer–Wigderson lower bound for the graph s-t connectivity function, based on the above proof overview.

DEFINITIONS. Let f be a monotone Boolean function acting on graphs with n ordinary vertices and two distinguished vertices. Say that f is an (n, l, α) *path acceptor* if it outputs 1 on a fraction at least α of the l-path graphs. Say that f is an (n, β) *cut rejector* if it outputs 0 on a fraction at least β of the cut graphs. Finally, say that f is an (n, l, α, β) *separator* if it is both an (n, l, α) path acceptor and an (n, β) cut rejector.

DEFINITION. Let f be a function acting on graphs with vertex set V, and let ρ be a mapping from V to V'. The *induced function* f_ρ is the following function acting on graphs with vertex set V'. Given a graph $G=(V', E')$, form the graph $G_\rho=(V, E)$ by setting $\{i, j\} \in E$ iff $\rho(i)=\rho(j)$ or $\{\rho(i), \rho(j)\} \in E'$. The value of f_ρ on G is assigned the value of f on G_ρ.

DEFINITION. Let f be a function acting on graphs with vertex set V, and let ρ be a mapping from V to V'. The *induced function* f_ρ is the following function acting on k-subset W of $V-\{s, t\}$. Second, place each vertex of W into either set W_s or set W_t, randomly and independently. Finally, form the mapping $\rho: V \to V - W$ by mapping all of W_s into s, mapping all of W_t into t, and mapping every other vertex to itself.

Our first lemma shows that the path-accepting density of a function is likely to increase greatly by switching to a randomly induced function.

5.1. LEMMA. *Let f be an (n, l, α) path acceptor, and let ρ be a random mapping from $\mathscr{R}_k$. Suppose that $100l/\alpha \leqslant k \leqslant n/100l$. Then with probability at least $\frac{3}{4}$, the induced function f_ρ is an $(n-k, l/2, \sqrt{\alpha/2})$ path acceptor.*

PROOF. Let V be the set of n ordinary vertices. Let P be the collection of l-path graphs that f accepts; we will identify each l-path graph with its sequence of intermediate vertices. Let $L=\{1, \ldots, l/2\}$ and $R=\{l/2+1, \ldots, l\}$. Given a subpath p in V^L (or V^R), an *extension* of p is a path in V^l that agrees with p on L (R). Define P_L (P_R) to be the collection of subpaths in V^L (V^R) with more than $\frac{1}{4}\alpha n^{l/2}$ extensions in P.

We first show that either P_L or P_R is large. Observe that every path in P is either: (a) an element of $P_L \times P_R$, or (b) an extension of some subpath in $V^L - P_L$, or (c) an extension of some subpath in $V^R - P_R$. The number of type-(a) paths in P is at most $|P_L| \cdot |P_R|$. The number of type-(b) paths in P is at most $\frac{1}{4}\alpha n^l$, since there are at most $n^{l/2}$ subpaths in $V^L - P_L$ each with at most $\frac{1}{4}\alpha n^{l/2}$ extensions in P. Similarly, the number of type-(c) paths in P is at most $\frac{1}{4}\alpha n^l$. In total we obtain $|P| \leqslant |P_L| \cdot |P_R| + \frac{1}{2}\alpha n^l$. On the other hand, by hypothesis we have $|P| \geqslant \alpha n^l$. Comparing the two bounds, it follows that $|P_L| \cdot |P_R| \geqslant \frac{1}{2}\alpha n^l$. Hence either P_L or P_R has size at least $\sqrt{\alpha/2}n^{l/2}$; without loss of generality, suppose it is P_L.

Next, consider the following collection of $l/2$-subpaths:

$$P_\rho = \{p \in P_L : p \in (V-W)^L \text{ and } \exists p' \in (W_t)^R \text{ such that } (p, p') \in P\}.$$

From the definition of f_ρ, it is straightforward to see that f_ρ accepts all paths in P_ρ. Thus we only need show that P_ρ is large with high probability.

Toward this goal, we estimate the probability that an arbitrary path p in P_L belongs to P_ρ. The probability that $p \notin (V-W)^L$ is at most $l/2 \cdot k/n$, since there are $l/2$ coordinates and each has probability k/n of being in W. This probability is at most $\frac{1}{200}$ by hypothesis.

We next estimate the probability that p fails the second condition of being in P_ρ. Since W_t is a random subset of W, it has size at least $k/4$ with exponentially high probability. In that case, let X be a random $k/4$-subset of W_t; observe that X has the distribution of a random $k/4$-subset of V. One way to form such a random $k/4$-subset is as follows: select $k/(2l)$ subpaths $p'_1, p'_2, \ldots, p'_{k/2l}$ randomly and independently from V^R. Obtain X by taking the union of the vertices of all these subpaths, and if necessary randomly adding new vertices until there are $k/4$ vertices. The probability that p fails the second condition of P_ρ is

$$\begin{aligned}\Pr[\nexists p' \in (W_t)^R \text{ s.t. } (p, p') \in P] &\leqslant \Pr[|W_t| < k/4] + \Pr[\nexists i \text{ s.t. } (p, p'_i) \in P]\\ &\leqslant 2^{-k/10} + \left(1 - \frac{\alpha}{4}\right)^{k/2l} \leqslant \frac{1}{200}.\end{aligned}$$

We now complete the proof. Combining the two estimates above shows that p belongs to P_ρ with probability at least $1 - \frac{1}{200} - \frac{1}{200} = \frac{99}{100}$. Hence the expected size of P_ρ is at least $\frac{99}{100}|P_L|$. On the other hand, we always have $|P_\rho| \leqslant |P_L|$. Thus with probability at least $\frac{3}{4}$ we have $|P_\rho| \geqslant \frac{24}{25}|P_L|$, which is greater than $\frac{1}{2}\sqrt{\alpha} n^{l/2}$ by our prior estimate of $|P_L|$. This is what we wanted to prove. □

Our second lemma shows that the cut-rejecting density of a function is likely to decrease only slightly by switching to a randomly induced function.

5.2. LEMMA. *Let f be an (n, β) cut rejector, and let ρ be a random mapping from $\mathcal{R}_k$. Suppose that $\beta \geqslant 2 \cdot (\frac{3}{4})^{n/k}$. Then with probability at least $\frac{3}{4}$, the induced function f_ρ is an $(n-k, \frac{\beta k}{2n})$ cut rejector.*

PROOF. Again let V be the set of n ordinary vertices. Let Q be the collection of cut graphs that f rejects; we will identify each cut graph with a vector in $\{s, t\}^V$. Given a k-subset X of V and a vector x in $\{s, t\}^X$, an *extension* of x is a cut in $\{s, t\}^V$ that agrees with x on X. Define $Q(X)$ to be those vectors in $\{s, t\}^X$ with more than $\beta k/(2n) \cdot 2^{n-k}$ extensions in Q.

Our first goal is to show that $|Q(X)|$ has large expectation for a random k-subset X. Consider an arbitrary partition of V into n/k subsets $X_1, \ldots, X_{n/k}$ each of size k. An argument similar to the first part of the proof of Lemma 5.1 will show that $|Q(X_1)| \cdots |Q(X_{n/k})| \geqslant \frac{1}{2}\beta 2^n$. From the arithmetic-geometric mean inequality, we obtain

$$\frac{|Q(X_1)| + \cdots + |Q(X_{n/k})|}{n/k} \geqslant (|Q(X_1)| \cdots |Q(X_{n/k})|)^{k/n} \geqslant \left(\frac{\beta}{2} 2^n\right)^{k/n} \geqslant \frac{3}{4} 2^k.$$

This bound holds for an arbitrary such partition, so by averaging over all such partitions, the expected size of $Q(X)$ is at least $\frac{3}{4}2^k$, where X is a random k-subset of V.

We use this estimate to complete the proof. Let x be the vector in $\{s,t\}^W$ that is s on W_s and t on W_t. Consider the set Q_ρ of subcuts q in $\{s,t\}^{V-W}$ such that (q,x) is in Q. From the definition of f_ρ, it is straightforward to see that f_ρ rejects all the cuts in Q_ρ. It thus suffices to show that Q_ρ is large with high probability. Notice that the event "$|Q_\rho|>\beta k/(2n)\cdot 2^{n-k}$" is identical to the event "$x\in Q(W)$". Thus, by the previous paragraph, we have

$$\Pr\left[|Q_\rho|>\frac{\beta k}{2n}\cdot 2^{n-k}\right]=\Pr[x\in Q(W)]=\frac{\mathrm{E}[Q(W)]}{2^k}\geqslant\frac{3}{4},$$

which completes the proof. □

Combining Lemmas 5.1 and 5.2 yields the following corollary showing how to improve a separator.

5.3. COROLLARY. *Let f be an (n,l,α,β) separator. Suppose that $100l/\alpha\leqslant k\leqslant n/(100l)$ and $\beta\geqslant 2\cdot(\frac{3}{4})^{n/k}$. Then for some mapping ρ in the support of $\mathscr{R}_k$ the induced function f_ρ is an $(n-k,l/2,\sqrt{\alpha}/2,\beta k/(2n))$ separator.*

We finally prove the superlogarithmic lower bound on the monotone depth complexity of the graph s-t connectivity function.

5.4. THEOREM. $\mathrm{CONNECT}_n$ *requires monotone circuits of depth* $\Omega((\log n)^2)$.

PROOF. Let C be a monotone circuit for $\mathrm{CONNECT}_n$, and assume it has depth at most $\frac{1}{25}(\log_2 n)^2$, for n sufficiently large. Adding dummy gates if necessary, we can suppose that the underlying graph of C is a complete binary tree of depth $\frac{1}{25}(\log_2 n)^2$. Set the path length $l=n^{1/4}$ and the path density $\alpha=\frac{1}{4}n^{-1/5}$.

We do a top-down exploration of the circuit C. The circuit C is an $(n,l,1,1)$ separator, and in particular is an $(n,l,\alpha,1)$ separator. Break C into $\frac{1}{5}\log_2 n$ blocks of $\frac{1}{5}\log_2 n$ gate levels each. By following C down one block as described in the proof overview, we obtain a subcircuit that is an $(n,l,\alpha n^{-1/5},n^{-1/5})$ separator. Applying Corollary 5.3 gives a new subcircuit that is an $(n-n^{2/3},l/2,\alpha,n^{-1})$ separator. Notice that the path-accepting density is back up to α.

Repeat the above process successively for each block of the circuit. Upon reaching the bottom of the circuit, we end up with a monotone depth-0 circuit that is an $(n',l',\alpha,n^{-\log_2 n})$ separator, where $n'=n-\frac{1}{5}n^{2/3}\log_2 n$ and $l'=l/n^{1/5}=n^{1/20}$.

A monotone depth-0 circuit is either a single input variable or a constant. An input variable $x_{i,j}$, for $\{i,j\}\cap\{s,t\}=\emptyset$, accepts a fraction at most $l'/\binom{n'}{2}$ of the l'-path graphs. An input variable of the form $x_{s,i}$ or $x_{i,t}$ accepts a fraction at most $1/n'$ of the l'-path graphs. A constant either rejects all l'-path graphs or accepts all cut graphs. None of these cases gives a good separator, so we have a contradiction. Thus the original circuit C must have had depth larger than $\frac{1}{25}(\log_2 n)^2$. □

5.3. *Nonmonotone formulas*

This subsection will prove the largest lower bounds known for nonmonotone formulas, the results of Andreev and Nečiporuk.

Let us start with lower bounds for DeMorgan formulas. Observe that negations in DeMorgan formulas may always be pushed down to the bottom level. Random restrictions, as they were for bounded-depth circuits (Section 3), will be useful here. The probability distribution $\mathscr{R}_k$ of restrictions used here is the following: randomly assign k variables to be $*$, and assign all other variables to be 0 or 1 randomly and independently. Intuitively, a random restriction should reduce considerably the size of a formula. The following lemma of Subbotovskaya [86] makes this intuition precise.

5.5. LEMMA. *Let f be a Boolean function of n variables, and let ρ be a random restriction from $\mathscr{R}_k$. Then with probability at least $\frac{3}{4}$,*

$$L_D(f|_\rho) \leqslant 4 \cdot \left(\frac{k}{n}\right)^{3/2} \cdot L_D(f).$$

PROOF. Let F be an optimal DeMorgan formula for the function f, of size $s = L_D(f)$. Construct the restriction ρ in $n-k$ stages as follows. At any stage, choose a variable randomly from the ones remaining, and assign it 0 or 1 randomly. We analyze the effect of this restriction stage-by-stage.

Suppose the first stage chooses the variable x_i. When the variable x_i is set, the literals x_i and $\bar{x}_i$ will disappear from the formula F. By averaging, the expected number of such literals is s/n.

In fact, the formula is likely to be reduced even further. For each of the literals x_i or $\bar{x}_i$, consider the gate which it feeds into. For example, suppose the gate is $x_i \wedge G$ for some subformula G. We may assume without loss of generality that G does not contain the literals x_i or $\bar{x}_i$. If the variable x_i is assigned 0, then the subformula G will disappear from the formula F, thereby erasing at least one more literal. Since x_i is assigned 0 or 1 randomly, we expect at least $\frac{1}{2} \cdot s/n$ literals to disappear because of these secondary effects. In total, we thus expect at least $s/n + \frac{1}{2} \cdot s/n = \frac{3}{2} \cdot s/n$ literals to disappear in the first stage, yielding a new formula with expected size at most $s \cdot (1 - 3/(2n)) \leqslant s \cdot (1 - 1/n)^{3/2}$.

The succeeding stages of the restriction can be analyzed in the same way. After each stage the number of variables decrements by one. Hence the expected size of the final formula is

$$\mathrm{E}[L_D(f|_\rho)] \leqslant s \cdot \left(1 - \frac{1}{n}\right)^{3/2} \cdot \left(1 - \frac{1}{n-1}\right)^{3/2} \cdot \ldots \cdot \left(1 - \frac{1}{k+1}\right)^{3/2} = s \cdot \left(\frac{k}{n}\right)^{3/2}.$$

The probability that $L_D(f|_\rho)$ is more than 4 times its expected value is less than $\frac{1}{4}$, which completes the proof. □

Lemma 5.5 is useful for proving lower bounds on formula size. For example, Subbotovskaya [86] applied it to show that the parity function of n variables requires

DeMorgan formulas of size $\Omega(n^{3/2})$. Khrapchenko [41] later improved the parity lower bound to $\Omega(n^2)$ using a method we describe in Subsection 5.4. For now we show how Andreev [9] used Lemma 5.5 to prove an $\Omega(n^{5/2-\varepsilon})$ lower bound for another explicit function.

First we define Andreev's function. Given a Boolean function f of b variables, let $f^{\oplus m}$ be the Boolean function of bm variables defined by

$$f^{\oplus m}(x_1,\ldots,x_{bm})=f(x_1\oplus\cdots\oplus x_m, x_{m+1}\oplus\cdots\oplus x_{2m},\ldots,x_{(b-1)m+1}\oplus\cdots\oplus x_{bm}).$$

The variables $x_{(i-1)m+1},\ldots,x_{im}$ are said to form the ith *block*. Andreev's function, denoted $A_{b,m}$, is a Boolean function of 2^b+bm variables. The first 2^b variables of $A_{b,m}$ will specify a Boolean function of b variables by listing its truth table; call the resulting function f. The value of $A_{b,m}$ is then obtained by applying the function $f^{\oplus m}$ to the last bm variables of $A_{b,m}$. The collection of functions $\{A_{b,m}\}$ is easy to compute in time polynomial in the number of variables. Andreev proved the following lower bound for $A_{b,m}$ over the DeMorgan basis.

5.6. THEOREM. *Let $b=\lfloor\log_2 n-1\rfloor$ and $m=\lfloor n/(2b)\rfloor$. Then the function $A_{b,m}$ of at most n variables requires DeMorgan formulas of size $\Omega(n^{5/2-\varepsilon})$ for every fixed $\varepsilon>0$.*

PROOF. Let f be the Boolean function of b variables that requires the largest DeMorgan formulas. An argument analogous to that of Theorem 2.4 will show that $L_D(f)$ is $\Theta(2^b/\log b)$. Since $f^{\oplus m}$ is a subfunction of $A_{b,m}$, we have $L_D(A_{b,m})\geqslant L_D(f^{\oplus m})$. Let ρ be a random restriction from $\mathcal{R}_k$ for $k=\lceil b\ln(4b)\rceil$.

Applying Lemma 5.5, with probability at least $\frac{3}{4}$ we have

$$L_D(f^{\oplus m}|_\rho)\leqslant 4\cdot\left(\frac{k}{bm}\right)^{3/2}\cdot L_D(f^{\oplus m}).$$

An easy probability calculation shows that, also with probability at least $\frac{3}{4}$, the restriction ρ assigns at least one star to each of the b blocks of variables. Some restriction ρ will thus satisfy both conditions. From the second condition, we have $L_D(f^{\oplus m}|_\rho)\geqslant L_D(f)$. Combining the above inequalities, we obtain

$$L_D(A_{b,m})\geqslant L_D(f^{\oplus m})\geqslant\frac{1}{4}\cdot\left(\frac{bm}{k}\right)^{3/2}\cdot L_D(f)=\Omega\left(\left(\frac{bm}{k}\right)^{3/2}\cdot\frac{2^b}{\log b}\right)=\Omega(n^{5/2-\varepsilon}).$$

This establishes the theorem. □

We next turn to lower bounds for binary formulas. Let f be a Boolean function on a variable set X. A *subfunction* of f on $Y\subseteq X$ is a function obtained from f by setting the variables of $X-Y$ to constants. Let $N_Y(f)$ be the number of different nonconstant subfunctions of f on Y. Intuitively, if f has many subfunctions, then it is complicated and hence should require large formulas. This intuition was made precise by Nečiporuk [54] who proved the following theorem with a weaker constant factor; the version below is due to Paterson (unpublished) and Zwick [98].

5.7. THEOREM. *Let f be a Boolean function on a variable set X, and let $Y_1, Y_2, \ldots, Y_k$ be disjoint subsets of X. Then*

$$L_B(f) \geqslant \sum_{i=1}^{k} \log_5[2N_{Y_i}(f)+1].$$

PROOF. Let $S_Y(f)$ be the collection of nonconstant functions g on a variable set Y for which either g or $\neg g$ is a subfunction of f. Clearly $N_Y(f) \leqslant |S_Y(f)|$. Let $size_Y(F)$ be the number of occurrences of variables of Y in a Boolean formula F. It is simple to see that the theorem follows from the claim below.

5.8. CLAIM. *For every binary formula F and every variable set Y, we have*

$$2|S_Y(F)|+1 \leqslant 5^{size_Y(F)}.$$

The claim is proved by induction on the size of F. The base case $F = x_i$ divides into two subcases ($i \in Y$ or $i \notin Y$); both subcases satisfy the claim. Assume by induction that $F = F_1 * F_2$, where F_1 and F_2 satisfy the claim and $*$ is a binary operator. For brevity let $S_1 = S_Y(F_1)$ and $S_2 = S_Y(F_2)$. Consider the following two collections of Boolean functions on Y:

$$T = \{g_1 * g_2 \colon g_1 \in S_1 \text{ and } g_2 \in S_2\}, \qquad \tilde{T} = \{\neg g \colon g \in T\}.$$

It is straightforward to check that

$$S_Y(F) \subseteq T \cup \tilde{T} \cup S_1 \cup S_2,$$

and therefore

$$\begin{aligned} 2|S_Y(F)|+1 &\leqslant 2\cdot(|T|+|\tilde{T}|+|S_1|+|S_2|)+1 \\ &\leqslant 4|S_1||S_2|+2|S_1|+2|S_2|+1 = (2|S_1|+1)\cdot(2|S_2|+1) \\ &\leqslant 5^{size_Y(F_1)} \cdot 5^{size_Y(F_2)} = 5^{size_Y(F)}. \end{aligned}$$

This completes the proof of the claim and hence the proof of the theorem. □

Theorem 5.7 can be used to prove lower bounds of size $\Omega(n^2/\log n)$ for many explicit functions. For instance, it applies to Andreev's function $A_{b,m}$, where each subset Y_i consists of one variable from every block. The theorem also applies to the element distinctness function defined in Section 6. Unfortunately the theorem is inherently unable to give a lower bound larger than $\Theta(n^2/\log n)$.

5.4. *Symmetric functions*

Symmetric functions are those Boolean functions whose value only depends on the number of variables that are 1. Being so natural, symmetric functions have been studied by many researchers in formula complexity. In this subsection, we present the best upper and lower bounds known on the formula complexity of symmetric functions.

We start with upper bounds. That all symmetric functions have polynomial-size formulas is implicit in the work of Ofman [56] and Wallace [93], who independently showed how to add n integers with n bits each in logarithmic depth. The best bounds known are due to Khrapchenko [42] and Peterson [62][2]. Khrapchenko showed that all symmetric functions have DeMorgan formulas of size $O(n^{4.93})$. Peterson, improving constructions of Pippenger [63] and Paterson [60], showed that all symmetric functions have binary formulas of size $O(n^{3.37})$.

For monotone symmetric functions (i.e., threshold functions), it seems more natural to use monotone formulas. Valiant [92], using an elegant argument, shows that all threshold functions have monotone formulas of size $O(n^{5.3})$. Boppana [17], improving results of Khasin [40] and Friedman [29], showed that the threshold function $TH_{k,n}$ has monotone formulas of size $O(k^{4.3} n \log n)$. Unfortunately all these results use probabilistic methods and hence do not explicitly construct the formulas. Ajtai, Komlós and Szemerédi [5] give a very clever explicit construction of polynomial-size monotone formulas for all threshold functions, where the degree of the polynomial is a large constant. Using their construction, Friedman [29] explicitly constructed monotone formulas for $TH_{k,n}$ of size $O(k^c n \log n)$ for a large constant c.

What about lower bounds for symmetric functions? Over the DeMorgan basis, Khrapchenko [41] presented a method for obtaining lower bounds of size $\Omega(n^2)$ for certain symmetric functions, which we describe next. Let A and B be two disjoint subsets of $\{0,1\}^n$. A Boolean formula F *separates* A and B if it outputs 0 for every input in A and outputs 1 for every input in B. Define the set

$$A \otimes B = \{(a,b) : a \in A \text{ and } b \in B \text{ and } a \sim b\},$$

where $a \sim b$ means that the inputs a and b differ on exactly one bit. Intuitively, if $A \otimes B$ is large, then every formula separating A and B should be large, since the formula must distinguish many pairs of adjacent inputs. The following theorem of Khrapchenko makes this intuition precise.

5.9. THEOREM. *Let F be a DeMorgan formula that separates A and B. Then*

$$size(F) \geqslant \frac{|A \otimes B|^2}{|A| \cdot |B|}.$$

PROOF (*due to Paterson*). The proof is by induction on the size of F. If the size of F is 1, then F is just a single literal. In that case, it is easy to see that $|A \otimes B| \leqslant |A|$ and $|A \otimes B| \leqslant |B|$, settling the base case of the induction.

Assume by induction the theorem holds for all formulas smaller than F, and suppose that $F = F_1 \wedge F_2$ (the case $F = F_1 \vee F_2$ is similar). Define

$$A_1 = \{a \in A : F_1(a) = 0\}, \qquad A_2 = A - A_1.$$

Notice that F_i is a separator of A_i and B for $i = 1,2$. Define $a_i = |A_i|$, $c_i = |A_i \otimes B|$, and $b = |B|$. Applying the induction hypothesis to the subformula F_i yields

$$size(F_i) \geqslant c_i^2 / (a_i \cdot b).$$

[2] In 1990, Paterson and Zwick [100] improved both of these bounds.

Thus we obtain

$$size(F)=size(F_1)+size(F_2)\geqslant\frac{c_1^2}{a_1\cdot b}+\frac{c_2^2}{a_2\cdot b}\geqslant\frac{(c_1+c_2)^2}{(a_1+a_2)\cdot b},$$

where the last inequality can be established by cross-multiplication. Since $c_1+c_2=|C|$ and $a_1+a_2=|A|$, this completes the induction step for F. □

Khrapchenko's theorem shows that the parity function of n variables requires DeMorgan formulas of size $\Omega(n^2)$, which is tight. It also shows that the threshold function $\mathrm{TH}_{k,n}$ requires DeMorgan formulas of size $\Omega(k\cdot(n-k+1))$. For $2\leqslant k\leqslant n-1$, Krichevskii [44] showed that $\mathrm{TH}_{k,n}$ requires DeMorgan formulas of size $\Omega(n\log n)$, which beats the Khrapchenko bound for very small or very large k.

Over the full binary basis, the best lower bounds known for symmetric functions are of size $\Omega(n\log n)$, due to Fischer, Meyer and Paterson [28]. Their bound applies to the majority function and many other symmetric functions. Pudlák [70], improving a result of Hodes and Specker [35], showed that all symmetric functions of n variables, except for 16 of them having linear-size binary formulas, require binary formulas of size $\Omega(n\log\log n)$. This bound applies for instance to the function $\mathrm{TH}_{k,n}$ for $2\leqslant k\leqslant n-1$.

6. Branching programs

The branching program model is useful for investigating the amount of space necessary to compute various functions.

Definitions. A *branching program* is a directed acyclic graph all of whose nodes of non-zero out-degree are labeled with a variable x_i and whose nodes of out-degree zero are labeled with an output value. The edges are labeled by 0 or 1. One of the nodes is designated the start node. A setting of the inputs determines a collection of paths from the start node to output nodes giving a collection of output values. The branching program is *deterministic* if every non-output node has exactly one 0 edge and one 1 edge leaving it. Otherwise it is *nondeterministic*. It *accepts* its input if at least one path leads to an accepting output node. The *size* of a branching program is the number of nodes. The branching program complexity of functions and languages is defined analogously with that of circuits. If the nodes are arranged into a sequence of levels with edges going only from one level to the next, then the *width* is the size of the largest level.

6.1. *Relationship with space complexity*

We select the Turing machine as a formal model to define space complexity. Designate one of the tapes to be a read-only input tape and call the others worktapes. The space used by this machine on a given input is defined to be the number of work-tape cells scanned by a head at any point in the computation. Define the deterministic, nondeterministic, and alternating space complexity classes by analogy with the time complexity classes. Of particular interest are

$$L=\mathrm{SPACE}(\log n), \qquad \mathrm{NL}=\mathrm{NSPACE}(\log n),$$

$$\mathrm{PSPACE}=\bigcup_k \mathrm{SPACE}(n^k).$$

The following connection between the size of branching programs and space complexity was first observed by Masek [48].

6.1. THEOREM. *For* $S(n)\geqslant \log n$, *if* $A\in \mathrm{SPACE}(S(n))$, *then A has branching program complexity at most* $c^{S(n)}$ *for some constant c.*

PROOF. The possible configurations of the machine on an input of length n form the nodes of a branching program operating on inputs of length n. Place an edge labeled 0 or 1 from a node a to node b if configuration a reading 0 or 1 respectively can yield configuration b. □

The above theorem also applies for nondeterministic space and nondeterministic branching programs.

6.2. THEOREM. *For* $S(n)\geqslant \log n$, *if* $A\in \mathrm{NSPACE}(S(n))$, *then A has nondeterministic branching program complexity at most* $c^{S(n)}$ *for some constant c.*

Thus every language in L has polynomial branching program complexity. We may obtain a converse by extending L to the nonuniform class L/poly.

6.3. THEOREM. $A\in$ L/poly *iff A has polynomial branching program complexity.*

6.4. THEOREM. $A\in$ NL/poly *iff A has polynomial nondeterministic branching program complexity.*

6.2. Bounds on size

There have been a number of results bounding the size of branching programs with various restrictions. None of these is yet strong enough to have any implications for space complexity. Just as for Boolean circuits, we know via a counting argument that most functions have exponential branching program complexity. The best lower bound for an explicit function is $\Omega(n^2/\log^2 n)$ due to Nečiporuk [54]. Using the same method Beame and Cook (unpublished) have observed that the following "element distinctness" function may be proved to require large branching programs. Consider the input to represent m strings $s_1,\ldots, s_m$ each of length $2\log m$ where $n=2m\log m$. Define the function so that it is 1 iff the s_i are all distinct.

6.5. THEOREM. *The element distinctness function requires branching programs of size* $\Omega(n^2/\log^2 n)$.

PROOF. We show that a function is hard if its inputs may be partitioned into blocks b_i so that there are a large number of functions on each block b_i obtained by setting the remaining variables. By letting b_i be the inputs for s_i we see that the given function has this property. For each b_i there are $\binom{m^2}{m-1}$ ways of setting the remaining s's distinctly and each way gives a different induced function. Each such setting also yields an induced branching program on the nodes labeled from b_i plus an accept and a reject node. Say there are h_i such nodes. The number of branching programs on h_i nodes is at most $n^{h_i}h_i^{2h_i}$. Thus $n^{h_i}h_i^{2h_i} \geqslant \binom{m^2}{m-1}$ and so $h_i \geqslant m/2$. The total size of the original branching program is $\Sigma(h_i-2)+2$ since the accept and reject nodes are common but the others are not. This is $\Omega(m^2)$ and hence $\Omega(n^2/\log^2 n)$. □

Since stronger lower bounds on general branching programs seem hard to come by, researchers have turned to restricted versions. Furst, Saxe and Sipser [30] and Borodin, Dolev, Fich and Paul [18] conjectured that the majority function required superpolynomial size if the width of the branching program is held fixed. Partial progress was made by Chandra, Furst and Lipton [19] who gave a superlinear lower bound. Pudlak [71] improved this by eliminating the width restriction. Then Barrington [12], with a surprising construction, disproved this conjecture. He showed that constant-width branching programs are unexpectedly powerful, being able to accept all NC^1-languages.

6.6. THEOREM. *A language has polynomial size, width 5 branching programs iff it is in nonuniform* NC^1.

PROOF. One direction ($\rightarrow$) is easy since a fixed amount of circuitry can compose two levels into one. Doing this in parallel across the branching program and repeating it $O(\log n)$ times yields the desired NC^1-circuit.

For the other direction we construct a special kind of width 5 branching program from the $O(\log n)$ depth circuit. All levels have 5 nodes and all nodes in a given level are labeled with the same variable. Additionally, at each level the 0 edges and the 1 edges going to the next level form permutations. Then an input setting yields a permutation which is the composition of the selected permutations at each level. Call such a branching program p a *permuting* branching program and let $p(x)$ be the resulting permutation on input x. For a language B and permutation σ say that branching program p *σ-accepts B* if for each string $x \in B$ we have $p(x)=\sigma$ and for each $x \notin B$ we have $p(x)=e$, the identity permutation.

A permutation is *cyclic* if it is composed of a single cycle on all of its elements. In the following lemmas let σ and τ be cyclic permutations, B and C be languages, and p and q be permuting branching programs.

6.7. LEMMA. If *p σ-accepts B, then there is a permuting branching program of the same size τ-accepting B.*

PROOF. Since σ and τ are cyclic we may write $\tau=\gamma\sigma\gamma^{-1}$ for some permutation γ. Then simply reorder the left and right nodes of p according to γ to obtain the τ-accepting branching program. □

6.8. LEMMA. *If p σ-accepts B, then there is a permuting branching program of the same size σ-accepting $\bar{B}$.*

PROOF. Use the previous lemma to obtain a σ^{-1}-acceptor for B. Then reorder the final level by σ so that it becomes a σ-acceptor for $\bar{B}$. □

6.9. LEMMA. *If p σ-accepts B and q τ-accepts C, then there is a permuting branching program $\sigma\tau\sigma^{-1}\tau^{-1}$-accepting $B \cap C$ of size $2(size(p)+size(q))$.*

PROOF. Use Lemma 6.7 to get a σ^{-1}-acceptor for B and a τ^{-1}-acceptor for C. Now compose these four acceptors in the order $\sigma, \tau, \sigma^{-1}, \tau^{-1}$. This has the desired effect because replacing either σ or τ by e in $\sigma\tau\sigma^{-1}\tau^{-1}$ yields e. □

6.10. LEMMA. *There are cyclic permutations σ and τ in S_5 such that $\sigma\tau\sigma^{-1}\tau^{-1}$ is cyclic.*

PROOF. (12345) (13542) (54321) (24531) = (13254). □

The above lemma is the only place where the value 5 is important.

PROOF OF THEOREM 6.6 (*continued*). Now we may finish the proof of the theorem. If a depth d circuit contains only NOT-gates and AND-gates of fan-in 2, then it is easily seen by the above lemmas that we may construct a width 5 branching program of length at most 4^d. Hence, if the depth is $O(\log n)$, then the branching program has polynomial size. □

7. Conclusion

Let us speculate on the future for research on lower bounds on the complexity of finite functions. Looking back, it is clear that the last decade has seen some important progress on restricted models of computation. It is hard to say whether this has any bearing on the unrestricted case. An optimist might argue that by considering models with weaker and weaker restrictions one may incrementally approach the unrestricted case. A very recent result of Razborov [75] indicates however that the unrestricted case may be qualitatively different from the cases that have been successfully treated so far. This result shows that the "approximation method", which lies at the heart of all of the strong lower bounds presented in this chapter except for the Karchmer–Wigderson monotone depth bound, cannot be used to prove superpolynomial lower bounds on the size of general circuits. It seems that a fundamentally different idea is needed.

A different approach was proposed by Sipser [80]. Using an analogy between countability and polynomiality one may derive a suggestive correspondence between finite complexity and definability in descriptive set theory (see Moschovakis [51] for a comprehensive treatment). In this way the class AC^0 corresponds to the class of Borel sets and the class NP corresponds to the class of analytic sets. The former correspondence played a critical role in the discovery of the proof of the theorem of Furst, Saxe and Sipser [30]. By first formulating and solving an infinite version of their conjecture about the parity function they were guided in the search for a solution to the conjecture.

The latter correspondence may have more direct bearing on the P=NP, or more precisely, the NP=co-NP problem. Sipser [82] gives a new proof of the classical theorem due to Lebesgue [47] stating that the analytic sets are not closed under complement. There is a possibility that this proof contains a hint of how to proceed with showing that NP is not closed under complement.

Acknowledgment

We wish to thank M. Karchmer, M. Paterson, N. Pippenger, A.A. Razborov, and U. Zwick for extensive comments on an earlier draft of this paper.

The work of R.B. Boppana was supported by an NSF Mathematical Sciences Postdoctoral Fellowship and NSF Grant CCR-8902522. The work of M. Sipser was supported by NSF Grant CCR-8912586 and Air Force Contract AFOSR-86-0078.

References

[1] ADLEMAN, L., Two theorems on random polynomial time, in: *Proc. 19th Ann. IEEE Symp. on Foundations of Computer Science* (1978) 75–83.

[2] AJTAI, M., Σ_1^1-formulae on finite structures, *Ann. Pure Appl. Logic* **24** (1983) 1–48.

[3] AJTAI, M. and M. BEN-OR, A theorem on probabilistic constant depth circuits, in: *Proc. 16th Ann. ACM Symp. on Theory of Computing* (1984) 471–474.

[4] AJTAI M. and Y. GUREVICH, Monotone versus positive, *J. Assoc. Comput. Mach.* **34**(4) (1987) 1004–1015.

[5] AJTAI, M., J. KOMLÓS and E. SZEMERÉDI, An $O(n \log n)$ sorting network, in: *Proc. 15th Ann. ACM Symp. on Theory of Computing* (1983) 1–9; revised version in *Combinatorica* **3**(1) (1983) 1–19.

[6] ALON, N. and R.B. BOPPANA, The monotone circuit complexity of Boolean functions, *Combinatorica* **7**(1) (1987) 1–22.

[7] ANDREEV, A.E., On a method for obtaining lower bounds for the complexity of individual monotone functions, *Dokl. Akad. Nauk SSSR* **282**(5) (1985) 1033–1037 (in Russian); English translation in: *Soviet Math. Dokl.* **31**(3) (1985) 530–534.

[8] ANDREEV, A.E., On a family of Boolean matrices, *Vestnik Moskov. Univ. Mat.* **41**(2) (1986) 97–100 (in Russian); English translation in: *Moscow Univ. Math. Bull.* **41**(2) (1986) 79–82.

[9] ANDREEV, A.E., On a method for obtaining more than quadratic effective lower bounds for the complexity of π-schemes, *Vestnik Moskov. Univ. Mat.* **42**(1) (1987) 70–73 (in Russian); English translation in: *Moscow Univ. Math. Bull.* **42**(1) (1987) 63–66.

[10] BAKER, T.P., J. GILL and R. SOLOVAY, Relativizations of the P=?NP question, *SIAM J. Comput.* **4**(4) (1975) 431–442.

[11] BAKER, T.P. and A.L. SELMAN, A second step toward the polynomial hierarchy, *Theoret. Comput. Sci.* **8** (1979) 177–187.

[12] BARRINGTON, D.A., Bounded-width polynomial-size branching programs recognize exactly those languages in NC^1, in: *Proc. 18th Ann. ACM Symp. on Theory of Computing* (1986) 1–5; revised version in: *Comput. System Sci.* **38**(1) (1989) 150–164.

[13] BENNETT, C. and J. GILL, Relative to a random oracle A, $P^A \neq NP^A \neq$co-NP^A with probability 1, *SIAM J. Comput.* **10** (1981) 96–113.

[14] BERKOWITZ, S.J., On some relationships between monotone and non-monotone circuit complexity, Tech. Report, Comput. Sci. Dept., Univ. of Toronto, 1982.

[15] BLUM, N., A Boolean function requiring $3n$ network size, *Theoret. Comput. Sci.* **28** (1984) 337–345.

[16] BOPPANA, R.B., Threshold functions and bounded depth monotone circuits, *J. Comput. System Sci.* **32**(2) (1986) 222–229.

[17] BOPPANA, R.B., Amplification of probabilistic Boolean formulas, in: S. Micali, ed., *Advances in Computer Research, Vol. 5: Randomness and Computation* (JAI Press, Greenwich, CT, to appear).

[18] BORODIN, A., D. DOLEV, F.E. FICH and W.J. PAUL, Bounds for width two branching programs, in: *Proc. 15th Ann. ACM Symp. on Theory of Computing* (1983) 87–93.

[19] CHANDRA, A.K., M.L. FURST and R.J. LIPTON, Multiparty protocols, in: *Proc. 15th Ann. ACM Symp. on Theory of Computing* (1983) 94–99.

[20] CHANDRA, A.K., D. KOZEN and L. STOCKMEYER, Alternation, *J. Assoc. Comput. Mach.* **28** (1981) 114–133.

[21] CHANDRA, A.K., L.J. STOCKMEYER and U. VISHKIN, Constant depth reducibility, *SIAM J. Comput.* **13**(2) (1984) 423–439.

[22] COOK, S.A., A taxonomy of problems with fast parallel algorithms, *Inform. and Control* **64** (1985) 2–22.

[23] COPPERSMITH, D. and S. WINOGRAD, Matrix multiplication via arithmetic progressions, in: *Proc. 19th Ann. ACM Symp. on Theory of Computing* (1987) 1–6.

[24] DUNNE, P.E., Lower bounds on the monotone network complexity of threshold functions, in: *Proc. 22nd Ann. Allerton Conf. on Communication, Control and Computing* (1984) 911–920.

[25] EDMONDS, J., Paths, trees, and flowers, *Canad. J. Math.* **17** (1965) 449–467.

[26] EMDE BOAS, P. VAN, Machine models and simulations, in: J. van Leewen, ed., *Handbook of Theoretical Computer Science Vol. A* (North-Holland, Amsterdam, 1990) 1–66.

[27] ERDÖS, P. and R. RADO, Intersection theorems for systems of sets, *J. London Math. Soc.* **35** (1960) 85–90.

[28] FISCHER, M.J., A.R. MEYER and M.S. PATERSON, $\Omega(n \log n)$ lower bounds on length of Boolean formulas, *SIAM J. Comput.* **11**(3) (1982) 416–427.

[29] FRIEDMAN, J., Constructing $O(n \log n)$ size monotone formulae for the k-th elementary symmetric polynomial of n Boolean variables, *SIAM J. Comput.* **15**(3) (1986) 641–654.

[30] FURST, M., J. SAXE and M. SIPSER, Parity, circuits and the polynomial time hierarchy, *Math. Systems Theory* **17** (1984) 13–27.

[31] GRAHAM, R.L., B.L. ROTHSCHILD and J.H. SPENCER, *Ramsey Theory* (Wiley, New York, 1980).

[32] HARPER, L.H. and J.E. SAVAGE, The complexity of the marriage problem, *Adv. in Math.* **9** (1972) 299–312.

[33] HARTMANIS, J. and R.E. STEARNS, On the computational complexity of algorithms, *Trans. Amer. Math. Soc.* **117** (1965) 285–306.

[34] HASTAD, J. Almost optimal lower bounds for small depth circuits, in: S. Micali, ed., *Advances in Computer Research, Vol. 5: Randomness and Computation* (JAI Press, Greenwich, CT, to appear); see also *Computational Limitations for Small Depth Circuits* (MIT Press, Cambridge, MA, 1986).

[35] HODES, L. and E. SPECKER, Length of formulas and elimination of quantifiers I, in: H.A. Schmidt, K. Schutte and H.-J. Thiele, eds., *Contributions to Mathematical Logic* (North-Holland, Amsterdam, 1968) 175–188.

[36] HOPCROFT, J.E. and J.D. ULLMAN, *Introduction to Automata Theory, Languages, and Computation* (Addison-Wesley, Reading, MA, 1979).

[37] IMMERMAN, N., Languages that capture complexity classes, *SIAM J. Comput.* **16**(4) (1987) 760–778.

[38] KARCHMER, M. and A. WIGDERSON, Monotone circuits for connectivity require super-logarithmic depth, in: *Proc. 20th Ann. ACM Symp. on Theory of Computing* (1988) 539–550.

[39] KARP, R.M. and R. LIPTON, Turing machines that take advice, *Enseign. Math.* **28** (1982) 191–209.

[40] KHASIN, L.S., Complexity bounds for the realization of monotone symmetrical functions by means of formulas in the basis $+, \cdot, -$, *Dokl. Akad. Nauk SSSR* **189**(4) (1969) 752–755 (in Russian); English translation in: *Soviet Phys. Dokl.* **14**(12) (1970) 1149–1151.

[41] KHRAPCHENKO, V.M., A method of determining lower bounds for the complexity of Π-schemes, *Mat. Zametki* **10**(1) (1971) 83–92 (in Russian); English translation in: *Math. Notes* **10**(1) (1971) 474–479.

[42] KHRAPCHENKO, V.M., The complexity of the realization of symmetrical functions by formulae, *Mat. Zametki* **11**(1) (1972) 109–120 (in Russian); English translation in: *Math. Notes* **11**(1) (1972) 70–76.

[43] KLAWE, M., W.J. PAUL, N. PIPPENGER and M. YANNAKAKIS, On monotone formulae with restricted depth, in: *Proc. 16th Ann. ACM Symp. on Theory of Computing* (1984) 480–487.

[44] KRICHEVSKII, R.E., Complexity of contact circuits realizing a function of logical algebra, *Dokl. Akad. Nauk SSSR* **151**(4) (1963) 803–806 (in Russian); English translation in: *Soviet Phys. Dokl.* **8**(8) (1964) 770–772.

[45] LAMAGNA, E.A., The complexity of monotone networks for certain bilinear forms, routing problems, sorting and merging, *IEEE Trans. Comput.* **28** (1979) 773–782.

[46] LAMAGNA, E.A. and J.E. SAVAGE, Combinational complexity of some monotone functions, in: *Proc. 15th Ann. IEEE Symp. on Switching and Automata Theory* (1974) 140–144.

[47] LEBESGUE, H., Sur les fonctions représentables analytiquement, *J. de Math. 6ᵉ serie* **1** (1905) 139–216 (in French).

[48] MASEK, W., A fast algorithm for the string editing problem and decision graph complexity, M.S. Thesis, Dept. Electrical Eng. Comput. Sci., Massachusetts Institute of Technology, Cambridge, MA, 1976.

[49] MEHLHORN, K., Some remarks on Boolean sums, *Acta Inform.* **12** (1979) 371–375.

[50] MEHLHORN, K. and Z. GALIL, Monotone switching circuits and Boolean matrix product, *Computing* **16** (1976) 99–111.

[51] MOSCHOVAKIS, Y.N., *Descriptive Set Theory* (North-Holland, Amsterdam, 1980).

[52] MULLER, D.E., Complexity in electronic switching circuits, *IRE Trans. Electronic Computers* **5** (1956) 15–19.

[53] MULLER, D.E. and F.P. PREPARATA, Bounds to complexities of networks for sorting and switching, *J. Assoc. Comput. Mach.* **22**(2) (1975) 195–201.

[54] NEČIPORUK, E.I., A Boolean function, *Dokl. Akad. Nauk SSSR* **169**(4) (1966) 765–766 (in Russian); English translation in: *Soviet Math. Dokl.* **7**(4) (1966) 999–1000.

[55] NEČIPORUK, E.I., On a Boolean matrix, *Problemy Kibernet.* **21** (1969) 237–240 (in Russian); English translation in: *Systems Theory Research* **21** (1971) 236–239.

[56] OFMAN, YU., On the algorithmic complexity of discrete functions, *Dokl. Akad. Nauk SSSR* **145**(1) (1962) 48–51 (in Russian); English translation in: *Soviet Phys. Dokl.* **7**(7) (1963) 589–591.

[57] OKOL'NISHNIKOVA, E.A., On the influence of negations on the complexity of a realization of monotone Boolean functions by formulas of bounded depth, *Metody Diskret. Analiz.* **38** (1982) 74–80 (in Russian).

[58] PATERSON, M.S., Complexity of monotone networks for Boolean matrix product, *Theoret. Comput. Sci.* **1**(1) (1975) 13–20.

[59] PATERSON, M.S., An introduction to Boolean function complexity, *Astérisque* **38–39** (1976) 183–201.

[60] PATERSON, M.S., New bounds on formula size, in: *Proc. 3rd GI Conf. on Theoretical Computer Science*, Lecture Notes in Computer Science, Vol. 48 (Springer, Berlin, 1977) 17–26.

[61] PAUL, W.J., A $2.5n$ lower bound on the combinational complexity of Boolean functions, *SIAM J. Comput.* **6**(3) (1977) 427–443.

[62] PETERSON, G.L., An upper bound on the size of formulae for symmetric Boolean functions, Tech. Report 78-03-01, Dept. Comput. Sci., Univ. Washington, 1978.

[63] PIPPENGER, N., Short formulae for symmetric functions, Research Report RC-5143, IBM T.J. Watson Research Center, Yorktown Heights, 1974.

[64] PIPPENGER, N., On simultaneous resource bounds, in: *Proc. 20th Ann. IEEE Symp. on Foundations of Computer Science* (1979) 307–311.

[65] PIPPENGER, N., On another Boolean matrix, *Theoret. Comput. Sci.* **11** (1980) 49–56.

[66] PIPPENGER, N. and M.J. FISCHER, Relations among complexity measures, *J. Assoc. Comput. Mach.* **26** (1979) 361–381.

[67] PIPPENGER, N. and L.G. VALIANT, Shifting graphs and their applications, *J. Assoc. Comput. Mach.* **23** (1976) 423–432.

[68] PRATT, V.R., The power of negative thinking in multiplying Boolean matrices, *SIAM J. Comput.* **4** (1974) 326–330.

[69] PRATT, V.R., The effect of basis on size of Boolean expressions, in: *Proc. 16th Ann. IEEE Symp. on Foundations of Computer Science* (1975) 1119–121.

[70] PUDLÁK, P., Bounds for Hodes–Specker theorem, in: *Logic and Machines: Decision Problems and Complexity*, Lecture Notes in Computer Science, Vol. 171 (Springer, Berlin, 1984) 421–445.

[71] PUDLÁK, P., A lower bound on complexity of branching programs, in: *Mathematical Foundations of Computer Science 1984*, Lecture Notes in Computer Science, Vol. 176 (Springer, Berlin, 1984) 480–489.

[72] RAZBOROV, A.A., Lower bounds on the monotone complexity of some Boolean functions, *Dokl. Akad. Nauk SSSR* **281**(4) (1985) 798–801 (in Russian); English translation in: *Soviet Math. Dokl.* **31** (1985) 354–357.

[73] RAZBOROV, A.A., A lower bound on the monotone network complexity of the logical permanent, *Mat. Zametki* **37**(6) (1985) 887–900 (in Russian); English translation in: *Math. Notes* **37**(6) (1985) 485–493.

[74] RAZBOROV, A.A., Lower bounds on the size of bounded depth networks over a complete basis with logical addition, *Mat. Zametki* **41**(4) (1987) 598–607 (in Russian); English translation in: *Math. Notes* **41**(4) (1987) 333–338.

[75] RAZBOROV, A.A., On the method of approximations, in: *Proc. 21st Ann. ACM Symp. on Theory of Computing* (1989) 167–176.

[76] RAZBOROV, A.A., Applications of matrix methods to the theory of lower bounds in computational complexity, *Combinatorica*, to appear.

[77] SAVAGE, J.E., Computational work and time on finite machines, *J. Assoc. Comput. Mach.* **19**(4) (1972) 660–674.

[78] SAVAGE, J.E., *The Complexity of Computing* (Wiley, New York, 1976).

[79] SHANNON, C.E., The synthesis of two-terminal switching circuits, *Bell Systems Tech. J.* **28**(1) (1949) 59–98.

[80] SIPSER, M., On polynomial vs. exponential growth, Unpublished manuscript, 1981.

[81] SIPSER, M., Borel sets and circuit complexity, in: *Proc. 15th Ann. ACM Symp. on Theory of Computing* (1983) 61–69.

[82] SIPSER, M., A topological view of some problems in complexity theory, in: L. Lovász and E. Szemerédi, eds., *Theory of Algorithms*, Colloquia Mathematica Societatis János Bolyai, Vol. 44 (1984) 387–391.

[83] SKYUM, S. and L.G. VALIANT, A complexity theory based on Boolean algebra, *J. Assoc. Comput. Mach.* **32**(2) (1985) 484–502.

[84] SMOLENSKY, R., Algebraic methods in the theory of lower bounds for Boolean circuit complexity, in: *Proc. 19th Ann. ACM Symp. on Theory of Computing* (1987) 77–82.

[85] SPIRA, P.M., On time-hardware complexity tradeoffs for Boolean functions, in: *Proc. 4th Hawaii Symp. on System Sciences* (Western Periodicals Company, North Hollywood, 1971) 525–527.

[86] SUBBOTOVSKAYA, B.A., Realizations of linear functions by formulas using $+, \cdot, -$, *Dok. Akad. Nauk SSSR* **136(3)** (1961) 553–555 (in Russian); English translation in: *Soviet Math. Dokl.* **2** (1961) 110–112.

[87] TARDOS, É., The gap between monotone and non-monotone circuit complexity is exponential, *Combinatorica* **8**(1) (1988) 141–142.

[88] TIEKENHEINRICH, J., A $4n$ lower bound on the monotone network complexity of a one-output Boolean function, *Inform. Process. Lett.* **18** (1984) 201–202.

[89] TURING, A.M., On computable numbers with an application to the Entscheidungsproblem, *Proc. London Math. Soc.* (2) **42** (1937) 230–265; correction, ibidem **43** (1936) 544–546.

[90] UGOL'NIKOV, A.B., Complexity and depth of formulas realizing functions from closed classes, in: *Fundamentals of Computation Theory*, Lecture Notes in Computer Science, Vol. 278 (Springer, Berlin, 1987) 456–461.

[91] VALIANT, L.G., Completeness classes in algebra, in: *Proc. 11th ACM Ann. Symp. on Theory of Computing* (1979) 249–261.

[92] VALIANT, L.G., Short monotone formulae for the majority function, *J. Algorithms* **5** (1984) 363–366.

[93] WALLACE, C.S., A suggestion for a fast multiplier, *IEEE Trans. Comput.* **3**(1) (1964) 14–17.

[94] WEGENER, I., Boolean functions whose monotone complexity is of size $n^2/\log n$, *Theoret. Comput. Sci.* **21** (1982) 213–224.

[95] WEGENER, I., Relating monotone formula size and monotone depth of Boolean functions, *Inform. Process. Lett.* **16** (1983) 41–42.

[96] WEGENER, I., *The Complexity of Boolean Functions*, Wiley-Teubner Series in Computer Science (Teubner, Stuttgart/Wiley, Chichester, 1987).

[97] YAO, A.C., Separating the polynomial-time hierarchy by oracles, in: *Proc. 26th Ann. IEEE Symp. on Foundations of Computer Science* (1985) 1–10.
[98] ZWICK, U., Optimizing Nečiporuk's theorem, Tech. Report 86/1987, Dept. Comput. Sci., Tel Aviv Univ., 1987.
[99] DUNNE, P.E., *The Complexity of Boolean Networks* (Academic Press, London, 1988).
[100] PATERSON, M.S. and U. ZWICK, Improved circuits and formulae for multiple addition, multiplication, and symmetric Boolean functions, Preprint, 1990.
[101] RAZ, R. and A. WIGDERSON, Monotone circuits for matching require linear depth, in: *Proc. 22nd Ann. ACM Symp. on Theory of Computing* (1990).

CHAPTER 15

Communication Networks

Nicholas PIPPENGER

Department of Computer Science, The University of British Columbia, 6356 Agricultural Road, Vancouver, Canada BC V6T 1W5

Contents

HANDBOOK OF THEORETICAL COMPUTER SCIENCE
Edited by J. van Leeuwen

1. Introduction

1.1. Communication networks and computation

Communication networks occur as components of computer systems, but their study has also been of use in deriving both upper and lower bounds to the complexity of computations in various models, and it is this latter aspect that is of greatest importance in theoretical computer science. This chapter surveys results concerning communication networks that have found application in this way, as well as such other results are necessary to make the picture cohere. It also includes a complete proof of the most celebrated and frequently applied of these results, that of Ajtai, Komlós and Szemerédi concerning the existence of sorting networks of logarithmic depth.

What is the role of communication networks in theoretical computer science? At least three answers can be given to this question. First, a computer (particularly, a parallel computer) may involve a communication network in its design, for communication among processors or between processors and memory devices. This role is the most obvious one, and some aspects of it are described in Chapter 18 of this Handbook, "General Purpose Parallel Architectures". Second, operations performed within a computer (by hardware or software) may involve communication to such an extent that it dominates other activities. This is the case with operations that shift, permute and replicate data in an oblivious fashion. In some contexts, it may also apply to operations such as sorting that depend on the data being manipulated, and to operations such as prefix iteration that alter the data in stylized ways. Thus the means for performing these operations may in many instances be modelled by communication networks (whether these means are networks realized by hardware or networks simulated by software). Third, it is sometimes possible to obtain lower bounds for the resources needed to solve computational problems from corresponding lower bounds for communication networks (even when the computational problems and the means used to solve them involve activities other than communication). Progress in exploiting this third role has been slow, but we shall describe a few successes that complement the methods of obtaining lower bounds described in Chapter 14 "The Complexity of Finite Functions" and Chapter 11 "Algebraic Complexity Theory". This chapter is organized around a discussion of the first role, with occasional digressions to deal with the second and third.

1.2. Modes of operation

Imagine n agents called "sources" and n other agents called "targets". Our interest is in networks that might be used to provide simultaneous communication between various combinations of sources and targets. (We emphasize the condition "simultaneous", for without it we would be interested in what might better be called "busses" than networks.) There are two principal ways in which we may envisage the communication to occur; they are usually referred to as "packet switching" and "circuit switching". In packet switching, "packets" or "messages" move through the network from their sources to their targets; they may be stored temporarily and later forwarded, or they may be discarded and later resent. In circuit switching, "routes" or "paths" are established through the network from their sources to their targets. Later these paths may be dissolved, and their constituent components used to establish other paths. The essence of the distinction is that the resources used to establish a path in a circuit switching network are committed simultaneously, and on a time scale long compared with that of the transmission or reception of a message. Contrastingly, the path followed by a message in a packet switching network may engage different resources at different times, on time scales short compared with that of the sojourn of a message through the network. Packet switching networks are dealt with in Chapter 18 "General Purpose Parallel Architectures"; the present chapter deals with circuit switching networks.

For circuit switching networks, there are two principal ways in which we may envisage the paths to be established and dissolved; they are usually referred to as "nonblocking" and "rearrangeable" operation. In nonblocking operation, a request to establish a path from a source to a target may arrive at any time, and must be satisfied without disturbing previously existing paths in the network; from the time it is satisfied until the time that its dissolution is requested, an arbitrary number of other requests to establish and dissolve paths may arrive and must be honored. In rearrangeable operation, requests to establish paths arrive in batches, so that all of the requests are known before any need be satisfied; the satisfying paths may then all be dissolved before any of the next batch of requests is considered. The essence of the distinction is that in rearrangeable operation each batch of requests is independent of others that precede or follow it, whereas with nonblocking operation the establishment of each path is constrained by previously established paths, and may constrain the paths used to satisfy future requests. Nonblocking operation epitomizes the activity of networks for human telephone communication; it was introduced by Clos [19] in 1953, and an account of subsequent work is given by Feldman, Friedman and Pippenger [26]. The present chapter deals with rearrangeable operation, which is more typical of communication within computers.

For rearrangeable operation of a circuit switching network, a basic problem is: given a batch of requests (called an "assignment"), find a batch of paths (called a "configuration") that simultaneously satisfy the requests (referred to as "realizing" the assignment). There are two principal ways in which we may envisage this problem being solved; they are usually referred to as "global" and "local" routing. With global routing, the problem is to be solved by an agent external to the network, with knowledge of the network and of the assignment to be realized. One may then study sequential and parallel algorithms for solving this problem; for a typical result of this form, see [40]. With local routing, the problem is to be solved by the network itself, with the requests blazing their own trails as they pass through the network. We shall see that the construction of "self-routing" networks that support local routing presents much more challenging problems than that of networks supporting global routing.

2. Communication problems

2.1. The connection problem

Suppose for the moment that each of the sources requests a path to one of the targets, and that these requests constitute a bijection between sources and targets. In this situation, a network that supports global routing will be called a "connector". To formulate the property more precisely, we shall model a network as an acyclic directed graph, with vertices and directed edges. The sources will be represented by n distinguished vertices called *inputs*, and the targets by n other distinguished vertices called *outputs*. Vertices other than inputs and outputs will be called *links*. Such a network will be called an *n-connector* if, for every assignment comprising n requests that establish a bijection between the inputs and the outputs, there exists a configuration comprising n vertex-disjoint paths joining each input to the request output.

There are two parameters of particular interest in comparing networks; they are the *size*, or number of edges in the network, and the *depth*, or the number of edges on the longest path from an input of the network to an output. Edges, or configurations of edges, correspond to switching components, so that size corresponds, at least in some approximate sense, to the cost of the network. Similarly, depth corresponds to the delay that signals suffer as they pass through the network. Other things being equal, it is usually preferable to have networks with smaller size and smaller depth.

There is a very natural construction, due to Beneš [12], for n-connectors with size $O(n \log n)$ and depth $O(\log n)$. Beneš ascribes the main idea in his construction to unpublished work of Slepian, Duguid and LeCorre during the 1950s. (It had been suggested earlier (without a correct proof) that it permutes, and it has been rediscovered many times since then.) A lower bound due to Shannon [70] shows that the size bound is, to within a constant factor, the smallest possible, but the depth bound can be decreased, at a cost of greater size. Indeed, it is clear that by taking an edge between each input and each output, the depth can be decreased to 1, at a cost of increasing the size to n^2. This question of the trade-off between depth and size will be one of the themes of this survey, with the case of fixed depth exhibiting the most striking results.

Another feature of the natural construction for connectors is that the edges can be partitioned into configurations of four edges, which we shall call *blocks*, with each configuration containing edges directed out of a pair of vertices called *inlets*, and directed into another pair of vertices called *outlets*, with one edge from each inlet to each outlet. Furthermore each input of the network is an inlet of just one block, each output is an outlet of just one block, and every other vertex is an inlet of exactly one block and an outlet of exactly one block. A network with these properties will be called *homogeneous*. Such a network may be viewed as an interconnection of components (the blocks) capable of routing signals from their inlets to their outlets in various possible ways; the interconnections are effected by wires (the links), each of which connects an outlet of one block to an inlet of another. A block is a 2-connector (which is called a *transposer* in this context). The question of whether or not a network is homogeneous gives insight into how it performs its function, and this question will be another of the themes of this survey.

When dealing with homogeneous networks it is possible and often convenient to use a more structured formulation than that of an acyclic directed graph. We shall present such a formulation in terms of "programs" consisting of "instructions" acting on "items" in "registers".

Let n be a nonnegative integer. Let A be a totally ordered set of n objects called *registers*. Let B be a totally ordered set of n objects called *items*. A bijection $f: A \to B$ will be called a *state*. The item $f(m)$ is called the *contents* of register m in state f.

By an *instruction* we shall mean an object $(i:j)$, where i and j are distinct registers. If f is a state and q is an instruction, the state fq, the *action* of q on f, is defined as follows. If $q=(i:j)$, then

$$(f(i:j))(m) = \begin{cases} f(i) & \text{if } m=j, \\ f(j) & \text{if } m=i, \\ f(m), & \text{otherwise.} \end{cases}$$

(Thus $(i:j)$ exchanges the contents of registers i and j.)

By a *program* we shall mean a sequence of instructions. There is a unique *empty* program, which contains no instructions and which will be denoted Π. Every other program Q has the form Pq, where P is a program and q is an instruction. We define the state fQ (the result of Q acting on f) to be f if $Q=\Pi$, and to be $(fP)q$ if $Q=Pq$.

By the *size* $C(Q)$ of a program Q we shall mean the number of instructions it contains. Thus $C(Q)$ is 0 if $Q=\Pi$ and $1+C(P)$ if $Q=Pq$. (We have defined the size of a program to be one-quarter of the size of the corresponding homogeneous network, since the latter is always a multiple of four.)

If Q is a program and m is a register, define $\Delta(m, Q)$, the *depth* of m in Q, to be 0 if $Q=\Pi$, and to be

$$\begin{cases} 1+\max\{\Delta(i, P), \Delta(j, P)\} & \text{if } m \in \{i, j\}, \\ \Delta(m, P), & \text{otherwise,} \end{cases}$$

if $Q=P(i:j)$. By the *depth* $D(Q)$ of Q we shall mean the maximum of $\Delta(m, Q)$ over all registers m. (The depth of a program equals that of the corresponding homogeneous network.)

Since each instruction increases the depth of two registers, and the maximum depth of a register is at least as large as the average, $D(Q) \geqslant 2C(Q)/n$.

If $(i:j)$ and $(i':j')$ are instructions involving disjoint sets $\{i, j\}$ and $\{i', j'\}$ of registers, then $(i:j)$ and $(i':j')$ commute, in the sense that the action on states, the size and the depth of a program are unaffected if the consecutive instructions $(i:j)\,(i':j')$ are replaced by their transposition $(i':j')\,(i:j)$. It is sometimes convenient to identify two programs if one can be obtained from the other by a sequence of transpositions of pairs of consecutive instructions involving disjoint pairs of registers. This convention abstracts away those properties of programs that we consider irrelevant to networks.

If f is a state and X is a set of registers, we shall write $f(X)$ for the set of items in registers of X in f. Say that a program Q is *supported* by X if $i, j \in X$ for every instruction $(i{:}j)$ in Q. If Q is supported by X, then $(fQ)(X) = f(X)$ for every state f. Say that a set of registers is *sorted* in a state f if $f(i) < f(j)$ whenever $i < j$.

Say a program Q is *regular* if every instruction in Q is of the form $(i{:}j)$, with $i < j$. Say that a regular program *connects* X if Q is supported by X and for every state f there is a subsequence P of Q such that X is sorted in fP. Say that Q *connects* (without further qualification) if Q connects A. Then a program connects if and only if the corresponding homogeneous network is an n-connector.

Returning from programs to networks, let us now consider the minimum possible size of n-connectors with fixed depth k. The case of $k = 2$ was investigated by de Bruijn et al. [24, 38], who gave a lower bound of $\Omega(n^{3/2})$. They also gave an upper bound of $O((n \log n)^{3/2})$, but only by nonconstructive methods. For larger k, a lower bound of $\Omega(n^{1+1/k})$ was given by Pippenger and A.C. Yao [66], who also gave the upper bound $O((n \log n)^{1+1/k})$, again by nonconstructive methods. The best explicit construction known (see [12]) gives a size of $O(n^{1+1/k})$ for depth $2k - 1$ rather than k. There is a "sporadic" explicit construction, based on combinatorial designs, with depth 2 and size $O(n^{5/3})$ (see [26, 69]); it can be used to extend the preceding explicit construction to even depths, giving size $O(n^{1+2/(3k-1)})$ for depth $2k$.

If we seek a connector that supports local routing, the most immediate formulation of the problem leads to the notion of a "sorter". To define this, we first augment our notion of a network by assuming that every vertex other than an input is provided with one of two labels, "min" or "max". Suppose then that n distinct and totally ordered values called *items* are associated in a one-to-one fashion with the inputs of the network. We may then associate an item with each other vertex in the network according to the following rule: suppose inductively that items have been associated with all of the immediate predecessors of the vertex v (these are the vertices u for which there is an edge (u, v) in the network); if v is labelled "min", associate the minimum of these items with v; if v is labelled "max", associate the maximum. In this way we arrive at an association of the items with the outputs of the network. Such a network is called an *n-sorter* if all $n!$ possible bijective associations of items with inputs lead to the same bijective association of items with outputs. It is easy to see that if a sorter is a homogeneous network, then each block has one of its outlets labelled "min" and one labelled "max", so that it may be regarded as a 2-sorter (which is called a *comparator* in this context).

To discuss homogeneous sorters in terms of programs, we augment our notion of program by allowing an instruction to be an object $[i{:}j]$, where i and j are registers. We define the action of these instructions as follows. If $q = [i{:}j]$, then

$$(f[i{:}j])(m) = \begin{cases} \min\{f(i), f(j)\} & \text{if } m = i, \\ \max\{f(i), f(j)\} & \text{if } m = j, \\ f(m), & \text{otherwise.} \end{cases}$$

(Thus $[i{:}j]$ conditionally exchanges the contents of registers i and j, putting the smaller item into i and the larger item into j.)

Say that a program Q *sorts* X if Q is supported by X and X is sorted in fQ for every state f. Say that QA *sorts* (without further qualification) if Q sorts A.

Say that a program Q is *standard* if every instruction in Q is of the form $[i:j]$, with $i<j$. If Q sorts, then there is a standard program P that also sorts, where P has the same size and depth as Q (see [35, Subsection 5.3.4, Example 16]).

The problem of constructing efficient sorting networks was raised in a U.S. Patent issued to O'Connor and Nelson [49]; they observed that size $\binom{n}{2}$ suffices for all n and gave smaller programs for $n \leqslant 8$. Bose and Nelson [15], in 1962, were the first to present a nontrivial result for large n; they constructed networks of size $O(n^{\log_2 3})$ ($\log_2 3 = 1.585\ldots$). Hibbard [32], in 1963, gave another construction satisfying a similar bound. A few years later, in 1967, Floyd and Knuth [28] announced a scheme with size $O(n^{1+1/\sqrt{\log_4 n}} \log n)$. Still later, in 1968, Batcher [11] constructed networks of size $(n/4)(\log_2 n)^2 + O(n \log n)$. (Indeed, Batcher gave two schemes; if $n = 2^\nu$, they have depth $\binom{\nu+1}{2}$.) Many variants have been proposed since then by other authors, but none of these has even reduced the coefficient $\frac{1}{4}$ in the leading term.

The problem then is to resolve the gap between the lower bound of $n \log_2 n + O(n)$ and the upper bound of $(n/4)(\log_2 n)^2 + O(n \log n)$. (See [29], [35, Subsection 5.3.4] for surveys as of 1973.) The resolution of this problem, due to Ajtai, Komlós and Szemerédi, will be presented in Section 3.

In a sorter, there is a natural total ordering of the outputs. To see how a sorter serves as a connector that supports local routing, assign "addresses" to the targets in a way that is compatible with the natural total ordering of the outputs representing them. Let each source send the address of the target to which it requests a path. Then (assuming for simplicity that the network is homogeneous) if each comparator can compare the two addresses it receives and can route them accordingly, the paths followed by these addresses will establish the requested connections.

2.2. *The generalized connection problem*

Now suppose that each of the sources may or may not request a path to one of the targets, and suppose further that two or more sources may request paths to the same target. (Thus the assignment constitutes a partial, many-to-one mapping rather than a one-to-one correspondence.) We shall assume that the paths connecting several sources to a common target should constitute a tree with the inputs as its leaves and the output as its root, and that this tree should be vertex-disjoint from the trees rooted at other outputs. In this way, operations such as the simultaneous addition by several processors (sources) to the contents of a memory cell (target) can be implemented by the network if the components corresponding to vertices can perform the indicated additions.

In this situation, a network that supports global routing is called a "generalized connector". (Many discussions of generalized connectors interchange the roles of inputs and outputs.) One particularly straightforward strategy for constructing a generalized connector is to obtain it as the "cascade" interconnection of a connector with another network called a "generalizer": the inputs of the connector are the inputs of the generalized connector, the outputs of the connector are identified in a one-to-one fashion with the inputs of the generalizer, and the outputs of the generalizer are the outputs of the generalized connector. For each output y of the generalized connector, if m inputs $x_1, \ldots, x_m$ of the generalized connector request paths to y, the generalizer provides paths (constituting a tree, vertex-disjoint from the trees provided for other outputs) from some set $v_1, \ldots, v_m$ of its inputs to y. The connector prefaced to the generalizer then provides paths from its inputs $x_1, \ldots, x_m$ to the inputs $v_1, \ldots, v_m$ of the generalizer, through which they proceed to y.

The first construction for a generalized connector was given by Ofman [51] in 1965. This network, like an improved version discovered independently by Garmash and Shor [31] and by Thompson [74], is homogeneous, and is the cascade of a connector and a generalizer.

Further reductions in the size of generalizers have been given by Chung and Wong [18] and by Dolev, Dwork, Pippenger and Wigderson [25]; in [57], it is shown that there are n-generalizers with size $O(n)$ (rather than $O(n \log n)$, as in all the other cited results). None of these generalizers is homogeneous, and it is not known if homogeneous generalizers with linear size exist.

Let us now consider the minimum possible size of generalized n-connectors with fixed depth k. We inherit the lower bound $\Omega(n^{1+1/k})$ from connectors. An upper bound of $O(n^{1+1/k}(\log n)^{1-1/k})$, nearly matching the lower bound, is known (see [25, 26]), but only by nonconstructive arguments. Efforts to find corresponding explicit constructions fall into three classes. First there is a recursive construction using "concentrators" (see [46, 55]). This construction does not give a very good result: to obtain a size of $O(n^{1+1/k})$ requires a depth on the order of k^2 rather than k. Second there is the decomposition into a connector followed by a generalizer. The best explicitly known generalizers for this purpose are described in [25]; they result in generalized connectors of depth $3k-2$ and size $O(n^{1+1/k})$, and appear to be the best for practical values of n. Finally there are direct explicit constructions for generalized connectors; an early example is a network of depth 3 and size $O(n^{5/3})$ described by Masson and Jordan [45]. A direct construction with depth $2k-1$ and size $O(n^{1+1/k}(\log n)^{(k-1)/2})$ has been given by Kirkpatrick, Klawe and Pippenger [34]; Spencer [7] has pointed out that this can be improved to $O(n^{1+1/k})$, thus at least matching the bound for explicitly constructed (ungeneralized) connectors. The explicit constructions for connectors of Richards and Hwang [69] and Feldman, Friedman and Pippenger [26] in fact give generalized connectors with depth 2 and size $O(n^{5/3})$, so the preceding bound can also be extended to even depths, giving size $O(n^{1+2/(3k-1)})$ for depth $2k$.

A network that supports local routing in this more general situation might reasonably be called a "generalized sorter", though in fact no name appears to be attached to it in the literature. The problem of constructing such a generalized sorter can, however, be reduced to that of constructing sorters and one other type of network (which we shall call a "prefixer"). The first phase of the reduction is to eliminate the need to provide paths from "groups" containing more than one input to a common output; this is done by electing a "leader" for each group, deleting the other requests, and providing paths whereby the inputs other than the leader in a group can share the connection satisfying the leader's request. The second phase is to eliminate the asymmetry whereby certain inputs request paths to outputs and others do not; this is done by creating "dummy" requests for those inputs that would not otherwise request a path, either because they were not in a group or because they were not elected leader of their group. These dummy requests are for paths to precisely those outputs to which no input or group of inputs would otherwise request a path; it is easy to see that there are exactly enough such outputs to provide a distinct one for each dummy request. After this reduction, we are in a situation in which each input requests a path to a different output, wherein a sorter suffices to support local routing.

The key element in both phases of the reduction (aside from sorters) is a type of network called a "prefixer". To define this, we augment our notion of network by assuming that the inputs and the outputs are each totally ordered, and furthermore that the immediate predecessors of each vertex are totally ordered. Let the inputs be $x_0, \ldots, x_{n-1}$ in increasing order, and let the outputs be $y_0, \ldots, y_{n-1}$ in increasing order. Let $\xi_0, \ldots, \xi_{n-1}$ be n indeterminates, and associate with each input x_i the indeterminate ξ_i. We may now associate with each vertex other than an input a product of indeterminates (an element of the semigroup freely generated by $\xi_0, \ldots, \xi_{n-1}$) as follows. Let the immediate predecessors of the vertex v be the vertices $u_1, \ldots, u_k$ in increasing order, and suppose inductively that the products $\eta_1, \ldots, \eta_k$ have been associated with $u_1, \ldots, u_k$, respectively; then associate the product $\eta_1 \cdots \eta_k$ with v. In this way we arrive at an association of products with the outputs of the network. Such a network is called an *n-prefixer* if the products $\xi_0, \xi_0\xi_1, \ldots, \xi_0\xi_1 \cdots \xi_{n-1}$ are associated with the outputs $y_0, \ldots, y_{n-1}$, respectively.

The notion of an n-prefixer, with vertices of in-degree 2 but unrestricted out-degree, is implicit in Ofman's paper [50] of 1963; he gave a construction with size O(n) and depth O(log n), and observed that this implied similar bounds for simulating finite automata with Boolean circuits. The notion was rediscovered by Ladner and Fischer [36], who gave two constructions, one of which is homogeneous, and thus has out-degree 2. (When discussing prefixers, it is convenient to define a homogeneous network in terms of blocks containing three edges rather than four, so that a block is a 2-prefixer. By the *in-* or *out-degree* of a vertex we mean the number of edges directed into or out of it, respectively. If the in-degree of a network is 2, the out-degree can also be reduced to 2 without affecting size or depth by more than constant factors, but the resulting network is not necessarily homogeneous; see [33].) The possible combinations of size and depth have not yet been completely resolved (see [27, 71]).

The question of what size is compatible with a fixed depth (with both in- and out-degree unrestricted) was raised by Chandra, Fortune and Lipton [16, 17]. To state their results we shall need to define some slowly growing functions. Define

$$\log^* n = \min\{l \geq 0: \underbrace{\log \ldots \log}_{l} n \leq 1\},$$

where the logarithms are to base 2. By induction on k, define

$$\log^{\overbrace{*\cdots*}^{k}} n = \min\{l \geq 0: \underbrace{\log^{\overbrace{*\cdots*}^{k-1}} \ldots \log^{\overbrace{*\cdots*}^{k-1}}}_{l} n \leq 1\}.$$

(From now on, this expression will be denoted by $\log^{*\cdots*^{(k)}} n$.)

Chandra, Fortune and Lipton showed that for an n-prefixer with depth $2k \geqslant 2$, size $O(n \log^{*\cdots*^{(k-1)}} n)$ is sufficient and depth $\Omega(n \log^{*\cdots*^{(2k-2)}} n)$ is necessary. These results have direct application to Boolean circuits with unbounded fan-in for "prefix iteration" of an operation, when the operation is one that can be performed by the gates (Boolean conjunction or disjunction, for example). The upper bound carries over from prefixers to circuits, and the lower bound carries over if the circuits are monotone. Chandra, Fortune and Lipton [17] showed, however, that prefix-disjunction or prefix-conjunction can be computed by nonmonotone circuits with fixed depth and linear size, so there is no hope of applying this lower bound to nonmonotone circuits.

An important situation in which the operation is not one that can be performed by the gates is the propagation of carries in addition (see [64] for a history and survey of this problem). For this situation, and in fact for the more general one in which the operation is that of a "group-free" finite semigroup, Chandra, Fortune and Lipton [16] showed that for circuits of depth $6k \geqslant 6$, size $O(n(\log^{*\cdots*^{(k-1)}} n)^2)$ is sufficient, a bound only slightly more nonlinear than that for prefixers. If the circuits are monotone, the lower bound given above again applies; but addition cannot be performed by monotone circuits, so this result is not very interesting. It turns out, however, that an even better lower bound can be obtained with the aid of another type of communication network. This bound will be described in Section 2.4.

2.3. *Shifting, merging and classifying networks*

In this section we shall consider some types of networks that satisfy weaker conditions than those we have studied so far. The first of these is called a "shifter". To define this, we augment our notion of network by assuming that the inputs and outputs are each totally ordered; let the inputs be $x_0, \ldots, x_{n-1}$ in increasing order, and let the outputs be $y_0, \ldots, y_{n-1}$ in increasing order. We shall say that such a network is an *n-shifter* if for every $0 \leqslant k \leqslant n-1$, there exists a configuration containing vertex-disjoint paths joining each input x_i to the output y_j, where j is congruent to $i+k$ modulo n. (We have defined "cyclic" shifters, but obvious variants without "wrap-around" are possible, and have similar properties.)

Let X and Y be disjoint sets of registers. Say that X and Y are *ordered* if $i<j$ whenever $i \in X$ and $j \in Y$. Say X and Y are *classified* in a state f if $f(i) < f(j)$ whenever $i \in X$ and $j \in Y$. Say that f is *shifted* if the registers can be partitioned into two ordered sets X and Y such that X is sorted, Y is sorted, and Y and X are classified. There are exactly n shifted states, which are "cyclically shifted" versions of the unique state in which A is sorted.

Say that a regular program Q *shifts* if for every shifted state f there is a subsequence P of Q such that A is sorted in fP.

Pippenger and Valiant [65] have shown that if Q shifts, then $C(Q) \geqslant (n/2)\log_2 n + O(n)$, and thus $D(Q) \geqslant \log_2 n + O(1)$. (The proof considers the associated network, and does not require that it is homogeneous. If $n=2^\nu$, these bounds can be achieved; see [65].) Thus, although shifting is superficially a much simpler problem than connecting, the size and depth required by networks are smaller only by a constant factor, rather than by having a different order of growth. This phenomenon will be encountered again later.

The principal motivation for studying shifters is the frequent occurrence of shifting operations in computer hardware. In addition to instructions for explicit shifting, floating-point arithmetic instructions may involve shifting for alignment of operands and normalization of results. Thus the $O(n \log n)$ upper bound was discovered by many designers, and it is difficult to trace its history. For computer systems in which many processors are interconnected with many memory units, the use of shifters has been proposed to implement data distribution patterns that avoid access conflicts (see [37]). (The patterns proposed actually require "multiplicative" rather than "additive" shifts, but since a finite commutative group is a direct sum of cyclic groups, the multiplicative shifters can be constructed by interconnecting additive ones; see [23].)

Since shifters provide a convenient abstraction of the $O(n \log n)$ upper bound for shifting by Boolean circuits, it is natural to ask if they may be used to establish a corresponding lower bound. The answer is: for monotone circuits, yes; for nonmonotone circuits, no. Consider a Boolean circuit that takes n Boolean values $x_1, \ldots, x_n$ as inputs and "computes" these same values as outputs: $y_1 = x_1, \ldots, y_n = x_n$. If the circuit is monotone, we may conclude that in the underlying graph there are n vertex-disjoint paths, one joining each input x_i to the corresponding output y_i. The argument, given by Pippenger and Valiant [65], can be used to infer lower bounds for monotone circuits from point-to-point connection properties (such as that of being a shifter) which guarantee the existence of paths from particular inputs to particular outputs. If the circuit is nonmonotone, however, we may conclude that there are n vertex-disjoint paths joining the set of inputs to the set of outputs, but not that any particular input is joined thereby to any particular output. (The simplest example computes $z = x_1 \oplus x_2$, $y_1 = x_2 \oplus z$ and $y_2 = x_1 \oplus z$; then $y_1 = x_1$ and $y_2 = x_2$, but the unique paths from x_1 to y_1 and from x_2 to y_2 intersect at z.) Thus to infer lower bounds for nonmonotone circuits we must use set-to-set connection properties (some of which we shall encounter below), and the results available are much weaker.

Let us now consider the depth of shifters. The shifters obtained from the homogeneous construction have vertices of in-degree 2 and depth $O(\log n)$, and it is clear that if the in-degree is bounded, then depth $\Omega(\log n)$ is necessary. If we relax the restriction on in-degree, we may reduce the depth to any value we please. Pippenger and A.C. Yao [66] adapted the argument of Pippenger and Valiant [65] to show that an n-shifter with depth k must have size at least $kn^{1+1/k}$. An upper bound not much larger than this is easily obtained by expressing the amount of the shift as a k-digit number in radix $\lceil n^{1/k} \rceil$, and interconnecting k networks of depth 1, with one network devoted to each of the digits.

We shall now turn to "mergers" and "classifiers". We shall confine our attention to programs, since it is in this context that they are most frequently discussed.

Let X and Y be disjoint sets of registers. Say that a program Q *merges* X and Y if Q is supported by $X \cup Y$ and $X \cup Y$ is sorted in fQ whenever X and Y are sorted in f.

A scheme for constructing merging networks yields one for constructing sorting networks in the following way. Partition the registers in A into two ordered sets X and Y. Let $Q(X)$ be a program that sorts X (constructed by recursive application of the scheme being described), let $Q(Y)$ be a program that sorts Y, and let $Q(X, Y)$ be a program that merges X and Y. Then the program $Q(X)Q(Y)Q(X, Y)$ sorts.

All of the bounds for sorting networks described above were obtained by first obtaining bounds for merging networks, then using this scheme for constructing sorting networks. In particular, Batcher constructed networks that merge two sets each containing m registers with size $m \log_2 m + O(m)$. The recursive scheme then yields networks that sort n registers with size $(n/4)(\log_2 n)^2 + O(n \log n)$. If the bound for merging networks could be reduced to $O(m)$, the recursive scheme would yield $O(n \log n)$. Floyd has shown, however (see [35, Subsection 5.3.4, Theorem F]), that networks that merge two sets each containing m registers must have size at least $(m/2)\log_2 m + O(m)$. Thus sorting networks of size $O(n \log n)$ cannot be obtained through this strategy.

Let X and Y be disjoint sets of registers. Say that a program Q *classifies* X and Y if Q is supported by $X \cup Y$ and X and Y are classified in fQ for every state f.

A scheme for constructing classifying networks yields one for constructing sorting networks in a way dual to that for merging networks. Partition the registers in A into two ordered sets X and Y. Let $Q(X)$ be a program that sorts X (constructed by recursive application of the scheme being described), let $Q(Y)$ be a program that sorts Y, and let $Q(X, Y)$ be a program that classifies X and Y. Then the program $Q(X, Y)Q(X)Q(Y)$ sorts.

Just as for merging, if it were possible to construct networks that classify two sets each containing m registers with size $O(m)$, this recursive scheme would yield networks that sort n registers with size $O(n \log n)$. Alekseev [6] has shown, however, that networks that classify two sets each containing m registers must have size at least $(m/2)\log_2 m + O(m)$. Thus sorting networks of size $O(n \log n)$ cannot be obtained through this strategy.

2.4. *Other problems*

All of the assignments we have considered thus far have specified connections between particular inputs and particular outputs. Contrasting with these are assignments for networks such as “concentrators”, “hyperconcentrators” and “superconcentrators”, which specify only sets of inputs and outputs. Of these, we shall consider only superconcentrators; the others have similar properties. A network is an *n-superconcentrator* if, for every $1 \leqslant k \leqslant n$, every set X of k inputs and every set Y of k outputs, there exists a configuration containing k vertex-disjoint paths joining the inputs of X to the outputs of Y. (Note that nothing is said about which input of X is joined to which output of Y.)

One motivation for studying superconcentrators and their kin is their ability to perform resource allocation in computer hardware. Suppose that the sources are n processors connected to the inputs of an n-superconcentrator, and the targets are n memory devices connected to the outputs. If at some time k processors are working and k memories are working, the superconcentrator will provide a path from each working processor to a working memory device; it is assumed that the working processors or the working memory devices are all equivalent, so that it does not matter which is connected to which.

The problem of determining the minimum possible size of an n-superconcentrator was posed by Aho, Hopcroft and Ullman ([1, Problem 12.37]), who ascribed it to discussions with Floyd. Their hope was (see their Problem 12.38) that these networks might play a role in establishing a nonlinear lower bound for the Boolean circuit complexity of multiplication (of integers in binary), by first establishing a nonlinear lower bound for the size of superconcentrators, then showing that every Boolean circuit for multiplication must "contain" a superconcentrator. This hope was dashed by Valiant [75], who gave an upper bound of $O(n)$ (and coined the name "superconcentrators"). Though superconcentrators did not fulfil this initial hope, they were central to a number of important developments concerning communication networks, and they eventually did play a role in establishing lower bounds for Boolean circuit complexity.

Valiant's upper bound for superconcentrators is based on a recursive construction using another type of network, called a "concentrator", as a building block, and it relied on a linear upper bound for the size of concentrators due to Pinsker [54]. Pinsker's upper bound in turn is established by a nonconstructive argument; it shows that a graph having a certain property exists by constructing a graph according to a certain random process, then showing that with probability greater than zero, the resulting graph has the property. (One could always "construct" such a graph by exhaustively searching through all graphs of a certain size for one having the property, but this procedure gives no insight into the structure of the resulting graph and would take too long to be used in practice.) As superconcentrators were applied in theoretical computer science (see [53] for an early example), their "nonconstructive existence" cast a disquieting shadow on their applications.

Margulis [43] gave an "explicit construction" for concentrators that did not require exhaustive searching, but which had a different nonconstructive aspect: the coefficient in the linear bound was shown to exist, but no upper bound for it emerged from the argument. Finally Gabber and Galil [30] gave an alternative proof of Margulis' construction that gave an explicit coefficient, although a very large one. Much effort has been devoted to obtaining a practical version of this result, and the latest development in this direction (due to Lubotzky, Phillips and Sarnak [41], and Margulis [44]) will be described later in connection with sorting networks.

The explicit constructions for concentrators mentioned above all proceed by way of an even more basic building block called an "expander", and once explicit constructions for expanders were available, many other applications were found for them in areas not involving networks; see [5, 22, 62] for a variety of examples.

Let us now consider the depth of superconcentrators; it is this aspect that eventually led to their use in lower bounds for Boolean circuit complexity. Valiant's argument gives n-superconcentrators with size $O(n)$ and depth $O((\log n)^2)$; a simplification by Pippenger [56] reduced the depth to $O(\log n)$. The question of what size is compatible with a fixed depth was raised by Pippenger [63], who showed that for depth 2, size $\Omega(n \log n)$ is necessary and $O(n(\log n)^2)$ is sufficient. (As usual, for depth 1, size n^2 is trivially necessary and sufficient.) Pippenger's result was extended to arbitrary fixed depth by Dolev, Dwork, Pippenger and Wigderson [25], who showed that for fixed $k \geqslant 2$, size $\Theta(n \log^{*\cdots *^{(k-1)}} n)$ is necessary and sufficient for n-superconcentrators with depth $2k$.

We now come to the network that yields nonlinear lower bounds for nonmonotone Boolean circuits with fixed depth. Consider an n-network in which the union of the set of input and the set of outputs is totally ordered, with inputs and outputs alternating in this order, and with an input coming first. Say that a set of designated inputs and a set of designated outputs is *interleaved* if inputs and outputs alternate, with an input coming first, in the order induced on the union of these sets. Say that the network is a *weak superconcentrator* if, for any equinumerous and interleaved sets of designated inputs and designated outputs, there is a set of vertex-disjoint paths joining the designated inputs to the designated outputs. Chandra, Fortune and Lipton [16] observed that a circuit for carry propagation or addition of n-bit integers in binary must contain a weak n-superconcentrator in its underlying graph. Dolev, Dwork, Pippenger and Wigderson [25] showed, on the other hand, that a weak n-superconcentrator with depth $2k \geqslant 2$ must have size at least $\Omega(n \log^{*\cdots*^{(k-1)}} n)$. Thus, to add longer and longer numbers, even with unbounded fan-in, either depth must grow unboundedly, or size must grow nonlinearly. This result, one of the few that apply to nonmonotone circuits, may be viewed as a belated fulfilment of the hope with which superconcentrators were introduced.

The method just described has recently been applied to some other problems of Boolean circuit complexity. Pudlák and Savický [67] have shown that even nonmonotone shifting circuits with fixed depth must have nonlinear size. Bilardi and Preparata [13] have determined which of the group-free semigroups have operations that (like conjunction and disjunction) have nonmonotone prefix circuits with fixed depth and linear size, and which (like carry propagation) do not. (For semigroups that are not group-free, Chandra, Fortune and Lipton [16] observed that the known methods for parity circuits (see [14]) apply to show that prefix circuits with fixed depth cannot have polynomial size.)

The question of the minimum possible size of a homogeneous n-superconcentrator remains virtually untouched. It is known that $n \log_2 n + O(n)$ instructions are sufficient (since a connector is a superconcentrator), and that $2n + O(n^{1/2})$ are necessary (see [39]), but even the rate of growth remains unknown. (A homogeneous weak n-superconcentrator with linear size can be derived from an n-prefixer.)

Before concluding this survey, we should mention a degenerate type of communication network in which there is a fixed assignment that specifies which input–output pairs are to be connected by paths, and which are not (there is no requirement in this case that the paths be disjoint in any way). The minimum possible size now depends of course on the assignment, and a natural question is: what is the maximum (over assignments) of this minimum (over networks)? This question was raised by Lupanov [42] in 1956, and solved by him for the case of equal numbers of inputs and outputs. The answer is, asymptotically, $n^2/2 \log_2 n$. The method of proof suggests the conjecture that, if there are n inputs and m outputs, and if $(\log_2 n)/m$ and $(\log_2 m)/n$ both tend to 0, then the answer is asymptotic to $nm/\log_2(nm)$. A key step toward the proof of this conjecture was taken by Nečiporuk [47], and a complete proof was given by Pippenger [58] (see [59] for an application in algebraic complexity theory).

Another application of these degenerate networks arises for certain fixed assignments, and yields lower bounds for the complexity of monotone circuits for computing sets of disjunctions (or, dually, conjunctions) of the arguments. Say that an assignment is *k-free* if there does not exist a set X of k inputs and a set Y of k outputs for which the assignment requests k^2 paths joining each input in X to each output in Y. An assignment that requests l paths but is k-free yields a set of disjunctions or conjunctions for which every monotone circuit must have size at least $l/2k^3$ (see [60] for the proof). This relationship leads to the search for explicitly constructed assignments requesting many paths but being k-free for a value of k as small as possible. (We are only interested in explicitly constructed assignments because a nonconstructive proof is known for the existence of sets of disjunction or conjunctions requiring monotone circuits of size asymptotic to $n^2/2\log_2 n$; this is closely related to the result of Lupanov [42] mentioned above.) Nečiporuk [48] constructed a 2-free assignment requesting $\Omega(n^{3/2})$ paths, where n is again both the number of inputs and the number of outputs. Pippenger [60] extended this to $\Omega(n^{5/3})$ paths in a 3-free request, and both of these results are best possible. Andreev [8] has shown that for every $\varepsilon > 0$ there exists a k such that k-free assignments requesting $\Omega(n^{2-\varepsilon})$ can be explicitly constructed, but the question of the best possible relationship between ε and k remains unresolved.

3. The Ajtai, Komlós and Szemerédi Theorem

3.1. *Applications of the theorem*

This was the state of affairs at the outset of the 1980s: several methods were known for constructing sorting networks with size $(n/4)(\log_2 n)^2 + O(n \log n)$ (see [81] for the sharpest of these results); but the best lower bound known was $n\log_2 n + (n/2)\log_2\log_2 n + O(n)$ (see [79, 80]). Furthermore, the two most plausible avenues for progress were known to be blocked. So unlikely did it seem that the lower bound had the right order of growth, that Knuth presented as an open problem [35, Subsection 5.3.4, Example 51] to "Prove that the asymptotic value of $\hat{S}(n)$ [the minimum possible size of a network that sorts n registers] is not $O(n \log n)$."

The reader will therefore appreciate the sensation that attended the announcement by Ajtai, Komlós and Szemerédi of the existence of sorting networks for n registers with depth $O(\log n)$ (and therefore size $O(n \log n)$). Not only did their result solve an important open problem; but it yielded as corollaries the solutions to several other open problems as well. We have seen, for example, that networks that classify two sets each containing m registers must have size at least $(m/2)\log_2 m + O(m)$, and therefore depth at least $\log_2 m + O(1)$. But only with the announcement of the Ajtai, Komlós and Szemerédi result was it known that they could have depth $O(\log m)$, or even size $O(m \log m)$. (Simpler proof of these results are now known, but they are obtained by "pruning" the arguments of the sorting result to achieve more modest goals.)

Only from their result is it known that a network of n processors, each connected to only a bounded number of other processors, can sort n items in time $O(\log n)$. (See [38] for this result. A randomized algorithm achieving expected time $O(\log n)$ has been obtained by other methods; see [68].)

Only from their result is it known that networks of gates, of size $O(n(\log n)^2)$ and depth $O(\log n)$, can sort n binary numbers of length $O(\log n)$. Indeed, only from their result was it known that networks of monotone gates, of size $O(n \log n)$ and depth $O(\log n)$, can sort n binary numbers of length 1. (This last result was long known for nonmonotone networks. For monotone networks, a simpler proof has been found for the depth bound (see [76]), but none has been found for the size bound.)

Only from their result was it known that a decision tree performing n comparisons at each node can sort n items in time $O(\log n)$. Indeed, the minimum possible time for a decision tree that performs at most p comparisons at each node to sort n items is $\Theta((\log n)/\log(1+p/n))$ (see [9]. This last result is obtained by simulating an Ajtai, Komlós and Szemerédi network if $p=n$, slowing down the simulation for smaller p, and speeding it up for larger p. (A simpler proof is now known for the case $p=n$ (see [21]). This proof is based on merging and is, in a sense, dual to the proof of the result given here.)

The arguments of Ajtai, Komlós and Szemerédi have also been used by Ajtai, Komlós, Steiger and Szemerédi [2] to prove that a decision tree that performs n comparisons at each node can select the median of n items in time $O(\log \log n)$. As a final example, the result of Ajtai, Komlós and Szemerédi has been used by Cole [20] to obtain results concerning serial algorithms. We shall conclude this section with an account of the strategy of the Ajtai, Komlós and Szemerédi Theorem. The details will be presented in the next section.

Let X and Y be disjoint sets of registers, let ε be a nonnegative real number and let f be a state. Say that X and Y are *approximately classified* with *tolerance* ε in f if, for every $1 \leqslant m \leqslant |X|$, at most εm of the m smallest items in $f(X \cup Y)$ are in $f(Y)$, and for every $1 \leqslant m \leqslant |Y|$, at most εm of the m largest items in $f(X \cup Y)$ are in $f(X)$. Say that a program Q *approximately classifies* X and Y with *tolerance* ε if Q is supported by $X \cup Y$ and if X and Y are approximately classified with tolerance ε in fQ, for every state f.

The notion "approximately classified with tolerance 0" coincides with the notion "classified", but for $\varepsilon > 0$ the approximate notion is more liberal, as it allows for a fraction ε of mistakes. The key to the Ajtai, Komlós and Szemerédi strategy is their discovery that for every $\varepsilon > 0$ there exists a constant c such that for every m there is a network that approximately classifies two sets each containing m registers with tolerance ε and depth c (and therefore size at most $cm/2$).

This fact is an almost immediate consequence of the existence of "expanding graphs", which have played an important role in many recent results concerning communication and computation. Ajtai, Komlós and Szemerédi then showed how approximately classifying networks could be assembled to construct sorting networks. This construction is more complicated than the construction given above from classifying networks, and it constitutes the bulk of the proof.

3.2. Approximate classifying networks

We shall construct approximate classifiers from expanding graphs. Let us review the best currently known explicit construction for expanding graphs. (See [41, 73] for more details.)

Let p and q be primes congruent to 1 modulo 4. Let $\mathscr{P}$ denote the set of quadruples (a, b, c, d) such that $p = a^2 + b^2 + c^2 + d^2$ is a representation of p as a sum of four squares with a odd and positive. There are $p+1$ quadruples in $\mathscr{P}$. Let $\mathscr{Q} = \{0, 1, \ldots, q-1, \infty\}$ denote the projective line over the field with q elements. There are $q+1$ points in $\mathscr{Q}$. Let i denote a square root of -1 modulo q. Associate with each quadruple $(a, b, c, d) \in \mathscr{P}$ the transformation $T_{a,b,c,d}(x) = ((a+b\mathrm{i})x + (c+d\mathrm{i}))/((-c+d\mathrm{i})x + (a-b\mathrm{i}))$ (with obvious conventions regarding ∞).

Let X and Y be disjoint sets, each containing $q+1$ registers and each indexed by $\mathscr{Q}$. We shall construct a program Q that approximately classifies X and Y with tolerance $\varepsilon = 2p^{1/2}/(1 + 2p^{1/2} + p)$. The program Q will consist of $p+1$ subprograms $Q_{a,b,c,d}$, where (a, b, c, d) runs through $\mathscr{P}$. Each subprogram $Q_{a,b,c,d}$ will consist of $q+1$ instructions $[x\colon T_{a,b,c,d}(x)]$, where x runs through $\mathscr{Q}$.

Since each map $T_{a,b,c,d}$ is a bijection, the $q+1$ instructions in each subprogram $Q_{a,b,c,d}$ involve disjoint pairs of registers. Thus each subprogram $Q_{a,b,c,d,}$ has depth 1, and the program Q has depth $p+1$.

Let M denote the matrix whose rows are indexed by the registers in X, whose columns are indexed by the registers in Y, and whose (x, y) entry is the number of quadruples (a, b, c, d) such that $T_{a,b,c,d}(x) = y$. The matrix M is real and symmetric, since the transformation $T_{a,-b,-c,-d}$ is the inverse of the transformation $T_{a,b,c,d}$, so that $(a, -b, -c, -d)$ appears in $\mathscr{P}$ if and only if (a, b, c, d) does. Thus the eigenvalues of M are real. According to Lubotzky, Phillips and Sarnak [41], M has a simple eigenvalue $\lambda = p+1$, and all of its other eigenvalues are at most $\mu = 2p^{1/2}$ in absolute value.

If Z is a set of registers in X, let $Q(Z)$ denote the set of registers y in Y such that Q includes the instruction $[x\colon y]$ for some register x in Z. For $0 \leqslant \xi \leqslant 1$, let $\varphi(\xi) = \xi/(\xi + (\mu/\lambda)^2(1-\xi))$. Then according to the result of Tanner [73], $|Q(Z)|/(q+1) \geqslant \varphi(|Z|/(q+1))$. In particular, since φ is increasing, if $|Z| \geqslant \varepsilon(q+1)$, then $|Q(Z)| \geqslant \varphi(\varepsilon)(q+1) = (1-\varepsilon)(q+1)$. And for $0 \leqslant \vartheta \leqslant 1$, since φ is concave, if $|Z| \geqslant \vartheta\varepsilon(q+1)$, then $|Q(Z)| \geqslant \vartheta\varphi(\varepsilon)(q+1) = \vartheta(1-\varepsilon)(q+1)$.

We shall show that Q approximately classifies X and Y with tolerance ε. Let f be a state, let $u = \vartheta(q+1)$ and let the set U comprise the smallest u items in $f(\mathrm{X} \cup Y)$. If Q contains the instruction $[x\colon y]$, then after the execution of this instuction, the contents of x are smaller than the contents of y. Since the contents of registers in X can only decrease during the execution of Q, and the contents of registers in Y can only increase, the contents of x are also smaller than the contents of y in fQ. If more than εu registers of X were to contain items of U in fQ, then at least $(1-\varepsilon)u$ registers of Y would contain items of U in fQ, contradicting the fact that U contains only u items. Thus $(fQ)(X)$ contains at most εu items of U. Similarly, by virtue of the symmetry of Q, $(fQ)(Y)$ contains at most εu of the largest u items of $f(X \cup Y)$. This completes the proof that Q approximately classifies X and Y with tolerance ε.

Two small extensions of this construction will be needed later. Firstly the cardinality of X and Y might not be of the form $q+1$ with q a prime congruent to 1 modulo 4. If this cardinality is $m \geqslant 13$, however, we can find a prime q congruent to 1 modulo 4 such that $m \leqslant q < 2m$. If we now let X and Y be indexed by $\mathscr{Q}$ modulo m, each register of X or Y will correspond to either 1 or 2 points of $\mathscr{Q}$. The depth of Q is thus increased by a factor of at most 4, and the tolerance with which it classifies X and Y is increased by a factor of at most 2.

Secondly the cardinalities of X and Y might not be equal. We shall need later the cases in which they differ by a factor of k, where $k \in \{3, 4, 15, 20\}$. If X has cardinality m and Y has cardinality km, let Y be partitioned into k sets $Y_1, \ldots, Y_k$, each containing m registers. Let the program Q consist of k subprograms $Q_1, \ldots, Q_k$, each having depth c, where Q_j approximately classifies X and Y_j with tolerance ε. Then Q has depth at most kc and approximately classifies X and Y with tolerance at most $k\varepsilon$.

We have presented the best currently known explicit construction for expanding graphs, which yields approximate classifiers with tolerance ε and depth $O(1/\varepsilon^2)$. If we do not insist on an explicit construction, and use random expanding graphs (as does Paterson [52]), then this depth may be reduced to $O((1/\varepsilon)\log(1/\varepsilon))$.

3.3. *Register movements*

Let us assume that the number n of registers (and thus also the number of items to be sorted) is an integral power of 16, so that $n = 2^\nu$ for some integer $\nu \geqslant 8$ with $\nu \equiv 0 \pmod 4$. Let the (names for the) registers be the binary strings of length ν. Let $I = \{0, 1\}^\nu$ denote the set of registers.

Let $J = \bigcup_{0 \leqslant \mu \leqslant \nu} \{0, 1\}^\mu$ denote a set of binary strings that will be called *nodes*. The set J has the structure of a complete binary tree of depth ν. The empty node, which will be denoted Λ, is the *root*. Every other node x has a *father* x'. The nodes of length ν are the *leaves*. Every other node x has a *left son* $x0$ and a *right son* $x1$. The *level* of a node is given by its length $\ell(x)$. If $r \leqslant \ell(x)$, we shall write $x^{(r)}$ for the rth *ancestor* of x, defined by $x^{(0)} = x$ and $x^{(r)} = x'^{(r-1)}$. We have $x0' = x1' = x$ if x is not a leaf, $x'0 = x$ if x is a left son and $x'1 = x$ if x is a right son.

Let $T = \{0, 1, \ldots, 3(\nu - 5)\}$ denote a set of integers that will be called *times*. We shall define a certain "movement" whereby registers in I move from node to node in J as time passes in T. We shall define this movement in terms of a function $\xi: I \times T \to J$. We shall say that a register i is at node $\xi(i, t)$ at time t. All registers will start at the root; thus $\xi(i, 0) = \Lambda$ for all $i \in I$. Each register will move up or down one level at each time; thus $\xi(i, t+1) \in \{\xi(i, t)', \xi(i, t)0, \xi(i, t)1\}$.

For $x \in J$ and $t \in T$, let $I(x, t) = \{i \in I: \xi(i, t) = x\}$ denote the set of registers at node x at time t. We have

$$I(x, 0) = \begin{cases} I & \text{if } x = \Lambda, \\ \emptyset & \text{otherwise.} \end{cases}$$

Furthermore $I(x, t) = \emptyset$, unless $\ell(x) \equiv t \pmod 2$.

Define $I'(x, t) = \{i \in I(x, t): \xi(i, t+1) = x'\}$, $I^0(x, t) = \{i \in I(x, t): \xi(i, t+1) = x0\}$ and $I^1(x, t) = \{i \in I(x, t): \xi(i, t+1) = x1\}$ to be the sets of registers at node x at time t that move to the father, the left son and the right son, respectively, at time $t+1$. Then we have

$$I(x, t) = I'(x, t) \cup I^0(x, t) \cup I^1(x, t)$$

and

$$I(x,t+1)=\begin{cases} I'(x0,t)\cup I'(x1,t) & \text{if } x=\Lambda,\\ I^{\sigma}(x',t) & \text{if } x \text{ is a leaf},\\ I'(x0,t)\cup I'(x1,t)\cup I^{\sigma}(x',t) & \text{otherwise},\end{cases}$$

where $x=x'\sigma$ (and the unions are disjoint).

Let $m(x,t)$, $m'(x,t)$, $m^0(x,t)$ and $m^1(x,t)$ denote the cardinalities of $I(x,t)$, $I'(x,t)$, $I^0(x,t)$ and $I^1(x,t)$, respectively. We shall begin by assigning these number so that

$$m(x,0)=\begin{cases} n & \text{if } x=\Lambda,\\ 0 & \text{otherwise},\end{cases}$$

$$m(x,t)=m'(x,t)+m^0(x,t)+m^1(x,t)$$

and

$$m(x,t+1)=\begin{cases} m'(x0,t)+m'(x1,t) & \text{if } x=\Lambda,\\ m^{\sigma}(x',t) & \text{if } x \text{ is a leaf},\\ m'(x0,t)+m'(x1,t)+m^{\sigma}(x',t) & \text{otherwise},\end{cases}$$

where $x=x'\sigma$.

For $x\in J$ and $t\in T$, let $n(x,t)=2^{\nu+2\ell(x)-t}$. Suppose that we were to have $m(x,t)=63\cdot 2^{-6}n(x,t)$ for every node x with $\ell(x)\equiv t \pmod 2$. Then we could set $m'(x,t)=3\cdot 2^{-7}n(x,t)$ and $m^0(x,t)=m^1(x,t)+30\cdot 2^{-7}n(x,t)$; we would then have $m(x,t+1)=63\cdot 2^{-7}n(x,t+1)$ for every node x with $\ell(x)\equiv t+1 \pmod 2$ that is not a root or a leaf, since $3n(x0,t)+3n(x1,t)+30n(x',t)=63n(x,t+1)$. This simple idea requires some modifications, not only for the boundary conditions (the initial time, and the root and leaves of the tree), but also to ensure that the numbers of registers moving from one node to another are always nonnegative integers (we shall, in fact, make them nonnegative even integers).

We shall associate with each time t a top level $a(t)$ and a bottom level $b(t)$ such that $a(t), b(t)\equiv t \pmod 2$. We shall set $m(x,t)=0$ if $\ell(x)<a(t)$ or $\ell(x)>b(t)$. Specifically, we shall take

$$a(t)=\left\lfloor\frac{t+1}{2}\right\rfloor-\left\lfloor\frac{t}{2}\right\rfloor,$$

if $0\leqslant t\leqslant \nu-3$, and

$$a(t)=\left\lfloor\frac{t-\nu-1}{4}\right\rfloor+\left\lfloor\frac{t-\nu-2}{4}\right\rfloor+\left\lfloor\frac{t-\nu-3}{4}\right\rfloor-\left\lfloor\frac{t-\nu-4}{4}\right\rfloor,$$

if $\nu-4\leqslant t\leqslant 3(\nu-5)$. The level $a(t)$ alternates between 0 and 1 until $t=\nu-4$, when it begins to descend in the pattern "down-down-down-up", with an average rate of descent of one level every two time steps. We shall take

$$b(t)=\left\lfloor\frac{t+2}{3}\right\rfloor+\left\lfloor\frac{t+1}{3}\right\rfloor-\left\lfloor\frac{t}{3}\right\rfloor.$$

The level $b(t)$ descends in the pattern "down-down-up", with an average rate of descent of one level every three time steps. (The expressions for $a(t)$ and $b(t)$ can be rewritten so as to contain at most two "floors" in each case; the expressions used here have been chosen to make the motion as clear as possible.)

Define

$$\alpha(x,t)=\begin{cases}1 & \text{if } \ell(x)=a(t),\\ 0 & \text{otherwise.}\end{cases}$$

Define

$$\beta(x,t)=\begin{cases}3 & \text{if } \ell(x)=b(t) \text{ and } t\equiv 2 \pmod 3,\\ 15 & \text{if } \ell(x)=b(t) \text{ and } t\equiv 1 \pmod 3,\\ 63 & \text{otherwise.}\end{cases}$$

We shall set

$$m(x,t)=\begin{cases}(\beta(x,t)+\alpha(x,t))2^{-6}n(x,t) & \text{if } a(t)\leqslant\ell(x)\leqslant b(t) \text{ and } \ell(x)\equiv t \pmod 2,\\ 0 & \text{otherwise.}\end{cases}$$

Define

$$\gamma(x,t)=\begin{cases}0 & \text{if } \ell(x)=a(t) \text{ and } a(t+1)=a(t)+1,\\ 4 & \text{if } \ell(x)=a(t) \text{ and } a(t+1)=a(t)-1,\\ 3 & \text{otherwise.}\end{cases}$$

We shall set

$$m'(x,t)=\begin{cases}\gamma(x,t)2^{-6}n(x,t) & \text{if } a(t)\leqslant\ell(x)\leqslant b(t) \text{ and } \ell(x)\equiv t \pmod 2,\\ 0 & \text{otherwise,}\end{cases}$$

and

$$m^0(x,t)=m^1(x,t)=2^{-1}(m(x,t)-m'(x,t)).$$

It is straightforward, though tedious, to verify from these definitions that $m(x,t)$, $m'(x,t)$, $m^0(x,t)$ and $m^1(x,t)$ are nonnegative even integers that satisfy the conditions recited previously. As an example, the complete table of $m(x,t)$ for $n=4096$ is shown in Fig. 1. Only the entry 252 at level 2 and time 8, and the entry 126 at level 3 and time 11 are "generic"; the others illustrate various special cases.

We shall need some additional properties of these numbers. Firstly since $m^0(x,t)=m^1(x,t)$, it follows that $m(x,t)$ depends on x only through $\ell(x)$, so that nodes on the same level will have the same number of registers. Secondly, we have

$$\sum_{0\leqslant r\leqslant\ell(x)} 2^{-r}m(x^{(r)},t)\leqslant n(x,t) \tag{3.1}$$

if $\ell(x)\equiv t \pmod 2$, as is easily verified from the definitions.

For $x\in J$, let $J(x)$ denote the subtree rooted at x, the nodes of which are the strings beginning with x. Let $M(x,t)=\sum_{y\in J(x)}m(y,t)$ denote the number of registers in this

	0	1	2	3	4	5	6	7	8	9	10	11	12	13	14	15	16	17	18	19	20	21
0	4096		1024		256		64		16													
1		2048		2048		512		128		32		8										
2			768		960		1008		252		64		16									
3						384		480		504		126		32		8						
4									192		240		252		64		16					
5												96		120		126		32		8		
6															48		60		64		16	
7																		24		30		32
8																					12	

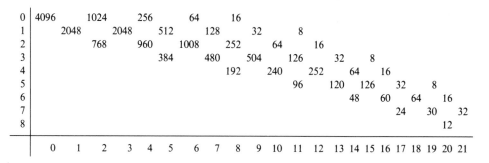

Fig. 1. $m(x, t)$ as a function of t (horizontally) and $\ell(x)$ (vertically), for $n=4096$ elements.

subtree at time t. Since $M(x,t)$ depends on x only through $\ell(x)$, we have $M(x,t) \leqslant 2^{\nu-\ell(x)}$, since at most 2^ν registers are distributed among $2^{\ell(x)}$ subtrees. We shall also need a converse inequality,

$$2^{\nu-\ell(x)} - M(x,t) \leqslant 2^{-1}m(x',t) + 2^{-7}n(x',t), \tag{3.2}$$

which holds if $\ell(x) \not\equiv t \pmod 2$. To see this, observe that

$$2^{\nu-\ell(x)} - M(x,t) = \sum_{1 \leqslant r \leqslant \ell(x)} 2^{-r} m(x^{(r)}, t),$$

since each side depends on x only through $\ell(x)$, and the sum of either side over all nodes x with $\ell(x)=q$ is the number of registers at nodes with levels at most $q-1$ at time t. The terms with $r \equiv 0 \pmod 2$ vanish, since $\ell(x) \not\equiv t \pmod 2$. Splitting off the term $r=1$ and applying (3.1) to the remaining terms yields (3.2).

Finally at time $t=3(\nu-5)$, we have $a(t)=b(t)=\nu-5$. Thus all registers are at level $\nu-5$, with $2^5=32$ registers at each node on this level.

We have specified the numbers $m(x,t)$, $m'(x,t)$, $m^0(x,t)$ and $m^1(x,t)$ for each node x and time t. We shall now specify the contents of the sets $I(x,t)$, $I'(x,t)$, $I^0(x,t)$ and $I^1(x,t)$. We shall do this by means of functions $j(x,t,p)$, $j'(x,t,p)$, $j^0(x,t,p)$ and $j^1(x,t,p)$ that assume registers as values and run through the sets $I(x,t)$, $I'(x,t)$, $I^0(x,t)$ and $I^1(x,t)$ as p runs through the sets $[m(x,t)]$, $[m'(x,t)]$, $[m^0(x,t)]$ and $[m^1(x,t)]$, respectively. Here, $[m]$ denotes the set $\{0, 1, \ldots, m-1\}$.

For later convenience, we shall set $m'_0(x,t)=m'_1(x,t)=2^{-1}m'(x,t)$ and write $I'(x,t)=I'_0(x,t) \cup I'_1(x,t)$, with $I'_0(x,t)$ and $I'_1(x,t)$ containing $m'_0(x,t)$ and $m'_1(x,t)$ registers, respectively. (Since $m'(x,t)$ is an even integer, $m'_0(x,t)$ and $m'_1(x,t)$ are integers.) We shall define $j'_0(x,t,p)$ and $j'_1(x,t,p)$ to run through $I'_0(x,t)$ and $I'_1(x,t)$, respectively, as p runs through $[m'_0(x,t)]$ and $[m'_1(x,t)]$, respectively.

We shall define these functions by induction on t. For time 0, we define $j(\Lambda,0,p)$ to run through $I=I(\Lambda,0)$ in increasing lexicographic order. Suppose now that $j(x,t,p)$ has been defined for all x and some t. We define $j'_0(x,t,p)$ to run through the first $m'_0(x,t)$ values assumed by $j(x,t,p)$, $j^0(x,t,p)$ to run through the next $m^0(x,t)$, $j^1(x,t,p)$ to run through the next $m^1(x,t)$ and $j'_1(x,t,p)$ to run through the last $m'_1(x,t)$, in each case in the same order as $j(x,t,p)$. We define $j'(x,t,p)$ to run first through the values assumed by $j'_0(x,t,p)$, then through the values assumed by $j'_1(x,t,p)$. Finally, we define $j(x,t+1,p)$ to run first through the values assumed by $j'(x0,t,p)$ (with this case omitted if x is a leaf), then through the values assumed by $j^{\sigma}(x',t,p)$ (where $x=x'\sigma$ and this case is omitted if $x=\Lambda$) and then through the values assumed by $j'(x1,t,p)$ (with this case omitted if x is a leaf).

For later convenience, we shall set $I_0(x,t)=I'_0(x,t)\cup I^0(x,t)$ and $I_1(x,t)=I'_1(x,t)\cup I^1(x,t)$, so that $I(x,t)=I_0(x,t)\cup I_1(x,t)$.

3.4. *Construction of the network*

The program Q that we construct will consist of $3(v-5)+1$ subprograms Q_t, where t runs through T. First consider $0\leqslant t\leqslant 3(v-5)-1$. Each subprogram Q_t will consist of $2^{v+1}-1$ further subprograms $Q_{t,x}$, where x runs through J. Each subprogram $Q_{t,x}$ will consist of two further subprograms $F_{t,x}$ and $G_{t,x}$. The subprogram $F_{t,x}$ will approximately classify the sets $I_0(x,t)$ and $I_1(x,t)$, with tolerance 2^{-7}. The subprogram $G_{t,x}$ will consist of two further subprograms $G_{t,x,0}$ and $G_{t,x,1}$. The subprogram $G_{t,x,0}$ will approximately classify the sets $I'_0(x,t)$ and $I^0(x,t)$, and the subprogram $G_{t,x,1}$ will approximately classify the sets $I'_1(x,t)$ and $I^1(x,t)$, in each case with tolerance 2^{-7}.

The subprogram $Q_{3(v-5)}$ will consist of 2^{v-5} further subprograms $Q_{3(v-5),x}$, where x runs through the nodes on level $v-5$. The subprogram $Q_{3(v-5),x}$ will sort the 32 items in registers at x.

Let us consider the depth of the program Q. Those subprograms $F_{t,x}$, $G_{t,x,0}$ and $G_{t,x,1}$ that approximately classify two sets have depths bounded by a constant, which will be denoted c. Since $G_{t,x,0}$ and $G_{t,x,1}$ involve disjoint sets of registers, $G_{t,x}$ also has depth at most c. Thus each subprogram $G_{t,x}$ has depth at most $2c$. Since for each $0\leqslant t\leqslant 3(v-5)-1$, the subprograms $G_{t,x}$ involve disjoint sets of registers, Q_t also has depth at most $2c$.

By the result of Batcher [11], each subprogram $Q_{3(v-5),x}$ can be constructed so as to have depth 15. Since these subprograms involve disjoint sets of registers, $Q_{3(v-5)}$ also has depth 15. Thus Q has depth at most $3(v-5)2c+15=O(v)=O(\log n)$.

It remains to prove that Q sorts.

3.5. *Local arguments*

Let us assume that the items to be sorted are, like the registers, the binary strings of length v. Say that an item y is *correct* for a node x if x is a prefix of y. Say that y is *wrong* for x if it is not correct for x. If y is wrong for x, then it is either too small for x or too large for x.

Let U be a set of u items at node x at time t. Say U is a *left segment* or *right segment* if it comprises the smallest or largest, respectively, u of the items at x at t. Say that a left segment or right segment is a *left fringe* or *right fringe*, respectively, if it contains at most $2^{-1}m'(x,t)$ items.

3.1. LEMMA. *Let U be a set containing at most $3\cdot 2^{-7}n(x,t)$ items at x at t. If U is a left segment whose items are too small for x or a right segment whose items are too large for x, then U is a left fringe or right fringe, respectively.*

PROOF. Let U contain u items. We must show that $u \leqslant 2^{-1}m'(x,t)$.

If $x=\Lambda$, then $u=0$, since no item is too small or too large for the root. If $n(x,t)\leqslant 32$, then again $u=0$, since $u\leqslant 3\cdot 2^{-7}n(x,t)\leqslant 3\cdot 2^{-2}<1$. Thus in either of these cases, $u\leqslant 2^{-1}m'(x,t)$. But otherwise we have $m'(x,t)\geqslant 3\cdot 2^{-6}n(x,t)$, and again $u\leqslant 2^{-1}m'(x,t)$. □

3.2. LEMMA. *If U is a left fringe of u items at x at t, then at most $2^{-7}u$ items of U move to $x0$ at $t+1$.*

PROOF. After execution of $F_{t,x}$, at most $2^{-7}u$ items of U are in registers of $I^0(x,t)$, and only items in these registers move from x to $x0$ at $t+1$. □

Say that a right segment is a *right range* if it contains at most $2^{-2}m(x,t)$ items. Since $m'(x,t)\leqslant m(x,t)$, a right fringe is a right range.

3.3. LEMMA. *If U is a right range of u items at x at t, then at most $2^{-7}u$ items of U move to $x0$ at $t+1$.*

PROOF. After execution of $G_{t,x,0}$, at most $2^{-7}u$ items of U are in registers of $I_0(x,t)$, and only items in these registers move from x to $x0$ at $t+1$. □

3.4. LEMMA. *Suppose U is a right segment containing u items. Then at most $2^{-7}\min\{u, 2^{-1}m(x,t)\}+\max\{0, u-2^{-1}m(x,t)\}$ items of U move to $x0$ at $t+1$.*

PROOF. Let U be the disjoint union of a right range V containing $\min\{u, 2^{-1}m(x,t)\}$ items and a set containing $\max\{0, u-2^{-1}m(x,t)\}$ items. Applying Lemma 3.3 to V, we reach the desired conclusion. □

3.5. LEMMA. *If U, containing at most u items, is the disjoint union of a left fringe and a right fringe, then at most $2^{-7}u$ items of U move to $x0$ at $t+1$.*

PROOF. Let U be the disjoint union of a left fringe V with v items and a right fringe W with w items, so that $v+w\leqslant u$. Applying Lemma 3.2 to V and Lemma 3.3 to W, we reach the desired conclusion. □

3.6. LEMMA. *If U, containing at most $2^{-1}m(x,t)+u$ items, is the disjoint union of a left fringe and a right segment, then at most $2^{-8}m(x,t)+u$ items of U move to $x0$ at $t+1$.*

PROOF. Let U be the disjoint union of a left fringe V with v items and a right segment W containing w items, so that $v+w\leqslant 2^{-1}m(x,t)+u$. Applying Lemma 3.2 to V and

Lemma 3.4 to W, we have that at most $2^{-7}v+2^{-7}\min\{w,2^{-1}m(x,t)\}+\max\{0, w-2^{-1}m(x,t)\}\leqslant 2^{-8}m(x,t)+u$ items of U move to $x0$ at $t+1$. □

3.6. The global argument

For $r\geqslant 1$, define $S_r(x,t)$ to be the number of items in (registers in) $J(x)$ at time t that are wrong for $x^{(r-1)}$ (or 0, if $r-1>\ell(x)$). The key to the theorem is the following estimate:

$$S_r(x,t)\leqslant 3\cdot 2^{-7r}n(x,t), \tag{3.3}$$

for all x such that $\ell(x)\equiv t \pmod 2$. If $\ell(x)\not\equiv t \pmod 2$, we have

$$\begin{aligned} S_r(x,t) &= S_{r+1}(x0,t)+S_{r+1}(x1,t) \\ &\leqslant 3\cdot 2^{-7(r+1)}(n(x0,t)+n(x1,t)) \\ &= 3\cdot 2^{-7r-4}n(x,t) \end{aligned} \tag{3.4}$$

as a corollary of (3.3).

We shall prove (3.3) by induction on t. The estimate (3.3) holds for $t=0$, since then all items are at the root, and no item is wrong for the root. To complete the induction, we shall assume (3.3) and prove

$$S_r(x,t+1)\leqslant 3\cdot 2^{-7r}n(x,t+1)$$

for $\ell(x)\equiv t+1 \pmod 2$. This estimate holds for $x=\Lambda$, since no item is wrong for the root. Thus we may assume that x is of the form $y0$ or $y1$. We shall prove

$$S_r(y0,t+1)\leqslant 3\cdot 2^{-7r}n(y0,t+1), \tag{3.5}$$

the case of $y1$ being analogous.

Consider the items in $J(y0)$ at time $t+1$ that are wrong for $y0^{(r-1)}$, where $\ell(y0)\equiv t+1 \pmod 2$. These items are of two types: those that were in $J(y0)$ at time t, and those that were at y at time t. The number $Q_r(y0,t+1)$ of items of the first type can be estimated by inductive hypothesis, applying (3.4) to $y0$:

$$\begin{aligned} Q_r(y0,t+1) &= S_r(y0,t)\leqslant 3\cdot 2^{-7r-4}n(y0,t) \\ &= 3\cdot 2^{-7r-3}n(y0,t+1). \end{aligned}$$

Below we shall show that the number $R_r(y0,t+1)$ of items of the second type satisfies

$$R_r(y0,t+1)\leqslant 21\cdot 2^{-7r-3}n(y0,t+1). \tag{3.6}$$

Adding the last two inequalities will then complete the proof of (3.5).

Say that an item is a *candidate* if it is at y at time t and is wrong for $y0^{(r-1)}$.

Consider first the case $r\geqslant 2$. Candidates must be in $J(y)$ at time t and be wrong for $y^{(r-2)}$. Thus the number $C_r(y0,t+1)$ of candidates can be estimated by inductive hypothesis, applying (3.3) to y:

$$C_r(y0,t+1)\leqslant S_{r-1}(y,t)\leqslant 3\cdot 2^{-7(r-1)}n(y,t).$$

Since $r\geqslant 2$, there are at most $3\cdot 2^{-7}n(y,t)$ candidates. Since candidates are wrong for y, they are either too small or too large for y. Those that are too small for y form a left segment and, by Lemma 3.1, a left fringe. Those that are too large for y form a right segment and, by Lemma 3.1, a right fringe. Thus, by Lemma 3.5, we have

$$\begin{aligned} R_r(y0, t+1) &\leqslant 2^{-7}C_r(y0, t+1) \leqslant 3 \cdot 2^{-7}n(y, t) \\ &= 3 \cdot 2^{-7r-1}n(y0, t+1). \end{aligned}$$

Since $3 \cdot 2^{-1} < 21 \cdot 2^{-3}$, this completes the proof of (3.6) for $r \geqslant 2$.

We consider now the case $r=1$. Candidates are of two types: those that are wrong for y, and those that are correct for y but wrong for $y0$. The number $A(y0, t+1)$ of candidates of the first type can be estimated by inductive hypothesis, applying (3.3) to y:

$$A(y0, t+1) \leqslant S_1(y, t) \leqslant 3 \cdot 2^{-7}n(y, t).$$

Candidates of the second type are correct for y but wrong for $y0$, and thus are correct for $y1$. The number of items that are correct for $y1$ is $2^{v-\ell(y1)}$. Of these $M(y1, t) - S_1(y1, t)$ are in $J(y1)$ at time t. Since they are not at y they are not candidates. Thus the number $B(y0, t+1)$ of candidates of the second type satisfies

$$B(y0, t+1) \leqslant 2^{v-\ell(y1)} - M(y1, t) + S_1(y1, t).$$

The difference between the first two terms can be estimated by applying (3.2) to $J(y1)$:

$$2^{v-\ell(y1)} - M(y1, t) \leqslant 2^{-1}m(y, t) + 2^{-7}n(y, t).$$

The remaining term can be estimated by inductive hypothesis, applying (3.4) to $y1$:

$$S_1(y1, t) \leqslant 3 \cdot 2^{-11}n(y1, t) = 3 \cdot 2^{-9}n(y, t).$$

Adding the estimates for the two types of candidates yields

$$C_1(y0, t+1) = A(y0, t+1) + B(y0, t+1) \leqslant 2^{-1}m(y, t) + 19 \cdot 2^{-9}n(y, t).$$

Candidates are either too small or too large for $y0$. Candidates that are too small for $y0$ are also too small for y. They form a left segment at y and are at most $3 \cdot 2^{-7}n(y, t)$ in number. Thus, by Lemma 3.1, they form a left fringe. Candidates that are too large for $y0$ form a right segment at y. Thus, using Lemma 3.6 and $m(y, t) \leqslant n(y, t)$, we have

$$\begin{aligned} R_1(y0, t+1) &\leqslant 2^{-8}m(y, t) + 19 \cdot 2^{-9}n(y, t) = 21 \cdot 2^{-9}n(y, t) \\ &= 21 \cdot 2^{-10}n(y0, t+1). \end{aligned}$$

This completes the proof of (3.6) for $r=1$, and thus completes the proof of (3.3).

Finally, we apply (3.3) for $t = 3(v-5)$. At this time, all items are at level $v-5$, with 32 items at each such node. Inequality (3.3) tells us that $S_1(x, 3(v-5)) \leqslant 3 \cdot 2^{-2} < 1$ for each such node x. Thus each item is correct for the node at which it resides, and all that remains is to sort the 32 items at each node. This is accomplished by the subprogram $Q_{3(v-5)}$. Thus the program Q sorts.

Acknowledgment

The author is indebted to Miklós Ajtai, János Komlós and Endre Szemerédi for showing that it could be done; to Michael Paterson, for showing how easily; and to Donald Knuth, for communicating the unpublished details of his work with Robert Floyd.

References

[1] AHO, A.V., J.E. HOPCROFT and J.D. ULLMAN, *The Design and Analysis of Computer Algorithms* (Addison-Wesley, Reading, MA, 1974).

[2] AJTAI, M., J. KOMLÓS, W.L. STEIGER and E. SZEMERÉDI, Deterministic selection in O(log log n) time, in: *Proc. 18th Ann. ACM Symp. on Theory of Computing* (1986) 188–195.

[3] AJTAI, M., J. KOMLÓS and E. SZEMERÉDI, Sorting in c log n parallel steps, *Combinatorica* **3** (1983) 1–19.

[4] AJTAI, M., J. KOMLÓS and E. SZEMERÉDI, An O(n log n) sorting network, in: *Proc. 15th Ann. ACM Symp. on Theory of Computing* (1983) 1–9.

[5] AJTAI, M., J. KOMLÓS and E. SZEMERÉDI, Deterministic simulation in LOGSPACE, in: *Proc. 17th Ann. ACM Symp. on Theory of Computing* (1987) 132–140.

[6] ALEKSEEV, V.E., Sorting algorithms with minimum memory, *Kibernetica* **5** (1969) 99–103.

[7] SPENCER, J.H., Personal communication, 1985.

[8] ANDREEV, A.E., On a family of Boolean matrices, *Vestnik Moskov. Univ. Mat. Mekh.* **2** (1986) 97–100.

[9] AZAR, Y. and U. VISHKIN, Tight comparison bounds on the complexity of parallel sorting, *SIAM J. Comput.* **16** (1987) 458–464.

[10] BASSALYGO, L.A., Asymptotically optimal switching circuits, *Problems Inform. Transmission* **17** (1981) 206–211.

[11] BATCHER, K.E., Sorting networks and their applications, in: *Proc. 32nd Ann. AFIPS Spring Joint Computer Conf.* (1968) 307–314.

[12] BENEŠ, V.E., Optimal rearrangeable multistage connecting networks, *Bell Systems Tech. J.* **43** (1964) 1641–1656.

[13] BILARDI, G. and F.P. PREPARATA, Characterization of associative operations with prefix circuits of constant depth and linear size, to appear.

[14] BOPPANA, R.B. and M. SIPSER, The complexity of finite functions, in: J. van Leeuwen, ed., *Handbook of Theoretical Computer Science, Vol. A* (North-Holland, Amsterdam, 1990) 757–804.

[15] BOSE, R.C. and R.J. NELSON, A sorting problem, *J. ACM* **9** (1962) 282–296.

[16] CHANDRA, A.K., S.J. FORTUNE and R.J. LIPTON, Unbounded fan-in circuits and associative functions, in: *Proc. 15th Ann. ACM Symp. on Theory of Computing* (1983) 52–60.

[17] CHANDRA, A.K., S.J. FORTUNE and R.J. LIPTON, Lower bounds for constant depth monotone circuits for prefix functions, in: *Proc. 10th Internat. Coll. on Automata, Languages and Programming*, Lecture Notes in Computer Science, Vol. 154 (Springer, Berlin, 1983) 109–117.

[18] CHUNG, K.M. and C.K. WONG, Construction of a generalized connector with $5.8n \log_2 n$ edges, *IEEE Trans. Comput.* **29** (1980) 1029–1032.

[19] CLOS, C., A study of non-blocking switching networks, *Bell Systems Tech. J.* **32** (1953) 406–424.

[20] COLE, R., Slowing down sorting networks to obtain faster sorting algorithms, *J. ACM* **34** (1987) 200–208.

[21] COLE, R., Parallel merge sort, *SIAM J. Comput.* **17** (1988) 770–785.

[22] COLE, R. and U. VISHKIN, Approximate and exact parallel scheduling with applications to list, tree and graph problems, in: *Proc. 27th Ann. IEEE Symp. on Foundations of Computer Science* (1986) 478–491.

[23] DAVIO, M. and C. RONSE, Rotator design, *Discr. Appl. Math.* **5** (1983) 253–277.

[24] DE BRUIJN, N.G., P. ERDŐS and J. SPENCER, Solution 350, *Nieuw Arch. Wisk.* **22** (1974) 94–109.

[25] DOLEV, D., C. DWORK, N. PIPPENGER and A. WIGDERSON, Superconcentrators, generalizers and

generalized connectors with limited depth, in: *Proc. 15th Ann. ACM Symp. on Theory of Computing* (1983) 42–51.

[26] FELDMAN, P., J. FRIEDMAN and N. PIPPENGER, Wide-sense nonblocking networks, *SIAM J. Discr. Math.* **1** (1988) 158–173.

[27] FICH, F.E., New bounds for parallel prefix circuits, in: *Proc. 15th Ann. ACM Symp. on Theory of Computing* (1983) 100–109.

[28] FLOYD, R.W. and D.E. KNUTH, Improved constructions for the Bose–Nelson sorting problem, *Notices AMS* **14** (1967) 283.

[29] FLOYD, R.W. and D.E. KNUTH, The Bose–Nelson sorting problem, in: J.N. Srivastava et al., eds., *A Survey of Combinatorial Theory* (North-Holland, Amsterdam, 1973) 163–172.

[30] GABBER, O. and Z. GALIL, Explicit construction of linear-sized superconcentrators, *J. Comput. System Sci.* **22** (1981) 407–420

[31] GARMASH, V.A. and L.A. SHOR, Multistage one-shot batch switching arrangements, *Problems Inform. Transmission* **13** (1978) 320–323.

[32] HIBBARD, T.N., A simple sorting algorithm, *J. ACM* **10** (1963) 142–150.

[33] HOOVER, J.H., M.M. KLAWE and N.J. PIPPENGER, Bounding fan-out in logical networks, *J. ACM* **31** (1984) 13–18.

[34] KIRKPATRICK, D.G., M. KLAWE and N. PIPPENGER, Some graph-colouring theorems with applications to generalized connection networks, *SIAM J. Algebraic Discrete Methods* **6** (1985) 576–582.

[35] KNUTH, D.E., *The Art of Computer Programming, Vol. 3: Sorting and Searching* (Addison-Wesley, Reading, MA, 1973).

[36] LADNER, R.E. and M.J. FISCHER, Parallel prefix computation, *J. ACM* **27** (1980) 831–838.

[37] LAWRIE, D.H., Access and alignment of data in an array processor, *IEEE Trans. Comput.* **24** (1975) 1145–1154.

[38] LEIGHTON, F.T., Tight bounds on the complexity of parallel sorting, *IEEE Trans. Comput.* **34** (1985) 344–354.

[39] LEV, G., Size bounds and parallel algorithms for networks, Ph. D. Thesis, Univ. of Edinburgh, 1980.

[40] LEV, G., N. PIPPENGER and L.G. VALIANT, A fast parallel algorithm for routing in permutation networks, *IEEE Trans. Comput.* **30** (1981) 93–100.

[41] LUBOTZKY, A., R. PHILLIPS and P. SARNAK, Ramanujan graphs, *Combinatorica* **8** (1988) 261–277.

[42] LUPANOV, O.B., On rectifier and contact-rectifier networks, *Dokl. Akad. Nauk SSSR* **111** (1956) 1171–1174.

[43] MARGULIS, G.A., Explicit constructions of concentrators, *Problems Inform. Transmission* **9** (1973) 325–332.

[44] MARGULIS, G.A., Explicit group-theoretic constructions for combinatorial designs with applications to expanders and concentrators, *Problems Inform. Transmission* **24** (1988) 39–46.

[45] MASSON, G.M. and B.W. JORDAN JR., Generalized multi-stage connection networks, *Networks* **2** (1972) 191–209.

[46] NASSIMI, D. and S. SAHNI, Parallel permutation and sorting algorithms and a new generalized connection network, *J. ACM* **29** (1982) 642–667.

[47] NEČIPORUK, E.I., Rectifier networks, *Soviet Phys. Dokl.* **8** (1963) 5–7.

[48] NEČIPORUK, E.I., On a Boolean matrix, *Systems Theory Res.* **21** (1971) 236–239.

[49] O'CONNOR, D.G. and R.J. NELSON, Sorting system with *n*-line sorting switch, U.S. Patent 3,029,413, April 10, 1962.

[50] OFMAN, YU.P., On the algorithmic complexity of discrete functions, *Soviet Phys. Dokl.* **7** (1963) 589–591.

[51] OFMAN, YU.P., A universal automaton, *Trans. Moscow Math. Soc.* **14** (1965) 200–215.

[52] PATERSON, M.S., Improved sorting networks with O(log *n*) depth, *Algorithmica*, to appear.

[53] PAUL, W.J., R.E. TARJAN and J.R. CELONI, Space bounds for a game on graphs, *Math. Systems Theory* **10** (1977) 239–251.

[54] PINSKER, M.S., On the complexity of a concentrator, in: *Proc. 7th Internat. Teletraffic Congr.* (1973) 318/1–4.

[55] PIPPENGER, N., The complexity theory of switching networks, Ph. D. Thesis, Massachusetts Institute of Technology, 1973.

[56] PIPPENGER, N., Superconcentrators, *SIAM J. Comput.* **6** (1977) 298–304.

[57] PIPPENGER, N., Generalized connectors, *SIAM J. Comput.* **7** (1978) 510–514.

[58] PIPPENGER, N., The minimum number of edges in graphs with prescribed paths, *Math. Systems Theory* **12** (1979) 325–346.

[59] PIPPENGER, N., On the evaluation of powers and monomials, *SIAM J. Comput.* **9** (1980) 230–250.

[60] PIPPENGER, N., On another Boolean matrix, *Theoret. Comput. Sci.* **11** (1980) 49–56.

[61] PIPPENGER, N., A new lower bound for the number of switches in rearrangeable networks, *SIAM J. Algebraic Discrete Methods* **1** (1980) 164–167.

[62] PIPPENGER, N., Pebbling with an auxiliary pushdown, *J. Comput. System Sci.* **23** (1981) 151–165.

[63] PIPPENGER, N., Superconcentrators of depth two, *J. Comput. System Sci.* **24** (1982) 82–90.

[64] PIPPENGER, N., The complexity of computations of networks, *IBM J. Res. Develop.* **31** (1987) 235–243.

[65] PIPPENGER, N. and L.G. VALIANT, Shifting graphs and their applications, *J. ACM* **23** (1976) 423–432.

[66] PIPPENGER, N. and A.C.-C. YAO, Rearrangeable networks with limited depth, *SIAM J. Algebraic Discrete Methods* **3** (1982) 411–417.

[67] PUDLÁK, P. and P. SAVICKÝ, On shifting networks, to appear.

[68] REIF, J.H. and L.G. VALIANT, Logarithmic time sort for linear size networks, *J. ACM* **34** (1987) 60–76.

[69] RICHARDS, G.W. and F.K. HWANG, A two-stage rearrangeable broadcast switching network, *IEEE Trans. Comm.* **33** (1985) 1025–1035.

[70] SHANNON, C.E., Memory requirements in a telephone exchange, *Bell Systems Tech. J.* **29** (1950) 343–349.

[71] SNIR, M., Depth-size trade-offs for parallel prefix computations, *J. Algorithms* **7** (1986) 185–201.

[72] STRASSEN, V., Algebraic complexity theory, in: J. van Leeuwen, ed., *Handbook of Theoretical Computer Science, Vol. A* (North-Holland, Amsterdam, 1990) 633–672.

[73] TANNER, R.M., Explicit construction of concentrators from generalized N-gons, *SIAM J. Algebraic Discrete Methods* **5** (1984) 287–293.

[74] THOMPSON, C.D., Generalized connection networks for parallel processor interconnection, *IEEE Trans. Comput.* **27** (1978) 1119–1125.

[75] VALIANT, L.G., Graph-theoretic properties in computational complexity, *J. Comput. System Sci.* **13** (1976) 278–285.

[76] VALIANT, L.G., Short monotone formulae for the majority function, *J. Algorithms* **5** (1984) 363–366.

[77] VALIANT, L.G., General purpose parallel architectures, in: J. van Leeuwen, ed., *Handbook of Theoretical Computer Science, Vol. A* (North-Holland, Amsterdam, 1990) 943–971.

[78] VAN LINT, J.H., Problem 350, *Nieuw Arch. Wisk.* **21** (1973) 179.

[79] VAN VOORHIS, D.C., An improved lower bound for sorting networks, *IEEE Trans. Comput.* **21** (1972) 612–613.

[80] VAN VOORHIS, D.C., Toward a lower bound for sorting networks, in: R.E. Miller and J.W. Thatcher, eds., *The Complexity of Computer Computations* (Plenum Press, New York, 1972) 119–129.

[81] VAN VOORHIS, D.C., An economical construction for sorting networks, in: *Proc. 43rd AFIPS Nat. Computer Conf.* (1974) 921–926.

[82] YAO, A.C.-C., Bounds on selection networks, *SIAM J. Comput.* **9** (1980) 566–582.

CHAPTER 16

VLSI Theory

Thomas LENGAUER

Fachbereich 10—Informatik, Universität Paderborn, Postfach 1621, D-4790 Paderborn, FRG

Contents

HANDBOOK OF THEORETICAL COMPUTER SCIENCE
Edited by J. van Leeuwen

1. Introduction

The rapid progress in microelectronics has a great impact on theoretical computer science. New fabrication technologies suggest different machine architectures and call for new design methodologies for integrated circuits and systems. In this chapter we will consider various theoretical research areas that have been motivated by the revolution in microelectronics. One of these research areas is discussed in detail. For the other areas references for further reading are given

The chapter is concerned with integrated circuit (or chip) technologies. The most sophisticated version of such technologies is commonly referred to as VLSI (Very Large Scale Integration). In VLSI technologies well over 100 000 transistors are laid down on a single *chip*. This can be done by using one of several fabrication technologies, such as MOS (metal-oxide semiconductor), bipolar technologies or others. Many of the theoretical results also apply to packing technologies for *printed circuit boards*, in which the active circuit components are chips which are wired together on the board. On an even higher level of hierarchy, printed circuit boards form the active components, and they are wired together on *backplanes*. For the purposes of theoretical research the details of the fabrication technology are not of importance. Rather only the following two characteristics common to all chip technologies (and met by most printed circuit board and many backplane technologies) are modeled.

Finite resolution

A component on the chip, i.e., a switching element (transistor) or a wire takes a finite amount of space and cannot be made arbitrarily small. The technological parameters that determine this *minimum feature size* are the resolution of the photo-lithographic fabrication process (i.e., the wavelength of the electromagnetic radiation used to define structures on the chip surface) and several statistical phenomena that introduce imperfections into the features on the chip.

Quasi planarity

The *layout* of a circuit made up of switching elements and wires on a chip can only use a small limited number of *wiring layers*. This number can vary from technology to technology. For simple chip fabrication technologies only two wiring layers are available. For highly sophisticated packing technologies the number of wiring layers can be as large as about thirty. However, in all technologies we will consider, and in most technologies that are in use today, the number of wiring layers *does not grow with the size of the circuit*. Such technologies are called *quasi planar* because they produce sandwiched structures of small depth on a planar surface.

Let us discuss which theoretical research problems are suggested by the presence of fabrication technologies with such characteristics. We group the respective problems into two categories.

The circuit design problem

Stated in the most general way the design problem receives as an input a (high-level) description of the functional behavior and performance constraints of a circuit and produces the *best* circuit complying with this specification. Here "best" is defined via a cost function involving all important performance parameters of the circuit, such as chip area, computation delay, power or energy consumption, etc. VLSI technologies are capable of fabricating such large circuits that the design problem becomes intractable. Formally, it is an NP-hard problem. In order to solve this problem at least approximate optimization techniques have to be employed. These techniques incorporate a decomposition of the design problem into subproblems such as hardware synthesis, simulation and verification, layout synthesis, and testing. Each of these subproblems is in turn a very hard problem and thus is decomposed further. For instance, the layout problem is decomposed into placement, routing, compaction, etc. Since the design problem does depend on the fabrication process, at least to some extent, the optimization techniques also must be matched with the fabrication process. In other words, a special *design technology* must complement the fabrication technology. The development of effective design technologies is a very important problem with many theoretical and practical facets. It is a generally accepted fact that the design problem dominates the fabrication problem. Put differently, the fabrication technology provides us with means to produce circuits that are so complex that we do not know how to design them effectively. This is the reason why circuit design is one of the most critical areas in computer science today.

The theoretical research problems related to the design problem are mostly NP-hard optimization problems. They are solved with methods from graph theory (see Chapter 10 on graph algorithms), combinatorial optimization (see [43, 26, 25]), or statistics [58]. Asymptotic results are not sufficient here, since the constant factors involved are usually impractically large. We will not discuss design problems in detail. They deserve a separate thorough treatment (see also [50, 57, 34, 33]).

The architecture problem

The high-level part of the design problem is to decide upon an architecture for the circuit or machine one wants to build. Since a chip allows for many *processing elements* (*p.e.s*) to be connected by wires such that they can compute *concurrently*, the von Neumann architecture is not the only architecture possible, and often it is not even a favorable architecture. Rather parallelism can be employed to greatly speed up the computation, at the expense of extra space for additional processing elements and connections between them. The introduction of parallelism introduces different cost measures than we are used to in the von Neumann context.

In a von Neumann machine the two main complexity measures are (processor) runtime and (memory) space. In a parallel environment the computation is distributed over many p.e.s and the memory may be segmented into chunks pertaining to each p.e. A p.e. can be a sophisticated (sequential) processor or just a simple switching element, such as a transistor or a logic gate. Thus, in VLSI circuits the following additional complexity measures come into play.

Communication time

The time to transfer data between p.e.s.

Wiring space
The space needed to put down the wires on the chip surface.
Switching energy
The energy spent by a computation.

These additional complexity measures impose on the architecture problem for VLSI circuits a completely different structure from what we are used to in sequential computation. In the remaining sections of this chapter we will discuss in detail what complexity measures can be used to assess VLSI computations and how one can find lower bounds for these complexity measures. Section 9 will refer to upper bounds but not discuss them in as much detail. All results we will discuss are asymptotic.

The architecture problem for VLSI computations receives as an input a functional description of a circuit. Often this is just a Boolean function, such as multiplication, sorting or others. But it can also be the description of a finite automaton or another kind of functional description. The architecture problem generates a network of processing elements and an accompanying (parallel) program that complies with the functional specification. The architecture problem of VLSI theory, in addition, has to provide a *quasi planar embedding of the processor network onto the chip surface* that optimizes certain complexity measures.

The collection of processor network, quasi planar layout, and program forms a solution of an instance of the architecture problem and thus an *upper bound* on the best possible solution w.r.t. the respective complexity measures. Such upper bounds can be matched with *lower bounds* on all possible circuits complying with the respective functional specification. It turns out that asymptotically tight lower bounds can often be found by disregarding the sequential computations done at each processor and just taking into account the complexity of the *communication* of data between the processing elements. This is impressive evidence of the fact that in VLSI computation the communication complexity dominates the complexity of the total computation.

So far we have discussed the architecture problem for *special-purpose* VLSI chips, i.e., chips that are tailored to one or a small class of specific Boolean functions or circuits. Equally interesting is the *general-purpose* architecture problem where we want to design VLSI systems that implement a large variety of functions efficiently. Understandably, this variant of the architecture problem is not as well suited for the development of good lower bounds. There are several interesting upper bounds, however, and we will detail them in Section 9.

2. VLSI complexity measures

The first author to propose a theoretical model for VLSI computations that encompasses the finite resolution and the quasi planarity properties was Clark Thompson. His Ph.D. Thesis [54] also started VLSI theory, in a sense. Since then several modifications and variations of the model have been proposed [9, 11, 36, 35, 56, 55]. Rather than discussing all variations we will present a representative version of a model for VLSI computations.

The main complexity measures studied are the chip area A and the computing time T. For A we measure the area on the chip surface that is taken up by switches and wires, the so-called *active area* [54, 35] or, alternatively, the area of the smallest rectangle enclosing the whole circuit, the so-called *total* area [9, 11, 36]. The first measure is directly related to the chip yield during fabrication. The second area measure accounts for the total size of the chip. For our purposes both area measures are equivalent [35]. The question how to measure time is more difficult to answer. Most authors only study synchronous circuits. Here a global clock counts the synchronous steps of the computation performed by the processing elements. Then T is interpreted as the number of clock cycles spent in the computation. Note that this is not a "physical" measure of time since the length of a clock cycle may itself vary with the size of the chip. As long as the delay of signal propagation along wires is not significantly longer than the delay of the processing elements, the number of clock cycles gives a good representation of the time spent, i.e., it is asymptotically accurate. As the number of transistors on a chip increases, this ceases to be the case. Chazelle and Monier [11] introduced a physical time measure T by measuring time in microseconds under the assumption that signals propagate along wires at a constant speed. (This is sometimes called the *resistive* model, because it accounts for resistances in wires.) They get dramatically different lower bounds on circuit complexity. Not only are their lower bounds larger, as we expect, but the optimal chips according to their complexity measures differ significantly from the optimal chips if T is measured in clock cycles. This is because if the time for propagating a signal along a wire depends on the length of the wire, one has to pay a penalty for long wires across which communication is expensive. Time measures such as the one discussed by Chazelle and Monier are becoming technologically significant as the scale of integration increases further. Eventually also *capacitive* and *inductive* phenomena will become important, leading to even larger, i.e., quadratic, delays along wires [4]. Such delays can be reduced by placing appropriate driver stages on wires [38]. We will not address these technological questions in detail.

There is a theoretical reason why synchronous models are interesting. While in all models of VLSI computations the complexity of communicating data along wires dominates the complexity of computations inside processing elements, in synchronous models lower bounds on the communication complexity can be derived by information flow arguments. In contrast, in other models of VLSI computations more trivial geometric arguments involving the distances between processors are applied.

In this chapter we will only consider synchronous models of VLSI computations. Since synchronous circuits can be pipelined, another complexity measure gains importance in this context, namely the *period* P. A pipelined circuit receives input to a new problem instance to be solved before it has finished solving the previous problem instance. P is the number of clock cycles between the first input cycle of two consecutive problem instances (see Fig. 1). When a circuit solves many problem instances, i.e., the number N of problem instances greatly exceeds T/P, then P assumes the role of T, since the time $T+(N-1)P$ to solve all N problem instances does not differ significantly from $N \cdot P$. We will not discuss P explicitly in this chapter but just mention that the lower bounds we prove even hold if P is substituted for T and the relevant definitions are modified accordingly.

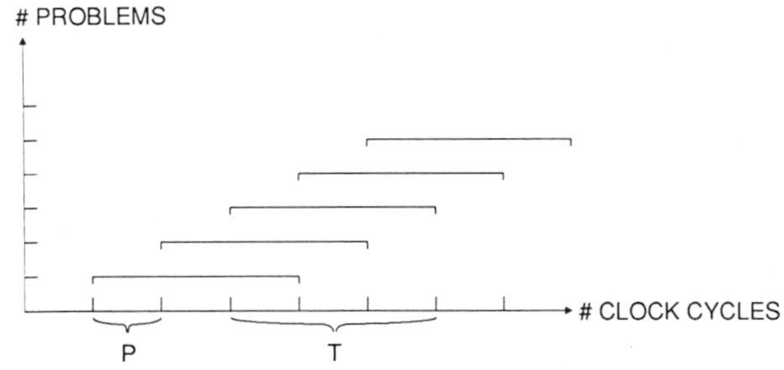

Fig. 1. Time vs. period.

Besides area, time and period, the switching energy is a complexity measure of interest in VLSI computations. Energy consumption is related to heat dissipation on chips. Therefore it is of interest from a technological standpoint. Furthermore it is of fundamental theoretical interest from an information theory point of view. In synchronous circuits the most natural definition of the switching energy E is based on the following assumption.

Uniswitch model: Every unit of active chip area consumes one unit of switching energy whenever it changes its state from 0 to 1 or vice versa in a clock cycle.

This definition is optimistic in the sense that it assumes that a state change in a clock cycle is achieved by a *clean* transition from 0 to 1. In fact, internal delays in the circuit can cause races that lead to oscillations on wires. In such a case the state changes back and forth several times before finally settling at its stable value at the end of a clock cycle. This, of course, increases the switching energy expended. The energy model that takes such races into account is called the *multiswitch model*. It is discussed in [1]. We do not consider it in detail, but mention that lower bounds on the energy in the uniswitch model also apply to the multiswitch model.

In technologies with high d.c. currents (such as NMOS [38]) the energy consumption is determined by the number of devices on the chip. In technologies without high a.c. currents (such as CMOS [60]) the energy consumption is dominated by the switching energy.

In this chapter we will place specific emphasis on developing lower bounds for the respective complexity measures. Lower bounds on A can be obtained in many cases [59, 35]. The respective arguments are based on measuring the storage capacity necessary for the computation and do not involve information flow arguments. We will not discuss them in detail here. Looking for lower bounds on T is the same problem as in parallel complexity theory. A lower bound on E will be discussed in Section 8. It is taken from [35].

The above complexity measures can be combined to form new interesting complexity measures that exhibit trade-offs. Clearly there is a trade-off between A and T, since more chip area allows for more concurrency. Thus the quantity AT is an interesting complexity measure. However, w.r.t. AT many circuits are optimal that we intuitively would not consider to be good VLSI circuits. (An example is the bit-serial adder). This is because the concurrency in the computation is not given enough priority. We can make the complexity measure more sensitive to concurrency by giving the T term a higher weight. This is one motivation for the investigation of AT^2 as a complexity measure. Another one—and the main motivation from a theoretical point of view—is that the AT^2-complexity measure can be analyzed using information flow arguments. Thompson [54] has presented a proof method for deriving lower bounds on the AT^2-complexity of VLSI chips, and later other authors have elaborated on it. This method will be the main subject of the chapter. The method measures the amount of information that has to be exchanged between different regions on the chip and relates it to the chip area and computing time. It only accounts for the cost of communication, not for the cost of computation inside a p.e., or of storage. The fact that many of the derived lower bounds are asymptotically tight shows that the AT^2-complexity of VLSI computations is dominated by the cost of communicating data across the chip.

In Section 3 we introduce the synchronous model for VLSI computations that we use in this chapter. In Section 4 we give an intuitive outline of the method for proving AT^2 lower bounds. Section 5 contains the geometric tools in the form of a separator theorem. Sections 6 and 7 provide the information theoretic tools for proving lower bounds on the AT^2-complexity of Boolean predicates respectively. Boolean functions with many outputs. In Section 8 we use the ideas developed in the previous sections to provide a lower bound on the switching energy spent by VLSI chips. Finally in Section 9 we survey results on upper bounds in the AT^2-measure. In essence we present good special and general purpose architectures in this section.

3. The VLSI model

Our VLSI model is based on Boolean circuits. This choice is adequate also for modeling more general "multidirectional" VLSI structures, e.g., buses [35].

1. Definition. A chip $\chi=(\Gamma, \Lambda, \Delta)$ consists of three structures.

(a) The *circuit* Γ: A synchronous Boolean circuit with feedback and unbounded fan-in. Formally, this is a directed bipartite graph $\Gamma=(V, E)$ where V is partitioned into a set S of *switches* and a set W of *wires*. Here $S=P\cup G$ where P is a set of *ports* labeled *in* or *out* and G is a set of *gates* labeled *and*, *or*, *nand* or *nor*. For $s\in S$ and $w\in W$, if $(s, w)\in E$, then w is called an *output of* s, if $(w, s)\in E$, then w is called an *input of* s. All gates have out-degree 1, all wires have in-degree 1. Each input port has one input and one output. Each output port has two inputs and no output. The "additional" input signal for the ports is an enable signal computed on the chip that activates the port.

(b) The *layout* Λ: The layout maps every vertex in the circuit into a compact connected region in the plane. Furthermore, each point in the region lies inside some square of side length $\lambda>0$ that is completely contained in the region. This provision models the finite resolution of the fabrication process for VLSI chips. Each point in the plane belongs to the interior of at most $\nu\geqslant 2$ regions representing vertices in the circuit. Since $\nu\geqslant 2$, also nonplanar circuits Γ can be laid out. The fact that ν is bounded models the quasi planarity of the fabrication process. Two regions touch exactly if the vertices they represent are neighbors in Γ. (We say that regions R_1 and R_2 *touch* if $R_1 \cap R_2 \neq \emptyset$ but $R_1^0 \cap R_2^0 = \emptyset$, where R_1^0, R_2^0 are the interiors of R_1 resp. R_2.)

(c) The *manual* Δ: The manual is a set of directions for the communication between the chip and its environment. For every input port it contains a sequence of numbers that identifies the input bits that enter through this port. The sequence also determines the order in which the input bits enter. When its enable signal is raised the port requests the next input bit in the sequence to enter. Analogously, for each output port the manual contains a sequence of numbers identifying the output bits produced at that port and their order. When its enable signal is raised the port produces the next output bit in the sequence.

After defining all components of the chip we can define the operation of the circuit. To this end we associate with each port a word $w \in B^*$, $B = \{0,1\}$, that we call its *history*. Furthermore, we label each wire with an initial Boolean value from $B \cup \{X\}$. (X stands for the undefined Boolean value, $0 \wedge X = 0$, $1 \vee X = 1$, $0 \vee X = 1 \wedge X = X$.) Such a labeling we call a *state of the circuit*. The initial state is most often the completely undefined state. In the ith cycle the circuit does the following. Each gate "reads" the values on all its input wires and computes the Boolean operation given by its label. The resulting value is put on its output wire. Each input port puts an X on its output wire if its enable signal is 0, otherwise it puts the next bit from its history on its output wire. Each output port checks if its enable signal is 1, and it so it puts its other input at the end of its history. Thus, input ports consume their histories and output ports produce them. All actions of the gates happen in parallel. Thus a new state is reached on which the $(i+1)$st cycle of the computation is started.

Finally, we discuss what A, T and E mean in this context. The area A is the area of the set M of all points in the plane that belong to at least one region in the layout. (Thus we measure active area.) The time T is the number of steps taken by the circuit to produce the desired outputs in its output histories. The switching energy E is based on the area of the regions in the layout.

There has been some confusion about different kinds of manuals in the past. Therefore we discuss the issue here at greater length.

The manual used in Definition 1 is called *strongly where-oblivious* in [35]. Many authors use manuals of a different kind to prove the same lower bounds. Lipton and Sedgewick [36] call a manual *where-oblivious* (*when-oblivious*) if the port (cycle) at which an input bit is requested, resp. an output bit is produced, is independent of the input. In this paper we will also consider manuals which are both where- or when-oblivious. Such manuals simply list for each I/O bit the time and port at which it crosses the chip boundary. These data are independent of the input. (Note that strongly where-oblivious manuals are in general not when-oblivious.) If we restrict ourselves to manuals which are where- and when-oblivious, then we do not have to require the I/O ports to be activated by enable signals which are produced by the circuit. We can instead assume that all ports are active all the time, and that they sometimes consume resp. produce bits which are insignificant for the problem to be solved. This assumption takes some computational burden off the chip, because it can "blindly" consume and produce bits without having to compute the knowledge on what data it is currently working. Even so we can still prove the same lower bounds as for strongly where-oblivious chips. This is, because if the manual is where- and when-oblivious, then even on "blind" chips, solutions to different problems are always represented by different I/O histories. If the manual is strongly where-oblivious, this is not the case however, and the chip has to "know" what bits cross its boundary when and where. If this knowledge were not computed by the chip but could be found inside the manual, every problem could be solved by a trivial chip with a manual that contains all the computation. (The chip takes no input, runs for two cycles and produces a 0 in the first and a 1 in the second cycle at its unique output port. One of these two values is the unique output and the manual says which one it is, depending on the input.)

We distinguish two kinds of complexity analysis: worst-case analysis and average-case analysis. In worst-case analysis we ask for the greatest value of a complexity measure (A, T, E) for all possible input sequences of the same length. In average-case analysis we average the complexity measure over all input sequences of the same length under the assumption that they are equally likely. For the area A, worst-case analysis and average-case analysis coincide. For the switching energy E they may differ. For T they coincide if the manual is where- and when-oblivious, and they may differ if the manual is strongly where-oblivious.

4. The basic lower bound argument

In this section we give an intuitive outline of the argument for proving lower bounds on the AT^2-complexity of VLSI chips.

Let us assume that we have a rectangular chip computing some Boolean predicate $p: B^n \to B$. The chip has side lengths a and b, $a \leqslant b$. Let us furthermore assume that the chip has n input ports and one output port, and that a unique I/O bit is assigned to each port. Clearly we can cut the chip into two halves L and R by a line C parallel to the sides of length a of the chip, such that about half the input ports lie on the side L resp. R. We first observe that the length of the cut is $\lg(C) = a \leqslant \sqrt{A}$. The chip is where-oblivious and therefore we can also assign to each input bit a side among L, R through which it enters the chip, independently of the input. This partitions the input bits into two classes X and Y corresponding to the sides L and R. Let us assume that the output port lies on side R. If we can now show that in order to compute the output bit we have to know many, e.g., ω of the input bits in X, we can argue as follows: since the chip is planar, the input bits in X can only be communicated to the side R across the cut C. By our layout model C can only be crossed by a number of wires that is proportional to its length. Thus during each cycle only $O(\lg(C))$ bits can be communicated across C. Since ω bits have to cross C this takes time $T = \Omega(\omega/\lg(C))$. By our bound on $\lg(C)$ we get $T = \Omega(\omega/\sqrt{A})$, or $AT^2 = \Omega(\omega^2)$.

Thus the argument is divided into two parts. The first part is the proof of a geometric separator theorem that says that any well behaved (e.g., rectangular) region in the plane with area A representing the chip layout can be cut by a nicely behaved (e.g. straight) line into two "balanced" halves (e.g., with respect to the number of input ports in them).

Furthermore the length of the cut line is small, i.e., $O(\sqrt{A})$. We call such a theorem a *separator theorem*. The notion of a separator is known from graph theory [37], and in nontrivial cases the methods from graph theory can be carried over to prove geometric separator theorems. We will prove a general separator theorem in Section 5 that allows the application of the lower bound argument for the case that we are interested in active area. The obvious separator theorem for rectangles contained in the above outline applies to the notion of total area.

The second part of the lower bound argument lies in the definition of a certain notion of "communication complexity" for the Boolean function p to be computed. Several different notions are possible here. We will discuss them in detail in Sections 6 and 7. Intuitively the communication complexity of a function p is the amount of information that has to be exchanged between two cooperating processors that each have a partial knowledge of the input and that want to aid each other in computing p. The communicating processors are, in this case, the two halves of the chip as divided by the cut C. The communication complexity of p then takes the part of the quantity ω in the above outline.

These ideas will be made more precise in the following sections.

5. A geometric separator theorem

This section gives a separator theorem for compact plane regions. The theorem is a special case of a more general separator theorem presented in [35]. The proof is patterned after the proof of a similar separator theorem for planar graphs given in [37].

2. THEOREM. *Let M be a compact plane region. Let $p_1, \ldots, p_n$ be n different points inside the interior of M of which n are specially "marked". Then M can be cut by a set C of three straight cuts into two compact sets M_l and M_r such that*

(1) $M_l \cup M_r = M$, $M_l \cap M_r = C$,

(2) $\lg(C) \leqslant 2\sqrt{A}$ *where A is the area of M,*

(3) *M_l, M_r both have at most $2n/3$ marked points of $p_1, \ldots, p_n$ in their interior and none of the points $p_1, \ldots, p_n$ on their border.*

PROOF. First we note that because the set $P = \{p_1, \ldots, p_n\}$ is finite M can be put into a Cartesian coordinate system such that no line parallel to the x- or y-axis contains more than one point in P. We devise the following procedure for cutting M.

W.l.o.g. assume that $\lceil n/2 \rceil$ of the marked points lie above the x-axis, $\lfloor n/2 \rfloor$ of the marked points lie below the x-axis, and no point in P lies on the x-axis. Let $M_+ = \{(x, y) \in M \mid y \geqslant 0\}$ and $M_- = \{(x, y) \in M \mid y \leqslant 0\}$. Let A_+ be the area of M_+ and A_- be the area of M_- ($A_+ + A_- = A$). The method for cutting M has two steps:

Step 1: For any real l, let $C_l = \{(x, y) \in M \mid y = l\}$. We choose two cuts of M parallel to the x-axis at distances $y = l_t > 0$ and $y = l_b < 0$. C_{l_t} and C_{l_b} cut M into three parts M_t, M_m, and M_b as suggested by Fig. 2. We choose l_t such that $\lg(C_{l_t}) + l_t \leqslant \sqrt{2A_+}$. Analogously we choose l_b such that $\lg(C_{l_b}) - l_b \leqslant \sqrt{2A_-}$. Such l_t and l_b can be found.

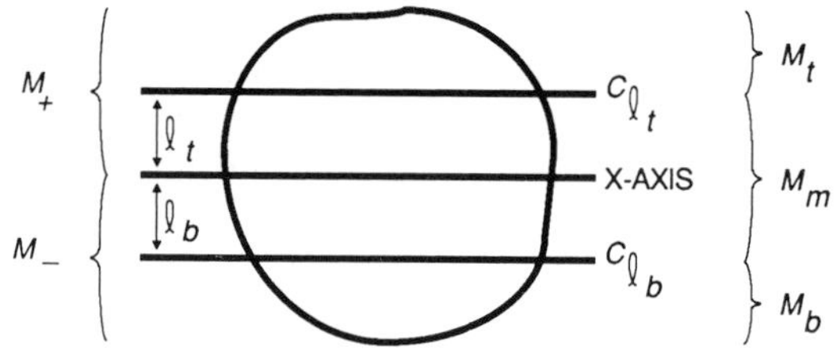

Fig. 2. The separator theorem.

For assume that such an l_t does not exist. Then

$$A_+ = \int_0^\infty \lg(C_l)\,\mathrm{d}l \geqslant \int_0^{\sqrt{2A_+}} \lg(C_l)\,\mathrm{d}l > \int_0^{\sqrt{2A_+}} (\sqrt{2A_+} - l)\,\mathrm{d}l = A_+,$$

which is a contradiction. In fact we can assume that no point of P lies on C_{l_t} since the set $\{l \mid l>0 \wedge \lg(C_l)+l \leqslant \sqrt{2A_+}\}$ must have a positive measure. An analogous argument applies to l_b.

Now, if M_m contains no more than $2n/3$ marked points of P, then we are done. Because M_t and M_b both have at most $n/2$ marked points of P we only have to choose M_l to be the one of the three sets with the most marked points of P in it. If, however, M_m has more than $2n/3$ marked points in it, then we have to continue cutting.

Step 2: We bisect M_m with a line $x=l_v$ such that half the marked points in M_m are on either side on the line. More precisely the cut consists of the segment $\{(x, y) \mid x=l_v$ and $l_t \geqslant y \geqslant l_b\}$. Now clearly the half of M_m containing the most marked points of P will serve as M_l.

The total length of the cut C is bounded by

$$\begin{aligned}\lg(C) &\leqslant (C_{l_t}) + \lg(C_{l_b}) + (l_t - l_b)\\ &\leqslant \sqrt{2A_+} + \sqrt{2A_-} \leqslant \sqrt{2A_+} + \sqrt{2(A-A_+)} \leqslant 2\sqrt{A},\end{aligned}$$

since the function $z(x)=\sqrt{x}+\sqrt{1-x}$ has a maximum of $\sqrt{2}$ at $x=\frac{1}{2}$. □

6. The communication complexity of Boolean predicates

As motivated in Section 4 we start from the following definition.

3. DEFINITION (Yao [61]). Let $p: X \times Y \to B$ be a Boolean predicate. We call X the set of *left inputs* and Y the set of *right inputs*. Let L and R be two computing agents, one of which knows $x \in X$ and the other of which knows $y \in Y$. L and R want to compute $p(x, y)$ by exchanging information between each other. We are primarily interested in deterministic computations.

A *deterministic algorithm* is given by two *response functions* $\rho_L: X \times B^* \to B$ and $\rho_R: Y \times B^* \to B$ and the *partial output function* $a: B^* \to B$. For computing $p(x, y)$ L starts sending the bit $w_1=\rho_L(x,\varepsilon)$ to R. R responds with $w_2=\rho_R(y,w_1)$; L returns $w_3=\rho_L(x,w_2)$ and so on, until $(w_1,\ldots,w_{k(x,y)})\in dom(a)$. At this point the computation stops with the result $a(w_1,\ldots,w_{k(x,y)})=p(x,y)$. $k(x,y)$ is the length of the computation.

The *deterministic two-way communication complexity* of p is given by

$$C_{\text{det}}(p,L\leftrightarrow R)=\min_{A}\max_{x\in X,y\in Y} k_A(x,y)$$

where A is any deterministic algorithm computing p and $k_A(x,y)$ is the length of the computation on input (x,y), if algorithm A is used.

For theoretical purposes it is often interesting to introduce nondeterminism into the model. In a *nondeterministic algorithm* we are allowed guesses as to the outcome of certain computations. These guesses then only have to be verified. W.l.o.g. we can assume for nondeterministic algorithms that information is only sent from L to R. (L could guess all of the information sent from R and R just has to verify these guesses.)

4. DEFINITION. A *nondeterministic (one-way) algorithm* is given by a set $\rho_L(x)\in B$ for every x and by a response function $b: Y\times B^*\to B$. On input (x,y), L sends some word $w\in\rho_L(X)$ to R and R outputs $b(y,w)$. In order for R to know when L has finished sending its information we assume $\rho_L(x)$ to be prefix-free. The nondeterministic algorithm computes p if $p(x,y)=1$ exactly if there is a $w\in\rho_L(x)$ such that $b(y,w)=1$. The *nondeterministic communication complexity* of p is

$$C_{\text{ndet}}(p,L\to R)=\min_{A}\max_{x\in X,y\in Y,p(x,y)=1}\min\{|w|\,|\,w\in\rho_{L,A}(x)\wedge b_A(y,w)=1\}.$$

A nondeterministic algorithm is unambiguous if the last minimum in the above equation is taken over singletons (and can thus be omitted). We define

$$C_{\text{unamb}}(p,L\to R)=\min_{A\text{ unamb}}\max_{x\in X,y\in Y,p(x,y)=1}|w(x,y)|$$

where $w(x,y)$ is the unique w such that $w\in\rho_{L,A}(x)$ and $b(y,w)=1$.

A third class of algorithms which is of interest is "Las Vegas" algorithms. *Las Vegas algorithms* use internal randomization (coin flipping) to determine the result. However, the randomization does not affect the outcome of the computation but just its complexity. Las Vegas algorithms not only are of theoretical interest but of practical significance since the realization of true coin flipping devices in VLSI technologies seems in the realm of the possible.

5. DEFINITION. A *Las Vegas algorithm with t coin tosses* or an *LV-t algorithm* is given by two response functions $\rho_L: X\times B^*\times B^t\to B$ and $\rho_R: Y\times B^*\times B^t\to B$ and the partial output function $a: B^*\to B$. We assume that both L and R first toss t coins to determine the third arguments t_L, t_R in the response functions, and then start a deterministic computation as in Definition 3. The computation ends when $(w_1,\ldots,$

$w_{k(x,y,t_L,t_R)}) \in dom(a)$. Its result is $a(w_1, \ldots, w_k)$. The *Las Vegas communication complexity* of p is

$$C_{\mathrm{LV}}(p, L \leftrightarrow R) = \lim_{t \to \infty} \min_{A\,\mathrm{LV}\text{-}t\,\mathrm{alg}} \sum_{t_L, t_R \in B^t} k(x, y, t_L, t_R)/2^{2t}.$$

We also define a deterministic average-case communication complexity of p just by averaging over all inputs instead of maximizing over them:

$$C_{\mathrm{ave}}(p, L \leftrightarrow R) = \min_{A\,\det} \sum_{x \in X, y \in Y} k_A(x, y)/(|X||Y|).$$

The relationship between the quantities defined above and the AT^2-complexity of chips will be discussed later in this section. However, they are also of use for concerns outside the VLSI area. For instance, they apply to problems in one-tape Turing machine complexity. Indeed they can be used as a basis for a theory of communication complexity, and as such we will consider them now. This kind of complexity theory has been extensively studied. For a survey see [46].

We now study methods for finding lower bounds on the communication complexity of Boolean predicates. The methods consider the Boolean predicate p as an $|X| \times |Y|$ matrix P of zeros and ones. The input x selects the row, y selects the column of the matrix. Thus $P_{xy} = p(x, y)$.

Method 1 (*crossing sequences*)

This method is an adaptation of the crossing sequence technique for showing lower time bounds for one-tape Turing machine complexity. It was first presented in our context in [51]. It is based on the following theorem.

6. THEOREM. *Consider the matrix P defining the Boolean predicate p. By permuting rows and columns independently put P into the form* $\binom{I\,B}{A\,C}$ *where I is an* $m \times m$ *identity matrix and m is as large as possible. Then* $C_{\det}(p, L \leftrightarrow R) \geq \log m$.

PROOF. The proof is an application of the Pigeon Hole Principle. Assume that $C_{\det}(p, L \leftrightarrow R) < \log m$. Then there is a deterministic algorithm computing p that always exchanges less than $\log m$ bits. Thus there are two inputs (x_1, y_1) and (x_2, y_2) corresponding to diagonal elements in the identity matrix I of P for which the same information is exchanged between L and R. By Definition 3 the same information is also exchanged on inputs (x_1, y_2) resp. (x_2, y_1). Thus the algorithm yields the same result for all four input values. Since (x_1, y_2) and (x_2, y_1) are off the diagonal in I the algorithm must therefore yield incorrect results on some of the four input values. □

7. COROLLARY. *The above theorem is still true if we substitute* C_{ndet} *or* C_{unamb} *for* $C_{\det}$.

PROOF. Analogous to the proof of Theorem 6. □

Brent and Goldschlager [8] use Method 1 to derive lower bounds on the AT^2-complexity of context-free language recognition, the evaluation of propostional formulae, and certain set problems. They also discuss (nonmatching) upper bounds.

Corollary 7 is the manifestation for communication complexity of the well-known fact that the crossing sequence technique is not able to differentiate between determinism and nondeterminism.

Note that Method 1 is not strong enough to show average-case lower bounds. This is because we bound the communication complexity of a Boolean predicate from below by showing that it entails an identity predicate $id(x, y) := (x = y)$ for strings of some length. But identity predicates are easy to compute on the average. In fact, most often the inputs (x, y) of an identity predicate will not agree and this can be found out very fast. (Discrepancies will occur in the first bit half of the time.) One easily shows that $C_{\text{ave}}(id, L \leftrightarrow R) \leqslant 2$.

Although Corollary 7 may at first sight seem to show a strength of Method 1, it really exhibits a weakness. Nondeterministic lower bounds are in general weaker than deterministic lower bounds because nondeterminism is a more powerful concept than determinism. Thus Method 1 will yield bad results for predicates that are difficult to compute deterministically but easy to compute nondeterministically. Sometimes one can get better results by complementing the predicate, i.e., by considering $\bar{p} = 1 - p$ instead of p. $\bar{p}$ has deterministically the same communication complexity as p but may be harder nondeterministically. An example for such a predicate is again the identity predicate. By Corollary 7 we have for the identity predicate on strings of length n, $C_{\text{ndet}}(id, L \leftrightarrow R) \geqslant n$. However we can show that $C_{\text{ndet}}(\overline{id}, L \leftrightarrow R) \leqslant \log n + 1$. This is because in order to show that $x \neq y$, L only has to guess one bit position in which x and y differ. It then sends the number of this position ($\log n$ bits) and the corresponding bit value of x (1 bit) to R. Thus $\overline{id}$ is easy to compute nondeterministically. However we can still show a good deterministic lower bound on $C_{\text{ndet}}(\overline{id}, L \leftrightarrow R)$ by applying Method 1 to its complement id, which is difficult to compute nondeterministically.

However, there are natural Boolean predicates p such that

$$C_{\text{ndet}}(p, L \rightarrow R),\ C_{\text{ndet}}(\bar{p}, L \rightarrow R) \ll C_{\text{det}}(p, L \leftrightarrow R).$$

For such predicates Method 1 fails to prove good deterministic lower bounds. For the purpose of proving lower bounds in such cases a second method is presented.

Method 2 (rank lower bound)

This method has been presented in [39].

8. DEFINITION. Let r be a ring and let $r^{(n,m)}$ be the set of $n \times m$ matrices over r. The *rank of* $A \in r^{(n,m)}$ over r is the minimum k such that A can be written as $A = C \cdot D$, where $C \in r^{(n,k)}$ and $D \in r^{(k,m)}$.

If r is a field, the above definition coincides with the definition of matrix rank known from linear algebra. Method 2 is based on the following theorem.

9. THEOREM. *Let p be a Boolean predicate, and let P be its associated matrix. Then $C_{\text{det}}(p, L \leftrightarrow R) \geqslant \log rank_N(P) \geqslant \log rank_r(P)$, where N is the ring of integers and r is any field.*

PROOF. The second inequality is known from algebra. For the proof of the first inequality we state the following lemma.

10. LEMMA. *Let r be a ring, $A \in r^{(n,m)}$, $B \in r^{(n,k)}$, and $C \in r^{(k,m)}$. Then*

$$rank_r((A\ B)) \leqslant rank_r(A) + rank_r(B),$$

$$rank_r\left(\begin{pmatrix} A \\ C \end{pmatrix}\right) \leqslant rank_r(A) + rank_r(C).$$

PROOF. If $A = D_1 \cdot E_1$ and $B = D_2 \cdot E_2$, then

$$(AB) = (D_1 D_2) \cdot \begin{pmatrix} E_1 & 0 \\ 0 & E_2 \end{pmatrix}.$$

The proof of the second inequality is analogous. □

PROOF OF THEOREM 9 (*continued*). Now consider any deterministic algorithm for computing p. Inductively on the length of $w \in B^*$ we define the matrix P_w as follows:

$|w| = 0$: $P_\varepsilon := P$.

$|w| > 0$: if $|w| = 2l$, then P_{w0} (P_{w1}) is obtained from P_w by selecting all rows x with $\rho_L(x, w) = 0$ ($\rho_L(x, w) = 1$). If $|w| = 2l+1$, then P_{w0} (P_{w1}) is obtained from P_w by selecting all columns y with $\rho_R(y, w) = 0$ ($\rho_R(y, w) = 1$).

By Lemma 10 we have $\max(rank_N(P_{w0}), rank_N(P_{w1})) \geqslant rank_N(P_w)/2$. Moreover, if $w \in dom(a)$, then P_w must be monochromatic, i.e., all of the entries in P_w must have the same value $a(w)$, and thus $rank_N(P_w) \leqslant 1$ must hold. Thus there is a $w \in dom(a)$ such that $|w| \geqslant \log rank_N(P)$. □

In [39] it is shown that in fact $C_{\text{unamb}}(p, L \to R) = \lceil \log rank_N(P) \rceil$. On the other hand $C_{\text{ndet}}(p, L \to R)$ can be much smaller, as the following example shows.

11. DEFINITION. Let $n \in N$ and $X = Y = [0:2^n - 1]^n$. For $x = (x_1, \ldots, x_n) \in X$ and $y = (y_1, \ldots, y_n) \in Y$ let

$$p_1(x, y) = \begin{cases} 1 & \text{if } x_i = y_i \text{ for some } i,\ 1 \leqslant i \leqslant n, \\ 0 & \text{otherwise.} \end{cases}$$

The above definition is the result of modifying the identity predicate such that it and its complement are nondeterministically about equally hard. Instead of checking the identity of one pair of n-bit words we now check the identity of n pairs of n-bit words. If any pair is identical we output 1 else 0.

12. THEOREM

(a) $C_{\text{det}}(p_1, L \leftrightarrow R) \geqslant n^2$.

(b) $C_{\text{ndet}}(p_1, L \to R) = O(n + \log n)$.

(c) $C_{\text{ndet}}(\overline{p_1}, L \to R) = O(n \log n)$.

(d) $C_{\text{LV}}(p_1, L \leftrightarrow R) = O(n)$.

(e) $C_{\text{ave}}(p_1, L \leftrightarrow R) = O(n)$.

PROOF. (a) Since $P_1 \in B^{2(n^2)}$ we only have to show that $rank_{\mathrm{GF}(2)} P_1 \geqslant 2^{(n^2)}$, where GF(2) is the field of characteristic 2. Let $\oplus$ denote addition modulo 2. We transform the matrix $\overline{P_1}$ associated with $\overline{p_1}$ into the identity matrix of size $2^{(n^2)} \times 2^{(n^2)}$ by means of linear transformations.

13. LEMMA. *Let $w_1, \ldots, w_n, y_1, \ldots, y_n \in [0\colon 2^n - 1]$. Define*

$$g(w_1, \ldots, w_n, y_1, \ldots, y_n) := \bigoplus_{x_1, x_1 \neq w_1} \cdots \bigoplus_{x_n, x_n \neq w_n} \overline{p_1}(x_1, \ldots, x_n, y_1, \ldots, y_n).$$

Then $g(w_1, \ldots, w_n, y_1, \ldots, y_n) = id(w_1, \ldots, w_n; y_1, \ldots, y_n)$.

PROOF. Omitted. □

PROOF OF THEOREM 12 (*continued*). (b) L guesses the number $i \in [1\colon n]$ such that $x_i = y_i$ and sends i and x_i to R. If $x_i = y_i$, then R outputs 1.

(c) For each $i \in [1\colon n]$ L guesses a position $j_i \in [1\colon n]$ such that x_i and y_i differ in position j_i. L sends j_i and the corresponding bit of x_i to R. R verifies that the bits indeed differ.

(d) See [14]. Since

$$C_{\mathrm{ndet}}(p_1, L \rightarrow R),\ C_{\mathrm{ndet}}(\overline{p_1}, L \rightarrow R) \leqslant C_{\mathrm{LV}}(p_1, L \leftrightarrow R).$$

Part (d) supersedes parts (b) and (c) but it is more difficult to prove. We have included parts (b) and (c) here for the sake of their simple proof.

(e) For $i = 1, \ldots, n$, L and R alternate sending bits of x_i resp. y_i until one bit is found in which x_i and y_i differ. Then L and R proceed with (x_{i+1}, y_{i+1}). If $x_i = y_i$ is verified, then the computation stops immediately. This deterministic algorithm computes n independent identity predicates. We already noted that the average-case communication complexity of the identity predicate is less than 2. Thus $C_{\mathrm{ave}}(p_1, L \leftrightarrow R) \leqslant 2n$. □

Part (e) of Theorem 12 shows that Method 2 is not strong enough to show average-case lower bounds either.

Theorem 12 exhibits a quadratic gap between the nondeterministic and the deterministic two-way communication complexity for a particular predicate. Aho et al. [2] prove that for no predicate this gap is larger than quadratic.

14. THEOREM (Aho, Ullman and Yannakakis [2]). *For any Boolean predicate p we have*

$$C_{\mathrm{det}}(p(x, y), L \leftrightarrow R) \leqslant \min\{C_{\mathrm{ndet}}(p(x, y), L \leftrightarrow R),\ C_{\mathrm{ndet}}(\bar{p}(x, y), L \leftrightarrow R)\}^2.$$

With the above two methods for showing lower bounds on the communication complexity of Boolean predicates we are now able to devise strategies for making statements about the AT^2-complexity of chips computing such predicates.

15. DEFINITION. Let p be a Boolean predicate with n "marked" input bits. p is called *ω-separable* if for every partition of the input bits into sets X, Y such that both X and

Y contain at most $2n/3$ of the marked input bits we have

$$C_{\det}(p(x,y), L\leftrightarrow R) \geqslant \omega.$$

The following theorem emerged over a series of papers [54, 9, 59, 51, 36, 22].

16. THEOREM. *Let p be an ω-separable Boolean predicate. Then $AT^2 = \Omega(\omega^2)$ for every (where- and when-oblivious or strongly where-oblivious) chip computing p.*

PROOF. Let $\chi = (\Gamma, \Lambda, \Delta)$ be a chip computing p. Let π be an input port and let R_π be its associated region in the layout Λ. Let the input bit x_i enter the chip through input port π. With each such x_i we associate a point p_i in the interior of R_π such that different points are associated with different input bits. For the purposes of the lower bound proof we will consider the bit x_i to enter the chip through point p_i. If x_i is a marked input bit we consider p_i a marked point.

We now apply Theorem 2. Thus we cut the layout area M into two halves M_{l} and M_{r}, and induce a partition of the input bits as follows:

$$X = \{x_i \mid p_i \in M_{\mathrm{l}}\}, \qquad Y = \{x_i \mid p_i \in M_{\mathrm{r}}\}.$$

Furthermore X, Y contain at most $2n/3$ marked inputs.

Since the cut C has a length of at most $2\sqrt{A}$, at most $h = \mathrm{O}(\sqrt{A})$ regions of Λ associated with components of Γ can intersect C. This is, because in the VLSI model only O(1) regions can intersect any square of unit area in Λ.

Now consider the computation of p by χ. At each cycle we associate two values with C, a left and a right "crossing" value. The left (right) crossing value $v^{\mathrm{l}} = (v_1^{\mathrm{l}}, \ldots, v_h^{\mathrm{l}})$ ($v^{\mathrm{r}} = (v_1^{\mathrm{r}}, \ldots, v_h^{\mathrm{r}})$) contains a component v_i^{l} (v_i^{r}) for each region R_i intersecting C. If R_i is a (*nand*, *nor*, *and*, *or*) gate, then v_i^{l} (v_i^{r}) is the (*nand*, *nor*, *and*, *or*) of all input values during the last cycle whose regions intersect M_{l} (M_{r}). If R_i is a wire, then v_i^{l} (v_i^{r}) is the above value for its input gate. Ports act as *and*-gates in this context.

With these definitions the computation of p by χ can be regarded as a deterministic algorithm in the sense of Definition 3. The information exchanged between L and R are the crossing values. The left crossing values are sent from L to R, and the right crossing values are sent from R to L. After the last cycle L and R exchange two more bits which make the result of the computation known globally.

If the length of the VLSI computation is T, the communication length of the associated deterministic algorithm is $\mathrm{O}(Th)$. We get $Th = T\sqrt{A} = \Omega(\omega)$, or $AT^2 = \Omega(\omega^2)$. □

Note that if we substitute C_{ave} for $C_{\det}$ in Definition 15, Theorem 16 yields an average-case lower bound. However, as mentioned before we have no methods of proving good average-case lower bounds on the communication complexity of Boolean predicates.

The identity predicate $id(x, y)$ that served as an example for the application of Method 1 is not $\Omega(n)$-separable because we can partition the input bits cleverly to assure that little information has to be exchanged between L and R. Specifically if we choose $X = \{x_1, \ldots, x_{n/2}, y_1, \ldots, y_{n/2}\}$ and Y accordingly, L has to send just one bit to

R, namely the bit $x_1 = y_2 \wedge x_2 = y_2 \wedge \cdots \wedge x_{n/2} = y_{n/2}$. We can modify the identity predicate to yield an $\Omega(n)$-separable Boolean predicate.

17. DEFINITION. Let $id_1(x, y, k)$ be a predicate of three inputs. x and y are n-bit strings ($n = 2^m$) and k is an m-bit string encoding a number $0 \leqslant k < n$. We define

$$id_1(x, y, k) = \begin{cases} 1 & \text{if for all } i = 0, \ldots, n-1, \text{ we have } x_i = y_{i+k} \bmod n, \\ 0 & \text{otherwise.} \end{cases}$$

The predicate id_1 tests whether y is the same as x rotated cyclically by k positions.

18. THEOREM. *With the choice of x and y as the marked input bits of id_1, id_1 is $\Omega(n)$-separable.*

PROOF. The proof is an application of Theorem 27 in Section 7 to the cyclic shift function defined in Section 7. □

By Theorem 16, for every chip computing id_1 we have $AT^2 = \Omega(n^2)$.

We can modify the predicate p_1 in a similar way to carry Theorem 12 over to VLSI. In particular, this yields a predicate p_2 of n inputs such that for every chip computing p_2 we have $AT^2 = \Omega(n^2)$. However, in a straightforward modification of the VLSI model to allow coin tossing on a chip there is a chip that computes p_2 with $AT^2 = O(n^{3/2})$. For details see [39, 14].

Hajnal et al. [19] use Method 2 to prove lower bounds on the communication complexity of several graph properties.

7. The communication complexity of Boolean functions with many outputs

Boolean functions with many outputs do not fit the framework outlined in Section 6, because just to put the output bits on the communication channel will cost a lot. But in VLSI an output bit does not have to be made known to the whole chip. Rather it just has to be known to the output port that produces it. Thus, in multiple output functions output bits may be computed "locally" and we have to measure how much information has to be gathered from far away to compute each output bit.

Therefore, when dividing the inputs between L and R we assign sides to the output bits as well. The corresponding definition is the following.

19. DEFINITION. Let $f: X \times Y \to B^m$. Let I, J be a partition of the output bits. We assume that L has to compute I, and R has to compute J.

A *deterministic algorithm* is again given by two response functions as in Definition 3. But now for each input $x \in X$ and output bit $i \in I$ we have a partial output function $a(x, i): B^* \to B$. Similarly, for each input $y \in Y$ and output bit $j \in J$ we have a partial output function $a(y, j): B^* \to B$. The computation proceeds as in Definition 3. To avoid conflicts we assume that if w is a prefix of w' and $a(x, i)(w)$ is defined, then $a(x, i)(w')$ is defined and $a(x,i)(w')=a(x,i)(w)$. (Similarly for $a(y,j)$.) The computation stops as soon as $a(x,i)(w)$ is defined for all $i\in I$ and $a(y,j)(w)$ is defined for all $j\in J$. $C_{\text{det}}(f, L\leftrightarrow R)$ is defined as in Definition 3.

Definition 19 differs from Definition 3 in that the partial inputs x resp. y can help in determining an output. If $m=1$, then Definition 19 yields a notion of communication complexity which is no greater and at most 2 less than the communication complexity defined in Definition 3. This is because once a side has computed the unique output bit, it can make it globally known by transmitting it to the other side.

The concept of nondeterminism does not make sense for multiple output functions. Las Vegas algorithms can be defined analogously to Definition 5.

There is one method for proving lower bounds on the communication complexity of multiple output functions.

Method 3 (crossing sequences for multiple output functions)

20. DEFINITION. Let $f: X\times Y\to B^m$ and let I, J be a partition of the output bits of f. f has *ω-flow* if there is a partial input $y\in Y$ such that f restricted to $X\times\{y\}$ in its domain and J in its range has more than $2^{\omega-1}$ different points in its range.

21. THEOREM. *If f has ω-flow, then $C_{\text{det}}(f, L\leftrightarrow R)\geqslant\omega$.*

PROOF. The proof is similar to the proof of Theorem 6. Assume that there is a deterministic algorithm computing f that has a communication length of less than ω. Then for two inputs (x_1, y), (x_2, y) generating different output configurations in J, the same communication sequence w is generated. Thus for some $j\in J$ $f_j(x_1, y)\neq f_j(x_2, y)$, but the algorithm computes $f_j(x_1, y)=a(y,j)(w)=f_j(x_2, y)$. Thus the algorithm does not compute f correctly, a contradiction. □

Theorem 21 can be applied to get lower bounds on the AT^2-complexity of chips computing multiple output functions. The corresponding definition and theorem are the following.

22. DEFINITION. Let $f: B^n\to B^m$. Let A be a subset of the input bits and B be a subset of the output bits containing the "marked" I/O bits. f is called *ω-separable* if for all partitions of the I/O bits into sets X, Y such that both X and Y contain at most 2/3 of all marked I/O bits we have

$$C_{\text{det}}(f, L\leftrightarrow R)\geqslant\omega.$$

23. THEOREM. *If $f: B^n\to B^m$ is ω-separable, then $AT^2=\Omega(\omega^2)$ for every chip computing f.*

PROOF. Analogous to the proof of Theorem 16. □

Vuillemin [59] gives an example of a class of functions to which Method 3 applies.

24. DEFINITION. Let $f(x_1,\ldots,x_n,s_1,\ldots s_m)=(y_1,\ldots,y_n)$ be a Boolean function. f computes a permutation group G on n elements if for all $g\in G$ there is an assignment $\alpha_1,\ldots,\alpha_m$ to the $s_1,\ldots,s_m$ such that $f(x_1,\ldots,x_n,\alpha_1,\ldots,\alpha_m)=(x_{g(1)},\ldots,x_{g(n)})$ for all choices of $x_1,\ldots,x_n$. We call $x_1,\ldots,x_n$ the *permutation inputs* and $s_1,\ldots,s_k$ the *control inputs*. f is called *transitive of degree n* if G is a transitive group, i.e., for all $i,j=1,\ldots,n$, there is a $g\in G$ such that $g(i)=j$.

The most straightforward example of a transitive function of degree n is the *cyclic shift* function $cs(x_1,\ldots,x_n,s_1,\ldots s_m)=(y_1,\ldots,y_n)$ where $n=2^m$ and the $s_1,\ldots,s_m$ encode a number k, $0\leqslant k<n$ and $y_i=x_{(i+k)\bmod n}$. cs computes the transitive group of cyclic permutations. Other examples of transitive functions of degree $\Omega(n)$ are the multiplication of n-bit integers, the multiplication of three $\sqrt{n}\times\sqrt{n}$ matrices, the sorting of n numbers between 0 and n, etc.

25. THEOREM. *Let $f: B^{n+m}\to B^n$ be transitive of degree n. Then f is $n/6$-separable.*

PROOF. Let G be the transitive group computed by f. The equivalence relation $g(i)=h(i)$ for fixed but arbitrary $i\in\{1,\ldots,n\}$ divides G into n equivalence classes of size $|G|/n$. Let A be the set of all permutation input bits and let B be the set of all output bits. Let X, Y be any partition of $A\cup B$ such that $|X|,|Y|\leqslant 4n/3$. For each input bit i in X and output bit j in Y there are $|G|/n$ group elements $g\in G$ such that $g(i)=j$. Let w.l.o.g. X be no greater than Y and assume that X contains at least as many input bits as output bits. (The other cases can be argued similarly.) Let S be the set of input bits in X and S' be the set of output bits in Y. Then

$$|S|\cdot|S'|\geqslant n/3\cdot n/2=n^2/6.$$

For each of the pairs $(i,j)\in S\times S'$ there are $|G|/n$ group elements matching them. Since there is only a total of $|G|$ group elements there must be one element $g_0\in G$ realizing at least $n/6$ matchings between inputs in S and outputs in S'. The partial input y realizing the flow sets the control input bits in Y such that together with appropriate assignments to the control input bits in X they encode this element g_0. The other input bits in Y are assigned arbitrarily. □

We can establish a relationship between Method 3 and Method 1.

26. DEFINITION. Let $f: X\to B^m$. The characteristic predicate of f is the predicate $p_f: X\times B^m\to B$ such that

$$p_f(x,y)=\begin{cases}1 & \text{if } y=f(x),\\ 0 & \text{otherwise.}\end{cases}$$

The characteristic predicate of the cyclic shift function cs is the predicate id_1 defined in Definition 17.

27. THEOREM. *Let $f: X\times Y\to B^m$. Let p_f be the characteristic predicate of f. Let I, J be a partition of the output bits of f, and consider the induced partition of the input of p_f. (An* *input bit of p_f corresponding to an output f of f enters on the side on which the output bit is produced.) If f has ω-flow, then p_f fulfils the premise of Theorem* 6 *with* $m=2^{\omega}$.

PROOF. The rows and columns of the identity matrix are defined by the more than $2^{\omega-1}$ input configurations of p_f realizing the flow and their respective output configurations. □

An application of Theorem 27 to the cyclic shift function *cs* proves Theorem 18.

Theorem 12 shows that, as Method 1, Method 3 is also rather weak: if we can prove a lower bound on the communication complexity of f, then we can prove the same nondeterministic lower bound on the communication complexity of p_f.

In general, proving average-case lower bounds on the AT^2-complexity of chips is hard. However there are examples where it can be done. Thompson [54] proves an average-case lower bound for the Discrete Fourier Transform and for sorting in a modified VLSI model. We will now prove an average-case lower bound for the cyclic shift function. The proof is based on two observations. First we note that for the cyclic shift function every assignment of the control input bits implements exactly one group element. Second we observe that the worst-case lower bound on the number of matchings realized simultaneously by a group element (see proof of Theorem 25) can be extended to an average-case lower bound.

28. LEMMA. *Let f be a transitive function of degree n computing the transitive group G. Then for each partition of the I/O bits of f as in Theorem* 25 *at least $|G|/11$ group elements realize each at least $n/12$ matchings across the cut.*

PROOF. Otherwise at most

$$(|G|/11-1)n+(10|G|/11+1)(n/12-1)<n|G|/6$$

matchings across the cut are realized in total. □

29. THEOREM. *For any chip computing the cyclic shift function cs we have $AT^2=\Omega(n^2)$ on the average.*

For the proof of Theorem 29 we need the following technical lemma.

30. LEMMA. *Let $\{w_1,\ldots,w_r\}$ be a set of r different bit strings. Then*

$$\sum_{1\leqslant j\leqslant r} |w_j| \geqslant (\lfloor \log r\rfloor - 2)r.$$

PROOF

$$\sum_{1\leqslant j\leqslant r} |w_j| \geqslant \sum_{1\leqslant j<\lfloor\log r\rfloor} j2^j+(r-2^{\lfloor\log r\rfloor})\lfloor\log r\rfloor \geqslant (\lfloor\log r\rfloor-2)r. \qquad \square$$

PROOF OF THEOREM 29. Let C be the cut bisecting the chip. Let $g \in G$ be one of the $n/11$ group elements that realizes $n/12$ matchings across the cut C. Consider the assignment to the control inputs that encodes g, and any assignment to the $11n/12$ permutation inputs that are not matched across C. By varying the $n/12$ permutation input bits that are matched across C we have to generate $2^{n/12}$ different sequences of bits across C. By Lemma 30 the average length of each bit sequence is thus at least $n/12-2$. Thus the average number of communicated bits for all inputs whose control part encodes g is at least $n/12-2$. This means that $C_{\text{ave}}(cs, L\leftrightarrow R) \geqslant (n/12-2)/11$. The rest follows as in the proof of Theorem 16. □

Note that by the argument known to us from Section 6 it can be shown that $C_{\text{ave}}(id_1, L\leftrightarrow R)=O(1)$ across any cut bisecting a chip computing id_1. Indeed, one can devise chips that compute id_1 with $AT^2=O(n \text{ polylog}(n))$ on the average.

Method 3 applies to many functions that are not transitive, e.g., matrix multiplication [51, 44].

Aho et al. [2] observed that Theorem 23 (and analogously, Theorem 16) can be strengthened in certain cases, by using different notions of communication complexity. Ullman [57] and Siegel [52] also discuss modifications of information flow arguments for proving lower bounds on the AT^2-complexity of VLSI circuits. In particular Ullman presents a notion of information transfer that allows to derive lower bounds on the AT^2-complexity of randomized chips that may make errors in rare cases. Siegel extends the rank lower bound to functions with multiple outputs.

Kedem and Zorat [21] and Kedem [20] extend the theory to chips which can read an input several times.

So far we have discussed quasi planar chips. The lower bound argument described in Section 4 generalizes to a three-dimensional VLSI model. Intuitively in such a model a layout is an embedding into three-dimensional space. The finite resolution property is maintained, but quasi planarity is given up. The lower bound argument goes through as in two-dimensional VLSI, with the modification that the length of a cut line is now replaced with the area of a cut plane. The separator theorem presented in Section 5 generalizes to showing that one can always find a separator whose size is $O(V^{2/3})$ where V is the active volume of the layout [35]. Substituting this into the lower bound argument yields $T=\Omega(\omega/V^{2/3})$ or $T^3V^2=\Omega(\omega^3)$. Three-dimensional VLSI models are discussed in [47, 30].

8. A lower bound on the switching energy of VLSI chips

In Section 7 we proved lower bounds on the AT^2-complexity of chips computing Boolean functions with multiple outputs by measuring the amount of information flow that has to flow across a cut bisecting the chip. We argued that each bit needs some time, namely one cycle, to cross the cut, and that because of the limited length of the cut only a limited number of bits can cross simultaneously. The switching energy essentially counts the number of state changes happening in the circuit. More precisely, the switching energy is defined by the following statement.

Uniswitch model

Each unit of chip area consumes one unit of switching energy every time that it changes its state from 0 to 1 or vice versa.

Therefore if we want to prove a lower bound on the switching energy, considering one cut is not enough. Many of the state changes could happen in circuit components that do not cross the cut. Therefore we must consider many cuts to cover a large part of the circuit. By proving a lower bound on the amount of information flowing across *each* cut, and summing over all cuts, we can account for many state changes happening in the circuit.

Since we have to consider many cuts it will not suffice to consider functions with a great communication complexity *in the worst case*. However a property such as the one proved for transitive functions in Lemma 28 will do.

For the purpose of proving the main theorem in this section let us assume that the chip under consideration fulfils the following additional requirements.

Restriction 1. It is when- and where-oblivious.

Restriction 2. Each bit is communicated through its own private port.

Restriction 3. The layout of the chip is done on a rectangular grid.

31. THEOREM (Lengauer and Mehlhorn [35]). *Let f be a transitive function of degree n. Let E be the worst-case switching energy consumed by any chip computing f that satisfies the above three requirements. Let A be the active chip area and let T be the worst-case computing time. Then*

$$c_1AT^2 \geqslant ET \geqslant \frac{c_2n^2}{\log\dfrac{c_3AT^2}{n^2}} \geqslant 0$$

for appropriate constants $c_1, c_2, c_3 > 0$.

PROOF. Let $\chi=(\Gamma, \Lambda, \Delta)$ be any chip computing f that fulfils the above three requirements. The upper bound on ET is trivial. We prove the lower bound in two steps. First we define a set of rectilinear cuts through the chip. Second we count state changes in the circuit components crossing the cuts.

In the first step we cut the chip according to a technique devised by Thompson [54]. To this end we overlay the chip with a square grid of mesh size λ. Then we define cuts $C_1, \ldots, C_Q$ bisecting the chip as shown in Fig. 3.

Hereby each cut consists of several sections as shown in Fig. 4.

The middle section of cut C_i has a length of at most $(2i-1)\lambda$. The vertical sections of all cuts C_i are disjoint. Each cut induces a partition of the permutation input bits A into sets X, Y and of the output bits B into sets I, J such that $|X|+|I|, |Y|+|J| \leqslant 4n/3$. We will count state changes happening on the vertical section of the cuts. Let L_i be the number of circuit components crossing C_i and let l_i be the number of circuit components crossing the vertical sections of C_i. Then $l_i \geqslant L_i - c_0 i$ for some appropriate constant $c_0 > 0$.

In the second step we start by noting that there must be a group element g_0 such that $\Omega(n)$ matchings are realized by g_0 across each of $\Omega(Q)$ cuts.

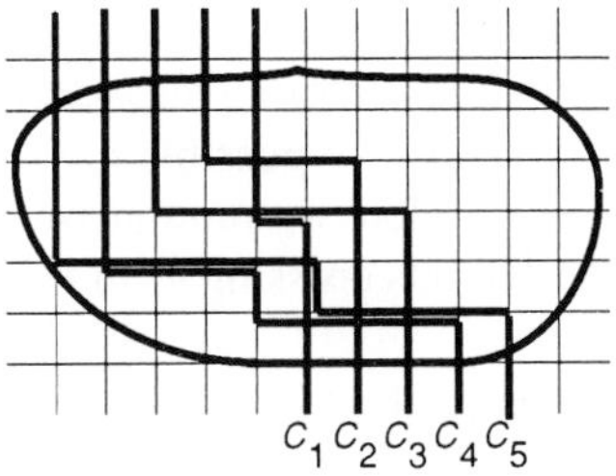

Fig. 3.

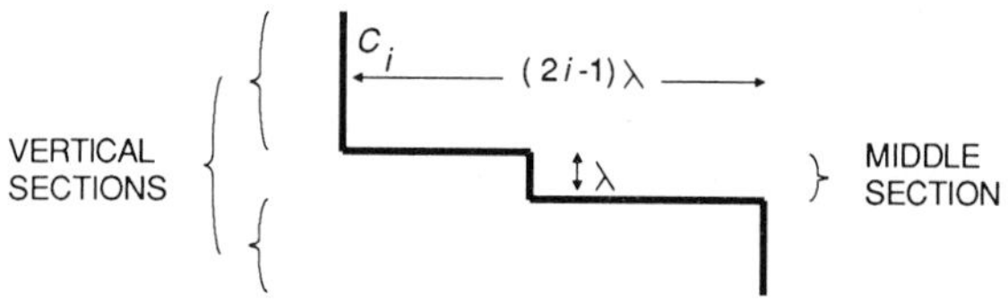

Fig. 4.

32. LEMMA. *There is a* $g_0 \in G$ *such that for each of* $Q/11$ *of the cuts* C_i *at least* $n/12$ *matchings are realized.*

PROOF. Analogous to Lemma 28. □

In the sequel we denote the set of cuts identified by Lemma 32 by Φ.

We will prove a lower bound on the average-case switching energy consumed for computing g_0 on any configuration of the permutation inputs. This is then also a lower bound for the worst-case energy for computing f on any input.

Before we start the actual argument we discuss an encoding that expresses bit strings in terms of the state changes happening in them.

33. DEFINITION. Let w be an arbitrary bit string $w=0^{l_1}1^{l_2}0^{l_3}\dots{}^{l_t}$.

(a) $s(w)=t$, i.e., $s(w)$ is the number of state changes in w *plus* 1.

(b) *bin*(w) is the bit string obtained from w after substituting each 0 with 00 and each 1 *with* 11.

(c) *compress*(w)$=0\ bin(l_1)\ 01\ bin(l_2)\ 01\ bin(l_3)\ 01 \dots 01\ bin(l_1)$. If the first bit of w is a 1 so is the first bit of *compress*(w).

34. EXAMPLE. *compress*(00011)$=$0 11 11 01 11 00.

35. LEMMA. $|compress(w)| \leqslant 4s(w)+2s(w)\log(|w|/s(w))$.

PROOF

$$|compress(w)| \leqslant 2s(w)+2\sum_{j=1}^{s(w)}(1+\log l_j) \leqslant 4s(w)+2\sum_{j=1}^{s(w)}\log l_j$$

$$\leqslant 4s(w)+2\log\left(\left(\left(\prod_{j=1}^{s(w)} l_j\right)^{1/s(w)}\right)^{s(w)}\right) \leqslant 4s(w)+2\log\left(\sum_{j=1}^{s(w)}\frac{l_j}{s(w)}\right)^{s(w)}$$

$$\leqslant 4s(w)+2s(w)\log\frac{|w|}{s(w)}.$$

Here the next-to-last inequality is derived using the well-known fact that the geometric mean is no greater than the arithmetic mean. □

We now consider an arbitrary but fixed cut C_i in Φ. We choose an arbitrary but fixed assignment of the $11n/12$ permutation inputs that are not matched across C_i by g_0 and let the $n/12$ permutation inputs matched across C_i by g_0 vary in all $2^{n/12}$ possible ways. Thus we generate $2^{n/12}$ different bit sequences (w_{ij}, h_{ij}) across C_i. Here w_{ij} summarizes all bits crossing C_i in the vertical sections and h_{ij} summarizes all bits crossing C_i in the middle section $(1 \leqslant j \leqslant 2^{n/12})$.

Using Lemma 30 we now compute an upper bound on the length of *compress*(w_{ij}) averaged over all $1 \leqslant j \leqslant 2^{n/12}$.

36. Lemma

$$\sum_{1 \leqslant j \leqslant 2^{n/12}} |compress(w_{ij})| \geqslant \left(\frac{n}{12} - c_0 iT - 3\right) 2^{n/12}.$$

Proof. Since the length of the middle section of cut C_i is at most $(2i-1)\lambda$ there are at most $c_0 i$ circuit components that intersect the middle section of cut C_i. Thus there are at most t, $1 \leqslant t \leqslant 2^{c_0 iT}$, different sequences h_{ij}. For each such sequence there is a number u_k of different sequences w_{ij} such that (w_{ij}, h_{ij}) is generated by some input as described above. We have $U := \sum_{1 \leqslant k \leqslant t} u_k \geqslant 2^{n/12}$. Since the mapping $w \to compress(w)$ is injective all strings *compress*(w_{ij}) are distinct. Thus by Lemma 30

$$\begin{aligned}\sum_{1 \leqslant j \leqslant 2^{n/12}} |compress(w_{ij})| &\geqslant \sum_{1 \leqslant k \leqslant t} (\log u_k - 3) u_k \\ &\geqslant U \log(U/t) - 3U \\ &\geqslant 2^{n/12}(n/12 - c_0 iT - 3).\end{aligned}$$

Here we used the fact that $\sum_{1 \leqslant k \leqslant t} u_k \log u_k$ is minimum if for all k, $u_k = U/t$. □

Using the upper bound on $|compress(w)|$ derived in Lemma 35 we can now give a lower bound on the average number of state changes in all sequences w_{ij}. Let $S_i = \sum_{1 \leqslant j \leqslant 2^{n/12}} s(w_{ij})/2^{n/12}$ be the average number of state changes occurring in any w_{ij} $(1 \leqslant j \leqslant 2^{n/12})$.

37. Lemma. $(n/12 - c_0 iT - 3) \leqslant 4S_i + 2S_i \log(Tl_i/S_i)$.

Proof. Apply Lemma 35 and Lemma 36 and observe that $\sum_{1 \leqslant j \leqslant 2^{n/12}} s(w_{ij}) \times \log(Tl_i/s(w_{ij}))$ is maximum if for all j, $s(w_{ij}) = S_i$. □

We now sum over all cuts C_i in Φ. Let $l := \sum_{i+1}^{Q/11} l_i$ and $S := \sum_{i=1}^{Q/11} S_i$. Thus S is the average number of state changes in the sequences w_{ij} across all cuts in Φ.

38. Lemma. $\sum_{1 \leqslant i \leqslant Q/11} (n/12 - c_0 QT - 3) \leqslant 4S + 2S \log(Tl/S)$.

Proof. We manipulate the formula in a way similar to the proof of Lemma 37. □

Choosing $Q = (n-36)/24c_0 T$ yields the next lemma.

39. LEMMA. *For suitable constants $c_2, c_3 > 0$ and sufficiently large n we get*

$$ST \geqslant \frac{c_2 n^2}{\log \dfrac{c_3 T^2 l}{n^2}}.$$

PROOF. Substituting the value for Q into Lemma 38 yields

$$\frac{1}{11}\left(\frac{n-36}{24}\right)^2 \frac{1}{c_0 t} \leqslant 4S\left(1+\tfrac{1}{2}\log\frac{Tl}{S}\right). \qquad (*)$$

Thus

$$\log S \geqslant \log\left(\left(\frac{n-36}{24}\right)^2 \frac{1}{4c_0 T}\right) - \log\left(1+\tfrac{1}{2}\log\frac{Tl}{S}\right).$$

Applying the inequality $\log(1+x) \leqslant x$ we get

$$\log S \geqslant 2\log\left(\left(\frac{n-36}{24}\right)^2 \frac{1}{4c_0 T}\right) - \log(Tl).$$

Substituting this into $(*)$ proves the lemma. □

PROOF OF THEOREM 31 (*continued*). We are left with relating S to the switching energy and bounding l from above. Both can be done with the following argument.

Let c be a component crossing the vertical section of cut C_i. It can be shown that we can charge to c an area of size $\Omega(1)$ inside the layout of component c such that areas charged to different components overlap at most ν-fold. From this it immediately follows that $l = O(A)$. Furthermore $S = O(E + A)$. (We have to add A here for the state changes in w_{ij} that can occur when switching from one component crossing C_i to the next. Those state changes in w_{ij} do not correspond to quanta of switching energy dissipated on the chip.) Thus

$$ET \geqslant \frac{c_2 n^2}{\log \dfrac{c_3 A T^2}{n^2}} - c_4 A.$$

If we assume that on each circuit component there will be at least one state change, say, during initialization, then the last term can be omitted proving the theorem. □

Note that if the chip is AT^2-optimal, i.e., if $AT^2 = O(n^2)$, then the above lower bound on E is tight up to a constant factor. This shows that AT^2-optimal chips use their computing resources to capacity (up to a constant factor). This means that in an AT^2-optimal chip on the average at each time a constant fraction of the chip area changes state.

In Theorem 31 we imposed three restrictions on the circuit computing the transitive Boolean function. Let us now remove the restrictions one by one.

Restriction 1. In Section 4 we remarked that the cutting argument generalizes to strongly where-oblivious chips.

Restriction 2. If we assume that at most cn I/O bits share a port, where $0<c<1$, then we can construct a sequence of cuts, in which each cut partitions the chips such that at most $(c+1)n/2$ bits are located on each side of the cut. The additional factor of $c+1$ in this formula decreases the constant factor in the lower bound on the switching energy, but the argument still goes through. Obviously if more than cn bits share a port, then $T>cn$. Since the permutation input bits and the corresponding output bits can assume any value, $E=\Omega(n)$ and thus $ET=\Omega(n^2)$ follows in this case.

Restriction 3. We can superimpose a grid of mesh size λ on an arbitrary chip layout satisfying the requirements of the model introduced in Section 3. Then L_i has to be defined as the number of circuit components that intersect C_i, and l_i is the number of circuit components that intersect the vertical sections of C_i.

The logarithmic factor in the denominator in Theorem 31 is an artifact of our analysis. It gets introduced in Lemma 35 where we bound the length of a coding of the sequence of state changes. This suggests that the denominator may be eliminated with a more sophisticated technique. Indeed Aggarwal et al. [1] show how to do so.

9. Upper bounds on the AT^2-complexity of VLSI chips

In many cases the lower bounds on the AT^2-complexity that can be obtained using the methods described in this chapter can be matched with tight upper bounds. This shows that indeed the communication complexity often dominates the AT^2-complexity of a VLSI chip.

In this section we give an overview of methods for developing tight upper bounds w.r.t. the AT^2-measure. Such bounds can also be developed for several other complexity measures. In fact, what we are doing is constructing (asymptotically) efficient VLSI chips.

This first step of this process is to find a network topology in which to arrange the processors such as to allow for efficient communication among them. It turns out that for many functions such topologies are highly regular and, in fact, a comparatively small number of topologies covers a wide range of applications [13]. Therefore solutions for instances of the special purpose architecture problem can also contribute to solving general purpose architecture problems.

The set of promising network topologies can be classified into ones with dense interconnections (*high bandwidth*) and ones with sparse interconnections (*low bandwidth*). For technical reasons one is limited to bounding the degree of the network by a small constant. (Ports are expensive in VLSI.) Thus the sparsity is not measured by the number of edges but by the *diameter* of the network. The diameter is the maximum distance between any pair of nodes of the network, measured in the number of edges that have to be traversed to get from one node to the other. Roughly the diameter of a high-bandwidth network with n nodes is about $O(\log n)$, whereas the diameter of a low-bandwidth network is $\Omega(\sqrt{n})$.

Popular instances of high-bandwidth networks are the *shuffle-exchange graph* (see

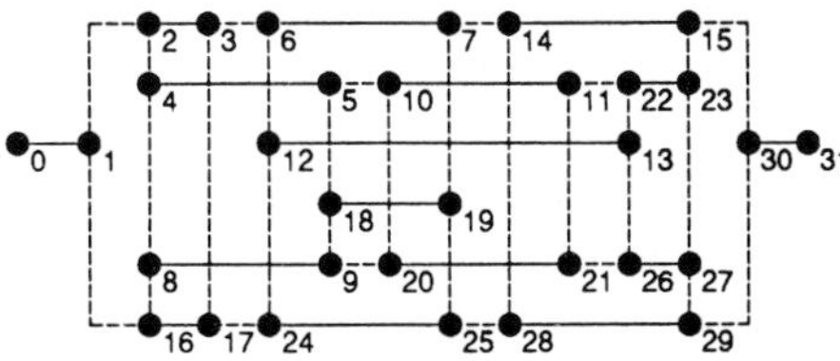

Fig. 5. An efficient layout of a small shuffle-exchange graph.

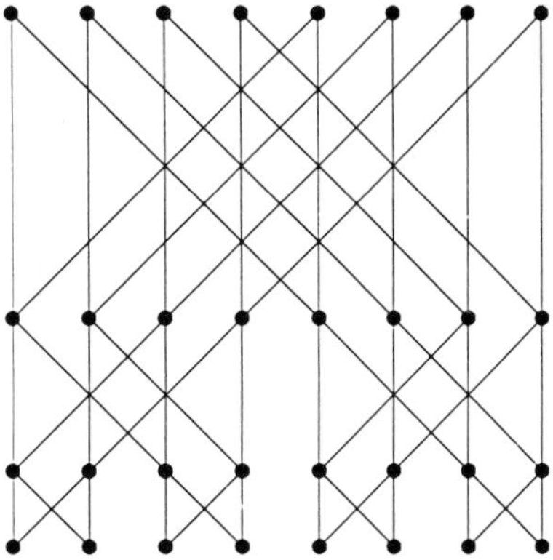

Fig. 6. A butterfly network.

Fig. 5) and its unfolded variant the *butterfly network* (see Fig. 6) as well as the *hypercube* and its bounded-degree version, the *cube-connected cycles* (see Fig. 7). These networks are all intimately related; in fact they can be transformed into each other by straightforward graph operations [57]. They all need large wiring area. Specifically an asymptotically tight bound for the area needed to layout a shuffle exchange graph resp. cube-connected cycle network with n nodes is $\Theta(n^2/(\log n)^2)$ [27, 45]. On the other hand, many functions can be computed quickly on these networks. Examples are sorting which can be done in $O((\log n)^2)$ cycles of arithmetic operations [53, 56], and the Discrete Fourier Transform which can be done in $O(\log n)$ cycles [45].

To map these results onto our VLSI model, one has to translate them from using *word* to using *bit* operations, because the length of the numbers involved contributes to the length of the input and to the complexity of the computation. This fact complicates the issue. If the input consists of n integers, each between $-n+1$ and $n-1$, then the length of the input is $N=\Theta(n \log n)$. A lower bound for the AT^2-complexity of sorting and Discrete Fourier Transform in our bit-oriented VLSI model can then be proved to be $AT^2=\Omega(n^2(\log n)^2)=\Omega(N^2)$ [29, 54]. A bit-serial implementation of the Discrete Fourier Transform on the cube-connected cycles takes area $\Theta(n^2/(\log n)^2)$ and time

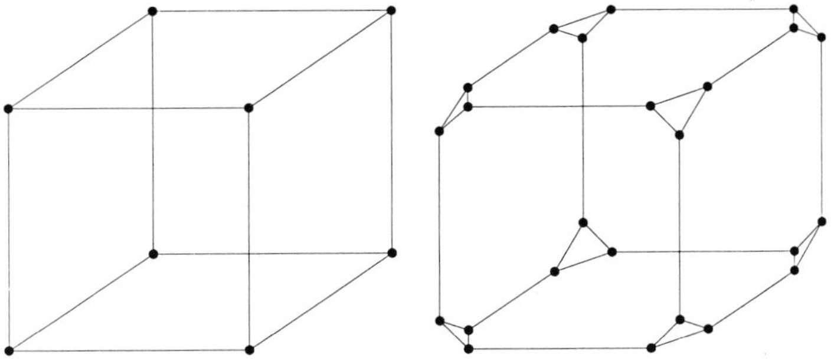

Fig. 7. Hypercube and cube-connected cycles.

$T=O((\log n)^2)$ and is thus AT^2-optimal also w.r.t. bit operations. Sorting on the cube-connected cycles leads to suboptimal chips because of the $O((\log n)^2)$ cycles of word operations needed. But with some modifications AT^2-optimal chips can be achieved for sorting, as well [5].

High-bandwidth networks are characterized by a high speed of operation and a high amount of wiring area. On the other hand, low-bandwidth networks operate more slowly but require much less area. Typical examples of low-bandwidth networks are meshes of various types (see Fig. 8). Meshes are the basis of many *systolic algorithms*. In

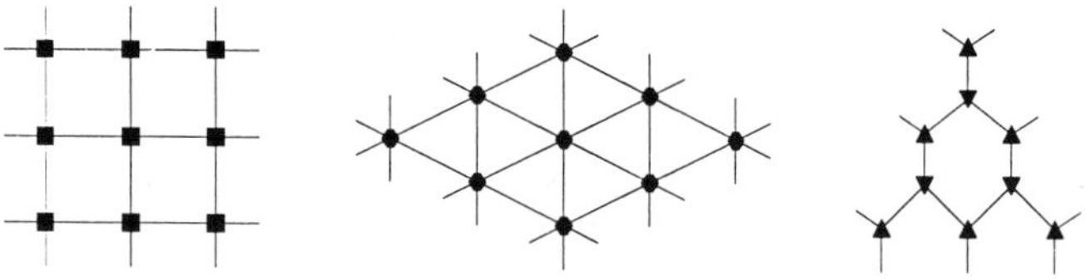

Fig. 8. Various types of meshes.

a systolic algorithm data are pulsed through the network in regular sequences, and the processors are data-driven, i.e., the operation that a processor has to perform is independent of the data [38, Ch. 8]. There are general purpose extensions to this concept which allow data to incorporate control information [23]. A large number of algorithms on meshes, especially of systolic algorithms has been reported for various problems including matrix problems, e.g., [10, 49, 12], graph problems, e.g., [17], sorting and permutation algorithms, e.g., [56, 24], and even data structures, e.g., [18]. Overviews on systolic algorithms can be found in [57]. A structural investigation of systolic systems is given in [16]. Leiserson and Saxe [32] present a general mechanism for optimizing the delay in systolic systems. As with high-bandwidth networks transformations have been identified that port systolic algorithms from one mesh type to the other [48].

The complete binary tree is not a popular processor network because it has a communication bottleneck at the root. In order to circumvent this problem two ways of merging the tree with a square mesh have been proposed, the *mesh of trees* (see Fig. 9) and the *tree of meshes* (see Fig. 10) [27]. The mesh of trees is a high-bandwidth network with diameter $O(\log n)$ that can be laid out in area $\Theta(n(\log n)^2)$ [27]. It has interesting applications [28, 42]. The tree of meshes can be laid out in area $\Theta(n \log n)$ [3]. It has been established as a suitable host graph in which one can efficiently embed any processor network [3]. This qualifies the tree of meshes as a processor network for general purpose applications [31].

The goal of theoretical research on upper bounds in VLSI theory is to find, for a given Boolean function, asymptotically optimal chips (w.r.t. the AT^2-measure or other complexity measures) for all eligible values of T. For instance the cube-connected cycles with adequate processor designs provide an AT^2-optimal chip for the Discrete Fourier Transform with $T=O((\log N)^2)$ bit operations. On the other hand, in principle, $AT^2=O(N^2)$ can be achieved for $\Omega(\log N)=T=O(\sqrt{N})$, since $T=\Omega(\log N)$ and

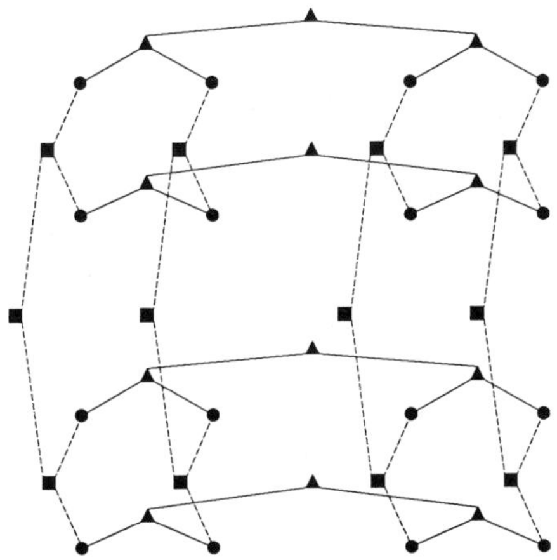

Fig. 9. A mesh of trees.

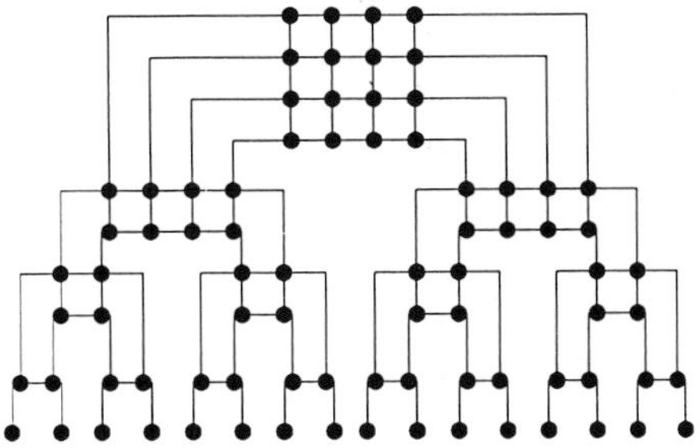

Fig. 10. A tree of meshes.

$A = \Omega(N)$ are trivial lower bounds, and $AT^2 = \Omega(n^2)$. By applying techniques such as pipelining and folding AT^2-optimal chips for the Discrete Fourier Transform have been found in the range of $\Omega((\log N)^2) = T = O(\sqrt{N})$ [55, 45]. Recently, Bilardi and Sarrafzadeh [7] have developed AT^2-optimal chips for the Discrete Fourier Transform in the whole range of interesting values for T, i.e., for $\Omega(\log N) = T = O(\sqrt{N})$. For sorting an analogous result has been proved in [6]. The same holds also for integer multiplication [41] and multiplication in Galois fields [15]. In the case of integer division AT^2-optimal chips are only known for $T = \Omega((\log n)^{1+\varepsilon})$ [40]. It should be noted that most of these AT^2-optimal designs are highly impractical in practice.

Several of the processor networks we mentioned, particularly the regular high-bandwidth networks and the tree of meshes are suitable network topologies for general purpose VLSI computers [13].

Acknowledgment

This chapter was written while the author was at XEROX PARC, 3333 Coyote Hill Rd., Palo Alto, CA 94304, USA.

References

[1] AGGARWAL, A., A.K. CHANDRA and P. RAGHAVAN, Energy consumption in VLSI circuits, in: *Proc. 20th Ann. ACM Symp. on Theory of Computing* (1988) 205–216.

[2] AHO, A.V., J.D. ULLMAN and M. YANNAKAKIS, On notions of information transfer in VLSI circuits, in: *Proc. 15th Ann. Symp. on Theory of Computing* (1983) 133–139.

[3] BHATT, S.N. and F.T. LEIGHTON, A framework for solving VLSI graph layout problems, *J. Comput. System Sci.* **28**(2) (1984) 300–343.

[4] BILARDI, G., M. PRACCHI and F.P. PREPARATA, A critique and appraisal of VLSI models of computation, *IEEE J. Solid-State Circuits* **17** (1982) 696–702.

[5] BILARDI, G. and F.P. PREPARATA, An architecture for bitonic sorting with optimal VLSI performance, *IEEE Trans. Comput.* **33**(7) (1984) 646–651.

[6] BILARDI, G. and F.P. PREPARATA, A minimum area VLSI network for O(log n) time sorting, *IEEE Trans. Comput.* **34**(4) (1985) 336–343.

[7] BILARDI, G. and M. SARRAFZADEH, Optimal VLSI circuits for discrete Fourier transform, in: F.P. Preparata, ed., *Advances in Computing Research, Vol. 4* (JAI Press, Greenwich, CT, 1987) 87–101.

[8] BRENT, R.P. and L.M. GOLDSCHLAGER, Some area-time tradeoffs in VLSI, *SIAM J. Comput.* **11**(4) (1982) 737–747.

[9] BRENT, R.P. and H.T. KUNG, The area-time complexity of binary multiplication, *J. Assoc. Comput. Mach.* **28**(3) (1981) 521–524.

[10] BRENT, R.P. and F.T. LUK, A systolic array for the linear time solution of Toeplitz systems and equations, *J. VLSI Systems Sci.* **1**(1) (1983) 1–22.

[11] CHAZELLE, B. and L. MONIER, A model of computation for VLSI with related complexity results, in: *Proc. 13th Ann. ACM Symp. on Theory of Computing* (1981) 318–325.

[12] DELOSME, J. and I.C.F. IPSEN, Parallel solution of symmetric positive definite systems with hyperbolic rotations, *Linear Algebra Appl.* **77** (1986) 75–111.

[13] FENG, T., A survey of interconnection networks, *Computer* **14**(11) (1981) 12–27.

[14] FÜRER, M., The power of randomness for communication complexity, in: *Proc. 19th Ann. ACM Symp. on Theory of Computing* (1987) 178–181.

[15] FÜRER, M. and K. MEHLHORN, AT^2-optimal Galois-field multiplier for VLSI, in: *VLSI Algorithms and Architectures, 3rd Aegaen Workshop*, Lecture Notes in Computer Science, Vol. 227 (Springer, Berlin, 1986) 217–225.

[16] GRUSKA, J., Systolic architectures, systems and computations, in: T. Lepistö and A. Salomaa, eds., *Proc. 15th Internat. Coll. on Automata, Languages, and Programming*, Lecture Notes in Computer Science, Vol. 317 (Springer, Berlin, 1988) 254–270.

[17] GUIBAS, L., H.T. KUNG and C.D. THOMPSON, Direct VLSI implementation of combinatorial algorithms, in: *Proc. Caltech Conf. on VLSI* (1979) 309–325.

[18] GUIBAS, L.J. and F.M. LIANG, Systolic stacks, queues and counters, in: P. Penfield Jr., ed., *Proc. MIT Conf. on Advanced Research in VLSI* (Artech House, Dedham, MA, 1982) 155–164.

[19] HAJNAL, A., W. MAASS and G. TURAN, On the communication complexity of graph properties, in: *Proc. 20th Ann. ACM Symp. on Theory of Computing* (1988) 186–191.

[20] KEDEM, Z., Optimal allocation of computational resources in VLSI, in: *Proc. 23rd Ann. IEEE Symp. on Foundations of Computer Science* (1982) 379–386.

[21] KEDEM, Z. and Z.A. ZORAT, On relations between input and communication/computation in VLSI, in: *Proc. 22nd Ann. IEEE Symp. on Foundations of Computer Science* (1981) 37–44.

[22] KOLLA, R., Where-oblivious is not sufficient, *Inform. Process. Lett.* **17**(5) (1983) 263–268.

[23] KUNDE, M., H. LANG, M. SCHIMMLER, H. SCHMECK and H. SCHRÖDER, The instruction systolic array and its relation to other models of parallel computers, *Parallel Comput.* **7**(1) (1988) 25–39.

[24] LANG, H., M. SCHIMMLER, H. SCHMECK and H. SCHRÖDER, Systolic sorting on a mesh-connected network, *IEEE Trans. Comput.* **34**(7) (1985) 652–658.

[25] LAWLER, E.L., *Combinatorial Optimization: Networks and Matroids* (Holt, Rinehart and Winston, New York, 1976).

[26] LAWLER, E.L., J.K. LENSTRA, A.H.G. RINNOOY KAN and D.B. SHMOYS, *The Traveling Salesman Problem: A Guided Tour of Combinatorial Optimization* (Wiley, New York, 1985).

[27] LEIGHTON, F.T., *Complexity Issues in VLSI* (MIT Press, Cambridge, MA, 1983).

[28] LEIGHTON, F.T., Parallel computation using meshes of trees, in: *Proc. 9th Internat. Workshop on Graphtheoretic Concepts in Computer Science* (Trauner, Linz, 1983) 200–218.

[29] LEIGHTON, F.T., Tight bounds on the complexity of parallel sorting, *IEEE Trans. Comput.* **34**(4) (1985) 344–354.

[30] LEIGHTON, F.T. and A.L. ROSENBERG, Three-dimensional circuit layouts, *SIAM J. Comput.* **15**(3) (1986) 793–813.

[31] LEISERSON, C.E., Fat-trees: universal network for hardware-efficient supercomputing, in: *Internat. Conf. on Parallel Processing* (1985) 393–402.

[32] LEISERSON, C.E. and J.B. SAXE, Retiming synchronous circuitry, Tech. Report, DEC System Research Center, Palo Alto, CA, 1986.

[33] LENGAUER, T., *Combinatorial Algorithms for Integrated Circuit Layout*, Teubner–Wiley Series of Applicable Theory in Computer Science (Teubner/Wiley, Stuttgart/New York, 1990).

[34] LENGAUER, T., The combinatorial complexity of layout problems, in: B.T. Preas and M.J. Lorenzetti, eds., *Physical Design Automation of VLSI Systems* (Benjamin/Cummings, Menlo Park, CA, 1988) Ch. 10, 461–497.

[35] LENGAUER, T. and K. MEHLHORN, Four results on the complexity of VLSI computations, in: F.P. Preparata, ed., *Advances in Computing Research, Vol. 2: VLSI Theory* (JAI Press, Greenwich, CT, 1984) 1–22.

[36] LIPTON, R.J. and R. SEDGEWICK, Lower bounds for VLSI, in: *Proc. 13th Ann. ACM Symp. on Theory of Computing* (1981) 300–307.

[37] LIPTON, R.J. and R.E. TARJAN, A separator theorem for planar graphs, *SIAM J. Appl. Math.* **36**(2) (1979) 177–189.

[38] MEAD, C. and L. CONWAY, *Introduction to VLSI Systems* (Addison-Wesley, Reading, MA, 1980).

[39] MEHLHORN, K. and E. MEINECKE-SCHMIDT, Las Vegas is better than determinism in VLSI and distributed computing, in: *Proc. 14th Ann. ACM Symp. on Theory of Computing* (1982) 330–337.

[40] MEHLHORN, K. and F.P. PREPARATA, Area-time optimal division for $T=\Omega((\log n)^{1+\varepsilon})$, *Inform. and Comput.* **62**(3) (1987) 270–282.

[41] MEHLHORN, K. and F.P. PREPARATA, Area-time optimal VLSI integer multiplier with minimum computation time, *Inform. and Control* **58** (1983) 137–156.

[42] NATH, D., S.N. MAHESHWARI and P.C.P. BHATT, Efficient VLSI networks for parallel processing based on orthogonal trees, *IEEE Trans. Comput.* **32**(6) (1983) 569–581.

[43] PAPADIMITRIOU, C.H. and K. STEIGLITZ, *Combinatorial Optimization: Algorithms and Complexity* (Prentice Hall, Englewood Cliffs, NJ, 1982).

[44] PREPARATA, F.P. and J. VUILLEMIN, Area-time optimal VLSI networks for multiplying matrices, *Inform. Process. Lett.* **11**(2) (1980) 77–80.

[45] PREPARATA, F.P. and J. VUILLEMIN, The cube-connected cycles: a versatile network for parallel computation, *Comm. ACM* **24**(5) (1981) 300–309.

[46] REISCHUK, R., Parallel machines and their communication theoretical limits, in: *Proc. STAC 86, 3rd Ann. Symp. on Theoretical Aspects of Computing*, Lecture Notes in Computer Science, Vol. 210 (Springer, Berlin, 1986) 359–368.

[47] ROSENBERG, A.L., Three-dimensional integrated circuitry, in: H.T. Kung, B. Sproull and G. Steele, eds., *Proc. CMU Conf. on VLSI Systems and Computations* (Computer Science Press, Rockville, MD, 1981) 69–80.

[48] ROTE, G., On the connection between hexagonal and unidirectional rectangular systolic arrays, in: F. Makedon, K. Mehlhorn, T. Papatheodorou and P. Spirakis, eds., *AWOC'86: VLSI Algorithms and Architectures*, Lecture Notes in Computer Science, Vol. 227 (Springer, Berlin, 1986) 70–83.

[49] ROTE, G., A systolic array algorithm for the algebraic path problem (shortest path, matrix inversion), *Computing* **34**(3) (1985) 191–219.

[50] SAHNI, S., A. BHATT and R. RAGHAVAN, The complexity of design automation problems, in: *Proc. Automation 86 High Technology Computer Conf.* (Van Nostrand-Reinhold, New York, 1986) 82–98.

[51] SAVAGE, J.E., Planar circuit complexity and the performance of VLSI algorithms, in: H.T. Kung, B. Sproull and G. Steele, eds., *CMU Conf. on VLSI Systems and Computations* (Computer Science Press, Rockville, MD, 1981) 61–68.

[52] SIEGEL, A.R., Aspects of information flow in VLSI circuits, in: *Proc. 18th Ann. ACM Symp. on Theory of Computing* (1986) 448–459.

[53] STONE, H.S., Parallel processing with the perfect shuffle, *IEEE Trans. Comput.* **20**(2) (1971) 153–161.

[54] THOMPSON, C.D., A complexity theory for VLSI, Ph.D. Thesis, Dept. Comput. Sci., Carnegie-Mellon Univ., Pittsburgh, PA, 1980.

[55] THOMPSON, C.D., Fourier transforms in VLSI, *IEEE Trans. Comput.* **32**(11) (1983) 1047–1057.

[56] THOMPSON, C.D., The VLSI complexity of sorting, *IEEE Trans. Comput.* **32**(12) (1983) 1171–1184.

[57] ULLMAN, J.D., *Computational Aspects of VLSI* (Computer Science Press, Rockville, MD, 1984).

[58] VAN LAARHOVEN, P.J.M. and E.H.L. AARTS, *Simulated Annealing: Theory and Applications* (Reidel, Boston, MA, 1987).

[59] VUILLEMIN, J., A combinatorial limit to the power of VLSI circuits, in: *Proc. 12th Ann. ACM Symp. on Theory of Computing* (1980) 294–300.

[60] WESTE, N.H.E. and K. ESHRAGIAN, *Principles of CMOS VLSI Design: A Systems Perspective* (Addison-Wesley, Reading, MA, 1985).

[61] YAO, A.C., Some complexity questions related to distributive computing, in: *Proc. 11th Ann. ACM Symp. on Theory of Computing* (1979) 209–213.

CHAPTER 17

Parallel Algorithms for Shared-Memory Machines

Richard M. KARP

Department of Computer Science, University of California at Berkeley, Berkeley, CA 94720, USA

Vijaya RAMACHANDRAN

Department of Computer Sciences, University of Texas at Austin, Austin, TX 78712, USA

Contents

HANDBOOK OF THEORETICAL COMPUTER SCIENCE
Edited by J. van Leeuwen

1. Introduction

Parallel computation is rapidly becoming a dominant theme in all areas of computer science and its applications. It is likely that, within a decade, virtually all developments in computer architecture, systems programming, computer applications and the design of algorithms will be taking place within the context of parallel computation.

In preparation for this revolution, theoretical computer scientists have begun to develop a body of theory centered around parallel algorithms and parallel architectures. Since there is no consensus yet on the appropriate logical organization of a massively parallel computer, and since the speed of parallel algorithms is constrained as much by limits on interprocessor communication as it is by purely computational issues, it is not surprising that a variety of abstract models of parallel computation have been pursued.

Closest to the hardware level are the VLSI models, which focus on the technological limits of today's chips, in which gates and wires are packed into a small number of planar layers. At the next level of abstraction are those models in which a parallel computer is viewed as a set of processors interconnected in a fixed pattern, with each processor having a small number of neighbors.

At one further remove from physical reality is the parallel random-access machine (PRAM), in which it is assumed that, in addition to the private memories of the processors, there is a shared memory, and that any processor can access any cell of that memory in unit time. The PRAM cannot be considered a physically realizable model, since, as the number of processors and the size of the global memory scales up, it quickly becomes impossible to provide a constant-length data path from any processor to any memory cell. Nevertheless, the PRAM has proved to be an extremely useful vehicle for studying the logical structure of parallel computation in a context divorced from issues of parallel communication, and it is the focus of attention in the present survey. Algorithms developed for other, more realistic, models are often based on algorithms originally designed for the PRAM.

Many studies of algorithms and complexity in the PRAM model focus on the trade-off between the time for a parallel computation and the number of processors required. In a practical situation the number of processors available is fixed, but our theoretical studies are enriched if we let the number of processors grow as a function of the size of the problem instance being solved. Of particular interest are so-called *efficient* algorithms, which run in *polylog time* (i.e., the parallel computation time is bounded by a fixed power of the logarithm of the size of the input), and in which the processor–time product exceeds the number of steps in an optimal sequential algorithm by at most a polylog factor. Section 2 surveys efficient PRAM algorithms for bookkeeping operations such as compacting an array by squeezing out its "dead" elements, for evaluating algebraic expressions, for searching a graph and decomposing it into various kinds of components, and for sorting, merging and selection. These efficient parallel algorithms are typically completely different from the best sequential algorithms for the same problems, and their discovery has required the creation of a new set of paradigms for the construction of parallel algorithms.

In the PRAM model there is the possibility of read- and write-conflicts, in which two or more processors try to read from or write into the same memory cell concurrently. Distinctions in the way these conflicts are handled lead to several different variants of the model. The weakest of these is the exclusive-read exclusive-write (EREW) PRAM, in which concurrent reading or writing are forbidden; of intermediate strength is the concurrent-read exclusive-write (CREW) PRAM, which allows concurrent reading but not concurrent writing; and strongest of all is the concurrent-read concurrent-write (CRCW) PRAM, which permits both kinds of concurrency. Several varieties of CRCW PRAMs have been defined; they differ in their method of resolving write conflicts. In Section 3 it is shown that, although these variants do not differ very greatly in computation speed, the CREW PRAM is strictly more powerful than the EREW PRAM and strictly less powerful than the CRCW PRAM. It is also shown that certain simple tasks, such as multiplying two n-bit numbers, inherently require time $\Omega(\log n/\log\log n)$ even on the strongest PRAM model provided the number of processors is polynomial-bounded in the size of the input.

Section 3 goes on to study the relationship between the PRAM and other abstract models of parallel computation, such as Boolean circuits, alternating Turing machines and vector machines. It turns out that all these models are equivalent in their ability to solve problems in polylog time using a polynomial-bounded number of computing elements (processors or gates). This motivates the definition of NC as the class of problems that can be solved within these resource bounds by deterministic algorithms. Two important refinements of this result are presented, each showing that certain parallel computation models are equivalent in their ability to solve problems in time $O(\log^k n)$, where k is a fixed positive integer, using a polynomial-bounded amount of hardware. For PRAMs the amount of hardware is measured by the number of processors, for Boolean circuits by the number of wires and for alternating Turing machines by the number of possible configurations, which is exponential in the space. The first refinement, by Ruzzo [203], states that alternating Turing machines are equivalent to bounded fan-in circuits; the second, by Stockmeyer & Vishkin [220], states that CRCW PRAMs are equivalent to unbounded fan-in circuits.

Section 4 gives a survey of important problems that lie in the class NC. Among these are the basic arithmetic operations, transitive closure and Boolean matrix multiplication, the computation of the determinant, the rank or the inverse of a matrix, the evaluation of certain classes of straight-line programs, and the construction of a maximal independent set of vertices in a graph. Section 4 also presents randomized algorithms that operate in polylog time using a polynomial-bounded number of processors for problems such as finding a maximum matching in a graph.

The study of parallel complexity within the PRAM model has led to some important negative results. Using a theory of reducibility analogous to the theory of NP-completeness, it has been possible to identify certain problems as P-complete; such problems are solvable sequentially in polynomial time, but do not lie in the class NC unless every problem solvable in sequential polynomial time lies in NC. This is evidence that the P-complete problems are inherently resistent to ultrafast parallel solution. Our survey concludes, in Section 4, by exploring this concept and deriving a number of examples of P-complete problems, including the maximum-flow problem and the problem of evaluating the output of a monotone Boolean circuit when all the inputs are fixed at constant values.

2. Efficient PRAM algorithms

2.1. The PRAM model and the concepts of efficient and optimal algorithms

The primary model of parallel computation that we will be working with is the PRAM (*P*arallel *R*andom-*A*ccess *M*achine) (Fortune & Wyllie [85]; Goldschlager [106]; Savitch & Stimson [206]). This is an idealized model, and can be viewed as the parallel analogue of the sequential RAM (Cook & Reckhow [64]). A PRAM consists of several independent sequential processors, each with its own private memory, communicating with one another through a global memory. In one unit of time, each processor can read one global or local memory location, execute a single RAM operation, and write into one gobal or local memory location.

PRAMs can be classified according to restrictions on global memory access. An Exclusive-Read Exclusive-Write (or EREW) PRAM is a PRAM for which simultaneous access to any memory location by different processors is forbidden for both reading and writing. In a Concurrent-Read Exclusive-Write (or CREW) PRAM simultaneous reads are allowed but no simultaneous writes. A Concurrent-Read Concurrent-Write (or CRCW) PRAM allows simultaneous reads and writes. In this case we have to specify how to resolve write conflicts. Some commonly used methods of resolving write conflicts are

(a) all processors writing into the same location must write the same value (the COMMON model);

(b) any one processor participating in a common write may succeed, and the algorithm should work correctly regardless of which one succeeds (the ARBITRARY model);

(c) there is a linear ordering on the processors, and the minimum numbered processor writes its value in a concurrent write (PRIORITY model).

Even though there is a variety of PRAM models, they do not differ very widely in their computational power. We show in Section 3 that any algorithm for a CRCW PRAM in the PRIORITY model can be simulated by an EREW PRAM with the same number of processors and with the parallel time increased by only a factor of $O(\log P)$, where P is the number of processors; further, any algorithm for a PRIORITY PRAM can be simulated by a COMMON PRAM with no loss in parallel time provided sufficiently many processors are available.

Define $polylog(n) = \bigcup_{k>0} O(\log^k n)$. Let S be a problem whose fastest sequential algorithm runs in time proportional to $T(n)$ (although, by Blum's Speed-up Theorem (cf. Seiferas [210]), there are pathological problems for which no fastest sequential algorithm exists, this phenomenon will not trouble us in practice). Many problems have the property that any algorithm that solves them would have to look at all of the input in the worst case, and hence $T(n) = \Omega(n)$ in such cases. A PRAM algorithm A for S, running in parallel time $t(n)$ with $p(n)$ processors is *optimal* if

(a) $t(n) = polylog(n)$; and

(b) the *work* $w(n) = p(n) \cdot t(n)$ is $O(T(n))$.

An optimal parallel algorithm achieves a high degree of parallelism in an optimal way. Analogously, we can define an *efficient* parallel algorithm for problem S as one for which the work $w(n)$ is $T(n) \cdot polylog(n)$ with the parallel time $t(n) = polylog(n)$; i.e., an efficient parallel algorithm is one that achieves a high degree of parallelism and comes within a polylog factor of optimal speed-up. A major goal in the design of parallel algorithms is to find optimal and efficient algorithms with $t(n)$ as small as possible. Clearly it is easier to design optimal algorithms on a CRCW PRAM than on a CREW or EREW PRAM. However, the simulations between the various PRAM models make the notion of an *efficient* algorithm invariant with respect to the particular PRAM model used. Thus this latter notion is more robust.

Consider a computation that can be done in t parallel steps with x_i primitive operations at step i. If we implement this computation directly on a PRAM to run in t parallel steps, the number of processors required would be $m = \max x_i$. If we have $p < m$ processors, we can still simulate this computation effectively by observing that the ith step can be simulated in time $\lceil x_i/p \rceil \leqslant (x_i/p) + 1$, and hence the total parallel time to simulate the computation with p processors is no more than $\lceil x/p \rceil + t$ where $x = \Sigma_i x_i$. This observation, known as *Brent's scheduling principle* (Brent [42]), is often used in the design of efficient parallel algorithms. It should be noted that this simulation assumes that *processor allocation* is not a problem, i.e., that it is possible for each of the p processors to determine, on-line, the steps it needs to stimulate. We will see below that this is sometimes a nontrivial task.

Brent's scheduling principle implies that, when processor allocation is not a problem, a parallel algorithm requiring work $w(n)$ and time $t(n)$ can be simulated using p processors in time $w(n)/p + t(n)$. In practice, p is often small compared to n, and in such cases the ratio $w(n)/p$ is typically much greater than $t(n)$. In view of this fact, Kruskal, Rudolph & Snir [245] suggest that it may be unnecessarily restrictive to concentrate on parallel algorithms with $t(n) = polylog(n)$, and they emphasize instead the concept of polynomial speed-up. For a problem requiring sequential time $T(n)$, a parallel algorithm running in time $t(n)$ is said to have *polynomial speed-up* if, for some $\varepsilon < 1$, $t(n) = O(T(n^{\varepsilon}))$. Algorithms with polynomial speed-up are further classified according to their *innefficiency*. For a problem whose best sequential algorithm runs in time $T(n)$, a parallel algorithm that performs an amount of work $w(n)$ is said to have *constant inefficiency* if $w(n) = O(T(n))$, *polylog inefficiency* if

$$w(n) = O(T(n)\ polylog(T(n)))$$

and *polynomial inefficiency* if $w(n)$ is bounded by a polynomial in $T(n)$. Kruskal, Rudolph & Snir argue that algorithms with polynomial speed-up and constant or polylog inefficiency may be extremely useful, even if they do not run in polylog time.

For PRAM algorithms we would like to have simple algorithms that are easy to specify and code. Most of the algorithms we describe will have this feature.

There are a few key methods that have emerged as fundamental subroutines in the design of efficient and optimal parallel algorithms. In the following subsections, we review these basic techniques and algorithms.

2.2. *Basic PRAM techniques*

2.2.1. *Prefix sums*

The first problem we consider is *prefix sums*. Let $*$ be an associative operation over a domain D. Given an array $[x_1, \ldots, x_n]$ of n elements from D, the prefix sums problem is to compute the n prefix sums $S_i = x_1 * x_2 * \cdots * x_i = \Sigma_{j=1}^{i} x_j$, $i = 1, \ldots, n$. This problem has several applications. For example, consider the problem of compacting a sparse array: given an array of n elements, many of which are zero, we wish to generate a new array containing the non-zero elements in their original order. We can compute the position of each non-zero element in the new array by assigning value 1 to the non-zero elements, and computing prefix sums with $*$ operating as regular addition. An application to the problem of adding two n-bit numbers is given in Section 4.2. Further, recognition of any regular language whose input size is restricted to n can be viewed as a prefix sums problem [150].

There is a simple sequential algorithm to solve the prefix sums problem using $n-1$ operations, by computing S_i incrementally from S_{i-1}, for $i = 2, \ldots, n$. Unfortunately this algorithm has no parallelism in it since one of the two operands for the ith $*$ operation is the result of the $(i-1)$st operation.

We now describe a simple parallel algorithm to compute prefix sums in parallel (Ladner & Fischer [150]). For simplicity we assume that n is a power of 2.

PARALLEL PREFIX ALGORITHM

Input: an array $[x_1, \ldots, x_n]$ of elements from domain D.

if $n = 1$ **then** $S_1 := x_1$ **else**

(1) **for** $i = 1, \ldots, n/2$ compute $y_1 := x_{2i-1} * x_{2i}$

(2) recursively compute prefix sums S_i, $i = 1, \ldots, n/2$, for the new array $[y_1, \ldots, y_{n/2}]$

(3) **for** even i in $\{1, 2, \ldots, n\}$ set $S_i := S_{i/2}$

(4) **for** odd i in $\{1, 2, \ldots, n\}$ set $S_i := S_{(i-1)/2} * x_i$.

This algorithm runs on an EREW PRAM since there are no conflicts in the memory accesses. The parallel time $t(n)$ satisfies the recurrence $t(n) = t(n/2) + O(1)$ with $t(1) = 0$, and the work satisfies the recurrence $w(n) = w(n/2) + O(n)$ with $w(1) = 0$. Thus $t(n) = O(\log n)$ and $w(n) = O(n)$. By invoking Brent's scheduling principle we see that this is an optimal EREW PRAM algorithm for $p(n) = O(n/\log n)$. Processor allocation is straightforward in this case and we illustrate it by a standard technique used to implement Brent's principle: Let the number of processors available be $p \leqslant n/\log n$, and let $q = [n/p]$. We first assign the ith processor to elements $x_{(i-1)\cdot q+1}, x_{(i-1)\cdot q+2}, \ldots, x_{i\cdot q}$, $i = 1, \ldots, p$. The ith processor stores these values in an array in its local memory, combines these q values (sequentially) using $*$ in $O(q)$ time and places the result in y_i. Now the array has only p elements in it, and the parallel prefix algorithm is used to compute these prefix sums in $O(\log n)$ time using p processors. Finally, in additional $O(n/p)$ time, the ith processor computes the prefix sums for its local array with S_{i-1} as the first element of the array. This algorithm runs in $O(n/p)$ time with p processors on an EREW PRAM for $p \leqslant n/\log n$.

On a CRCW PRAM, the above algorithm can be modified to run optimally in $O(\log n/\log\log n)$ time when the x_i are $O(\log n)$-bit numbers and $*$ is ordinary addition (Reif [196]; Cole & Vishkin [56]).

Since the prefix sums problem is an important one, much attention has been given to fine-tuning the constants in the time and processor bounds (see, e.g., [150, 80]).

2.2.2. *List ranking*

A problem closely related to the prefix sums problem is the *list ranking problem*: Given a linked list of n elements, compute the suffix sums of the last i elements of the list, $i=1, \ldots, n$. This is a variant of prefix sums, in which the ordered sequence of elements is given in the form of a linked list rather than an array, and the sums are computed from the end, rather than from the beginning. The term "list ranking" is usually applied to the special case of this problem in which all elements have value 1, and $*$ stands for addition (and hence the result of the list ranking computation is to obtain, for each element, the number of elements ahead of it in the list, i.e., its rank in the list); however, the technique we shall present easily adapts to our generalization.

We assume that the linked list is represented by a contents array $c[1, \ldots, n]$ and a successor array $s[1, \ldots, n]$: here, $c(i)$ gives the value of the element at location i, and $s(i)$ gives the location j of the successor to $c(i)$ on the list. Without loss of generality we assume that our domain has a zero element z, that the last element on the list $c(i_l)$ has value z, and that $s(i_l)=i_l$. The following simple algorithm solves the list ranking problem on an EREW PRAM in $O(\log n)$ time with n processors (see, e.g., [242]).

BASIC LIST RANKING ALGORITHM

for $\lceil \log n \rceil$ iterations **repeat**
  in parallel for $i=1, \ldots, n$ **do**
    $c(i) := c(i) * c(s(i))$;
    $s(i) := s(s(i))$;
for $i=1, 2, \ldots, n$ output $c(i)$ as the rank of element i.

The operation used in this algorithm of replacing each pointer $s(i)$ by the pointer's pointer $s(s(i))$ is called *pointer jumping*, and is a fundamental technique in parallel algorithm design. Let the *rank* $r(i)$ of the element in location i be the distance of this element from the end of the input linked list. The correctness of this algorithm follows from the observation that at the start of each step, $c(i)$ equals the sum of elements in the input list with ranks $r(i), r(i)-1, \ldots, r(s(i))+1$ for the current $s(i)$; and after $\lceil \log n \rceil$ iterations, $s(i)=i_l$ for all i. By assigning a processor to each location i, we obtain an n-processor $O(\log n)$ time parallel algorithm. Observe, however, that the work done by this algorithm is $\Theta(n \log n)$ and hence this algorithm as it stands does not lead to an optimal parallel algorithm (since there is a simple linear-time sequential algorithm for the list ranking problem).

The list ranking problem is similar to the prefix sums problem, which has a simple optimal parallel algorithm. The optimal parallel prefix algorithm reduces the problem of computing prefix sums on n elements to one of computing prefix sums on the $n/2$ elements at even positions on the array. This reduction is done in constant time and is data-independent in the sense that the locations of the $n/2$ elements in the reduction list are predetermined. If we try to implement a list ranking algorithm with this property, we run into the problem that a given element has no way of knowing whether it is at an odd or even position on the list. Except for the beginning and end elements, there is no way to determine this information from the local environment of an element.

In order to overcome this problem, we note that we need not necessarily locate the elements at even positions. It suffices to construct a set S of no more than $c \cdot n$ elements in the list, with $c < 1$, such that the distance in the list between any two consecutive elements of S is small. The list ranking problem can then be solved as follows:

(i) *List contraction:* Create a contracted list composed of the elements of S, in which each element of S has as its successor the first element of S that follows it in the original list, and a value equal to its own value in the original list, plus the sum of the values of the elements that lie between it and this successor;

(ii) Recursively, solve the list ranking problem for the contracted list. The suffix sum for each element in the contracted list is the same as its suffix sum in the original list;

(iii) Extend this solution to all elements of the original list. The time to do this is proportional to the maximum distance between two elements of S in the original list, and the work is proportional to the length of the original list.

We shall present an optimal $O(\log n)$ time randomized list ranking algorithm that takes this approach, but with the following exception: once a contracted list of length less than $n/\log n$ is obtained, list contraction is no longer used; instead, the list ranking problem for this contracted list is solved using the Basic List Ranking Algorithm. This requires time $O(\log n)$ using $n/\log n$ processors.

It is necessary to specify how the list contraction step is carried out. This entails giving a method for choosing the set S, and a method for the compaction process needed to place the elements of S in consecutive locations, in preparation for the recursive solution of the list ranking problem on the compacted list.

We can construct S by the following simple randomized algorithm (see e.g., Vishkin [235]; Miller & Reif [171]) called the *random mate algorithm.* Each element chooses a gender, female or male, with equal probability. An element is in set S if and only if it is female or has a male predecessor. It is easy to see that, with probability $1-o(1)$, the size of S is not more than $15n/16$, and each element in S can find its successor in S in constant time, since the distance to its successor is at most 2. With random mating each list contraction tends to shrink the length of the list by a constant factor, and thus the number of contractions needed to pass from the original list of length n to a list of length less than $n/\log n$ is $O(\log\log n)$.

The process of compacting S into consecutive locations could be done by the $O(\log n)$ time optimal method for prefix sums that we saw earlier, but this method would lead to a list ranking algorithm, running in time $O(\log n \log\log n)$, since $\log\log n$ list contractions need to be performed. Instead, we can use either an optimal $O(\log\log n)$ time randomized algorithm on an ARBITRARY CRCW PRAM that approximately compacts an array (Miller & Reif [171]), or the optimal $O(\log n/\log\log n)$ time deterministic prefix sums algorithm of Cole & Vishkin [56], which runs on a COMMON PRAM. Either of these leads to a method that, with high probability, solves the list ranking problem in time $O(\log n)$ and work $O(n)$. Thus, using Brent's scheduling principle, one obtains an optimal $O(\log n)$ time randomized list ranking algorithm using $n/\log n$ processors.

In order to obtain an optimal $O(\log n)$ time deterministic list ranking algorithm it is necessary to replace the random mating procedure by a deterministic method of isolating a contracted set of elements S. A symmetry breaking technique known as *deterministic coin tossing* can be used for this purpose (Cole & Vishkin [55]). The technique is based on the concept of an *r-ruling set* [55]. Given an n-element list, a subset S of these elements is an r-ruling set if S contains no two adjacent elements of the list, and every element not in S is at a distance no more than $c \cdot r$ on the list from an element in S, where c is a suitable constant.

Define $\log^{(k)} n$ to be the log function iterated k times, and let $r = \log^{(k)} n$. The following algorithm finds an r-ruling set in an n-element linked list in $O(k)$ time using n processors [55] (see also Goldberg, Plotkin & Shannon [102]). We assume that the linked list is doubly linked with successor pointer $s(i)$ and predecessor pointer $p(i)$.

RULING SET ALGORITHM

Input: n-element linked list with successor pointers $s(i)$ and predecessor pointers $p(i)$; integer k to set $r = log^{(k)} n$.

(1) **for** $i = 1, \ldots, n$ initialize $c(i) := i$.

(2) **for** k iterations **do**

  in parallel for each i **do**

   determine the rightmost bit position q such that the qth bit of $c(i)$ differs from the qth bit of $c(s(i))$; let b be the qth bit of $c(i)$; $c(i) := b$ concatenated with the binary representation of q;

(3) **in parallel for** each i **do**

  if $c(p(i)) \leqslant c(i)$ **and** $c(s(i)) \leqslant c(i)$ **then** assign i to the ruling set.

It is straightforward to verify that in this algorithm $c(i) \neq c(s(i))$ at every iteration. Further, the maximum value of any $c(i)$ at the end of the jth iteration is $B_j = O(\log^{(j)} n)$. Finally, the distance between two local maxima at the end of the jth iteration is no more than $2 \cdot B_j$, and hence at the end of the algorithm, any element on the list is within distance $O(\log^{(k)} n)$ of an element in the ruling set.

Two special cases of the ruling set algorithm deserve special attention. When $k = c$, a constant, the algorithm obtains an $O(\log^{(k)} n)$ ruling set in constant time using n processors. Define $\log^* n$ as the minimum value of k such that $\log^{(k)} n \leqslant 3$; $\log^* n$ is a very slowly growing function of n. Using the ruling set algorithm we can obtain an $O(1)$ ruling set in $O(\log^* n)$ time with n processors. Since no two elements in a ruling set are adjacent, the size of any r-ruling set is at most one more than half the number of elements in the list. In additional $O(r)$ time, each element in the ruling set can locate its successor in the ruling set by following the successor pointers in the linked list, thus forming a contracted list.

The ruling set algorithm with appropriately chosen values of k has been used in rather elaborate procedures to obtain optimal O(log n) deterministic EREW PRAM algorithms for list ranking (Cole & Vishkin [58]; Anderson & Miller [12]).

2.2.3. Tree contraction

There are several applications that require computation on a rooted tree. One such problem is the *expression evaluation problem*: Given a parenthesized arithmetic expression (using + and · operations) with values assigned to the variables, evaluate E and all subexpressions of E. Note that the prefix sums problem is the expression evaluation problem on the parenthesized expression $(\ldots(x_1+x_2)+x_3\ldots)+x_n)$.

Associated with a parenthesized expression is a binary tree with n leaves that specifies the parenthesization. As in the prefix sums problem there is a simple linear-time sequential algorithm for the expression evaluation problem: compute the values on the internal nodes of the expression tree from the leaves upward to the root. The value at the root gives the value of the expression, and the value at each internal node is the value of the subexpression rooted at that node. However, if the tree is highly imbalanced, i.e., its height is large in relation to its size, then this method performs poorly in parallel.

Tree contraction is a method of evaluating expression trees efficiently in parallel. The method transforms the input tree in stages using local operations in such a way that an n-node tree is contracted into a single node in O(log n) stages, each of which takes constant time on a PRAM. An efficient method for tree contraction was first introduced by Miller & Reif [171]. There are several optimal tree contraction algorithms that run in O(log n) time on EREW PRAM (Gibbons & Rytter [98]; Cole & Vishkin [57]; Abrahamson, Dadoun, Kirkpatrick & Przytycka [1]; Kosaraju & Delcher [145]; Gazit, Miller & Teng [97]). We describe the method reported independently in [1] and [145].

The tree contraction algorithm works on a rooted, ordered binary tree, i.e., a rooted ordered tree in which every vertex is either a *leaf*, having no children, or an *internal node*, having exactly two children, and each arc points from a child to its parent. Let l be a leaf in an n-leaf binary tree T. The SHUNT operation applied to l results in a contracted tree T' in which l and $p(l)$, the parent of l in T, are deleted, and the other child l' of $p(l)$ has the parent of $p(l)$ as its parent, while leaving the relative ordering of the remaining leaves unchanged (see Fig. 1).

We now describe the tree contraction algorithm.

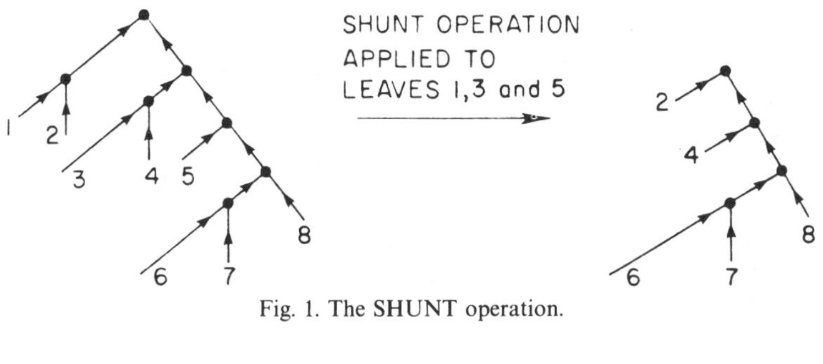

Fig. 1. The SHUNT operation.

TREE CONTRACTION ALGORITHM

Input: a rooted, ordered, binary n-leaf tree T.

(1) *Preprocess*: label the leaves in order from left to right as $1,\ldots, n$.

(2) **for** $\lceil \log n \rceil$ iterations **do**

   (a) apply SHUNT in parallel to all odd numbered leaves that are the left child of their parent;

   (b) apply SHUNT in parallel to all odd numbered leaves that are the right child of their parent;

   (c) shift out the rightmost bit in the labels of all remaining leaves.

It is straightforward to see that the operations in this algorithm can be implemented on an EREW PRAM. After each iteration, half of the leaves are deleted from the current tree, and no new leaves are created. Hence after $\lceil \log n \rceil$ iterations, the tree is contracted to a single node.

Step 1 can be implemented optimally in $O(\log n)$ time on an EREW PRAM using the *Euler tour technique* (Tarjan & Vishkin [221]), which we describe in Section 2.3. Then, in constant time, the leaves can be placed in an array A in the order in which they will be processed. The total work done in step 2 is

$$O\left(\sum_{i=1}^{\lceil \log n \rceil} n/2^i\right) = O(n)$$

and processor allocation is no problem since we have the array A. Thus this gives an optimal $O(\log n)$ tree contraction algorithm on an EREW PRAM.

By associating appropriate computations with the SHUNT operation, we can evaluate an arithmetic expression while performing tree contraction on its associated tree. We associate with each arch (u, v) an ordered pair of values (a, b) with the interpretation that if the value of u is x, then the operand supplied to v along arc (u, v) is $a \cdot x + b$. Initially, every arc has the ordered pair $(1, 0)$. Thus, initially, the value of each node in the tree is exactly the value of its subexpression.

Now consider a SHUNT operation on a leaf l with parent p, sibling s and grandparent q. Let the value of leaf l be v, and let arc (l, p) have value (a_1, b_1), arc (s, p) have value (a_2, b_2) and arc (p, q) have value (a_3, b_3). In the contracted tree, all three of these arcs are deleted and replaced by the arc (s, q). Let a and b be constants such that

$$a \cdot y + b = a_3 \cdot ((a_1 \cdot v + b_1) * (a_2 \cdot y + b_2)) + b_3,$$

where $*$ represents the operation at p and y is an indeterminate. Then it is easy to see that assigning the value (a, b) to the newly introduced arc (s, q) leaves the values of the vertices remaining in the new tree unaltered. Thus we obtain an optimal $O(\log n)$ time EREW PRAM algorithm for expression evaluation. All subexpressions in the expression tree can be evaluated within the same bounds by having an expansion phase at the end, similar to that in the parallel prefix algorithm. This technique works for the evaluation of expressions over a semiring, ring or field (in the case of a field, the value on each edge is an ordered set of four values, to represent the ratio of two linear forms). If the input is in the form of a parenthesized expression, the tree form can be extracted from it optimally in $O(\log n)$ time using an algorithm by Bar-On & Vishkin [24].

In addition to expression evaluation, tree contraction has been applied to a wide variety of problems. The technique easily generalizes to arbitrary (nonbinary) trees, and has been used to derive parallel algorithms for various graph-theoretic computations on trees such as maximum matching, minimum vertex cover and maximum independent set [115]. Other applications of tree contraction and its variants can be found in [101, 171, 189]. A generalization of the SHUNT operation is used in the more general problem of straight-line program evaluation, or evaluation of a DAG (see Section 4); in fact, the SHUNT operation was first introduced in this more general setting [170].

2.2.4. *Conclusion*

We have described optimal parallel algorithms for three basic problems: prefix sums, list ranking and expression evaluation.

In Section 3 we show that it requires $\Omega(\log n)$ time to compute the OR of n bits on a CREW PRAM, regardless of the number of processors available. Since the three problems we considered in this section are at least as difficult as the OR function, the lower bound applies for these problems as well. Thus the results we have given are optimal with maximum possible speed-up.

A lower bound of $\Omega(\log n/\log\log n)$ time holds for computing the parity of n bits on a PRIORITY CRCW PRAM with a polynomial number of processors (Beame & Hastad [29]). Since we could solve the parity problem if we could solve an arbitrary prefix sums, list ranking or tree contraction problem, this lower bound applies to these problems as well. While an optimal $O(\log n/\log\log n)$ time CRCW PRAM algorithm is known for the prefix sums problem under certain conditions, it is not known if the list ranking and tree contraction problems have sublogarithmic time algorithms on a CRCW PRAM with a polynomial number of processors.

2.3. *Efficient graph algorithms*

Graphs play an important role in modeling real-world problems, and sequential algorithms for graph problems have been studied extensively. Almost without exception, all of the optimal (i.e., linear-time) sequential algorithms for these problems use one of two methods to search a graph: *depth-first search* or *breadth-first search*. At present, neither of these techniques has an efficient parallel algorithm. The best polylog-time PRAM algorithm known for breadth-first search of an n-node graph uses $M(n)$ processors, where $M(n)$ is the number of processors required to multiply two $n \times n$ matrices over a general ring in $O(\log n)$ time. The best upper bound presently known for $M(n)$ is $O(n^{2.376})$ (Coppersmith & Winograd [66]). For depth-first search there is currently no deterministic polylog-time parallel algorithm known that uses a polynomial number of processors. For more on parallel breadth-first search and depth-first search, see Section 4.

The early work on parallel graph algorithms [48, 76, 192, 205, 221, 224] used various methods to circumvent the lack of an efficient parallel method of searching a graph. More recently, a new efficient graph searching technique called *ear decomposition* [241, 156, 168, 162] has been developed for undirected graphs. Using this technique, efficient parallel algorithms for several graph problems including strong orientation, biconnectivity, triconnectivity, four-connectivity, and s-t numbering have been developed. Several of these algorithms also have optimal sequential implementations, thus giving us new algorithms for the sequential case. This is an example of a new emerging discipline enriching an existing one. We briefly survey the results in the following, where we assume that the input graph G has n vertices and m edges.

2.3.1. *Connected components*

The problem of computing connected components is often considered the most basic graph problem. While it is not known how to apply depth-first search or breadth-first search to obtain an efficient parallel connected-components algorithm, the following approach (Hirschberg, Chandra & Sarwate [118]) does give an efficient parallel algorithm for the problem. The algorithm works in $O(\log n)$ stages. At each stage, the vertices of G are organized in a forest of directed trees, with a directed arc from each vertex to its parent in the tree. All vertices in any given tree in the forest are known to be in the same connected component of G. In the first stage of the algorithm, each vertex is in a tree by itself, and at the end of the last stage, all vertices in a connected component are in a tree of height 1. In going from stage i to stage $i+1$, some of the trees containing adjacent vertices in G are linked by a *hooking* process and then the heights of the resulting new trees are compressed by pointer jumping, i.e., each vertex that is not a root or a child of a root in the new tree chooses the parent of its parent to be its new parent. The hooking process has to be implemented properly in order to maintain the tree structure and to guarantee termination of the algorithm in $O(\log n)$ stages. Various implementations of the above basic idea for the CREW PRAM [118], EREW PRAM (Nath & Maheshwari [176]) and CRCW PRAM (Awerbuch & Shiloach [21]; Shiloach & Vishkin [212]) are available. The implementation bound is $O(\log n)$ time using $O(m+n)$ processors on an ARBITRARY CRCW PRAM (recall that by the simulations between the PRAMs, this implies an $O(\log^2 n)$ time algorithm using $O(m+n)$ processors on an EREW PRAM). This algorithm is easily extended to obtain a spanning tree for G with the same time and processor bounds.

By applying more elaborate techniques based on those for optimal list ranking to the basic algorithm we have outlined, the connected components of a graph can be determined on an ARBITRARY CRCW PRAM in $O(\log n)$ time with

$$O((m+n)\alpha(m,n)/\log n)$$

processors (Cole & Vishkin [56]), where $\alpha(m,n)$ is the inverse Ackermann function, which is an extremely slowly growing function of m and n. There is also a randomized optimal $O(\log n)$ algorithm for finding connected components on an ARBITRARY CRCW PRAM (Gazit [94]).

2.3.2. *Euler tour technique*

The starting point for most other graph algorithms is the construction of a rooted spanning tree T, and the computation of simple tree functions such as pre- and postorder numbering of vertices in the tree, the level and height of each vertex in the tree, and the number of descendants of each vertex in the tree. For this, we can use the

Euler tour technique on trees (Tarjan & Vishkin [221]), which we describe below. This method works by reducing the computation of these tree functions to list ranking.

Given an unrooted tree T we can convert it into an Eulerian directed graph D by replacing each edge $\{u, v\}$ by two directed arcs, one from u to v and the other from v to u. Let $E=\langle e_1, e_2, \ldots \rangle$ be an Euler tour of D, with e_1 being the directed arc from u to v. Then it is easy to see that E represents a depth-first search traversal of T with u as the root. Each undirected edge $\{x, y\}$ appears once on the list as the directed arc (x, y), and once as the directed arc (y, x). If (x, y) appears before (y, x) in E, then x is the parent of y in T rooted at u, since in this case, (x, y) represents the forward traversal of the edge in the depth-first search and (y, x) represents the backtracking along the edge in the depth-first search. Thus parent–child relationships in T rooted at u can be determined once we have ranked the elements in list E.

Given a tree T, finding an Euler tour E for its directed Eulerian version is very simple. We assume that the tree is specified by an adjacency list for each vertex, which can be interpreted as a list of outgoing arcs from that vertex. This automatically gives us two arcs directed in opposite directions for each edge. We assume that there is a pointer from each edge to its reversal. With this representation we can obtain an Euler tour E in constant time with n processors by specifying, for each edge e, its next edge on E as the edge next to $\bar{e}$, the reversal of e, in the adjacency list containing $\bar{e}$ (if $\bar{e}$ is the last edge on its adjacency list, its next edge is the first edge on this list). To obtain a depth-first search from a given root u, we simply pick any arc (u, v) as the starting arc of the tour. Now we can use list ranking to determine the position of each arc in E, and hence the parent–child relation. Other tree functions such as preorder number, level, and number of descendants of each vertex can be determined by list ranking using appropriate initial weights. For instance, to compute preorder numbers, we can assign a weight of 1 to forward arcs and a weight of 0 to back arcs. Then, for each forward arc (u, v), the preorder number of v is $n-1-$(the weighted rank of (u, v) in the list). The optimal list ranking algorithm implies optimal $O(\log n)$ time EREW PRAM algorithms for these problems.

We can also use the Euler tour technique on trees to implement the preprocessing step in the Tree Contraction Algorithm of the previous section optimally. For this we merely need to give a weight of 1to leaves and a weight of 0 to internal nodes, and then compute the weighted rank of each leaf in the Euler tour.

The Euler tour technique on trees generalizes to finding Euler tours in general Eulerian graphs with the same complexity bounds as the connected-components algorithm (Atallah & Vishkin [18]; Awerbuch, Israeli & Shiloach [20]). For this, we construct the tour E as above by specifying for each edge, the edge that follows it in E. For a general Eulerian graph, this results in a collection of edge disjoint (possibly nonsimple) cycles. Two cycles having a common vertex u can be "stitched together" by swapping the successor edges of the two incoming arcs to u in the Euler tour E. The algorithm obtains an Euler tour for the graph by stitching all the cycles together into a single connected structure through an appropriate choice of such swaps.

2.3.3. Ear decomposition

An *ear decomposition* $D=[P_0, \ldots, P_{r-1}]$ of an undirected graph $G=(V, E)$ is a partition of E into an ordered collection of edge-disjoint simple paths $P_0, \ldots, P_{r-1}$

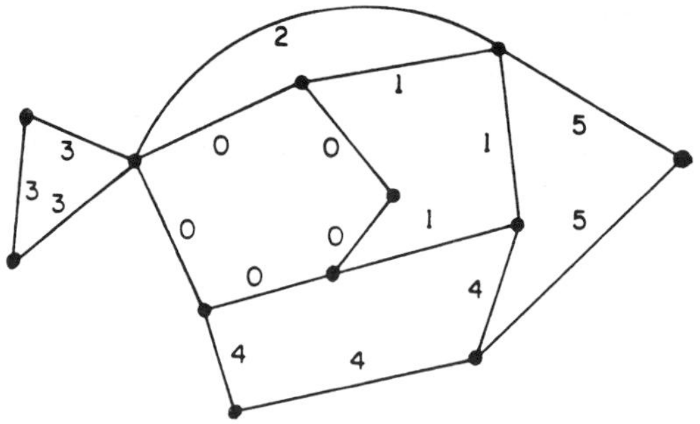

Fig. 2. An ear decomposition.

called *ears*, such that P_0 is a simple cycle, and for $i>0, P_i$ is a simple path (possibly a simple cycle) with each endpoint belonging to a lower-numbered ear, and with no internal vertices belonging to lower-numbered ears (see Fig. 2). An ear with no internal vertex is called a *trivial ear*.

An *open ear decomposition* is an ear decomposition in which none of the $P_i, i>0$, is a simple cycle.

It is known that a graph has an ear decomposition if and only if it is 2-edge connected and a graph has an open ear decomposition if and only if it is biconnected, i.e., 2-vertex connected (Whitney [241]).

Let T be a spanning tree of an undirected, 2-edge connected graph $G=(V, E)$, and let N be the set of non-tree edges in G, i.e., the edges in $G-T$. Then each edge e in N completes exactly one cycle in $T\cup\{e\}$, called a *fundamental cycle of G with respect to T*. It is easy to see that the number of ears in any ear decomposition or open ear decomposition of an n-vertex, m-edge graph is $m-n+1$, which is also the number of fundamental cycles in the graph. In the ear decomposition algorithm presented below (Maon, Schieber & Vishkin [162]; Miller & Ramachandran [168]), each ear is generated as part of a fundamental cycle of G with respect to T, and contains the non-tree edge in that fundamental cycle.

EAR DECOMPOSITION ALGORITHM

Input: Undirected, 2-edge connected graph $G=(V, E)$.

(1) *Preprocess G*:
   (a) Find a spanning tree T for G;
   (b) Root T and number the vertices in preorder;
   (c) label each non-tree edge by the least common ancestor (lca) of its endpoints in T.

(2) *Assign ear numbers to non-tree edges*: number non-tree edges from 0 to $r-1$ in nondecreasing order of their lca labels.

(3) *Assign ear numbers to tree edges*: number each tree edge with the number of the minimum-numbered non-tree edge whose fundamental cycle it belongs to.

The correctness of this algorithm follows from a straightforward induction on ear number. It is easy to see that the edges with ear number 0 form a simple cycle passing through the root of T, and if we assume inductively that edges with ear numbers 0 to i satisfy the definition of an ear decomposition, it is not difficult to show that the edges with ear number $i+1$ have the desired properties as well.

All of the steps in the algorithm can be implemented using the Euler tour algorithm, together with efficient parallel algorithms for finding a spanning tree, sorting, prefix sums, and finding lca's. The algorithm runs in $O(\log n)$ time while performing $O(m+n\log n)$ work. If consecutive ear numbers are not required, but only distinct labels from a totally ordered set, then the parallel sorting algorithm is not required and the algorithm can be implemented to run in $O(\log n)$ time while performing the same amount of work as the connected-components algorithm, i.e., $O((m+n)\alpha(m,n))$ work, by using an optimal $O(\log n)$ time parallel algorithm to find lca's (Schieber & Vishkin [207]). A randomized version of the ear decomposition algorithm with consecutive ear numbers runs in $O(\log n)$ time while performing $O(m+n)$ work, by using optimal $O(\log n)$ time randomized parallel algorithms for connectivity (Gazit [94]) and for sorting integers in the range 1 to n (Reif [196]).

The ear decomposition algorithm, in general, does not give an open ear decomposition, but can be modified to do so (Maon, Schieber & Vishkin [162]; Miller & Ramachandran [168]) with the same parallel complexity by refining the numbering in step 2 so that non-tree edges with the same lca labels are further ordered in such a way that the resulting ear numbers give an open ear decomposition.

2.3.4. *Applications of ear decomposition*

The efficient parallel ear decomposition algorithm implies efficient parallel algorithms for 2-edge connectivity and biconnectivity. Other efficient parallel algorithms for biconnectivity are known [48, 221].

We describe two other applications of ear decomposition. A *strong orientation* of an undirected graph $G=(V,E)$ is an assignment of a direction to each edge of G such that the resulting directed graph is strongly connected. A graph has a strong orientation if and only if it is 2-edge connected. To strongly orient a 2-edge connected graph, we find an ear decomposition for it, and then orient the edges in each ear from one (arbitrary) endpoint of the ear to the other. Parallel algorithms for strong orientation are reported in [13, 236].

We now briefly describe how to use open ear decomposition to obtain an efficient parallel algorithm to test triconnectivity, and to find the triconnected components of a biconnected graph (Miller & Ramachandran [169]). Let $D=[P_0,\ldots,P_{r-1}]$ be an open ear decomposition of a biconnected graph $G=(V,E)$. If G is not triconnected, then it contains a pair of vertices x, y whose removal separates the graph into two or more pieces. For such a pair of vertices x, y, it is easy to see that there must be some ear P_i in D that contains both of them, such that the portion of P_i between x and y is separated from the rest of P_i when x and y are removed from G. The triconnectivity algorithm tests if such a separating pair of vertices exists by looking for them in each ear in parallel. It does so by constructing, for each nontrivial ear, its *ear graph*, which is a graph derived from the input graph that contains the necessary information about separating pairs of vertices, if any, on the ear. The algorithm then further processes the ear graph to obtain its *coalesced graph*, using which all separating pairs on the ear can be determined quickly. Efficient parallel algorithms for finding the ear graphs of all nontrivial ears and the coalesced graph of each ear graph are given in [169, 190], leading to a parallel graph triconnectivity algorithm that runs in $O(\log n)$ time with $O(m \cdot \log^2 n)$ work on a CRCW PRAM. These ideas generalize to efficient parallel algorithms for finding all separating pairs of vertices in a biconnected graph and for finding the triconnected (or Tutte) components of a biconnected graph. These ideas also generalize to testing for graph four-connectivity (Kanevsky & Ramachandran [129]) giving a parallel algorithm that runs in $O(\log^2 n)$ time with $O(n^2)$ processors on an ARBITRARY CRCW PRAM. This approach also gives an $O(n^2)$ sequential algorithm for the problem, which is an improvement over previously known sequential algorithms for the problem.

Open ear decomposition has been used to obtain a parallel algorithm for finding an s-t numbering in a biconnected graph (Maon, Schieber & Vishkin [162]) with the same complexity as the connected-components algorithm. This efficient parallel s-t numbering algorithm has been used in conjunction with an efficient parallel implementation of PQ trees, to obtain a parallel planarity algorithm (Klein & Reif [143]) that runs in $O(\log^2 n)$ time using $O(m+n)$ processors on a CREW PRAM.

At present there is no efficient graph search technique known for directed graphs. Thus for most problems on directed graphs, including the basic one of testing if one vertex is reachable from another in a directed graph, the best polylog-time parallel algorithm currently known needs on the order of $M(n)$ processors.

2.3.5. *Conclusion*

In this section we have presented efficient parallel algorithms for several problems on undirected graphs.

Using the fact that graph problems are typically at least as hard as computing the OR of n bits or the parity of n bits, it follows from the lower bounds of Section 3 that graph problems need $\Omega(\log n)$ parallel time on a CREW or EREW PRAM with no restriction on the number of processors, and $\Omega(\log n/\log \log n)$ parallel time on a CRCW PRAM with a polynomial number of processors. In practice, graph problems seem to need at least $\log^2 n$ time on a CREW and EREW PRAM and $\log n$ time on a CRCW PRAM. We have stated many of the results in this section only for CRCW PRAMs, when it is the case that, by the simulations between the various types of PRAMs, this gives the best bounds known for CREW and EREW PRAMs as well.

2.4. *Sorting, merging and selection*

In this section we discuss parallel algorithms for which the input is an array of n elements from a linearly ordered set. In the sorting problem, the task is to rearrange the elements into nondecreasing order. In the merging problem, the input array is partitioned into two subarrays, each of which is known to be in nondecreasing order, and the task is to rearrange the entire array into nondecreasing order. In the selection problem an integer k between 1 and n is given, and the task is to find the kth-smallest element of the array. Except for a brief remark in Section 2.4.3. we restrict attention to comparison algorithms, in which the only means of gathering information about the elements is through pairwise comparisons (i.e., tests of the form "is x less than y?", where x and y are elements of the array). For convenience we also assume that the n elements are distinct.

Valiant [229] proposed a model of parallel comparison algorithms in which, at each step, p comparisons are performed, where p is a parameter that we shall call the *degree of parallelism*. It is not required that the p pairs of elements compared at a given step be disjoint. The choice of the p comparisons to be performed at a given step can depend in an arbitrary manner on n, the number of elements, and on the outcomes of previous comparisons. The algorithm terminates when it has acquired enough information about the input to specify the answer (the identity of the kth smallest element in the case of selection, and the permutation required to put the elements in increasing order, in the case of sorting or merging). The execution time of an algorithm is the number of steps performed. This model is called the *parallel comparison model.*

A second model, the *comparator network*, is a restriction of the parallel comparison model. The basic operation in a comparator network is the i–j comparison-exchange operation, where i and j are distinct integers between 1 and n. Such an operation has the following interpretation: compare the contents of location i with the contents of location j; store the smaller of the two values in location i and the larger of the two in location j. If a comparator network has degree of parallelism p, then each of its steps is specified by at most p disjoint i–j pairs, and consists of the simultaneous execution of the comparison-exchange operations corresponding to these pairs. The network is oblivious, in the sense that the set of pairs of locations compared in any given step does not depend on the outcomes of the comparisons performed at earlier steps. A comparator network can be represented by a diagram reminiscent of a sheet of music paper, in which each memory cell is represented by a horizontal line and each i–j comparison-exchange operation is represented by a vertical line segment between horizontal lines i and j, with an arrow directed toward i, the line that will receive the smaller of the two values being compared. The order in which the comparison-exchange operations affecting a given cell are executed is given by the left-to-right order of the corresponding vertical line segments. From this description it should be clear that, in contrast to the more general parallel comparison model, comparator networks can easily be realized in hardware.

A comparator network is called a *sorting network* if it is guaranteed to rearrange the contents of locations 1 to n into nondecreasing order. Figure 3 depicts an 8-element

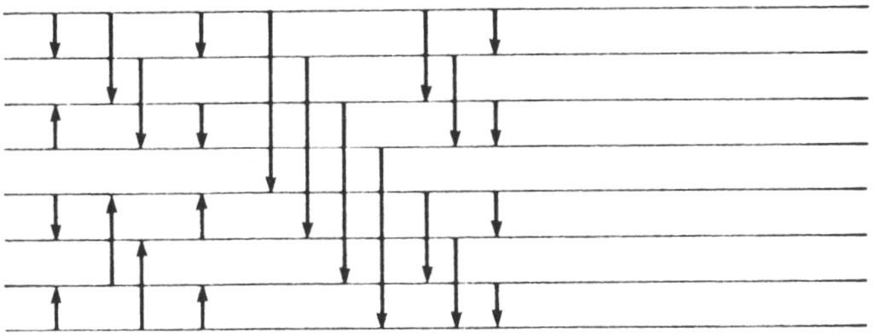

Fig. 3. A sorting network.

comparator network with degree of parallelism 4 and execution time 6. Our discussion of bitonic sort in Section 2.4.2 will establish that this network is a sorting network.

A third model is the comparison PRAM. This is a PRAM for which the input is an array of n elements from a linearly ordered set. In addition to its usual instruction set, the comparison PRAM is provided with instructions for comparing the elements of the input array and for moving them around in memory. Elements of the input array have no interpretation as bit strings or integers, and thus cannot be used as addresses or as arguments for the arithmetic or shift instructions of the PRAM. The comparison PRAM is a more realistic model than the parallel comparison model, because the time required to move data around, as well as the time required to decide which comparisons are to be performed and which processors will perform them, are counted in the cost of computation along with the cost of the comparisons themselves.

Since the parallel comparison model is more permissive than the comparator network or comparison PRAM, lower bounds on the complexity of comparison problems within that model are valid for the other models as well.

2.4.1. *The complexity of finding the maximum and merging*

In this section we derive tight upper and lower bounds on the complexity of two problems: finding the largest element in an array and merging two sorted arrays. Our lower bounds are for the parallel comparison model, and our upper bounds are for the comparison PRAM. Much of our discussion is based on [229].

We begin by presenting an algorithm for finding the maximum of n elements on a COMMON CRCW comparison PRAM with p processors. At a general step within the algorithm, the search for the maximum will have been narrowed down to a set S of elements from the input array. Initially, S consists of all n elements, and the computation terminates when S becomes a singleton set.

Let x be the smallest integer such that, when S is partitioned into x blocks of size $\lfloor |S|/x \rfloor$ or $\lceil |S|/x \rceil$, p comparisons suffice for the direct comparison of each pair of elements that lie in a common block. Then, in one step, the algorithm determines the maximum element in each block of such a partition, and thus eliminates all but x elements of S.

It can be shown that, when $n/2 \leqslant p \leqslant \binom{n}{2}$, the number of iterations required to reduce the cardinality of S from n to 1 is $\log\log n - \log\log(2p/n) + O(1)$. Also, the computations required at each step for each of the p processors to determine which pair of elements to compare and for the set of block maxima to be assembled into an array can be carried out in constant time. This establishes that the complexity of finding the maximum on a p-processor COMMON CRCW comparison PRAM is $O(\log\log n - \log\log(2p/n))$.

To prove a lower bound in the parallel comparison model we use a graph-theoretic argument. At the beginning of any step of any comparison-based algorithm, there will remain a set S of candidates for the maximum; an element will lie in S if it has never been compared with a larger element. At a general step, the comparisons performed between candidate elements can be represented by the edges of a graph G whose vertices are the candidate elements. An adversary can choose any set S' which is independent in G (i.e.. no two elements of S' are compared with each other in the step), and can consistently specify the outcomes of comparisons so that all the elements of S' remain candidates. A theorem of Turan in extremal graph theory shows that, in any graph G having v vertices and e edges, there is an independent set of size at least $v^2/(v+2e)$. Thus, the adversary can ensure that the candidate set S' is of size at least $|S|^2/(|S|+2p)$. It follows that, against such an adversary strategy, any algorithm requires $\log\log n - \log\log(2p/n)+O(1)$ steps when $n/2 \leqslant p \leqslant \binom{n}{2}$. Hence, for this range of values of p, the algorithm described above is optimal in the parallel comparison model and optimal within a constant factor on a COMMON CRCW comparison PRAM. Interestingly, on a CREW comparison PRAM, finding the maximum requires time $\Omega(\log n)$, even when no limit is placed on the number of processors, and there is a simple optimal $O(\log n)$ time EREW PRAM algorithm for the problem.

For the problem of selecting the median of n elements with degree of parallelism n, a lower bound of $\Omega(\log\log n)$ is implied by the lower bound for the maximum problem. A matching upper bound for the comparison model has been given by Ajtai, Komlós, Steiger & Szemerédi [6] building on earlier work by Cole & Yap [59]. It follows from [29] (Beame & Hastad) that finding the median on a CRCW comparison PRAM using a polynomial-bounded number of processors requires time $\Omega(\log n/\log\log n)$.

We now turn to the problem of merging nondecreasing sequences of lengths n and m, where $n \leqslant m$. We begin by giving an algorithm within the parallel comparison model that has degree of parallelism $n+m$ and execution time $O(\log\log n)$. Let the two sequences be A and B, where $A=(a_1, a_2, \ldots, a_n)$ and $B=(b_1, b_2, \ldots, b_m)$. The algorithm is as follows.

(1) Divide A into $\sqrt{n}$ blocks of length $\sqrt{n}$, and B into $\sqrt{m}$ blocks of length $\sqrt{m}$ (here we ignore the simple modifications needed when the length of A or the length of B is not a perfect square).

(2) Let α_i be the first element of the ith block of A, and β_j the first element of the jth block of B. **In parallel**, compare each α_i with each β_j (the number of processors required is $\sqrt{n}\sqrt{m}$, which is at most $(m+n)/2$).

(3) **In parallel for** each α_i **do**
  let $j(i)$ be the unique index such that $\beta_{j(i)} < \alpha_i < \beta_{j(i)+1}$ (here we use the convention that $\beta_0 = -\infty$ and $\beta_{\sqrt{m+1}} = +\infty$).

(4) Compare α_i with each element of the block starting with $\beta_{j(i)}$.

At this point, the algorithm has determined where each α_i fits into B, and thus the problem has been reduced to a set of disjoint merging problems, each of which involves merging a block of length $\sqrt{n}$ from A with some consecutive subsequence of B. Recursively, solve each of these subproblems, using a degree of parallelism equal to the number of elements.

Let $T(n)$ be the time required by this algorithm to merge a sequence of length n with a sequence of length m, where $n \leqslant m$, using $n+m$ processors. Since the parallel computation model charges only for the comparison steps, and not for the bookkeeping involved in keeping track of the results of comparisons and determining which comparisons to perform next, we have $T(1)=1$ and $T(n) \leqslant 2+T(\sqrt{n})$. This gives $T(n)=O(\log\log n)$.

This algorithm can be implemented to run in time O(log log n) time with $n+m$ processors on a CREW PRAM, although the processor allocation problems are not entirely trivial (Borodin & Hopcroft [41]).

Following Borodin & Hopcroft we now prove that, in the parallel comparison model, the time required to merge two sequences of length n using $2n$ processors is $\Omega(\log\log n)$. In order to prove the lower bound we consider the following generalization of the problem: given ordered sequences $A_1, B_1, A_2, B_2, \ldots, A_k, B_k$, each of length s, and the information that, for $i=1,2,\ldots,k-1$, each element of $A_i \cup B_i$ is less than each element of $A_{i+1} \cup B_{i+1}$, merge each of the pairs A_i, B_i using cks processors. We refer to this problem as the c, k, s problem.

Let the worst-case time to solve this problem in the parallel comparison model be $t(c,k,s)$, and let $T(c,s)=\min_k t(c,k,s)$ where k ranges over the positive integers. We shall use an adversary argument to prove that $T(c,s) \geqslant 1+T(16c/7, s')$ where $s'=\sqrt{s/8c}$. Consider the first step in solving a c,k,s problem. We have $ks \times s$ merging problems and cks processors. For $i=1,2,\ldots,k$, partition A_i into $2\sqrt{2cs}$ blocks $A_i^1, A_i^2, \ldots, A_i^{2\sqrt{2cs}}$, each of size s', partition B_i similarly, form a $2\sqrt{2cs} \times 2\sqrt{2cs}$ matrix M_i, and mark the $[a,b]$ cell of this matrix if some element of A_i^a is compared with some element of B_i^b at the first step. Each of the M_i, $i=1,\ldots,k$ has $8cs$ cells, so that the total number of cells in all matrices is $8cks$. At most cks of these cells are marked. Also, each matrix has $4\sqrt{2cs}-1$ diagonals parallel to the main diagonal. By a simple averaging argument, it is possible to choose a diagonal in each of these matrices, such that the number of unmarked cells in the chosen diagonals is at least $7k\sqrt{2cs}/8$. An adversary can specify the outcomes of these comparisons so that each unmarked cell on the selected diagonals corresponds to an independent $s' \times s'$ merging problem at the next step. It follows that the adversary can leave the algorithm with k' independent merging problems to solve, each of which is $s' \times s'$, where $k'=7k\sqrt{2cs}/8$. The number of processors is cks, which is equal to $16ck's'/7$. This leads to the inequality $T(c,s) \geqslant 1+T(16c/7, s')$. It follows that $T(c,s) \geqslant d \log\log_c cs$, where d is a certain positive constant. In particular, the complexity of merging two ordered sequences of length n with $2n$ processors is at least $T(2,n)$ which is $\Omega(\log\log n)$. This establishes that the algorithm we have given is optimal up to a constant factor, both for the parallel comparison model and the CREW comparison PRAM.

2.4.2. Sorting networks

One of the classic parallel sorting methods is Batcher's bitonic sorting network [26]. This network is based on the properties of certain sequences. Let $A=(a_1, a_2, \ldots, a_n)$ be a sequence of distinct elements of a linearly ordered set. For $i=2,3,\ldots,n-1$, call element a_i a *local minimum* if both a_{i-1} and a_{i+1} are greater than a_i, and a *local maximum* if both a_{i-1} and a_{i+1} are less than a_i. The sequence A is called *unimodal* if it has at most one element that is a local minimum or local maximum, and *bitonic* if it is a cyclic shift of a unimodal sequence. The following lemma is due to Batcher.

LEMMA. *Let $A=(a_1, a_2, \ldots, a_{2N})$ be a bitonic sequence of even length. Define the sequence $L(A)$ and $R(A)$ as follows:*

$$L(A)=(\min(a_1, a_{N+1}), \min(a_2, a_{N+2}), \ldots, \min(a_N, a_{2N}))$$

and

$$R(A)=(\max(a_1, a_{N+1}), \max(a_2, a_{N+2}), \ldots, \max(a_N, a_{2N})).$$

Then the sequences $L(A)$ *and* $R(A)$ *are bitonic, and each element of* $L(A)$ *is less than each element of* $R(A)$.

The lemma suggests a parallel method of sorting a bitonic sequence A whose length n is a power of 2: if A is of length 1 then halt; else compare corresponding elements of the left and right half of A and form the arrays $L(A)$ and $R(A)$; in parallel, sort $L(A)$ and $R(A)$. This algorithm can be realized by a comparator network with degree of parallelism $n/2$ and execution time $\log n$.

Batcher's bitonic sorting network is built upon this algorithm for sorting a bitonic sequence. Starting with an unsorted n-element array, which can be regarded as a list of $n/2$ bitonic sequences of length 2, the algorithm constructs a list of $n/4$ bitonic sequences of length 4, then a list of $n/8$ bitonic sequences of length 8, and so forth until the array has been transformed into a single bitonic sequence of length n, which may then be sorted. The algorithm exploits the fact that the concatenation of an increasing sequence with a decreasing sequence is a bitonic sequence. Thus, to convert a list of $n/2^i$ bitonic sequences of length 2^i into a list of $n/2^{i+1}$ bitonic sequences of length 2^{i+1}, it suffices to sort the sequences of length 2^i alternately into increasing and decreasing order, using the algorithm for sorting a bitonic sequence. Figure 3 shows a bitonic sorting network for 8 elements.

We now analyze the execution time of the bitonic sorting algorithm. Let $B(n)$ be the time required to sort a bitonic sequence of length n using degree of parallelism $n/2$, and let $S(n)$ be the time required to sort an array of length n, again using degree of parallelism $n/2$. Then $B(2^k)=k$, and

$$S(2^k)=\sum_{i=1}^{k} B(2^i)=\binom{k+1}{2}$$

Thus bitonic sort requires time $O(\log^2 n)$ to sort n elements using degree of parallelism $n/2$. Bitonic sort is also easily implemented on an EREW comparison PRAM. Again, it runs in time $O(\log^2 n)$ using $n/2$ processors. As an aside, we mention that bitonic sort can also be implemented very neatly to run within these time and processor bounds on certain fixed-degree networks, such as the butterfly and shuffle-exchange network.

Batcher's bitonic sorting network requires $\theta(n \log^2 n)$ comparators. Since there exist sequential sorting algorithms that require only $O(n \log n)$ comparisons in the worst case, it was natural to ask whether one could improve on Batcher's construction by exhibiting sorting networks requiring only $O(n \log n)$ comparators. In 1983 this question was answered affirmatively by Ajtai, Komlós & Szemerédi [7]. The family of sorting networks given by these three authors also has theoretical advantages for parallel computation, since they execute in time $O(\log n)$ using degree of parallelism $n/2$. However, despite the substantial improvements obtained by later researchers (Paterson [183]), the constant implied by the "big O" is so large that bitonic sort is preferable for all practical values of n. Since the Ajtai–Komlós–Szemerédi networks are presented in detail in [186], we shall not attempt to describe them here.

2.4.3. Sorting on a PRAM

Since the sequential complexity of sorting by comparisons is $\Theta(n \log n)$, it is of interest to investigate methods for sorting on a comparison PRAM that run in polylog time and have a processor–time product that is $O(n \log n)$. Randomized methods that use $O(n)$ processors and run in $O(\log n)$ time with high probability are given in [201, 200]. The first deterministic method to achieve such performance was an EREW comparison PRAM algorithm based on the Ajtai–Komlós–Szemerédi sorting network. It achieves an execution time of $O(\log n)$ using $O(n)$ processors; however, the constant factor in the time bound is so large as to render the method impractical. Using a new version of bitonic merging, Bilardi & Nicolau [35] give a sorting algorithm for the EREW comparison PRAM that achieves a processor–time product of $O(n \log n)$ using $O(n/\log n)$ processors. Moreover, the constant factor in the time bound is small, so that the method is attractive for practical use. Cole [53] has given a practical deterministic method of sorting on an EREW comparison PRAM in time $O(\log n)$ using $O(n)$ processors. His algorithm can be viewed as a pipelined version of merge sort. Finally, we point out that when the elements to be sorted are integers in a limited range, better bounds are achievable by using bucket sorting methods rather than comparison algorithms [196, 111].

The rest of this section is devoted to a presentation of Cole's sorting algorithm. For simplicity, we confine ourselves to describing a version of the algorithm that runs on a CREW comparison PRAM. We assume for convenience that the number of elements to be sorted is a power of 2. Let the 2^k elements to be sorted be placed in correspondence with the leaves of a complete binary tree T of height k; hereafter, we make no distinction between a leaf and the corresponding element. Each internal node v within T is the root of a subtree; let T_v be the set of leaves of that subtree. Then the task of node v is to arrange the elements of T_v into a sorted list.

An obvious method of creating the required lists would be to move up the tree level-by-level from the leaves to the root, using the merging algorithm of Section 2.4.1 to create the list for each node by merging the lists for its two children. Using the fact that the time to merge two sorted lists of length t using $2t$ processors is $O(\log \log t)$, a simple analysis shows that this obvious method runs in time $O(\log n \log \log n)$ using $O(n)$ processors. Cole improves on this approach by having the algorithm work on many levels of the tree at once, creating successively more refined approximations to the lists that the nodes must eventually produce. The method of approximation is chosen so that each approximation to the final list for a node can be obtained from the preceding approximation in constant time.

Associated with any time step s and internal node v is a list $LIST_v(s)$; this list is an increasing sequence of elements drawn from T_v. We say v is *finished* at time s if $LIST_v(s)$ contains all the elements of T_v. Node v is a *frontier node* when v is finished but its parent is not. Initially, the leaves of T are its frontier nodes. At any time s, all the frontier nodes are at the same distance from the root of T.

At every step, each frontier node or unfinished node passes a subsequence of its list up to its parent. For node v at step s we call this subsequence $UP_v(s)$. Node v forms $LIST_v(s)$ by merging $UP_x(s)$ with $UP_y(s)$, where x and y are the children of v. If x is unfinished at step $s-1$ then $UP_x(s)$ consists of every fourth element of $LIST_x(s-1)$; i.e., elements $1, 5, 9, 13, \ldots$ If x becomes finished at step s then $UP_x(s+1)$ consists of every fourth element of $LIST_x(s)$, $UP_x(s+2)$ consists of every second element of $LIST_x(s)$ and $UP_x(s+3)$ is equal to $LIST_x(s)$. It follows that a node becomes finished three steps after its children do. The sorting process is completed after $3k$ steps, when the root becomes finished.

It remains to show that each step can be completed in constant time on a CREW PRAM using $O(n)$ processors. At each step the sum of the lengths of all the lists associated with unfinished or frontier nodes is $O(n)$. Thus it suffices to give a method of merging the lists $UP_x(s)$ and $UP_y(s)$ in constant time, using one processor per list element. The key idea is to use information from previous steps. Given two ordered lists A and B, define a *cross-link array* from A into B to be an array of pointers from A into B such that each element a in A points to the least element in B that is greater than a (or, if no such element exists, to a sentinel placed at the end of the list B). If the ordered lists A and B are disjoint then, given cross-links from A into B and from B into A, each element can quickly calculate its rank order in the list that results from merging A and B, and thus A and B can be merged in constant time. Cole's algorithm maintains certain cross-link arrays in order to speed up the merging process. If nodes x and y are siblings in T then, upon the completion of step $s-1$, the algorithm maintains cross-links from $UP_x(s-1)$ into $UP_y(s-1)$ and $LIST_x(s-1)$, and, symmetrically, from $UP_y(s-1)$ into $UP_x(s-1)$ and $LIST_y(s-1)$. Given the cross-links from $UP_x(s-1)$ into $LIST_x(s-1)$ it is easy, in constant time, to create cross-links from $UP_x(s-1)$ into $UP_x(s)$. Similarly, cross-links can be constructed in constant time from $UP_y(s-1)$ into $UP_y(s)$. Moreover, it can be shown that at most four elements of $UP_x(s-1)$ point to any element of $UP_x(s)$. Given all this cross-link information, it is possible in constant time on a CREW comparison PRAM to create cross-links between $UP_x(s)$ and $UP_y(s)$, to merge those two sequences, and to create the cross-links required for the next step. It follows that each step can be executed in constant time using $O(n)$ processors on a CREW comparison PRAM, and thus Cole's algorithm sorts in time $O(\log n)$ using $O(n)$ processors on a CREW comparison PRAM.

2.5. *Further topics*

Efficient parallel algorithms have been developed for a number of other combinatorial problems, including string matching [88, 137, 151, 237, 33], set manipulation [17], computational geometry [4, 14, 15, 16, 51, 54, 108, 198] and graph algorithms [52, 112, 142]. In [78, 99] extensive surveys of efficient parallel algorithms are given including the topics of Sections 2.1–2.3.

Some of the challenges that have thus far eluded researchers in the area of efficient and optimal combinatorial parallel algorithms are the following:

(i) The construction of an efficient parallel $O(\log n/\log\log n)$ time CRCW PRAM algorithm for the n-element list ranking problem. At present it is not known how to achieve this time bound with any polynomial-bounded number of processors.

(ii) The construction of an efficient parallel algorithm for breadth-first search in an undirected graph. This problem can be solved on a CRCW PRAM in O(log n) time using a polynomial-bounded number of processors (see the discussion of the "transitive closure bottleneck" in Section 4.8.1).

(iii) The construction of efficient parallel algorithms for various reachability problems on directed graphs, including reachability from one given vertex to another, breadth-first search, strong connectedness and topological sorting of a directed acyclic graph. These problems can be solved on a CRCW PRAM in O(log n) time using a polynomial-bounded number of processors (see Section 4.8.1).

(iv) The construction of efficient algorithms to find the connected components of an undirected graph in time $o(\log^2 n)$ on a CREW PRAM or time $o(\log n)$ on a CRCW PRAM. The known lower bounds are $\Omega(\log n)$ and $\Omega(\log n/\log\log n)$ respectively.

(v) The construction of n-element sorting networks with degree of parallelism at most Cn and execution time $D \log n$, where C and D are moderate constants (rather than the very large constants achieved in the construction of Ajtai, Komlós, & Szemerédi [7]).

(vi) The construction of an optimal deterministic algorithm for sorting n integers in the range $[1, \ldots, n^k]$, for fixed k, on an EREW PRAM in time O(log n). Reif [196] gives an optimal randomized O(log n) time algorithm on an ARBITRARY CRCW PRAM to sort n integers in the rage $[1, \ldots, n]$, and Hagerup [111] gives a deterministic algorithm to sort n integers from $[1, \ldots, n^k]$ in O(log n) time using $O(n \log\log n/\log n)$ processors on a PRIORITY CRCW PRAM; this algorithm requires space $\Omega(n^{1+\varepsilon})$, where ε is an arbitrarily small positive constant.

3. Models of parallel computation

3.1. Relations between PRAM models

Our primary model for parallel computation is the PRAM family [85, 106, 206]. In Section 2.1 PRAMs were classified according to restriction on global memory access as EREW, CREW or CRCW, and CRCW PRAMs were further classified as COMMON, ARBITRARY or PRIORITY. It should be noted that this listing represents the PRAM models in increasing order of their power. Thus, any algorithm that works on an EREW PRAM works on a CREW PRAM, and in turn, any algorithm on a CREW PRAM works on a COMMON CRCW PRAM, and so on. The most powerful model in this spectrum is the PRIORITY CRCW PRAM.

We now show that any algorithm for a PRIORITY CRCW PRAM can be simulated by an EREW PRAM with the same number of processors and with the parallel time increased by only a factor of O(log P), where P is the number of processors [77, 234]. This is done as follows: Let $P_1, \ldots, P_r$ be the processors and $M_1, \ldots, M_s$ be the memory locations used by the PRIORITY algorithm. The simulating EREW PRAM uses r auxiliary memory locations $N_1, \ldots, N_r$ for simulating a write or read step. If processor P_i needs to access location M_j in the PRIORITY algorithm, then it writes the ordered pair (j, i) in location N_i. The array N_j, $j = 1, \ldots, r$ is then sorted in lexicographically increasing order in O(log r) time using the r processors [53]. Then, by reading adjacent entries in this sorted array, the highest priority processor accessing any given location can be determined in constant time. For a write instruction, these processors then execute the write as specified in the PRIORITY algorithm. For a read instruction, the processors read the specified locations, and then in additional O(log r) time, duplicate the value read so that there are enough copies of each value for all the processors that need to read it.

We also have the result that any algorithm for a PRIORITY PRAM can be simulated by a COMMON PRAM with no loss in parallel time provided sufficiently many processors are available [148]: Let $P_1, \ldots, P_r$ be the processors used by the PRIORITY algorithm. The simulating COMMON algorithm uses auxiliary processors $P_{i,j}$ and memory locations M_j, $1 \leqslant i, j \leqslant r$. The locations M_j are initialized to 0. Processor $P_{i,j}$, $i<j$ determines the memory addresses m_i and m_j that processors P_i and P_j were to access in the PRIORITY algorithm and writes a 1 in location M_j if $m_i = m_j$. Now P_j can ascertain if it is the lowest-numbered processor that needs to write into m_j by testing if M_j is still zero. If so, it writes into m_j the value it was supposed to write by the PRIORITY algorithm.

3.2. *Lower bounds for PRAMs*

There is a substantial body of literature which explores lower bounds on the time required by EREW, CREW or CRCW PRAMs to perform simple computational tasks. For the purpose of proving such lower bounds it is customary to adopt a model called the ideal PRAM, in which no limits are placed on the computational power of individual processors or on the capacity of a cell in the shared memory. Each processor in a PRAM can compute an arbitrary function of the values in its private memory at each step, and thus the ideal PRAM is much more powerful than the ordinary PRAM, whose processors are restricted to executing conventional RAM instructions. Lower bounds proved within such a powerful model of computation have great generality because they do not depend on particular assumptions about the instruction set or internal structure of the individual processors. Such lower bounds capture the intrinsic limitations of global memory as a means of communication between processors, and demonstrate clear distinctions in power among the various concurrent-read and concurrent-write arbitration mechanisms.

An *ideal PRAM* consists of processors which communicate through a global memory divided into cells of unbounded storage capacity. Each processor has a private memory of unbounded size and the ability to compute in unit time any function of the contents of its private memory. The input data are assumed to be stored in locations $M_1, M_2, \ldots, M_n$ of the global memory, and the computation is required to terminate with its output in location M_1. The computation proceeds in steps, with each step consisting of a read phase, a compute phase and a write phase. In the read phase, each processor reads into its private memory the contents of one cell in the gobal memory. In the compute phase, each processor computes some function of the contents of its private memory. In the write phase, each processor stores a value dependent on the contents of its private memory in some cell of global memory. An ideal PRAM is designated as EREW, CREW or CRCW according to whether concurrent reading and/or concurrent writing are permitted, and concurrent-write ideal PRAMs are further classified as COMMON, ARBITRARY, PRIORITY, etc. according to the method of write-conflict resolution. Even the weakest of these models, the EREW ideal PRAM, is so powerful that any function of n variables can be computed in time $O(\log n)$ using n processors, simply by assembling all the input data together in the private memory of one processor, which can then use its unlimited computation power to determine the output and store it in the global memory.

In the paper by Cook, Dwork & Reischuk [63] it is shown that the OR function requires time $\Omega(\log n)$ on an ideal CREW PRAM, no matter how many processors or memory cells are used. Here each input cell contains a bit. The output is 0 if all the input bits are zero, and 1 otherwise. Since this function can be computed in constant time by a COMMON CRCW PRAM with n processors and a very limited instruction set, this result clearly demonstrates that the concurrent-write mechanism is strictly more powerful than exclusive-write. This lower bound may appear obvious, since, at first sight, there seems to be no method of solving the problem on a CREW PRAM better than halving the number of inputs at each step by "OR"ring them together in pairs. But an alternate method can be given with runs in time $\log_{2.618} n + O(1)$ on a CREW PRAM, and thus beats the obvious halving method.

The lower bound proof requires the following definitions. Let us say that input bit i *affects* processor P at time t if the contents of the processor's private memory at time t differ according to whether input i is 1 or 0, when the other input bits are fixed at 0. Similarly, we can speak of an input bit affecting global memory cell M at time t. By induction on t, one can prove that the number of input bits affecting any processor or memory cell at time t is at most c^t, where c is a suitable constant. Since every input bit must affect the output cell at the end of the computation, it then follows that the computation time is at least $\log_c n$.

Extending the work of Cook, Dwork & Reischuk, Nisan [177] has recently given the following very precise characterization of the time required by an ideal CREW PRAM to compute a finite function f with domain B^n, where $B = \{0, 1\}$. For any $w \in B^n$ and any subset S of the index set $\{1, 2, \ldots, n\}$, let us say that f is *sensitive* to S at w if the value of f changes when w is changed by flipping those input bits with indices in S. We say that f is *k-block sensitive* at w if f is sensitive at w to each of k disjoint index sets. The *block-sensitivity* $bs(f)$ is defined as the largest k such that, for some w, f is k-block sensitive at w. Then the time required to compute f on an ideal CREW PRAM is bounded both above and below by bounds of the form $c \log(bs(f)) + d$, where c and d are constants.

The paper by Snir [217] studies the complexity of solving the following table look-up problem using p processors: given an array of distinct integers $\langle x_1, x_2, \ldots, x_n, y\rangle$ where $x_1 < x_2 < \cdots < x_n$, find the index i such that $x_i < y < x_{i+1}$ (by definition, $x_0 = -\infty$ and $x_{n+1} = \infty$). The problem can be solved on a CREW PRAM with a conventional instruction set in time $O(\log_{p+1} n)$ using a variant of binary search, and a simple adversary argument shows that $\log_{p+1} n$ is also a lower bound for the problem on an ideal CREW PRAM. Snir proves that the problem requires time $\Omega(\log n - \log p)$ on an ideal EREW PRAM, thereby showing that the concurrent-read PRAM is strictly more powerful than the exclusive-read PRAM.

One component of Snir's proof is a Ramsey-theoretic argument showing that, when the x_i are allowed to be arbitrarily large integers, one can restrict attention to algorithms in which the only information gathered about the x_i is obtained through comparing the x_i directly with y; i.e., for each algorithm that violates this restriction, there is another algorithm that respects the restriction and has equally good worst-case behavior. In view of this restriction one can rephrase the problem as the following *zero-counting problem*: given an array of m 0s followed by $n-m$ 1s, determine m. Here a 0 in position i means that $x_i<y$, and a 1 in position i means that $x_i<y$. At the end of the computation, cell M_m is to contain a 1 if the answer is m, and a 0 otherwise.

We sketch Snir's proof of a lower bound on the time required to solve the zero-counting problem on an ideal EREW PRAM with p processors. For any processor P and time t, say that input coordinate m *affects* P at time t if the contents of the private memory of P at time t is different on input $0^{m-1}1^{n-m+1}$ than it is on input $0^m 1^{n-m}$. Let $P(t)$ be the set of input coordinates that affect P at time t. Similarly, let $M(t)$ be the set of input coordinates that affect global memory cell M at time t. As a measure of the progress of the computation, define

$$c(t)=\sum_P |P(t)|+\sum_M \max(0,|M(t)|-1),$$

where the first summation is over all processors, and the second is over all global memory cells. Then $c(0)=0$, and if the computation halts at time T then $c(T)\geqslant n+1$. Snir proves the inequality $c(t+1)\leqslant 4c(t)+p$, and the lower bound $T=\Omega(\log n-\log p)$ follows. By a similar argument Snir proves that when the number of processors is unlimited, there is a lower bound of $\Omega(\sqrt{\log n})$ on the parallel time for the table look-up problem.

The paper by Beame & Hastad [29], improving an earlier result in Beame [27], proves that, on an ideal PRIORITY CRCW PRAM, the number of processors required to compute the parity of n bits in time T is at least $^{\lfloor (n^{1/T}/96)-2\rfloor}$. It follows that the time required to compute the parity of n bits of an ideal CRCW PRAM using a polynomial-bounded number of processors in $\Omega(\log n/\log\log n)$.

Many further lower bounds for ideal PRAMs can be cited. The papers [36, 81, 82, 110, 153, 166, 238] study the effects of the size of global memory and the choice of a write-conflict mechanism on the time required to solve problems on an ideal CRCW PRAM. These results show that, in various settings, the ARBITRARY model is strictly more powerful than the COMMON model, but strictly less powerful than the PRIORITY model.

3.3. Circuits

So far we have mainly considered the PRAM model. There are several other models of parallel computation, and of these, the *circuit model* has emerged as an important medium for defining parallel complexity classes. By a *circuit* we mean a bounded fan-in combinational Boolean circuit. More formally, a *circuit* is a labeled directed acyclic graph (DAG). Nodes are labeled as *input*, *constant*, *AND*, *OR*, *NOT*, or *output* nodes. Input and constant nodes have zero fan-in, AND and OR nodes have fan-in of 2, NOT and output nodes have fan-in of 1. Output nodes have fan-out 0.

Let $B=\{0,1\}$. A circuit with n input nodes and m output nodes computes a Boolean function $f: B^n \to B^m$, where we assume the input nodes to be ordered as $\langle x_1, \ldots, x_n \rangle$ and similarly the output nodes as $\langle y_1, \ldots, y_m \rangle$. The *size* of a circuit is the number of edges in the circuit. The *depth* of the circuit is the length of a longest path from some input node to some output node. We note that the size of a circuit is a measure of its hardware content and its depth measures the time required to compute the output, assuming unit delay at each gate.

A rather general formulation of a *problem* is as a *transducer of strings over B*: i.e., as a function from B^* to B^*. By using a suitable encoding scheme, we can assume, without loss of generality, that the size n of the input string determines the size $l(n)$ of the output string. Let $\boldsymbol{C}=\{C_i\}$, $i=1,2,\ldots$ be a family of circuits for which C_i has i input bits and $l(i)$ output bits. Then the family of circuits $\boldsymbol{C}$ *solves* a problem P if the function computed by C_i defines precisely the string transduction required by P for inputs of length i.

When using a family of circuits as a model of computation, it is necessary to introduce some notion of *uniformity* if we wish to correlate the size and depth of a family $\boldsymbol{C}$ that solves a problem P with the parallel time and hardware complexity of P. If not, we could construct a family of circuits of constant size and depth to solve an undecidable problem (for which all inputs of a given length have the same one-bit output; it is well-known that nonrecursive sets with this property exist). A notion of uniformity that is commonly accepted for parallel computation is *logspace uniformity* [203, 62]: A family of circuits $\boldsymbol{C}$ is logspace uniform if the description of the nth circuit C_n can be generated by a Turing machine using $O(\log n)$ workspace. By the description of a circuit, we mean a listing of its nodes, together with their type, followed by a listing of the inputs to each node. For $C(n) \geqslant n$ and $D(n) \geqslant \log n$, a problem P is said to be in the class CKT($C(n)$, $D(n)$) if there is a logspace uniform family of circuits $\boldsymbol{C}=\{C_n\}$ solving P such that C_n is of size $O(C(n))$ and depth $O(D(n))$.

The class NC^k, $k>1$, is the class of all problems that are solvable in CKT($poly(n)$, $O(\log^k n)$), where $poly(n)=\bigcup_{k\geqslant 1} O(n^k)$. For technical reasons we define NC^1 to be the class of problems solvable by alternating Turing machines in time $O(\log n)$ (see Section 3.5). The class $NC=\bigcup_{k\geqslant 1} NC^k$ is generally accepted as a characterization of the class of problems that can be solved with a high degree of parallelism using a feasible amount of hardware (Cook [61], Pippenger [184]). As mentioned in Section 2, we refer to the quantity $\bigcup_{k\geqslant 1} O(\log^k n)$ as *polylog*(n). Thus NC = CKT(*poly*(n), *polylog*(n)). For many commonly used models of parallel computation, the class of problems that can be solved in polylog time with a polynomial-bounded amount of hardware coincides with NC.

If we remove the fan-in restriction on AND and OR gates in the circuit model, we obtain the *unbounded fan-in circuit* model, where, as before, the size of the circuit is the number of edges in the circuit, and the depth is the length of a longest path from an input node to an output node in the unbounded fan-in circuit [47, 62, 86, 220].) A family of unbounded fan-in circuits is logspace uniform if the description of the ith circuit can be generated in logspace. By analogy with the bounded fan-in case, the class UCKT($C(n)$, $D(n)$), where $C(n) \geqslant n$, is defined as the class of problems solvable by logspace-uniform circuit families whose nth element has size $O(C(n)$ and depth $O(D(n))$. For $k \geqslant 1$, *the class* UCKT(*poly*(n), $\log^k n$) is denoted AC^k. The class $AC = \bigcup_{k\geqslant 1} AC^k$.

Since any gate in an unbounded fan-in circuit of size $p(n)$ can have fan-in at most $p(n)$, each such gate can be converted into a tree of gates of the same type with fan-in 2, such that the output gate computes the same function as the original gate. By applying this transformation to each gate in an unbounded fan-in circuit in UCKT$(p(n), d(n))$ we obtain a bouned fan-in circuit in

$$\mathrm{CKT}(\mathrm{O}(p(n)), \mathrm{O}(d(n)\cdot \log p(n))).$$

It is straightforward to see that this transformation is in logspace. Thus $\mathrm{AC}^k \subseteq \mathrm{NC}^{k+1}$ for $k \geqslant 1$. Clearly $\mathrm{NC}^k \subseteq \mathrm{AC}^k$. Thus we conclude that AC = NC. We also note that we can always compress O(log log n) levels of a bounded fan-in circuit into two levels of a polynomial-size, unbounded fan-in circuit [47] and hence

$$\mathrm{CKT}(poly(n), \log^k n) \subseteq \mathrm{UCKT}(poly(n), \log^k n/\log\log n).$$

3.4. Relations between circuits and PRAMS

If we assume a bounded amount of local computation per processor per unit time, we can establish a strong correspondence between the computational power of unbounded fan-in circuits and that of CRCW PRAMs [220]. More specifically, assume the PRIORITY write model, and that the instruction set of the individual RAMs consists of binary addition and subtraction of poly-size numbers, binary Boolean operations, left and right shifts, and conditional branching on zero, and that indirect memory access is allowed. Each instruction is assumed to execute in unit time. The input to a problem of size n is specified by n n-bit numbers. Let CRCW$(P(n), T(n))$ be the class of problems solvable on such a PRAM in O$(T(n))$ time with O$(P(n))$ processors.

Given any unbounded fan-in circuit in UCKT$(S(n), D(n))$ we can simulate it on a CRCW PRAM in time O$(D(n))$ with $S(n)$ processors and $n+M(n)$ memory locations, where $M(n)$ is the number of gates in the circuit. Any bounded fan-in or unbounded fan-in circuit can be converted in logspace into an equivalent circuit of comparable size and depth for which negations occur only at the inputs. These negations can now be removed by supplying the complement value as input. So we can assume that our input circuit has no negations. In the simulation, each processor is assigned to an edge in the circuit and each memory location to a gate. Initially, the inputs are in memory locations 1 through n, and memory location $n+i$ is assigned to gate i in the circuit, $i = 1, \ldots, M(n)$; these latter locations are initialized to 0 for OR gates and to 1 for AND gates.

Each step in the simulation consists of performing the following three substeps in sequence:

(a) Each processor p determines the current value c on its edge $e = (u, v)$ by reading memory location $n+u$.

(b) If $c = 0$ and v is an OR gate, or if $c = 1$ and v is an AND gate then p writes value c into location $n+v$.

(c) If $c = 1$ and v is an OR gate, or if $c = 0$ and v is an AND gate then p writes c into location $n+v$.

Clearly, after $D(n)$ steps, memory location $n+i$ has the value of gate i for $i = 1, \ldots, M(n)$. Thus

$$\mathrm{UCKT}(S(n), D(n)) \subseteq \mathrm{CRCW}(S(n), D(n)).$$

For the reverse part, we note that each of the binary operations in the instruction set of the PRAM can be implemented by constant-depth, polynomial-size, unbounded fan-in circuits, as shown in Section 4. It is also fairly easy to implement the conditional branching by such circuits, by suitably updating the program counter. The nontrivial part of the simulation of a CRCW PRAM by an unbounded fan-in circuit lies in simulating the memory accesses. Since a combinational circuit has no memory, the simulation retains all values that are written into a given memory location during the computation, and has a bit associated with each such value that indicates whether the value is current or not. With this system, it is not difficult to construct a constant-depth unbounded fan-in circuit to implement reads and writes into memory. Thus a single step of the PRIORITY CRCW PRAM can be simulated by an unbounded fan-in circuit of polynomial size and constant depth, and it follows that

$$\mathrm{CRCW}(P(n), T(n)) \subseteq \mathrm{UCKT}(poly(P(n)), T(n)).$$

Let us call a PRAM algorithm *logspace uniform* if there is a logspace Turing machine that, on input n, generates the program executed by each processor on inputs of size n. (We note that all of the PRAM algorithms we describe have programs that are parametrized by n and the processor number, and such algorithms are clearly logspace uniform.) Define CRCW^k as the class of problems solvable by logspace uniform CRCW PRAM algorithms in time $O(\log^k n)$ using a polynomial-bounded number of processors; let CREW^k and EREW^k be the analogous classes for CREW PRAMs and EREW PRAMs. We will see in Section 3.9 that NL (nondeterministic logspace) is in CRCW^1, and that L (deterministic logspace) is in EREW^1. Thus, by the results of Stockmeyer & Vishkin we have just described [220], we have $\mathrm{CRCW}^k = \mathrm{AC}^k$ for $k \geqslant 1$, and

$$\bigcup_{k>0} \mathrm{CRCW}(poly(n), \log^k n) = \mathrm{NC}.$$

Since we also noted earlier that simulations between the various types of PRAM result in only an $O(\log P(n))$ increase in time and a squaring of the processor count, we have

$$\mathrm{PRAM}(poly(n), polylog(n)) = \mathrm{NC},$$

where the PRAM processor and time bounds can refer to any of the PRAM models.

It is shown in Hoover, Klawe & Pippenger [119] that any bounded fan-in circuit of size $S(n)$ and depth $D(n)$ can be converted into an equivalent circuit of size $O(S(n))$ and depth $O(D(n))$ having both bounded fan-in and bounded fan-out. Using this result, it is easy to see that $\mathrm{NC}^k \subseteq \mathrm{EREW}^k$. Thus we have the following chain of inclusions for $k \geqslant 1$:

$$\mathrm{NC}^k \subseteq \mathrm{EREW}^k \subseteq \mathrm{CREW}^k \subseteq \mathrm{CRCW}^k = \mathrm{AC}^k \subseteq \mathrm{NC}^{k+1}.$$

As noted earlier, we also have

$$\mathrm{NC}^k \subseteq \mathrm{UCKT}(poly(n), \log^k n/\log\log n).$$

Earlier in this section, we referred to the lower bound of $\Omega(\log n/\log\log n)$ for computing parity on an ideal PRIORITY CRCW PRAM with a polynomial number of processors [29]. This immediately gives the same lower bound for computing parity with an unbounded fan-in circuit of polynomial size. Historically this lower bound was first developed for the case of unbounded fan-in circuits in a sequence of papers [86, 5,

243, 113, 114, 216], thus implying the lower bound for CRCW PRAMs with restricted instruction set. The extension of the result to ideal PRAMs requires a substantial refinement of the probabilistic restriction techniques used to obtain the lower bounds for circuits. For bounded fan-in circuits, a lower bound of $\Omega(\log n)$ for the depth required to compute any function whose value depends on all n input bits is easily established by a simple fan-in argument.

3.5. Alternating Turing machines

Another important model of parallel computation is the *alternating Turing machine* (*ATM*) [46, 203]. An ATM is a generalization of a nondeterministic Turing machine whose states can be either existential or universal. Some of the states are *accepting* states. As with regular Turing machines, we can represent the *configuration* α of an ATM at a given stage in the computation by the current state, together with the current contents of worktapes and the current positions of the read and worktape heads. A configuration is *accepting* if it contains an accepting state. The space required to encode a configuration is proportional to the space used by the computation on the worktapes. Configuration β *succeeds* configuration α if α can change to β in one move of the ATM. For convenience we assume that an accepting configuration has transitions only to itself, and that each nonaccepting configuration can have at most two different configurations that succeed it.

We define the concept of an *accepting computation* from a given initial configuration α. If α is an accepting configuration, then α itself comprises an accepting computation. Otherwise, if α is existential, then there is an accepting computation from α if and only if there is an accepting computation from some configuration β that succeeds α; and if α is universal, then there is an accepting computation from α if and only if there is an accepting computation from every configuration β that succeeds α. Thus we can represent an accepting computation of an ATM on input x as a rooted tree whose root is the initial configuration, whose leaves are accepting configurations, and whose internal nodes are configurations such that if c is a node in the tree representing a configuration with an existential state, then c has one child in the tree which is a configuration that succeeds it, and if c represents a configuration with a universal state, then the children of c are all the configurations that succeed it. An ATM *accepts* input x if it has an accepting computation on input x. A node α at depth t in the computation tree represents the event that the ATM can reach configuration α after t steps of the computation, and will be denoted by the ordered pair (α, t). The *computation DAG* of an accepting computation of an ATM on input x is the DAG derived from the computation tree by identifying together all nodes in the tree that represent the same ordered pair.

Alternating Turing machines are generally defined as acceptors of sets. We can view them as transducers of strings by considering the input as an ordered pair $\langle w, i\rangle$ and the output bit as specifying the ith bit of the output string when the input string is w [61]. Here we assume that the length of the output is at most exponential in the input length.

An ATM M operates in time $T(n)$ if, for every accepted input of length n, M has an accepting computation of depth $O(T(n))$. Similarly, an ATM operates in space $S(n)$ if, for every accepted input of length n, M has an accepting computation in which the configurations require at most $O(S(n))$ space. We define $ATM(S(n), T(n))$ to be the class of languages that are accepted by ATMs operating simultaneously in time $O(T(n))$ and space $O(S(n))$. Also, let $ATIME(T(n))$ denote the class of languages accepted by ATMs operating in time $O(T(n))$.

In order to allow sublinear computation times, we use a *random-access model* to read the input tape: when in a specified *read* state, we allow the ATM to write a number in binary which is then interpreted as the address of a location on the input tape, whose symbol is then read onto the worktape in unit time. Thus $\log n$ time suffices to read any input symbol.

We shall now show that, for $k \geqslant 1$, $ATM(\log n, \log^k n) = NC^k$ (Ruzzo [203]). This can be seen as follows. Consider a language accepted by an ATM M in $ATM(S(n), T(n))$, with $S(n) = \Omega(\log n)$. The *full computation DAG D* of M for inputs of length n is obtained by having a vertex for each pair (α, t), where α is a configuration of the ATM in space $O(S(n))$ and $0 \leqslant t \leqslant T(n)$, and an arc from vertex (α, t) to vertex $(\beta, t-1)$ if α succeeds β. We can construct a circuit from D by replacing each existential node in D by an OR gate, each universal node by an AND gate, each accepting leaf by a constant input 1, each nonaccepting leaf by a constant input 0, and each node in which the state is a read state by the corresponding input. The output of the circuit corresponds to vertex $(\alpha_0, 0)$, where α_0 is the initial configuration. It is not difficult to see that the output of this circuit is 1 if and only if the input is accepted by the ATM in the prescribed time and space bounds. Further the depth of this circuit is $O(T(n))$ and its size is $O(c^{S(n)})$ for a suitable constant c. Thus,

$$ATM(S(n), T(n)) \subseteq CKT(c^{S(n)}, T(n)).$$

Also, since the resulting family of circuits is easily shown to be logspace uniform, we have

$$ATM(\log n, \log^k n) \subseteq NC^k \quad \text{for } k \geqslant 2, \quad \text{and} \quad ATM(\log n, polylog(n)) \subseteq NC.$$

For the reverse, consider any logspace uniform circuit family with size $O(S(n))$ and depth $O(T(n))$. We may assume that all NOT gates have been pushed back to the inputs and eliminated. Let p be a sequence of Ls and Rs, where L stands for left input and R stands for right input. Given two gates g and h in the circuit C_n, we say that $h = g(p)$ if h is the gate reached starting from g and following the sequence p. The ATM simulating C_n computes a function $cv(n, g, p)$, which evaluates the output of $g(p)$ by guessing $g(p)$, recursively evaluating $cv(n, g, p\mathrm{L})$ and $cv(n, g, p\mathrm{R})$, and combining the results appropriately, depending on the gate type of $g(p)$, to obtain $cv(n, g, p)$. When the length of p grows longer than $\log n$, g is replaced by $g(p)$ and p is truncated to ε. The procedure for computing $cv(n, g, p)$ is as follows.

PROCEDURE $CV(n, g, p)$
(1) Guess $g(p) = h$ (using existential states);
(2) **In parallel** (using universal states)
    verify that $g(p) = h$;
    if $|p| = \lceil \log n \rceil$
    then return $CV(n, h, \varepsilon)$
    else return $CV(n, h, p\mathrm{L}) * CV(n, h, p\mathrm{R})$, where $*$ is the type of gate h

The initial call to the procedure is CV(n, *outputgate*, ε). Because $|p|$ is always at most $\lceil \log n \rceil$ and our circuit family is logspace uniform, the step which verifies that $g(p)=h$ can be performed in logspace, and hence in ATM($\log n$, $\log^2 n$) [46]. Hence a logspace uniform circuit family of size $S(n)$ and depth $T(n) \geqslant \log^2 n$ can be simulated in ATIME(O($T(n)$)) using space O($\log S(n)$). In particular, $\mathrm{NC}^k \subseteq \mathrm{ATM}(poly(n), \log^k n)$ for $k \geqslant 2$. For $k=1$ a stronger notion of uniformity is required for the above inclusion to hold [203]; alternatively, following [43, 44] we can *define* NC^1 to be ATM($\log n$, $\log n$) (which is the same as ATIME($\log n$) or *alternating logtime*). Let $\mathrm{ATM}^k = \mathrm{ATM}(\log n, \log^k n)$. The two results outlined above relating ATM computations with uniform families of circuits establish that $\mathrm{ATM}^k = \mathrm{NC}^k$ and $\mathrm{ATM}(\log n, polylog(n)) = \mathrm{NC}$.

Computation on ATMs can also be related to unbounded fan-in circuits. A computation on an ATM is said to be in ALT($S(n)$, $f(n)$) if the configurations require O($S(n)$) space, and if any path in an accepting computation DAG has at most $f(n)$ alternations between existential and universal states. The result

$$\mathrm{UCKT}(c^{S(n)}, D(n)) = \mathrm{ALT}(S(n), D(n))$$

is readily established by observing that the unbounded fan-in circuit can be converted into a bounded fan-in circuit with the same number of alternations between AND and OR gates, which can then be simulated in a manner similar to the bounded fan-in case; and conversely the full computation DAG of such a resource-bounded ATM has O($c^{S(n)}$) nodes, c a constant. Thus, defining ALT^k to be ALT($\log n$, $\log^k n$), we have

$$\mathrm{ALT}^k = \mathrm{AC}^k \quad \text{for } k \geqslant 1, \quad \text{and} \quad \mathrm{ALT}(\log n, polylog(n)) = \mathrm{NC}.$$

3.6. *Vector machines*

The last model of parallel computation that we consider is the *vector machine* [187, 214], which consists of a collection of bit processors together with a collection of registers that can hold bit vectors. All processors contain the same program whose instruction set consists of binary Boolean operations on the registers, complementation of the contents of a register, conditional jump on zero, the right or left shift of the contents of a register by a shift parameter specified in a register, and a mask instruction that inhibits some processors from executing the next instruction. Some of the registers are identified as *input* or *output* registers. The inputs to the computation are supplied in the input registers. At a given instant of time, the ith processor reads the ith bit of the operands (if any) specified in the current instruction, executes the prescribed instruction and writes the result in the ith position of the output vector. In the case of a shift operation, the processor writes in the appropriately shifted bit position; this is the means by which interprocessor communication takes place. When the computation terminates, the results of the computation are available in the output registers. A vector machine algorithm is logspace uniform if there is a deterministic Turing machine operating in space $\log n$ that, on input n, generates the program for inputs of length n. As in the case of PRAMs, in practice, we would expect vector machine code for a problem to be fixed, with n as a parameter.

Let VM($S(n)$, $T(n)$) be the class of problems that can be solved on a vector machine with $O(S(n))$ processors (and hence with vectors of length $O(S(n))$) in $O(T(n))$ time, and let VM^k be the class of problems that have logspace uniform vector machine algorithms in $VM(poly(n), \log^k n)$. It is readily seen that

$$VM(S(n), T(n)) \subseteq UCKT(poly(S(n)), T(n))$$

since the instruction set of vector machines can be simulated in constant depth and polynomial size by unbounded fan-in circuits. It can also be shown that

$$CKT(S(n), T(n)) \subseteq VM(poly(S(n)), T(n))$$

provided $T(n) \geqslant \log^2 S(n)$ [100, 222]. Thus, for $k \geqslant 2$, we have

$$NC^k \subseteq VM^k \subseteq AC^k \quad \text{and} \quad VM(poly(n), polylog(n)) = NC.$$

Several other parallel models and complexity results can be found in [25, 61, 62, 73, 74, 182, 184, 202, 222, 223, 233] (See also Chapter 1 on machine models and simulations in this Handbook).

3.7. *Randomized complexity classes*

In discussing randomized algorithms, we limit ourselves to problems defined in terms of a binary input–output relation $S(x, y)$. On input x, the task is to find a y satisfying $S(x, y)$, if such a y exists. A randomized algorithm will output one of the following three answers:

(a) a suitable value for y;

(b) an indication that no suitable y exists;

(c) an indication of failure, i.e., inability to determine if a suitable y exists or not.

We distinguish between zero-error algorithms (also known as Las Vegas algorithms) and algorithms with one-sided error (also known as Monte Carlo algorithms). If, on input x, $S(x, y)$ holds for some y, then the two types of algorithms act alike: each type produces a suitable y with probability greater than $\frac{1}{2}$, and otherwise reports failure. On an input x such that there is no y satisfying $S(x, y)$, the two types of algorithms behave differently. A zero-error algorithm reports "No suitable y exists" with probability greater than $\frac{1}{2}$, and otherwise reports failure, but a one-sided-error algorithm always reports failure.

Each of the complexity classes we have defined has its zero-error and one-sided-error randomized counterparts, indicated by the prefixes Z and R respectively. These are the classes of problems that have randomized algorithms with the corresponding hardware and time bounds. For randomized algorithms on PRAMs each processor is assumed to have the capability to generate random ($\log n$)-bit numbers; for randomized computation on circuits, an n-input circuit is allowed $poly(n)$ additional random input bits. Thus, for example, a problem is in $ZCREW^k$ if it is solvable by a zero-error randomized algorithm that runs in time $O(\log^k n)$ using $poly(n)$ processors on a CREW PRAM in which each processor can generate random ($\log n$)-bit numbers, and a problem is in RNC^k if it is solvable with one-sided error by a logspace-uniform family of bounded fan-in circuits which receive, in addition to the problem input, $poly(n)$ random input bits.

3.8. *Arithmetic models*

For computation involving elements from an arbitrary domain D, it is convenient to assume that certain specified binary operations on the elements take unit time. This leads to the definition of *arithmetic PRAMs*, *arithmetic circuits* and *arithmetic Boolean circuits* (von zur Gathen [93]). An arithmetic PRAM is a regular PRAM which in addition can execute certain binary operations on elements over D in unit time. Sometimes D is specified not as a specific structure but according to axiomatic properties; for example, as an abstract ring. An arithmetic Boolean circuit is analogous to a Boolean circuit except that the set of gates is augmented to include gates that compute the specified arithmetic operations on elements over D in unit time. The conversion from arithmetic to Boolean values is performed by gates that test for zero, detect the sign, etc. In the reverse direction, Boolean selection circuits are used to select an output from among several arithmetic inputs. A special case of an arithmetic Boolean circuit is an *arithmetic circuit*, which has no Boolean gates. For *unbounded fan-in arithmetic Boolean circuits* we allow the Boolean gates to have unbounded fan-in while the arithmetic gates continue to have fan-in 2. The complexity classes *arithmetic* NC^k and *arithmetic* AC^k are analogues of NC^k and AC^k for arithmetic Boolean circuits and unbounded fan-in arithmetic Boolean circuits respectively. Similarly we have the analogous PRAM classes *arithmetic* EREW^k, *arithmetic* CREW^k, and *arithmetic* CRCW^k. It should be noted that other authors (cf. [93]) define arithmetic NC^k in terms of arithmetic circuits (without Boolean gates) with a further restriction to polynomials or rational functions of polynomial-bounded degree. This restriction avoids certain trivial separations between complexity classes; for instance, without it, the function $x^{2^{\log^2 n}}$ would lie in arithmetic NC^2 but not arithmetic NC^1.

When dealing with arithmetic computation involving addition, subtraction and multiplication of numbers represented in binary, there is an obvious conversion from arithmetic circuits in Boolean ones that maintains size to within a polynomial, and increases depth by a factor of $\log\log s$, where s is a bound on the size of the arithmetic operands (and thus the number of bits needed to represent any operand is no more than $\lceil \log s \rceil$). The bounds on size and depth follow from well-known poly-size log-depth circuits for addition, subtraction and multiplication of two n-bit numbers (see Section 4.2).

3.9. *Parallel Computation Thesis*

Finally, an important connection between sequential and parallel computation is highlighted in the *Parallel Computation Thesis* [38, 46, 85, 106, 187]. Let us say that two functions are polynomially equivalent if each is bounded above by a polynomial function of the other. The Parallel Computation Thesis states that parallel time is polynomially equivalent to sequential space. This relationship has been established in many forms. Parallel time on a vector machine is polynomially equivalent to sequential space [187, 214]. An ATM can be viewed as a parallel machine, and the result follows since alternating time is polynomially equivalent to sequential space [46]. Computation on a PRAM can be simulated by a Turing machine with space polynomially bounded in the parallel time, and conversely, provided the number of processors is no more than an exponential in the parallel time [85, 106]. Finally any computation in nondeterministic space $S(n)$ can be simulated by a circuit of depth $S(n)^2$ and any circuit of depth $D(n)$ can be simulated in deterministic space $D(n)$ [38]. We conclude this section by illustrating the Parallel Computation Thesis for some of the models we have considered.

We first relate nondeterministic space to parallel time. Consider the computation of any nondeterministic $S(n)$ space-bounded Turing machine M. Given an input x to M, we can formulate the acceptance problem as a reachability problem on a directed graph G, whose vertices are the configurations of M, and for which there is an arc from vertex u to vertex v if and only if the configuration presented by v can be reached in a single step from the configuration represented by u. There is also a dummy vertex z with an arc into it from every vertex corresponding to an accepting configuration. If s is the vertex corresponding to the initial configuration of M on input x, then M accepts x if and only if z is reachable from s in G. The size of G is $O(c^{S(n)})$ for some constant c, since the number of different configurations of an $S(n)$ space bounded Turing machine is no more than an exponential in $S(n)$. The reachability problem can be solved in AC^1 (see Section 4) by finding the transitive closure of the adjacency matrix of G. Since this construction is logspace uniform, it follows that any computation in nondeterministic space $S(n)$ is in $CKT(c^{S(n)}, S^2(n))$ and in $UCKT(c^{S(n)}, S(n))$, and also in $CRCW(c^{S(n)}, S(n))$ and $EREW(c^{S(n)}, S^2(n))$. Thus, in these parallel models, nondeterministic space-bounded computation can be parallelized to run in time at most the square of the space bound, and using hardware at most exponential in the space bound.

A similar technique shows that deterministic space $S(n)$ is in $EREW(c^{S(n)}, S(n))$. In this case the computation of the space-bounded Turing machine can be modeled as a directed tree, and an input is accepted if and only if the final configuration is reachable from the initial one. Since reachability in a directed tree is a special case of expression evaluation, the tree contraction algorithm of Section 2.2 gives the required result.

For the reverse, consider a circuit in $CKT(S(n), D(n))$. Given a description of this circuit together with an input to it, a deterministic Turing machine can compute its output in $O(D(n))$ space by starting at the output and working its way back to the input, while using a stack to keep track of the path taken (using L for left input and R for right input), first testing left inputs and then the right ones. Since the depth of the circuit is $D(n)$, the stack has at most $D(n)$ entries, each of which is a constant. Hence the value computed by the circuit can be evaluated in $O(D(n))$ space. By using the results we described earlier in this section relating parallel time on various models to depth on a corresponding uniform family of bounded fan-in circuits, it is easy to verify that a parallel algorithm on all of the models we have considered implies a deterministic sequential algorithm that uses space at most the square of the parallel time.

4. NC-algorithms and P-complete problems

4.1. Introduction

In this section we locate several important problems within the hierarchy $\{NC^k\}$. We restrict ourselves to problems that are of central importance because they can be used as subroutines in the solution of a wide range of other problems. Among these are the basic arithmetic operations, Boolean matrix multiplication and transitive closure, the computation of the inverse and the rank of a matrix, the evaluation of straight-line programs, and the computation of a maximal independent set in a graph. We also introduce the concept of P-completeness, state the evidence that the P-complete problems are unlikely to lie in NC, and give several examples of P-complete problems.

The algorithms we give are not necessarily efficient, in the sense of Section 2, since their processor–time product may be asymptotically much larger than the execution time of the best sequential algorithm for the same problem. In this sense, our level of aspiration is lower than in Section 2, where we concerned ourselves with efficient and optimal algorithms.

4.2. *NC-circuits for arithmetic operations*

This section is concerned with Boolean circuits for addition, subtraction, multiplication and division of integers. We shall show that addition and subtraction are in AC^0, that multiplication is in NC^1 but not AC^0, and that division is realizable by a family of bounded fan-in circuits that is of logarithmic depth and polynomial size, but does not appear to be logspace uniform. Further information on circuits for arithmetic operations can be found in [9, 185, 204].

4.2.1. *Addition and subtraction*

The addition problem takes as input two n-bit binary numbers and produces as output their $(n+1)$-bit sum. We represent numbers as tuples in binary notation. Let the input numbers be $(x_{n-1}, x_{n-2}, \ldots, x_0)$ and $(y_{n-1}, y_{n-2}, \ldots, y_0)$, and let the output be $(z_n, z_{n-1}, \ldots, z_0)$. Let c_j denote the carry out of the jth position. Let $g_j = x_j y_j$ and $p_j = x_j \vee y_j$; g_j is called the jth *carry generate bit* and p_j is called the jth *carry propagate bit*. Then, letting $\oplus$ represent "exclusive or" and letting $c_{-1} = 0$, we have the recurrences $c_j = g_j \vee p_j c_{j-1}$ and $z_j = x_j \oplus y_j \oplus c_{j-1}$ for $j = 0, 1, \ldots, n$. It follows that $c_j = \bigvee_{i \leqslant j} g_i p_{i+1} \cdots p_j$. These formulas yield a polynomial-size constant-depth circuit for addition:

Stage 1: compute all carry generate bits g_j and carry propagate bits p_j.
Stage 2: compute all products of the form $g_i p_{i+1} \cdots p_j$.
Stage 3: compute all c_j as the OR of products computed in Stage 2.
Stage 4: compute each output bit z_j from x_j, y_j and c_{j-1}.

This establishes that addition is in AC^0, and hence, also in NC^1.

A more economical NC^1-circuit for addition can be obtained using the prefix sums algorithm of Section 2.2. That algorithm takes as input an array $(a_1, a_2, \ldots, a_n)$ and produces the array $(a_1, a_1 * a_2, \ldots, a_1 * a_2 * \cdots * a_n)$, where $*$ is an associative binary operation. Since the memory accesses are data independent in the PRAM algorithm for prefix sums in Section 2.2, that algorithm can also be represented as a circuit of depth $O(\log n)$ and size $O(n)$, with gates that compute $*$. To apply this construction to addition, let T_j be the transformation that computes carry bit c_j from carry bit c_{j-1}. This affine Boolean transformation is specified by the equation $c_j = g_j \vee p_j c_{j-1}$. Then c_j

is obtained by evaluating the transformation $T_j * T_{j-1} * \cdots * T_0$ at the point 0; here the associative operation $*$ is the composition of affine Boolean transformations. Using a parallel prefix circuit to compute the necessary compositions, one obtains a bounded fan-in Boolean circuit of size O(n) and depth O(log n) which computes the carry bits c_j, and then the output bits z_j. A variant of this construction gives bounded fan-in circuits of linear size and logarithmic depth for subtraction in either the 1's complement or 2's complement representation.

In [45] it is shown that, if the semigroup defined by $*$ does not contain a nontrivial group as a subset then, for any strictly increasing primitive recursive function f, the prefix sums problem can be solved by an unbounded fan-in circuit family of constant depth and size $O(nf^{-1}(n))$. Each gate in such a circuit computes the semigroup product of its inputs. Dolev, Dwork, Pippenger & Wigderson [70] show that addition cannot be performed by unbounded fan-in circuits of constant depth and linear size.

4.2.2. Multiplication

The multiplication problem takes as input two n-bit binary numbers and produces as output their $2n$-bit product. Let the inputs be $(x_{n-1}, x_{n-2}, \ldots, x_0)$ and $(y_{n-1}, y_{n-2}, \ldots, y_0)$, and let the output be $(z_{2n-1}, z_{2n-2}, \ldots, z_0)$. The standard shift-and-add algorithm for multiplication can be implemented in the time required to add n binary numbers, each of length $2n$. This latter problem can be solved by a bounded fan-in circuit of depth O(log n) and polynomial size using the following *three-for-two trick* which reduces the problem of adding three n-bit numbers to the problem of adding two $(n+1)$-bit numbers [179, 240]. Let $a=(a_{n-1}, a_{n-2}, \ldots, a_0)$, $b=(b_{n-1}, b_{n-2}, \ldots, b_0)$ and $c=(c_{n-1}, c_{n-2}, \ldots, c_0)$ be the three n-bit numbers to be added. If we add the three bits a_i, b_i and c_i for $i=0, 1, \ldots, n-1$, we will obtain for each i a two-bit number whose upper and lower bits we shall denote by u_i and v_i. The u_i and v_i can be generated by a linear-size constant-depth circuit that, for each i, adds the three 1-bit numbers a_i, b_i and c_i. Then $a+b+c=u+v$, where $u=(u_{n-1}, u_{n-2}, \ldots, u_0, 0)$ and $v=(v_{n-1}, v_{n-2}, \ldots, v_0)$. The addition of n numbers, each of length $2n$, can be achieved by applying O(log n) iterations of the three-for-two trick, each of which reduces the number of numbers by a factor of $\frac{2}{3}$, followed by a final stage of adding two O(n)-bit numbers. This establishes that multiplication is in NC^1.

The best bounded fan-in circuit known for multiplication achieves O(log n) depth with O(n log n log log n) size [208]. The construction is logspace-uniform; it is based on circuits for computing the Discrete Fourier Transform over certain finite rings.

The paper by Furst, Saxe & Sipser [86] was the first to establish that multiplication is not in AC^0. This was done by showing that the n-input parity function, which is equal to 1 if and only if an odd number of its inputs are 1, does not lie in AC^0, and then showing that if multiplication were in AC^0, then parity would also lie in AC^0. These results are presented in [37]. As mentioned in Section 3, parity, and hence multiplication, requires unbounded fan-in circuits of depth $\Omega(\log n/\log\log n)$ if the size is to be polynomial. The circuit of Schönhage & Strassen [208] can be converted into an $O(\log n/\log\log n)$ depth unbounded fan-in circuit for multiplication of size $O(n^{1+\varepsilon})$ for any $\varepsilon>0$, using a standard technique of compressing O(log log n) levels of a bounded fan-in circuit [47].

4.2.3. Division

The input to the division problem is a pair of n-bit binary strings representing integers x and y with $y \neq 0$. The output is the n-bit binary representation of the integer part of x/y. We present a simplified version of a construction due to Beame, Cook & Hoover [28] which yields a family of bounded fan-in division circuits of polynomial size and logarithmic depth. The construction appears not to be logspace-uniform, and thus does not establish that division lies in NC^1.

To describe the construction we require the concept of NC^1-reducibility. Let A and B be functions from $\{0,1\}^*$ into $\{0,1\}^*$, each having the property that the length of the output is determined by the length of the input. For each positive integer n we may derive from B a function B_n from $\{0,1\}^n$ into $\{0,1\}^m$, where m is the length of the string $B(x)$ whenever the string x is of length n. Similarly, we may derive from A a family of functions A_n. The function A is said to be NC^1-reducible to B (denoted $A <_{NC^1} B$) if the family of functions $\{A_n\}$ can be realized by a logspace-uniform family of bounded fan-in circuits of polynomial size and logarithmic depth, together with "oracle gates" realizing functions in the family $\{B_n\}$, subject to the restriction that no two oracle gates lie on the same input-output path. It follows from this definition that if $A <_{NC^1} B$ and B is realizable by a bounded fan-in circuit family of polynomial size and logarithmic depth, then A is also realizable by such a circuit family; moreover, if the circuit family for B is logspace-uniform, then the circuit family for A will also be logspace-uniform.

We shall show that

$$\text{DIVISION} <_{NC^1} \text{RECIPROCAL} <_{NC^1} \text{POWERING} <_{NC^1} \text{ITERATED PRODUCT},$$

where the reciprocal, powering and iterated product problems are defined as follows.

RECIPROCAL

Input: A non-zero n-bit integer y.

Output: An n-bit binary function $\tilde{y}^{-1}$ such that $y^{-1} - 2^{-n} \leqslant \tilde{y}^{-1} \leqslant y^{-1}$.

POWERING

Input: An n-bit integer x and an integer i, where $1 \leqslant i \leqslant n$.

Output: x^i expressed as an n^2-bit integer.

ITERATED PRODUCT

Input: n-bit binary integers $w_1, w_2, \ldots, w_n$.

Output: The n^2-bit product $w_1 w_2 \cdots w_n$.

The reductions are as follows.

DIVISION $<_{NC^1}$ RECIPROCAL. It suffices to treat the case $y > 0$:

(1) using an oracle gate for RECIPROCAL, compute $\tilde{y}^{-1}$;

(2) using an NC^1-circuit, compute the product $x\tilde{y}^{-1}$;

(3) **if** $(x\tilde{y}^{-1})y \leqslant x$

  then $\lfloor xy^{-1} \rfloor = x\tilde{y}^{-1}$

  else $\lfloor xy^{-1} \rfloor = x\tilde{y}^{-1} - 1$.

RECIPROCAL $<_{NC^1}$ POWERING
(1) let $y=2^j w$, where $\frac{1}{2}\leqslant w<1$, and let $t=1-w$;
(2) **in parallel**, using oracle gates for powering, compute $t^2, t^3, \ldots, t^n$;
(3) using an NC^1-circuit for iterated addition, compute $1+t+t^2+\cdots+t^n$;
(4) $\tilde{y}^{-1}=2^{-j}(1+t+t^2+\cdots+t^n)$, rounded to n-bit precision.

POWERING $<_{NC^1}$ ITERATED PRODUCT. This is immediate, since powering is a special case of iterated product.

Given these reductions, the task remaining is to give a polynomial-size logarithmic-depth bounded fan-in circuit for iterated product. The method used by Beame, Cook & Hoover [28] for constructing this circuit depends on the following ancient theorem.

CHINESE REMAINDER THEOREM. *Let $c_1, c_2, \ldots, c_m$ be pairwise relatively prime and let $c=\Pi_{i=1}^m c_i$. Then*

(i) *there exist integers $v_1, v_2, \ldots, v_m$ such that, for $i=1,2,\ldots,m$ and $j=1,2,\ldots,m$,*

$$v_i \bmod c_j = \begin{cases} 1 & \text{if } i=j, \\ 0 & \text{if } i\neq j; \end{cases}$$

(ii) *for any integer u, $u \bmod c = \Sigma_{i=1}^m u_i v_i \bmod c$, where $u_i = u \bmod c_i$.*

The circuit for computing the product $w_1 w_w \cdots w_n$, where the w_i are n-bit integers, is as follows:

(i) Let $c_1, c_2, \ldots, c_m$ be distinct primes less than n^2 whose product is greater than 2^{n^2} (for all $n>5$ such a set of primes exists). Let $c=\Pi_{j=1}^m c_j$.
(ii) For $i=1,2,\ldots,n$ and $j=1,2,\ldots,m$ compute $w_i \bmod c_j$.
(iii) For $j=1,2,\ldots,n$ compute $w_1 w_2 \cdots w_n \bmod c_j$.
(iv) Using the Chinese Remainder Theorem, compute $w_1 w_2 \cdots w_n \bmod c$.

Each of the steps is carried out by a polynomial-size, logarithmic-depth circuit which has access to certain precomputed constants. We illustrate by describing step (iii), in which the essential task is to multiply together n elements x_i, each of which is a residue modulo a prime p which is less than n^2 (here p corresponds to c_j, and x_i to $w_i \bmod c_j$). Let $\mathbf{Z}_p^*$ be the field of non-zero residues modulo p. Then $\mathbf{Z}_p^*$ has a *primitive root*: i.e., an element g such that every element of $\mathbf{Z}_p^*$ is a power of g. If $x=g^y$ then y is called the (discrete) logarithm of x, and x is the antilogarithm of y. With the help of circuits of depth $O(\log n)$ for computing logarithms and antilogarithms, we can replace iterated product mod p by iterated summation mod$(p-1)$. The circuit for computing $x_1 x_2 \cdots x_n$ mod p has the following parts:

(i) For $i=1,2,\ldots,n$ compute y_i, the logarithm of x_i;
(ii) Using an NC^1-circuit for iterated addition, compute $y=y_1+y_2+\cdots+y_n$ mod$(p-1)$.
(iii) Compute x, the antilogarithm of y. Then $x=x_1 x_2 \cdots x_n \bmod p$.

This completes our description of a polynomial-size, logarithmic-depth bounded fan-in circuit for division. Because of the precomputed constants $v_1, v_2, \ldots, v_m$ and $\Pi_{j=1}^m c_j$, the circuit appears not to be logspace-uniform, and it remains an open question whether DIVISION is in NC^1. Reif [197] has shown that division is "almost" in NC^1 by giving a logspace-uniform family of division circuits of polynomial size and depth $O(\log n \log\log n)$. Shankar & Ramachandran [211] refined the constructions of Beame, Cook & Hoover and Reif [28, 197] by showing that in each case the size can be reduced to $O(n^{1+\delta})$, where δ is an arbitrarily small positive constant. Reif & Tate [199] improve upon the result of Shankar & Ramachandran by giving logspace-uniform division circuits having size $O(n \log n \log\log n)$ and depth $O(\log n \log\log n)$.

4.3. *Circuits for expression evaluation*

In Section 2.2.3 we described a tree contraction algorithm and used it to derive a simple, optimal $O(\log n)$ time EREW PRAM algorithm for the expression evaluation problem. Since the expression evaluation problem is an important one, considerable work has been done on solving the problem efficiently for a different model of parallel computation, the arithmetic network.

The problem has both a static version and a dynamic version. In the static version we are given an n-variable expression over an algebraic structure, and our task is to construct an arithmetic circuit of small size and depth for that particular expression. The inputs to the circuit will be the values of the n variables in the expression, and the output will be the value of the expression. In the more general dynamic version we wish to construct an arithmetic Boolean circuit that takes as inputs an n-variable arithmetic expression, presented as a well-formed string of operators, operands and left and right parentheses, together with the values of the variables, and produces the value of the expression.

The early circuits for static expression evaluation are based on methods of decomposing a binary tree into two subtrees of approximately equal size by deleting a single edge. Let T be a tree representing an expression on n variables over a ring. Then T has $m=2n-1$ vertices. It is easy to see that T has an edge e whose removal breaks it into two subtrees, T_1 and T_2, each of which has no more than $2n/3$ vertices. Let e be directed from node u to its parent v. Then the subtree rooted at u is one of the two operands for the operator at v. Let T_1 be the subtree containing v and T_2, the subtree containing u. Then we can write the value of the expression computed by T_1 as $Ax+B$, where x is the value of u, which is also the value of the expression computed by T_2. The method of Brent [42] recursively computes the coefficients A and B for T_1 and the value x for T_2, and, using a constant-size circuit, combines the outputs of the two subtrees to obtain $Ax+B$, the value of the expression. This leads to a log-depth linear-size arithmetic circuit for the static expression evaluation problem. The tree contraction algorithm of Section 2.2.3 gives another optimal $O(\log n)$ depth circuit for the same problem. Both of these methods are easily extended to handle division.

Brent's method leads to an NC^2 arithmetic Boolean circuit for the dynamic expression evaluation problem since, at each recursive stage, the separating edge for each subtree at that level of recursion can be found in log-depth by a fairly easy construction. Similarly, since $EREW^1 \subseteq NC^2$, the tree contraction algorithm gives another NC^2 arithmetic Boolean circuit for the dynamic expression evaluation problem.

In [43, 44] an NC^1 arithmetic Boolean circuit is given for the problem. As in Brent [42], the method is based on recursively computing the value of the expression by decomposing the tree into subtrees; a factor of $\log n$ in depth is saved by computing all the decompositions simultaneously, instead of doing it separately at each of the $\log n$ recursive stages. The algorithm first transforms the expression into a postfix expression with the property that, for each operator, the length of the second operand is no greater than that of the first. This expression E is then recursively segmented into three equal-sized, overlapping strings of half its length, consisting of the first half of E, the middle half of E and the last half of E. Each of these strings represents a collection of subtrees of E, and the algorithm chooses the roots of the first two subtrees in each string to be the positions at which the expression is to be decomposed. It can be shown that this method of decomposition leads to a log-depth recursive algorithm for evaluating E. The method has the advantage that the positions where decomposition takes place at all levels of recursion can be determined in logarithmic depth by suitably parsing the string representing the expression E, and this leads to an NC^1 arithmetic Boolean circuit for the dynamic expression evaluation problem.

The NC^1 circuits for static and dynamic expression evaluation work for expressions over a field, ring or semiring. In particular, the Boolean expression evaluation problem, in which each input value is 0 or 1, is complete for NC^1 under deterministic log-time reductions, and can be considered, in some sense, to be the canonical problem for NC^1 [43].

4.4. *Boolean matrix multiplication and transitive closure*

We show that the Boolean matrix multiplication problem is in AC^0 and the transitive closure problem is in AC^1. Let $A=(a_{ij})$ and $B=(b_{ij})$ be $n \times n$ matrices of zeros and ones. Then the (Boolean) product of A and B is the $n \times n$ matrix $C=(c_{ij})$, where $c_{ij}=\bigvee_{k=1}^{n} a_{ik} \cdot b_{kj}$. The Boolean product of A and B can be computed by a two-level unbounded fan-in circuit. At the first level there are n^3 AND-gates, each of which computes one of the products $a_{ik} \cdot b_{kj}$. At the second level there are n^2 n-input OR-gates, each of which computes $\bigvee_{k=1}^{n} a_{ik} \cdot b_{kj}$ for a specific i and j. The specification of these circuits is clearly logspace-uniform. Thus Boolean matrix multiplication is in AC^0, and hence in NC^1. By imbedding the problem in an appropriate ring and applying fast matrix multiplication, the number of gates in the bounded fan-in circuit can be reduced asymptotically to $O(M(n))$, where $M(n)$ is the number of processors required by an arithmetic PRAM to perform $n \times n$ matrix multiplication over a general ring in time $O(\log n)$. It is known that $M(n)=O(n^{2.376})$ (Coppersmith & Winograd [66]).

Let $A=(a_{ij})$ be an $n \times n$ matrix of zeros and ones. Let I denote the $n \times n$ identity matrix. For $k=0, 1, 2, \wedge$ let A^k, the kth power of A, be defined inductively as follows: $A^0=I$; for $k=1, 2, \ldots, A^k=A^{k-1} \cdot A$. Define A^*, the reflexive transitive closure of A as follows: $A^*=\bigvee_{k=0}^{\infty} A^k$. It is not difficult to show that $A^*=(I \vee A)^{2\lceil \log_2 n \rceil}$. Hence A^* can be computed by initializing the matrix B to the value $I \vee A$, and successively squaring the matrix B $\lceil \log_2 n \rceil$ times. Since matrix multiplication is in AC^0, it follows that transitive closure lies in AC^1, and hence in NC^2.

4.5. *Matrix computations*

In discussing matrix computations, we work with arithmetic circuits as defined in Section 3: acyclic circuits in which the inputs are elements of a field **F** and the gates perform the four field operations $+$, $-$, $*$ and $/$. It is clear from its defining formula that matrix multiplication is in arithmetic NC^1. Similarly, the problem of computing A^n is in arithmetic NC^2, since it is possible to compute A^n in at most $2\lceil \log_2 n \rceil$ matrix multiplications by first computing $A^2, A^4, A^8, \ldots$ via repeated squaring, and then multiplying together appropriate matrices of the form A^{2^k}.

The problem of computing the inverse of a lower triangular matrix illustrates a divide-and-conquer technique that is often used for the construction of parallel algorithms. Let A be a square nonsingular lower triangular matrix whose number of rows is a power of 2 (square matrices of other dimensions can easily be handled by adding additional rows and columns, whose entries are 0 except for 1s on the diagonal, to pad the dimension out to a power of 2). Then we can block-decompose A into four matrices of half its dimension, as follows:

$$A = \left[\begin{array}{c|c} A_{11} & 0 \\ \hline A_{21} & A_{22} \end{array}\right]$$

Because A is lower triangular, it follows that the upper right-hand submatrix is the zero matrix, and that A_{11} and A_{22} are lower triangular. The inverse of A is given by

$$A^{-1} = \left(\begin{array}{c|c} A_{11}^{-1} & 0 \\ \hline -A_{22}^{-1} A_{21} A_{11}^{-1} & A_{22}^{-1} \end{array}\right)$$

This formula for the inverse suggests a recursive parallel algorithm for computing A^{-1}. First, A_{11}^{-1} and A_{22}^{-1} are computed recursively in parallel. Then $-A_{22}^{-1}A_{21}A_{11}^{-1}$ is computed via two matrix multiplications. This recursive algorithm leads to a uniform family of arithmetic circuits with $O(M(n))$ gates and depth $O(\log^2 n)$; hence the problem of inverting a lower triangular matrix lies in arithmetic NC^2.

4.5.1. *Computing the determinant*

We next turn to the problem of computing the determinant of an $n \times n$ matrix. This problem presents an interesting challenge because Gaussian elimination, the standard sequential algorithm for this problem, does not lead to an NC-algorithm. Gaussian elimination computes the determinant in a series of n stages, each of which transforms the given matrix without changing its determinant. Each stage requires $O(n^2)$ operations which can be performed in parallel, but it appears that the stages must be performed in sequence, so that the running time cannot be reduced below $\Theta(n)$.

The characteristic polynomial of a square matrix A is the nth-degree polynomial $\det(A - xI)$. In 1976 Csanky [67] showed that, for matrices over a field of characteristic zero, the problem of computing the characteristic polynomial lies in arithmetic NC^2. The result was extended to matrices over an arbitrary ring in [40] (see also [34]) and in [49]. Let the characteristic polynomial be $\Sigma_{i=0}^{n} c_i x^i$. Then $\det A = c_0$; and since, by the

Cayley–Hamilton Theorem, a matrix satisfies its own characteristic equation,

$$A^{-1}=\frac{-(c_1+c_2A+\cdots+c_nA^{n-1})}{c_0}.$$

This equation yields an arithmetic NC^2-algorithm for computing A^{-1} and det A from the characteristic polynomial.

We present Chistov's algorithm [49] for computing the characteristic polynomial. The following observation will be required. Let B be an $n\times n$ matrix. Let B_k be the $k\times k$ matrix in the lower right-hand corner of B; i.e., $(B_k)_{ij}=(B)_{n-k+i,n-k+j}$. Assume that, for $k=1,2,\ldots,n$, B_k is nonsingular. Then

$$\frac{1}{\det B}=\prod_{k=1}^{n}(B_k^{-1})_{11}.$$

Chistov's algorithm computes the polynomial $\det(I-xA)$ in time $O(\log^2 n)$ using $n^2M(n)$ processors. The coefficients of the polynomial $\det(I-xA)$ are those of the characteristic polynomial, but in reverse order and, if n is odd, with their signs reversed. The algorithm performs computations on power series in x; however, since the final result of the computation is a polynomial of degree n, these power series can be computed modulo n^{n+1}; i.e., they can be truncated to polynomials of degree n. We will use the fact that if $A(x)$ is a square matrix in which each element is a power series in x with constant term zero, then

$$(I-A(x))^{-1}\bmod x^{n+1}=\sum_{i=0}^{n}A(x)^i \bmod x^{n+1}.$$

By using repeated squaring, the powers of $A(x)$ up to the nth can be computed in time $O(\log^2 n)$ using $O(nM(n))$ processors, and hence $(I-A(x))^{-1}\bmod x^{n+1}$ can be computed in time $O(\log^2 n)$ using $O(nM(n))$ processors.

Our goal is to compute det B, where $B=I-xA$. Henceforth, all the objects computed are power series modulo x^{n+1} or matrices whose elements are such power series. For $k=1,2,\ldots,n$, B_k is a nonsingular matrix of the form I_k-xA_k, where I_k is the $k\times k$ identity matrix. Then $B_k^{-1}=(I_k-xA_k)^{-1}$; and, modulo x^{n+1}, this quantity is equal to $\Sigma_{i=0}^{n}(xA_k)^i$.

Chistov's algorithm proceeds as follows:

(i) for $k=1,2,\ldots,n$ compute B_k^{-1} and extract the element $(B_k^{-1})_{11}$;

(ii) multiply together the n elements extracted in step (i); the resulting product is $1/\det B$;

(iii) compute det B by taking the inverse of the power series obtained in step (ii).

Chistov's algorithm requires $O(n^2M(n))$ processors. For fields of characteristic zero, a variant of Csanky's algorithm due to Preparata & Sarwate [188] computes the characteristic polynomial in time $O(\log^2 n)$ using $O(n^{3.5})$ processors. Even this algorithm is not efficient, since, over a field of characteristic zero, the characteristic polynomial can be calculated sequentially in time $O(n^{2.376})$, where the constant implied by the "big O" is enormous, or, more practically, in $O(n^3)$ steps, where the implied constant is small.

Borodin, Cook & Pippenger [39] have shown that, for each fixed k, the following problem is in NC^2: compute the determinant of a matrix A whose entries are rational functions of the k variables $x_1, x_2, \ldots, x_k$, where each entry is presented as a pair of polynomials of degree at most n in which each coefficient is an n-bit integer. The same problem, but with the coefficients taken from an arbitrary field, is in arithmetic NC^2.

4.5.2. *Computing the rank*

The problem of computing the rank of a matrix A with elements in a field $\mathbf{F}$ has been treated in [121, 40, 49, 174]. If A is diagonalizable then its rank can be read off from the characteristic polynomial: it is simply $n-m$, where x^m is the highest power of x that divides the characteristic polynomial. If A is not diagonalizable, however, then this rule does not correctly give the rank. For example, the matrix $\left[\begin{smallmatrix}0 & 1\\ 0 & 0\end{smallmatrix}\right]$ has rank 1, but its characteristic polynomial is x^2. In [121], it is shown that, if $\mathbf{F}$ is a subfield of the reals, then $\operatorname{rank}(AA^{\mathrm{T}}) = \operatorname{rank}(A)$ and the matrix AA^{T} is diagonalizable. Hence, in this case, the rank may be read off from the characteristic polynomial.

In [174] it is shown that the problem of computing the rank of matrix A over an arbitrary field $\mathbf{F}$ is in arithmetic NC^2. Mulmuley's algorithm is remarkably simple. Assume that A is square and symmetric; for, if not, one may work instead with the square, symmetric matrix

$$\begin{bmatrix} 0 & A \\ A^{\mathrm{T}} & 0 \end{bmatrix}$$

whose rank is twice the rank of A. Let Z be a diagonal matrix in an indeterminate z such that $Z_{ii} = z^{i-1}$. Using the algorithm of Borodin, Cook & Pippenger [39], compute $Q(x) = \det(xI - ZA)$, the characteristic polynomial of the matrix ZA. Then the rank of A is $n-m$, where x^m is the highest power of x that divides the characteristic polynomial $Q(x)$.

4.6. *Dynamic evaluation of straight-line code*

A commutative semiring $(R, +, *, 0, 1)$ is an arithmetic structure with domain R and two commutative, associative binary operations $+$ and $*$, such that 0 is an additive identity, 1 is a multiplicative identity, and the distributive law $a*(b+c) = a*b + a*c$ holds. In conformity with the definition of Section 3.8, an arithmetic circuit over this semiring is an acyclic connection of input nodes of in-degree 0 and addition and multiplication nodes of in-degree 2, together with an assignment to each input node of a value from the domain R. The execution of the indicated multiplication and addition operations assigns a value from R to each node of the circuit, and the problem of computing these values is known as the evaluation problem. It is equivalent to the problem of evaluating straight-line programs with operations from a commutative semiring.

The performance of algorithms for this problem is stated in terms of two parameters: n, the size of the circuit C, and d, the (formal algebraic) degree of C. Each node in C can be viewed as computing a polynomial expression in the inputs, and the degree of C is the maximum of the formal degrees of these expressions. More precisely, the degrees of the nodes are defined inductively: the degree of an input node is 1, the degree of an addition node is the larger of the degrees of its two inputs, and the degree of a multiplication node is the sum of the degrees of its two inputs. The paper by Valiant, Skyum, Berkowitz & Rackoff [231] gives a method of converting any arithmetic circuit of size n and degree d into one of size $poly(n)$ and depth $O(\log n \cdot \log d)$. The paper by Miller, Ramachandran & Kaltofen [170] considers the more general dynamic version of the problem, in which the input to the evaluation algorithm consists of the arithmetic circuit C and the values of its inputs; thus no preprocessing based on C alone is allowed. The algorithm of [170] runs in time $O(\log n \log(nd))$ using $M'(n)$ processors, where $M'(n)$ is the number of processors required by an arithmetic PRAM to perform $n \times n$ matrix multiplication over a general semiring in time $O(\log n)$. It is evident that $M'(n) = O(n^3)$.

In order to describe the algorithm of Miller, Ramachandran & Kaltofen, we require the concept of a weighted arithmetic circuit over the semiring R. Such a circuit is a directed acyclic graph in which each edge (u, v) has a weight $W(u, v)$ which is an element of R. The nodes are of three types: leaves, multiplication nodes and addition nodes. A leaf has in-degree 0, a multiplication node has in-degree 2, and an addition node has in-degree greater than 0. Associated with each leaf is a value in R. It is required that no edge be directed from a multiplication node to a multiplication node.

Associated with each node v in a weighted arithmetic circuit is a value $VAL(v)$. The values of the leaves are given as part of the specification. If an addition node v has edges directed into it from nodes $v_1, v_2, \ldots, v_h$, then

$$VAL(v) = \sum_{i=1}^{h} VAL(v_i) * W(v_i, v).$$

If a multiplication node has edges directed into it from nodes v_1 and v_2 then

$$VAL(v) = VAL(v_1) * W(v_1, v) * VAL(v_2) * W(v_2, v).$$

We see that an ordinary arithmetic circuit over R can be viewed as a weighted arithmetic circuit in which every weight is equal to 1. The requirement that no edge runs from one multiplication node to another is easily met by inserting extra addition nodes of in-degree 1. Starting with the given arithmetic circuit, the algorithm constructs a sequence of weighted arithmetic circuits, all of which have the same set of nodes. Moreover, the iteration that produces each weighted arithmetic circuit from its predecessor in the sequence preserves the values of all nodes.

The iteration is accomplished in three steps:

(1) (*MM*): compress subcircuits consisting of additions only; the nature of this operation is indicated in Fig. 4;

(2) (*Rake*): simultaneously evaluate every node for which all direct predecessors are leaves, and delete the edges directed into those nodes;

(3) (*Shunt*): simultaneously bypass every multiplication node having a leaf as a direct predecessor; the nature of this operation is indicated in Fig. 5.

The cost of executing this iteration is dominated by the first step, which can be

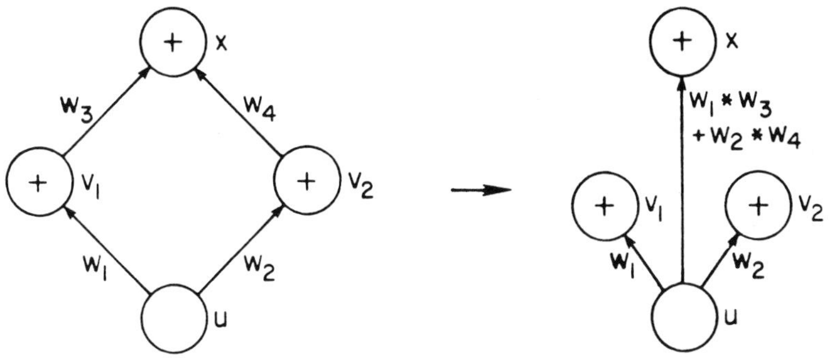

Fig. 4. The MM operation.

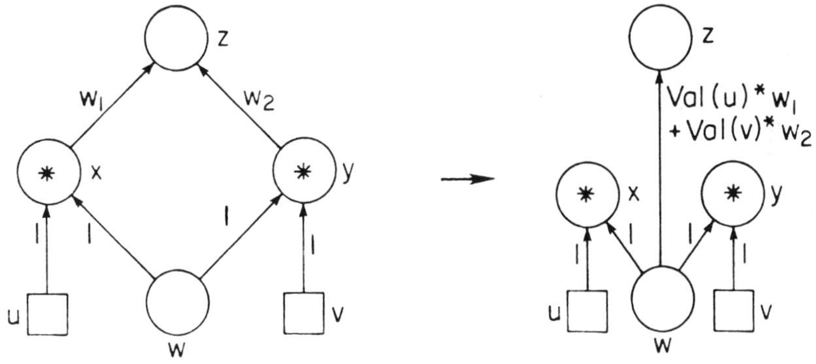

Fig. 5. The Shunt operation.

organized as a matrix multiplication. Thus each iteration can be performed in time $O(\log n)$ using $M'(n)$ processors. It can be shown that all nodes get evaluated within $O(\log(nd))$ iterations of this transformation.

The paper by Miller & Teng [172] extends the result of [170] to a wider class of algebraic structures, and Kaltofen [126] gives a randomized algorithm for the static problem in the case where division is allowed.

4.7. *The maximal independent set problem*

A set of vertices S in a graph G is called *independent* if no vertices in S are adjacent, and *maximal independent* if it is independent and not properly contained in any independent set. Noting that the most obvious methods of creating a maximal independent set do not parallelize, the paper [230] suggested the possibility that the problem of constructing such a set might be inherently resistant to solution in parallel. This was shown to be false in 1985, when Karp & Wigderson [140] showed that the problem of constructing a maximal independent set of vertices in a graph is in NC^4.

Soon thereafter, effecient randomized parallel algorithms for the problem were presented in [157] and [8]. On a graph with n vertices and m edges, these algorithms run in time $O(\log^2 n)$ on an EREW PRAM, and require $O(m+n)$ processors. Goldberg & Spencer [104] give a deterministic algorithm for constructing a maximal independent set that runs on an EREW PRAM in time $O(\log^4 n)$ using $O(m+n)$ processors.

All of these algorithms for constructing a maximal independent set in a graph $G'=(V', E')$ have the following overall structure.

```
begin
  I←∅: V→V';
  while V≠∅ do
  begin
    G←the subgraph of G' induced by the vertex set V;
    S←IN(G);
    V←V−(S∪N(S))
    I←I∪S
  end
end.
```

Here $IN(G)$ is an independent set in the graph G, and $N(S)$ denotes the set of vertices in V' that are adjacent to one or more vertices in S. It is easy to check that, upon termination of this algorithm, S is a maximal independent set in G.

The algorithms differ in the way they construct the independent set $IN(G)$. Luby's randomized method [157] is as follows. Let $d_G(v)$ denote the degree of vertex v in G. Then $IN(G)$ is constructed as follows:

```
X←∅
in parallel for all v∈V do
  insert v into X with probability 1/(2d_G(v))
in parallel for all 2-element sets {u, v}⊆X such that {u, v} is an edge of G do
  if d_G(u)<d_G(v) then delete u from X;
  if d_G(v)<d_G(u) then delete v from X;
  if d_G(u)=d_G(v) then randomly choose either u or v and delete it from X.
IN(G)←X.
```

The crucial lemma in Luby's analysis of his algorithm is as follows.

LEMMA. *Let E be the edge set of G. Then the expected number of edges of G incident with vertices in $IN(G)\cup N(IN(G))$ is at least $\frac{1}{8}|E|$.*

From the lemma, it easily follows that the expected number of calls on procedure $IN(G)$ required to construct a maximal independent set in the original graph G' is $O(\log n)$, where n is the number of vertices in G'. As pointed out in [157], it is sufficient for the random trials in the algorithm to be pairwise independent rather than mutually independent. This observation permits the trials to be taken from a probability space small enough to be searched exhaustively, yielding a deterministic NC-algorithm for the MIS problem. Luby [158] gives a more refined method of searching through this probability space, leading to a deterministic NC-algorithm with a linear number of processors.

The algorithm of Goldberg & Spencer [104] is based on a direct deterministic method of constructing an independent set S in a graph $G=(V,E)$. Let $|V|=p$ and $|E|=q$. Then, on an EREW PRAM their method runs in time $O(\log^2 p)$ using $O(p+q)$ processors, and guarantees that $|S \cup N(S)| \geqslant p/3 \log p$. If procedure IN is implemented using their method then the total number of calls on the procedure will be $O(\log^2 n)$, and the overall algorithm will require $n+m$ processors and run in time $O(\log^4 n)$ on an EREW PRAM.

Thus we see that there is a polylog-time deterministic algorithm to construct a maximal independent set using a linear number of processors; perhaps, in future work, the power of $\log n$ in the time bound will be reduced below 4.

4.8. *Applications*

The algorithms given above for computing the transitive closure of a Boolean matrix, for computing the characteristic polynomial and the rank of a matrix with entries drawn from a field, and for the fast parallel evaluation of functions specified by straight-line programs, provide fundamental tools for placing problems in NC. In this section we describe several of these applications.

4.8.1. *Applications of transitive closure techniques*

The following is a list of six basic computational problems regarding digraphs:

(i) computing the strong components of a digraph G;

(ii) determining whether G is acyclic;

(iii) constructing a tree rooted at a given vertex v of G and containing all vertices reachable from v;

(iv) constructing a breadth-first search tree rooted at a given vertex v;

(v) constructing a topological ordering of the vertices of an acyclic digraph G; i.e., a bijection h from the set of n vertices onto the integers $1, 2, \ldots, n$ such that, for every directed edge (u, v), $h(u) < h(v)$;

(vi) computing shortest paths from a given root vertex v to all other vertices, in a digraph G whose edges have nonnegative weights.

All of these problems can be placed in NC^2 using techniques related to transitive closure. The solutions of (i) and (ii) can be read off directly from the transitive closure. An NC^2-algorithm for problem (vi) is easily constructed, based on the well-known technique of iterated min/+ matrix multiplication. Problem (iv) is a special case of problem (vi), and a solution to (iv) also solves (iii). Problem (v) can be solved by computing the length of a longest path to each vertex by iterated max/+ matrix multiplication, sorting the vertices in increasing order of their longest path lengths (ties being broken arbitrarily), and then assigning each vertex a number equal to its rank in this sorted order.

Using techniques related to transitive closure, we have exhibited NC^2-algorithms for six elementary problems related to digraphs. Unfortunately, none of these algorithms are efficient. On a graph with n vertices and m edges, problem (vi) can be solved sequentially in time $O(n \log n + m)$, and each of the other problems can be solved sequentially in time $O(n+m)$. By contrast, our parallel algorithms run in time $O(\log^2 n)$ and require $M(n)$ processors in the case of problems (i)–(v), and $M'(n)$ processors in case (vi) (see [96]). Thus, we see that the processor–time product for each of our algorithms is far in excess of the time required to solve the same problem sequentially. In order to construct efficient parallel algorithms for these problems, it will be necessary to avoid the use of matrix powering or transitive closure as a subroutine; our inability to do so is sometimes called the *transitive closure bottleneck.*

4.8.2. Problems reducible to the solution of linear equations

We have shown that the problems of inverting a matrix and computing the rank of a matrix are in arithmetic NC^2. It follows easily that the problem of solving the linear system $Ax=b$ (where A is not necessarily of full rank) is also in arithmetic NC^2.

Many problems can be reduced to the solution of a system of linear equations. As an example, we consider the problem of computing the greatest common divisor (gcd) of two univariate polynomials f and g of degree n [40]. It can be shown that $\gcd(f, g)$ is of degree $\leqslant i$ if and only if there exist polynomials s and t of degree less than $n-i$ such that $sf+tg$ is of degree i. The condition that such polynomials exist is expressed by a nonsingular system of $2n-2i$ linear equations in which the unknowns are the $2n-2i$ coefficients of s and t. Each entry in the matrix of this system is either 0, a coefficient of f, or a coefficient of g.

As a second example, consider the problem of drawing a 3-connected planar graph in the plane without crossing edges, in such a way that all the edges are line segments. Call a cycle C of G a bounding cycle if there exists a plane imbedding of G in which the cycle C bounds a face. If T is any spanning tree of G then each edge e not in T forms a unique "fundamental cycle" when added to T; at least one of these fundamental cycles is a bounding cycle. In an interesting paper entitled "How to Draw a Graph", Tutte [225] gives the following result: Let G be a 3-connected planar graph, and let C be a bounding cycle. Let the successive vertices of C in cyclic order be $v_1, v_2, \ldots, v_k$. Let $p_1, p_2 \ldots, p_k$ be points in the plane such that the line segments $\overline{p_1 p_2}, \overline{p_2 p_3}, \ldots, \overline{p_k p_1}$ determine a convex polygon. Let the vertices of G be placed in the plane so that v_i is placed at p_i, $i=1,2,\ldots,k$, and each vertex not on C is located at the center of gravity of the vertices adjacent to it in G. For each edge $\{u, v\}$, let the points corresponding to u and v be joined by a line segment. Then no two of these lines segments will intersect except at a common endpoint; i.e., the process produces a straight-line imbedding of G. Thus, as Ja'Ja' & Simon [123] have observed, a straight-line imbedding of G can be constructed by identifying a bounding cycle C, and then constructing the associated placement. Once the vertices of C have been placed at the vertices of a convex polygon, the placement of the remaining points can be obtained by solving a nonsingular system of linear equations. The unknowns are the x- and y-coordinates of the vertices not on C, and the linear equations express the condition that each vertex lies at the center of gravity of its neighbors. This approach leads to an NC-algorithm for constructing a straight-line imbedding.

4.8.3. Applications of Mulmuley's rank algorithm

Mulmuley's algorithm [174] for computing the rank of a matrix over an arbitrary field is also a powerful tool for placing problems in NC. Among these are the problem of solving a (possibly singular) system of linear equations and the problem of factoring polynomials over finite fields. Mulmuley's algorithm also enables a number of problems about permutation groups presented by generators to be placed in NC [23]. These include the problems of determining the order, finding the derived series or a composition series, testing membership, testing if the permutation group is solvable, finding a central composition series, and finding pointwise stabilizers of sets when the permutation group is nilpotent. In each case Mulmuley's algorithm permits an earlier randomized algorithm to be made deterministic.

4.8.4. Context-free recognition—an application of the technique of Miller, Ramachandran and Kaltofen

The technique of Miller, Ramachandran & Kaltofen [170] for parallel evaluation of straight-line code enables several particular problems to be placed in NC. It is especially applicable to problems that can be solved in polynomial sequential time using dynamic programming. We shall illustrate the approach by giving an AC^1-algorithm for the problem of deciding whether a given string is in the language generated by a given context-free grammar in Chomsky Normal Form.

Our starting point is the well-known Cocke–Kasami–Younger dynamic programming algorithm for context-free recognition [141, 244]. Let (V, Σ, P, S) be a context-free grammar, where V denotes the set of symbols, Σ denotes the set of terminal symbols, P denotes the set of productions, and S denotes the initial symbol. Since the grammar is in Chomsky Normal Form, each production is of the form $A \to BC$ or $A \to a$, where A, B and C denote nonterminal symbols and a denotes a terminal symbol. For $A \in V - \Sigma$ and $w \in \Sigma^*$, we say that $A \overset{*}{\to} w$ (*A derives w*) if the nonterminal symbol A can generate the string w.

Let $x = a_1 a_2 \dots a_n$ be the input string, and let $x(i,j)$ be the substring $a_1 a_{i+1} \dots a_j$. For $1 \leqslant i \leqslant j \leqslant n$, and for every nonterminal symbol A, let the predicate $A(i,j)$ be true if $A \overset{*}{\to} x(i,j)$. The Cocke–Kasami–Younger algorithm evaluates all such predicates; the string x is accepted if and only if the predicate $S(1, n)$ is true.

The evaluation of these predicates is based on the following rules:

(i) for $i = 1, 2, \dots, n$, $A(i, i)$ is true if and only if there is a production of the form $A \to a_i$;

(ii) for $1 \leqslant i < j \leqslant n$, $A(i,j)$ is true if and only if there exists a k, $i \leqslant k < j$, and a production $A \to BC$, such that $B(i, k)$ and $C(k+1, j)$ are true.

The resulting algorithm can be represented as an arithmetic circuit over the semiring $(\{0, 1\}, +, *, 0, 1)$, where $+$ is Boolean OR and $*$ is Boolean AND. The circuit contains $O(n^3)$ nodes, and the node that computes the predicate $A(i,j)$ is of degree $j - i + 1$. Thus d, the maximum degree of any node, is $n + 1$, and it follows that the number of iterations required by the parallel evaluation algorithm of [170] is $O(\log n)$. The time for each iteration is dominated by the matrix multiplication required in the MM step of that algorithm. Since the semiring in this case is Boolean, matrix multiplication is in AC^0, and thus the time for the entire algorithm is $O(\log n)$, using a polynomial-bounded number of processors. It follows that the problem of deciding whether a given string is in the language generated by a given context-free grammar in Chomsky Normal Form is in AC^1. In [202], Ruzzo proved the static version of this result in a different way by showing that every context-free language can be recognized by an alternating Turing machine operating within space $\log n$ and using $\log n$ alternations. It follows that every context-free language is in AC^1.

4.9. *Randomized NC-algorithms*

4.9.1. *Testing whether a symbolic determinant is nonzero*

The following lemma is an important tool for the design of randomized parallel algorithms.

LEMMA. *Let $F(x_1, x_2, \ldots, x_n)$ be a polynomial of degree d. If F is not identically zero then, for at least half of the $(2d+1)^n$ n-tuples in which each component is an integer between $-d$ and d, $F(a_1, a_2, \ldots, a_n) \neq 0$.*

We omit the proof, which goes by induction on n, with the case $n=1$ corresponding to the Fundamental Theorem of Algebra (see, e.g., [209]).

The lemma suggests a randomized algorithm for testing whether a polynomial of degree d is not identically zero: simply substitute independent random integers between d and $-d$ for the variables. If the polynomial is not identically zero then, with probability greater than $\frac{1}{2}$, a non-zero value will result. For any class of polynomial expressions that can be evaluated by an NC-algorithm, we obtain in this way an RNC-algorithm for testing whether a polynomial in that class is not identically zero.

4.9.2. *The matching problem*

We give an RNC^2-algorithm for recognizing the graphs that possess perfect matchings (recall that a perfect matching in a graph G is a set of edges M, such that each vertex of G is incident with exactly one vertex of M). The algorithm is based on the following theorem attributed to Tutte.

THEOREM. *Let G be a simple graph (no loops or multiple edges) with vertex set $\{1, 2, \ldots, n\}$ and edge set E. Let $A=(a_{ij})$ be the following $n \times n$ matrix, in which the variables x_{ij} are indeterminates:*

$$a_{ij} = \begin{cases} x_{ij} & \text{if } \{i,j\} \in E \text{ and } i<j, \\ -x_{ji} & \text{if } \{i,j\} \in E \text{ and } i>j, \\ 0 & \text{if } \{i,j\} \notin E. \end{cases}$$

Then G has a perfect matching if and only if the determinant of A is not identically 0.

The matrix A is called the Tutte matrix of the graph G. The determinant of A is a polynomial of degree n in the indeterminates x_{ij}. Combining Tutte's Theorem, the lemma, and the existence of an NC^2-algorithm for computing the determinant of an $n \times n$ matrix with integer entries in $[-n, n]$, we obtain the desired one-sided-error algorithm for deciding whether a graph has a perfect matching.

This result does not directly yield an RNC^2-algorithm for constructing a perfect matching when one exists. Such an algorithm was first provided in [138]. We present here a particularly elegant RNC^2-algorithm for the problem, due to Mulmuley, Vazirani & Vazirani [175]. Their algorithm is based on the following probabilistic lemma.

LEMMA. *Let $\boldsymbol{C}$ be any nonempty collection of subsets of $\{1, 2, \ldots, N\}$. Let $w(1), w(2), \ldots, w(N)$ be independent random variables, each with the uniform distribution over $\{0, 1, 2, \ldots, 2N\}$. Associate with each set $S \subseteq \{1, 2, \ldots, N\}$ a weight $w(S) = \Sigma_{i \in S} w(i)$. Then, with probability greater than $\frac{1}{2}$, the family $\boldsymbol{C}$ contains a unique set of minimum weight.*

The lemma can be applied to the matching problem by taking $\boldsymbol{C}$ to be the set of perfect matchings in a graph with N edges. If each edge is given a weight drawn from the uniform distribution over $\{0, 1, \ldots, 2N\}$, and the weight of a matching is the sum of the weights of its edges, then, with probability $> \frac{1}{2}$, there will be a unique perfect matching of minimum weight.

Mulmuley, Vazirani & Vazirani go on to show that when there is a unique perfect matching of minimum weight, it can be constructed at the cost of a single matrix inversion by the following algorithm.

(1) For each edge $\{i, j\}$, draw a weight w_{ij} from the uniform distribution over $\{0, 1, \ldots, 2|E|\}$.
(2) Form the Tutte matrix of G and, for each indeterminate x_{ij} occurring in the Tutte matrix, substitute the constant $2^{w_{ij}}$; let the resulting matrix be B.
(3) Using a parallel matrix inversion algorithm that yields the determinant and the adjoint, such as the one in [180], compute $|B|$ and adj(B). The (i, j) entry of adj(B) is the minor $|B_{ij}|$.
(4) In parallel, for all edges $\{i, j\}$, compute $|B_{ij}| 2^{w_{ij}}/2^{2w}$. Let M be the set of edges for which this quantity is odd.
(5) If M is a perfect matching then output M.

Whenever the weights w_{ij} are such that there is a unique perfect matching of minimum weight, the algorithm will produce this matching. Thus each execution of the algorithm produces a perfect matching with probability $> \frac{1}{2}$, provided that a perfect matching exists. The algorithm runs in $O(\log^2 n)$ time using a polynomial-bounded number of processors. Thus we can conclude that the problem of constructing a perfect matching is in RNC^2. Further results by Karloff [134] establish that the problem lies in ZNC^2.

We have discussed two problems related to perfect matchings: the *decision problem*, in which the task is to decide whether a perfect matching exists, and the *search problem*, in which the task is to construct a perfect matching when one exists. Karp, Upfal & Wigderson [139] have studied the general question of how to construct a parallel algorithm for a search problem, given a subroutine for the corresponding decision problem.

4.9.3. Applications of matching

Karp, Upfal & Wigderson [138] have shown that the following problems related to matching and network flows are in RNC:

(i) constructing a perfect matching of maximum weight in a graph whose edge weights are given in unary notation;

(ii) constructing a matching of maximum cardinality;

(iii) constructing a matching that covers a set of vertices of maximum weight in a graph whose vertex weights are given in binary;

(iv) constructing a maximum s-t flow in a directed or undirected network whose edge weights are given in unary notation.

Each of these results is obtained by a reduction of the given problem to the problem of constructing a perfect matching, or by a closely related algorithmic technique. In view of the above result of Mulmuley, Vazirani & Vazirani [175], these four problems lie in RNC^2.

4.9.4. Depth-first search

Let $G=(V, E)$ be a connected graph. Let T be a spanning tree of G rooted at r. Then T is called a *depth-first-search tree* if, for every edge $e \in E$, one of the two endpoints of e is an ancestor of the other.

There is a sequential algorithm running in time $O(|E|)$ for the construction of a depth-first search tree rooted at a given vertex. This algorithm is the backbone of linear-time sequential algorithms for testing whether a graph is planar, computing the biconnected components of a graph, and many other important problems (see, e.g., [79].

Although the goal of finding an efficient parallel algorithm for depth-first search has not been reached, some progress has been made. Define a *splitting path* as a simple path having the root as an endpoint, with the property that its deletion breaks the remaining vertices into connected components of size less than or equal to $c|V|$, for a suitable constant c. Once such a splitting path is found, a depth-first search tree for the overall graph can be constructed recursively out of the splitting path together with depth-first search trees for these connected components. This approach is used in [215] to solve the depth-first search problem in the case of planar graphs (see also [102, 116, 122] for efficient implementations of this algorithm), and by Ramachandran [189] in the case of directed acyclic graphs and reducible flow graphs.

It remained an open question whether the depth-first search problem for general graphs was in ZNC. Aggarwal & Anderson [2] settled this question by giving a ZNC^5-algorithm that constructs a depth-first search tree in an n-vertex undirected graph. The algorithm has $O(\log n)$ levels of recursion, in each of which it constructs a splitting path; in order to construct such a path, it makes $O(\log^2 n)$ successive calls to a subroutine for the following problem: given a bipartite graph in which each edge has weight 0 or 1, find a perfect matching of minimum weight. Each of the bipartite graphs presented to the subroutine has at most n vertices. Since this matching problem is in

ZNC^2, it follows that the depth-first search problem is in ZNC^5. The algorithm of Aggarwal & Anderson is quite intricate, and we shall not attempt to describe it here. Building on results of Kao [131] for planar directed graphs, this result is extended to yield a ZNC algorithm for depth-first search in directed graphs in [3].

4.10. *Further results*

Our treatment of NC-algorithms has focused on methods and results that appear to be of general utility for placing problems in the classes NC, RNC or ZNC, or for locating them in the hierarchies such as $\{AC^k\}$ and $\{NC^k\}$. Many further problems have been analyzed from this point of view. The following references give a representative sample of such results.

Graph theory: Babai [22]; Coppersmith, Raghavan & Tompa [65]; Dahlhaus & Karpinski [68]; Galil & Pan [89]; Gazit & Miller [95]; Gibbons, Karp, Miller & Soroker [101]; Grigoriev & Karpinski [109]; Johnson [124]; Karloff [134]; Karloff & Shmoys [136]; Klein & Reif [143]; Kozen, Vazirani & Vazirani [146]; Lev, Pippenger & Valiant [152]; Lingas & Karpinski [154]; Lovász [156]; Nisan & Soroker [178]; Reif [193, 194]; Savage & Ja'Ja' [205]; Soroker [218, 219], Tsin & Chin [224]; Vazirani [232].

Scheduling theory: Hembold & Mayr [117].

Algebra: Babai, Luks & Seress [23]; Ben-Or & Tiwari [32]; Borodin, von zur Gathen & Hopcroft [40]; Eberly [75]; Fich & Tompa [84]; Galil & Pan [89]; von zur Gathen [90, 91, 92]; Ibarra, Moran & Rosier [121]; Kaltofen, Krishnamoorthy & Saunders [127]; Kaltofen & Saunders [128]; Kannan, Miller & Rudolph [130]; Litow & Davida [155]; Lueker, Megiddo & Ramachandran [159]; Luks [160]; Luks & McKenzie [161]; McKenzie & Cook [165]; Pan [180]; Pan & Reif [181]; Reif [197].

Language theory: Ibarra, Jiang, Ravikumar & Chang [120]; Klein & Reif [144].

Logic programming: Mayr [163]; Ullman & van Gelder [226].

Analysis: Ben-Or, Feig, Kozen & Tiwari [30]; Ben-Or, Kozen & Reif [31]; Kozen & Yap [147].

In addition to the work on NC, there is an important body of research concerned with algorithms that achieve significant speed-up, even if they do not run in polylog time. A prominent example is Gaussian elimination, which solves a system of n linear equations in time $O(n)$ using n^2 processors. Further examples include the algorithm due to Eckstein [76] for performing depth-first search in an n-node, m-edge graph in time $O(m/p+n\log p)$, using p processors; the algorithm of Gabow & Tarjan [87] for the assignment problem on a weighted bipartite graph with n nodes, m edges and maximum edge weight N that runs in

$$O(\sqrt{n}\, m \log(nN)\, (\log 2p)/p)$$

time with $p \leqslant m/(\sqrt{n}\log^2 n)$ processors; and the algorithms of Goldberg, Plotkin & Vaidya [103] for performing depth-first search on a graph with n vertices and m edges in time $O(\sqrt{n}\, polylog(n))$ using $n+m$ processors, and for computing a maximum matching in an n-node bipartite graph in time $O(n^{2/3} polylog(n))$ using $M(n)$

processors. In the terminology of Kruskal, Rudolph & Snir (cf. Section 2.1), Eckstein's algorithm achieves polynomial speed-up for dense graphs with constant inefficiency, the algorithms of Gabow & Tarjan achieve polynomial speed-up with logarithmic inefficiency, and the algorithms of Goldberg, Plotkin & Vaidya achieve polynomial speed-up with polynomial inefficiency.

4.11. P-*complete problems*

Let P denote the set of decision problems solvable by deterministic Turing machines in polynomial time. Every decision problem in NC lies in P. A fundamental open question is whether every problem in P lies in NC. If this were so, it would mean, roughly speaking, that every problem that is efficiently solvable in a sequential model of computation can be solved very fast in parallel, using a polynomial-bounded number of processors (this interpretation is disputed in the interesting paper by Vitter & Simons [239]). Using a reducibility technique, we shall identify a set of problems within P called the P-complete problems; a P-complete problem lies in NC if and only if P = NC. Thus the P-complete problems can be viewed as the problems in P most resistant to parallelization.

We adopt the usual convention of representing a decision problem as a subset of $\{0,1\}^*$. Decision problem A is said to be *logspace-reducible* to decision problem B if there is a function $f:\{0,1\}^* \to \{0,1\}^*$ such that f is computable by a logspace Turing machine and, for all $x \in \{0,1\}^*$, $x \in A$ if and only if $f(x) \in B$. A decision problem in P is called *P-complete* if every problem in P is logspace-reducible to it. The relation of logspace-reducibility is transitive and has the following further property: if A is logspace-reducible to B and B is in NC^k, where $k \geqslant 2$, then A is in NC^k. Our interest in P-completeness stems from the following consequences of this observation: let A be a P-complete problem; then $P = NC$ if and only if A lies in NC, and, for $k \geqslant 2$, $P \subseteq NC^k$ if and only if A lies in NC^k.

Our official definition of a P-complete problem requires that it be specified as a function from strings over one alphabet to strings over another. In practice, the inputs and outputs to a problem are often other kinds of symbolic objects: graphs, formulas, circuits, grammars, families of sets and the like. In describing such problems and proving them P-complete, we will often not specify how their inputs and outputs are encoded as strings, since the details of that encoding are seldom of interest.

The usual method of proving a problem P-complete is to show that it lies in P and that some standard P-complete problem is logspace-reducible to it. The standard problem most often used for this purpose is the monotone circuit value problem (MCVP) (Goldschlager [105]). Informally, the input to this problem is a single-output fan-in-2 Boolean circuit without NOT-gates, together with an assignment of a constant value (0 or 1) to each input line. More precisely, the input is given as a sequence of equations, specifying the outputs of the gates in a monotone circuit. The first two equations are $g_0 = 0$ and $g_1 = 1$, and for each i, $i = 2, 3, \ldots, n$, there is either an equation of the form $g_i = g_j \vee g_k$ or an equation of the form $g_i = g_j \wedge g_k$, where j and k are nonnegative indices less than i. The first of these two equations corresponds to the case where gate i is an OR-gate with inputs from gates j and k; similarly, the second equation corresponds to the case where gate i is an AND-gate. The output is the value of g_n in the unique solution of this system of equations.

We sketch the proof that MCVP is P-complete. Let A be a decision problem in P. Then A is accepted by a deterministic one-tape Turing machine M which has a unique accepting state q^* and operates within the polynomial time bound $T(n)$. For a given input string x, let $T = T(|x|)$. Then T is an upper bound on the execution time of M on x, and the only tape squares whose contents can change during the computation are within distance T of the home square. Introduce the following Boolean variables to represent the computation of M on input x:

- $h(i,t)$, meaning "the head is on tape square i at time t";
- $a(i,t)$, meaning "the symbol on square i at time t is a";
- $q(t)$, meaning "the state at time t is q".

Here a ranges over the tape alphabet of M, q ranges over the state set of M, $-T \leqslant i \leqslant T$ and $0 \leqslant t \leqslant T$. The input x is accepted if and only if $q^*(T) = 1$.

Given input x and the description of the transition function of M, there is a logspace algorithm to generate a system of monotone Boolean equations, in the format of the MCVP problem, specifying the values of these variables. This is the required logspace reduction from problem A to the MCVP. Since A is an arbitrary problem in P, it follows that MCVP is P-complete.

We shall give two further examples of P-completeness proofs, each involving a reduction from MCVP. The first example involves a certain "greedy" sequential algorithm for constructing a maximal independent set of vertices in a graph. Let G be a graph with vertex set V, and let a linear ordering of V be given. The following algorithm constructs a maximal independent set in G.

GREEDY INDEPENDENT SET ALGORITHM
$S \leftarrow \emptyset$;
for $i = 1, 2, \ldots, |V|$ **do**
begin
  let v be the ith element in the linear ordering of V;
  if v is adjacent to no vertex in S
  then $S \leftarrow S \cup \{v\}$
end.

The following decision problem related to this algorithm is P-complete (Cook [61]).

GREEDY INDEPENDENT SET
Input: Graph $G = (V, E)$ where V is linearly ordered,
Property: The last vertex in the linear ordering of V is in the independent set constructed by the greedy independent set algorithm.

We give a reduction from MCVP. Let an instance of MCVP be specified, as described above, by equations for the gate outputs $g_0, g_1, \ldots, g_n$. The reduction will produce a graph G with vertex set $V = \{v_0, v_1, \ldots, v_n\} \cup \{w_0, w_1, \ldots, w_n)$, together with a linear ordering of V. The linear ordering is such that, whenever $i < j$, v_i and w_i precede

v_j and w_j. The relative ordering of v_i and w_i and the specification of the edges incident with v_i and w_i depend on the nature of the equation for g_i as follows: w_0 precedes v_0; v_1 precedes w_1; if the equation for g_i is $g_i = g_j \vee g_k$ then w_i precedes v_i and the graph contains the edges $\{v_j, w_i\}$ and $\{v_k, w_i\}$; if the equation is $g_i = g_j \wedge g_k$ then v_i precedes w_i and the graph contains the edges $\{w_j, v_i\}$ and $\{w_k, v_i\}$. In addition, all edges $\{v_i, w_i\}$ are present.

The construction of the graph from the circuit can be performed by a logspace algorithm. It is easy to prove by induction on n that v_i lies in the greedy independent set if and only if the output of g_i is 1, and w_i lies in the greedy independent set if and only if the output of g_i is 0. Thus we have a logspace-reduction from the monotone circuit value problem to the lexicographically-first-independent-set problem, showing that the latter problem is P-complete. The P-completeness of this problem stands in contrast to the result that the problem of constructing some maximal independent set, not necessarily the one produced by the greedy algorithm, is in EREW^4.

As a final example, we show that the following problem is P-complete [107].

MAX-FLOW

Input: A directed graph $G=(V, E)$, a pair of distinct vertices s and t called the source and sink respectively, and a function c from E into the nonnegative integers, assigning to each edge e a capacity $c(e)$.

Property: The value of the maximum flow from s to t is an odd integer.

We give a reduction from the MCVP to MAX-FLOW. Let C be a monotone circuit. We may assume without loss of generality that the following properties hold:

(i) each gate has fan-out at most 2 (i.e., each variable g_i occurs on the right-hand side of at most two equations);

(ii) g_n is the output of an OR-gate.

We refer to the edges of the circuit as wires in order to distinguish them from the edges of the network produced by the reduction. The network has vertex set $\{s, t, v_0, v_1, \ldots, v_n\}$, and its edges and capacities are specified as follows:

(i) for each wire of C connecting the output of g_j to the input of g_i, there is an edge of capacity 2^{n-j} from v_j to v_i;

(ii) there is an edge of capacity 2^{n+1} from s to v_n;

(iii) there is an edge of capacity 1 from v_n to t;

(iv) if g_i is an OR-gate then there is an edge from v_i to s;

(v) if g_i is an AND-gate, there is an edge from v_i to t and

(vi) the capacities of the edges specified in (iv) and (v) are such that the sum of the capacities of the edges directed into any vertex v_i is equal to the sum of the capacities of the edges directed out of that vertex.

It is easy to prove by induction on n that there is a maximum flow in G having the following properties:

(i) if a wire in C carries the signal 0 then the flow in the corresponding edge of G is 0;

(ii) if a wire in C carries the signal 1 then the flow in the corresponding edge of G is equal to the capacity of that edge;

(iii) with the exception of the edges from v_n to s and t, the flow in every edge of G is an even integer;

(iv) The flow in the edge from v_n to t is 0 if the output of C is 0, and 1 if the output of C is 1.

It follows that the value of a maximum flow in G is odd if and only if the output of C is 1. This establishes that the reduction is correct.

Thus we see that the max-flow problem is P-complete. In contrast to this is the result given in Section 4.9.3 that when the capacities of the edges are bounded by a polynomial in the number of vertices, the max-flow problem is in ZNC^2.

The seminal results on P-complete problems are given in [60, 125, 149]. More recent P-completeness results can be found in [11, 19, 69, 71, 135, 195].

4.12. Open problems

The following is a list of problems, solvable in sequential polynomial time, whose parallel complexity remains unknown despite the efforts of many researchers. It is not known whether these problems lie in ZNC, and they have not been proven to be P-complete. One resolution of these questions would be a proof that P = NC, but this seems unlikely to be true and, in any case, unlikely to be settled by known proof methods.

1. EXISTENCE OF A PERFECT MATCHING
Input: A graph G.
Question: Does G have a perfect matching?

This problem is in ZNC, as is the related problem of constructing a perfect matching when one exists (see Section 4.9.2). In the bipartite case, an algorithm running in $O(n^{2/3}\mathit{polylog}(n))$ time using a polynomial-bounded number of processors is known [103].

2. UNDIRECTED DEPTH-FIRST SEARCH
Input: A connected graph G and a vertex v.
Output: A spanning tree T of G, rooted at v and having the following property: for each non-tree edge $\{u, w\}$, u is either an ancestor or a descendant of w in T.

This problem is in RNC (see Section 4.9.4). A $\sqrt{n}\,\mathit{polylog}(n)$ time algorithm using $\mathit{poly}(n)$ processors is known [103].

3. DIRECTED DEPTH-FIRST SEARCH
Input: A digraph G and a vertex v from which all vertices of G are reachable.
Output: An oriented tree T, rooted at v and containing a directed path from v to each vertex of G, such that T is a subgraph of G and has the following directed depth-first search property: there is a preorder numbering of G such that, if (u, w) is an edge of $G-T$ then either w precedes u in the preorder numbering, or w is a descendant of u in T.

The problem is in RNC [3].

4. WEIGHTED MATCHING

Input: A graph G with n vertices and, for each edge e, a positive integer weight $w(e)$.

Output: A matching of maximum total weight.

When the weights are polynomial in n this problem is in ZNC [138].

5. MAXIMAL INDEPENDENT SET IN A HYPERGRAPH

Input: A collection C of subsets of a finite set X.

Output: A set $X' \subseteq X$ which is maximal with respect to the property that it does not contain any set in the collection C.

The case where C is a collection of 2-element sets is the maximal independent set problem for graphs, which is in NC (see Section 4.7).

6. TWO-VARIABLE LINEAR PROGRAMMING

Input: A linear system of inequalities $Ax \leqslant b$ over the rationals, such that each row of A has at most two non-zero elements.

Output: A feasible solution, if one exists.

This problem has a polylog-time parallel algorithm with $n^{polylog(n)}$ processors [159].

7. INTEGER GCD

Input: Integers a and b.

Output: The greatest common divisor of a and b.

A sublinear-time algorithm using a polynomial-bounded number of processors is given in [130], and subsequently improved in [50]. The problem of computing the greatest common divisor of two polynomials is in NC (see Section 4.8.2).

8. MODULAR INTEGER EXPONENTIATION

Input: n-bit integers a, b and m.

Output: $a^b \bmod m$.

9. MODULAR POLYNOMIAL EXPONENTIATION

Input: Polynomials $a(x)$ and $m(x)$ with coefficient from a ring, and a positive integer e.

Output: $a(x)^e \bmod m(x)$.

Problem 8 is a special case. Over fields of some fixed characteristic the problem is in arithmetic NC^2, provided that unit-time operations over the base field are assumed [84].

10. POLYNOMIAL ROOT APPROXIMATION

Input: An nth-degree polynomial $P(x)$ with integer coefficients, and a positive error bound ε.

Output: For each (real or complex) root ξ of P, an approximation ξ' such that $|\xi - \xi'| < \varepsilon$.

When all roots are real the problem is in NC [30].

11. COMPARATOR CIRCUIT VALUE
Input: A comparator network (see Section 2.4) together with values for its inputs.
Output: The value of a specified output.

This problem is complete in the complexity class CC defined in [164].

5. Conclusion

The examples we have given in this survey of parallel algorithms are of interest not only for their theoretical significance, but also as illustrations of typical methods of exploiting the parallelism inherent in problems. Basic algorithms for such problems as prefix sums, list ranking, sorting, and graph searching can serve as fundamental building blocks for further algorithm construction. As parallel computation grows in importance, such algorithms will find their way into the undergraduate textbooks, and will become part of the general lore of computer science.

We have seen that, within many different abstract models, the class NC represents the collection of problems solvable in polylog time with a polynomial-bounded number of computing elements, and we have identified many basic problems as lying in NC. The robustness of this class suggests that it is of fundamental importance, and lends interest to the question of whether NC coincides with the familiar complexity class P. We have also seen the usefulness of the concept of P-completeness in identifying the problems in P least likely to lie in NC.

So far, the studies of complexity within the PRAM model have been somewhat unrealistic because of the assumption that the number of processors is allowed to grow as a function of the size of the input. It would be useful to complement this line of research with studies in which the number of processors remains fixed as the input size grows; this assumption is closer to the typical situation in practice.

Although the PRAM model neglects communication issues, it is a very convenient vehicle for the logical design of parallel algorithms. For this reason, there has been intense interest in the simulation of PRAMs on more feasible models of parallel computation. The chief components of such a simulation are the choice of a processor interconnection pattern; the mapping of the address space of the PRAM onto the set of memory cells of the simulating machine; and the algorithm for routing read and write requests, and the replies to read requests, through the network of processors.

A series of more and more refined PRAM simulations [166, 227, 228, 133, 10] has culminated in Ranade's efficient randomized simulation of an n-processor CRCW PRAM on an n-processor butterfly network [191]. The simulation time per PRAM step is only logarithmic in the number of processors. This simulation opens the way for a programming environment in which algorithms are designed within the convenient

PRAM model and then simulated efficiently on a network of processors, and thus underscores the value of the PRAM model, and the applicability of algorithms designed within it.

Acknowledgment

Many colleagues have provided valuable comments on earlier drafts of this chapter. We would like to express our thanks to Richard Anderson, Laszlo Babai, David Barrington, Robert Boyer, Joachim von zur Gathen, Hillel Gazit, Mark Goldberg, Michael Goodrich, Erich Kaltofen, János Komlós, Jan van Leeuwen, Michael Loui, Eugene Luks, Jay Misra, Ketan Mulmuley, Steve Omohundro, Victor Pan, James Renegar, John Reif, Wojciech Rytter, John Savage, Marc Snir, Robert Tarjan, Prasoon Tiwari, Martin Tompa, Stephen Vavasis, Jeffrey Vitter, Yaacov Yesha and, especially, Torben Hagerup and Larry Stockmeyer. Finally, we owe an enormous debt of gratitude to Ann DiFruscia, who assisted us with great dedication and skill in the preparation of this manuscript.

The research of the first author was supported by the International Computer Science Institute, Berkeley, CA, and NSF Grant Nos. DCR-8411954 and CCR-8612563. The research of the second author was supported by the International Computer Science Institute, Berkeley, CA, and by Joint Services Electronics Program NOOO14-84-C-0149 while she was with the University of Illinois, Urbana, IL.

References

[1] ABRAHAMSON, K., N. DADOUN, D.A. KIRKPATRICK and T. PRZYTYCKA, A simple parallel tree contraction algorithm, in: *Proc. 25th Ann. Allerton Conf. on Communication, Control and Computing* (1987) 624–633.

[2] AGGARWAL, A. and R.J. ANDERSON, A random NC algorithm for depth first search, in: *Proc. 19th Ann. ACM Symp. on Theory of Computing* (1987) 325–334; revised version, *Combinatorica* **8** (1988) 1–12.

[3] AGGARWAL, A., R.J. ANDERSON and M. KAO, Parallel depth-first search in general directed graphs, in: *Proc. 21st Ann. ACM Symp. on Theory of Computing* (1989) 297–308.

[4] AGGARWAL, A., B. CHAZELLE, L. GUIBAS, C. O'DUNLAING and C. YAP, Parallel computational geometry, in: *Proc. 26th Ann. IEEE Symp. on Foundations of Computer Science* (1985) 468–477.

[5] AJTAI, M., Σ_1^1-formulae on finite structures, *Ann. Pure Appl. Logic* **24** (1983) 1–48.

[6] AJTAI, M., J. KOMLÓS, W.L. STEIGER and E. SZEMERÉDI, Deterministic selection in $O(\log\log n)$ parallel time, in: *Proc. 18th Ann. ACM Symp. on Theory of Computing* (1986) 188–195.

[7] AJTAI, M., J. KOMLÓS and E. SZEMERÉDI, Sorting in $c \log n$ parallel steps, *Combinatorica* **3** (1983) 1–19.

[8] ALON, N., L. BABAI and A. ITAI, A fast and simple randomized parallel algorithm for the maximal independent set problem, *J. Algorithms* **7** (1986) 567–583.

[9] ALT, H., Comparison of arithmetic functions with respect to Boolean circuit depth, in: *Proc. 16th Ann. ACM Symp. on Theory of Computing* (1984) 466–470.

[10] ALT, H., T. HAGERUP, K. MEHLHORN and F.P. PREPARATA, Deterministic simulation of idealized parallel computers on more realistic ones, *SIAM J. Comput* **16** (1987) 808–835.

[11] ANDERSON, R.J. and E.W. MAYR, Parallelism and greedy algorithms, in: Advances in Computing Research, Vol. 24 (JAI Press, Greenwich, CT, 1987) 17–38.

[12] ANDERSON, R.J. and G.L. MILLER, Deterministic parallel list ranking, in: *VLSI Algorithms and*

Architectures, Proc. 3rd Aegean Workshop on Computing, Lecture Notes in Computer Science, Vol. 319 (Springer, Berlin, 1988) 81–90.

[13] ATALLAH, M.J., Parallel strong orientation of an undirected graph, *Inf. Proc. Letters* **18** (1984) 37–39.

[14] ATALLAH, M.J., R. COLE and M.T. GOODRICH, Cascading divide-and-conquer: a technique for designing parallel algorithms, in: *Proc. 28th Ann. IEEE Symp. on Foundations of Computer Science* (1987) 151–160.

[15] ATALLAH, M.J. and M.T. GOODRICH, Efficient parallel solutions to some geometric problems, *J. Parallel Distr. Comput.* **3** (1986) 492–507.

[16] ATALLAH, M.J. and M.T. GOODRICH, Parallel algorithms for some functions of two convex polygons, *Algorithmica* **3** (1988) 535–548.

[17] ATALLAH, M.J., M.T. GOODRICH and S.R. KOSARAJU, Parallel algorithms for evaluating sequences of set manipulation operations, in: *VLSI Algorithms and Architectures, Proc. 3rd Aegean Workshop on Computing*, Lecture Notes in Computer Science, Vol. 319 (Springer, Berlin, 1988) 1–10.

[18] ATALLAH, M. and U. VISHKIN, Finding Euler tours in parallel, *J. Comput. System Sci.* **29** (1984) 330–337.

[19] AVENHAUS J. and K. MADLENER, The Nielsen reduction and P-complete problems in free groups, *Theoret. Comput. Sci.* **32** (1984) 61–76.

[20] AWERBUCH, B., A. ISRAELI and Y. SHILOACH, Finding Euler circuits in logarithmic parallel time, in: *Proc. 16th Ann. ACM Symp. on Theory of Computing* (1984) 249–257.

[21] AWERBUCH, B. and Y. SHILOACH, New connectivity and MSF algorithms for shuffle-exchange network and PRAM, *IEEE Trans. Comput.* **36** (1987) 1258–1263.

[22] BABAI, L., A Las-Vegas NC algorithm for isomorphism of graphs with bounded multiplicity of eigenvalues, in: *Proc. 27th Ann. IEEE Symp. on Foundations of Computer Science* (1986) 303–312.

[23] BABAI, L., E.M. LUKS and A. SERESS, Permutation groups in NC, in: *Proc.19th Ann. ACM Symp. on Theory of Computing* (1987) 409–420.

[24] BAR-ON, I. and U. VISHKIN, Optimal parallel generation of a computation tree form, *ACM Trans. Programming Languages and Systems* **7** (1985) 348–357.

[25] BARRINGTON, D.A., Bounded-width polynomial-size branching programs recognize exactly those languages in NC^1, in: *Proc 18th Ann. ACM Symp. on Theory of Computing* (1986) 1–5.

[26] BATCHER, K.E., Sorting networks and their applications, in: *Proc. AFIPS Spring Joint Summer Computer Conf., Vol. 32* (1968) 307–314.

[27] BEAME, P.W., Limits on the power of concurrent-write parallel machines, *Inform. and Comput.* **76** (1988) 13–28.

[28] BEAME, P.W., S.A. COOK and H.J. HOOVER, Log depth circuits for division and related problems, *SIAM J. Comput.* **15** (1986) 994–1003.

[29] BEAME, P. and J. HASTAD, Optimal bounds for decision problems on the CRCW PRAM, in: *Proc. 19th Ann. ACM Symp. on Theory of Computing* (1987) 83–93.

[30] BEN-OR, M., E. FEIG, D. KOZEN and P. TIWARI, A fast parallel algorithm for determining all roots of a polynomial with real roots, *SIAM J. Comput.* **17** (1988) 1081–1092.

[31] BEN-OR, M., D. KOZEN and J. REIF, The complexity of elementary algebra and geometry, in: *Proc. 16th Ann. ACM Symp. on Theory of Computing* (1984) 457–464.

[32] BEN-OR, M. and P. TIWARI, A deterministic algorithm for sparse multivariate polynomial interpolation, in: *Proc. 20th Ann. ACM Symp. on Theory of Computing* (1988) 301–309.

[33] BERKMAN, O., D. BRESLAUER, Z. GALIL, B. SCHIEBER and U. VISHKIN, Highly parallelizable problems, in: *Proc. 21st Ann. ACM Symp. on Theory of Computing* (1989).

[34] BERKOWITZ, S.J., On computing the determinant in small parallel time using a small number of processors, *Inf. Proc. Letters* **18** (1984) 147–150.

[35] BILARDI, G. and A. NICOLAU, Bitonic sorting with $O(N \log N)$ comparisons, in: *Proc. 20th Ann. Conf. on Information Science and Systems*, Princeton Univ., Princeton, NJ (1986) 309–319.

[36] BOPPANA, R., Optimal separations between concurrent-write parallel machines, in: *Proc. 21st Ann. ACM Symp. on Theory of Computing* (1989) 320–326.

[37] BOPPANA, R. and M. SIPSER, The complexity of finite functions, in: J. van Leeuwen, ed., *Handbook of Theoretical Computer Science, Vol. A* (North-Holland, Amsterdam, 1990) 757–804.

[38] BORODIN, A., On relating time and space to size and depth, *SIAM J. Comput.* **6** (1977) 733–744.

[39] BORODIN, A., S.A. COOK and N. PIPPENGER, Parallel computation for well-endowed rings and space-bounded probabilistic machines, *Inform. and Control* **58** (1983) 113–136.

[40] BORODIN, A., J. VON ZUR GATHEN and J.E. HOPCROFT, Fast parallel matrix and GCD computations, in: *Proc. 23rd Ann. IEEE Symp. on Foundations of Computer Science* (1982) 65–71.

[41] BORODIN, A. and J.E. HOPCROFT, Routing, merging, and sorting on parallel models of computation, *J. Comput. System Sci.* **30** (1985) 130–145.

[42] BRENT, R.P., The parallel evaluation of general arithmetic expressions, *J. ACM* **21** (1974) 201–206.

[43] BUSS, S.R., The Boolean formula value problem is in ALOGTIME, in: *Proc. 19th Ann. ACM Symp. on Theory of Computing* (1987) 123–131.

[44] BUSS, S.R., S.A. COOK, A. GUPTA and V. RAMACHANDRAN, An optimal parallel algorithm for fomula evaluation, Manuscript, Toronto, Canada, 1989.

[45] CHANDRA, A.K., S. FORTUNE, R.J. LIPTON, Unbounded fan-in circuits and associative functions, *J. Comput. System Sci.* **30** (1985) 222–234.

[46] CHANDRA, A.K, D.C. KOZEN and L.J. STOCKMEYER, Alternation, *J. ACM* **28** (1981) 114–133.

[47] CHANDRA, A.K., L. STOCKMEYER and U. VISHKIN, Constant depth reducibility, *SIAM J. Comput.* **13** (1984) 423–439.

[48] CHIN, F.Y., J. LAM and I. CHEN, Efficient parallel algorithms for some graph problems, *Comm. ACM* **25** (1982) 659–665.

[49] CHISTOV, A.L., Fast parallel calculation of the rank of matrices over a field of arbitrary characteristic, in: *Fundamentals of Computation Theory, FCT'85*, Lecture Notes in Computer Science, Vol. 199 (Springer, Berlin, 1985) 63–79.

[50] CHOR, B. and O. GOLDREICH, An improved parallel algorithm for integer gcd, Manuscript, 1985.

[51] CHOW, A., Parallel algorithms for geometric problems, Dissertation, Computer Science Dept., Univ. of Illinois at Urbana-Champaign, 1980.

[52] CHROBAK, M. and M. YUNG, Fast parallel and sequential algorithms for edge -coloring planar graphs, in: *VLSI Algorithms and Architectures, Proc. 3rd Aegean Workshop on Computing*, Lecture Notes in Computer Science, Vol. 319 (Springer, Berlin, 1988) 11–23.

[53] COLE, R., Parallel merge sort, *SIAM J. Comput.* **17** (1988) 770–785.

[54] COLE, R. and M.T. GOODRICH, Optimal parallel algorithms for polygon and point-set problems, in: *Proc. 4th Ann. ACM Conf. on Computational Geometry* (1988) 201–210.

[55] COLE, R. and U. VISHKIN, Deterministic coin tossing with applications to optimal parallel list ranking, *Inform. and Control* **70** (1986) 32–53.

[56] COLE, R. and U. VISHKIN, Approximate and exact parallel scheduling with applications to list, tree and graph problems, in: *Proc. 27th Ann. IEEE Symp. on Foundations of Computer Science* (1986) 478–491.

[57] COLE, R. and U. VISHKIN, The accelerated centroid decomposition technique for optimal parallel tree evaluation in logarithmic time, *Algorithmica* **3** (1988) 329–346.

[58] COLE, R. and U. VISHKIN, Approximate parallel scheduling, part I: the basic technique with applications to optimal parallel list ranking in logarithmic time, *SIAM J. Comput* **17** (1988) 128–142.

[59] COLE, R. and C.K. YAP, A parallel median algorithm, *Inf. Proc. Letters* **20** (1985) 137–139.

[60] COOK, S.A., An observation on time-storage trade-off, *J. Comput. System Sci.* **9** (1974) 308–316.

[61] COOK, S.A., Towards a complexity theory of synchronous parallel computation, *Enseign. Math.* **27** (1981) 99–124.

[62] COOK, S.A., A taxonomy of problems with fast parallel algorithms, *Inform. and Control* **64** (1985) 2–22.

[63] COOK, S.A., C. DWORK and R. REISCHUK, Upper and lower time bounds for parallel random access machines without simultaneous writes, *SIAM J. Comput.* **15** (1986) 87–97.

[64] COOK, S.A. and R.A. RECKHOW, Time-bounded random access machines, *J. Comput. System Sci.* **7** (1973) 354–375.

[65] COPPERSMITH, D., P. RAGHAVAN and M. TOMPA, Parallel graph algorithms that are efficient on the average, in: *Proc. 28th Ann. IEEE Symp. on Foundations of Computer Science* (1987) 260–270.

[66] COPPERSMITH, D. and S. WINOGRAD, Matrix multiplication via arithmetic progressions, in: *Proc. 19th Ann. ACM Symp. on Theory of Computing* (1987) 1–6.

[67] CSANKY, L., Fast parallel matrix inversion algorithms, *SIAM J. Comput.* **5** (1976) 618–623.

[68] DAHLHAUS, E. and M. KARPINSKI, The matching problem for strongly chordal graphs is in NC, Tech. Report 855-CS, Institut für Informatik, Universität Bonn, 1986.

[69] DOBKIN, D., R.J. LIPTON and S. REISS, Linear programming is log-space hard for P, *Inf. Proc. Letters* **8** (1979) 96–97.
[70] DOLEV, D., C. DWORK, N. PIPPENGER and A. WIGDERSON, Superconcentrators, generalizers and generalized connectors with limited depth, in: *Proc: 15th Ann. ACM Symp. on Theory of Computing* (1983) 42–51.
[71] DWORK, C., P.C. KANELLAKIS and J.C. MITCHELL, On the sequential nature of unification, *J. Logic Programming* **1** (1984) 35–50.
[72] DWORK, C., P.C. KANELLAKIS and L. STOCKMEYER, Parallel algorithms for term matching, *SIAM J. Comput* **17** (1988) 711–731.
[73] DYMOND, P.W. and S.A. COOK, Hardware complexity and parallel complexity, in: *Proc. 21st Ann. IEEE Symp. on Foundations of Computer Science* (1980) 360–372.
[74] DYMOND, P.W. and W.R. RUZZO, Parallel RAMs with owned global memory and deterministic context-free language recognition, in: *Proc. 13th Internat. Coll. on Automata Languages and Programming* (1986) 95–104.
[75] EBERLY W., Very fast parallel matrix and polynomial arithmetic, in: *Proc. 25th Ann. IEEE Symp. on Foundations of Computer Science* (1984) 21–30.
[76] ECKSTEIN, D.M., Parallel processing using depth-first search and breadth-first search, Ph.D. Thesis, Dept. of Computer Science, Univ. of Iowa, Iowa City, Iowa, 1977.
[77] ECKSTEIN, D.M., Simultaneous memory access. Tech. Report TR-79-6, Computer Science Dept., Iowa State Univ., Ames, IA, 1979.
[78] EPPSTEIN, D. and Z. GALIL, Parallel algorithmic techniques for combinatorial computation, Manuscript, Dept. of Computer Science, Columbia Univ., New York, 1988.
[79] EVEN, S., *Graph Algorithms* (Computer Science Press, Potomac, MD, 1979).
[80] FICH, F.E., New bounds for parallel prefix circuits, in: *Proc. 15th Ann. ACM Symp. on Theory of Computing* (1983) 27–36.
[81] FICH, F.E., F. MEYER AUF DER HEIDE and A. WIGDERSON, Lower bounds for parallel random-access machines with unbounded shared memory, Advances in Computing Research, Vol. 4 (JAI Press, Greenwich, CT, 1987) 1–15.
[82] FICH, F.E., P. RAGDE and A. WIGDERSON, Simulations among concurrent-write models of parallel computation, *Algorithmica* **3** (1988) 43–51.
[83] FICH, F.E., P. RAGDE and A. WIGDERSON, Relations between concurrent-write models of parallel computation, *SIAM J. Comput.* **17** (1988) 606–627.
[84] FICH, F.E. and M. TOMPA, The parallel complexity of exponentiating polynomials over finite fields, *J. ACM* **35** (1988) 651–667.
[85] FORTUNE, S. and J. WYLLIE, Parallelism in random access machines, in: *Proc. 10th Ann. ACM Symp. on Theory of Computing* (1978) 114–118.
[86] FURST, M., J.B. SAXE and M. SIPSER, Parity, circuits and the polynomial time hierarchy, *Math. Systems Theory* **17** (1984) 13–28.
[87] GABOW, H.N. and R.E. TARJAN, Almost-optimum speed-ups of algorithms for bipartite matching and related problems, in: *Proc. 20th Ann. ACM Symp. on Theory of Computing* (1988) 514–527.
[88] GALIL, Z., Optimal parallel algorithms for string matching, *Inform. and Control* **67** (1985) 144–157.
[89] GALIL, Z. and V. PAN, Improved processor bounds for algebraic and combinatorial problems in RNC, in: *Proc. 26th Ann. IEEE Symp. on Foundations of Computer Science* (1985) 490–495.
[90] VON ZUR GATHEN, J., Parallel algorithms for algebraic problems, *SIAM J. Comput.* **13** (1984) 802–824.
[91] VON ZUR GATHEN, J., Parallel arithmetic computations: a survey, in: *Proc. 12th. Internat. Symp. on Mathematical Foundations of Computer Science*, Lecture Notes in Computer Science, Vol. 233 (Springer, Berlin, 1986) 93–113.
[92] VON ZUR GATHEN, J., Computing powers in parallel, *SIAM J. Comput.* **16** (1987) 930–945.
[93] VON ZUR GATHEN, J., Algebraic complexity theory, *Ann. Rev. Comp. Sci.* **3** (1988) 317–347.
[94] GAZIT, H., An optimal randomized parallel algorithm for finding connected components in a graph, in: *Proc. 27th Ann. IEEE Symp. on Foundations of Computer Science* (1986) 492–501.
[95] GAZIT, H. and G.L. MILLER, A parallel algorithm for finding a separator in planar graphs, in: *Proc. 28th Ann. IEEE Symp. on Foundations of Computer Science* (1987) 238–248.
[96] GAZIT, H. and G.L. MILLER, An improved parallel algorithm that computes the bfs numbering of a directed graph, *Inf. Proc. Letters* **28** (1988) 61–65.

[97] GAZIT, H., G.L. MILLER and S.H. TENG, Optimal tree contraction in the EREW model, in: S.K. Tewksbury, B.W. Dickinson and S.C. Schwartz, eds., *Concurrent Computations: Algorithms, Architecture, and Technology* (Plenum, New York, 1988) 139–156.

[98] GIBBONS, A.M. and W. RYTTER, An optimal parallel algorithm for dynamic expression evaluation and its applications, in: *Proc. 6th Conf. on Foundations of Software Technology and Theoretical Computer Science*, Lecture Notes in Computer Science, Vol. 241 (Springer, Berlin, 1986) 453–469.

[99] GIBBONS, A.M. and W. RYTTER, *Efficient Parallel Algorithms* (Cambridge Univ. Press, Cambridge, UK, 1988).

[100] GIBBONS, P., Personal communication, 1987.

[101] GIBBONS, P., R.M. KARP, G. MILLER and D. SOROKER, Subtree isomorphism is in Random NC, in: *VLSI Algorithms and Architectures, Proc. 3rd Aegean Workshop on Computing*, Lecture Notes in Computer Science, Vol. 319 (Springer, Berlin, 1988) 43–52.

[102] GOLDBERG, A.V., S.A. PLOTKIN and G.E. SHANNON, Parallel symmetry-breaking in sparse graphs, in: *Proc. 19th Ann. ACM Symp. on Theory of Computing* (1987) 315–324.

[103] GOLDBERG, A.V., S.A. PLOTKIN and P.M. VAIDYA, Sublinear-time parallel algorithms for matching and related problems, in: *Proc. 29th Ann. IEEE Symp. on Foundations of Computer Science* (1988) 174–185.

[104] GOLDBERG, M. and T. SPENCER, A new parallel algorithm for the maximal independent set problem, in: *Proc. 28th Ann. IEEE Symp. on Foundations of Computer Science* (1987) 161–165.

[105] GOLDSCHLAGER, L.M., The monotone and planar circuit value problems are log space complete for P, *SIGACT News* **9** (1977) 25–29.

[106] GOLDSCHLAGER, L.M., A unified approach to models of synchronous parallel machines, *J. ACM* **29** (1982) 1073–1086.

[107] GOLDSCHLAGER, L.M., R.A. SHAW and J. STAPLES, The maximum flow problem is log space complete for P, *Theoret. Comput. Sci.* **21** (1982) 105–111.

[108] GOODRICH, M.T., Triangulating a polygon in parallel, *J. Algorithms* **10** (1989) 327–351.

[109] GRIGORIEV, D.Y. and M. KARPINSKI, The matching problem for bipartite graphs with polynomially bounded permanents is in NC, in: *Proc. 28th Ann. IEEE Symp. on Foundations of Computer Science* (1987) 166–172.

[110] GROLMUSZ, V. and P. RAGDE, Incomparability in parallel computation, in: *Proc: 28th Ann. IEEE Symp. on Foundations of Computer Science* (1987) 89–98.

[111] HAGERUP, T., Towards optimal parallel bucket sorting, *Inform. and Comput.* **75** (1987) 39–51.

[112] HAGERUP, T., Optimal parallel algorithms on planar graphs, in: *VLSI Algorithms and Architectures, Proc. 3rd Aegean Workshop on Computing*, Lecture Notes in Computer Science, Vol. 319 (Springer, Berlin, 1988) 24–32.

[113] HASTAD, J., Almost optimal lower bounds for small depth circuits, in: *Proc. 18th Ann. ACM Symp. on Theory of Computing* (1986) 6–20.

[114] HASTAD, J., *Computational Limitations for Small Depth Circuits* (MIT Press, Cambridge, MA, 1986).

[115] HE, X. and Y. YESHA, Binary tree algebraic computation and parallel algorithms for simple graphs, *J. Algorithms* **9** (1988) 92–113.

[116] HE, X. and Y. YESHA, A nearly optimal parallel algorithm for constructing depth-first spanning trees in planar graphs, *SIAM J. Comput.* **17** (1988) 486–492.

[117] HELMBOLD, D. and E. MAYR, Two-processor scheduling is in NC, in: *VLSI Algorithms and Architectures, Proc. Aegean Workshop on Computing*, Lecture Notes in Computer Science, Vol. 227 (Springer, Berlin, 1986) 12–25.

[118] HIRSCHBERG, D.S., A.K. CHANDRA and D.V. SARWATE, Computing connected components on parallel computers, *Comm. ACM* **22** (1979) 461–464.

[119] HOOVER, H.J., M.M. KLAWE and N.J. PIPPENGER, Bounding fan-out in logical networks, *J. ACM* **31** (1984) 13–18.

[120] IBARRA, O.H., T. JIANG, B. RAVIKUMAR and J.H. CHANG, On some languages in NC^1, in: *VLSI Algorithms and Architectures, Proc. 3rd Aegean Workshop on Computing*, Lecture Notes in Computer Science, Vol. 319 (Springer, Berlin, 1988) 64–73.

[121] IBARRA, O.H., S. MORAN and L.E. ROSIER, A note on the parallel complexity of computing the rank of order n matrices, *Inf. Proc. Letters* **11** (1980) 162.

[122] JA'JA', J. and S.R. KOSARAJU, Parallel algorithms for planar graph isomorphism and related problems, *IEEE Trans. on Circuits and Systems*, **35** (1988) 304–311.

[123] JA'JA', J. and J. SIMON, Parallel algorithms in graph theory: planarity testing, *SIAM J. Comput.* **11** (1982) 314–328.

[124] JOHNSON, D.B., Parallel algorithms for minimum cuts and maximum flows in planar networks, *J. ACM* **34** (1987) 950–967.

[125] JONES, N.D. and W.T. LAASER, Complete problems for deterministic polynomial time, *Theoret. Comput. Sci.* **3** (1977) 105–117.

[126] KALTOFEN, E., Uniform closure properties of P-computable functions, in: *Proc. 18th Ann. ACM Symp. on Theory of Computing* (1986) 330–337.

[127] KALTOFEN, E., M.S. KRISHNAMOORTHY and B.D. SAUNDERS, Fast parallel computation of Hermite and Smith forms of polynomial matrices, *SIAM J. Alg. Discrete Methods* **8** (1987) 683–690.

[128] KALTOFEN, E. and B.D. SAUNDERS, Parallel algorithms for matrix normal forms, Report No. 88-6, Dept. of Computer Science, RPI, Troy, NY, 1988.

[129] KANEVSKY, A. and V. RAMACHANDRAN, Improved algorithms for graph four-connectivity, in: *Proc. 28th Ann. IEEE Symp. on Foundations of Computer Science* (1987) 252–259.

[130] KANNAN, R., G.L. MILLER and L. RUDOLPH, Sublinear parallel algorithm for computing the greatest common division of two integers, in: *Proc. 25th Ann. IEEE Symp. on Foundations of Computer Science* (1984) 7–11.

[131] KAO, M., All graphs have cycle separators and planar directed depth-first search is in DNC, in: *VLSI Algorithms and Architectures, Proc. 3rd Aegean Workshop on Computing*, Lecture Notes in Computer Science, Vol. 319 (Springer, Berlin, 1988) 53–63.

[132] KAO, M. and G. SHANNON, Local reorientation, global order, and planar topology, in: *Proc. 21st Ann. ACM Symp. on Theory of Computing* (1989) 286–296.

[133] KARLIN, A.R. and E. UPFAL, Parallel hashing—an efficient implementation of shared memory in: *Proc. 18th Ann. ACM Symp. on Theory of Computing* (1986) 160–168.

[134] KARLOFF, H.J., A Las-Vegas RNC algorithm for maximum matching, *Combinatorica* **6** (1986) 387–392.

[135] KARLOFF, H.J. and W.L. RUZZO, The iterated mod problem, *Inform. and Comput.* **80** (1989) 193–204.

[136] KARLOFF, H.J. and D.B. SHMOYS, Efficient parallel algorithms for edge coloring problems, *J. Algorithms* **8** (1987) 39–52.

[137] KARP, R.M. and M.O. RABIN, Efficient randomized pattern-matching algorithms, *IBM J. Res. Dev.* **31** (1987) 249–260.

[138] KARP, R.M., E. UPFAL and A. WIGDERSON, Constructing a perfect matching is in random NC, *Combinatorica* **6** (1986) 35–48.

[139] KARP, R.M., E. UPFAL and A. WIGDERSON, The complexity of parallel search, *J. Comput. System. Sci.* **36**(2) (1988) 225–253.

[140] KARP, R.M. and A. WIGDERSON, A fast parallel algorithm for the maximal independent set problem, *J. ACM* **32** (1985) 762–773.

[141] KASAMI, T., An efficient recognition and syntax—analysis algorithm for context-free languages, Science Report AFCRL-65-758, Air Force Cambridge Research Lab., Bedford, MA, 1965.

[142] KLEIN, P.N., Efficient parallel algorithms for chordal graphs, in: *Proc. 29th Ann. IEEE Symp. on Foundations of Computer Science* (1988) 150–161.

[143] KLEIN, P.N. and J.H. REIF, An efficient parallel algorithm for planarity, in: *Proc. 27th Ann. IEEE Symp. on Foundations of Computer Science* (1986) 465–477.

[144] KLEIN, P.N. and J.H. REIF, Parallel time $O(\log n)$ acceptance of deterministic CFLs on an exclusive-write P-RAM, *SIAM J. Comput.* **17** (1988) 463–485.

[145] KOSARAJU, S.R. and A.L. DELCHER, Optimal parallel evaluation of tree-structured computations by raking, in: *VLSIAlgorithms and Architectures, Proc. 3rd Aegean Workshop on Computing*, Lecture Notes in Computer Science, Vol. 319 (Springer, Berlin, 1988) 101–110.

[146] KOZEN, D., U.V. VAZIRANI and V.V. VAZIRANI, NC algorithms for comparability graphs, interval graphs, and testing for unique perfect matching, in: *Proc. 5th Conf. on Foundations of Software Technology and Theoretical Computer Science*, Lecture Notes in Computer Science, Vol. 206 (Springer, Berlin, 1985) 496–503.

[147] KOZEN, D. and C.-K. YAP, Algebraic cell decomposition in NC, in: *Proc. 26th Ann. IEEE Symp. on Foundations of Computer Science* (1985) 515–521.

[148] KUČERA, L., Parallel computation and conflicts in memory access, *Inf. Proc. Letters* **14** (1982) 93–96.

[149] LADNER, R.E., The circuit value problem is log space complete for P, *SIGACT News* 7(2) (1975) 18–20.

[150] LADNER, R.E. and M.J. FISCHER, Parallel prefix computation, *J. ACM* **27** (1980) 831–838.

[151] LANDAU, G.M. and U. VISHKIN, Introducing efficient parallelism into approximate string matching and a new serial algorithm, in: *Proc. 18th Ann ACM Symp. on Theory of Computing* (1986) 220–230.

[152] LEV, G.F., N. PIPPENGER and L.G. VALIANT, A fast algorithm for routing in permutation networks, *IEEE Trans. Comput.* **30** (1981) 93–100.

[153] LI, M. and Y. YESHA, New lower bounds for parallel computation, *Inform. and Comput.* **73** (1987) 102–128.

[154] LINGAS, A. and M. KARPINSKI, Subtree isomorphism and bipartite perfect matching are mutually NC-reducible, Report No. 856-CS, Institut für Informatik, Universität Bonn, 1986.

[155] LITOW, B.E. and G.I. DAVIDA, O(log(n)) parallel time finite field inversion, in: *VLSI Algorithms and Architectures, Proc. 3rd Aegean Workshop on Computing*, Lecture Notes in Computer Science, Vol. 319 (Springer, Berlin, 1988) 74–80.

[156] LOVÁSZ, L., Computing ears and branchings in parallel, in: *Proc. 26th Ann. IEEE Symp. on Foundations of Computer Science* (1985) 464–467.

[157] LUBY, M., A simple parallel algorithm for the maximal independent set problem, *SIAM J. Comput.* **15** (1986) 1036–1053.

[158] LUBY, M., Removing randomness in parallel computation without a processor penalty, in: *Proc. 29th Ann. IEEE Symp. on Foundations of Computer Science* (1988) 162–173.

[159] LUEKER, G.S., N. MEGIDDO and V. RAMACHANDRAN, Linear programming with two variables per inequality in poly-log time, in: *Proc. 18th Ann. ACM Symp. on Theory of Computing* (1986) 196–205.

[160] LUKS, E.M., Parallel algorithms for permutation groups and graph isomorphism, in: *Proc. 27th Ann. IEEE Symp. on Foundations of Computer Science* (1986) 292–302.

[161] LUKS, E.M. and P. MCKENZIE, Fast parallel computation with permutation groups, in: *Proc. 26th Ann. IEEE Symp. on Foundations of Computer Science* (1985) 505–514.

[162] MAON, Y., B. SCHIEBER and U. VISHKIN, Parallel ear decomposition search (EDS) and *st*-numbering in graphs, *Theoret. Comput. Sci.* **47** (1986) 277–298.

[163] MAYR, E.W., The dynamic tree expression problem, in: S.K. Tewksbury, B.W. Dickinson and S.C. Schwartz, eds. *Concurrent Computations: Algorithms, Architecture, and Technology* (Plenum, New York, 1988) 157–180.

[164] MAYR, E.W. and A. SUBRAMANIAN, The complexity of circuit value and network stability, in: *Proc. 4th Ann. Conf. on Structure in Complexity Theory* (1989) 114–123.

[165] MCKENZIE, P. and S.A. COOK, The parallel complexity of the abelian permutation group membership problem, in: *Proc. 24th Ann. IEEE Symp. on Foundations of Computer Science* (1983) 154–161.

[166] MEHLHORN, K. and U. VISHKIN, Randomized and deterministic simulation of PRAMs by parallel machines with restricted granularity of parallel memories, *Acta Inform.* **21** (1984) 339–374.

[167] MEYER AUF DER HEIDE, F. and A. WIGDERSON, The complexity of parallel sorting, *SIAM J. Comput.* **16** (1987) 100–107.

[168] MILLER, G.L. and V. RAMACHANDRAN, Efficient parallel ear decomposition with applications, Manuscript, MSRI, Berkeley, CA, 1986.

[169] MILLER, G.L. and V. RAMACHANDRAN, A new graph triconnectivity algorithm and its parallelization, in: *Proc. 19th Ann. ACM Symp. on Theory of Computing* (1987) 335–344.

[170] MILLER, G.L., V. RAMACHANDRAN and E. KALTOFEN, Efficient parallel evaluation of straight-line code and arithmetic circuits, *SIAM J. Comput.* **17** (1988) 687–695.

[171] MILLER, G.L. and J.H. REIF, Parallel tree contraction and its application, in: *Proc. 26th Ann. IEEE Symp. on Foundations of Computer Science* (1985) 478–489.

[172] MILLER, G.L. and S. TENG, Dynamic parallel complexity of computational circuits, in: *Proc. 19th Ann. ACM Symp. on Theory of Computing* (1987) 254–263.

[173] MIYANO, S., The lexicographically first maximal subgraph problems: P-completeness and NC algorithms, Manuscript, Universität Paderborn, Paderborn, Fed. Rep. Germany, 1986.

[174] MULMULEY, K., A fast parallel algorithm to compute the rank of a matrix over an arbitrary field, *Combinatorica* **7** (1987) 101–104.

[175] MULMULEY, K., U.V. VAZIRANI and V.V. VAZIRANI, Matching is as easy as matrix inversion, in: *Proc. 19th Ann. ACM Symp. on Theory of Computing* (1987) 345–354.

[176] NATH, D. and S.N. MAHESHWARI, Parallel algorithms for the connected components and minimal spanning tree problems, *Inf. Proc. Letters* **14** (1982) 7–11.

[177] NISAN, N., CREW PRAMs and decision trees, in: *Proc. 21st Ann. ACM Symp. on Theory of Computing* (1989) 327–335.

[178] NISAN, N. and D. SOROKER, Parallel algorithms for zero-one supply-demand problems, *SIAM J. Discrete Math.* **2** (1989) 108–125.

[179] OFMAN, YU., On the algorithmic complexity of discrete functions, English translation in: *Sov. Phys. Dokl.* **7** (1963) 589–591; original in: *Dokl. Akad. Nauk SSSR* **145** (1963) 48–51.

[180] PAN, V., Fast and efficient algorithms for the exact inversion of integer matrices, in: *Proc. 5th Ann. Conf. on Foundations of Software Technology and Theoretical Computer Science*, Lecture Notes in Computer Science, Vol. 206 (Springer, Berlin, 1985) 504–521.

[181] PAN, V. and J. REIF, Efficient parallel solution of linear systems, in: *Proc. 17th Ann. ACM Symp. on Theory of Computing* (1985) 143–152.

[182] PARBERRY, I., *Parallel Complexity Theory* (Pitman, London, 1987).

[183] PATERSON, M.S., Improved sorting networks with O(log N) depth, Research Report 89, Dept. of Computer Science, Univ. of Warwick, Coventry, 1987.

[184] PIPPENGER, N., On simultaneous resource bounds, in: *Proc. 20th Ann. IEEE Symp. on Foundations of Computer Science* (1979) 307–311.

[185] PIPPENGER, N., The complexity of computations by networks, *IBM J. Res. Dev.* **31** (1987) 235–243.

[186] PIPPENGER, N., Communication networks, in: J. van Leeuwen, ed., *Handbook of Theoretical Computer Science, Vol. A* (North-Holland, Amsterdam, 1990) 805–833.

[187] PRATT, V.R. and L.J. STOCKMEYER, A characterization of the power of vector machines, *J. Comput. System Sci.* **12** (1976) 198–221.

[188] PREPARATA, F.P. and D.V. SARWATE, An improved parallel processor bound in fast matrix inversion, *Inf. Proc. Letters* **7** (1978) 148–150.

[189] RAMACHANDRAN, V., Fast parallel algorithms for reducible flow graphs, in: S.K. Tewksbury, B.W. Dickinson and S.C. Schwartz, eds., *Concurrent Computations: Algorithms, Architecture, and Technology* (Plenum, New York, 1988) 117–138.

[190] RAMACHANDRAN, V. and U. VISHKIN, Efficient parallel triconnectivity in logarithmic time, in: *VLSI Algorithms and Architectures, Proc. 3rd Aegean Workshop on Computing*, Lecture Notes in Computer Science, Vol. 319 (Springer, Berlin, 1988) 33–42.

[191] RANADE, A.G., How to emulate shared memory, in: *Proc. 28th Ann. IEEE Symp. on Foundations of Computer Science* (1987) 185–194.

[192] REGHBATI (ARJOMANDI), E. and D.G. CORNEIL, Parallel computations in graph theory, *SIAM J. Comput.* **7** (1978) 230–237.

[193] REIF, J.H., Parallel algorithms for graph isomorphism, Tech. Report TR-14-83, Aiken Computation Laboratory, Harvard. Univ. Cambridge, MA, 1983.

[194] REIF, J.H., Symmetric complementation, *J. ACM* **31** (1984) 401–421.

[195] REIF, J.H., Depth-first search is inherently sequential, *Inf. Proc. Letters* **20** (1985) 229–234.

[196] REIF, J.H., An optimal parallel algorithm for integer sorting, in: *Proc. 26th Ann. IEEE Symp. on Foundations of Computer Science* (1985) 496–504.

[197] REIF, J.H., Logarithmic depth circuits for algebraic functions, *SIAM J. Comput.* **15** (1986) 231–242.

[198] REIF, J. and S. SEN, Polling: a new randomized sampling technique for computational geometry, in: *Proc. 21st Ann. ACM Symp. on Theory of Computing* (1989) 394–404.

[199] REIF, J. and S. TATE, Optimal size integer division circuits, in: *Proc. 21st Ann. ACM Symp. on Theory of Computing* (1989) 264–273.

[200] REIF, J.H. and L.G. VALIANT, A logarithmic time sort for linear size networks, *J. ACM* **34** (1987) 60–76.

[201] REISCHUK, R., A fast probabilistic sorting algorithm, *SIAM J. Comput.* **14** (1985) 396–409.

[202] RUZZO, W.L., Tree-size bounded alternation, *J. Comput. System Sci.* **21** (1980) 218–235.
[203] RUZZO, W.L., On uniform circuit complexity, *J. Comput. System. Sci.* **22** (1981) 365–383.
[204] SAVAGE, J.E., *The Complexity of Computing* (Wiley, New York, 1976).
[205] SAVAGE, C. and J. JA'JA', Fast, efficient parallel algorithms for some graph problems, *SIAM J. Comput.* **10** (1981) 682–691.
[206] SAVITCH, W.J. and M.J. STIMSON, Time bounded random access machines with parallel processing, *J. ACM* **26** (1979) 103–118.
[207] SCHIEBER B. and U. VISHKIN, On finding lowest common ancestors: simplification and parallelization, in: *VLSI Algorithms and Architectures, Proc. 3rd Aegean Workshop on Computing*, Lecture Notes in Computer Science, Vol. 319 (Springer, Berlin, 1988) 111–123.
[208] SCHÖNHAGE, A. and V. STRASSEN, Schnelle Multiplikation grosser Zahlen, *Comput.* **7** (1971) 281–292.
[209] SCHWARTZ, J.T., Fast probabilistic algorithms for verification of polynomial identities, *J. ACM* **27** (1980) 701–717.
[210] SEIFERAS, J., Machine-independent complexity theory, in: J. van Leeuwen, ed., *Handbook of Theoretical Computer Science, Vol. A* (North-Holland, Amsterdam, 1990).
[211] SHANKAR, N. and V. RAMACHANDRAN, Efficient parallel circuits and algorithms for division, *Inf. Proc. Letters* **29** (1988) 307–313.
[212] SHILOACH, Y. and U. VISHKIN, Finding the maximum, merging, and sorting in a parallel computation model, *J. Algorithms* **2** (1981) 88–102.
[213] SHILOACH, Y. and U. VISHKIN, An O(log *n*) parallel connectivity algorithm, *J. Algorithms* **3** (1982) 57–67.
[214] SIMON, J., On feasible numbers, in: *Proc. 9th Ann. ACM Symp. on Theory of Computing* (1977) 195–207.
[215] SMITH, J.R., Parallel algorithms for depth-first searches: I. planar graphs, *SIAM J. Comput.* **15** (1986) 814–830.
[216] SMOLENSKY, R., Algebraic methods in the theory of lower bounds for Boolean circuit complexity, in: *Proc. 19th Ann. ACM Symp. on Theory of Computing* (1987) 77–82.
[217] SNIR, M., On parallel searching, *SIAM J. Comput.* **14** (1985) 688–708.
[218] SOROKER, D., Fast parallel algorithms for finding hamiltonian paths an cycles in tournaments, *J. Algorithms* **9** (1988) 276–286.
[219] SOROKER, D., Fast parallel strong orientation of mixed graphs and related augmentation problems, *J. Algorithms* **9** (1988) 205–223.
[220] STOCKMEYER, L. and U. VISHKIN, Simulation of parallel random access machines by circuits, *SIAM J. Comput.* **13** (1984) 409–422.
[221] TARJAN, R.E. and U. VISHKIN, An efficient parallel biconnectivity algorithm, *SIAM J. Comput.* **14** (1985) 862–874.
[222] TRAHAN, J., Instruction sets for parallel random access machines, Ph. D. Thesis, Coordinated Science Lab., Univ. of Illinois, Urbana, IL, 1988.
[223] TRAHAN, J., V. RAMACHANDRAN and M.C. LOUI, The power of parallel random access machines with augmented instruction sets, in: *Proc. Fourth Ann. Conf. on Structure in Complexity Theory* (1989) 97–103.
[224] TSIN, Y.H. and F.Y. CHIN, Efficient parallel algorithms for a class of graph theoretic problems, *SIAM J. Comput.* **13** (1984) 580–598.
[225] TUTTE, W.T., How to draw a graph, *Proc. London Math. Soc.* **13** (1963) 743–767.
[226] ULLMAN, J.D. and A. VAN GELDER, Parallel complexity of logical query programs, *Algorithmica* **3** (1988) 5–42.
[227] UPFAL, E., A probabilistic relation between desirable and feasible models of parallel computation, in: *Proc. 16th Ann. ACM Symp. on Theory of Computing* (1984) 258–1265.
[228] UPFAL, E. and A. WIGDERSON, How to share memory in a distributed system, *J. ACM* **34** (1987) 116–127.
[229] VALIANT, L.G., Parallelism in comparison problems, *SIAM J. Comput.* **4** (1975) 348–355.
[230] VALIANT, L.G., Parallel computation, in: *Proc. 7th IBM Symp. on Mathematical Foundations of Computer Science* (1982) 173–189.

[231] VALIANT, L.G., S. SKYUM, S. BERKOWITZ and C. RACKOFF, Fast parallel computation of polynomials using few processors, *SIAM J. Comput.* **12** (1983) 641–644.

[232] VAZIRANI, V.V., NC algorithms for computing the number of perfect matchings in $K_{3,3}$-free graphs and related problems, Computer Science Dept., Cornell Univ., Ithaca, NY, 1987.

[233] VENKATESWARAN, H. and M. TOMPA, A new pebble game that characterizes parallel complexity classes, in: *Proc. 27th Ann. IEEE Symp. on Foundations of Computer Science* (1986) 348–360.

[234] VISHKIN, U., Implementation of simultaneous memory address access in models that forbid it, *J. Algorithms* **4** (1983) 45–50.

[235] VISHKIN, U., Randomized speed-ups in parallel computation, in: *Proc. 16th Ann. ACM Symp. on Theory of Computing* (1984) 230–239.

[236] VISHKIN, U., On efficient parallel strong orientation, *Inf. Proc. Letters* **20** (1985) 235–240.

[237] VISHKIN, U., Optimal parallel pattern matching in strings, *Inform. and Control* **67** (1985) 91–113.

[238] VISHKIN, U. and A. WIGDERSON, Trade-offs between depth and width in parallel computation, *SIAM J. Comput.* **14** (1985) 303–314.

[239] VITTER, J.S. and R.A. SIMONS, New classes for parallel complexity: a study of unification and other complete problems for P, *IEEE Trans. on Computers* **35** (1986) 403–418.

[240] WALLACE, C.S., A suggestion for a fast multiplier, *IEEE Trans. Comput.* **13** (1964) 14–17.

[241] WHITNEY, H., Non-separable and planar graphs, *Trans. Amer. Math. Soc.* **34** (1932) 339–362.

[242] WYLLIE, J.C., The complexity of parallel computations, Ph.D. Dissertation, Computer Science Dept., Cornell Univ., Ithaca, NY, 1981.

[243] YAO, A.C., Separating the polynomial-time hierarchy by oracles; Part I, in: *Proc. 26th Ann. IEEE Symp. on Foundations of Computer Science* (1985) 1–10.

[244] YOUNGER, D.H., Recognition and parsing of context-free languages in time n^3, *Inform. and Control* **10** (1967) 189–208.

[245] KRUSKAL, C.P., L. RUDOLPH and M. SNIR, A complexity theory of efficient parallel algorithms, RC 13572 (#60702), IBM Research Division, 1988.

CHAPTER 18

General Purpose Parallel Architectures

L.G. VALIANT

Aiken Computation Laboratory, Harvard University, Cambridge, MA 02138, USA

Contents

HANDBOOK OF THEORETICAL COMPUTER SCIENCE
Edited by J. van Leeuwen

1. Introduction

It is already explicit in Turing's fundamental work [68] that a computer may be viewed both as a special purpose device executing a particular program, as well as a universal device capable of simulating all programs. The importance of this duality in sequential computation is enhanced by the fact that universality can be made efficient. This means that special purpose machines have no major advantage since general purpose machines can perform the same functions almost as fast. Efficient universality also makes possible general purpose high-level languages and transportable software, both of which contribute decisively to the ease of use of computers.

The question of whether efficient universality can be found, and the accompanying benefits reaped, in the context of parallel computing, is a central problem for contemporary computer science. In this chapter we review some of the theoretical results that relate to these issues, but do not attempt an exhaustive survey. Our central theme is that a theory is now available in the light of which parallel computation looks as efficiently universal as does the sequential case.

To develop the theory we have to discuss both *realistic* parallel computers, those that are to be built, as well as *idealized* models that describe what a general purpose machine is expected to be able to do. To reflect current technologies we will take as our main model of a realistic machine that of a set of sequential processors with local memories connected by a sparse message-passing communication network. The theory of this model is concerned essentially with the communication of information in the network. To reflect widespread usage in the literature on parallel algorithms we will take as our main model of idealized machines varieties of the PRAM, where storage management and communication issues are hidden from the programmer. The theory of this model deals with how its various varieties can simulate each other. At the intersection of these two theories is the XPRAM, a model we introduce here that can be implemented fast on networks and on which higher-level PRAMs can be simulated efficiently.

For a program running on a p-processor machine in time T we will call the pT product the operation count. If this program is simulated on a p'-processor machine in time T' we will say that the efficiency of the simulation is E if E is a lower bound on the ratio $pT/p'T'$. We will aim for efficiencies that are constant and independent of p and T. To achieve this we will need to allow some parallel *slackness*, namely to have p' smaller than p, for example $p/\log p$. This will correspond to the situation of having a program written for a certain number p of processors, and its being efficiently executable on any universal parallel machine with fewer than $p/\log p$ processors.

There is essentially one class of networks that has been extensively studied and proved to have good communication capabilities. This is the family related to the binary n-dimensional cube and also containing the butterfly and shuffle-exchange graph. These also have the additional advantage that they can perform several important special purpose functions especially efficiently. We shall restrict attention exclusively to these. As far as communication capabilities alone is concerned, there is a second class, that of expander graphs, that have desirable properties but currently they do not offer a practical alternative.

Our network model attempts to capture some of the basic phenomena relating to large systems of sequential computers working in parallel. For simplicity we have to ignore numerous parameters that may have crucial relevance to the performance of any actual machine. Among our assumptions are that the system is globally synchronized, error-free, communicates by packet-switching and has fixed transmission time for all connections. Each of these assumptions can be relaxed somewhat, however, without affecting the conclusions. Also, we do not consider physical packaging and network realizability as a separate parameter. The VLSI literature addresses these issues more directly [32, 38, 41, 69]. The networks we consider do have some desirable features from this viewpoint, as is evidenced by the fact that several machines based on them have been built [23, 63]. A further significant parameter, relating to granularity, is the ratio of the time of a basic transmission step to that of a basic processor operation. The theory presented is most relevant when these two times are or can be viewed as roughly comparable.

The organization of this chapter is as follows. First we shall describe some interconnection patterns related to the n-dimensional cube, and briefly discuss their basic properties and suitability for certain particular algorithms. In Section 3 we go on to analyze the problem of parallel communication or routing for these networks. The main result, which we give in detail, is that even at maximal packet densities, permutation routing can be done in a sparse p-node network in $O(\log p)$ time by a distributed algorithm. In Section 4 we formalize various notions of idealized parallel computers. We show that the exclusive-read exclusive-write PRAM can be simulated optimally given logarithmic parallel slackness and fast implementations of certain hash functions. For concurrent-read concurrent-write PRAMs the same can be achieved given p^{ε} slackness for any $\varepsilon > 0$. The simulations can be carried out either on the network model or on a very different one inspired by optical communication. Our simulations all go through the XPRAM model which is further discussed in [87].

In computer science several kinds of fundamental impediments to efficient computation have been uncovered such as noncomputability. The main conclusion of the results reviewed here is that there is no comparable fundamental impediment at all to the construction of efficient general purpose computers in the sense considered. A major feature of our proposal is that as the machines scale up, communication facilities grow slightly faster than computational facilities. The proofs of these results give clues about how these machines might be built. Their very positive nature suggests that there may be a rich space of good solutions. Hence there is reason for optimism about the eventual success of such machines.

2. Some networks

2.1. *Introduction*

Our basic model of a realistic parallel computer consists of p processors each located at a distinct node of a p-node graph G. Each processor is a universal sequential computer with some local memory and the capability of sending packets of information directly to processors at neighboring nodes of the graph. We assume that each edge can transmit one packet of information in unit time and has a queue for storing packets that have to be transmitted along it. Instead of a fixed graph we shall consider families of them parametrized by their size. To reflect physical packaging constraints we insist that the degree d of G, the maximum number of edges from any node, is small. It is either independent of p or grows slowly, for example, as the logarithm of p.

There are several variations of the model for which similar analysis is possible. We may, for example, distinguish three kinds of nodes, one having only a processor, one only a memory unit and a third having only switching capabilities.

Among the graph properties that are desirable for such computers some are self-evident. The *diameter* of a graph is the minimal-length path between the pair of nodes for which this length is maximal. If a graph has p nodes and degree d, then there are at most $(d-1)^r$ nodes at distance r from any fixed node and it follows that the diameter must be at least about $\log_{d-1} r$. Graphs of small diameter are desirable because they enable any single packet to be sent fast between any two nodes. Graphs with optimal diameter for given p and d are called Moore graphs and have been studied extensively [11, 24]. In parallel computers where many packets are in transit simultaneously we need much stronger graph properties that enable a large number, such as p, of packets to be transmitted simultaneously and all delivered fast.

Other desirable features include homogeneity. If the graph looks isomorphic when viewed from any part of the graph, the resulting machine should be easier to program. Also, recursive decomposability, the possibility of constructing the graph from a large number of relatively few kinds of components, greatly helps manufacturing.

2.2. *The hypercube family*

We now define the connection patterns that are the subject of our study. A *directed graph* G is denoted by the pair (V, E) where V is the set of nodes and E, a set of pairs of nodes, the set of edges. A directed edge $(u, v) \in E$ denotes an edge from node u to node v.

For simplicity we define the families of graphs below as *undirected* (or bidirectional) graphs. When the graph is regarded, as in Section 3, as a communication scheme, each undirected edge (u, v) has to be interpreted as a pair of directed edges (u, v) and (v, u).

Each graph family will be parametrized by an integral power of 2: $N = 2^n$. Thus $n = \log N$, all logarithms being assumed to be to the base 2. For integer x we denote the integers $\{0, 1, \ldots, x-1\}$ by $\langle x \rangle$. For integer x we refer to the $(i+1)$st least significant bit in its binary representation $(i = 0, 1, 2, \ldots)$ as the "ith bit" of x. The integer obtained by changing the ith bit of x is denoted by x_i. (For instance, if $x = 5$, then $x_0 = 4$, $x_1 = 7$ and $x_2 = 1$).

The *N-cube* is (V, E) where $V = \langle N \rangle$ and $E = \{(w, w_i) \mid w \in V,\ i \in \langle n \rangle\}$. It is also known as the binary n-dimensional hypercube. For each i the edges (w, w_i) are called the edges of dimension i. The graph has N nodes and degree and diameter both equal to n. Note that for this degree the diameter does not achieve the Moore bound of about $n/\log n$ but experience suggests that this may be compensated for by the cube's rich communication properties. The graph families below can be viewed as constant-degree derivatives of it that each sacrifice some capability in order to achieve the lower degree.

The *N-butterfly* is (V, E) where $V = \langle N \rangle \times \langle n+1 \rangle$ and

$$E = \{([w, t], [w, t+1]), ([w, t], [w_t, t+1]) \mid w \in \langle N \rangle,\ t \in \langle n \rangle\}.$$

It can be regarded as a graph with $n+1$ *levels* of nodes with N nodes at each level. It has $N(\log N+1)$ nodes, degree 4 and diameter $2\log N$. It is closely related to the N-cube, for if for each w the node set $\{[w,t] \mid t \in \langle n+1\rangle\}$ were identified as a single node, then it would become exactly the N-cube.

The *N-CCC* or *cube-connected-cycles* is defined here exactly as the N-butterfly except that for each $w \in \langle N\rangle$ the nodes $[w,0]$ and $[w,n]$ are identified as one node. It can be visualized as an N-cube in which each node has been cut off so as to become a ring of n nodes, each adjacent to an original cube edge of a different dimension. The original definition given in [52] is closely related to our one here (see [69]).

The *N-shuffle-exchange* is (V,E) where $V=\langle N\rangle$ and

$$E=\{[w,w+1] \mid w \in V,\ w \text{ even}\} \tag{1}$$

$$\cup\{[w,2w \bmod (N-1)] \mid w \in \langle N-1\rangle\} \cup \{[N-1,N-1]\}. \tag{2}$$

Here there are two kinds of edges. The *exchange* edges connect each node w to w_0, the node differing in the least significant bit. The *shuffle* edges connect each w to w' obtained by cyclically shifting the binary representation of w. (For instance, if $N=8$ and $w=5$, then the edges incident to w are (5, 4), (5, 6) and (3, 5), the first being an exchange edge and the last two shuffle edges.) This graph has N nodes, degree 3 and diameter $2\log N$.

The important commonality among the N-CCC and N-butterfly is that they can both efficiently simulate any N-cube algorithm that is *normal* [69]. Normal here means that at each time step the edges used by the algorithm are all of the same dimension i in the cube, and that at the following time step the edges used must all use either dimension $i-1$ or i or $i+1$ (modulo n). The simulation is obvious for the CCC and applies also to the N-butterfly if the modulo n moves between dimensions 0 and n are forbidden. To simulate a normal algorithm on the N-shuffle-exchange we use the exchange edges for any dimension i and use the shuffle edges to right shift (left shift) if the dimension is to be changed to $i+1$ ($i-1$), respectively.

The above families of graphs have been suggested numerous times as appropriate for multiprocessor computers (e.g., [66]). These and related networks have been extensively studied [64, 65]. Among these others, not mentioned so far, the de Bruijn graph is particularly simple. It is like the shuffle-exchange except that the exchange edges are replaced by "exchange-shuffle" edges (i.e., all edges do a right shift and either change the 0th bit or they do not). This can be generalized to k-ary representations of integers. Then each edge from a node w corresponds to choosing an integer in $\langle k\rangle$, changing the 0th bit of w into it, and cyclically right shifting.

In conclusion we remark that aside from the cube there is another family of graphs with remarkable communication properties. These are the expander graphs which form the basis of many efficient asymptotic constructions in computer science [3, 48, 49, 72, 74, 84]. Unfortunately even using the best bounds known on their size from existential proofs, they are not currently competitive in a quantitative sense. Their main theoretical advantage is that they allow for $O(\log N)$ deterministic routing and sorting whereas for the hypercube family, the only algorithms known for achieving these bounds use randomization.

2.3. Special purpose applications

The main reason given for the attractiveness of the hypercube family in the early literature is their suitability for several specific important algorithms. Some of the most important examples of this can be seen by considering the following single schema. Let $\mathrm{OPER}_{t,j}$ be any operation with two input values and two output values. For $N=2^n$ consider the following parallel algorithm acting on data held in a linear array $A[0,\ldots,N-1]$.

for $t:=0$ **to** $n-1$ **do**
  for $j\in\langle N\rangle$
    pardo if tth bit of j is 0
      then $A[j],\ A[j+2^t]:=\mathrm{OPER}_{t,j}(A[j],\ A[j+2^t])$

Several important algorithms such as the Fast Fourier Transform [2, 18] and odd–even merge [7, 28] have exactly this structure. In the former case OPER computes linear forms. In the latter it does a comparison on the inputs, and outputs them in order of magnitude. The important point is that any algorithm with this structure can be implemented directly on the N-cube, by using the dimension-t edges to implement the data transfers at the tth cycle. Note that since this algorithm is normal it can be implemented directly on the N-CCC and N-shuffle-exchange, and also on the N-butterfly since the connections between level 0 and n are not required. For further details see [51, 52, 66].

We note that this simulation is not quite optimal in the case of the N-CCC and the N-butterfly since these networks have $N\log N$ processors, of which only N are being used at any time. As shown by Preparata and Vuillemin [52], however, these networks can actually implement instances of the parallel algorithm for array size as large as $N\log N$ rather than N by assigning segments of length $\log N$ of the array A to each CCC-cycle, performing the first $\log N$ parallel operations locally on the cycles, and using pipelining. See [51, 52] for further details of this and of numerous other algorithms besides the FFT and odd–even merge that either have this structure or can be made up of combinations of them. Note also that the complexity of such problems can be nailed down more easily for the cube than for many other models of computation. The results in [74] about the BRAM model imply, for example, $\Omega(N\log N)$ lower bounds on the total number of communication steps needed for computing the N-point discrete Fourier transform on the N-cube if the inputs, as well as the outputs, are at distinct nodes.

Graphs such as grid-graphs and binary trees also reflect information flows in important algorithms. A great advantage of the cube, not apparently shared by its three relations, is that both of the above classes of graphs can be embedded into it efficiently. For example, the $2^{d_1}\times 2^{d_2}\times\cdots\times 2^{d_r}$ grid is defined as having node set $\{1,2,3,\ldots,2^{d_1}\}\times\cdots\times\{1,2,\ldots,2^{d_r}\}$ and edges between pairs of nodes that differ in only one coordinate and there by exactly one unit. By a simple recursive construction these can be embedded into a Σd_i-dimensional cube, such that nodes are mapped to distinct nodes and adjacent nodes in the grid map to adjacent nodes in the cube. As shown by Bhatt et al. [9] arbitrary binary trees can be also embedded without increasing the number of nodes, except now adjacent nodes may map to nodes of distance up to a constant in the cube. For each tree edge a path in the cube can be assigned so that only a constant number of such paths share any cube edge.

Besides grids and trees several other connection patterns can be embedded efficiently into the cube [43]. One important such example is the mesh of trees, which is an efficient host for numerous computational problems [34].

3. Routing

3.1. Parallel communication schemes

A basic difference between single and multiprocessor computers is that in the latter there is potentially a very large overhead in interprocessor communication. In this section we shall analyze in detail the routing problem, which is an abstraction of this communication overhead.

A solution to the routing problem can be formalized as a *parallel communication scheme* (PCS). Adapting from [56] and [76] we shall define a PCS here as a triple (G, R, Q) where $G=(V, E)$ is a directed graph, R is a *routing algorithm* and Q a *queueing discipline*. An *initialized* PCS is defined as a quintuple (G, R, Q, I, D) where I is the *input specification*, a mapping from $X \to V$ that specifies for each packet in the packet set X, the node at which it is located initially. D is the *output specification*, a mapping from $X \to V$, specifying for each $x \in X$ the desired destination address. At each integral time instant $t=0, 1, \ldots$ the routing algorithm R decides for each packet at each node to which adjacent node it is to be sent, if possible, during the next unit time interval. Each directed edge $e \in E$ has a queue q_e where all the packets competing to travel down that edge during the next interval wait. The queueing discipline Q determines which one of the packets will be actually sent at the next interval. We assume that each packet takes unit time to traverse an edge and that all other computational operations are instantaneous (or at least dominated by this). We note that once G, R, Q, I and D are defined, the run of the associated initialized PCS is completely specified to within any randomization allowed.

The paradigmatic communication pattern (I, D) is that of a *permutation*. Here $|X| = |V|$, and I and D are both one-to-one mappings. If $|X| < |V|$ we have a *partial permutation*. If $|X| \leqslant h|V|$ and for all $v \in V$ at most h packets start at v and are destined for v, then we have an *h-relation*.

For universal parallel computers of the kind considered in this paper we need routing algorithms that are totally *distributed*. This means that the routing algorithm R can make its decisions at node v at time t based only on information available in packets that have passed through this node by this time. This information is typically $O(\log|V|)$ bits, and certainly not more than polynomial in this. This possibility that a complete specification of I and D ($V \log V$ bits) is available at each node is definitely excluded.

In the *nondistributed* case when global information is everywhere available, permutation routing is easier and well understood. The Beneš construction [8, 81], consisting of two butterfly networks back to back, can realize permutations in $2 \log N$

steps without queueing. It can be readily adapted to the cube, CCC and shuffle-exchange by the remarks of the previous section. If the same communication pattern is to be used for a very long sequence of packets, then it becomes worthwhile to precompute the routes and distribute this information. Parallel algorithms for doing this precomputation have been devised [39, 45, 62] and their consequences for parallel computers investigated [21].

The remainder of this section is devoted therefore to distributed routing algorithms for the hypercube family. Among *deterministic* algorithms the fastest known is based on Batcher's comparator networks for merging [7, 28] and takes time $(\log N)^2$ for permuting $N=|V|$ packets. (See [82] for a recent asymptotic improvement.) It is also known that if we further restrict deterministic algorithms to be *oblivious* (i.e., the path of each packet is determined by its source and destination only), then no routing algorithm can do better than $\Omega(N^{1/2}/d)$ for any degree d graph [12, 83]. Among examples of oblivious algorithms are the *greedy* algorithms. For each member of the hypercube family one can define a natural greedy algorithm that routes each packet by the shortest route. In each case simple permutations can be found that force the N^{α} $(\alpha>0)$ behavior as predicted by Borodin and Hopcroft [12]. In the case of the cube there are exponentially many shortest routes, but even randomizing among these still gives N^{α} behavior [76].

That distributed routing is possible on a sparse network in time $O(\log N)$ was discovered in 1980 [75, 76, 79]. The two-phase randomized algorithm proposed was first shown to work in the required time for the cube. Subsequent work using innovative techniques extended it to the CCC, butterfly and shuffle-exchange [4, 70]. The subject of this section is a proof of a strong version of these results which incorporates techniques from all these sources as well as from [56] and [60].

The two-phase randomized algorithm works as follows. In phase A each packet chooses a random destination independently and executes an appropriate greedy algorithm to get it there. Phase B executes a greedy algorithm to take each packet from the random destination of the first phase to the actual intended destination. Note that even when a permutation is to be realized, the result of the first phase is a random mapping rather than a random permutation. (In 1987 Ranade [55] showed that on the butterfly a random choice from a certain subset of permutations suffices and can be realized without queueing in the first phase.)

Besides specifying the greedy algorithm it remains only to describe the queueing discipline Q. For ease of proof we shall use random priorities [56, 60]; each packet is initially assigned a random priority π in the range $\langle n\rangle$. In any queue at any time a packet with lower-numbered priority will leave ahead of any higher-numbered ones. As we shall see later the runtime of the algorithm is invariant under a wide class of disciplines that includes first-in-first-out and last-in-first-out [56].

Below we prove a result (Theorem 3.1) from which it can be deduced easily that the cube, butterfly and CCC can all support optimal traffic density to within a constant factor. In particular, in $O(\log N)$ time the N-cube can realize $O(\log N)$-relations, while the constant degree butterfly and CCC can realize $O(1)$-relations. These are clearly optimal if each packet has typical path length $\log N$. Also these results can be translated into statements about continuous packet traffic [15].

In the suboptimal traffic density case of realizing permutations on the N-cube, the following result of Tsantilas [86] based on techniques due to Ranade [56] shows that queueing delays are, in fact, $o(n)$.

3.1. THEOREM. *For all $\alpha>1$ there is an $\eta>0$ such that if T is the runtime of either phase of the two-phase algorithm on an N-cube when realizing a permutation, then*

$$\Pr(T>n+\alpha n/\log n)=O(N^{-\eta}).$$

For the maximal traffic density case the queueing delays still contribute no more than an $O(n)$ term. The proof of Theorem 3.3 in Section 3.2 will establish for the appropriate two-phase algorithms the following theorem, where T is the runtime of either phase.

3.2. THEOREM. *There exist α_0, β, $\delta>0$ such that for all $\gamma>0$ and all $\alpha>\alpha_0$,*

(1) *for realizing a γ-relation on an n-CCC or N-butterfly, or*

(2) *for realizing a γn-relation on an N-cube,*

$$\Pr(T>\alpha\gamma n)\leqslant N^{-\alpha\beta\gamma+\delta}.$$

This result upper bounds the probability of exceeding arbitrarily large runtimes in terms of this runtime. The proof also supports various other quantifications of the parameters.

We shall consider the *directed butterfly* which is exactly as the undirected case except that each edge $([w,t],[w',t+1])$ is considered as directed in the direction of increasing t. We consider the level $t=0$ as *input* nodes and the level $t=n$ nodes as *output* nodes. The *greedy* algorithm for the directed butterfly is the following: if a packet at node $[w,0]$ is destined for $[w',n]$, then in each of the steps $i\in\langle n\rangle$ it goes from the current $[w'',i]$ to $[w'',i+1]$ or $[w''_i,i+1]$ according to whether the ith bit of w is the same as that of w'' or not. Phase A for an h-relation is an initialization (I,D) where initially there are at most h packets at each input node, each with a randomly chosen output node as destination address. Phase B is an (I,D) where initially each packet is placed at a random input node and at most h altogether share a common output destination.

3.3. THEOREM. *There exist α_0, β, $\delta>0$ such that, for all $\gamma>0$ and all $\alpha>\alpha_0$, for an initialized PCS (G,R,Q,I,D) where G is the directed N-butterfly, R is the greedy algorithm, Q is random priority based, and T is the runtime, when (I,D) is either Phase* A *or Phase* B *for a γn-relation, then*

$$\Pr(T>\alpha\gamma n)\leqslant N^{-\alpha\beta\gamma+\delta}.$$

Theorem 3.2 follows easily from Theorem 3.3. To simulate the CCC or butterfly by the directed case we send the packets to level-zero nodes initially, then do the directed butterfly algorithm, and finally send the packets that accumulate at level n to the correct levels. To simulate the cube we just simulate the CCC and identify all the nodes $[w,t]$ of the latter with node w of the former.

3.2. *The proof for the directed butterfly*

The probabilistic arguments in routing results can be presented in various styles. These include the use of well-known bounds on the tails of distributions as in [4, 70, 76, 79], the use of Kolmogorov complexity or counting as in [56, 61], or the use of generating functions as suggested by Leighton. They have different didactic advantages. Here the probabilistic arguments will be encapsulated into a single proof, that of Claim 3, and will combine elementary use of the first and third styles. The proof is given for Phase A. Essentially the same argument can be used also for Phase B.

Following ideas in [4, 70] we define a directed *task graph* H with respect to the directed butterfly graph $G=(V, E)$, to have node set $\{\tau=[u, \pi] \mid u \in V, \pi \in \langle n \rangle\}$ and edge set $\{([u, \pi], [v, \rho]) \mid \text{either } u=v \text{ and } \rho=\pi+1 \text{ or } (u, v) \in E \text{ and } \pi=\rho\}$. The two kinds of edges represent the two ways in which the completion of a task $[v, \rho]$ may have to await the completion of other tasks. First at node v packets with priorities lower than ρ will have precedence. Second at node v the packets of priority ρ can be processed only once they have arrived from a neighboring node.

A node (u, π) in H will denote the task of transmitting through $u \in V$ all packets of priority π. A *delay sequence* D is any sequence of tasks $(\tau_1, \ldots, \tau_d)$ along a directed path in H where $\tau_1=(u, \pi)$ and u has level zero in G and $\pi=0$. Clearly the longest directed path in G has length less than $2n$. Since there are 2^n starting points and the out-degree of H is three, the number of distinct delay sequences is less than $2^n \cdot 3^{2n}=18^n$.

In a particular execution of Phase A the execution of the various tasks may overlap temporarily in complex ways. For a particular delay sequence $D=\{\tau_1, \ldots, \tau_d\}$ and particular execution we define f_i and t_i, for $1 \leqslant i \leqslant d$, as follows. Let f_i be the total number of packet transmissions involved in task $\tau_i=(u, \pi)$, and let the *completion time* t_i be the time at which the last packet of priority at most π has been transmitted from node u. Denote the sum of all the f_i by $\mathrm{T}(D)$.

1. CLAIM. *If an execution of the algorithm takes T steps, then there is some delay sequence D such that* $\mathrm{T}(D) \geqslant T$.

PROOF. We consider a *critical* delay sequence constructed as follows. Choose τ_{2n} to be one of the tasks with maximum completion time. Having chosen τ_j choose τ_{j-1} to be task τ with maximum completion time t amongst those tasks τ such that (τ, τ_j) is an edge of H. The construction stops when we reach a task with priority zero and node level zero in G. The indices of the τ sequence are finally shifted down to start from zero.

Observe that for this sequence $t_i - t_{i-1} \leqslant f_i$ for every $i \geqslant 1$ (where $t_0=0$). For by construction the three tasks that are predecessors of t_i in H all have completion times at most t_{i-1}. Furthermore, once these tasks are completed nothing prevents $\tau_i=(u, \pi)$ from being completed except for queueing at node u of the at most f_i packets of priority π that go through u. Consequently

$$T=\sum_{i=1}^{d} t_i - t_{i-1} \leqslant \sum_{i=1}^{d} f_i = \mathrm{T}(D) \qquad \square.$$

2. CLAIM. *Under the conditions of Theorem* 3.3 *there exist* $\alpha_0, \beta>0$ *such that for all*

$\gamma>0$ and $\alpha>\alpha_0$ for any delay sequence D for Phase A *or Phase* B,

$$\Pr(\mathrm{T}(D)\geqslant\alpha\gamma n)\leqslant N^{-\alpha\beta\gamma}.$$

PROOF. Consider the set X_u of packets that can intersect D for the first time at $\tau=(u,\pi)$ for some π where $u=[w,i]$. These are the packets whose initial addresses agree with w in the most significant $n-i$ bits (but not in the next most significant bit) and their number is at most $2^i\gamma n$. Each of these packets will go through τ according to the random choices of priority and destination. An important point is that the X_u are disjoint for the set of nodes u in D. This ensures that the probabilistic events analyzed in Claim 3 are probabilisticly independent. Also if some $x\in X_u$ does indeed intersect D, then it will go through node u.

The probability that any fixed $x\in X_u$ has priority π is n^{-1}. The probability that it goes through u is 2^{-i}, the probability that the last i bits (but not in the next most significant bit) of the random destination coincide with those of u. Hence the actual probability of $x\in X_u$ intersecting D at (u,π) is $(n2^i)^{-1}$. Suppose there are $\bar{K}_i$ tasks in D with nodes at level i in G. This corresponds to tasks (u,π) all with the same u, but with K_i consecutive values of π. Then the probability that $x\in X_u$ intersects D for the first time at (u,π) for some π is $\bar{K}_i(n2^i)^{-1}$. If x receives a higher priority, it can intersect D for the first time later. The probability that this happens at level $i+j$ is $\bar{K}_{i+j}(n2^{i+j})^{-1}$. Hence the probability that x intersects D is $K_i(n2^i)^{-1}$, where $K_i=\sum_{j=0}^{n-i-1}\bar{K}_{i+j}2^{-j}$. Since $\sum\bar{K}_i\leqslant 2n$, it follows that $\sum K_i=4n$.

Now we compute for $x\in X_u$ the conditional probability that x intersects D at least $j+1$ times given that it intersects it $j\geqslant1$ times. Since the priority of a packet, once chosen, is unchanged, repeated intersections with D can only occur with a subsequence $\tau_r,\tau_{r+1},\ldots,\tau_{r+m}$ all of which have the same priority, in other words a sequence corresponding to a path in G. But if a packet intersects D at (u,π), then the probability that it will intersect it at an adjacent node (v,π) is exactly $\frac{1}{2}$ since the choice between the outgoing edges of u is made with equal probability by the routing algorithm. (N.B. By construction, once a packet departs from D it cannot intersect it again.)

Now for each $x\in X_u$ let y be a random variable that has value z with the same probability that packet x has of intersecting D at least z times. Let the set of all such random variables y for X_u be Y_u. Denote by S the sum $\sum y$ over all Y_u and u. But the sum of the values of all the y is just T(D) by definition, since counting the intersections with multiplicity of all the γN packets with D is the same as counting the total number of packet transmissions in D. Hence by Claim 3 (to follow), putting $Y_i=\{Y_u\mid u=(w,i)$ for some $w\}$, there exist $c_1,c_2>1$ such that if $s=\alpha h=\alpha\gamma n$, then for $\alpha>c_2$

$$\Pr(\mathrm{T}(D)\geqslant s)\leqslant c_1^{-\alpha\gamma n}\leqslant N^{-\alpha\beta\gamma}$$

for $\beta=\log_2 c_1>0$. □

3. CLAIM. *Let $Y_0,Y_1,\ldots,Y_n$ be sets of random variables, all mutually independent and taking nonnegative integral values, such that $|Y_i|=h2^i$,*

$$\Pr(y\geqslant1\mid y\in Y_i)\leqslant K_i(n2^i)^{-1}$$

and

$$\Pr(y \geqslant j+1 \mid y \in Y_i, y \geqslant j \geqslant 1) \leqslant \tfrac{1}{2},$$

where $\sum_{i=0}^{n} K_i \leqslant 4n$. *If R denotes the number of non-zero y's and S the sum of the values of all the y's (in* $Y_0 \cup Y_1 \cup \cdots \cup Y_n$*), then there exist* $c_1, c_2 > 1$ *such that*

$$\Pr(R \geqslant r) \leqslant (4he/r)^r \quad \textit{if } r \geqslant 8h \tag{3}$$

and

$$\Pr(S \geqslant s) \leqslant c_1^{-s} \qquad \textit{if } s > c_2 h. \tag{4}$$

PROOF. (i) Let $A_i(x)$ be the ordinary generating function [40] in which the coefficient of x^r is the probability that the number of non-zero y's in Y_i is r. Then by definition,

$$A_i(x) \leqslant \sum_{r=0}^{\infty} \left(\frac{K_i}{n2^i}\right)^r \binom{h2^i}{r} x^r$$

$$\leqslant \sum_{r=0}^{\infty} \left(\frac{K_i}{n2^i}\right)^r (h2^i)^r \frac{x^r}{r!} \quad \text{since } \binom{B}{C} \leqslant \frac{B^C}{C!} \text{ for } B, C \geqslant 1$$

$$= e^{K_i hx/n} \qquad \text{by definition.}$$

Hence if $A(x)$ is the ordinary generating function in which the coefficient of x^r is the probability that $R = r$ then, by the multiplication rule,

$$A(x) = \prod_{i=0}^{n} A_i(x) \leqslant e^{(\sum K_i)hx/n} \leqslant e^{4hx} \quad \text{since } \textstyle\sum K_i = 4n.$$

Hence

$$\Pr(R \geqslant r) = \sum_{t=r}^{\infty} \frac{(2h)^t}{t!} \leqslant \frac{(4h)^r}{r!} \quad \text{if } r \geqslant 8h$$

$$\leqslant \left(\frac{4he}{r}\right)^r \qquad \text{since } r! \geqslant \left(\frac{r}{e}\right)^r.$$

(ii) If $R = r$ then for any choice of the r non-zero variables the probability that $S = s$ given this choice is $2^{-(s-r)}$ times the number of ways of distributing the $s - r$ excess ones among the r variables. Hence,

$$\Pr(S = s \mid R = r) \leqslant 2^{r-s} \binom{s-1}{r-1}$$

$$\leqslant 2^{r-s} \left(\frac{e(s-1)}{r-1}\right)^r \quad \text{since } \binom{B}{C} \leqslant \left(\frac{eB}{C}\right)^C \text{ for } B, C \geqslant 1.$$

Then for large enough c there exists a $k_1 < 1$ such that if $s \geqslant cr$, then $\Pr(S = s \mid R = r) < k_1^s$. Summation gives that there exists $k_2 < 1$ such that $\Pr(S \geqslant s \mid R = r) \leqslant k_2^s$. It follows that

$$\Pr(S \geqslant s) \leqslant \Pr(R \leqslant s/c) \cdot \Pr(S \geqslant s \mid R \leqslant s/c) + \Pr(R > s/c)$$
$$\leqslant k_2^s + (4hec/s)^{s/c} \leqslant c_1^{-s} \quad \text{if } s > c_2 h \text{ for appropriate } c_2 > 0. \quad \Box$$

PROOF OF THEOREM 3.3. Since there are at most 18^n choices of D, by Claim 2 the probability that for any choice of D T(D) exceeds $\alpha\gamma n$ is less than $18^n \cdot N^{-\alpha\beta\gamma+\delta}$. □

We note that while it is convenient to present the proofs of the two phases separately, in practice it seems advantageous for each packet to start Phase B as soon as Phase A has been completed. The above proof technique can be applied to this algorithm (on say the cube) if, for example, the priorities assigned in Phase B are all higher-numbered than those of Phase A.

While the above results do not imply O(log N) routing for the shuffle-exchange, the analysis of Aleliunas [4] does establish this. Also, two-phase routing has been shown by Peleg and Upfal [47] to work for a large class of expander graphs. The amount of randomization required for routing is analyzed in [26] and [29]. For routing in $n \times n$ grids a related three-phase algorithm achieves $2n + \mathrm{O}(\log n)$ time with constant buffers [30]. In 1989 this bound has been improved to the exactly optimal $2n-2$ bound by a deterministic algorithm [37] of greater intricacy. Greenberg and Leiserson [22] have also used randomization for routing but in a different way.

3.3. *Invariance to queueing discipline*

Ranade [56] showed that a wide class of queueing disciplines have identical probabilistic behaviors for a class of routing algorithms that includes the one analyzed above. The *state* of a PCS at time t is $S_t : X \to V$, the mapping that determines for each packet its location at time t. He defines a queueing discipline to be *nonpredictive* if it is deterministic and makes its selection among a set of packets at time t only on the basis of S_u for $u \leqslant t$ and of information carried with packets that do not depend on their destination. Examples of nonpredictive disciplines are first-in-first-out, last-in-first-out and any priority-based scheme as described in Section 3.2 where equal priorities are resolved nonpredictively. Ranade's surprising discovery is that the runtime behavior of each phase of the two-phase algorithm is the same for all nonpredictive schemes. An example of a non-nonpredictive discipline is "farthest to go first out". Experimental evidence suggests that in at least one context (for the greedy routing algorithm for the cube if the order of dimensions is randomized) this is slightly faster than certain nonpredictive ones [13, 75].

Given a graph $G=(V, E)$ and a routing algorithm R we define the *resource graph* of G to be $G'=(V \cup E, E')$ where E' is the set of directed edges (g, g') such that one of the following is allowed by R:

(i) $g \in V$, and $g' \in E$ is an edge along which some message originating at g may depart.

(ii) $g, g' \in E$ and g, g' are consecutive edges on some possible path of a message.

(iii) $g \in E, g' \in V$, and g is the last edge on some possible path of a message destined for g'.

Now define a routing algorithm R for G to be *tree-based* if for every $v, w \in V$ there is exactly one path from v to w in its resource graph G' that passes only through nodes in E. This is a very strong restriction meaning as it does that R allows only one path from v to w even if packets could "exchange identity" when they meet in a queue. The greedy routing algorithm described for the butterfly is, however, tree-based.

Finally we define the *population descriptor* P of an initialized PCS to describe the pattern of package transmissions over time in a way that is blind to their identities. Thus:

$$P: V' \times \{1, 2, \ldots\} \to V' \cup \{\perp\},$$

where $P(g, t)$ tells where the packet at g at time t will go at the next time unit, and is undefined if there is no packet there.

1. CLAIM. *For any G, R, I and any nonpredictive Q, any pair of distinct destination settings gives rise to distinct population descriptors.*

PROOF. Consider the earliest time t at which some packet is sent to different places in G'. Then at time $t-1$ they would have been at the same node. Since Q depends only on the previous states and other information independent of the destinations, Q would select this packet to be sent in both cases. Hence transmission at time t from this queue would be to distinct places in G', making the descriptors distinct. □

2. CLAIM. *If R is tree-based and Q_1, Q_2 are nonpredictive, then for every I, D_1 there is a unique D_2 such that (G, R, I, Q_2, D_2) and (G, R, I, Q_1, D_1) have the same population descriptors.*

PROOF. Clearly it is easy to construct the population descriptor P of (G, R, I, Q_1, D_1) for successive time steps $t = 1, 2, \ldots$ by simulation. We fix P in this way. Changing Q at each step without changing P means that we are selecting different packets from the queues and sending them as before. This amounts to confusing the identities of the packets and hence their destinations. The only issue is whether the routes constructed are valid routes for R for some D_2. This is ensured, however, by the definition of tree-based. The nonpredictiveness of Q_2 ensures that D_2 is unique. □

3.4. THEOREM. *If R is tree-based and D is chosen from the uniform distribution of destination settings, then the probability that (G, R, I, Q, D) has runtime T is the same for all nonpredictive Q.*

PROOF. For each Q, D the runtime is determined by its population descriptor. The previous claim shows that changing Q merely permutes the descriptors among the choices of D. Since D is chosen from the uniform distribution, the result follows. □

This establishes the invariance of Phase A under changes in Q. Essentially the same argument proves it also for Phase B.

3.4. *Testability*

A feature of Phases A and B when considered separately is that their runtime behavior is independent of the permutation being realized. This is simply because Phase A is totally independent of the destinations, while phase B when run on a random mapping as input (with each queue order initially randomized) is independent of the original sources. Hence the exact probabilistic behavior of each phase can be tested experimentally by simulation on any permutation, such as the identity. Such experiments prove that even for modest values of N two-phase routing is very efficient on the N-cube, and give accurate comparisons among the various networks [75, 79].

Experiments often show that the behavior of randomized routing is even better than could be expected from existing analysis. Experiments by Brebner [13] suggest that for h-relations on cubes each phase takes expected time less than $h+\log N$. See [86] for the best current analysis.

Note that this testability is independent of the queueing disciplines, the greedy algorithm or any quantitative features of the hardware model. Hence it holds for any implementation in physical hardware.

3.5. *Path length*

One immediate question is whether the factor of 2 in the packet route length introduced by the two-phase method is unavoidable. In [77] it is shown that for the directed de Bruijn graph of any degree k, this factor is essential in the following sense: for any algorithm where the distribution of routes taken by a packet from source u to destination v is dependent only on u and v, if the mean route length is less than $(2-\varepsilon)$ times the diameter for any $\varepsilon>0$, the runtime will be N^{η} for some $\eta>0$. The tightness of this result is related to the optimality of the directed de Bruijn graph with respect to diameter among directed graphs of this degree and number of nodes.

On the N-cube the factor of 2 is not, however, unavoidable. For permutations the algorithm suggested in [77] has route length $n+o(n)$ and runtime $O(n)$. For each packet starting at u and destined for v the routing algorithm makes random guesses $w_1, \ldots, w_r \in \langle N \rangle$ and chooses the w_i such that the sum of the shortest-length paths from u to w_i and w_i to v is minimized over the r choices of w. The packet then takes the greedy algorithm first from u to w_i and then from w_i to u. If $r=\sqrt{n}$, say, then the probability that no w_i is within distance $n(1+\varepsilon)/2$ from both u and v is less than $e^{-\omega(1)n}$ if $\varepsilon=n^{-1/3}$ for example. Hence with probability $1-N^{-\omega(1)}$ the route length will be less than $n+n^{4/5}$. To show that the runtime is $O(n)$ it suffices to observe that even if copies of the packets were sent by all the r routes $w_1, \ldots, w_r$, all of them could be delivered in time $O(n)$ since this is an $O(n)$-relation.

3.6. *Bounded buffers*

The routing algorithms considered so far require queues of potentially unbounded size. While the queue sizes grow to at most $O(\log N)$ with high probability, it would still be desirable to eliminate them, if possible, in favor of fixed-size buffers. Three different ingenious approaches to this have been suggested which open up interesting new areas. We can comment on them only briefly here.

Pippenger [50] was the first to prove that even in the constant-degree case of the CCC, fixed-size buffers suffice for permutation routing. Both the algorithm and the proof are somewhat intricate.

A subsequent solution by Ranade [57] eliminates queueing without substantially complicating the packet routing. It does use however auxiliary control operations including the sending of non-packet "ghost" messages. The solution does have the additional significant advantage of also solving the many-to-one routing problem as described in the next section. Some extensions can be found in [36, 58].

Rabin [53] uses a different approach based on an error-resilient information dispersal system of wider applicability. His scheme, proved for permutations in the N-cube, encodes each packet of m bits redundantly as n minipackets of $(1+\varepsilon)m/n$ bits each and sends these in two-phase fashion. Minipackets that would cause the constant buffers to overflow are just discarded. With large probability, however, enough of them arrive at their destinations such that each original packet can be successfully reconstructed.

Testability suggests that one can determine the behavior of routing with bounded buffers experimentally. It turns out that the simplest adaptation of two-phase routing to fixed buffer size k degrades performance very little. In the case of a directed graph of in-degree d each edge has a buffer of size k. Whenever such a buffer has more than $k-d$ packets, it sends a "buffer full" message to the d predecessor nodes so that these will not forward packets destined for that buffer at the next time unit. Experiments by Tsantilas [86] show that for permutation routing on the N-directed butterfly for values of N up to 2^{14} the runtime of each phase is the same, about $1.4 \log_2 N$, for $k=4$ as it is for $k=\infty$. Also, it degrades for $k=3$ only by less than 3%. For $\log_2 N$-relations buffers of size $k=4$ cause degradation, compared with $k=\infty$, of only a few percent for these values of N.

4. Universality

4.1. General purpose parallel computation

The phenomenon of general purpose computation in the sequential domain is both specific and powerful. We take the view here that it can be characterized by the simultaneous satisfaction of three requirements. First there exists at least one high-level language U that humans find satisfactory for expressing arbitrary complex algorithms. Second there exists a machine architecture M, in this case the von Neumann model, that can be implemented efficiently in existing technologies. Finally there exists a compilation algorithm A that can transform arbitrary programs in U to run on M efficiently.

We hypothesize that parallel computation will have only limited applicability until analogous such triples (U, M, A) are found and exploited in the parallel domain also. The purpose of this chapter is to show, using various recent insights, that at least in the area of large granularity programs for problems with large inherent parallelism, such triples do now exist at the theoretical level.

This area can be characterized as the one where the actual machine M has, say, p processors, while the high-level parallel language U expresses an algorithm for v virtual processors with $v \geqslant p$. The extra *slackness* in parallelism made available by allowing v to be larger than p gives the compiler more flexibility in scheduling communication and computation (e.g., [80]). Recently it has become clear that such parallel slackness can make possible compilers that produce code optimal to within constant factors in a range of situations [31, 78]. This section describes and extends such results.

The theory developed here deals not just with two levels of description U and M but with several levels. We show how each of these several levels can be implemented at a lower level efficiently given some parallel slackness. In sequential computation high-level languages invariably automate memory management. In the parallel case automating communication becomes an additional primary objective. The differences among the models we define below relate to the extent to which they do automate these two tasks.

4.2. PRAM *models*

The most elegant formalization of the von Neumann model that takes into account efficiency considerations is the Random-Access Machine (RAM) of Cook and Reckhow [17]. Numerous attempts have been made to generalize this to the parallel case and these have given rise to a plethora of models (e.g., [1, 20, 39, 42]). Reviews and comparisons of some of these can be found, for example, in [19, 27, 31].

Our approach here will be to emphasize the view that the PRAM models are not just theoretical constructs, but are potentially the basis of parallel programming languages and machine designs. For this reason, at the expense of introducing new terminology, we will define the various versions we deal with in a unified notation.

All the versions consist of a number of processors, each one attached to a memory unit local only to itself. Each processor is a universal sequential machine with its local instruction set which can access words only from the local memory. In addition, there is a set of global parallel instructions that allow accesses by processors to the memories of other processors. The main global instructions will be reads and writes, where the processors simultaneously read (write) from (to) places in the whole memory space. We assume that each local operation takes unit time while, typically, (in each case (i)–(iii)) each global operation takes time $g \geqslant 1$. We assume that each processor executes its own, possibly unique, program. (The SIMD model, where one insists on identical programs, gives an equivalent model to within constant factors.)

We shall describe our models in terms of their global operations.

(i) A *concurrent PRAM* or *CPRAM* allows unrestricted reads and writes. Thus for a read instruction each of the p processors can issue simultaneously an instruction to read from an arbitrary address in the memory space, and these p words need not be distinct. Similarly in a write the p addresses need not be distinct. In the latter case, if there is multiple access to a single word, some convention has to resolve which one of the writes succeeds (see [27]). One such convention is that the processor with lowest identifier succeeds.

(ii) An *exclusive PRAM* or *EPRAM* is defined as a CPRAM except that two processors are not allowed to access any one word simultaneously.

(iii) A *seclusive PRAM* or *SPRAM* is defined as an EPRAM except that now two processors are not allowed to access words in the same memory unit simultaneously. An S* PRAM is a stronger version where a program is allowed to make simultaneous accesses but if that happens, then at most one of the processors accessing that unit succeeds. In constant time all the processors requesting accesses are informed whether they have succeeded.

(iv) An *XPRAM* with p processors executes operations in *supersteps*. In each superstep each processor executes a number of local instructions and sends or receives some messages that implement global read or write instructions. In an execution of a superstep suppose that processor i performs a_i local operations, sends b_i messages and receives c_i messages from other processors. Let $r_i = g^{-1}a_i + b_i + c_i$. Then if $t = \max\{r_i \mid i \in \langle p \rangle\}$, we consider the runtime of this superstep to be that of $(\lceil t/L \rceil)L$ standard global operations, or time $(\lceil t/L \rceil)Lg$. All the processors know whether a superstep is complete or not at the end of each period of L global operations. Within each such period, however, there is no synchronization among the processors. In this paper we will take the parameter L to have value $\log p$.

We note that the CPRAM and EPRAM are identical to the standard CRCW and EREW models described, for example, in [27], in the case that g is unity. Since in current technologies g is often larger, it is reasonable to incorporate this feature into the model since a programmer may choose to exploit the strict form of locality where accesses by processors to nonlocal memory are minimized. Aggarwal and Chandra [1] have shown that for several problems such as sorting and the Discrete Fourier Transform, the number of global accesses need not grow as fast as the total operation count, provided the problem size suitably majorizes the number of processors. For such programs it is relevant to express complexity in terms of the parameter g. On the other hand the simulations described below (Theorems 4.5 and 4.6) that use hashing do not preserve this locality, and it remains an open question whether this problem can be alleviated. Since the remainder of this chapter consists of such simulations we shall assume from now on that g is some constant independent of p.

The SPRAM (called a Direct Connection Machine in [31]) is much closer to realistic technologies but, because of its deterministic nature it appears too restrictive to serve as a host for executing compiled code from the more general models. For pragmatic reasons the XPRAM appears to be the ideal candidate for implementation. It embodies the principle of *bulk-synchrony*, that is, the processors should be barrier synchronized at regular time intervals long enough for several packets to be transmitted between a pair of nodes. As we shall see CPRAMs and EPRAMs can be implemented optimally on it given sufficient slackness. Furthermore it can be implemented optimally not only in the packet-routing setting described in Section 3 of this chapter, but also on the S*PRAM, which is a recently suggested model inspired by optical technology [6, 59].

For each kind of PRAM we will denote by x-PRAM the instance with x processors. Also we shall assume in general that the memory space is shared uniformly among the processors and that inputs to programs are also so shared.

4.3. The simulations

We will say that a p-processor machine M simulates a p'-processor machine M' with *efficiency* E if there is a compilation algorithm that can transform any time-T' program on M' to a time-T program on M where $(Tp)/(T'p') \geq E$.

We will denote p' by v to suggest that M' is a *virtual* machine that is being simulated. E will be a function of v and p therefore. We aim for *optimal* efficiency, where E is a constant independent of p and v. We shall say that a simulation has *expected* efficiency E if there is randomization in the compiled algorithm and E is a lower bound on the ratio of the probabilistic expectations of Tp and $T'v$ for all programs and inputs. The analyses of the simulations support stronger statements, saying that substantial departures from the expected behavior are very unlikely. For brevity we will omit discussion of these here.

In simulation results we always assume that the local instruction sets of both models are identical and, therefore, need not be specified.

We shall first give expected optimal simulations of EPRAMs and CPRAMs by XPRAMs. They both depend on the hypothesis that certain "good" hash functions can be implemented in time proportional to g, the basic global operation time. Mehlhorn and Vishkin [42] have shown that such good hash functions always exist, but their construction involves evaluating polynomials of degree $\log p$. Thus we can either take the view that special purpose hardware is made available to compute this function in time g. This has some plausibility since no memory accesses are required. Alternatively we can hypothesize that some hash function requiring a constant number of local processor operations will one day be verified to be good.

Suppose we have a v-processor EPRAM M'. We regard its memory space to be the addresses $\langle m \rangle$ distributed as equally as possible among the v processors. Suppose that we wish to simulate this on a p-processor XPRAM M with addresses $\langle m \rangle$ also. Suppose also that for each $i \in \langle p \rangle$ the memory of the ith processor of M contains all the words $j \in \langle m \rangle$ such that $j \bmod p = i$.

For each program execution we shall choose a hash function h randomly from a class H of universal hash functions [14]. Each h maps $\langle m \rangle$ to itself. The intention is that in simulating M' machine M will represent address j of M' in memory unit $h(j) \bmod p$. The purpose of h is to spread out the memory accesses as uniformly as possible among the memory units of M even if the memory accesses of M' are arbitrarily nonuniform. Specifically the property of H that we require is that for any set S of up to v addresses (i.e., the addresses chosen by a global read or write operation of M') $h(S)$ should be spread evenly among the p processors of M, with high probability. Formally let

$$R(p, S \mid h) = \max_{i \in \langle p \rangle} |\{ j \in S \mid h(j) \bmod p = i \}|$$

and let

$$\hat{R}(p, H, v) = \max_{S, |S| = v} \left\{ \frac{1}{|H|} \sum_{h \in H} R(p, S, h) \right\}.$$

This is the expected number of references to the memory module of M that has the longest such number, for the worst case S.

Mehlhorn and Vishkin [42] introduced the class H of hash functions h of the form

$$h(j)=\left(\sum_{i\in\langle k\rangle} a_i j^i \bmod m\right).$$

They proved the following lemma.

4.1. LEMMA. *If m is prime, then for all S with $|S|=v$*

$$\Pr\{R(p,S,h)\geqslant k \mid h\in H\}\leqslant\binom{v}{k}p^{1-k}\,\mathrm{e}^{kp/m}.$$

PROOF. Let $H'=\{h \bmod p \mid h\in H\}$. For any $j_1,\ldots,j_k,l_1,\ldots,l_k\in\langle m\rangle$ there is at most one polynomial

$$h(j)=\sum_{i\in\langle k\rangle} a_i j^i \bmod m$$

such that $h(j_r)=l_r$ for all $r=1,\ldots,k$. Hence there are at most $(m/p+1)^k$ functions $h'\in H'$ with the same property if now instead $l_1,\ldots,l_k\in\langle p\rangle$.

Hence the probability that any fixed set of k of the v accesses (i.e., values of $j\in\langle m\rangle$) are mapped under h to the same value in $\langle p\rangle$ is $(m/p+1)^k/|H|=(1+p/m)^k p^{-k}<\mathrm{e}^{kp/m}p^{-k}$. Hence the probability that some set of k of the v accesses are all mapped under h to the same one of the p possible values is less than

$$\binom{v}{k}\mathrm{e}^{kp/m}p^{1-k}. \qquad \square$$

We shall define a family $\mathscr{H}$ of hash function classes to be *good* if for all p and all $m\geqslant p\log p$ there is an $H\in\mathscr{H}$ of functions $\{h:\langle m\rangle\rightarrow\langle m\rangle\}$ such that

$$\hat{R}(p,H,p\log p)=\mathrm{O}(\log p).$$

4.2. COROLLARY. *There exists a good family of hash function classes.*

PROOF. If m is prime we substitute $k=4\log p$ and $v=p\log p$ in Lemma 4.1. This gives that for all S of size $p\log p$

$$\Pr\{R(p,S,h)\geqslant 4\log p\}=\mathrm{o}(p^{-1}).$$

For m not prime we define the corresponding H as above but for m' rather than m, where m' is the smallest prime larger than m. $\square$

We shall say that a family of PRAMs parametrized by p, the number of processors, and m, the size of the total memory space, has a *fast good hash function* (*f.g.h.f.*) if it has implemented on it a good hash function that takes time αg where α is a constant independent of m and p.

Our first simulation result can be stated as follows.

4.3. THEOREM. *A p-XPRAM with f.g.h.f. can simulate a v-EPRAM with expected optimal efficiency if* $v \geqslant p \log p$.

PROOF. For each execution of a program on an input on the EPRAM M' we shall randomly choose a hash function $h \in H \in \mathscr{H}$ with appropriate parameters (p, m) and will represent word j in the memory of M' by a word in memory unit $h(j) \bmod p$ of the XPRAM M. We shall distribute the tasks of the v processors of M' arbitrarily so that each processor of M simulates at most $\lceil v/p \rceil$. In simulating one step of M' we will execute $\lceil v/p \log p \rceil$ supersteps of M. In each superstep for each processor we will select randomly a set of $\log p$ accesses from the up to $\lceil v/p \rceil$ that have still to be made. By the choice of $\mathscr{H}$ the expectation of the largest number of accesses made to any memory unit in a superstep is $O(\log p)$, and hence the time for M to execute a superstep will have the same expectation. Summing expectations over the supersteps of M and the steps of M' gives an expected runtime of M of $O(v/p)$ as required.

We note that as long as the number of words of the input data is no more than the total number of computational operations performed by the M' program, and as long as these data are distributed uniformly among the processors of M, all the data can be initially rehashed according to h in the same $O(v/p)$ parallel time. □

In stating and proving the above result we have neglected to say how the accesses to the various addresses in each memory unit are implemented. If h is a permutation, then there is no problem since we can allocate address $h(j) \in \langle m \rangle$ in M to address j of M' without causing conflicts. For the specific H of Corollary 4.2, however, the hash functions are not one-to-one, and up to $k = 4 \log p$ addresses may map to one. We therefore need to say how the address set $A_i \subseteq \langle m \rangle$ that is mapped by h to the ith unit is to be allocated among the addresses of the latter. To illustrate the possibilities we shall give a scheme using a randomly chosen hash function h_i^* for each i that has a simple analysis for m large enough and with M having memory space larger than that of M' by a constant factor.

For each processor i we choose a hash function h_i^* randomly from a class H^* such that, even if linear chaining is used in this second hashing, the probability that the up to $k = 4 \log p$ requests to the ith unit require more than $\beta \log p$ steps, for a certain constant β, is less than $(p \log p)^{-1}$. We can then deduce that all p processors finish a superstep in time $\beta \log p$, except with probability $(\log p)^{-1}$. If they do not finish then the worst case, $O(\log p)^2$ steps, is taken, giving an expected time of $O(\log p)$.

We define $h_i^* = h \bmod \lceil m/p \rceil$ where h is drawn randomly from the class H defined above but with $k^* = \beta \log p$ for some $\beta > 4$. Then it is easy to show that, if $m > 16p^2(\log p)^3$, then any $4 \log p$ elements of A_i will be mapped by h^* to distinct points, except with probability $(p \log p)^{-1}$. It then remains to show that the number x of points in A_i that are mapped by h^* to any set of $4 \log p$ distinct points is less than $\beta \log p$, except with probability $(p \log p)^{-1}$. Clearly x upper bounds the cost of the $4 \log p$ accesses even if linear chaining is used in implementing the h^* hash table. But the probability that x exceeds $\beta \log p$ is no more than

$$\binom{|A_i|}{\beta \log p} \left(\frac{4 \log p}{m/p} \right)^{\beta \log p}.$$

This is $O(p^{-2})$ if β is large enough as long as $|A_i| = O(m/p)$, which is guaranteed, with high probability, by Corollary 4.2.

We now go on to consider the case of concurrent or CPRAMS where arbitrary patterns of simultaneous reads and writes have to be implemented. One approach is to allow the communications hardware to contain certain "combining" capabilities to combine requests to one location as they meet in the network [57]. This is generally considered to complicate the switching hardware significantly. The point of our next theorem is to show that given enough slackness the combining capability is unnecessary for optimality and can be simulated by an XPRAM, which totally separates out communication from other functions.

One technique for simulating concurrent PRAMs is to first sort the simultaneous requests according to the destination addresses so that access requests to the same location in memory will be adjacent in the sorted array. This technique has been widely used [12, 29, 44, 46, 69, 80]. No parallel sorting algorithm is known, however, that would achieve our optimal simulation for all memory sizes m. We shall define the weaker operation of *semisorting* an array of keys as that of rearranging the items so that any subset of identical keys will be adjacent in the array (but these segments need not be ordered).

4.4. LEMMA. *For some integers p, m and v, let B be an array of length v of items from $\langle m\rangle$, possibly with repetitions. Then there is a p-XPRAM algorithm for semisorting B in expected time* $O((v/p)(\log v)/\log(v/p))$ *if* $v > p(\log p)^2$.

PROOF. As observed by Kruskal et al. [31] a stable sorting algorithm of Reif (see [54]) can sort v items from any $\langle r\rangle$ on a p-SPRAM in $(\log r)/\log(v/p)$ phases of stable sorting, each phase taking time $O(v/p)$, provided $v > p \log p$. The algorithm views the $\log r$ bits of the items as $(\log r)/\log(v/p)$ subsequences each of $\log(v/p)$ bits. The ith phase sorts on the ith least significant such subsequence leaving the relative ordering on the less significant bits unchanged whenever the bits of the ith subsequence are equal.

Each phase assigns v/p consecutive items in the current array to each of the p processors and can be viewed as sorting items in the range $\langle v/p\rangle$. Each processor $k \in \langle p\rangle$ then sequentially completes column k of the $v/p \times p$ array C where C_{jk} is the number of items assigned to processor k having value j. Then the position in the sorted array of the first item j' that was assigned to processor k' is given by

$$D_{j'k'} = 1 + \sum_{j,k} \{C_{jk} \mid (j<j') \text{ or } (j=j' \text{ and } k<k')\}.$$

Also if each processor k' knows the value of $D_{j'k'}$ for all j', then it can place all its elements in the sorted output of this phase in time $O(v/p)$.

To complete the description of each phase of stable sorting we need to describe how D is computed from C. This is done using the parallel prefix operation of Ladner and Fischer [33]. This is defined for an arbitrary array $x_1, \ldots, x_v$ of elements and an arbitrary associative operation $\circ$ on them. The outputs of the operation are the v values $y_i = x_1 \circ x_2 \circ \cdots \circ x_i$ for $i = 1, \ldots, v$. This can be computed in $O(v/p)$ steps on a p-SPRAM if $v \geqslant p \log p$ [27] and in the same time on a p-XPRAM if $v \geqslant p(\log p)^2$. To compute D we do a parallel prefix addition operation on the one-dimensional array obtained by traversing $\{C_{jk}\}$ along each row in the direction of increasing k, successively for increasing j.

Finally we want to remove the dependency of our algorithm on the memory size m. To do this we shall choose a function h randomly from a universal hash function class H mapping $\langle m\rangle$ to $\langle 2v^2\rangle$. Carter and Wegman [14] give classes of universal hash functions requiring only a constant number of arithmetic operations such that $\forall S\subseteq\langle m\rangle$ such that $|S|=v$

$$\Pr\{\exists b_1, b_2\in S, b_1\neq b_2, h(b_1)=h(b_2)\}\leqslant\tfrac{1}{2}.$$

Hence to semisort B we choose $h\in H$ randomly and perform the above-described stable sort on $\{h(b)|b\in B\}$ in the desired $O((v/p)(\log v)/\log(v/p))$ steps. This will be considered to be successful if h does map distinct elements of S to distinct numbers, which will happen with probability at least a half. In case of failure the procedure is repeated until success in reached. The expected number of trials required is clearly a constant. Failure can be detected in $O(v/p)$ steps by detecting locally whether there is a pair of adjacent items with $b_1\neq b_2$ but $h(b_1)=h(b_2)$ and performing a parallel disjunction on these. □

The simulation result now follows easily.

4.5. THEOREM. *A p-XPRAM with f.g.h.f. can simulate a v-CPRAM with expected optimal efficiency if* $v\geqslant p^{1+\varepsilon}$ *for some* $\varepsilon>0$.

PROOF. The simulation is similar to that for EPRAMs. In performing a superstep, however, the XPRAM first semisorts the v requests using Lemma 4.4. Once we have semisorted the v requests into an array, the read and write operations can be completed using parallel prefix operations. We distribute the semisorted array among the p processors giving each one a v/p length segment. For a read instruction, for each word to be accessed the first request to it in the array is selected by a parallel prefix operation and the processor to which it belongs performs the access. The word read is then distributed to the other requests in the array by another parallel prefix. Each processor then returns the result of each access to the processor that requested it originally. A write instruction is implemented similarly. □

The significance of the above result is that it shows that our most powerful PRAM model can be implemented optimally on a communication model that assumes permutation routing but no combining capabilities. We note that recent analysis of polynomial hash functions of fixed degree [31] shows that the f.g.h.f. hypothesis is redundant here (see also [85]).

It turns out that the XPRAM can be implemented optimally, in turn, on at least two models of computation that are inspired by very different technological assumptions. The first is packet-switching network technology and the second optically-switched crossbars.

Let the *cube* and *directed butterfly* be communication networks as defined in previous sections, with a processor and memory unit attached to each node in the cube, and to each level zero node in the directed butterfly. Suppose a message can be sent from a node to any of its neighbors in time g. Then Theorems 3.2 and 3.3 establish.

4.6. THEOREM. *A p-cube or p-directed butterfly can simulate a p-XPRAM with expected optimal efficiency.*

An entirely distinct model of computation suggested by optical switching [6, 59] is the S*PRAM we defined earlier. The idea is that if a processor directs a beam of light to another, then it will be correctly received if no others are directed to the same recipient. Correct receipt is acknowledged by returning the message to the sender. If transmission fails because of a concurrent access attempt, then this will be discovered by the sender.

We can see that a superstep of a p-XPRAM, in which at most $\log p$ messages are sent to any processor, can be simulated by a p-S*PRAM in $O(\log p)$ steps as follows: Anderson and Miller [6] gave a solution to this message passing problem in the special case that for every node the set of messages destined for it have distinct labels from $1, 2, \ldots, \log p$. We shall use the first phase of their algorithm that does not use these labels. In this phase each processor repeats the following steps for $c_1 \log p$ steps: Choose a random message from those still to be sent and attempt to send it with a certain probability. If a processor fails to keep up with a specified level of successful transmissions, it is considered oversize and it attempts no more transmissions. They show that for a suitable constant c_1, only cp packets still need to be transmitted at the end of this phase, for some constant c. We now observe that the simulation can be completed in $O(\log p)$ steps, by performing, for example, the following three stages. (i) First the packets are redistributed without changing their overall order so that each node gets at most $\lceil c \rceil$. To do this a parallel prefix operation computes the required addresses. Redistribution can be done deterministically on an SPRAM in $O(\log p)$ steps. (ii) Second the packets are sorted by some $O(\log p)$ time SPRAM algorithm (e.g., [60]). (iii) Lastly since the packets destined for any one processor will have been sorted into adjacent places in adjacently numbered processors, they can be sent to the correct destination without further contention by an $O(\log p)$ SPRAM algorithm.

We can extend the above to show that if at most $r > \log p$ packets are destined to any processor, then $O(r)$ time suffices. Hence we conclude the following.

4.7. THEOREM. *A p-S*PRAM can simulate a p-XPRAM with expected optimal efficiency.*

Putting together Theorems 4.3, 4.5–4.7 gives the following summary of the simulations given, all of which go through the XPRAM model.

4.8. THEOREM. *A p-cube, p-directed butterfly or a p-S*PRAM, each with a f.g.h.f., can simulate a v-EPRAM (v-CPRAM) with expected optimal efficiency if* $v \geqslant p \log p$ ($v \geqslant p^{1+\varepsilon}$ *for some* $\varepsilon > 0$).

As historical background we note that efficient PRAM simulations were first obtained without parallel slackness [5, 39, 71, 73]. In that setting an optimal result was first obtained by Karlin and Upfal [25] who gave a simulation on a p-butterfly of a p-EPRAM with efficiency $\Omega(1/\log p)$. Their result, when stated for this efficiency, does not require the f.g.h.f. assumption. Methodologically our treatment differs from theirs also in that it separates out the hashing and routing aspects.

For simulating CPRAMs in this same setting $\Omega(1/\log p)$ efficiency can be obtained by extending the EPRAM simulation using an appropriate sorting algorithm (e.g., [60]). More directly it can be obtained using the simulation of Ranade [57].

All these previous simulations can be adapted to obtain optimal efficiency by allowing logarithmic parallel slackness [78]. A substantial difference does arise, however, in the case of the CPRAM simulation since the resulting p-processor simulation requires the network to have $p \log p$ elements that perform some computational function such as combining messages or supporting sorting. The advantage of Theorem 4.5 is that it achieves an optimal simulation on a pure message-passing model.

The constant factors implied by the general CPRAM simulation may grossly overestimate what can be obtained for any specific parameters. For example there already exists a rich variety of ideas for parallel sorting (e.g., [17, 35]) in addition to the ones already discussed and we expect that substantial improvements are possible. For the EPRAM simulation, the algorithm given is already practical. As previously mentioned, experimental results give small constants for the routing, while the good hash function discussed provably contributes small constants too. We note also that many algorithms written for CPRAMs can be implemented as efficiently on an EPRAM enhanced with a constant-time parallel prefix operation [10]. As already observed, however, a parallel prefix of size $p \log p$ can be implemented optimally on an ordinary p-EPRAM. Hence the enhancement is actually redundant when computing problems of larger size.

4.4. Conclusion

We have considered multiprocessor parallel computation in the context of problems with large inherent parallelism. The results suggest that the XPRAM may turn out to be an appropriate aspiration for the parallel computer architect much as the von Neumann model is in the sequential case. It is conceptually simple, capturing as it does the idea of message passing synchronized in bulk. It is a suitable host on to which higher-level communication and storage management functions can be efficiently compiled. Furthermore it is a promising candidate for efficient implementation in foreseeable technologies.

Acknowledgment

The author is grateful to Nick Littlestone, Thanasis Tsantilas and Eli Upfal for their comments on this chapter. The preparation of this chapter was supported in part by the National Science Foundation under Grant DCR-86-00379 and by the Office of Naval Research under Grant ONR-N00014-85-K-0445.

References

[1] AGGARWAL, A. and A.K. CHANDRA, Communication complexity of PRAMs, in: *Proc. Internat. Coll. on Automata, Languages and Programming*, Lecture Notes in Computer Science, Vol. 317 (Springer, Berlin, 1988) 1–17.

[2] AHO, A.V., J.E. HOPCROFT and J.D. ULLMAN, *The Design and Analysis of Computer Algorithms* (Addison-Wesley, Reading, MA, 1974).

[3] AJTAI, M., J. KOMLÓS and E. SZEMERÉDI, Sorting in $c \log n$ steps, *Combinatorica* **3** (1983) 1–19.

[4] ALELIUNAS, R., Randomized parallel communication, in: *Proc. ACM Symp. on Principles of Distributed Computing* (1982) 60–72.

[5] ALT, H., T. HAGERUP, K. MEHLHORN and F.P. PREPARATA, Deterministic simulation of idealized parallel computers on more realistic ones, *SIAM J. Comput.* **16**(5) (1987) 808–835.

[6] ANDERSON, R.J. and G.L. MILLER, Optimal communication for pointer based algorithms, Tech. Report, CRI 88-14, Comput. Sci. Dept., Univ. of Southern California, Los Angeles, 1988.

[7] BATCHER, K., Sorting networks and their applications, in: *Proc. AFIPS Spring Joint Comp. Conf.* (1968) 307–314.

[8] BENEŠ, V.E., *Mathematical Theory of Connecting Networks and Telephone Traffic* (Academic Press, New York, 1965).

[9] BHATT, S., F. CHUNG, T. LEIGHTON and A. ROSENBERG, Optical simulations of tree machines, in: *Proc. 27th Ann. IEEE Symp. on Foundations of Computer Science* (1986) 274–282.

[10] BLELLOCH, G., Scans as primitive parallel operations, *IEEE Trans. Comput.* **38**(11) (1989) 1526–1538.

[11] BOLLOBÁS, B., *Random Graphs* (Academic Press, London, 1985).

[12] BORODIN, A. and J.E. HOPCROFT, Routing, merging and sorting on parallel models of computation, *J. Comput. System Sci.* **30** (1985) 130–145.

[13] BREBNER, G.J., Ph.D. Thesis, Comput. Sci. Dept., Edinburgh Univ. (1983).

[14] CARTER, J.L. and M.N. WEGMAN, Universal classes of hash functions, *J. Comput. System Sci.* **18** (1979) 143–154.

[15] CHANG, Y. and J. SIMON, Continuous routing and batch routing on the hypercube, *5th Ann. ACM Symp. on Principles of Distributed Computing* (1986) 272–281.

[16] COLE, R., Parallel merge sort, *SIAM J. Comput.* **17**(4) (1988) 770–785.

[17] COOK, S.A. and R.A. RECKHOW, Time bounded random access machines, *J. Comput. System Sci.* **7** (1973) 354–375.

[18] COOLEY, J.M. and J.W. TUKEY, An algorithm for the machine calculation of complex Fourier series, *Math. Comp.* **19** (1965) 297–301.

[19] EPPSTEIN, D. and Z. GALIL, Parallel algorithmic techniques for combinatorial computation, *Ann. Rev. Comput. Sci.* **3** (1988) 233–283.

[20] FORTUNE, S. and J. WYLLIE, Parallelism in random access machines, in: *Proc. 10th Ann. ACM Symp. on Theory of Computing* (1978) 114–118.

[21] GALIL, Z. and W.J. PAUL, An efficient general-purpose parallel computer, *J. ACM* **30**(2) (1983) 360–387.

[22] GREENBERG, R.I. and C.E. LEISERSON, Randomized routing on fat-trees, in: *Proc. 26th Ann. IEEE Symp. on Foundations of Computer Science* (1985) 241–249.

[23] HILLIS, W.D., *The Connection Machine* (MIT Press, Cambridge, MA, 1985).

[24] JERRUM, M.R. and S. SKYUM, Families of fixed degree graphs for processor interconnection, *IEEE Trans. Comput.* **33** (1984) 190–194.

[25] KARLIN, A. and E. UPFAL, Parallel hashing—an efficient implementation of shared memory, *J. ACM* **35**(4) (1988) 876–892.

[26] KARLOFF, H.J. and P. RAGHAVAN, Randomized algorithms and pseudo-random numbers, in: *Proc. 20th Ann. ACM Symp. on Theory of Computing* (1988) 310–321.

[27] KARP, R.M. and V. RAMACHANDRAN, Parallel algorithms for shared-memory machines, in: J. van Leeuwen, ed., *Handbook of Theoretical Computer Science, Vol. A* (North-Holland, Amsterdam, 1990).

[28] KNUTH, D.E., *The Art of Computer Programming, Vol. 3: Sorting and Searching* (Addison-Wesley, Reading, MA, 1973).

[29] KRIZANC, D., D. PELEG and E. UPFAL, A time-randomness tradeoff for oblivious routing, in: *Proc. 20th Ann. ACM Symp. of Theory of Computing* (1988) 93–102.

[30] KRIZANC, D., S. RAJASEKARAN and A. TSANTILAS, Optimal routing algorithms for mesh-connected processor arrays, in: *VLSI Algorithms and Architectures*, Lecture Notes in Computer Science, Vol. 319 (Springer, Berlin, 1988) 411–422.

[31] KRUSKAL, C.P., L. RUDOLPH and M. SNIR, A complexity theory of efficient parallel algorithms, *Theoret. Comput. Sci.* **71** (1990) 95–132.

[32] KUNG, H.T. and C.E. LEISERSON, Systolic arrays (for VLSI), in: I.S. Duff and G.W. Stewart, eds., *Sparse Matrix Proceedings 1978* (SIAM, Philadelphia, PA, 1979) 256–282.

[33] LADNER, R.E. and M.J. FISCHER, Parallel prefix computation, *J. ACM* **27** (1980) 831–838.

[34] LEIGHTON, F.T., Parallel computation using meshes of trees, in: M. Nagl and J. Perl, eds., *Graph Theoretic Concepts in Computer Science* (Trauner, Linz, 1983) 200–218.

[35] LEIGHTON, F.T., Tight bounds on the complexity of sorting, *IEEE Trans. Comput.* **34**(4) (1985) 344–354.

[36] LEIGHTON, F.T., B. MAGGS and S. RAO, Universal packet routing algorithms, in: *Proc. 29th Ann. IEEE Symp. on Foundations of Computer Science* (1988) 256–269.

[37] LEIGHTON, F.T., F. MAKEDON and I. TOLLIS, A $2n-2$ steps algorithm for routing in an $n \times n$ array with constant-size queues, in: *Proc. 1989 ACM Symp. on Parallel Algorithms and Architectures* (1989) 328–335.

[38] LEISERSON, C.E., Fat-trees: universal networks for hardware- efficient supercomputing, *IEEE Trans. Comput.* **34**(10) (1985) 892–901.

[39] LEV, G., N.J. PIPPENGER and L.G. VALIANT, A fast parallel algorithm for routing in permutation networks, *IEEE Trans. Comput.* **30**(2) (1981) 93–100.

[40] LIU, C.L., *Introduction to Combinatorial Mathematics* (McGraw-Hill, New York, 1968).

[41] MEAD, C.A. and L.A. CONWAY, *Introduction to VLSI Systems* (Addison-Wesley, Reading, MA, 1979).

[42] MEHLHORN, K. and U. VISHKIN, Randomized and deterministic simulations of PRAMs by parallel machines with restricted granularity of parallel memories, *Acto Inform.* **21** (1984) 339–374.

[43] MONIEN, B. and H. SUDBOROUGH, Comparing interconnection networks, in: *Math. Foundations of Comput. Sci.*, Lecture Notes in Computer Science, Vol. 324 (Springer, Berlin, 1988) 138–153.

[44] NASSIMI, D. and S. SAHNI, Data broadcasting in SIMD computers, *IEEE Trans. Comput.* **30** (1981) 101–107.

[45] NASSIMI, D. and S. SAHNI, Parallel algorithms to set-up the Beneš permutation networks, *IEEE Trans. Comput.* **31**(2) (1982) 148–154.

[46] PARBERRY, I., *Parallel Complexity Theory* (Pitman, London, 1987).

[47] PELEG, D. and E. UPFAL, The token distribution problem, in: *Proc. 27th Ann. IEEE Symp. on Foundations of Computer Science* (1986) 418–427.

[48] PINSKER, M.S., On the complexity of a concentrator, in: *Proc. of 7th Internat. Teletraffic Congress, Stockholm* (1973) 318.

[49] PIPPENGER, N.J., Superconcentrators, *SIAM J. Comput.* **6**(2) (1977) 298–304.

[50] PIPPENGER, N.J., Parallel communication with bounded buffers, in: *Proc. 25th Ann. IEEE Symp. on Foundations of Computer Science* (1984) 127–136.

[51] PREPARATA, F.P., VLSI algorithms and architectures, in: *Math. Foundations of Computer Science*, Lecture Notes in Computer Science, Vol. 176 (Springer, Berlin, 1984) 149–161.

[52] PREPARATA, F.P. and J. VUILLEMIN, The cube-connected-cycles: a versatile network for parallel computation, *Comm. ACM* **24**(5) (1981) 300–309.

[53] RABIN, M.O., Efficient dispersal of information for security, load balancing and fault tolerance, *J. ACM* **36** (1989) 335–348.

[54] RAJASEKARAN, S. and J.H. REIF, Optimal and sublogarithmic time randomized parallel sorting algorithms, *SIAM J. Comput.* **18** (1989) 594–607.

[55] RANADE, A.G., Constrained randomization for parallel communication, Tech. Report YALEU/DCS/TR-511, Dept. Comput. Sci., Yale University, New Haven, CT, 1987.

[56] RANADE, A.G., Equivalence of message scheduling algorithms for parallel communication, Tech. Report YALEU/DCS/TR-512, Dept. Comput. Sci., Yale University, New Haven, CT, 1987.

[57] RANADE, A.G., How to emulate shared memory, in: *Proc. 28th Ann. IEEE Symp. on Foundations of Computer Science* (1987) 185-194.

[58] RANADE, A.G., Fluent parallel computation, Ph.D. Thesis, Dept. Comput. Sci., Yale University, New Haven, CT, 1988.

[59] REIF, J.H. and K. JOHNSON, Constant time holographic message routing using $O(n \log n)$ switches, Manuscript, 1989; see also: E.S. Maniloff, K.M. Johnson and J. Reif, Halographic routing network for parallel processing machines, in: *Proc. EPS/EUROPTICA/SDIE Internat. Congress on Optical Science and Engineering* (1989).

[60] REIF, J.H. and L.G. VALIANT, A logarithmic time sort on linear size networks, *J. ACM* **34**(1) (1987) 60–76.

[61] REISCH, S. and G. SCHNITGER, Three applications of Kolmogorov complexity, in: *Proc. 23rd. Ann. IEEE Symp. on Foundations of Computer Science* (1982) 45–52.

[62] SCHWARTZ, J.T., Ultracomputers, *ACM TOPLAS* **2** (1980) 484–521.

[63] SEITZ, C., The cosmic cube, *Comm. ACM* **28** (1985) 22–33.

[64] SIEGEL, H.J., Interconnection networks for SIMD machines, *Computer* **12**(6) (1979) 57–65.

[65] SIEGEL, H.J., *Interconnection Networks for Large-scale Parallel Processing* (Heath, Lexington, MA, 1985).

[66] STONE, H., Parallel processing with the perfect shuffle, *IEEE Trans. Comput.* **20**(2) (1971) 153–161.

[67] SULLIVAN, H. and T.R. BASHKOW, A large scale homogeneous machine, in: *Proc. 4th Ann. ACM Symp. on Computer Architecture* (1977) 105–124.

[68] TURING, A.M., On computable numbers with an application to the Entscheidungs problem, *Proc. London Math. Soc. Ser. 2* **42** (1936) 230–265; correction, *ibidem* **43** (1937) 544–546.

[69] ULLMAN, J.D., *Computational Aspects of VLSI* (Computer Science Press, Rockville, MD, 1984.

[70] UPFAL, E., Efficient schemes for parallel communication, *J. ACM* **31**(3) (1984) 507–517.

[71] UPFAL, E., A probabilistic relation between desirable and feasible models of parallel computation, in: *Proc. 16th Ann. ACM Symp. on Theory of Computing* (1984) 258–265.

[72] UPFAL, E., An $O(\log N)$ deterministic packet routing scheme, in: *Proc. 21st ACM Symp. on Theory of Computing* (1989) 241–250.

[73] UPFAL, E. and A. WIGDERSON, How to share memory in a distributed system, *J. ACM* **34**(1) (1987) 116–127.

[74] VALIANT, L.G., Graph-theoretic properties in computational complexity, *J. Comput. System Sci.* **13** (1976) 278–285.

[75] VALIANT, L.G., Experiments with a parallel communication scheme, in: *Proc. 18th Allerton Conf. on Communication, Control and Computing* (1980) 802–811.

[76] VALIANT, L.G., A scheme for fast parallel communication, *SIAM J. Comput.* **11** (1982) 350–361.

[77] VALIANT, L.G., Optimality of a two-phase strategy for routing in interconnection networks, *IEEE Trans. Comput.* **32**(9) (1983) 861–863.

[78] VALIANT, L.G., Optimally universal parallel computers, *Phil. Trans. Roy. Soc. Lond. Ser. A* **326** (1988) 373–376.

[79] VALIANT, L.G. and G.J. BREBNER, Universal schemes for parallel communication, in: *Proc. 13th Ann. ACM Symp. on Theory of Computing* (1981) 263–277.

[80] VISHKIN, U., A parallel-design distributed implementation (PDDI) general-purpose computer, *Theoret. Comput. Sci.* **13** (1984) 157–172.

[81] WAXMAN, A., A permutation network, *J. ACM* **15** (1968) 159–163 and *J. ACM* **15** (1968) 340.

[82] CYPHER, R. and C.G. PLAXTON, Deterministic sorting in nearly logarithmic time on the hypercube and related computers, in: *Proc. 22nd Ann. ACM Symp. on Theory of Computing* (1990).

[83] KAKLAMANIS, C., D. KRIZANC and A. TSANTILAS, Tight bounds for oblivious routing in the hypercube, Tech. Report TR-23-89, Aiken Computation Lab., Harvard Univ., 1989.

[84] LEIGHTON, T. and B. MAGGS, Expanders might be practical, in: *Proc. 30th Ann. IEEE Symp. on Foundations of Computer Science* (1989) 384–389.

[85] SIEGEL, A., On universal classes of fast high performance hash functions, in: *Proc. 30th Ann. IEEE Symp. on Foundations of Computer Science* (1989) 20–25.

[86] TSANTILAS, A.M., A refined analysis of the Valiant–Brebner algorithm, Tech. Report TR-82-89, Aiken Computation Lab., Harvard Univ., 1989.

[87] VALIANT, L.G., Bulk-synchronous parallel computers, in: M. Reeve and S.E. Zenith, eds., *Parallel Processing and Artificial Intelligence* (Wiley, Chichester, UK, 1989); see also: A bridging model for parallel computation, *Comm. ACM*, to appear.

Subject Index